Springer Series in Astrophysics and Cosmology

The series covers all areas of astrophysics and cosmology, including theory, observations, and instrumentation. It publishes monographs and edited volumes. All books are authored or edited by leading experts in the field and are primarily intended for researchers and graduate students.

Cosimo Bambi · Yosuke Mizuno ·
Swarnim Shashank · Feng Yuan
Editors

New Frontiers in GRMHD Simulations

Springer

Editors
Cosimo Bambi
Department of Physics
Fudan University
Shanghai, China

Yosuke Mizuno
Tsung-Dao Lee Institute
Shanghai Jiao Tong University
Shanghai, China

Swarnim Shashank
Department of Physics
Fudan University
Shanghai, China

Feng Yuan
Department of Physics
Fudan University
Shanghai, China

ISSN 2731-734X ISSN 2731-7358 (electronic)
Springer Series in Astrophysics and Cosmology
ISBN 978-981-97-8521-6 ISBN 978-981-97-8522-3 (eBook)
https://doi.org/10.1007/978-981-97-8522-3

This Springer imprint is published by the registered company Springer Nature Singapore Pte Ltd.
The registered company address is: 152 Beach Road, #21-01/04 Gateway East, Singapore 189721, Singapore

Preface

Astrophysical systems are primarily studied through observations of electromagnetic radiation. This electromagnetic radiation arises from fluids/plasmas present in these sources. To study these systems, we need to model the dynamics of these fluids and their interactions with radiation and electromagnetic fields. In the case of compact objects like black holes and neutron stars, fluids and radiation interact in a strong gravitational potential, necessitating the study of dynamics in curved spacetime as described by general relativity. One particularly important class of phenomena to study is the accretion process.

Accretion onto a compact object (black hole or neutron star) is a common phenomenon in the Universe and an extremely efficient mechanism to release energy. In the case of a black hole accreting from a geometrically thin and optically thick disk, the accretion process can convert up to 5.7% of the rest-mass of the accreting material into radiation if the black hole is nonrotating and up to 42% in the case of an extremal Kerr black hole with spin parameter $a_* = 1$. If we consider nuclear reactions inside stars, the efficiency is much lower and we can convert not more than 0.7% of the rest-mass of the particles involved in the fusion process into energy (and only a fraction of this energy can be used to heat the interior of the star). Analytical and semianalytical studies of the accretion process are possible only in exceptional cases. Numerical studies are crucial to make progress in this field.

General relativistic magnetohydrodynamic (GRMHD) simulations are a fundamental tool for studying accretion processes onto black holes and neutron stars, neutron star mergers, core-collapse supernovae, etc. The magnetorotational instability is a crucial ingredient to transport angular momentum in the disk and implies that the essential dynamics of accreting systems can be derived from the GRMHD equations. The first GRMHD simulations of the accretion process onto a black hole were run by Jim Wilson in the late 1970s. At that time, the computing power was a serious limitation to make progress in the field. The past 20–30 years have seen remarkable advancements in the field and GRMHD codes have been developed by several groups. The increase in computing power in the last two decades has permitted us to develop even GRMHD codes in a dynamical spacetime, in which both the magnetized fluid and the spacetime metric evolve in time. With the recent detection of gravitational waves, multi-messenger astronomy, and observations of black hole shadows, we can compare the theoretical predictions of these simulations with real astrophysical observations. Thus, this field is bright and still rapidly progressing.

This book aims to review the state-of-the-art in GRMHD simulations, as we realized that there is not a similar review in the literature at present. We organized the book into four parts. Part I is *Methods*, where we review the basic formalism to calculate how a magnetized fluid and the background spacetime metric evolve with time. Part II is *Codes* and we present an overview of several GRMHD codes including GR-Athena++, KHARMA, Einstein Toolkit, and SPHINCS_BSSN. Part III is *Non-Self-Gravitating Fluids* and is dedicated to the astrophysics of accretion processes onto black holes, neutron stars, and exotic compact objects assuming a fixed background metric. Part IV is *Self-Gravitating Fluids*, where both the magnetized fluid and the spacetime metric evolve in time. For every chapter, we invited experts on the topic. We are extremely grateful to all authors who enthusiastically agreed to contribute to this volume. Their dedication and hard work have been pivotal in making this project successful. We value their expertise and are grateful for their generous contribution of knowledge and time.

We hope that the present book can become a valuable reference work for the community working on GRMHD simulations and be useful both to researchers and graduate students. We are aware that some parts of the book may become out-of-date quickly, as this is a rapidly evolving field. However, we think it is important to review the current state-of-the-art in the field in a single work and we promise to include progress in the field in future editions.

Shanghai, China
August 2024

Cosimo Bambi
Yosuke Mizuno
Swarnim Shashank
Feng Yuan

Contents

Part I
Methods

Chapter 1
General-Relativistic Magnetohydrodynamic Equations: The Bare Essential

Yosuke Mizuno and Luciano Rezzolla

Abstract Recent years have seen a significant progress in the development of general relativistic codes for the numerical solution of the equations of magnetohydrodynamics in spacetimes with high and dynamical curvature. These codes are valuable tools to explore the large-scale plasma dynamics such as that takes place when two neutron stars collide or when matter accretes onto a supermassive black hole. This chapter is meant to provide a very brief but complete overview of the set of equations that are normally solved in modern numerical codes after they are cast into a conservative formulation within a 3+1 split of spacetime.

1.1 Introduction

Relativistic astrophysics studies the most energetic and violent astrophysical processes that are characterized by very high speeds, strong gravitational fields, very high temperatures and ultra-intense magnetic fields. Under these conditions, which are normally met near neutron stars and black holes, a fully general relativistic treatment is necessary for an accurate description of the physical conditions.

In this context, relativistic magnetohydrodynamics (MHD) represents a very effective framework to describe the dynamics of macroscopic plasma in a relativistic

Y. Mizuno
Tsung-Dao Lee Institute, Shanghai Jiao Tong University, 520 Shengrong Road, Shanghai 201210, China

School of Physics and Astronomy, Shanghai Jiao Tong University, 800 Dongchuan Road, Shanghai 200240, China

L. Rezzolla (✉)
Institute for Theoretical Physics, Goethe University Frankfurt, Max-von-Laue-Strasse 1, 60438 Frankfurt am Main, Germany
e-mail: rezzolla@itp.uni-frankfurt.de

Frankfurt Institute for Advanced Studies, 60438 Frankfurt am Main, Germany

School of Mathematics Trinity College, Dublin 2, Ireland

C. Bambi et al. (eds.), *New Frontiers in GRMHD Simulations*, Springer Series in Astrophysics and Cosmology, https://doi.org/10.1007/978-981-97-8522-3_1

regime, also when considering non-astrophysical scenarios, such as the collision of heavy ions (see, e.g., [73, 81]). It is in fact important to note that plasma is the most diffused state of matter in the Universe and that most plasmas are electrically conducting. In the MHD approximation, the plasma is treated as a macroscopic fluid that is coupled with electromagnetic fields that it produces with its dynamics or that may present from external sources. In addition, the collision times between particles in astrophysical fluids are usually smaller than the other relevant timescales of the systems. As a result, the mean free paths of the plasma constituents are shortened, causing particles to interact on much smaller spatial scales than those of the underlying macroscopic system. As a result, astrophysical plasmas represent what is referred to as collisional systems so that, despite being composed of particles at a microscopic level, the plasmas can be described as a continuous medium with well-defined macroscopic average quantities such as velocity, density, and pressure.

Relativistic MHD is often employed to study the dynamics of relativistic plasma, such as the collision of two magnetized neutron stars and the ensuing gamma-ray burst [3, 17, 34, 43, 50, 60, 63, 64, 69, 77], or the accretion and outflows onto a central compact object [19, 40, 45, 48, 54, 58, 59, 62, 87, 89, 91]. Under these conditions, general relativistic effects become important if not dominant and, hence, the solution of the full set of general-relativistic MHD (GRMHD) equations represent the only avenue to obtain an accurate and physically consistent description of the system. Numerical simulations have proven to be crucial in modern theoretical astrophysics, enhancing our understanding of the dynamics of astrophysical systems in highly nonlinear regimes. From the point of view of GRMHD, over the past few decades, many GRMHD codes have been developed [4, 6–8, 11, 13, 16, 18, 21, 23, 26, 28, 31, 32, 37, 41, 45, 49, 57, 66, 72, 83, 90, 92] employing the 3+1 decomposition of spacetime and conservative 'Godunov' schemes based on approximate Riemann solvers [29, 33, 52, 78]. These codes are utilized to study various high-energy astrophysical phenomena. Some of these GRMHD codes incorporate radiation [55, 82, 86], and/or non-ideal MHD processes [12, 14, 15, 22, 24, 25, 74, 76, 80]. State-of-the-art GRMHD codes implement full adaptive mesh refinement [49, 66, 72, 85, 92], which is useful for obtaining higher spatial resolution in particularly interesting regions such as strong shocks, turbulence, and shear regions.

Chapter 1 of this book provides an overview of some essential properties of GRMHD equations. The structure of the chapter is as follows. Section 1.2 introduces the covariant GRMHD equations, while Sect. 1.3 introduces the basic decomposition of a four-dimensional spacetime into a timelike time-line and space-like hypersurface. Section 1.4 finally provides the form of the GRMHD equations when cast in a conservative form within a 3+1 split spacetime. A brief summary is contained in Sect. 1.5.

Before concluding, two important remarks should be made. First, we will not discuss here the numerical methods that are normally employed to solve the equations of GRMHD. This is because they are often complex, involving a variety of Riemann solvers and reconstruction methods, and hence needing an independent discussion; the interested reader can find an introduction to such methods in a number of textbooks (see, e.g., [9, 47, 52, 79, 88]) and more code-specific details in Part II of this book. Second, in this chapter, we will not discuss the solution to the Einstein equations, which are needed to account for the evolution of spacetime and will be covered in Chap. 2. Throughout the text, we adopt units where the speed of light, $c = 1$, and the gravitational constant, $G = 1$ and we absorb a factor of $\sqrt{4\pi}$ of the magnetic field four-vector, b^{μ}. We also use the index notation for contracted indices, employ Greek (Latin) letters for indices running between 0 (1) and 3, and a metric with signature $(-, +, +, +)$.

1.2 Covariant General-Relativistic Magnetohydrodynamic Equations

Hereafter, we will generally consider an ideal fluid endowed with electromagnetic fields whose equations of motion,—that is, the general-relativistic magnetohydrodynamic (GRMHD) equations—can be derived after imposing the local conservation laws of rest-mass (the continuity equation) and of the energy-momentum tensor, $T_{\mu\nu}$ (the Bianchi identities):

$$\nabla_{\mu} \tilde{J}^{\mu} = \nabla_{\mu} \rho u^{\mu} = 0 \,, \tag{1.1}$$

$$\nabla_{\mu} T^{\mu\nu} = 0 \,, \tag{1.2}$$

where ∇_{μ} is the covariant derivative associated with the four-dimensional spacetime metric $g_{\mu\nu}$, $\tilde{J}^{\mu}$ is the rest-mass density current and $T^{\mu\nu}$ is the energy-momentum tensor of the plasma.

Equation (1.1) represents the well-known mass-conservation law, where u^{μ} is the fluid four-velocity and ρ is the proper rest-mass density. After introducing the projector operator orthogonal to u_{μ}, i.e.,

$$h_{\mu\nu} := u_{\mu} u_{\nu} + g_{\mu\nu} \,, \tag{1.3}$$

and such that $h_{\mu\nu} u^{\mu} = 0$, it is possible to realise that Eq. (1.2) are four distinct equations representing respectively the conservation of energy

$$u_{\nu} \nabla_{\mu} T^{\mu\nu} = 0 \,, \tag{1.4}$$

and of four-momentum

$$h_{\alpha\nu} \nabla_{\mu} T^{\mu\nu} = 0 \,. \tag{1.5}$$

Note that the total energy-momentum is the linear combination of the contributions coming from the matter and from the electromagnetic fields, i.e., $T^{\mu\nu} := T^{\mu\nu}_{\rm m} + T^{\mu\nu}_{\rm f}$, where

$$T^{\mu\nu}_{\rm m} := \rho h u^{\mu} u^{\nu} + p g^{\mu\nu} , \tag{1.6}$$

and

$$T_{\rm f}^{\mu\nu} := F^{\mu}_{\lambda} F^{\lambda\nu} - \frac{1}{4}(F^{\lambda\delta} F_{\lambda\delta}) g^{\mu\nu} \,. \tag{1.7}$$

In the expressions above,

$$h := 1 + \varepsilon + \frac{p}{\rho} \,, \tag{1.8}$$

is the specific enthalpy, ε is specific internal energy, p is the fluid pressure, and $F^{\mu\nu}$ is the Faraday tensor.

The presence of electric and magnetic fields in the GRMHD equations requires the solution of additional equations expressing the corresponding conservation laws, namely, the Maxwell equations

$$\nabla_{\mu} F^{\mu\nu} = \mathcal{J}^{\mu} \,, \tag{1.9}$$

$$\nabla_{\mu} {}^{*}F^{\mu\nu} = 0 \,, \tag{1.10}$$

where $\mathcal{J}^{\mu}$ is the charge current density, and ${}^{*}F^{\mu\nu}$ is the dual of the Faraday tensor. The Faraday tensor $F^{\mu\nu}$ is constructed from the electric and magnetic fields, E^{α} and B^{α}, as measured in the generic frame having U^{α} as tangent vector, i.e.,

$$F^{\mu\nu} = U^{\mu} E^{\nu} - U^{\nu} E^{\mu} - \sqrt{-g}\eta^{\mu\nu\lambda\delta} U_{\lambda} B_{\delta} \,, \tag{1.11}$$

where $\eta^{\mu\nu\lambda\delta}$ is the fully-antisymmetric symbol (see, e.g., [79]) and g is the determinant of the spacetime four-metric. The dual Faraday tensor

$${}^{*}F^{\mu\nu} := \sqrt{-g}\eta^{\mu\nu\lambda\delta} F_{\lambda\delta} \,, \tag{1.12}$$

is written as

$${}^{*}F^{\mu\nu} = U^{\mu} B^{\nu} - U^{\nu} B^{\mu} + \sqrt{-g}\eta^{\mu\nu\lambda\delta} U_{\lambda} E_{\delta} \,. \tag{1.13}$$

Most of the GRMHD simulations to date have explored scenarios within the so-called "ideal MHD limit", that is, a limit in which the electrical conductivity is assumed to be infinite. This limit represents a rather good first approximation in astrophysical plasmas, where the conductivity is actually very large.[1] Under these

[1] To be more precise, in geometrized units, as the one adopted here, the induction equation reveals that the scalar component of the electrical conductivity tensor $\sigma_{ij} = \sigma\delta_{ij}$ can be expressed as the ratio between the Ohmic diffusion timescale $\tau_{\rm diff}$ and the (square of the) dynamical timescale $\tau_{\rm dyn}$ of the magnetic field, i.e., $\sigma \simeq \tau_{\rm diff}/\tau^2_{\rm dyn}$ [36]. In typical astrophysical plasmas, the dynamical timescales of the system are much shorter than the timescale associated with the diffusion of the the magnetic field so that it is reasonable to assume that the conductivity is actually infinite.

conditions, the electric charges are "infinitely effective" in canceling any electric field, which are therefore zero in the frame comoving with the fluid u^μ, i.e.,

$$F^{\mu\nu}u_\nu = 0\,. \tag{1.14}$$

The main consequence of the condition (1.14) is that the electric fields cease to be independent vector fields and can be obtained from simple algebraic expressions involving the fluid four-velocity and the magnetic fields. In particular, after defining the electric and magnetic four-vectors in the fluid frame as

$$e^\mu := F^{\mu\nu}u_\nu\,, \tag{1.15}$$
$$b^\mu :=^* F^{\mu\nu}u_\nu\,, \tag{1.16}$$

with the constraints that

$$e^\mu = 0\,, \tag{1.17}$$

and that the comoving magnetic field is fully spatial

$$u_\mu b^\mu = 0\,. \tag{1.18}$$

Under these conditions, the Faraday tensor can then be rewritten as

$$F^{\mu\nu} = -\sqrt{-g}\eta^{\mu\nu\lambda\delta}u_\lambda b_\delta\,, \tag{1.19}$$
$${}^*F^{\mu\nu} = b^\mu u^\nu - b^\nu u^\mu\,. \tag{1.20}$$

We can write the total energy-momentum tensor in terms of the vectors u^μ and b^μ [5] as

$$T^{\mu\nu} = \rho h_{\rm tot} u^\mu u^\nu + p_{\rm tot} g^{\mu\nu} - b^\mu b^\nu\,, \tag{1.21}$$

where we introduced total pressure

$$p_{\rm tot} := p + \frac{b^2}{2}\,, \tag{1.22}$$

which now includes the magnetic pressure

$$p_{\rm mag} := \frac{1}{2}b^\mu b_\mu =: \frac{1}{2}b^2\,, \tag{1.23}$$

while the total specific enthalpy is given by

$$h_{\rm tot} := h + \frac{b^2}{\rho}\,. \tag{1.24}$$

Note that the square of the magnetic-field strength in the fluid frame also satisfies the following identity $b^2 = B^2 - E^2 > 0$, which is sometimes used as a physical constraint in numerical evolutions [2, 68]. Finally, it is useful to express the current density $\mathscr{J}^\mu$ into two components, i.e., a "convection" (or advection) term and a "conduction" term

$$\mathscr{J}^\mu = qn^\mu + J^\mu, \tag{1.25}$$

where $q = -\mathscr{J}^\mu n_\mu$ is the charge density, qn^μ is the convection current and J^μ is conduction current, that is, the current density measured by the Eulerian observer and such that $J^\mu n_\mu = 0$.

In summary, the set of covariant GRMHD equations consists of the coupled system of two conversation laws (1.1)–(1.2) and of the two sets of Maxwell equations (1.9) and (1.10). However, as such, the corresponding system of equations is not closed and needs to be complemented with an equation of state that normally prescribes the behavior of the pressure as a function of the rest-mass density, specific internal energy (temperature) and particle abundances. The complexity of such equations of state varies enormously, from very simple and analytic ones—as those employed in simulations of accretion onto supermassive black holes—to very sophisticated and tabulated ones—as those employed in simulations of binary neutron stars. Chapter 3 of this book will be dedicated to a detailed discussion of the equations of state employed in modern numerical simulations.

However, more importantly, as presented here, Eqs. (1.1)–(1.2) and (1.9)–(1.10) are not particularly useful for a numerical solution in a simulation code; rather, they first need to be cast within a 3+1 decomposition of spacetime and then expressed in a conservative formulation, as we will discuss in the following two sections.

1.3 The 3+1 Decomposition of Spacetime

The intrinsically "covariant view" of Einstein's theory of general relativity is based on the concept that all coordinates are equivalent and, hence, the distinction between spatial and time coordinates is more an organizational matter than a strict requirement of the theory. Yet, our experience, and the laws of physics on sufficiently large scales, do suggest that a distinction between the time coordinate from the spatial ones is the most natural one in describing physical processes. Furthermore, such a distinction of time and space is the simplest way to exploit a large literature on the numerical solution of hyperbolic partial differential equations as those of relativistic MHD. Adopting this principle, and following closely the presentation already offered in Ref. [79], we "foliate" spacetime in terms of a set of non-intersecting spacelike hypersurfaces $\Sigma := \Sigma(t)$, each of which is parameterized by a constant value of the coordinate t. In this way, the three spatial coordinates are split from the one temporal coordinate and the resulting construction is called the 3+1 decomposition of spacetime [56].

Given one such constant-time hypersurface, Σ_t, belonging to the foliation Σ, we can introduce a timelike four-vector $\boldsymbol{n}$ normal to the hypersurface at each event in the spacetime and such that its dual one-form $\boldsymbol{\Omega} := \boldsymbol{\nabla} t$ is parallel to the gradient of the coordinate t, i.e.,

$$n_\mu = A\Omega_\mu = A\nabla_\mu t\,, \tag{1.26}$$

with $n_\mu = \{A, 0, 0, 0\}$ and A a constant to be determined. If we now require that the four-vector $\boldsymbol{n}$ defines an observer and thus that it measures the corresponding four-velocity, then from the normalization condition on timelike four-vectors, $n^\mu n_\mu = -1$, we find that

$$n^\mu n_\mu = g^{\mu\nu} n_\mu n_\nu = g^{tt} A^2 = -\frac{1}{\alpha^2} A^2 = -1\,, \tag{1.27}$$

where we have defined $\alpha^2 := -1/g^{tt}$. From the last equality in expression (1.27) it follows that $A = \pm\alpha$ and we will select $A = -\alpha$, such that the associated vector field n^μ is future directed. The quantity α is commonly referred to as the *lapse* function, it measures the rate of change of the coordinate time along the vector n^μ (see Fig. 1.1), and will be a building block of the metric in a 3+1 decomposition [cf., Eq. (1.37)].

The specification of the normal vector $\boldsymbol{n}$ allows us to define the metric associated to each hypersurface, i.e.,

$$\gamma_{\mu\nu} := g_{\mu\nu} + n_\mu n_\nu\,, \qquad\qquad \gamma^{\mu\nu} := g^{\mu\nu} + n^\mu n^\nu\,, \tag{1.28}$$

where $\gamma^{0\mu} = 0$, $\gamma_{ij} = g_{ij}$, but in general $\gamma^{ij} \neq g^{ij}$. Also note that $\gamma^{ik}\gamma_{kj} = \delta^i_{\ j}$, that is, γ^{ij} and γ_{ij} are the inverse of each other, so that the spatial metric $\boldsymbol{\gamma}$ can be used for raising and lowering the indices of purely spatial vectors and tensors.

The tensors $\boldsymbol{n}$ and $\boldsymbol{\gamma}$ provide us with two useful tools to decompose any four-dimensional tensor into a purely spatial part (hence contained in the hypersurface Σ_t) and a purely timelike part (hence orthogonal to Σ_t and aligned with $\boldsymbol{n}$). Not surprisingly, the spatial part is readily obtained after contracting with the *spatial projection operator* (or *spatial projection tensor*)

$$\gamma^\mu_{\ \nu} := g^{\mu\alpha}\gamma_{\alpha\nu} = g^\mu_{\ \nu} + n^\mu n_\nu = \delta^\mu_{\ \nu} + n^\mu n_\nu\,, \tag{1.29}$$

while the timelike part is obtained after contracting with the *time projection operator* (or *time projection tensor*)

$$N^\mu_{\ \nu} := -n^\mu n_\nu\,, \tag{1.30}$$

and where the two projectors are obviously orthogonal, i.e.,

$$\gamma^\alpha_{\ \mu} N^\mu_{\ \nu} = 0\,. \tag{1.31}$$

Hence, a generic four-vector $\boldsymbol{U}$ can be decomposed as

$$U^\mu = \gamma^\mu_{\ \nu} U^\nu + N^\mu_{\ \nu} U^\nu\,, \tag{1.32}$$

where the *purely spatial* part $\gamma^{\mu}_{\ \nu} U^{\nu} = V^{\mu}$ is still a four-vector that, by construction, has a zero contravariant time component, i.e., $V^t = 0$, whereas it has the covariant time component, $V_t = g_{\mu t} V^{\mu}$, which is nonzero in general. Analogous considerations can be done about tensors of any rank.

We have already seen in Eq. (1.26) that the unit normal $\boldsymbol{n}$ to a spacelike hypersurface Σ_t does not represent the direction along which the time coordinate changes, that is, it is not the direction of the time derivative. Indeed, if we compute the contraction of the two tensors we obtain

$$n^{\mu} \Omega_{\mu} = \frac{1}{A} n^{\mu} n_{\mu} = \frac{1}{\alpha} \neq 1 \,. \tag{1.33}$$

We can therefore introduce a new vector, $\boldsymbol{t}$, along which to carry out the time evolutions and that is dual to the surface one-form $\boldsymbol{\Omega}$. Such a vector is just the time-coordinate basis vector and is defined as the linear superposition of a purely temporal part (parallel to $\boldsymbol{n}$) and of a purely spatial one (orthogonal to $\boldsymbol{n}$), namely

$$\boldsymbol{t} = \boldsymbol{e}_t = \partial_t := \alpha \boldsymbol{n} + \boldsymbol{\beta} \,. \tag{1.34}$$

The purely spatial vector $\boldsymbol{\beta}$ [i.e., $\beta^{\mu} = (0, \beta^i)$] is usually referred to as the shift vector and will be another building block of the metric in a 3+1 decomposition [cf., Eq. (1.37)]. The decomposition of the vector $\boldsymbol{t}$ into a timelike component $\alpha \boldsymbol{n}$ and a spatial component $\boldsymbol{\beta}$ is shown in Fig. 1.1 (note that $\alpha = 1$, $\beta^i = 0$ in special relativity).

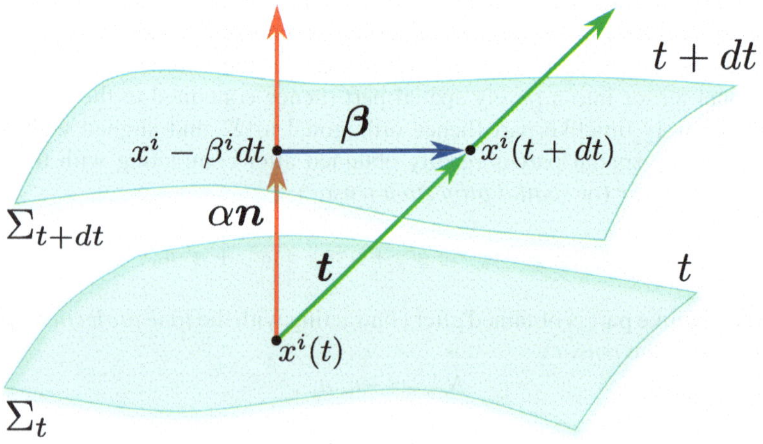

Fig. 1.1 Schematic representation of the 3+1 decomposition of spacetime with hypersurfaces of constant time coordinate Σ_t and Σ_{t+dt} foliating the spacetime. The four-vector $\boldsymbol{t}$ represents the direction of evolution of the time coordinate t and can be split into a timelike component $\alpha\boldsymbol{n}$, where $\boldsymbol{n}$ is a timelike unit normal to the hypersurface, and into a spacelike component, represented by the spacelike four-vector $\boldsymbol{\beta}$. The function α is the "lapse" and measures the proper time between adjacent hypersurfaces, while the components of the "shift" vector β^i measure the change of coordinates from one hypersurface to the subsequent one. Figure reproduced from [79]

We can check that $\boldsymbol{t}$ is a coordinate basis vector by verifying that

$$t^{\mu}\Omega_{\mu} = \alpha n^{\mu}\Omega_{\mu} + \beta^{\mu}\Omega_{\mu} = \frac{\alpha}{\alpha} = 1\,, \tag{1.35}$$

from which it follows that the vector $\boldsymbol{t}$ is effectively dual to the one-form $\boldsymbol{\Omega}$. This guarantees that the integral curves of t^{μ} are naturally parameterized by the time coordinate. As a result, all infinitesimal vectors t^{μ} originating on one hypersurface Σ_t would end up on the same hypersurface Σ_{t+dt}. Note that this is not guaranteed for translations along Ω_{μ} and that since $t^{\mu}t_{\mu} = g_{tt} = -\alpha^2 + \beta^{\mu}\beta_{\mu}$, the vector $\boldsymbol{t}$ is not necessarily timelike (the shift can in fact be superluminal).

Using the components of $\boldsymbol{n}$

$$n_{\mu} = (-\alpha, 0, 0, 0)\,, \qquad n^{\mu} = \frac{1}{\alpha}\left(1, -\beta^i\right)\,, \tag{1.36}$$

we can now express the generic *line element* in a 3+1 decomposition as

$$ds^2 = -(\alpha^2 - \beta_i\beta^i)dt^2 + 2\beta_i dx^i dt + \gamma_{ij}dx^i dx^j\,. \tag{1.37}$$

Expression (1.37) clearly emphasises that when $\beta^i = 0 = dx^i$, the lapse measures the proper time, $d\tau$, between two adjacent hypersurfaces, i.e.,

$$d\tau^2 = \alpha^2(t, x^j)dt^2\,, \tag{1.38}$$

while the shift vector measures the change of coordinates of a point from the hypersurface Σ_t to the hypersurface Σ_{t+dt}, i.e.,

$$x^i_{t+dt} = x^i_t - \beta^i(t, x^j)dt\,. \tag{1.39}$$

Similarly, the covariant and contravariant components of the metric (1.37) can be written explicitly as

$$g_{\mu\nu} = \begin{pmatrix} -\alpha^2 + \beta_i\beta^i & \beta_i \\ \beta_i & \gamma_{ij} \end{pmatrix}\,, \qquad g^{\mu\nu} = \begin{pmatrix} -1/\alpha^2 & \beta^i/\alpha^2 \\ \beta^i/\alpha^2 & \gamma^{ij} - \beta^i\beta^j/\alpha^2 \end{pmatrix}\,, \tag{1.40}$$

from which it is easy to obtain an important identity which will be used extensively hereafter, i.e.,

$$\sqrt{-g} = \alpha\sqrt{\gamma}\,, \tag{1.41}$$

where $g := \det(g_{\mu\nu})$ and $\gamma := \det(\gamma_{ij})$.

When defining the unit timelike normal $\boldsymbol{n}$ in Eq. (1.27), we have mentioned that it can be associated to the four-velocity of a special class of observers, which are referred to as *normal* or *Eulerian observers*. Although this denomination is somewhat confusing, since such observers are not at rest with respect to infinity but have a coordinate velocity $dx^i/dt = n^i = -\beta^i/\alpha$, we will adopt this traditional nomenclature also in the following and thus take an "Eulerian observer" as one with four-velocity given by (1.36).

When considering a fluid with four-velocity $\boldsymbol{u}$, the spatial four-velocity $\boldsymbol{v}$ measured by an Eulerian observer will be given by the ratio between the projection of $\boldsymbol{u}$ in the space orthogonal to $\boldsymbol{n}$, i.e., $\gamma^i{}_\mu u^\mu = u^i$, and the Lorentz factor of $\boldsymbol{u}$ as measured by $\boldsymbol{n}$ [20]

$$-n_\mu u^\mu = \alpha u^t \,. \tag{1.42}$$

As a result, the spatial four-velocity of a fluid as measured by an Eulerian observer will be given by

$$\boldsymbol{v} := \frac{\boldsymbol{\gamma} \cdot \boldsymbol{u}}{-\boldsymbol{n} \cdot \boldsymbol{u}}\,, \tag{1.43}$$

or, in component form, by

$$v^t = 0\,, \qquad v^i = \frac{\gamma^i{}_\mu u^\mu}{\alpha u^t} = \frac{1}{\alpha}\left(\frac{u^i}{u^t} + \beta^i\right), \tag{1.44}$$

$$v_t = \beta_i v^i\,, \qquad v_i = \frac{\gamma_{i\mu} u^\mu}{\alpha u^t} = \frac{u_i}{\alpha u^t} = \frac{\gamma_{ij}}{\alpha}\left(\frac{u^j}{u^t} + \beta^j\right). \tag{1.45}$$

Using now the normalisation condition $u^\mu u_\mu = -1$ and indicating as usual with W the Lorentz factor, we obtain

$$\alpha u^t = -\boldsymbol{n} \cdot \boldsymbol{u} = \frac{1}{\sqrt{1 - v^i v_i}} = W\,, \qquad u_t = W(-\alpha + \beta_i v^i)\,, \tag{1.46}$$

so that the components (1.44)–(1.45) can finally, be written as

$$v^i = \frac{u^i}{W} + \frac{\beta^i}{\alpha} = \frac{1}{\alpha}\left(\frac{u^i}{u^t} + \beta^i\right), \qquad v_i = \frac{u_i}{W} = \frac{u_i}{\alpha u^t}\,, \tag{1.47}$$

where in the last equality we have exploited the fact that $\gamma_{ij} u^j = u_i - \beta_i W/\alpha$. Finally, using expressions (1.43) and (1.46), it is also possible to write the fluid four-velocity as

$$u^\mu = W(n^\mu + v^\mu)\,, \tag{1.48}$$

which highlights the split of $\boldsymbol{u}$ into a temporal and a spatial part.

The three different unit four-vectors in a 3+1 decomposition of spacetime are shown in Fig. 1.2, which should be compared with Fig. 1.1. The four-vectors $\boldsymbol{n}$, $\boldsymbol{t}$ and $\boldsymbol{u}$ represent the unit timelike normal, the time-coordinate basis vector and the fluid four-velocity, respectively. Also shown are the associated worldlines, namely, the

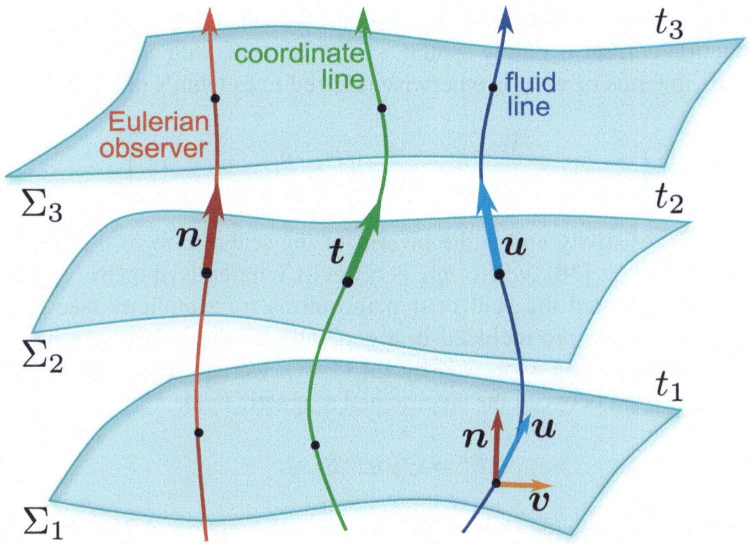

Fig. 1.2 Schematic representation of the different unit vectors in a 3+1 decomposition of spacetime (see Fig. 1.1). The four-vectors $\boldsymbol{n}$, $\boldsymbol{t}$, and $\boldsymbol{u}$ represent the unit timelike normal, the time-coordinate basis vector and the fluid four-velocity, respectively. It is shown the associated worldlines, namely, the normal line, the coordinate line, and the fluidline. The spatial projection $\boldsymbol{v}$ of the fluid four-velocity $\boldsymbol{u}$ as measured by the (Eulerian) normal observer $\boldsymbol{n}$ are also shown. Figure reproduced from [79]

normal line representing the worldline of an Eulerian observer, the coordinate line representing the worldline of a coordinate element, and the fluidline. The figure also reports the spatial projection $\boldsymbol{v}$ of the fluid four-velocity $\boldsymbol{u}$, thus highlighting that $\boldsymbol{v}$ is the three-velocity as measured by the normal observer.

What remains to be done at this point is to apply the 3+1 decomposition to the relevant vector fields that appear in the GRMHD equations. In particular, we start by expressing the Faraday tensor and the dual of the Faraday tensor in Maxwell's equations (1.9) and (1.10) respectively as

$$F^{\mu\nu} = n^{\mu} E^{\nu} - n^{\nu} E^{\mu} - \sqrt{-g}\eta^{\mu\nu\lambda\delta} n_{\lambda} B_{\delta} \,, \tag{1.49}$$

$$^{*}F^{\mu\nu} = n^{\mu} B^{\nu} - n^{\nu} B^{\mu} + \sqrt{-g}\eta^{\mu\nu\lambda\delta} n_{\lambda} E_{\delta} \,, \tag{1.50}$$

where the spatial components of the electric and magnetic fields measured by the Eulerian observer are given as

$$E^{i} := F^{i\nu} n_{\nu} = \alpha F^{it} \,, \qquad B^{i} :=^{*} F^{i\nu} n_{\nu} = \alpha^{*} F^{it} \,. \tag{1.51}$$

We should note that our definition of the electric and magnetic fields differs by a factor α from the corresponding definition used in Refs. [31, 44].

Going back to the definition of the total current density (1.25) and recalling that the conduction current is purely spatial, i.e., $J^{\mu}n_{\mu} = 0$, we can express its spatial components in terms of what is otherwise referred to as Ohm's law [12, 70],

$$J^i = qv^i + \frac{W}{\eta}\left[E^i + \frac{1}{\sqrt{\gamma}}\eta^{ijk}v_j B_k - (v_k E^k)v^i\right], \tag{1.52}$$

where η is the resistivity and is the inverse of the scalar term of the conductivity tensor, i.e., $\eta := 1/\sigma$ [36], where η_{ijk} is a Levi-Civita antisymmetric symbol, and where we have ignored the Hall or dynamo terms for simplicity (see Refs. [12, 70] where these terms are included in a generalized Ohm's law). Assuming now the ideal MHD condition expressed by Eq. (1.14), we can obtain the explicit and algebraic expression between the electric and magnetic fields

$$E^i = \sqrt{\gamma}\eta^{ijk}B_j v_k . \tag{1.53}$$

This result, which coincides with the equivalent expression in Newtonian MHD, underlines the passive role of dependent quantity for the electric field in the ideal-MHD equations.

Finally, using Eq. (1.20) together with (1.51), we can obtain the transformation between magnetic four-vector field in the fluid frame b^{μ} and the magnetic four-vector field in the Eulerian frame B^{μ} as

$$b^t = \frac{W}{\alpha}(B^i v_i), \qquad b^i = \frac{1}{W}(B^i + \alpha b^t u^i), \tag{1.54}$$

which allows us to express the dual Faraday tensor (1.20) in terms of the magnetic four-vector field in the Eulerian frame as

$${}^*F^{\mu\nu} = \frac{1}{W}(B^{\mu}u^{\nu} - B^{\nu}u^{\mu}), \tag{1.55}$$

and calculate the scalar b^2 as

$$b^2 = \frac{B^2 + \alpha^2(b^t)^2}{W^2} = \frac{B^2}{W^2} + (B^i v_i)^2, \tag{1.56}$$

where $B^2 := B^i B_i$.

1.4 Formulation of the GRMHD Equations for Numerical Simulations

1.4.1 Conservative Formulations

The equations of (relativistic) hydrodynamics and MHD can be written in the generic first-order-in-time form

$$\partial_t \boldsymbol{U} + \boldsymbol{A} \cdot \nabla \boldsymbol{U} = \boldsymbol{S}\,, \tag{1.57}$$

and the system above is said to be *hyperbolic* if the matrix of coefficients $\boldsymbol{A}$ is diagonalisable with a set of real eigenvalues, or *eigenspeeds*, $\lambda_1, \ldots, \lambda_N$ and a corresponding set of N linearly independent *right eigenvectors* $\boldsymbol{R}^{(1)}, \ldots, \boldsymbol{R}^{(m)}$, such that $\boldsymbol{A}\boldsymbol{R}^{(i)} = \lambda_i \boldsymbol{R}^{(i)}$, $\boldsymbol{\Lambda} := \boldsymbol{R}^{-1}\boldsymbol{A}\boldsymbol{R} = \mathrm{diag}(\lambda_1, \ldots, \lambda_N)$ is the diagonal matrix of eigenvalues and $\boldsymbol{R}$ the matrix of right eigenvectors.

The most important property of *hyperbolic equations* is that they *are well-posed*, hence suitable for numerical solution. Moreover, if the matrix $\boldsymbol{A}(\boldsymbol{U})$ is the Jacobian of a *flux vector* $\boldsymbol{F}(\boldsymbol{U})$ with respect to the state vector $\boldsymbol{U}$, namely if $\boldsymbol{A}(\boldsymbol{U}) := \partial \boldsymbol{F}/\partial \boldsymbol{U}$, then the homogeneous version of the system (1.57) can be written in *conservative form* as

$$\partial_t \boldsymbol{U} + \nabla \boldsymbol{F}(\boldsymbol{U}) = 0\,, \tag{1.58}$$

where $\boldsymbol{U}$ is therefore called the vector of *conserved variables*. With this definition in mind, we can now discuss two theorems underlining the importance of a conservative formulation. The first one loosely speaking states that: conservative numerical schemes,—that is, a numerical scheme based on the conservative formulation of the equations—if convergent, do converge to the weak solution of the problem [46]. The second theorem states instead that: *non-conservative schemes, i.e., schemes in which the equations are not written in the conservative form* (1.58)*, do not converge to the correct solution if a shock wave is present in the flow* [38]. In other words, the two theorems above state that if a conservative formulation *is used*, then we are guaranteed that the numerical solution will converge to the correct one, while if a conservative formulation *is not used*, we are guaranteed to converge to the incorrect solution in the likely event in which the flow develops a discontinuity.

1.4.2 The 3+1 Valencia Formulation(s)

Given the importance of a conservative formulation to obtain a well-posed system of equations, more than 30 years ago the group of Valencia started to develop Eulerian 3+1 formulations of the relativistic hydrodynamic and MHD equations written in conservative forms. Because of this, these formulations are often referred to as the "Valencia formulations". The first step in this direction was taken by considering the equations of special-relativistic hydrodynamics, which were cast into a conservative formulation [51] and solved in conjunction with High-Resolution Shock-Capturing (HRSC) methods (see, e.g., [79] for an introduction to HRSC methods). The second step was taken a few years later, when the equations of genera-relativistic hydrodynamics were cast into a conservative formulation [10] and have been employed in a variety of numerical simulations starting from Refs. [8, 30]. A final step, which is very relevant to the content of this chapter, was taken when the equations of GRMHD were cast in an Eulerian conservative formulation [7].

In what follows, we detail the steps that are needed in order to cast the covariant set ideal-GRMHD equations (1.1), (1.2), (1.9), (1.10) into the following 3+1 and Eulerian conservative form

$$\partial_t \left(\sqrt{\gamma}\boldsymbol{U}\right) + \partial_i \left(\sqrt{\gamma}\boldsymbol{F}^i\right) = \sqrt{\gamma}\boldsymbol{S}\,, \tag{1.59}$$

where $\boldsymbol{U}$ is the vector of conserved variables, $\boldsymbol{F}^i$ are the flux vectors (or "fluxes"), and $\boldsymbol{S}$ is the vector of source terms.

We can start from the covariant continuity equation (1.1), which can be written as

$$\begin{aligned}\nabla_\mu \left(\rho u^\mu\right) &= \frac{1}{\sqrt{-g}}\partial_\mu \left(\sqrt{-g}\rho u^\mu\right)\\ &= \frac{1}{\sqrt{-g}}\left[\partial_t \left(\sqrt{-g}\rho u^t\right) + \partial_i \left(\sqrt{-g}\rho u^i\right)\right] = 0\,,\end{aligned} \tag{1.60}$$

which, after introducing the conserved rest-mass density in the Eulerian frame,

$$D := \rho u^\mu n_\mu = \rho\alpha u^t = \rho W\,, \tag{1.61}$$

can be rewritten in the conservative form

$$\partial_t \left(\sqrt{\gamma}D\right) + \partial_i \left[\sqrt{\gamma}D\left(\alpha v^i - \beta^i\right)\right] = 0\,. \tag{1.62}$$

Equivalently, after defining the transport velocity $\mathcal{V}^i := \alpha v^i - \beta^i$, Eq. (1.62) can be further rewritten as

$$\partial_t \left(\sqrt{\gamma}D\right) + \partial_i \left[\sqrt{\gamma}D\mathcal{V}^i\right] = 0\,, \tag{1.63}$$

which represents the 3+1 conservative form of the continuity equation in a generic (curved) spacetime.

Before considering the 3+1 conservative form of the energy-momentum equations, it is useful to write the energy-momentum tensor in terms of quantities measured by the Eulerian observer. In particular, we can define the conserved total energy density $\mathcal{U}$ as the full projection of the energy-momentum tensor $T^{\mu\nu}$ along the unit normal $\boldsymbol{n}$ to the spatial hypersurface Σ_t, i.e.,

$$
\begin{aligned}
\mathscr{U} :=& n_\mu n_\nu T^{\mu\nu} \\
=& \rho h W^2 - p + \frac{1}{2}\left[B^2\left(1+v^2\right) - \left(B^j v_j\right)^2\right] .
\end{aligned} \tag{1.64}
$$

Similarly, the three-momentum density measured by the Eulerian observer is defined as the mixed parallel-transverse component of the energy-momentum tensor

$$
\begin{aligned}
S_i :=& \gamma^\mu_{\ i} n^\nu T_{\nu\mu} \\
=& \rho h W^2 v_i + B^2 v_i - (B^j v_j) B_i .
\end{aligned} \tag{1.65}
$$

while the purely spatial part of the energy-momentum tensor is given by

$$
\begin{aligned}
W^{ij} :=& \gamma^i_{\ \mu} \gamma^j_{\ \nu} T^{\mu\nu} \\
=& S^i v^j + p_{\rm tot} \gamma^{ij} - \frac{B^i B^j}{W^2} - (B^k v_k) v^i B^j .
\end{aligned} \tag{1.66}
$$

The corresponding four-dimensional definitions are given respectively by

$$
W^{\alpha\beta} := \gamma^\alpha_{\ \mu} \gamma^\beta_{\ \nu} T^{\mu\nu} , \quad \text{and} \quad S_\alpha := \gamma^\mu_{\ \alpha} n^\nu T_{\nu\mu} , \tag{1.67}
$$

and allow us to rewrite the energy-momentum tensor in its generic 3+1 decomposition

$$
T^{\mu\nu} = \mathscr{U} n^\mu n^\nu + S^\mu n^\nu + S^\nu n^\mu + W^{\mu\nu} . \tag{1.68}
$$

Next, recalling that the four-divergence of a symmetric rank-2 tensor is given by

$$
\nabla_\mu T^{\mu\nu} = g^{\nu\lambda}\left[\frac{1}{\sqrt{-g}} \partial_\mu \left(\sqrt{-g} T^\mu_{\ \lambda}\right) - \frac{1}{2} T^{\alpha\beta} \partial_\lambda g_{\alpha\beta}\right] , \tag{1.69}
$$

we can express the conservation of energy-momentum tensor (1.2) as

$$
\frac{1}{\sqrt{-g}} \partial_\mu \left(\sqrt{-g} T^\mu_{\ \nu}\right) = \frac{1}{2} T^{\mu\lambda} \partial_\nu g_{\mu\lambda} , \tag{1.70}
$$

and thus, using expression (1.68), obtain the 3+1 conservative form of the momentum-conservation equation in a curved spacetime

$$
\partial_t \left(\sqrt{\gamma} S_j\right) + \partial_i \left[\sqrt{\gamma}\left(\alpha W^i_j - \beta^i S_j\right)\right] = \frac{1}{2} \sqrt{-g} T^{\mu\nu} \partial_j T_{\mu\nu} . \tag{1.71}
$$

In a similar way, we can combine Eqs. (1.2) and (1.4) as

$$
\nabla_\mu \left(T^{\mu\nu} n_\nu\right) - T^{\mu\nu} \nabla_\mu n_\nu = 0 , \tag{1.72}
$$

and replacing $T^{\mu\nu}$ with its decomposed form (1.68) so that, after some algebra, we obtain the 3+1 conservative form of the energy-conservation equation in curved spacetime

$$\partial_t\left(\sqrt{\gamma}\,\mathscr{U}\right)+\partial_i\left[\sqrt{\gamma}\left(\alpha S^i-\mathscr{U}\beta^i\right)\right]=-\sqrt{-g}\,T^{\mu\nu}\nabla_\mu n_\nu\,. \tag{1.73}$$

Equations (1.71) and (1.73) both involve source terms on their right-hand sides. In particular, the source term of the momentum-conservation equation (1.71) can be written more explicitly as

$$\begin{aligned}\frac{1}{2}\sqrt{-g}\,T^{\mu\nu}\partial_j g_{\mu\nu}&=\sqrt{-g}\left(\frac{1}{2}W^{ik}\partial_j\gamma_{ik}+S^\mu n^\nu\partial_j g_{\mu\nu}+\frac{1}{2}\mathscr{U}n^\mu n^\nu\partial_j g_{\mu\nu}\right)\\&=\sqrt{-g}\left(\frac{1}{2}W^{ik}\partial_j\gamma_{ik}+\frac{1}{\alpha}S_i\partial_j\beta^i-\mathscr{U}\partial_j\ln\alpha\right),\end{aligned} \tag{1.74}$$

where we have used the following identities

$$\partial_j g_{\mu\nu}=\Gamma^\kappa_{\ j\nu}g_{\mu\kappa}+\Gamma^\kappa_{\ j\nu}g_{\kappa\nu}\,. \tag{1.75}$$

and

$$n^\mu\nabla_j n_\mu=0\,, \tag{1.76}$$

where $\Gamma^\alpha_{\ \beta\gamma}$ are the Christoffel symbols. Similarly, the source terms of the energy-conservation equation (1.73) can be expressed explicitly as

$$-\sqrt{-g}\,T^{\mu\nu}\nabla_\mu n_\nu=\sqrt{-g}\left(K_{ij}W^{ij}-S^i\partial_i\ln\alpha\right), \tag{1.77}$$

where $K_{\mu\nu}$ is the extrinsic curvature [1, 35, 79]. Expression (1.77) can be further simplified if the spacetime is stationary, in which case the term $\alpha K_{ik}W^{ik}$ reduces to

$$\alpha K_{ik}W^{ik}=\frac{1}{2}W^{ik}\beta^j\partial_j\gamma_{ik}+W^j_{\ i}\partial_j\beta^i\,. \tag{1.78}$$

We should note that the expressions for the right-hand side of the energy and momentum conservation equations are due to Ref. [23] and do not correspond to the ones originally presented in Ref. [10], which expressed the four divergences of the energy-momentum tensor as

$$\nabla_\mu T^{\mu\nu}=\frac{1}{\sqrt{-g}}\partial_\mu\left(\sqrt{-g}\,T^{\mu\nu}\right)+\Gamma^\nu_{\ \mu\lambda}T^{\mu\lambda}\,. \tag{1.79}$$

While this expression is mathematically equivalent to Eq. (1.69), it leads to more complex expressions for the right-hand sides when expressed in a 3+1 decomposition.

The final equation needed to complete the set of GRMHD equations in 3+1 conservative form is relative to the electromagnetic sector and, in particular, Faraday's law of induction. Using Eq. (1.10) and the definition of the dual Faraday tensor, it is possible to rewrite Eq. (1.20) in the 3+1 equivalent form

$$\partial_t \sqrt{\gamma} B^j + \partial_i \left[\sqrt{\gamma} B^j \left(\alpha v^i - \beta^i\right) - B^i \left(\alpha v^j - \beta^j\right)\right] = 0\,, \tag{1.80}$$

or, using the transport velocity, as

$$\partial_t \sqrt{\gamma} B^j + \partial_i \left[\sqrt{\gamma} \left(B^j \mathscr{V}^i - B^i \mathscr{V}^j\right)\right] = 0\,. \tag{1.81}$$

Equation (1.81) is also known as the general-relativistic induction equation and highlights how the evolution of the magnetic field is directly related to the curl of the vector product of the three-velocity and of the magnetic field, as in Newtonian MHD. Once the magnetic field is known, the electric field follows trivially from Eq. (1.53).

At this point, it is possible to collect Eqs. (1.62), (1.71), (1.73), and (1.80), so as to build the state vector, the fluxes and the source vector and hence write the ideal-GRMHD equations in the conservative Valencia formulation (1.59) [7]. More specifically, after some simple algebra, it is not difficult to derive the following expressions for the state and flux vectors

$$\boldsymbol{U} = \begin{pmatrix} D \\ \\ S_j \\ \\ \mathscr{U} \\ \\ B^j \end{pmatrix}, \qquad \boldsymbol{F}^i = \begin{pmatrix} D\mathscr{V}^i \\ \\ \alpha W^i{}_j - \beta^i S_j \\ \\ \alpha S^i - \beta^i \mathscr{U} \\ \\ B^j \mathscr{V}^i - B^i \mathscr{V}^j \end{pmatrix}, \tag{1.82}$$

while the source vector has components

$$\boldsymbol{S} = \begin{pmatrix} 0 \\ \\ \frac{1}{2}\alpha W^{ik} \partial_j \gamma_{ik} + S_i \partial_j \beta^i - \mathscr{U} \partial_j \alpha \\ \\ \frac{1}{2} W^{ik} \beta^j \partial_j \gamma_{ik} + W_i^j \partial_j \beta^i - S^j \partial_j \alpha \\ \\ 0 \end{pmatrix}. \tag{1.83}$$

We note that since the linear combination of a conserved variable is still a conserved variable, the state vector of conserved variables (1.82) is not unique and different codes implement different definitions. A particularly common choice is to replace the conserved total energy density $\mathscr{U}$ with its equivalent

$$\tau := \mathscr{U} - D = \rho W(hW - 1) - p + \frac{1}{2}[B^2(1 + v^2) - (B_j v^j)^2]\,. \tag{1.84}$$

so that the corresponding state and flux vectors are given by

$$\boldsymbol{U} = \begin{pmatrix} D \\ S_j \\ \tau \\ B^j \end{pmatrix}, \qquad \boldsymbol{F}^i = \begin{pmatrix} D\mathcal{V}^i \\ \alpha W^i_j - \beta^i S_j \\ \alpha(S^i - v^i D) - \beta^i \tau \\ B^j \mathcal{V}^i - B^i \mathcal{V}^j \end{pmatrix}, \tag{1.85}$$

while the source terms are not changed. Obviously, expressions (1.82) and (1.85) are mathematically equivalent, but their numerical implementation has shown that the latter systematically leads to more accurate evolutions.

A few remarks before concluding this section. First, in stark contrast with what happens for the conservative formulation of the Newtonian MHD equations (i.e., when $v^2 \ll 1$, $p \ll \rho$, and $E^2 \ll B^2 \ll \rho$), in GRMHD (but already in general-relativistic hydrodynamics) the relation between the primitive variables, i.e., ρ, v^i, ε (or p), and B^i, and the conserved variables, i.e., D, S_j, τ, and B^j, is not analytic. In other words, the calculation of the primitive variables from the conserved one, i.e., what is normally referred to as the "primitive recovery" procedure, cannot be done analytically but requires the use of a multidimensional root-finding approach. The latter needs to be implemented at each numerical cell of the computational domain and can be particularly complex when tabulated equations of state are employed. Over the years, various algorithms have been developed to ensure an accurate, efficient, and stable primitive recovery, aiming at minimizing error accumulation during the matter evolution. For compactness, we will not discuss here the feature of such algorithms, but detailed discussions and comparisons of different algorithms have been studied in Refs. [27, 42, 65, 84] with specific focus on their accuracy and robustness. Second, the set of evolution equations (1.59) does not include another important equation that needs to be solved together such a set and that represents one of the most important ones in the actual solution of the GRMHD equations, namely, the divergence-free condition (1.90). Also in this case, we will not discuss here the numerous numerical approaches that are possible to limit or prevent the growth of the violations of the divergence-free condition—from the divergence-cleaning methods, over to the evolution of the vector potential and up to the sophisticated constrained-transport methods—and refer the interested reader to Ref. [66] for a recent review of the various methods, and to the following chapters in this book. Finally, in the absence of gravity and thus in flat spacetimes, the corresponding form of the special relativistic MHD equations can be obtained trivially after setting $\alpha = 1$, $\beta^i = 0$, and $\sqrt{\gamma} = 1$. The corresponding set of equations is widely employed in particle physics to simulate the dynamics of the collisions of heavy ions and some representative examples can be found in Refs. [39, 53].

1.4.3 General-Relativistic Resistive MHD Equations

As mentioned in the Introduction, the electrical conductivity in astrophysical plasmas is extremely high and the ideal-MHD condition of infinite conductivity represents a very good approximation. In this case, the magnetic flux is conserved and the magnetic field is frozen in the fluid, being simply advected with it, and the electric field is trivially obtained from expression (1.53). By construction, therefore, the solution of the ideal-GRMHD equations neglects any effect of resistivity on the dynamics. In practice, however, even in astrophysical plasmas there will be spatial regions with very high temperatures where the electrical conductivity is finite and the resistive effects, most notably, the creation of current sheets and the consequent reconnection, will play a role. Such effects are expected to take place, for example, during the merger of two magnetized neutron stars or in accretion disks onto supermassive black holes, and could provide an important contribution to the energy losses from the system.

In all of these scenarios, the ideal-MHD limit may not be sufficient to study those physical processes that involve reconnection or the presence of anisotropic resistivities. This has motivated the derivation of a more extended set of GRMHD equations that accounts for resistive effects and that, in practice is augmented by an evolution equation for the electric field and by an extended Ohm's law. The corresponding system is normally referred to as the set of general-relativistic resistive MHD (GRRMHD) equations, and we refer to Ref. [70], where this system was first presented.

Since the changes with respect to the GRMHD equations take place only in the electromagnetic sector, the conservation equations of rest-mass (1.62), momentum (1.71), and energy (1.73) remain unchanged. On the other hand, the terms involved in the basic decomposition of the energy-momentum tensor (1.68) are modified because of the explicitly appearance of the electric fields and take the form

$$\mathscr{U} := n_\mu n_\nu T^{\mu\nu} = \rho h W^2 - p + \frac{1}{2}(E^2 + B^2)\,, \tag{1.86}$$

$$S_i := \gamma^\mu_{\ i} n^\alpha T_{\alpha\mu} = \rho h W^2 v_i + \sqrt{\gamma}\eta_{ijk} E^k B^k\,, \tag{1.87}$$

$$W^{ij} := \gamma^i_{\ \mu}\gamma^j_{\ \nu} T^{\mu\nu} = \rho h W^2 v^i v^j - E^i E^j - B^i B^j + \left[p + \frac{1}{2}(E^2 + B^2)\right]\gamma^{ij}\,. \tag{1.88}$$

The new expressions for the electromagnetic sector involve the evolution equation for the magnetic field (1.10) (Faraday's induction law), whose modified expression takes into account also the contributions from the electric fields

$$\partial_t \sqrt{\gamma} B^j + \partial_i \left[\sqrt{\gamma}\left(\beta^j B^i - \beta^i B^j + \frac{1}{\sqrt{\gamma}}\eta^{ijk}\alpha E_k\right)\right] = 0\,. \tag{1.89}$$

and the corresponding divergence-free constraint following from the temporal component of Eq. (1.10)

$$\frac{1}{\sqrt{\gamma}}\partial_i\sqrt{\gamma}B^i = 0\,. \tag{1.90}$$

In addition, Ampere's law for the evolution of the electric field follows from the spatial part of the second couple of Maxwell equations (1.10) and is given by

$$\partial_t\sqrt{\gamma}E^j + \partial_i\left[\sqrt{\gamma}\left(\beta^j E^i - \beta^i E^j - \frac{1}{\sqrt{\gamma}}\eta^{ijk}\alpha B_k\right)\right] = -\sqrt{\gamma}(\alpha J^i - q\beta^j)\,, \tag{1.91}$$

while the temporal component of (1.10) expresses the charge density in terms of the divergence of the electric field, i.e.,

$$\frac{1}{\sqrt{\gamma}}\partial_i\left(\sqrt{\gamma}E^i\right) = q\,. \tag{1.92}$$

Substituting now Eqs. (1.52) and (1.92) in Ampere's law (1.91), we can rewrite it as

$$\begin{aligned}\partial_t\sqrt{\gamma}E^j &+ \partial_i\left[\sqrt{\gamma}\left(\beta^j E^i - \beta^i E^j - \frac{1}{\sqrt{\gamma}}\eta^{ijk}\alpha B_k\right)\right]\\ &= -\sqrt{\gamma}\frac{\alpha W}{\eta}\left[E^i + \frac{1}{\sqrt{\gamma}}\eta^{ijk}v_j B_k - \left(v_k E^k\right)v^i\right] - \left(\alpha v^j - \beta^j\right)\partial_j\left(\sqrt{\gamma}E^j\right)\,.\end{aligned} \tag{1.93}$$

At this point, we can finally collect Eqs. (1.62), (1.71), (1.73), (1.89), and (1.91), so as to obtain the conservative form (1.59) of the GRRMHD equations. The corresponding augmented state and flux vectors are given by

$$\boldsymbol{U} = \begin{pmatrix} D \\ S_j \\ \mathcal{U} \\ B^j \\ E^j \end{pmatrix}, \qquad \boldsymbol{F}^i = \begin{pmatrix} D\mathcal{V}^i \\ \alpha W^i_j - \beta^i S_j \\ \alpha S^i - \beta^i\mathcal{U} \\ \beta^j B^i - \beta^i B^j + \alpha\eta^{ijk}E_k/\sqrt{\gamma} \\ \beta^j E^i - \beta^i E^j - \alpha\eta^{ijk}B_k/\sqrt{\gamma} \end{pmatrix}, \tag{1.94}$$

while the source terms are

$$
S = \begin{pmatrix} 0 \\ \frac{1}{2}\alpha W^{ik}\partial_j \gamma_{ik} + S_i \partial_j \beta^i - \mathscr{U}\partial_j \alpha \\ \frac{1}{2}W^{ik}\beta^j \partial_j \gamma_{ik} + W_i^j \partial_j \beta^i - S^j \partial_j \alpha \\ 0 \\ -\alpha J^j + \beta^j q \end{pmatrix}. \tag{1.95}
$$

Note that, as for the set of GRMHD equations, also the set of GRRMHD equations is not closed and requires the specification of an equation of state. Similarly, a complex primitive-recovery approach is needed to compute at each numerical cell the primitive variables from the conserved ones (see, e.g., Ref. [80]). What is new, however, is that Ohm's law (1.52) does not provide any information on the properties of the resistivity, which will in general be a function of space and time, i.e., $\eta = \eta(x^i, t)$. In principle, the information about the properties of the resistivity should follow from microphysical considerations and hence be calculated under specific physical conditions (see, e.g., [36] in the case of neutron-star matter). In practice, however, given the poor knowledge of the resistive aspects of astrophysical plasmas, much cruder choices are made that model the resistivity as a constant or as a simple function of the rest-mass density (see, e.g., [25, 67]). Finally, it should be mentioned that the solution of the GRRMHD equations is considerably more challenging that those of the GRMHD equations. For the former set, in fact, the equations become mixed hyperbolic-parabolic in Newtonian physics or hyperbolic with stiff relaxation terms in special relativity. The appearance of stiff terms in the equations follows from the fact that the diffusive effects take place on timescales that are intrinsically larger than the dynamical one or, in other words, from the fact the relaxation terms dominate over the purely hyperbolic ones, posing severe constraints on the timestep for the evolution (similar difficulties arise when considering the inclusion of radiative effects; see, e.g., Refs. [61, 75, 93]). Overall, the presence of stiff terms forces the use of specially designed numerical methods such as the implicit-explicit Runge–Kutta time-integration schemes or RK-IMEX [71], and hence a significant extension of the numerical infrastructure (see, e.g., Refs. [67, 80]).

1.5 Summary

The equations of MHD represent a very effective tool to describe the dynamics of astrophysical plasmas and because of this they are employed in a variety of scenarios in astrophysics and cosmology. When applied in their GRMHD form to study astrophysical compact objects, they are the most accurate and powerful tool to explore the properties of compact black holes and neutron stars in fully dynamical and nonlinear regimes. In this chapter, we have presented the MHD equations in generic and curved spacetimes, either in the ideal-MHD limit (GRMHD equations) or in the presence of resistivity (GRRMHD equations). We have also discussed the basic aspects of a 3+1 decomposition of spacetime and the importance of writing the MHD equations in a conservative form.

The purpose of this chapter—together with Chap. 2, which covers the coupling of the MHD equations with the Einstein equations, and of Chap. 3, which reviews the properties of modern equations of state—is to lay out the mathematical foundations of the set of equations that will be employed extensively in the rest of the book and which have a more applied nature. In particular, after an introduction of some representative GRMHD codes in Part II, Part III and IV are dedicated to the study of the dynamics of a variety of astrophysical plasmas, either when the spacetime is held fixed because the plasmas are not self-gravitating (Part III) or when the spacetime is dynamically evolved because the plasmas are self-gravitating (Part IV).

Acknowledgements Y. M. is supported by the Shanghai Municipality orientation program of Basic Research for International Scientists (Grant No. 22JC1410600), the National Natural Science Foundation of China (Grant No. 12273022), and the National Key R&D Program of China (Grant No. 2023YFE0101200). L.R. is supported in part from the State of Hesse within the Research Cluster ELEMENTS (Project ID 500/10.006), by the ERC Advanced Grant "JETSET: Launching, propagation and emission of relativistic jets from binary mergers and across mass scales" (Grant No. 884631), and by the Walter Greiner Gesellschaft zur Förderung der physikalischen Grundlagenforschung e.V. through the Carl W. Fueck Laureatus Chair.

References

1. M. Alcubierre, *Introduction to* 3 + 1 *Numerical Relativity* (Oxford University Press, Oxford, UK, 2008)
2. D. Alic, P. Moesta, L. Rezzolla, O. Zanotti, J.L. Jaramillo, Accurate simulations of binary black hole mergers in force-free electrodynamics. Astrophys. J. **754**, 36 (2012)
3. M. Anderson, E.W. Hirschmann, L. Lehner, S.L. Liebling, P.M. Motl, D. Neilsen, C. Palenzuela, J.E. Tohline, Magnetized neutron-star mergers and gravitational-wave signals. Phys. Rev. Lett. **100**(19), 191101 (2008)
4. M. Anderson, E.W. Hirschmann, S.L. Liebling, D. Neilsen, Relativistic MHD with adaptive mesh refinement. Class. Quantum Grav. **23**, 6503–6524 (2006)
5. A.M. Anile, *Relativistic Fluids and Magneto-fluids* (Cambridge University Press, Cambridge, UK, 1990)
6. P. Anninos, P.C. Fragile, J.D. Salmonson, Cosmos++: relativistic magnetohydrodynamics on unstructured grids with local adaptive refinement. Astrophys. J. **635**, 723 (2005)
7. L. Antón, O. Zanotti, J.A. Miralles, J.M. Martí, J.M. Ibáñez, J.A. Font, J.A. Pons, Numerical 3+1 general relativistic magnetohydrodynamics: a local characteristic approach. Astrophys. J. **637**, 296–312 (2006)
8. L. Baiotti, I. Hawke, P.J. Montero, F. Löffler, L. Rezzolla, N. Stergioulas, J.A. Font, E. Seidel, Three-dimensional relativistic simulations of rotating neutron-star collapse to a Kerr black hole. Phys. Rev. D **71**(2), 024035 (2005)
9. D.S. Balsara, Higher-order accurate space-time schemes for computational astrophysics–Part I: finite volume methods. Living Rev. Comput. Astrophys. **3**(1), 2 (2017)

10. F. Banyuls, J.A. Font, J.M. Ibáñez, J.M. Martí, J.A. Miralles, Numerical 3+1 general-relativistic hydrodynamics: a local characteristic approach. Astrophys. J. **476**, 221 (1997)
11. D. Bégué, A. Pe'er, G.Q. Zhang, B.B. Zhang, B. Pevzner, cuHARM: a New GPU-accelerated GRMHD code and its application to ADAF disks. Astrophys. J. Supp. **264**(2), 32 (2023)
12. N. Bucciantini, L. Del Zanna, A fully covariant mean-field dynamo closure for numerical 3 + 1 resistive GRMHD. Mon. Not. R. Astron. Soc. **428**, 71–85 (2013)
13. P. Cerdá-Durán, J.A. Font, L. Antón, E. Müller, A new general relativistic magnetohydrodynamics code for dynamical spacetimes. Astron. Astrophys. **492**, 937–953 (2008)
14. M. Chandra, F. Foucart, C.F. Gammie, grim: a Flexible, Conservative scheme for relativistic fluid theories. Astrophys. J. **837**(1), 92 (2017)
15. M. Chandra, C.F. Gammie, F. Foucart, E. Quataert, An extended magnetohydrodynamics model for relativistic weakly collisional plasmas. Astrophys. J. **810**(2), 162 (2015)
16. P.C.-K. Cheong, A.T.-L. Lam, H.H.-Y. Ng, T.G.F. Li, Gmunu: paralleled, grid-adaptive, general-relativistic magnetohydrodynamics in curvilinear geometries in dynamical spacetimes. Mon. Not. R. Astron. Soc. **508**(2), 2279–2301 (2021)
17. R. Ciolfi, The key role of magnetic fields in binary neutron star mergers. Gen. Relativ. Gravit. **52**(6), 59 (2020)
18. F. Cipolletta, J.V. Kalinani, B. Giacomazzo, R. Ciolfi, Spritz: a new fully general-relativistic magnetohydrodynamic code. Class. Quantum Grav. **37**(13), 135010 (2020)
19. A. Cruz-Osorio, C.M. Fromm, Y. Mizuno, A. Nathanail, Z. Younsi, O. Porth, J. Davelaar, H. Falcke, M. Kramer, L. Rezzolla, State-of-the-art energetic and morphological modelling of the launching site of the M87 jet. Nat. Astron. **6**, 103–108 (2022)
20. F. de Felice, C.J.S. Clarke, *Relativity on Curved Manifolds* (Cambridge University Press, 1990)
21. J.-P. De Villiers, J.F. Hawley, A numerical method for general relativistic magnetohydrodynamics. Astrophys. J. **589**, 458–480 (2003)
22. L. Del Zanna, N. Bucciantini, Covariant and 3 + 1 equations for dynamo-chiral general relativistic magnetohydrodynamics. Mon. Not. R. Astron. Soc. **479**(1), 657–666 (2018)
23. L. Del Zanna, O. Zanotti, N. Bucciantini, P. Londrillo, ECHO: a Eulerian conservative high-order scheme for general relativistic magnetohydrodynamics and magnetodynamics. Astron. Astrophys. **473**, 11–30 (2007)
24. K. Dionysopoulou, D. Alic, C. Palenzuela, L. Rezzolla, B. Giacomazzo, General-relativistic resistive magnetohydrodynamics in three dimensions: formulation and tests. Phys. Rev. D **88**(4), 044020 (2013)
25. K. Dionysopoulou, D. Alic, C. Palenzuela, L. Rezzolla, B. Giacomazzo, General-relativistic resistive magnetohydrodynamics in three dimensions: Formulation and tests. Phys. Rev. D **88**(4), 044020 (2013)
26. M.D. Duez, Y.T. Liu, S.L. Shapiro, B.C. Stephens, Relativistic magnetohydrodynamics in dynamical spacetimes: numerical methods and tests. Phys. Rev. D **72**(2), 024028 (2005)
27. P.L. Espino, G. Bozzola, V. Paschalidis, Quantifying uncertainties in general relativistic magnetohydrodynamic codes (2022). arXiv:2210.13481
28. Z.B. Etienne, V. Paschalidis, R. Haas, P. Mösta, S.L. Shapiro, IllinoisGRMHD: an open-source, user-friendly GRMHD code for dynamical spacetimes. Class. Quantum Grav. **32**(17), 175009 (2015)
29. J.A. Font, Numerical hydrodynamics in general relativity. Living Rev. Relativ. **6**, 4 (2003)
30. J.A. Font, T. Goodale, S. Iyer, M. Miller, L. Rezzolla, E. Seidel, N. Stergioulas, W.-M. Suen, M. Tobias, Three-dimensional numerical general relativistic hydrodynamics. II. Long-term dynamics of single relativistic stars. Phys. Rev. D **65**(8), 084024 (2002)
31. C.F. Gammie, J.C. McKinney, G. Tóth, HARM: a numerical scheme for general relativistic magnetohydrodynamics. Astrophys. J. **589**(1), 444–457 (2003)
32. B. Giacomazzo, L. Rezzolla, WhiskyMHD: a new numerical code for general relativistic magnetohydrodynamics. Class. Quantum Grav. **24**, 235 (2007)
33. B. Giacomazzo, L. Rezzolla, The exact solution of the Riemann problem in relativistic MHD. J. Fluid Mech. **562**, 223–259 (2006)

34. O. Gottlieb, S. Moseley, T. Ramirez-Aguilar, A. Murguia-Berthier, M. Liska, A. Tchekhovskoy, On the Jet-Ejecta interaction in 3D GRMHD simulations of a binary neutron star merger aftermath. Astrophys. J. Lett. **933**(1), L2 (2022)
35. E. Gourgoulhon, *3+1 Formalism in General Relativity*, Lecture Notes in Physics, vol. 846. (Springer, Berlin, 2012)
36. A. Harutyunyan, A. Nathanail, L. Rezzolla, A. Sedrakian, Electrical resistivity and Hall effect in binary neutron star mergers. Eur. Phys. J. A **54**(11), 191 (2018)
37. J.F. Hawley, L.L. Smarr, J.R. Wilson, A numerical study of nonspherical black hole accretion. I Equations and test problems. Astrophys. J. **277**, 296–311 (1984)
38. T.Y. Hou, P.G. LeFloch, Why nonconservative schemes converge to wrong solutions: error analysis. Math. Comp. **62**, 497–530 (1994)
39. G. Inghirami, L. Del Zanna, A. Beraudo, M. Haddadi Moghaddam, F. Becattini, M. Bleicher, Magneto-hydrodynamic simulations of Heavy Ion Collisions with ECHO-QGP. J. Phys. Conf. Ser. **1024**, 012043 (2018)
40. B. James, A. Janiuk, F.H. Nouri, Modeling the Gamma-Ray burst jet properties with 3D general relativistic simulations of magnetically arrested accretion flows. Astrophys. J. **935**(2), 176 (2022)
41. A. Janiuk, The r-process nucleosynthesis in the outflows from short GRB accretion disks. Astrophys. J. **882**(2), 163 (2019)
42. W. Kastaun, J.V. Kalinani, R. Ciolfi, Robust recovery of primitive variables in relativistic ideal magnetohydrodynamics. Phys. Rev. D **103**(2), 023018 (2021)
43. K. Kiuchi, Y. Sekiguchi, K. Kyutoku, M. Shibata, K. Taniguchi, T. Wada, High resolution magnetohydrodynamic simulation of black hole-neutron star merger: mass ejection and short gamma ray bursts. Phys. Rev. D **92**(6), 064034 (2015)
44. S.S. Komissarov, A Godunov-type scheme for relativistic magnetohydrodynamics. Mon. Not. R. Astron. Soc. **303**, 343–366 (1999)
45. S. Koide, D.L. Meier, K. Shibata, T. Kudoh, General relativistic simulation of early jet formation in a rapidly rotating black hole magnetosphere. Astrophys. J. **536**, 668–674 (2000)
46. P.D. Lax, B. Wendroff, Systems of conservation laws. Commun. Pure Appl. Math. **13**, 217–237 (1960)
47. R.J. Leveque, *Finite Volume Methods for Hyperbolic Problems* (Cambridge University Press, New York, 2002)
48. M. Liska, C. Hesp, A. Tchekhovskoy, A. Ingram, M. van der Klis, S. Markoff, Formation of precessing jets by tilted black hole discs in 3d general relativistic mhd simulations. Mon. Not. R. Astron. Soc. **474**(1), L81–L85 (2018)
49. M. Liska, K. Chatterjee, A. Tchekhovskoy, D. Yoon, D. van Eijnatten, C. Hesp, S. Markoff, A. Ingram, M. van der Klis, H-AMR: a new GPU-accelerated GRMHD code for exascale computing with 3D adaptive mesh refinement and local adaptive time-stepping (2019). arXiv:1912.10192
50. Y.T. Liu, S.L. Shapiro, Z.B. Etienne, K. Taniguchi, General relativistic simulations of magnetized binary neutron star mergers. Phys. Rev. D **78**, 024012 (2008)
51. J.M. Martí, J.M. Ibáñez, J.A. Miralles, Numerical relativistic hydrodynamics: local characteristic approach. Phys. Rev. D **43**, 3794 (1991)
52. J.M. Martí, E. Müller, Grid-based methods in relativistic hydrodynamics and magnetohydrodynamics. Living Rev. Comput. Astrophys. 1 (2015)
53. M. Mayer, L. Rezzolla, H. Elfner, G. Inghirami, D.H. Rischke, BHAC-QGP: three-dimensional MHD simulations of relativistic heavy-ion collisions, II. Application to Au-Au collisions (2024). arXiv:2403.08669
54. J.C. McKinney, General relativistic magnetohydrodynamic simulations of the jet formation and large-scale propagation from black hole accretion systems. Mon. Not. R. Astron. Soc. **368**, 1561–1582 (2006)
55. J.C. McKinney, A. Tchekhovskoy, A. Sadowski, R. Narayan, Three-dimensional general relativistic radiation magnetohydrodynamical simulation of super-Eddington accretion, using a new code HARMRAD with M1 closure. Mon. Not. R. Astron. Soc. **441**, 3177–3208 (2014)

56. C.W. Misner, K.S. Thorne, J.A. Wheeler, *Gravitation* (W. H. Freeman, San Francisco, 1973)
57. Y. Mizuno, K.-I. Nishikawa, S. Koide, P. Hardee, G.J. Fishman, RAISHIN: a high-resolution three-dimensional general relativistic magnetohydrodynamics code. *ArXiv Astrophysics e-prints* (2006)
58. Y. Mizuno, S. Yamada, S. Koide, K. Shibata, General relativistic magnetohydrodynamic simulations of collapsars: rotating black hole cases. Astrophys. J. **606**, 395 (2004)
59. Y. Mizuno, Z. Younsi, C.M. Fromm, O. Porth, M. De Laurentis, H. Olivares, H. Falcke, M. Kramer, L. Rezzolla, The current ability to test theories of gravity with black hole shadows. Nat. Astron. **2**, 585–590 (2018)
60. A. Murguia-Berthier, E. Ramirez-Ruiz, G. Montes, F. De Colle, L. Rezzolla, S. Rosswog, K. Takami, A. Perego, W.H. Lee, The properties of short Gamma-Ray burst jets triggered by neutron star mergers. Astrophys. J. Lett. **835**, L34 (2016)
61. C. Musolino, L. Rezzolla, A practical guide to a moment approach for neutrino transport in numerical relativity (2023). arXiv:2304.09168
62. R. Narayan, A. Sądowski, R.F. Penna, A.K. Kulkarni, GRMHD simulations of magnetized advection-dominated accretion on a non-spinning black hole: role of outflows. Mon. Not. R. Astron. Soc. **426**, 3241–3259 (2012)
63. A. Nathanail, R. Gill, O. Porth, C.M. Fromm, L. Rezzolla, On the opening angle of magnetized jets from neutron-star mergers: the case of GRB170817A. Mon. Not. R. Astron. Soc. **495**(4), 3780–3787 (2020)
64. A. Nathanail, R. Gill, O. Porth, C.M. Fromm, L. Rezzolla, 3D magnetized jet break-out from neutron-star binary merger ejecta: afterglow emission from the jet and the ejecta. Mon. Not. R. Astron. Soc. **502**(2), 1843–1855 (2021)
65. H.H.-Y. Ng, J.-L. Jiang, C. Musolino, C. Ecker, S.D. Tootle, L. Rezzolla, Hybrid approach to long-term binary neutron-star simulations. Phys. Rev. D **109**(6), 064061 (2024)
66. H. Olivares, O. Porth, J. Davelaar, E.R. Most, C.M. Fromm, Y. Mizuno, Z. Younsi, L. Rezzolla, Constrained transport and adaptive mesh refinement in the Black Hole Accretion Code. Astron. Astrophys. **629**, A61 (2019)
67. C. Palenzuela, Modelling magnetized neutron stars using resistive magnetohydrodynamics. Mon. Not. R. Astron. Soc. **431**, 1853–1865 (2013)
68. C. Palenzuela, T. Garrett, L. Lehner, S.L. Liebling, Magnetospheres of black hole systems in force-free plasma. Phys. Rev. D **82**(4), 044045 (2010)
69. C. Palenzuela, L. Lehner, M. Ponce, S.L. Liebling, M. Anderson, D. Neilsen, P. Motl, Electromagnetic and gravitational outputs from binary-neutron-star coalescence. Phys. Rev. Lett. **111**(6), 061105 (2013)
70. C. Palenzuela, L. Lehner, O. Reula, L. Rezzolla, Beyond ideal MHD: towards a more realistic modelling of relativistic astrophysical plasmas. Mon. Not. R. Astron. Soc. **394**, 1727–1740 (2009)
71. L. Pareschi, G. Russo, Implicit-explicit runge-kutta schemes and applications to hyperbolic systems with relaxation. J. Sci. Comput. **25**(1), 129 (2005)
72. O. Porth, H. Olivares, Y. Mizuno, Z. Younsi, L. Rezzolla, M. Moscibrodzka, H. Falcke, M. Kramer, The black hole accretion code. Comput. Astrophys. Cosmol. **4**, 1 (2017)
73. P. Shi, V. Roy, L. Rezzolla, D.H. Rischke, Bjorken flow in one-dimensional relativistic magnetohydrodynamics with magnetization. Phys. Rev. D **93**(7), 074022 (2016)
74. Q. Qian, C. Fendt, S. Noble, M. Bugli, rHARM: accretion and ejection in resistive GR-MHD. Astrophys. J. **834**, 29 (2017)
75. D. Radice, S. Bernuzzi, A. Perego, R. Haas, A new moment-based general-relativistic neutrino-radiation transport code: methods and first applications to neutron star mergers. Mon. Not. Roy. Astron. Soc. **512**(1), 1499–1521 (2022)
76. S.M. Ressler, A. Tchekhovskoy, E. Quataert, M. Chandra, C.F. Gammie, Electron thermodynamics in GRMHD simulations of low-luminosity black hole accretion. Mon. Not. R. Astron. Soc. **454**(2), 1848–1870 (2015)
77. L. Rezzolla, B. Giacomazzo, L. Baiotti, J. Granot, C. Kouveliotou, M.A. Aloy, The missing link: merging neutron stars naturally produce jet-like structures and can power short Gamma-ray bursts. Astrophys. J. Letters **732**, L6 (2011)

78. L. Rezzolla, K. Takami, Black-hole production from ultrarelativistic collisions. Class. Quantum Grav. **30**(1), 012001 (2013)
79. L. Rezzolla, O. Zanotti, *Relativistic Hydrodynamics* (Oxford University Press, Oxford, UK, 2013)
80. B. Ripperda, F. Bacchini, O. Porth, E.R. Most, H. Olivares, A. Nathanail, L. Rezzolla, J. Teunissen, R. Keppens, General-relativistic resistive magnetohydrodynamics with robust primitive-variable recovery for accretion disk simulations. Astrophys. J. Supp. **244**(1), 10 (2019)
81. V. Roy, P. Shi, L. Rezzolla, D. Rischke, Analytic Bjorken flow in one-dimensional relativistic magnetohydrodynamics. Phys. Lett. B **750**, 45–52 (2015)
82. A. Sądowski, R. Narayan, A. Tchekhovskoy, Y. Zhu, Semi-implicit scheme for treating radiation under M1 closure in general relativistic conservative fluid dynamics codes. Mon. Not. R. Astron. Soc. **429**, 3533–3550 (2013)
83. M. Shibata, Y.-I. Sekiguchi, Magnetohydrodynamics in full general relativity: formulation and tests. Phys. Rev. D **72**(4), 044014 (2005)
84. D.M. Siegel, P. Mösta, D. Desai, W. Samantha, Recovery schemes for primitive variables in general-relativistic Magnetohydrodynamics. Astrophys. J. **859**(1), 71 (2018)
85. J.M. Stone, K. Tomida, C.J. White, K.G. Felker, The Athena++ adaptive mesh refinement framework: design and magnetohydrodynamic solvers. Astrophys. J. Supp. **249**(1), 4 (2020)
86. H.R. Takahashi, K. Ohsuga, T. Kawashima, Y. Sekiguchi, Formation of overheated regions and truncated disks around black holes: three-dimensional general relativistic radiation-magnetohydrodynamics simulations. Astrophys. J. **826**, 23 (2016)
87. A. Tchekhovskoy, R. Narayan, J.C. McKinney, Efficient generation of jets from magnetically arrested accretion on a rapidly spinning black hole. Mon. Not. R. Astron. Soc. **418**, L79–L83 (2011)
88. E.F. Toro, *Riemann Solvers and Numerical Methods for Fluid Dynamics*, 3rd edn. (Springer, 2009)
89. J.P. De Villiers, J.F. Hawley, Global general relativistic magnetohydrodynamic simulations of accretion tori. Astrophys. J. **592**, 1060 (2003)
90. C.J. White, J.M. Stone, C.F. Gammie, An Extension of the Athena++ code framework for GRMHD based on advanced Riemann solvers and staggered-mesh constrained transport. Astrophys. J. **225**, 22 (2016)
91. C.J. White, E. Quataert, C.F. Gammie, The structure of radiatively inefficient black hole accretion flows. Astrophys. J. **891**(1), 63 (2020)
92. O. Zanotti, M. Dumbser, A high order special relativistic hydrodynamic and magnetohydrodynamic code with space-time adaptive mesh refinement. Comput. Phys. Commun. **188**, 110–127 (2015)
93. O. Zanotti, C. Roedig, L. Rezzolla, L. Del Zanna, General relativistic radiation hydrodynamics of accretion flows - I. Bondi-Hoyle accretion. Mon. Not. R. Astron. Soc. **417**, 2899–2915 (2011)

Chapter 2
Solving the Einstein Equations Numerically

David Hilditch

Abstract There are many complementary approaches to the construction of solutions to the field equations of general relativity. Among these, numerical approximation offers the only possibility to compute a variety of dynamical spacetimes, and so has come to play an important role for theory and experiment alike. Presently we give a brief introduction to this, the science of numerical relativity. We discuss the freedom in formulating general relativity as an initial (boundary) value problem. We touch on the fundamental concepts of well-posedness and gauge freedom and review the standard computational methods employed in the field. We discuss the physical interpretation of numerical spacetime data and end with an overview of a number of 3d codes that are either in use or under active development.

2.1 Overview

This book is primarily concerned with advanced computational methods for general relativistic magnetohydrodynamics (GRMHD) and the solutions they produce. It is therefore assumed at the outset that the reader already has a solid grasp of the fundamentals of general relativity (GR) [1]. There are, however, many practitioners working on GRMHD who enter the subject through computational astrophysics, where it is common to solve the field equations of GRMHD with a given, fixed, spacetime metric. Following on from the previous chapter [2], in which a concise overview to GRMHD was given, our purpose here is to give a brief introduction to the field of numerical relativity (NR), where the metric becomes an additional dependent variable. There are excellent textbooks [3–7], which are highly recommended for those seeking instead a complete introduction to the topic.

Starting with the basics, we recall that in GR, gravitation is described geometrically within the spacetime manifold M as the curvature of the Levi-Civita covariant

D. Hilditch (✉)
Departamento de Física, CENTRA, Instituto Superior Técnico - IST, Universidade de Lisboa - UL, Avenida Rovisco Pais 1, 1049-001 Lisboa, Portugal
e-mail: david.hilditch@tecnico.ulisboa.pt

C. Bambi et al. (eds.), *New Frontiers in GRMHD Simulations*, Springer Series in Astrophysics and Cosmology, https://doi.org/10.1007/978-981-97-8522-3_2

derivative ∇_a associated with the Lorentzian metric g_{ab}, which we take to have mostly plus signature. The goal of the relativist is then to understand the solution space of Einstein's equations, the field equations of the theory,

$$ {}^{(4)}G_{ab} \equiv {}^{(4)}R_{ab} - \tfrac{1}{2} g_{ab} {}^{(4)}R = 8\pi T_{ab} \,, \tag{2.1} $$

expressed here and throughout in units with $G = c = 1$. Latin indices starting from the beginning of the alphabet will be abstract, and we shall work in the physical $3 + 1$ dimensions. The label (4) denotes then the spacetime curvature, distinct from the spatial curvature introduced below. The generalization of the results presented in this chapter to higher or lower spacetime dimensions is straightforward. The Ricci tensor and scalar are given, respectively by,

$$ {}^{(4)}R_{ab} = {}^{(4)}R_{acb}{}^{c} \,, \qquad {}^{(4)}R = g^{ab}\, {}^{(4)}R_{ab} \,, \tag{2.2} $$

with the inverse metric denoted as usual by g^{ab}. Introducing coordinates x^{μ} and writing out the field equations in the associated coordinate basis we obtain a system of coupled nonlinear partial differential equations (PDEs) whose solution space we can first try to understand with various simplifying strategies. Under strong symmetry assumptions, for instance, we may be able to find exact analytic solutions. The most important of these for the astrophysics of black holes is the Kerr solution. If we are interested in physical scenarios that are close to a known solution, we may linearize about that background and solve these simplified equations, whose solutions nevertheless accurately describe the physics of interest. This approach is fundamental to understand the propagation of weak gravitational waves far from their original source. More generally, when we encounter a problem with two or more very different scales we can use the smaller of the two as an expansion parameter. This strategy is used in the self-force treatment of extreme mass-ratio compact binary systems and in the Post-Newtonian expansion of the field equations, which has proven highly successful in the computation of the inspiral motion of compact binary systems. Turning instead to produce and understand solutions to the full field equations without simplifying assumptions, methods of pure mathematical analysis are absolutely fundamental to obtain a rigorous picture. Yet the general message we must absorb from the PDE literature is that such pure analysis will not provide solutions *in hand*. In many physical applications however, we do need the solution itself. To overcome this we turn to NR: the use of numerical methods to find approximate solutions to the field equations that converge to solutions of the continuum equations in the limit of infinite resolution. We will see that the requirement of convergence guides much of the theory in the topic. The most important application of NR to astrophysics is in the computation of the late stages of compact binary inspiral and merger.

In the following we give an overview of the basics of NR. We begin in Sect. 2.2 by reviewing the formulation of GR as an initial value and initial boundary value problem. We discuss well-posedness and the most popular free-evolution formulations of GR. In Sect. 2.3 the post-processing used to give physical interpretation to numerically constructed spacetime data is discussed. Finally, in Sect. 2.4 we give an introduction to the standard methods used in NR, and finally give an overview of the 3d codes in use.

2.2 Formulations of GR

Perhaps the single most important insight of special and general relativity is that space and time should be viewed as a unified entity. Using this insight to express field theories of interest results in a minimal and elegant formulation of their field equations. Yet such a spacetime formulation is not optimally adapted to the treatment of the initial and initial boundary value problems that necessarily underpin any understanding of dynamics within the theory. Textbook numerical methods for time evolution moreover rely on a clear separation of temporal and spatial derivatives. In this section, focusing primarily on the evolution of the metric in GR, we review the $3 + 1$ decomposition, which allows this separation. Afterwards we move on to discuss free-evolution, well-posedness and gauge freedom as they are usually framed in the NR literature.

2.2.1 The 3+1 Split

Suppose that we are given a solution to Einstein's equations (2.1) expressed in coordinates x^μ in a coordinate patch. Then we may map from an abstract tensor $T_{a_1\ldots a_k}{}^{b_1\ldots b_l}$ to the components $T_{\mu_1\ldots\mu_k}{}^{\nu_1\ldots\nu_l}$ in the coordinate basis defined by ∂^a_α and $\nabla_a x^\mu$ simply by contracting on every abstract index with elements of the basis,

$$T_{\mu_1\ldots\mu_k}{}^{\nu_1\ldots\nu_l} = T_{a_1\ldots a_k}{}^{b_1\ldots b_l}(\partial^{a_1}_{\mu_1}\ldots\partial^{a_k}_{\mu_k})(\nabla_{b_1}x^{\nu_1}\ldots\nabla_{b_l}x^{\nu_l})\,. \tag{2.3}$$

Likewise the tensor itself can be reconstructed from the components using

$$T_{a_1\ldots a_k}{}^{b_1\ldots b_l} = T_{\mu_1\ldots\mu_k}{}^{\nu_1\ldots\nu_l}(\nabla_{a_1}x^{\mu_1}\ldots\nabla_{a_l}x^{\mu_l})(\partial^{b_1}_{\nu_1}\ldots\partial^{b_k}_{\nu_k})\,. \tag{2.4}$$

We suppose now that $x^\mu = (t, x^i)$ with level sets Σ_t of t spacelike, in the sense that vectors tangent to Σ_t are spacelike. We then refer to t as a time coordinate. The collection of level sets Σ_t is called a foliation of the spacetime. Individual level sets are equivalently referred to as either instants of time or slices. We define the lapse function α by,

$$\alpha^{-2} = -g^{ab}\nabla_a t\nabla_b t\,. \tag{2.5}$$

We may now define the future directed unit normal to slices of constant t by

$$n^a = -\alpha \nabla^a t \,. \tag{2.6}$$

We define a projection operator through

$$\perp^a{}_b = \delta^a{}_b + n^a n_b \,. \tag{2.7}$$

The lapse measures the proper time elapsed between neighboring slices of the foliation. Direct computation reveals that $\perp^a{}_c \perp^c{}_b = \perp^a{}_b$, so that this indeed defines a projection. The core idea of the $3+1$ decomposition is that we should never deal directly with general tensors. Instead, we decompose all tensors into a set of objects that exist naturally within the foliation. For example, given a general vector field v^a, we may define,

$$v^n \equiv v^a n_a \,, \qquad (\perp v)^a \equiv \perp^a{}_b v^b \,. \tag{2.8}$$

The vector field can be reconstructed from its normal v^n and spatial $(\perp v)^a$ components

$$v^a = (\perp v)^a - n^a v^n \,. \tag{2.9}$$

The spatial part of the vector is orthogonal to the normal vector $(\perp v)^a n_a = 0$. The generalization of this decomposition to higher rank tensors follows the obvious pattern. Any tensor which is orthogonal to n^a on every index is called spatial. Since the normal vector depends upon the time coordinate, this definition is made with respect to a foliation. Consequently every spatial vector is spacelike, but not vice-versa. Our task now is to apply the idea of decomposition to the objects that appear in GR.

These objects may be naturally organized in a hierarchy according to the number of derivatives that they (implicitly) contain. At zeroth order we have the coordinates $x^\mu = (t, x^i)$. Then at first order we have the metric g_{ab}, the coordinate basis vectors $\partial^a_\alpha = (\partial^a_t \equiv t^a, \partial^a_i \equiv \partial^a_{x^i})$ and the basis one-forms dx^μ (which can also be written $\nabla_a x^\mu$). Recall that, by definition, we have the relationships

$$\delta^a{}_b = \partial^a_\alpha \nabla_b x^\alpha \,, \qquad \delta^\alpha{}_\beta = \partial^a_\beta \nabla_a x^\alpha = \partial_\beta x^\alpha \,, \tag{2.10}$$

which we use without further comment in the manipulations below. Next are the connection coefficients ${}^{(4)}\Gamma^\alpha{}_{\mu\nu}$, and at third order, or equivalently with two derivatives of the metric, the curvature ${}^{(4)}R_{abcd}$. Finally for our present needs, the second Bianchi identities

$$\nabla_{[a}{}^{(4)}R_{bc]de} = 0 \,, \tag{2.11}$$

and any contractions thereof lie at the fourth level. We treat each level of this hierarchy in turn.

There is nothing do to treat the coordinates $x^\mu = (t, x^i)$ as they constitute a collection of scalar fields rather than a tensor.

Moving on the first level, the basis vectors decompose as,

$$t^a n_a = -\alpha\,, \qquad \perp t^a \equiv \beta^a\,,$$
$$\partial_i^a n_a = 0\,, \qquad \perp \partial_i^a = \partial_i^a\,, \tag{2.12}$$

where here and in what follows we adopt the convention that the appearance of the projection operator without indices means projection on all open abstract indices on the object directly to the right of it. We have defined the shift vector β^a as the spatial part of the time vector t^a. It encodes the difference between the integral curves of the time vector and the normal vector. Rearranging, we have

$$t^a = \alpha n^a + \beta^a\,, \qquad \partial_i^a = \perp \partial_i^a \tag{2.13}$$

and, unsurprisingly, we see that the 'spatial derivatives' ∂_i are spatial according to our definition without projection. The basis one-forms dx^μ are then given by,

$$n^a \nabla_a t = \alpha^{-1}\,, \qquad \perp \nabla_a t = 0\,,$$
$$n^a \nabla_a x^i = -\alpha^{-1}\beta^i\,, \qquad \perp \nabla_a x^i = \gamma_a{}^b \nabla_a x^i = \gamma_a{}^i\,. \tag{2.14}$$

These relationships may be rewritten as,

$$\nabla_a t = -\alpha^{-1} n_a\,, \quad \nabla_a x^i = \alpha^{-1}\beta^i n_a + \perp \nabla_a x^i\,. \tag{2.15}$$

The geometry of the foliation and the relationship between the normal and time vectors is sketched in Fig. 2.1. The metric decomposes as

$$g_{nn} = -1\,, \qquad \perp g_{an} = 0\,, \qquad \gamma_{ab} = \perp g_{ab} = \perp_{ab} = g_{ab} + n_a n_b\,. \tag{2.16}$$

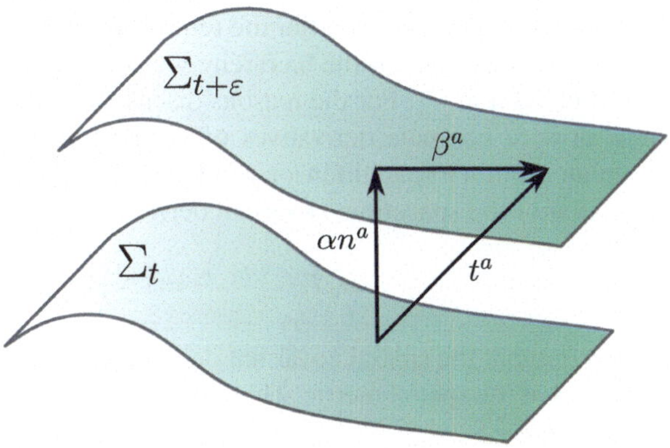

Fig. 2.1 Level sets Σ_t and neighbor $\Sigma_{t+\varepsilon}$ of the foliation, together with a depiction of the relationship between the time vector t^a, the lapse function α, the shift vector β^a, and the future pointing unit normal to slices of the foliation n^a.

In the last expressions here we define the spatial metric, which may be thought of geometrically as a metric on slices of the foliation induced by g_{ab}. Clearly the spatial metric is positive definite when acting on spatial vectors $\gamma_{ab}v^a v^b \geq 0$. Observe that within the $3+1$ split, the metric is special because it was used to define the unit normal vector n^a. Consequently rearranging (2.16), we arrive at a particularly simple expression

$$g_{ab} = \gamma_{ab} - n_a n_b \,. \tag{2.17}$$

Nevertheless, observe that the decomposition of the metric itself used the approach sketched above in a completely boiler-plate manner.

If we work in the coordinate basis ∂^a_α, spatial tensors take a particularly simple form. Consider once more an arbitrary spatial vector v^a. Expanding out the orthogonality condition $v^a n_a = 0$, we find,

$$0 = v^a n_a = -v^0 \alpha \,, \tag{2.18}$$

which means that the time component of a spatial vector must vanish $v^0 = 0$ at any point where the foliation has non-vanishing lapse. For an arbitrary spatial covector, we instead have

$$0 = n^a v_a = \alpha^{-1} v_0 - \alpha^{-1} \beta^i v_i \,, \tag{2.19}$$

which means that the time component is given by the inner product with the shift vector, which is itself spatial, so that so long as the lapse is bounded we have,

$$v_0 = \beta^i v_i \,. \tag{2.20}$$

The generalization of this result to tensors of arbitrary valence is straightforward. We conclude that, in a regular foliation with finite non-vanishing lapse, if we know the shift vector, then the spatial components of an arbitrary spatial tensor suffice to reconstruct the full tensor. We will use this to express the full Einstein equations in our coordinate basis below after the remainder of the geometric decomposition.

In the second level of the hierarchy we need to obtain a complete representation of the connection ∇_a. For the reasons discussed in the last paragraph, we need only know how to compute derivatives of spatial tensors within the $3+1$ formalism. Given an arbitrary spatial tensor we define the spatial covariant derivative by total projection of the spacetime covariant derivative

$$D_a T_{a_1\ldots a_k}{}^{b_1\ldots b_l} = \perp \nabla_a T_{a_1\ldots a_k}{}^{b_1\ldots b_l} \,. \tag{2.21}$$

It turns out that the spatial covariant derivative is nothing more than the Levi-Civita derivative of the spatial metric. This can be verified by verifying first that it is torsion free

$$2D_{[a}D_{b]}\phi = 2\perp\nabla_{[a}\left(\nabla_{b]}\phi + n_{b]}\mathscr{L}_n\phi\right) = 2\perp\nabla_{[a}n_{b]}\mathscr{L}_n\phi = 0\,. \tag{2.22}$$

Here the last term vanishes by Frobenius theorem because n^a is hypersurface orthogonal, square parentheses [ ] on indices denote antisymmetrization with the usual normalization of [1], and we employ the standard notation for the Lie-derivative. Second, we see that it is compatible with the spatial metric,

$$D_a\gamma_{bc} \equiv\perp\nabla_a(g_{bc} + n_bn_c) = 0\,. \tag{2.23}$$

Observe that with this definition the spatial coordinate basis from (2.15) can be rewritten as $\nabla_a x^i = \alpha^{-1}\beta^i n_a + D_a x^i$. To capture pieces of the projected covariant derivative $\perp^c{}_a\nabla_c T_{a_1\ldots a_k}{}^{b_1\ldots b_l}$ normal to the level set Σ_t we need a representation of derivatives of n_a. For this we define the acceleration vector a_a and the extrinsic curvature K_{ab}, which are given, respectively, by

$$a_a = n^b\nabla_b n_a\,, \qquad K_{ab} = -\gamma^c{}_a\nabla_c n_b\,. \tag{2.24}$$

Both are spatial by virtue of the unit normalization of n^a. One should be aware that in the mathematics literature the extrinsic curvature is occasionally defined with the opposite sign convention. As in Eq. (2.22), hypersurface orthogonality of n_a simplifies matters, allowing us to rewrite $a_a = D_a\ln\alpha$. The covariant derivative of n_a is completely represented through these variables and can now be expressed as

$$\nabla_a n_b = -K_{ab} - n_a a_b\,. \tag{2.25}$$

The same argument given below Eq. (2.22) shows that the extrinsic curvature is symmetric. Using orthogonality of n^a with spatial $T_{a_1\ldots a_k}{}^{b_1\ldots b_l}$ it follows that

$$n^{a_1}\perp^c{}_a\nabla_c T_{a_1\ldots a_k}{}^{b_1\ldots b_l} = -(\perp^c{}_a\nabla_c n^{a_1})T_{a_1\ldots a_k}{}^{b_1\ldots b_l} = K_a{}^{a_1}T_{a_1\ldots a_k}{}^{b_1\ldots b_l}\,, \tag{2.26}$$

and likewise with the contraction with n^a on any slot within the derivative, so that we now have a complete representation of any covariant derivative of a spatial tensor projected with $\perp$ on the derivative index itself. The extrinsic curvature plays yet another role, however, as it can be viewed morally as a 'time derivative' of the spatial metric, through the relation

$$K_{ab} = -\tfrac{1}{2}\mathscr{L}_n\gamma_{ab}\,. \tag{2.27}$$

There is variability in the literature in the representation of the 'time derivative' of an arbitrary spatial tensor under the $3+1$ split that depends on the specific equations under consideration, but this serves as the canonical example. The core idea is always to introduce the Lie derivative along a vector field with non-vanishing contraction with n^a as a dependent variable, and then project the resulting object on every remaining open index. For instance for $T_{a_1\ldots a_k}{}^{b_1\ldots b_l}$, we might introduce

$$S_{a_1\ldots a_k}{}^{b_1\ldots b_l} = \perp \mathcal{L}_n T_{a_1\ldots a_k}{}^{b_1\ldots b_l}\,. \tag{2.28}$$

The remainder of the Lie-derivative of $T_{a_1\ldots a_k}{}^{b_1\ldots b_l}$ along n^a is given by relationships akin to Eq. (2.26) but with the acceleration a_a playing the role of the extrinsic curvature. Again contracting on the first index of a given type as a representative example, we have,

$$n^{a_1}\mathcal{L}_n T_{a_1\ldots a_k}{}^{b_1\ldots b_l} = 0\,, \tag{2.29}$$

when the contraction is made with any index downstairs, and

$$n_{b_1}\mathcal{L}_n T_{a_1\ldots a_k}{}^{b_1\ldots b_l} = -a_{b_1} T_{a_1\ldots a_k}{}^{b_1\ldots b_l}\,. \tag{2.30}$$

otherwise. In fact we have now introduced enough variables to achieve a complete description of the connection coefficients in our coordinate basis. To verify this we simply use the general definition,

$${}^{(4)}\Gamma^{\alpha}{}_{\mu\nu} = \partial^a_\mu\,(\nabla_a\partial^b_\nu)\,\nabla_b x^\alpha\,, \tag{2.31}$$

express our basis vectors as in (2.13), and (2.15), and collect all possible contractions and projections, which results in

$$\begin{aligned}
{}^{(4)}\Gamma^{n}{}_{nn} &= -\partial_n \ln\alpha\,,\\
{}^{(4)}\Gamma^{n}{}_{in} &= -a_i = -\partial_i \ln\alpha\,,\\
{}^{(4)}\Gamma^{\alpha}{}_{nn}\perp^i{}_\alpha &= \alpha^{-1}\partial_n\beta^i + a^i\,,\\
{}^{(4)}\Gamma^{\alpha}{}_{in}\perp^j{}_\alpha &= \alpha^{-1}\partial_i\beta^j - K_i{}^j\,,\\
{}^{(4)}\Gamma^{n}{}_{ij} &= K_{ij}\,,\\
{}^{(4)}\Gamma^{\alpha}{}_{ij}\perp^k{}_\alpha &= \partial^a_i(D_a\partial^b_j)D_b x^k \equiv \Gamma^k{}_{ij}\,.
\end{aligned} \tag{2.32}$$

Observe that, despite the name, the extrinsic curvature really belongs to the connection. We realized already above that the spatial covariant derivative is the Levi-Civita covariant derivative of the spatial metric. It follows that the spatial connection coefficients $\Gamma^k{}_{ij}$ are just the associated Christoffel symbols,

$$\Gamma^k{}_{ij} = \tfrac{1}{2}\gamma^{kl}(\partial_i\gamma_{jl} + \partial_j\gamma_{il} - \partial_l\gamma_{ij})\,. \tag{2.33}$$

In a coordinate basis the spacetime connection coefficients ${}^{(4)}\Gamma^{\alpha}{}_{\mu\nu}$ have 40 algebraically independent components. Taking into account the symmetry of K_{ij} and $\Gamma^k{}_{ij}$ in the downstairs indices and adding the number of components of each line of (2.32), we have $1+3+3+9+6+18=40$, confirming that we have captured the spacetime object completely.

We now move on to the third level in our derivative hierarchy, the geometric decomposition of curvature quantities. In $3+1$ dimensions there are 20 algebraically independent components of the spacetime Riemann curvature ${}^{(4)}R_{abcd}$, and we therefore seek 20 algebraically independent equations upon decomposition. In other words; symmetries of the Riemann tensor vastly reduce the number of different contractions and projections we must consider. We start by projecting on every index, which reveals,

$$\perp {}^{(4)}R_{abcd} = R_{abcd} + 2K_{a[c}K_{d]b} \,. \tag{2.34}$$

Here we have introduced the spatial Riemann tensor R_{abcd}, the curvature associated with the spatial covariant derivative, defined as usual through commutation,

$$2D_{[a}D_{b]}w_c = R_{abc}{}^{d}w_d \,, \tag{2.35}$$

with w_a an arbitrary spatial one-form. Equation (2.34) is known as the Gauss equation. It is straightforwardly derived by contracting ${}^{(4)}R_{abcd}$ on the last index with an arbitrary spatial vector, projecting on the remaining indices and using the definition of the spatial covariant derivative D_a and extrinsic curvature K_{ab}. Recall that, although we are using abstract spacetime indices, since we work with spatial tensors we need only count their spatial components. In 3 dimensions the Riemann tensor has 6 algebraically independent components. A useful expression of this fact is that in the $3+1$ setting the spatial Riemann tensor R_{abcd} is determined entirely by the spatial Ricci tensor $R_{ab} = R_{acb}{}^{c}$ and its trace, the spatial Ricci scalar $R = \gamma^{ab}R_{ab}$,

$$R_{abcd} = 2\left(\gamma_{a[c}R_{d]b} - \gamma_{b[c}R_{d]a}\right) - R\gamma_{a[c}\gamma_{d]b} \,, \tag{2.36}$$

which can also be viewed as the fact that in 3 dimensions the Weyl tensor vanishes identically. Next we have the Codazzi–Mainardi equations, which are built by contracting once with the normal vector n^a before projecting on the remaining indices, giving

$$\perp {}^{(4)}R_{abcd}n^d = -2D_{[a}K_{b]c} \,. \tag{2.37}$$

The derivation of this result follows the exact same pattern sketched above for the Gauss equation. Counting here is a little more subtle. For each of the 3 independent components of the index without symmetrization, there are 3 equations associated with the antisymmetrized indices, giving a total of 9 equations. But the totally antisymmetric part also vanishes by the first spacetime Bianchi identity ${}^{(4)}R_{[abc]d} = 0$. In 3 dimensions this amounts to one more equation that we must subtract, giving 8 algebraically independent components in total in (2.37). The same combination of contractions and projections can in fact be used together with the Bianchi identity to efficiently obtain an evolution equation for the spatial connection coefficients, but this not needed presently. Finally, the Ricci equations are obtained by contracting twice with the normal vector and projecting on the remaining indices, which gives

$$\perp {}^{(4)}R_{abcd}n^b n^d = \mathcal{L}_n K_{ab} + \alpha^{-1} D_a D_b \alpha + K^c{}_a K_{bc}\,, \tag{2.38}$$

again by following the procedure sketched above. The symmetries of the Riemann tensor actually make the projection superfluous, but maintaining it makes counting more transparent. Since we have now a symmetric spatial tensor, there are 6 algebraically independent components here. Double-checking, from (2.34), (2.37) and (2.38) we have a total of $6 + 8 + 6 = 20$ algebraically independent components, as required.

Moving now to the fourth level of the hierarchy, the highest that we require here, the decomposition of the second Bianchi identities (2.11) proves useful. These correspond to algebraically 20 independent equations, with four independent contractions of interest. We begin by projecting on every index, which reveals the second Bianchi identity for D_a,

$$0 = \perp \nabla_{[a}{}^{(4)}R_{bc]de} = D_{[a} R_{bc]de}\,, \tag{2.39}$$

which could also be viewed as a consequence of the fact that the spatial covariant derivative is itself torsion-free, again as a result of the fact that n^a is hypersurface orthogonal. These correspond to 3 equations, consistent with the number of contracted Bianchi identities in 3 dimensions, which can be seen by substitution of (2.36) into (2.39). Next we contract with the normal vector on one of the non-explicitly antisymmetrized indices and project the rest. This gives

$$0 = -\perp \nabla_{[a}{}^{(4)}R_{bc]de}n^e = 2D_{[a} D_b K_{c]d} - K^e{}_{[a} R_{bc]de}\,, \tag{2.40}$$

which contains another 3 algebraically independent equations. Viewed from the perspective of the spatial covariant derivative, this identity holds simply by the definition of the spatial Riemann curvature and the Bianchi identity (2.39). Contracting instead with the normal vector on the antisymmetrized indices and projecting, we have

$$\begin{aligned}
0 &= 3 \perp \nabla_{[a}{}^{(4)}R_{bc]de}n^a \\
&= \perp \mathcal{L}_n\big(\perp {}^{(4)}R_{bcde}\big) - 2 \perp {}^{(4)}R_{bca[d}K_{e]}{}^a + 2\gamma^{fg} \perp {}^{(4)}R_{feg[b}K_{c]d} - 2\gamma^{fg} \perp {}^{(4)}R_{fdg[b}K_{c]e} \\
&\quad - 2D_b\big(D_{[d}K_{e]c}\big) + 2D_c\big(D_{[d}K_{e]b}\big) + 4\big(D_{[d}K_{e][b}\big)a_{c]} + 4\big(D_{[b}K_{c][d}\big)a_{e]} \\
&\quad + 2 \perp {}^{(4)}R_{d[b}K_{c]e} - 2 \perp {}^{(4)}R_{e[b}K_{c]d}\,,
\end{aligned} \tag{2.41}$$

which encodes 6 equations. In this equation, for brevity, we have not expanded out $\perp {}^{(4)}R_{abcd}$ from (2.34). Finally, we contract once with the unit normal in each of the two blocks of indices,

$$
\begin{aligned}
0 &= 3 \perp \nabla_{[a}{}^{(4)}R_{bc]de}n^a n^e \\
&= -2 \perp \mathcal{L}_n (D_{[b}K_{c]}{}^e)\gamma_{de} + 2K D_{[b}K_{c]d} + K_c{}^e \left(2D_{[d}K_{e]b} - D_b K_{de} + D_b(\gamma_{de}K)\right) \\
&\quad - 2K_b{}^e D_{[d}K_{e]c} + 2D_{[b}R_{c]d} + K_{eb}D_c K_{de} - K_{bd}D_c K + 2a_{[b}\left(R_{c]d} + K_{c]d}K - K_{c]}{}^e K_{de}\right) \\
&\quad - a^e\left(R_{bcde} + 2K_{b[d}K_{e]c}\right) - 2^{-1}D_{[b}(\perp{}^{(4)}R_{c]d}) - 2a_{[b}\perp{}^{(4)}R_{c]d}\,,
\end{aligned}
\tag{2.42}
$$

which results in the remaining 8 algebraically independent equations.

To this point our discussion was entirely geometrical. Given a spacetime manifold M, Lorentzian metric g_{ab} with an associated Levi-Civita connection ∇_a, coordinates $x^\mu = (t, x^i)$ whose time coordinate t has spacelike level sets and associated coordinate basis vectors ∂_α, and one-forms dx^μ, we have understood how to rewrite the geometry in a $3+1$ decomposed form. We now use this formalism to reformulate Einstein's equations (2.1). For this we could apply our recipe above in a purely agnostic fashion to the field equations. But to streamline the derivation it is convenient to begin with the observation that in $3+1$ dimensions the Einstein tensor can be decomposed in the form

$$
G_{ab} = G_{nn}(n_a n_b + \gamma_{ab}) + 2n_{(a} \perp G_{b)n} + \perp{}^{(4)}R_{ab} - \gamma_{ab}{}^{(4)}R_{cd}\gamma^{cd}\,, \tag{2.43}
$$

where round parentheses ( ) on indices denote symmetrization with the usual normalization of [1]. Thus we may consider the three pieces G_{nn}, $\perp G_{an}$ and $\perp{}^{(4)}R_{ab}$. We define variables for the decomposition of the stress-energy tensor, namely

$$
\rho = T_{ab}n^a n^b\,, \quad j_a = -\perp T_{ab}n^b\,, \quad s_{ab} = \perp T_{ab}\,. \tag{2.44}
$$

the energy density, momentum density and spatial stress tensors, respectively, as viewed by normal observers who travels along integral curves of n^a. Straightforward application of the $3+1$ decomposition on G_{nn} then reveals the first of the field equations as

$$
H = 2G_{nn} - 16\pi T_{nn} = R + K^2 - K^{ab}K_{ab} - 16\pi\rho = 0\,. \tag{2.45}
$$

Observe that no explicit time derivatives occur here (in this formalism these would occur in the form $\mathcal{L}_n$). For this reason, this equation constraints the permissible data at any instant of time Σ_t. It is called the Hamiltonian constraint. Taking now the $\perp G_{an}$ and using a trace of the Codazzi–Mainardi formula (2.37) we obtain,

$$
M_a = -\perp G_{an} = D^b K_{ab} - D_a K - 8\pi j_a = 0\,. \tag{2.46}
$$

Again, there are no explicit time derivatives and so these equations restrict permissible data at all times. This is known as the momentum constraint. For the last piece $\perp{}^{(4)}R_{ab}$, we project the trace-reversed Einstein equations, obtaining

$$\mathcal{L}_n K_{ab} = -\alpha^{-1} D_a D_b \alpha + R_{ab} - 2K^c{}_a K_{bc} + K K_{ab} \\ + 8\pi \left[s_{ab} - \tfrac{1}{2}\gamma_{ab} s + \tfrac{1}{2}\gamma_{ab}\rho \right] , \qquad (2.47)$$

These are now true evolution equations which, in some sense, encode the dynamical behavior of GR.

It remains to offer a little tidying up. Observe that the Einstein equations do not determine the time evolution of γ_{ab} but rather that of the extrinsic curvature. This is no problem, however, since (2.27) already told us that the extrinsic curvature can be viewed as the time derivative of the metric. For practical numerical purposes we need to express the field equations not as Lie derivatives of abstract index tensors but directly with partial derivatives of their components. This is easily achieved by using the relation (2.13) to replace the normal vector with the time vector t^a and expanding out the equations in our coordinate basis. Doing so, we finally arrive at the Arnowitt–Deser–Misner (ADM), or York, evolution equations,

$$\partial_t \gamma_{ij} = -2\alpha K_{ij} + \mathcal{L}_\beta \gamma_{ij} , \\ \partial_t K_{ij} = -D_i D_j \alpha + \alpha \left[R_{ij} - 2K^k{}_i K_{jk} + K K_{ij} \right] \\ + 4\pi\alpha \left[\gamma_{ij}(s-\rho) - 2s_{ij} \right] + \mathcal{L}_\beta K_{ij} . \qquad (2.48)$$

We make the usual abuse of notation when expressing derivatives, in a basis, writing

$$D_i D_j \alpha \equiv (\partial_i^a \partial_j^b) D_a D_b \alpha , \quad \mathcal{L}_\beta \gamma_{ij} \equiv (\partial_i^a \partial_j^b) \mathcal{L}_\beta \gamma_{ab} , \qquad (2.49)$$

and so forth. If we take a divergence of the Einstein equations, the left-hand side vanishes identically, giving $\nabla^b T_{ab} = 0$, which can be interpreted as the local conservation of energy and momentum. Under the $3+1$ decomposition these four equations can be written as

$$\partial_t \rho = \alpha[D^i j_i + 2a^i j_i - K\rho - K^{ij} s_{ij}] + \mathcal{L}_\beta \rho , \\ \partial_t j_i = -\alpha[D^j s_{ij} + a^j s_{ij} - K j_i + a_i \rho] + \mathcal{L}_\beta j_i . \qquad (2.50)$$

Observe that the lapse and shift are completely absent from the constraints, and furthermore that they have no equations of motion. Instead, they encode the freedom in the choice of coordinates. In NR this is usually viewed as the gauge freedom of the theory. We asserted above that the Hamiltonian (2.45) and momentum (2.46) constraints restrict the permissible data at each instant of time. This can only be self-consistent if the evolution equations (2.48) respect the constraints, in the sense that the time development of constraint-satisfying initial data remains constraint-satisfying. It turns out that, under the evolution equations (2.48), the constraints evolve according to the formal evolution system

$$
\begin{aligned}
\partial_t H &= -2\alpha D^i M_i - 4M^i D_i\alpha + 2\alpha K H + \mathcal{L}_\beta H \,, \\
\partial_t M_i &= -\tfrac{1}{2}\alpha D_i H + \alpha K M_i - (D_i\alpha) H + \mathcal{L}_\beta M_i \,.
\end{aligned} \tag{2.51}
$$

These equations can be obtained, for instance, by taking suitable traces of (2.41) and (2.42), plugging in the Einstein equations and using (2.50), or equivalently by 3 + 1 decomposition of the spacetime contracted Bianchi identities. Obviously $H = M_a = 0$ is a solution to these equations. Therefore given constraint satisfying initial data that can be evolved in time, there ought to be at least one solution that is constraint satisfying.

Putting aside for now any subtleties associated with the stress-energy tensor, we can now give a complete geometric formulation of the initial value (Cauchy) problem for GR. At an instant of time Σ_t we choose (γ_{ab}, K_{ab}) satisfying the Hamiltonian and momentum constraints, which then evolve according to the evolution system (2.48). This leads to the concept of free-evolution in NR, in which the idea is to solve the constraints just once for initial data and then evolve those data in time, relying on closure of the constraint subsystem to guarantee they are satisfied at later times. In practice numerical error violates the constraints, but these violations should converge away with increasing resolution. Strategies to obtain initial data of interest are discussed in the Sect. 2.2.2 next. To be sure that initial data give rise to a time development that can be reliably calculated by numerical methods, we must understand whether or not the evolution PDE we are solving is well-posed. Well-posedness is the requirement that a PDE problem admits a unique solution that depends continuously on the given data. This is discussed for evolution problems in Sect. 2.2.3. The choice of gauge has a direct effect on well-posedness, and so is fundamental in the construction of the free-evolution formulations of GR. This freedom is discussed below in Sect. 2.2.4.

2.2.2 *Initial Data*

Our aim in the construction of initial data is to find a configuration of the variables that models a physical scenario of interest whilst simultaneously solving the Hamiltonian and momentum constraints. To fulfill the former condition it is often advantageous to append additional equations to the constraints to impose physical restrictions on the data. To keep the discussion here brief, we review only the most common approach.

As we saw above, the Hamiltonian and momentum constraints

$$
\begin{aligned}
R + K^2 - K^{ab} K_{ab} &= 16\pi\rho \,, \\
D^b K_{ab} - D_a K &= 8\pi j_a \,,
\end{aligned} \tag{2.52}
$$

restrict the permissible data at each instant of time Σ_t. They are a collection of four equations for the twelve components of the spatial metric, extrinsic curvature and four components of the stress-energy tensor. Since this system of equations is highly under-determined and does not have a fixed PDE character, many different approaches may be taken to find solutions. What all strategies have in common is that decisions must be made about what part of the initial data will be specified, what 'dependent variables' will be introduced and solved for, what data will be given to those variables, and where it should be given.

In asymptotically flat spacetimes, the class most prevalent in NR, the metric and extrinsic curvature reduce to the flat-metric far away from a gravitating source. An attractive approach is thus to specify data for the dependent variables at spatial infinity, where they ought to be simple. In practice this means that we would like to solve a boundary value problem. The canonical PDE with a well-posed boundary value problem is the Laplace equation, which is of elliptic type. For these reasons it is popular to formulate the constraints as a system of second order elliptic PDEs and solve a boundary value problem. There are various different approaches even to this, but we shall focus on the archetype, the conformal transverse-traceless decomposition.

We begin by making a conformal decomposition of the metric,

$$\gamma_{ij} = \psi^4 \tilde{\gamma}_{ij} \,. \tag{2.53}$$

We use the common convention to raise and lower the indices of objects with a tilde using the inverse conformal metric $\gamma^{ij} = \psi^{-4}\tilde{\gamma}^{ij}$. We denote by $\tilde{D}_i$ the Levi-Civita connection of $\tilde{\gamma}_{ij}$, write the associated curvature as $\tilde{R}_{ijkl}$, and denote contractions similarly. Straightforward calculations relate the spatial and conformal curvature scalars

$$R = \psi^{-4}\tilde{R} - 8\psi^{-5}\tilde{\Delta}\psi \,, \tag{2.54}$$

where the conformal Laplace operator is $\tilde{\Delta} \equiv \tilde{D}^i \tilde{D}_i$. Plugging this into the Hamiltonian constraint, we obtain an elliptic equation for the dependent variable ψ. Besides the fact that the conformal decomposition naturally introduces an elliptic operator on the conformal factor, it is aesthetically pleasing that in this construction no single metric component is singled out for special treatment.

The momentum constraint contains only one derivative of the extrinsic curvature, and so to render it a second order equation the obvious maneuver is to introduce a potential. This is the transverse-traceless part of the setup. We write the extrinsic curvature as

$$K^{ij} = \psi^{-10}\tilde{A}^{ij} + \tfrac{1}{3}\gamma^{ij}K \,, \tag{2.55}$$

and place the ansatz

$$\tilde{A}^{ij} = \tilde{D}^i\tilde{v}^j + \tilde{D}^j\tilde{v}^i - \tfrac{2}{3}\tilde{\gamma}^{ij}\tilde{D}_k\tilde{v}^k + \tilde{M}^{ij} \,, \tag{2.56}$$

on the conformal trace-free part $\tilde{A}_{ij}$, with $\tilde{M}^{ij}$ an arbitrary symmetric trace-free tensor. Plugging this into the momentum constraint, we again obtain a second order elliptic equation, this time for the vector potential $\tilde{v}^i$, albeit with a more complicated elliptic operator

$$\tilde{\Delta}_{\mathrm{v}}\tilde{v}^i = \tilde{\Delta}\tilde{v}^i + \tfrac{1}{3}\tilde{D}^i(\tilde{D}_j\tilde{v}^j) + \tilde{R}^i{}_j v^j\,, \tag{2.57}$$

than for the conformal factor.

Putting all of this together, the conformal transverse-traceless decomposition gives the constraints the form

$$\begin{aligned} 8\tilde{\Delta}\psi - \tilde{R}\psi + \psi^{-7}\tilde{A}_{ij}\tilde{A}^{ij} - \tfrac{2}{3}\psi^5 K^2 + 16\pi\psi^5\rho &= 0\,,\\ \tilde{\Delta}_{\mathrm{v}}\tilde{v}^i + \tilde{D}_j\tilde{M}^{ij} - \tfrac{2}{3}\psi^6\tilde{D}^i K - 8\pi\psi^{10}j^i &= 0\,. \end{aligned} \tag{2.58}$$

We choose data for the conformal metric $\tilde{\gamma}_{ij}$, the trace of the extrinsic curvature K, the symmetric trace-free tensor $\tilde{M}^{ij}$ and pick a matter distribution, which will generate ρ and j^i. As mentioned above, in the routine asymptotically flat setting we impose boundary conditions $\psi = 1$ and $v^i = 0$ at spatial infinity and then solve for (ψ, v^i) everywhere on Σ_t. Yet another beautiful property of elliptic equations is that existence theorems, when they can be applied, guarantee that the solution exists on the entire domain.

Once we have a concrete formulation of the constraints we can turn our attention to making a choice of data that represents the scenario we are interested in studying. In doing so there is frequently a trade-off between the mathematical simplicity and accurately capturing this scenario. For instance, working in vacuum $\rho = j_i = 0$, taking the conformal metric flat $\tilde{\gamma}_{ij} = \delta_{ij}$ and choosing $K_{ij} = 0$, the Hamiltonian constraint reduces to the flat-space Laplace equation and the momentum constraint is trivially fulfilled. The Hamiltonian constraint is then easily solved. This choice of initial data, called Brill–Lindquist data, represents a number of superposed black holes instantaneously at rest. These are evidently modeled after the Schwarzschild solution in isotropic coordinates, which we recall has spatial metric

$$\gamma_{ij}dx^i dx^j = \psi^4(dx^2 + dy^2 + dz^2) = \left(1 + \tfrac{M}{2r}\right)^4(dx^2 + dy^2 + dz^2)\,, \tag{2.59}$$

with $r^2 = x^2 + y^2 + z^2$ as usual. Since we are most often interested in black holes in orbit around each other, this is too restrictive. Working in vacuum with the same conformally flat metric, choosing $K = 0$, called maximal slicing, and $\tilde{M}^{ij} = 0$, the momentum constraint decouples from the conformal factor and admits an exact solution, called the Bowen–York solution. This can be used to give linear momentum and spin to the black holes, with the price that the Hamiltonian constraint remains a nonlinear PDE. To solve it we turn then to numerical methods. These data are called 'puncture-initial data'. They are among the most used binary black hole configurations. Thus, Schwarzschild written in isotropic coordinates serves as the prototype for one of the two main classes of black hole initial data. Again the assumptions placed to simplify the constraints are a direct shortcoming, since neither instants of time in the Schwarzschild spacetime that carry linear momentum nor time slices of the Kerr spacetime can be conformally flat. The prototype for the second class of black hole initial data, called excision data, is the Schwarzschild solution written in horizon penetrating coordinates. The basic idea for the time-evolution of this data is that, since the interior of the black hole can not effect what happens outside, we may cut the black hole region out of the computational domain. See below for a short discussion of what this entails. For excision initial data it is more common to drop the assumptions of conformal flatness and maximal slicing. Likewise when using more sophisticated physical input such as that from Post-Newtonian theory [8], simplifying mathematical assumptions must be abandoned. In that case, the system (2.58) remains coupled and so requires a general purpose numerical solver for elliptic systems. For an up-to-date review of what is known rigorously for this complete system, see [9].

Setting up initial data for systems including fluid matter poses its own complications. Initial data that are co-rotational or irrotational can be constructed using the clean procedure as presented in [4]. But for generic binary configurations it is common to boost numerically constructed Tolman-Oppenheimer-Volkoff (TOV) configurations to help construct the given data for whatever formulation of the constraints (most often the extended-conformal-thin-sandwich system) is being used. Another subtlety lies at the treatment of the stellar surface, where the Euler equations become formally singular. Further discussion of this can be found in the review article [10].

2.2.3 Well-Posedness of Time Evolution PDEs

The complete specification of a PDE problem consists of the domain on which the problem is to be solved, the data that is given and of course the PDEs themselves. If a PDE problem admits a unique solution that depends continuously on the given data in some norm we say that the problem is well-posed (with respect to that norm). Well-posedness is a fundamental requirement on a PDE problem, and is of practical concern for numerical approximation. If a PDE is ill-posed, that is, if there exists no norm in which it is well-posed, then solutions of any approximation scheme will fail to converge to the continuum solution (if it even exists) in the limit of infinite resolution.

In this subsection we state without proof the most relevant well-posedness results for the initial value problem and initial boundary value problem. For proofs, together with a numerical perspective, we recommend the classic textbooks [11, 12]. For a discussion closest to the language and needs of NR, we suggest [13, 14], the fascinating review [15], and for a pure mathematical treatment [16]. For an introduction to the long-term behavior of nonlinear wave equations, we also recommend [17].

Consider the first order linear constant coefficient evolution system

$$\partial_t \mathbf{u} = \mathbf{A}^p \partial_p \mathbf{u} + \mathbf{s} \,. \tag{2.60}$$

The principal part of the system consists just of the terms involving the highest derivatives. We assume that the components of the state vector $\mathbf{u}$, the principal part matrices $\mathbf{A}^p$ and the source terms $\mathbf{s}$ are real. Observe that the source terms could be of the form $\mathbf{s} = \mathbf{B}\mathbf{u} + \mathbf{s}'$. The initial value problem consists of setting data,

$$\mathbf{u}(0, x) = \mathbf{f}(x), \tag{2.61}$$

and requesting a solution at times $t > 0$. In this context, well-posedness of the initial value problem in a norm $||\cdot||$ can be characterized by the existence of positive constants α, K such that an estimate

$$||\mathbf{u}(t, \cdot)|| \leq K e^{\alpha t} ||\mathbf{f}(\cdot)|| \tag{2.62}$$

holds regardless of the choice of initial data.

Let s_p denote an arbitrary unit spatial covector. The principal symbol $\mathbf{P}^s$ in the s_p direction is defined by $\mathbf{P}^s = \mathbf{A}^s \equiv \mathbf{A}^p s_p$. The system (2.60) is called weakly hyperbolic if, for every such covector, the principal symbol has real eigenvalues. These can be interpreted as the speeds of propagation of the system. If, furthermore, for every s_p the principal symbol has a complete set of eigenvectors and the bound

$$|\mathbf{T}_s| + |\mathbf{T}_s^{-1}| \leq K, \tag{2.63}$$

holds uniformly in the direction s_p for some $K > 0$, then we say that the system is strongly hyperbolic. Here $\mathbf{T}_s$ is the matrix that takes the right eigenvectors as columns and where $|\cdot|$ denotes here the spectral norm of a matrix. If there exists a symmetrizer $\mathbf{H}$ for the system, a symmetric positive definite matrix such that $\mathbf{H}\mathbf{A}^p$ is symmetric for each p, we say that the system is symmetric hyperbolic. Symmetric hyperbolicity is yet a stronger condition than strong hyperbolicity.

The first, crucial theorem is that the initial value problem for the system (2.60) is well-posed in L^2

$$||\mathbf{u}(t, \cdot)||^2_{L^2} = \int_{\Sigma_t} \mathrm{d}x\, \mathbf{u}(t, x)^T \mathbf{u}(t, x) \tag{2.64}$$

if and only if it is strongly hyperbolic. We denote the transpose by superscript T. Interestingly, if the system is only weakly hyperbolic, it may be well-posed in a norm that contains specific derivatives of the certain components of the state vector. In contrast to L^2 well-posedness for strongly hyperbolic systems however, the weakly hyperbolic system is sensitive to changes in the source terms $\mathbf{s}$, and well-posedness is therefore delicate to verify in practice. The second important result is that, for the initial boundary value problem, which consists of specifying initial data on a partial Cauchy slice at $t = 0$ together with boundary conditions at the boundary of the domain at each time, there is a simple prescription to choose boundary conditions such that the problem becomes well-posed when the system (2.60) is symmetric hyperbolic. This straightforward approach is made possible by the fact that symmetric hyperbolic systems admit energy estimates like (2.62) by simple physical space arguments involving the divergence theorem, without having to use the Fourier-transform. We demonstrate this below for the wave equation. In GR the application of this result for the initial boundary value problem is greatly complicated by the constraints. The reason for this is that, just as we need to choose initial data that satisfies the constraints, so do we have to choose boundary conditions that are constraint preserving. Likewise, boundary conditions need to be chosen so that they model the physics we are interested in studying. For instance, reflecting boundary conditions are less favored than those which can absorb outgoing waves.

There are obvious differences between the simple system considered in this section and the field equations of GR. First, the formulations of GR we have seen are not first order in all derivatives but rather, first order in time and second order in space. This is purely a technical detail, however, which can be overcome introducing a first order reduction and then applying the definitions as given here. An elementary example of this is given already by the wave equation, discussed momentarily, for which it is easy to see that there is a symmetric hyperbolic first order reduction. Second, the Einstein equations are not linear. This can be surmounted by linearizing about an arbitrary background and then working in a small neighborhood of a point, where to a good approximation the coefficients in the linearized problem will be constant, and we can apply the definitions given in this section. This is known as the localization principle. With additional smoothness assumptions, strong hyperbolicity still guarantees *local well-posedness* in a certain norm. The price we pay for linearization and working just in the neighborhood of each point is that, without investing much more work, we can only guarantee existence of the solution for a short time. Naturally, long-term behavior of solutions is a relevant question of great interest, and one on which much progress is being made. But from a pure NR perspective, it is the property of local well-posedness that determines whether or not we can hope to accurately approximate solutions computationally. There is much to say about all this, but to avoid a lengthy discussion, we direct the interested reader to the references highlighted above.

The canonical example of a PDE with a well-posed Cauchy problem is the wave equation in flat-space, which we now consider as a simple example. The wave equation is given by

$$\Box_\eta \phi = \eta^{ab}\nabla_a\nabla_b\phi = -\partial_t^2\phi + \partial^i\partial_i\phi = 0\,, \tag{2.65}$$

where here $\Box_\eta$ is the d'Alembert operator associated with the Minkowski metric, ∇_a is the Levi-Civita derivative of η_{ab} and in the second equality we assume that the coordinates are global inertial. We can draw a crude analogy between GR and the wave equation in which this is identified with the fully second order form of the field equations (2.1). Introducing the time reduction variable $\pi = -\partial_t\phi$, we can write the system in first-order-in-time-second-order-in-space form

$$\partial_t\phi = -\pi\,, \quad \partial_t\pi = -\partial^i\partial_i\phi\,. \tag{2.66}$$

In analogy to our treatment of GR, this is like the evolution equations (2.48). We can then reduce the system to first order by introducing the spatial reduction variable Φ_i and the reduction constraint

$$C_i = \partial_i \phi - \Phi_i = 0\,. \tag{2.67}$$

Even in the linear constant coefficient setting, not every PDE will admit a first order reduction of the type (2.60). Such PDEs are not hyperbolic. The canonical example of this is the heat equation $\partial_t \phi = \partial^i \partial_i \phi$. In first order form, the wave equation can be rewritten as

$$\partial_t \phi = -\pi\,, \quad \partial_t \Phi_i = -\partial_i \pi\,, \quad \partial_t \pi = -\partial^i \Phi_i\,. \tag{2.68}$$

Solutions of the reduction agree with those of the wave equation whenever the reduction constraint (2.67) is satisfied. We did not introduce a first order reduction of GR, but can perfectly easily do so. In common with what we saw above for GR however, if the reduction constraint is initially satisfied, it will remain so, since we can easily show that $\partial_t C_i = 0$. Taking the state vector to be $\mathbf{u} = (\phi, \Phi_i, \pi)^T$, in the notation of (2.60), we have

$$\mathbf{A}^p = \begin{pmatrix} 0 & 0 & 0 \\ 0 & 0 & -\delta_i{}^p \\ 0 & -\delta^{jp} & 0 \end{pmatrix}\,, \quad \mathbf{s} = \begin{pmatrix} -\pi \\ 0 \\ 0 \end{pmatrix}\,. \tag{2.69}$$

The principal symbol $\mathbf{P}^s$ is then

$$\mathbf{P}^s = \mathbf{A}^p s_p = \begin{pmatrix} 0 & 0 & 0 \\ 0 & 0 & -s_i \\ 0 & -s^j & 0 \end{pmatrix}\,. \tag{2.70}$$

It has real eigenvalues $0, \pm 1$, and so the reduction is at least weakly hyperbolic. The eigenvectors of $\mathbf{P}^s$ are complete. They are given by

$$\mathbf{u}_0 = (1, 0_j, 0)^T\,, \quad \mathbf{u}_A = (0, v^A_j, 0)^T\,, \quad \mathbf{u}_\pm = 1/\sqrt{2}(0, s_j, \pm 1)^T\,, \tag{2.71}$$

where we have introduced two orthogonal unit covectors v^A_j for $A = 1, 2$ that are orthogonal to s_i. The uniformity condition (2.63) is satisfied with $K = 1$. In this case the principal matrices are already symmetric, so we can choose the symmetrizer $\mathbf{H} = \mathbf{1}$, the 5×5 identity.

Consider the following norm for our first order reduction of the wave equation

$$E(t, \mathbf{u}) = \int_{\Sigma_t} \mathrm{d}x\, \mathbf{u}(t, x)^T \mathbf{H} \mathbf{u}(t, x)\,. \tag{2.72}$$

In the present context this corresponds almost exactly to the physical energy of the solution. For this reason estimates that hold for such norms are often referred to as energy estimates. Taking a time derivative of (2.72), substituting the equations of motion and rearranging, we obtain

$$\partial_t E(t, \mathbf{u}) \leq E(t, \mathbf{u}) + \int \mathrm{d}x\, \partial_p \left(\mathbf{u}(t, x)^T \mathbf{H}\mathbf{A}^p \mathbf{u}(t, x)\right) . \tag{2.73}$$

Using the divergence theorem to remove the second term, and a simple application of Grönwall's lemma, we conclude with a well-posedness estimate

$$E(t, \mathbf{u})^{1/2} \leq e^{t/2} E(0, \mathbf{f})^{1/2} , \tag{2.74}$$

with initial data $\mathbf{u}(0, x) = \mathbf{f}(x)$. This is precisely of the form (2.62) in the L^2 norm. The computations to obtain estimates for an arbitrary symmetric hyperbolic PDE are identical in spirit.

There are various PDE setups of interest in NR besides the initial and initial boundary value problems. Most important are the characteristic initial and characteristic initial boundary value problems, and the hyperboloidal initial value problem. For an introduction to these, see [18–20].

In the following section, besides presenting the equations themselves, we give a summary of the status of the initial and initial boundary value problems for the most popular formulations of GR in use in NR.

2.2.4 Free-Evolution Formulations of GR

We saw in Sect. 2.2.1 that the constraint subsystem for the ADM evolution equations (2.48) is closed. If we modify the evolution equations by adding combinations of the constraints, this property still holds. We also saw that the field equations of GR do not by themselves impose a choice for the lapse and shift. Thus there is a large freedom in the choice of equations to be solved. A concrete choice for these is known as a (free-evolution) formulation of GR. We observed in Sect. 2.2.3 that for sufficiently smooth initial data, strong hyperbolicity guarantees local well-posedness of the Cauchy problem. Taking these considerations together, we are guided when constructing formulations of GR to impose *at least* strong hyperbolicity. Such a formulation is often called a hyperbolic reduction. But there are issues besides this. It is highly desirable to obtain gauge conditions that are flexible enough to reliably treat a large range of scenarios of interest. A clever choice of evolved variables can moreover help to reduce numerical error in applications, or even to make it possible to treat data otherwise out of reach.

The basic ingredients in the construction of free-evolution formulations are first, to make a choice of gauge, second, the definition of auxiliary constraints, and finally the choice of coupling between the field equations and the complete set of constraints. These are already well-illustrated by the archetypal example of a hyperbolic formulation of GR, the generalized harmonic gauge formulation. To construct this, let us begin with the trace-reversed Einstein equations written out, in coordinates x^μ defined as in Sect. 2.2.1, in the form,

$$\begin{aligned} R_{\alpha\beta} &= -\tfrac{1}{2} g^{\mu\nu} \partial_\mu \partial_\nu g_{\alpha\beta} + \partial_{(\alpha} {}^{(4)}\Gamma_{\beta)} - {}^{(4)}\Gamma^\gamma{}_{\alpha\beta} {}^{(4)}\Gamma_\gamma \\ &\quad + g^{\mu\nu} g^{\delta\varepsilon} \left(\partial_\delta g_{\mu\alpha} \partial_\varepsilon g_{\nu\beta} - {}^{(4)}\Gamma_{\alpha\mu\delta} {}^{(4)}\Gamma_{\beta\nu\varepsilon} \right) \\ &= 8\pi \left(T_{\alpha\beta} - \tfrac{1}{2} g_{\alpha\beta} T \right), \end{aligned} \tag{2.75}$$

where we introduce the contracted Christoffel symbols ${}^{(4)}\Gamma_\alpha = g^{\mu\nu\,(4)}\Gamma_{\alpha\mu\nu}$. Comparing the principal part with the wave equation (2.65), we see that there would be a strong structural similarity if only the second term in the right-hand side, the gradient of the contracted Christoffel symbol were absent. Of course we are not free to simply change the equations to be solved. But looking more closely we realize that the contracted Christoffel symbol can be written as,

$${}^{(4)}\Gamma_\alpha = -g_{\alpha\beta} \Box_g x^\beta \,, \tag{2.76}$$

with $\Box_g = \nabla^a \nabla_a$ here the curved space d'Alembertian. Therefore, we might work with generalized harmonic, also called generalized wave, coordinates

$$\Box_g x^\alpha = H^\alpha(x, g) \,. \tag{2.77}$$

When the gauge source functions $H^\alpha(x, g)$ vanish, these are called harmonic coordinates or wave coordinates. Substituting this expression back into (2.75) directly and using the fact that the gauge source functions depend upon the metric, but not partial derivatives of the metric components, the principal part of the equations indeed agrees with that of the curved space wave equation. Formally this is achieved by defining the auxiliary Harmonic constraints

$$C_\alpha = {}^{(4)}\Gamma_\alpha + H_\alpha = 0 \,, \tag{2.78}$$

and replacing the Einstein equations by the reduced field equations

$$\begin{aligned} \mathscr{R}_{\alpha\beta} &= R_{\alpha\beta} - \nabla_{(\alpha} C_{\beta)} \\ &= -\tfrac{1}{2} g^{\mu\nu} \partial_\mu \partial_\nu g_{\alpha\beta} + \nabla_{(\alpha} H_{\beta)} \\ &\quad + g^{\mu\nu} g^{\delta\varepsilon} \left(\partial_\delta g_{\mu\alpha} \partial_\varepsilon g_{\nu\beta} - {}^{(4)}\Gamma_{\alpha\mu\delta} {}^{(4)}\Gamma_{\beta\nu\varepsilon} \right) \\ &= 8\pi \left(T_{\alpha\beta} - \tfrac{1}{2} g_{\alpha\beta} T \right), \end{aligned} \tag{2.79}$$

where $\mathscr{R}_{\alpha\beta}$ is called the reduced Ricci tensor. This is called the generalized harmonic gauge (GHG) formulation of GR. Due to the similarity with the wave equation discussed in the last section, it is not surprising that the GHG formulation admits a symmetric hyperbolic first order reduction, and thus has a well-posed Cauchy problem. The terminology 'reduced' arose as an abbreviation of hyperbolic reduction. To see that the constraint subsystem remains closed, we reverse the trace once more to obtain

$$G_{ab} - \nabla_{(a}C_{b)} + \tfrac{1}{2}g_{ab}(\nabla_c C^c) = 8\pi T_{ab} \,. \tag{2.80}$$

Contracting then with the normal vector n^a and employing the language of the $3+1$ decomposition once more, a short calculation gives

$$\begin{aligned} \mathcal{L}_t C_n &= \alpha\big[H - D_a(\perp C^a) + a^a(\perp C_a) - 2KC_n\big] + \mathcal{L}_\beta C_n \,, \\ \perp\mathcal{L}_t(\perp C)_a &= \alpha[2M_a - D_a C_n + a_a C_n - 2K_a{}^b(\perp C)_b] + \perp\mathcal{L}_\beta(\perp C)_a \,, \end{aligned} \tag{2.81}$$

for $C_n = C_a n^a$ and $\perp C_a$, so we see that time derivatives of the harmonic constraints are essentially given by the Hamiltonian and momentum constraints we met in Sect. 2.2.2. Taking the divergence of (2.80) and using local stress-energy conservation we have

$$\Box C_a + R_{ab}C^b = 0 \,, \tag{2.82}$$

where we can now substitute back from (2.79) to remove the Ricci tensor. This means that the harmonic constraints satisfy a second order nonlinear wave equation in which every term appears with at least one constraint. Combining this with (2.81) we can conclude that if we start with constraint satisfying initial data, the time development, which is guaranteed to exist, at least locally, by well-posedness, will satisfy the constraints too.

We have presented GHG formulation itself by returning to the spacetime view, but this was simply because the structure in the equations is more evident in that notation. We can equivalently work with the $3+1$ language. The harmonic constraints, for instance, can be rewritten as

$$\begin{aligned} \partial_t \alpha &= -\alpha^2(K + C_n + H_n) + \mathcal{L}_\beta \alpha \,, \\ \partial_t \beta^i &= \alpha^2(\Gamma^i + (\perp C)^i + (\perp H)^i) - \alpha D^i\alpha + \beta^j\partial_j\beta^i \,. \end{aligned} \tag{2.83}$$

So it turns out that a choice of gauge as a second order wave equation on the coordinates corresponds to a choice of first order equations of motion for the lapse and shift in the $3+1$ setting, consistent with the interpretation we gave earlier that the lapse and shift encode the freedom in choosing coordinates.

To build the reduced Einstein equations (2.79), we chose a specific combination of derivatives of constraints to add to the Einstein equations. As we have seen, local well-posedness of hyperbolic PDEs only depends upon the principal part. Therefore one may wonder why we chose to add a symmetrized covariant derivative, which contains non-principal constraint additions. The answer in this case is that the choice we made presently results in the simplest form of the equations, which we hope makes the general strategy more transparent. More generally, one might feel that the use of the Levi-Civita derivative is preferred as it is covariant, but if we take seriously the point of view of [1] then the same applies to partial derivatives. Nevertheless, it is perfectly legitimate to ask whether one non-principal constraint addition or another is to be preferred. In numerical applications, error will generically violate the constraints. Although such violations should converge to zero with increasing resolution, any strategy to suppress them at finite resolution is welcome. The behavior of numerical error reflects to some extent the behavior of the continuum equations. Motivated by this type of reasoning, a crucial adjustment to the formulation (2.79) for numerical applications was obtained. The idea is to carefully make non-principal constraint additions,

$$\mathscr{R}_{\alpha\beta} = R_{\alpha\beta} - \nabla_{(\alpha}C_{\beta)} + \gamma_0\left(n_{(a}C_{b)} - \tfrac{1}{2}g_{ab}C_c n^c\right) = 8\pi\left(T_{\alpha\beta} - \tfrac{1}{2}g_{\alpha\beta}T\right). \quad (2.84)$$

Linearizing, it can be seen that choosing γ_0 positive leads to the exponential suppression of small, high-frequency constraint violations. Combined with the wavelike propagation of the harmonic constraints, this idea is essentially the same as that of hyperbolic divergence cleaning in GRMHD. The use of such constraint damping schemes is now standard with various formulations of GR. Besides this constraint damping scheme, there are various different choices of non-principal constraint addition in use in the NR literature, giving rise to slightly different flavors of the GHG formulation.

The GHG formulation as presented here is often subjected to a first order in time (or even fully first order) reduction when implemented in practice. Together these different flavors of GHG constitute one of the two main types of formulations of GR in regular use for astrophysical work in NR. The GHG formulation is symmetric hyperbolic. As discussed above, this allows for much more straightforward PDE analysis. In particular, this has led to a proof of well-posedness of the initial boundary value problem with boundary conditions that are constraint preserving and gravitational radiation controlling. The GHG formulation is the only formulation of GR with this property in frequent use in NR simulations that actually uses these boundary conditions in applications.

We now come to the second class of formulations of GR in frequent use for astrophysical NR. In the literature these are referred to as the BSSN (or BSSNOK), CCZ4 and Z4c formulations. All three are constructed following the general template laid out above. They are furthermore closely related and indeed, the latter two are essentially variations of each other, as they share principal part and constraint damping terms. In particular, all three formulations share in common the use of a conformal decomposition in the choice of variables. We will try to give a unified presentation that covers all three simultaneously. We begin with the trace-reversed Einstein equations and, as above, define auxiliary constraints, this time called Z_a. We then replace the Einstein equations by the Z4 equations of motion

$$R_{ab} + 2\nabla_{(a}Z_{b)} - \kappa_1\left[2n_{(a}Z_{b)} - (1+\kappa_2)g_{ab}n_c Z^c\right] + W_{ab}(Z) = 8\pi\left(T_{ab} - \tfrac{1}{2}g_{ab}T\right), \quad (2.85)$$

where $W_{ab}(Z)$ is a symmetric tensor that vanishes when $Z_a = 0$, but which is otherwise arbitrary. This tensor parametrizes the different flavors of the formulation. As in the GHG formulation, we have introduced constraint damping terms with parameters κ_1 and κ_2. Solutions of these equations agree with those of GR when $Z_a = 0$.

Comparing the Z4 equations of motion with (2.84), there is an obvious similarity. In fact, choosing

$$Z_a = -\tfrac{1}{2}C_a\,, \quad \kappa_1 = \gamma_0\,, \quad \kappa_2 = 0\,, \tag{2.86}$$

and taking $W_{ab} = 0$, the expressions align perfectly. The important difference, however, is that in the Z4 setup we have not yet chosen an equation of motion for the lapse and shift. Morally, therefore, we may think of Z4 as a formulation with the same constraint subsystem as GHG, but without imposing generalized harmonic coordinates.

We will follow the common presentation of these conformal variable formulations in using the $3+1$ language and working in adapted coordinates, writing the components of the equations of motion. We begin by decomposing the Z4 constraint, introducing $\Theta = -n_a Z^a$, and the spatial part Z_i. Geometrically this is defined by application of the projection operator, but in adapted coordinates we can use the components downstairs. The definition of the extrinsic curvature remains unchanged, so we have

$$\partial_t \gamma_{ij} = -2\alpha K_{ij} + \mathcal{L}_\beta \gamma_{ij}\,, \tag{2.87}$$

as in the ADM setup above. For the extrinsic curvature itself we simply take the spatial part of (2.85). This of course gives equations that look like the ADM evolution equations for K_{ij}, plus constraint additions,

$$\begin{aligned}\partial_t K_{ij} = & -D_i D_j \alpha + \alpha\left[R_{ij} - 2K^k{}_i K_{jk} + K K_{ij} + 2D_{(a}Z_{b)} - 2K_{ij}\Theta + W_{ij}\right] \\ & - \alpha\kappa_1(1+\kappa_2)\gamma_{ij}\Theta + 4\pi\alpha\left[\gamma_{ij}(s-\rho) - 2s_{ij}\right] + \mathcal{L}_\beta K_{ij}\,.\end{aligned} \tag{2.88}$$

For the constraints, we trace-reverse (2.85) and contract with n^a twice (for the Θ equation of motion) and likewise contract once with n^a and once with the projection operator (for the Z_i equation). This gives,

$$\begin{aligned}\partial_t \Theta &= \alpha\left[\tfrac{1}{2}H + D^i Z_i - a^i Z_i - K\Theta - \kappa_1(2+\kappa_2)\Theta + \hat{W}_{nn}\right] + \mathcal{L}_\beta \Theta\,, \\ \partial_t Z_i &= \alpha\left[M_i + D_i\Theta - a_i\Theta - 2K_i{}^j Z_j - \kappa_1 Z_i - W_{ni}\right] + \mathcal{L}_\beta Z_i\,.\end{aligned} \tag{2.89}$$

Where we denote $\hat{W}_{nn} = W_{nn} - W/2$. The complete collection of the constraints is then given by

$$\Theta = 0\,, \quad Z_i = 0\,, \quad H = 0\,, \quad M_i = 0\,, \tag{2.90}$$

with the Hamiltonian and momentum constraints defined exactly as earlier in (2.45) and (2.46).

To derive the formal evolution equations for the Hamiltonian and momentum constraints, we trace-reverse (2.85) and take the divergence with ∇_a, then $3+1$ decompose. Although straightforward, this is considerably lengthier than for the ADM equations. This reveals

$$\begin{aligned}\partial_t H &= -2\alpha D^i M_i - 4M_i D^i\alpha + 2\alpha K H + 4\alpha(K\gamma^{ij} - K^{ij})(D_i Z_j - K_{ij}\Theta + \tfrac{1}{2}W_{ij})\\ &\quad - 4\alpha\kappa_1(1+\kappa_2)K\Theta + \mathcal{L}_\beta H\,,\\ \partial_t M_i &= -\tfrac{1}{2}\alpha D_i H + \alpha K M_i - (D_i\alpha)H + D^j\big(\alpha(2D_{(i}Z_{j)} - 2K_{ij}\Theta + W_{ij})\big)\\ &\quad - D_i\big(\alpha(2D_j Z^j - 2K\Theta + \gamma^{jk}W_{jk})\big) + 2D_i\big(\alpha\kappa_1(1+\kappa_2)\Theta\big) + \mathcal{L}_\beta M_i\,. \end{aligned} \tag{2.91}$$

As discussed above, the actual evolved variables involve a conformal decomposition not dissimilar to, and historically inspired by, that employed in the treatment of the constraints as outlined in Sect. 2.2.2. The variables themselves are defined through

$$\begin{aligned} &\chi = \gamma^{-1/3}\,, && \tilde{\gamma}_{ij} = \gamma^{-1/3}\gamma_{ij}\,, && \hat{K} = \gamma^{ij}K_{ij} - 2\Theta\,,\\ &\tilde{A}_{ij} = \gamma^{-1/3}(K_{ij} - \tfrac{1}{3}\gamma_{ij}K)\,, && \tilde{\Gamma}^i = 2\tilde{\gamma}^{ij}Z_j + \tilde{\Gamma}_{\mathrm{d}}{}^i\,, && \tilde{\Gamma}_{\mathrm{d}}{}^i = \tilde{\gamma}^{jk}\tilde{\Gamma}^i{}_{jk} = \tilde{\gamma}^{ij}\tilde{\gamma}^{kl}\partial_k\tilde{\gamma}_{lj}\,. \end{aligned} \tag{2.92}$$

There are minor changes in the choice of variables from one formulation to another, and indeed across different implementations. The key benefit of this change of variables is that they help in the treatment of data in which the metric components blow-up in an isotropic manner. Consider for instance the prototype for puncture-type initial data, the Schwarzschild metric with the usual time coordinate and isotropic coordinates. In that case, all non-vanishing component of the spatial metric (2.59) diverge as $r \to 0$, but the evolved metric remains trivial. One would therefore expect that perturbations would give a conformal metric that could also be inverted without problem. This turns out to be the case in a large variety of situations of interest. Nevertheless, since the determinant of the metric diverges we must still manage one singular variable. In practice this is managed by evolving χ, a negative power of the determinant of the spatial metric, see the definition in (2.92). The definition of χ thus guarantees that the evolved variable remains finite, although keeping in mind (2.59) as a model, we may be concerned about the degree of smoothness of solutions when χ vanishes, and moreover any inverse power of χ that appears in the equations of motion. The remaining variables are chosen so that the complete set of equations of motion appear as regular as possible. (There is no proof that this is the optimal choice, but many different choices have been investigated. With carefully chosen numerical methods, even less regular choices of variable may be managed, but the conformal decomposition remains the most prevalent choice to manage this type of data). We see that under this change we have apparently increased the number of evolved variables, but we have also introduced the algebraic constraints

$$\det \tilde{\gamma}_{ij} = \tilde{\gamma} = 1\,, \qquad \tilde{\gamma}^{ij}\tilde{A}_{ij} = \tilde{A} = 0\,. \tag{2.93}$$

When these constraints are explicitly enforced, which is typically done in practice, the solution spaces of the evolution equations with either the ADM or the conformal variables are isomorphic and they therefore have shared PDE properties.

Pushing the evolution equations (2.87), through the change of variables, we obtain

$$\begin{aligned} \partial_t \chi &= \frac{2}{3}\chi\big[\alpha(\hat{K} + 2\Theta) - D_i\beta^i\big]\,, \\ \partial_t \tilde{\gamma}_{ij} &= -2\alpha\tilde{A}_{ij} + \beta^k\partial_k\tilde{\gamma}_{ij} + 2\tilde{\gamma}_{k(i}\partial_{j)}\beta^k - \frac{2}{3}\tilde{\gamma}_{ij}\partial_k\beta^k\,, \end{aligned} \tag{2.94}$$

Likewise from (2.89) we have

$$\partial_t\Theta = \alpha\big[\tfrac{1}{2}H + D^iZ_i - a^iZ_i - K\Theta - \kappa_1(2+\kappa_2)\Theta + \hat{W}_{nn}\big] + \mathcal{L}_\beta\Theta\,, \tag{2.95}$$

for the Θ constraint and the complicated expression

$$\begin{aligned} \partial_t\tilde{\Gamma}^i = {}& 2\alpha\big[\tilde{\Gamma}^i{}_{jk}\tilde{A}^{jk} - \tfrac{3}{2}\tilde{A}^{ij}\partial_j\ln\chi - \tilde{A}^{ij}a_j - 4\pi\tilde{\gamma}^{ij}s_j - \tfrac{1}{3}\tilde{\gamma}^{ij}\partial_j(2\hat{K}+\Theta) \\ & - \tfrac{1}{3}(\hat{K}+2\Theta)(\tilde{\Gamma}^i - \tilde{\Gamma}_{\rm d}{}^i) - \tilde{\gamma}^{ij}a_j\Theta - \tilde{\gamma}^{ij}W_{jn}\big] - \alpha\kappa_1(\tilde{\Gamma}^i - \tilde{\Gamma}_{\rm d}{}^i) \\ & + \tilde{\gamma}^{jk}\partial_j\partial_k\beta^i + \tfrac{1}{3}\tilde{\gamma}^{ij}\partial_j\partial_k\beta^k + \beta^j\partial_j\tilde{\Gamma}^i - \tilde{\Gamma}^j\partial_j\beta^i + \tfrac{2}{3}\tilde{\Gamma}^i\partial_j\beta^j\,, \end{aligned} \tag{2.96}$$

when we combine derivatives of (2.94) with (2.89). Finally, doing the same for (2.88), we arrive at

$$\begin{aligned} \partial_t\hat{K} = {}& -D^iD_i\alpha + \alpha[\tilde{A}^{ij}\tilde{A}_{ij} + \tfrac{1}{3}(\hat{K}+2\Theta)^2] + 4\pi\alpha[s+\rho] \\ & + \alpha[2a^iZ_i + K\Theta + 2\kappa_1(1+\kappa_2)\Theta - W_{nn}] + \mathcal{L}_\beta\hat{K}\,, \end{aligned} \tag{2.97}$$

for the trace variable, and

$$\begin{aligned} \partial_t\tilde{A}_{ij} = {}& \chi\big[-D_iD_j\alpha + \alpha(R_{ij} + 2D_{(i}Z_{j)} + W_{ij} - 8\pi s_{ij})\big]^{\rm tf} + \alpha\big[\hat{K}\tilde{A}_{ij} - 2\tilde{A}^k{}_i\tilde{A}_{jk}\big] \\ & + \beta^k\partial_k\tilde{A}_{ij} + 2\tilde{A}_{k(i}\partial_{j)}\beta^k - \tfrac{2}{3}\tilde{A}_{ij}\partial_k\beta^k\,, \end{aligned} \tag{2.98}$$

for the trace-free part of the extrinsic curvature. Here we have employed a number of auxiliary definitions and shorthands. The superscript tf denotes the trace-free part as usual. The spatial Christoffel symbols can be expressed in terms of the conformal connection via,

$$\Gamma^k{}_{ij} = \tilde{\Gamma}^k{}_{ij} - \delta^k{}_{(i}\partial_{j)}\ln\chi + \tfrac{1}{2}\tilde{\gamma}_{ij}\tilde{\gamma}^{kl}\partial_l\ln\chi\,. \tag{2.99}$$

In terms of the conformal variables, the spatial Ricci tensor can be expressed as

$$
\begin{aligned}
R_{ij} &= \tilde{R}^{\chi}_{ij} + \tilde{R}_{ij}\,,\\
\tilde{R}^{\chi}_{ij} &= \tfrac{1}{2\chi}\tilde{D}_i\tilde{D}_j\chi + \tfrac{1}{2\chi}\tilde{\gamma}_{ij}\tilde{D}^k\tilde{D}_k\chi - \tfrac{1}{4\chi^2}\tilde{D}_i\chi\tilde{D}_j\chi - \tfrac{3}{4\chi^2}\tilde{\gamma}_{ij}\tilde{D}^k\chi\tilde{D}_k\chi\,,\\
\tilde{R}_{ij} &= -\tfrac{1}{2}\tilde{\gamma}^{kl}\partial_k\partial_l\tilde{\gamma}_{ij} + \tilde{\gamma}_{k(i}\partial_{j)}\tilde{\Gamma}_{\mathrm{d}}{}^k + \tilde{\Gamma}_{\mathrm{d}}{}^k\tilde{\Gamma}_{(ij)k} + \tilde{\gamma}^{lm}\left(2\tilde{\Gamma}^k{}_{l(i}\tilde{\Gamma}_{j)km} + \tilde{\Gamma}^k{}_{im}\tilde{\Gamma}_{klj}\right),
\end{aligned}
\tag{2.100}
$$

To summarize, in terms of the conformal variables the Z4 constraints are

$$
\Theta = 0\,, \qquad 2Z_i = \tilde{\gamma}_{ij}(\tilde{\Gamma}^j - \tilde{\Gamma}_{\mathrm{d}}{}^j)\,. \tag{2.101}
$$

The Hamiltonian and momentum constraints are

$$
\begin{aligned}
H &= R - \tilde{A}^{ij}\tilde{A}_{ij} + \tfrac{2}{3}(\hat{K} + 2\Theta)^2 - 16\pi\rho = 0\,,\\
\tilde{M}^i &= \tilde{\gamma}^{ij}M_j = \tilde{D}_j\tilde{A}^{ij} - \tfrac{2}{3}\tilde{D}^i(\hat{K} + 2\Theta) - \tfrac{3}{2}\tilde{A}^{ij}\tilde{D}_j\ln\chi - 4\pi\tilde{\gamma}^{ij}s_j = 0\,,
\end{aligned}
\tag{2.102}
$$

in addition to which we have the algebraic constraints (2.93), which are assumed to be imposed for the reasons discussed above.

The most popular gauge condition in use with these conformal formulations can be written

$$
\begin{aligned}
\partial_t\alpha &= -2\alpha\hat{K} + \mathscr{L}_\beta\alpha\,,\\
\partial_t\beta^i &= \mu_S\tilde{\Gamma}^i - \eta\beta^i + \beta^j\partial_j\beta^i\,.
\end{aligned}
\tag{2.103}
$$

This is known as the moving-puncture gauge. Again, there are various different flavors in use, and here we have selected just the simplest. Observe that just as in the GHG formulation, we choose here evolution equations for the lapse and shift, and in fact these look structurally similar to the generalized harmonic choice. The lapse condition is referred to as the '$1 + \log$' variant of the Bona-Massó slicing condition. The shift condition is known as 'Gamma-driver shift'. In GHG, all of the speeds of propagation associated with the gauge choice coincide with the speed of light. That is not the case in moving puncture gauge, which instead has various superluminal speeds.

Moving now to the constraint addition tensor W_{ab}, observe that we have

$$
W_{ab} = n_a n_b W_{nn} - 2n_{(a}\perp W_{b)n} + \tfrac{1}{3}\gamma_{ab}[2\hat{W}_{nn} - W_{nn}] + \perp W^{\mathrm{tf}}_{ab}\,. \tag{2.104}
$$

The first of the three common flavors of this 'conformal decomposition' setup, which we give in reverse historical order of their appearance, is called CCZ4. In CCZ4 the only non-trivial piece of the constraint addition tensor W_{ab} is taken to be

$$W_{in} = \alpha^{-1}(\kappa_3 - 1)\big(\tfrac{2}{3}Z_i\partial_k\beta^k - \gamma_{ij}\gamma^{kl}Z_k\partial_l\beta^j\big)\,, \tag{2.105}$$

often with $\kappa_3 = 1$, but occasionally with $\kappa_3 = 1/2$. One needs to check the literature for details in different cases. CCZ4 is often used with $\chi^{1/2}$ in place of χ as an evolved variable. The second variation is called Z4c, in which the constraint addition tensor is

$$\begin{aligned}
W_{nn} &= 2a^iZ_i + K\Theta + \kappa_1(1+\kappa_2)\Theta\,,\\
\hat{W}_{nn} &= a^iZ_i + K\Theta + \Gamma^iZ_i - \tfrac{2}{3}\Gamma^i{}_{ij}Z^j\,,\\
\tilde{\gamma}^{ij}W_{jn} &= \tfrac{1}{3}(\hat{K}+2\Theta)(\tilde{\Gamma}^i - \tilde{\Gamma}_{\rm d}{}^i) + \tilde{\gamma}^{ij}a_j - \alpha^{-1}\tilde{\gamma}^{jk}Z_k\partial_j\beta^i + \tfrac{2}{3}\alpha^{-1}\tilde{\gamma}^{ij}Z_j\partial_k\beta^k\,,\\
W_{ij}^{\rm tf} &= 2\Theta K_{ij}^{\rm tf} + 2\big[\Gamma^k{}_{ij}Z_k - \tfrac{2}{3}\Gamma^k{}_{k(i}Z_{j)}\big]^{\rm tf}\,.
\end{aligned} \tag{2.106}$$

The reason for these choices was to make the equations of motion as close as possible to those of the following formulation. To obtain the the third popular option, the BSSN formulation (called on occasion BSSNOK), one need only take the Z4c equations of motion, and set

$$\Theta = 0\,, \tag{2.107}$$

discarding it as an evolved variable, and set all of the constraint damping parameters κ_1, κ_2 to zero. Further variations, including first order reductions and slightly different choices of variable have been investigated. Observe that here we have sacrificed fidelity to the notation used in the literature to present all three formulations together simultaneously.

With suitable choices for the gauge parameter μ_S in (2.103) all three variants are strongly hyperbolic and therefore as discussed in Sect. 2.2.3 admit a well-posed initial value problem. For the moving puncture gauge none are symmetric hyperbolic, and so well-posedness results for the initial boundary value problem are technically harder to come by. Such results are well summarized in [13] (see also [21] for a proof of well-posedness for a system close to those presented here). Consequently most implementations working with puncture gauge employ a naive outer boundary treatment, but try to place the boundary sufficiently far away from the region of interest that it has a negligible effect on the physics of the system.

In the principal part the relationship between gauge, constraints and the gravitational wave degrees of freedom can be understood straightforwardly. It turns out that hyperbolicity of gauge choices can be understood independently of the formulation of GR, and that strong hyperbolicity of a gauge choice is a necessary condition that a formulation built with that gauge will itself be strongly hyperbolic. The same holds for the constraint subsystem.

Supposing we want to tackle a particular physical problem, what principles can we use to choose between different formulations, both here and more generally? In accordance with our discussion above, a first requirement is that of well-posedness of the initial boundary value problem, ideally with boundary conditions that can be implemented in practice. From this point of view the GHG formulation is preferred over those which use moving puncture gauge. On the other hand, as discussed momentarily, the use of the conformally decomposed variables does give a clear advantage in managing the puncture representation of black holes. As we have seen, there is little difference between the different flavors of conformal decomposition of Z4. Both have a constraint subsystem that looks like the wave equation in the principal part, and accept the same constraint damping scheme as the GHG formulation. Non-principal differences in the constraint subsystem do make a difference to the long-term behavior of constraint violations and can induce exponential growth or worse. The analysis required to understand this for a particular constraint addition is, unfortunately, technically involved, and so there is no systematic understanding of what choices are optimal in general. The wavelike nature and damping scheme for the Z4 constraint subsystem has been shown to give advantages over BSSN computations in certain scenarios. The reason for this is that the BSSN constraint subsystem contains slow speeds, which result in constraint violation that sits stationary on the numerical domain and grows.

A crucial feature of formulations of NR is in how black holes are to be managed. We saw in our discussion of initial data that binary black hole initial data can be modeled on the Schwarzschild solution in isotropic coordinates. These are called puncture initial data. The crucial strength of working with the conformally decomposed variables (2.92) and the moving-puncture gauge (2.103) is that the points at which the conformal factor χ vanishes (the puncture) is naturally advected around by the shift condition. In practice this results in a highly versatile treatment with many different physical configurations that requires little fine-tuning of the numerical algorithm. The disadvantage of the moving-puncture approach is the limited regularity of solutions at the puncture itself. Excision, in which the black hole region is cut out of the computational domain explicitly, results in smooth solutions but poses a much greater technical obstacle, since we have to monitor the solution, either actively or passively, to be sure that the excision boundary does not require boundary conditions and that (some notion of) the black hole horizon remains within the computational domain.

2.3 The Physical Interpretation of Numerical Spacetimes

In most cases the raw solution to the Einstein equations provided by the metric over an extended region of spacetime does not correspond directly to the physical quantity of interest. Instead the data is subjected to post-processing. The key examples of this are black hole horizons and gravitational waves. For brevity, we give only an informal overview of each.

Asymptotically flat spacetimes can be characterized by the existence of a certain region in which the metric can be represented as a perturbation of the Minkowski metric in global inertial coordinates. In these spacetimes there is a natural notion of future null-infinity, the final destination for outgoing gravitational waves emitted from the strong-field region. The black hole region, if it is non-empty, is defined as the complement of the causal past of future null-infinity. The boundary of the black hole region is called the event horizon. This definition relies on global structure and is therefore not applicable if we are given only a subset of the spacetime without future null-infinity. Provided a large enough region of data, we can nevertheless compute an accurate approximation to the position of the event horizon. In practice this is done by integrating null-geodesics backwards in time to find the points in spacetime around which they accumulate. Since we compute spacetimes in NR forward in time, this procedure has to be done by post-processing data.

To understand the physics during the evolution itself it is often helpful to have a proxy for the event horizon. For this we build on the concept of a trapped surface. A trapped surface within a foliation is a closed surface at a fixed instant of time Σ_t whose area form would decrease from any point on the surface if we dragged that point infinitesimally to the future along an outgoing null-geodesic starting at said point. In case the area form would remain constant under such a deformation, the surface is called marginally trapped. The outermost marginally trapped surface is called the apparent horizon. Trapped surfaces play an important role in the singularity theorems. For the purposes of NR, the most relevant fact is that if we have a spacetime that does not contain naked singularities, as is expected generically in black hole spacetimes, then any trapped surface must lie inside a cross-section of the event horizon. The definition of the apparent horizon refers to instants of time and so appears in different positions in different foliations, but since it is defined locally in time can be used to characterize black holes during a numerical evolution itself. For details of numerical methods for apparent and event horizon searches, see [22]. For details of additional quasilocal quantities that may be used to interpret spacetimes, see [23].

The primary deliverable of NR for gravitational wave astronomy is the gravitational waveform emitted from a compact binary system. The natural mathematical idealization of a distant observer in an asymptotically flat spacetime, which serves as a model observer for experiments, lies at future null-infinity. From the waveform one can compute the total energy and linear momentum emitted from a system. As is the case with the event horizon, since the computational domain does not, in most cases, include future null-infinity, we can not extract this signal from numerical data. Nevertheless, if a sufficiently large domain is chosen an accurate approximation to the signal at infinity can be computed by extrapolation. This wave extraction procedure can be performed in multiple ways but the most popular approach employs the Newman–Penrose formalism. Besides textbook treatments, the review article [24] contains a complete overview. There is an ongoing effort to include future null-infinity in the computational domain [18, 25–27] in generic asymptotically flat systems. Even with such an improvement, calculating and extracting gravitational waveforms with NR will remain too expensive, and too slow, to make an extremely dense sampling of the solution space. For applications to gravitational wave astronomy, accurate models that aggregate Post-Newtonian, perturbative and NR information are constructed. For textbook introductions to gravitational wave science and modeling, see [28–31].

2.4 Numerical Methods, Implementation and Codes

Numerical methods and code engineering for NR is similar to that in other subfields of computational physics, in particular those involving nonlinear PDEs such as computational fluid dynamics and the like.

For time evolution, most common is the use of the method of lines with explicit integrators. This means discretizing in space to obtain a large set of nonlinear ODEs, with one equation per variable for each position on the grid, and the time coordinate as the independent variable. These ODEs are then typically solved with an explicit integration algorithm such as a Runge–Kutta method. The coupling between these different equations is determined by whatever approximation to the spatial derivative is made. For this the most common options are either high-order finite-differencing or pseudospectral methods, in which the grid-function is represented by a basis of orthogonal polynomials. See [33–36] for textbook introductions to the latter. When working with pseudospectral methods the spatial domain is typically split up into small subdomains, called spectral elements, each of which has an associated initial boundary value problem, with data being communicated from one element to neighbors using a penalty method. The closely related discontinuous Galerkin method [37] is also becoming more popular in the field.

A qualitative difference between these methods and those used for hydrodynamics is that here a high degree of differentiability is assumed and so there is no mention of shocks or shock capturing. The reason for this is that for systems of nonlinear wave equations, the most natural model for formulations of GR, one expects regularity of solutions to persist unless the solution itself breaks down. Regularity of solutions is of course limited when the field equations are coupled to fluid matter which can form shocks. An introduction to numerical methods to manage such irregular solutions can be found in the textbooks [38–41]. Numerical simulations employing such methods are discussed later in this volume.

Adaptive-mesh-refinement (AMR) is a staple for applications in 3d NR. The basic idea is to introduce a measure of numerical error and use that to steer the algorithm to place finer grids where they are needed. The most influential method for finite difference codes is the Berger-Oliger method or derivatives thereof. For compact binary systems it is common to use a simplified 'moving-box' restriction of this approach. Pseudospectral codes employ instead a hybrid approach in which both the spectral elements are made smaller and the number of basis polynomials is increased. For an illustration of such a refined mesh, see Fig. 2.2. Although still somewhat niche in NR, impressive progress has been made in using wavelets for AMR [42].

For elliptic boundary value problems such as those we saw emanating from the constraint equations, many of the methods just mentioned for approximating derivatives are employed within iterative schemes that should converge to solutions of the discretized equations at finite resolution. Nonlinearities may be treated by explicit linearization, with preconditioning for the linear solve, under a Newton-Raphson root finding method, and for efficiency multigrid or AMR is again in widespread use.

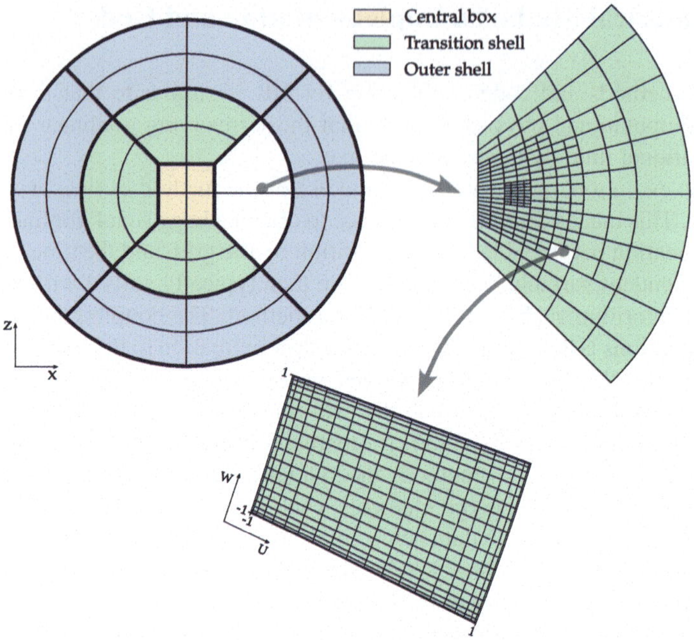

Fig. 2.2 A simple example of a refined grid structure from a pseudospectral code. Figure taken from [32].

From an engineering point of view, 3d NR codes are fairly complicated. They are generally designed to be run on supercomputers. Low-level parts of the code are written in a variety of different programming languages including `c`, `c++`, `Fortran` and `Julia`, with code generated using computer algebra packages such as `Mathematica` or `Mathics`. Job sizes vary greatly, but computation on several thousand cores is common. There are numerous paradigms to effectively achieve parallelism, but ultimately to do so the computational domain has to be divided up and split across different processors. To share data between different processing cores the message-passing standard MPI is usually used. For shared memory architectures the openMP standard is popular. Post-processing, data analysis and the creation of figures is often performed in `Julia`, `Mathematica` or `Python`.

Earlier in the text we asserted that well-posedness of a PDE problem is a necessary condition for an approximation scheme to converge to the continuum solution in the limit of infinite resolution. But what of sufficient conditions? The fact is that in NR there is a considerable gap between what has been proven and that which is implemented and used in practice. The state-of-the-art in this direction is for linear symmetric hyperbolic systems with variable coefficients. There is no proof of convergence for the nonlinear equations for any of the formulations or discretizations discussed above, even without AMR. See [43] for a textbook introduction to numerical analysis, and [13] for a summary of relevant work in NR. For lack of this we have to rely on convergence series. Numerical results are obtained for the exact same physical configuration but at different resolutions. These data are then compared to establish consistency with convergence to the continuum solution, ideally with an expected rate. With this established, the same data can be used to make error-estimates.

Table 2.1 A collection of 3d NR codes in current use, listed together with their developers and with links to technical descriptions, useful introductions and/or their webpages. We restrict here to tools that solve for the metric. The Einstein Toolkit is a broad infrastructure under which we have placed (indented) important individual projects. The ExaHype code likewise provides infrastructure, but to the best of our knowledge there is only one NR module, which we point to here

Code	Developers	Resources
AMSS-NCKU	Beijing Normal University	[44]
bam	CoRe Collaboration	[45]
bamps	Jena/Lisbon	[46, 47]
COFFEE	Canterbury/Otago	[48, 49]
Dendro-GR	Brigham-Young/Utah	[42, 50, 51]
Einstein Toolkit	The ET Consortium	[52–54]
Antelope (FIL)	Frankfurt/Princeton	[55, 56]
Canuda	Aveiro/Illinois	[57–59]
LazEv	Rochester	[60, 61]
LEAN	Cambridge	[62]
MAYA	Austin	[63, 64]
McLachlan	Lousianna-State/Perimeter	[65, 66]
prague	Charles University Prague	[67, 68]
SphericalNR	Bowdoin/Rochester/West Virginia	[69, 70]
Elliptica	Florida-Atlantic	[71]
ExaHyPE	Frankfurt/Trento	[72, 73]
FUKA	Frankfurt/Paris Observatory/Princeton	[56, 74]
GR-Athena++	Jena/Penn-State/Princeton	[75]
GRCHOMBO	GRTL Collaboration	[76–78]
HAD	Long-Island	[79–81]
KADATH	Paris Observatory	[82, 83]
LORENE		[84]
MHDueT	ICE-CSIC/Hamburg/Long-Island/Nottingham/UIB	[85, 86]
NMESH	Florida-Atlantic	[87]
gh3d2m	Perimeter/Princeton	[88, 89]
SACRA	Illinois/Kyoto/Okinawa/Tsukuba/Toho/Tokyo/Wako	[90]
SENR/NRpy+	University of Idaho	[91]
SpEC	SxS Collaboration	[92]
SpECTRE		[93, 94]
SphGR	Bowdoin	[95, 96]
SPHINCS_BSSN	Hamburg-Louisiana-Stockholm	[97, 98]

In Table 2.1 we give a brief overview of the 3d NR codes currently in use. We include here only codes that solve for the metric variables, and only those designed to treat the initial or initial boundary value problems. A similar summary, which includes also tools for treating fluid matter rather than just focusing on the metric can be found in [99]. For an overview of codes written to solve the characteristic initial boundary value problem for astrophysical applications, see [18, 26, 27]. Tools to manage, post-process and plot data produced by each of these tools are described either within the references or at the associated webpages.

Acknowledgements I am grateful to Jorge Expósito Patiño, Tomas Ledvinka, Christian Peterson Bórquez, Maciej Maliborski, Krinio Marouda, Hannes Rüter, Alex Vañó-Viñuales and Miguel Zilhão for helpful discussions and/or feedback on the text. This work was supported in part by FCT (Portugal) Project No. UIDB/00099/2020.

References

1. R.M. Wald, *General Relativity* (The University of Chicago Press, Chicago, 1984)
2. Y. Mizuno, L. Rezzolla, General-Relativistic Magnetohydrodynamic Equations: the bare essential (2024)
3. M. Alcubierre, *Introduction to 3+1 Numerical Relativity* (Oxford University Press, Oxford, 2008)
4. T.W. Baumgarte, S.L. Shapiro, N. Relativity, *Solving Einstein's Equations on the Computer* (Cambridge University Press, Cambridge, 2010)
5. E. Gourgoulhon, *3+1 Formalism in General Relativity* (Springer, Berlin, 2012)
6. M. Shibata, *Numerical Relativity* (World Scientific, Singapore, 2016)
7. T.W. Baumgarte, S.L. Shapiro. *Numerical Relativity: Starting from Scratch* (Cambridge University Press, 2021)
8. E. Poisson, C.M. Will, *Gravity: Newtonian, Post-Newtonian, Relativistic* (Cambridge University Press, Cambridge, England, 2014)
9. A. Carlotto, The general relativistic constraint equations. Living Rev. Relativ. **24**, 2 (2021)
10. W. Tichy, The initial value problem as it relates to numerical relativity. Rept. Prog. Phys. **80**(2), 026901 (2017)
11. H.-O. Kreiss, J. Lorenz, *Initial-Boundary Value Problems and the Navier-Stokes Equations* (Academic, New York, 1989)
12. B. Gustafsson, H.-O. Kreiss, J. Oliger, *Time Dependent Problems and Difference Methods* (Wiley, New York, 1995)
13. O. Sarbach, M. Tiglio, Continuum and discrete initial-boundary value problems and Einstein's field equations. Living Rev. Relativ. **15**(9) (2012)
14. D. Hilditch, An introduction to well-posedness and free-evolution. Int. J. Mod. Phys. A **28**, 1340015 (2013)
15. H. Friedrich, A.D. Rendall, The Cauchy problem for the Einstein equations. Lect. Notes Phys. **540**, 127–224 (2000)
16. H. Ringström, *The Cauchy Problem in General Relativity* (European Mathematical Society, 2009)
17. C.D. Sogge. *Lectures on nonlinear wave equations*. Number Bd. 2 in Monographs in analysis (International Press, 1995)
18. J. Winicour, Characteristic evolution and matching. Living Rev. Relativ. **8**, 10 (2005). [Online article]
19. J. Frauendiener, Conformal infinity. Living Rev. Relativ. **7**(1) (2004)

20. J.-A. Valiente-Kroon, *Conformal Methods in General Relativity* (Cambridge University Press, Cambridge, 2016)
21. D. Hilditch, M. Ruiz, The initial boundary value problem for free-evolution formulations of General Relativity (2016)
22. J. Thornburg, Event and apparent horizon finders for 3 + 1 numerical relativity. Living Rev. Relativ. (2006). [Online article]
23. L.B. Szabados, Quasi-local energy-momentum and angular momentum in GR: a review article. Living Rev. Relativ. **12**, 4 (2009)
24. N.T. Bishop, L. Rezzolla, Extraction of gravitational waves in numerical relativity. Living Rev. Relativ. **19**(1), 2 (2016)
25. The Hyperboloidal Research Network: https://hyperboloid.al/
26. C. Reisswig, N.T. Bishop, D. Pollney, B. Szilagyi, Unambiguous determination of gravitational waveforms from binary black hole mergers. Phys. Rev. Lett. **103**, 221101 (2009)
27. S. Ma et al., Fully relativistic three-dimensional Cauchy-characteristic matching (2023)
28. M. Maggiore. *Gravitational Waves. Vol. 1: Theory and Experiments* (Oxford University Press, Oxford, 2007)
29. J.D.E. Creighton, W.G. Anderson, Gravitational-wave physics and astronomy: an introduction to theory, experiment and data analysis (2011)
30. M. Maggiore. *Gravitational Waves. Vol. 2: Astrophysics and Cosmology* (Oxford University Press, 2018)
31. N. Andersson, *Gravitational-Wave Astronomy: Exploring the Dark Side of the Universe*. Oxford Graduate Texts (Oxford University Press, 2019)
32. D. Cors, S. Renkhoff, H.R. Rüter, D. Hilditch, B. Brügmann, Formulation improvements for critical collapse simulations. Phys. Rev. D **108**(12), 124021 (2023)
33. L.N. Trefethen, *Spectral Methods in MATLAB* (SIAM, Philadelphia, 2000)
34. J.P. Boyd, *Chebyshev and Fourier Spectral Methods* (2nd edn., Revised) (Dover Publications, New York, 2001)
35. J.S. Hesthaven, S. Gottlieb, D. Gottlieb, *Spectral Methods for Time-Dependent Problems* (Cambridge University Press, Cambridge, 2007)
36. D.A. Kopriva, *Implementing Spectral Methods for Partial Differential Equations* (Springer, New York, 2009)
37. J.S. Hesthaven, T. Warburton, *Nodal Discontinuous Galerkin Methods* (Springer, New York, 2008)
38. R.J. Leveque, *Numerical Methods for Conservation Laws* (Birkhauser Verlag, Basel, 1992)
39. R.J. LeVeque. *Finite Volume Methods for Hyperbolic Problems* (Cambridge University Press, 2002)
40. J.R. Wilson, G.J. Mathews, *Relativistic Numerical Hydrodynamics* (Cambridge University Press, 2003)
41. L. Rezzolla, O. Zanotti, *Relativistic Hydrodynamics* (Oxford University Press, Oxford, 2013)
42. M. Fernando, D. Neilsen, Y. Zlochower, E.W. Hirschmann, H. Sundar, Massively parallel simulations of binary black holes with adaptive wavelet multiresolution. Phys. Rev. D **107**(6), 064035 (2023)
43. J.W. Thomas, *Numerical Partial Differential Equations: Finite Difference Methods*. Texts in Applied Mathematics (Springer, New York, 1998)
44. Z.-J. Cao, H.-. Yo, Y. Jui-Ping, A reinvestigation of moving punctured black holes with a new code. Phys. Rev. D **78**, 124011 (2008)
45. Core collaboration. http://www.computational-relativity.org/
46. D. Hilditch, A. Weyhausen, B. Brügmann, Pseudospectral method for gravitational wave collapse. Phys. Rev. D **93**(6), 063006 (2016)
47. S. Renkhoff, D. Cors, D. Hilditch, B. Brügmann, Adaptive hp refinement for spectral elements in numerical relativity. Phys. Rev. D **107**, 104043 (2023)
48. J. Frauendiener, C. Stevens, The non-linear perturbation of a black hole by gravitational waves. I. The Bondi–Sachs mass loss. Class. Quant. Grav. **38**(19), 194002 (2021)

49. G. Doulis, J. Frauendiener, C. Stevens, B. Whale, COFFEE - an MPI-parallelized Python package for the numerical evolution of differential equations. SoftwareX **10**, 100283 (2019)
50. M. Fernando, D. Neilsen, H. Lim, E. Hirschmann, H. Sundar, Massively parallel simulations of Binary Black Hole intermediate-mass-ratio inspirals. SIAM J. Sci. Comput. **41**(2), C97–C138 (2019)
51. Dendro-gr website. https://paralab.github.io/Dendro-GR/
52. F. Löffler, J. Faber, E. Bentivegna, T. Bode, P. Diener, R. Haas, I. Hinder, B.C. Mundim, C.D. Ott, E. Schnetter, G.A. Allen, M. Campanelli, P. Laguna, The Einstein Toolkit: a community computational infrastructure for relativistic astrophysics. Class. Quant. Grav. **29**, 115001 (2012)
53. M. Zilhão, F. Löffler, An Introduction to the Einstein Toolkit. Int. J. Mod. Phys. A **28**, 1340014 (2013)
54. Einstein Toolkit
55. E.R. Most, L.J. Papenfort, L. Rezzolla, Beyond second-order convergence in simulations of magnetized binary neutron stars with realistic microphysics. Mon. Not. Roy. Astron. Soc. **490**(3), 3588–3600 (2019)
56. L.J. Papenfort, S.D. Tootle, P. Grandclément, E.R. Most, L. Rezzolla, New public code for initial data of unequal-mass, spinning compact-object binaries. Phys. Rev. D **104**(2), 024057 (2021)
57. H. Okawa, H. Witek, V. Cardoso, Black holes and fundamental fields in Numerical Relativity: initial data construction and evolution of bound states. Phys. Rev. D **89**(10), 104032 (2014)
58. M. Zilhão, H. Witek, V. Cardoso, Nonlinear interactions between black holes and Proca fields. Class. Quant. Grav. **32**, 234003 (2015)
59. Canuda, https://bitbucket.org/canuda/
60. M. Campanelli, C.O. Lousto, P. Marronetti, Y. Zlochower, Accurate evolutions of orbiting black-hole binaries without excision. Phys. Rev. Lett. **96**, 111101 (2006)
61. Y. Zlochower, J.G. Baker, M. Campanelli, C.O. Lousto, Accurate black hole evolutions by fourth-order numerical relativity. Phys. Rev. D **72**, 024021 (2005). (arXiv:gr-qc/0505055)
62. U. Sperhake, Binary black-hole evolutions of excision and puncture data. Phys. Rev. D **76**, 104015 (2007)
63. K. Jani, J. Healy, J.A. Clark, L. London, P. Laguna, D. Shoemaker, Georgia tech catalog of gravitational waveforms. Class. Quant. Grav. **33**(20), 204001 (2016)
64. D. Ferguson, et al., Second MAYA catalog of binary black hole numerical relativity waveforms (2023)
65. J.D. Brown, P. Diener, O. Sarbach, E. Schnetter, M. Tiglio, Turduckening black holes: an analytical and computational study. Phys. Rev. D **79**, 044023 (2009)
66. C. Reisswig, C.D. Ott, U. Sperhake, E. Schnetter, Gravitational wave extraction in simulations of rotating Stellar core collapse. Phys. Rev. D **83**, 064008 (2011)
67. A. Khirnov, T. Ledvinka, Slicing conditions for axisymmetric gravitational collapse of Brill waves. Class. Quant. Grav. **35**(21), 215003 (2018)
68. T. Ledvinka, A. Khirnov, Universality of curvature invariants in critical vacuum gravitational collapse. Phys. Rev. Lett. **127**, 011104 (2021)
69. V. Mewes, Y. Zlochower, M. Campanelli, I. Ruchlin, Z.B. Etienne, T.W. Baumgarte, Numerical relativity in spherical coordinates with the Einstein Toolkit. Phys. Rev. D **97**(8), 084059 (2018)
70. V. Mewes, Y. Zlochower, M. Campanelli, T.W. Baumgarte, Z.B. Etienne, F.G. Lopez Armengol, F. Cipolletta, Numerical relativity in spherical coordinates: a new dynamical spacetime and general relativistic MHD evolution framework for the Einstein Toolkit. Phys. Rev. D **101**(10), 104007 (2020)
71. A. Rashti, F.M. Fabbri, B. Brügmann, S.V. Chaurasia, T. Dietrich, M. Ujevic, W. Tichy, New pseudospectral code for the construction of initial data. Phys. Rev. D **105**(10), 104027 (2022)
72. M. Dumbser, F. Guercilena, S. Köppel, L. Rezzolla, O. Zanotti, Conformal and covariant Z4 formulation of the Einstein equations: strongly hyperbolic first-order reduction and solution with discontinuous Galerkin schemes. Phys. Rev. D **97**(8), 084053 (2018)
73. www.exahype.org/
74. https://bitbucket.org/fukaws/workspace/repositories/

75. B. Daszuta, F. Zappa, W. Cook, D. Radice, S. Bernuzzi, V. Morozova, GR-Athena++: puncture evolutions on vertex-centered oct-tree adaptive mesh refinement. Astrophys. J. Supp. **257**(2), 25 (2021)
76. K. Clough, P. Figueras, H. Finkel, M. Kunesch, E.A. Lim, S. Tunyasuvunakool, GRChombo: numerical relativity with adaptive mesh refinement. Class. Quant. Grav. **32**(24), 245011 (2015)
77. T. Andrade et al., GRChombo: an adaptable numerical relativity code for fundamental physics. J. Open Source Softw. **6**(68), 3703 (2021)
78. Grtl collaboration. https://www.grchombo.org/
79. S.L. Liebling, The singularity threshold of the nonlinear sigma model using 3D adaptive mesh refinement. Phys. Rev. D **66**, 041703(R) (2002)
80. L. Lehner, S.L. Liebling, O. Reula, AMR, stability and higher accuracy (2005)
81. http://had.liu.edu/
82. P.G. Kadath, A spectral solver for theoretical physics. J. Comput. Phys. **229**, 3334–3357 (2010)
83. https://kadath.obspm.fr/
84. E. Gourgoulhon, P. Grandclément, J.-A. Marck, J. Novak, K. Taniguchi. http://www.lorene.obspm.fr
85. C. Palenzuela, B. Miñano, D. Viganò, A. Arbona, C. Bona-Casas, A. Rigo, M. Bezares, C. Bona, Joan Massó, A Simflowny-based finite difference code for high-performance computing in numerical relativity. Class. Quant. Grav. **35**(18), 185007 (2018)
86. http://mhduet.liu.edu/
87. W. Tichy, L. Ji, A. Adhikari, A. Rashti, M. Pirog, The new discontinuous Galerkin methods based numerical relativity program Nmesh. Class. Quant. Grav. **40**(2), 025004 (2023)
88. F. Pretorius, Numerical relativity using a generalized harmonic decomposition. Class. Quant. Grav. **22**, 425–451 (2005)
89. W.E. East, F. Pretorius, B.C. Stephens, Hydrodynamics in full general relativity with conservative AMR. Phys. Rev. D **85**, 124010 (2012)
90. T. Yamamoto, M. Shibata, K. Taniguchi, Simulating coalescing compact binaries by a new code SACRA. Phys. Rev. D **78**, 064054 (2008)
91. I. Ruchlin, Z.B. Etienne, T.W. Baumgarte, SENR/NRPy+: numerical relativity in singular curvilinear coordinate systems. Phys. Rev. D **97**(6), 064036 (2018)
92. SpEC - Spectral Einstein Code. http://www.black-holes.org/SpEC.html
93. L.E. Kidder et al., SpECTRE: a Task-based discontinuous Galerkin code for relativistic astrophysics. J. Comput. Phys. **335**, 84–114 (2017)
94. https://spectre-code.org/
95. T.W. Baumgarte, P.J. Montero, I. Cordero-Carrion, E. Muller, Numerical relativity in spherical polar coordinates: evolution calculations with the BSSN formulation. Phys. Rev. D **87**(4), 044026 (2013)
96. T.W. Baumgarte, P.J. Montero, E. Müller, Numerical relativity in spherical polar coordinates: off-center simulations. Phys. Rev. D **91**(6), 064035 (2015)
97. S. Rosswog, P. Diener, SPHINCS_BSSN: a general relativistic Smooth Particle Hydrodynamics code for dynamical spacetimes. Class. Quant. Grav. **38**(11), 115002 (2021)
98. S. Rosswog, F. Torsello, P. Diener, The Lagrangian numerical relativity code SPHINCS_BSSN_v1.0. 6 (2023)
99. N. Afshordi et al., Waveform modelling for the laser interferometer space antenna (2023)

Chapter 3
Equations of State for Neutron Star Simulations

Debarati Chatterjee

Abstract This chapter provides an introduction to the techniques for modelling fluids in numerical simulations in general relativity. It provides a broadbrush picture of the commonly used descriptions of equations of state in GRMHD simulations. It also provides a summary of the state-of-the-art microphysical descriptions of compact objects, compatible with recent constraints from multi-disciplinary physics. This would be useful to students and researchers outside the nuclear astrophysics community who would like to have some knowledge of realistic equations of state that are currently being implemented in tabulated form into such simulations.

3.1 Background

With recent advances in observational astronomical techniques, a large pool of electromagnetic multi-wavelength data for compact objects, such as neutron stars (NSs) and black holes (BHs), has become accessible. Further, the recent direct detection of Gravitational Waves (GWs) from compact binary mergers [1] has also opened a new window of possibility to study their interior properties directly [2–5]. For correct interpretation of multi-messenger data and for improving our understanding of dynamical astrophysical systems, numerical simulations are indispensable.

General Relativity (GR) is an essential ingredient to describe astrophysical scenarios involving strongly relativistic compact objects. In systems such as binary neutron star mergers or astrophysical jets, effects of electromagnetic field also need to be accounted for. In general, to describe the dynamics of relativistic astrophysical systems, one needs to solve coupled equations of time-dependent partial differential equations, comprising the general-relativistic hydrodynamic and magnetohydrodynamic (GRMHD) equations along with the Einstein equations for the gravitational field. These equations must be supplemented as necessary with further

D. Chatterjee (✉)
Inter-University Centre for Astronomy and Astrophysics, Post Bag 4, Ganeshkhind, Pune University Campus, Pune 411007, India
e-mail: debarati@iucaa.in

C. Bambi et al. (eds.), *New Frontiers in GRMHD Simulations*, Springer Series in Astrophysics and Cosmology, https://doi.org/10.1007/978-981-97-8522-3_3

inputs on nonadiabatic processes, e.g. transport processes, radiative transfer or the microphysical description of the compact star interior.

As numerical simulations are extremely demanding in terms of computational costs, one must introduce simplifications wherever possible. In several astrophysical scenarios of interest (e.g. accretion of matter onto compact objects, relativistic outflows and jets, or stellar oscillations), an ideal-fluid and ideal-MHD approximation (neglecting dissipation) is good enough to provide a relatively accurate description of the underlying dynamics. However, in the recent past, rapid advances in computational power and supercomputing techniques is making large-scale numerical simulations in relativistic framework possible, not only in two-dimensional but also in three-dimensional systems. It has become possible to progressively include complexities in these simulations, such as realistic microphysical descriptions of neutron star interior including electromagnetic fields and thermal effects.

This chapter aims to provide the readers a broadbrush picture of the commonly used techniques for modelling fluids in numerical simulations. It also summarizes the state-of-the-art in microphysical descriptions of compact objects, compatible with recent constraints from multi-disciplinary physics. The article is organised as follows: in Sect. 3.2 a simplified description of fluids applicable for GRMHD simulations is outlined. In Sect. 3.3, microphysical descriptions of Neutron Stars is described in detail. In Sect. 3.4, further advanced descriptions of NS models are described. In Sect. 3.5, parametric and non-parametric models are discussed. Section 3.6 elaborates on the state-of-the-art constraints from multi-disciplinary physics on the NS EoS. Finally, Sect. 3.7 discusses some ways forward.

3.2 Modelling Fluids: The Simplified Picture

Laws of Thermodynamics govern the state of the matter in a given fluid element. All thermodynamic properties of a system are determined completely if a thermodynamic potential is known as a function of its state variables, e.g. baryon number density n, mass density ρ, energy density ε, pressure p, temperature T, specific entropy s, etc. However, not all the thermodynamic variables are independent; they are connected via the so-called Equation of State (EoS). NS matter has three degrees of freedom, and can be parametrized in terms of baryon number density n, temperature T and electron fraction Y_e, i.e. $p(n, T, Y_e)$. In general for cold neutron stars, thermal effects can be neglected and for the barotropic EoS the pressure can be described as function of the density as $p = p(n)$.

The EoS is therefore a key link between the microphysics that governs the local fluid behaviour and global quantities (such as the mass and radius of a star) that can be derived from astrophysical observations. In numerical simulations, the equations of motion and the continuity equation must be supplemented by an EoS relating some fundamental thermodynamic quantities. For a more systematic discussion of the EoS from a kinetic point of view, the reader may refer to the book "Relativistic Hydrodynamics" by Rezzolla and Zanotti [6].

3.2.1 Free Fermi Gas

Relativistic fluids in general can be modelled as non-degenerate and degenerate fluids, composed of bosons or fermions. Compact stars, such as white dwarfs and neutron stars, are primarily composed of degenerate Fermions (electrons and neutrons respectively) which obey Fermi-Dirac statistics. We therefore examine the general case of degenerate Fermi gas in further detail below [7].

For a degenerate gas of Fermions, the number of states available at momentum k per unit volume is

$$dn = \frac{d^3k}{(2\pi\hbar)^3} = \frac{4\pi k^2 dk}{(2\pi\hbar)^3}.$$

The number density is given by

$$n = \frac{8\pi}{(2\pi\hbar)^3}\int_0^{k_F} k^2 dk = \frac{{k_F}^3}{3\pi^2\hbar^3}, \tag{3.1}$$

where k_F is the Fermi momentum. Integrating the energy of Fermions over the Fermi sphere, one can obtain the energy density

$$\begin{aligned}\varepsilon(k_F) &= \frac{8\pi}{(2\pi\hbar)^3}\int_0^{k_F} (k^2c^2 + m^2c^4)^{1/2}k^2 dk \\ &= \varepsilon_0 \int_0^{k_F/mc} (u^2+1)^{1/2}u^2 du \\ &= \frac{\varepsilon_0}{8}\left[(2x^3+x)(1+x^2)^{1/2} - \sinh^{-1}(x)\right],\end{aligned}$$

defining dimensionless variables $u = k/mc$ and $x = k_F/mc$, with $\varepsilon_0 = \frac{m^4c^5}{\pi^2\hbar^3}$ (in dyne/cm^2). Similarly, one obtains the pressure from an isotropic distribution of momenta (v being the velocity):

$$\begin{aligned}p &= \frac{1}{3}\frac{8\pi}{(2\pi\hbar)^3}\int_0^{k_F} kvk^2 dk \\ &= \frac{8\pi}{3(2\pi\hbar)^3}\int_0^{k_F} (k^2c^2 + m^2c^4)^{-1/2}c^2k^4 dk \\ &= \frac{\varepsilon_0}{3}\int_0^{k_F/mc} (u^2+1)^{-1/2}u^4 du \\ &= \frac{\varepsilon_0}{24}\left[(2x^3-3x)(1+x^2)^{1/2} + 3\sinh^{-1}(x)\right].\end{aligned}$$

The relation between the pressure and energy density gives the EoS. For the non-relativistic case, $x \ll 1$ ($k_F \ll mc$), we get

$$\varepsilon(x) \simeq \frac{\varepsilon_0}{3} x^3,$$

$$p(x) \simeq \frac{\varepsilon_0}{15} x^5.$$

Defining mass density $\rho = mn$, $p = K_{NR}\rho^{5/3}$ where $K_{NR} = \frac{\hbar^2}{5}\left(\frac{3\pi^2}{m^4}\right)^{2/3}$.

An important application of this result is in the case of white dwarfs, composed of electrons and atoms. At a temperature 10^7 K, the electrons behave as a cold relativistic Fermi gas, while atoms are non-relativistic. In 1931 Chandrasekhar used the properties of degenerate relativistic Fermi fluid to describe white dwarfs and to infer the maximum mass limit [8].

In the ultra-relativistic case, $x \gg 1$ ($k_F \gg mc$), we get

$$\varepsilon(x) \simeq \frac{\varepsilon_0}{3} x^4,$$

$$p(x) \simeq \frac{\varepsilon_0}{12} x^4.$$

Then $p \simeq K_R \rho^{4/3}$ where $K_R = \frac{\hbar c}{4}\left(\frac{3\pi^2}{m^4}\right)^{1/3}$, which implies $p = \varepsilon/3$.

Unlike the case of white dwarfs, the NS interior cannot be described in terms of relativistic fluid consisting only of free neutrons. It was shown by Oppenheimer and Volkoff [9] by solving the hydrostatic equilibrium equations (see Sect. 3.3.4) that such a system supports a maximum mass of only $\sim 0.7 M_\odot$, far below observed NS masses. It was therefore evident that strong nuclear forces should be accounted for in order to explain astrophysical observations of NSs [10]. The microphysical neutron star models describing nucleon-nucleon interaction in dense matter will be discussed in Sect. 3.3.

3.2.2 *Polytropic EoS*

The results for the degenerate non-relativistic and relativistic free Fermi gas can be generalised to parametrization of the EoS of the form, known as a *Polytropic EoS* [11]

$$p = K\rho^\Gamma$$

where $\Gamma = 1 + \frac{1}{n}$ is the polytropic exponent and n the polytropic index. Γ takes the value $\frac{5}{3}$ in the non-relativistic limit and $\frac{4}{3}$ in the relativistic limit and K is a constant. The polytropic exponent Γ can also be defined by the relation

$$\Gamma = \frac{\partial \log p}{\partial \log \rho}. \tag{3.2}$$

The energy density of a polytrope is given by

$$\varepsilon(\rho) = \frac{K}{\Gamma - 1} \cdot \rho^{\Gamma} \tag{3.3}$$

for particle of mass m. One must note that polytropes demonstrate acausal behaviour at high densities. The squared speed of sound at constant entropy

$$c_s^2 = \left.\frac{\partial p}{\partial \varepsilon}\right|_s \tag{3.4}$$

becomes unity in the limiting case $\Gamma = 2$, corresponding to the stiffest possible EoS $p = \varepsilon$, also referred to as the *causal limit*. This implies polytropes become acausal for $\Gamma > 2$.

3.2.3 Ideal Fluid

For solving the hydrostatic equilibrium equations, one may also define a relation between the pressure and density of the form

$$p \propto \rho^{\gamma},$$

where γ is the adiabatic index for an ideal fluid EoS. One must note that although γ values coincide with those of Γ for non-relativistic and relativistic cases for a composite Fermion gas, in the ultra-relativistic limit for a single Fermion gas, one has $p = \frac{1}{3}\varepsilon$, i.e. $\gamma = 1$ while $\Gamma = \frac{4}{3}$.

One may note that the EoS of a classical monatomic fluid, found to be valid also for non-degenerate fluids, can be written in terms of the adiabatic index as

$$p = \rho\varepsilon(\gamma - 1). \tag{3.5}$$

Due to its simplicity, the ideal fluid EoS is the most commonly adopted EoS in numerical simulations and referred to as the *ideal fluid* or Γ-law EoS

$$p = (\Gamma - 1)\rho\varepsilon,$$

where Γ is the adiabatic index.

3.3 Modelling Neutron Stars: The Microphysical Picture

For compact stars, one may go beyond the simple descriptions mentioned in this chapter to employ microphysical EoSs to model the complex interior. There are a large number of available EoSs that include state-of-the-art information from nuclear physics, astrophysics, condense matter physics etc. This will be elaborated in the next section. Although there are uncertainties in such theoretical models as they extrapolate known laboratory physics to unknown regimes of temperature and densities, such EoSs are being increasingly used in the numerical modelling of relativistic stars.

3.3.1 What We Know from Astrophysical Observations

With the diverse ground- and space-based telescopes available today that scan the skies at multiple wavelengths, NSs are observed across the electromagnetic spectrum, from γ-rays to X-rays all the way to radio frequencies. The vast pool of observational data provides insight into a number of global properties of these objects [12–14]. In pulsars, one can accurately measure their spin frequencies, as well as their spin derivatives, from which one can derive other properties such as their magnetic fields, characteristic age or braking index. For NSs in binary systems with another NS or a companion star (such as an ordinary star or a white dwarf), one can measure precisely relativistic effects beyond the usual Keplerian orbital properties, and these post-Keplerian parameters allow the determination of the component masses in the binary to high accuracy. Although most of the NSs are found close to a canonical mass of around 1.4 $M_{\odot}$, we are particularly interested in the highest observed NS masses that test the limit of the theoretical models that can explain such an observation. In the past decade, several NSs have been found with masses close to 2 $M_{\odot}$ [15–17]. The heaviest known pulsar PSR J0740+6620 has been observed with a mass $2.08^{+0.07}_{-0.07}$ $M_{\odot}$ [18, 19].

The determination of NS radii is more challenging. The most reliable NS radii are derived from observations of thermal emission from NS surfaces. Radius estimates from isolated neutron stars are affected by large distance uncertainties [20–24]. Pulsar distances are obtained from dispersion measures, which have large ambiguities (e.g. in the absorption of X-rays in the interstellar medium, atmospheric composition and magnetic field). Measurements of radii from NSs in globular clusters might be relatively accurate if distance measurements are improved. If the atmosphere and distance are well known, one can obtain radius measurements. One may also obtain NS radius measurements by combining observation of NS mass M with another astrophysical observable, such as its moment of inertia $I \propto MR^2$ (derived from NS precession) or its compactness $C = M/R$ (obtained from X-ray spectra). Most observations indicate NSs to have radii ∼9–15 km. With the recent launch of the Neutron Star Interior Composition ExploreR (NICER) [25] mission from NASA, the accuracy in the measurements of NS radii have drastically improved to a few per cent. By employing the novel technique of Pulse Profile modelling, NS radii can be inferred from general relativistic effects on hotspots in accreting NSs. In particular, two NICER measurements of pulsars PSR J0030+0451 [26, 27] and PSR J0740+6620 [28, 29] have resulted in improved radius estimates.

A recent breakthrough in NS radius measurements also came from the direct detection of GWs from the merger of NSs. Neutron stars, being strongly relativistic, are important sources of GWs, both in isolated or binary systems. The LIGO-Virgo Collaboration in 2017 detected GWs from a NS merger event GW170817 [1]. During the inspiral phase of the merger, the tidal deformation of each of the NSs in the gravitational field of the other could be accurately determined, which is correlated with the radius of the NSs [30]. Therefore the analysis of the GW170817 data provided important constraints on NS radii [2, 4, 5]. It is however important to note that these measurements indicate NS radii ∼10–15 km, which is in slight tension with the NICER measurements. Therefore combining information from GW data with NICER measurements as well as those from the electromagnetic counterparts of the event provide even more stringent constraints on the NS radius.

These observations of NS masses and radii are consistent with those from the theories of stellar evolution and the calculations/simulations of core-collapse supernova. These indicate that NSs have masses $\sim 1-2M_{\odot}$, confined within a very small radius ∼10 km. A simple estimate indicates that the central densities of such objects could be as large as $\sim 2-10$ times those of nuclear densities.

3.3.2 What We Know from Terrestrial Experiments

In order to understand the fundamental constituents of atoms and to probe the behaviour of dense matter, we have at our disposal several terrestrial experiments, such as nuclear and heavy-ion experiments. **Nuclear experiments** allow investigation of the behaviour of properties of matter close to saturation nuclear densities ($n_{sat} \sim 10^{14}$ g/cm^3) [31]. However most of these experiments probe properties of nuclei close to the stability line, with almost equal number of neutron and protons (i.e. “isospin symmetric”), while only a few (e.g. ISOLDE, SIDDHARTA) targeted experiments probe properties of neutron-rich nuclei. Neutron stars, on the other hand, are composed almost entirely of neutrons, and are therefore highly isospin asymmetric [13].

Heavy-ion collision experiments in particle accelerators, e.g. CERN, GANIL, LHC investigate the properties of hot and dense neutron star matter, exceeding densities of ∼2–3 n_{sat} [32, 33]. Although this is the density regime relevant to the NS core, the matter in heavy-ion collisions is isospin symmetric and at high temperature. Therefore any information from heavy-ion data would require extrapolation to the unknown regimes of low temperature and finite isospin asymmetries. Further, the analysis of heavy-ion data is itself strongly dependent on the models employed and comparison with astrophysical data therefore requires caution.

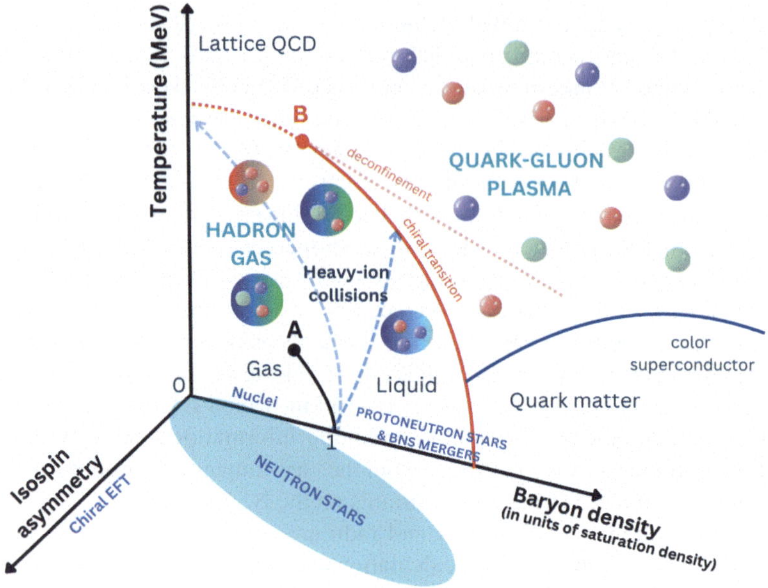

Fig. 3.1 Neutron stars on the QCD phase diagram. [*Credit* Debarati Chatterjee]

The differences between terrestrial experiments and compact stars can be understood from the phase diagram of Quantum Chromodynamics (QCD), the theory of strong interactions [34, 35]. Referring to Fig. 3.1, we see that the two regimes on the diagram are distinct and complementary. Due to the nature of the QCD coupling constant (decreasing or "running" of the coupling at high densities), perturbative approach to the theory (pQCD) is applicable only at very high densities (close to $\sim 50n_{sat}$). At high temperature and low baryon density, which is relevant for the early Universe and accessible to particle accelerators, the non-perturbative nature of QCD is described using simulations of QCD on the lattice (Lattice QCD). Nuclear experiments correspond to the low temperature and baryon density axis of the phase diagram. The coexistence curve corresponding to the liquid-gas phase transition (marked by critical point A) demarcates the distinct phases of finite nuclei and nuclear matter. It is now established that the nature of the phase transition close to the critical point B close to $T_c \sim 120$ MeV is a crossover. The coexistence curve differentiates the hadronic phase of matter from the deconfined quark-gluon plasma phase at high baryon densities and temperatures. It is conjectured that a zoo of color superfluid phases (2SC, CFL, LOFF etc.) may appear at high densities and low temperatures due to Cooper pairing of the deconfined fermionic quarks (see Sect. 3.4.1.2).

Neutron stars span a wide range of densities from 1 to 10 n_{sat} close to the low temperature axis, and therefore encounter a wide range of dense matter phases. This is described in detail in the next Sect. 3.3.3. In contrast with the other terrestrial experiments, neutron stars lie along a different axis in the figure, corresponding to finite isospin asymmetry. Newly born neutron stars also extend to the finite temperature region of the diagram. It is therefore clear that although the different first-principle calculations and experiments provide important information at other conditions of temperature and densities, to describe the dense matter properties in neutron stars one must resort to theoretical models in order to extrapolate the information to unknown regimes of temperature and densities.

3.3.3 NS Interior Structure

In the previous section, we outlined the different phases of matter that appear across the phase diagram at different densities and temperatures. As neutron stars span a wide range of densities (1–10 n_{sat}) increasing from the surface to the interior, they encounter different forms of matter [34, 36–39]. The atmosphere and envelope play a negligible role in the structure properties of the NS, so we will refrain from their detailed description. The outermost layer (the *outer crust*) of the NS matter comprise of a lattice of nuclei, close to the iron (Fe) family, embedded in a sea of free electrons. This is expected from the formation of the remnant proto-neutron star in a core-collapse supernova explosion. As the density increases towards the centre, one may encounter more and more neutron rich nuclei. Beyond $4\times\ 10^{11}$ g cm^{-3} (known as *neutron drip density*), the chemical potential of neutrons inside nuclei become equal to those outside, and free neutrons along with electrons are encountered along with the nuclei. This region is also known as the *inner crust*. As the density and therefore pressure increases further, the nuclei start to get deformed into complex irregular shapes resembling spaghetti and lasagna, therefore popularly known as *pasta phases*. Figure 3.2 provides a schematic outline of the NS interior structure.

Further increase in densities results in the pasta structures dissolving into a homogeneous liquid in the *outer core*, mainly composed of neutrons. A small fraction of protons and electrons are also present to ensure charge neutrality and chemical equilibrium, which are requirements for stability of the star. This is expected as the density range allows for a liquid-gas phase transition at the crust-core interface. While the crust is only $\sim$500 m in thickness, it plays an important role in governing surface properties as well as transport properties. However it is the core that comprises the majority of the stellar interior and determines global properties such as its mass.

Heavy-ion experiments indicate the appearance of strangeness containing exotic particles, such as hyperons (strange baryons) or kaons (strange mesons) at high densities beyond 2–3 n_{sat} [40–42]. Although these appear for a very brief time in particle accelerators, the high density in the neutron star *inner core* provides a natural environment for such particles to appear as stable components. Although current experiments cannot provide information for densities in the range 3–10 n_{sat}, QCD calculations even suggest the appearance of deconfined quark matter in the inner NS core. Such possibilities and their effects on NS global properties will be discussed in the subsequent sections.

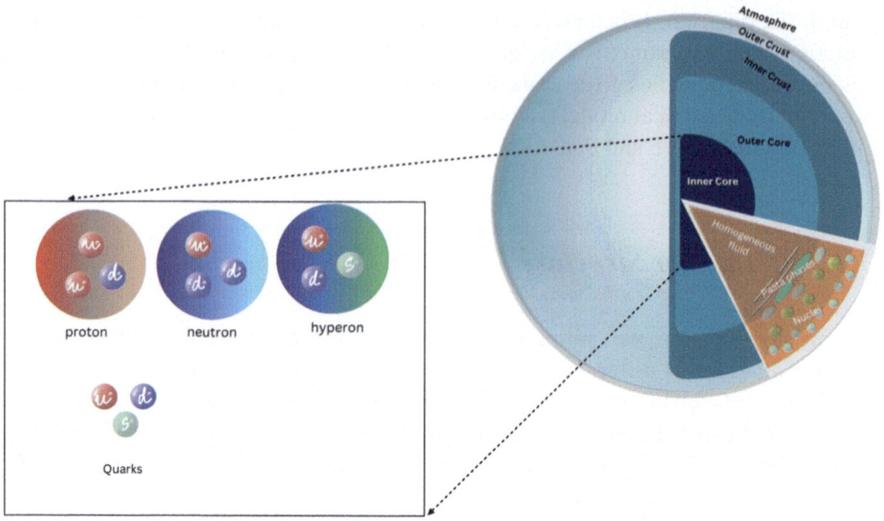

Fig. 3.2 Interior structure of a Neutron Star. [*Credit* Debarati Chatterjee]

3.3.4 Macroscopic NS Structure

Assuming a compact star to be a homogeneous fluid sphere, one may calculate the macroscopic properties, e.g., mass or radius by solving equations of hydrostatic equilibrium [10]. This is done by balancing the pressure $p(r)$ per unit area of a spherical shell of radius r having mass $m(r)$ with the gravitational force. Then the equations of hydrostatic equilibrium for the Newtonian (non-relativistic case) can be written as (in natural units $G = c = 1$):

$$\frac{dm(r)}{dr} = 4\pi r^2 \rho(r),$$
$$\frac{dp(r)}{dr} = -\frac{m(r)\rho(r)}{r^2}. \tag{3.6}$$

The boundary conditions are that the pressure at the centre is $p(0)$ is finite and vanishes at the surface $p(r = R) = 0$.

The Newtonian equation of hydrostatic equilibrium can be generalised to the relativistic case. Using spherical coordinates (t, r, θ, ϕ) for time and space, where θ and ϕ represent spherical angular coordinates, the background spacetime is characterized by the line element:

$$ds^2 = -e^{2\Phi(r)}dt^2 + e^{2\Lambda(r)}dr^2 + r^2 d\theta^2 + r^2 \sin(\theta) d\phi^2, \tag{3.7}$$

where Φ, Λ are metric functions with respect to r. The equilibrium configurations of non-rotating relativistic stars can be obtained by solving the Tolman–Oppenheimer–Volkoff (TOV) equations of hydrostatic equilibrium [10]

$$\begin{aligned}\frac{dm(r)}{dr} &= 4\pi\varepsilon(r)r^2,\\ \frac{dp(r)}{dr} &= -\frac{[p(r)+\varepsilon(r)][m(r)+4\pi r^3 p(r)]}{r(r-2m(r))},\\ \frac{d\Phi(r)}{dr} &= -\frac{1}{(\varepsilon+p)}\frac{dp}{dr}.\end{aligned} \tag{3.8}$$

For both the Newtonian and the relativistic cases, one requires an EoS (see Sect. 3.2) to close the system of equations. To obtain the global NS structure, i.e. mass (M), radius (R) and compactness ($C = M/R$), one needs to integrate the TOV equations from the centre of the star to the surface. The boundary conditions that must be satisfied are a vanishing mass ($m|_{r=0} = 0$) at the centre of the star, and a vanishing pressure at the surface ($p|_{r=R} = 0$), along with the metric function ($\Phi|_{r=R} = \frac{1}{2}\ln(1-2M/R)$) at the surface of the star. These can then be compared to the state-of-the-art limits derived from astrophysical data. The tidal deformability (Λ) can be obtained by solving a set of differential equations coupled with the TOV equations [30, 43–45]. The tidal deformability parameter Λ depends on the mass and radius of the NS (and hence its EoS) [46]

$$\Lambda = \frac{2}{3}k_2\left(\frac{R}{M}\right)^5, \tag{3.9}$$

where k_2 is the $l = 2$ dimensionless tidal Love number.

One may also rewrite the Newtonian hydrostatic equilibrium equations (see Eq. 3.6) in an analytical form using polytropic EoSs (see Sect. 3.2.2) [11]. To redefine the equations in dimensionless units, one can define a dimensionless radial coordinate ξ such that

$$r = \xi r_c, \tag{3.10}$$

in terms of the characteristic scale r_c. Similarly a dimensionless density coordinate θ is defined by

$$\rho(\xi) = \theta(\xi)^{1/(\Gamma-1)}\cdot\rho_c = \theta(\xi)^n\cdot\rho_c, \tag{3.11}$$

where ρ_c is the characteristic scale for mass density. The polytropic index n is related to Γ by $\Gamma = 1+\frac{1}{n}$. Substituting in Eq. 3.6, we get the dimensionless *Lane Emden equation*

$$\frac{1}{\xi^2}\cdot\frac{d}{d\xi}\left(\xi^2\frac{d\theta(\xi)}{d\xi}\right) = -\theta(\xi)^n. \tag{3.12}$$

This analytical equation can be solved subject to some boundary conditions: (i) vanishing density at the surface of the star $\theta_n(\xi = \xi_n) = 0$, and (ii) vanishing derivative of the scaling function at the centre $\theta'(\xi = 0) = \left.\frac{d\theta(\xi)}{d\xi}\right|_{\xi=0} = 0$.

One can also derive useful information about the stability of compact stars by using the polytropic EoS in the hydrostatic equilibrium equations [11]. Assuming a constant average mass density $\bar{\rho} = \frac{3M}{4\pi R^3}$, one can approximate the first equation by

$$\frac{dP}{dr} = \frac{-P_c}{R} = G\frac{M\bar{\rho}}{R^2}.$$

relating the central pressure to the mass and radius. Inserting the polytropic EoS relation, one arrives at an equation relating the mass and radius

$$M^{2-\Gamma} \cdot R^{3\Gamma-4} = const. \tag{3.13}$$

This relation has important consequences on the stability depending on the value of Γ, for example:

1. $\Gamma < \frac{4}{3}$: the star collapses to a black hole
2. $\Gamma = \frac{4}{3}$: is independent of R as the relation (3.13) reduces to $M = const$, corresponds to the case of limiting stability.
3. $\Gamma = \frac{5}{3}$: $M \cdot R^3 = const$, corresponds to the non-relativistic limit of Fermi gas EoS.

3.3.5 Realistic EoS Models: Crust EoS

The crust forms the outermost layer of NS matter, and comprises of inhomogeneous matter at subsaturation densities. It extends to about 1 km along the radius and is composed of crystalline matter (light and heavy nuclei). The properties of the NS crust-core interface play an important role in determining NS radii as well as the correct interpretation of astrophysical phenomena such as glitches and X-ray bursts. However, modelling neutron stars across their entire range of densities within the same theoretical framework is a major challenge, as different types of matter appear at different density regimes. Often different theoretical models are employed separately for the description of homogeneous matter in the core and the clusterized matter in the crust.

The composition of the **outer crust** for isolated NSs is given by the ground state of matter below the neutron drip density (roughly $\simeq 4 \times 10^{11}$ g cm^{-3}). According to the cold catalyzed matter hypothesis, matter is in full thermodynamic equilibrium at zero temperature. The outer crust is composed of perfect crystals with a representative nuclear species at the lattice sites (bcc), embedded in a sea of electrons. A single lattice volume is represented by a Wigner-Seitz cell, assumed to be charge neutral and in chemical equilibrium. Among the most commonly used models to describe the composition of the outer crust is the *BPS model* by Baym et al. [47]. In the BPS model, the equilibrium sequence of nuclei in the ground state is determined by minimising the Gibbs free energy per baryon with respect to mass number A and atomic number Z. The determination of the composition of the outer crust is largely sensitive to the experimentally determined nuclear masses [47]. Experimental nuclear masses from the atomic mass tables [48] are extrapolated theoretically to higher densities. The DD2 EoS [49, 50] is an improved version of the BPS model [47].

Beyond the neutron drip density, the **inner crust** consists of neutron-rich clusters, embedded in a neutron gas in a background of uniform electron gas. While nuclear mass data may help to constrain the composition of the subsaturation matter in the outer crust, in the inner crust the neutron-rich nuclei are far from the valley of stability and therefore beyond the reach of nuclear experiments. Therefore for the description of the inner crust, one needs to resort to theoretical models for the extrapolation to higher densities. Some commonly used techniques are Compressible Liquid Drop model [51, 52], Hartree–Fock/Hartree–Fock Bogoliubov [49, 53], Extended Thomas-Fermi approximation [54]. The effect of the free neutron gas was included in the calculations by Baym, Bethe and Pethick *(BBP)* [51], and further improved in subsequent models by Negele and Vautherin [53] by treatment of the many-body nuclear matter within an energy density functional. Other calculations include the Thomas-Fermi approximation [55] and extended Nuclear Statistical Equilibrium (NSE) models by Hempel and Schaffner-Bielich 2010, Rüster et al. [56].

It has been found that under high densities nuclei undergo transformation to unusual shapes called *nuclear pasta* phases, before melting into a homogeneous crust made of neutrons, protons and electrons. This has also been supported by molecular dynamic simulations of the crust [57]. In the case of "non-unified" models (where the crust and core composition are calculated using different approaches) usually the crust-core matching is done in a way that the pressure is always an monotonic function of density. However it has been demonstrated [58] that the use of non-unified models at the **crust-core interface** may leads to an uncertainty in the crust thickness of upto 30 % and upto 4 % for the estimation of the NS radius. Further the non-unified models may show no or spurious correlations with experimentally determined observables such as symmetry energy and its derivatives, as demonstrated in the works by [59, 60]. One of the main challenges for nuclear theory is therefore to develop unified models that are able to reproduce both clusterized matter in nuclei in the crust on one hand, and homogeneous matter in the core on the other, within the same formalism.

A more consistent approach is to obtain "unified" EoS models by applying the nuclear energy density functional (EDF) method [61, 62]. There has been considerable success in the application of the density functional theory (DFT) for the description of non-relativistic many particle systems. DFT calculations involve solving a system of non-interacting particles, which interact through a self-consistent effective potential. The parameters of the density functionals are optimized to selected experimental data and known properties of homogeneous nuclear matter. Different calculations and numerical techniques can be used to investigate the crust-core interface region, involving binodal sections, thermodynamical and dynamical spinodals [63] and the pasta phase transition to homogeneous phase [64, 65], obtaining comparable values for crust-core transition densities.

3.3.6 Realistic EoS Models: Core EoS

As described in Sect. 3.3.2, to describe the NS EoS, dense matter properties calibrated close to nuclear saturation density need to be extrapolated from nuclear saturation density to higher densities and from symmetric (SNM) to asymmetric nuclear matter (ANM) via theoretical models. The energy per nucleon can be expanded around the nuclear saturation density as (upto fourth order),

$$e(x,\delta) = e_{is}(x) + \delta^2 e_{iv}(x) \tag{3.14}$$

where $x = (n - n_{sat})/3n_{sat}$ is the dimensionless density and $\delta = (n_n - n_p)/(n_n + n_p)$ is the isospin asymmetry parameter. Here $e_{is}(x)$ and $e_{iv}(x)$ are the isoscalar and isovector contributions to the energy per nucleon, respectively. $e_{is}(x)$ and $e_{iv}(x)$ can be expanded around saturation density ($x = 0$) as,

$$e_{is}(x) = E_{sat} + \frac{1}{2!}K_{sat}x^2 + \frac{1}{3!}Q_{sat}x^3 + \frac{1}{4!}Z_{sat}x^4 + \cdots \tag{3.15}$$

$$e_{iv}(x) = J_{sym} + L_{sym}x + \frac{1}{2!}K_{sym}x^2 + \frac{1}{3!}Q_{sym}x^3 + \frac{1}{4!}Z_{sym}x^4 + \cdots \tag{3.16}$$

where

$$\begin{aligned}
K_{sat} &= 9n^2\frac{d^2e_{is}}{dn^2}|_{n=n_{sat}} & L_{sym} &= 3n\frac{de_{iv}}{dn}|_{n=n_{sat}} \\
Q_{sat} &= 27n^3\frac{d^3e_{is}}{dn^3}|_{n=n_{sat}} & K_{sym} &= 9n^2\frac{d^2e_{iv}}{dn^2}|_{n=n_{sat}} \\
J_{sym} &= e_{iv}(x=0) & Q_{sym} &= 27n^3\frac{d^3e_{iv}}{dn^3}|_{n=n_{sat}}
\end{aligned} \tag{3.17}$$

Here $n_{sat}, E_{sat}, K_{sat}, J_{sym}, L_{sym}, K_{sym}, Q_{sat}, Q_{sym}, Z_{sat}, Z_{sym}$ and m^*/m are empirical nuclear parameters, measured in terrestrial experiments. Nuclear saturation density (n_{sat}) is the density at which the energy per nucleon will be minimum for nuclear matter. E_{sat}, K_{sat}, J_{sym}, and L_{sym} are the energy per nucleon (or binding energy per nucleon), the compressibility factor, the symmetry energy and its slope parameters respectively, all calculated at saturation density. The parameter m^*/m is the effective nucleon mass at saturation. Q_{sat}, Q_{sym}, Z_{sat}, and Z_{sym} are isoscalar skewness, isovector skewness, isoscalar kurtosis, and isovector kurtosis, respectively. The empirical parameters at saturation density are determined from neutron skin thickness measurements of ^{208}Pb, electric dipole polarizability α_D, and various other experiments.

The description of nucleon-nucleon interaction in the dense NS environment suffers from uncertainties in the knowledge of interactions and unknown treatment of the many-body problem. There are different approaches to deal with the many-body problem of strongly interacting matter in neutron stars. These approaches can be broadly categorised into ab-initio or *microscopic* and *phenomenological* schemes (see [66] for a detailed review). These are described in the following Sects. 3.3.6.1 and 3.3.6.2.

3.3.6.1 Microscopic Approaches

In microscopic or ab-initio models, the many-body interaction problem for nuclear matter is solved starting from a Hamiltonian with a two-nucleon potential that fits nucleon-nucleon (NN) scattering data in vacuum and properties of deuteron. The realistic interactions are based on meson-exchange [67] or chiral perturbation theory [68]. Examples of such models are non-relativistic Brueckner-Hartree Fock (BHF) approximation [69] of the Brueckner-Bethe-Goldstone (BBG) theory, the relativistic Dirac-Brueckner-Hartree Fock (DBHF) models [70, 71] or the variational method [72].

At low energies, the ground state of nuclear matter made of nucleons and mesons is highly non-perturbative, and therefore perturbative techniques are not applicable. One may therefore resort to Effective Field Theory (EFT) techniques in this regime. Chiral symmetry is an approximate symmetry of low-energy QCD. Therefore in this regime, **Chiral Effective Field Theory (CEFT)** is a useful tool based on the symmetries of low-energy QCD, and therefore its domain of applicability is limited to close to saturation point [68]. The chiral symmetry breaking scale, $\Lambda_\chi \approx 1$ GeV, sets a limit to the momentum or energy domains where pions and nucleons are the degrees of freedom. It is a powerful and reliable model-independent framework for studying nuclear matter properties at low-density regimes based on order-to-order nucleon-nucleon interaction for modelling NS matter. The main advantage of this microscopic CEFT approach over the phenomenological approach is that theoretical uncertainties can be quantified by analyzing the order-by-order convergence of the expansion.

The main challenge in nuclear many-body theory in the realistic nucleon-nucleon description is the treatment of the short-range repulsive core interaction. Perturbative approaches cannot be applied, since the matrix element of the interaction is large. One way to circumvent this problem is the non-relativistic **Brueckner–Hartree–Fock (BHF) model**, based on the Brueckner theory of in-medium NN interaction and Hartree–Fock (HF) mean-field approximation to extract the effective single-particle potential [73]. According to the Brueckner theory, the bare nucleon-nucleon interaction (in vacuum) is not directly applied to the in-medium two-body nucleon-nucleon (NN) interaction due to its strong repulsive behaviour at short distances, but the bare nucleon interaction potential matrix is replaced by the Brueckner reaction matrix G. It represents the effective two-body interaction in the presence of the other nucleons and is derived perturbatively by summation of ladder diagrams [73]. In the BHF approximation, an auxiliary potential U can be introduced by splitting the Hamiltonian H of the two-particle interaction as

$$H = T + V = T + U + (V - U) = H_0' + V', \tag{3.18}$$

in terms of the kinetic energy T and NN interaction V, with the new single particle Hamiltonian H_0' and the new interaction potential V'. U can then be determined self-consistently such that the expansion in V' converges faster.

The BHF model however fails to reproduce the empirical properties of symmetric nuclear matter (SNM) at saturation density without adding three-body forces (TBF). The relativistic extension of this model is known as the **Dirac–Brueckner–Hartree–Fock (DBHF)** [70, 71]. The saturation SNM properties calculated within the DBHF formalism agree with empirical values satisfactorily [74]. In the DBHF approach, the Dirac equation is used to describe single particle motion of nucleons in nuclear matter. Different approaches may be used to calculate the single-particle potential in the DBHF theory. The free nucleon-nucleon interaction is via a one boson exchange, while the nucleon self-energy is composed of attractive scalar and repulsive vector fields. The excitation of nucleon anti-nucleon pairs captures the essence of three-body forces in the relativistic formalism.

Another alternative formulation of many-body interaction in nuclear matter is using the non-relativistic **variational technique** developed by Pandharipande and Friedman [75, 76]. In this method, phenomenological nucleon-nucleon interactions were used, including short and intermediate-range components as well as pion-exchange. The three-nucleon interactions were determined by reproducing the nuclear matter saturation density, binding energy, and incompressibility, using variational calculations. Nucleon-nucleon scattering cross-sections and nuclear matter properties were satisfactorily explained by this model. The Akmal–Pandharipande–Ravenhall (APR) EoS [72] is the most widely used example of variational EoSs.

3.3.6.2 Phenomenological Approaches

In phenomenological models, the model parameters are adjusted within their allowed uncertainties so as to reproduce nuclear empirical observables at saturation density. Examples of this type of models are non-relativistic Skyrme or Gogny models or relativistic density functional models. Although these models are able to satisfy current constraints within uncertainties of the couplings, they lack predictive power and are less robust.

The **Skyrme model** is among the most widely applied non-relativistic phenomenological models [77, 78]. In this model, the interaction occurs through the zero-range force, which can be considered as effective parametrization of the G-matrix but with only a few free adjustable parameters. The model parameters are fitted so as to reproduce the saturation nuclear properties such as E_{sat}, K_{sat}, J_{sym}, L_{sym} and $m^{SNM}_{\text{eff,N}}$ for symmetric matter, and $(E/A)_{PNM}$, and $m^{PNM}_{\text{eff,N}}$ for pure neutron matter. There are different Skyrme parameterizations (MSk7, UNEDF0, UNEDF1, SLy4, SLy5, SkI2, SkI5, etc.), that can cover the entire range of nuclear saturation properties like $220 \leq K_{sat} \leq 270$ MeV, $26 \leq J_{sym} \leq 37$ MeV, and the slope $9 \leq L \leq 130$ MeV coming from different experiments.

The **Gogny interaction model** is a finite-range non-relativistic phenomenological model [79] to simultaneously describe the mean and pairing fields of the nuclear interaction. Among the different Gogny parametrizations, the D1 family (i.e. D1S, D1N, and D1M) is able to successfully reproduce the ground state finite nuclei properties, but cannot explain the maximum observed mass ($2M_{\odot}$) of neutron stars due to small values of the symmetry energy slope. However other parametrizations of the D1M interaction, namely D1M* and D1M** are able to satisfy the NS properties as well as the finite nuclei properties. The recently proposed extended analytical form of the Gogny interaction, called D2 [80] Gogny interaction, can easily reproduce maximum NSs masses beyond $2M_{\odot}$.

The **Relativistic Mean Field (RMF) model**, is based on Quantum Hydrodynamics (QHD) introduced by J. D. Walecka in 1974 [81]. According to QHD, also known as the $\sigma - \omega$ model, baryons in nuclei and the nuclear matter interact through the exchange of isoscalar sigma σ and isoscalar vector ω mesons. The effective nucleon-nucleon potential of QHD in terms of Yukawa potentials for σ and ω mesons is given by

$$V_{eff} = \frac{g_\omega^2}{4\pi}\frac{e^{-m_\omega r}}{r} - \frac{g_\sigma^2}{4\pi}\frac{e^{-m_\sigma r}}{r}, \tag{3.19}$$

where m_σ and m_ω represent the masses of the σ and ω mesons respectively, and g_σ and g_ω are the scalar and vector coupling constants. While the σ meson is responsible for the long-range attractive force the ω meson reproduces the short-range repulsive force of the nuclear potential. The Lagrangian density describing the exchange of σ and ω mesons can be written as

$$\begin{aligned}\mathscr{L} &= \sum_N \bar{\psi}_B(i\gamma^\mu \partial_\mu - m + g_\sigma \sigma - g_\omega \gamma_\mu \omega^\mu)\psi_B \\ &+ \frac{1}{2}(\partial_\mu \sigma \partial^\mu \sigma - m_\sigma^2 \sigma^2) \\ &+ \frac{1}{2} m_\omega^2 \omega_\mu \omega^\mu - \frac{1}{4}\omega_{\mu\nu}\omega^{\mu\nu},\end{aligned} \tag{3.20}$$

where ψ_B is the baryon spinor, m is the vacuum nucleon mass, while γ^μ are the Dirac matrices. The effective in-medium nucleon mass is defined as $m^* = m - g_\sigma \sigma$ and $\omega_{\mu\nu} = \partial_\mu \omega_\nu - \partial_\nu \omega_\mu$. From the Lagrangian, using Euler-Lagrange equations the field equations for σ and ω can be obtained. The equations of motion are non-linear, coupled equations that are difficult to solve, but they can be solved within the mean-field approximation, in which the meson fields are replaced by the classical fields i.e., their ground state expectation values. From the Lagrangian one can also derive the energy momentum tensor, whose components are the energy density and pressure defining the equation of state. To describe the properties of asymmetric nuclear matter, the σ-ω model was extended to include the isovector-vector meson ρ to account for isospin asymmetry while the isovector-scalar meson δ takes care of mass asymmetry [82]. The coupling constants of mesons with the baryons are determined by fitting them to saturation nuclear properties. Currently a large number of QHD parameter sets exist, which differ from each other depending on the meson fields considered and different couplings between the fields [83].

For better reproduction of properties of nuclei, in the **non-linear (NL-RMF) models** several other mesons along with the self-and cross-coupling terms were added. The Lagrangian density describing the interaction between baryons through the exchange of mesons: the scalar-isoscalar (σ), vector-isoscalar (ω), vector-isovector (ρ) mesons is given in Eq. (3.21). In addition, the Lagrangian density takes into account the scalar and vector self-interactions, while also including contributions from possible mixed interactions between the mesons upto quartic order:

$$
\begin{aligned}
\mathcal{L} = & \sum_B \bar{\psi}_B(i\gamma^\mu\partial_\mu - m + g_{\sigma B}\sigma - g_{\omega B}\gamma_\mu\omega^\mu - \frac{g_{\rho B}}{2}\gamma_\mu \boldsymbol{\tau}\cdot\boldsymbol{\rho}^\mu)\psi_B \\
& + \frac{1}{2}(\partial_\mu\sigma\partial^\mu\sigma - m_\sigma^2\sigma^2) - \frac{1}{3}bm(g_{\sigma B}\sigma)^3 + \frac{1}{4}c(g_{\sigma B}\sigma)^4 \\
& + \frac{1}{2}m_\omega^2\omega_\mu\omega^\mu - \frac{1}{4}\omega_{\mu\nu}\omega^{\mu\nu} + \frac{\zeta}{4!}\left(g_{\omega B}^2\omega_\mu\omega^\mu\right)^2 \\
& - \frac{1}{4}(\boldsymbol{\rho}_{\mu\nu}.\boldsymbol{\rho}^{\mu\nu} - 2m_\rho^2\boldsymbol{\rho}_\mu\boldsymbol{\rho}^\mu) + \Lambda_\omega(g_{\rho B}^2\boldsymbol{\rho}_\mu\boldsymbol{\rho}^\mu)\,(g_{\omega B}^2\omega_\mu\omega^\mu)
\end{aligned}
\tag{3.21}
$$

where $\boldsymbol{\tau}$ are the Pauli matrices. σ, ω, ρ denote the meson fields, with isoscalar coupling constants g_σ, g_ω, isovector coupling g_ρ and mixed $\omega - \rho$ coupling Λ_ω. b and c represent the scalar meson self interaction, while the coupling ζ represents the quartic vector self-interaction.

Charge neutrality and chemical equilibrium in neutron star matter requires the inclusion of leptons (e.g. electrons e^- and muons μ^-). The Lagrangian for the leptons can be written as:

$$
\mathcal{L}_\ell = \sum_{\ell=\{e^-,\ \mu^-\}} \bar{\psi}_\ell(i\gamma^\mu\partial_\mu - m_\ell)\psi_\ell, \tag{3.22}
$$

where ψ_ℓ is the lepton spinor and m_ℓ lepton mass.

As before from the given Lagrangian density, the equations of motion for the nucleons and mesons can be obtained in the mean-field approximation. The energy density can be derived from the energy-momentum tensor:

$$
\varepsilon = \sum_B \frac{1}{8\pi^2}\left[k_{F_B}E_{F_B}^3 + k_{F_B}^3 E_{F_B} - m^{*4}\ \ln\left(\frac{k_{F_B}+E_{F_B}}{m^*}\right)\right] + \frac{1}{2}m_\sigma^2\bar{\sigma}^2 + \frac{1}{2}m_\omega^2\bar{\omega}^2
$$

$$+\frac{1}{2}m_\rho^2\bar{\rho}^2+\frac{1}{3}bm(g_\sigma\bar{\sigma})^3+\frac{1}{4}c(g_\sigma\bar{\sigma})^4+\frac{\zeta}{8}(g_\omega\bar{\omega})^4+3\Lambda_\omega(g_\rho g_\omega\bar{\rho}\bar{\omega})^2, \quad (3.23)$$

while the pressure can be obtained from the energy density ε via the Gibbs–Duhem relation,

$$p=\sum_B \mu_B n_B-\varepsilon. \quad (3.24)$$

where baryon chemical potential μ_B is defined as

$$\mu_B=E_{F_N}+g_{\omega B}\bar{\omega}+\frac{g_{\rho B}}{2}\tau_{3B}\bar{\rho}. \quad (3.25)$$

Here, k_{F_B} is the Fermi momentum of the baryon, and Fermi energy is given by $E_{F_B}=\sqrt{k_{F_B}^2+m_B^{*\,2}}$.

The isoscalar set of couplings $g_{\sigma B}$, $g_{\omega B}$, b and c are determined by fixing the saturation density n_{sat}, binding energy per nucleon E_{sat}, the incompressibility coefficient K_{sat} and the effective nucleon mass m^*/m at saturation. On the other hand, isovector couplings $g_{\rho B}$ and Λ_ω are determined as a function of the symmetry energy E_{sym} and the slope of the symmetry energy at saturation L_{sym}. An example of NL-RMF parametrization [84] is shown in Fig. 3.3.

In the NL-RMF models, the interaction strengths are calculated at nuclear saturation density, and used throughout the density range. This is the main drawback of this model, as the interaction strength depends on density. The self- and cross-coupling of various mesons in the NL-RMF model are replaced by the density-dependent baryon-meson coupling constants in the relativistic **density-dependent (DD-RMF) models**, to address the high density behaviour of the EoS. A rearrangement self-energy term, which physically accounts for the effects of static polarisation in the

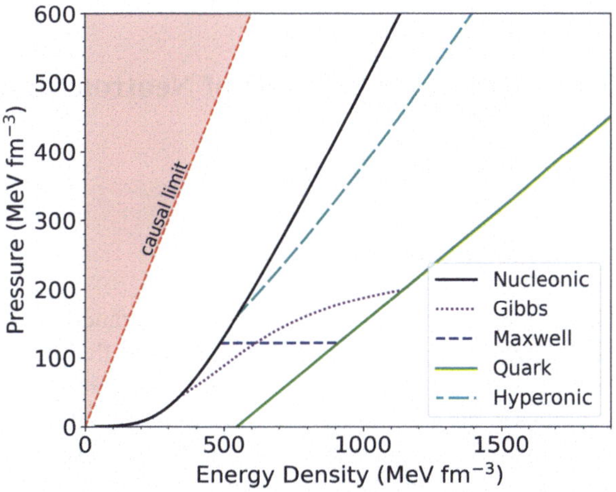

Fig. 3.3 Comparison of EoSs for different NS core composition. [*Credit* Swarnim Shirke (IUCAA)]

nuclear medium, is added to the DD-RMF field equations. The density-dependent coupling constants can be expressed as [50, 85]:

$$g_i(n_B) = g_i(n_{sat}) f_i(x), \tag{3.26}$$

where

$$f_i = a_i \frac{1 + b_i(x + d_i)^2}{1 + c_i(x + d_i)^2}, \; i = \sigma, \omega, \tag{3.27}$$

is a rational function of $x = n_B/n_{sat}$ for the isoscalar mesons. For the isovector mesons, it is replaced by an exponential function

$$f_i(n_B) = \exp[-a_i(x - 1)]. \tag{3.28}$$

Due to the density dependent coupling constants, a rearrangement term $\sum_R$ is introduced in the equation of motion of the mesons. As before, the energy-momentum tensor obtained from the Lagrangian gives the equation of state.

For nucleonic matter, thermodynamic stability is ensured by certain conservation conditions. **Chemical equilibrium** is maintained by neutrons, protons and leptons (electrons and muons), which may be expressed by a balance of the chemical potentials

$$\mu_n = \mu_p + \mu_e,$$

$$\mu_\mu = \mu_e.$$

Charge neutrality may be ensured by demanding the same number of oppositely charged protons, electrons and muons, i.e.

$$n_p = n_e + n_\mu.$$

3.4 Advanced Modelling of Neutron Stars

3.4.1 Including Strangeness

3.4.1.1 Hyperon Matter

Heavy-ion experiments indicate the appearance of strange baryons or *hyperons* at high densities. It was therefore conjectured that hyperons should be present in the high density environment of the neutron star core [86]. The effect of the presence of hyperons in the neutron star core has been studied extensively within microscopic and phenomenological models (see e.g. [42, 87] for recent reviews). There is a consensus in these models that hyperons may appear at densities ~ 2–$3 n_{sat}$ in the inner core, when the nucleon chemical potential exceeds that of hyperons, making it energetically favourable. However the appearance of hyperons results in a strong softening of the EoS, leading to maximum neutron star masses not compatible with observations (see Sect. 3.6). The situation can be remedied by the inclusion of hyperonic three-body forces [88] involving one or more hyperons (NNY, NYY, YYY) which may provide additional repulsion making the EoS stiffer. In DBHF models, three-nucleon forces are included via nucleon-antinucleon excitations due to the dressed Dirac spinor in the medium. Some efforts have been made recently to extend the DBHF formalism to the hyperonic sector [89, 90].

Several works used phenomenological EoSs including hyperonic degrees of freedom, which rely on nuclear experimental data to calibrate the NN interaction and hypernuclear data for the hyperon-nucleon (YN) and hyperon-hyperon (YY) interactions. In the previous section it was described how beta-equilibrated and charge-neutral hadronic matter composed of baryons (all baryons in baryon octet, including hyperons) and leptons (e^- and μ^-) can be described within the RMF model. In this model, hyperon-nucleon interaction is mediated by the exchange of scalar (σ), vector (ω) and iso-vector (ρ) mesons as described by the Lagrangian density for the hadronic phase in Eq. 3.21. Within the RMF model, hyperon-hyperon interaction is mediated by two additional mesons, strange scalar (σ^*) and strange vector (ϕ) mesons [41]:

$$\begin{aligned}\mathscr{L}_{YY} &= \sum_Y \bar{\psi}_Y (g_{\sigma_Y^*}\sigma^* - g_{\phi_Y}\gamma_\mu\phi^\mu)\psi_Y + \frac{1}{2}m_\phi^2\phi_\mu\phi^\mu - \frac{1}{4}\phi_{\mu\nu}\phi^{\mu\nu}\\ &+\frac{1}{2}(\partial_\mu\sigma^*\partial^\mu\sigma^* - m_{\sigma^*}^2\sigma^{*2}).\end{aligned} \tag{3.29}$$

The vector-hyperon coupling constants ($g_{\omega_Y}, g_{\rho_Y}, g_{\phi_Y}$) are fixed to their theoretical values from SU(6) quark model [91].

$$\begin{aligned} g_{\omega\Lambda} = g_{\omega\Sigma} = 2g_{\omega\Xi} &= \frac{2}{3}g_{\omega N}\\ g_{\rho\Lambda} &= 0\\ g_{\rho\Sigma} = g_{\rho\Xi} &= g_{\rho N}\\ 2g_{\phi\Lambda} = 2g_{\phi\Sigma} = g_{\phi\Xi} &= \frac{-2\sqrt{2}}{3}g_{\omega N}. \end{aligned} \tag{3.30}$$

The scalar-hyperon coupling constants are obtained by fixing the hyperon-nucleon ($U_Y^{(N)}(n_0)$) depth at normal nuclear matter [41, 91].

$$U_Y^N(n_0) = -g_{\sigma_Y}\sigma + g_{\omega_Y}\omega. \tag{3.31}$$

Among the nucleon-hyperon potential depths, the best known is the hyperon Λ about $U_\Lambda^N = -30$ MeV, and there are larger uncertainties for the Σ and Ξ potentials. State-of-the-art hypernuclear experiments indicate Σ and Ξ potential depths of +30

MeV and −18 MeV, respectively. The effect of uncertainties of hyperon potential depth on maximum neutron star mass was studied systematically in [92]. Figure 3.3 shows an example of a hyperonic EoS with NL-RMF parametrization [93].

The hyperon-σ^* coupling constants may be estimated by fixing the potential depth ($U_Y^{Y'}$), potential of a hyperon (Y) in a bath of the other hyperon (Y') at normal nuclear matter density obtained from double hypernuclei data [40, 41, 94]. These are highly uncertain due to lack of sufficient experimental data [37]. Usually one adapts following values

$$U_{\Xi}^{\Xi} = 2U_{\Lambda}^{\Lambda} = 2U_{\Xi}^{\Lambda} = U_{\Lambda}^{\Xi} = -40\,\text{MeV}. \tag{3.32}$$

The system maintains chemical equilibrium via weak processes of the type $B_1 \rightarrow B_2 + \ell + \bar{\nu_\ell}$ and $B_2 + \ell \rightarrow B_1 + \nu_\ell$ where B is a baryon, ℓ is lepton and ν_ℓ ($\bar{\nu_\ell}$) is corresponding neutrino (anti-neutrino). For evolved neutron stars transparent to neutrinos, the neutrino chemical potential set to 0. For a generalized system characterised by two conserved charges q_i (electric charge) and b_i (baryon number), chemical equilibrium condition may be summarized to a generic equation:

$$\mu_i = b_i \mu_n - q_i \mu_e \tag{3.33}$$

where μ_i , μ_n and μ_e are chemical potentials of ith species, neutron and electron respectively, similarly b_i and q_i are baryonic and electric charge of species 'i' respectively. Charge neutrality of hadronic phase requires total charge should be 0, i.e:

$$\sum_B q_B n_B - n_e - n_\mu = 0 \tag{3.34}$$

where q_B is electric charge of baryon B and n_B, n_e, n_μ are number densities of baryon B, electron and muon respectively.

3.4.1.2 Quark Matter

According to QCD, the "running" coupling constant for strong interactions decreases in strength with increasing interaction distance, making it impossible to isolate a colored quark. However at high densities (∼2–10 n_{sat}) relevant to the inner core of neutron stars, the nuclear boundaries begin to dissolve and quarks approach each other, weakening the strong forces and making the quarks practically free. In this scenario one may consider the possibility of having deconfined quarks. There are different possibilities of having compact stars with quark matter. *Hybrid stars* have a quark matter core, in pure quark matter phase or in a mixed phase of quarks and hadrons, surrounded by a shell of hadronic matter. The description of hybrid stars requires a hadronic and quark matter EoS, while the hadron-quark phase transition may occur via *Maxwell* (sharp phase transition with an energy density jump) or *Gibbs* (via a mixed phase) construction (see illustration in Fig. 3.5). Under certain

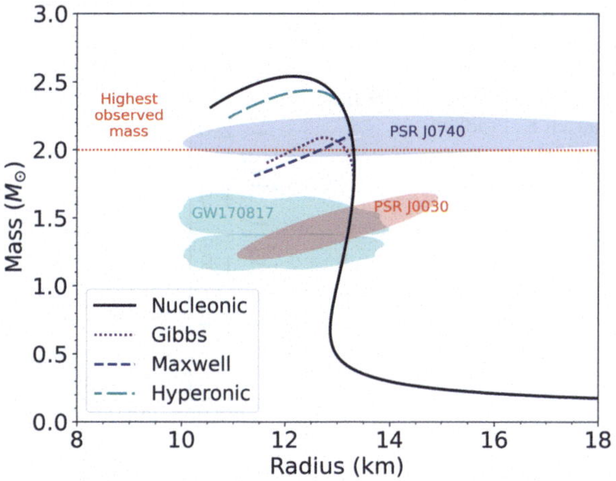

Fig. 3.4 Comparison of mass-radius relations for the different NS core composition in Fig. 3.3. [*Credit* Swarnim Shirke (IUCAA)]

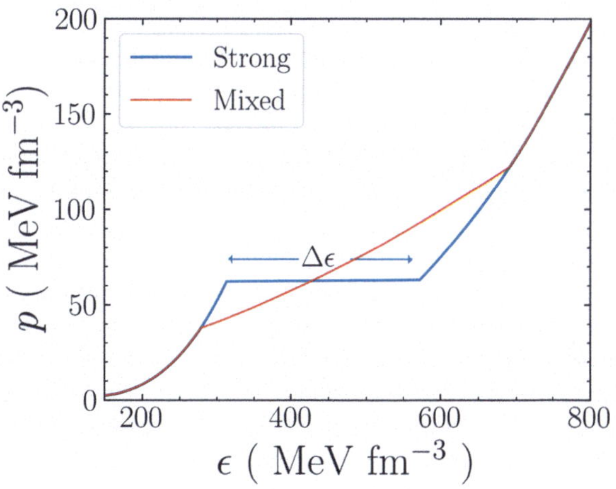

Fig. 3.5 Comparison of a Maxwell (strong) hadron-quark phase transition vs Gibbs (mixed). [*Credit* Bikram K. Pradhan (IUCAA)]

conditions (when Seidov criterion is satisfied [95]), a Maxwell phase transition may result in the formation of a *twin star*, i.e. a stable compact object with the same mass as a neutron star but with a different radius (see Fig. 3.6).

At large densities or temperatures ($T \gtrsim 300$ MeV, $\mu_B = 0$ or $n_B \gtrsim 40\, n_{sat}$, $T = 0$), the coupling strength of QCD decreases logarithmically, and quarks can be considered asymptotically free. In this regime, properties of quark matter can be calculated using analytical techniques of **perturbative QCD** (pQCD) upto second order in strong coupling constant ($\mathcal{O}(\alpha_s^2)$) [96–98] and more recently partially to ($\mathcal{O}(\alpha_s^3)$) [99–101].

First-principle calculations are not available to describe strongly interacting matter in the non-perturbative regime of QCD, except at very large densities $\sim 50 n_{sat}$. Such high densities are not directly relevant in neutron stars, therefore one needs to rely on effective theories to capture major features of QCD, such as confinement and

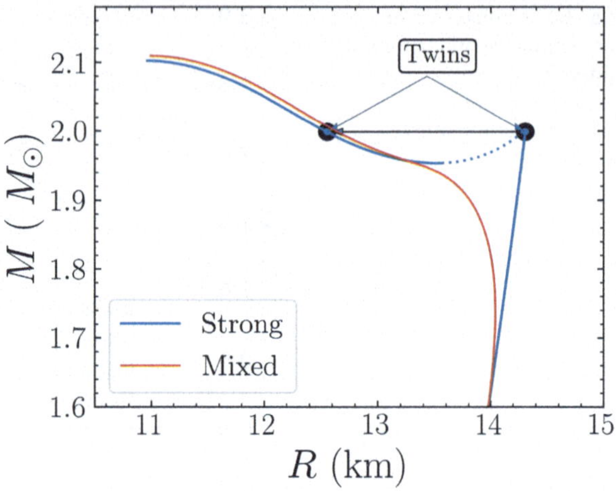

Fig. 3.6 Mass-radius relations corresponding to the hadron-quark phase transitions in Fig. 3.5 [*Credit* Bikram K. Pradhan (IUCAA)]

restoration of chiral symmetry at high densities. Therefore theoretical descriptions of quark matter rely on phenomenological models such as the MIT Bag model or the Nambu–Jona–Lasinio (NJL) model.

The **MIT Bag Model** [102–104] is one of the simplest phenomenological models that takes into account confinement and asymptotic freedom. In this model, at low densities and temperatures quarks are strongly interacting and confined in colorless hadrons or "bags" that are embedded in the non-perturbative QCD vacuum. Inside these bags, quarks have their usual masses and interact weakly, while outside they have infinite mass. The energy density of the quark gas in the deconfined phase is given by free fermi-gas expressions, with quarks having their bare masses, with an additional "Bag constant" $B > 0$ (in units of energy density) that defines the energy difference between the perturbative and true vacuum. Its effect is to minimize the volume and hence contributes negatively to the pressure. At high temperatures and densities, the bag radius diverges, leading to the deconfinement transition. The equations for pressure and energy density are:

$$P_{\text{Bag}} = \sum_{f=u,d(,s)} P_{\text{FG},f}(\mu_f, m_f) - B \tag{3.35}$$

$$\varepsilon_{\text{Bag}} = \sum_{f=u,d(,s)} \varepsilon_{\text{FG},f}(\mu_f, m_f) + B \tag{3.36}$$

where m is the bare quark mass, μ is the chemical potential, P_{FG} and ε_{FG} stands for the pressure and energy density of Fermi-gas. The chemical potential at which P_{Bag} becomes non-negative defines the critical point for deconfinement. For a relativistic Fermi-gas, we have $P_{FG,f} = \varepsilon_{FG,f}/3$. Using this and Eqs. 3.35, 3.36, we obtain the EoS of QM as

$$\varepsilon_{\text{Bag}} = 3P_{\text{Bag}} + 4B. \tag{3.37}$$

Although the simple bag model explains several features of QCD, this model results in a rather soft EoS that cannot explain observations of massive NSs. This problem can be resolved by adding perturbative corrections in powers of the strong running coupling constant α_s, since α_s although small at high density, is non-zero. Upto first order corrections, this results in the addition of a term in the grand potential density [105] ($\Omega = -P$) and becomes

$$\Omega = \sum_{i=u,d,s,e} \Omega_{i,free} + \frac{3\mu^4}{4\pi^2}(1 - a_4) + B \tag{3.38}$$

where $a_4 = 1 - 2\alpha_s/\pi$ and $\mu = (\mu_u + \mu_d + \mu_s)/3 = \mu_B/3$ is the quark chemical potential. Considering massless up and down quarks, and expanding the pressure to first order in mass of strange quark ($m_s = 100$ MeV), we obtain [106]

$$P = \frac{3\mu^4}{4\pi^2}a_4 - \frac{3\mu^2}{4\pi^2}m_s^2 - B. \tag{3.39}$$

Figure 3.3 illustrates an example of a quark EoS described by MIT Bag model, with a mixed phase via Gibbs and Maxwell construction. The corresponding mass-radius relations for the hybrid star are displayed in Fig. 3.4.

Although simple and applicable, the Bag model lacks non-perturbative vector interactions, which are important for explaining large observed NS masses. It also lacks the QCD feature of chiral symmetry restoration. As the Bag parameter B is independent of flavour, the critical point for chiral symmetry restoration in this model is not flavour-dependent. These shortcomings are taken care of in the vector-enhanced (vMIT) Bag model [107].

Quark matter in the inner core may also be in **color superconducting** states [108]. This is possible if quarks near the Fermi surface form Cooper pairs via the attractive quark-quark interaction in the antisymmetric color channel. The most energetically favoured state at very high densities is the color-flavor-locked (CFL) phase, where massless quarks with all the three flavors and colors are paired. Some of the other possible states are the two-flavor color superconducting state (2SC) where only light u and d quarks are paired.

The SU(3) Nambu–Jona–Lasinio **NJL model** [109] is another effective theory that captures several basic aspects of QCD. In this model, the constituent quark mass is connected with the restoration and breaking of chiral symmetry, while the origin of the nucleon mass is based on analogy with the BCS theory of superconductivity. This model is efficient for studying the formation of quark pair condensates via spontaneous breaking of chiral symmetry and dynamical generation of quark masses. However, due to its simplicity, the model lacks some features of QCD such as local color gauge symmetry.

There are also other novel approaches such as **Holographic QCD** which use gauge/gravity duality to model dense nuclear and quark matter in neutron stars [110, 111]. Some of the recent results within holographic framework also include theoretical constraints both at low and high densities and temperatures such as lattice QCD results, chiral effective field theory, QCD perturbation theory and statistical modelling [112] and have been applied to study hot and dense QCD matter relevant for neutron star mergers [113]. However, there still remain several open questions within this approach e.g. flavor dependence, magnetic field dependence at large densities of nuclear or quark matter, transport in the presence of magnetic field, both for nuclear and quark matter etc. which remain to be investigated.

According to the Bodmer-Witten hypothesis [114, 115] which suggests *strange quark matter* (SQM) to be the true ground state of strongly interacting matter, another class of self-bound compact stars completely made up of deconfined u, d, s quarks, called *strange stars* or *quark stars*, may exist. These objects follow a completely different mass-radius behaviour as compared to neutron stars.

3.4.2 Effect of Temperature

For NS matter, we consider cold ($T = 0$), β-equilibrated and neutrino transparent matter. At low temperature, the mean free path of neutrinos (and anti-neutrinos) is large, and the star is transparent to neutrinos, which stream out freely. At zero temperature, all the nucleonic and other fermionic degrees of freedom occupy states up to their respective Fermi levels. Due to loss of neutrinos, the charge fraction (Y_Q) of the system changes with density.

However, in astrophysical scenarios such as core-collapse supernovae, newly born proto-neutron stars (PNS) and BNS mergers, the temperature can exceed several tens of MeV and the thermal effect on the EoS can no longer be ignored. Supernova simulations indicate that during post-bounce evolution, the EoS is governed by thermal properties of the out-of-β equilibrium system i.e. entropy per baryon (S/A) and/or lepton fractions (in case of neutrino trapping) [23]. From an initial hot (high entropy) state with low electric charge or lepton fraction, the star evolves to a low entropy state with increase in charge/lepton fractions until it attains β equilibrium. At temperatures $T \geq 5$ MeV, neutrinos are trapped inside the star and affect the EoS and macroscopic properties non-negligibly. This is illustrated in Fig. 3.7, where change in entropy affects the mass-radius relations in neutron stars. Anti-particles can be thermally excited and exotic degrees of freedom can also appear at low densities (sub-saturation densities and even in the crust). For $T \leq 5$ MeV, the neutrinos are free streaming and the EoS behaviour resembles that of cold matter.

Currently there exist only a limited number of fully temperature-dependent EoS models, and the number of studies of NS mergers or core-collapse supernovae where such EoS tables have been applied are also few. For systems at finite temperature, general purpose EoSs that are available include Lattimer and Swesty (LS) [116] and Shen et al. (STOS) [117]. Both models employ the single nucleus approximation (SNA) and all light nuclei except alpha particles are ignored. Alternatively, there are models based on nuclear statistical equilibrium (NSE), which incorporate realistic nuclear masses and go beyond the single-nucleus approximation, e.g. the HS

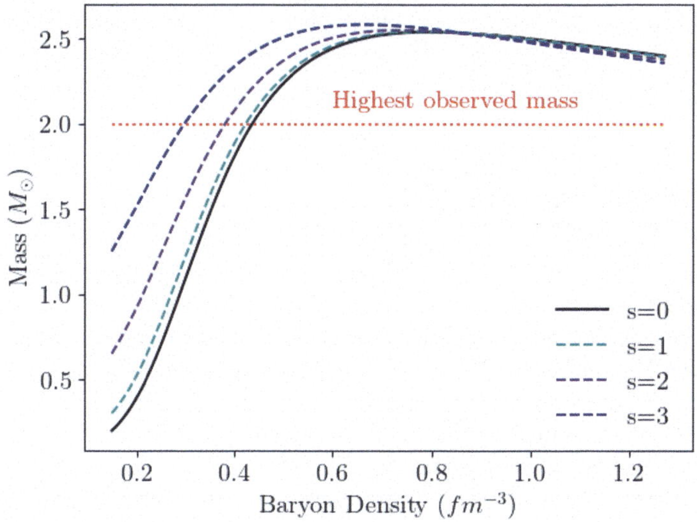

Fig. 3.7 Finite entropy affects the mass-radius relation in neutron stars [*Credit* Nilaksha Barman (IUCAA)]

EoS model of Hempel and Schaffner-Bielich [118]. This EoS is often employed in RMF EoSs for matching at low densities. Many of the available finite temperature EoS models are based on the RMF parameterization with density dependent (DD) meson couplings [119], e.g. DD/DD2 [50] and DDME1/DDME2 [85]. Subsequently these have been extended to finite temperature (e.g. HS(DD2)) [50, 120]; as well as non-nucleonic degrees of freedom, such as hyperons: BHB (DD2$\Lambda\phi$) [121], OMHN (DD2Y) [122]; delta baryons: R (DD2YDelta) [123]; kaons (Malik et al. [124]) and quarks (Bastian et al. [125]). More recent EoSs include finite temperature effects in non-linear RMF models, e.g. SFHo and SFHx [23] and the QMC-RMF model [126], which also includes constraints from CEFT and maximum NS mass. There are available finite temperature hybrid EoSs based on the Chiral Mean Field (CMF) model [127, 128] which includes scalar mesons (σ, δ, ζ, χ) and vector mesons (ω, ρ, ϕ) to describe interactions among baryons. The additional scalar-isoscalar field (χ) is used as an effective gluon condensate while σ and ζ are responsible for generating baryon masses [127].

Due to the dearth of EoSs for different NS composition spanning the required range of temperature, density and electron fraction, finite temperature EoS effects have been incorporated in numerical simulations for over two decades by supplementing a cold EoS with ideal-gas thermal contributions using the so-called Γ-law [129–132], with Γ values in the range 1.5–2. However, recent studies have demonstrated that when non-nucleonic degrees of freedom are taken into account, Γ values exhibit a wide variation with density and hence a simple Γ-law introduces errors in numerical simulations [123]. For quark-hadron phase transition, it has been shown [133] that the Γ value changes from ($\Gamma \sim 1.75$) in hadronic phase to $\Gamma \sim 4/3$ in the quark phase, and therefore the transitions to mixed phase and the pure quark phase occur at lower densities compared to the cold case, since the Γ law overestimates transition densities.

3.4.3 Effect of Magnetic Fields

Neutron stars are associated with phenomena involving strong electromagnetic fields. Typical pulsars are known to possess magnetic fields as high as $10^9 - 10^{12}$ G. Observations indicate the existence of a subclass of neutron stars called **magnetars**, detected as Anomalous X-ray Pulsars (AXPs) and Soft Gamma ray Repeaters (SGRs), with ultra-strong surface magnetic fields $10^{15}G$ [134, 135]. According to the scalar Virial theorem, magnetic fields in the interior of a magnetar can be even higher, reaching upto $10^{18}G$ in the core.

Ultrastrong interior magnetic field strengths may affect compact stars mainly in two ways:

1. The magnetic field interacts with the particles in the stellar interior, thus affecting the Equation of State (EoS)
2. Strong magnetic fields result in a modification of the energy-momentum tensor, breaking of the spherical symmetry, and therefore affect the external structure of the compact star.

In presence of a magnetic field, the motion of charged particles becomes confined to quantized Landau levels [136] and the EoS is affected by the Landau quantization of the charged particles (e.g. protons, electrons). Further, the interaction of the magnetic moments, including the anomalous magnetic moments of the neutral particles (e.g. neutrons, Λ-hyperons) with the magnetic field also influences the EoS. The critical field b_c for Landau quantization of a charged particle is defined as the value where the cyclotron quantum is equal to or above the rest energy , i.e. $b_c = m^2/|q|$, where m and q are its mass and electric charge. This value is $b_c^e \sim 4.4 \times 10^{13}$ G for electrons and $b_c^p \sim 10^{20}$ G for protons. At magnetic fields below this critical field, there exists a continuum of energy states. The energy of a charged particle changes significantly in the quantum limit when the magnetic field is comparable to or above the critical value, and the quantum effects are most pronounced when the particle is confined to the lowest Landau level. The effect of quantizing magnetic fields on the EoS for different models of NS matter has been widely investigated, see e.g. [137–141]. In presence of a uniform external magnetic field in the z direction, the transverse momenta of particles with an electric charge q are restricted to discrete Landau levels with momenta $k_\perp^2 = 2\nu q b$ where $\nu \geq 0$ is the Landau quantum number. For spin-1/2 particles, ν is related to the orbital angular momentum n by

$$\nu = n + \frac{1}{2} - \frac{s}{2}\frac{q}{|q|},$$

where $s \pm 1$ is the spin projection of the particle in the direction of the magnetic field. The total energy of a charged particle becomes quantized as:

$$E = \sqrt{k_z^2 + \left[(m^2 + 2\nu|q|b)^{1/2} - s\kappa b\right]^2} = \sqrt{k_z^2 + \bar{m}^2(\nu)}, \tag{3.40}$$

k_z being the momenta in the z-direction and κ the anomalous magnetic moment [140], while

$$\bar{m}^2(\nu) = (\sqrt{m^2 + 2\nu|q|b} - s\kappa b)^2. \tag{3.41}$$

If one considers the simplest case with $\kappa = 0$, the maximum momentum k_z in terms of the chemical potential μ is defined as:

$$k_{z,F} = \sqrt{\mu^2 - 2\nu|q|b - m^2}. \tag{3.42}$$

To ensure that the quantity under the square root is positive, one may impose the condition:

$$\nu \leq \nu_{max} = \frac{\mu^2 - m^2}{2|q|b}. \tag{3.43}$$

A magnetic field strong enough such that only the lowest Landau level is occupied is in general called a strongly quantising field. Using the above relations, one can write down the expression for the number density as

$$n = \frac{|q|b}{2\pi^2} \sum_{s=\pm 1} \sum_{n=0}^{\nu_{max}} k_{z,F}(\nu), \tag{3.44}$$

and the energy density

$$\varepsilon = \frac{|q|b}{2\pi^2} \sum_{s=\pm 1} \sum_{\nu=0}^{\nu_{max}} \int_0^{k_{z,F}} \sqrt{k_z^2 + \bar{m}^2(\nu)}. \tag{3.45}$$

From the energy density, one may derive the pressure using the Gibbs–Duhem relation, and therefore obtain the EoS. In presence of non-zero anomalous magnetic moment [140], the full expression in Eq. (3.41) must be used.

In the **neutron star crust**, due to the magnetic field electron motion is quantised into Landau levels. In a typical magnetar with $b \gg b_c^e$, the crust composition depends on the magnetic field strength with typical quantum oscillations observed in the transition density from one nuclear cluster to another. If the magnetic field strength exceeds the quantising value [142], this transition density increases linearly with the magnetic field strength, leading to a less neutron rich crust and an increased density for neutron drip, i.e. shifting the transition from the outer to the inner crust to higher densities. For magnetic fields in the crust above a few times 10^{17} G, nuclear binding energies start to become significantly altered, modifying additionally the structure of the crust [143]. A review of crust properties under strong magnetic fields can be found in [144].

The **crust-core region** of magnetars has also recently studied within the relativistic mean field model framework [145, 146]. In these works, the effects of strong magnetic fields ($10^{15} - 10^{17}$ G) on the instability region was investigated with the help of the Vlasov formalism used to determine the dynamical spinodals. It was found that, since the inner crust becomes less neutron rich, the transition to homogeneous matter is shifted to higher densities.

Magnetic field influence on **homogeneous hadronic matter**, be it purely nucleonic or including other baryons such as hyperons or Δ-baryons (see e.g. Refs. [32, 139, 147]) can be studied using different models for the underlying hadronic interaction. Estimates of the critical magnetic field for the different baryons and the energy of the interaction of the respective magnetic moments with the magnetic field show that below a field strength of $\sim 10^{18}$ G the influence on the EoS is negligible. Magnetic field also influences the **hadron-quark phase transition** [147, 148] A strong magnetic field shifts the transition to quark matter to higher densities [148, 149].

The influence of the magnetic field on **quark matter** has been investigated with much interest, in compact stars and also the complex structure of the QCD phase diagram and phenomena such as magnetic catalysis [141, 150–152]. The Nambu–Jona–Lasinio (NJL) model without pairing interaction [153] has been considered, as well as magnetic colour-flavor locked state (mCFL) [141].

It has been shown that spherical TOV equations cannot be applied to describe magnetar structure, since magnetars are highly deformed from their spherical shape. Bocquet et al. [154] were the first to include magnetic fields to rotating neutron star models, by solving coupled Einstein-Maxwell equations. Subsequently fully relativistic numerical computations of the magnetar structure were performed by others taking into account the anisotropy of the stress-energy tensor [155–159] assuming a perfect fluid, polytrope or a realistic EoS, but not taking into account the magnetic field modifications to the EoS due the quantizing magnetic field. Solutions to coupled Einstein-Maxwell equations along with a magnetic field dependent EoS to obtain the global quantities, has been addressed in a few recent works [160, 161].

3.4.4 *Effect of Rotation*

Electromagnetic observations of neutron stars indicate that most of them are rotating slowly. The fastest spinning neutron star PSR J1748-2446 observed yet has a frequency 716 Hz [162]. On the other hand, newly born neutron stars rotate rapidly. In hydrostatic equilibrium, the structure of a spherically symmetric, static relativistic NS is given by Tolman–Oppenheimer–Volkoff (TOV) Eq. 3.8, using the Schwarzschild metric. Given a barotropic EoS (of the form of $\varepsilon = \varepsilon(n_b)$ and $P = P(n_b)$), the stellar structure can be computed by numerically integrating the TOV equations with suitable boundary conditions. However rotation results in deformation of the star from spherical symmetry. The metric for a rigidly rotating neutron star in axial symmetry is defined by the following line element:

$$g_{\mu\nu}dx^{\mu}dx^{\nu} = e^{2\nu(r,\theta)}dt^2 - e^{2\lambda(r,\theta)}dr^2 - e^{2\mu(r,\theta)}\left[r^2d\theta^2 + r^2\sin\theta^2\theta(d\phi - \omega(r,\theta)dt^2\right] \tag{3.46}$$

where ν, λ, μ are metric potentials and $\omega(r,\theta)$ is the angular frequency of the local inertial frame. In the case of a **slowly rotating** neutron star, the modification to the metric potentials can be considered as a small perturbation to the non-rotating structure, as proposed first by Hartle and Thorne [163]. The metric coefficients can be expanded in terms of Legendre polynomials, upto second order as

$$\begin{aligned} e^{2\nu(r,\theta)} &= e^{2\nu}\left[1 + 2(h_0 + h_2P_2(\cos\theta))\right] \\ e^{2\lambda(r,\theta)} &= e^{2}\lambda\left[1 + 2\frac{(m_0 + m_2P_2(\cos\theta)))}{r - 2M(r)}\right] \\ e^{2\mu(r,\theta)} &= \left[1 + 2(v_2 - h_2)P_2(\cos\theta)\right], \end{aligned} \tag{3.47}$$

where m_0, m_2, h_0, h_2, v_2 are obtained from Einstein field equations and are functions of r. The angular velocity of the fluid element with respect to the local inertial frame is defined as

$$\bar{\omega}(r,\theta) = \Omega - \omega(r,\theta). \tag{3.48}$$

Angular momentum can be written in terms of $\bar{\omega}$ as

$$J = -\frac{2}{3}\int_0^R r^3\frac{dj}{dr}\bar{\omega}dr. \tag{3.49}$$

The moment of inertia I can be obtained from the relation $J = I\Omega$ and is given by

$$I = \frac{8\pi}{3}\int_0^R drr^4e^{-\nu}\frac{[\varepsilon(r) + P(r)]}{\sqrt{(1 - 2M(r)/r)}}\frac{[\Omega - \omega(r)]}{\Omega}. \tag{3.50}$$

In general, equilibrium configurations of **rapidly rotating** neutron stars can be calculated by solving Einstein equations, with the assumptions of stationarity, axisymmetry and circularity (absence of meridional convective currents). An EoS is required to close the system of equations. For finite temperature, the EoS depends on temperature as well as on the particle number densities. The partial differential Einstein equations can be solved using a multidomain spectral method [164]. A rapidly rotating star can be modelled in general relativity within a 3 + 1 dimensional space plus time decomposition of Einstein equation [164], where spacetime is sliced by a family of spacelike hypersurfaces Σ_t, labeled by the time coordinate t. The line element can be written as

$$ds^2 = -N^2dt^2 + \gamma_{ij}(dx^i + \beta^idt)(dx^j + \beta^jdt). \tag{3.51}$$

Here x^i is coordinate on the hyperspace, N is the proportionality coefficient of two collinear vectors ∇t and $\mathbf{n}$ in $\mathbf{n} = -N\nabla t$, called lapse function, and β^i is the shift vector and γ_{ij} the 3-metric on each hypersurface [164]. Since each hypersurface Σ_t is assumed to be spacelike, the components of γ are given in terms of the components of the normal via $\gamma_{\mu\nu} = g_{\mu\nu} + n_\mu n_\nu$.
Spacetime can be stationary and axisymmetric at the same time which implies the existence of timelike Killing vector field $\boldsymbol{\zeta} = (\partial_t)$ and a spacelike Killing field $\boldsymbol{\xi} = \partial_\phi$ respectively. These two commuting Killing vectors implies that we can choose spherical polar coordinates ($x^0 = t$, x^1 =r , $x^2 = \phi$, $x^3 = \theta$). Furthermore for an asymptotically flat spacetime $\beta^r = \beta^\phi = 0$, $\gamma_{r\phi} = \gamma_{\theta\phi} = 0$. Hence, the line element for a rotating star in quasi-isotropic coordinates, which additionally gives $\gamma_{r\theta}$ =0, is given by

$$g_{\alpha\beta}dx^\alpha dx^\beta = -N^2dt^2 + A^2\left(dr^2 + r^2d\theta^2\right) + B^2r^2sin^2\theta(d\phi - \omega dt)^2. \quad (3.52)$$

where N, A, B and ω are four functions of (r, θ). Four Einstein equations for the rotating star involving these four metric potentials reduce to partial differential equations with the matter source terms as well as the gravitational field on their right hand side.

For a perfect fluid, the stress-energy tensor is given by, $T^{\mu\nu} = (\varepsilon + p)u^\mu u^\nu + pg^{\mu\nu}$, where ε is energy density, p is pressure, both in the fluid frame and u^μ is the fluid 4-velocity. Due to the circularity hypothesis, the fluid 4-velocity with respect to ZAMO (zero angular momentum observer) can be written as, $\mathbf{u} = \Gamma(\mathbf{n} + \mathbf{U})$. where, $\mathbf{n}$ are the 4-velocity of the ZAMO, $\Gamma = (1 - U^2)^{-1/2}$ and $\mathbf{U}$ are the Lorentz factor and 3-velocity of the fluid with respect to ZAMO respectively. The equations of motion follow from the energy-momentum conservation law, $\nabla^\mu T_{\mu\nu} = 0$ and the baryon number conservation law $\nabla^\mu(n_b u^\mu) = 0$. First integral of motion is expressed as [164]

$$\nabla(H + \ln N - \ln\Gamma) + F(\Omega)\nabla\Omega = 0, \quad (3.53)$$

where $H := \ln\frac{h}{m_B}$, m_B being the mean baryon mass: $m_B \simeq 1.66\times10^{27}$ kg. For **rigid rotation,** Ω = constant and Eq. 3.53 becomes

$$H + \ln N - \ln\Gamma = const.$$

In case of **differential rotation,** $\nabla\Omega \neq 0$ and the first integral of motion is given by

$$H + \ln N - \ln\Gamma + \int_0^\Omega F(\Omega')d\Omega' = const. \quad (3.54)$$

For hot stars, equation of fluid motion is given by

$$\partial_i(H + \ln N - \ln\Gamma) = \frac{Te^{-H}}{m_B}\partial_i s_B - F\partial_i\Omega. \quad (3.55)$$

This numerical method gives a unique solution with the input parameters (H, Ω) and a fixed EoS for a rotation law. In order to investigate the role of differential rotation, one may employ the Komatsu–Eriguchi–Hachisu (KEH) scheme [165] or j-constant rotation law defined by the velocity profile:

$$F(\Omega) = R_0^2\,(\Omega_c - \Omega)$$

where Ω_c is the central angular frequency and R_0 is a free parameter with dimensions of length that determines the degree of differential rotation [166]. We consider the dimensionless parameter $a = R_e/R_0$, where R_e is the equatorial NS radius. Thus the limit of uniform rotation is obtained when $a \to 0$ and increasing a denotes increasing degree of differential rotation. The advantage of the j-constant law is that it approximately reproduces the rotation-profile obtained in 2D-simulations, and is a "simple" law, with the minimum number of free parameters. Although the j-constant rotation law is the most widely used, alternative rotation laws have been discussed in the literature [167, 168].

Differential rotation may support hot hypermassive NS merger remnants against collapse. There exist several numerical schemes that compute equilibrium solutions of uniform and differentially rotating cold NSs, see e.g. the numerical library LORENE (https://lorene.obspm.fr/). Within this scheme, calculations are performed solving general relativistic equations of hydrostatic equilibrium of rotating, axially symmetric stars. The first attempts towards equilibrium models including thermal effects in uniformly and differentially rotating NSs were introduced by Goussard et al. [169, 170] for realistic EoSs. It was shown that for finite temperature, the integrability of the equation of stationary motion requires an isentropic (constant entropy) or isothermal (constant temperature) solution.

3.5 Parametric and Non-parametric EoS

The many-body nucleon-nucleon interaction in dense matter is modelled using different schemes, as described in Sect. 3.3.6. However among the different realistic nuclear models, there are uncertainties and unresolved differences to extract model-independent information. An alternative model-independent way of constructing EoSs is where a number of free parameters are fitted directly to observational data. In the past decade, there have been many studies using such phenomenological **parametric models** to impose constraints on the nuclear EoS within a statistical Bayesian scheme [171–179], starting from the pioneering work by Steiner et al. [180]. The idea of such a scheme is to construct posterior functions, with the low density EoS constrained by theoretical and experimental nuclear physics, and the parametrized high density EoS satisfying gravitational wave and electromagnetic data [181–185]. Usually EoSs based on different parametrization schemes such as *piecewise polytropes* [5, 186–188], *spectral representation* [189–191] or *speed-of-sound parametrization* [182, 192–194] have been used for such studies, convenient for numerical relativity simulations and parameter estimation from gravitational waveforms. Parametric models rely on some functional forms to reproduce a large array of EoSs, but this is limited as any finite-parameter model cannot accurately approximate all possible realistic EoSs. **Non-parametric inference schemes** overcome this by increasing the freedom in the parameter space of their model, at the cost of increased complexity. They attempt to combine information from recent astrophysical observations while avoiding assumption of modelling systematics inherent to parametric inference schemes, via nonparametric representation of the EoS based on Gaussian processes conditioned on nuclear theory models. Several such schemes have also been developed recently [193, 195] and even combined with CEFT [196] and pQCD [197–200] results.

The main issue with such studies is that the non-nucleonic degrees of freedom (e.g. hyperons or quark matter) at intermediate or high densities are overlooked and the EoS curves are modelled by chosen functional forms or considering general physical conditions such as causality. The role of the underlying nuclear physics, such as the key nuclear saturation parameters that control the behaviour of NS global observables, are not apparent from such studies. To remedy this problem, a few recent studies have used Bayesian scheme within the RMF model [201] or hybrid (nuclear + piecewise polytope) parametrizations [173] to obtain posterior distributions of empirical parameters, but no clear correlation between the nuclear empirical and astrophysical observables could be established in such studies. An alternative approach is to employ the non-relativistic *meta-modelling technique* [202, 203], where the energy per particle is directly expressed in terms of the empirical isoscalar and isovector parameters. Some of the techniques described above are elaborated in the following Sects. 3.5.1 and 3.5.2.

3.5.1 Piecewise Polytropes

In this method polytropic EoSs with different indices are used in different density regimes [187]. The EoSs are assumed to be of the form

$$P_i = K_i n^{\Gamma_i}$$

where n is the baryon density and i refers to the density regime. Using the relation $n = \partial P/\partial\mu$, one can write pressure as a function of chemical potential as,

$$P_i = K_i \left(n_i^{\Gamma_i - 1} + \frac{\Gamma_i - 1}{K_i \Gamma_i} (\mu - \mu_i) \right)^{\frac{\Gamma_i}{\Gamma_i - 1}},$$

where μ_i and n_i refer to the baryon chemical potential and number density respectively at the starting point of the interval. This ensures that the EoS is continuous. The curves may also be smoothened at these critical points so as to mimic a smooth phase transition. Using the first law of thermodynamics $d(\varepsilon/n) = -Pd(1/n)$ one gets,

$$\varepsilon_i = c_i n + \frac{K_i}{\Gamma_i - 1} n^{\Gamma_i},$$

where $c_i = \frac{\varepsilon_{i-1}}{\rho_{i-1}} - \frac{K_i}{\Gamma_i - 1} n_{i-1}^{\Gamma_i - 1}$ wherein $i - 1$ refers to the end point of the previous interval. The corresponding speed of sound ($c_s = \sqrt{dp/d\varepsilon}$) is given by

$$c_s = \sqrt{\frac{\Gamma_i p}{\varepsilon + p}}.$$

3.5.2 Meta-Modelling Scheme

In several studies involving different nuclear models, certain correlations are found to exist between the nuclear empirical parameters and the NS observables for some of the nuclear models. To investigate whether these correlations are model-dependent, a non-relativistic **meta-modelling (MM) scheme** was introduced based on the Taylor-Series expansion, assuming no apriori underlying nuclear correlations [202, 203]. In the MM approach, the energy per particle is defined as:

$$e^N(n, n_1) = t^{FG*}(n, n_1) + V^N(n, n_1). \tag{3.56}$$

within a non-relativistic approach. The kinetic energy $t^{FG*}(n, n_1)$ and the potential energy $V^N(n, n_1)$ terms are defined as:

$$t^{FG*}(n, n_1) = \frac{t_{sat}^{FG}}{2} \left(\frac{n}{n_{sat}}\right)^{2/3} \left[\left(1 + \kappa_{sat} \frac{n}{n_{sat}}\right) f_1(\delta) + \kappa_{sym} \frac{n}{n_{sat}} f_2(\delta)\right] \tag{3.57}$$

$$V^N(n, n_1) = \sum_{\alpha \geq 0}^{N} \frac{1}{\alpha!} (v_\alpha^{is} + v_\alpha^{iv} \delta^2) x^\alpha u_\alpha^N(x) \tag{3.58}$$

where asymmetry coefficient is $\delta = [0, 1]$. A correction factor

$$u_\alpha^N(x) = 1 - (-3x)^{N+1-\alpha} e^{-\frac{bn}{n_{sat}}}$$

is introduced to reproduce the correct behavior of energy density at zero density (model *ELFc*) [203], where parameter $b = 10 ln\ 2$. The functions f_1 and f_2 are defined as follows,

$$f_1(\delta) = (1 + \delta)^{5/3} + (1 - \delta)^{5/3} \tag{3.59}$$

$$f_2(\delta) = \delta[(1 + \delta)^{5/3} - (1 - \delta)^{5/3}]. \tag{3.60}$$

The Landau effective nucleon mass (m^*_N) in MM is defined as [204]

$$\frac{m}{m^*_N} = 1 + (\kappa_{sat} + \tau_{3N}\kappa_{sym}\delta)\frac{n}{n_{sat}}. \tag{3.61}$$

Here τ_3 = +1 and -1 for neutrons and protons, respectively. The parameters κ_{sat} and κ_{sym} are functions of Landau effective mass at saturation for SNM (m^*_{sat}), and isospin splitting at saturation (Δm^*_{sat}) for pure neutron matter (PNM) [205, 206]. These can be written as follows,

$$\begin{aligned} \kappa_{sat} &= \frac{m}{m^*_{sat}} - 1 \\ \kappa_{sym} &= \frac{1 - \sqrt{1 + \left(\frac{\Delta m^*_{sat}}{m^*_{sat}}\right)^2}}{\frac{\Delta m^*_{sat}}{m}}. \end{aligned} \tag{3.62}$$

Now, one can derive the relations between the model and nuclear empirical parameters at saturation. For isoscalar model parameters,

$$\begin{aligned} v_0^{is} &= E_{sat} - t_{sat}^{FG}(1 + \kappa_{sat}) \\ v_1^{is} &= -t^{FG}sat(2 + 5\kappa_{sat}) \\ v_2^{is} &= K_{sat} - 2t_{sat}^{FG}(-1 + 5\kappa_{sat}) \\ v_3^{is} &= Q_{sat} - 2t_{sat}^{FG}(4 - 5\kappa_{sat}) \\ v_4^{is} &= Z_{sat} - 8t_{sat}^{FG}(-7 + 5\kappa_{sat}). \end{aligned} \tag{3.63}$$

Similarly, for isovector parameters,

$$\begin{aligned} v_0^{iv} &= J_{sym} - \frac{5}{9}t_{sat}^{FG}[1 + (\kappa_{sat} + 3\kappa_{sym})] \\ v_1^{iv} &= L_{sym} - \frac{5}{9}t^{FG}sat[2 + 5(\kappa_{sat} + 3\kappa_{sym})] \\ v_2^{iv} &= K_{sym} - \frac{10}{9}t_{sat}^{FG}[-1 + 5(\kappa_{sat} + 3\kappa_{sym})] \\ v_3^{iv} &= Q_{sym} - \frac{10}{9}t_{sat}^{FG}[4 - 5(\kappa_{sat} + 3\kappa_{sym})] \\ v_4^{iv} &= Z_{sym} - \frac{40}{9}t_{sat}^{FG}[-7 + 5(\kappa_{sat} + 3\kappa_{sym})]. \end{aligned} \tag{3.64}$$

It is enough to consider up to 4th order in the Taylor series for the meta-model to reproduce the EoS and its derivatives of the different models.

One may note that as the higher-order nuclear parameters have large uncertainties, the MM has large uncertainties at high density, which poses a problem when estimating NS maximum mass using MM EoS. Further, since it is a non-relativistic model, causality has to be imposed by hand, and it is not straight-forward to include other non-nucleonic degrees of freedom such as hyperons or mesons. Recently, a meta-modelling technique based on a relativistic density functional with density-dependent couplings has been proposed [207].

3.6 Multidisciplinary Constraints on the Equation of State

The description of nuclear many-body interactions in dense matter using different EoS models suffers from uncertainties. In order to reduce the parameter space of uncertainties, information can be combined from multi-disciplinary physics. Recent progresses in theoretical physics that provide important insights at asymptotic regions of the QCD phase diagram i.e., very low or very high densities, will be described in Sect. 3.6.1. Important constraints also come from terrestrial experiments, studying properties of nuclei or hypernuclei, as well as from heavy-ion experiments. This is described in Sect. 3.6.2. Finally, recent breakthrough in multi-messenger astrophysical data in electromagnetic and gravitational waves, and their implications for NS EoS, will be summarised in Sect. 3.6.3.

3.6.1 Theoretical Constraints

3.6.1.1 Chiral Effective Field Theory

Effective and reliable constraints on the EoS of NS matter at low baryon densities $n_b \sim 0.5 - 1.4 n_{sat}$ come from microscopic ab-initio calculations of pure neutron matter (PNM) using CEFT (see Sect. 3.3.6.1) [68, 186, 208]. CEFT starts from a low momentum expansion of nuclear forces consistent with the symmetries of QCD. It accounts for many-body interactions among nucleons using order-by-order expansions in terms of long-range pion exchange interactions and contact interactions to describe the short-range. The EoSs of PNM from phenomenological calculations can be compared with the latest CEFT calculation results [182] within Bayesian statistical methods. The CEFT formalism is currently also being extended to include hyperonic degrees of freedom [209] and to asymmetric nuclear matter at finite temperature [210].

3.6.1.2 Perturbative QCD

As discussed in Sect. 3.4.1.2, pQCD calculations provide reliable constraints at large densities or temperatures ($T \gtrsim 300$ MeV, $\mu_B = 0$ or $n_B \gtrsim 40 n_{sat}$, $T = 0$), where quarks can be treated as asymptotically free. It was shown that causality and stability bounds allow pQCD to be applied at lower densities [211]. It has further been demonstrated that combining low energy constraints from CEFT and very high density constraints from pQCD can provide important insights into the behaviour of dense matter EoS, and may even indicate a phase transition to quark matter (see [5, 194, 199, 212]).

3.6.1.3 Lattice QCD

Recent progress in lattice QCD [213–215] are also providing reliable constraints on the properties of matter at high temperature ($T \gtrsim 135$ MeV) and vanishing densities. Lattice QCD calculations are based on solving QCD numerically on a very large grid. It has been established that the phase transition from a hadron resonance gas (HRG) at low temperatures into a quark-gluon plasma at high temperatures occurs via a smooth crossover. Recent results by the HALQCD group extracted hyperon potentials at almost physical quark masses and used it to estimate the hyperon potentials in pure neutron matter and in nuclear matter by using the Brueckner–Hartree–Fock approximation [213, 214]. They find that the hyperon potentials in nuclear matter are $U_\Lambda = -28$ MeV, $U_\Sigma = +15$MeV and $U_\Xi = -4$ MeV. Despite these successes, Lattice QCD EoS calculations cannot be extended to temperatures and densities relevant to low-energy heavy-ion collisions or neutron stars due to the sign problem.

3.6.2 *Experimental Constraints*

3.6.2.1 Nuclear Physics

Low energy nuclear physics experiments ($T \to 0$) probe the properties of nuclei near the stability line ($Y_Q \sim 0.5$) close to saturation density n_{sat}. In phenomenological EoS models, the symmetric matter parameters are fitted to nuclear empirical observables measured at saturation density (e.g. binding energy per nucleon, compressibility, effective nucleon mass). While the binding energy at saturation is relatively well constrained, the compressibility and effective nucleon mass have large uncertainties. There are a few experiments that study neutron rich heavy nuclei and unstable nuclei close to neutron drip line. However this information cannot be directly applied in neutron stars, and extrapolation in symmetry energy (difference between energy of symmetric nuclear matter and pure neutron matter) becomes inevitable. In phenomenological models, the asymmetric matter parameters are adjusted to reproduce symmetry energy and its slope, which are determined from the measurements of neutron skin thickness of ^{208}Pb or ^{48}Ca [216–218], from electric dipole polarizability α_D, from isoscalar giant monopole and dipole resonances [219, 220], nuclear masses [221–224] and from isobaric analog states [225].

3.6.2.2 Hypernuclear Physics

Compared to the nucleon-nucleon (NN) interaction, which is fairly well known from existing scattering data and measured properties of nuclei, YN and YY interactions are poorly constrained. The number of ΛN and ΣN scattering events are limited to ≈ 600 due to the short lifetime of hyperons and low intensity beam fluxes, while that of ΞN events are even fewer [226–230]. In the case of the YY interaction there is no available scattering data.

Alternative information about YN and YY interactions can be obtained from the study of hypernuclei, i.e. bound systems composed of nucleons and one or more hyperons. High energy accelerators as well as modern electronic counters have led to the identification of more than 40 single Λ and few double-Λ hypernuclei. As there are no unambiguous confirmations of the existence of single Σ-hypernuclei, it is suggested that ΣN interaction is repulsive [42]. The values of the hyperon potentials in nuclear matter compatible from such experiments are $U_\Lambda = -30$ MeV, $U_\Sigma = +30$MeV and $U_\Xi = -18$ MeV.

3.6.2.3 Heavy-Ion Collisions

While nuclear and hypernuclear physics gives us important information at low energies, heavy-ion collisions provide information on compressed baryonic matter at intermediate energies in the range $1-3n_{sat}$, which are relevant densities for the neutron star interior where exotic particles may appear. There are several heavy-ion collision experiments, which provide important information on the nuclear EoS in the density regime $\sim 1-3n_{sat}$:

1. **KaoS experiment** Studies of K^+ meson production in nuclear collisions at subthreshold energies were performed with the Kaon Spectrometer (KaoS experiment) at GSI, Darmstadt [231]. Kaon multiplicity in Au+Au and C+C collisions is an indicator of the compressibility of dense matter at baryon densities in the range $\sim 2-3n_{sat}$, which is in turn controlled by the stiffness of nuclear matter via the nucleon potential U_N. The more attractive the nucleon potential U_N, the higher is the K^+ meson abundance, which therefore serves as a probe for the stiffness of nuclear matter. The analysis of the experimental results using (isospin dependent quantum molecular dynamics or IQMD) transport models indicate that the EoS in this density range is soft, using a simple Skyrme ansatz with an incompressibility $K_{sat} \sim 200$ MeV or less [231, 232]. These results have implications for the maximum neutron star mass [233]. It was shown that the heavy-ion data can be reconciled with NS maximum mass observations only if the EoS has a transition from a soft EoS (compatible with heavy-ion data from KaoS experiment at 2–3 n_{sat}) to a stiff EoS (the maximum value is limited by causality) beyond a critical density.

2. **FOPI experiment** Elliptic flow data from Au+Au collisions within 0.4 - 1.5A GeV from the FOPI collaboration [234] can also provide constraints on the symmetric nuclear matter (SNM) EoS. Using the IQMD transport code, an observable is introduced to describe the evolution of the size of the elliptic flow as a function of rapidity, which is sensitive to the nuclear EoS and therefore a robust tool to constrain the incompressibility of nuclear matter up to 2 n_{sat}. Using the FOPI data, a constraint for the binding energy of SNM in the density range $\sim 1.4 - 2n_{sat}$ [234] can be obtained.
3. **ASY-EOS experiment** In the ASY-EOS experiment at GSI in Germany [235], directed and elliptic flows of neutrons and light charged particles were measured for the reaction ^{197}Au+^{197}Au at 400 MeV/nucleon incident energy. By comparing the ratio of the elliptic flow of neutrons to charged particles with predictions from Ultra relativistic Quantum Molecular Dynamics or UrQMD calculations, the density dependence of the potential part in the parametrization of the symmetry energy can be derived. This provides a more stringent constraint on the symmetry energy for Asymmetric Nuclear Matter (ANM) for supra-saturation density $\sim 2n_{sat}$.
   It should be noted that although the constraints from heavy-ion collisions provide important and interesting information, the results are model-dependent. The heavy-ion data are confronted by transport codes such as IQMD [231, 236] or UrQMD using different phenomenological EoSs to conclude which EoS is most compatible with the data.

3.6.3 Astrophysical Constraints

Any theoretical EoS is required to predict a maximum neutron star mass, M_{max}, compatible with observed maximum NS mass observations. GR effects in pulsars in binaries can lead to a very accurate determination of NS masses. Until the last decade, the most precise neutron star **mass** observations clustered around a "canonical" value close 1.4 $M_{\odot}$ [24]. This constraint was not stringent enough to discriminate between the different EoS models. A breakthrough came with the precise measurements of the NSs with masses as high as $2M_{\odot}$ [15–17]. Recent observations of the heaviest known pulsar PSR J0740+6620 indicate that the highest maximum mass of the neutron star observed is $2.08^{+0.07}_{-0.07}$ $M_{\odot}$ [18, 19]. These measurements impose a stronger constraint on the EoS models, such that any reliable one should predict $M_{max} \geq 2M_{\odot}$.

It is well known that the presence of **hyperons** in the neutron star interior, which is energetically favorable for densities above 2–3 n_{sat}, leads to the softening of the neutron star EoS and to the reduction of the maximum mass to values that can be smaller than 2 $M_{\odot}$. The discovery of neutron stars with masses of $\sim 2M_{\odot}$ then lead to the so-called "hyperon puzzle", and reconciling this constraint with the presence of hyperons in neutron stars required the conditions for appearance of hyperons in the neutron star core to be scrutinized in great detail [42].

In addition to the maximum mass, the measurement of **radii** can also serve to constrain the EoS and estimate the strangeness content in neutron stars. NS radii may be determined using hot spots on NS surface, offset from the rotating pole modulated by stellar rotation [20–24]. However, the analysis of observations from quiescent low-mass X-ray binaries suffered from large uncertainties. The recently launched Neutron star Interior Composition ExploreR (NICER) [25] mission from NASA is providing highly improved estimates of the mass and radius measurements for pulsars e.g. data for PSR J0030+0451 [26, 27] and PSR J0740+6620 [28, 29]. The Bayesian parameter estimation of the mass and equatorial radius of the millisecond pulsar PSR J0030+0451 [26] yield $M = 1.44^{+0.15}_{-0.14}\ \mathrm{M}_{\odot}$ and $R = 13.02^{+1.24}_{-1.06}$ km, consistent with the independent analysis by [27]. The more recent data for PSR J0740+6620 yields an equatorial circumferential radius of $13.7^{+2.6}_{-1.5}$ km [28] (68%).

Thermal evolution in neutron stars can also be very sensitive to their composition. The presence of strangeness in their core can affect neutrino emissivities allowing for fast and more efficient cooling mechanisms and therefore affect the cooling properties. Measurements of the surface temperatures of neutron stars, with satellite-based X-ray observatories, allow us to determine the role of the strange components of dense matter in their cooling behaviour. A well-known example is the compact star in Cassiopeia A (Cas A) [237, 238] whose thermal evolution studied over the last 10 years suggests fast cooling mechanisms to explain the data, such as via phase transitions among different superconducting phases of quark matter or Cooper pair-breaking processes [239, 240]. Cooling studies of other neutron stars can serve in general to rule out slow cooling via Urca (direct and modified) processes involving exotic matter, and constrain the maximum strangeness fraction in the neutron star core [241].

The most exciting and revolutionary astrophysical discovery in recent times has been the direct detection of **Gravitational Waves** (GW) from NSs [1]. The detection of the NS merger event GW170817 [1] is particularly unique because this event seen in gravitational waves and its subsequent followup in electromagnetic counterparts by telescopes such as Fermi, Integral and Chandra opened up a new era of Multi-Messenger (MM) astronomy. The GW170817 event puts a tight constraint on the upper bound of the dimensionless tidal deformability $\Lambda < 720$ [3, 242]. As tidal deformability depends on the compactness and radius, this result leads to a constraint on the mass-radius relation, imposing that NSs should have $R_{1.4M_{\odot}} < 13.5$ km [1, 4, 5]. The role of dense matter EoS is crucial in the tidal response of NSs during late stage of binary inspiral and post-merger remnant oscillations [43].

Unstable oscillation modes (e.g. f-,p-,g-modes) in isolated neutron stars may also emit GW, which contains signature of its internal composition [243–246]. The emission of GW in hot and rapidly rotating newly born neutron stars due to r-mode instability can also be affected by the presence of strangeness [42, 94, 247], since bulk viscosity of neutron star matter is dominated by the contribution of the strange components as soon as they appear in the neutron star interior.

The combined constraints at low density from CEFT, combined with multi-messenger astrophysical observations or pQCD at high density, as well as heavy-ion physics at intermediate densities, can provide important information about the nuclear EoS [248–252].

3.7 Looking Ahead

There are several new hypernuclear facilities planned or under construction e.g. in GSI in Germany, JLAB in USA and J-PARC in Japan, which may provide more precise updates on the properties of hyperon- nucleon and hyperon-hyperon interactions [253–255]. Experimental hypernuclear physics is still an active field of research. In the next years, lattice QCD calculations may be able to provide crucial information on ΛN and ΛNN interactions [256–258]. Recent lattice simulations of the binding energy of dibaryons indicate the existence of hyperons. Compatibility with the maximum mass constraint requies several of the hyperon EoSs to go beyond the SU(6) symmetry towards a universal hyperon-meson coupling. Whether such approaches are justified can be tested with experimental data of hyperon matter at large densities, such as with possible future measurements of multi-hypernuclei at FAIR in GSI. The analysis of hyperon-hyperon correlation in heavy-ion collisions may provide further understanding of hyperons. The Compressed Baryonic Matter (CBM) experiment at FAIR (GSI, Germany) is a future experimental facility dedicated to constrain the EoS at moderate temperatures and high densities $\sim 4-5 n_{sat}$ [259]. However extraction of data from heavy-ion collisions will require the development of more sophisticated dynamical models.

With the advent of the LIGO-Virgo-Kagra (LVK) [260–263] global network of gravitational wave detectors, a new window of astrophysics has emerged, and the multi-messenger data is expected to be further enriched with the future generations of detectors such as the Cosmic Explorer [264], Einstein telescope [265] or LIGO-India [266]. Finally, new accurate measurements of masses and radii of neutron stars with future missions such as Strobe-X [267] will play an important role in pinning down the EoS of nuclear matter at high densities. In addition, observational information such as cooling studies, pulsar back-bending, quasi-normal modes and gravitational waves would be useful tools to probe the existence of exotic matter in neutron stars.

Acknowledgements I am grateful to Prof Luciano Rezzolla for reading the article and providing valuable guidance. I also acknowledge Christian Ecker and Tyler Gorda for their useful comments and pointing to important references. I am thankful to my institute IUCAA for providing invaluable support, and also to the students in my NSGW research group, Bikram, Swarnim, Rajesh and Nilaksha, for helping with the writeup and preparing the figures for this chapter. I also acknowledge Ashwin Shirke, masters student at University of Potsdam, for reading the manuscript and providing useful feedback.

References

1. B.P. Abbott, R. Abbott, T.D. Abbott, F. Acernese, K. Ackley, C. Adams, T. Adams, P. Addesso, R.X. Adhikari, V.B. Adya, et al., Gw170817: observation of gravitational waves from a binary neutron star inspiral. Phys. Rev. Lett. **119**(16) (2017)
2. B.P. Abbott, R. Abbott, T.D. Abbott, F. Acernese, K. Ackley, C. Adams, T. Adams, P. Addesso, R.X. Adhikari, V.B. Adya, et al., Gw170817: measurements of neutron star radii and equation of state. Phys. Rev. Lett. **121**(16) (2018)
3. B.P. Abbott, R. Abbott, T.D. Abbott, F. Acernese, K. Ackley, C. Adams, T. Adams, P. Addesso, R.X. Adhikari, V.B. Adya et al., Properties of the binary neutron star merger gw170817. Phys. Rev. X **9**, 011001 (2019)
4. E.R. Most, L.R. Weih, L. Rezzolla, J. Schaffner-Bielich, New constraints on radii and tidal deformabilities of neutron stars from gw170817. Phys. Rev. Lett. **120**(26) (2018)
5. E. Annala, T. Gorda, A. Kurkela, A. Vuorinen, Gravitational-wave constraints on the neutron-star-matter equation of state. Phys. Rev. Lett. **120**, 172703 (2018)
6. Copyright Page, in *Relativistic Hydrodynamics* (Oxford University Press, 2013)
7. R.R. Silbar, S. Reddy, Neutron stars for undergraduates. Amer. J. Phys. **72**(7), 892–905 (2004)
8. S. Chandrasekhar, The Maximum Mass of Ideal White Dwarfs. Astrophys. J. **74**, 81 (1931)
9. J.R. Oppenheimer, G.M. Volkoff, On massive neutron cores. Phys. Rev. **55**, 374–381 (1939)
10. N.K. Glendenning, *Compact Stars: Nuclear Physics, Particle Physics and General Relativity*. Astronomy and Astrophysics Library (Springer, New York, 2012)
11. J. Schaffner-Bielich, *Compact Star Physics* (Cambridge University Press, 2020)
12. J.M. Lattimer, The nuclear equation of state and neutron star masses. Ann. Rev. Nucl. Part. Sci. **62**(1), 485–515 (2012)
13. J.M. Lattimer, Introduction to neutron stars. AIP Conf. Proc. **1645**(1), 61–78 (2015)
14. F. Özel, P. Freire, Masses, radii, and the equation of state of neutron stars. Ann. Rev. Astron. Astrophys. **54**(1), 401–440 (2016)
15. P.B. Demorest, T. Pennucci, S.M. Ransom, M.S.E. Roberts, J.W.T. Hessels, A two-solar-mass neutron star measured using shapiro delay. Nature **467**(7319), 1081–1083 (2010)
16. J. Antoniadis, P.C.C. Freire, N. Wex, T.M. Tauris, R.S. Lynch, M.H. van Kerkwijk, M. Kramer, C. Bassa, V.S. Dhillon, T. Driebe et al., A massive pulsar in a compact relativistic binary. Science **340**(6131), 1233232–1233232 (2013)
17. E. Fonseca, T.T. Pennucci, J.A. Ellis, I.H. Stairs, D.J. Nice, S.M. Ransom, P.B. Demorest, Z. Arzoumanian, K. Crowter, T. Dolch et al., The nanograv nine-year data set: mass and geometric measurements of binary millisecond pulsars. Astrophys. J. **832**(2), 167 (2016)
18. H.T. Cromartie, E. Fonseca, S.M. Ransom, P.B. Demorest, Z. Arzoumanian, H. Blumer, P.R. Brook, M.E. DeCesar, T. Dolch, J.A. Ellis et al., Relativistic shapiro delay measurements of an extremely massive millisecond pulsar. Nat. Astron. **4**(1), 72–76 (2019)
19. E. Fonseca, H.T. Cromartie, T.T. Pennucci, P.S. Ray, A.Y. Kirichenko, S.M. Ransom, P.B. Demorest, I.H. Stairs, Z. Arzoumanian, L. Guillemot, A. Parthasarathy, M. Kerr, I. Cognard, P.T. Baker, H. Blumer, P.R. Brook, M. DeCesar, T. Dolch, F.A. Dong, E.C. Ferrara, W. Fiore, N. Garver-Daniels, D.C. Good, R. Jennings, M.L. Jones, V.M. Kaspi, M.T. Lam, D.R. Lorimer, J. Luo, A. McEwen, J.W. McKee, M.A. McLaughlin, N. McMann, B.W. Meyers, A. Naidu, C. Ng, D.J. Nice, N. Pol, H.A. Radovan, B. Shapiro-Albert, C.M. Tan, S.P. Tendulkar, J.K. Swiggum, H.M. Wahl, W. Zhu, Refined mass and geometric measurements of the high-mass psr j0740+6620 (2021)
20. F. Özel, G. Baym, T. Güver, Astrophysical measurement of the equation of state of neutron star matter. Phys. Rev. D **82**, 101301 (2010)
21. S. Guillot, M. Servillat, N.A. Webb, R.E. Rutledge, Measurement of the radius of neutron stars with high signal-to-noise quiescent low-mass x-ray binaries in globular clusters. Astrophys. J. **772**(1), 7 (2013)
22. S. Guillot, R.E. Rutledge, Rejecting proposed dense matter equations of state with quiescent low-mass x-ray binaries. Astrophys. J. **796**(1), L3 (2014)

23. A.W. Steiner, M. Hempel, T. Fischer, Core-collapse supernova equations of state based on neutron star observations. Astrophys. J. **774**(1), 17 (2013)
24. J.M. Lattimer, A.W. Steiner, Neutron star masses and radii from quiescent low-mass x-ray binaries. Astrophys. J. **784**(2), 123 (2014)
25. Z. Arzoumanian, K.C. Gendreau, C.L. Baker, T. Cazeau, P. Hestnes, J. W. Kellogg, S.J. Kenyon, R.P. Kozon, K.C. Liu, S.S. Manthripragada, C.B. Markwardt, A.L. Mitchell, J.W. Mitchell, C.A. Monroe, T. Okajima, S.E. Pollard, D.F. Powers, B.J. Savadkin, L.B. Winternitz, P.T. Chen, M.R. Wright, R. Foster, G. Prigozhin, R. Remillard, J. Doty, The neutron star interior composition explorer (NICER): mission definition, in *Space Telescopes and Instrumentation 2014: Ultraviolet to Gamma Ray*, ed. by T. Takahashi, J.-W.A. den Herder, M. Bautz. *Society of Photo-Optical Instrumentation Engineers (SPIE) Conference Series*, vol. 9144 (2014), pp. 914420
26. M.C. Miller, F.K. Lamb, A.J. Dittmann, S. Bogdanov, Z. Arzoumanian, K.C. Gendreau, S. Guillot, A.K. Harding, W.C.G. Ho, J.M. Lattimer et al., Psr j0030+0451 mass and radius from nicer data and implications for the properties of neutron star matter. Astrophys. J. **887**(1), L24 (2019)
27. T.E. Riley, A.L. Watts, S. Bogdanov, P.S. Ray, R.M. Ludlam, S. Guillot, Z. Arzoumanian, C.L. Baker, A.V. Bilous, D. Chakrabarty et al., A nicer view of psr j0030+0451: Millisecond pulsar parameter estimation. Astrophys. J. **887**(1), L21 (2019)
28. M.C. Miller, F.K. Lamb, A.J. Dittmann, S. Bogdanov, Z. Arzoumanian, K.C. Gendreau, S. Guillot, W.C.G. Ho, J.M. Lattimer, M. Loewenstein, S.M. Morsink, P.S. Ray, M.T. Wolff, C.L. Baker, T. Cazeau, S. Manthripragada, C.B. Markwardt, T. Okajima, S. Pollard, I. Cognard, H.T. Cromartie, E. Fonseca, L. Guillemot, M. Kerr, A. Parthasarathy, T.T. Pennucci, S. Ransom, I. Stairs, The radius of psr j0740+6620 from nicer and xmm-newton data (2021)
29. T.E. Riley, A.L. Watts, P.S. Ray, S. Bogdanov, S. Guillot, S.M. Morsink, A.V. Bilous, Z. Arzoumanian, D. Choudhury, J.S. Deneva, K.C. Gendreau, A.K. Harding, W.C.G. Ho, J.M. Lattimer, M. Loewenstein, R.M. Ludlam, C.B. Markwardt, T. Okajima, C. Prescod-Weinstein, R.A. Remillard, M.T. Wolff, E. Fonseca, H. Thankful Cromartie, M. Kerr, T.T. Pennucci, A. Parthasarathy, S. Ransom, I. Stairs, L. Guillemot, I. Cognard, A nicer view of the massive pulsar psr j0740+6620 informed by radio timing and xmm-newton spectroscopy (2021)
30. T. Hinderer, Tidal love numbers of neutron stars. Astrophys. J. **677**(2), 1216–1220 (2008)
31. S. Gandolfi, A. Gezerlis, J. Carlson, Neutron matter from low to high density. Ann. Rev. Nucl. Part. Sci. **65**(1), 303–328 (2015)
32. V. Dexheimer, J. Noronha, J. Noronha-Hostler, N. Yunes, C. Ratti, Future physics perspectives on the equation of state from heavy ion collisions to neutron stars. J. Phys. G: Nuclear Part. Phys. (2021)
33. C.Y. Tsang, M.B. Tsang, P. Danielewicz, W.G. Lynch, F.J. Fattoyev, Constraining neutron-star equation of state using heavy-ion collisions (2018)
34. G. Baym, T. Hatsuda, T. Kojo, P.D. Powell, Y. Song, T. Takatsuka, From hadrons to quarks in neutron stars: a review. Rep. Progr. Phys. **81**(5), 056902 (2018)
35. B. David, A. Alexander, F. Alexandra, G. Hovik, Universe (2020)
36. J.M. Lattimer, M. Prakash, The physics of neutron stars. Science **304**(5670), 536–542 (2004)
37. I. Vidana, A short walk through the physics of neutron stars (2018)
38. J.M. Lattimer, M. Prakash, Neutron star observations: prognosis for equation of state constraints. Phys. Rep. **442**(1–6), 109–165 (2007)
39. H. Heiselberg, V. Pandharipande, Recent progress in neutron star theory. Ann. Rev. Nucl. Part. Sci. **50**(1), 481–524 (2000)
40. J. Schaffner, C.B. Dover, A. Gal, C. Greiner, H. Stöcker, Strange hadronic matter. Phys. Rev. Lett. **71**, 1328–1331 (1993)
41. J. Schaffner, I.N. Mishustin, Hyperon-rich matter in neutron stars. Phys. Rev. C **53**, 1416–1429 (1996)
42. D. Chatterjee, I. Vidaña, Do hyperons exist in the interior of neutron stars? Eur. Phys. J. A **52**(2), 29 (2016)

43. É.É. Flanagan, T. Hinderer, Constraining neutron-star tidal love numbers with gravitational-wave detectors. Phys. Rev. D **77**, 021502 (2008)
44. T. Damour, A. Nagar, Relativistic tidal properties of neutron stars. Phys. Rev. D **80**(8) (2009)
45. K. Yagi, N. Yunes, I-love-q relations in neutron stars and their applications to astrophysics, gravitational waves, and fundamental physics. Phys. Rev. D **88**, 023009 (2013)
46. T. Hinderer, B.D. Lackey, R.N. Lang, J.S. Read, Tidal deformability of neutron stars with realistic equations of state and their gravitational wave signatures in binary inspiral. Phys. Rev. D **81**, 123016 (2010)
47. G. Baym, C. Pethick, P. Sutherland, The ground state of matter at high densities: equation of state and stellar models. Astrophys. J. **170**, 299 (1971)
48. W.J. Huang, Meng Wang, F.G. Kondev, G. Audi, S. Naimi, The ame 2020 atomic mass evaluation (i). Evaluation of input data, and adjustment procedures*. Chin. Phys. C **45**(3), 030002 (2021)
49. F. Grill, H. Pais, C. Providência, I. Vidaña, S.S. Avancini, Equation of state and thickness of the inner crust of neutron stars. Phys. Rev. C **90**, 045803 (2014)
50. S. Typel, G. Röpke, T. Klähn, D. Blaschke, H.H. Wolter, Composition and thermodynamics of nuclear matter with light clusters. Phys. Rev. C **81**, 015803 (2010)
51. G. Baym, H.A. Bethe, C.J. Pethick, Neutron star matter. Nucl. Phys. A **175**(2), 225–271 (1971)
52. F. Douchin, P. Haensel, A unified equation of state of dense matter and neutron star structure. Astron. Astrophys. **380**(1), 151–167 (2001)
53. J.W. Negele, D. Vautherin, Neutron star matter at sub-nuclear densities. Nucl. Phys. A **207**(2), 298–320 (1973)
54. M. Onsi, H. Przysiezniak, J.M. Pearson, Equation of state of stellar nuclear matter in the temperature-dependent extended thomas-fermi formalism. Phys. Rev. C **55**, 3139–3148 (1997)
55. A.R. Raduta, F. Gulminelli, Statistical description of complex nuclear phases in supernovae and proto-neutron stars. Phys. Rev. C **82**, 065801 (2010)
56. S.B. Rüster, M. Hempel, J. Schaffner-Bielich, Outer crust of nonaccreting cold neutron stars. Phys Rev. C **73**(3), 035804 (2006)
57. M.E. Caplan, A.S. Schneider, C.J. Horowitz, D.K. Berry, Pasta nucleosynthesis: molecular dynamics simulations of nuclear statistical equilibrium. Phys. Rev. C **91**, 065802 (2015)
58. M. Fortin, C. Providência, A.R. Raduta, F. Gulminelli, J.L. Zdunik, P. Haensel, M. Bejger, Neutron star radii and crusts: uncertainties and unified equations of state. Phys. Rev. C **94**, 035804 (2016)
59. E. Khan, J. Margueron, Determination of the density dependence of the nuclear incompressibility. Phys. Rev. C **88**, 034319 (2013)
60. C. Ducoin, J. Margueron, C. Providência, I. Vidaña, Core-crust transition in neutron stars: predictivity of density developments. Phys. Rev. C **83**, 045810 (2011)
61. M. Bender, P.-H. Heenen, P.-G. Reinhard, Self-consistent mean-field models for nuclear structure. Rev. Mod. Phys. **75**, 121–180 (2003)
62. J.R. Stone, P.-G. Reinhard, The skyrme interaction in finite nuclei and nuclear matter. Progr. Part. Nucl. Phys. **58**(2), 587–657 (2007)
63. C. Ducoin, J. Margueron, C. Providência, Nuclear symmetry energy and core-crust transition in neutron stars: a critical study. EPL (Europhysics Letters) **91**(3), 32001 (2010)
64. P. Débora, Menezes, Constança Providência, Hyperons in the nuclear pasta phase. Phys. Rev. C **96**(4), 045803 (2017)
65. S.S. Avancini, D.P. Menezes, M.D. Alloy, J.R. Marinelli, M.M.W. Moraes, C. Providência, Warm and cold pasta phase in relativistic mean field theory. Phys. Rev. C **78**, 015802 (2008)
66. M. Oertel, M. Hempel, T. Klähn, S. Typel, Equations of state for supernovae and compact stars. Rev. Mod. Phys. **89**(1), 015007 (2017)
67. R. Machleidt, *The Meson Theory of Nuclear Forces and Nuclear Structure* (Springer, US, Boston, MA, 1989), pp.189–376
68. C. Drischler, J.W. Holt, C. Wellenhofer, Ann. Rev. Nucl. Part Sci. (2021)

69. M. Baldo, G.F. Burgio, H.Q. Song, F. Weber, Neutron star properties with nuclear bhf equations of state (1998)
70. R. Brockmann, R. Machleidt, Relativistic nuclear structure. i. nuclear matter. Phys. Rev. C **42**, 1965–1980 (1990)
71. H. Müther, F. Sammarruca, Z. Ma, Relativistic effects and three-nucleon forces in nuclear matter and nuclei. Int. J. Mod. Phys. E **26**(03), 1730001 (2017)
72. A. Akmal, V.R. Pandharipande, D.G. Ravenhall, Equation of state of nucleon matter and neutron star structure. Phys. Rev. C **58**, 1804–1828 (1998)
73. M. Baldo, G.F. Burgio, M. Centelles, B.K. Sharma, X. Viñas, From the crust to the core of neutron stars on a microscopic basis. Phys. Atom. Nucl. **77**(9), 1157–1165 (2014)
74. H. Tong, C. Wang, S. Wang, Nuclear matter and neutron stars from relativistic brueckner-hartree-fock theory. Astrophys. J. **930**(2), 137 (2022)
75. B. Friedman, V.R. Pandharipande, Hot and cold, nuclear and neutron matter. Nucl. Phys. A **361**, 502–520 (1981)
76. V.R. Pandharipande, Hyperonic matter. Nucl. Phys. A **178**, 123–144 (1971)
77. T.H.R. Skyrme, A unified field theory of mesons and baryons. Nucl. Phys. **31**, 556–569 (1962)
78. E. Chabanat, P. Bonche, P. Haensel, J. Meyer, R. Schaeffer, A skyrme parametrization from subnuclear to neutron star densities part ii. Nuclei far from stabilities. Nucl. Phys. A **635**(1), 231–256 (1998)
79. D. Gogny, *Self-consistent Pairing Calculations* (North-Holland Publishing Company, Netherlands, 1975)
80. F. Chappert, N. Pillet, M. Girod, J.-F. Berger, Gogny force with a finite-range density dependence. Phys. Rev. C **91**, 034312 (2015)
81. J.D. Walecka, A theory of highly condensed matter. Ann. Phys. **83**(2), 491–529 (1974)
82. B.D. Serot, J.D. Walecka, *Relativistic Nuclear Many-Body Theory* (Springer, US, Boston, MA, 1992), pp.49–92
83. M. Dutra, O. Lourenço, S.S. Avancini, B.V. Carlson, A. Delfino, D.P. Menezes, C. Providência, S. Typel, J.R. Stone, Relativistic mean-field hadronic models under nuclear matter constraints. Phys. Rev. C **90**, 055203 (2014)
84. N. Hornick, L. Tolos, A. Zacchi, J.-E. Christian, J. Schaffner-Bielich, Relativistic parameterizations of neutron matter and implications for neutron stars. Phys. Rev. C **98**, 065804 (2018)
85. G.A. Lalazissis, T. Nikšić, D. Vretenar, P. Ring, New relativistic mean-field interaction with density-dependent meson-nucleon couplings. Phys. Rev. C **71**, 024312 (2005)
86. V.A. Ambartsumyan, G.S. Saakyan, The Degenerate Superdense Gas of Elementary Particles. Soviet Astron. **4**, 187 (1960)
87. D. Logoteta, Hyperons in neutron stars. Universe **7**(11) (2021)
88. Y. Yamamoto, T. Furumoto, N. Yasutake, T.A. Rijken, Hyperon-mixed neutron star with universal many-body repulsion. Eur. Phys. J. A **52**(2), 19 (2016)
89. F. Sammarruca, L. Coraggio, J.W. Holt, N. Itaco, R. Machleidt, L.E. Marcucci, Toward order-by-order calculations of the nuclear and neutron matter equations of state in chiral effective field theory. Phys. Rev. C **91**, 054311 (2015)
90. T. Katayama, K. Saito, Properties of dense, asymmetric nuclear matter in dirac-brueckner-hartree-fock approach. Phys. Rev. C **88**, 035805 (2013)
91. S. Weissenborn, D. Chatterjee, J. Schaffner-Bielich, Hyperons and massive neutron stars: Vector repulsion and su(3) symmetry. Phys. Rev. C **85**, 065802 (2012)
92. S. Weissenborn, D. Chatterjee, J. Schaffner-Bielich, Hyperons and massive neutron stars: the role of hyperon potentials. Nucl. Phys. A **881**, 62–77 (2012). Progress in Strangeness Nuclear Physics
93. B.K. Pradhan, D. Chatterjee, R. Gandhi, J. Schaffner-Bielich, Role of vector self-interaction in neutron star properties. Nucl. Phys. A **1030**, 122578 (2023)
94. D. Chatterjee, D. Bandyopadhyay, Effect of hyperon-hyperon interaction on bulk viscosity and r-mode instability in neutron stars. Phys. Rev. D **74**, 023003 (2006)
95. Z.F. Seidov, Stability of stellar cores. Sov. Astron. Lett. **4**, 144–147 (1978)

96. A. Kurkela, P. Romatschke, A. Vuorinen, Cold quark matter. Phys. Rev. D **81**(10) (2010)
97. A. Kurkela, A. Vuorinen, Cool quark matter. Phys. Rev. Lett. **117**(4) (2016)
98. T. Gorda, S. Säppi, Cool quark matter with perturbative quark masses. Phys. Rev. D **105**(11) (2022)
99. T. Gorda, A. Kurkela, P. Romatschke, M. Säppi, A. Vuorinen, Next-to-next-to-next-to-leading order pressure of cold quark matter: leading logarithm. Phys. Rev. Lett. **121**(20) (2018)
100. T. Gorda, A. Kurkela, R. Paatelainen, S. Säppi, A. Vuorinen, Cold quark matter at n3lo: soft contributions. Phys. Rev. D **104**(7) (2021)
101. T. Gorda, R. Paatelainen, S. Säppi, K. Seppänen, Equation of state of cold quark matter to $\mathcal{O}(\alpha_s^3 \ln \alpha_s)$. Phys. Rev. Lett. **131**(18) (2023)
102. A. Chodos, R.L. Jaffe, K. Johnson, C.B. Thorn, Baryon structure in the bag theory. Phys. Rev. D **10**(8), 2599–2604 (1974)
103. A. Chodos, R.L. Jaffe, K. Johnson, C.B. Thorn, V.F. Weisskopf, New extended model of hadrons. Phys. Rev. D **9**(12), 3471–3495 (1974)
104. E. Farhi, R.L. Jaffe, Strange matter. Phys. Rev. D **30**, 2379–2390 (1984)
105. Simon Weissenborn, Irina Sagert, Giuseppe Pagliara, Matthias Hempel, Jürgen. Schaffner-Bielich, Quark matter in massive compact stars. Astrophys. J. **740**(1), L14 (2011)
106. A. Schmitt, *Dense Matter in Compact Stars: A Pedagogical Introduction*, vol. 811 (Springer, 2010)
107. T. Klähn, T. Fischer, Vector interaction enhanced bag model for astrophysical applications. Astrophys. J. **810**(2), 134 (2015)
108. M.G. Alford, A. Schmitt, K. Rajagopal, T. Schäfer, Color superconductivity in dense quark matter. Rev. Mod. Phys. **80**(4), 1455–1515 (2008)
109. K. Schertler, S. Leupold, J. Schaffner-Bielich, Neutron stars and quark phases in the nambu-jona-lasinio model. Phys. Rev. C **60**(2) (1999)
110. N. Kovensky, A. Poole, A. Schmitt, Building a realistic neutron star from holography. Phys. Rev. D **105**(3) (2022)
111. M. Järvinen, Holographic modeling of nuclear matter and neutron stars. Eur. Phys. J. C **82**(4) (2022)
112. T. Demircik, C. Ecker, M. Järvinen, Dense and hot qcd at strong coupling. Phys. Rev. X **12**(4) (2022)
113. S. Tootle, C. Ecker, K. Topolski, T. Demircik, M. Järvinen, L. Rezzolla, Quark formation and phenomenology in binary neutron-star mergers using v-qcd. SciPost Physics **13**(5) (2022)
114. A.R. Bodmer, Collapsed nuclei. Phys. Rev. D **4**(6), 1601 (1971)
115. E. Witten, Cosmic separation of phases. Phys. Rev. D **30**, 272–285 (1984)
116. J.M. Lattimer, D.F. Swesty, A generalized equation of state for hot, dense matter. Nuclear Phsyics A **535**(2), 331–376 (1991)
117. H. Shen, H. Toki, K. Oyamatsu, K. Sumiyoshi, Relativistic equation of state of nuclear matter for supernova and neutron star. Nuclear Phsyics A **637**(3), 435–450 (1998)
118. M. Hempel, T. Fischer, J. Schaffner-Bielich, M. Liebendörfer, New equations of state in simulations of core-collapse supernovae. Apj **748**(1), 70 (2012)
119. S. Typel, H.H. Wolter, Relativistic mean field calculations with density-dependent meson-nucleon coupling. Nucl. Phys. A **656**(3), 331–364 (1999)
120. A.R. Raduta, M. Oertel, A. Sedrakian, Proto-neutron stars with heavy baryons and universal relations. Mon. Not. R. Astron. Soc. **499**(1), 914–931 (2020)
121. S. Banik, M. Hempel, D. Bandyopadhyay, New hyperon equations of state for supernovae and neutron stars in density-dependent hadron field theory. Astrophys. J. Suppl. Ser. **214**(2), 22 (2014)
122. M. Marques, M. Oertel, M. Hempel, J. Novak, New temperature dependent hyperonic equation of state: Application to rotating neutron star models and i-q relations. Phys. Rev. C **96**(4), 045806 (2017)
123. A.R. Raduta, Equations of state for hot neutron stars–ii. the role of exotic particle degrees of freedom. Eur. Phys. J. A **58**(115), 1–28 (2022)

124. T. Malik, S. Banik, D. Bandyopadhyay, New equation of state involving bose-einstein condensate of antikaon for supernova and neutron star merger simulations. Eur. Phys. J. ST **230**(2) (2021)
125. F. Niels-Uwe, Bastian, Phenomenological quark-hadron equations of state with first-order phase transitions for astrophysical applications. Phys. Rev. D **103**(2), 023001 (2021)
126. M.G. Alford, L. Brodie, A. Haber, I. Tews, Tabulated equations of state from models informed by chiral effective field theory. Phys. Scr. **98**(12), 125302 (2023)
127. P. Papazoglou, D. Zschiesche, S. Schramm, J. Schaffner-Bielich, H. Stöcker, W. Greiner, Nuclei in a chiral su(3) model. Phys. Rev. C **59**(1), 411–427 (1999)
128. V. Dexheimer, S. Schramm, Proto-neutron and neutron stars in a chiral su(3) model. Astrophys. J. **683**(2), 943 (2008)
129. A. Bauswein, H.-T. Janka, R. Oechslin, Testing approximations of thermal effects in neutron star merger simulations. Phys. Rev. D **82**(8), 084043 (2010)
130. K. Hotokezaka, K. Kiuchi, K. Kyutoku, T. Muranushi, Y. Sekiguchi, M. Shibata, K. Taniguchi, Remnant massive neutron stars of binary neutron star mergers: Evolution process and gravitational waveform. Phys. Rev. D **88**(4), 044026 (2013)
131. A. Endrizzi, D. Logoteta, B. Giacomazzo, I. Bombaci, W. Kastaun, R. Ciolfi, Effects of chiral effective field theory equation of state on binary neutron star mergers. Phys. Rev. D **98**(4), 043015 (2018)
132. L.R. Weih, M. Hanauske, L. Rezzolla, Postmerger gravitational-wave signatures of phase transitions in binary mergers. Phys. Rev. Lett. **124**(17), 171103 (2020)
133. S. Blacker, A. Bauswein, S. Typel, Exploring thermal effects of the hadron-quark matter transition in neutron star mergers (2023)
134. V.M. Kaspi, A.M. Beloborodov, Magnetars. Annu. Rev. Astron. Astrophys. **55**(1), 261–301 (2017)
135. P. Esposito, N. Rea, G.L. Israel, *Magnetars: A Short Review and Some Sparse Considerations* (Springer, Berlin, Heidelberg, 2021), pp.97–142
136. L.D. Landau, E.M. Lifshitz, *Electrodynamics of Continuous Media* (Pergamon, New York, 1984)
137. D. Bandyopadhyay, S. Chakrabarty, S. Pal, Quantizing magnetic field and quark-hadron phase transition in a neutron star. Phys. Rev. Lett. **79**, 2176–2179 (1997)
138. S. Chakrabarty, D. Bandyopadhyay, S. Pal, Dense nuclear matter in a strong magnetic field. Phys. Rev. Lett. **78**, 2898–2901 (1997)
139. A. Broderick, M. Prakash, J.M. Lattimer, The equation of state of neutron star matter in strong magnetic fields. Astrophys. J. **537**(1), 351 (2000)
140. M. Strickland, V. Dexheimer, D.P. Menezes, Bulk properties of a fermi gas in a magnetic field. Phys. Rev. D **86**, 125032 (2012)
141. J.L. Noronha, I.A. Shovkovy, Color-flavor locked superconductor in a magnetic field. Phys. Rev. D **76**, 105030 (2007)
142. N. Chamel, Z.K. Stoyanov, L.M. Mihailov, Y.D. Mutafchieva, R.L. Pavlov, C.J. Velchev, Role of landau quantization on the neutron-drip transition in magnetar crusts. Phys. Rev. C **91**, 065801 (2015)
143. D. Peña Arteaga, M. Grasso, E. Khan, P. Ring, Nuclear structure in strong magnetic fields: nuclei in the crust of a magnetar. Phys. Rev. C **84**, 045806 (2011)
144. D. Blaschke, N. Chamel, Phases of dense matter in compact stars, in *Astrophysics and Space Science Library*, ed. by L. Rezzolla, P. Pizzochero, D.I. Jones, N. Rea, I. Vidaña. Astrophysics and Space Science Library, vol. 457, p. 337 (2018)
145. J. Fang, H. Pais, S. Avancini, C. Providência, Larger and more heterogeneous neutron star crusts: a result of strong magnetic fields. Phys. Rev. C **94**(6) (2016)
146. J. Fang, H. Pais, S. Pratapsi, S. Avancini, J. Li, C. Providência, Effect of strong magnetic fields on the crust-core transition and inner crust of neutron stars. Phys. Rev. C **95**(4) (2017)
147. A. Rabhi, C. Providência, J. Da Providência, Stellar matter with a strong magnetic field within density-dependent relativistic models. J. Phys. G: Nuclear Part. Phys. **35**(12), 125201 (2008)

148. V. Dexheimer, R. Negreiros, S. Schramm, Hybrid stars in a strong magnetic field. Eur. Phys. J. A **48**(12), 189 (2012)
149. B.C.T. Backes, K.D. Marquezb, D.P. Menezes, Effects of strong magnetic fields on the hadron-quark deconfinement transition. Eur. Phys. J. A **57**(7) (2021)
150. *Strongly Interacting Matter in Magnetic Fields* (Springer, Berlin, Heidelberg, 2013)
151. E.J. Ferrer, V. de la Incera, Exploring dense and cold qcd in magnetic fields. Eur. Phys. J. A **52**(8), 266 (2016)
152. E.J. Ferrer, A. Hackebill, Hadron-quark phase transition at finite density in the presence of a magnetic field: Anisotropic approach (2022)
153. S.S. Avancini, V. Dexheimer, R.L.S. Farias, V.S. Timóteo, Anisotropy in the equation of state of strongly magnetized quark matter within the nambu-jona-lasinio model. Phys. Rev. C **97**, 035207 (2018)
154. M. Bocquet, S. Bonazzola, E. Gourgoulhon, J. Novak, Rotating neutron star models with magnetic field (1995)
155. C.Y. Cardall, M. Prakash, J.M. Lattimer, Effects of strong magnetic fields on neutron star structure. Astrophys. J. **554**(1), 322 (2001)
156. K. Kiuchi, S. Yoshida, Relativistic stars with purely toroidal magnetic fields. Phys. Rev. D **78**, 044045 (2008)
157. K. Kiuchi, S. Yoshida, Relativistic stars with purely toroidal magnetic fields. Phys. Rev. D **78**, 044045 (2008)
158. S. Yoshida, K. Kiuchi, M. Shibata, Stably stratified magnetized stars in general relativity. Phys. Rev. D **86**, 044012 (2012)
159. A.G. Pili, N. Bucciantini, L. Del Zanna, Axisymmetric equilibrium models for magnetized neutron stars in General Relativity under the Conformally Flat Condition. Mon. Not. R. Astron. Soc. **439**(4), 3541–3563 (2014)
160. D. Chatterjee, T. Elghozi, J. Novak, M. Oertel, Consistent neutron star models with magnetic-field-dependent equations of state. Mon. Not. R. Astron. Soc. **447**(4), 3785–3796 (2015)
161. D. Chatterjee, J. Novak, M. Oertel, Structure of ultra-magnetised neutron stars. Eur. Phys. J. A **57**(8) (2021)
162. J.W.T. Hessels, S.M. Ransom, I.H. Stairs, P.C.C. Freire, V.M. Kaspi, F. Camilo, A radio pulsar spinning at 716 hz. Science **311**(5769), 1901–1904 (2006)
163. J.B. Hartle, K.S. Thorne, Slowly rotating relativistic stars. II. Models for neutron stars and supermassive stars. Astrophys. J. **153**, 807 (1968)
164. S. Bonazzola, E. Gourgoulhon, M. Salgado, J.A. Marck, Axisymmetric rotating relativistic bodies: a new numerical approach for "exact" solutions. Astron. Astrophys. **278**(2), 421–443 (1993)
165. H. Komatsu, Y. Eriguchi, I. Hachisu, Rapidly rotating general relativistic stars. I - numerical method and its application to uniformly rotating polytropes. Mon. Not. R. Astron. Soc. **237**, 355–379 (1989)
166. T.W. Baumgarte, S.L. Shapiro, M. Shibata, On the maximum mass of differentially rotating neutron stars. Astrophys. J. **528**(1), L29 (1999)
167. G. Bozzola, N. Stergioulas, A. Bauswein, Universal relations for differentially rotating relativistic stars at the threshold to collapse. Mon. Not. Roy. Astron. Soc. **474**(3), 3557–3564 (2018)
168. K. Uryū, A. Tsokaros, L. Baiotti, F. Galeazzi, K. Taniguchi, S. Yoshida, Modeling differential rotations of compact stars in equilibriums. Phys. Rev. D **96**, 103011 (2017)
169. J.O. Goussard, P. Haensel, J.L. Zdunik, Rapid differential rotation of protoneutron stars and constraints on radio pulsars periods (1997)
170. J.O. Goussard, P. Haensel, J.L. Zdunik, Rapid uniform rotation of protoneutron stars (1996)
171. P.T.H. Pang, I. Tews, M.W. Coughlin, M. Bulla, C. Van Den Broeck, T. Dietrich, Nuclear-physics multi-messenger astrophysics constraints on the neutron-star equation of state: adding nicer's psr j0740+6620 measurement (2021)
172. M.W. Coughlin, T. Dietrich, B. Margalit, B.D. Metzger, Multimessenger bayesian parameter inference of a binary neutron star merger. Mon. Not. R. Astron. Soc.: Lett. **489**(1), L91–L96 (2019)

173. B. Biswas, P. Char, R. Nandi, S. Bose, Towards mitigation of apparent tension between nuclear physics and astrophysical observations by improved modeling of neutron star matter. Phys. Rev. D **103**(10) (2021)
174. B. Biswas, Impact of prex-ii and combined radio/nicer/xmm-newton's mass-radius measurement of psrj0740+6620 on the dense matter equation of state (2021)
175. B. Biswas, Bayesian model-selection of neutron star equation of state using multi-messenger observations (2021)
176. T. Dietrich, M.W. Coughlin, P.T.H. Pang, M. Bulla, J. Heinzel, L. Issa, I. Tews, S. Antier, Multimessenger constraints on the neutron-star equation of state and the hubble constant. Science **370**(6523), 1450–1453 (2020)
177. M.F. O'Boyle, C. Markakis, N. Stergioulas, J.S. Read, Parametrized equation of state for neutron star matter with continuous sound speed. Phys. Rev. D **102**(8) (2020)
178. S. Altiparmak, C. Ecker, L. Rezzolla, On the sound speed in neutron stars. Astrophys. J. Lett. **939**(2), L34 (2022)
179. C. Ecker, L. Rezzolla, Impact of large-mass constraints on the properties of neutron stars. Mon. Not. R. Astron. Soc. **519**(2), 2615–2622 (2022)
180. A.W. Steiner, J.M. Lattimer, E.F. Brown, The equation of state from observed masses and radii of neutron stars. Astrophys. J. **722**(1), 33–54 (2010)
181. C.D. Capano, I. Tews, S.M. Brown, B. Margalit, S. De, S. Kumar, D.A. Brown, B. Krishnan, S. Reddy, Stringent constraints on neutron-star radii from multimessenger observations and nuclear theory. Nat. Astron. **4**(6), 625–632 (2020)
182. I. Tews, J. Margueron, S. Reddy, Critical examination of constraints on the equation of state of dense matter obtained from gw170817. Phys. Rev. C **98**(4) (2018)
183. I. Tews, J. Margueron, S. Reddy, Constraining the properties of dense matter and neutron stars by combining nuclear physics and gravitational waves from gw170817. *XIAMEN-Custipen Workshop on the Equation of State of Dense Neutron-Rich Matter in the Era of Gravitational Wave Astronomy* (2019)
184. I. Tews, J. Margueron, S. Reddy, Confronting gravitational-wave observations with modern nuclear physics constraints. Eur. Phys. J. A **55**(6) (2019)
185. S. Gandolfi, J. Lippuner, A.W. Steiner, I. Tews, X. Du, M. Al-Mamun, From the microscopic to the macroscopic world: from nucleons to neutron stars. J. Phys. G: Nuclear Part. Phys. **46**(10), 103001 (2019)
186. K. Hebeler, J.M. Lattimer, C.J. Pethick, A. Schwenk, Equation of state and neutron star properties constrained by nuclear physics and observation. Astrophys. J. **773**(1), 11 (2013)
187. J.S. Read, B.D. Lackey, B.J. Owen, J.L. Friedman, Constraints on a phenomenologically parametrized neutron-star equation of state. Phys. Rev. D **79**(12) (2009)
188. R. Gamba, J.S. Read, L.E. Wade, The impact of the crust equation of state on the analysis of GW170817. Class. Q. Grav. **37**(2), 025008 (2019)
189. M. Fasano, T. Abdelsalhin, A. Maselli, V. Ferrari, Constraining the neutron star equation of state using multiband independent measurements of radii and tidal deformabilities. Phys. Rev. Lett. **123**, 141101 (2019)
190. L. Lindblom, Causal representations of neutron-star equations of state. Phys. Rev. D **97**, 123019 (2018)
191. L. Lindblom, N.M. Indik, Spectral approach to the relativistic inverse stellar structure problem. Phys. Rev. D **86**, 084003 (2012)
192. S.K. Greif, G. Raaijmakers, K. Hebeler, A. Schwenk, A.L. Watts, Equation of state sensitivities when inferring neutron star and dense matter properties. Mon. Not. R. Astron. Soc. **485**(4), 5363–5376 (2019)
193. P. Landry, R. Essick, K. Chatziioannou, Nonparametric constraints on neutron star matter with existing and upcoming gravitational wave and pulsar observations. Phys. Rev. D **101**, 123007 (2020)
194. E. Annala, T. Gorda, A. Kurkela, J. Nättilä, A. Vuorinen, Evidence for quark-matter cores in massive neutron stars. Nat. Phys. **16**(9), 907–910 (2020)

195. I. Legred, K. Chatziioannou, R. Essick, S. Han, P. Landry, Impact of the psr J0740 + 6620 radius constraint on the properties of high-density matter. Phys. Rev. D **104**, 063003 (2021)
196. R. Essick, I. Tews, P. Landry, S. Reddy, D.E. Holz, Direct astrophysical tests of chiral effective field theory at supranuclear densities. Phys. Rev. C **102**, 055803 (2020)
197. T. Gorda, O. Komoltsev, A. Kurkela, Ab-initio qcd calculations impact the inference of the neutron-star-matter equation of state. Astrophys. J. **950**(2), 107 (2023)
198. T. Gorda, O. Komoltsev, A. Kurkela, A. Mazeliauskas, Bayesian uncertainty quantification of perturbative qcd input to the neutron-star equation of state. J. High Energy Phys. **2023**(6) (2023)
199. E. Annala, T. Gorda, J. Hirvonen, O. Komoltsev, A. Kurkela, J. Nättilä, A. Vuorinen, Strongly interacting matter exhibits deconfined behavior in massive neutron stars. Nat. Commun. **14**(1) (2023)
200. O. Komoltsev, R. Somasundaram, T. Gorda, A. Kurkela, J. Margueron, I. Tews, Equation of state at neutron-star densities and beyond from perturbative qcd (2024)
201. S. Traversi, P. Char, G. Pagliara, Bayesian inference of dense matter equation of state within relativistic mean field models using astrophysical measurements. Astrophys. J. **897**(2), 165 (2020)
202. J. Margueron, R. Hoffmann Casali, F. Gulminelli, Equation of state for dense nucleonic matter from metamodeling. i. foundational aspects. Phys. Rev. C **97**, 025805 (2018)
203. J. Margueron, R. Hoffmann Casali, F. Gulminelli. Equation of state for dense nucleonic matter from metamodeling. ii. predictions for neutron star properties. Phys. Rev. C **97**, 025806 (2018)
204. A.R. Raduta, F. Nacu, M. Oertel, Equations of state for hot neutron stars. Eur. Phys. J. A **57**(12), 329 (2021)
205. D. Alvarez-Castillo, A. Ayriyan, G.G. Barnaföldi, H. Grigorian, P. Pósfay, Studying the parameters of the extended σ-ω model for neutron star matter. Eur. Phys. J. Spec. Top. **229**(22–23), 3615–3628 (2020)
206. D. Alvarez-Castillo, A. Ayriyan, G.G. Barnaföldi, P. Pósfay, Studying the landau mass parameter of the extended σ-ω model for neutron star matter. Phys. Part. Nucl. **51**(4), 725–729 (2020)
207. P. Char, C. Mondal, F. Gulminelli, M. Oertel. Phys. Rev. D (2023)
208. A. Carbone, A. Cipollone, C. Barbieri, A. Rios, A. Polls, Self-consistent green's functions formalism with three-body interactions. Phys. Rev. C **88**, 054326 (2013)
209. J. Haidenbauer, U.-G. Meißner, A. Nogga, H. Le, Hyperon-nucleon interaction in chiral effective field theory at next-to-next-to-leading order. Eur. Phys. J. A **59**(3) (2023)
210. J. Keller, K. Hebeler, A. Schwenk, Nuclear equation of state for arbitrary proton fraction and temperature based on chiral effective field theory and a gaussian process emulator. Phys. Rev. Lett. **130**(7) (2023)
211. O. Komoltsev, A. Kurkela, How perturbative qcd constrains the equation of state at neutron-star densities. Phys. Rev. Lett. **128**(20) (2022)
212. A. Kurkela, E.S. Fraga, J. Schaffner-Bielich, A. Vuorinen, Constraining neutron star matter with quantum chromodynamics. Astrophys. J. **789**(2), 127 (2014)
213. Takashi Inoue and for HAL QCD Collaboration. Hyperon single-particle potentials from qcd on lattice (2017)
214. Strange nuclear physics from qcd on lattice, in *Central European Symposium on Thermophysics 2019 (CEST)* (AIP Publishing, 2019)
215. L. Fabbietti, V.M. Sarti, O.V. Doce, Study of the strong interaction among hadrons with correlations at the lhc. Ann. Rev. Nucl. Part. Sci. **71**(1), 377–402 (2021)
216. X. Roca-Maza, M. Centelles, X. Viñas, M. Warda, Neutron skin of ^{208}Pb, nuclear symmetry energy, and the parity radius experiment. Phys. Rev. Lett. **106**, 252501 (2011)
217. M. Centelles, X. Roca-Maza, X. Viñas, M. Warda, Nuclear symmetry energy probed by neutron skin thickness of nuclei. Phys. Rev. Lett. **102**, 122502 (2009)
218. M. Warda, X. Viñas, X. Roca-Maza, M. Centelles, Neutron skin thickness in the droplet model with surface width dependence: indications of softness of the nuclear symmetry energy. Phys. Rev. C **80**, 024316 (2009)

219. S. Shlomo, V.M. Kolomietz, G. Colò, Deducing the nuclear-matter incompressibility coefficient from data on isoscalar compression modes. Eur. Phys. J. A **30**(1), 23–30 (2006)
220. U. Garg, T. Li, S. Okumura, H. Akimune, M. Fujiwara, M.N. Harakeh, H. Hashimoto, M. Itoh, Y. Iwao, T. Kawabata, K. Kawase, Y. Liu, R. Marks, T. Murakami, K. Nakanishi, B.K. Nayak, P.V. Madhusudhana Rao, H. Sakaguchi, Y. Terashima, M. Uchida, Y. Yasuda, M. Yosoi, J. Zenihiro, The giant monopole resonance in the sn isotopes: why is tin so "fluffy"? Nucl. Phys. A **788**(1), 36–43 (2007)
221. M.C. Atkinson, W.H. Dickhoff, M. Piarulli, A. Rios, R.B. Wiringa, Reexamining the relation between the binding energy of finite nuclei and the equation of state of infinite nuclear matter. Phys. Rev. C **102**, 044333 (2020)
222. V. Kejzlar, L. Neufcourt, W. Nazarewicz, P.-G. Reinhard, Statistical aspects of nuclear mass models. J. Phys. G: Nuclear Part. Phys. **47**(9), 094001 (2020)
223. M. Wang, G. Audi, A.H. Wapstra, F.G. Kondev, M. MacCormick, X. Xu, B. Pfeiffer, The Ame 2012 atomic mass evaluation. Chin. Phys. C **36**(12), 1603–2014 (2012)
224. I. Angeli, K.P. Marinova, Table of experimental nuclear ground state charge radii: an update. Atom. Data Nucl. Data Tables **99**(1), 69–95 (2013)
225. P. Danielewicz, J. Lee, Symmetry energy i: semi-infinite matter. Nucl. Phys. A **818**(1), 36–96 (2009)
226. R. Engelmann, H. Filthuth, V. Hepp, E. Kluge, Inelastic Σ −p-interactions at low momenta. Phys. Lett. **21**(5), 587–589 (1966)
227. G. Alexander, U. Karshon, A. Shapira, G. Yekutieli, R. Engelmann, H. Filthuth, W. Lughofer, Study of the lambda-n system in low-energy lambda-p elastic scattering. Phys. Rev. **173**, 1452–1460 (1968)
228. B. Sechi-Zorn, B. Kehoe, J. Twitty, R.A. Burnstein, Low-energy lambda-proton elastic scattering. Phys. Rev. **175**, 1735–1740 (1968)
229. J.A. Kadyk, G. Alexander, J.H. Chan, P. Gaposchkin, G.H. Trilling, λ p interactions in momentum range 300 to 1500 mev/c. Nucl. Phys. B **27**(1), 13–22 (1971)
230. F. Eisele, H. Filthuth, W. Föhlisch, V. Hepp, G. Zech, Elastic $\sigma - \pm$ p scattering at low energies. Phys. Lett. B **37**(2), 204–206 (1971)
231. C. Hartnack, R.K. Puri, J. Aichelin, J. Konopka, S.A. Bass, H. Stöcker, W. Greiner, Modelling the many-body dynamics of heavy ion collisions: present status and future perspective. Eur. Phys. J. A **1**(2), 151–169 (1998)
232. C. Fuchs, A. Faessler, E. Zabrodin, Y.-M. Zheng, Probing the nuclear equation of state by k^+ production in heavy-ion collisions. Phys. Rev. Lett. **86**, 1974–1977 (2001)
233. I. Sagert, L. Tolos, D. Chatterjee, J. Schaffner-Bielich, C. Sturm, Soft nuclear equation-of-state from heavy-ion data and implications for compact stars. Phys. Rev. C **86**, 045802 (2012)
234. A. Le Fèvre, Y. Leifels, W. Reisdorf, J. Aichelin, Ch. Hartnack, Constraining the nuclear matter equation of state around twice saturation density. Nucl. Phys. A **945**, 112–133 (2016)
235. P. Russotto, S. Gannon, S. Kupny, P. Lasko, L. Acosta, M. Adamczyk, A. Al-Ajlan, M. Al-Garawi, S. Al-Homaidhi, F. Amorini, et al., Results of the asy-eos experiment at gsi: the symmetry energy at suprasaturation density. Phys. Rev. C **94**(3) (2016)
236. J. Aichelin, "quantum" molecular dynamics-a dynamical microscopic n-body approach to investigate fragment formation and the nuclear equation of state in heavy ion collisions. Phys. Rep. **202**(5), 233–360 (1991)
237. W.C.G. Ho, C.O. Heinke, A neutron star with a carbon atmosphere in the cassiopeia a supernova remnant. Nature **462**(7269), 71–73 (2009)
238. W.C.G. Ho, M.J.P. Wijngaarden, P. Chang, C.O. Heinke, D. Page, M. Beznogov, D.J. Patnaude, Cooling of the cassiopeia a neutron star and the effect of diffusive nuclear burning, in *XIAMEN-CUSTIPEN Workshop on the Equation of State of Dense Neutron-Rich Matter in the Era of Gravitational Wave Astronomy* (AIP Publishing, 2019)
239. D. Page, M. Prakash, J.M. Lattimer, A.W. Steiner, Rapid cooling of the neutron star in cassiopeia a triggered by neutron superfluidity in dense matter. Phys. Rev. Lett. **106**, 081101 (2011)

240. P.S. Shternin, D.G. Yakovlev, C.O. Heinke, W.C.G. Ho, D.J. Patnaude, Cooling neutron star in the cassiopeia a supernova remnant: evidence for superfluidity in the core. Mon. Not. R. Astron. Soc.: Lett. **412**(1), L108–L112 (2011)
241. D. Page, *Neutron Star Cooling: I* (Springer, Berlin, Heidelberg, 2009), pp.247–288
242. H. Tong, P. Zhao, J. Meng, Symmetry energy at supra-saturation densities via the gravitational waves from gw170817. Phys. Rev. C **101**, 035802 (2020)
243. C.J. Krüger, K.D. Kokkotas, P. Manoharan, S.H. Völkel, Fast rotating neutron stars: oscillations and instabilities. Front. Astron. Space Sci. **8** (2021)
244. B.K. Pradhan, D. Chatterjee, M. Lanoye, P. Jaikumar, General relativistic treatment of f-mode oscillations of hyperonic stars. Phys. Rev. C **106**(1) (2022)
245. B.K. Pradhan, D. Pathak, D. Chatterjee, Constraining nuclear parameters using gravitational waves from f-mode oscillations in neutron stars. Astrophys. J. **956**(1), 38 (2023)
246. V. Tran, S. Ghosh, N. Lozano, D. Chatterjee, P. Jaikumar. g-mode oscillations in neutron stars with hyperons. Phys. Rev. C **108**(1) (2023)
247. B. Haskell, R-modes in neutron stars: Theory and observations. Int. J. Modern Phys. E **24**(09), 1541007 (2015)
248. S. Ghosh, D. Chatterjee, J. Schaffner-Bielich, Imposing multi-physics constraints at different densities on the neutron star equation of state. Eur. Phys. J. A **58**(3) (2022)
249. S. Ghosh, B.K. Pradhan, D. Chatterjee, J. Schaffner-Bielich, Multi-physics constraints at different densities to probe nuclear symmetry energy in hyperonic neutron stars. Front. Astron. Space Sci. **9** (2022)
250. S. Shirke, S. Ghosh, D. Chatterjee, Constraining the equation of state of hybrid stars using recent information from multidisciplinary physics. Astrophys. J. **944**(1), 7 (2023)
251. S. Huth et al., Constraining neutron-star matter with microscopic and macroscopic collisions. Nature **606**, 276–280 (2022)
252. E. Annala, T. Gorda, E. Katerini, A. Kurkela, J. Nättilä, V. Paschalidis, A. Vuorinen, Multimessenger constraints for ultradense matter. Phys. Rev. X **12**(1), 011058 (2022)
253. J. Pochodzalla, Future experiments on hypernuclei and hyperatoms. Nucl. Instrum. Meth. B **214**, 149–152 (2004)
254. S.N. Nakamura, Future hypernuclear experiments at jlab, in *Proceedings of The IX International Conference on Hypernuclear and Strange Particle Physics*, ed. by J. Pochodzalla, T. Walcher. (Springer, Berlin, Heidelberg, 2007), pp. 51–56
255. T. Saito, Past and future of hypernuclei research using heavy ion beams: from gsi to fair/hiaf. Genshikaku Kenkyu **64**(2), 75–89 (2020). Nuclear Physics and Radiation Physics
256. P.F. Bedaque, A.W. Steiner, Hypernuclei and the hyperon problem in neutron stars. Phys. Rev. C **92**(2) (2015)
257. N. Miller, G. Bradley, M.A. Clark, B. Hörz, D. Howarth, M. Lazarow, H. Monge-Camacho, A. Nicholson, E. Rinaldi, P. Vranas, A. Walker-Loud, The hyperon spectrum from lattice qcd (2022)
258. H. Nemura, Hyperon-nucleon interaction from lattice QCD at $(m_\pi, m_K) \approx (146, 525)$ MeV. AIP Conf. Proc. **2130**(1), 040005 (2019)
259. K. Agarwal, The compressed baryonic matter (cbm) experiment at fair - physics, status and prospects. Phys. Scr. **98**, 02 (2023)
260. J. Aasi, B.P. Abbott, R. Abbott, T. Abbott, M.R. Abernathy, K. Ackley, C. Adams, T. Adams, P. Addesso et al., Advanced ligo. Class. Q. Grav. **32**(7), 074001 (2015)
261. B.P. Abbott, R. Abbott, T.D. Abbott, M.R. Abernathy, F. Acernese, K. Ackley, C. Adams, T. Adams, P. Addesso, R.X. Adhikari et al., Gw150914: the advanced ligo detectors in the era of first discoveries. Phys. Rev. Lett. **116**, 131103 (2016)
262. F. Acernese, M. Agathos, K. Agatsuma, D. Aisa, N. Allemandou, A. Allocca, J. Amarni, P. Astone, G. Balestri, G. Ballardin et al., Advanced virgo: a second-generation interferometric gravitational wave detector. Class. Q. Grav. **32**(2), 024001 (2014)
263. T. Akutsu, M. Ando, K. Arai, Y. Arai, S. Araki, A. Araya, N. Aritomi, Y. Aso, S.W. Bae, Y.B. Bae et al., Overview of KAGRA: detector design and construction history. Progr. Theor. Exper. Phys. **2021**(5), 05A101 (2021)

264. M. Evans, A. Corsi, C. Afle, A. Ananyeva, K.G. Arun, S. Ballmer, A. Bandopadhyay, L. Barsotti, M. Baryakhtar, E. Berger, E. Berti, S. Biscoveanu, S. Borhanian, F. Broekgaarden, D.A. Brown, C. Cahillane, L. Campbell, H.-Y. Chen, K.J. Daniel, A. Dhani, J.C. Driggers, A. Effler, R. Eisenstein, S. Fairhurst, J. Feicht, P. Fritschel, P. Fulda, I. Gupta, E.D. Hall, G. Hammond, O.A. Hannuksela, H. Hansen, C.-J. Haster, K. Kacanja, B. Kamai, R. Kashyap, J.S. Key, S. Khadkikar, A. Kontos, K. Kuns, M. Landry, P. Landry, B. Lantz, T.G.F. Li, G. Lovelace, V. Mandic, G.L. Mansell, D. Martynov, L. McCuller, A.L. Miller, A.H. Nitz, B.J. Owen, C. Palomba, J. Read, H. Phurailatpam, S. Reddy, J. Richardson, J. Rollins, J.D. Romano, B.S. Sathyaprakash, R. Schofield, D.H. Shoemaker, D. Sigg, D. Singh, B. Slagmolen, P. Sledge, J. Smith, M. Soares-Santos, A. Strunk, L. Sun, D. Tanner, L.A.C. van Son, S. Vitale, B. Willke, H. Yamamoto, M. Zucker, Cosmic explorer: a submission to the nsf mpsac nggw subcommittee (2023)
265. M. Maggiore, C. Van Den Broeck, N. Bartolo, E. Belgacem, D. Bertacca, M.A. Bizouard, M. Branchesi, S. Clesse, S. Foffa, J. García-Bellido, S. Grimm, J. Harms, T. Hinderer, S. Matarrese, C. Palomba, M. Peloso, A. Ricciardone, M. Sakellariadou, Science case for the einstein telescope. J. Cosmol. Astropart. Phys. **2020**(03), 050–050 (2020)
266. M. Saleem, J. Rana, V. Gayathri, A. Vijaykumar, S. Goyal, S. Sachdev, J. Suresh, S. Sudhagar, A. Mukherjee, G. Gaur, B. Sathyaprakash, A. Pai, R.X. Adhikari, P. Ajith, S. Bose, The science case for ligo-india. Class. Q. Grav. **39**(2), 025004 (2021)
267. P.S. Ray, Z. Arzoumanian, D. Ballantyne, E. Bozzo, S. Brandt, L. Brenneman, D. Chakrabarty, M. Christophersen, A. DeRosa, M. Feroci, K. Gendreau, A. Goldstein, D. Hartmann, M. Hernanz, P. Jenke, E. Kara, T. Maccarone, M. McDonald, M. Nowak, B. Phlips, R. Remillard, A. Stevens, J. Tomsick, A. Watts, C. Wilson-Hodge, K. Wood, S. Zane, M. Ajello, W. Alston, D. Altamirano, V. Antoniou, K. Arur, D. Ashton, K. Auchettl, T. Ayres, M. Bachetti, M. Balokovic, M. Baring, A. Baykal, M. Begelman, N. Bhat, S. Bogdanov, M. Briggs, E. Bulbul, P. Bult, E. Burns, E. Cackett, R. Campana, A. Caspi, Y. Cavecchi, J. Chenevez, M. Cherry, R. Corbet, M. Corcoran, A. Corsi, N. Degenaar, J. Drake, S. Eikenberry, T. Enoto, C. Fragile, F. Fuerst, P. Gandhi, J. Garcia, A. Goldstein, A. Gonzalez, B. Grefenstette, V. Grinberg, B. Grossan, S. Guillot, T. Guver, D. Haggard, C. Heinke, S. Heinz, P. Hemphill, J. Homan, M. Hui, D. Huppenkothen, A. Ingram, J. Irwin, G. Jaisawal, A. Jaodand, E. Kalemci, D. Kaplan, L. Keek, J. Kennea, M. Kerr, M. van der Klis, D. Kocevski, M. Koss, A. Kowalski, D. Lai, F. Lamb, S. Laycock, J. Lazio, D. Lazzati, D. Longcope, M. Loewenstein, D. Maitra, W. Majid, W.P. Maksym, C. Malacaria, R. Margutti, A. Martindale, I. McHardy, M. Meyer, M. Middleton, J. Miller, C. Miller, S. Motta, J. Neilsen, T. Nelson, S. Noble, P. O'Brien, J. Osborne, R. Osten, F. Ozel, N. Palliyaguru, D. Pasham, A. Patruno, V. Pelassa, M. Petropoulou, M. Pilia, M. Pohl, D. Pooley, C. Prescod-Weinstein, D. Psaltis, G. Raaijmakers, C. Reynolds, T.E. Riley, G. Salvesen, A. Santangelo, S. Scaringi, S. Schanne, J. Schnittman, D. Smith, K.L. Smith, B. Snios, A. Steiner, J. Steiner, L. Stella, T. Strohmayer, M. Sun, T. Tauris, C. Taylor, A. Tohuvavohu, A. Vacchi, G. Vasilopoulos, A. Veledina, J. Walsh, N. Weinberg, D. Wilkins, R. Willingale, J. Wilms, L. Winter, M. Wolff, J. in 't Zand, A. Zezas, B. Zhang, A. Zoghbi, Strobe-x: x-ray timing and spectroscopy on dynamical timescales from microseconds to years (2019)

Part II
Codes

Chapter 4
GR-Athena++: Magnetohydrodynamical Evolution with Dynamical Space-Time

Boris Daszuta and William Cook

Abstract We present a self-contained overview of GR-Athena++, a general-relativistic magnetohydrodynamics (GRMHD) code, that incorporates treatment of dynamical space-time, based on the recent work of Daszuta+ [49] and Cook+ [45]. General aspects of the Athena++ framework we build upon, such as oct-tree based, adaptive mesh refinement (AMR) and constrained transport, together with our modifications, incorporating the Z4c formulation of numerical relativity, judiciously coupled, enables GRMHD with dynamical space-times. Initial verification testing of GR-Athena++ is performed through benchmark problems that involve isolated and binary neutron star space-times. This leads to stable and convergent results. Gravitational collapse of a rapidly rotating star through black hole formation is shown to be correctly handled. In the case of non-rotating stars, magnetic field instabilities are demonstrated to be correctly captured with total relative violation of the divergence-free constraint remaining near machine precision. The use of AMR is show-cased through investigation of the Kelvin-Helmholtz instability which is resolved at the collisional interface in a merger of magnetised binary neutron stars. The underlying task-based computational model enables GR-Athena++ to achieve strong scaling efficiencies above 80% in excess of 10^5 CPU cores and excellent weak scaling up to $\sim 5 \times 10^5$ CPU cores in a realistic production setup. GR-Athena++ thus provides a viable path towards robust simulation of GRMHD flows in strong and dynamical gravity with exascale high performance computational infrastructure.

4.1 Introduction

Embedding the techniques of numerical relativity (NR) within robust, highly-performant simulation infrastructure provides an indispensable tool for understanding the phenomenology of astrophysical inspiral and merger events. This is especially

B. Daszuta (✉) · W. Cook
Theoretisch-Physikalisches Institut, Friedrich-Schiller-Universität Jena, 07743 Jena, Germany
e-mail: boris.daszuta@uni-jena.de

W. Cook
e-mail: william.cook@uni-jena.de

C. Bambi et al. (eds.), *New Frontiers in GRMHD Simulations*, Springer Series in Astrophysics and Cosmology, https://doi.org/10.1007/978-981-97-8522-3_4

pertinent due to the arrival of multimessenger astronomy observations. Indeed recent, near-simultaneous detections of the gravitational wave signal GW170817, and associated electromagnetic counterpart GRB 170817A, alongside a kilonova, resulting from the merger of a Binary Neutron Star (BNS) system, have been made [2, 72, 128]. These two events provide insight on the astrophysical origins of short Gamma Ray bursts (SGRB) as ensuing from BNS progenitors [1].

The end products of BNS mergers have long been viewed as candidates for the launching of relativistic jets giving rise to SGRBs [25, 58, 73, 106, 107] however the precise mechanism remains an open question. The presence of large magnetic fields in a post-merger remnant may play a role in jet formation [38, 91, 113]. Such fields may arise as the result of magnetohydrodynamic (MHD) instabilities during the late inspiral such as the Kelvin-Helmholtz instability (KHI) [83, 117, 124], magnetic winding, and the Magneto-Rotational Instability (MRI) [15]. The configuration of magnetic fields for a BNS system during the early inspiral is an open question. Already in the case of an isolated neutron star an initially poloidal field can give rise to instability [63, 99, 100, 141, 142, 154], and long-term simulations involving isolated stars suggest that higher resolution is required to understand reasonable initial configurations for the magnetic fields [39, 40, 88, 92, 111, 112, 140], with thermal effects also playing a key role in the evolution of the field [115].

The distribution and nature of matter outflow from a BNS merger may also be influenced by the presence of magnetic fields [43, 44, 47, 50, 84, 87, 105, 132, 133]. These outflows determine the nature of the long lived electromagnetic signal following merger, such as AT2017gfo which accompanied GW170817 [3, 10, 46, 135], known as the kilonova [90, 93, 102], as well as the r-process nucleosynthesis responsible for heavy element production occurring in ejecta [80, 110]. It is clear that understanding the evolution of a BNS system from its late inspiral, through merger, to the post-merger is a formidable endeavour. A broad range of physical processes must be modeled consistently that include general relativity (GR), MHD, weak nuclear processes leading to neutrino emission and reabsorption, and the finite temperature behaviour of dense nuclear matter.

For the solution of the Einstein field equations, two commonly utilized classes of free evolution formulations of NR are: those based on moving puncture gauge conditions [14, 30, 34], such as BSSN [17, 130], Z4c [22, 78, 126, 151], and CCZ4 formulations [7, 8]; and the Generalised Harmonic Gauge approach [116]. These may be coupled to MHD so as to provide a system of GRMHD evolution equations. In this context the ideal MHD limit of a non-resistive and non-viscous fluid is typically assumed. The aforementioned are written as a system of balance laws, which is key for shock capture and mass conservation [64]. In order to preserve the magnetic field divergence-free condition a variety of approaches have been considered: Constrained Transport (CT) [61, 98, 127]; Flux-CT [147]; vector potential formulation [60]; and divergence cleaning [51]. Extensive effort has been invested into constructing codes for GRMHD simulation involving such techniques, a (non-exhaustive) list of examples include [9, 13, 36, 41, 55–57, 59, 66, 70, 82, 94, 97, 104, 109, 121, 129, 131, 143, 149]. Aside from the complexity of the underlying physics modeled, a further important concern for NR based investigations is efficient resolution of solution features that develop over widely disparate scales that scale when simulations utilize high performance computing (HPC) infrastructure.

To this end we overview and demonstrate the recently developed GRMHD capabilities of our code `GR-Athena++` [45, 49]. We build upon a public version of `Athena++` [138] which is an astrophysical (radiation), GRMHD, `c++` code for stationary space-times that features a task-based computational model built to exploit hybrid parallelism through dual use of message passing interface `MPI` and threading via `OpenMP`. Our efforts embed the Z4c formulation suitably coupled to GRMHD thereby removing the stationarity restriction. This allows `GR-Athena++` to evolve magnetised BNS. To this end we leverage the extant `Athena++` infrastructure and methods such as CT. We inherit block-based, adaptive mesh refinement (AMR) capabilities which exploit (binary, quad, oct)-tree structure with synchronisation restricted to layers at block boundaries. This can be contrasted against other codes utilizing AMR with fully-overlapping, nested grids that rely on the Berger-Oliger [19, 20] time subcycling algorithm.

On account of this `GR-Athena++` can efficiently resolve features developing during the BNS merger, such as small scale structures in the magnetic field instabilities, without requiring large scale refinement of coarser features. Performance scaling is demonstrated to $\mathcal{O}(10^5)$ CPU cores, showing efficient utilization of exascale HPC architecture.

4.2 Code Overview

`GR-Athena++` builds upon `Athena++` thus in order to specify nomenclature, provide a self-contained description, and explain our extensions, we first briefly recount some details of the framework (see also [62, 138, 153]).

In `Athena++`, overall details about the domain Ω over which a problem is formulated are abstracted from the salient physics and contained within a class called the `Mesh`. Within the `Mesh` an overall representation of the domain as a logical n-rectangle is stored, together with details of coordinatization type (Cartesian or more generally curvilinear), number of points along each dimension for the coarsest sampling $N_M = (N_{M_1}, \ldots, N_{M_d})$, and physical boundary conditions on $\partial\Omega$. Next, we would like to decompose the domain into smaller blocks, described by so-called `MeshBlock` objects. In order to partition the domain we first fix a choice $N_B = (N_{B_1}, \ldots, N_{B_d})$ where each element of N_B must divide each element of N_M component-wise. Then Ω is domain-decomposed through rectilinear sub-division into a family of n-rectangles satisfying $\Omega = \sqcup_{i\in Z}\Omega_i$, where Z is the set of `MeshBlock` indices, corresponding to the ordering described in Sect. 4.2.1. Nearest-neighbor elements are constrained to only differ by a single sub-division at most. The `MeshBlock` class stores properties of an element Ω_i of the sub-division. In particular the number of points in the sampling of Ω_i is controlled through the choice of N_B. For purposes of communication of data between nearest neighbor `MeshBlock` objects the sampling over Ω_i is extended by a thin layer of so-called "ghost nodes" in each direction. Furthermore the local values (with respect to the chosen, extended sampling on Ω_i) of any discretized, dependent field variables of interest are stored within the `MeshBlock`.

In both uniform grid $(\forall i \in Z)$ $\mathrm{vol}(\Omega_i) = C$ and refined meshes $(\exists i,\ j \in Z)$ $\mathrm{vol}(\Omega_i) \neq \mathrm{vol}(\Omega_j)$ it is crucial to arrange inter-`MeshBlock` communication efficiently – to this end the relationships between differing `MeshBlock` objects are arranged in a tree data structure, to which we now turn.

4.2.1 *Tree Structure of the* `Mesh`

For the sake of exposition here and convenience in later sections we now particularize to a Cartesian coordinatization though we emphasize that the general picture (and our implementation) of the discussions here and in Sect. 4.2.2 carry over to the curvilinear context with only minor modification.

`(GR-)Athena++` stores the logical relationship between the `MeshBlock` objects (i.e. Ω_i) involved in description of a domain Ω within a tree data structure. A binary-tree, quad-tree or oct-tree is utilized when $d := \dim(\Omega) = 1, 2, 3$ respectively. The relevant tree is then constructed by first selecting the minimum N such that 2^N exceeds the largest number of Ω_i along any dimension. The root of the tree is assigned a logical level of zero and the tree terminates at level N with every `MeshBlock` assigned to an appropriate leaf, though some leaves and nodes of the tree may remain empty. Data locality is enhanced, as references to `MeshBlock` objects are stored such that a post-order, depth-first traversal of the tree follows Morton order (also termed Z-order) [103]. This order can be used to encode multi-dimensional coordinates into a linear index parametrizing a Z-shaped, space-filling curve where small changes in the parameter imply spatial coordinates that are close in a suitable sense [33].

As an example we consider a three-dimensional `Mesh` described by $(N_x, N_y, N_z) = (2, 5, 3)$ `MeshBlock` objects in each direction at fixed physical level in Fig. 4.1.

4.2.1.1 `Mesh Refinement`

In order to resolve solution features over widely varying spatial (temporal) scales `(GR-)Athena++` implements block-based adaptive mesh refinement (AMR) [139]. This naturally fits into the domain-decomposition structure previously described as each physical position on a computational domain is covered by one and only one level. Salient field data need only be synchronized at block (`MeshBlock`) boundaries. The logical arrangement into a (binary,quad,oct)-tree structure has the added benefit of improving computational efficiency through preservation of data locality in memory. Similarly the underlying task-based parallelism (see Sect. 4.3.6)

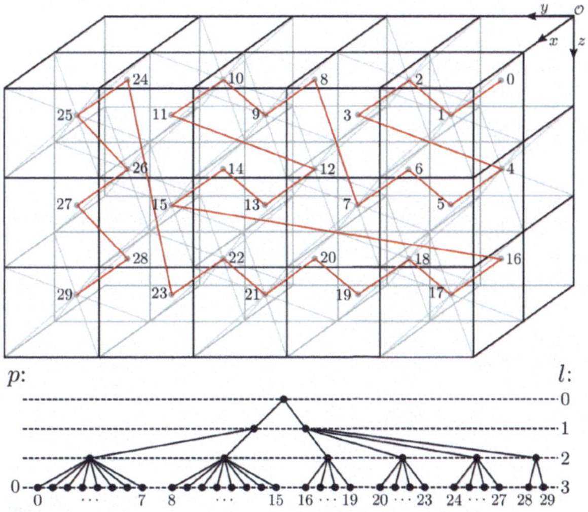

Fig. 4.1 Example of `Mesh` partitioned uniformly by `MeshBlock` objects indexed via Z-order and traced in red through constituent geometric centroids. The logical relationship between Ω_i is stored in an oct-tree. Empty leaves are suppressed though each populated node above logical level three has eight children. Notice that physical level p and logical level l are distinct. (Figure adapted from [49])

as the computational model greatly facilitates the overlap of communication and computation.

Consider now a `Mesh` with refinement. The number of physical refinement levels added to a uniform level, domain-decomposed Ω is controlled by the parameter N_L. By convention N_L starts at zero. Subject to satisfaction of problem-dependent, user-prescribed, refinement criteria, there may exist physical levels at $0, \ldots, N_L$. When a given `MeshBlock` is refined (coarsened) 2^d `MeshBlock` objects are constructed (destroyed). This is constrained to satisfy a $2:1$ refinement ratio where nearest-neighbor `MeshBlock` objects can differ by at most one physical level. Function data at a fixed physical level is transferred one level finer through use of a prolongation operator $\mathscr{P}$; dually, function data may be coarsened by one physical level through restriction $\mathscr{R}$. The details of the $\mathscr{P}$ and $\mathscr{R}$ operators depend on the underlying field treated, together with the selection of discretization.

A concrete example of a potential overall structure is provided in Fig. 4.2 where we consider a non-periodic Ω described by $N_x = N_y = N_z = 2$ `MeshBlock` objects with $N_L = 3$ selected with refinement introduced at the corner $x_{\max}$, $z_{\max}$. If periodicity conditions are imposed on $\partial\Omega$ then additional refinement may be required for boundary intersecting `MeshBlock` objects so as to maintain the aforementioned inter-`MeshBlock` $2:1$ refinement ratio.

Two main strategies for control over refinement are offered. One may pose that a fixed, region contained within the `Mesh` is at a desired level N_L. Alternatively, on individual `MeshBlock` objects a refinement condition may be imposed. This latter is quite flexible and allows for e.g. time-dependent testing as to whether dynamical fields satisfy some desired criteria. In particular, in the binary-black-hole context, recent work with `GR-Athena++` [123] has compared "box-in-box", "sphere-in-sphere", and truncation error estimation as ingredients for the aforementioned condition and the effect on gravitational wave mismatch.

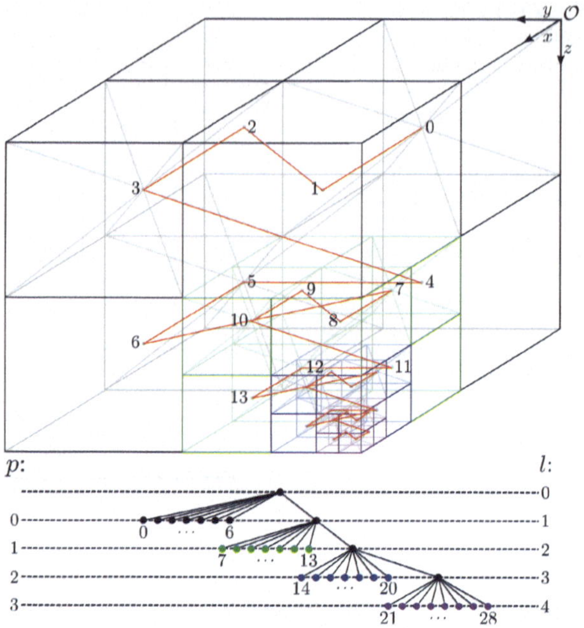

Fig. 4.2 Example of `Mesh` partitioned and refined by `MeshBlock` objects indexed via Z-order and traced in red through constituent geometric centroids. The logical relationship between ${}^{p}\Omega_i$ and neighbors is stored in an oct-tree. There are no unpopulated leaves. Notice that physical level p and logical level l are distinct; coloring corresponds to physical level: $p = 0$ in black, $p = 1$ in dark green, $p = 2$ in blue, and $p = 3$ in purple. (Figure adapted from [49])

4.2.2 Field Discretization

A variety of discretizations of dynamical field components are required to facilitate the evolution of the GRMHD system embedded (Sect. 4.3.1) within `GR-Athena++`. Natively `Athena++` supports cell-centered (CC), face-centered (FC), and edge-centered (EC) grids. Thus volume averaged, conserved (primitive) quantities of the Euler equations are taken on CC and suitable mapping to FC (during calculation of e.g. fluxes) leads to a shock-capturing, conservative, Godunov-style finite volume method. When magnetic (electric) fields are present then EC is employed so as to enable use of the Contrained Transport algorithm (see Sect. 4.3.3.2) which preserves the divergence free condition on magnetic fields.

To make the connection to a `MeshBlock` object concrete, we can first consider the one-dimensional domain $\Omega_I := [a, b]$. Grids for uniformly sampled CC and vertex-centers (VC) respectively are:

$$
\begin{aligned}
\mathscr{G}_{\mathrm{CC}} &:= \left\{ a + \left(n + \frac{1}{2}\right)\frac{b-a}{N_B} \,\middle|\, n \in \{0, \ldots N_B - 1\} \right\}, \\
\mathscr{G}_{\mathrm{VC}} &:= \left\{ a + n\frac{b-a}{N_B} \,\middle|\, n \in \{0, \ldots N_B\} \right\};
\end{aligned}
\tag{4.1}
$$

where $\delta x := (b - a)/N_B$ is the grid spacing and N_B the sampling parameter. To facilitate communication between `MeshBlock` objects a thin layer of N_g "ghost nodes" (also with spacing δx) is extended in the direction of each nearest-neighbor `MeshBlock`. Products of combinations of such grids allow for straight-forward extension to 2^{d} and 3^{d}, together with construction of e.g. FC sampling.

For the evolution of purely geometric quantities (induced spatial metric, extrinsic curvature, etc.) `GR-Athena++` extends the framework by incorporating infrastructure that allows for point-wise description of fields on VC. Our primary motivation for this is one of computational efficiency which may be understood through the following. Consider splitting a fixed $\mathscr{G}_{\mathrm{VC}}$ exactly in half into two new grids each with the same sampling parameter as the parent. This doubles the resolution. We also find that the coarsely spaced parent nodes exactly coincide with interspersed nodes of the children. This behaviour occurs at refinement boundaries (in ghost-layers) for VC sampling due to the $2:1$ ratio nearest-neighbor `MeshBlock` objects must obey (see Sect. 4.2.1.1). For refinement based on polynomial interpolation this reduces the restriction operation for VC to an injection (copy), and, simplifies prolongation. For a full discussion of our high-order $\mathscr{R}$ and $\mathscr{P}$ for VC see [49].

4.2.2.1 Geodesic Spheres

In `GR-Athena++`, during simulations, evaluation of diagnostic quantities that are naturally defined as integrals over spherical surfaces is often required. As a particular example we extract gravitational radiation (Sect. 4.3.5.2) through evaluation of numerical quadrature over triangulated geodesic spheres. Using a geodesic grid ensures more even tiling of the sphere compared to the uniform latitude-longitude grid of similar resolution.

A geodesic sphere of radius R (denoted Q_R) may be viewed as the boundary of a convex polyhedron embedded in $\mathbb{R}^3$ with triangular faces, i.e., a simplicial 2-sphere that is homeomorphic to the 2-sphere of radius R. To construct the geodesic grid we start from a regular icosahedron with 12 vertices and 20 plane equilateral triangular faces, embedded in a unit sphere. This is refined using the so called "non-recursive" approach [150]. This leads to a polyhedron with $10n_Q^2 + 2$ vertices where the refinement parameter n_Q controls the so-called grid level. In Fig. 4.3 we show an example of two such grids.

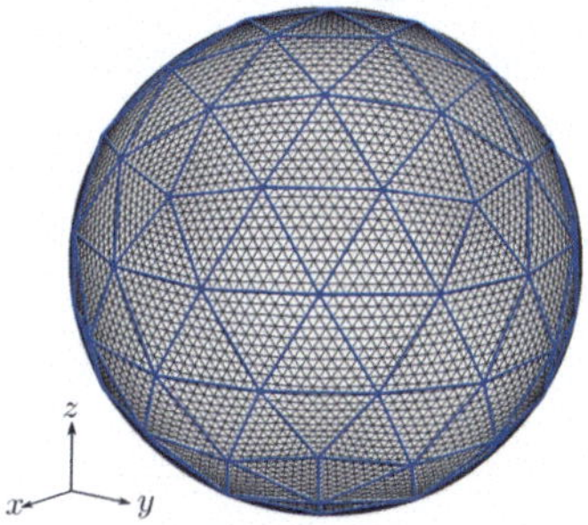

Fig. 4.3 Structure of the geodesic grid used by `GR-Athena++`. Depicted in light-blue is a grid in consisting of 92 vertices ($n_Q = 3$). In black we show a grid of 9002 vertices ($n_Q = 30$); a typical sampling associated with gravitational wave extraction in production simulations

Note that field data over the `Mesh` (i.e. local to `MeshBlock` objects) is treated as independent from a geodesic sphere in our implementation, with transfer achieved via standard Lagrange interpolation.

For evaluation of numerical quadratures we associate to each grid point of a geodesic sphere a solid angle by connecting the circumcenters of any pair of triangular faces that share a common edge. The solid angles subtended by the cells at the center of the sphere are used as weighting coefficients when computing the averages. The logical connection between the neighboring cells is implemented as described in [122].

4.3 GRMHD

4.3.1 Dynamical System

Our treatment of GRMHD within `GR-Athena++` involves the Z4c system (Sect. 4.3.1.1) coupled to the equations of ideal MHD through matter sources (Sect. 4.3.1.2). The setting of the formulation utilizes the standard geometric view (and definitions) of ADM decomposition [11] whereby a space-time manifold endowed with metric $(\mathscr{M}, g)$ is viewed as foliated by a family of non-intersecting, spatial hypersurfaces with members Σ_t. Fixing notation, recall that one introduces a future-directed t^a satisfying $t^a \nabla_a[t] = 1$ and considers $t^a = \alpha n^a + \beta^a$ where n^a is a future-directed, time-like, unit normal n^a to each member of the foliation Σ_t, α is the lapse and β^a the shift. Projections of ambient fields over $\mathscr{M}$ to (various products of) the tangent and normal bundles of Σ leads to evolution, and constraint equations. Here we summarise the resulting equations[1] as we have implemented them.

4.3.1.1 Z4c System

In brief, to stabilize the time-evolution problem, the Z4 formulation [26] directly augments the Einstein field equations via suitable introduction of an auxiliary, dynamical vector field Z^a. This allows for derivation of evolution equations for the induced metric γ_{ij} on Σ, the extrinsic curvature $K_{ij} := -\frac{1}{2}\pounds_{\mathbf{n}}[\gamma_{ij}]$ (with $\pounds_{\mathbf{n}}[\cdot]$ the Lie derivative along n^a), and the projected auxiliary quantities $\Theta := -n_a Z^a$ and $\check{Z}_i := \perp_i^a Z_a$ (with $\perp_b^a := g^a{}_b + n^a n_b$).

At its core Z4c [22, 78] is a conformal reformulation featuring constraint damping [77, 151]. A further, spatial conformal degree of freedom is factored out:

$$\tilde{\gamma}_{ij} := \psi^{-4}\gamma_{ij}, \qquad \tilde{A}_{ij} := \psi^{-4}\Big(K_{ij} - \frac{1}{3}K\gamma_{ij}\Big); \tag{4.2}$$

[1] Ambient quantities take indices $a, b, \ldots$ and $\mu, \nu, \ldots$ whereas spatial quantities take $i, j, \ldots$.

with the trace defined as $K := K_{ij}\gamma^{ij}$, and conformal factor $\psi := (\gamma/f)^{1/12}$ where γ and f are determinants of γ_{ij} and some spatial reference metric f_{ij} respectively. It is assumed that f_{ij} is flat and in Cartesian coordinates which immediately yields the *algebraic constraints*[2]:

$$\mathscr{C}_A := \left(\ln(\tilde{\gamma}),\ \tilde{\gamma}^{ij}\tilde{A}_{ij}\right) = 0. \tag{4.3}$$

Additionally we introduce the definitions:

$$\chi := \gamma^{-1/3}, \quad \hat{K} := K - 2\Theta; \quad \tilde{\Gamma}^i := 2\tilde{\gamma}^{ij}\check{Z}_j + \tilde{\gamma}^{ij}\tilde{\gamma}^{kl}\partial_l[\tilde{\gamma}_{jk}], \quad \widehat{\Gamma}^i := \tilde{\gamma}^{jk}\tilde{\Gamma}^i{}_{jk}. \tag{4.4}$$

The dynamical variables $\left(\chi,\ \tilde{\gamma}_{ij},\ \hat{K}, \tilde{A}_{ij},\ \Theta,\ \tilde{\Gamma}^i\right)$ obey the *evolution equations*:

$$\partial_t[\chi] = \frac{2}{3}\chi\left(\alpha(\hat{K}+2\Theta) - \partial_i[\beta^i]\right) + \beta^i\partial_i[\chi], \tag{4.5}$$

$$\partial_t[\tilde{\gamma}_{ij}] = -2\alpha\tilde{A}_{ij} + \beta^k\partial_k[\tilde{\gamma}_{ij}] - \frac{2}{3}\tilde{\gamma}_{ij}\partial_k[\beta^k] + 2\tilde{\gamma}_{k(i}\partial_{j)}[\beta^k]. \tag{4.6}$$

$$\begin{aligned}\partial_t[\hat{K}] =& -\mathrm{D}^i[\mathrm{D}_i[\alpha]] + \alpha\left[\tilde{A}_{ij}\tilde{A}^{ij} + \frac{1}{3}(\hat{K}+2\Theta)^2\right]\\ &+ \beta^i\partial_i[\hat{K}] + \alpha\kappa_1(1-\kappa_2)\Theta + 4\pi\alpha[\,{}^AS + {}^A\rho],\end{aligned} \tag{4.7}$$

$$\begin{aligned}\partial_t[\tilde{A}_{ij}] =& \chi[-\mathrm{D}_i[\mathrm{D}_j[\alpha]] + \alpha(R_{ij} - 8\pi\,{}^AS_{ij})]^{\mathrm{tf}} + \alpha[(\hat{K}+2\Theta)\tilde{A}_{ij} - 2\tilde{A}^k{}_i\tilde{A}_{kj}]\\ &+ \beta^k\partial_k[\tilde{A}_{ij}] + 2\tilde{A}_{k(i}\partial_{j)}[\beta^k] - \frac{2}{3}\tilde{A}_{ij}\partial_k[\beta^k],\end{aligned} \tag{4.8}$$

$$\partial_t[\Theta] = \frac{\alpha}{2}\left[\tilde{\mathscr{H}} - 2\kappa_1(2+\kappa_2)\Theta\right] + \beta^i\partial_i[\Theta], \tag{4.9}$$

$$\begin{aligned}\partial_t[\tilde{\Gamma}^i] =& -2\tilde{A}^{ij}\partial_j[\alpha] + 2\alpha\Big[\tilde{\Gamma}^i{}_{jk}\tilde{A}^{jk} - \frac{3}{2}\tilde{A}^{ij}\partial_j[\ln(\chi)] - \kappa_1(\tilde{\Gamma}^i - \widehat{\Gamma}^i)\\ &- \frac{1}{3}\tilde{\gamma}^{ij}\partial_j[2\hat{K}+\Theta] - 8\pi\tilde{\gamma}^{ij}\,{}^AS_j\Big] + \tilde{\gamma}^{jk}\partial_k[\partial_j[\beta^i]] + \frac{1}{3}\tilde{\gamma}^{ij}\partial_j[\partial_k[\beta^k]]\\ &+ \beta^j\partial_j[\tilde{\Gamma}^i] - \widehat{\Gamma}^j\partial_j[\beta^i] + \frac{2}{3}\widehat{\Gamma}^i\partial_j[\beta^j];\end{aligned} \tag{4.10}$$

where in Eq. (4.8) the trace-free (tf) operation is computed with respect to γ_{ij} and $\tilde{\mathscr{H}}$ is defined in Eq. (4.11). The scalars κ_1 and κ_2 are freely chosen, damping parameters. Definitions of matter fields are standard and based on projections of the decomposed

[2] We enforce $\mathscr{C}_A = 0$ during numerical evolution to ensure consistency [35].

space-time, energy-momentum-stress tensor $T_{ab} = {}^{A}\rho n_a n_b + 2{}^{A}S_{(a}n_{b)} + {}^{A}S_{ab}$ in terms of the energy density ${}^{A}\rho := T_{ab}n^a n^b$, momentum ${}^{A}S_i = -T_{bc}n^b \perp^c_i$, and spatial stress ${}^{A}S_{ij} := T_{cd} \perp^c_i \perp^d_j$. The intrinsic curvature R_{ij} appearing in Eq. (4.8) is decomposed according to $R_{ij} = \tilde{R}^{\chi}{}_{ij} + \tilde{R}_{ij}$ where in terms of the connection $\tilde{\mathrm{D}}_i$ compatible with $\tilde{\gamma}_{jk}$:

$$\tilde{R}^{\chi}{}_{ij} = \frac{1}{2\chi}\left[\tilde{\mathrm{D}}_i[\tilde{\mathrm{D}}_j[\chi]] + \tilde{\gamma}_{ij}\tilde{\mathrm{D}}^l[\tilde{\mathrm{D}}_l[\chi]] - \frac{1}{2\chi}\tilde{\mathrm{D}}_i[\chi]\tilde{\mathrm{D}}_j[\chi]\right] - \frac{3}{4\chi^2}\tilde{\mathrm{D}}^l[\chi]\tilde{\mathrm{D}}_l[\chi]\tilde{\gamma}_{ij},$$

and:

$$\tilde{R}_{ij} = -\frac{1}{2}\tilde{\gamma}^{lm}\partial_l[\partial_m[\tilde{\gamma}_{ij}]] + \tilde{\gamma}_{k(i}\partial_{j)}[\tilde{\Gamma}^k] + \hat{\Gamma}^k\tilde{\Gamma}_{(ij)k} + \tilde{\gamma}^{lm}(2\tilde{\Gamma}^k{}_{l(i}\tilde{\Gamma}_{j)km} + \tilde{\Gamma}^k{}_{im}\tilde{\Gamma}_{klj}).$$

The *dynamical constraints* in terms of transformed variables ($\tilde{\mathscr{H}}$, $\tilde{\mathscr{M}}_i$, Θ, $\check{Z}^i$) may be monitored[3] to assess the quality of a numerical calculation:

$$\tilde{\mathscr{H}} := R - \tilde{A}_{ij}\tilde{A}^{ij} + \frac{2}{3}(\hat{K} + 2\Theta)^2 - 16\pi\ {}^{A}\rho = 0, \tag{4.11}$$

$$\tilde{\mathscr{M}}_j := \tilde{\mathrm{D}}_i[\tilde{A}^i{}_j] - \frac{3}{2}\tilde{A}^i{}_j\partial_i[\ln(\chi)] - \frac{2}{3}\partial_j[\hat{K} + 2\Theta] - 8\pi\ {}^{A}S_j = 0, \tag{4.12}$$

$$\Theta = 0, \qquad \check{Z}^i = \tilde{\Gamma}^i - \hat{\Gamma}^i = 0. \tag{4.13}$$

In order to close the Z4c system we must supplement it with gauge conditions. These are conditions on (α, β^i) that in `GR-Athena++` are based on Bona-Másso lapse [27] and the gamma-driver shift [6]:

$$\partial_t[\alpha] = -\mu_L\alpha^2\hat{K} + \beta^i\partial_i[\alpha], \qquad \partial_t[\beta^i] = \mu_S\alpha^2\tilde{\Gamma}^i - \eta\beta^i + \beta^j\partial_j[\beta^i]. \tag{4.14}$$

In specification of Eq. (4.14) we employ the $1 + \log$ lapse variant $\mu_L = 2/\alpha$ together with $\mu_S = 1/\alpha^2$. The η term serves as a free damping parameter.

4.3.1.2 MHD and Coupling

In order to couple to the Z4c system of Sect. 4.3.1.1 consider the energy-momentum-stress (EMS) tensor factored into perfect fluid and electromagnetic parts:

$$T^{\mu\nu} = \left[\rho h u^\mu u^\nu + p g^{\mu\nu}\right] + \left(b^2 u^\mu u^\nu + \frac{1}{2}b^2 g^{\mu\nu} - b^\mu b^\nu\right). \tag{4.15}$$

[3] Alternatively these may be assessed mapping back from the conformal variables.

Focusing on the perfect fluid sector (first bracketed term) we have fluid quantities u^μ, ρ, p, $h := 1 + \epsilon + p/\rho$, and ϵ, which are respectively a time-like velocity 4-vector, rest mass density, pressure, relativistic specific enthalpy, and specific internal energy. Write $u^\mu = (u^0, u^i)$ and define spatial components of the fluid 4-velocity by forming the projection to Σ_t as:

$$\tilde{u}^i = \perp^i{}_\mu u^\mu = u^i + \frac{W\beta^i}{\alpha}, \tag{4.16}$$

where $W := -n_\mu u^\mu$ is the Lorentz factor between Eulerian and comoving fluid observers. This leads to the fluid 3-velocity $v^i := \tilde{u}^i/W$ and $\tilde{v}^i := v^i - \beta^i/\alpha$. The second bracketed term of Eq. (4.15) carries the electromagnetic contribution. The terms b^μ and b^2 appearing there describe the magnetic field and arise as projections of the dual Faraday tensor along u^μ. Recall that with respect to Eulerian observers the magnetic $\mathscr{B}^\mu$ and electric field $\mathscr{E}^\mu$ can be used to express the Faraday tensor as:

$$F^{\mu\nu} = n^\mu \mathscr{E}^\nu - \mathscr{E}^\mu n^\nu - \epsilon^{\mu\nu\rho\sigma} n_\rho \mathscr{B}_\sigma,$$

and $\epsilon^{\mu\nu\rho\sigma}$ is the 4 dimensional Levi-Civita pseudo-tensor. Define the dual tensor ${}^*F^{\mu\nu} := \frac{1}{2}\epsilon^{\mu\nu\rho\sigma} F_{\rho\sigma}$. Projecting as $b^\mu = u_\nu {}^*F^{\mu\nu}$ leads to[4, 5]:

$$b^0 = \frac{W\mathscr{B}^i v_i}{\alpha}, \quad b^i = \frac{\mathscr{B}^i + \alpha b^0 W \tilde{v}^i}{W}, \quad b^2 = g_{\mu\nu} b^\mu b^\nu = \frac{(\alpha b^0)^2 + \mathscr{B}^i \mathscr{B}^j \gamma_{ij}}{W^2}.$$

We form evolution equations for MHD as a balance (conservation) law as in [16]:

$$\partial_t \mathbf{q} + \partial_i \mathbf{F}^i = \mathbf{s}, \tag{4.17}$$

where the conservative variables $\mathbf{q}$ are associated to the Eulerian observer. In our convention $\mathbf{q} := (D, S_j, \tau, B^k)$. Note that we take as primitive variables $\mathbf{w} = (\rho, \tilde{u}^i, p, B^i)$ and may map to $\mathbf{q}$ through:

$$\mathbf{q} = \sqrt{\gamma}\left(\rho W, (\rho h + b^2)W\tilde{u}_j - \alpha b^0 b_j, (\rho h + b^2)W^2 - \rho W - \left(p + \frac{b^2}{2}\right) - (\alpha b^0)^2, \mathscr{B}^k\right).$$

The inversion of this map, where required, is evaluated numerically (Sect. 4.3.4).

In order to determine the flux vector $\mathbf{F}^i$, and source $\mathbf{s}$ of Eq. (4.17) one considers the conservation of rest mass density $\nabla_\mu[\rho u^\mu] = 0$, the Bianchi identity $\nabla_\mu[T^{\mu\nu}] = 0$, and the Maxwell relation $\nabla_\mu[{}^*F^{\mu\nu}] = 0$. The former two equations are projected onto and normal to Σ_t whereas the latter is projected onto Σ_t. This results in:

[4] Assuming ideal MHD, in the comoving fluid-frame, the electric field vanishes: $u^\mu F_{\mu\nu} = 0$.

[5] A factor of $\sqrt{4\pi}$ has been absorbed into the magnetic field definition throughout.

$$\mathbf{F}^i = \begin{bmatrix} D\alpha\tilde{v}^i \\ S_j\alpha\tilde{v}^i + \delta^i_j\left(p+\frac{b^2}{2}\right)\alpha\sqrt{\gamma} - \frac{\alpha\sqrt{\gamma}b_j\mathscr{B}^i}{W} \\ \tau\alpha\tilde{v}^i + \alpha\sqrt{\gamma}\left(\left(p+\frac{b^2}{2}\right)v^i - \frac{\alpha b^0\mathscr{B}^i}{W}\right) \\ \alpha(B^k\tilde{v}^i - B^i\tilde{v}^k) \end{bmatrix}. \tag{4.18}$$

and:

$$\mathbf{s} = \alpha\sqrt{\gamma}\begin{bmatrix} 0 \\ T^{00}\left(\frac{1}{2}\beta^i\beta^k\partial_j\gamma_{ik} - \alpha\partial_j\alpha\right) + T^{0i}\beta^k\partial_j\gamma_{ik} + T^0_i\partial_j\beta^i + \frac{1}{2}T^{ik}\partial_j\gamma_{ik} \\ T^{00}\left(\beta^i\beta^jK_{ij} - \beta^i\partial_i\alpha\right) + T^{0i}\left(2\beta^jK_{ij} - \partial_i\alpha\right) + T^{ij}K_{ij} \\ 0^k \end{bmatrix}.$$

One further projection can be formed. The Maxwell relation when projected in the normal direction to Σ_t gives rise to the divergence-free constraint on B^i, that is $\partial_i[B^i] = 0$. This will be automatically enforced by the discretisation of the equations (Sect. 4.3.3.2). Finally, the MHD equations are coupled to the Z4c system (4.3.1.1) through the the ADM sources (here prefixed with superscript A):

$$\begin{aligned} {}^A\rho &= (\rho h + b^2)W^2 - \left(p + \frac{b^2}{2}\right) - (\alpha b^0)^2, \\ {}^AS_i &= (\rho h + b^2)W^2 v_i - b^0 b_i\alpha, \\ {}^AS_{ij} &= (\rho h + b^2)W^2 v_i v_j + \left(p + \frac{b^2}{2}\right)\gamma_{ij} - b_i b_j. \end{aligned} \tag{4.19}$$

4.3.2 Equation of State

The GRMHD system of conservation laws together with the magnetic field divergence-free condition restricts 8 degrees of freedom for the 9 primitive variables $(\rho, p, \epsilon, \tilde{u}^i, B^i)$. The role of the equation of state (EOS) is to reduce this underdeterminedness through imposition of a thermodynamical relationship.

Here we highlight two classes of EOS used in `GR-Athena++`:

(i) Ideal gas with Gamma-law: A relation between (ρ, p, ϵ) is formed with $p = \rho\epsilon(\Gamma - 1)$ imposed where $\Gamma = 1 + 1/n$ and n is the adiabatic index of the fluid.[6] During initialisation this is supplemented by the barotropic relation $p = K\rho^\Gamma$ where K serves as a problem-dependent, fluid mass-scale.

(ii) Tabulated, 3-dimensional finite-temperature: First, species fractions Y_i that track fluid composition are introduced (CMA scheme [114]). For simplicity, consider only the electron fraction Y_e. This extra variable is incorporated into the GRMHD system as an additional evolved variable DY_e with zero source. Overall this

[6] Henceforth we restrict to the $\Gamma = 2$ case.

allows us to write the relation $p = p(\rho, T, Y_e)$. Quantities ρ and T are tabulated in logarithmic space, and, together with Y_e are linearly interpolated so as to provide $\log p$ when required.

Our implementation supports tables in the form used by the CompOSE database [146] processed to also include the sound speed by the PyCompOSE tool.[7]

4.3.3 Numerical Technique

In treatment of the GRMHD system in `GR-Athena++` the method of lines approach is taken. In this overview attention is restricted to time-evolution based on the third order, three stage, low-storage SSPRK(3, 3) method of [74].

Field components involved in the Z4c sub-system (Sect. 4.3.1.1) are sampled on vertex-centers (VC). Spatial derivatives on the interior of the computational domain $\mathrm{int}(\Omega)$ (i.e. in the bulk away from $\partial\Omega$) are computed with high-order, centered, finite difference (FD) stencils with one-point lop-siding for advective terms (see e.g.[32]). The ghost-layer extent (choice of N_g) fixes the overall, formal FD approximation order as $2(N_g - 1)$ throughout $\mathrm{int}(\Omega)$. The interpolation order during calculation of $\mathscr{P}$ under mesh refinement is consistently selected to match that of the FD; whereas $\mathscr{R}$ is exact for VC.

To fix physical boundary conditions, within every time-integrator sub-step, an initial Lagrange extrapolation (involving $N_g + 1$ points) is performed so as to populate a ghost-layer at $\partial\Omega$. The dynamical Z4c equations then furnish data for $\{\chi, \tilde{\gamma}_{ij}, \alpha, \beta^i\}$ at $\partial\Omega$ whereas values there for $\{\hat{K}, \tilde{\Gamma}, \Theta, \tilde{A}_{ij}\}$ are based on the Sommerfeld prescription of [78].

During evolution the Courant-Friedrich-Lewy (CFL) condition must be satisfied. In the context of `(GR-)Athena++` grid-spacing on the most refined level of a refinement hierarchy determines the global time-step that is applied on each `MeshBlock`. Finally for the evolved Z4c state-vector, high-order Kreiss-Oliger (KO) dissipation [89], of degree $2N_g$, with uniform factor σ_D is applied.

In the case of the MHD sub-system (Sect. 4.3.1.2) a standard conservative, Godunov-style finite volume method is employed. Thus, practically, when the conserved variables $\mathbf{q}$ are to be evolved, we require fluxes (Eq. (4.18)) evaluated at cell interfaces. Interface data is prepared based on reconstruction of known cell-averaged quantities to the left and right of an interface. Once prepared we can utilize e.g. Local-Lax-Friedrichs (LLF):

$$\hat{\mathbf{F}}_{i+1/2} = \frac{1}{2}\left(\mathbf{F}[\mathbf{q}^-_{i+1/2}] + \mathbf{F}[\mathbf{q}^+_{i-1/2}] - \alpha(\mathbf{q}^+_{i-1/2} - \mathbf{q}^-_{i+1/2})\right),$$

where α is taken as the maximum over characteristic speeds, which, are computed from locally reconstructed variables, while we note the availability of more sophis-

[7] https://bitbucket.org/dradice/pycompose/.

ticated Riemann solvers for magnetic fields such as those presented in [85]. In the absence of magnetic fields exact characteristic speeds are provided in [16] whereas for MHD we follow the approximate prescription of [67]. In the MHD context preservation of the magnetic field divergence-free condition requires some care. This is achieved through a constrained transport scheme which we return to in Sect. 4.3.3.2.

4.3.3.1 Intergrid Transfer, and Reconstruction

In `(GR-)Athena++`, data transfer between differing discretizations for the evolved field components of GR and MHD is required at various points of the time-evolution algorithm. Indeed CC grids form the primary representation for the evolved, conserved quantities $\mathbf{q}$, furthermore the FC and EC grids are utilized for constrained transport. The (smooth) geometric data is evolved on VC. For simplicity, this latter, is transferred to the other grids, as required, at linear order.

In order to evaluate the numerical fluxes at the cell interfaces, we reconstruct the primitive variables $\mathbf{w}$ from CC to FC. Reconstruction avoids the increase of total variation, providing a limited interpolation strategy appropriate for functions of locally reduced smoothness. To this end the Piecewise Linear Method (PLM) [148], and Piecewise Parabolic Method (PPM) [42, 101], as implemented in `Athena++`, are further supplemented in `GR-Athena++` by WENO5 [79] and WENOZ [28].

4.3.3.2 Constrained Transport

The ideal MHD approximation reduces the Maxwell equations to a hyperbolic equation for the magnetic field components, the induction equation, and an elliptic equation, Gauss's Law. The solution of this elliptic equation at each time-step in a numerical simulation through e.g. a relaxation method would be incredibly costly and so various schemes have been developed to avoid this requirement, as discussed above. `Athena++` employs an implementation of the constrained transport approach to ensure divergence free evolution of the magnetic field, following the original proposal of [61], with the development of the specific implementation in `Athena++` following [18, 68, 69, 137, 153].

Here the magnetic field is discretised, not as a cell average as for the other hydrodynamical variables, or as in other implementations such as the Flux-CT implementation of [147], but as face averages:

$$B^x_{i-1/2,j,k} = \frac{1}{\Delta A} \int_{A_{i-1/2,j,k}} B^x \mathrm{d}y\mathrm{d}z, \tag{4.20}$$

with the other components B^y, B^z discretised on, respectively, the y and z faces of the computational cell $A_{i,j-1/2,k}$, $A_{i,j,k-1/2}$, with ΔA the area of the face being integrated over.

Noting that the electric field is given by the cross product of the magnetic field with the fluid velocity $\alpha\tilde{v}$, integrating the induction equation over a face of the computational cell gives an evolution equation for the face averaged magnetic field in terms of the integrated curl of the electric field over the face. By applying Stokes theorem, the face centred magnetic field is updated by evaluation of the integral of the edge centred, time averaged, electric field around the face in question:

$$\begin{aligned}
\int_{t_n}^{t_{n+1}} \partial_t B^x_{i-1/2,j,k} \mathrm{d}t &= \frac{1}{\Delta A}\int_{t^n}^{t^{n+1}} \mathrm{d}t \int_{A_{i-1/2,j,k}} \Big\{ - \partial_y(\alpha(B^x\tilde{v}^y - B^y\tilde{v}^x)) \\
&\quad -\partial_z(\alpha(B^x\tilde{v}^z - B^z\tilde{v}^x))\mathrm{d}y\mathrm{d}z\Big\}, \\
&= \frac{1}{\Delta A}\int_{t^n}^{t^{n+1}} \mathrm{d}t \int_{A_{i-1/2,j,k}} -\partial_y(E^z) + \partial_z(E^y)\mathrm{d}y\mathrm{d}z, \\
B^{x,n+1}_{i-1/2,j,k} &= B^{x,n}_{i-1/2,j,k} - \frac{\Delta t}{\Delta A}\int_{\partial A_{i-1/2,j,k}} \mathbf{E}^{n+1/2} \cdot \mathbf{dl},
\end{aligned} \tag{4.21}$$

where $\mathbf{dl}$ is the line element around the edge of face $A_{i-1/2,j,k}$, and $\mathbf{E}^{n+1/2}$ the time averaged electric field between t_n and t_{n+1}.

It is clear that this discretisation of the field automatically preserves the divergence of B between time-steps; when evaluating the divergence of B^{n+1} using the above expression, contributions from the electric field will cancel due to shared edges between faces having opposite signs due to the orientations of the integrals around the face edges. The divergence of B^{n+1} must then equal the divergence of B^n.

The final requirement for evaluating the magnetic field is to calculate the time averaged, edge centred, electric fields.

Following the standard Godunov approach, the time averaged face centred electric field can be calculated as the solution to the Riemann problem at the cell interface, which is calculated along with the hydrodynamical fluxes. Then the cell centred electric field can be explicitly constructed as the cross product of the magnetic field and fluid 3-velocity. The derivative of the electric field from the cell centre to face centre can be calculated, and then upwinded to the cell face in the upwind direction as determined by the sign of the mass flux. This gradient is then used to integrate the time averaged face centred electric field to the cell edge.

A detailed description of this procedure may be found in [153], though we note that the implementation for dynamical space-times requires the appropriate densitisation of variables by the time dependent metric determinant, which can no longer be factored out of time derivatives.

For simulations with AMR new `MeshBlocks` will be created dynamically, on which data will be populated through interpolation from the coarser (finer) `MeshBlock(s)` that they replace. Naively, this interpolation will not respect the divergence free constraint on B, and so `Athena++` implements the restriction and prolongation operators of [145] designed to preserve the curl and divergence of interpolated vector fields.

At the interface between refinement levels, faces and edges of `MeshBlocks` of different levels overlap. To ensure a conservative evolution the face centred hydrodynamical fluxes are kept consistent, by replacing the face centred flux on the coarser block with the area weighted sum of the fluxes on the finer blocks. Similarly for the magnetic field evolution, the edge centred electric field on the coarser block is replaced with the length weighted sum of the edge centred fields on the finer block. To ameliorate other effects such as the non-deterministic order of `MPI` processes, shared edge centred electric fields are replaced by their averages also in the case of neighboring `MeshBlocks` on the same refinement level.

4.3.4 Conservative-To-Primitive Variable Inversion

In order to evaluate the fluxes in Eq. (4.17) we convert the conservative variables, $\mathbf{q}$ to their primitive representation $\mathbf{w}$ by inverting the definitions of $\mathbf{q}$ in Eq. (4.18). This involves solution of a non-linear system for the primitive variables.[8] To this end `GR-Athena++` features a custom, error-policy based implementation `PrimitiveSolver`. We have also coupled the external library `RePrimAnd`[9] of [81]. Here we note some additional considerations related to the latter method. Generically the conservative-to-primitive inversion is evaluated pointwise, where required, over a `MeshBlock`, though is not subject to the underlying particulars in choice of discretization.

Physically we expect a neutron-star (NS) to have a well defined, sharp surface. Consider a radial slice of the fluid density from the NS center: a non-negative profile that drops to zero is expected. Resolving such sharp features can lead to numerical issues during the variable conversion process. In order to address this we follow the standard technique of introducing a tenuous atmosphere region where primitive variables are set according to:

$$\rho_{\mathrm{atm}} = f_{\mathrm{atm}}\rho_{\mathrm{max}}, \qquad p_{\mathrm{atm}} = p(\rho_{\mathrm{atm}}), \qquad \tilde{u}^i_{\mathrm{atm}} = 0; \tag{4.22}$$

with ρ_{max} the maximum value of the fluid density in the initial data, and f_{atm} a problem-specific parameter. The magnetic field is left unaltered in the atmosphere. During the course of the evolution, physical processes of the star, or numerical dissipation, may drive the fluid into the atmosphere region. On account of this we impose an additional criterion where if $\rho < \rho_{\mathrm{thr}} = f_{\mathrm{thr}}\rho_{\mathrm{atm}}$ the cell is reset to atmosphere.[10] Similarly cells are reset to atmosphere if conversion from conservative to primitive variables fails to converge. When atmosphere values are set, we recalculate the conservative variables from the atmosphere primitives and metric variables.

[8] A variety of strategies are summarised and compared in [134].

[9] We find robust performance when employing a tolerance of 10^{-10} in the bracketing algorithm and a ceiling of 20 on the velocity variable $\tilde{u}$.

[10] Typically we set $f_{\mathrm{thr}} = 1$ with $f_{\mathrm{atm}} \in [10^{-13}, 10^{-18}]$.

4.3.5 Diagnostic Quantities

For later convenience common scalar diagnostic quantities computed during simulations are collected here.

4.3.5.1 Energetics

The baryon mass is a conserved quantity defined as:

$$M_b = \int_{\Sigma_t} D\mathrm{d}^3x; \tag{4.23}$$

where the integral is evaluated over the entire computational domain. Various energy contributions to the GRMHD system are similarly evaluated:

$$E_{\mathrm{Kin}} = \frac{1}{2}\int_{\Sigma_t} \frac{S_i S^i}{D}\mathrm{d}^3x, \quad E_B = \frac{1}{2}\int_{\Sigma_t} \sqrt{\gamma} W b^2 \mathrm{d}^3x, \quad E_{\mathrm{int}} = \int_{\Sigma_t} D\epsilon\mathrm{d}^3x. \tag{4.24}$$

4.3.5.2 Gravitational Wave Extraction

To obtain the gravitational wave (GW) content of the space-time, we calculate the Weyl scalar Ψ_4, the projection of the Weyl tensor onto an appropriately chosen null tetrad $k, l, m, \bar{m}$ with the conventions of [32]. Multipolar decomposition onto spin-weighted spherical harmonics[11] ${}_sY_{lm}$ of spin-weight $s = -2$ is evaluated through:

$$\psi_{\ell m} = \int_0^{2\pi}\int_0^{\pi} \Psi_4 \, {}_{-2}\bar{Y}_{\ell m}(\vartheta, \varphi) \sin\vartheta \mathrm{d}\vartheta \mathrm{d}\varphi. \tag{4.25}$$

Thus Ψ_4 is first calculated at all grid points throughout the `Mesh` whereupon data is interpolated onto a set of geodesic spheres Q_R (see Sect. 4.2.2.1) at given extraction radii R_Q. Subsequently the numerical quadrature of Eq. (4.25) may be evaluated. In order to select the grid level parameter n_Q that controls the number of samples on Q_R we may match the area element to that of any intersected `MeshBlock` objects through $n_Q = \left\lceil \sqrt{\frac{\pi R_Q^2}{\delta x^2}} - 2 \right\rceil$. The modes of the GW strain h may be computed from the projected Weyl scalar by integrating twice in time $\psi_{\ell m} = \ddot{h}_{\ell m}$. The strain is then given by the mode-sum:

$$h_+ - ih_\times = \frac{G}{D_L}\sum_{\ell=2}^{\infty}\sum_{m=-\ell}^{\ell} h_{\ell m}(t) \, {}_{-2}Y_{\ell m}(\vartheta, \varphi), \tag{4.26}$$

[11] Convention here is that of [71] up to a Condon-Shortley phase factor of $(-1)^m$.

where $D_L = (1 + z)R$ is the luminosity distance of a source located at distance R to the observer and at redshift z. Phase and frequency conventions follow the LIGO algorithms library [95].

4.3.6 Task-List Execution Model

The individual components of a calculation during a time-evolution sub-step that must be executed must satisfy a particular order of execution. The task-based execution model employed in `(GR-)Athena++` may be viewed as assembling this information in a directed, acyclic graph (DAG). A task specifies a given operation that must be performed and may feature dependency on the prior completion of other tasks. These may be assembled so as to represent a more complicated calculation into a so-called task-list. This allows to neatly encapsulate the calculation steps involved in GRMHD in Fig. 4.4. One advantage of this form of calculation assembly

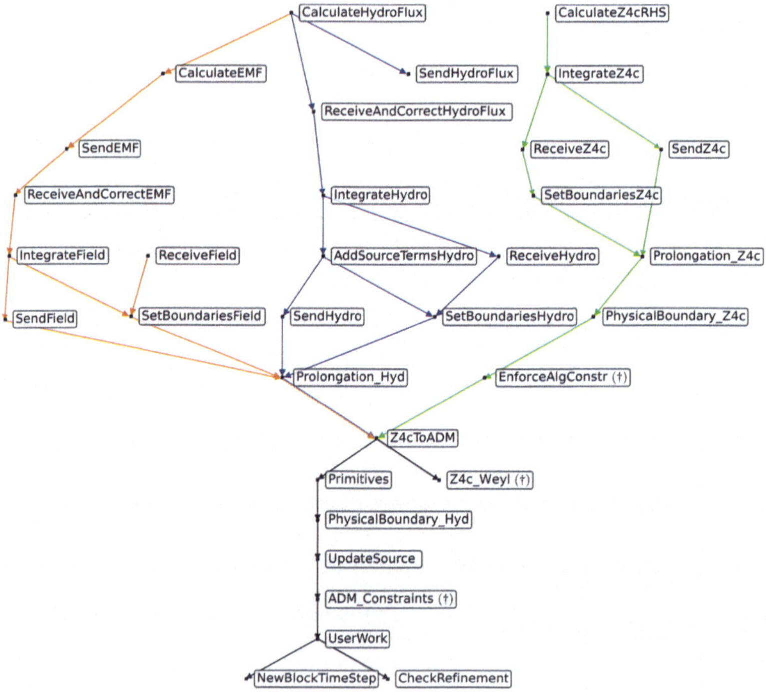

Fig. 4.4 Simplified DAG of GRMHD task-list in `GR-Athena++` executed on each `MeshBlock` during a time-integrator sub-step. Tasks are boxed and arrows between nodes indicate direction of dependency. In the absence of a dependency, remaining tasks may be executed in parallel. Labels indicate operation being performed. Arrows are color-coded according to type of field data manipulated where: (orange) corresponds to magnetic field; (blue) fluid; (green) Z4c; (black) general operations. Tasks labeled with a (†) are only executed on the final integrator sub-step

is that wait time during buffer-communication can be masked as e.g. system calculations proceed in parallel, this feature improves efficiency. Thus, for example, as shown in Fig. 4.4 once the `CalculateZ4cRHS` and `IntegrateZ4c` tasks are completed on the current `MeshBlock` a suitable subset of this data may be sent via communication buffers (`SendZ4c` task) to a nearest-neighbor `MeshBlock` so as to populate its ghost-layer (and vice-versa through the `RecvZ4c` task). Once these tasks complete, boundary data on the current `MeshBlock` may be set as represented by `SetBoundariesZ4c` and prolongated where required. As a final note, operations involving global data reduction (e.g. numerical quadrature over geodesic sphere in gravitational wave calculation Sect. 4.2.2.1) are performed external to the task-list.

4.4 Applications

We now demonstrate the performance and accuracy of `GR-Athena++` through solution of problems in dynamical space-times with GRMHD. A subset of tests based on our work [45] is summarised for NS space-times, followed by a set of first applications, in the long term study of magnetic field development in an isolated NS, and in using AMR to efficiently study the Kelvin-Helmholtz instability during a BNS merger.

Throughout this discussion $N_g = 4$ is selected which sets 6th order accurate finite differencing in treatment of Z4c. Damping parameters are taken as $\kappa_1 = 0.02$, and $\kappa_2 = 0$. Kreiss-Oliger dissipation is fixed at $\sigma_D = 0.5$.

4.4.1 Single Star Spacetimes

4.4.1.1 GRHD Evolution of Static Neutron Star

Our first test is of the evolution of a static NS model known as A0 in the convention of [53], commonly used as a code benchmark, e.g. [65, 143]. The star has baryon mass $M_b = 1.506\mathrm{M}_\odot$ and gravitational mass $M = 1.4\mathrm{M}_\odot$, with initial data calculated by solving the Tolman-Oppenheimer-Volkoff (TOV) equations using a central rest-mass density of $\rho_c = 7.905 \times 10^{14}\,\mathrm{g\,cm^{-3}}$, with an ideal gas EOS with $\Gamma = 2$, and initial pressure set using the polytropic EOS with $K = 100$.

Our grid configuration has outer boundaries at $[\pm 378.0, \pm 378.0, \pm 378.0]$ km, with 6 levels of static mesh refinement fully covering the star, with the innermost level covering the region $[\pm 14.8, \pm 14.8, \pm 14.8]$ km. The grid resolution is set through the `Mesh` parameter $N_M = 32, 48, 64, 128$, which gives a grid spacing of $[369, 246, 185, 92.3]$ m on the finest level.

We first demonstrate the evolution of the central rest mass density of the star $\rho_c(t)$ as a function of resolution and for a variety of reconstruction schemes in Fig. 4.5.

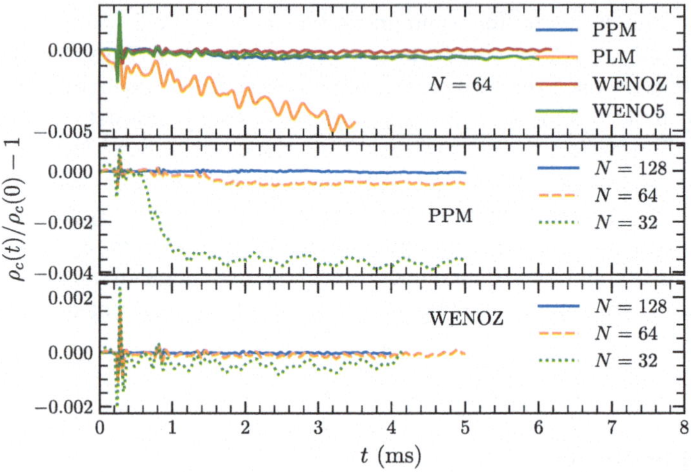

Fig. 4.5 GRHD Evolution of central rest-mass density of the static star A0 model. Top: varying reconstruction at constant resolution $N_M = 64$. Central, Bottom: varying resolution for PPM and WENOZ reconstruction respectively. Note: Some data has been truncated to 5 ms for ease of visualisation. (Figure adapted from [45])

The star is perturbed by the presence of truncation errors, leading to the excitation of characteristic oscillations in the central density, the frequencies of which match those predicted by perturbation theory for the first three radial modes, at frequencies of $\nu_F = 1462$, 3938 and 5928 Hz [12], and the relative amplitudes of which remain below 10^{-2}, and converge away as resolution increases. We find that, as expected, PLM is the least accurate scheme, with PPM performing poorly at low resolutions, but performing consistently with WENO schemes at higher resolutions.

In Fig. 4.6 we demonstrate a relative conservation of baryon mass at the level of 10^{-11} consistently across schemes and resolutions, with violations occuring due the setting of atmosphere levels and matter leaving the computational domain. The best performance is found for PPM, which preserves the star surface most accurately.

The convergence of three, spatially integrated, quantities is demonstrated for this model: the absolute difference between the density profile at time t and the initial density profile; and the Hamiltonian and Momentum constraints (Sect. 4.3.1.1). We show the former in Fig. 4.7, with dashed lines showing linear extrapolations from the first two data points, the slopes of which indicate the order of convergence. Approximately $2nd$ convergence is found, most clearly seen in the WENO(5,Z) schemes, while at the resolutions shown the PPM scheme has the lowest overall error. In Fig. 4.8, the constraints are shown to converge at an order between 3.51 and 5.46 depending on the reconstruction scheme used.

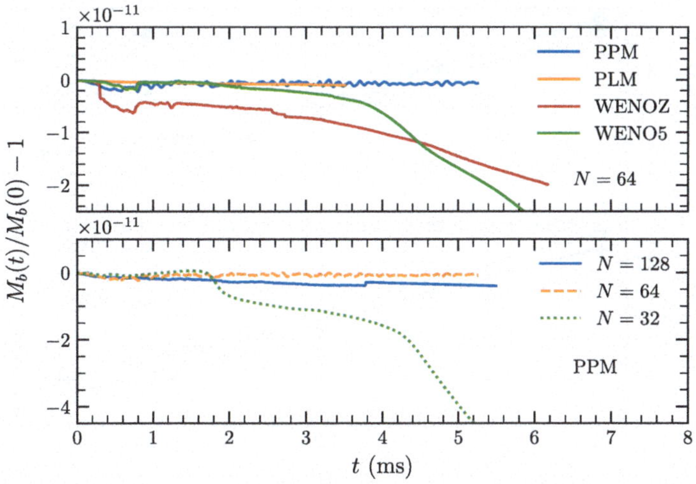

Fig. 4.6 Baryon mass conservation during evolution of the static star A0 model. The mass conservation can be violated due to the artificial low-density atmosphere as the fluid expands beyond the computational domain. (Figure adapted from [45])

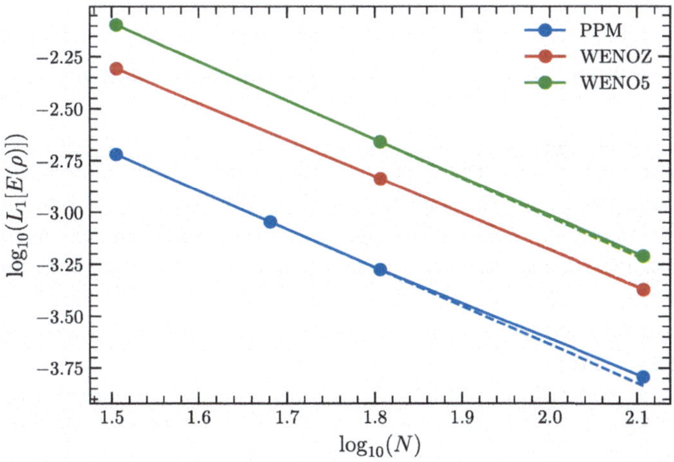

Fig. 4.7 Convergence of $L_1[E(\rho)] := |\rho(\mathbf{x}, t) - \rho(\mathbf{x}, 0)|$ for static star model A0 at $t = 2.96$ ms. Order of convergence is 1.85, 1.76, 1.87 for PPM, WENOZ, WENO5 reconstruction schemes respectively. (Figure adapted from [45])

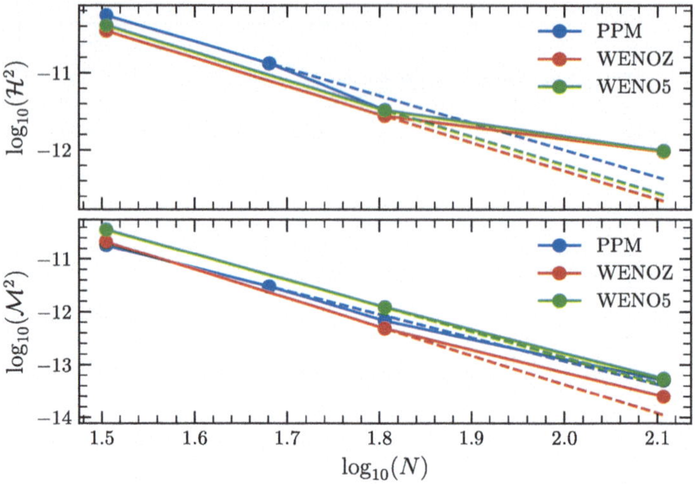

Fig. 4.8 Hamiltonian (top) and momentum (bottom) constraints for static star model A0 at $t = 2.96$ ms. Order of convergence in the Hamiltonian (momentum) constraint is 3.51(4.41), 3.65(5.46), 3.64(4.88) for PPM, WENOZ, WENO5 respectively. (Figure adapted from [45])

4.4.1.2 Long-Term Magnetic Field Dynamics in Static Neutron Star

We now present an application of `GR-Athena++` to the long term evolution of magnetic fields within NSs. The existence of a magnetic field configuration that is stable over long timescales within a NS is an open problem, with analytic and numerical studies showing the instability of purely poloidal fields on Alfven timescales, and suggesting the need for the development of a toroidal component to stabilise the field [29, 63, 99, 100, 141, 142, 154], with long term numerical relativity evolutions necessary to understand the rearrangement of the field profile and energy distribution [39, 88, 92, 112, 140].

We evolve an isolated star given by model A0, using WENOZ reconstruction, the grid set up of Sect. 4.4.1.1 and resolutions $N_M = 32, 64, 128$, with a purely poloidal field given by [96]:

$$(A_x, A_y, A_z) = (-yA_\varphi, xA_\varphi, 0), \tag{4.27a}$$

$$A_\varphi := A_b \max(p - 0.04\, p_{\max}, 0). \tag{4.27b}$$

where $p_{\max}$ is the maximum value of the pressure within the star.

The magnetic field is given by $B = \nabla \times A$, with A_b a free parameter scaling the overall strength of the field. Below we set this so that the maximum value of the magnetic field inside the star is 1.84×10^{16} G, approximately the magnetic field strength expected in magnetars.

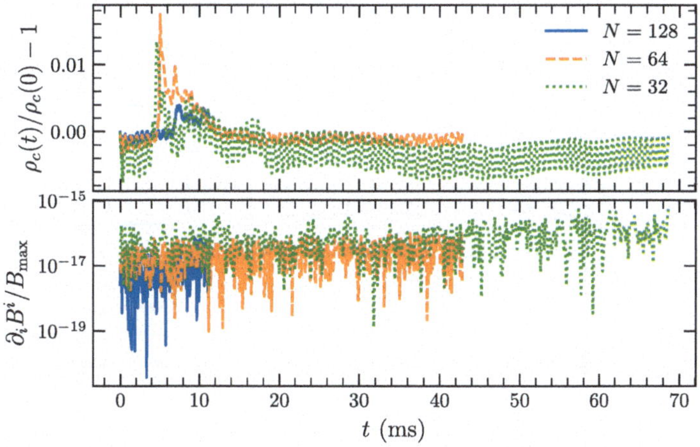

Fig. 4.9 Top: Evolution of central rest-mass density of the static star A0 model with superposed poloidal magnetic field, and WENOZ reconstruction. Bottom: $\partial_i B^i$ integrated over full computational domain Ω and normalised by $B_{\max} := \max_{\Omega} B$. (Figure adapted from [45])

We compare our results with our earlier simulations of a similar system, performed in the Cowling approximation in [140].

We first demonstrate the oscillations of the central density of the star in the upper panel of Fig. 4.9. In comparison to the case in the absence of magnetic fields (Fig. 4.5) we see larger oscillations, driven by the superposed B-field. We see the peak in oscillations corresponds to the saturation time of the toroidal field component growth. Throughout the evolution the constrained transport algorithm maintains the divergence free condition on the magnetic field with a relative violation of $\sim 10^{-16}$ shown in the lower panel of Fig. 4.9.

As expected we see a growth of a toroidal field component to stabilise the poloidal component, shown in the upper panel of Fig. 4.10. After $\sim$ 16 ms, the ratio of toroidal to total magnetic field energy $b^{\phi} b_{\phi} / b^{\mu} b_{\mu}$ grows to $\sim 10\%$, with the saturation of this growth delayed as resolution increases. This result is in line with the Cowling simulations demonstrated in Fig. 3 of [140].

We present the energy budget of the space-time in the lower panel of Fig. 4.10 with energies defined in Sect. 4.3.5.1. As the magnetic field profile rearranges it drives fluid motion leading to a growth in kinetic energy, with local maxima in the kinetic energy profile tracking the maxima within the toroidal field strength. In agreement with Fig. 5 of [140] we see a loss in overall magnetic field energy, though we see a much improved retention of internal energy, which we attribute to an improved atmosphere treatment, and the increased grid size in comparison to the previous simulations.

Through 3D visualisations (Fig. 4.11) of the magnetic field lines seeded on a circle in the equatorial plane of radius 5.91 km, we observe the onset of the "varicose" and "kink" instabilities of [141]. In a cylindrical system these correspond to, respectively,

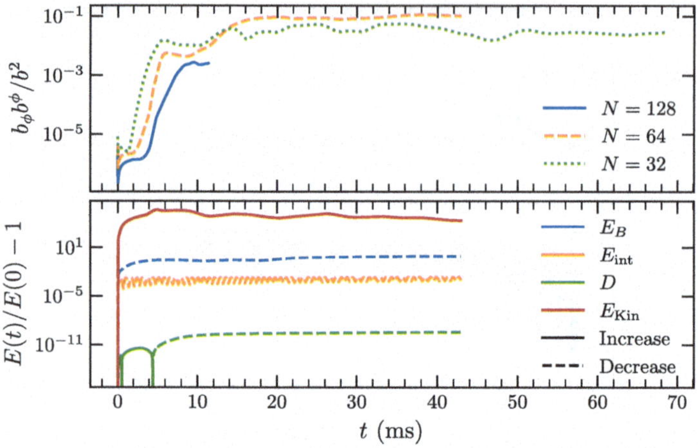

Fig. 4.10 Top: Growth of the toroidal component of magnetic field energy as a fraction of the total for model shown in Fig. 4.9. Bottom: relative change in magnetic energy, internal energy, mass and kinetic energy for $N_M = 64$. Solid lines denote a positive change in energy, dashed lines denote a negative change. (Figure adapted from [45])

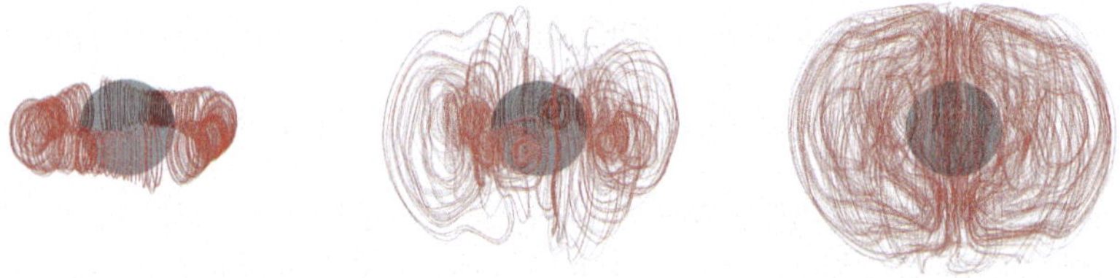

Fig. 4.11 Evolution of magnetic streamlines seeded in equatorial plane at radius 5.91 km. Gray isocontour shows surface of $\rho = 6.18 \times 10^{14}\,\mathrm{g\,cm^{-3}}$. Left: Varicose instability at $t = 4.93$ ms. Middle: Kink instability at $t = 12.8$ ms. Right: Late time non-linear field arrangement $t = 44.3$ ms. (Figure adapted from [45])

an $m = 0$ and $m = 1$ perturbation. The first of these breaks the rotational invariance of the fieldlines, with the initially constant cross-sectional area now varying as a function of the azimuthal angle around the star, with the latter displacing the fieldlines in the direction perpendicular to the gravitational field.

After 4.93 ms (left panel) the cross sectional area of the streamlines is no longer rotationally invariant, a sign of the onset of the varicose instability, with the kink instability disrupting streamlines orthogonal to the equatorial plane by 12.8 ms (middle panel), coinciding with the saturation of the growth of the toroidal field component. After 44.3 ms (right panel) we see an overall poloidal structure with clear toroidal contributions after the non linear dynamical evolution of the star.

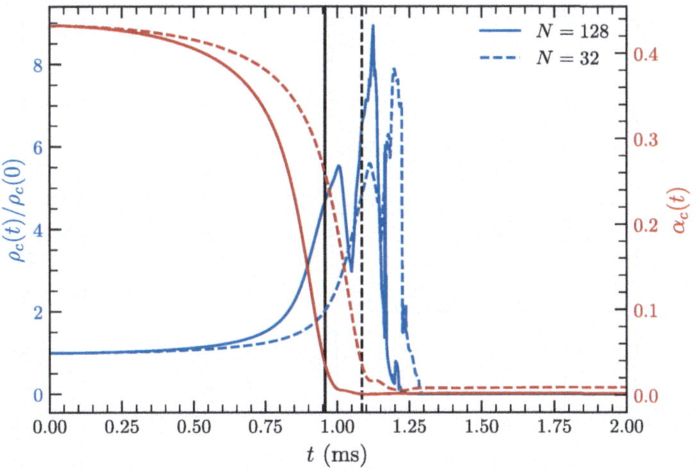

Fig. 4.12 Collapsing D4 model for simulations wth PPM reconstruction. Evolution of central rest-mass density and lapse of at different resolutions. Vertical black lines denote collapse time for each simulation. Oscillatory behaviour occurs within the horizon, after collapse. (Figure adapted from [45])

4.4.2 Gravitational Collapse of Rotating Neutron Star

The ability of `GR-Athena++` to handle gravitational collapse is confirmed, by simulating the unstable, uniformly rotating, equilibrium model D4 of [13, 52, 125]. This model is triggered to collapse by the introduction of a perturbation to the pressure field of 0.5%, following [52]. The initial data is calculated with the `RNS` code of [136] with a central density of $\rho_c = 1.924388 \times 10^{15}\,\mathrm{g\,cm^{-3}}$ and a polar-to-equatorial coordinate axis ratio of $r_p/r_e = 0.65$. This leads to a star with rotational frequency of 1276 Hz, baryon mass of $M_b = 2.0443\mathrm{M}_\odot$ and gravitational mass $M = 1.8604\mathrm{M}_\odot$.

Our grid consists of 8 refinement levels, 3 of which lie within the star radius, with the outer boundaries the same as for Sect. 4.4.1.1, without imposed grid symmetries. We perform simulations with a base level resolution $N_M = 32, 64, 128$ corresponding to finest grid spacings of 92.2, 46.1, 23.1 m respectively, using PPM reconstruction. The evolution successfully passes through gravitational collapse and horizon formation without the need for excision.

We first monitor the central density and lapse as indicators of the star collapse (Fig. 4.12). As the density and hence curvature increase at the centre of the star, the moving puncture gauge conditions used here [144], cause the lapse to collapse at the origin, and handle black hole formation.

Using an apparent horizon finder, employing the spectral fast-flow algorithm of [5, 76] we locate an apparent horizon (AH) after one ms, at which point we see losses of baryon mass from the grid (Fig. 4.13) as the determinant of the spatial metric becomes very large at the centre of collapse. Prior to this we see mass conservation with relative error 10^{-13} despite the presence of refinement boundaries within the star,

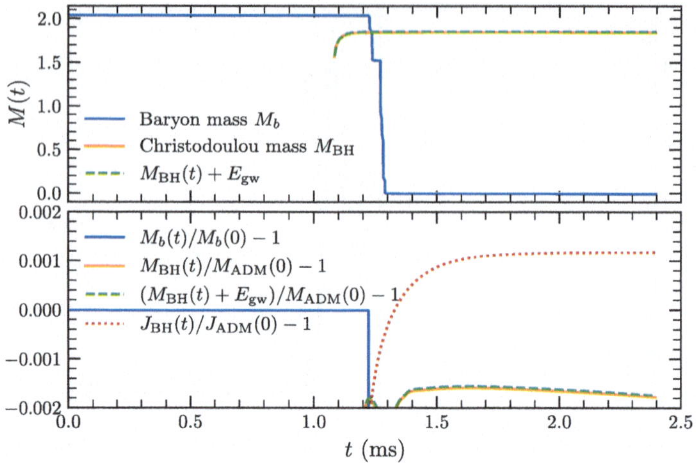

Fig. 4.13 Collapsing D4 model for simulations wth PPM reconstruction. Top: Evolution of baryon mass and black hole (Christodoulou) mass for $N_M = 32$. The dashed line shows the sum of the black hole mass and the gravitational-wave energy in the (2, 0) mode extracted at the coordinate sphere. Bottom: Baryon mass and angular momentum conservation and relative differences of Christodoulou mass with and without gravitational-wave energy added with respect to the initial ADM mass. (Figure adapted from [45])

providing a robust test of the flux correction of `Athena++`. From the AH shape we calculate the Christodoulou mass [37] of $M_{\rm BH} = 1.857{\rm M}_\odot$ and angular momentum $J_{\rm BH} = 1.884{\rm M}_\odot{}^2{\rm G/c}$ of the horizon. These values differ from the expected ADM mass and angular momentum of the space-time by a relative error of order 10^{-3} for the lowest resolution simulation $N_M = 32$, when the energy loss through GWs is included, which improves as a function of resolution. These results favourably compare to previous tests for this model in [52, 125].

Gravitational waveforms (Sect. 4.3.5.2) are extracted using geodesic spheres at radius $R = 221$ km, and the dominant $(\ell, m) = (2, 0)$ of the strain is computed integrating in time and adjusting for drift [12]. In Fig. 4.14 we demonstrate these as a function of the retarded time to the extraction spheres, given by $t - r_*$ with $r_* = r + 2M \log(r/2M - 1)$, and $r(R)$ the areal radius of the spheres of coordinate radius R (the isotropic Schwarzschild radius).

The waveform can be characterised by an initial burst of unphysical radiation after which we see the "precursor-burst-ringdown" behaviour expected for gravitational collapse. At $t - r_* \simeq 1$ ms for run $N_M = 128$ we see a local maximum in GW amplitude, corresponding to the formation of the AH, followed shortly after by the global maximum amplitude after collapse, with the quasi-normal mode oscillations of a perturbed Kerr BH forming the final part of the wave signal.

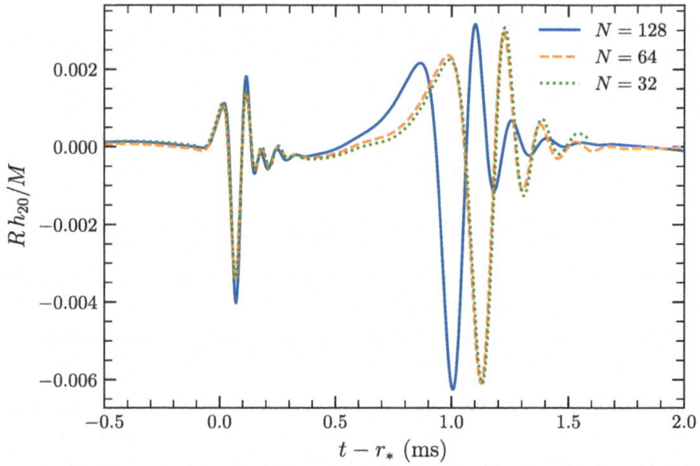

Fig. 4.14 Dominant (2, 0) multipole of the gravitational strain for the collapsing D4 model for simulations with PPM reconstruction. At $N_M = 32$ the resolution is too low to observe convergent behaviour, though at higher resolutions we do see convergence in the waveforms at approximate first order. Note usage of geometric units on the vertical axis. (Figure adapted from [45])

4.4.3 Binary Neutron Star Space-Times

The evolution of BNS space-times, and the extraction of highly accurate gravitational waveforms is a key target for `GR-Athena++`. We demonstrate this capability through direct, cross-code comparison of BNS evolution, without magnetic fields (GRHD), against the `BAM` code [31, 32, 143]. Self consistency of `GR-Athena++` is established through self-convergence of gravitational waveforms, together with good mass conservation for both GRHD and GRMHD simulations.

The initial data evolved is irrotational and constraint satisfying, as generated by the `Lorene` library [75], describing a quasi-circular equal-mass merger with baryon mass $M_b = 3.2500\mathrm{M}_\odot$ and gravitational mass $M = 3.0297\mathrm{M}_\odot$ at an initial separation of 45 km. The ADM mass of the binary is $M_{\mathrm{ADM}} = 2.9984\mathrm{M}_\odot$, the angular momentum $J_{\mathrm{ADM}} = 8.83542{\mathrm{M}_\odot}^2\mathrm{G/c}$, and the initial orbital frequency is $f_0 \simeq 294$ Hz.

An ideal gas EOS is used with $K = 124$ and, in the case in which magnetic fields are present they are initialised as in the single star case with the vector potential in Eq. (4.27b), with A_b chosen to give a maximum value of $1.77 \times 10^{15} G$.

For runs with magnetic fields the grid outer boundaries are set at [$\pm$2268, $\pm$2268, $\pm$2268] km with no symmetry imposed. A refined static `Mesh` is defined with 7 levels of refinement, with the innermost grid covering both stars, from [$\pm$37, $\pm$37, $\pm$37] km. For runs without magnetic fields the grid is identical except for the imposition of bitant symmetry across the plane $z = 0$. The `Mesh` parameters are $N_M = 64, 96, 128$, corresponding to a grid spacing on the finest refinement level of (554, 369, 277) m respectively. Simulations are performed with WENOZ reconstruction, and hydrodynamical variables are excised within the apparent horizon for runs with GRMHD.

4.4.3.1 Benchmark Against BAM

From the beginning of the simulation with GRHD, the binary inspiral lasts for ~ 3 orbits before merging to form a massive remnant star that undergoes gravitational collapse at $\sim$19 ms for the lowest resolution simulation. In the case of GRMHD we see similar behaviour in the inspiral phase and visualise the the pre-merger, merger and pre-collapse phases in Fig. 4.15.

We now directly compare to evolutions in BAM. GR-Athena++ and BAM employ slightly different hydrodynamical evolution schemes. For instance, the latter reconstructs the primitives using the fluid internal energy, rather than pressure as in GR-Athena++. BAM also lacks the intergrid operators of GR-Athena++ (Sect. 4.3.3.1), and uses a box-in-box style refinement strategy with Berger-Oliger time sub-cycling. For the sake of comparison, we perform two BAM simulations with differing hydrodynamical schemes, one using the LLF flux and WENOZ reconstruction similar to GR-Athena++, and one using the entropy flux-limited (EFL) scheme of [54], which is a fifth-order accurate scheme. The BAM grid uses six refinement levels, two of which move and contain the stars. The finest refinement level entirely covers the star at a resolution of $\sim$461 m, that is comparable to the GR-Athena++ grid configuration with $N_M = 64$.

The conservation of baryon mass is depicted in Fig. 4.16 for the GRHD and GRMHD evolutions at different resolutions. The maximum violation of the relative conservation $\sim 10^{-6}$ occurs at the lowest resolution. This error, while larger than for the single star, is sufficiently low to accurately study ejecta properties, and is consistent with other state-of-the-art Eulerian codes at the considered resolutions, e.g. [119]. It was found that through varying atmosphere parameter settings those leading to a mass increase provide the best mass conservation.

To compare evolutions in a gauge invariant manner, curves of binding energy against angular momentum [23, 48], are considered in Fig. 4.17. Close agreement up to merger at relatively low resolutions is found, with differences more pronounced post-merger. We note that the differences between GR-Athena++ and BAM are consistent with the internal differences between evolution schemes in BAM.

In Fig. 4.18, direct comparison of h_{22} between the three runs is made. The top panel compares the real part of the gravitational wave strain, where we see amplitudes closely matching, and the main difference coming in the phasing. The earlier merger time[12] of GR-Athena++ suggests that this run is slightly more dissipative than its BAM counterpart. Collapse time of runs is consistent to within 2.5 ms.

The middle panel demonstrates the instantaneous gravitational wave frequency, which closely match at the moment of merger, $\omega_{22} \simeq 1340$ Hz for GR-Athena++

[12] Merger time is signalled by the amplitude peak of h_{22}.

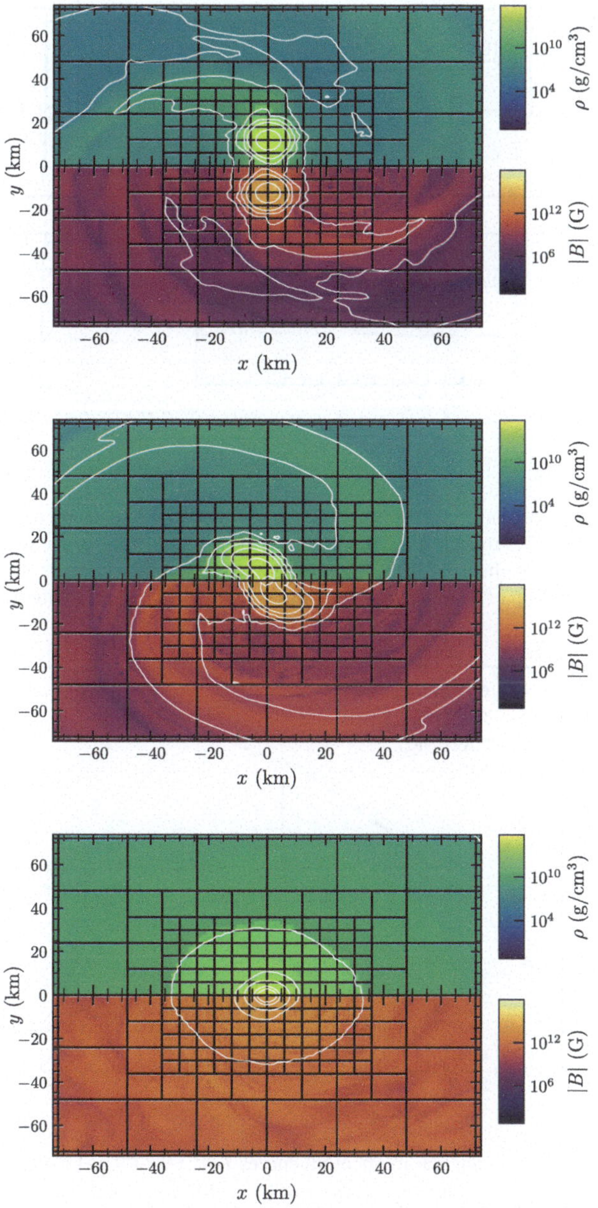

Fig. 4.15 Snapshots of rest-mass density (upper half) and magnetic field strength (lower half) in the orbital plane. (Top, middle, bottom) correspond to respectively the final orbit ($t = 5.96$ms), the moment of merger ($t = 7.83$ms), and a late stage shortly before gravitational collapse $t = 16$ms. White contours are lines of constant density at values $\rho = (6.18 \times 10^{14}, 3.09 \times 10^{14}, 6.18 \times 10^{13}, 6.18 \times 10^{12}, 6.18 \times 10^{10}, 6.18 \times 10^{8}, 6.18 \times 10^{6}, 6.18 \times 10^{4})\,\mathrm{g\,cm^{-3}}$. (Figure adapted from [45])

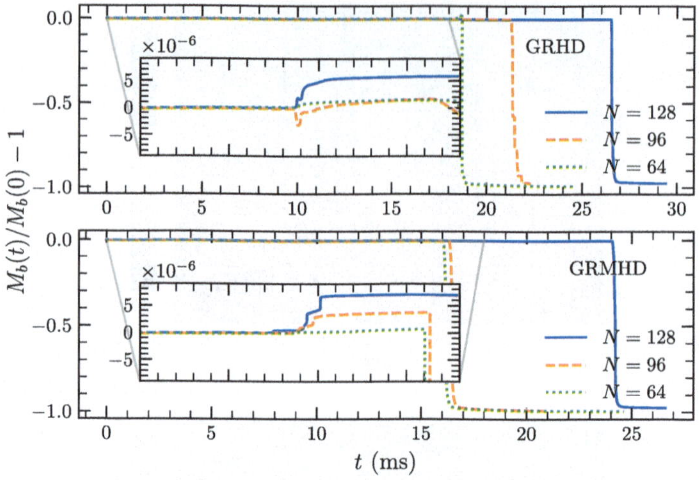

Fig. 4.16 Baryon mass conservation for GRHD and GRMHD evolutions of the BNS. To the moment of merger, $t \simeq 7.5$ ms, the maximum relative violation is of the order 10^{-7}. After merger, the violation increases to the 10^{-6} level. Note the mass typically increases due to atmosphere accretion. The abrupt drop of the mass at late times is due to black hole formation. (Figure adapted from [45])

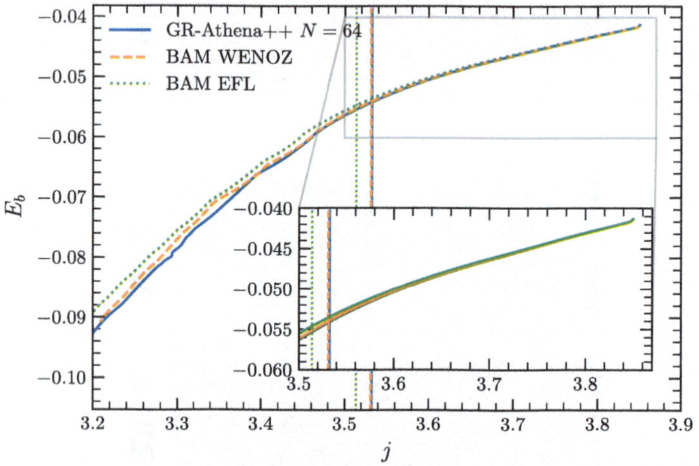

Fig. 4.17 Binding energy *vs* angular momentum curves $E_b(j)$ for GR-Athena++ and BAM simulations. These quantities are computed following [23, 48]; $E_b := (M_{\mathrm{ADM}} - E_{\mathrm{GW}})\nu/(M-1)$, $j := (J_{\mathrm{ADM}} - J_{\mathrm{GW}})/(M^2\nu)$, with ν the symmetric mass ratio, $\nu := q/(1+q)^2$, $q > 1$ the mass ratio and quantities suffixed with GW the contributions radiated away in gravitational waves. The moment of merger of each dataset is shown as vertical line. The inset captures energetics during the orbital phase, prior to the moment of merger. (Figure adapted from [45])

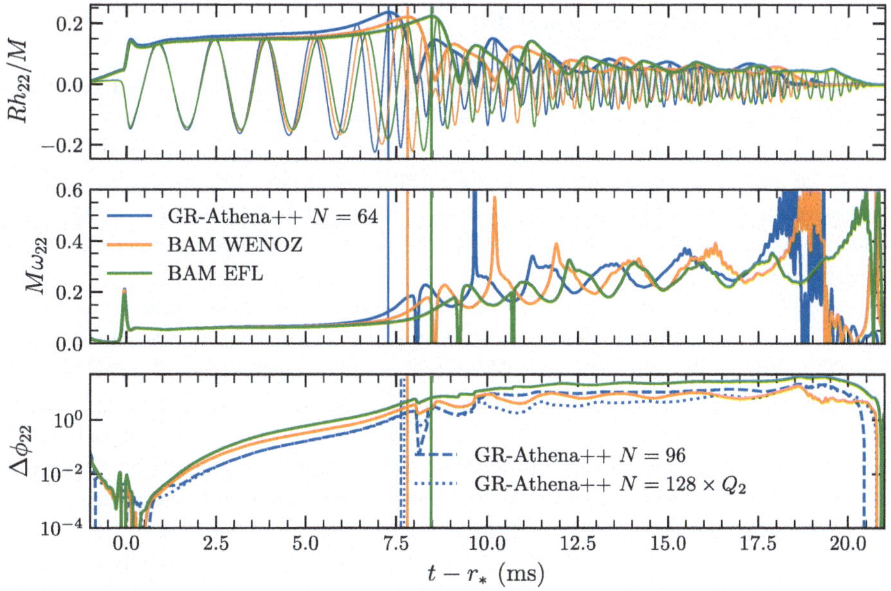

Fig. 4.18 Waveform analysis and comparison with BAM. (Top): Real part and amplitude of strain mode h_{22}. We show: GR-Athena++ data with WENOZ reconstruction and $N_M = 64$; BAM data with WENOZ reconstruction, and the EFL scheme at a similar resolution. (Middle): Instantaneous frequency of the waves for the three datasets. Note that in the top and middle panels the vertical axis is in geometric units. (Bottom): Phase differences between the GR-Athena++ data at $N_M = 64$ and: $N_M = 96$ (dashed); $N_M = 128$ (dotted lines); and BAM data. GR-Athena++ phase difference data at $N_M = 128$ is rescaled assuming $2nd$ convergence, matching with the $N_M = 96$ data up to merger. The moment of merger of each dataset is indicated with a vertical line. (Figure adapted from [45])

and BAM WENOZ and $\omega_{22} \simeq 1360$ Hz for BAM EFL. Similar consistency between runs is found in the frequency evolution of the post-merger remnant up to collapse, after which the quasi-normal modes of the remnant BH are insufficiently resolved, but are compatible with the fundamental mode frequency $\nu_{\mathrm{QNM}} \sim 6.5$ kHz ($M\omega_{22} \simeq 0.57$ in geometric units).

In the bottom panel of Fig. 4.18, phase differences of the GR-Athena++ $N_M = 64$ run with respect to the BAM runs and the GR-Athena++ runs at higher resolutions are quantified. The difference between the $N_M = 64$ and $N_M = 128$ runs is rescaled by Q_2 to compare to the difference between the $N_M = 64$ and $N_M = 96$ run assuming $2nd$ order convergence.[13] These curves closely match up to merger suggesting $2nd$ order convergence during the inspiral phase. Further, the difference to the BAM runs significantly reduces at higher resolutions.

---

[13] Here $Q_n := (\delta x_c^n - \delta x_m^n)/(\delta x_c^n - \delta x_f^n)$ where δx_c, δx_m, and δx_f are coarse, medium, and fine grid-spacings.

4.4.3.2 Kelvin-Helmholtz Instability

The Kelvin-Helmholtz instability (KHI) arises over a shearing interface in a fluid with a discontinuous velocity profile. This interface is unstable to perturbation, breaking down and forming small scale vortex like structures. It is suggested [117] that this should be expected during the merger of a BNS, as the two stars first make contact they shear past each other, creating an unstable interface. In the presence of magnetic fields numerical studies [4, 83, 86, 108] support that KHI leads to magnetic field amplification.

As small scale vortices must be resolved, efficiency considerations motivate the use of targeted AMR. Here, two AMR criteria are demonstrated to capture B-amplification. We perform two SMR based simulations denoted SMR64 and SMR128, with the same grid configuration as those in Sect. 4.4.3 utilising `Mesh` sampling $N_M = 64$ and $N_M = 128$. The AMRB and AMRσ runs are initialised from SMR64 after $t = 4.93$ms, allowing an extra refinement level to be generated according to the appropriate criterion, bringing the maximum refinement up to that of run SMR128. Grid resolution is bounded below by that of SMR64.

In the case of AMRB, a `MeshBlock` (MB) is refined if its maximum value of $|B|$ exceeds 8.35×10^{13}G and derefined if lower than this value. For AMRσ, a MB is refined if $\max_{\Omega_{i\in Z}} \sigma > 4.50 \times 10^9$cm/s, and derefined if $\max_{\Omega_{i\in Z}} \sigma < 2.25 \times 10^9$cm/s, where $\sigma := \sqrt{((\Delta_x \tilde{u}^y)^2 + (\Delta_y \tilde{u}^x)^2)}$ represents the shear of the fluid velocity and Δ_i is the undivided difference operator in the ith direction.

The grid structure of run AMRσ at $t = 7.88$ ms, shortly before merger, is shown in Fig. 4.19, where small MBs generated at the star interface can be seen.

The performance of an AMR criterion is judged on its ability to capture B-amplification, the #MB generated (as a measure of efficiency), and the violation of the divergence free constraint of the magnetic field introduced through MB creation and destruction operations, demonstrated in the upper, middle and lower panels of Fig. 4.20 respectively. In the upper panel we see that in SMR128 the energy is amplified by a factor of over 25 after the moment of merger, at $t \sim 10.2$ ms. The AMR simulations capture less of this amplification, with an AMRB amplification factor of 7.85, while AMRσ only captures a factor of 3.26. In [83], at considerably higher resolutions of 17.5m a much larger amplification, of 6 orders of magnitude, from a much weaker initial magnetic field profile, of initial strength order 10^{13}G is observed. Our expectation is that through addition of further levels of mesh refinement we should be able to capture such larger amplifications.

In the middle panel, we see that up to the moment of the peak magnetic field energy the run AMRB generates less than 10% more MBs than SMR64, a factor of 1.79 fewer than SMR128 and a factor 1.91 fewer at the moment of merger. This supports the idea that targeted use of AMR can resolve magnetic field amplification. AMRσ clearly however generates more MB than SMR128. This is due to MBs being generated at the star surface and in the region of unphysical ejecta from the merger. This criterion performs worse at capturing B-amplification, but shows improved mass conservation, which may have applications in ejecta tracking for simulations featuring physical ejecta.

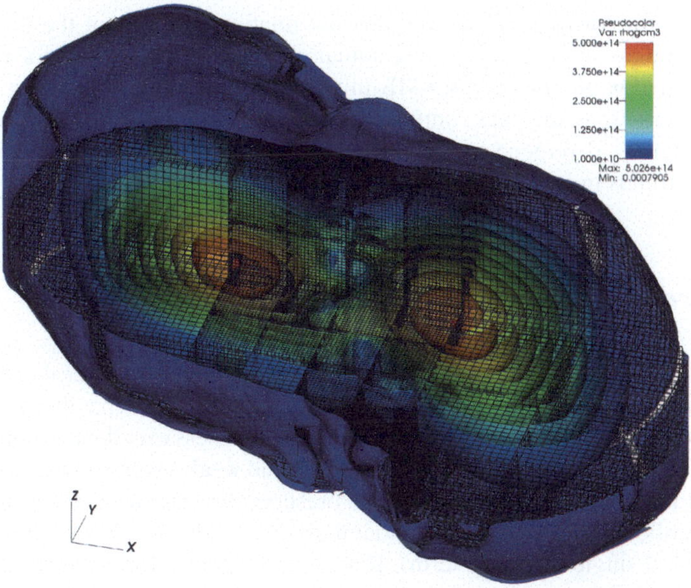

Fig. 4.19 `Mesh` structure at $t = 7.88$ms shortly before merger in the AMRσ run. Isocontours of density are shown for $\rho > 10^{11}\,\mathrm{g\,cm^{-3}}$. To reveal internal structure the $z > 0,\ y < 0$ quadrant is suppressed. (Figure adapted from [45])

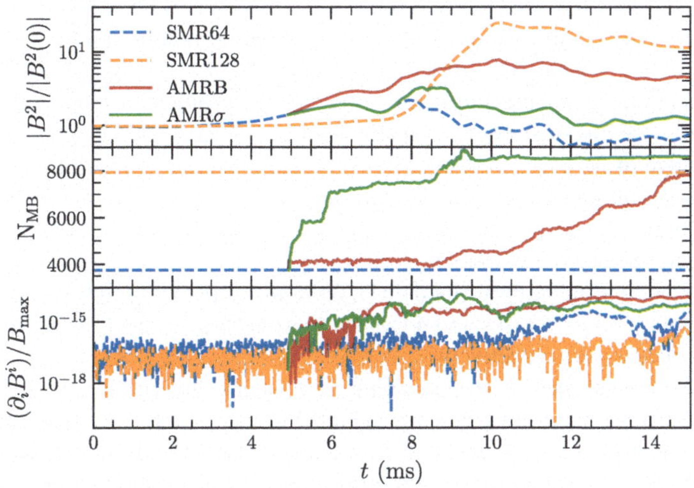

Fig. 4.20 Top: Amplification of total magnetic field strength $|B^2|$ in BNS merger for SMR, AMRB, and AMRσ refinement. Middle: Total #MB present in each run. Bottom: $\partial_i B^i$ integrated over full computational domain Ω and normalised by $B_{\max} := \max_{\Omega} B$. (Figure adapted from [45])

Finally in the lower panel we see that the relative violation of the divergence-free constraint, as integrated over the domain, remains below 10^{-14} for the AMR runs. Note that, while the overall #MB in the middle panel appears approximately constant, oscillations indicate continual destruction and creation of `MeshBlock` objects. Crucially, divergence preserving interpolation [145], ensures an absence of any secular trends in $\partial_i B^i$ post AMR activation.

4.4.4 Scaling Tests

The (binary, quad, oct)-tree based grid (Sect. 4.2.1), adaptive mesh refinement (Sect. 4.2.1.1), and parallel task-based execution featuring hybrid parallelism through `MPI` and `OpenMP` (Sect. 4.3.6) lead `Athena++` to possess excellent scaling properties [138]. Here we confirm that, based on strong and weak scaling tests,[14] extensions introduced in `GR-Athena++` continue to preserve this. Our tests are performed in the context of evolution, with and without magnetic fields, of: BNS inspiral through to merger, and single NS (long term). Robust performance for demanding production grids, attained over multiple machines that involve a variety of architectures, is crucial for opening up the feasibility of very high resolution studies of these and related problems.

Scaling tests are performed on three clusters with different architectures: HLRS-HAWK[15]; SUPERMUC-NG[16]; and Frontera.[17] The physical and grid configurations employed here match those described for the relevant problem in prior sections.

Strong scaling tests are performed by fixing the `Mesh` sampling N_M of the unrefined grid and increasing the number of cores that the problem is solved on. For each choice of N_M the physical extent of the innermost refined region is tuned to achieve a similar range of `MeshBlock`/core ratios across all series of resolutions.

The strong scaling efficiency is defined as $\eta_S := 1 - \frac{t - t_{\mathrm{ideal}}}{t_{\mathrm{ideal}}}$ where t is the elapsed wallclock time and t_{ideal} is the time expected for perfect strong scaling, halving exactly when resources are doubled.

Strong scaling tests are performed on HAWK and SuperMUC, where a given node is saturated with 32 `MPI` tasks and 4 `OpenMP` threads/task, and 12 `MPI` tasks and 4 `OpenMP` threads/task respectively.

In Fig. 4.21 we show the strong scaling performance for BNS with GRMHD. In the top panel, demonstrating efficiency against number of cores, we find efficiencies in excess of 80% up to $\sim 10^5$ cores (gray and teal lines) for both SuperMUC (dashed lines) and HAWK (solid lines). An important aspect of maintaining high efficiencies is sufficiently saturating computational load [49]. This can be seen in Fig. 4.21 (lower) where the efficiency on SuperMUC is $\gtrsim 90\%$ as long as `MeshBlock/core` $\gtrsim 6$,

---

[14] In all the scaling tests presented $N_B = 16$, and we evolve for 20 time-steps.

[15] 2 AMD EPYC 7742 CPUs/node, totalling 128 cores/node.

[16] 2 Intel Skylake Xeon Platinum 8174 CPUs/node, totalling, 48 cores/node.

[17] 2 Intel 8280 Cascade Lake CPUs/node, totalling 56 cores/node.

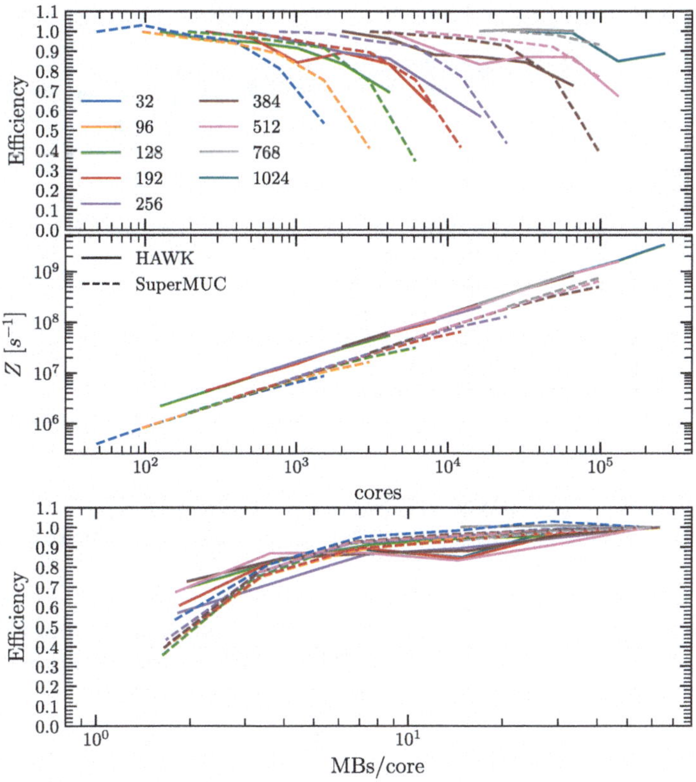

Fig. 4.21 Strong scaling efficiency η_S for the binary neutron star configuration with MHD. Results for HAWK and SuperMUC are included, up to $\mathcal{O}(10^5)$ cores. Top: efficiency versus number of cores. Middle: zone cycles per second (Z) versus number of cores. Bottom: efficiency in terms of the `MeshBlock` to core ratio. (Figure adapted from [45])

and the drop in efficiency on HAWK corresponding to this ratio dropping below 4. The differences between machines are attributed to varied machine architectures. This causes a difference in the raw performance in terms of zone cycles per second (Z) (middle panel), where we observe runs on HAWK executing $\gtrsim 2$ faster than on SuperMUC. Comparable results found for simulations with magnetic fields disabled, and single NS configurations, have been omitted for brevity.

For weak scaling the executed zone cycles per second, Z, are measured. The expectation is that for perfect scaling, Z should double every time the computational load and resources are concurrently doubled. The weak efficiency measure $\eta_W := 1 - \frac{Z - Z_{\rm ideal}}{Z_{\rm ideal}}$ captures this.

In Fig. 4.22 we demonstrate this weak scaling performance on the full range of machines and problems discussed above. Observe that scaling is maintained up to $\sim 5 \times 10^5$ CPU cores for the single star test at 89% efficiency on Frontera, with efficiencies dropping to, at worst 72% on HAWK. For binary tests we see an efficiency

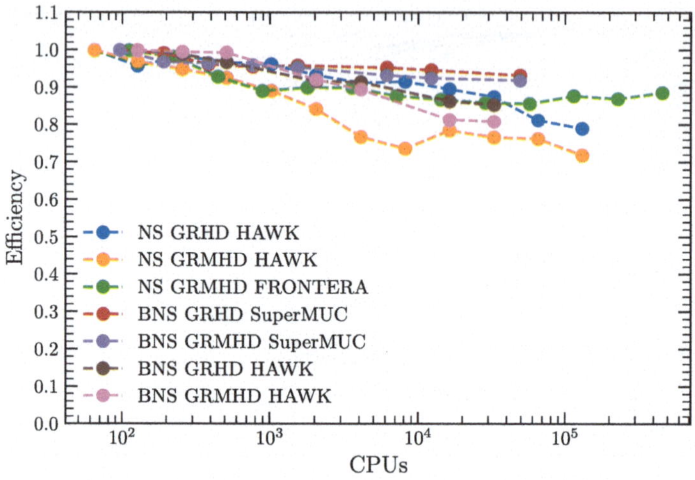

Fig. 4.22 Weak scaling η_W for SMR grids, over different target machines, for NS and BNS configurations with (GRMHD), and without (GRHD) magnetic fields. (Figure adapted from [45])

$\gtrsim 90\%$ up to $\sim 5 \times 10^4$ cores on SuperMUC-NG, with slightly lower efficiencies seen on HAWK for the same problem. This is also consistent with our previous vacuum sector tests [49] and those for stationary space-times [138].

4.5 Summary and Outlook

Our overview of `GR-Athena++` for GRMHD simulations of astrophysical flows on dynamical space-times described several novel aspects we have introduced with respect to the original `Athena++`. In particular our treatment of: Z4c coupling to GRMHD; equation of state; discretizations over differing grids including reconstruction; the constrained transport algorithm; conservative-to-primitive variable inversion; extraction of gravitational waves utilizing geodesic grids. When operation of these features is interwoven during GRMHD simulation, robust performance characteristics on multiple machines was demonstrated. Evolution on production grids in the absence of symmetries shows efficient use of exascale HPC architecture where: strong scaling efficiency in excess of 80% up to 10^5 CPU cores, and weak scaling efficiency of 89% for $\sim 5 \times 10^5$ CPU cores is demonstrated.

Representative of the modeling challenges posed by astrophysical applications, benchmark problems were utilized to establish simulation quality of `GR-Athena++` with focus on the absence of any simplifying symmetry reductions. This involved evolution of: isolated equilibrium NS over the long-term; unstable, rotating NS through gravitational collapse and black hole remnant; BNS inspiral through merger. Gravitational waveforms are shown to be convergent as resolution is increased, and consistent in cross-code comparison with BAM.

Anticipating full-scale simulations that seek to start tackling questions surrounding the open problems outlined in the introduction we presented novel simulations involving magnetic field instabilities. Long-term evolution of isolated magnetised NS with initially poloidal fields up to 68.7 ms extend our previous simulations [140] to include a dynamical space-time. Similar results for the growth of a toroidal field component were found, saturating at approximately 10% of the total magnetic field energy. In contrast, with the current GRMHD treatment of GR-Athena++, we find superior conservation of the internal energy of the NS, and total relative violations of the magnetic field divergence-free condition of the order of machine round-off. Similarly, we show that the KHI may be efficiently resolved through our AMR infrastructure. Indeed even initial low resolutions yield an amplification of a factor ~ 8 with a single, additional level of refinement whilst utilizing approximately half the number of MeshBlock objects at the moment of merger when contrasted with a comparable static mesh refinement approach.

Active work on GR-Athena++ is underway so as to incorporate high-order schemes [21, 54, 120] essential for further improving waveform convergence and quality. The neutrino transport scheme developed by [118] is presently being ported together with an improved treatment of weak reactions and reaction rates. In future we also plan to couple the recently developed radiation solvers of [24, 152]. Once these developments are mature, it is our intention to make GR-Athena++ publicly available.

Acknowledgements BD acknowledges funding from the EU H2020 under ERC Starting Grant, no. BinGraSp-714626, and from the EU Horizon under ERC Consolidator Grant, no. InspiReM-101043372.

References

1. B.P. Abbott et al., Gravitational waves and gamma-rays from a binary neutron star merger: GW170817 and GRB 170817A. Astrophys. J. **848**(2), L13 (2017). 1710.05834
2. B.P. Abbott et al., GW170817: observation of gravitational waves from a binary neutron star inspiral. Phys. Rev. Lett. **119**(16), 161101 (2017). 1710.05832
3. B.P. Abbott et al., Multi-messenger observations of a binary neutron star merger. Astrophys. J. **848**(2), L12 (2017). 1710.05833
4. R. Aguilera-Miret, C. Palenzuela, F. Carrasco, D. Viganò, The role of turbulence and winding in the development of large-scale, strong magnetic fields in long-lived remnants of binary neutron star mergers. 2307.04837
5. M. Alcubierre, S. Brandt, B. Bruegmann, C. Gundlach, J. Masso, E. Seidel, P. Walker, Test beds and applications for apparent horizon finders in numerical relativity. Class. Quant. Grav. **17**, 2159–2190 (2000). ((gr-qc/9809004))
6. M. Alcubierre, B. Brügmann, P. Diener, M. Koppitz, D. Pollney et al., Gauge conditions for long term numerical black hole evolutions without excision. Phys. Rev. D **67**, 084023 (2003). ((gr-qc/0206072))

7. D. Alic, C. Bona-Casas, C. Bona, L. Rezzolla, C. Palenzuela, Conformal and covariant formulation of the Z4 system with constraint-violation damping. Phys. Rev. D **85**, 064040 (2012). ((1106.2254))
8. D. Alic, W. Kastaun, L. Rezzolla, Constraint damping of the conformal and covariant formulation of the Z4 system in simulations of binary neutron stars. Phys. Rev. D **88**(6), 064049 (2013). ((1307.7391))
9. M. Anderson, E. Hirschmann, S.L. Liebling, D. Neilsen, Relativistic MHD with adaptive mesh refinement. Class. Quant. Grav. **23**, 6503–6524 (2006). ((gr-qc/0605102))
10. I. Arcavi et al., Optical emission from a kilonova following a gravitational-wave-detected neutron-star merger. Nature **551**, 64 (2017). 1710.05843
11. R.L. Arnowitt, S. Deser, C.W. Misner, Dynamical structure and definition of energy in general relativity. Phys. Rev. **116**, 1322–1330 (1959)
12. L. Baiotti, S. Bernuzzi, G. Corvino, R. De Pietri, A. Nagar, Gravitational-wave extraction from neutron stars oscillations: comparing linear and nonlinear techniques. Phys. Rev. D **79**, 024002 (2009). ((0808.4002))
13. L. Baiotti, I. Hawke, P.J. Montero, F. Loffler, L. Rezzolla et al., Three-dimensional relativistic simulations of rotating neutron star collapse to a Kerr black hole. Phys. Rev. D **71**, 024035 (2005). ((gr-qc/0403029))
14. J.G. Baker, J. Centrella, D.-I. Choi, M. Koppitz, J. van Meter, Gravitational wave extraction from an inspiraling configuration of merging black holes. Phys. Rev. Lett. **96**, 111102 (2006). ((gr-qc/0511103))
15. S.A. Balbus, J.F. Hawley, A powerful local shear instability in weakly magnetized disks. I. Linear Analysis. Astrophys. J. **376**, 214 (1991)
16. F. Banyuls, J.A. Font, J.M.A. Ibanez, J.M.A. Marti, J.A. Miralles, Numerical 3+1 general relativistic hydrodynamics: a local characteristic approach. Astrophys. J. **476**, 221 (1997)
17. T.W. Baumgarte, S.L. Shapiro, On the numerical integration of Einstein's field equations. Phys. Rev. D **59**, 024007 (1999). ((gr-qc/9810065))
18. K. Beckwith, J.M. Stone, A second order Godunov method for multidimensional relativistic magnetohydrodynamics. Astrophys. J. Suppl. **193**, 6 (2011). ((1101.3573))
19. M.J. Berger, P. Colella, Local adaptive mesh refinement for shock hydrodynamics. J. Comput. Phys. **82**, 64–84 (1989)
20. M.J. Berger, J. Oliger, Adaptive mesh refinement for hyperbolic partial differential equations. J. Comput. Phys. **53**, 484 (1984)
21. S. Bernuzzi, T. Dietrich, Gravitational waveforms from binary neutron star mergers with high-order weighted-essentially-nonoscillatory schemes in numerical relativity. Phys. Rev. **D94**(6), 064062 (2016). 1604.07999
22. S. Bernuzzi, D. Hilditch, Constraint violation in free evolution schemes: comparing BSSNOK with a conformal decomposition of Z4. Phys. Rev. D **81**, 084003 (2010). ((0912.2920))
23. S. Bernuzzi, A. Nagar, M. Thierfelder, B. Brügmann, Tidal effects in binary neutron star coalescence. Phys. Rev. D **86**, 044030 (2012). ((1205.3403))
24. M.K. Bhattacharyya, D. Radice, A finite element method for angular discretization of the radiation transport equation on spherical geodesic grids. 2212.01409
25. S.I. Blinnikov, I.D. Novikov, T.V. Perevodchikova, A.G. Polnarev, Exploding neutron stars in close binaries. Sov. Astron. Lett. **10**, 177–179 (1984). 1808.05287
26. C. Bona, T. Ledvinka, C. Palenzuela, M. Zacek, General-covariant evolution formalism for numerical relativity. Phys. Rev. D **67**, 104005 (2003). ((gr-qc/0302083))
27. C. Bona, J. Massó, E. Seidel, J. Stela, New formalism for numerical relativity. Phys. Rev. Lett. **75**, 600–603 (1995). ((gr-qc/9412071))
28. R. Borges, M. Carmona, B. Costa, W.S. Don, An improved weighted essentially non-oscillatory scheme for hyperbolic conservation laws. J. Comput. Phys. **227**(6), 3191–3211 (2008)
29. J. Braithwaite, H.C. Spruit, Evolution of the magnetic field in magnetars. Astron. Astrophys. **450**, 1097 (2006). ((astro-ph/0510287))

30. S. Brandt, B. Brügmann, A simple construction of initial data for multiple black holes. Phys. Rev. Lett. **78**, 3606–3609 (1997). ((gr-qc/9703066))
31. B. Brügmann, Adaptive mesh and geodesically sliced Schwarzschild spacetime in 3+1 dimensions. Phys. Rev. D **54**, 7361–7372 (1996). ((gr-qc/9608050))
32. B. Brügmann, J.A. Gonzalez, M. Hannam, S. Husa, U. Sperhake et al., Calibration of moving puncture simulations. Phys. Rev. D **77**, 024027 (2008). ((gr-qc/0610128))
33. C. Burstedde, J. Holke, T. Isaac, On the number of face-connected components of Morton-type space-filling curves. Found. Comput. Math. **19**(4), 843–868 (2019)
34. M. Campanelli, C.O. Lousto, P. Marronetti, Y. Zlochower, Accurate evolutions of orbiting black-hole binaries without excision. Phys. Rev. Lett. **96**, 111101 (2006). ((gr-qc/0511048))
35. Z. Cao, D. Hilditch, Numerical stability of the Z4c formulation of general relativity. Phys. Rev. D **85**, 124032 (2012). ((1111.2177))
36. P.C.-K. Cheong, A.T.-L. Lam, H.H.-Y. Ng, T.G.F. Li, Gmunu: paralleled, grid-adaptive, general-relativistic magnetohydrodynamics in curvilinear geometries in dynamical space–times. Mon. Not. Roy. Astron. Soc. **508**(2), 2279–2301 (2021). 2012.07322
37. D. Christodoulou, Reversible and irreversible transformations in black hole physics. Phys. Rev. Lett. **25**, 1596–1597 (1970)
38. R. Ciolfi, Short gamma-ray burst central engines. Int. J. Mod. Phys. D **27**(13), 1842004 (2018). 1804.03684
39. R. Ciolfi, S.K. Lander, G.M. Manca, L. Rezzolla, Instability-driven evolution of poloidal magnetic fields in relativistic stars. Astrophys. J. **736**, L6 (2011). ((1105.3971))
40. R. Ciolfi, L. Rezzolla, Twisted-torus configurations with large toroidal magnetic fields in relativistic stars. Mon. Not. Roy. Astron. Soc. **435**, L43–L47 (2013). ((1306.2803))
41. F. Cipolletta, J.V. Kalinani, B. Giacomazzo, B., and Ciolfi, R. Spritz: a new fully general-relativistic magnetohydrodynamic code. Class. Quant. Grav. 37, 13 (2020), 135010, 1912.04794
42. P. Colella, M.R. Dorr, J.A.F. Hittinger, D.F. Martin, High-order, finite-volume methods in mapped coordinates. J. Comput. Phys. **230**(8), 2952–2976 (2011)
43. L. Combi, D.M. Siegel, GRMHD simulations of neutron-star mergers with weak interactions: r-process nucleosynthesis and electromagnetic signatures of dynamical ejecta. Astrophys. J. **944**(1), 28 (2023). 2206.03618
44. L. Combi, D.M. Siegel, Jets from neutron-star merger remnants and massive blue kilonovae. Phys. Rev. Lett. **131**, 23, 231402 (2023). 2303.12284
45. W. Cook, B. Daszuta, J. Fields, P. Hammond, S. Albanesi, F. Zappa, S. Bernuzzi, D. Radice, GR-Athena++: general-relativistic magnetohydrodynamics simulations of neutron star spacetimes. 2311.04989
46. D.A. Coulter, et al., Swope Supernova Survey 2017a (SSS17a), the optical counterpart to a gravitational wave source. Science (2017), 1710.05452. [Science358,1556(2017)]
47. S. Curtis, P. Mösta, Z. Wu, D. Radice, L. Roberts, G. Ricigliano, A. Perego r-process nucleosynthesis and kilonovae from hypermassive neutron star post-merger remnants. Mon. Not. Roy. Astron. Soc. **518**(4), 5313–5322 (2022). 2112.00772
48. T. Damour, A. Nagar, D. Pollney, C. Reisswig, Energy versus angular momentum in black hole binaries. Phys. Rev. Lett. **108**, 131101 (2012). ((1110.2938))
49. B. Daszuta, F. Zappa, W. Cook, D. Radice, S. Bernuzzi, V. Morozova, GR-Athena++: puncture evolutions on vertex-centered oct-tree adaptive mesh refinement. Astrophys. J. Supp. **257**(2), 25 (2021). 2101.08289
50. S. de Haas, P. Bosch, Mösta, S. Curtis, N. Schut, Magnetic field effects on nucleosynthesis and kilonovae from neutron star merger remnants. 2208.05330
51. A. Dedner, F. Kemm, D. Kröner, C.-D. Munz, T. Schnitzer, M. Wesenberg, Hyperbolic divergence cleaning for the MHD equations. J. Comput. Phys. **175**, 645–673 (2002)
52. T. Dietrich, S. Bernuzzi, Simulations of rotating neutron star collapse with the puncture gauge: end state and gravitational waveforms. Phys. Rev. D **91**(4), 044039 (2015). ((1412.5499))
53. H. Dimmelmeier, N. Stergioulas, J.A. Font, Non-linear axisymmetric pulsations of rotating relativistic stars in the conformal flatness approximation. Mon. Not. Roy. Astron. Soc. **368**, 1609–1630 (2006). ((astro-ph/0511394))

54. G. Doulis, F. Atteneder, S. Bernuzzi, B. Brügmann, Entropy-limited higher-order central scheme for neutron star merger simulations. Phys. Rev. D **106**(2), 024001 (2022). 2202.08839
55. M.D. Duez, F. Foucart, L.E. Kidder, H.P. Pfeiffer, M.A. Scheel, S.A. Teukolsky, Evolving black hole-neutron star binaries in general relativity using pseudospectral and finite difference methods. Phys. Rev. D **78**, 104015 (2008). ((0809.0002))
56. M.D. Duez, Y.T. Liu, S.L. Shapiro, B.C. Stephens, Relativistic magnetohydrodynamics in dynamical spacetimes: numerical methods and tests. Phys. Rev. D **72**, 024028 (2005). ((astro-ph/0503420))
57. W.E. East, F. Pretorius, B.C. Stephens, Hydrodynamics in full general relativity with conservative AMR. Phys. Rev. D **85**, 124010 (2012). ((1112.3094))
58. D. Eichler, M. Livio, T. Piran, D.N. Schramm, Nucleosynthesis, neutrino bursts and gamma-rays from coalescing neutron stars. Nature **340**, 126–128 (1989)
59. Z.B. Etienne, V. Paschalidis, R. Haas, P. Mösta, S.L. Shapiro, IllinoisGRMHD: an open-source, user-friendly GRMHD code for dynamical spacetimes. Class. Quant. Grav. **32**, 175009 (2015). 1501.07276
60. Z.B. Etienne, V. Paschalidis, Y.T. Liu, S.L. Shapiro, Relativistic MHD in dynamical spacetimes: improved EM gauge condition for AMR grids. Phys. Rev. D **85**, 024013 (2012). ((1110.4633))
61. C.R. Evans, J.F. Hawley, Simulation of magnetohydrodynamic flows - A constrained transport method. Astrophys. J. **332**, 659–677 (1988)
62. K.G. Felker, J.M. Stone, A fourth-order accurate finite volume method for ideal MHD via upwind constrained transport. J. Comput. Phys. **375**, 1365–1400 (2018)
63. E. Flowers, M.A. Ruderman, Evolution of pulsar magnetic fields. Astrophys. J. **215**, 302–310 (1977)
64. J.A. Font, Numerical hydrodynamics and magnetohydrodynamics in general relativity. Living Rev. Rel. **11**, 7 (2007)
65. J.A. Font, et al. Three-dimensional general relativistic hydrodynamics. II: Long-term dynamics of single relativistic stars. Phys. Rev. D65 (2002), 084024, gr-qc/0110047
66. F. Foucart, M.D. Duez, L.E. Kidder, S.A. Teukolsky, Black hole-neutron star mergers: effects of the orientation of the black hole spin. Phys. Rev. D **83**, 024005 (2011). ((1007.4203))
67. C.F. Gammie, J.C. McKinney, G. Toth, HARM: A Numerical scheme for general relativistic magnetohydrodynamics. Astrophys. J. **589**, 444–457 (2003). ((astro-ph/0301509))
68. T.A. Gardiner, J.M. Stone, An unsplit Godunov method for ideal MHD via constrained transport. J. Comput. Phys. **205**(2), 509–539 (2005). ((astro-ph/0501557))
69. T.A. Gardiner, J.M. Stone, An unsplit godunov method for ideal MHD via constrained transport in three dimensions. J. Comput. Phys. **227**, 4123–4141 (2008). ((0712.2634))
70. B. Giacomazzo, L. Rezzolla, WhiskyMHD: a new numerical code for general relativistic magnetohydrodynamics. Class. Quant. Grav. **24**, S235–S258 (2007). ((gr-qc/0701109))
71. J.N. Goldberg, A.J. MacFarlane, E.T. Newman, F. Rohrlich, E.C.G. Sudarshan, Spin s spherical harmonics and edth. J. Math. Phys. **8**, 2155 (1967)
72. A. Goldstein, et al., An ordinary short gamma-ray burst with extraordinary implications: fermi-GBM detection of GRB 170817A. Astrophys. J. **848**(2), L14 (2017). 1710.05446
73. J. Goodman, Are gamma-ray bursts optically thick? Astrophys. J. Lett. **308**, L47 (1986)
74. S. Gottlieb, D.I. Ketcheson, C.-W. Shu, High order strong stability preserving time discretizations. J. Sci. Comput. **38**(3), 251–289 (2009)
75. E. Gourgoulhon, P. Grandclement, K. Taniguchi, J.-A. Marck, S. Bonazzola, Quasiequilibrium sequences of synchronized and irrotational binary neutron stars in general relativity: 1. Method and tests. Phys. Rev. D **63**, 064029 (2001). ((gr-qc/0007028))
76. C. Gundlach, Pseudospectral apparent horizon finders: An Efficient new algorithm. Phys. Rev. D **57**, 863–875 (1998). ((gr-qc/9707050))
77. C. Gundlach, J.M. Martin-Garcia, G. Calabrese, I. Hinder, Constraint damping in the Z4 formulation and harmonic gauge. Class. Quant. Grav. **22**, 3767–3774 (2005). ((gr-qc/0504114))
78. D. Hilditch, S. Bernuzzi, M. Thierfelder, Z. Cao, W. Tichy, B. Bruegmann, Compact binary evolutions with the Z4c formulation. Phys. Rev. D **88**, 084057 (2013). ((1212.2901))

79. G.-S. Jiang, C.-W. Shu, Efficient implementation of weighted ENO schemes. J. Comput. Phys. **126**(1), 202–228 (1996)
80. D. Kasen, B. Metzger, J. Barnes, E. Quataert, E. Ramirez-Ruiz, Origin of the heavy elements in binary neutron-star mergers from a gravitational wave event. Nature (2017), 1710.05463. [Nature551,80(2017)]
81. W. Kastaun, J.V. Kalinani, R. Ciolfi, Robust recovery of primitive variables in relativistic ideal magnetohydrodynamics. Phys. Rev. D **103**(2), 023018 (2021). 2005.01821
82. L.E. Kidder, et al., SpECTRE: a task-based discontinuous galerkin code for relativistic astrophysics. J. Comput. Phys. **335**, 84–114 (2017). 1609.00098
83. K. Kiuchi, P. Cerdá-Durán, K. Kyutoku, Y. Sekiguchi, M. Shibata, Efficient magnetic-field amplification due to the Kelvin-Helmholtz instability in binary neutron star mergers. Phys. Rev. D **92**(12), 124034 (2015). 1509.09205
84. K. Kiuchi, S. Fujibayashi, K. Hayashi, K. Kyutoku, Y. Sekiguchi, M. Shibata, Self-consistent picture of the mass ejection from a one second long binary neutron star merger leaving a short-lived remnant in a general-relativistic neutrino-radiation magnetohydrodynamic simulation. Phys. Rev. Lett. **131**(1), 011401 (2023). 2211.07637
85. K. Kiuchi, L.E. Held, Y. Sekiguchi, M. Shibata, Implementation of advanced Riemann solvers in a neutrino-radiation magnetohydrodynamics code in numerical relativity and its application to a binary neutron star merger. Phys. Rev. D **106**(12), 124041 (2022). 2205.04487
86. K. Kiuchi, K. Kyutoku, Y. Sekiguchi, M. Shibata, Global simulations of strongly magnetized remnant massive neutron stars formed in binary neutron star mergers. Phys. Rev. D **97**(12), 124039 (2018). 1710.01311
87. K. Kiuchi, K. Kyutoku, Y. Sekiguchi, M. Shibata, T. Wada, High resolution numerical-relativity simulations for the merger of binary magnetized neutron stars. Phys. Rev. D **90**(4), 041502 (2014). ((1407.2660))
88. K. Kiuchi, M. Shibata, S. Yoshida, Evolution of neutron stars with toroidal magnetic fields: Axisymmetric simulation in full general relativity. Phys. Rev. D **78**, 024029 (2008). ((0805.2712))
89. H.O. Kreiss, J. Oliger, *Methods for the approximate solution of time dependent problems, in International Council of Scientific Unions* (World Meteorological Organization, Geneva, 1973)
90. S.R. Kulkarni Modeling supernova-like explosions associated with gamma-ray bursts with short durations. astro-ph/0510256
91. P. Kumar, B. Zhang, The physics of gamma-ray bursts and relativistic jets. Phys. Rept. **561**, 1–109 (2014). ((1410.0679))
92. P.D. Lasky, B. Zink, K.D. Kokkotas, K. Glampedakis, Hydromagnetic instabilities in neutron stars. Astrophys. J. Lett. **735**, L20 (2011). ((1105.1895))
93. L.-X. Li, B. Paczynski, Transient events from neutron star mergers. Astrophys. J. **507**, L59 (1998). ((astro-ph/9807272))
94. S.L. Liebling, L. Lehner, D. Neilsen, C. Palenzuela, Evolutions of magnetized and rotating neutron stars. Phys. Rev. D **81**, 124023 (2010). ((1001.0575))
95. LIGO Scientific Collaboration. LIGO Algorithm Library - LALSuite. free software (GPL) (2018)
96. Y.T. Liu, S.L. Shapiro, Z.B. Etienne, K. Taniguchi, General relativistic simulations of magnetized binary neutron star mergers. Phys. Rev. D **78**, 024012 (2008). ((0803.4193))
97. F. Löffler et al., The Einstein toolkit: a community computational infrastructure for relativistic astrophysics. Class. Quant. Grav. **29**, 115001 (2012). ((1111.3344))
98. P. Londrillo, L. Del Zanna, On the divergence-free condition in godunov-type schemes for ideal magnetohydrodynamics: the upwind constrained transport method. J. Comput. Phys. **195**, 17–48 (2004). ((astro-ph/0310183))
99. P. Markey, R.J. Tayler, The adiabatic stability of stars containing magnetic fields. II. Poloidal fields. Mon. Not. R. Astron. Soc. **163**, 77–91 (1973)
100. P. Markey, R.J. Tayler, The adiabatic stability of stars containing magnetic fields-III. Additional results for poloidal fields. Mon. Not. R. Astron. Soc. **168**, 505–514 (1974)

101. P. McCorquodale, M.R. Dorr, J.A.F. Hittinger, P. Colella, High-order finite-volume methods for hyperbolic conservation laws on mapped multiblock grids. J. Comput. Phys. **288**, 181–195 (2015)
102. B. Metzger, G. Martinez-Pinedo, S. Darbha, E. Quataert, A. Arcones et al., Electromagnetic Counterparts of Compact Object Mergers Powered by the Radioactive Decay of R-process Nuclei. Mon. Not. Roy. Astron. Soc. **406**, 2650 (2010). ((1001.5029))
103. G.M. Morton, A computer oriented geodetic data base and a new technique in file sequencing. Tech. rep. (1966)
104. P. Mösta, B.C. Mundim, J.A. Faber, R. Haas, S.C. Noble et al., GRHydro: a new open source general-relativistic magnetohydrodynamics code for the Einstein Toolkit. Class. Quant. Grav. **31**, 015005 (2014). ((1304.5544))
105. P. Mösta, D. Radice, R. Haas, E. Schnetter, S. Bernuzzi, A magnetar engine for short GRBs and kilonovae. Astrophys. J. Lett. **901**, L37 (2020). 2003.06043
106. R. Narayan, B. Paczynski, T. Piran, Gamma-ray bursts as the death throes of massive binary stars. Astrophys. J. **395**, L83–L86 (1992). ((astro-ph/9204001))
107. B. Paczynski, Gamma-ray bursters at cosmological distances. Astrophys. J. **308**, L43–L46 (1986)
108. C. Palenzuela, R. Aguilera-Miret, F. Carrasco, R. Ciolfi, J.V. Kalinani, W. Kastaun, B. Miñano, D. Viganò, Turbulent magnetic field amplification in binary neutron star mergers. Phys. Rev. D **106**(2), 023013 (2022). 2112.08413
109. C. Palenzuela, B. Miñano,D. Viganò, A. Arbona, C. Bona-Casas, A. Rigo, M. Bezares, C. Bona, J. Massó, A simflowny-based finite-difference code for high-performance computing in numerical relativity. Class. Quant. Grav. **35**(18), 185007 (2018). 1806.04182
110. E. Pian, et al. Spectroscopic identification of r-process nucleosynthesis in a double neutron star merger. Nature (2017), 1710.05858
111. A.G. Pili, N. Bucciantini, L. Del Zanna, Axisymmetric equilibrium models for magnetized neutron stars in General Relativity under the Conformally Flat Condition. Mon. Not. Roy. Astron. Soc. **439**, 3541–3563 (2014). ((1401.4308))
112. A.G. Pili, N. Bucciantini, L. Del Zanna, General relativistic models for rotating magnetized neutron stars in conformally flat space–time. Mon. Not. Roy. Astron. Soc. **470**(2), 2469–2493 (2017). 1705.03795
113. T. Piran, The physics of gamma-ray bursts. Rev. Mod. Phys. **76**, 1143–1210 (2004). ((astro-ph/0405503))
114. T. Plewa, E. Mueller, The consistent multi-fluid advection method. Astron. Astrophys. **342**, 179 (1999). ((astro-ph/9807241))
115. J.A. Pons, D. Viganò, Magnetic, thermal and rotational evolution of isolated neutron stars. 1911.03095
116. F. Pretorius, Evolution of binary black hole spacetimes. Phys. Rev. Lett. **95**, 121101 (2005). ((gr-qc/0507014))
117. D. Price, S. Rosswog, Producing ultra-strong magnetic fields in neutron star mergers. Science **312**, 719 (2006). ((astro-ph/0603845))
118. D. Radice, S. Bernuzzi, A. Perego, and R. Haas A new moment-based general-relativistic neutrino-radiation transport code: Methods and first applications to neutron star mergers. Mon. Not. Roy. Astron. Soc. **512**(1), 1499–1521 (2022). 2111.14858
119. D. Radice, A. Perego, K. Hotokezaka, S.A. Fromm, S. Bernuzzi, L.F. Roberts, Binary neutron star mergers: mass ejection, electromagnetic counterparts and nucleosynthesis. Astrophys. J. **869**(2), 130 (2018). 1809.11161
120. D. Radice, L. Rezzolla, F. Galeazzi, Beyond second-order convergence in simulations of binary neutron stars in full general-relativity. Mon. Not. Roy. Astron. Soc. **437**, L46–L50 (2014). ((1306.6052))
121. D. Radice, L. Rezzolla, F. Galeazzi, High-order fully general-relativistic hydrodynamics: new approaches and tests. Class. Quant. Grav. **31**, 075012 (2014). ((1312.5004))
122. D.A. Randall, T.D. Ringler, R. Heikes, P. Jones, J. Baumgardner, Climate modeling with spherical geodesic grids. Comput. Sci. Eng. **4**, 32–41 (2002)

123. A. Rashti, M. Bhattacharyya, D. Radice, B. Daszuta, W. Cook, S. Bernuzzi, Adaptive mesh refinement in binary black holes simulations **2312**, 05438 (2023)
124. F.A. Rasio, S.L. Shapiro, Topical review: coalescing binary neutron stars. Class. Quantum Gravity **16**(6), R1–R29 (1999). ((gr-qc/9902019))
125. C. Reisswig, R. Haas, C.D. Ott, E. Abdikamalov, P. Mösta, D. Pollney, E. Schnetter, Three-dimensional general-relativistic hydrodynamic simulations of binary neutron star coalescence and stellar collapse with multipatch grids. Phys. Rev. D **87**(6), 064023 (2013). ((1212.1191))
126. M. Ruiz, D. Hilditch, S. Bernuzzi, Constraint preserving boundary conditions for the Z4c formulation of general relativity. Phys. Rev. D **83**, 024025 (2011). ((1010.0523))
127. D. Ryu, F. Miniati, T.W. Jones, A. Frank, A divergence-free upwind code for multidimensional magnetohydrodynamic flows. Astrophys. J. **509**, 244 (1998). ((astro-ph/9807228))
128. V. Savchenko, et al., INTEGRAL detection of the first prompt gamma-ray signal coincident with the gravitational-wave event GW170817. Astrophys. J. **848**(2), L15 (2017). 1710.05449
129. S. Shankar, P. Mösta, S.R. Brandt, R. Haas, E. Schnetter, Y. de Graaf, GRaM-X: a new GPU-accelerated dynamical spacetime GRMHD code for exascale computing with the Einstein Toolkit. Class. Quant. Grav. **40**(20), 205009 (2023). 2210.17509
130. M. Shibata, T. Nakamura, Evolution of three-dimensional gravitational waves: Harmonic slicing case. Phys. Rev. D **52**, 5428–5444 (1995)
131. M. Shibata, Y.-I. Sekiguchi, Magnetohydrodynamics in full general relativity: formulation and tests. Phys. Rev. D **72**, 044014 (2005). ((astro-ph/0507383))
132. D.M. Siegel, R. Ciolfi, L. Rezzolla, Magnetically driven winds from differentially rotating neutron stars and X-ray afterglows of short gamma-ray bursts. Astrophys. J. **785**, L6 (2014). ((1401.4544))
133. D.M. Siegel, B.D. Metzger, Three-dimensional general-relativistic magnetohydrodynamic simulations of remnant accretion disks from neutron star mergers: outflows and r-process nucleosynthesis. Phys. Rev. Lett. **119**(23), 231102 (2017). 1705.05473
134. D.M. Siegel, P. Mösta, D. Desai, S. Wu, Recovery schemes for primitive variables in general-relativistic magnetohydrodynamics. Astrophys. J. **859**(1), 71 (2018). 1712.07538
135. M. Soares-Santos, et al., The electromagnetic counterpart of the binary neutron star merger LIGO/Virgo GW170817. I. Dark energy camera discovery of the optical counterpart. Astrophys. J. **848**(2), L16 (2017). 1710.05459
136. N. Stergioulas, J.L. Friedman, Comparing models of rapidly rotating relativistic stars constructed by two numerical methods. Astrophys. J. **444**, 306 (1995). ((astro-ph/9411032))
137. J.M. Stone, T.A. Gardiner, P. Teuben, J.F. Hawley, J.B. Simon, Athena: a new code for astrophysical MHD. Astrophys. J. Suppl. **178**, 137 (2008). ((0804.0402))
138. J.M. Stone, K. Tomida, C.J. White, K.G. Felker, The Athena++ adaptive mesh refinement framework: design and magnetohydrodynamic solvers. Astrophys. J. Suppl. **249**(1), 4 (2020). 2005.06651
139. Q.F. Stout, D.L. De Zeeuw, T.I. Gombosi, C.P. Groth, H.G. Marshall, K.G. Powell, Adaptive blocks: a high performance data structure, in *Proceedings of the 1997 ACM/IEEE Conference on Supercomputing, SC '97* (Association for Computing Machinery, New York, 1997), pp. 1–10
140. A. Sur, W. Cook, D. Radice, B. Haskell, S. Bernuzzi, Long-term general relativistic magnetohydrodynamics simulations of magnetic field in isolated neutron stars. Mon. Not. Roy. Astron. Soc. **511**(3), 3983–3993 (2022). 2108.11858
141. R.J. Tayler, Hydromagnetic instabilities of an ideally conducting fluid. Proc. Phys. Soc. B **70**(1), 31–48 (1957)
142. R.J. Tayler The adiabatic stability of stars containing magnetic fields-I. Toroidal fields. Mon. Not. R. Astron. Soc. **161**, 365 (1973)
143. M. Thierfelder, S. Bernuzzi, B. Brügmann, Numerical relativity simulations of binary neutron stars. Phys. Rev. D **84**, 044012 (2011). ((1104.4751))
144. M. Thierfelder, S. Bernuzzi, D. Hilditch, B. Brügmann, L. Rezzolla, The trumpet solution from spherical gravitational collapse with puncture gauges. Phys. Rev. D **83**, 064022 (2011). ((1012.3703))

145. G. Tóth, P.L. Roe, Divergence- and curl-preserving prolongation and restriction formulas. J. Comput. Phys. **180**(2), 736–750 (2002)
146. S. Typel, M. Oertel, T. Klähn, CompOSE CompStar online supernova equations of state harmonising the concert of nuclear physics and astrophysics compose.obspm.fr. Phys. Part. Nucl. **46**(4), 633–664 (2015). 1307.5715
147. G. Tóth, The $\nabla \cdot B = 0$ constraint in shock-capturing magnetohydrodynamics codes. J. Comput. Phys. **161**(2), 605–652 (2000)
148. B. van Leer, Towards the ultimate conservation difference scheme. II. Monotonicity and conservation combined in a second-order scheme. J. Comput. Phys. **14**(4), 361–370 (1974)
149. D. Viganò, D. Martínez-Gómez, J.A. Pons, C. Palenzuela, F. Carrasco, B. Miñano, A. Arbona, C. Bona, J. Massó, A simflowny-based high-performance 3D code for the generalized induction equation. 1811.08198
150. N. Wang, J.-L. Lee, Geometric properties of the icosahedral-hexagonal grid on the two-sphere. SIAM J. Sci. Comput. **33**(5), 2536–2559 (2011)
151. A. Weyhausen, S. Bernuzzi, D. Hilditch, Constraint damping for the Z4c formulation of general relativity. Phys. Rev. D **85**, 024038 (2012). ((1107.5539))
152. C.J. White, P.D. Mullen, Jiang, Y.-F., S.W. Davis, J.M. Stone, V. Morozova, L. Zhang, An extension of the Athena++ code framework for radiation-magnetohydrodynamics in general relativity using a finite-solid-angle discretization. Astrophys. J. **949**(2), 103 (2023). 2302.04283
153. C.J. White, J.M. Stone, C.F. Gammie, An extension of the Athena++ code framework for GRMHD based on advanced riemann solvers and staggered-mesh constrained transport. Astrophys. J. Suppl. **225**(2), 22 (2016). 1511.00943
154. G.A.E. Wright, Pinch instabilities in magnetic stars. Mon. Not. R. Astron. Soc. **162**, 339–358 (1973)

Chapter 5
KHARMA: Flexible, Portable Performance for GRMHD

Cora Prather

Abstract KHARMA (an acronym for "Kokkos-based High-Accuracy Relativistic Magnetohydrodynamics with Adaptive mesh refinement") is a new open-source code for conducting general-relativistic magnetohydrodynamic simulations in stationary spacetimes, primarily of accretion systems. It implements among other options the High-Accuracy Relativistic Magnetohydrodynamics (HARM) scheme, but is written from scratch in C++ with the Kokkos programming model in order to run efficiently on both CPUs and GPUs. In addition to being fast, KHARMA is written to be readable, modular, and extensible, separating functionality into "packages," representing, e.g., algorithmic components or physics extensions. Components of the core ideal GRMHD scheme can be swapped at runtime, and additional packages are included to simulate electron temperature evolution, viscous hydrodynamics, and for designing chained multi-scale "bridged" simulations. This chapter presents the computational environment and requirements for KHARMA, features and design which meet these requirements, and finally, validation and performance data.

KHARMA[1] (an acronym for "Kokkos-based High-Accuracy Relativistic Magnetohydrodynamics with Adaptive mesh refinement") is a new open-source performance-portable code for performing general-relativistic magnetohydrodynamic (GRMHD) simulations, targeting stationary spacetimes such as black hole accretion systems. It implements the High-Accuracy Relativistic Magnetohydrodynamics (HARM) scheme [14], written from scratch in C++ 17 to leverage the Kokkos programming model [57] in order to run efficiently on various computer and accelerator architectures using the same source code. KHARMA also uses the Parthenon adaptive mesh refinement (AMR) framework [17] to support static and adaptive mesh refinement and to provide a consistent overall code structure.

KHARMA is in production use as one of the primary codes used for GRMHD simulations in the Event Horizon Telescope Collaboration (EHTC), furnishing many of

[1] https://github.com/AFD-Illinois/kharma.

C. Prather (✉)
Los Alamos National Laboratory, Los Alamos PO Box 1663, NM, USA
e-mail: cprather@fas.harvard.edu

C. Bambi et al. (eds.), *New Frontiers in GRMHD Simulations*, Springer Series in Astrophysics and Cosmology,https://doi.org/10.1007/978-981-97-8522-3_5

the GRMHD simulations used in official papers such as [48, 49] and collaboration projects such as [12, 41]. In addition to being fast, KHARMA has been designed from the ground up to be modular and extensible through the use of packages, designed to each contain a single algorithmic component or physics extension. This modularity extends to all features, such as the magnetic field transport and primitive variable inversion, in addition to the usual modular components such as reconstruction scheme and Riemann solver. Various additional physics packages have been developed to treat new systems and processes beyond ideal GRMHD, including an approximate viscosity treatment useful for very low accretion-rate systems as described in [7], and a co-evolved electron temperature treatment for two-temperature plasmas as described in [39], as well as a mode for running chained "multi-scale" simulations as described in [8]. An extension is also being developed for simulating radiative effects, which are extremely important at high accretion rates and increasingly computationally feasible for GPU-based codes on modern supercomputers.

This chapter provides a brief case study of KHARMA, beginning with the hardware trends and community requirements which led to its development, then covering the code's scheme, software design, and features, and ending with basic validation and performance data.

5.1 Necessity of Performance Portability

Perhaps the single choice that most dictates KHARMA's code structure is to be "performance-portable," targeting efficient operation on general-purpose graphics processing units (GPUs) as well as other accelerator architectures and traditional central processing unit (CPU) cores. Adding performance portability to an existing code often requires substantially re-writing it to leverage new loop ordering and data structures, and this difficulty was a primary motivator for developing KHARMA from scratch.

In order to understand the trends in computing which made performance portability so important as to abandon an existing fast C implementation of HARM [37], it's useful to provide a bit of history and context around high-performance computing (HPC). This context is unnecessary for understanding KHARMA itself; readers already convinced of the universal need for performance portability can skip to KHARMA's more unique code requirements in Sect. 5.2.

5.1.1 Performance and Power

In 1965, Gordon Moore predicted that the number of transistors on a given integrated circuit would double each year. In 1975, he revised this to predict a slower doubling rate of 2 years from 1980 onward [29, 30]. Whether by prescience or self-fulfilling prophecy, this doubling rate has held for nearly all of the ensuing 4½ decades. Usually, this increase is interpreted as computer chips getting "faster" or "more capable."

However, over the years, the purpose of these transistors—that is, how new transistors are "spent" to improve a chip's capabilities, has shifted. For the first decades of chip development, from the 1970s to 2000s, extra transistors were used to increase the clock speed and instruction processing rate of otherwise similar chips—that is, to make existing programs run faster. However, power constraints cut short the increasing clock rates in the mid-2000s, necessitating that designers find other ways to improve chips—thus, improvements to CPUs since that time have taken the form of adding more processing cores to a single chip, operating at the same frequency but processing more data (so long as programs could take advantage of the parallelism). This much is well-known in the supercomputing community.

Parallel to this more obvious trend of multiple cores, we see a trend toward not just duplicating existing processors, but specializing them for particular kinds of work—seen in GPUs, as well as vector-optimized or "many-core" CPUs. Specialization involves taking out functionality like branch prediction, out-of-order execution, and large instruction sets, and using their area and transistor count to further boost the amount of data a chip can process at once, through additional parallel execution units. At the far end of specialization, chips capable of providing only one function (encryption, say, or video encoding) can provide orders of magnitude faster processing than software implementations.

Both trends are illustrated in parallel in Fig. 5.1. Plotted in blue and gold are CPU scores in the PassMark general benchmark,[2] for single cores and full sockets [35]. Single-core performance stagnates with the power limits to clock rate in 2008, and has grown at a rate of only $2^{0.12t}$ (where t is in years) since then, mostly consisting of hard-won efficiency improvements in the rate of instructions cleared per clock cycle (IPC). Instead, performance benefits are realized by increasing the core count; parallel workloads have still seen a $2^{0.40t}$ improvement as core counts have increased over the last decade. Plotted in red are GPU performance numbers in floating-point operations per second (FLOPS/s) [47]. Note this is a significantly different measurement from a PassMark score, evaluating GPUs' effectiveness at number crunching specifically, not general workloads (which they are generally unsuited for anyway). This means that GPU performance is plotted on a *different axis* from CPU performance. However, both axes span exactly four decades, to compare relative improvement, where GPUs clearly have an advantage, especially in recent years. Over the same measured span 2006–2023, GPUs grow in floating-point performance at $2^{0.49t}$, paying forward the full benefits of their increasing transistor count.

It should not be surprising that special-purpose chips can improve at their intended task faster than general-purpose chips can improve at a broad range of tasks; they trade generality for efficiency. As power constraints and the end of Moore's law force the use of more creative methods to improve performance, the pressure toward specialization will likely increase.

---

[2] https://www.passmark.com/.

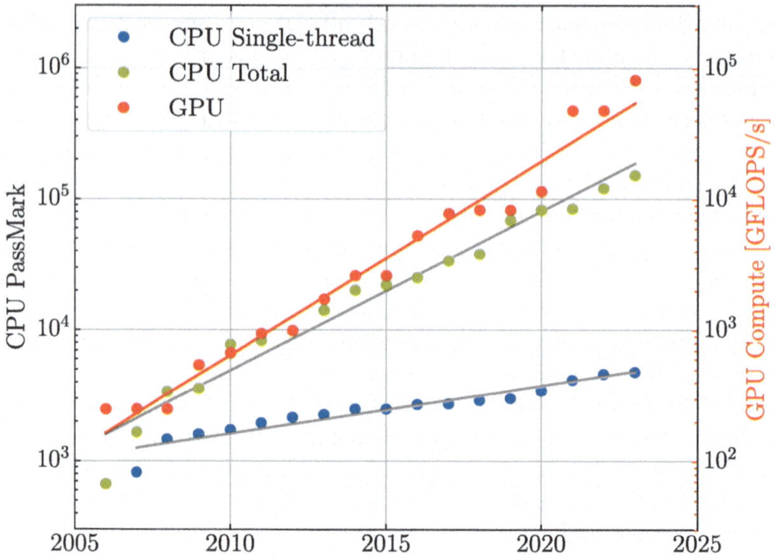

Fig. 5.1 CPU single-core and total performance development, plotted against GPU theoretical peak performance performing floating-point operations. Scales for CPU and GPU metrics are independent, and are adjusted so that metrics coincide in 2006, but all scales span exactly 4 decades to accurately depict relative improvement. Fitting functions reflect exponential growth of $2^{0.12t}$, $2^{0.40t}$, and $2^{0.49t}$ for CPU single, full, and GPU performance respectively, with t measured in years. CPU performance is measured via public listed results of the PassMark combined CPU benchmark software. GPU performance is taken from combinations of manufacturer-listed and reviewer-tested floating point operations per second (FLOPS/s) at double precision (not AI-specific packed operations marketed as "matrix" or "tensor" cores). CPU scores from [35], GPU performance from [47], figure originally from [38]

The benefits of specialization are nowhere more evident than in Artificial Intelligence (AI). AI research now drives the development of its own class of specialized chips, designed even more narrowly than GPUs for processing the large matrix multiplications used in training and inference. The clear efficiency benefits of specialized hardware have driven a rush of new architectures, such as Google's Tensor Processing Units [20], Cerebras systems' Wafer-Scale Engine [23] and Tesla's many-core "Dojo" architecture [46], which have become increasingly widespread. Even architectures which remain general in principle are gaining subsystems targeted at AI calculations specifically (Nvidia's Tensor Cores, Intel's AMX Engines, etc.).

5.1.2 Specialization in Supercomputing

Though more recently than the frequency limit, supercomputing has similarly hit a hard limit: power draw. Machines consuming more than about 1MW can cost more in electricity than hardware, so the top machines cannot continue to draw more power without incurring spiraling costs. To illustrate trends in supercomputing, we use the open top 500 supercomputers list compiled semi-annually [28]. While not a complete list of every machine, is still indicative of overall trends in supercomputers both public and private. As shown at left of Fig. 5.2, from 2008 (when recording of power usage began) to 2013, the average power draw of the reporting portion of the top 500 machines grew by nearly an order of magnitude in 5 years. Since then, however, it has grown by only another factor of 2, and appears to be slowing further still. This is good news for global energy usage, of course, but it has visibly slowed the performance development of the top 500 machines, as shown at right of Fig. 5.2. Before the increasing power usage slowed in 2013, machines doubled in performance every year; afterward, every two years.

Even discounting the very top machines, where exotic architectures have always been common, the closing power envelope is pushing HPC systems toward special-purpose chips, mostly GPUs. Indeed, the number of machines in the top 500 list powered primarily by GPUs has increased substantially: in 2013, under 10% of systems used some kind of GPU accelerator—as of November 2023, the share is just

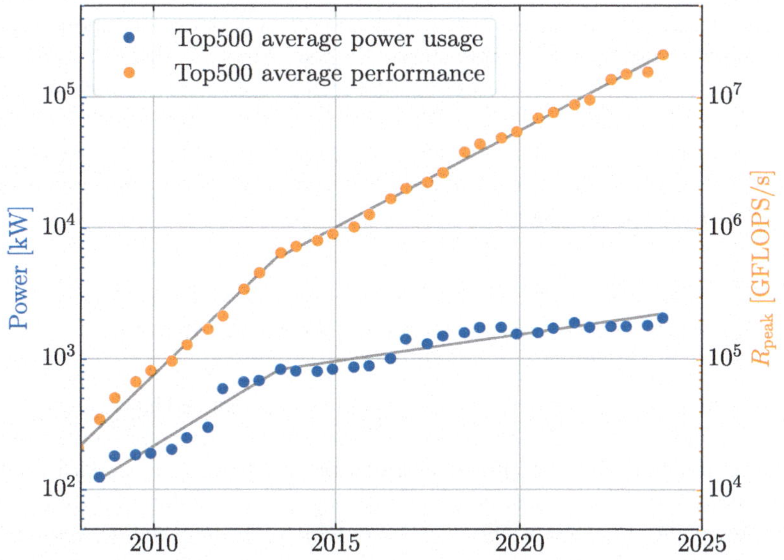

Fig. 5.2 Average power draw of the top 500 machines, plotted with average performance development, showing a clear slowing of power usage increases in about 2013, when average machine power usage began to exceed 1 MW. Average power usage of the reporting fraction of the top 500 machines shows exponential growth at $2^{0.56t}$ from 2008 (when reporting began) to 2013, and $2^{0.14t}$ thereafter. Average peak performance R_{peak} in GFLOPS/s of all top 500 machines shows development at $2^{0.88t}$ before 2013, $2^{0.49t}$ thereafter. Data from [28], figure originally from [38]

about 30%, together providing half of the list's total computing power [28].[3] While initial adoption of accelerators was slow due to programming challenges, maturing software (AI frameworks, programming models, etc.) and the closing power envelope have made GPUs an ever clearer choice for new systems.

Even the word GPU in the HPC context has shifted to mean something more specialized than it did a decade ago: back then, top machines such Oak Ridge National Laboratory's Titan were powered by what were effectively consumer graphics chips—the GK110 chips used in that machine were also sold to consumers as the Titan GPU. Even when the die was not exactly the same, the technologies were: many supercomputers over the course of the 2000s were built with technologies sold concurrently (or even primarily) to consumers. Modern "GPUs," in contrast, are specialized computational designs, increasingly distinct in architecture from their consumer counterparts (e.g., CDNA vs. RDNA from AMD, XeHPC vs. XeLP from Intel, Hopper and Blackwell vs. Lovelace from Nvidia).

The sudden specialization has a clear cause: the massive hardware market for running artificial intelligence models. As the large tech companies build AI training supercomputers on the scale of cloud server farms, they finally represent the kind of large and stable market that will drive stiff competition, and it seems likely that chip architectures will continue to shift to serve the needs of the larger AI market, rather than traditional HPC use cases. The current generation of supercomputers uses three substantially new architectures (Fujitsu A64FX, AMD CDNA2, Intel XeHPC), and as AI drives further architecture changes, the next generation may use substantially different architectures still. Yanked out of its complacency by a strong AI hardware market and increasing power efficiency concerns, not to mention the end of Moore's law, HPC faces a decade of either rapid architectural changes or complete performance stagnation.

5.1.3 Specialized Hardware Requires Portable Software

It should be no surprise that in places where complex specialized hardware has found success, it has generally included an easy and standard software interface, to abstract away the complexity. Take graphics acceleration: calls using a standard interface such as OpenGL or DirectX can be accelerated by a huge number of different hardware implementations, each providing a driver which translates the standard calls to specific instructions for the accelerator hardware. Thus innovations to the hardware do not require changing existing applications at all. A similar separation of standard interface layer and driver implementation layer has found success in enabling many other specialized hardware implementations, for video encoding/decoding, networking operations such as encryption, etc.

[3] This number excludes specialized CPUs such as the Fujitsu A64FX and Intel Knights Landing architectures. Nevertheless, these architectures present many of the same challenges and opportunities as accelerators.

There is no single standard language of this kind for HPC as there is for other accelerators. This is not for lack of trying, however, and there are promising modern contenders: projects within HPC such as Kokkos [57], OCCA [27], RAJA [2], as well as industry standards such as OpenCL,[4] SYCL,[5] OpenACC,[6] and efforts within the C++ standard under the namespace `std::parallel`. All of these attempt to express accelerator programming in a hardware-independent way, striking a balance between providing an easy-to-use interface and allowing low-level access and flexible programming.

While the lack of a standard is disappointing, it is not reason to avoid any of the existing options. Nearly all of the listed performance portability options involve abstractions which simplify device programming, and provide convenient frameworks to simplify host programming as well. Many of them share a similar overall structure for expressing compute kernels, so porting between them may often be easier than adapting "landlocked" pure C code of any complexity.

5.2 Other Requirements

Besides performance portability, there were several requirements that shaped KHARMA's design. KHARMA was designed primarily as a code for black hole accretion simulations, specifically of low-luminosity active galactic nuclei (LLAGN), that is, very low-accretion-rate supermassive black hole (SMBH) accretion systems where radiation does not play a primary role in the fluid dynamics. The primary targets were M87* and Sagittarius A*, in order to understand and utilize the images from the Event Horizon Telescope (EHT). This provided a simple and useful target for early development, though more recently the code has been extended well beyond this original purpose.

The following sections give background on the scientific requirements which shaped KHARMA's development.

5.2.1 Simulation Libraries

The first black hole (BH) images of M87* by the EHT Collaboration (EHTC) [50] presented a unique opportunity to put accretion theories and simulations directly to the test, by comparing concrete predictions from ray-traced simulations to the EHT image properties and other measurements of M87*. Within the collaboration, a parameter study or "library" of GRMHD simulations was created. The core of this library consisted of 10 simulations performed with the HARM-based GRMHD code

[4] https://www.khronos.org/opencl/.

[5] https://www.khronos.org/sycl/.

[6] https://www.openacc.org/.

`iharm3D` [37]. Simulations were performed on Stampede2 at the Texas Advanced Computing Center (TACC), where a set of just 10 simulations used a significant allocation of CPU-based resources.

From this core library of GRMHD simulations, images could be created via general-relativistic ray-traced volume rendering of the diffuse optically thin emission, generating a prediction for emission intensity at EHT frequencies [62]. With these images, especially when used in combination with other measurements (e.g., estimates of the jet power), the collaboration was able to significantly narrow the space of realistic accretion possibilities, and thereby understand the salient image features in terms of the accretion state around the event horizon (EH) [52]. Many research projects within the collaboration have made use of this standard library, from validation and calibration for official results [51, 53] to interpretation of polarized results [54], and many individual studies [16, 19, 24, 34, 42].

However, running libraries of simulations is expensive, and the conclusions are limited to motivating or demotivating particular models within a set of options necessarily limited in both its coverage of the available parameters, and its included physics. With better data, some clear and independent observational signatures by which to measure particular system parameters may yet be forthcoming—for example, measurement of BH spin by directly measuring the shape of the photon ring [19]. Likewise, sufficiently accurate modeling may reveal reliable trends under certain circumstances—for example, the β_2 parameter describing rotational symmetry of the resolved linear polarization, which can reliably distinguish between different magnetic field strengths in simulations, and provides a rough indicator of BH spin if accretion is viewed nearly face-on [34].

However, in general, changes to BH system parameters cause complex and overlapping changes in the observable image and other measurements, which are difficult to disentangle into clear relationships between individual parameters and measured values. Thus, simulation libraries are still the most reliable way to understand the constraints that new data provide, as well as the most useful current tool available to develop better techniques. Simulations retain an important advantage over phenomenological models in particular, as predictions can be made about many disparate measurements based on the same underlying simulated model, and thus it can be evaluated against all available measurements of a source in concert.

5.2.2 Many Long Simulations

A relatively short and sparse parameter study sufficed in order to understand the basic features of M87*—however, in order to evaluate the much faster-moving Sagittarius A* (Sgr A*) in detail, longer simulations were required. Ideal GRMHD simulations can be rescaled to the system's characteristic timescale, but the relevant scale is much shorter for Sgr A* than M87*, requiring much longer simulations in order to cover just a few days of real time. Longer simulations also allowed better characterizing the variability of the light-curve, as well as providing more snapshots to compare against the data—useful considering the less definitive reconstructions of the Sgr A* image. This need provided the original motivation for writing KHARMA.

Notably, in this case and generally, the advantage to simulation libraries and parameter studies is to produce *many long* simulations, rather than necessarily the *largest* possible single examples. In GRMHD in particular, so long as basic resolution requirements are met (such as resolving the magnetorotational instability) simulations tend to vary little with resolution—at least, on the scales at which it is feasible to run an entire simulation library. Thus a code must produce simulations at the existing resolutions more quickly—it must use each processor to its full potential (good base performance) and scale efficiently with the addition of more processors to a problem of the same size (good strong scaling).

In particular, the requirement of good base performance also motivates performance portability, as single GPU chips or nodes can vastly outperform single CPU nodes—thus, the difficult or unavoidable performance penalties of strong scaling can be partially recovered simply because fewer nodes need to communicate in order to achieve the same performance. There is also great utility in porting simulations which might have required 10–20 CPU nodes onto a single GPU, as single machines are much more likely to be available as a part of campus computing resources, or even within the means of an average research budget.

5.2.3 Mesh Refinement

Simulation efficiency benefits from larger time steps, which in turn require the largest possible zones, placing resolution only where it's needed. This is a universal problem with Logically Cartesian grids in spherical coordinate systems: the allowable simulation timestep set by the Courant-Friedrichs-Lewy (CFL) condition is much smaller than it needs to be, set by the many small zones next to the coordinate pole. In a single regular mesh, increasing the ϕ-resolution necessarily arranges more zones next to the coordinate pole, and increasing the θ-resolution causes all zone centers to grow closer to one another as they approach the pole. Both effects reduce the width, and therefore the CFL limit, of these zones, and waste computation time with needlessly small timesteps (Fig. 5.3).

This can be mitigated somewhat by the use of coordinate-based refinement; specially designed coordinate transforms like "Funky" Modified Kerr-Schild coordinates [62], which increase the θ-widths of the circumpolar zones at the coordinate level, so that zones equally spaced in the native coordinate system nevertheless grow in the base spherical coordinate system. At nominal resolutions, this provides from a 2–8x greater timestep than spacing zones evenly in θ.

However, really addressing this problem requires mesh refinement, which can widen zones in ϕ without losing θ-resolution near each coordinate pole. Successive derefinement of blocks approaching the axis can reduce almost arbitrarily high resolutions in the midplane down to a manageable number of axis zones in the coarsest, polar blocks. Internal refinement can further reduce the number of zones surrounding

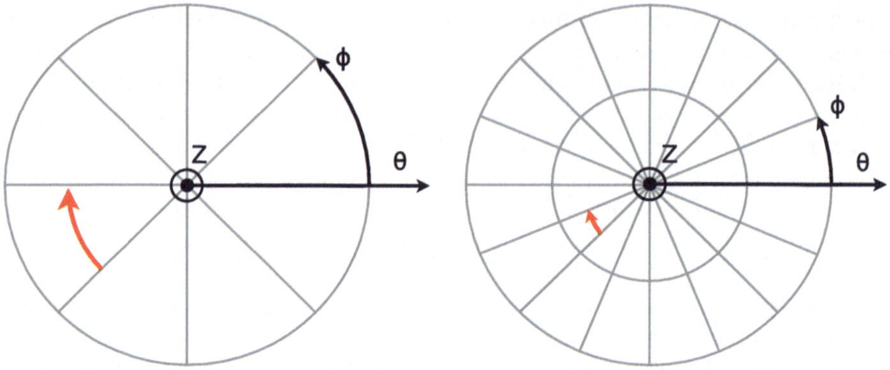

Fig. 5.3 Diagram illustrating the timestep problem faced by non-refined logically Cartesian meshes in spherical polar coordinates. The illustration represents a slice at constant radius, arranged around the coordinate axis (i.e., z-axis) of a spherical polar coordinate grid, as marked (note that exactly the same problem exists in cylindrical or two-dimensional polar coordinates as well, with θ replaced by the radial coordinate). As the resolution in the θ or ϕ direction is increased, zones next to the axis bunch together, with smaller zone widths (e.g., those marked in red above). For solvers with a maximum timestep limited to the signal crossing time of the smallest zone (i.e., subject to the Courant-Friedrichs-Lewy condition) this can cause the stability-limited timestep to be much smaller than what would be required to maintain good accuracy, wasting significant computation time

the axis to just 4–8 total zones, orders of magnitude fewer zones than an unrefined system, with a proportional speedup.

Other solutions exist for the timestep problem, e.g., overstepping the Courant condition of the circumpolar zones without modifying the mesh geometry [18]. However, these are complex to implement when compared to mesh refinement, and there is no need for them in a flexible self-contained code.

5.2.4 Extensibility

Perhaps the strictest requirement for KHARMA short of performance portability was that it be extensible, able to accommodate user changes and new physics onto its fast and scalable platform. Due to the broad range of physics at play over different scales and accretion rates, greater understanding will require treatment of new phenomena, not just modifications to the problem setup.

It is well known that ideal GRMHD does not capture all the physical effects which operate in most black hole accretion systems, and indeed the most exciting opportunities in really understanding these systems lie in treating physics beyond ideal GRMHD at resolution and accuracy scales where the observational consequences of new physics can be effectively studied.

One target of particular interest was dissipative effects, which may become important at very low accretion rates such as observed in Sgr A*. Even the simplest and most streamlined treatment for Sgr A* [7] uses a locally semi-implicit scheme as it must treat time-dependent and stiff source terms. Thus costs from 3 to 10 times more than fully-explicit ideal GRMHD evolution. If a more detailed treatment is ever required, it would cost even more. Resistivity may also play a role at small length scales close to the horizon, which can affect variability and observational appearance—a resistive treatment requires a similar implicit scheme to dissipative effects, with similar cost.

At the other end of the range of accretion rates, radiative effects quickly become dominant, at first indirectly through radiative cooling, and then directly as radiation pressure. Radiative effects dictate the evolution of high-accretion-rate systems, which are characterized by bright X-ray emission. Due to the difficulty of simulating accretion disks with such high luminosity, existing models for these systems are structured around "thin-disk" analytic solutions as in e.g., [32]. These models have proven tremendously useful, but are limited in their applicability due to the assumptions involved. Algorithmic treatments of radiative GRMHD are well-understood [43, 61]. However, global treatments are computationally expensive—computational power is only now reaching the point that full thin-disk radiative simulations at high accretion rate are possible, potentially revolutionizing the field.

As mentioned, algorithms exist for all of these extensions. However, generally they are expensive, ranging from 3 to 20 times the base computational cost of ideal GRMHD simulations. Thus, they have traditionally been simulated at low resolutions, over short timeframes, and only with isolated simulations (not parameter studies). With a performance-portable framework, implementations of these extra effects can finally be incorporated into libraries of longer simulations, dramatically increasing their utility in comparing to observational data by parameterizing over the system's unknown incidental features.

In short, KHARMA's performance portability finally makes feasible a number of exciting simulations treating computationally expensive new physics, by making available the increased computational power of GPUs and the exascale machines which rely on them. Providing easy interfaces for these new features was therefore a primary goal for the code.

5.2.5 Readability

Finally, any code which expects to see adoption in academia must be understandable to a domain scientist graduate student using the code for a particular project, and not primarily tasked with its development. Large, complex and multipurpose scientific codebases have a place in laboratory science and engineering in industry, but codes in academia cannot rely on the same level of long-term staffing and funding, and must be less complex to be applied productively. This is not a strict requirement of simplicity, but rather a dual requirement of modularity and documentation: KHARMA cannot assume a high level of institutional knowledge, and must be possible to modify without deep knowledge of its various components.

Many academic codes take a strict simplicity-first approach to this problem, restricting themselves to features and implementations which will be understandable in full to a new student, without requiring that months be devoted to simply learning the code. It is arguable whether this approach is strictly necessary, and KHARMA was designed under the assumption that it is not, as the other desired features and dependencies are simply not compatible with a minimal codebase. KHARMA instead takes a documentation- and standards-based approach to the problem: the hope is that with a proper introduction, new developers need understand only the package interface and a few relevant packages in order to get started. However, while incredibly successful in most software, physics simulation codes provide something of a worst case for this approach, since debugging numerical instabilities can involve understanding the interactions between different packages modifying the fluid state. KHARMA's long-term success in taking this alternative approach remains to be seen.

5.3 Design for Requirements

To summarize the previous sections, KHARMA was required to:

- Run simulation steps quickly by leveraging specialized architectures
- Simulate the same time periods with fewer steps using mesh refinement
- Easily support additional physics in combination with ideal GRMHD, such as viscosity and radiative processes
- Still be understandable to the average graduate student not primarily tasked with its development

KHARMA's major design choices are very explicitly geared toward meeting these four requirements, so the design is presented here in reference to how it helped to address each requirement.

5.3.1 Performance Portability

Performance portability was the core requirement, for reasons exhaustively catalogued in Sect. 5.1, and is provided by Kokkos[7] [4, 56, 57]. The Kokkos programming model enables performance portability by providing a vendor-neutral library, compiled alongside an otherwise standard C++ program. It was chosen as the programming model for Parthenon, and therefore KHARMA, due to its simple syntax, universal support, and high performance ceiling relative to native performance.

Kokkos is an effort primarily of Sandia National Laboratory, and supports the major CPU and GPU architectures used in HPC. In the Kokkos context, "performance portability" means enabling the use of a single primary source code base to generate

[7] https://github.com/kokkos/kokkos.

efficient compiled code for many different HPC architectures. Kokkos is designed to be compiled into C++ programs: calls into the library from user C++ code cause the compiler to transform certain user functions into computational kernels in a number of different languages and language extensions, including Nvidia's CUDA, Khronos SYCL (used for Intel GPUs), and AMD HIP. The transformation happens entirely at compile-time, without wasting cycles performing just-in-time (JIT) compilation when the program is run. As it supports the use of any C++17-compliant compiler, most vendor programming models, and nearly all accelerator hardware in use in HPC, Kokkos remains almost universally portable yet entirely vendor-neutral.

While this choice was made in 2020, before the universal availability of OpenMP offload implementations, OpenMP offload still does not meet our requirements of near-native performance or broad flexibility. Besides OpenMP, the competitor technologies were C-like systems such as OpenCL, as well as various "turnkey" JIT-compiled solutions. Modern options such as SYCL and `std::parallel` are also worth considering. In general, however, the existing options are all designed either for convenience when porting existing code, or for writing specifically GPU-first code. As a single programming model for CPUs and GPUs, Kokkos abstracts away the divide between CPU and GPU code entirely, enabling performance portability as well as a few convenience features at little overall cost to complexity.

The primary changes to user code under Kokkos are the abstraction of looping and memory access: the user writes parallel code once in an abstract way, and the memory layout and loop ordering are decided at compile time based on the compile target architecture. In part because of this heavy abstraction, Kokkos provides an easy and natural GPU programming model for the beginner. Take a normal one-dimensional loop in C:

```
for (int64_t i = 0; i < N; ++i) {
    c[i] = a[i] + b[i];
}
```

In Kokkos it is written in a similar way, albeit with very different results:

```
Kokkos::parallel_for(N,
    KOKKOS_LAMBDA (const int& i) {
        c(i) = a(i) + b(i);
    }
);
```

The latter snippet replaces the loop declaration with a function call into the Kokkos library, and the loop body with the definition of a C++ lambda function to be run once with each value of the index as its argument. Use of this construct is indication from the user that there are no loop dependencies, so Kokkos will map loop iterations to the available compute units on the target device in parallel as efficiently as possible, be it CPU, GPU, etc. The lambda function uses capture by value, so it operates almost as transparently as a normal loop body.

Multi-dimensional arrays (in Kokkos, `Views`) are also accessed in a straightforward way, substituting an inlined function call `()` for pointer-based array access

`[]`. Multi-dimensional and nested loops are a bit more complex, but all of Kokkos's loop formulations are ultimately much more usable than manually specifying thread and block geometry in CUDA, for example. While the user does eventually have to understand that the host and device do not share a memory space, most host-to-device copies are automatic through the lambda capture, allowing the user to write code nearly identical to what would be written in a host-only `for` loop. It's important to note this design enables performance portability without obvious GPU-first design overhead such as dictating block/thread breakdowns, and without separating the CPU and GPU codepaths.

In fact, by providing integrated bounds checking in debug builds, automatic memory management tied to the lifetime of the `View` object, and transparent multidimensional array handling, Kokkos can make some aspects of programming and debugging significantly *easier* for the beginner than C, especially when debugging common code errors such as indexing and memory initialization. After an initial learning curve understanding the syntax, development using the Kokkos abstractions is efficient, and the added features save time. Code can be developed and debugged locally on the CPU, and only once working compiled for GPU and checked for any (rare) GPU-specific bugs.

It is worth noting a Kokkos-based implementation is not forever locked into Kokkos, either. The basic lambda-argument interface for kernels is now quite common, shared by many performance portability tools. With a simple substitution of syntax, the above example could easily be rewritten with C++ `std::parallel`, or using SYCL calls. As mentioned, Kokkos has a backend for generating SYCL code automatically: it is perhaps indicative that despite having feature parity, this backend is only about half the size of the translation layer for CUDA (4442 non-comment lines vs 7562 as of February 2024). Especially for KHARMA, which mostly uses simple loop kernels, transitioning to e.g. SYCL would be straightforward.

5.3.2 Mesh Refinement

Importantly for KHARMA's design, Adaptive Mesh Refinement (AMR) cannot easily be added to a code not designed to accommodate it: keeping track of meshblocks at different refinement levels in a structured-mesh code requires a number of invasive changes to the systems handling coordinates, domain boundary communications, file operations, and (if the mesh layout is to change during the run) also the memory backing and load-balancing. Thus if AMR is desired, it should be planned into a code from the beginning, if possible.

Parthenon[8] [17] was selected as a dependency partially because it was designed to support AMR—it makes the construction of an AMR code feasible for a small team by handling nearly all aspects of mesh refinement, allowing the developer to focus on the algorithm as if writing for a single block. This abstraction breaks down

[8] https://github.com/lanl/parthenon.

in places, but overall makes the experience of writing and maintaining AMR code much more similar to any traditional multi-block code.

KHARMA has carefully preserved AMR support for unmagnetized runs since its inception, regularly testing simple unmagnetized AMR operations to ensure no single-level assumptions crept into the code. Due to this planning, when a face-centered constrained magnetic field transport was added last year (finally allowing a divergence-preserving prolongation operator, and thus prolongation/restriction of the B field), KHARMA could immediately support SMR and AMR meshes for all runs.

In addition to AMR, Parthenon provides tools for nearly all other aspects of an Eulerian structured-mesh code except the physics: parameter handling, file I/O, dynamic tasking, MPI communication, and some basic code organization tools are all provided.

5.3.3 Extensibility

Even the fastest code cannot be restricted to performing only ideal GRMHD simulations if it is to be useful going forward. To be useful, modification of KHARMA should be as simple as possible, in order to run new problems, add new features, and simulate new physics beyond ideal GRMHD.

Parthenon provides a pattern for separating code functionality into "packages." Packages do not have strict interface requirements or traditional class inheritance, but rather consist of:

1. A C++ namespace for all of a package's unique functions and classes
2. A `Params` map object (much like a `dictionary` in Python) for storing non-grid-based information such as run-time parameters.
3. A list of `Fields` (and `Sparse Fields` and `Swarms`) indicating variables which should be allocated on each new `MeshBlock`. `Fields` can be tied to the size of the `MeshBlock` (i.e., number of cells, faces, edges...), or independent of it.
4. Some number of "callbacks"—functions implemented by the package which are called at specified points by the code (e.g. a source term, or diagnostics to be printed after every step)

Each package thereby encapsulates code with its relevant data (parameters and fields) similarly to an object in Object-Oriented Programming, with the advantage of translating neatly to templated, device-first implementations. However, packages usually contain more functionality than individual objects, and are only instantiated once to handle large pieces of program functionality, rather than many times to handle specific instances of data.

To elaborate: in Parthenon, the data itself (particular arrays corresponding to fields) is handled by traditionally object-oriented code in `MeshBlock`, `Mesh`, and `View`. A package registers fields to be added automatically to each new `MeshBlock`, and then registers callbacks to manipulate that data at particular points during the run (e.g., before a step, when exchanging boundaries, before output, etc.). This design ensures that packages are neutral to the mesh structure and decisions, but rather represent processes and steps which modify variables on the mesh.

The encapsulation of packages has allowed KHARMA to accommodate several different extensions and projects within the same codebase, sharing features upstream in a way that does not impinge on other uses of the code. Rather than forking for each new project, and eventually implementing overlapping sets of features in several fundamentally incompatible downstream codes, KHARMA's package structure encourages centralizing changes, allowing everyone to benefit quickly from improvements. Two particular elements of the package structure enable this:

1. Disabled packages are *disabled*. Invoking code in a package which is not loaded will nearly always crash rather than producing subtle errors. Nearly all additional features peculiar to a package can be encapsulated within its own namespace and code, hooking into the main loop with callbacks or behind checks, reducing feature conflicts as much as possible.
2. Package-based implementations rarely modify the same code and cause conflicts, allowing more developers to work on forks independently for extended times, while still allowing the completed features to be brought back upstream later.

Extensibility also motivates developing KHARMA as a GRMHD-first code, as opposed to a general-relativistic extension to an existing MHD framework. By incorporating differential geometry from the ground up, KHARMA can easily run simulations in any theory with a metric, not just General Relativity, simply by implementing the metric once. Additionally, it is easy to implement new coordinate transformations, which can be used to compress and expand resolution in a more granular way than block-based mesh refinement.

5.3.4 Simplicity

The modular package structure necessarily makes for a more complex piece of software than a traditional C-based GRMHD code. At several points, these design considerations have required sacrificing strict simplicity in order to introduce features. This is not to say, however, that simplicity is abandoned as a requirement. Rather, it is treated as a matter of managing an increased scope and ambition in a way that preserves the goals of a simple code. Generally, simplicity accomplishes two goals, which remain central constraints in KHARMA:

1. A new user can learn to modify the code with a minimum of required reading, ideally of documentation and comments rather than source.
2. Modification of the code is as unlikely as possible to introduce regressions in existing functionality, and any introduced are as easy as possible to find.

These are ultimately not constraints on how big a code is (though both encourage smaller codes, of course). Rather, they are constraints on how someone interacts with and modifies the code, and thus on its structure and documentation.

KHARMA is designed and (especially) documented with these requirements in mind. As described in Sect. 5.3.3, modular code is the key to writing compatible implementations of complex physics. A modular structure also attempts to enforce that familiarity be required with as little of the code as possible in order to begin modifying it—implementations which conform to interfaces can (ideally) be assumed work within those interface parameters, reducing the amount of searching and background information necessary to modify another particular module. Care must be taken to ensure the maintenance of this design, and of course cross-package bugs are still a distinct possibility, but a modular approach pays off when managing and adding many different features, potentially at the same time.

Modularity also allows easily trimming code which is no longer useful. Several older optional pieces of KHARMA's scheme have been superseded by new components and are now rarely used. When no longer needed for backward compatibility, these can be easily removed to reduce line count and confusion, without any major disruption to the overall code structure. Similarly, there is no excuse to hold on to features in the conviction that they might one day be useful: the same package code can simply be reintroduced later, modulo any updates to the package interface.

5.4 Code Structure

The following sections describe KHARMA's structure as a code, including describing the primary algorithm, philosophy, and breakdown into packages.

5.4.1 Algorithm

At its base, KHARMA implements the algorithm used in HARM, described in [14], though it evolves an amended version of the 3-velocity as described in [26]. For primitive variable inversion, it has traditionally used the $1D_w$ scheme from [31], with the "Flux-CT" divergence-preserving magnetic field transport described in [55].

However, alternatives to nearly every part of the algorithm are available as different packages, which can be enabled to construct a particular algorithm, or emulate another code's choices for comparisons.

For example, four transport schemes are available for the magnetic field: in addition to "Flux-CT," KHARMA implements the face-centered constrained transport scheme outlined in [44], including a number of schemes for determining the EMFs along zone edges with and without upwinding. Also, though these are rarely used now, it implements a constraint-damping scheme similar to [11] usable in flat spacetimes, as well as a projection operator approach (described in [55]) using a stabilized

Biconjugate Gradient (BiCGStab) solver [58]. These schemes are separated into different packages and made mutually exclusive at run-time so as not to interfere with one another.

In addition to the $1D_w$ primitive variable recovery noted above, KHARMA now implements the robust primitive variable recovery of [21], which greatly improves stability.

KHARMA implements a first-order flux correction (FOFC) scheme [3], which speculatively determines which zones will encounter floors or inversion failures, and pre-emptively uses first order, Local Lax-Friedrichs fluxes with Donor Cell reconstruction to update just those potentially problematic cells, providing extra stability and less injection of floor material, at the cost of using a less accurate scheme in just those zones which would have required intervention. Combined with reliable variable inversion, this allows significantly more stable performance when integrating shocks at high magnetization, allowing for the introduction of fewer stability floors in the simulation.

KHARMA also implements a large number of different variable reconstruction schemes (WENO5 [25], WENOZ [5] including a linearized robust form,[9] MP5 [45], PPM [10], PPMX [9], linear with Van Leer and Monotonized Central limiters [59, 60], and Donor Cell [15]). Other reconstruction schemes can be added easily with just a single-zone implementation in C or C++.

For time integration, KHARMA supports not only the traditional predictor-corrector update used in most HARM codes, but RK1, RK2, and SSPRK(5)4 schemes as well—the KHARMA sub-step is implemented so as to be independent of the integrator used, so any integrator implemented in Parthenon can be immediately used in KHARMA.

When the density or internal energy fall below predetermined values, or become too small relative to the magnetic pressure, extra material and energy are directly injected to keep stability. KHARMA supports injecting floor material in any of a number of different frames, the most common of which is the drift frame (as tested in [40]).

KHARMA also supports theories with arbitrary implicit source terms, such as the "Extended GRMHD" theory of [7] treating viscosity and heat conduction in Coulomb-collisionless plasmas (see Sect. 5.5.1 for details). When evolving a semi-implicit theory, KHARMA supports only a predictor-corrector scheme for integration, as it must be augmented by a nonlinear solve for the new state after each sub-step. In addition, the implicit solver doubles as the primitive variable recovery, so no traditional primitive recovery solver is used.

---

[9] To be characterized in Barker et al. [1] (in preparation).

5.4.2 Package Structure

KHARMA relies on Parthenon as its single direct dependency, and this dictates much of its overall code structure and conventions. Parthenon provides interfaces to make use of MPI communication, HDF5 parallel file operations, and Kokkos loops and `Views`—KHARMA then uses these features almost exclusively through Parthenon. Parthenon additionally provides a framework for code modularization, "packages," which KHARMA uses heavily. In fact, perhaps the best introduction to the code structure is to look at the internal dependency structure between KHARMA's many packages.

A full diagram of KHARMA's package structure is provided in Fig. 5.4. Starting from the bottom row, it shows system dependencies of KHARMA: MPI, HDF5, and an accelerator framework: usually OpenMP, CUDA, HIP, or SYCL. (KHARMA obviously also requires a C++17-compliant compiler.) Next are the dependencies compiled alongside KHARMA: Kokkos and Parthenon. Finally, items within the box are KHARMA components, almost universally packages.

KHARMA's components are effectively split into four levels, detailed in the following four sub-sections.

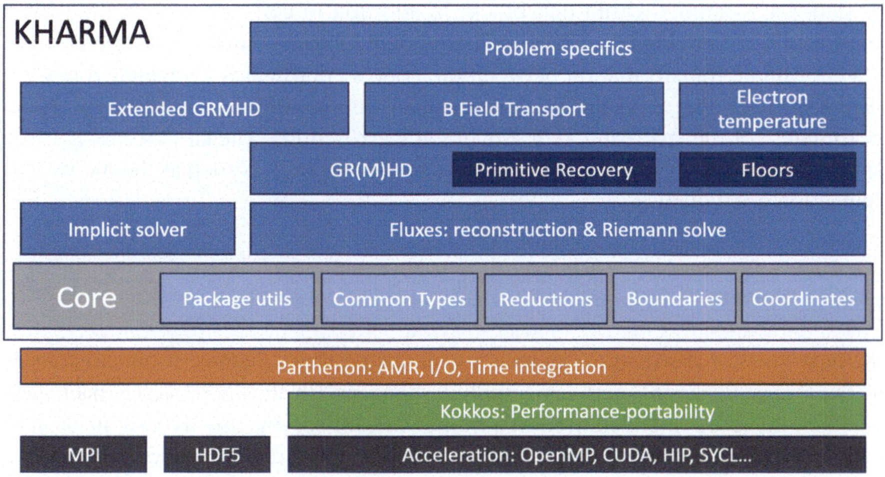

Fig. 5.4 Diagram of KHARMA's structure and dependencies. Dark grey boxes at bottom are system dependencies, above those are the bundled dependencies Kokkos and Parthenon. The large outline marks KHARMA proper. The lightest blue boxes within the grey outline denote core KHARMA features on which all packages depend. Darker blue boxes are KHARMA packages, with the darkest boxes indicating fully-interdependent "sub-packages." Placement of a module above another module or group denotes a direct dependency

5.4.2.1 Core

The core level consists of mostly convenience features, parameters, and common datatypes—anything common to the whole code, which could be used in any other package. Among these only `Globals`, `Reductions`, and `Boundaries` are formal packages.

The `Package` core module/namespace defines the standard interface for packages in KHARMA, which adds features to the Parthenon definition for convenience. KHARMA provides several new callbacks for its packages, structured around common points in a GRMHD update step; e.g., for adding source terms, operator-split physics, or conversions between primitive and conserved variables.

Other core modules house common data structures and code for easily calculating index ranges, as well as code for performing reductions among MPI ranks, used for diagnostics and summary statistics by several different packages. The core modules also include a `Globals` package, which provides more universal access to a couple of useful variables, such as the current simulation time and state (initializing vs. stepping).

Finally, the core includes a differential-geometry-aware coordinates object, `GRCoordinates`, which takes the place of Parthenon's built-in coordinates. A new `GRCoordinates` object is created automatically with each `MeshBlock`, providing data for locations within that block. In addition to cell locations in the code's coordinates, as provided by Parthenon, `GRCoordinates` introduces the concept of a coordinate transformation between the code's coordinates (in which zones are evenly spaced) and a system of "embedding" coordinates representing the basic coordinates for the spacetime (e.g. Spherical Kerr-Schild). The `GRCoordinates` object also carries caches of expensive-to-calculate quantities such as the metric and connection coefficients.

5.4.2.2 Solvers

The next level of abstraction implements reconstruction schemes, Riemann solvers (in the `Flux` package), and semi-implicit stepping (in the `Implicit` package). Together, these are intended to form a framework for solving any hyperbolic system in curved spacetime via finite volumes, allowing for the same code to be shared between different schemes with or without magnetic fields, viscous effects, radiative feedback, etc.

This separation is imperfect still, as the Riemann solver requires determining sound speeds and states of conserved variables from primitive variables, both of which are theory-dependent but must live in the same kernels as the Riemann solve itself, to maintain performance. However, the distinction is enough that KHARMA supports a few theories (hydrodynamics, magnetohydrodynamics, and extended MHD) in a relatively natural way with a minimum of code duplication.

The `Flux` package also handles correcting any fluxes which will lead to floor hits or primitive recovery failures (i.e., first-order flux corrections, see Sect. 5.4.1).

The implicit solver is written to be general, allowing any variables marked with the tag `Metadata::Implicit`, to be evolved semi-implicitly. Any time-dependent source terms acting on the new variable must still be added manually, however. The implicit solver is described in Sect. 5.5.1.

5.4.2.3 Physics

The `GRMHD` package handles updating the primitive hydrodynamic fluid variables associated with a HARM scheme—density, internal energy density, and velocity. It does not evolve the magnetic field components, however. Thus, while it optionally uses a magnetic field to compute conserved variables like stress-energy tensor (and the signal speeds, etc.), it is agnostic to the magnetic field transport scheme, and supports several different transports implemented in different packages.

When evolving the GRMHD variables explicitly, the `Inverter` package handles the numerical inversion of conserved to primitive fluid quantities (when evolved implicitly, the variable inversion is done automatically by the nonlinear solver). As they share substantial structure and code, this package implements both of the available primitive variable recovery schemes (see Sect. 5.4.1), which can be selected at runtime.

The `Floors` package, implementing various density and internal energy floors and injection frames, as well as a number of variable limits, also lives alongside/within the GRMHD algorithm.

5.4.2.4 Extras

There are a few additional packages which enhance the main GRMHD package. Generally, these implement any additional physics beyond evolving the core GRHD variables: electron temperatures, dissipative effects, and so on.

These packages also include the magnetic field transports, solutions for evolving the magnetic field while preserving low divergence. KHARMA supports several different magnetic field transport schemes, but modern simulations generally use the face-centered constrained transport scheme unless another package is required for backward compatibility, testing, or speed.

5.4.2.5 Glue

Time-stepping is orchestrated by the `Driver`, which constructs a list of tasks constituting a single simulation sub-step. KHARMA implements three different time-steps, which produce near-identical results but differ in which variables they synchronize, and the order they update: the first type mirrors most HARM codes in treating the primitive variable state as the fundamental representation, exchanging primitive variable values at boundaries. The second treats conserved state as fundamental and exchanges conserved variables, allowing consistent evolution with just one synchronization per sub-step in some circumstances (generally, if primitive variable recovery failures may need to be fixed, HARM requires two synchronizations). The last step type is a simplified algorithm for debugging and testing, without the additional tasks needed for AMR or face-centered magentic fields.

Fluid initialization code can also take advantage of any package: different problems can require or make optional whatever fields make sense in context. Where there is a common, standard initialization which makes sense (i.e. a nominal "cold" value for electron temperatures which will be heated to equilibrium by the fluid) this is applied if the package is loaded but the field is not otherwise initialized. Initialization routines which require fields not enabled by the user generate errors.

Since different magnetic field transports preserve different representations of the magnetic field divergence, initialization of the magnetic field is dependent on what field transport will be used, and thus is handled separately from problem initialization. Where problems require a specific magnetic field configuration, they can set parameters for this later initialization, so as not to have to implement two magnetic field initializations per problem.

While it contains quite a few different modules, KHARMA is not (yet) a terribly complicated code overall—Fig. 5.4 lays out almost all of the code modules and internal dependencies. Freed from implementing the extensive bookkeeping of mesh refinement or specialized syntax of device code, KHARMA consists almost exclusively of physics-related code, which makes up 3 of the 4 main layers.

5.5 Code Features

5.5.1 Implicit Solver and Extended Theory

KHARMA allows for semi-implicit time-stepping in limited instances, using a non-linear solver to obtain field primitive values at the end of a substep after any conserved fluxes have been applied explicitly. This operation is applied separately to each zone, and necessarily after the fluxes have been applied, allowing KHARMA to accommodate theories with implicit source terms.

The solver itself is a multi-dimensional Newton-Raphson root finder that minimizes a vector-valued loss function over the primitive variables, representing their state at the end of the timestep. From an initial guess consisting of primitive variables at the beginning of the timestep, it computes a Jacobian via the secant method, and evaluates a correction via linear matrix solve (we use a pivoted version of the serial QR decomposition from the `kokkos-kernels`). The process is iterated to desired accuracy, optionally with a backtracking line search for robustness.

The package is able to evolve an arbitrary subset of the overall variables implicitly, from a single-variable solve as in moment closure or constraint-damping schemes, to the semi-implicit evolution of all fluid variables. The latter is required for the "Extended GRMHD" scheme of [7], which KHARMA fully supports.

5.5.1.1 Extended GRMHD

The Extended GRMHD (EGRMHD) theory adapts the equations of ideal GRMHD to include shear viscosity and heat conduction, in a simple way suitable for Coulomb-collisionless systems such as the accretion of material onto Sagittarius A*. In such systems, the length scale for Coulomb collisions $\lambda_{\rm mfp}$ is much longer than the dynamical length scale r_g, which is longer than the Larmor radius ρ_c. This implies collisionality perpendicular to the magnetic field, but collisionless kinetic behavior parallel to it. This separation of directions forms the main assumption underlying the simplifications in EGRMHD.

EGRMHD treats only two additional variables, beyond the eight in ideal GRMHD. The first is ΔP, the pressure anisotropy with respect to the local magnetic field (since isotropy will be maintained in directions perpendicular to the field). The second is q, the scalar heat flux along magnetic field lines.

Figure 5.5 illustrates the scheme by modeling the anisotropic heat conduction on an artificial background magnetic field.

We provide a brief description of the EGRMHD equations following [13]. In contrast to the ideal GRMHD stress-energy tensor, the now-familiar

$$T^{\mu\nu}_{\rm ideal} = (\rho + u + P + b^2) u^\mu u^\nu + (P + b^2/2) g^{\mu\nu} - b^\mu b^\nu \tag{5.1}$$

EGRMHD evolves a version with additional contributions,

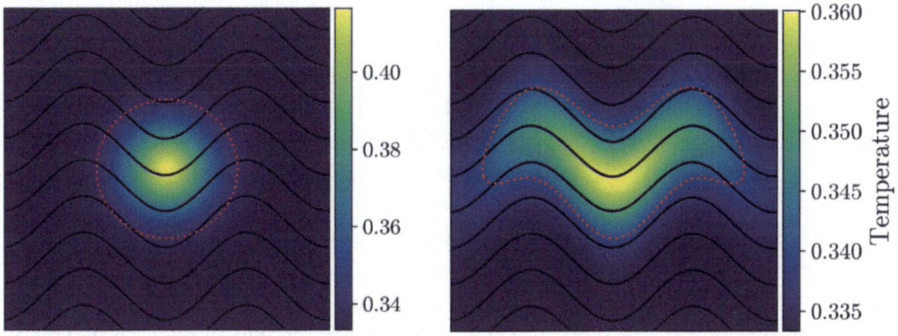

Fig. 5.5 Anisotropic conduction of heat in strong magnetic field, simulated in KHARMA by solving the equations of Extended GRMHD using a semi-implicit scheme. At left is a plot of the temperature in code units at initialization, with a single circular hotspot initialized in the center. The dotted red contour denotes the extent of non-background temperature, i.e. $> 3\%$ above the background value. At right is a later snapshot, showing the thermal energy diffused via conduction along the field lines

$$T^{\mu\nu} = T^{\mu\nu}_{\text{ideal}} + q^{\mu}u^{\nu} + q^{\nu}u^{\mu} + \Pi^{\mu\nu} \tag{5.2}$$

where $q^{\mu} = q\hat{b}^{\mu}$ is the heat flux, $\hat{b}^{\mu} = b^{\mu}/b$ is the unit vector along b^{μ},

$$\Pi^{\mu\nu} = -\Delta P\left(\hat{b}^{\mu}\hat{b}^{\nu} - \frac{1}{3}h^{\mu\nu}\right) \tag{5.3}$$

is the viscous stress-energy tensor, and

$$h^{\mu\nu} = g^{\mu\nu} + u^{\mu}u^{\nu} \tag{5.4}$$

is the projection on a hypersurface orthogonal to the 4-velocity u^{μ}. The total stress-energy tensor is evolved according to the Bianchi identity as usual, and the evolution equations for density and magnetic field are unchanged from ideal GRMHD. This leaves the evolution equations for the heat flux and pressure anisotropy. These are written with respect to rescaled variables

$$\tilde{q} = q\left(\frac{\tau_R}{\chi\rho\Theta^2}\right)^{1/2}, \tag{5.5}$$

$$\Delta\tilde{P} = \Delta P\left(\frac{\tau_R}{\nu\rho\Theta}\right)^{1/2}, \tag{5.6}$$

as

$$\nabla_{\mu}(\tilde{q}u^{\mu}) = -\frac{\tilde{q} - \tilde{q}_0}{\tau_R} + \frac{\tilde{q}}{2}\nabla_{\mu}u^{\mu}, \tag{5.7}$$

$$\nabla_{\mu}(\Delta\tilde{P}u^{\mu}) = -\frac{\Delta\tilde{P} - \Delta\tilde{P}_0}{\tau_R} + \frac{\Delta\tilde{P}}{2}\nabla_{\mu}u^{\mu}. \tag{5.8}$$

That is, $\tilde{q}$ and $\Delta\tilde{P}$ are damped towards their target value $\tilde{q}_0$ and $\Delta\tilde{P}_0$ over a timescale τ_R. The evolution equations follow from the requirement that the model satisfies the second law of thermodynamics, as detailed in [7], and reflect the change of variable introduced in [13]. Both papers additionally detail the closure model used for setting τ_R, which we will not describe in detail here.

In the context of implementation, it suffices for us to note that due to the covariant derivatives on the right-hand side of (5.7) and (5.8), the evolution of q and ΔP must be done locally implicitly, by solving a system of equations involving the time-dependent source terms in order to obtain the state at the end of each step. This is done using KHARMA's implicit solver, by marking the GRMHD and EGRMHD variables (though not the magnetic field!) as implicitly evolved, adding them to the solver's update step.

EGRMHD is a promising avenue for future research, especially in understanding the time variability of sparse accretion systems like Sagittarius A*. Further details on KHARMA's implementation, e.g. simulation floors and stability techniques, will be provided along with the first physics results from KHARMA simulations using the scheme.

5.5.2 *Simulation Re-gridding*

One of the greatest time investments in running high-resolution simulations is the evolution from initial conditions to the steady state. With modern simulations using physically-larger disk sizes which take longer to equilibrate, coupled with greater accuracy requirements necessitating an accurate idea of the variability after erasing initial conditions, just erasing the initial conditions can amount to millions of steps of wasted evolution, costing many GPU hours and days of real time.

It is much more efficient to restart a simulation which was initialized at low resolution, and simply interpolate its state onto a finer grid as desired. KHARMA can linearly interpolate between arbitrary grids when restarting from an existing state, allowing simulations to be arbitrarily refined or de-refined when they are continued. This is automatic—restarting with files of the correct format, if the specified grid is a different size from the snapshot's grid, KHARMA will automatically interpolate to the new grid size and shape.

Linear interpolation of the magnetic field, however, does not preserve the divergence. Thus, KHARMA also automatically cleans the magnetic field of divergence when interpolating, by finding the projection operator to the closest divergence-free field to the interpolated state. KHARMA includes a stabilized biconjugate gradient solver for eliminating magnetic field divergence; this is applied to produce an almost arbitrarily small field divergence, which is then preserved in the subsequent run by any of KHARMA's magnetic field transports.

5.5.3 *Multi-Zone Operation*

One pressing problem in understanding large-scale galactic and cosmic evolution is that of supermassive black hole (SMBH) feedback. Large-scale galactic or cosmic simulations cannot capture individual SMBHs, but must incorporate their accretion and feedback for accuracy. Thus when simulating, they are modeled with simple sub-grid accretion models which make many unverified assumptions.

Ongoing work with KHARMA [8] aims to bridge these scales, conducting chained GRMHD simulations from the event horizon scale all the way out to the galactic scales 10^7 times greater. In these "multi-zone" simulations, the total domain is split into 8 annuli, each spanning two orders of magnitude and overlapping by half. Simulations are conducted of each annulus for a portion of the characteristic timescale, and the fluid state of the simulation's inner portion becomes the outer portion of the next simulation, which is continued for its own characteristic time, and so on. The resulting chain of simulations is then run repeatedly, allowing material and magnetic field to inflow and escape, eventually reaching a steady-state solution.

Using multi-zone simulations, we hope to dramatically improve the feedback models used in large-scale simulations of galactic and cosmic development, both by producing better semi-analytic characterizations usable in galactic simulations, but eventually also by self-consistently alternating between galactic and SMBH-scale simulations, treating the feedback in full 3D.

5.5.4 Run-Time Options

All of KHARMA's features, including all problem setups and the selection of any spacetime geometry, are available at run-time. This means that once compiled, the single resulting binary is capable of doing anything implemented in the code. Those compile-time flags which do exist are almost purely for changing performance characteristics—splitting loops for faster operation on GPUs, or using constants for faster operation in flat spacetimes.

Since compile-time options are so limited, they need only be modified when porting KHARMA to a new machine, not when running a new problem, test, etc. Once a particular compiler stack works, it can be codified as a "machine script" with the module environment, compile options, etc. Further compilations of the code can then use a simple invocation of the top-level script `make.sh`, with just a few optional arguments for specifying the type of build (e.g., debugging vs production).

One other advantage to run-time options is to allow binary and container distribution. This doesn't often matter for KHARMA yet, as most users are also developers, and generally require builds for GPUs and specialized hardware. However, containerized KHARMA has proven useful a few times already, and may prove more useful in the future as container tools become more universal.

5.6 Readiness and Testing

At its core, KHARMA is an implementation of the well-understood HARM scheme for GRMHD. It was designed first to closely match its predecessor code `iharm3D`, which was validated as a part of the EHT collaboration GRMHD code comparison [36]. Where the scheme contains new elements, such as more modern magnetic field transport and primitive variable recovery schemes, these are tested to be drop-in replacements for existing functionality in existing problems. A particular boon for debugging is that KHARMA has enough scheme options that it can usually closely emulate the schemes of other GRMHD codes, to cut down on variables when understanding differences between codes, or specific quirks of behavior.

Each time new code is committed to the main KHARMA repository, a number of regression tests are run: convergence on GRMHD and Extended GRMHD problems (three linear modes and one Bondi accretion problem each) is tested using as many different schemes and features as possible—at current count, 46 ideal MHD linear modes convergence tests are performed, not counting Extended MHD modes or mesh-refined versions of the problem. Figure 5.6 shows these. The L1 norm is checked to match second-order convergence very closely, usually to scale with resolution at a power between -1.9 and -2.1. While KHARMA does not have separate unit tests, nearly all relevant codepaths for physics are tested against alternatives as a part of the regression tests, which isolates bugs in most subsystems. Regression tests are currently run on CPUs automatically, and on GPUs upon manual invocation.

In addition to convergence tests, checks are also performed for determinism upon multiple invocations, and for consistency after restarting from a checkpoint file. The initialization and (short) evolution of a torus problem is also tested, for error and basic consistency metrics such as the magnetic field divergence. Both test types include versions with and without mesh refinement. Another test is run verifying that none of the default parameter files shipped with the code generate errors (due to a change in parameter names or defaults, for example).

Several other well-known standard test problems are implemented which may be run manually, such as the Orszag-Tang vortex [33], the magnetized explosion problem of [22], and others. These are useful for gauging relative performance and stability of different algorithmic choices available in KHARMA, or characterizing KHARMA's performance against other MHD codes.

Finally, KHARMA includes several tests of the electron temperature tracking scheme, adapted from tests of the original implementation in [39], and for the extended GRMHD implementation (see Sect. 5.5.1), adapted from [6]. These will be detailed more closely in a future work, when physics results from these modules are first published.

5.7 Performance

Scaling tests of KHARMA have been performed on many different machines: Stampede2, Frontera, and the Frontera GPU subsystem, Longhorn, at the Texas Advanced Computing Center (TACC), Summit and Frontier at the Oak Ridge Leadership Computing Facility (OLCF), Delta at the National Center for Supercomputing Applications (NCSA) at the University of Illinois, and ThetaGPU at the Argonne Leadership Computing Facility (ALCF). Results here reflect performance running a torus accretion problem in general relativity using ideal GRMHD, in spherical coordinates, with the $1D_w$ inverter and Flux-CT magnetic field transport, for consistency between different KHARMA versions.

KHARMA was initially designed and tested for the Nvidia V100 GPU and IBM AC922 architecture used by Summit and Longhorn, where it achieves about 21 million zones by a cycle (ZCPS) per V100 GPU. This weakly scaled near-perfectly

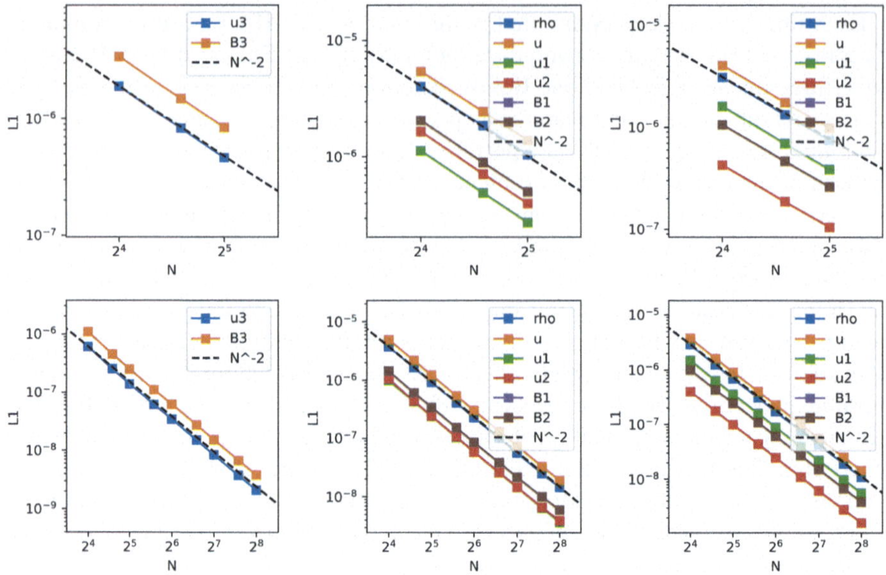

Fig. 5.6 Convergence figures from some of the tests performed on KHARMA after every code commit, illustrating convergence when propagating (from left to right) the Alfvén, fast, and slow modes in 3D (top) and 2D (bottom). Each panel plots the L1 norm of each primitive variable when compared to the analytic solution, versus the simulation resolution. In a second-order scheme, the L1 norm should decrease as the inverse square of resolution, as demonstrated in each test. These plots are taken directly from the most recent automated tests at time of writing (hence the independent legends and axes); these and nearly a hundred other figures corresponding to every test configuration are generated automatically on each code commit and available publicly. A total of 46 linear modes tests are performed, covering all KHARMA's available sub-step types, integrators, magnetic field transports, mesh refinements, and so on. In addition to convergence on the linear modes problems, another 14 different types of correctness and stability tests are performed—where applicable, these each cover a wide array of KHARMA's different configurations

to arbitrary problem sizes/node counts, but more importantly, strongly scaled to problem sizes of interest (e.g., 256^3 total zones) across enough nodes to produce efficient turnaround times of a few days for runs of 10,000 t_g. Numbers from these machines necessarily reflect earlier versions of the code, as KHARMA has not been run on these machines recently (indeed, Longhorn has been retired). Due to a number of new features in the modern code, KHARMA versions from this timeframe are generally slightly faster than the most recent commits.

More recently, KHARMA has primarily targeted A100 GPUs of Chicoma and Delta, which use the now-common architecture pairing AMD EPYC CPUs with Nvidia A100 GPUs (also used by e.g., Polaris at ALCF and other machines). Base performance with A100s is promising: running production problems of 128^3 is nearly twice as fast as on V100 GPUs, or about 35 million Zone-cycles per second per card. Thus on one node of Detla (4 GPUs) KHARMA achieves the performance of more than 50 nodes of Stampede2. KHARMA has not yet been tested at scale on the new

H100 GPUs, which promise some potentially relevant HPC-targeted features as well as higher memory bandwidth than Ampere-generation GPUs.

The current target for performance tuning of KHARMA is the AMD MI250X GPUs of Frontier, at Oak Ridge Leadership Computing Facility (OLCF). Due to register pressure constraints of its complex compute kernels, KHARMA achieves only about 100M ZCPS per node of Frontier—about 25M ZCPS per MI250X card, or 12.5M ZCPS per GPU die. This is only similar to performance on a node of Summit, despite the increased memory bandwidth and compute available on Frontier—thus if KHARMA were fully memory-bandwidth constrained, performance on Frontier could theoretically improve by a factor of 2.

Finally, it is important to note that performance of KHARMA is widely variable, depending on the individual components used. Different primitive variable solvers, magnetic field transports, spacetime geometries, and stability strategies (floors, corrections, etc.), and the presence of SMR/AMR will significantly affect performance. The configuration used for these tests is production-ready but relatively simple compared to KHARMA's modern feature set, as the hope is to be as directly comparable as possible across many different performance tests over years of development. More advanced features can run 10–30% more slowly than the base configuration here.

One feature bears special mention: the semi-implicit mode for running "Extended GRMHD" problems (see Sect. 5.5.1) is very costly. On CPUs, KHARMA runs EGRMHD problems about five times more slowly than ideal problems, and on GPUs the factor is even worse, between 10 and 20 times depending on the hardware. The semi-implicit evolution of EGRMHD problems is more computationally complex than explicit evolution, but performance on GPUs appears to be a factor of 2-4x below what might be expected given CPU performance. This is because the complex GPU kernels of the nonlinear solver require many registers for each thread, severely reducing thread occupancy by filling each block's available registers well before the

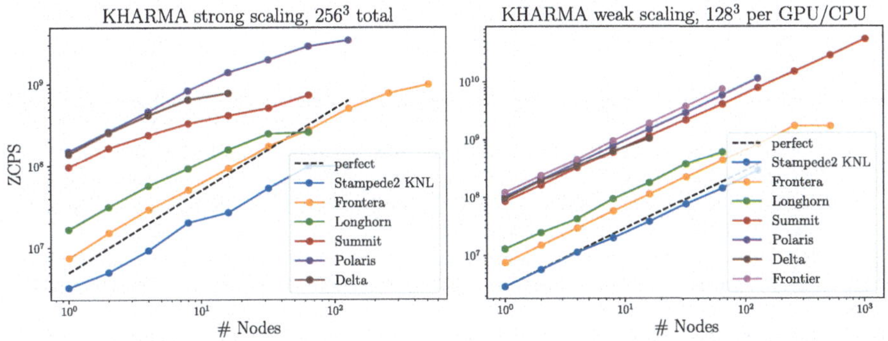

Fig. 5.7 (Left) Strong scaling plot showing KHARMA's performance in ZCPS plotted against the number of GPUs or CPU nodes used, when run on a problem of constant size, 256^3 zones (scaling results on Delta use a slightly larger problem, 384^3 zones). (Right) Weak scaling performance on problems of increasing size, allocating 128^3 zones per GPU. Scaling results from [38] and personal testing. Results on Delta courtesy of Vedant Dhruv (private communication)

available execution units. Work is ongoing to solve this, but must balance the thread occupancy with increased memory bandwidth requirements from splitting kernels, requiring use of global memory to store intermediate results (Fig. 5.7).

Acknowledgements Many thanks to all of the KHARMA developers and users, especially Vedant Dhruv, Hyerin Cho, and César Díaz-Blanco, and to Charles Gammie for useful advice, and of course for HARM. BP gratefully acknowledges support from the Los Alamos National Laboratory Metropolis Fellowship. Computing time for developing, testing, and applying KHARMA has been provided by Oak Ridge Leadership Computing Facility (OLCF) and Argonne Leadership Computing Facility (ALCF) through director's discretionary allocations and through the INCITE program, the National Center for Supercomputing Applications through the Advanced Cyberinfrastructure Coordination Ecosystem: Services & Support (ACCESS) program, the Texas Advanced Computing Center through the Extreme Science and Engineering Development Environment (XSEDE) and Frontera Large-Scale Community Partnerships (LSCP) programs, and by Los Alamos National Laboratory Institutional Computing.

References

1. B. Barker, M. Gogilashvili, J. Rodriguez-Bueno, C. Fields, J. Dolence, J. Miller, J. Murphy, L. Roberts, B. Ryan, *Phoebus: performance portable GRRMHD for relativistic astrophysics* (2024). https://arxiv.org/abs/2410.09146
2. D.A. Beckingsale, T.R. Scogland, J. Burmark, R. Hornung, H. Jones, W. Kil lian, A.J. Kunen, O. Pearce, P. Robinson, B.S. Ryujin, RAJA: portable performance for large-scale scientific applications, in *2019 IEEE/ACM In ternational Workshop on Performance, Portability and Productivity in HPC (P3HPC)* (IEEE, Denver, CO, USA, 2019), pp. 71–81, ISBN: 978-1-72816-003-0. https://doi.org/10.1109/P3HPC49587.2019.00012
3. K. Beckwith, J.M. Stone, A second-order godunov method for multi-dimensional relativistic magnetohydrodynamics. Astrophys. J. Suppl. Ser. **193**(1), 6 (2011). ISSN: 0067-0049. https://doi.org/10.1088/0067-0049/193/1/6
4. H. Carter Edwards, C.R. Trott, D. Sunderland, Kokkos: enabling many core performance portability through polymorphic memory access patterns. J. Parallel Distrib. Comput. Domain-Specific Languages and High-Level Frameworks for High-Performance Computing **74**(12), 3202–3216 (2014). ISSN: 0743-7315. https://doi.org/10.1016/j.jpdc.2014.07.003. Accessed from 1 Dec 2014
5. M. Castro, B. Costa, W.S. Don, High order weighted essentially non oscillatory WENO-Z schemes for hyperbolic conservation laws. J. Comput. Phys. **230**(5), 1766–1792 (2011), ISSN: 0021-9991. https://doi.org/10.1016/j.jcp.2010.11.028. Accessed from 1 Mar 2011
6. M. Chandra, F. Foucart, C.F. Gammie, Grim: a flexible, conservative scheme for relativistic fluid theories. Astrophys. J. **837**, p. 92 (2017). ISSN: 0004-637X. https://doi.org/10.3847/1538-4357/aa5f55. Accessed from 1 Mar 2017
7. M. Chandra, C.F. Gammie, F. Foucart, E. Quataert, an extended mag netohydrodynamics model for relativistic weakly collisional plasmas. Astrophys. J. **810**, 162 (2015)/ ISSN: 0004-637X. https://doi.org/10.1088/0004-637X/810/2/162. Accessed from 1 Sept 2015
8. H. Cho, C. Prather, R. Narayan, P. Natarajan, K.-Y. Su, A. Ricarte, K. Chatterjee, Bridging scales in black hole accretion and feedback: magnetized bondi accretion in 3D GRMHD. Astrophys. J. Lett. **959**(2), L22 (2023), ISSN: 2041-8205. https://doi.org/10.3847/2041-8213/ad1048
9. P. Colella, M.D. Sekora, A limiter for PPM that preserves accuracy at smooth extrema. J. Comput. Phys. **227**, 7069- 7076 (2008). ISSN: 0021-9991. https://doi.org/10.1016/j.jcp.2008.03.034. Accessed from 1 June 2008

10. P. Colella, P.R. Woodward, The piecewise parabolic method (ppm) for gas-dynamical simulations. J. Comput. Phys. **54**(1), 174–201 (1984). ISSN: 0021-9991. https://doi.org/10.1016/0021-9991(84)90143-8. Accessed from 1 Apr 1984
11. A. Dedner, F. Kemm, D. Kroner, C. .-.-D. Munz, T. Schnitzer, M. Wesenberg, Hyperbolic divergence cleaning for the MHD equations. J. Comput. Phys. **175**(2), 645–673 (2002), ISSN: 0021-9991. https://doi.org/10.1006/jcph.2001.6961. Accessed from 20 Jan 2002
12. R. Emami, A. Ricarte, G. N. Wong, D. Palumbo, D. Chang, S.S. Doeleman, A.E. Broderick, R. Narayan, M. Wielgus, L. Blackburn, C. Prather, A.A. Chael, R. Anantua, K. Chatterjee, I. Marti-Vidal, J.L. Gomez, K. Akiyama, M. Liska, L. Hernquist, G. Tremblay, M. Vogelsberger, C. Alcock, R. Smith, J. Steiner, P. Tiede, F. Roelofs, Unraveling twisty linear polarization morphologies in black hole images. Astrophys. J. **950**(1), 38. ISSN: 0004-637X. https://doi.org/10.3847/1538-4357/acc8cd
13. F. Foucart, M. Chandra, C.F. Gammie, E. Quataert, Evolution of accretion discs around a kerr black hole using extended magnetohydrodynamics. Mon. Not. Roy. Astron. Soc. **456**, 1332–1345 (2016). ISSN: 0035-8711. https://doi.org/10.1093/mnras/stv2687. Accessed from 1 Feb 2016
14. C.F. Gammie, J.C. McKinney, G. Toth, HARM: a numerical scheme for general relativistic magnetohydrodynamics. Astrophys. J. **589**(1), 444 (2003), ISSN: 0004-637X. https://doi.org/10.1086/374594. Accessed from 20 May 2003
15. S.K. Godunov, A difference method for numerical calculation of discontinuous solutions of the equations of hydrodynamics. Matematicheskii? Sbornik. Novaya Seriya, **47**(89), 271–306 (1959). ISSN: 0368-8666
16. R. Gold, A. E. Broderick, Z. Younsi, C. M. Fromm, C. F. Gammie, M. Moscibrodzka, H.-Y. Pu, T. Bronzwaer, J. Davelaar, J. Dexter, D. Ball, C.-k. Chan, T. Kawashima, Y. Mizuno, B. Ripperda, K. Akiyama, A. Alberdi, W. Alef, K. Asada, R. Azulay, A.-K. Baczko, M. Balokovic, J. Barrett, D. Bintley, L. Blackburn, W. Boland, K. L. Bouman, G. C. Bower, M. Bremer, C. D. Brinkerink, R. Brissenden, S. Britzen, D. Broguiere, D.-Y. Byun, J. E. Carl strom, A. Chael, K. Chatterjee, S. Chatterjee, M.-T. Chen, Y. Chen, I. Cho, P. Christian, J. E. Conway, J. M. Cordes, G. B. Crew, Y. Cui, M. De Lau rentis, R. Deane, J. Dempsey, G. Desvignes, S. S. Doeleman, R. P. Eatough, H. Falcke, V. L. Fish, E. Fomalont, R. Fraga-Encinas, B. Freeman, P. Friberg, J. L. Gomez, P. Galison, R. García, O. Gentaz, B. Georgiev, C. Goddi, M. Gu, M. Gurwell, K. Hada, M. H. Hecht, R. Hesper, L. C. Ho, P. Ho, M. Honma, C.-W. L. Huang, L. Huang, D. H. Hughes, M. Inoue, S. Issaoun, D. J. James, B. T. Jannuzi, M. Janssen, B. Jeter, W. Jiang, A. Jimenez-Rosales, M. D. Johnson, S. Jorstad, T. Jung, M. Karami, R. Karuppusamy, G. K. Keating, M. Kettenis, J.-Y. Kim, J. Kim, J. Kim, M. Kino, J. Y. Koay, P. M. Koch, S. Koyama, M. Kramer, C. Kramer, T. P. Krichbaum, C.-Y. Kuo, T. R. Lauer, S.-S. Lee, Y.-R. Li, Z. Li, R. Lico, M. Lindqvist, K. Liu, E. Liuzzo, W.-P. Lo, A. P. Lobanov, L. Loinard, C. Lonsdale, R.-S. Lu, N. R. MacDonald, S. Markoff, J. Mao, D. P. Marrone, A. P. Marscher, I. Martí-Vidal, S. Mat sushita, L. D. Matthews, L. Medeiros, K. M. Menten, I. Mizuno, J. M. Moran, K. Moriyama, C. Muller, H. Nagai, M. Nakamura, N. M. Nagar, R. Narayan, G. Narayanan, I. Natarajan, R. Neri, C. Ni, A. Noutsos, H. Okino, G. N. Ortiz-Leon, T. Oyama, F. Ozel, D. C. M. Palumbo, J. Park, N. Patel, U.-L. Pen, D. W. Pesce, R. Plambeck, V. Pietu, A. PopStefanija, O. Porth, J. A. Preciado-Lopez, D. Psaltis, V. Ramakrishnan, R. Rao, M. G. Rawlings, A. W. Raymond, L. Rezzolla, F. Roelofs, A. Rogers, E. Ros, M. Rose, A. Rosha nineshat, H. Rottmann, A. L. Roy, C. Ruszczyk, K. L. J. Rygl, S. Sanchez, D. Sanchez-Arguelles, M. Sasada, T. Savolainen, K.-F. Schuster, F. P. Schloerb, L. Shao, Z. Shen, D. Small, B. W. Sohn, J. SooHoo, P. Tiede, F. Tazaki, R. P. J. Tilanus, M. Titus, K. Toma, P. Torne, T. Trent, T. Traianou, S. Trippe, S. Tsuda, H. J. van Langevelde, I. van Bemmel, D. R. van Rossum, J. Wag ner, J. Wardle, N. Wex, J. Weintroub, R. Wharton, M. Wielgus, G. N. Wong, Q. Wu, D. Yoon, K. Young, A. Young, F. Yuan, Y.-F. Yuan, J. A. Zensus, G. Zhao, S.-S. Zhao, Z. Zhu, and Event Horizon Telescope Collaboration, Veri fication of Radiative Transfer Schemes for the EHT. Astrophys. J. **897**, 148 (2020). ISSN: 0004-637X. https://doi.org/10.3847/1538-4357/ab96c6
17. P. Grete, J.C. Dolence, J.M. Miller, J. Brown, B. Ryan, A. Gaspar, F. Glines, S. Swaminarayan, J. Lippuner, C.J. Solomon, G. Shipman, C. Junghans, D. Holladay, J.M. Stone, Parthenon - a

performance portable block structured adaptive mesh refinement framework (2022). Accessed from 1 Feb 2022

18. L. Ji, V. Mewes, Y. Zlochower, L. Ennoggi, F.G.L. Armengol, M. Campan elli, F. Cipolletta, Z.B. Etienne, Ameliorating the Courant-Friedrichs Lewy condition in spherical coordinates: a double FFT filter method for gen eral relativistic MHD in dynamical spacetimes. Phys. Rev. D **108**(10), 104 005 (2023). https://doi.org/10.1103/PhysRevD.108.104005
19. M.D. Johnson, A. Lupsasca, A. Strominger, G.N. Wong, S. Hadar, D. Kapec, R. Narayan, A. Chael, C.F. Gammie, P. Galison, D.C.M. Palumbo, S.S. Doeleman, L. Blackburn, M. Wielgus, D.W. Pesce, J.R. Farah, J.M. Moran, Universal interferometric signatures of a black hole's photon ring. Sci. Adv. **6**(12), eaaz1310 (2020). ISSN: 2375-2548. https://doi.org/10.1126/sciadv.aaz1310
20. N. P. Jouppi, C. Young, N. Patil, D. Patterson, G. Agrawal, R. Bajwa, S. Bates, S. Bhatia, N. Boden, A. Borchers, R. Boyle, P.-l. Cantin, C. Chao, C. Clark, J. Coriell, M. Daley, M. Dau, J. Dean, B. Gelb, T. V. Ghaemmaghami, R. Gottipati, W. Gulland, R. Hagmann, C. R. Ho, D. Hogberg, J. Hu, R. Hundt, D. Hurt, J. Ibarz, A. Jaffey, A. Jaworski, A. Kaplan, H. Khaitan, D. Killebrew, A. Koch, N. Kumar, S. Lacy, J. Laudon, J. Law, D. Le, C. Leary, Z. Liu, K. Lucke, A. Lundin, G. MacKean, A. Maggiore, M. Mahony, K. Miller, R. Nagarajan, R. Narayanaswami, R. Ni, K. Nix, T. Norrie, M. Omer nick, N. Penukonda, A. Phelps, J. Ross, M. Ross, A. Salek, E. Samadiani, C. Severn, G. Sizikov, M. Snelham, J. Souter, D. Steinberg, A. Swing, M. Tan, G. Thorson, B. Tian, H. Toma, E. Tuttle, V. Vasudevan, R. Walter, W. Wang, E. Wilcox, D. H. Yoon, In-datacenter performance analysis of a tensor processing unit, in *Proceedings of the 44th Annual International Symposium on Computer Architecture, ser. ISCA '17* (Association for Computing Machinery, New York, 2017), pp. 1-12, ISBN: 978- 1-4503-4892-8. https://doi.org/10.1145/3079856.3080246
21. W. Kastaun, J. Vijay Kalinani, R. Ciolfi, *Robust Recovery of Primi tive Variables in Relativistic Ideal Magnetohydrodynamics*. arXiv e-prints, vol. 2005, arXiv:2005.01821, May 1, 2020
22. S.S. Komissarov, A Godunov-type scheme for relativistic magnetohydro dynamics. Monthly Notices R. Astron. Soc. **303**, 343–366, (1999). ISSN: 0035-8711. https://doi.org/10.1046/j.1365-8711.1999.02244.x
23. G. Lauterbach, The path to successful wafer-scale integration: the cere bras story. IEEE Micro **41**(6), 52–57 (2021). ISSN: 0272- 1732, 1937-4143. https://doi.org/10.1109/MM.2021.3112025
24. J.Y.-Y. Lin, G.N. Wong, C. Prather, C.F. Gammie, Feature Ex traction on Synthetic Black Hole Images. arXiv e-prints, arXiv:2007.00794, Jul. 1, 2020
25. X.-D. Liu, S. Osher, T. Chan, Weighted essentially non-oscillatory schemes. J. Comput. Phys. **115**(1), 200–212 (1994). ISSN: 0021-9991. https://doi.org/10.1006/jcph.1994.1187
26. J.C. McKinney, C.F. Gammie, A Measurement of the Electromagnetic Luminosity of a Kerr Black Hole. Astrophys. J. **611**(2), 977 (2004). ISSN: 0004-637X. https://doi.org/10.1086/422244
27. D.S. Medina, A. St-Cyr, T. Warburton, OCCA: A unified approach to multi-threading languages (2014). arXiv: 1403.0968 [cs], preprint
28. H. Meuer, E. Strohmaier, J. Dongarra, H. Simon, M. Meuer. (Nov. 2023). Top500 Supercomputers List
29. G. Moore, Progress In Digital Integrated Electronics. Technical Digest 1975. International Electron Devices Meeting (1975), pp. 11–13. https://doi.org/10.1109/N-SSC.2006.4804410
30. G.E. Moore, Cramming more components onto integrated circuits. Electronics **38**(8), 114 (1965). ISSN: 1098-4232. https://doi.org/10.1109/N-SSC.2006.4785860. Accessed from 19 Apr 1965
31. S.C. Noble, C.F. Gammie, J.C. McKinney, L. Del Zanna, Primitive variable solvers for conservative general relativistic magnetohydrodynam ics. Astrophys. J. **641**, 626–637 (2006). ISSN: 0004-637X. https://doi.org/10.1086/500349
32. I.D. Novikov, K.S. Thorne, Astrophysics of black holes, presented at the Black Holes (Les Astres Occlus), 1973, pp. 343-450

33. S.A. Orszag, C.-M. Tang, Small-scale structure of two-dimensional magnetohydrodynamic turbulence. J. Fluid Mech. **90**(1), 129–143 (1979). ISSN: 1469-7645, 0022-1120. https://doi.org/10.1017/S002211207900210X
34. D.C. M. Palumbo, G.N. Wong, C. Prather, Discriminating Accretion States via Rotational Symmetry in Simulated Polarimetric Images of M87. Astrophys. J. **894**, 156 (2020). ISSN: 0004-637X. https://doi.org/10.3847/1538-4357/ab86ac
35. PassMark Software (2022). PassMark - CPU Benchmarks - CPU Mega Page - Detailed List of Benchmarked CPUs
36. O. Porth, K. Chatterjee, R. Narayan, C.F. Gammie, Y. Mizuno, P. Anni nos, J.G. Baker, M. Bugli, C.-k. Chan, J. Davelaar, L. Del Zanna, Z.B. Etienne, P.C. Fragile, B.J. Kelly, M. Liska, S. Markoff, J.C. McKinney, B. Mishra, S.C. Noble, H. Olivares, C. Prather, L. Rezzolla, B.R. Ryan, J.M. Stone, N. Tomei, C. J. White, Z. Younsi, K. Akiyama, A. Alberdi, W. Alef, K. Asada, R. Azulay, A.-K. Baczko, D. Ball, M. Balokovic, J. Barrett, D. Bintley, L. Blackburn, W. Boland, K. L. Bouman, G. C. Bower, M. Bremer, C. D. Brinkerink, R. Brissenden, S. Britzen, A. E. Broderick, D. Broguiere, T. Bronzwaer, D.-Y. Byun, J. E. Carlstrom, A. Chael, S. Chatterjee, M.-T. Chen, Y. Chen, I. Cho, P. Christian, J. E. Conway, J. M. Cordes, Geoffrey, B. Crew, Y. Cui, M. De Laurentis, R. Deane, J. Dempsey, G. Desvignes, S. S. Doele man, R. P. Eatough, H. Falcke, V. L. Fish, E. Fomalont, R. Fraga-Encinas, B. Freeman, P. Friberg, C. M. Fromm, J. L. Gomez, P. Galison, R. García, O. Gentaz, B. Georgiev, C. Goddi, R. Gold, M. Gu, M. Gurwell, K. Hada, M. H. Hecht, R. Hesper, L. C. Ho, P. Ho, M. Honma, C.-W. L. Huang, L. Huang, D. H. Hughes, S. Ikeda, M. Inoue, S. Issaoun, D. J. James, B. T. Jan nuzi, M. Janssen, B. Jeter, W. Jiang, M. D. Johnson, S. Jorstad, T. Jung, M. Karami, R. Karuppusamy, T. Kawashima, G. K. Keating, M. Kettenis, J.-Y. Kim, J. Kim, J. Kim, M. Kino, J. Y. Koay, Patrick, M. Koch, S. Koyama, M. Kramer, C. Kramer, T. P. Krichbaum, C.-Y. Kuo, T. R. Lauer, S.-S. Lee, Y.-R. Li, Z. Li, M. Lindqvist, K. Liu, E. Liuzzo, W.-P. Lo, A. P. Lobanov, L. Loinard, C. Lonsdale, R.-S. Lu, N. R. MacDonald, J. Mao, D. P. Marrone, A. P. Marscher, I. Martí-Vidal, S. Matsushita, L. D. Matthews, L. Medeiros, K. M. Menten, I. Mizuno, J. M. Moran, K. Moriyama, M. Moscibrodzka, C. Muller, H. Nagai, N. M. Nagar, M. Nakamura, G. Narayanan, I. Natarajan, R. Neri, C. Ni, A. Noutsos, H. Okino, T. Oyama, F. Ozel, D. C. M. Palumbo, N. Patel, U.-L. Pen, D. W. Pesce, V. Pietu, R. Plambeck, A. PopStefanija, J. A. Preciado-Lopez, D. Psaltis, H.-Y. Pu, V. Ramakrishnan, R. Rao, M. G. Rawlings, A. W. Raymond, B. Ripperda, F. Roelofs, A. Rogers, E. Ros, M. Rose, A. Roshanineshat, H. Rottmann, A. L. Roy, C. Ruszczyk, K. L. J. Rygl, S. Sanchez, D. Sánchez-Arguelles, M. Sasada, T. Savolainen, F. P. Schloerb, K.-F. Schuster, L. Shao, Z. Shen, D. Small, B. W. Sohn, J. SooHoo, F. Tazaki, P. Tiede, R. P. J. Tilanus, M. Titus, K. Toma, P. Torne, T. Trent, S. Trippe, S. Tsuda, I. van Bemmel, H. J. van Langevelde, D. R. van Rossum, J. Wagner, J. Wardle, J. Weintroub, N. Wex, R. Wharton, M. Wielgus, G. N. Wong, Q. Wu, K. Young, A. Young, F. Yuan, Y.-F. Yuan, J. A. Zensus, G. Zhao, S.-S. Zhao, Z. Zhu, and Event Horizon Telescope Collaboration, The Event Hori zon General Relativistic Magnetohydrodynamic Code Comparison Project. Astrophys. J. Suppl. Ser. **243**, 26 (2019). ISSN: 0067-0049. https://doi.org/10.3847/1538-4365/ab29fd
37. C. Prather, G.N. Wong, V. Dhruv, B.R. Ryan, J.C. Dolence, S.M. Ressler, C.F. Gammie, iharm3D: vectorized general relativistic magnetohy drodynamics. J. Open Sourc. Softw. **6**(66), 3336 (2021). ISSN: 2475-9066. https://doi.org/10.21105/joss.03336
38. C. Prather, GRMHD simulations and analysis of polarized emission from black hole accretion systems. Thesis, University of Illinois at Urbana-Champaign, Sep. 8, 2022
39. S.M. Ressler, A. Tchekhovskoy, E. Quataert, M. Chandra, C.F. Gammie, Electron thermodynamics in GRMHD simulations of low-luminosity black hole accretion. Monthly Notices R. Astron. Soc. **454**, 1848–1870 (2015). ISSN: 0035-8711. https://doi.org/10.1093/mnras/stv2084
40. S.M. Ressler, A. Tchekhovskoy, E. Quataert, C.F. Gammie, The disc jet symbiosis emerges: modelling the emission of Sagittarius A* with elec tron thermodynamics. Mon. Not. Roy. Astron. Soc. **467**, 3604–3619 (2017). ISSN: 0035-8711. https://doi.org/10.1093/mnras/stx364. Accessed from 1 May 2017

41. A. Ricarte, C. Gammie, R. Narayan, C. Prather, Probing plasma physics with spectral index maps of accreting black holes on event horizon scales. Monthly Notices R. Astron. Soc. **519**(3), 4203–4220 (2023). ISSN: 0035-8711. https://doi.org/10.1093/mnras/stac3796
42. A. Ricarte, R. Qiu, R. Narayan, Black hole magnetic fields and their imprint on circular polarization images. Monthly Notices R. Astron. Soc. **505**, 523–539 (2021). ISSN: 0035-8711. https://doi.org/10.1093/mnras/stab1289
43. B.R. Ryan, J.C. Dolence, C.F. Gammie, Bhlight: general relativistic radiation magnetohydrodynamics with Monte Carlo transport. Astrophys. J. **807**(1), 31 (2015). ISSN: 0004-637X. https://doi.org/10.1088/0004-637X/807/1/31
44. J.M. Stone, T. Gardiner, A simple unsplit Godunov method for multidi mensional MHD. New Astron. **14**(2), 139–148 (2009). ISSN: 1384-1076. https://doi.org/10.1016/j.newast.2008.06.003
45. A. Suresh, H.T. Huynh, Accurate monotonicity-preserving schemes with Runge-Kutta time stepping. J. Comput. Phys. **136**(1), 83–99 (1997). ISSN: 0021-9991. https://doi.org/10.1006/jcph.1997.5745
46. E. Talpes, D.D. Sarma, D. Williams, S. Arora, T. Kunjan, B. Floering, A. Jalote, C. Hsiong, C. Poorna, V. Samant, J. Sicilia, A.K. Nivarti, R. Ra machandran, T. Fischer, B. Herzberg, B. McGee, G. Venkataramanan, P. Banon, The microarchitecture of DOJO, Tesla's Exa-scale computer. IEEE Micro **43**(3), 31–39 (2023). ISSN: 0272-1732, 1937- 4143. https://doi.org/10.1109/MM.2023.3258906
47. TechPowerUp. (Mar. 2024). TechPowerUp GPU Specs Database, TechPowerUp
48. The Event Horizon Telescope Collaboration et al., First Sagittarius A* event horizon telescope results. VIII. Physical interpretation of the polarized ring. Astrophys. J. Lett. **964**(2), L26 (2024), ISSN: 2041-8205. https://doi.org/10.3847/2041-8213/ad2df1
49. The Event Horizon Telescope Collaboration et al., First Sagittarius A* event horizon telescope results. V. testing astrophysical models of the galactic center black hole. Astrophys. J. Lett. **930**(2), L16 (2022). ISSN: 2041-8205. https://doi.org/10.3847/2041-8213/ac6672. Accessed from 12 May 2022
50. The Event Horizon Telescope Collaboration et al., First M87 event horizon telescope results. I. The shadow of the super massive black hole. Astrophys. J. **875**, L1 (2019). ISSN: 0004-637X. https://doi.org/10.3847/2041-8213/ab0ec7 Accessed from 1 Apr 2019
51. The Event Horizon Telescope Collaboration et al., First M87 Event Horizon Telescope Results. IV. Imaging the central supermassive black hole. Astrophys. J. **875**, L4, Apr. 1, 2019, ISSN: 0004-637X. https://doi.org/10.3847/2041-8213/ab0e85
52. The Event Horizon Telescope Collaboration et al., First M87 Event Horizon Telescope Results. V. Physical Origin of the Asymmetric Ring. Astrophys. J. **875**, L5, Apr. 1, 2019, ISSN: 0004-637X. https://doi.org/10.3847/2041-8213/ab0f43
53. The Event Horizon Telescope Collaboration et al., First M87 Event Horizon Telescope Results. VI. The Shadow and Mass of the Central Black Hole. Astrophys. J. **875**, L6 (2019). ISSN: 0004-637X. https://doi.org/10.3847/2041-8213/ab1141
54. The Event Horizon Telescope Collaboration et al., First M87 Event Horizon Tele scope Results. VIII. Magnetic Field Structure near The Event Horizon. Astrophys. J. **910**, L13, (2021). ISSN: 0004-637X. https://doi.org/10.3847/2041-8213/abe4de
55. G. Toth, The $\nabla \cdot B=0$ constraint in shock-capturing magnetohydrodynam ics codes. J. Comput. Phys. **161**(2), 605–652 (2000). ISSN: 0021-9991. https://doi.org/10.1006/jcph.2000.6519
56. C.R. Trott, D. Lebrun-Grandie, D. Arndt, J. Ciesko, V. Dang, N. Elling-wood, R. Gayatri, E. Harvey, D.S. Hollman, D. Ibanez, N. Liber, J. Madsen, J. Miles, D. Poliakoff, A. Powell, S. Rajamanickam, M. Simberg, D. Sunder land, B. Turcksin, J. Wilke, Kokkos 3: programming model extensions for the exascale era. IEEE Trans. Parallel Distrib. Syst. **33**(4), 805–817 (20222). ISSN: 1558-2183. https://doi.org/10.1109/TPDS.2021.3097283
57. C. Trott, L. Berger-Vergiat, D. Poliakoff, S. Rajamanickam, D. Lebrun Grandie, J. Madsen, N. Al Awar, M. Gligoric, G. Shipman, G. Womel dorff, The Kokkos EcoSystem: comprehensive performance portability for high performance computing. Comput. Sci. Engin. **23**(5), 10–18 (2021). ISSN: 1558-366X. https://doi.org/10.1109/MCSE.2021.3098509

58. H.A. van der Vorst, Bi-CGSTAB: a fast and smoothly converging variant of Bi-CG for the solution of nonsymmetric linear systems. SIAM J. Sci. Stat. Comput. **13**(2), 631–644 (1992). ISSN: 0196-5204. https://doi.org/10.1137/0913035
59. B. van Leer, Towards the ultimate conservative difference scheme. II. Mono tonicity and conservation combined in a second-order scheme. J. Comput. Phys. **14**(4), 361–370 (1974). ISSN: 0021-9991. https://doi.org/10.1016/0021-9991(74)90019-9
60. B. van Leer, Towards the ultimate conservative difference scheme. III. Upstream centered finite-difference schemes for ideal compressible flow. J. Comput. Phys. **23**, 263–275 (1977). ISSN: 0021- 9991. https://doi.org/10.1016/0021-9991(77)90094-8
61. C.J. White, P.D. Mullen, Y.-F. Jiang, S.W. Davis, J.M. Stone, V. Mo rozova, L. Zhang, An extension of the Athena++ code framework for radiation-magnetohydrodynamics in general relativity using a finite-solid angle discretization. Astrophys. J. **949**(2), 103 (2023). ISSN: 0004-637X. https://doi.org/10.3847/1538-4357/acc8cf
62. G.N. Wong, C. Prather, V. Dhruv, B.R. Ryan, M. Moscibrodzka, C.-k. Chan, A.V. Joshi, R. Yarza, A. Ricarte, H. Shiokawa, J.C. Dolence, S.C. Noble, J.C. McKinney, C.F. Gammie, PATOKA: simulating elec tromagnetic observables of black hole accretion. Astrophys. Suppl. Ser. **259**, 64 (2022). ISSN: 0067-0049. https://doi.org/10.3847/1538-4365/ac582e

Chapter 6
Einstein Toolkit

Samuel Cupp, Steven R. Brandt, Lucas T. Sanches, Roland Haas, Leonardo R. Werneck, and Philipp Mösta

Abstract The Einstein Toolkit is one of the oldest open-source, freely-accessible, community-driven code frameworks for studying fully general relativistic systems such as black holes, neutron stars, and supernovae. The toolkit was developed by an international collaboration of students and professors working at many institutions. The toolkit is built in layers, offering generic computational infrastructure components at its lowest level—e.g., adaptive mesh refinement, method of lines, and I/O—and base modules which provide commonly-defined variables for an evolution system at a higher level. The base level provides a communication interface for more specific spacetime formulations, hydrodynamical schemes, and analysis tools like horizon finders, gravitational wave extractors, ejecta trackers, and others.

S. Cupp
Department of Physics, University of Idaho, Moscow, USA
e-mail: scupp1@my.apsu.edu

S. R. Brandt
Center for Computation and Technology, Luoisiana State Univesity, Baton Rouge, USA
e-mail: sbrandt@cct.lsu.edu

L. T. Sanches
Programa De Engenharia De Sistemas E Computação, Universidade Federal Do Rio De Janeiro (UFRJ), Rio De Janeiro, Brazil

R. Haas (✉)
National Center for Supercomputing Applications, Univesity of Illinois, Champaign, USA
e-mail: rhaas@illinois.edu

L. R. Werneck
Department of Physics, University of Idaho, Moscow, USA

P. Mösta
GRAPPA, Anton Pannekoek Institute for Astronomy and Institute of High-Energy Physics, University of Amsterdam, Science Park 904, 1098 XH Amsterdam, The Netherlands
e-mail: p.moesta@uva.nl

C. Bambi et al. (eds.), *New Frontiers in GRMHD Simulations*, Springer Series in Astrophysics and Cosmology, https://doi.org/10.1007/978-981-97-8522-3_6

6.1 Cactus/Carpet

The Einstein Toolkit (ET) [37] is based on the Cactus code [25, 49], originally developed at the NCSA and the Max Planck Institute for Gravitational Physics by Paul Walker and Joan Massó. Cactus has been used on high-performance computing (HPC) systems for nearly three decades while being continuously updated to support new hardware technologies [89]. During all this time, it has been regularly used for production-level science involving complex astrophysical simulations of colliding black hole (BH) [44, 53, 54], colliding neutron star (NS) [27, 48, 66, 84, 88, 106, 108, 109], supernovae (SN) [74], and cosmological simulations [17, 65] on some of the largest supercomputers in the world [72].

6.1.1 Basic Infrastructure

At its core, Cactus is a code framework designed to facilitate solving of partial differential equations (PDE) using finite difference and finite volume methods. Cactus itself only implements the connective tissue required to support these features, with additional modules implementing all other features. It provides a minimal infrastructure called the *flesh*, which acts like a kind of glue to bind the various computational infrastructure and science modules together. Following the biological metaphor provided by its name, these modules are called *thorns*. Indeed, the flesh does not even provide memory allocation for its arrays, delegating that task to a "driver thorn." A key benefit to the interfaces established by Cactus is that its library of science thorns was developed independently of numerical and infrastructure thorns (examples include time integration via the MoL thorn [99], HDF5 output, and checkpointing).

One of the key functionalities provided by the flesh is a basic workflow scheduler. Looking at the scheduler from a high level, Cactus "schedules" C, C++, and Fortran routines in specific "scheduling bins" that represent different phases in the simulation workflow. Cactus begins by running routines in setup and initial data bins, continues by running the scheduled functions in its evolution bins in a loop until a terminating condition is reached (such as number of iterations, final simulation time or compute walltime), and finally executes the scheduled functions in its shutdown and analysis bins. By default, Cactus provides a standard set of schedule bins, but new bins can be defined by thorns to be used by other thorns.

Cactus also provides users with an "inheritance" mechanism. A thorn's variables and functions can be marked as publicly available. Other thorns can then import these variables and routines from their parent thorns, using and modifying them. This mechanism allows for the implementation of "interface-like" thorns, that provide a common way of doing a computation or storing variables.

Since 2003, Carpet [91] has been the principle driver used by Cactus. It provides access to full Berger-Oliger adaptive mesh refinement (AMR), data distribution, communication through MPI, and memory management.

6.1.2 Reproducibility of Results and Regression Testing

Whenever new thorns are added to the ET, tests are provided to ensure the reproducibility of results obtained. These tests are run and checked rigorously as part of the regular release schedule of the ET (every six months, typically in May and November). Because reproducibility is such a high priority for this framework, thorn Formaline is typically compiled into each executable. Formaline captures all details about a simulation, including local edits in the source code that developers may have forgotten to commit and push to a repository and embeds them as a tar-archive in the executable. Thus, simulations performed with the ET retain a snapshot of the source code and compilation options used. Formaline also collects runtime information and records all runtime parameters used during the simulation.

6.2 Thorn Library Overview

The ET contains a wide variety of scientific modules. Support for curvilinear coordinates is provided by the Llama [81] code, general relativistic magnetohydrodynamics (GRMHD) by either the GRHydro [61, 71] or IllinoisGRMHD [39], and spacetime evolution by McLachlan [24], Baikal [87], BaikalVacuum [87], and lean_public [95, 107] thorns. McLachlan uses code automatically generated and optimized by the Kranc [55] code generator which includes loop-tiling, vectorization, and kernel splitting to ensure that compute kernels fit into L1 cache. Similarly, Baikal and BaikalVacuum use code generated by the NRPy+ [86] code generator, which takes symbolic expressions in Python and converts them to highly optimized C code using SIMD, common subexpression elimination (CSE), and other techniques. lean_public uses manually coded Fortran subroutines, providing a very straightforward implementation of the evolution equations.

GRHydro and IllinoisGRMHD support a variety of tabulated and analytical equations of state. These are implemented by the EOS_Omni and GRHayLib thorns, respectively. Both thorns support a tabulated equation of state (EOS), which imports a table of data for the different hydrodynamic quantities like pressure and specific internal energy, among others. Usually, these quantities are known for some values of the fluid density, temperature, and electron fraction, but one can determine them at any point that is within the bounds of the table data via interpolation. The analytic EOS approximates these tables by defining the pressure as a function of the density. While a simple polytropic EOS can be used, both thorns also provide implementations of the more accurate piecewise polytropic EOS, which more closely matches the table data at fixed temperature and electron fraction.

Despite its imperfections, Cactus has had enormous success in uniting the numerical relativity community in achieving science goals for several decades. It has won awards [2], been the basis of SPEC [1] benchmarks, and spawned countless Ph.D. theses and scientific papers.

Detailed descriptions and documentation for each module are included in the source distribution [30, 61].

6.2.1 Attribution of Credit

One difficulty of open-source projects is the proper attribution of credit, particularly with regard to citations. Instead of relying on a static publication for giving credit to the people who contribute to the ET, a Zenodo DOI is provided. With each release of the ET, new authors and contributors are added, and the Zenodo reference is updated. As of this writing, the current DOI is `doi:10.5281/zenodo.10380404`. This allows for newer contributors to be added to the ET's author list, so that all thorn authors are properly acknowledged for their contributions.

6.2.2 Base Modules

ET thorn variables are documented to be used within a system of units such that the solar mass, the speed of light, and the gravitational constant are all set to one (i.e., $G = c = M_{\odot} = 1$).

The ADMBase thorn provides grid functions for the spatial 3-metric γ_{ij}, the unit normal of the spacelike slices of the 4D spacetime manifold $n^{\mu} = (\alpha^{-1}, \beta^i/\alpha)$, the lapse function α, the shift vector β^i, and the extrinsic curvature associated with the embedding k_{ij} in the Arnowitt-Deser-Misner (ADM) formulation [12]. In addition, it also provides a grid function for the determinant of the spatial metric $\gamma = \det(\gamma_{ij})$.

Similar to ADMBase, the HydroBase thorn provides a definition for the fluid rest-mass density ρ, the pressure P, the specific internal energy density ϵ, the contravariant three-velocity in the fluid rest frame v^i, the electron fraction Y_e, the temperature T, the specific entropy s, and the contravariant magnetic field vector B^i.

Finally, the TmunuBase thorn defines the components of the 4D energy-momentum tensor $T_{\mu\nu}$, as well as scheduling bins to determine when and how to calculate $T_{\mu\nu}$.

These "Base" modules make it possible to couple various independently coded initial data (ID) schemes, spacetime solvers, and hydrodynamical schemes, providing a great deal of generality and computational flexibility.

6.2.3 CarpetRegrid2

Carpet is capable of full Berger-Oliger structure mesh refinement with arbitrary shapes for the refined regions and subcycling in time. For compact object merger simulations however, the regions of the simulation domain that require high resolution, can be computed based on the known location of the objects themselves. This results in a simpler regridding scheme, less reliant on ad-hoc prescriptions for a refinement criterion, a result also reported in [9]. To provide this "steered" mesh refinement, Carpet provides the CarpetRegrid2 thorn, which allows users to define sets of nested refined regions that can dynamically change in size, location and depth of nesting during the evolution. Coupled with object tracking capabilities provided by PunctureTracker, which tracks black holes, and VolumeIntegrals_GRMHD, which tracks NS and other material objects this allows for efficient refinement around the region of interest while ensuring simple geometries with low surface to volume ratio, thus limiting communication overhead.

6.2.4 Coordinate Systems/Llama

One of the other major concerns for any general relativistic code is the coordinate system employed for discretizing and evolving a set of PDEs. Depending on the needs of the programmer and the physical problem in question, one may use either 2 or 3 dimensions. In the former case, Cactus offers support for using 3D Cartesian Coordinates in "2D" through a thorn called Cartoon2D. Based on a concept originally proposed by Paul Walker, Cartoon2D fills in ghost zone values by rotations (see [8] for further information).

For large scale 3D simulations, the use of non-Cartesian coordinate systems is often employed. Such systems can offer advantages over Cartesian ones. Certain problems are naturally better-suited for certain coordinate systems. In the early days of numerical relativity, Čadež coordinates [105], which naturally described two spheroidal throats and a spheroidal outer boundary, helped the first collisions of black holes on a computer to succeed. Similarly, fisheye coordinates were used by the early inspiral black hole codes (before AMR was used).

An advantage of non-Cartesian coordinate systems becomes apparent when one desires to obtain the gravitational wave (GW) signal output from a numerical simulation. In order to do so, it is necessary to extract the wave signal very far away from the physical system producing the spacetime perturbations. In practice, this amounts to placing the outer boundaries of the simulation domain very far away from the center of interest in order to minimize the influence from the artificial outer boundary conditions that must be imposed in any simulation on the extracted wave signal. Using a spherical coordinate system, for instance, disentangles the angular and radial resolutions of a domain, allowing them to extend further away with smaller memory requirements when compared to their Cartesian counterpart.

Implementing non-Cartesian coordinate systems without introducing new coordinate singularities is paramount for any numerical simulation code, otherwise all potential advantages derived from such coordinate systems may nullified (but see [69] for a scheme to handle certain types of coordinate singularities in the simulation grid). These coordinate singularities can be avoided by decomposing the coordinate system in question into multiple coordinate "patches," each covering a piece of the complete domain of interest. As an example, let us consider spherical coordinates, which are known to be singular at the poles. It is possible to remove these singularities and cover the entire surface of the sphere by utilizing 6 coordinate patches, each covering a section of the whole domain. To visualize how such a system can be constructed, picture a malleable cube in 3D which is uniformly filled with air. Each of the six faces of the cube will round as the cube inflates and turn into spherical sections. The union of these 6 sections can uniquely describe the surface of a sphere without any coordinate singularities. Given the nature of the theory of General Relativity, decomposing a spacetime (a 4D manifold) into multiple sub-regions, each described by their own coordinate systems is mathematically very sound and can be easily done.

In the ET, when utilizing the Carpet driver, the Llama [81, 84] thorn allows users to utilize multiple coordinate patches for achieving non-Cartesian simulation domains. Most often, Llama is used on spherical domains, both with either spherical hole or Cartesian "core."

Figure 6.1 is a schematic representation of the currently supported multipatch coordinate systems. The figure represents a constant 2D slice of the simulation domain where each patch is colored differently. The Llama thorn supports spherical coordinates with a spherical core (left panel) and spherical coordinates with a Cartesian core (right panel). Llama handles the spherical-Cartesian inner boundary by making sure that the Cartesian core and the spherical patches overlap. Ghost zones are filled by interpolating values between patches.

From the user's perspective, using the aforementioned thorns involves simply modifying the derivative operators used (due to technical details on how tensor components are stored). Generally speaking, users interested in coding a spacetime or hydrodynamics formulation need not be concerned about multiple patches, AMR regions, or parallelization. In each of these cases, the Cactus infrastructure calls their code for a logical box, a 2 or 3-dimensional array with start and end indexes,

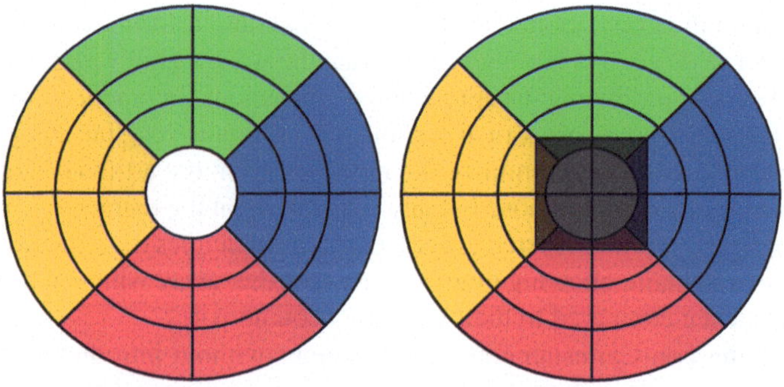

Fig. 6.1 Multipatch coordinates currently supported in the ET. Left panel: Spherical patches with spherical inner boundary. Right panel: Spherical patches with spherical inner boundary and an overlapping Cartesian core

grid spacings and other required information. For an example of GRMHD simulations utilizing multiple grid patches with Llama, see for instance [84].

6.2.5 *Time Stepping*

The actual time integration (e.g., RK4 and other Runge-Kutta schemes, Iterative Crank-Nicholson, Adams-Bashforth, multi-rate Runge-Kutta schemes, etc.) of any Cactus simulation is implemented by the MoL (Method of Lines) thorn. To use MoL, the user writes a small C routine which registers the variables to be evolved and the corresponding time-derivative of that value (which MoL calls the RHS value). One then schedules a routine to run in the MoL_CalcRHS bin to fill in the RHS values.

Therefore, new formulations of the hydrodynamical quantities can be created by (1) implementing a thorn which schedules a new function in MoL_CalcRHS, and (2) providing routines in the MoL_PostStep bin to communicate needed information between the thorn-specific evolution variables and the variables in ADMBase, HydroBase, and TmunuBase.

To implement a new spacetime module, one would need (in addition to writing and scheduling the RHS function) to provide a way to convert any new variables to the ADM variables.

Users choose which timestepping method to use via runtime parameters, based on the specific needs of the simulated system.

In either case, users benefit from Cactus's time integration, I/O modules, checkpointing, and AMR capabilities. They do not need to constantly re-implement these core functionalities, but can instead rely on the already existing, well-tested, and optimized implementations. Similarly, all modules using MoL immediately have access to any new time stepping scheme implemented in MoL, showcased, for example, by the introduction of the Adams-Bashforth multistep method, or, with some extra declarations, the multi-rate Runge-Kutta schemes.

6.2.6 *Spacetime Evolution*

The ET contains multiple alternative modules to evolve spacetime using the Baumgarte-Shapiro-Shibata-Nakamura (BSSN) and ccZ4 formulation of numerical relativity. McLachlan [24], Baikal [87], and lean_public [95, 107] are included in ET by default, and users can switch between them at runtime. The modular nature of the ET makes it possible to use other custom spacetime evolution modules. For example, the CTGamma [81] thorn is optimized to use curvilinear, multi-patch coordinates employing the Z4c formulation using Llama.

McLachlan, and lean_bssn both support curvilinear coordinates provided by Llama, and all codes use NewRad to implement approximate radiative boundary conditions at the outer boundary of the simulation domain.

Typically ID for spacetime metric and gauge quantities is either provided along with ID for fluid quantities, or, in the case a vacuum binary black hole (BBH) simulations by TwoPunctures [10], which constructs ID for a BBH system with arbitrary spin using the Brandt-Bruegmann method [21] in the Bowen-York formulation [20].

To support the modularity of the ET, ID is provided through a set of spacetime variables set based on the ADM decomposition, which can then be used by any of these evolution thorns regardless of the specific formulation. Similarly, these ADM variables are used to communicate between the evolution thorns and any analysis and fluid evolution thorns.

6.3 Hydrodynamics

Following the Valencia Formulation [11, 16, 67], all the relativistic hydrodynamic thorns use a combination of the primitive variables

$$\boldsymbol{P} = \begin{bmatrix} \rho \\ \varepsilon \\ T \\ P \\ v^i \\ B^i \\ Y_\mathrm{e} \\ S \end{bmatrix}, \tag{6.1}$$

and conserved variables

$$\boldsymbol{C} = \begin{bmatrix} \tilde{D} \\ \tilde{\tau} \\ \tilde{S}_i \\ \tilde{B}^i \\ \tilde{Y}_\mathrm{e} \\ \tilde{S} \end{bmatrix} \equiv \sqrt{\gamma} \begin{bmatrix} D \\ \tau \\ S_i \\ B^i \\ DY_\mathrm{e} \\ WS \end{bmatrix} \equiv \sqrt{\gamma} \begin{bmatrix} W\rho \\ n_\mu n_\nu T^{\mu\nu} - D \\ -n_\mu T^\mu_{\ i} \\ B^i \\ DY_\mathrm{e} \\ WS \end{bmatrix} = \sqrt{\gamma} \begin{bmatrix} W\rho \\ \alpha^2 T^{00} - D \\ \alpha T^0_{\ i} \\ B^i \\ DY_\mathrm{e} \\ WS \end{bmatrix}, \tag{6.2}$$

to evolve the system. The conserved variables $\boldsymbol{C}$ are an alternative representation of the primitive variables $\boldsymbol{P}$ that are better-suited for hyperbolic evolution.

In (6.1), ρ is the baryon density, P is the pressure, ε is the specific internal energy, T is the temperature, v^i is the fluid 3-velocity, Y_e is the electron fraction (possibly many such composition variables exist for more complex EOS), S is the entropy per baryon, and B^i are the spatial components of the magnetic field (B^μ). The fluid 3-velocity is related to the 4-velocity u^μ by

$$v^i = \frac{u^i}{W} + \frac{\beta^i}{\alpha}.$$

The frame of B^μ is that of normal (or Eulerian) observers with 4-velocity n^μ and is normal to the spatial hypersurface ($B^\mu n_\mu = 0$). The set of primitive variables is not fully independent, being related to each other through the EOS, for example, $P = P(\rho, \varepsilon)$ or $P = P(\rho, T, Y_e)$ depending on the EOS. Different codes may make different choices as to which primitive variables are considered independent and which ones are considered dependent variables, possibly even based on the temperature, density or magnetization regime the fluid is in.

In (6.2), D is the rest-mass density, τ is the energy variable, and $W = \alpha u^0 = \left(1 - \gamma_{ij} v^i v^j\right)^{-1/2}$ is the Lorentz factor. The GRMHD energy-momentum tensor, $T^{\mu\nu}$, is given by

$$T^{\mu\nu} = \left(\rho h + b^2\right) u^\mu u^\nu + \left(P + P_{\text{mag}}\right) g^{\mu\nu} - b^\mu b^\nu , \tag{6.3}$$

where $h = 1 + \epsilon + P/\rho$ is the specific enthalpy, and $P_{\text{mag}} = b^2/2$, $b^2 = g_{\mu\nu} b^\mu b^\nu$. For the magnetic field in the lab frame associated with unit normal n^μ, we choose the definition

$$B^\mu = \frac{1}{\sqrt{4\pi}} n_\nu {}^{*}F^{\mu\nu} . \tag{6.4}$$

where ${}^{*}F^{\mu\nu}$ is the dual of the Faraday tensor. Using this definition, $b^\mu = \frac{1}{\sqrt{4\pi}} u_\nu {}^{*}F^{\mu\nu}$ is the magnetic field in the co-moving frame u_μ. This gives the relation

$$b^0 = \frac{u_i B^i}{\alpha} , \tag{6.5}$$

$$b^i = \frac{B^i + B^j u_j u^i}{W} , \tag{6.6}$$

which completes our prescription for the magnetohydrodynamics (MHD) fields. The evolution equations are then given by the equation of rest-mass conservation, conservation of energy-momentum, and Maxwell's equation:

$$\begin{aligned} (D u^\mu)_{;\mu} &= 0 \\ T_\mu{}^\nu{}_{;\nu} &= 0 \\ {}^{*}F^{\mu\nu}{}_{;\nu} &= 0, \end{aligned} \tag{6.7}$$

where a semicolon indicates covariant derivatives with respect to $g_{\mu\nu}$.

At this point, we note that some of the evolution equations are optional and dependent on the EOS chosen to model nuclear matter. The electron fraction is only used when using a finite-temperature EOS, and $\tilde{Y}_e$ is not evolved when using a polytropic hybrid or simple gamma-law EOS. Similarly, entropy evolution is entirely optional and usually only used for some conservative-to-primitive recovery schemes.

6.3.1 Initial Data

To begin a simulation, we first initialize $\boldsymbol{P}$ using one of the ID thorns. Many options exist for this in the ET depending on the system of interest. Most ID solvers are outside of and independent of the Cactus framework. They are standalone programs for which specialized Cactus importer thorns exist.

Cactus typically is able to install a version of the initial data tool using its "External Library" mechanism. This is something like a very simple package manager designed to help install (or link to) Cactus dependencies.

6.3.1.1 TOVSolver

The Tolman-Oppenheimer-Volkoff (TOV) star [77, 103] is a spherically symmetric solution to the Einstein Equations that has, historically, been used to constrain the possible equations of state for NSs. It remains useful as a test of the various 3D machinery required to evolve more complex systems, and as a toy problem for students and workshops. Our standard tutorial used in our portal [23] and workshops involves a short simulation of a TOV star at low resolution.

6.3.1.2 Hydro_RNSID

The Rotating Neutron Star ID thorn uses the Komatsu-Eriguch-Hachisu (KEH) method [60] to construct a NS with a single killing vector. As such, this thorn generates 2D rotationally symmetric data. The RNSID solver is, therefore, a slightly more complex data set than the TOV star and is also useful for testing. Beyond testing evolution codes, single rotating NS are also used to study EOS, and to serve as a building block to construct NS-disk ID. This solver can be run within Cactus, or as a standalone utility that generates data files to be imported by Cactus at a later time.

6.3.1.3 FishboneMoncriefID

The Fishbone-Moncrief Initial Data creates ID for a black hole surrounded by a polytropic disk. Input parameters to the solution include the radius at which the density of the disk is at its maximum as well as the polytropic parameters. The code for this dataset is generated using NRPy, a Python based code generator. It also has a random noise parameter for use in testing.

The solver follows the prescription outlined in [45], describing a way to produce axisymmetric, isentropic disks with constant angular momentum per baryon.

6.3.1.4 LORENE/Meudon

LORENE is perhaps the oldest tool used to generate ID for BBH or binary neutron star (BNS) mergers [18]. LORENE uses spectral methods to generate data files which ET is able to import using the Meudon thorns Meudon_Bin_NS, Meudon_Bin_BH, and Meudon_Mag_NS.

Several options exist to initialize magnetic fields (or the vector potential) for a given hydrodynamic setup. The Meudon_Mag_NS thorn can directly set up both the hydrodynamic and magnetic variables for a single star. Seed_Magnetic_Fields provides several vector potential ID configurations for both single and binary NSs.

LORENE generally supplies a cold equation of state. This is fine for the ID when the NSs are widely separated. Later, during the merger, thermal effects can become important.

6.3.1.5 SGRID/DNSdata

SGRID is a similar to LORENE in purpose, but it can generate data for more extreme corners of the NS collision parameter space, e.g. total mass of $\approx 3.4 M_{\odot}$, highly spinning and precessing systems close to breakup, high mass ratio systems for soft EOSs, elliptical orbits, etc. SGRID [101] solves the conformal thin-sandwich equations using pseudo-spectral methods, so it has exponential convergence on its patchwork of grids.

The DNSdata thorn was designed by the creators of SGRID to read SGRID-generated ID files into Cactus and facilitate the use of SGRID for BNS collisions in Cactus.

6.3.1.6 KadathThorn/KadathImporter

The Kadath thorns are an interface for using files generated by the FUKA ID solver [50, 79]. The Frankfurt University/Kadath solver is capable not only of generating BNS data, but also black hole/neutron star (BHNS) binaries. In all these cases, Kadath can, like SGRID, solves for NSs with tabulated equations of state and spins near the mass-shedding limit.

A limitation is that the Kadath thorns require a strict z-symmetry, so spins can only be aligned or anti-aligned with the orbital axis.

6.3.1.7 FLRWSolver

FLRWSolver [64] constructs ID suitable for a dust-filled spatially flat Friedmann-Lemaître-Roberston-Walker (FLRW) cosmological spacetime evolved in full general relativity. Optionally, a linear perturbation can be seeded to study its linear and non-linear growth.

6.3.1.8 Vacuum Initial Data

A number of analytic spacetime solutions are provided by the Exact, and EinsteinExact thorns, useful to both test evolution codes and as building blocks for BH-disk, or similar, systems. TwoPunctures [10] and NRPyEllipticET [13] construct ID for BBH, with arbitrary spin and linear momentum values. The former also accepts a matter source to construct constraint satisfying ID data in presence of (prescribed) matter. Finally, NRPyEllipticPN [51] implements post-Newtownian (PN) expressions to compute parameters resulting in low-eccentricity BBH inspiral simulations.

6.3.2 Equation of State

The thorn EOS_Omni can be used to implement the equation of state. It is capable of modeling cold or hot EOSs. The thorn assumes nuclear statistical equilibrium with rest-mass density ρ, specific internal energy ϵ (alternatively, temperature T), and electron fraction Y_e as independent variables. The equation of state type can be specified as a polytropic, Γ-law, piecewise-polytropic with thermal part, cold tabulated with thermal part, hot tabulated, and barotropic tabulated EOS. Tables of some EOS are available at [6].

GRHayL contains its own independent EOS routines supporting Γ-law, piecewise-polytropic with thermal part, and hot tabulated EOS using the same tables as EOS_Omni. The details of this library and its implementation in the ET are included in the description of GRHayL and the new GRHayL-based IllinoisGRMHD in Sect. 6.6.

RePrimAnd [59] provides a complete EOS framework and conservative to primitive inversion method with strong guarantees on the existence and validity of solutions. A version is included in ET but is not currently in use due to various technical issues. It supports non-magnetized fluids (a version suitable for magnetized fluids exists [57]) described by a Γ-law, or cold tabulated EOS with a thermal part.

One can use these in conjunction with GRHydro or GRHayL to continue the evolution of NS ID created using such tools as LORENE, SGRID, or FUKA.

6.3.3 GRMHD Evolution

After initializing the primitives and spacetime quantities, the conserved variables are calculated. During each step of the evolution, the primitive variables are used to compute the RHS values used to numerically integrate the evolution equations.

The basic workflow for computing the time derivatives of the evolved quantities is depicted in Fig. 6.2 (adapted from [4]).

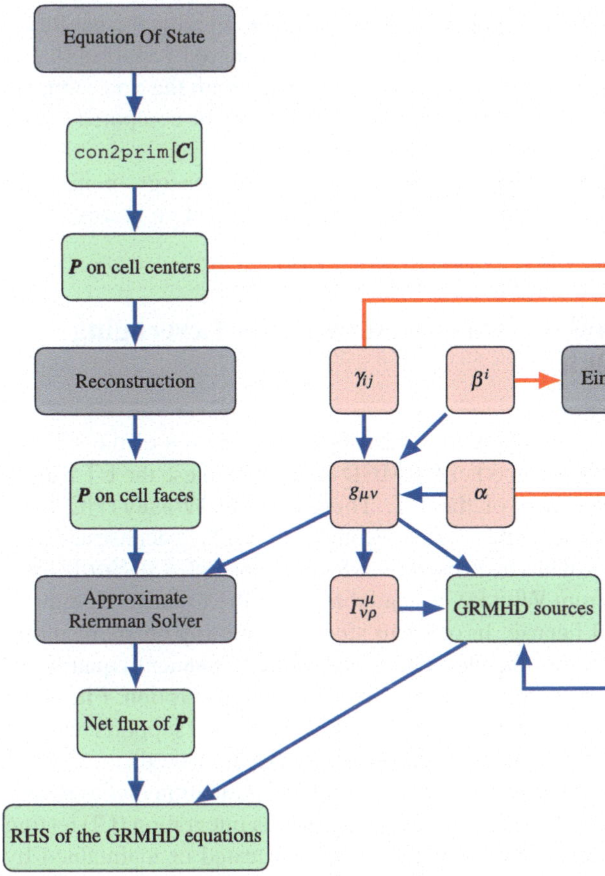

Fig. 6.2 Diagram depicting the conceptual workflow of a GRMHD code. In the figure, elements outlined by green boxes denote GRMHD variables, red boxes denote Einstein's equations' components required for the GRMHD evolution and gray boxes denote functional steps in the GRMHD workflow. Concurrently, blue arrows denote data flow for computing the GRMHD RHS while those depicted by red boxes and arrows pertain to relevant segments of the spacetime evolution scheme

In an initial step, the force terms appearing in the 3+1 decomposed energy-momentum conservation equation (6.7) are computed from the spacetime metric and Christoffel symbols, setting the stage to compute the numerical fluxes for the fluid variables. The primitive variables $\boldsymbol{P}$ are then computed from the conserved variables $\boldsymbol{C}$, which generally involves EOS-specific methods using iterative solvers. This step is in general only second order convergent, since all GRMHD codes in ET (except WhiskyTHC and Sprtiz which employ higher order finite *difference* schemes, but are not shipped as part of ET) do not distinguish between the cell-centered value of $\boldsymbol{P}$ and their cell averages. After primitive recovery, the values of the recovered primitive variables are then reconstructed to cell faces, where they form the input to a Riemann problem that is used to compute numerical fluxes. Finally, the net fluxes are used to compute an average rate of change for the conserved variables that is passed to the time stepper. It is worth noting that, when used with the time steppers provided by thorn MoL, this averaged rate of change needs only be computed to first order to achieve the full convergence order of the time stepper.

Simultaneously the fluid variables are used to compute the energy-momentum tensor $T_{\mu\nu}$, which is used as a source term by the spacetime evolution thorns.

6.4 Non-Einstein Toolkit Hydrodynamic Codes Leveraging the Einstein Toolkit

Before describing the tools available within the official distribution of the ET, we would like to point out that there are other GRMHD codes which use the ET framework but have never become part of the ET. These include WhiskyTHC [82], Spritz [28] and FIL [70]. WhiskyTHC is available under the GPLv3 license and provides advanced microphysics. It has been used for NS collisions [82, 83]. Spritz [46], which borrows some ideas from Whisky, is available under the Creative Commons Attribution 4.0 International license. In order to study gamma-ray bursts resulting from BNS, this code employs a tabulated and temperature-dependent equation of state to describe NS matter as well as neutrino emission and absorption. FIL is the private "Frankfurt Illinois" code derived from IllinoisGRMHD and supports realistic microphysics and neutrino cooling using high-order numerical methods.

As these exist outside the ET umbrella, their testing and validation methods vary from group to group and do not leverage the ET's continuous integration (CI) testing framework. That does not mean that these codes are not tested or maintained by their original developers. However, the standards of support and testing of these other codes depend on the needs of their developers for their own published research projects. The existence of these codes helps to illustrate the strength of the open source software model for science in general as well as the vibrancy of the ET community in particular.

The creation of these codes is facilitated by the open source license of the ET. When creating codes based on the ET, researchers may opt to (and this is frequently the case) create a closed-source code that leverages the Toolkit, keeping it private until they have validated and published results using it. At this stage, researchers then may decide to move their code to open source and petition for inclusion in the ET. It is very important to point out, however, that developers are never *required* to do this.

6.5 GRHydro

The original support for hydrodynamics within the ET was the GRHydro thorn [71], which itself descends from the EU Network on Sources of Gravitational Radiation code, Whisky [14]. It was also the first publicly available, fully GRMHD code [46]. Initially, GRHydro used a relativistic polytropic or Γ-law gas equation of state, with support for tabulated EOS added via EOS_Omni later. In [71], MHD was added to the capabilities of GRHydro. In order to maintain the divergence-free character of the magnetic field, GRHydro implements both constrained transport [47] and hyperbolic divergence cleaning [80] schemes.

The current, open-sourced GRhydro [14, 71] implements GRMHD with a high-resolution, shock-capturing, finite-volume scheme with (e)PPM [29, 68], (W)ENO(-Z) [19, 94, 98], and MP5 [96] reconstruction schemes and HLLE [36, 52], Roe [85], and Marquina [31] approximate Riemann solvers. Conservative to primitive conversion uses a non-linear Newton-Raphson iteration scheme based on [75], though specific applications, for example, SN simulations [72, 73], have instead explored use of the method of [26], and other variations may exist. In combination with the Llama [81] infrastructure, it supports curvilinear coordinates allowing it to handle large simulation volumes at moderate cost [84].

GRHydro evolves the set of conserved variables $\boldsymbol{C}$ in (6.2) using the Valencia formulation. Similar to most existing codes, it usually reconstructs the primitive variables (6.1) to cell interfaces when setting up the Riemann problem computing numerical fluxes, resulting in a second order accurate evolution scheme. By default GRHydro will reconstruct the 3-velocity v^i and specific internal energy ϵ (in addition to rest mass density ρ). For a hot tabulated EOS it may be more appropriate to reconstruct temperature T, and Wv^i instead of v^i to guarantee that the reconstructed speed remains subluminal. Both variants are offered by GRHydro as runtime options.

GRHydro implements the "flux-CT" scheme of [47, 104], which employs a *cell* averaged magnetic field B^i rather than the face-averaged field used in [41]. As a consequence of this choice, the (exactly) conserved divergence operator is an edge-centered operator instead of a cell-centered operator. This choice simplifies implementation in Carpet which lacks full support for face-centered variables. As of this writing, GRHydro does not include methods to preserve the divergence-free condition across mesh refinement boundaries which results in a layer of constraint violating data near the edges of mesh-refinement levels. Due to the constraint preserving nature of the evolution equation, this data remains fixed in space and does not propagate. For SN simulations, where the grid does not move, this is sufficient to ensure that simulations succeed.

As an alternative, GRHydro also implements a hyperbolic divergence cleaning scheme following [80], which introduces an additional dynamic variable ψ that is driven to 0 to transport and damp away any violation of the divergence-free condition. This method is well-suited for AMR simulations since any constraint violation is damped away no matter whether it is sourced by finite resolution effects or by AMR data movement. This method, however, has proven difficult to control in curved spacetime simulations beyond simple TOV star and Bondi accretion test cases. Flat spacetime simulations, which are where the method originates from, are successful.

Beyond compact object simulations, GRHydro has recently been used for cosmological simulations [64] and has been regularly used for core-collapse SNe simulations [73, 74, 78].

6.6 GRHayL/IllinoisGRMHD

IllinoisGRMHD [39] is an open-source GRMHD code originating from the closed-source code developed by the Illinois numerical relativity group [35]. Several modified versions of this code also exist [38, 70, 106] outside of the ET, as different research groups have extended the original code to facilitate their own research interests and needs. Notably, it was the first public code to use the vector potential of the magnetic field as an evolution variable. The original release implements the piecewise parabolic method (PPM) [29], hybrid EOS, and HLLE [36, 52] Riemann solver. The IllinoisGRMHD family of thorns has been used by many groups, primarily for BNS simulations [33–35, 38, 39, 62, 70, 106, 108, 109]. It has also been used for circumbinary disk evolution [42, 43]. A recent development is to base IllinoisGRMHD on a common library of modules (GRHayL) that provides re-usable building blocks for GRMHD codes. In cases where this difference is relevant the newer version of IllinoisGRMHD is dubbed GRHIGM below.

One notable difference between IllinoisGRMHD and other codes in the ET is that it chooses a different primitive velocity. The more common choice is the Valencia 3-velocity v^i, used by HydroBase, but IllinoisGRMHD instead uses the 3-velocity appearing in the induction equation, i.e.,

$$\tilde{v}^i = \frac{u^i}{u^0} = \alpha v^i - \beta^i, \tag{6.8}$$

as its primitive variable. IllinoisGRMHD also evolves the vector potential A_i instead of directly evolving the magnetic field B^i. This has the advantage of ensuring that $\nabla \cdot B^i = 0$ (i.e. no magnetic monopoles) automatically. Additionally, IllinoisGRMHD uses a different definition of B^i than what is given in (6.4). However, GRHIGM uses the definition of (6.4), making the various ET codes now agree in their definition of B^i. IllinoisGRMHD's vector potential evolution method is particularly well suited for simulations involving moving meshes, as happens for example in magnetized BNS merger simulations, for which constraint transport methods that fail to preserve the divergence-free condition across mesh refinement boundaries can rapidly fill the simulation domain with constraint violating data [40].

6.6.1 GRHayL

IllinoisGRMHD's popularity has led to a number of forks implementing advanced functionality, e.g. in the FIL [70] code. However, these improvements often led to mutually incompatible code variants that made community collaboration more difficult. To better support community efforts to improve and extend the code, the core algorithms of IllinoisGRMHD were converted into a C library called the General Relativistic Hydrodynamics Library (GRHayL). GRHayL provides an infrastructure-agnostic, extensible, open-source library of building blocks to simplify the task of extending or improving IllinoisGRMHD's core features. As discussed in Sect. 6.3, there exist several discrete steps to integrate the evolution equation. These steps can be broken into independent operations which are used to compartmentalize the internal features of GRHayL. As such, GRHayL divides its features into Atmosphere, Con2Prim, EOS, Flux_Source, Induction, Neutrinos, and Reconstruction modules.

GRHayL provides all the features of IllinoisGRMHD through functions acting on pointwise or stencil-wise data within its modules. Most modules refer to a specific stage of the evolution procedure and do not depend on each other, allowing users to freely pick and choose which modules to use. However, EOS contains all the code implementing features for specific choices of the EOS and is used by the other modules. Similar to RePrimAnd, it defines a common application programming interface (API) to access the EOS. Codes wishing to use new EOS implement that API, potentially re-using functionality provided by existing EOS implementations. This allows for modular and customizable code behavior.

Starting with the ET_2024_05 "Lev Landau" release of ET, IllinoisGRMHD uses GRHayL, which itself is also included.

6.7 Data Analysis

The ET has a number of built-in thorns for diagnostics and analysis. The most basic of these are the I/O thorns, which include not only 1D, 2D, and 3D data output but also scalar reductions such as the norm, average, maximum, etc. As there are many analysis and diagnostics thorns in the ET, we do not provide an exhaustive list of these thorns but instead provide descriptions of several widely-used analysis thorns. Some of these thorns—such as AHFinderDirect—are focused on black hole physics, but they are also frequently used during matter simulations that can lead to the formation of a black hole, such as BNS mergers.

6.7.1 AHFinderDirect

The AHFinderDirect [100] thorn provides a way to search for trapped surfaces, 2-surfaces within a spacetime that obey the equation:

$$0 = D_i \Sigma^i + K_{ij} \Sigma^i \Sigma^j - K, \tag{6.9}$$

where Σ^i is an outgoing vector on a 2D surface within a 3D slice of the spacetime, and D_i is the covariant derivative associated with γ_{ij}.

These surfaces correspond to regions in the spacetime where outgoing null rays do not expand. They are often used as proxies for event horizons because they are easier to compute, since they can be computed locally rather than starting from future null infinity and working backward. Apparent horizons, if they are present, will be coincident with or inside event horizons, so their appearance provides evidence that a black hole has formed.

Towards the end of any numerical simulation, apparent horizons are expected to settle down to being the same as the event horizon (which is true for the stationary Kerr solution).

Unlike event horizons, e.g. during a black hole collision, apparent horizons can suddenly appear during the course of a simulation. This means that to find them, one most provide some initial guess for their location.

AHFinderDirect discretizes (6.9) on a star-shaped trial-surface directly solving the resulting systems of non-linear equations for the trapped surface's shape using Newton's method starting from the initial guess. This is in contrast to curvature flow based methods [102], which solve a parabolic equation whose steady state solution is the trapped horizon surface. For BBH simulations, initial guesses for the trapped surfaces are provided by users based on initial location, mass and spins of the initial BHs, while initial guesses for parameters for a BH that forms during a collapse are estimated based on the expected mass of the BH and its location at the center of mass of the system.

6.7.2 QuasiLocalMeasures

This module implements the calculation of mass and spin multipoles from the isolated and dynamical horizon formalism [32, 92], as well as a number of other proposed formulae for quasilocal mass, linear momentum, angular momentum, etc. Despite their approximate nature, they have been surprisingly helpful in numerical simulations (see, e.g., [63]), and are therefore an indispensable tool in numerical relativity. QuasiLocalMeasures takes as input a horizon surface (or any other surface that the user specifies, e.g. a large coordinate sphere) and calculates quantities such as the Weyl or Ricci scalars, the 3-volume element of the horizon world tube, as well as other physical observables such as mass and momenta. While initially designed for

BH simulations, it was found in [97] that the methods used by QuasiLocalMeasures can also be used to estimate spins of NSs.

6.7.3 WeylScal4

A common output produced by compact object merger simulations is data about the GWs produced during the event. For this the Newman-Penrose scalar Ψ_4

$$\Psi_4 = C_{\rho\sigma\mu\nu}\tilde{m}^{\rho}\tilde{n}^{\sigma}\tilde{m}^{\mu}\tilde{n}^{\nu}, \tag{6.10}$$

which encodes outgoing radiation contained in the Weyl tensor $C_{\rho\sigma\mu\nu}$ is often convenient. WeylScal4 computes this quantity, as well as the remaining Newman-Penrose components of the Weyl tensor and the curvature invariants. Ψ_4 is computed by separating time (0 index) and space (latin indices)

$$\begin{aligned}\Psi_4 = {} & R_{ijkl}n^i\tilde{m}^j n^k\tilde{m}^l + 2R_{0jkl}(n^0\tilde{m}^j\tilde{n}^k\tilde{m}^l - \tilde{m}^0 n^j\tilde{n}^k\tilde{m}^l) \\ & + R_{0j0l}(n^0\tilde{m}^j\tilde{n}^0\tilde{m}^l + \tilde{m}^0\tilde{n}^j\tilde{m}^0\tilde{\tilde{n}}^l - 2\tilde{n}^0\tilde{m}^j\tilde{m}^0\tilde{n}^l),\end{aligned} \tag{6.11}$$

where $R_{\rho\sigma\mu\nu}$ is the spacetime Riemmann tensor and $\tilde{n}^{\mu}$, $\tilde{m}^{\mu}$ are part of a complex valued fiducial null-tedrad defined in [15]. WeylScal4 uses Kranc [55], which generates thorns from Mathematica input, reducing development effort and error rate when transcribing expressions into code.

6.7.4 Multipole

Multipole is responsible for performing a mode decomposition of Cactus grid functions via a projection onto the spin weighted spherical harmonics ${}_sY_{\ell m}$. This is extremely useful while extracting GW components of Ψ_4, whose dominant modes are the quadrupolar $\ell = |m| = 2$ spin-weight $s = -2$ modes. Multipole produces output as either ASCII or more compact HDF5 files.

6.7.5 VolumeIntegrals

The VolumeIntegrals_GRMHD and VolumeIntegrals_vacuum thorns compute a number of volume integrals of GRMHD and spacetime quantities, respectively. The thorns accept a number of integration volumes consisting of spherical shells, including balls and hollowed out balls. For integrals that compute a location, the integration volume can be set to itself track this location and also track it with a refined region defined in CarpetRegrid2. In this manner, results from some of these integrals can be used to track NSs and BHs, for example.

VolumeIntegrals_GRMHD computes the center of mass

$$X^i = \frac{\int Dx^i dV}{\int DdV}, \tag{6.12}$$

as well as rest mass

$$M_0 = \int DdV \tag{6.13}$$

for each region.

VolumeIntegral_vacuum and ADMMass calculate volume integrals for the ADM mass using

$$M_{ADM} = \frac{1}{16\pi} \int \left(\alpha\sqrt{\gamma}\gamma^{ij}\gamma^{kl}(\gamma_{ik,j} - \gamma_{ij,k})\right)_{,l} dV, \tag{6.14}$$

which follows its definition using a surface integral [76], assuming $\alpha \to 1$ at infinity. VolumeIntegral_vacuum also computes a quantity

$$C^i_\alpha = \int \left[(1-\alpha)^{80}\right] dV \tag{6.15}$$

dubbed "center of lapse", which, for material objects, has qualitatively the same result as the center of mass.

VolumeIntegral_vacuum and VolumeIntegrals_GRMHD permit regions of the simulation to be excluded from the integral, which allows, e.g., the interior of BHs being masked when computing rest mass.

6.7.6 *Outflow*

Outflow calculates the flow of rest mass density across a fixed, star-shaped surface, e.g. an apparent horizon or a sphere at fixed radius. Starting from (6.7) the evolution equation for D is integrated over the enclosed volume V and transformed into a surface integral to yield

$$\dot{M}_0 = -\int_{\partial V} \sqrt{\gamma}\alpha D(v^i - \beta^i/\alpha) d\sigma_i, \tag{6.16}$$

where σ_i is the outward-pointing flat-space surface element of the enclosing surface. Additionally, conditions can be used to count only unbound fluid elements that will eventually escape gravity and be ejected from the system. This provides a convenient way to monitor ejecta during BNS mergers, for example.

6.7.7 *Particle_tracerET*

particle_tracerET tracks the motion of test particles that comove with fluid flows. Designed to visualize the evolution of magnetic field lines in ideal fluid simulations, tracers can also be used to create evolution traces of ejecta particles. It evolves each particle's location according to

$$\frac{dX^i}{dt} = \alpha v^i - \beta^i, \tag{6.17}$$

seeding initial particle locations such as to preferentially sample high density regions of the simulation grid.

6.7.8 *SmallbPoynET*

The smallbPoynET thorn computes the fluid four-velocity u^i, the magnetic field measured by a co-moving observer b^μ, and the Poynting vector S^i. While a relatively simple thorn, this provides useful diagnostic and analysis information about the magnetic evolution of the system, and possible regions of jet formation. Currently, this thorn only supports the old IllinoisGRMHD definition of B^i, though this is planned to change in future releases.

6.8 Current and Future Development

The ET has been very successful in enabling numerical astrophysics, resulting in close to 200 publications within the last 4 years. It has provided a stable basis for research by scientists all over the world, incorporating newly developed methods while providing a backwards compatible core set of functionality that allows users to maintain their own codes with minimal effort. Over the past decade, however, computational research has shifted from single socket scalar architectures (what Cactus was originally designed for) to NUMA architectures with vectorization, and GPUs. With the end of Dennard scaling increased parallelization at all levels of the software stack and support for accelerator architectures is crucial to tackle immensely resource intensive GRMHD simulations. These include highly turbulent fluid motion in large, asymmetric domains and realistic treatment of the EOS, and microphysical interaction such as neutrino transport, and nuclear reactions. All of these are required to properly model the physical processes involved in NS collisions such as the "kilonova event" GW170817/GRB 170817A observed in [7].

While there had been attempts to incorporate support for GPUs into ET in the past [89], these have been hampered by the need to maintain full backwards compatibility with existing code and need to restructure the underlying driver thorn Carpet to adequately support GPUs. More recently, efforts have been underway to develop a new driver, CarpetX [90], that is built on the AMReX [110] mesh refinement library developed by the DOE as part of the Exascale Computing Project. AMReX provides the necessary abstraction layers to support both CPU and GPU resources, providing a portability layer for CUDA, RoCM, and oneAPI.

As of this writing, CarpetX is mature enough to be used by early adopters to port codes and benchmark its algorithms on a variety of large CPU and GPU clusters. While functionality that is not immediately required, such as subcycling in time (local time-stepping) are still being developed, the framework is functional for a number of scientific applications such as SN explosion and BNS merger simulations.

There are now 3 independently developed GRMHD codes based on CarpetX [3, 58, 93] that are gearing up for their first production simulations. Each targets different problems and prioritizes different areas of the physics. GRaM-X is designed to simulate core-collapse SNe, targeting very large simulations, with fully energy dependent neutrino transport being among the implementation goals, while dealing with a fairly straightforward geometry requiring no advanced regridding schemes or complicated domain decomposition. AsterX, on the other hand, targets BNS merger simulations (for which an energy averaged neutrino transport scheme may be sufficient) but has to contend with two moving objects and ejecta whose geometry cannot be predicted in advance. AsterX implements the advanced primitive variable recovery scheme of RePrimAnd [59]. Finally, GRHayL, which currently only supports CPU based evolution, provides a method for preserving core code for reuse in new infrastructures, providing infrastructure-agnostic building blocks that incorporate scientific code modules and can be combined with architecture-dependent infrastructure modules. This also simplifies "swapping out" of individual modules for more advanced ones, without having to re-design the evolution code from scratch.

Major improvements [56] in the microphysics support in CarpetX-based codes will be implemented in the upcoming years, along with support for more realistic equations of state [5], and more initial data readers for advanced initial data solvers.

The ET has heavily relied on automated code generation [55, 87] for the Einstein field equations and other tensorial expressions. CarpetX provides enhanced correctness checking capabilities that a CarpetX aware version of the NRPy+ code generator can target to catch common errors and mistakes [22], beyond what is possible in Carpet.

In the ET the Llama thorns provide support for multi-patch domains with curvilinear coordinates used in each patch. They currently only support overlapping patch boundaries, which can be a source of numerical noise and complicate implementing refluxing schemes that guarantee conservation of rest mass and other evolved quantities between patches. For CarpetX, the CapyrX thorn is being developed to provide this functionality while also simplifying the process of adding new types of patch sets. Once fully implemented, CapyrX will support both interpatch boundary communication via interpolation or via characteristic variables exchanges.

While the list of thorns supporting CarpetX is currently much shorter than those using the Carpet driver, their number is steadily growing and the transition to a new driver also offers an opportunity to correct past design decisions that are hindering current research due to changing architecture constraints.

In conclusion, the ET provides a rich, mature software ecosystem for astrophysical GRMHD simulations that has stood the test of time. With CarpetX there is a leap forward in performance and level of support for modern GPU accelerators, ensuring continued high performance of ET-based simulations on the largest compute cluster and targeting simulations at the forefront of scientific research.

Acknowledgements The ET has been supported by NSF 2004157/2004044/2004311/2227105/ 2004879/2003893/2114582. Any opinions, findings, and conclusions or recommendations expressed in this material are those of the author(s) and do not necessarily reflect the views of the National Science Foundation.

References

1. 507.cactuBSSN_r. SPEC CPU®2017 benchmark description. https://www.spec.org/cpu2017/Docs/benchmarks/507.cactuBSSN_r.html
2. Awards won by cactus and the Einstein toolkit. https://www.cactuscode.org/media/prizes/index.html
3. GRHayL/GRHayLET: Einstein Toolkit thorns which require GRHayL. https://github.com/GRHayL/GRHayLET.git
4. GRMHD workflow: 10.5281/zenodo.10988388. https://zenodo.org/records/10988388
5. Modular unified solver of the equation of state. https://musesframework.io/
6. stellarcollapse.org. https://stellarcollapse.org/
7. B.P. Abbott, R. Abbott, T.D. Abbott, F. Acernese, K. Ackley, C. Adams, T. Adams, P. Addesso, R.X. Adhikari, V.B. Adya, C. Affeldt, M. Afrough, B. Agarwal, M. Agathos, K. Agatsuma, N. Aggarwal, O.D. Aguiar, L. Aiello, A. Ain, P. Ajith, B. Allen, G. Allen, A. Allocca, P.A. Altin, A. Amato, A. Ananyeva, S.B. Anderson, W.G. Anderson, S.V. Angelova, S. Antier, S. Appert, K. Arai, M.C. Araya, J.S. Areeda, N. Arnaud, K.G. Arun, S. Ascenzi, G. Ashton, M. Ast, S. M. Aston, P. Astone, D.V. Atallah, P. Aufmuth, C. Aulbert, K. AultONeal, C. Austin, A. Avila-Alvarez, S. Babak, P. Bacon, M.K.M. Bader, S. Bae, M. Bailes, P.T. Baker, F. Baldaccini, G. Ballardin, S.W. Ballmer, S. Banagiri, J.C. Barayoga, S.E. Barclay, B.C. Barish, D. Barker, K. Barkett, F. Barone, B. Barr, L. Barsotti, M. Barsuglia, D. Barta, S.D. Barthelmy, J. Bartlett, I. Bartos, R. Bassiri, A. Basti, J.C. Batch, M. Bawaj, J.C. Bayley, M. Bazzan, B. Bécsy, C. Beer, M. Bejger, I. Belahcene, A.S. Bell, B.K. Berger, G. Bergmann, S. Bernuzzi, J.J. Bero, C.P.L. Berry, D. Bersanetti, A. Bertolini, J. Betzwieser, S. Bhagwat, R. Bhandare, I.A. Bilenko, G. Billingsley, C.R. Billman, J. Birch, R. Birney, O. Birnholtz, S. Biscans, S. Biscoveanu, A. Bisht, M. Bitossi, C. Biwer, M.A. Bizouard, J.K. Blackburn, J. Blackman, C.D. Blair, D.G. Blair, R.M. Blair, S. Bloemen, O. Bock, N. Bode, M. Boer, G. Bogaert, A. Bohe, F. Bondu, E. Bonilla, R. Bonnand, B.A. Boom, R. Bork, V. Boschi, S. Bose, K. Bossie, Y. Bouffanais, A. Bozzi, C. Bradaschia, P.R. Brady, M. Branchesi, J.E. Brau, T. Briant, A. Brillet, M. Brinkmann, V. Brisson, P. Brockill, J.E. Broida, A.F. Brooks, D.A. Brown, D.D. Brown, S. Brunett, C.C. Buchanan, A. Buikema, T. Bulik, H.J. Bulten, A. Buonanno, D. Buskulic, C. Buy, R.L. Byer, M. Cabero, L. Cadonati, G. Cagnoli, C. Cahillane, J. Calderón Bustillo, T.A. Callister, E. Calloni, J.B. Camp, M. Canepa, P. Canizares, K.C. Cannon, H. Cao, J. Cao, C.D. Capano, E. Capocasa, F. Carbognani, S. Caride, M.F. Carney, G. Carullo, J. Casanueva Diaz, C. Casentini, S. Caudill, M. Cavaglià, F. Cavalier, R. Cavalieri, G. Cella, C.B. Cepeda,

P. Cerdá-Durán, G. Cerretani, E. Cesarini, S.J. Chamberlin, M. Chan, S. Chao, P. Charlton, E. Chase, E. Chassande-Mottin, D. Chatterjee, K. Chatziioannou, B.D. Cheeseboro, H.Y. Chen, X. Chen, Y. Chen, H.P. Cheng, H. Chia, A. Chincarini, A. Chiummo, T. Chmiel, H.S. Cho, M. Cho, J.H. Chow, N. Christensen, Q. Chu, A.J.K. Chua, S. Chua, A.K.W. Chung, S. Chung, G. Ciani, R. Ciolfi, C.E. Cirelli, A. Cirone, F. Clara, J.A. Clark, P. Clearwater, F. Cleva, C. Cocchieri, E. Coccia, P.F. Cohadon, D. Cohen, A. Colla, C.G. Collette, L.R. Cominsky, M. Constancio, L. Conti, S.J. Cooper, P. Corban, T.R. Corbitt, I. Cordero-Carrión, K.R. Corley, N. Cornish, A. Corsi, S. Cortese, C.A. Costa, M.W. Coughlin, S.B. Coughlin, J.P. Coulon, S.T. Countryman, P. Couvares, P.B. Covas, E.E. Cowan, D.M. Coward, M.J. Cowart, D.C. Coyne, R. Coyne, J.D.E. Creighton, T.D. Creighton, J. Cripe, S.G. Crowder, T.J. Cullen, A. Cumming, L. Cunningham, E. Cuoco, T. Dal Canton, G. Dálya, S.L. Danilishin, S. D'Antonio, K. Danzmann, A. Dasgupta, C.F. Da Silva Costa, V. Dattilo, I. Dave, M. Davier, D. Davis, E.J. Daw, B. Day, S. De, D. DeBra, J. Degallaix, M. De Laurentis, S. Deléglise, W. Del Pozzo, N. Demos, T. Denker, T. Dent, R. De Pietri, V. Dergachev, R. De Rosa, R.T. DeRosa, C. De Rossi, R. DeSalvo, O. de Varona, J. Devenson, S. Dhurandhar, M.C. Díaz, T. Dietrich, L. Di Fiore, M. Di Giovanni, T. Di Girolamo, A. Di Lieto, S. Di Pace, I. Di Palma, F. Di Renzo, Z. Doctor, V. Dolique, F. Donovan, K.L. Dooley, S. Doravari, I. Dorrington, R. Douglas, M. Dovale Álvarez, T.P. Downes, M. Drago, C. Dreissigacker, J.C. Driggers, Z. Du, M. Ducrot, R. Dudi, P. Dupej, S.E. Dwyer, T.B. Edo, M.C. Edwards, A. Effler, H.B. Eggenstein, P. Ehrens, J. Eichholz, S.S. Eikenberry, R.A. Eisenstein, R.C. Essick, D. Estevez, Z.B. Etienne, T. Etzel, M. Evans, T.M. Evans, M. Factourovich, V. Fafone, H. Fair, S. Fairhurst, X. Fan, S. Farinon, B. Farr, W.M. Farr, E.J. Fauchon-Jones, M. Favata, M. Fays, C. Fee, H. Fehrmann, J. Feicht, M.M. Fejer, A. Fernandez-Galiana, I. Ferrante, E.C. Ferreira, F. Ferrini, F. Fidecaro, D. Finstad, I. Fiori, D. Fiorucci, M. Fishbach, R.P. Fisher, M. Fitz-Axen, R. Flaminio, M. Fletcher, H. Fong, J.A. Font, P.W.F. Forsyth, S.S. Forsyth, J.D. Fournier, S. Frasca, F. Frasconi, Z. Frei, A. Freise, R. Frey, V. Frey, E.M. Fries, P. Fritschel, V.V. Frolov, P. Fulda, M. Fyffe, H. Gabbard, B.U. Gadre, S.M. Gaebel, J.R. Gair, L. Gammaitoni, M.R. Ganija, S.G. Gaonkar, C. Garcia-Quiros, F. Garufi, B. Gateley, S. Gaudio, G. Gaur, V. Gayathri, N. Gehrels, G. Gemme, E. Genin, A. Gennai, D. George, J. George, L. Gergely, V. Germain, S. Ghonge, A. Ghosh, A. Ghosh, S. Ghosh, J.A. Giaime, K.D. Giardina, A. Giazotto, K. Gill, L. Glover, E. Goetz, R. Goetz, S. Gomes, B. Goncharov, G. González, J.M. Gonzalez Castro, A. Gopakumar, M.L. Gorodetsky, S.E. Gossan, M. Gosselin, R. Gouaty, A. Grado, C. Graef, M. Granata, A. Grant, S. Gras, C. Gray, G. Greco, A.C. Green, E.M. Gretarsson, P. Groot, H. Grote, S. Grunewald, P. Gruning, G.M. Guidi, X. Guo, A. Gupta, M.K. Gupta, K.E. Gushwa, E.K. Gustafson, R. Gustafson, O. Halim, B.R. Hall, E.D. Hall, E.Z. Hamilton, G. Hammond, M. Haney, M. M. Hanke, J. Hanks, C. Hanna, M.D. Hannam, O.A. Hannuksela, J. Hanson, T. Hardwick, J. Harms, G.M. Harry, I.W. Harry, M.J. Hart, C.J. Haster, K. Haughian, J. Healy, A. Heidmann, M.C. Heintze, H. Heitmann, P. Hello, G. Hemming, M. Hendry, I.S. Heng, J. Hennig, A.W. Heptonstall, M. Heurs, S. Hild, T. Hinderer, W.C.G. Ho, D. Hoak, D. Hofman, K. Holt, D.E. Holz, P. Hopkins, C. Horst, J. Hough, E.A. Houston, E.J. Howell, A. Hreibi, Y.M. Hu, E.A. Huerta, D. Huet, B. Hughey, S. Husa, S.H. Huttner, T. Huynh-Dinh, N. Indik, R. Inta, G. Intini, H.N. Isa, J.M. Isac, M. Isi, B.R. Iyer, K. Izumi, T. Jacqmin, K. Jani, P. Jaranowski, S. Jawahar, F. Jiménez-Forteza, W.W. Johnson, N.K. Johnson-McDaniel, D.I. Jones, R. Jones, R.J.G. Jonker, L. Ju, J. Junker, C.V. Kalaghatgi, V. Kalogera, B. Kamai, S. Kand hasamy, G. Kang, J. B. Kanner, S.J. Kapadia, S. Karki, K.S. Karvinen, M. Kasprzack, W. Kastaun, M. Katolik, E. Katsavounidis, W. Katzman, S. Kaufer, K. Kawabe, F. Kéfélian, D. Keitel, A.J. Kemball, R. Kennedy, C. Kent, J.S. Key, F.Y. Khalili, I. Khan, S. Khan, Z. Khan, E.A. Khazanov, N. Kijbunchoo, C. Kim, J.C. Kim, K. Kim, W. Kim, W.S. Kim, Y.M. Kim, S.J. Kimbrell, E.J. King, P.J. King, M. Kinley-Hanlon, R. Kirchhoff, J.S. Kissel, L. Kleybolte, S. Klimenko, T.D. Knowles, P. Koch, S.M. Koehlenbeck, S. Koley, V. Kondrashov, A. Kontos, M. Korobko, W.Z. Korth, I. Kowalska, D.B. Kozak, C. Krämer, V. Kringel, B. Krishnan, A. Królak, G. Kuehn, P. Kumar, R. Kumar, S. Kumar, L. Kuo, A. Kutynia, S. Kwang, B.D. Lackey, K. H. Lai, M. Landry, R.N. Lang,

J. Lange, B. Lantz, R.K. Lanza, S.L. Larson, A. Lartaux-Vollard, P.D. Lasky, M. Laxen, A. Lazzarini, C. Lazzaro, P. Leaci, S. Leavey, C.H. Lee, H.K. Lee, H.M. Lee, H.W. Lee, K. Lee, J. Lehmann, A. Lenon, E. Leon, M. Leonardi, N. Leroy, N. Letendre, Y. Levin, T.G.F. Li, S.D. Linker, T.B. Littenberg, J. Liu, X. Liu, R.K.L. Lo, N.A. Lockerbie, L.T. London, J.E. Lord, M. Lorenzini, V. Loriette, M. Lormand, G. Losurdo, J.D. Lough, C.O. Lousto, G. Lovelace, H. Lück, D. Lumaca, A.P. Lundgren, R. Lynch, Y. Ma, R. Macas, S. Macfoy, B. Machenschalk, M. MacInnis, D.M. Macleod, I. Magaña Hernandez, F. Magaña-Sandoval, L. Magaña Zertuche, R.M. Magee, E. Majorana, I. Maksimovic, N. Man, V. Mandic, V. Mangano, G.L. Mansell, M. Manske, M. Mantovani, F. Marchesoni, F. Marion, S. Márka, Z. Márka, C. Markakis, A.S. Markosyan, A. Markowitz, E. Maros, A. Marquina, P. Marsh, F. Martelli, L. Martellini, I.W. Martin, R.M. Martin, D.V. Martynov, J.N. Marx, K. Mason, E. Massera, A. Masserot, T.J. Massinger, M. Masso-Reid, S. Mastrogiovanni, A. Matas, F. Matichard, L. Matone, N. Mavalvala, N. Mazumder, R. McCarthy, D.E. McClelland, S. McCormick, L. McCuller, S.C. McGuire, G. McIntyre, J. McIver, D.J. McManus, L. McNeill, T. McRae, S.T. McWilliams, D. Meacher, G.D. Meadors, M. Mehmet, J. Meidam, E. Mejuto-Villa, A. Melatos, G. Mendell, R.A. Mercer, E.L. Merilh, M. Merzougui, S. Meshkov, C. Messenger, C. Messick, R. Metzdorff, P.M. Meyers, H. Miao, C. Michel, H. Middleton, E.E. Mikhailov, L. Milano, A.L. Miller, B.B. Miller, J. Miller, M. Millhouse, M.C. Milovich-Goff, O. Minazzoli, Y. Minenkov, J. Ming, C. Mishra, S. Mitra, V.P. Mitrofanov, G. Mitselmakher, R. Mittleman, D. Moffa, A. Moggi, K. Mogushi, M. Mohan, S.R.P. Mohapatra, I. Molina, M. Montani, C.J. Moore, D. Moraru, G. Moreno, S. Morisaki, S.R. Morriss, B. Mours, C.M. Mow-Lowry, G. Mueller, A.W. Muir, Arunava Mukherjee, D. Mukherjee, S. Mukherjee, N. Mukund, A. Mullavey, J. Munch, E.A. Muñiz, M. Muratore, P.G. Murray, A. Nagar, K. Napier, I. Nardecchia, L. Naticchioni, R.K. Nayak, J. Neilson, G. Nelemans, T.J.N. Nelson, M. Nery, A. Neunzert, L. Nevin, J.M. Newport, G. Newton, K.K.Y. Ng, P. Nguyen, T.T. Nguyen, D. Nichols, A.B. Nielsen, S. Nissanke, A. Nitz, A. Noack, F. Nocera, D. Nolting, C. North, L.K. Nuttall, J. Oberling, G.D. O'Dea, G.H. Ogin, J.J. Oh, S.H. Oh, F. Ohme, M.A. Okada, M. Oliver, P. Oppermann, R.J. Oram, B. O'Reilly, R. Ormiston, L.F. Ortega, R. O'Shaughnessy, S. Ossokine, D.J. Ottaway, H. Overmier, B.J. Owen, A.E. Pace, J. Page, M.A. Page, A. Pai, S.A. Pai, J.R. Palamos, O. Palashov, C. Palomba, A. Pal-Singh, H. Pan, H.-W. Pan, B. Pang, P.T.H. Pang, C. Pankow, F. Pannarale, B.C. Pant, F. Paoletti, A. Paoli, M.A. Papa, A. Parida, W. Parker, D. Pascucci, A. Pasqualetti, R. Passaquieti, D. Passuello, M. Patil, B. Patricelli, B.L. Pearlstone, M. Pedraza, R. Pedurand, L. Pekowsky, A. Pele, S. Penn, C.J. Perez, A. Perreca, L.M. Perri, H.P. Pfeiffer, M. Phelps, O.J. Piccinni, M. Pichot, F. Piergiovanni, V. Pierro, G. Pillant, L. Pinard, I.M. Pinto, M. Pirello, M. Pitkin, M. Poe, R. Poggiani, P. Popolizio, E.K. Porter, A. Post, J. Powell, J. Prasad, J.W.W. Pratt, G. Pratten, V. Predoi, T. Prestegard, M. Prijatelj, M. Principe, S. Privitera, R. Prix, G.A. Prodi, L.G. Prokhorov, O. Puncken, M. Punturo, P. Puppo, M. Pürrer, H. Qi, V. Quetschke, E.A. Quintero, R. Quitzow-James, F.J. Raab, D.S. Rabeling, H. Radkins, P. Raffai, S. Raja, C. Rajan, B. Rajbhandari, M. Rakhmanov, K.E. Ramirez, A. Ramos-Buades, P. Rapagnani, V. Raymond, M. Razzano, J. Read, T. Regimbau, L. Rei, S. Reid, D.H. Reitze, W. Ren, S.D. Reyes, F. Ricci, P.M. Ricker, S. Rieger, K. Riles, M. Rizzo, N.A. Robertson, R. Robie, F. Robinet, A. Rocchi, L. Rolland, J.G. Rollins, V.J. Roma, J.D. Romano, R. Romano, C.L. Romel, J.H. Romie, D. Rosińska, M.P. Ross, S. Rowan, A. Rüdiger, P. Ruggi, G. Rutins, K. Ryan, S. Sachdev, T. Sadecki, L. Sadeghian, M. Sakellariadou, L. Salconi, M. Saleem, F. Salemi, A. Samajdar, L. Sammut, L.M. Sampson, E.J. Sanchez, L.E. Sanchez, N. Sanchis-Gual, V. Sand berg, J.R. Sanders, B. Sassolas, B.S. Sathyaprakash, P.R. Saulson, O. Sauter, R.L. Savage, A. Sawadsky, P. Schale, M. Scheel, J. Scheuer, J. Schmidt, P. Schmidt, R. Schnabel, R.M.S. Schofield, A. Schönbeck, E. Schreiber, D. Schuette, B.W. Schulte, B.F. Schutz, S.G. Schwalbe, J. Scott, S. M. Scott, E. Seidel, D. Sellers, A.S. Sengupta, D. Sentenac, V. Sequino, A. Sergeev, D.A. Shaddock, T.J. Shaffer, A.A. Shah, M.S. Shahriar, M.B. Shaner, L. Shao, B. Shapiro, P. Shawhan, A. Sheperd, D.H. Shoemaker, D.M. Shoemaker, K. Siellez, X. Siemens, M. Sieniawska, D. Sigg, A.D. Silva, L.P. Singer, A. Singh, A. Singhal, A.M. Sintes, B.J.J. Slagmolen, B. Smith, J.R. Smith,

R.J.E. Smith, S. Somala, E.J. Son, J.A. Sonnenberg, B. Sorazu, F. Sorrentino, T. Souradeep, A.P. Spencer, A.K. Srivastava, K. Staats, A. Staley, M. Steinke, J. Steinlechner, S. Steinlechner, D. Steinmeyer, S.P. Stevenson, R. Stone, D.J. Stops, K.A. Strain, G. Stratta, S.E. Strigin, A. Strunk, R. Sturani, A.L. Stuver, T.Z. Summerscales, L. Sun, S. Sunil, J. Suresh, P.J. Sutton, B.L. Swinkels, M.J. Szczepańczyk, M. Tacca, S.C. Tait, C. Talbot, D. Talukder, D.B. Tanner, M. Tápai, A. Taracchini, J.D. Tasson, J.A. Taylor, R. Taylor, S.V. Tewari, T. Theeg, F. Thies, E.G. Thomas, M. Thomas, P. Thomas, K.A. Thorne, K.S. Thorne, E. Thrane, S. Tiwari, V. Tiwari, K.V. Tokmakov, K. Toland, M. Tonelli, Z. Tornasi, A. Torres-Forné, C.I. Torrie, D. Töyrä, F. Travasso, G. Traylor, J. Trinastic, M.C. Tringali, L. Trozzo, K.W. Tsang, M. Tse, R. Tso, L. Tsukada, D. Tsuna, D. Tuyenbayev, K. Ueno, D. Ugolini, C.S. Unnikrishnan, A.L. Urban, S.A. Usman, H. Vahlbruch, G. Vajente, G. Valdes, M. Vallisneri, N. van Bakel, M. van Beuzekom, J.F.J. van den Brand, C. Van Den Broeck, D.C. Vand er-Hyde, L. van der Schaaf, J.V. van Heijningen, A.A. van Veggel, M. Vardaro, V. Varma, S. Vass, M. Vasúth, A. Vecchio, G. Vedovato, J. Veitch, P.J. Veitch, K. Venkateswara, G. Venugopalan, D. Verkindt, F. Vetrano, A. Viceré, A.D. Viets, S. Vinciguerra, D.J. Vine, J.Y. Vinet, S. Vitale, T. Vo, H. Vocca, C. Vorvick, S.P. Vyatchanin, A.R. Wade, L.E. Wade, M. Wade, R. Walet, M. Walker, L. Wallace, S. Walsh, G. Wang, H. Wang, J.Z. Wang, W.H. Wang, Y.F. Wang, R.L. Ward, J. Warner, M. Was, J. Watchi, B. Weaver, L. W. Wei, M. Weinert, A.J. Weinstein, R. Weiss, L. Wen, E.K. Wessel, P. Weßels, J. Westerweck, T. Westphal, K. Wette, J.T. Whelan, S.E. Whitcomb, B.F. Whiting, C. Whittle, D. Wilken, D. Williams, R.D. Williams, A.R. Williamson, J.L. Willis, B. Willke, M.H. Wimmer, W. Winkler, C.C. Wipf, H. Wittel, G. Woan, J. Woehler, J. Wofford, K.W.K. Wong, J. Worden, J.L. Wright, D.S. Wu, D.M. Wysocki, S. Xiao, H. Yamamoto, C.C. Yancey, L. Yang, M.J. Yap, M. Yazback, H. Yu, H. Yu, M. Yvert, A. ZadroŻny, M. Zanolin, T. Zelenova, J.P. Zendri, M. Zevin, L. Zhang, M. Zhang, T. Zhang, Y.H. Zhang, C. Zhao, M. Zhou, Z. Zhou, S.J. Zhu, X.J. Zhu, A.B. Zimmerman, M.E. Zucker, J. Zweizig, LIGO Scientific Collaboration, and Virgo Collaboration. GW170817: observation of gravitational waves from a binary neutron star inspiral. Phys. Rev. Lett. **119**(16), 161101 (2017)

8. M. Alcubierre, S. Brandt, B. Brügmann, D. Holz, E. Seidel, R. Takahashi, J. Thornburg, Symmetry without symmetry: numerical simulation of axisymmetric systems using cartesian grids. Int. J. Mod. Phys. D **10**, 273–290 (2001)
9. T. Andrade, L.A. Salo, J.C. Aurrekoetxea, J. Bamber, K. Clough, R. Croft, E. de Jong, A. Drew, A. Duran, P.G. Ferreira, P. Figueras, H. Finkel, T. França, B.-X. Ge, C. Gu, T. Helfer, J. Jäykkä, C. Joana, M. Kunesch, K. Kornet, E.A. Lim, F. Muia, Z. Nazari, M. Radia, J. Ripley, P. Shellard, U. Sperhake, D. Traykova, S. Tunyasuvunakool, Z. Wang, J.Y. Widdicombe, K. Wong, GRChombo: an adaptable numerical relativity code for fundamental physics. J. Open Source Softw. **6**(68), 3703 (2021)
10. M. Ansorg, B. Brügmann, W. Tichy, A single-domain spectral method for black hole puncture data. Phys. Rev. D **70**, 064011 (2004)
11. L. Antón, O. Zanotti, J.A. Miralles, J.M. Martí, J.M. Ibáñez, J.A. Font, J.A. Pons, Numerical 3+1 general relativistic magnetohydrodynamics: a Local characteristic approach. Astrophys. J. **637**, 296–312 (2006)
12. R.L. Arnowitt, S. Deser, C.W. Misner, The dynamics of general relativity. Gen. Relativ. Gravit. **40**, 1997–2027 (2008)
13. T. Assumpcao, L.R. Werneck, T.P. Jacques, Z.B. Etienne, Fast hyperbolic relaxation elliptic solver for numerical relativity: conformally flat, binary puncture initial data. Phys. Rev. D **105**(10), 104037 (2022)
14. L. Baiotti, I. Hawke, P.J. Montero, F. Löffler, L. Rezzolla, N. Stergioulas, J.A. Font, E. Seidel, Three-dimensional relativistic simulations of rotating neutron star collapse to a Kerr black hole. Phys. Rev. D **71**, 024035 (2005)
15. J.G. Baker, M. Campanelli, C.O. Lousto, The Lazarus project: A Pragmatic approach to binary black hole evolutions. Phys. Rev. D **65**, 044001 (2002)
16. F. Banyuls, J.A. Font, J.M.A. Ibanez, J.M.A. Marti, J.A. Miralles, Numerical 3+1 general relativistic hydrodynamics: a local characteristic approach. Astrophys. J. **476**, 221 (1997)

17. E. Bentivegna, M. Korzynski, Evolution of a periodic eight-black-hole lattice in numerical relativity. Class. Quant. Grav. **29**, 165007 (2012)
18. S. Bonazzola, E. Gourgoulhon, J.-A. Marck, Numerical approach for high precision 3d relativistic star models. Phys. Rev. D **58**(10), 104020 (1998)
19. R. Borges, M. Carmona, B. Costa, W.S. Don, An improved weighted essentially non-oscillatory scheme for hyperbolic conservation laws. J. Comp. Phys. **227**, 3191–3211 (2008)
20. J.M. Bowen, J.W. York Jr., Time asymmetric initial data for black holes and black hole collisions. Phys. Rev. D **21**, 2047–2056 (1980)
21. S. Brandt, B. Brügmann, Black hole punctures as initial data for general relativity. Phys. Rev. Lett. **78**, 3606–3609 (1997)
22. S.R. Brandt, Automatic generation of Carpet/CarpetX thorns with NRPy+, 2022. https://youtu.be/d6VZYiAKvXc
23. S.R. Brandt, H. Roland, The einstein toolkit tutorial server, in *Science Gateways 2023 Annual Conference* (2023)
24. J. David Brown, P. Diener, O. Sarbach, E. Schnetter, M. Tiglio, Turduckening black holes: an analytical and computational study. Phys. Rev. D **79**, 044023 (2009)
25. Cactus Computational Toolkit
26. P. Cerda-Duran, J.A. Font, L. Anton, E. Muller, A new general relativistic magnetohydrodynamics code for dynamical spacetimes. Astron. Astrophys. **492**, 937 (2008)
27. R. Ciolfi, W. Kastaun, B. Giacomazzo, A. Endrizzi, D.M. Siegel, R. Perna, General relativistic magnetohydrodynamic simulations of binary neutron star mergers forming a long-lived neutron star. Phys. Rev. D **95**(6), 063016 (2017)
28. F. Cipolletta, J.V. Kalinani, B. Giacomazzo, R. Ciolfi, Spritz: a new fully general-relativistic magnetohydrodynamic code. Class Quantum Gravity **37**(13), 135010 (2020)
29. P. Colella, P.R. Woodward, The Piecewise Parabolic Method (PPM) for gas dynamical simulations. J. Comp. Phys. **54**, 174–201 (1984)
30. S. Cupp, S.R. Brandt, G. Bozzola, P. Diener, Z. Etienne, D. Ferguson, R. Haas, L. Ji, H. Park, F.G. Lopez Armengol, M. Chabanov, C.-H. Cheng, A. Dima, J. Doherty, L. Ennoggi, T.P. Jacques, J. Kalinani, P. Moesta, M. Pirog, L.T. Sanches, E. Schnetter, S. Shankar, W. Tichy, L. Werneck, M. Alcubierre, D. Alic, G. Allen, M. Ansorg, M. Babiuc-Hamilton, L. Baiotti, W. Benger, E. Bentivegna, S. Bernuzzi, T. Bode, B. Brendal, B. Bruegmann, M. Campanelli, F. Cipolletta, G. Corvino, R. De Pietri, H. Dimmelmeier, R. Dooley, N. Dorband, M. Elley, Y. El Khamra, J. Faber, G. Ficarra, T. Font, J. Frieben, B. Giacomazzo, T. Goodale, C. Gundlach, I. Hawke, S. Hawley, I. Hinder, E.A. Huerta, S. Husa, T. Ikeda, S. Iyer, D. Johnson, A.V. Joshi, A. Kankani, W. Kastaun, T. Kellermann, A. Knapp, M. Koppitz, P. Laguna, G. Lanferman, P. Lasky, F. Löffler, H. Macpherson, J. Masso, L. Menger, A. Merzky, J.M. Miller, M. Miller, P. Montero, B. Mundim, P. Nelson, A. Nerozzi, S.C. Noble, C. Ott, L.J. Papenfort, R. Paruchuri, D. Pollney, D. Price, D. Radice, T. Radke, C. Reisswig, L. Rezzolla, C.B. Richards, D. Rideout, M. Ripeanu, L. Sala, J.A. Schewtschenko, B. Schutz, E. Seidel, E. Seidel, J. Shalf, K. Sible, U. Sperhake, N. Stergioulas, W.-M. Suen, B. Szilagyi, R. Takahashi, M. Thomas, J. Thornburg, C. Tian, M. Tobias, A. Tonita, S. Tootle, P. Walker, M.-B. Wan, B. Wardell, H. Witek, M. Zilhão, B. Zink, Y. Zlochower, The Einstein toolkit (2023)
31. R. Donat, A. Marquina, Capturing shock reflections: an improved flux formula. J. Comp. Phys. **125**, 42–58 (1996)
32. O. Dreyer, B. Krishnan, D. Shoemaker, E. Schnetter, Introduction to isolated horizons in numerical relativity. Phys. Rev. D **67**, 024018 (2003)
33. M.D. Duez, F. Foucart, L.E. Kidder, H.P. Pfeiffer, M.A. Scheel, S.A. Teukolsky, Evolving black hole-neutron star binaries in general relativity using pseudospectral and finite difference methods. Phys. Rev. D **78**, 104015 (2008)
34. M.D. Duez, Y.T. Liu, S.L. Shapiro, M. Shibata, B.C. Stephens, Collapse of magnetized hypermassive neutron stars in general relativity. Phys. Rev. Lett. **96**, 031101 (2006)
35. M.D. Duez, Y.T. Liu, S.L. Shapiro, B.C. Stephens, Relativistic magnetohydrodynamics in dynamical spacetimes: numerical methods and tests. Phys. Rev. D **72**, 024028 (2005)

36. B. Einfeldt, On Godunov-type methods for gas dynamics. SIAM J. Numer. Anal. **25**(2), 294–318 (1988)
37. Einstein Toolkit: Open software for relativistic astrophysics
38. P.L. Espino, G. Bozzola, V. Paschalidis, Quantifying uncertainties in general relativistic magnetohydrodynamic codes. Phys. Rev. D **107**(10), 104059 (2023)
39. Z.B. Etienne, V. Paschalidis, R. Haas, P. Mösta, S.L. Shapiro, IllinoisGRMHD: an open-source, user-friendly GRMHD code for dynamical spacetimes. Class. Quant. Grav. **32**(17), 175009 (2015)
40. Z.B. Etienne, V. Paschalidis, Y.T. Liu, S.L. Shapiro, Relativistic MHD in dynamical spacetimes: improved EM gauge condition for AMR grids. Phys. Rev. D **85**, 024013 (2012)
41. C.R. Evans, J.F. Hawley, Simulation of magnetohydrodynamic flows: a Constrained transport method. Astrophys. J. **332**, 659–677 (1988)
42. B.D. Farris, R. Gold, V. Paschalidis, Z.B. Etienne, S.L. Shapiro, Binary black-hole mergers in magnetized disks: simulations in full general relativity. Phys. Rev. Lett. **109**, 221102 (2012)
43. B.D. Farris, Y.T. Liu, S.L. Shapiro, Binary black hole mergers in gaseous disks: Simulations in general relativity. Phys. Rev. D **84**, 024024 (2011)
44. D. Ferguson et al., Second MAYA catalog of binary black hole numerical relativity waveforms 9 (2023)
45. L.G. Fishbone, V. Moncrief, Relativistic fluid disks in orbit around kerr black holes. Astrophys. J. **207**, 962–976 (1976)
46. B. Giacomazzo, Numerical simulations of coalescing neutron star binaries part i-numerical relativity (2023)
47. B. Giacomazzo, L. Rezzolla, WhiskyMHD: a new numerical code for general relativistic magnetohydrodynamics. Class. Quantum Grav. **24**, S235–S258 (2007)
48. R. Gold, V. Paschalidis, Z.B. Etienne, S.L. Shapiro, H.P. Pfeiffer, Accretion disks around binary black holes of unequal mass: General relativistic magnetohydrodynamic simulations near decoupling. Phys. Rev. D **89**(6), 064060 (2014)
49. T. Goodale, G. Allen, G. Lanfermann, J. Massó, T. Radke, E. Seidel, J. Shalf, *The Cactus framework and toolkit: design and applications, in Vector and Parallel Processing - VECPAR'2002, 5th International Conference* Lecture Notes in Computer Science. (Springer, Berlin, 2003)
50. P. Grandclement, Kadath: a spectral solver for theoretical physics. J. Comput. Phys. **229**, 3334–3357 (2010)
51. S. Habib, A. Ramos-Buades, E.A. Huerta, S. Husa, R. Haas, Z. Etienne, Initial data and eccentricity reduction toolkit for binary black hole numerical relativity waveforms 11 (2020)
52. A. Harten, High resolution schemes for hyperbolic conservation laws. J. Comput. Phys. **49**, 357–393 (1983)
53. J. Healy, C.O. Lousto, Fourth RIT binary black hole simulations catalog: extension to eccentric orbits. Phys. Rev. D **105**(12), 124010 (2022)
54. E.A. Huerta, R. Haas, S. Habib, A. Gupta, A. Rebei, V. Chavva, D. Johnson, S. Rosofsky, E. Wessel, B. Agarwal, D. Luo, W. Ren, Physics of eccentric binary black hole mergers: A numerical relativity perspective. Phys. Rev. D **100**(6), 064003 (2019)
55. S. Husa, I. Hinder, C. Lechner, Kranc: a Mathematica application to generate numerical codes for tensorial evolution equations. Comput. Phys. Commun. **174**, 983–1004 (2006)
56. M.R. Izquierdo, J. Fernando Abalos, C. Palenzuela, Guided moments formalism: a new efficient full-neutrino treatment for astrophysical simulations. Phys. Rev. D **109**(4), 043044 (2024)
57. J.V. Kalinani, R. Ciolfi, W. Kastaun, B. Giacomazzo, F. Cipolletta, L. Ennoggi, Implementing a new recovery scheme for primitive variables in the general relativistic magnetohydrodynamic code Spritz. Phys. Rev. D **105**(10), 103031 (2022)
58. J.V. Kalinani et al., AsterX: a new open-source GPU-accelerated GRMHD code for dynamical spacetimes 6 (2024)
59. W. Kastaun, J.V. Kalinani, R. Ciolfi, Robust recovery of primitive variables in relativistic ideal magnetohydrodynamics. Phys. Rev. D **103**(2), 023018 (2021)

60. H. Komatsu, Y. Eriguchi, I. Hachisu, Rapidly rotating general relativistic stars. I - Numerical method and its application to uniformly rotating polytropes. Mon. Not. Roy. Astron. Soc. **237**, 355–379 (1989)
61. F. Löffler, J. Faber, E. Bentivegna, T. Bode, P. Diener, R. Haas, I. Hinder, B.C. Mundim, C.D. Ott, E. Schnetter, G. Allen, M. Campanelli, P. Laguna, The Einstein toolkit: a community computational infrastructure for relativistic astrophysics. Class. Quantum Grav. **29**(11), 115001 (2012)
62. F.G. Lopez Armengol, Z.B. Etienne, S.C. Noble, B.J. Kelly, L.R. Werneck, B. Drachler, M. Campanelli, F. Cipolletta, Y. Zlochower, A. Murguia-Berthier, L. Ennoggi, M. Avara, R. Ciolfi, J. Faber, G. Fiacco, B. Giacomazzo, T. Gupte, T. Ha, J.H. Krolik, V. Mewes, R. O'Shaughnessy, J.M. Rueda-Becerril, J. Schnittman, Handing off the outcome of binary neutron star mergers for accurate and long-term postmerger simulations. Phys. Rev. D **106**, 083015 (2022)
63. G. Lovelace, Y. Chen, M. Cohen, J.D. Kaplan, D. Keppel, K.D. Matthews, D.A. Nichols, M.A. Scheel, U. Sperhake, Momentum flow in black-hole binaries. II. Numerical simulations of equal-mass, head-on mergers with antiparallel spins. Phys. Rev. D **82**, 064031 (2010)
64. H.J. Macpherson, P.D. Lasky, D.J. Price, Inhomogeneous cosmology with numerical relativity. Phys. Rev. D **95**(6), 064028 (2017)
65. H.J. Macpherson, P.D. Lasky, D.J. Price, The trouble with Hubble: local versus global expansion rates in inhomogeneous cosmological simulations with numerical relativity. Astrophys. J. Lett. **865**(1), L4 (2018)
66. F. Maione, R. De Pietri, A. Feo, F. Löffler, Binary neutron star merger simulations with different initial orbital frequency and equation of state. Class. Quant. Grav. **33**(17), 175009 (2016)
67. J.M. Martí, J.M. Ibáñez, J.M. Miralles, Numerical relativistic hydrodynamics: local characteristic approach. Phys. Rev. D **43**, 3794–3801 (1991)
68. P. McCorquodale, P. Colella, A high-order finite-volume method for conservation laws on locally refined grids. Comm. Appl. Math. Comput. Sci. **6**, 1 (2011)
69. V. Mewes, Y. Zlochower, M. Campanelli, I. Ruchlin, Z.B. Etienne, T.W. Baumgarte, Numerical relativity in spherical coordinates with the Einstein Toolkit. Phys. Rev. D **97**(8), 084059 (2018)
70. E.R. Most, L. Jens Papenfort, L. Rezzolla, Beyond second-order convergence in simulations of magnetized binary neutron stars with realistic microphysics. Mon. Not. Roy. Astron. Soc. **490**(3), 3588–3600 (2019)
71. P. Mösta, B.C. Mundim, J.A. Faber, R. Haas, S.C. Noble, T. Bode, F. Löffler, C.D. Ott, C. Reisswig, E. Schnetter, GRHydro: a new open source general-relativistic magnetohydrodynamics code for the Einstein Toolkit. Class. Quantum Gravity **31**(1), 015005 (2014)
72. P. Mösta, C.D. Ott, D. Radice, L.F. Roberts, E. Schnetter, R. Haas, A large scale dynamo and magnetoturbulence in rapidly rotating core-collapse supernovae. Nature **528**, 376 (2015)
73. P. Mösta, S. Richers, C.D. Ott, R. Haas, A.L. Piro, K. Boydstun, E. Abdikamalov, C. Reisswig, E. Schnetter, Magnetorotational core-collapse supernovae in three dimensions. Astrophys. J. Lett. **785**, L29 (2014)
74. P. Mösta, L.F. Roberts, G. Halevi, C.D. Ott, J. Lippuner, R. Haas, E. Schnetter, R-process nucleosynthesis from three-dimensional magnetorotational core-collapse supernovae. Astrophys. J. **864**(2), 171 (2018)
75. S.C. Noble, C.F. Gammie, J.C. McKinney, L. Del Zanna, Primitive variable solvers for conservative general relativistic magnetohydrodynamics. Astrophys. J. **641**, 626–637 (2006)
76. N.O. Murchadha, J.W. York, Gravitational energy. Phys. Rev. D **10**, 2345–2357 (1974)
77. J.R. Oppenheimer, G.M. Volkoff, On Massive Neutron Cores. Phys. Rev. **55**, 374–381 (1939)
78. C.D. Ott, E. Abdikamalov, P. Mösta, R. Haas, S. Drasco, E.P. O'Connor, C. Reisswig, C.A. Meakin, E. Schnetter, General-relativistic simulations of three-dimensional core-collapse supernovae. Astrophys. J. **768**, 115 (2013)
79. L. Jens Papenfort, S.D. Tootle, P. Grandclément, E.R. Most, L. Rezzolla, New public code for initial data of unequal-mass, spinning compact-object binaries. Phys. Rev. D **104**(2), 024057 (2021)

80. A.J. Penner, General relativistic magnetohydrodynamic Bondi-Hoyle accretion. Mon. Not. Roy. Astron. Soc. **414**, 1467–1482 (2011)
81. D. Pollney, C. Reisswig, E. Schnetter, N. Dorband, P. Diener, High accuracy binary black hole simulations with an extended wave zone. Phys. Rev. D **83**, 044045 (2011)
82. D. Radice, L. Rezzolla, F. Galeazzi, High-order fully general-relativistic hydrodynamics: new approaches and tests. Class. Quant. Grav. **31**, 075012 (2014)
83. D. Radice, L. Rezzolla, F. Galeazzi, High-order numerical-relativity simulations of binary neutron stars. ASP Conf. Ser. **498**, 121–126 (2015)
84. C. Reisswig, R. Haas, C.D. Ott, E. Abdikamalov, P. Mösta, D. Pollney, E. Schnetter, Three-dimensional general-relativistic hydrodynamic simulations of binary neutron star coalescence and stellar collapse with multipatch grids. Phys. Rev. D **87**, 064023 (2013)
85. P.L. Roe, Characteristic-based schemes for the Euler equations. Ann. Rev. Fluid Mech. **18**, 337–365 (1986)
86. I. Ruchlin, Z.B. Etienne, T.W. Baumgarte, SENR /NRPy +ÃƒÂ¢Ã¢âŁšÂ¬Ã‚Â¯: Numerical relativity in singular curvilinear coordinate systems. Phys. Rev. D **97**(6), 064036 (2018)
87. I. Ruchlin, Z.B. Etienne, T.W. Baumgarte, SENR/NRPy+: numerical relativity in singular curvilinear coordinate systems. Phys. Rev. D **97**(6), 064036 (2018)
88. M. Ruiz, A. Tsokaros, S.L. Shapiro, Jet launching from merging magnetized binary neutron stars with realistic equations of state. Phys. Rev. D **104**(12), 124049 (2021)
89. E. Schnetter, M. Blazewicz, S.R. Brandt, D.M. Koppelman, F. Löffler, *Chemora: a PDE solving framework for modern HPC architectures* (Comput. Sci, Engin, 2014). ((arXiv:1410.1764))
90. E. Schnetter, S. Brandt, S. Cupp, R. Haas, P. Mösta, S. Shankar, Carpetx, February 2022. Grants: 10.13039/100000001::2004157–10.13039/100000001::2004879
91. E. Schnetter, S.H. Hawley, I. Hawke, Evolutions in 3-D numerical relativity using fixed mesh refinement. Class. Quantum Grav. **21**, 1465–1488 (2004)
92. E. Schnetter, B. Krishnan, F. Beyer, Introduction to dynamical horizons in numerical relativity. Phys. Rev. D **74**, 024028 (2006)
93. S. Shankar, P. Mösta, S.R. Brandt, R. Haas, E. Schnetter, Y. de Graaf, GRaM-X: a new GPU-accelerated dynamical spacetime GRMHD code for exascale computing with the Einstein toolkit. Class Quantum Gravity **40**(20), 205009 (2023)
94. C.W. Shu, High order ENO and WENO schemes for computational fluid dynamics, in *High Order Methods for Computational Physics*. ed. by J. Timothy, H. Barth (Array Deconinck (Springer, New York, 1999), pp.439–582
95. U. Sperhake, Binary black-hole evolutions of excision and puncture data. Phys. Rev. D **76**, 104015 (2007)
96. A. Suresh, H.T. Huynh, Accurate monotonicity-preserving schemes with Runge-Kutta time stepping. J. Comput. Phys. **136**(1), 83–99 (1997)
97. N. Tacik, F. Foucart, H.P. Pfeiffer, R. Haas, S. Ossokine, J. Kaplan, C. Muhlberger, M.D. Duez, L.E. Kidder, M.A. Scheel, B. Szilágyi, Binary neutron stars with arbitrary spins in numerical relativity. Phys. Rev. D **92**(12), 124012 (2015). [Erratum: Phys. Rev. D **94**, 049903 (2016)]
98. A. Tchekhovskoy, J.C. McKinney, R. Narayan, WHAM: a WENO-based general relativistic numerical scheme I: Hydrodynamics. Mon. Not. Roy. Astron. Soc. **379**, 469–497 (2007)
99. J. Thornburg, A 3+ 1 computational scheme for dynamic spherically symmetric black hole spacetimes–ii: time evolution (1999). arXiv preprint gr-qc/9906022
100. J. Thornburg, A fast apparent-horizon finder for 3-dimensional cartesian grids in numerical relativity. Class. Quantum Grav. **21**, 743–766 (2004)
101. W. Tichy, A. Rashti, T. Dietrich, R. Dudi, B. Brügmann, Constructing binary neutron star initial data with high spins, high compactnesses, and high mass ratios. Phys. Rev. D **100**(12), 124046 (2019)
102. K.P. Tod, Looking for marginally trapped surfaces. Class. Quantum Gravity **8**(5), L115 (1991)
103. R.C. Tolman, Static solutions of Einstein's field equations for spheres of fluid. Phys. Rev. **55**, 364–373 (1939)

104. G. Tóth. The $\nabla.B = 0$ constraint in shock-capturing magnetohydrodynamics codes. J. Comput. Phys. **161**, 605–652 (2000)
105. A. Čadež, Ph.d. thesis, University of North Carolina at Chapel Hill, 1971
106. L.R. Werneck et al., Addition of tabulated equation of state and neutrino leakage support to illinoisgrmhd. Phys. Rev. D **107**(4), 044037 (2023)
107. H. Witek, M. Zilhao, G. Bozzola, M. Elley, G. Ficarra, T. Ikeda, N. Sanchis-Gual, H. Silva, Canuda: a public numerical relativity library to probe fundamental physics, October 2021
108. Y. Zenati, J. Krolik, L. Werneck, Z. Etienne, S. Noble, A. Murguia-Berthier, J. Schnittman, The dynamics of debris disk creation in neutron star mergers 4 (2024)
109. Y. Zenati, J.H. Krolik, L.R. Werneck, A. Murguia-Berthier, Z.B. Etienne, S.C. Noble, T. Piran, Bound debris expulsion from neutron star merger remnants. Astrophys. J. **958**(2), 161 (2023)
110. W. Zhang, A. Almgren, V. Beckner, J. Bell, J. Blaschke, C. Chan, M. Day, B. Friesen, K. Gott, D. Graves, M. Katz, A. Myers, T. Nguyen, A. Nonaka, M. Rosso, S. Williams, M. Zingale, AMReX: a framework for block-structured adaptive mesh refinement. J. Open Sour. Softw. **4**(37), 1370 (2019)

Chapter 7
SPHINCS_BSSN: Numerical Relativity with Particles

Stephan Rosswog and Peter Diener

Abstract In this chapter we describe the *Lagrangian* numerical relativity code `SPHINCS_BSSN`. This code evolves spacetimes in full General Relativity by integrating the BSSN equations on structured meshes with a simple dynamical mesh refinement strategy. The fluid is evolved by means of freely moving Lagrangian particles that are evolved using a modern Smooth Particle Hydrodynamics (SPH) formulation. To robustly and accurately capture shocks, our code uses artificial dissipation terms, but similar to Finite Volume schemes, we apply a slope-limited reconstruction within the dissipative terms and in addition we use time-dependent dissipation parameters, so that dissipation is only applied where needed. The technically most complicated, but absolutely crucial part of our methodology, is the coupling between the particles and the mesh. For the mapping of the energy-momentum tensor $T_{\mu\nu}$ from the particles to the mesh, we use a sophisticated combination of "Local Regression Estimate" (LRE) method and a "multi-dimensional optimal order detection" (MOOD) approach which we describe in some detail. The mapping of the metric quantities from the grid to the particles is achieved by a quintic Hermite interpolation. Apart from giving an introduction to our numerical methods, we demonstrate the accurate working of our code by presenting a set of representative relativistic hydrodynamics tests. We begin with a relativistic shock tube test, then compare the frequencies of a fully relativistic neutron star with reference values from the literature and, finally, we present full-blown merger simulations of irrotational binary systems, one case where a central remnant survives and another where a black hole forms, and of a binary where only one of the stars is rapidly spinning.

S. Rosswog (✉)
Sternwarte Hamburg, Gojenbergsweg 112, 21029 Hamburg, Germany
e-mail: stephan.rosswog@uni-hamburg.de

The Oskar Klein Centre, Department of Astronomy, Stockholm University, Stockholm, Sweden

P. Diener
Center for Computation and Technology, Louisiana State University, Baton Rouge, LA 70803, USA
e-mail: diener@cct.lsu.edu

Department of Physics and Astronomy, Louisiana State University, Baton Rouge, LA 70803, USA

C. Bambi et al. (eds.), *New Frontiers in GRMHD Simulations*, Springer Series in Astrophysics and Cosmology, https://doi.org/10.1007/978-981-97-8522-3_7

7.1 Introduction

Despite their early postulation by Albert Einstein in 1916 [1, 2], gravitational waves (GWs) remained a topic of sustained debate for more than half a century [3]. Their physical reality was only firmly established after the discovery of the Hulse-Taylor pulsar in 1974 [4]: the secular orbital decay of this neutron star binary system is in nearly perfect agreement with the prediction from Einstein's theory and this finally brought the decades' long controversy to an end. Still, the first *direct* detection of GWs had to wait until 2015 when both LIGO detectors recorded the last 0.2 s of the inspiral and the subsequent merger of two $\sim$30 $M_\odot$ black holes [5].

The first detection of a GW source that actually involved *matter* (rather than "just black holes"), followed soon: in August 2017 a $\sim$minute-long GW-signal from a merging neutron star binary (GW170817) was detected. Most excitingly, this event was also observed in electromagnetic (EM) waves, first as a short gamma-ray burst (GRB) that followed the peak of the GW emission with a delay of 1.7 s, later ($\sim$days) as a radioactively powered thermal transient peaking at optical/near-infrared wavelengths frequently called a "kilonova", and much later ($\sim$months) also in radio wavelengths, see [6] for a recent review.

This event meant a huge leap forward for many long-standing (astro-)physics questions. For example, the electromagnetic follow-up observations of GW170817 revealed the actual host galaxy where the merger took place and this placed the event in an astronomical environment and connected it to potential stellar evolution paths [7]. Since the distance to the host galaxy is known, it was possible to use the 1.7 s delay between GWs and the GRB to show that GWs propagate to a relative accuracy of 10^{-15} at the speed of light [8]. This, in turn, eliminated families of alternative theories of gravity where $c_{\rm GW} \neq c$. GW170817 also demonstrated the exciting possibility to constrain the properties of cold nuclear matter via the effects of tidal deformability and thereby ruled out some very stiff equations of state [9]. The event further showed that neutron star mergers are, as expected on theoretical grounds [10–14], major cosmic production sites of heavy "r-process" elements [15], and that they are useful probes of the cosmic expansion [16].

All of these breakthroughs were enabled by the *combined* detection of the same event via different messengers, here GWs and photons. This detection splendidly illustrated the huge discovery potential of multi-messenger astrophysics, but it also underlined the enormous modelling challenge: in addition to strong-field gravity, relativistic fluid dynamics, neutrino transport, magnetic fields, nuclear reactions and atomic physics, also (non-equilibrium) radiative transfer should be included in theoretical models in order to make the connection to the EM observations.

Numerical simulations, especially numerical relativity simulations [17–20], play a very crucial role in connecting the physical processes during merger to potentially observable signals. Such simulations have increasingly become tools for exploring signatures from both established and more speculative parts of fundamental physics. Multi-physics numerical relativity simulations are extremely complex and time consuming: codes typically evolve several dozen variables per spacetime (grid-) point and, depending on the simulated physics, many more variables may be needed. For example, neutrino transport is needed to make detailed predictions for the neutron-to-proton-ratio in the ejecta, which, in turn, determines which elements are formed and therefore also how the electromagnetic signal looks like. But evolving several neutrino flavours, with many energy groups for each of them, becomes a serious extra computational burden [21–23]. To make things worse, the spacetime needs to be evolved with very small numerical time steps that are limited by the Courant criterion [24]. The "signal speed" that enters this criterion is the speed of light, which limits the admissible time steps to $\Delta_t \approx 10^{-7}\, s\, \left(\frac{\Delta x}{100\text{m}}\right)$, where Δx is the spatial length to be resolved.

To date, essentially all numerical relativity codes that solve the full set of Einstein equations are *Eulerian*, i.e. they solve the fluid equations on computational (usually structured or adaptive) grids. The only exception is our code SPHINCS_BSSN [25–28], which is the main topic of this book chapter. Using an alternative methodology makes it possible to independently cross-validate the longer established Eulerian methods, but it also comes with distinctive advantages for multi-messenger astrophysics. One of its salient features is that–contrary to Eulerian methods–vacuum does not need to be modelled hydrodynamically: it is simply characterized by the absence of computational elements (here: particles) and no computational resources need to be spent to evolve it. Moreover, the neutron star surface, which is an eternal source of numerical trouble for Eulerian methods, is very well-behaved and does not need any particular treatment compared to the rest of the fluid.[1] The ejecta of a neutron star merger correspond to only $\sim 1\%$ of the binary mass, but they are responsible for essentially all of the EM emission. Therefore, the accurate prediction of the properties of this small amount of matter is of paramount importance for multi-messenger astrophysics. This is, in fact, one of the key strengths of a code like SPHINCS_BSSN: advection is exact (i.e. not depending on the resolution) and the properties of a particle, say its electron fraction Y_e, are simply transported by the particle unless they are changed by physical processes (e.g. by an electron capture reaction), but no purely numerical "diffusive" processes occur.

In the rest of this chapter, we will describe the methodology behind our Lagrangian Numerical Relativity code SPHINCS_BSSN and we will present some of the first applications. In Sect. 7.2, we will describe how we evolve fluids and, since in a numerical relativity context the use of particles is not widespread, we start with some basics of the SPH method. We further outline how the spacetime is evolved and how the fluid particles and the grid cells for the spacetime evolution exchange information. In Sect. 7.3, we will show some applications of the code including some standard benchmark tests, but also some of the first astrophysical simulations. We finally will summarize this chapter in Sect. 7.4.

[1] This is illustrated later, see Fig. 7.6.

7.2 Methodology

Here, we will briefly summarize the main methodological elements that enter in SPHINCS_BSSN. We will focus here on the main ideas and in some cases we will refer to our more technical papers [25–28] for the full details.

7.2.1 *Relativistic Smooth Particle Hydrodynamics*

Our main aim here is to describe how we evolve matter by means of a modern, general relativistic Smooth Particle Hydrodynamics (SPH) code. But before we do so, we will summarize some basic SPH techniques and we will briefly discuss the (simpler to understand) Newtonian SPH approach. Once equipped with this basic knowledge, the general relativistic version should be straight forward to grasp.

This introduction to the topic is by no means meant as a comprehensive SPH review, for a more complete introduction to the SPH-method, its variants and subtleties and its various applications in astrophysics and cosmology, we refer to extensive reviews that exist in the literature [29–35]. For relativistic formulations of SPH, we specifically want to point to [25–28, 31, 36–41]. Among these, we will frequently refer to [31] since here many results are derived very explicitly in a step-by-step fashion.

7.2.1.1 A Primer on Basic SPH-Techniques

The main idea of SPH is to model a fluid via spherical particles of finite size and overlapping support. These particles move with the local fluid velocity, i.e. SPH is a fully *Lagrangian* method, as opposed to Eulerian or Adaptive Lagrangian Eulerian, or "ALE" methodologies. The overlapping nature of the particles distinguishes them, for example, from (usually quasi-) Lagrangian Moving Mesh methods, see e.g. [42, 43], which instead tessellate space into non-overlapping cells which have a sharp interface surface between them. The task is now to translate the continuum equations of Lagrangian fluid dynamics into evolution equations for the particles, ideally in a way that ensures that nature's conservation laws are enforced on the level of the particle evolution equations. In the simplest case of ideal, non-relativistic fluids, the equations read

$$\frac{d\rho}{dt} = -\rho \nabla \cdot \mathbf{v} \tag{7.1}$$

$$\frac{d\mathbf{v}}{dt} = -\frac{\nabla P}{\rho} + \mathbf{f} \tag{7.2}$$

$$\frac{du}{dt} = \frac{dq}{dt} + \frac{P}{\rho^2}\frac{d\rho}{dt}. \tag{7.3}$$

Here, ρ is the mass density, $\mathbf{v}$ the fluid velocity, P the pressure, $\mathbf{f}$ denote other "body forces" e.g. from gravity, u is the specific internal energy (i.e. energy per mass) and dq/dt denotes potential sources of heating and/or cooling. The Lagrangian time derivative, d/dt, is related to the Eulerian time derivative $\partial/\partial t$ by

$$\frac{d}{dt}(.) = \left(\frac{\partial}{\partial t} + \mathbf{v} \cdot \nabla\right)(.). \tag{7.4}$$

Note that Eq. (7.3) is simply the first law of thermodynamics written "on a per mass basis", i.e. the involved quantities are e.g. energy per mass or volume per mass (= 1/density).

At the heart of the SPH method is kernel interpolation. Assuming that some quantity f is known at particle locations b, a smoothed version of the quantity f reads

$$\langle f\rangle(\mathbf{r}) = \int f(\mathbf{r}')\, W_h(\mathbf{r}' - \mathbf{r})\, d^3x' \approx \sum_b V_b f_b W_h(\mathbf{r_b} - \mathbf{r}) = \sum_b \frac{m_b}{\rho_b}\, f_b\, W_h(\mathbf{r_b} - \mathbf{r}). \tag{7.5}$$

Here W_h is a smoothing kernel, see below, whose support size is determined by the length scale h, which, in an SPH context, is usually referred to as the "smoothing length". In the approximation we have applied a one-point quadrature. In SPH, one usually chooses the particle volume as m_b/ρ_b, where m_b and ρ_b are the particle mass and mass density at the position of particle b, but other choices are possible, see e.g. [34, 44, 45] for alternatives. One simple application of Eq. (7.5) is to choose the density ρ for the quantity f, which yields

$$\langle\rho\rangle_a = \sum_b m_b W_{ab}(h_a), \tag{7.6}$$

where $W_{ab}(h_a)$ is a short-hand for $W((\mathbf{r}_a - \mathbf{r}_b)/h_a)$. In other words: the density can be calculated by summing over the local neighbourhood and weighting each particle's mass according to how far away it is from particle a. If one keeps each particle's mass fixed in time, one has enforced *exact* mass conservation and one does not have to solve the mass conservation equation (7.1) explicitly, but one can do so if this is desired.

Clearly, for the smoothing procedure in Eq. (7.5) to make sense, the smoothing kernel must have certain properties:

- it needs to be normalized, $\int W_h(\mathbf{r}')d^3r' = 1$,
- obviously, the dimension of the kernel has to be "1/volume",
- it has to have the "delta-property", so as to reproduce the original function in the limit of vanishing support size,

$$\lim_{h\to 0} \int W_h(\mathbf{r}' - \mathbf{r})\; f(r')\; d^3r' = f(r) \tag{7.7}$$

- and it should (at least) have a continuous second derivative.

A number of other properties are desirable, for example, the kernel should be "radial", i.e. $W_h(\mathbf{x}-\mathbf{x}') = W_h(|\mathbf{x}-\mathbf{x}')|)$, to easily allow for the conservation of momentum and angular momentum, and its Fourier transform should fall of rapidly with wave number [46]. As a side remark: it may seem appealing to use an oblate kernel to model, say, an accretion disk, but this approach would sacrifice one of SPH's most salient features: its exact angular momentum conservation. The latter can be easily enforced when radial kernels are used, see Sect. 2.4 in [31] for an explicit demonstration how radial kernels lead to exact angular momentum conservation.

For a gradient estimate, one can straight forwardly apply the nabla operator to Eq. (7.5)[2]

$$\nabla f(\mathbf{r}) = \sum_b \frac{m_b}{\rho_b} f_b \, \nabla_r W_h(\mathbf{r_b} - \mathbf{r}), \tag{7.8}$$

where the index r at the nabla operator indicates that the derivative is taken with respect to the variable $\mathbf{r}$ (rather than $\mathbf{r}_b$). The derivatives are usually needed at the position of a particle (here labelled a) and abbreviated as

$$(\nabla f)_a = \sum_b \frac{m_b}{\rho_b} f_b \nabla_a W_{ab}(h_a), \tag{7.9}$$

where the label at the nabla operator is again a reminder that the derivative is taken with respect to $\mathbf{r}_a$ (and not $\mathbf{r}_b$). From Eq. (7.5) it is obvious that one would want to have the "partition of unity" property in a discrete form at any position $\mathbf{r}$

$$1 = \sum_b \frac{m_b}{\rho_b} W_h(\mathbf{r_b} - \mathbf{r}) \tag{7.10}$$

$$0 = \nabla_r(1) = \sum_b \frac{m_b}{\rho_b} \nabla_r W_h(\mathbf{r_b} - \mathbf{r}), \tag{7.11}$$

where the second equation is simply the result of applying the nabla operator to the first one. In standard SPH these equations are only approximately fulfilled and the accuracy depends on the kernel function used, its support size (or the number of contributing particles) and on the exact particle distribution. This has the effect, that when applying Eq. (7.8) to a constant function $f = f_0$, one only approximately recovers the theoretical value of zero

$$\nabla_r \left(\sum_b \frac{m_b}{\rho_b} f_b W_h(\mathbf{r_b} - \mathbf{r}) \right) = f_0 \sum_b \frac{m_b}{\rho_b} \nabla_r W_h(\mathbf{r_b} - \mathbf{r}) \approx 0. \tag{7.12}$$

This, however, can be easily corrected by simply "subtracting a numerical zero", so that a better gradient estimate reads

---

[2] From now on we drop the distinction between a function and its numerical approximation.

$$(\nabla f)_a^{(1)} = \sum_b \frac{m_b}{\rho_b} f_b \nabla_a W_{ab}(h_a) - \left(f_a \sum_b \frac{m_b}{\rho_b} \nabla_a W_{ab}(h_a) \right)$$
$$= \sum_b \frac{m_b}{\rho_b} (f_b - f_a) \nabla_a W_{ab}(h_a), \tag{7.13}$$

and this expression vanishes by construction when all f_k are the same. As an example, we can apply this gradient prescription to find an estimate for the velocity divergence. This quantity plays an important role in fluid dynamics, since $\nabla \cdot \mathbf{v} = 0$ defines "incompressibility", while $\nabla \cdot \mathbf{v} > 0$ indicates local expansion and $\nabla \cdot \mathbf{v} < 0$ means local compression of the fluid. The resulting expression is

$$(\nabla \cdot \mathbf{v})_a = \sum_b \frac{m_b}{\rho_b} (\mathbf{v}_b - \mathbf{v}_a) \cdot \nabla_a W_{ab}(h_a). \tag{7.14}$$

An alternative discretization can be found by taking Eq. (7.1) and calculating the Lagrangian time derivative of Eq. (7.6) directly

$$(\nabla \cdot \mathbf{v})_a = -\frac{1}{\rho_a} \frac{d\rho_a}{dt} = -\frac{1}{\rho_a} \sum_b m_b \frac{dW_{ab}}{dt} = \frac{1}{\rho_a} \sum_b m_b (\mathbf{v}_b - \mathbf{v}_a) \cdot \nabla_a W_{ab}(h_a), \tag{7.15}$$

where we have taken the straight-forward Lagrangian time derivative of the kernel function (see the kernel table in Sect. 2.3 in [31] where this is shown explicitly). Thus, we have derived an alternative numerical discretization for $\nabla \cdot \mathbf{v}$, where the difference between the two estimates is whether the local density ("ρ_a") or the weighted nearby densities ("ρ_b") are used. Both are equally valid and both vanish for the uniform velocity case.

For easy later reference, we explicitly write down once more the Lagrangian time derivative of the density in Eq. (7.6) that we have just found

$$\frac{d\rho_a}{dt} = \sum_b m_b \frac{d}{dt} [W_{ab}(h_a)] = \sum_b m_b (\mathbf{v}_a - \mathbf{v}_b) \cdot \nabla_a W_{ab}(h_a) \tag{7.16}$$

and we also note that

$$\nabla_a \rho_b = \sum_k m_k \nabla_a W_{bk}(h_b) \tag{7.17}$$

(the detailed steps can be found in Sect. 3.2 of [31]). Note that we have neglected small terms, usually called "grad-h" terms, that result from taking derivatives of the kernel with respect to the smoothing lengths. These terms have been derived for Newtonian hydrodynamics by [47, 48], for special-relativistic SPH in [40] and for the General Relativistic case in [39]. We had explored the effects of these grad-h terms in [49], but found them to be too close to unity to warrant the extra effort of an additional iteration (ρ depends on h and h depends on ρ, so one needs an iteration for consistent values). Therefore, we will ignore these "grad-h"-terms for the rest of this chapter.

7.2.1.2 Newtonian SPH

We have already seen in Eq. (7.6) that the continuity equation does not need to be solved explicitly, and if we keep the particle masses fixed, we have enforced exact mass conservation. It now turns out that by insightful choices (see the discussion in Sect. 2.3–2.5 in [31]), one can also bring the energy and momentum equation into discrete forms so that they conserve the total energy and momentum exactly. If a radial kernel is used, then the total angular momentum is conserved exactly by construction[3] as well. Keep in mind that this exact conservation even holds at finite resolution, which is not the case for Eulerian methods. Moving-mesh methods, see e.g. [42, 43], usually show good overall conservation, but angular momentum is not conserved exactly.

One can, however, also achieve exact conservation in a more elegant way: by starting from a Lagrangian and by applying a variational principle [31, 33, 38, 47, 50, 51] one finds a fully conservative discrete form of the equations without ambiguities.

As before, we will start here with the simplest case of ideal, Newtonian hydrodynamics, but one can adhere to this strategy also for cases with additional physics, e.g. gravity [52]. One can start from a discretized Lagrangian of an ideal fluid

$$L = \sum_b m_b \left(\frac{v_b^2}{2} - u_b \right) \tag{7.18}$$

and apply the Euler-Lagrange equations

$$\frac{d}{dt}\frac{\partial L}{\partial \mathbf{v}_a} - \frac{\partial L}{\partial \mathbf{r}_a} = 0 \tag{7.19}$$

to find the momentum evolution equation of each particle, where $\partial L/\partial \mathbf{v}_a$ is the canonical momentum, which for Eq. (7.18) simply becomes $m_a \mathbf{v}_a$. We can use again fixed particle masses and the adiabatic first law of thermodynamics, $du = -Pd(1/\rho)$, which yields

$$\left(\frac{\partial u}{\partial \rho} \right)_s = \frac{P}{\rho^2}, \tag{7.20}$$

where s denotes the specific entropy. This yields

$$\begin{aligned} \frac{d\mathbf{v}_a}{dt} &= \frac{1}{m_a}\frac{\partial L}{\partial \mathbf{r}_a} = -\frac{1}{m_a}\sum_b m_b \frac{\partial u_b}{\partial \rho_b}\frac{\partial \rho_b}{\partial \mathbf{r}_a} = -\frac{1}{m_a}\sum_b m_b \frac{P_b}{\rho_b^2}\left[\sum_k m_k \nabla_a W_{bk}(h_b)\right] \\ &= -\sum_b m_b \left(\frac{P_a}{\rho_a^2}\nabla_a W_{ab}(h_a) + \frac{P_b}{\rho_b^2}\nabla_a W_{ab}(h_b) \right), \end{aligned} \tag{7.21}$$

---

[3] This statement assumes that forces and time integration have a zero/negligible error. Obviously, time integration or approximate forces, e.g. from a gravity solver, can in practice introduce violations of *exact* conservation.

where we have used $\nabla_a W_{bk} = \nabla_b W_{kb}(\delta_{ba} - \delta_{ka})$, see Sect. 2.3 in [31]. For the energy equation, a straight forward insertion of Eq. (7.15) into Eq. (7.3) (without the heating/cooling term) yields

$$\frac{du_a}{dt} = \frac{P_a}{\rho_a^2} \sum_b m_b (\mathbf{v}_b - \mathbf{v}_a) \nabla_a W_{ab}(h_a), \tag{7.22}$$

and with Eqs. (7.6), (7.21) and (7.22) we now have a complete set of discrete hydrodynamics equations. For practical simulations one would still need to choose an equation of state so that one can calculate pressures and one has to add dissipative mechanisms, e.g. a Riemann solver or artificial viscosity, so that shocks can be handled robustly.

7.2.1.3 General Relativistic SPH

Special- and general-relativistic versions of SPH can be derived similarly to the Newtonian case [31, 38–40]. We use geometric units with $G = c = 1$ and adopt (-, +, +, +) for the signature of the metric. We reserve greek letters for space-time indices from 0...3 with 0 being the temporal component and we use i and j for spatial components and SPH particles are labeled by a, b and k. Contravariant spatial indices of a vector quantity $\mathbf{V}$ at particle a are denoted as V_a^i, while covariant ones will be written as $(V_i)_a$. The line element and proper time are given by $ds^2 = g_{\mu\nu}\, dx^\mu\, dx^\nu$ and $d\tau^2 = -ds^2$. The proper time is related to the coordinate time t by

$$\Theta d\tau = dt, \tag{7.23}$$

where we have introduced a generalization of the Lorentz-factor

$$\Theta \equiv \frac{1}{\sqrt{-g_{\mu\nu} v^\mu v^\nu}}. \tag{7.24}$$

The coordinate velocity components are given by

$$v^\mu = \frac{dx^\mu}{dt} = \frac{dx^\mu}{d\tau}\frac{d\tau}{dt} = \frac{U^\mu}{\Theta} = \frac{U^\mu}{U^0}, \tag{7.25}$$

where we have used Eq. (7.23) and U^μ is the four-velocity which is normalized,

$$U^\mu U_\mu = -1. \tag{7.26}$$

The Lagrangian of a relativistic fluid is given by [53]

$$L = -\int T^{\mu\nu} U_\mu U_\nu \sqrt{-g}\, dV, \tag{7.27}$$

where $g = \det(g_{\mu\nu})$ and $T^{\mu\nu}$ denotes the energy-momentum tensor which, for an ideal fluid, reads

$$T^{\mu\nu} = (\tilde{\rho} + P)U^{\mu}U^{\nu} + Pg^{\mu\nu}. \tag{7.28}$$

Here $\tilde{\rho}$ is the energy density measured in the local rest frame of the fluid and P is its pressure. For clarity, we will write out explicit factors of c in the following lines. The energy density possesses a contribution from the rest mass and one from the thermal energy:

$$\tilde{\rho} = \rho + u\rho/c^2 = nm_0c^2(1 + u/c^2). \tag{7.29}$$

Here n is the baryon number density in the local fluid rest frame, m_0 is the baryon mass and $u = u(n, s)$ the specific energy, with s being the specific entropy.[4] In SPHINCS_BSSN we measure all energies in units of m_0c^2. With this convention (and now using again $c = 1$) the energy momentum tensor reads

$$T^{\mu\nu} = \{n(1 + u) + P\}\, U^{\mu}U^{\nu} + Pg^{\mu\nu}. \tag{7.30}$$

To perform practical simulations, we give up general covariance and choose a particular frame ("computing frame") in which the simulations are performed. Like in special-relativistic hydrodynamics, one must clearly distinguish between quantities that are measured in the computing frame and those measured in the local rest frame of the fluid. With the normalization of the four-velocity, Eq. (7.26), the Lagrangian can be written as

$$L = -\int n(1 + u)\sqrt{-g}dV. \tag{7.31}$$

Local baryon number conservation, $(U^{\mu}n)_{;\mu} = 0$, can be expressed as

$$\frac{1}{\sqrt{-g}}\partial_{\mu}(\sqrt{-g}U^{\mu}n) = 0, \tag{7.32}$$

where we have used an identity for the covariant derivative [54]. More explicitly, we can write

$$\partial_t(N) + \partial_i(Nv^i) = 0, \tag{7.33}$$

where we have made use of Eq. (7.25) and have introduced the computing frame baryon number density

$$N = \sqrt{-g}\Theta n. \tag{7.34}$$

In the special relativistic limit $g_{\mu\nu} \to \eta_{\mu\nu}$ (Minkowski), Θ reduces to the special relativistic Lorentz factor γ and $\sqrt{-g}$ becomes unity. In this limit, the computing frame density $N_{\rm sr}$ is equal to the local rest frame density n multiplied by the Lorentz

---

[4] Here we simply use $m_0 = m_u$ for the average baryon mass, where m_u is the atomic mass unit. In reality, m_0 depends on the exact composition, but even in extreme cases the deviations from m_u are only a small fraction of a percent. See Sect. 2.1 in [26] for a more detailed discussion.

factor γ between the fluid rest frame and the computing frame. This, of course, makes perfect sense: think of a box in the rest frame of the fluid that contains a certain number of baryons. If this box moves with respect to the computing frame along a coordinate axis, the corresponding box edge appears Lorentz contracted in the computing frame, so that an observer in the computing frame would conclude that the density is

$$N_{\rm sr} = \gamma n. \tag{7.35}$$

The total conserved baryon number, $\mathcal{N}$, can then be expressed as a sum over fluid parcels with volume ΔV_b located at $\mathbf{r}_b$, where each parcel carries a baryon number[5] ν_b

$$\mathcal{N} = \int N dV \simeq \sum_b N_b \Delta V_b = \sum_b \nu_b, \tag{7.36}$$

therefore the particle volume in the computing frame reads $\Delta V_b = \nu_b / N_b$. Equation (7.33) looks like the Newtonian continuity equation and we will use it for the SPH discretization process. Similar to Eq. (7.5), a quantity f can now be approximated by

$$\langle f \rangle(\mathbf{r}) \simeq \sum_b \frac{\nu_b}{N_b} f_b \, W(\mathbf{r} - \mathbf{r}_b, h), \tag{7.37}$$

where the particle's baryon number ν_b replaces the mass and the Newtonian mass density ρ is replaced by the computing frame baryon number density N. The latter can be calculated, very similarly to the Newtonian case, as

$$N_a = \sum_b \nu_b W_{ab}(h_a). \tag{7.38}$$

If each particle keeps its baryon number constant, we have exact baryon number conservation. The fluid Lagrangian, Eq. (7.31), then becomes

$$L = -\int \frac{1+u}{\Theta} N \, dV \approx -\sum_b \left(\frac{1+u}{\Theta}\right)_b N_b \Delta V_b = -\sum_b \nu_b \left(\frac{1+u}{\Theta}\right)_b, \tag{7.39}$$

where in the last step we have made use of the volume element $\Delta V_b = \nu_b / N_b$, as suggested by Eq. (7.36).

Again, as in the Newtonian case, one can use the Euler-Lagrange equations

$$\frac{d}{dt}\frac{\partial L}{\partial v_a^i} - \frac{\partial L}{\partial x_a^i} = 0, \tag{7.40}$$

[5] Be careful not to confuse the baryon number ν_b with the velocity component of a particle v_b^i.

to derive an evolution equation for the canonical momentum. One can use the Lagrangian to define a canonical momentum

$$(p_i)_a \equiv \frac{\partial L}{\partial v_a^i} = -\frac{\partial}{\partial v_a^i} \sum_b \nu_b \left(\frac{1+u}{\Theta}\right)_b. \tag{7.41}$$

Note that here $\partial u_b/\partial v_a^i$ has a non-zero value, since

$$\frac{\partial u_b}{\partial v_a^i} = \frac{\partial u_b}{\partial n_b}\frac{\partial n_b}{\partial v_a^i} = \frac{P_b}{n_b^2}\frac{\partial}{\partial v_a^i}\left(\frac{N_b}{\sqrt{-g}\,\Theta_b}\right), \tag{7.42}$$

where we have used the first law of thermodynamics and the relation Eq. (7.34) and the velocity dependence comes in via Θ, see Eq. (7.24) and one finds [39]

$$\frac{\partial}{\partial v_a^i}\left(\frac{1}{\Theta_b}\right) = -\Theta_b (g_{i\mu} v^\mu)_a \delta_{ab}. \tag{7.43}$$

The *canonical momentum per baryon* reads

$$(S_i)_a \equiv \frac{1}{\nu_a}\frac{\partial L}{\partial v_a^i} = \Theta_a\left(1 + u_a + \frac{P_a}{n_a}\right)(g_{i\mu}v^\mu)_a = \left(1 + u_a + \frac{P_a}{n_a}\right)(U_i)_a. \tag{7.44}$$

Similarly, one can use the canonical energy

$$E \equiv \sum_a \frac{\partial L}{\partial v_a^i} v_a^i - L = \sum_a \nu_a \left(v_a^i (S_i)_a + \frac{1+u_a}{\Theta_a}\right) \tag{7.45}$$

to identify the *canonical energy per baryon*

$$e_a \equiv v_a^i (S_i)_a + \frac{1+u_a}{\Theta_a}. \tag{7.46}$$

The evolution equation for the canonical momentum per baryon follows, according to the Euler-Lagrange equations (7.40), from $\frac{d(S_i)_a}{dt} = \frac{1}{\nu_a}\frac{\partial L}{\partial x_a^i}$, and after some algebra [39] one finds[6]

$$\frac{d(S_i)_a}{dt} = -\sum_b \nu_b \left\{\frac{P_a}{N_a^2} D_i^a + \frac{P_b}{N_b^2} D_i^b\right\} + \left(\frac{\sqrt{-g}}{2N} T^{\mu\nu} \frac{\partial g_{\mu\nu}}{\partial x^i}\right)_a, \tag{7.47}$$

[6] Note that here we are omitting the "grad-h terms", see [39] for the explicit expressions including these small correction terms.

where we have introduced the abbreviations

$$D_i^a \equiv \sqrt{-g_a}\,\frac{\partial W_{ab}(h_a)}{\partial x_a^i} \quad \text{and} \quad D_i^b \equiv \sqrt{-g_b}\,\frac{\partial W_{ab}(h_b)}{\partial x_a^i}. \tag{7.48}$$

The first term in Eq. (7.47) (including the summation) is due to hydrodynamic accelerations and it is formally very similar to the Newtonian momentum equation (7.21), although the involved quantities have a different meaning. The second term represents the metric accelerations from the spacetime curvature and it depends on the spatial derivatives and the determinant of the metric.

The evolution equation of canonical energy per baryon follows from straight forwardly taking the Lagrangian time derivative of Eq. (7.46), applying the first law of thermodynamics and after some algebra [39] one obtains

$$\frac{de_a}{dt} = -\sum_b \nu_b \left\{ \frac{P_a v_b^i}{N_a^2}\, D_i^a + \frac{P_b v_a^i}{N_b^2}\, D_i^b \right\} - \left(\frac{\sqrt{-g}}{2N} T^{\mu\nu} \partial_t g_{\mu\nu} \right)_a. \tag{7.49}$$

Similar to the momentum equation, there is a hydrodynamic contribution (involving the summation over neighbour particles) and a gravitational contribution that is proportional to the time derivative of the metric tensor.

The attentive reader may wonder how one finds the physical variables (i.e. v^i, n, u, P) that are needed in the RHSs of Eqs. (7.47) and (7.49) from the quantities that are actually updated during the integration process, i.e. from N, S_i and e. This is actually a not entirely trivial problem (sometimes called "recovery problem", "recovery" or "conservative-to-primitive transformation") that requires a numerical root finding algorithm. The exact algorithm depends on which equations are evolved and on the equation of state that is used. Our strategy in `SPHINCS_BSSN` is to express n and u in terms of the evolved variables N, S_i and e and the pressure P, substitute them into the (polytropic or piecewise polytropic) equation of state and use numerical rootfinding (we use Ridders' method [24]) to find the pressure value that solves this equation. Once this pressure value is found, all physical quantities can be obtained by a simple backsubstitution. See Sect. 2.2.4 in [25] and Appendix A in [27] for a detailed description of the polytropic and the piecewise polytropic case, respectively.

7.2.1.4 Dissipative Terms

So far, we have dealt exclusively with strictly non-dissipative hydrodynamics. Most astrophysical gas flows are overall very well described by such ideal fluids, but in reality shocks occur and, in a "microscopic" zoom into the shock front dissipative processes *do* occur which lead to an increase in entropy. Numerical methods need to mimick this behaviour, i.e. in smooth parts of the flow they should (ideally) be dissipationless, but they need to have a built-in mechanism which produces entropy in a shock front. This can be achieved by either some form of artificial viscosity or an (approximate or exact) Riemann solver. While both are valid possibilities

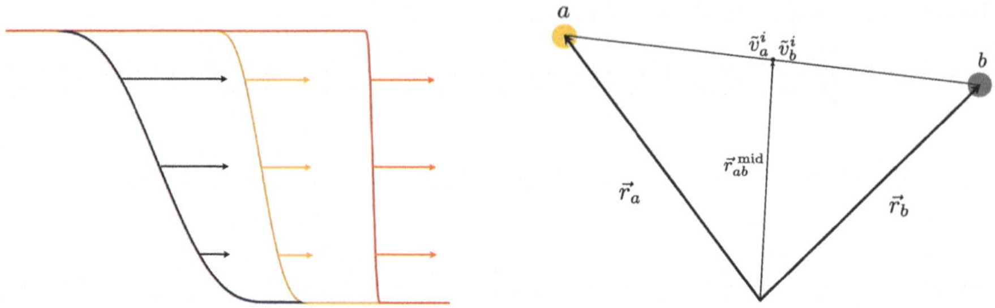

Fig. 7.1 Left: sound wave steepening into a shock. Right: reconstruction of the particle velocities to the mid-point between two particles

[29, 55], we have decided to use a modern form of artificial viscosity that shares features with what is usually done in a finite volume scheme [56], in particular, our artificial viscosity treatment makes use of slope-limited reconstruction. This new approach to artificial viscosity has been extensively tested in our Newtonian SPH code `MAGMA2` [57] which served as a technology testbed for `SPHINCS_BSSN`.
Artificial viscosity is actually one of the oldest techniques of computational physics [58], and while having had a questionable reputation for some time, it seems that it is becoming more popular again, though in more sophisticated forms than originally suggested [59–64]. It is further worth pointing out that modern artificial viscosities share many similarities with approximate Riemann solvers [65].
Seemingly harmless sound waves can steepen into shocks during the hydrodynamic evolution, because the high-density parts try to catch up with the slower low-density parts, see the sketch in the left panel of Fig. 7.1. Without any dissipation this process will keep going until the wave has steepened into a mathematical discontinuity where physical quantities change abruptly. In nature, dissipation becomes active on some small scale, so that the transition would be very sharp, usually much sharper than the lengths that can be numerically resolved, but not strictly discontinuous. The main idea of artificial viscosity is to add an "artificial pressure" q to the physical pressure P (wherever it occurs)

$$P \rightarrow P + q \tag{7.50}$$

in order to prevent the wave from becoming infinitely steep and instead keep it at a numerically treatable level. Or, in the words of John von Neumann and Robert Richtmyer [58]: "*Our idea is to introduce (artificial) dissipative terms into the equations so as to give shocks a thickness comparable to (but preferentially somewhat larger than) the spacing of the points of the network. Then the differential equations (more accurately, the corresponding difference equations) can be used for the entire calculation, just as if there were no shocks at all.*"
We will not go into the explicit expressions for the artificial pressure q that we use, for this we refer to Sect. 2.1.1 in [27], we do, however, want to describe the basic

ideas behind some of the recent artificial viscosity improvements. "Old style" SPH has been criticized for being overly dissipative. It is worth stating, however, that the SPH method as derived from the Lagrangian is completely inviscid, and in early SPH implementations the viscous terms were always switched on. For pedagogical reasons, we will describe the improvements at the example of Newtonian SPH where the expressions are simplest, but all the ideas on how to improve artificial dissipation translate straight forwardly to the GR case [27]. In the Newtonian case, the viscous pressure typically has the form

$$q_a \propto \alpha_v \, \Delta v_{ab} \tag{7.51}$$

where α_v is a numerical parameter that needs to be of order unity at a shock (but not elsewhere) while Δv_{ab} quantifies the "velocity jump" between particle a and a neighbouring particle b (different expressions for the velocity jump are possible). The major concern is usually to not have dissipation where it is not needed. There are essentially three approaches to this. The first one is to multiply the quantity α_v with a "limiter" that is smart enough to decide, based on the local fluid properties, whether a particle is in a shock or not and, in the latter case, the limiter should have a very small or zero value to avoid dissipation. Different versions of such limiters have been suggested [35, 59, 66]. The second approach is to make α_v time-dependent [67], so that it decays exponentially, when it is not needed, to some small, possibly zero value, α_0. This approach needs a "trigger" at each particle location that indicates whether more dissipation is needed. There are several possibilities for such triggers:

- The original suggestion of Morris and Monaghan [67] was to use $-\nabla \cdot \mathbf{v}$ as a trigger so that dissipation switches on during compression. This, however, does not necessarily indicate a shock: gas can also be *adiabatically* compressed in which case no dissipation is wanted.
- Cullen and Dehnen [59] suggested (apart from some other improvements) to construct a trigger that is instead based on the time derivative $d(-\nabla \cdot \mathbf{v})/dt$. This quantity indicates a steepening flow convergence which is characteristic for a particle moving into a shock.
- Rosswog [34] suggested to use in addition to the *shock* trigger of Cullen and Dehnen [59], a *noise* trigger that switches on (to a smaller amount), if the flow becomes numerically noisy, even if there is no shock. Here, the main idea is to measure sign fluctuations of $\nabla \cdot \mathbf{v}$ in the neighborhood of a given particle: if there is a clean compression or expansion, all nearby particles have the same sign of $\nabla \cdot \mathbf{v}$, but a large number of positive and negative signs, instead, indicates noise.
- Rosswog [62] suggested to measure how the entropy of a particle evolves in time. Since we are modelling an ideal fluid, the entropy should be strictly conserved, this, however, is not numerically enforced by construction. That means that we can monitor the entropy conservation as a measure for the numerical quality of the flow and entropy non-conservation (either via shock or numerical noise) indicates that dissipation should be applied.

The third approach to avoid excessive dissipation [57], is actually very similar to the reconstruction procedure in Finite Volume schemes [56]. Instead of using the velocity difference between two particles for the velocity jump, $\Delta v_{ab} = \mathbf{v}_a - \mathbf{v}_b$, one uses instead the differences of the slope-limited velocities *reconstructed to the midpoint* between two particles, $\Delta \tilde{v}_{ab} = \tilde{\mathbf{v}}_a - \tilde{\mathbf{v}}_b$, as sketched in Fig. 7.1, right panel. The slope-limited, first order reconstruction from the a-side reads

$$\tilde{v}_a^i = v_a^i + \Phi_{ab}(\partial_j v_a^i)\ \frac{1}{2}(r_b - r_a)^j \tag{7.52}$$

and correspondingly for the reconstruction from the b-side. For the slope limiter Φ_{ab} any of the standard slope limiters can be used in principle. The reconstruction can be extended to higher order, for example in the MAGMA2 code [57] we use a quadratic reconstruction.

In SPHINCS_BSSN we have implemented GR-versions of the second and the third of the above ideas. We also evolve our dissipation parameter triggered by a shock trigger similar to Cullen and Dehnen [59], enhanced by noise triggers as in Rosswog [34]. In addition, we perform linear reconstruction to the particle midpoint in which we use a minmod [68] slope limiter. Please consult Sect. 2.1.1 in [27] for the explicit expressions of the artificial viscosity terms used in SPHINCS_BSSN.

7.2.2 Spacetime Evolution

In general relativity, gravity is described as curvature of spacetime and the fundamental object is the spacetime metric, $g_{\mu\nu}$, from which the spacetime interval between infinitesimally close points can be calculated via $ds^2 = g_{\mu\nu}dx^\mu dx^\nu$. From the metric tensor we can define the *Christoffel symbols*

$$\Gamma^\lambda_{\mu\nu} = \frac{g^{\lambda\sigma}}{2}\left[\frac{\partial g_{\mu\sigma}}{\partial x^\nu} + \frac{\partial g_{\nu\sigma}}{\partial x^\mu} - \frac{\partial g_{\mu\nu}}{\partial x^\sigma}\right], \tag{7.53}$$

from which we can calculate the *Riemann curvature tensor*

$$R^\lambda{}_{\sigma\mu\nu} = \frac{\partial \Gamma^\lambda_{\sigma\nu}}{\partial x^\mu} - \frac{\partial \Gamma^\lambda_{\sigma\mu}}{\partial x^\nu} + \Gamma^\lambda_{\rho\mu}\Gamma^\rho_{\sigma\nu} - \Gamma^\lambda_{\rho\nu}\Gamma^\rho_{\sigma\mu}. \tag{7.54}$$

Contracting the first and third index on the Riemann curvature tensor delivers the *Ricci tensor*, $R_{\mu\nu} = R^\lambda{}_{\mu\lambda\nu}$ and by contracting the two indices on the Ricci tensor one finds the *Ricci scalar* $R = g^{\mu\nu}R_{\mu\nu}$. These combine to define the *Einstein tensor* that is related to the stress-energy tensor in the full covariant 4-dimensional set of Einstein field equations

$$G_{\mu\nu} = R_{\mu\nu} - \frac{1}{2}g_{\mu\nu}R = 8\pi T_{\mu\nu}, \tag{7.55}$$

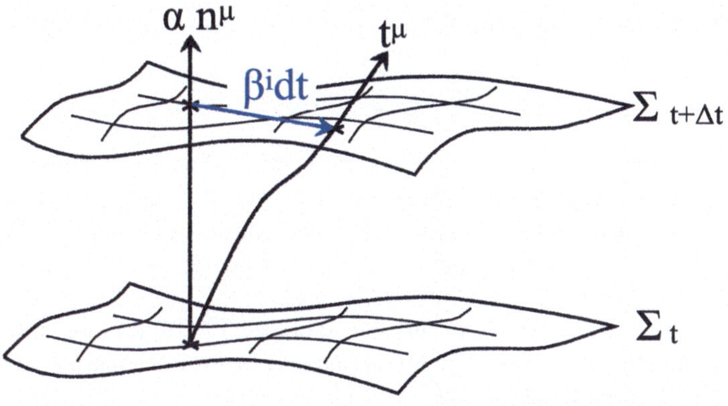

Fig. 7.2 3+1 foliation of the spacetime. The spacetime is sliced into spacelike hypersurfaces which are integrated forward in time. The lapse function α measures the proper time along the normal n^μ, the shift vector β^i measures the displacement, on consecutive hypersurfaces, between the observer time lines t^μ and the normal lines n^μ

which, in turn, is the starting point for any spacetime evolution code.

These equations are 4-dimensional and it is, therefore, not possible to evolve them like a normal initial value problem, where one provides 3-dimensional data at a given time and evolves it forward to the next time. What we can do to achieve this is to split the spacetime into 3+1 dimensions,[7] that is we write the line element as

$$ds^2 = g_{\mu\nu}dx^\mu dx^\nu = (-\alpha^2 + \beta_i\beta^i)dt^2 + 2\beta_i dt dx^i + \gamma_{ij}dx^i dx^j, \tag{7.56}$$

where we have introduced the lapse function, α, the shift vector, β^i, and the spatial 3-metric, γ_{ij}. This foliates spacetime into non-intersecting spacelike hypersurfaces Σ labeled by the coordinate time, t, see the sketch in Fig. 7.2. Each spatial slice has a future pointing normal vector, n^μ, whose coordinates are $n^\mu = (1/\alpha, -\beta^i/\alpha)$. In this picture the lapse α measures the elapsed proper time between two nearby spatial slices (in the normal direction), the shift vector, β^i, is the relative velocity of Eulerian observers and the lines of constant spatial coordinates and the spatial metric, γ_{ij}, measure proper distance within the slices.

The curvature that is intrinsic to the slices is, of course, given by the 3-dimensional Riemann tensor defined in terms of the 3-metric γ_{ij}. However, there is also an extrinsic curvature associated with how the 3-dimensional slices are embedded in the overall 4-dimensional spacetime. Remarkably, this can be described in terms of another 3-dimensional tensor, K_{ij}, that can be defined as the projection of the gradient of the normal vector onto the spatial slice

$$K_{\mu\nu} := -P_\mu^\lambda n_{\nu;\lambda}, \tag{7.57}$$

[7] This is certainly not the only possible way to convert the Einstein equations into an initial value problem, but it is a very common approach and the one we have chosen here.

where P^{λ}_{μ} is the projection operator,

$$P^{\lambda}_{\mu} := \delta^{\lambda}_{\mu} + n^{\lambda} n_{\mu}. \tag{7.58}$$

Note that even though the extrinsic curvature is defined as a 4-dimensional object in Eq. (7.57), it is indeed a 3-dimensional object with $K^{00} = K^{0i} = 0$ and can be referred to as K^{ij}. K_{00} and K_{0i} on the other hand are not zero, but the components of K_{ij} can still be obtained by using only the spatial metric, γ_{ij}, to lower the indices of K^{ij}.

Leaving out the details (see the excellent text books [17, 18] for more information) it is then possible to split the 10 original second order Einstein equations into 12 first order hyperbolic evolution equations

$$\partial_t \gamma_{ij} = -2\alpha K_{ij} + D_i \beta_j + D_j \beta_i, \tag{7.59}$$

$$\begin{aligned}\partial_t K_{ij} = & -D_i D_j \alpha + \alpha \left[{}^{(3)}R_{ij} + K K_{ij} - 2 K_{ik} K^k_j \right] \\ & + 4\pi\alpha[\gamma_{ij}(S - \rho) - 2S_{ij}] + \beta^k \partial_k K_{ij} + K_{ki} \partial_j \beta^k + K_{kj} \partial_i \beta^k,\end{aligned} \tag{7.60}$$

where D is the 3-dimensional covariant derivative related to γ_{ij}, $\rho = n^{\mu} n^{\nu} T_{\mu\nu}$, $S_{\mu\nu} = P^{\lambda}_{\mu} P^{\sigma}_{\nu} T_{\lambda\sigma}$ and $S = S^{\mu}_{\mu}$ and 4 elliptical constraint equations

$${}^{(3)}R + K^2 - K_{ij} K^{ij} = 16\pi\rho \tag{7.61}$$

$$D_j (K^{ij} - \gamma^{ij} K) = 8\pi j^i, \tag{7.62}$$

where $j^{\mu} = -P^{\mu\lambda} n^{\nu} T_{\lambda\nu}$. Equations (7.59) and (7.60), as written here, are due to York [69] and are known as the ADM equations after Arnowitt, Deser and Misner. Given initial data for γ_{ij} and K_{ij} on a 3-dimensional spatial slice, the ADM equations can in principle be used to evolve the spacetime forward in time. Equation (7.61) is the *Hamiltonian or energy constraint* and Equation (7.62) is the so-called *momentum constraint*. These do not involve any time derivatives but rather provide a set of equations that any physical data has to satisfy at any point in time. This is similar, for example, to the case of magneto-hydrodynamics, where one has evolution equations that have to fulfill the $\nabla \cdot \mathbf{B} = 0$-constraint at any point in time.

Note that the York form of the ADM equations, also known as standard ADM, differs from the original ADM equations [70] by a term proportional to the Hamiltonian constraint. Hence they are physically equivalent but not mathematically or numerically equivalent. This is because the Hamiltonian constraints contains second derivatives that changes the principal part and therefore the mathematical properties of the equations.

There is a lot of freedom in choosing which of the 12 components of γ_{ij} and K_{ij} to provide initial data for and which of the components to solve for using the constraint equations. Luckily, once constraint satisfying initial data are set up, evolution of that data using the ADM evolution equations will guarantee, in the absence of numerical errors, that the evolved data remains constraint satisfying. Of course numerical errors are unavoidable, which means that any numerical evolution will always have some constraint violations. In that case the constraint violations can be used to monitor the numerical quality of the simulation.

The only problem with the ADM equations is that they do not actually work in practice. They can be shown (see for example [17]) to be only *weakly hyperbolic* and hence they are not a well posed formulation of the Einstein equation. An alternative formulation due to Baumgarte, Shapiro, Shibata and Nakamura, the so-called BSSN formulation [18, 20], has proven to be a robust choice for many codes and has therefore become rather popular. In the BSSN formulation a conformal rescaling is introduced

$$\tilde{\gamma}_{ij} = e^{-4\phi}\gamma_{ij}, \tag{7.63}$$

where ϕ is defined in terms of the determinant of the physical three-metric, γ, as $\phi = \frac{1}{12}\log\gamma$. With this choice the determinant of the conformal metric becomes $\tilde{\gamma} = 1$. In addition, the extrinsic curvature is separated into its trace and its tracefree part and conformally rescaled

$$\tilde{A}_{ij} = e^{-4\phi}A_{ij} = e^{-4\phi}(K_{ij} - \frac{1}{3}\gamma_{ij}K). \tag{7.64}$$

The final addition is to also evolve the three quantities

$$\tilde{\Gamma}^i = \tilde{\gamma}^{jk}\tilde{\Gamma}^i_{jk} = -\partial_j\tilde{\gamma}^{ij}, \tag{7.65}$$

where $\tilde{\Gamma}^i_{jk}$ is the Christoffel symbols related to the conformal metric. That is, the BSSN evolution variables are ϕ, $\tilde{\gamma}$, K, $\tilde{A}$ and Γ^i, and result in the following set of evolution equations

$$\partial_t\phi = -\frac{1}{6}\left(\alpha K - \partial_i\beta^i\right) + \beta^i\bar{\partial}_i\phi, \tag{7.66}$$

$$\partial_t\tilde{\gamma}_{ij} = -2\alpha\tilde{A}_{ij} + \tilde{\gamma}_{ik}\partial_j\beta^k + \tilde{\gamma}_{jk}\partial_i\beta^k - \frac{2}{3}\tilde{\gamma}_{ij}\partial_k\beta^k + \beta^k\bar{\partial}_k\tilde{\gamma}_{ij}, \tag{7.67}$$

$$\begin{aligned}\partial_t K &= -e^{-4\phi}\left(\tilde{\gamma}^{ij}\left[\partial_i\partial_j\alpha + 2\partial_i\phi\,\partial_j\alpha\right] - \tilde{\Gamma}^i_{(n)}\partial_i\alpha\right)\\ &\quad + \alpha\left(\tilde{A}^i_j\tilde{A}^j_i + \frac{1}{3}K^2\right) + \beta^i\bar{\partial}_i K + 4\pi\alpha(\rho + s),\end{aligned} \tag{7.68}$$

$$\begin{aligned}\partial_t\tilde{A}_{ij} &= e^{-4\phi}\left[-\partial_i\partial_j\alpha + \tilde{\Gamma}^k_{ij}\partial_k\alpha + 2\left(\partial_i\alpha\,\partial_j\phi + \partial_j\alpha\,\partial_i\phi\right) + \alpha R_{ij}\right]^{\mathrm{TF}}\\ &\quad + \alpha(K\tilde{A}_{ij} - 2\tilde{A}_{ik}\tilde{A}^k_j) + \tilde{A}_{ik}\partial_j\beta^k + \tilde{A}_{jk}\partial_i\beta^k - \frac{2}{3}\tilde{A}_{ij}\partial_k\beta^k\\ &\quad + \beta^k\bar{\partial}_k\tilde{A}_{ij} - e^{-4\phi}\alpha 8\pi\left(T_{ij} - \frac{1}{3}\gamma_{ij}s\right),\end{aligned} \tag{7.69}$$

$$\partial_t \tilde{\Gamma}^i = -2\tilde{A}^{ij}\partial_j\alpha + 2\alpha\left(\tilde{\Gamma}^i_{jk}\tilde{A}^{jk} - \frac{2}{3}\tilde{\gamma}^{ij}\partial_j K + 6\tilde{A}^{ij}\partial_j\phi\right)$$
$$+ \tilde{\gamma}^{jk}\partial_j\partial_k\beta^i + \frac{1}{3}\tilde{\gamma}^{ij}\partial_j\partial_k\beta^k - \tilde{\Gamma}^j_{(n)}\partial_j\beta^i + \frac{2}{3}\tilde{\Gamma}^i_{(n)}\partial_j\beta^j$$
$$+ \beta^j\bar{\partial}_j\tilde{\Gamma}^i - 16\pi\alpha\tilde{\gamma}^{ij}s_j, \tag{7.70}$$

where

$$\rho = \frac{1}{\alpha^2}(T_{00} - 2\beta^i T_{0i} + \beta^i\beta^j T_{ij}), \tag{7.71}$$
$$s = \gamma^{ij}T_{ij}, \tag{7.72}$$
$$s_i = -\frac{1}{\alpha}(T_{0i} - \beta^j T_{ij}), \tag{7.73}$$

and $\beta^i\bar{\partial}_i$ denote partial derivatives that are "upwinded" based on the shift vector. This means that the stencil used for finite differencing is shifted by one in the direction of the shift. As an example of this, look at second order finite differencing, where a derivative in the x-direction at grid point x_i would normally be approximated by the centered finite difference

$$\left.\frac{\partial f}{\partial x}\right|_{x=x_i} \approx \frac{-f_{i-1} + f_{i+1}}{2\Delta x} + O\left(\Delta x^2\right). \tag{7.74}$$

If the shift is positive we instead use the "upwinded" finite difference approximation

$$\left.\frac{\bar{\partial} f}{\partial x}\right|_{x=x_i} \approx \frac{-3f_i + 4f_{i+1} - f_{i+2}}{2\Delta x} + O\left(\Delta x^2\right), \tag{7.75}$$

whereas if the shift is negative we use

$$\left.\frac{\bar{\partial} f}{\partial x}\right|_{x=x_i} \approx \frac{f_{i-2} - 4f_{i-1} + 3f_i}{2\Delta x} + O\left(\Delta x^2\right). \tag{7.76}$$

The superscript "TF" in the evolution equation of $\tilde{A}_{ij}$ denotes the trace-free part of the bracketed term. Note that there is a slight subtlety to the treatment of $\tilde{\Gamma}^i$. We introduce

$$\tilde{\Gamma}^i_{(n)} = \tilde{\gamma}^{jk}\tilde{\Gamma}^i_{jk}, \tag{7.77}$$

i.e. a numerical recalculation of the contracted conformal Christoffel symbols from the current conformal metric. This is used instead of $\tilde{\Gamma}^i$ whenever derivatives of $\tilde{\Gamma}^i$ are not needed. In all other places, i.e. when finite differences are needed, the evolved variables, $\tilde{\Gamma}^i$, are used directly. This helps with numerical stability and makes the constraint $\tilde{\Gamma}^i = -\partial_j\tilde{\gamma}^{ij}$ better behaved. Finally $R_{ij} = \tilde{R}_{ij} + R^{\phi}_{ij}$, where

$$\tilde{\Gamma}_{ijk} = \frac{1}{2}\left(\partial_k\tilde{\gamma}_{ij} + \partial_j\tilde{\gamma}_{ik} - \partial_i\tilde{\gamma}_{jk}\right), \tag{7.78}$$

$$\tilde{\Gamma}_{ij}{}^{k} = \tilde{\gamma}^{kl}\tilde{\Gamma}_{ijl}, \tag{7.79}$$

$$\tilde{\Gamma}^{i}_{jk} = \tilde{\gamma}^{il}\tilde{\Gamma}_{ljk}, \tag{7.80}$$

$$\tilde{\Gamma}^{i}_{(n)} = \tilde{\gamma}^{jk}\tilde{\Gamma}^{i}_{jk}, \tag{7.81}$$

$$\begin{aligned}\tilde{R}_{ij} = & -\frac{1}{2}\tilde{\gamma}^{kl}\partial_k\partial_l\tilde{\gamma}_{ij} + \tilde{\gamma}_{k(i}\partial_{j)}\tilde{\Gamma}^k + \tilde{\Gamma}^k_{(n)}\tilde{\Gamma}_{(ij)k} + \tilde{\Gamma}^k_{il}\tilde{\Gamma}_{jk}{}^{l} \\ & + \tilde{\Gamma}^k_{jl}\tilde{\Gamma}_{ik}{}^{l} + \tilde{\Gamma}^k_{il}\tilde{\Gamma}_{kj}{}^{l},\end{aligned} \tag{7.82}$$

$$\begin{aligned}R^{\phi}_{ij} = & -2\left(\partial_i\partial_j\phi - \tilde{\Gamma}^k_{ij}\partial_k\phi\right) - 2\tilde{\gamma}_{ij}\tilde{\gamma}^{kl}\left(\partial_k\partial_l\phi - \tilde{\Gamma}^m_{kl}\partial_m\phi\right) \\ & + 4\partial_i\phi\partial_j\phi - 4\tilde{\gamma}_{ij}\tilde{\gamma}^{kl}\partial_k\phi\partial_l\phi,\end{aligned} \tag{7.83}$$

where the standard notation for symmetrization, $A_{(ij)} = (A_{ij} + A_{ji})/2$, has been used. The additional constraints associated with the conformal metric, $1 - \tilde{\gamma} = 0$, and conformal traceless part of the extrinsic curvature, $\tilde{A} = \tilde{\gamma}^{ij}\tilde{A}_{ij} = 0$, are enforced at every substep of the time integrator.

Note that a slight variation of BSSN, where $W = \gamma^{-1/6}$ is used as an evolution variable instead of ϕ is implemented in the `SPHINCS_BSSN` code as well. This variant seems to be better behaved near the puncture of a black hole, but is not expected to have a significant effect for binary neutron star mergers.

The BSSN evolution equations above do not give a prescription for how to choose the gauge variables, α and β^i. In order to complete the system, we add the simplest form of the so-called *moving puncture gauges*, i.e.

$$\partial_t\alpha = -2\alpha K \tag{7.84}$$

and

$$\partial_t\beta^i = \frac{3}{4}\left(\tilde{\Gamma}^i - \eta\beta^i\right). \tag{7.85}$$

7.2.3 Particle-Mesh Interaction

Since we are solving the hydrodynamic equations by means of Lagrangian particles, but evolve the spacetime on a mesh, the particles and the mesh need to continuously exchange information. As one sees from Eqs. (7.47) and (7.49), the particle evolution equations need the metric and its derivatives, which are known at grid points, while the spacetime/BSSN-evolution equations, see Eqs. (7.68), (7.69) and (7.70), need the matter's energy momentum tensor, Eq. (7.30), which is known at the particle positions. In other words, the metric and its derivatives need to be interpolated–at every Runge-Kutta substep–to the particle positions ("mesh-to-particle"-, or M2P-step) and the energy momentum tensor needs to be mapped to grid points based on the values at the surrounding particles ("particle-to-mesh"-, or P2M-step).

The simpler of the two is the M2P-step. In `SPHINCS_BSSN` it is performed via a 5th order Hermite interpolation, that we have developed in [25] extending the work of [71]. Contrary to a standard Lagrange polynomial interpolation, the Hermite interpolation guarantees that the metric remains twice differentiable as particles pass from one grid cell to another and therefore this sophisticated interpolation avoids the introduction of additional noise. The details of our approach are explained in Sect. 2.4 of [25] to which we refer the interested reader.

The P2M-step is more complicated since the particle distribution is continuously changing and the particles are in particular not arranged on a regular lattice. Our original approach [25–27] was to use sophisticated kernels that we borrowed from vortex methods [72], but we have recently [28] further refined our P2M-step. Our new approach is based on a "local regression estimate" (LRE), where for a given polynomial order, we determine the $T_{\mu\nu}$ value on the grid points by the requirement that it is the best fit (according to a suitable error measure) to the surrounding particle values, see Fig. 7.3. The remaining question is then: which polynomial order should be used? Also here we follow an optimization approach: we perform trial LRE-mappings for all orders up to a maximal order (= 4), and out of those trial mappings we select the one that agrees best with the surrounding particles. This selection process for the polynomial order is called "Multidimensional Optimal Order Detection" (MOOD), and we therefore call our P2M-approach "LRE-MOOD".

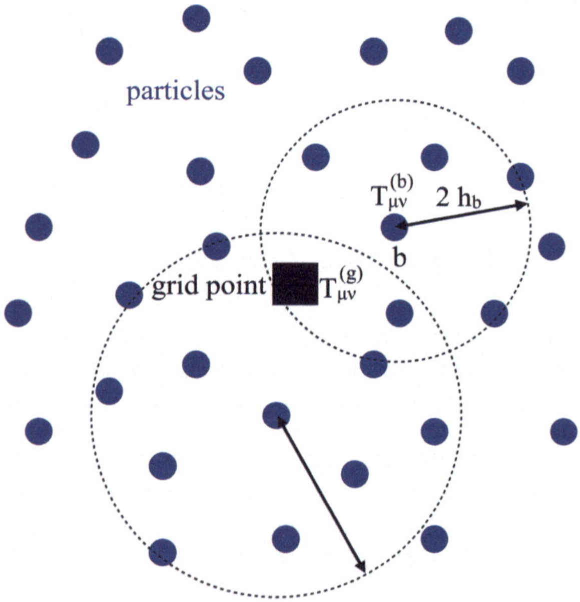

Fig. 7.3 Particle-to-mesh step: the energy momentum tensor $T_{\mu\nu}$ is naturally known at the particle positions (blue circles), since it is needed as a source of the spacetime evolution equations, the particle values need to be transferred to the grid points (black square). This is achieved by an LRE-MOOD approach, see the main text for an explanation

7.2.3.1 LRE-Mapping

The basic idea is to assume that the particle values are given by some unknown function f, which can be expanded in a Taylor series around a grid point located at $\mathbf{r}_{\rm G}$

$$f(\mathbf{r}) = f^G + (\partial_i f)^G\,(\mathbf{r}-\mathbf{r}^G)^i + \frac{1}{2}(\partial_{ij} f)^G\,(\mathbf{r}-\mathbf{r}^G)^i(\mathbf{r}-\mathbf{r}^G)^j + \text{h.o.t.} \tag{7.86}$$

The terms like $(\mathbf{r}-\mathbf{r}^G)^i$ can be interpreted as a polynomial basis P_l, shifted to the grid point $\mathbf{r}_{\rm G}$, with coefficients α_l that contain the derivatives of f. The approximation, "optimized at the grid point G", can then be written as

$$\tilde{f}^G(\mathbf{r}) = \sum_{l=0}^{l_{\max}} \alpha_l^G P_l^G(\mathbf{r}). \tag{7.87}$$

Here $l_{\max}$ is the number of degrees of freedom and for d spatial dimensions and a maximum polynomial order m, it is given by

$$l_{\max} = \frac{(d+m)!}{d!m!}. \tag{7.88}$$

That is we have 1, 4, 10, 20 and 35 degrees of freedom for constant, linear, quadratic, cubic and quartic polynomials. We want the α_l to be an optimal fit to the surrounding particles, therefore we minimize at each grid point the error measure

$$\epsilon^G \equiv \sum_p \left[f_p - \tilde{f}^G(\mathbf{r}_p)\right]^2\, W_{pG}, \tag{7.89}$$

where the f_p are the function values at the particle positions and W_{pG} is a smooth, positive definite function (e.g. a typical SPH kernel) that gives particles that are closer to the grid point a larger weight than to further away particles. The optimal coefficients, $\alpha^{\rm G}$, are then found from the condition

$$\left(\frac{\partial \epsilon^G}{\partial \alpha_i}\right)_G \stackrel{!}{=} 0, \tag{7.90}$$

which yields [28]

$$\alpha_i^G = (M_{ik})^{-1}\, B_k, \tag{7.91}$$

where the "moment matrix" reads

$$M_{ik} = \sum_p P_i^G(\mathbf{r}_p)\; P_k^G(\mathbf{r}_p)\; W_{pG} \tag{7.92}$$

and the vector that contains the information about the function values reads

$$B_k = \sum_p f_p \, P_k^G(\mathbf{r}_p) W_{pG}. \tag{7.93}$$

The moment matrix can be close to singular, therefore we solve for the α_i by means of a singular value decomposition [24]. For more details and some examples of function approximations, we refer to the appendix of our recent paper [28].

7.2.3.2 Multi-dimensional Optimal Order Detection (MOOD)

While we have just addressed how to determine the best coefficients for a given polynomial order, we still need to discuss how we decide which polynomial order is chosen. Here we follow an approach that is sometimes used in computational hydrodynamics [73], where the basic idea is to try a number of possibilities, dismiss not-admissible solutions (e.g. unphysical ones with $\rho < 0$) and pick the most accurate out of the admissible solutions. To take this decision, we use the following error measure:

$$\begin{aligned} E^{\mathrm{G},m} &\equiv \sum_p W_{pG} \left[\sum_{\mu,\nu} \left\{ \tilde{T}_{\mu\nu}^{\mathrm{G},m}(\mathbf{r}_p) - T_{\mu\nu,p} \right\}^2 \right] \\ &= \sum_p W_{pG} \left[\sum_{\mu,\nu} \{ (\boldsymbol{\beta}_{\mu,\nu}^{G,m} \cdot P^G(\mathbf{r}_p)) - T_{\mu\nu,p} \}^2 \right], \end{aligned} \tag{7.94}$$

where W_{pG} is, as before, a smooth positive definite function. In other words, for each polynomial order m, we use the optimal coefficients at the grid point to estimate the function values *at each particle position* for those particles that contribute to the grid point and from the difference between these estimates and the real function values at the particle positions we calculate an error measure. The aim is then to select the polynomial order with the smallest error measure that is physically admissible. There is, however, one more subtlety....

Identifying the stellar surface
We have tested the just described approach extensively and we noted when simulating neutron stars that in very few cases, when particles at the stellar surface contribute to grid cells lying just beneath the surface, the lowest error result can be an outlier value compared to the surrounding grid points. Therefore, we treat surface grid points differently: here we only apply the lowest polynomial order. This, however, requires the identification of which grids points are close to the stellar surface. We do this by checking which grid cells have many contributions from "surface particles", so we need to answer the question: *How can we identify particles at the stellar surface?* The decisive insight to identify the surface is that at this place "particles are missing

on one side" and that this will deteriorate standard SPH-approximations such as Eq. (7.13). If a particle is well engulfed in all directions by other particles, we will find a very good numerical approximation to an analytical result and–vice versa–a bad approximation indicates that the particle is located near a surface. In practice, we calculate a numerical approximation to $\nabla \cdot \mathbf{r} = 3$ and we use the gradient estimate Eq. (7.13) for the numerical approximation

$$(\nabla \cdot \mathbf{r})_a = \sum_b \frac{\nu_b}{N_b} (\mathbf{r}_b - \mathbf{r}_a) \cdot \nabla_a W_{ab}(h_a). \tag{7.95}$$

From the relative error

$$\delta_a \equiv \frac{|(\nabla \cdot \mathbf{r})_a - 3|}{3} \tag{7.96}$$

we calculate the average deviation $\langle \delta \rangle_G = (\sum_{b=1}^{n} \delta_b)/n$, where n is the number of particles that contribute to the error measure Eq. (7.94). We use $\langle \delta_G \rangle > 0.05$ as an accurate and robust identification of a grid point near the fluid surface. For such grid points we only use polynomial order $m = 0$.

7.3 Tests and First Applications

We show here some of the test cases to verify the correct implementation of the above described methods, more tests can be found in [25, 27]. We further show some of the first applications to neutron star mergers.

7.3.1 3D Shock Tube

With exactly known solutions, shock tubes are excellent tests to verify a correct implementation of the hydrodynamic equations. We perform here a 3D relativistic version of "Sod's shocktube" [74] which has become a widespread benchmark for relativistic hydrodynamics codes [37, 75–78]. The test uses a polytropic exponent $\Gamma = 5/3$ and as initial conditions

$$[N, P] = \begin{cases} \left[10, \frac{40}{3}\right], & \text{for } x < 0 \\ \left[1, 10^{-6}\right] & \text{for } x \geq 0, \end{cases} \tag{7.97}$$

with velocities initially being zero everywhere. We place particles with equal baryon numbers on close-packed lattices as described in [34], so that on the left side the particle spacing is $\Delta x_L = 0.0005$ and we have 12 particles in both y- and z-

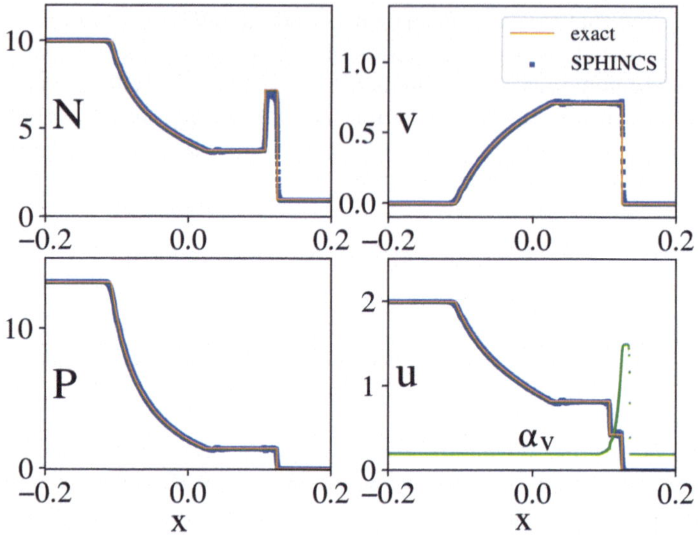

Fig. 7.4 3D, relativistic shocktube test. The exact solution is shown as orange lines, while the numerical results for the computing frame baryon number density N, the velocity v, the pressure P and the specific internal energy u are shown as blue squares. All particles are shown. We also show in last panel the dissipation parameter α (in green): it switches on just directly ahead of the shock, and then very quickly decays to the floor level value

direction.[8] This test is performed with the full 3+1 dimensional code, but using a fixed Minkowski metric. The result at $t = 0.15$ is shown in Fig. 7.4 with the `SPHINCS_BSSN` results marked with blue squares and the exact solution [78] with the orange line. In the last panel we also show the dissipation parameter α_v (in green): it switches on sharply just *ahead* of the shock front and then, once the shock wave has passed, very rapidly decays to the chosen floor value. Overall there is very good agreement with practically no spurious oscillations. There is only a small amount of "noise" directly after the shock front. This is to some extent unavoidable, since here the particles have to re-arrange themselves from their pre-shock close-packed configuration into their preferred post-shock configuration which goes along with some mild sideways motion.

7.3.2 Oscillating Neutron Stars

In order to test the full code, i.e. the relativistic hydrodynamics, the spacetime evolution and their mutual coupling via the mappings from the grid to the particles and from the particles to the grid, we have performed simulations of single oscillating

[8] The particle spacing on the right side is then given by the density jump and the requirement that particles have equal baryon numbers.

neutron stars. For this problem oscillation frequencies are known from independent, linear perturbation approaches [79, 80] which serve as a measure of the accuracy of our approach. We solve the Tolman-Oppenheimer-Volkoff (TOV) equations [81, 82] for a star in equilibrium which is then used to set up initial data for the particles and the spacetime grid. Due to truncation error the star does not remain in exact equilibrium and different eigenmodes of the star are excited at eigenfrequencies that depend on the equation of state and the mass of the star. With the choice of $K = 100$ and a central density of $\rho_c = 1.28 \times 10^{-3} = 7.91 \times 10^{14}\text{g/cm}^3$ we get a star with a gravitational mass of 1.4 $M_\odot$ and a baryonic mass of 1.506 $M_\odot$.

We performed 3 simulations with varying resolution of both the number of particles and the grid. For the particles we used 500k (low), 1M (medium) and 2M (high) particles. In all cases we used a grid with outer boundaries (in all directions) at 160 in code units, corresponding to 236 km. We used 4 levels of refinement with the star being completely contained within the finest grid. The resolution on the finest grid was 0.4≈590 m (low), 0.317≈468 m (medium) and 0.25≈369 m (high). The smallest smoothing length in each case was about 310 m (low), 255 m (medium) and 208 m (high). We evolved the stars up to physical times of almost 30 ms.

The results are shown in Fig. 7.5. In the top plot we show the relative difference between the maximal density, ρ_{max}, and the initial maximal density, ρ_0, as a function of time for the three runs. Here light blue is low, orange is medium and dark blue is high resolution. As expected, the amplitudes of the oscillations decrease with higher resolution (≈0.002 for 2M particles) and multiple modes at different frequencies are excited. Taking a Fourier transform of $(\rho_{\text{max}} - \rho_0)/\rho_0$ results in the spectrum shown in the bottom plot of Fig. 7.5. We see one weak peak as well as five strong peaks, all at frequencies that agree very well with the frequencies of the fundamental mode and the five first harmonics calculated[9] with the linear perturbation code written by Panagiota Kolitsidou [80]. The reference values for the known frequencies are indicated with black vertical lines in the plot. Quantitative error measures for the known frequencies are provided in Table 7.1. Note that even at the lowest resolution, the frequencies agree to better than 1% with the reference solution and they further improve with resolution.

In Fig. 7.6 we show a projection to the xy-plane of the particle distribution at the end of the simulation (left plot) and a comparison between the initial (black) and final (red) density as a function of radius (right plot). The final time, $t = 29.55$ ms, corresponds to more than 42 oscillation periods at the fundamental mode frequency. Note that, despite the long evolution time, the surface of the TOV star is extremely well behaved with no "outlier" particles (the plot shows all the particles). In contrast, the neutron star surface in *Eulerian* simulations is a perpetual source of numerical trouble where often (completely spurious) "winds" are driven away from the star. From the right plot it is clear that the density distribution has changed very little over the duration of the simulation. The main difference is that particles near the surface

[9] Kindly provided to us by Kostas Kokkotas and Nick Stergioulas.

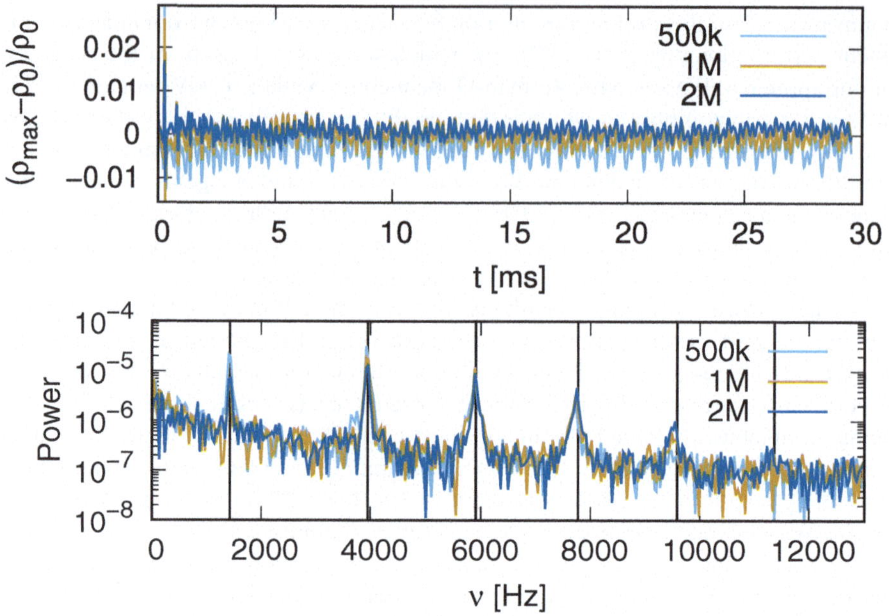

Fig. 7.5 Top plot: The relative difference between the maximal density ρ_{max} and the initial maximal density, ρ_0, for the TOV star as function of time in units of ms. Bottom plot: The power in the Fourier spectrum as function of frequency in units of Hz. In both plots the light blue line is for 500k, the orange is for 1M and the dark blue is for 2M particles. In the bottom plot, the vertical lines indicate (from left to right) the position of the expected frequency for the fundamental mode at 1.442 kHz, the first harmonic at 3.952 kHz, the second harmonic at 5.913 kHz, the third harmonic at 7.771 kHz, the fourth harmonic at 9.584 kHz and the fifth harmonic at 11.373 kHz. Even though the fifth harmonic is barely visible, we can still extract the frequency to better than 1% accuracy (see Table 7.1) from the simulation data

Table 7.1 Oscillation frequencies: fundamental mode (F), first, second, third, fourth and fifth harmonic (H1, H2, H3, H4 and H5) frequencies (from [80]) are given in Hz in the "Reference" column. For each resolution we list the extracted frequency of all modes (number before ±), as well as an estimate of the accuracy of its extraction (number after ±). Finally the relative error between the extracted frequency and the reference frequency is given in the parenthesis. Note that even at the lowest resolution the agreement of `SPHINCS_BSSN` results with the reference solution is better than 1%

Mode	Reference (Hz)	500k (Hz)	1M (Hz)	2M (Hz)
F	1441.9	1435.9 ± 2.5 (0.4)	1441.5 ± 6.1 (0.03)	1439.0 ± 4.0 (0.2)
H1	3952.4	3937.0 ± 3.3 (0.4)	3942.1 ± 5.5 (0.3)	3945.4 ± 4.2 (0.2)
H2	5912.5	5890.1 ± 2.7 (0.4)	5895.7 ± 5.9 (0.3)	5901.8 ± 4.4 (0.2)
H3	7770.6	7713.4 ± 5.0 (0.7)	7733.7 ± 6.1 (0.5)	7750.4 ± 4.0 (0.3)
H4	9583.9	9571.5 ± 32 (0.1)	9532.4 ± 52 (0.5)	9537.8 ± 21 (0.5)
H5	11373.1	11345.9 ± 19 (0.2)	11357.5 ± 31 (0.1)	11326.3 ± 36 (0.4)

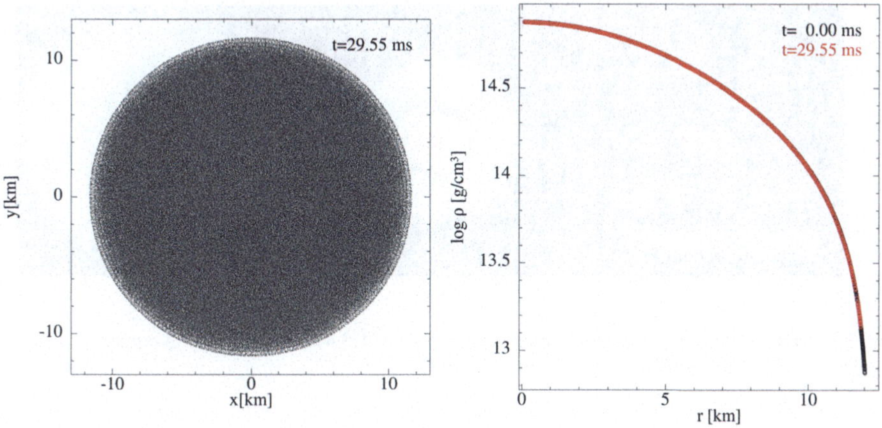

Fig. 7.6 The left plot shows a projection of the particle distribution of a 2M particle simulation (all particles are shown) at the final time, $t = 29.55$ ms. This time corresponds to more then 42 oscillation periods of the fundamental mode. Note that the neutron star surface has remained perfectly well behaved and the density distribution at late time (red) is nearly identical to the initial condition (black)

have moved inward by a very small amount compared to the starting position, but overall the density distribution has remained very well-behaved and is essentially identical to the initial condition.

7.3.3 Neutron Star Mergers

One of the key motivations behind the `SPHINCS_BSSN` development efforts are the multi-messenger signatures of merging neutron star binaries. We will show here two examples, one merger where the remnant remains stable until the end of the simulation and another one where the encounter results in a prompt collapse to a black hole.

7.3.3.1 Merger with Stable Remnant

As an example of a neutron star merger that results in a surviving remnant, we show a binary of $2\times 1.3\, M_\odot$ with the piecewise polytropic representation [83] of the APR3 equation of state [84], where we add a thermal pressure/internal energy component with a thermal polytropic exponent $\Gamma_{\rm th} = 1.75$, see Fig. 7.7. The spacetime/matter initial conditions for this binary system were calculated using the FUKA code [85], subsequently the particles were placed according to the "artificial pressure method" [28, 57] with the code SPHINCS_ID. We start from an initial separation a_0= 45 km

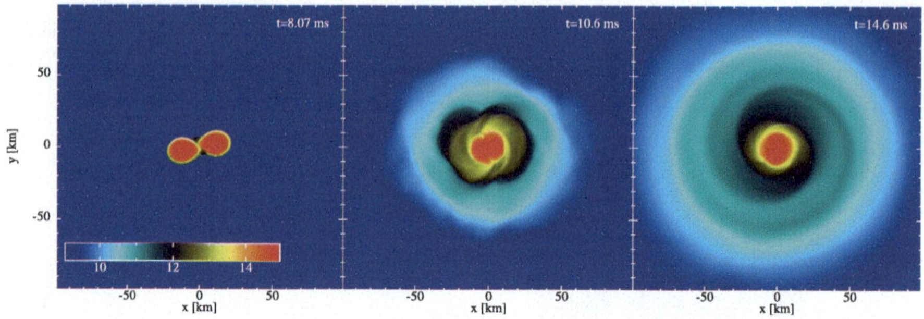

Fig. 7.7 Merger of an irrotational binary system with $2 \times 1.3\ M_{\odot}$ with the APR3 equation of state. Shown is the density distribution in the orbital plane

and use 2 million SPH particles to model the matter evolution while for the spacetime evolution we initially employ seven levels of mesh refinement with 193 points in each direction. During the merger and the subsequent evolution, the spacetime becomes more extreme and `SPHINCS_BSSN` automatically increases the number of refinement levels to eight, see Sect. 2.5 in [28] for more information on our mesh refinement algorithm.

Figure 7.7 shows the density evolution (in the orbital plane) and in the top of Fig. 7.8 we show how the maximum density and the minimum value of the lapse function evolve. As the stars approach each other, the lapse continuously decreases, while the peak density stays essentially constant. The merger results in a strong initial compression (by ~25% in the density), a subsequent "bounce back" during which the peak density drops below the initial single-star density, and then a continuous increase which levels off to a constant value at ~15 ms after the merger. The evolution of the lapse function during the merger is anti-correlated with the peak density: during maximum compression the laps drops to a minimum value of 0.42 and reaches after ~15 ms an asymptotic value of 0.45.

In the bottom plot of Fig. 7.8 we show, in light blue, the plus- and, in orange, the cross-polarization of the strain times the distance to the source, r. The gravitational waves has been extracted from the simulation via the Newman-Penrose Weyl scalar Ψ_4 at an extraction sphere of $R_{\rm extr} = 150 \approx 221.5$ km, see the appendix A of [26] for more details and we have used kuibit [86] for their analysis. At the earliest stages ($t < -7$ ms; $t = 0$ marking the peak GW emission) there is a small spurious transient introduced by the initial conditions, subsequently one recognizes the expected "chirp" signal with a long ring-down phase.

7.3.3.2 Merger with Prompt Black Hole Formation

As an example of a system that undergoes prompt collapse to a black hole, we show the merger of a $2\times1.5\ M_{\odot}$ binary with the same equation of state, initial separation, particle number and initial grid setup as in Sect. 7.3.3.1. As this system

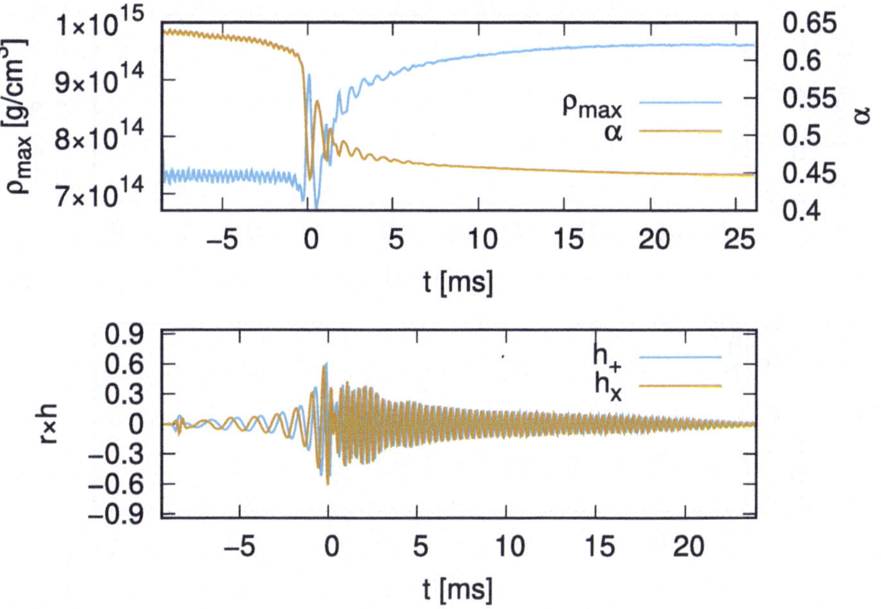

Fig. 7.8 Top: evolution of the peak density and minimum lapse value during the merger of the 2×1.3 $M_\odot$ merger with APR3 equation of state. Bottom: corresponding gravitational wave amplitudes (multiplied with the distance to the source)

is massive enough to undergo prompt collapse to a black hole, more refinement levels are added automatically as the collapse proceeds until, at the end, we have 11 levels of refinement. Again see Sect. 2.5 in [28] for the details. In the top plot of Fig. 7.9 we show the maximal density, ρ_{max}, (light blue, left axis), and the lapse, α, (orange, right axis) as function of time. As can be seen the lapse starts collapsing right after the merger while the maximal density starts to increase dramatically. The collapsing remnant requires a rapid decrease in the numerical time step as described in Sect. 2.5 in [28]. Once the lapse value of a particle decreases below $\alpha_{dust} = 0.05$, we convert it to "dust", which means that we ignore the particle's internal energy and pressure contribution. This makes the conservative-to-primitive recovery trivial. Below $\alpha_{cut} = 0.02$ we remove particles from the simulations. As can be double-checked by calculating a posteriori the apparent horizon, this procedure is safe and does not lead to notable artifacts. During the collapse we reach a maximal density of $\rho_{max} = 5 \times 10^{16}$ g/cm^3 before the removal of particles kicks in. At the end of the simulation about 7×10^{-4} $M_\odot$ of material are still outside the black hole out of which about 2×10^{-4} $M_\odot$ are unbound, and some fraction of which has speeds exceeding $0.7c$. The remaining bound material is expected to fall into the black hole eventually.

In the bottom plot of Fig. 7.9 we show, in light blue, the plus- and, in orange, the cross-polarization of r times the strain extracted from the simulation at an extraction

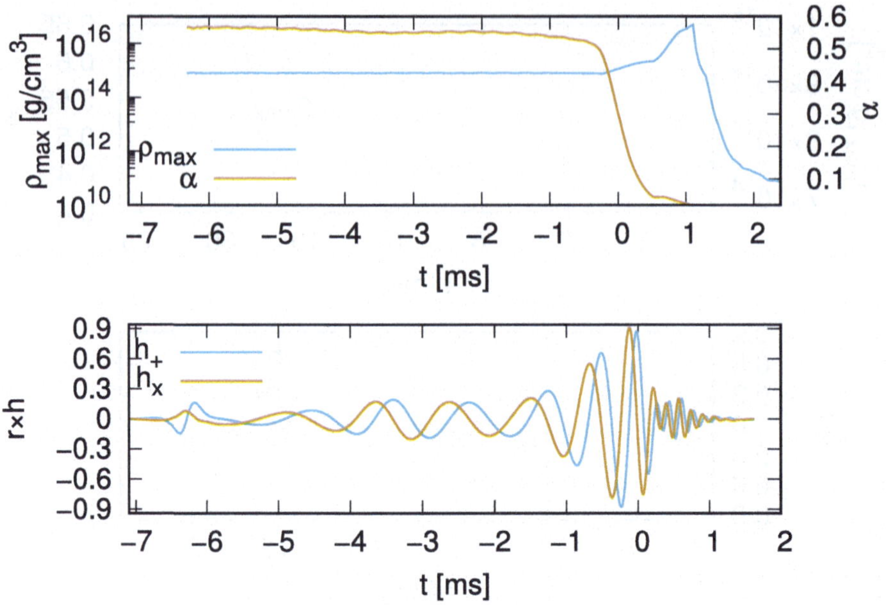

Fig. 7.9 Top plot: we show in blue (left axis) the maximal density, ρ_{max}, and in orange (right axis) the lapse, α, as function of time for the system with two 1.5 $M_\odot$ neutron stars with the APR3 equation of state. Bottom plot: we show in blue the plus-polarization, h_+, and in orange the cross-polarization, $h_\times$, of r times the gravitational wave strain

sphere of $R_{extr} = 150 \approx 221.5$ km. Note that, in both the bottom and the top plot, the data is shifted so that the time of maximal gravitational wave power corresponds to $t = 0$. The waveform initially (-7 ms $< t < -5.5$ ms) shows some spurious waves from the initial data, but then shows a clean "chirping" inspiral, the merger (-0.5 ms $< t < 0.6$ ms) and, finally, the black hole ringdown ($t > 0.6$ ms).

We have written a reader for spacetime data for the Einstein Toolkit [87] to analyze the properties of the black hole that forms. We first used `AHFinderDirect`[88, 89] to find the apparent horizon and, once found, `QuasiLocalMeasures`[90, 91] to calculate the spin and total mass of the black holes. The result is shown in Fig. 7.10. Here we show the total mass of the black hole for the 2M particle run in orange, the result of a lower resolution 1M particle run is shown in light blue. As can be seen the mass of the black hole grows rapidly at first then slows down and eventually levels off. The dark blue line shows the ADM mass of the initial data, while the black and red lines shows the initial ADM mass[10] minus the energy radiated away in gravitational waves as calculated from the 2M (black) and 1M (red) runs. This is the expected final mass, M_{final}, of the black hole. Both cases produce final black hole

[10] In General Relativity, energy and momentum are non-local quantities and difficult to define since the spacetime itself contributes [92]. There are various definitions of "mass", the ADM mass applies to asymptotically flat spacetimes and can be found by integrating over the spacetime.

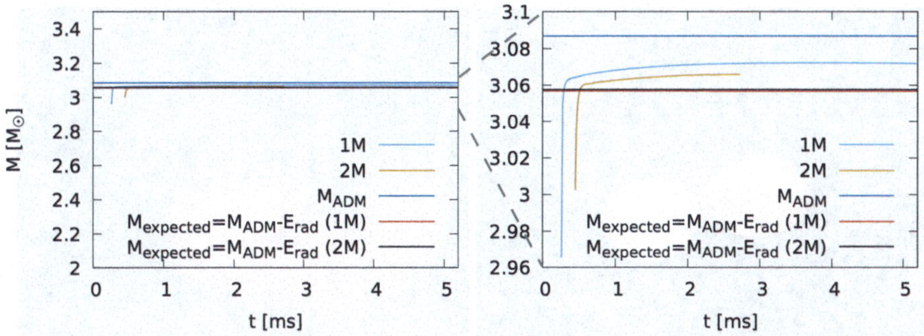

Fig. 7.10 The mass of the black hole (including rotational energy) as function of time (the time has been shifted so zero corresponds to the time of maximal gravitational wave power; left panel is a zoom-in of right panel). The orange curve is for the 2M particle runs mentioned earlier, while the light blue curve is for a lower resolution run with 1M particles. The dark blue curve shows the initial ADM mass of the system, while the black and red curves shows the initial ADM mass minus the radiated gravitational wave energy, i.e. the expected final mass of the black hole, as determined by the 2M and 1M runs, respectively

masses very close to this value with the 2M run reproducing it to within 0.3 %. For even better agreement, higher resolution would be needed. In the higher resolution run, the final dimensionless spin parameter is $a/M \approx 0.75$.

7.4 Summary and Outlook

We have presented here the main methodological elements of the *Lagrangian* numerical relativity code `SPHINCS_BSSN`. This code evolves the spacetime in a very similar fashion to standard *Eulerian* codes, in particular by solving the BSSN equations [17, 18] on an adaptive mesh. The major difference to more conventional numerical relativity codes is that the fluid is evolved by means of Lagrangian particles. This has major advantages for tracking and evolving the matter that is ejected in a neutron star merger. This matter, although only a relatively small fraction ($\sim$1%) of the binary mass, is responsible for all the electromagnetic emission and therefore of paramount importance for predicting the multi-messenger signatures on compact binary mergers. This mixed methodology with both particles and adaptive mesh, however, requires an accurate information exchange between the particles and the grid points and this has been one of the major challenges of the development process. As described in Sect. 7.2.3, we have found a way that produces accurate and robust results.

To demonstrate the proper functioning of the hydrodynamic part of `SPHINCS_BSSN` we have performed 3D shock tube tests which show very good agreement with the exact solution and thereby demonstrate the accurate functioning of the hydrodynamic part. Neutron star oscillations where the full spacetime is

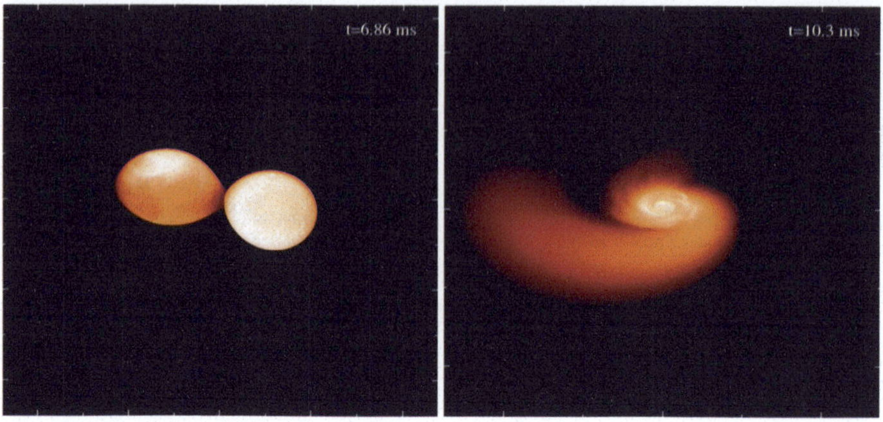

Fig. 7.11 Volume rendering of an equal mass ($2 \times 1.3\ M_\odot$) binary merger where one of the stars (the left star in the left panel) is rapidly spinning ($\chi_1 = 0.5$) while the other is not ($\chi_2 = 0$). The multi-messenger signatures of such systems are explored in detail in [93]

evolved are a good test for scrutinizing both the spacetime evolution and the coupling of spacetime and hydrodynamics, since measured oscillation frequencies can be compared against accurately known values from independent approaches. We have evolved a neutron star for ~30 ms and measured its oscillation frequencies. We find excellent agreement with the results of [79]: even at low resolution all frequencies agree already to better than 1% and become increasingly more accurate with higher resolution. The neutron star remains essentially perfectly in its initial state and its surface remains very well-behaved without any artifacts. This is a major advantage of `SPHINCS_BSSN` compared to Eulerian codes. We have further performed two binary neutron star simulations, one where a stable remnant survives and another one where a black hole forms promptly.

After an extensive phase of scrutinizing and further improving our methodology, we have recently begun to simulate astrophysical systems [94]. Figure 7.11 shows an example of a neutron star binary ($2 \times 1.3\ M_\odot$, APR3 EOS) where one star is rapidly spinning with dimensionless spin parameter $\chi_1 = 0.5$ while the other star has no spin ($\chi_2 = 0$). Such binaries with only one millisecond neutron star are not just a mere academic possibility, but they likely form in dense stellar environments such as globular clusters via binary-single exchange encounters [93]. The large majority of neutron stars observed in globular clusters are rapidly spinning[11] and encounters between a binary, consisting of a spun-up neutron star and a low-mass companion, and another neutron star can easily lead to the exchange of the low mass companion with the neutron star. According to our estimates [93], such binaries may account for approximately ~5% of the merging neutron star systems. Such mergers show a number of distinct signatures in their multi-messenger signals: they dynamically eject approximately an order of magnitude more matter and therefore produce particularly

[11] See P. Freire's website https://www3.mpifrbonn.mpg.de/staff/pfreire/NS_masses.html.

bright kilonovae. Since the collision itself is less violent they produce a smaller amount of fast ejecta ($> 0.5c$) and therefore weaker blue/UV kilonova precursors [95]. The GW-emission is substantially less efficient and therefore one can expect a longer lived central remnant and one may speculate that such mergers may be related to long GRBs that produce kilonova emission such as GRB 211211A [96] and GRB 230307A [97].

Future development efforts of `SPHINCS_BSSN` include improvements of its parallelization and the implementation of further physics ingredients such as, for example, more realistic nuclear matter equations of state and neutrino transport.

Acknowledgements It is a great pleasure to acknowledge interesting discussions with Sam Tootle and we are very grateful for his generous help with the FUKA library. We are further grateful to Kostas Kokkotas and Nick Stergioulas for sharing their insights into oscillating stars and we would also like to thank Diego Calderon, Jan-Erik Christian, Pia Jakobus, Laurens Paulsen, Lukas Schnabel and Wasif Shaquil for their careful reading of an earlier version of this text. SR has been supported by the Swedish Research Council (VR) under grant number 2020-05044, by the research environment grant "Gravitational Radiation and Electromagnetic Astrophysical Transients" (GREAT) funded by the Swedish Research Council (VR) under Dnr 2016-06012, by the Knut and Alice Wallenberg Foundation under grant Dnr. KAW 2019.0112, by the Deutsche Forschungsgemeinschaft (DFG, German Research Foundation) under Germany's Excellence Strategy - EXC 2121 "Quantum Universe"—390833306 and by the European Research Council (ERC) Advanced Grant INSPIRATION under the European Union's Horizon 2020 research and innovation programme (Grant agreement No. 101053985). Parts of the simulations for this chapter have been performed on the facilities of North-German Supercomputing Alliance (HLRN), and at the SUNRISE HPC facility supported by the Technical Division at the Department of Physics, Stockholm University. Special thanks go to Holger Motzkau and Mikica Kocic for their excellent support in upgrading and maintaining SUNRISE. Portions of this research were conducted with high performance computing resources provided by Louisiana State University (http://www.hpc.lsu.edu).

References

1. A. Einstein, Sitzungsberichte der Königlich Preußischen Akademie der Wissenschaften (Berlin). Seite **688–696**, 688 (1916)
2. A. Einstein, Sitzungsberichte der Königlich Preußischen Akademie der Wissenschaften (Berlin). Seite **154–167**, 154 (1918)
3. D. Kennefick, *Traveling at the Speed of Thought: Einstein and the Quest for Gravitational Waves* (2007). https://doi.org/10.1515/9781400882748
4. R.A. Hulse, J.H. Taylor, APJL **195**, L51 (1975). https://doi.org/10.1086/181708
5. B.P. Abbott, R. Abbott, T.D. Abbott, M.R. Abernathy, F. Acernese, K. Ackley, C. Adams, T. Adams, P. Addesso, R.X. Adhikari et al., Phys. Rev. Lett. **116**, 061102 (2016). https://doi.org/10.1103/PhysRevLett.116.061102. arXiv:1602.03837 [gr-qc]
6. R. Margutti, R. Chornock, Ann. Rev. Astron. Astrophys. **59**, 155 (2021). https://doi.org/10.1146/annurev-astro-112420-030742. arXiv:2012.04810 [astro-ph.HE]
7. A.J. Levan, J.D. Lyman, N.R. Tanvir, J. Hjorth, I. Mandel, E.R. Stanway, D. Steeghs, A.S. Fruchter, E. Troja, S.L. Schrøder, K. Wiersema, APJL **848**, L28 (2017). https://doi.org/10.3847/2041-8213/aa905f. arXiv:1710.05444 [astro-ph.HE]

8. B.P. Abbott, R. Abbott, T.D. Abbott, F. Acernese, K. Ackley, C. Adams, T. Adams, P. Addesso, R.X. Adhikari, V.B. Adya et al., APJL **848**, L13 (2017). https://doi.org/10.3847/2041-8213/aa920c. arXiv:1710.05834 [astro-ph.HE]
9. B.P. Abbott, R. Abbott, T.D. Abbott, F. Acernese, K. Ackley, C. Adams, T. Adams, P. Addesso, R.X. Adhikari, V.B. Adya et al., APJL **848**, L12 (2017). https://doi.org/10.3847/2041-8213/aa91c9. arXiv:1710.05833 [astro-ph.HE]
10. J.M. Lattimer, F. Mackie, D.G. Ravenhall, D.N. Schramm, APJ **213**, 225 (1977). https://doi.org/10.1086/155148
11. E. Symbalisty, D.N. Schramm, Astrophys. Lett. **22**, 143 (1982)
12. D. Eichler, M. Livio, T. Piran, D.N. Schramm, Nature **340**, 126 (1989)
13. S. Rosswog, M. Liebendörfer, F.-K. Thielemann, M. Davies, W. Benz, T. Piran, A & A **341**, 499 (1999)
14. C. Freiburghaus, S. Rosswog, F.-K. Thielemann, APJ **525**, L121 (1999)
15. J.J. Cowan, C. Sneden, J.E. Lawler, A. Aprahamian, M. Wiescher, K. Langanke, G. Martinez-Pinedo, F.-K. Thielemann, Rev. Mod. Phys. **93**, 015002 (2021). https://doi.org/10.1103/RevModPhys.93.015002. arXiv:1901.01410 [astro-ph.HE]
16. B.P. Abbott, R. Abbott, T.D. Abbott, F. Acernese, K. Ackley, C. Adams, T. Adams, P. Addesso, R.X. Adhikari, V.B. Adya et al., Nature **551**, 85 (2017). https://doi.org/10.1038/nature24471. arXiv:1710.05835
17. M. Alcubierre, *Introduction to 3+1 Numerical Relativity* (Oxford University Press, 2008)
18. T.W. Baumgarte, S.L. Shapiro, *Numerical Relativity: Solving Einstein's Equations on the Computer*, ed. by T.W. Baumgarte, S.L. Shapiro (Cambridge University Press, Cambridge, 2010). ISBN: 9780521514071
19. L. Rezzolla, O. Zanotti, *Relativistic Hydrodynamics* (Oxford University Press, 2013). ISBN-10: 0198528906; ISBN-13: 978-0198528906
20. M. Shibata, *Numerical Relativity* (World Scientific, 2016). https://doi.org/10.1142/9692
21. A. Mezzacappa, E. Endeve, O.E.B. Messer, S.W. Bruenn, Living Rev. Comput. Astrophys. **6**, 4 (2020). https://doi.org/10.1007/s41115-020-00010-8. arXiv:2010.09013 [astro-ph.HE]
22. D. Radice, S. Bernuzzi, A. Perego, R. Haas, MNRAS (2022). https://doi.org/10.1093/mnras/stac589. arXiv:2111.14858 [astro-ph.HE]
23. F. Foucart, Living Rev. Comput. Astrophys. **9**, 1 (2023). https://doi.org/10.1007/s41115-023-00016-y. arXiv:2209.02538 [astro-ph.HE]
24. W.H. Press, B.P. Flannery, S.A. Teukolsky, W.T. Vetterling, *Numerical Recipes* (Cambridge University Press, New York, 1992)
25. S. Rosswog, P. Diener, Class. Quantum Grav. **38**, 115002 (2021). https://doi.org/10.1088/1361-6382/abee65. arXiv:2012.13954 [gr-qc]
26. P. Diener, S. Rosswog, F. Torsello, Eur. Phys. J. A **58**, 74 (2022). https://doi.org/10.1140/epja/s10050-022-00725-7. arXiv:2203.06478 [astro-ph.HE]
27. S. Rosswog, P. Diener, F. Torsello, Symmetry **14**, 1280 (2022). https://doi.org/10.3390/sym14061280. arXiv:2205.08130 [gr-qc]
28. S. Rosswog, F. Torsello, P. Diener, Front. Appl. Math. Stat. **9**, (2023). https://doi.org/10.48550/arXiv.2306.06226. arXiv:2306.06226 [gr-qc]
29. J.J. Monaghan, Ann. Rev. Astron. Astrophys. **30**, 543 (1992)
30. J.J. Monaghan, Rep. Prog. Phys. **68**, 1703 (2005)
31. S. Rosswog, New Astron. Rev. **53**, 78 (2009)
32. V. Springel, ARAA **48**, 391 (2010). https://doi.org/10.1146/annurev-astro-081309-130914. arXiv:1109.2219 [astro-ph.CO]
33. D.J. Price, J. Comput. Phys. **231**, 759 (2012). https://doi.org/10.1016/j.jcp.2010.12.011. arXiv:1012.1885 [astro-ph.IM]
34. S. Rosswog, MNRAS **448**, 3628 (2015). https://doi.org/10.1093/mnras/stv225. arXiv:1405.6034 [astro-ph.IM]
35. S. Rosswog, Living Rev. Comput. Astrophys. **1**, (2015b). https://doi.org/10.1007/lrca-2015-1. arXiv:1406.4224 [astro-ph.IM]
36. P. Laguna, W.A. Miller, W.H. Zurek, APJ **404**, 678 (1993). https://doi.org/10.1086/172321

37. S. Siegler, H. Riffert, APJ **531**, 1053 (2000). https://doi.org/10.1086/308482. arXiv:astro-ph/9904070
38. J.J. Monaghan, D.J. Price, MNRAS **328**, 381 (2001). https://doi.org/10.1046/j.1365-8711.2001.04742.x
39. S. Rosswog, Class. Quantum Grav. **27**, 114108 (2010). https://doi.org/10.1088/0264-9381/27/11/114108
40. S. Rosswog, J. Comp. Phys. **229**, 8591 (2010). arXiv:0907.4890
41. S. Rosswog, *Meshfree Methods for Partial Differential Equations V*, ed. by M. Griebel, M.A. Schweitzer. Springer Lecture Notes in Computational Science and Engineering (Heidelberg, 2011), pp. 89–103
42. V. Springel, MNRAS **401**, 791 (2010). https://doi.org/10.1111/j.1365-2966.2009.15715.x. arXiv:0901.4107 [astro-ph.CO]
43. P.C. Duffell, A.I. MacFadyen, APJS **197**, 15 (2011). https://doi.org/10.1088/0067-0049/197/2/15. arXiv:1104.3562 [astro-ph.HE]
44. P.F. Hopkins, MNRAS **428**, 2840 (2013). https://doi.org/10.1093/mnras/sts210. arXiv:1206.5006 [astro-ph.IM]
45. R.M. Cabezon, D. Garcia-Senz, J. Figueira, A & A **606**, A78 (2017). https://doi.org/10.1051/0004-6361/201630208. arXiv:1607.01698 [astro-ph.IM]
46. J.J. Monaghan, J. Comput. Phys. **60**, 253 (1985)
47. V. Springel, L. Hernquist, MNRAS **333**, 649 (2002). https://doi.org/10.1046/j.1365-8711.2002.05445.x
48. J.J. Monaghan, MNRAS **335**, 843 (2002). https://doi.org/10.1046/j.1365-8711.2002.05678.x
49. S. Rosswog, D. Price, MNRAS **379**, 915 (2007)
50. R.A. Gingold, J.J. Monaghan, J. Comput. Phys. **46**, 429 (1982)
51. R. Speith, Untersuchung von smoothed particle hydrodynamics anhand astrophysikalischer beispiele. Ph.D. thesis, School Eberhard-Karls-Universität Tübingen, 1998
52. D. Price, J. Monaghan, MNRAS **374**, 1347 (2007)
53. V. Fock, *Theory of Space, Time and Gravitation* (Pergamon, Oxford, 1964)
54. B. Schutz, *A First Course in General Relativity*, 1st edn. (Cambridge University Press, Cambridge, 1989)
55. S.-I. Inutsuka, J. Comput. Phys. **179**, 238 (2002). arXiv:astro-ph/0206401
56. E. Toro, *Riemann Solvers and Numerical Methods for Fluid Dynamics* (Springer, Berlin, 1999)
57. S. Rosswog, MNRAS **498**, 4230 (2020). https://doi.org/10.1093/mnras/staa2591. arXiv:1911.13093 [astro-ph.IM]
58. J. von Neumann, R.D. Richtmyer, J. Appl. Phys. **21**, 232 (1950)
59. L. Cullen, W. Dehnen, MNRAS **408**, 669 (2010). https://doi.org/10.1111/j.1365-2966.2010.17158.x. arXiv:1006.1524 [astro-ph.IM]
60. J.-L. Guermond, R. Pasquetti, B. Popov, J. Comput. Phys. **230**, 4248 (2011). https://doi.org/10.1016/j.jcp.2010.11.043
61. J.-L. Guermond, B. Popov, V. Tomov, Comput. Methods Appl. Mech. Eng. **300**, 402 (2016). https://doi.org/10.1016/j.cma.2015.11.009
62. S. Rosswog, APJ **898**, 60 (2020). https://doi.org/10.3847/1538-4357/ab9a2e. arXiv:1912.01095 [astro-ph.IM]
63. M. Cernetic, V. Springel, T. Guillet, R. Pakmor, MNRAS **522**, 982 (2023). https://doi.org/10.1093/mnras/stad1043. arXiv:2208.11131 [astro-ph.IM]
64. M. Caldana, P.F. Antonietti, L. Dede', arXiv e-prints (2024). https://doi.org/10.48550/arXiv.2402.16517. arXiv:2402.16517 [math.NA]
65. J.J. Monaghan, J. Comput. Phys. **136**, 298 (1997)
66. D. Balsara, J. Comput. Phys. **121**, 357 (1995)
67. J. Morris, J. Monaghan, J. Comp. Phys. **136**, 41 (1997)
68. P.L. Roe, Annu. Rev. Fluid Mech. **18**, 337 (1986). https://doi.org/10.1146/annurev.fl.18.010186.002005

69. J.W. York, *Sources of Gravitational Radiation*, ed. by L.L. Smarr (Cambridge University Press, Cambridge, UK, 1979), pp. 83–126
70. R. Arnowitt, S. Deser, C.W. Misner, *Gravitation: An Introduction to Current Research*, ed. by L. Witten (Wiley, New York, 1962), pp. 227–265
71. F.X. Timmes, F.D. Swesty, APJS **126**, 501 (2000)
72. G.-H. Cottet, P.D. Koumoutsakos, *Vortex Methods* (2000)
73. S. Diot, R. Loubere, S. Clain, Int. J. Numer. Methods Fluids **73**, 362 (2013)
74. G. Sod, J. Comput. Phys. **43**, 1 (1978)
75. J. Marti, E. Müller, J. Comp. Phys. **123**, 1 (1996)
76. J.E. Chow, J. Monaghan, J. Comput. Phys. **134**, 296 (1997)
77. L. Del Zanna, N. Bucciantini, A & A **390**, 1177 (2002). https://doi.org/10.1051/0004-6361:20020776. arXiv:astro-ph/0205290
78. J.M. Marti, E. Müller, Living Rev. Relativ. **6**, 7 (2003)
79. J.A. Font, T. Goodale, S. Iyer, M. Miller, L. Rezzolla, E. Seidel, N. Stergioulas, W.-M. Suen, M. Tobias, Phys. Rev. D **65**, 084024 (2002). https://doi.org/10.1103/PhysRevD.65.084024
80. Radial oscillations of relativistic stars with piecewise polytropic equations of state. https://ikee.lib.auth.gr/record/303530/files/Thesis_P_Kolitsidou.pdf. Accessed 20th Sept 2024
81. R.C. Tolman, Phys. Rev. **55**, 364 (1939). https://doi.org/10.1103/PhysRev.55.364
82. J.R. Oppenheimer, G.M. Volkoff, Phys. Rev. **55**, 374 (1939). https://doi.org/10.1103/PhysRev.55.374
83. J.S. Read, B.D. Lackey, B.J. Owen, J.L. Friedman, Phys. Rev. D **79**, 124032 (2009). https://doi.org/10.1103/PhysRevD.79.124032. arXiv:0812.2163 [astro-ph]
84. A. Akmal, V.R. Pandharipande, D.G. Ravenhall, Phys. Rev. C **58**, 1804 (1998). https://doi.org/10.1103/PhysRevC.58.1804. hep-ph/9804388
85. L.J. Papenfort, S.D. Tootle, P. Grandclément, E.R. Most, L. Rezzolla, Phys. Rev. D **104**, 024057 (2021). https://doi.org/10.1103/PhysRevD.104.024057. arXiv:2103.09911 [gr-qc]
86. G. Bozzola, J. Open Source Softw. **6**, 3099 (2021). https://doi.org/10.21105/joss.03099. arXiv:2104.06376 [gr-qc]
87. F. Löffler, J. Faber, E. Bentivegna, T. Bode, P. Diener, R. Haas, I. Hinder, B.C. Mundim, C.D. Ott, E. Schnetter, G. Allen, M. Campanelli, P. Laguna, Class. Quantum Grav. **29**, 115001 (2012). https://doi.org/10.1088/0264-9381/29/11/115001. arXiv:1111.3344 [gr-qc]
88. J. Thornburg, Phys. Rev. D **54**, 4899 (1996). (gr-qc/9508014)
89. J. Thornburg, Class. Quantum Grav. **21**, 743 (2004). https://doi.org/10.1088/0264-9381/21/2/026. http://arxiv.org/abs/gr-qc/0306056gr-qc/0306056
90. O. Dreyer, B. Krishnan, D. Shoemaker, E. Schnetter, Phys. Rev. D **67**, 024018 (2003). https://doi.org/10.1103/PhysRevD.67.024018. arXiv:gr-qc/0206008
91. E. Schnetter, B. Krishnan, F. Beyer, Phys. Rev. D **74**, 024028 (2006). https://doi.org/10.1103/PhysRevD.74.024028. arXiv:gr-qc/0604015
92. C. Misner, K. Thorne, J. Wheeler, *Gravitation* (Freeman, New York, 1973)
93. S. Rosswog, P. Diener, F. Torsello, T. Tauris, N. Sarin, arXiv e-prints (2023). https://doi.org/10.48550/arXiv.2310.15920. arXiv:2310.15920 [astro-ph.HE]
94. S. Rosswog, O. Korobkin, Annalen der Physik **536**, 2200306 (2024). https://doi.org/10.1002/andp.202200306. arXiv:2208.14026 [astro-ph.HE]
95. B.D. Metzger, A. Bauswein, S. Goriely, D. Kasen, MNRAS **446**, 1115 (2015). https://doi.org/10.1093/mnras/stu2225. arXiv:1409.0544 [astro-ph.HE]
96. J.C. Rastinejad, B.P. Gompertz, A.J. Levan, W.-F. Fong, M. Nicholl, G.P. Lamb, D.B. Malesani, A.E. Nugent et al., Nature **612**, 223 (2022). https://doi.org/10.1038/s41586-022-05390-w. arXiv:2204.10864 [astro-ph.HE]

97. A.J. Levan, B.P. Gompertz, O.S. Salafia, M. Bulla, E. Burns, K. Hotokezaka, L. Izzo, G.P. Lamb, D.B. Malesani, S.R. Oates, M.E. Ravasio, A. Rouco Escorial, B. Schneider, N. Sarin, S. Schulze, N.R. Tanvir, K. Ackley, G. Anderson, G.B. Brammer, L. Christensen, V.S. Dhillon, P.A. Evans, M. Fausnaugh, W.-F. Fong, A.S. Fruchter, C. Fryer, J.P.U. Fynbo, N. Gaspari, K.E. Heintz, J. Hjorth, J.A. Kennea, M.R. Kennedy, T. Laskar, G. Leloudas, I. Mandel, A. Martin-Carrillo, B.D. Metzger, M. Nicholl, A. Nugent, J.T. Palmerio, G. Pugliese, J. Rastinejad, L. Rhodes, A. Rossi, A. Saccardi, S.J. Smartt, H.F. Stevance, A. Tohuvavohu, A. van der Horst, S.D. Vergani, D. Watson, T. Barclay, K. Bhirombhakdi, E. Breedt, A.A. Breeveld, A.J. Brown, S. Campana, A.A. Chrimes, P. D'Avanzo, V. D'Elia, M. De Pasquale, M.J. Dyer, D.K. Galloway, J.A. Garbutt, M.J. Green, D.H. Hartmann, P. Jakobsson, P. Kerry, C. Kouveliotou, D. Langeroodi, E. Le Floc'h, J.K. Leung, S.P. Littlefair, J. Munday, P. O'Brien, S.G. Parsons, I. Pelisoli, D.I. Sahman, R. Salvaterra, B. Sbarufatti, D. Steeghs, G. Tagliaferri, C.C. Thöne, A. de Ugarte Postigo, D.A. Kann, Nature **626**, 737 (2024). https://doi.org/10.1038/s41586-023-06759-1. arXiv:2307.02098 [astro-ph.HE]

Part III
Non-Self-Gravitating Fluids

Chapter 8
Wind from the Hot Accretion Flow and Super-Eddington Accretion Flow

Hai Yang and Feng Yuan

Abstract Wind is believed to be widespread in various black hole accretion flows. However, unlike the wind from thin disks, which have substantial observational evidence, the wind from hot accretion flows is difficult to observe due to the extremely high temperatures causing the gas to be almost fully ionized. Its existence was controversial until recent theoretical work demonstrated its presence and strength, which was subsequently confirmed by observations. Although there have been some new observations recently, the main progress still comes from theoretical studies. These studies investigate the effects of different magnetic fields and black hole spins on the wind, providing insights into properties such as mass flux and wind velocity. Wind is typically produced locally within the Bondi radius, and even wind generated on a small scale can propagate far beyond this radius. The situation with super-Eddington wind is similar, despite some recent observations, the main advances rely on theoretical studies. Recent research comparing the momentum and energy of wind and jets suggests that wind plays a more crucial role in active galactic nuclei feedback than jets, whether the wind originates from hot accretion flows or super-Eddington accretion flows.

8.1 Introduction

Black hole accretion is a prevalent and fundamental high-energy astrophysical phenomenon in the Universe and is generally regarded as the primary energy source for many observed phenomena, such as active galactic nuclei (AGNs), black hole

H. Yang (✉)
Tsung-Dao Lee Institute, Shanghai Jiao Tong University, 1 Lisuo Road,
Pudong New Area 201210, Shanghai, China
e-mail: hai-yang@sjtu.edu.cn

F. Yuan
Center for Astronomy and Astrophysics and Department of Physics, Fudan University,
Shanghai 200438, China
e-mail: fyuan@fudan.edu.cn

C. Bambi et al. (eds.), *New Frontiers in GRMHD Simulations*, Springer Series
in Astrophysics and Cosmology, https://doi.org/10.1007/978-981-97-8522-3_8

X-ray binaries (BHBs), and gamma-ray bursts. Depending on the temperature of the accretion flow, black hole accretion is commonly categorized into two modes: cold and hot modes. These classification criteria are based on the mass accretion rate of the black hole. In the cold mode, the accretion rate exceeds approximately 2% $\dot{M}_{\rm Edd}$, where $\dot{M}_{\rm Edd} = 10\, L_{\rm Edd}/c^2$ represents the Eddington accretion rate and $L_{\rm Edd}$ is the Eddington luminosity. Conversely, in the hot mode, the accretion rate is lower than about 2% $\dot{M}_{\rm Edd}$ [114].

In cold accretion modes, the accretion flow is optically thick, making radiation a crucial factor. Conversely, hot modes are typically optically thin, resulting in radiation being of lesser importance. Cold mode further segregates into two accretion disk models. Firstly, there is the standard thin disk [77, 84] characterized by accretion rates lower than $\dot{M}_{\rm Edd}$. This disk is geometrically thin, emitting thermal blackbody-like radiation that successfully describes luminous AGNs [3, 51, 77] and BHBs in their thermal state [55].

Secondly, there's the super-Eddington accretion flow [1, 46, 57, 68, 81] characterized by accretion rates surpassing $\dot{M}_{\rm Edd}$. In this scenario, the accretion disk is geometrically thick, with the gas being too optically thick to radiate all dissipated energy locally. Consequently, radiation becomes trapped and inwardly advected by the accretion flow. This model is employed to elucidate certain observations like tidal disruption events (TDEs) [21], SS433 [29], and ultraluminous X-ray sources [99].

The hot modes usually correspond to an advection-dominated accretion flow (ADAF), also known as hot accretion flow [63, 114]. In this scenario, the accretion disk is geometrically thick and the gas is optically thin, meaning that energy is accreted into the black hole before it can be cooled by radiation. This model has been applied to a wide variety of accretion systems, including the supermassive black holes (SMBHs) Sagittarius A* (Sgr A*), M87, low-luminosity AGNs, which are thought to be prevalent in most galaxies, and the hard/quiescent states of black hole binaries.

8.2 Theoretical Study of Wind from Hot Accretion Flows

AGN feedback is thought to play a key role in the formation and evolution of galaxies [28, 52, 61]. The wind from hot accretion flow may be one of the most important feedback media [101, 108, 116]. Recent IllustrisTNG cosmological simulations [101] research has found that winds from hot accretion flow are needed to solve some problems encountered in galaxy formation, such as the interaction of the wind with the interstellar medium at the galaxy scale is required to reduce the star formation efficiency in the most massive halos. Yuan et al. [116] studied the feedback effects involving wind and radiation from AGNs and found that wind plays a dominant role in controlling star formation and black hole mass growth.

In this section, we primarily introduce the latest research on winds in hot accretion flows. For a more comprehensive review of hot accretion flows and the driving mechanisms of winds from hot accretion flows, refer to [114].

The study of winds in accretion flows has long been a crucial research topic in this field and represents significant progress. Its importance and attraction stem mainly from two aspects. Firstly, winds constitute a fundamental component of black hole accretion flows. Research indicates that a substantial portion of gas flowing in from large radii is not accreted by the black hole but instead is predominantly converted into wind. Such powerful winds profoundly impact the dynamic properties of the accretion flow, including angular momentum transfer, mass accretion rate, temperature, density, and consequently influence radiation emitted from the accretion flow.

Secondly, AGN feedback is believed to play a pivotal role in the formation and evolution of galaxies [28, 52, 61]. Winds from hot accretion flows may constitute one of the most significant feedback mechanisms [101, 108, 116]. Recent cosmological simulations with IllustrisTNG [101] have revealed that winds originating from hot accretion flows are crucial for addressing certain challenges in galaxy formation. For instance, the interaction of these winds with the interstellar medium on galactic scales is necessary to reduce the star formation efficiency in the most massive halos. Yuan et al. [116] investigated feedback effects involving winds and radiation from AGNs, highlighting winds' dominant role in regulating star formation and the growth of black hole masses.

Narayan's earliest analytical self-similar solution to hot accretion flows, as proposed in [63], suggested the presence of strong winds. In 1999, a milestone in wind studies was achieved when Stone et al. [88] conducted the first comprehensive global numerical simulation of hot accretion flows. Their study calculated the inflow, outflow, and net accretion rate of mass flux.

$$\dot{M}_{\rm in} = -2\pi r^2 \int_0^{\pi} \rho \min(\mathrm{v_r}, 0) \sin\theta d\theta \tag{8.1}$$

$$\dot{M}_{\rm out} = 2\pi r^2 \int_0^{\pi} \rho \max(\mathrm{v_r}, 0) \sin\theta d\theta \tag{8.2}$$

$$\dot{M}_{\rm net} = \dot{M}_{\rm in} - \dot{M}_{\rm out} \tag{8.3}$$

As shown in Fig. 8.1, Stone et al. [88] found that both the inflow rate $\dot{M}_{\rm in}$ and outflow rate $\dot{M}_{\rm out}$ decrease as the radius decreases, following a power-law relationship,

$$\dot{M}_{\rm in}(r) = \dot{M}_{\rm in}(r_{\rm out}) \left(\frac{r}{r_{\rm out}}\right)^s, \quad s > 0 \tag{8.4}$$

where $\dot{M}_{\rm in}(r_{\rm out})$ is the mass flux of inflow at outer boundary $r_{\rm out}$. This result is surprising because it is generally believed that the accretion rate remains constant and does not vary with radius. Early analytical studies often assumed a constant accretion rate independent of radius. The radial distribution of density typically follows $\rho \propto$

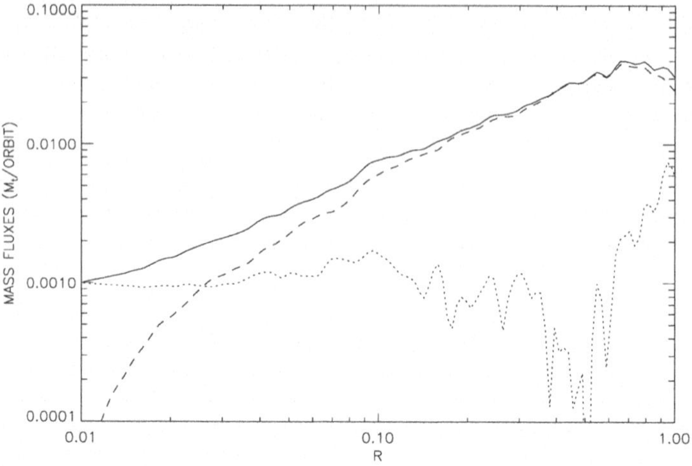

Fig. 8.1 Radial distribution of different mass accretion rates. The solid, dashed, and dotted lines represent inflow, outflow, and net accretion flow, respectively. Figure adapted from Stone et al. [88]

$r^{-3/2}$ [63]. However, subsequent numerical simulations, both hydrodynamic (HD) [41, 88, 111, 113] and magnetohydrodynamic (MHD) [36, 44, 48, 56, 66, 72, 74, 89, 112], have shown that the mass accretion rate decreases towards the black hole.

Due to technical constraints, most simulations only achieve equilibrium over two orders of magnitude in radius, potentially affected by boundary conditions and thus reliability. Yuan et al. [112] extended the equilibrium range to four orders of magnitude using a "two-zone" approach, from approximately $2\,r_g$ to $80{,}000\,r_g$, where $r_g = GM/c^2$ is the gravitational radius. Their work supports the earlier finding of a power-law distribution of inflow rate and density. They consolidated their findings with previous HD and MHD simulations, whether 2D or 3D, demonstrating a consistent density distribution $\rho \propto r^{-(0.5-1)}$. Variations in the exponent may arise from differences in parameters such as α, magnetic field strength, gravitational potential of the black hole, and initial simulation conditions.

To explain the radial variations in inflow and outflow observed in numerical simulations, two main competing models have been proposed. One is the ADIOS (Adiabatic Inflow-Outflow Solution) model [7, 11, 12] proposed by Blandford et al. This model posits that the presence of wind causes a decrease in the inward accretion rate $\dot{M}_{in}$. However, it does not specify the mechanism driving the wind, assuming instead that some mechanism extracts energy from the accretion flow to generate the wind. Self-similar solutions were derived for 1D and 2D cases, showing an index $0 < s < 1$.

Another model is the CDAF (Convection-Dominated Accretion Flow) model [2, 43, 64, 78] proposed by Narayan et al., which is based on the assumption that hot accretion flows are convectively unstable [63]. In the CDAF model, inward angular momentum transfer by convection balances outward transfer by viscous stresses, leading to the cancellation of these effects. This results in convective eddies around the black hole where gas continually convects inward and outward. The gas trapped in these eddies does not accrete onto the black hole, causing an inward decrease in the accretion rate.

The debate over which model is correct has persisted for over a decade and remains unresolved. Resolving this debate hinges on determining the extent to which the outflow in Eq. 8.2 represents actual outflow. Addressing this, Yuan et al. [112, 113] conducted MHD numerical simulations. In Eq. 8.2, any $v_r > 0$ denotes outflow; however, in accretion flows, both real outflow and the outward component of convective eddies contribute. Yuan et al. [113] thus compared various physical properties of inflows and outflows, including temperature, angular momentum, and velocity. They found systematic and significant differences between inflows and outflows. Analysis of angular momentum distribution in the θ direction indicated weaker inflow angular momentum where outflow was stronger. In HD simulations, outflow temperatures were consistently higher than inflow temperatures, suggesting buoyancy-driven by convective instabilities in HD accretion flows. In MHD simulations, outflow angular momentum approached Keplerian levels near the equatorial plane, contrasting sharply with low inflow angular momentum. These differences challenge the CDAF model's ability to explain these observations and provide strong evidence that both inflow and outflow mass fluxes are, if not dominant, at least significant.

Additionally, they examined the convective stability of accretion flows, finding that while HD flows are convectively unstable, the addition of a magnetic field stabilizes convection. Figure 8.2 illustrates these findings, showing that much of the accretion flow areas are not red, indicating convective stability in MHD flows. This result corroborates findings by Narayan et al. [66].

However, the observed inward decrease in accretion rate $\dot{M}_{\rm in}$ is evident in both HD and MHD simulations, suggesting the CDAF model's inadequacy in explaining these

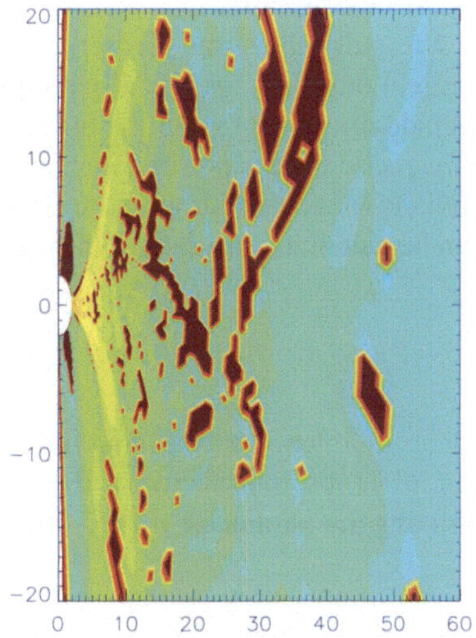

Fig. 8.2 Results of convective stability analysis of MHD accretion flow. Regions colored in red indicate instability. Adapted from Yuan et al. [113]

phenomena. Consequently, these results strongly support the existence of winds, aligning more closely with the ADIOS model.

While the aforementioned studies indicate the presence of winds, none quantifies the true strength of these winds-specifically, the mass flux of the wind. The method outlined in Eq. 8.2 for calculating outflow inevitably overestimates the magnitude of the wind mass flow due to its inclusion of turbulent components. Gas may momentarily move outward only to reverse direction and flow inward moments later, limiting accurate measurement to an upper bound. To address this issue, a method to account for turbulence effects is necessary.

Narayan et al. [66] proposed the use of a time-averaged "streamline method" to determine the real outflow. Initially, they conducted time and azimuthal averaging of pertinent physical quantities-notably density ρ, velocity v_r, and $\rho u_{\rm t}$, where v_i $(i = r,\ \theta,\ \varphi)$and u_ν $(\nu = t,\ r,\ \theta,\ \varphi)$ represent the 3-velocity and 4-velocity of the fluid respectively. This averaging effectively cancels out turbulent fluctuations within the flow's equilibrium region, facilitating a focus on its averaged properties. They defined $\langle v_r \rangle_{t\varphi} > 0$ as the criterion for genuine outflow and the opposite as inflow.

The outflow mass flux is then calculated using the formula,

$$\dot{M}_{\rm out} = 2\pi r^2 \int_0^\pi \max(\langle \rho {\rm v_r} \rangle_{\rm t\varphi}, 0) \sin\theta {\rm d}\theta \tag{8.5}$$

Although this approach effectively removes the effects of turbulence-induced oscillations, it tends to underestimate the mass flux of real outflows. This is because outflows are inherently intermittent and can wander in three-dimensional space. Even at the same spatial location, the velocity v_r in the accretion flow may be positive (indicating wind) at one moment and negative (indicating inflow) at the next. Therefore, if a time average is applied first, real outflows may be averaged out or diminished. Consequently, using $\langle v_r \rangle_{t\varphi} > 0$ as the criterion to identify outflows risks underestimating their true magnitude.

Secondly, they used the Bernoulli parameter as a necessary condition for identifying real outflows, specifically requiring $Be > 0$. The Bernoulli parameter typically includes the sum of gravitational potential energy, kinetic energy, and enthalpy. When a magnetic field is present, its contribution must also be considered. In General Relativistic Magnetohydrodynamic (GRMHD) accretion flows, the Bernoulli parameter can be calculated using the following formula [76],

$$Be = -\frac{\langle \rho u_t \rangle + \Gamma \langle u u_t \rangle + \langle b^2 u_t \rangle}{\langle \rho \rangle} - 1 \tag{8.6}$$

where $\langle ... \rangle$ denotes averaging over time and azimuth, u is the internal energy of the gas, b^ν is the four-vector of the fluid frame magnetic field. There is also research utilizing the parameter μ [66, 80], which is given by,

$$\mu = \frac{\langle T_t^p \rangle}{\langle \rho u^p \rangle} - 1 \tag{8.7}$$

In a steady-state axisymmetric ideal MHD flow, μ is conserved along the streamlines, where the index p represents "poloidal", T^{α}_{β} is the stress-energy tensor. Therefore, μ was used as a criterion for identifying real outflows, indicating that the gas can escape to infinity. This criterion helps to filter out a significant amount of turbulent outflow to a certain extent. However, in a non-steady-state accretion flow, the Bernoulli parameter is not constant along streamlines and may increase outward. This implies that at a certain radius, even if $Be < 0$, gas can still escape to infinity.

Indeed, Yuan et al. [115] found that the Bernoulli parameter Be of the wind can take any value-positive or negative-at any radius. Consequently, Be is not a reliable criterion for real outflow in accretion flows. Using Be as a criterion is likely to filter out some real outflows, thus causing an underestimation of the outflow mass flux. This explains why Narayan et al. [66] found the wind to be very weak, as shown in Fig. 8.3, and only becoming significant beyond approximately $100\,r_g$.

Therefore, the actual wind mass flux should be between the red solid line (calculated by Eq. 8.2) and the red dotted line (calculated by Eq. 8.5) in Fig. 8.3.

To determine the real wind mass flux, Yuan et al. [115] introduced the "trajectory" method in the analysis of accretion flows. Trajectories relate to the Lagrangian description of fluid motion, obtained by tracking the continuous motion of fluid elements over time. In contrast, streamlines relate to the Eulerian description of the fluid and are derived by connecting the velocity fields of adjacent fluid elements at a given moment. Although calculating trajectories is more complex than calculating streamlines, it provides a more realistic depiction of fluid element movements.

To obtain the trajectory line, a series of "test particles" are placed at a certain radius in space at a given time. These test particles are not real particles but represent spatial coordinate sets that serve as the starting points for trajectory lines. Their velocities

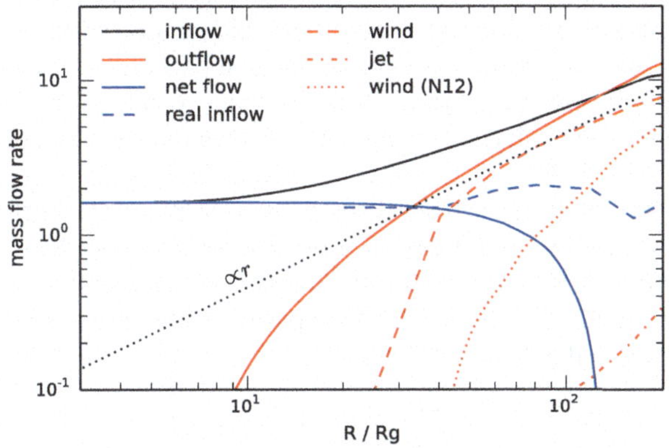

Fig. 8.3 Radial distribution of different mass accretion rates. The black, red, and blue solid lines represent the total inflow, outflow, and net flow calculated based on Eqs. 8.1–8.3, respectively. The red and blue dashed lines represent the real outflow and inflow, respectively, using the 'trajectory method'. The red-dotted line indicates the mass flux of the disk-jet, while the red-dashed line shows the actual outflow obtained from the 'streamline method'. Adapted from Yuan et al. [115]

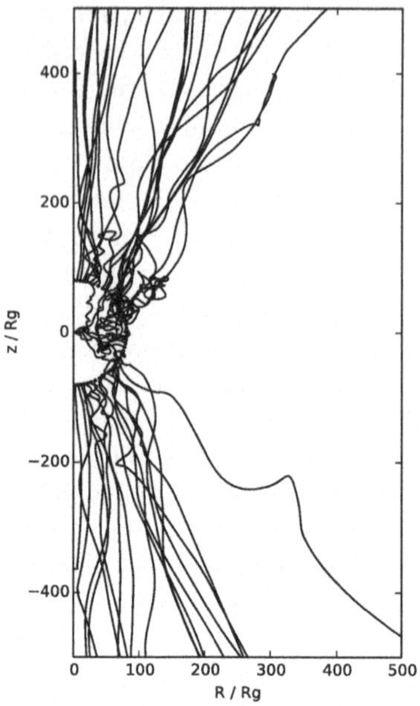

Fig. 8.4 The solid lines in the figure depict trajectories of various gas elements, with (0, 0) indicating the location of the black hole. Adapted from Yuan et al. [115]

are then obtained through interpolation, allowing the determination of their positions at the next time step $t + \delta t$. This process continues until the simulation ends or the test particle exits the simulation area.

A critical aspect of this calculation is ensuring that δt is sufficiently small to achieve convergence, depending on the rate at which the particle changes its speed as it moves. Typically, the time scale for δt is the Keplerian timescale.

Based on the trajectory method, Yuan et al. [115] analyzed the simulation data of 3D GRMHD. The trajectories of some of the test particles are shown in Fig. 8.4. These trajectories reveal the existence of significant real outflows in the coronal region, providing direct evidence of real outflows rather than relying solely on indirect evidence [113].

To perform this analysis, test particles were first placed at different θ and φ angles with a fixed radius, and their trajectories were obtained. Next, particles that maintained continuous outward motion and did not cross the radius r were identified as real outflows. Finally, the mass flux of the particles identified as real outflow was summed to determine the "real outflow rate",

$$\dot{M}_{\rm wind}(r) = \sum_i \rho_i(r) v_{r,i}(r) r^2 \sin\theta \delta\theta_i \delta\varphi_i \tag{8.8}$$

where $\rho_i(r)$ and $v_{r,i}(r)$ are the mass density and radial velocity of the test particle i at the initial position, respectively. $\delta\theta_i$ and $\delta\varphi_i$ represent the ranges of θ and φ occupied by the test particle i, respectively. From Eq. 8.8, we can derive the real outflow mass flux, which is shown as the red dashed line in Fig. 8.3. It is evident that the real outflow mass flux obtained using the trajectory method (Eq. 8.8) is significantly larger than the result obtained using the streamline method (Eq. 8.5). The mass flux of the real outflow is comparable to the accretion rate at the black hole horizon, $\dot{M}_{\rm BH}$, at approximately 40 $r_{\rm g}$. In contrast, Narayan et al. [66] found this value to be around 100 $r_{\rm g}$.

Additionally, they found that the mass flux of the wind can be well expressed by:

$$\dot{M}_{\rm wind}(r) \approx \dot{M}_{\rm BH} \left(\frac{r}{20\, r_{\rm s}}\right)^{s}, \quad s \approx 1, \tag{8.9}$$

where $r_{\rm s} \equiv 2\, r_{\rm g}$. Furthermore, the power-law index s in Eq. 8.9 aligns closely with the values obtained from the analytical work of Begelman et al. [7]. They also calculated the mass flux-weighted poloidal velocity ($v_p = \sqrt{v_r^2 + v_\theta^2}$) of the wind and found that it can be well expressed by the following formula,

$$v_{p,\rm wind}(r) \approx 0.21\, v_k(r), \tag{8.10}$$

where $v_k(r) \equiv \sqrt{GM/r}$ is the Keplerian velocity at radius r. It should be noted that this formula indicates the poloidal velocity of the wind at radius r. Considering that the wind is constantly generated locally, this is more likely to reflect the poloidal velocity of the wind generated at a radius close to r. Additionally, they found that the poloidal velocity of the test particle was either constant along the trajectory or slightly accelerated outward. Therefore, they provided the terminal poloidal velocity of the wind originating at radius r,

$$v_{p,\rm term}(r) \approx (0.2 \sim 0.4)\, v_{\rm k}(r), \tag{8.11}$$

Figure 8.5 illustrates the force analysis of the accretion flow, displaying gravity, gas pressure gradient, Lorentz force, and centrifugal force at typical locations. It is evident that the wind is primarily driven by centrifugal force and the magnetic pressure gradient. Additionally, the existence of a "disk-jet" was also discovered in their simulation. For more detailed information, please refer to Yuan et al. [115].

Yuan et al. work [115] focused on the standard and normal evolution (SANE) case [66] without considering black hole spin. However, strong magnetic field environments, such as magnetically arrested disks (MAD) [65], may exist in the real Universe. Recent studies on M87 and the SMBH at the center of the Milky Way have indicated that their accretion flows may be in a MAD state [25, 27, 117].

Sadowski et al. [80] have examined high-spin black holes and MAD situations. However, they still used the traditional streamline method to estimate the real outflow, resulting in a very weak wind mass flux. As shown in Yuan et al.'s work, they could only provide a lower limit on the mass flux of the wind.

To address this limitation, Yang et al. [104] extended Yuan et al.'s approach using the trajectory method to study the effects of magnetic fields and black hole spin on the wind. They utilized the Athena++ code [90, 102] to run three 3D GRMHD

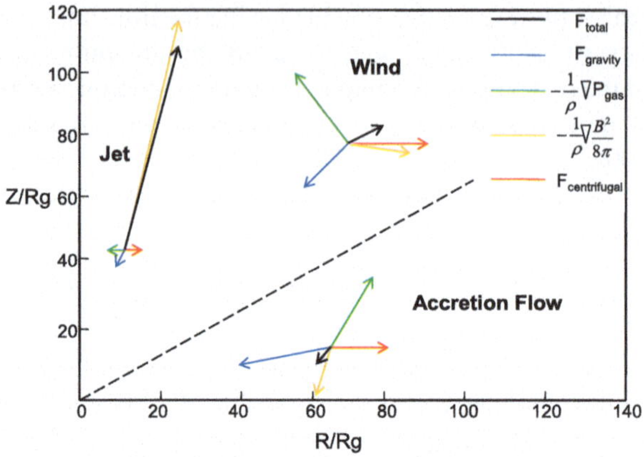

Fig. 8.5 Analysis of forces at three representative locations: jet, wind, and disk. Arrows indicate direction, and lengths represent force magnitude. Adapted from Yuan et al. [115]

simulations: MAD00, MAD98, and SANE98, where "MAD98" signifies a MAD accretion flow with a black hole spin of 0.98.

MAD and SANE represent two distinct accretion flow states characterized by different magnetic field configurations. MAD typically begins with a poloidal magnetic loop threading through the gas torus, facilitating rapid accretion. However, as magnetic flux accumulates near the black hole, it can eventually impede further accretion, leading to a saturated state [56, 65, 80].

In contrast, SANE initially comprises multiple magnetic field loops with opposing polarities. This setup promotes frequent magnetic reconnection during the accretion process, maintaining a weaker magnetic field configuration. This prevents excessive accumulation of magnetic flux around the black hole, allowing the accretion flow to remain in the SANE state [66].

The determination of whether the accretion flow transitions into the MAD or SANE state is typically based on the normalized averaged magnetic flux ϕ threading the hemisphere of the black hole event horizon [66, 92, 104].

$$\phi(t) = \frac{1}{2\sqrt{\dot{M}_{\rm BH}}} \int_\theta \int_\varphi |B^r(r_{\rm H}, t)| \sqrt{-g} d\theta d\varphi, \tag{8.12}$$

where B^r denotes the radial component of the magnetic field, $r_{\rm H} = r_{\rm g} + \sqrt{1 - a^2} r_{\rm g}$ is the radius of the event horizon, a is the dimensionless spin parameter of the black hole, and g represents the determinant of the metric. The accretion flow transitions to the MAD state once the normalized averaged magnetic flux ϕ exceeds a critical value of approximately 50 [92]. Figure 8.6 illustrates the evolution of ϕ over time. The green and red solid lines depict the accretion flow in the MAD state, while the solid blue line, appearing after $t \approx 50,000\, r_{\rm g}/c$, indicates the accretion flow in the SANE state.

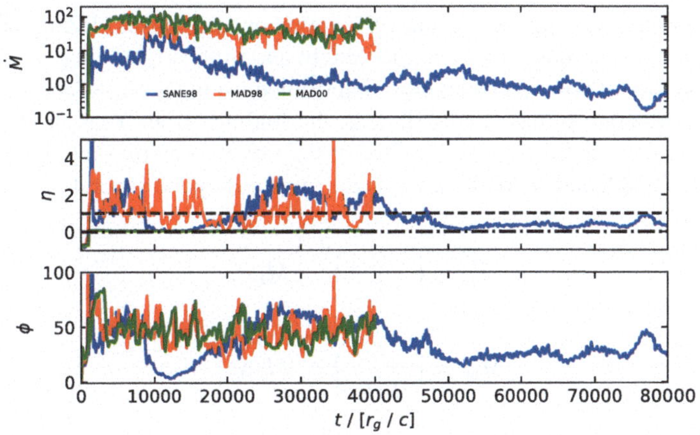

Fig. 8.6 Time evolution of the mass flux, energy extraction efficiency, and dimensionless magnetic flux ϕ. The blue, red, and green represent the SANE98, MAD98, and MAD00, respectively. Adapted from Yang et al. [104]

The plot shows that ϕ exhibits significant variation over time in the MAD state, with magnitudes of change much larger compared to the relatively stable and low values observed in the SANE state.

The presence of black hole spin can lead to the generation of relativistic outflows, known as jets, driven by magnetic fields. When a large-scale poloidal magnetic field crosses the ergosphere and threads the event horizon of a spinning black hole, it can extract rotational energy from the black hole, producing what is known as a Blandford-Znajek (BZ) jet [10]. The power of this jet is described by the BZ model [10, 92]:

$$P_{\rm BZ} = \frac{\kappa}{4\pi c}\Phi_{\rm H}^2\Omega_{\rm H}^2, \tag{8.13}$$

where κ is a numerical coefficient, typically around 0.5. $\Phi_{\rm H}$ represents the magnetic flux threading the event horizon, and $\Omega_{\rm H} = \frac{ac}{4r_{\rm H}}$ denotes the angular velocity of the event horizon. Equation 8.13 illustrates that the energy of the BZ jet increases with both the magnetic flux threading the event horizon and the spin of the black hole.

Winds and jets coexist in real outflows, necessitating a method to distinguish between them. Some works define the jet region as where the magnetization parameter $\sigma(= B^2/\rho) > 1$ [56]. Alternatively, Yang et al. [104] use magnetic field lines to delineate jets, defining them as regions where magnetic field lines connect to the ergosphere of the black hole, with the outer boundary anchored at the ergosphere with $\theta = 90°$. Consequently, the outflow beyond the jet region is classified as wind. This distinction is justified by Yang et al.'s study of the M87 jet, demonstrating that it is driven by the BZ mechanism [106]. They modeled the BZ jet's image and compared it with observed radio images of the M87 jet, showing that the BZ jet accurately reproduces its morphology, including its length and limb-brightening.

The mass flux of wind and jet can be determined using the trajectory method and Eq. 8.8. By integrating these methods with the findings from Yuan et al. [115] on SANE00, we can establish radial profiles of the inflow accretion rate for SANE00, SANE98, MAD00, and MAD98 at various distances beyond several r_g.

$$\dot{M}_{\rm in} = \left(\frac{r}{4.8r_s}\right)^{0.54} \dot{M}_{\rm BH}, \tag{8.14}$$

$$\dot{M}_{\rm in} = \left(\frac{r}{0.7r_s}\right)^{0.91} \dot{M}_{\rm BH}, \tag{8.15}$$

$$\dot{M}_{\rm in} = \left(\frac{r}{2.8r_s}\right)^{0.18} \dot{M}_{\rm BH}, \tag{8.16}$$

$$\dot{M}_{\rm in} = \left(\frac{r}{2.5r_s}\right)^{0.42} \dot{M}_{\rm BH}, \tag{8.17}$$

Similarly, the radial profiles of the mass flux of wind are,

$$\dot{M}_{\rm wind} = \left(\frac{r}{20r_s}\right)^{1.0} \dot{M}_{\rm BH}, \tag{8.18}$$

$$\dot{M}_{\rm wind} = \left(\frac{r}{15r_s}\right)^{1.16} \dot{M}_{\rm BH}, \tag{8.19}$$

$$\dot{M}_{\rm wind} = \left(\frac{r}{55r_s}\right)^{1.54} \dot{M}_{\rm BH}, \tag{8.20}$$

$$\dot{M}_{\rm wind} = \left(\frac{r}{30r_s}\right)^{1.26} \dot{M}_{\rm BH}, \tag{8.21}$$

respectively. By comparing SANE00 and MAD00, it is shown that near the black hole where the magnetic field is stronger, the wind is weaker. This phenomenon is likely due to the role of turbulence in wind generation, with SANE turbulence attributed to magneto-rotational instability (MRI), whereas in MAD, MRI is suppressed leading to the absence of significant turbulence [56, 65]. However, in regions farther from the black hole where the magnetic field strength diminishes, both SANE and MAD exhibit MRI which aids in wind production. Consequently, the power-law index of MAD00 (1.54) exceeds that of SANE00 (1.0). Similar trends are evident when comparing SANE98 and MAD98.

When comparing SAN00 with SANE98, it becomes evident that at smaller radii, higher spin enhances wind flux. This pattern is also observed in the comparison of MAD00 and MAD98, indicating that spin plays a crucial role in wind generation.

Notably, the power-law index from Eqs. 8.18–8.21 exceeds that from Eqs. 8.14–8.17, suggesting that beyond specific radii ($214\, r_g$ for SANE00, $10^6\, r_g$ for SANE98, $164\, r_g$ for MAD00, and $210\, r_g$ for MAD98), the mass flux of the outflow surpasses that of the inflow, which is unphysical. Therefore, the mass flux of the wind equals the inflow at these radii. Furthermore, Eq. 8.19 should only be applied within the Bondi radius, the outer limit of the hot accretion flow, typically located at $10^5\, r_g$.

As the magnetic field weakens at large radii, the wind velocity is primarily governed by the poloidal components [19, 20]. Therefore, particular attention is given to the poloidal velocity of the wind in these cases. For SANE00, MAD00, SANE98, and MAD98,

$$v_{\rm p}(r) = 0.21\, v_{\rm k}(r), \tag{8.22}$$

$$v_{\rm p}(r) = 0.66\, v_{\rm k}(r), \tag{8.23}$$

$$v_{\rm p}(r) = 0.24\, v_{\rm k}(r), \tag{8.24}$$

$$v_{\rm p}(r) = 0.64\, v_{\rm k}(r), \tag{8.25}$$

respectively. In the MAD model, the poloidal velocity of the wind is significantly higher than that in the SANE model. This disparity arises because the Poynting flux of the wind is converted into kinetic energy, consistent with findings by Yuan et al. [115] indicating that the Lorentz force plays a major role in accelerating the wind. The influence of spin on wind velocity is relatively minor, as the wind originates in regions at large radii where relativistic effects due to spin are less pronounced. Additionally, it has been noted that the rotational velocity of the MAD wind is much lower than the Keplerian velocity, which can be attributed to the strong outward magnetic pressure gradient force prevalent in the MAD scenario.

We can obtain the radial profiles of the flux of the momentum and kinetic energy of the wind and jet by the following formula,

$$\dot{E}_{\rm jet(wind)} = \frac{1}{2}\int \gamma\rho(r,\theta,\varphi)v_r^3(r,\theta,\varphi) \times (r^2 + a^2\cos\theta^2)\sin\theta d\theta d\varphi, \tag{8.26}$$

$$\dot{P}_{\rm jet(wind)} = \frac{1}{2}\int \gamma\rho(r,\theta,\varphi)v_r^2(r,\theta,\varphi) \times (r^2 + a^2\cos\theta^2)\sin\theta d\theta d\varphi, \tag{8.27}$$

Here $\gamma = 1/\sqrt{1 - v_r^2}$, v_r represents the radial three-velocity component in the Locally Non-Rotating Frame (LNRF). Since $v_\theta \ll v_r$, only v_r is considered here. Further details can be found in [104]. The results indicate that both black hole spin and magnetic field enhance the momentum and kinetic energy fluxes of the wind and jet. For non-spinning black holes, Yuan et al. [115] studied SANE00 and found that

$$\dot{E}_{\rm wind} \approx \frac{1}{1000}\, \dot{M}_{\rm BH} c^2, \tag{8.28}$$

Its energy flux corresponds closely with the requirements for large-scale AGN feedback simulations [18, 31]. Moreover, to be consistent with observations of isolated galaxies and galaxy clusters, the necessary "mechanical feedback efficiency" should fall within the range of approximately 10^3-10^4. Equation 8.28 naturally explains this required efficiency. Comparing the momentum and kinetic energy of the "disk jet" with that of the wind reveals that

$$\dot{P}_{\rm wind} \approx 15\, \dot{P}_{\rm disk\ jet}, \tag{8.29}$$

and,

$$\dot{E}_{\rm wind} \approx 3\, \dot{E}_{\rm disk\ jet}, \tag{8.30}$$

Whether considering momentum or kinetic energy, the flux of the wind is significantly greater than that of the disk jet. This finding is consistent for a non-spinning black hole, where the density of the disk jet is very low, with $\dot{M}_{\rm jet} \approx 1/20\, \dot{M}_{\rm wind}$, and velocities are only sub-relativistic, around $\sim 0.3-0.4\, c$. These results imply that the wind plays a more crucial role than the jet in AGN feedback. However, it is important to note that the black hole discussed here is non-spinning.

When a black hole has spin, a much more powerful BZ jet emerges. Its velocity is relativistic, reaching up to $0.9\, c$ even in the inner region. This jet carries significant energy, capable of extending to galactic scales, such as in M87, where it spans about 6,500 light-years. Due to the immense power of such jets, most AGN feedback studies focus solely on the jet, often assuming that the wind is insignificant. However, Yang et al. [104]'s research on high-spin black holes revealed surprising findings. They compared the momentum and kinetic energy fluxes of MAD98 and SANE98 at $r = 200\, r_g$. For SANE98, there are,

$$\dot{P}_{\rm wind} \approx 1.5\, \dot{P}_{\rm jet}, \tag{8.31}$$

and,

$$\dot{E}_{\rm jet} \approx 3.5\, \dot{E}_{\rm wind}, \tag{8.32}$$

For MAD98, there are

$$\dot{P}_{\rm wind} \approx \dot{P}_{\rm jet}, \tag{8.33}$$

and,

$$\dot{E}_{\rm jet} \approx 4\, \dot{E}_{\rm wind}, \tag{8.34}$$

These results show that even for rapidly rotating black holes, where the jet energy is the strongest, the kinetic flux of the jet is only about four times larger than that of the wind, whether in the MAD or SANE model. When considering the total energy flux (the sum of kinetic energy, Poynting flux, and enthalpy), the flux of the jet is only seven and ten times larger than that of the wind for SANE and MAD98, respectively. Moreover, the momentum of the wind exceeds that of the jet. In AGN feedback studies, momentum feedback may play a more critical role than energy feedback in several key aspects, such as the growth of black hole mass, AGN mass accretion rate, and suppression of the star formation rate in the central region of the galaxy [71].

Additionally, the wind's larger opening angle compared to the jet allows it to more effectively deposit energy into the interstellar medium. These findings suggest that wind plays a more significant role in AGN feedback than jets and should be considered in AGN feedback models rather than focusing solely on jets, as has been common in most previous studies.

However, due to the limitations of GRMHD simulations, these results are only applicable at $r = 200\,r_g$, making it uncertain if they hold at larger radii. Nonetheless, from a momentum perspective, the findings indicate that the momentum of the wind continues to grow as it propagates outward, whereas the momentum of the jet reaches saturation.

Equations 8.18–8.21 show that as the wind moves outward, the newly generated local wind continues to join. This raises the question: will this continuous generation of new wind persist indefinitely, or will it cease beyond a certain radius? To address this, Bu et al. [14, 15] conducted a series of 2D HD and MHD simulations to study accretion flow at very large radii. In their simulations, they considered the gravitational potential of both the black hole and the nuclear star cluster. They assumed that the velocity dispersion of the nuclear stars is constant with radius, leading to the potential of the star clusters:

$$\psi_{\rm star} = \sigma^2 In(r) + C, \tag{8.35}$$

where σ is the velocity dispersion of the stars, and C is a constant. Typical values for σ range from $100-400\,{\rm km\ s^{-1}}$ [52], thus $R_{\rm A}$, the radius where the gravity of the black hole equals that of the star cluster, can be defined as:

$$R_{\rm A} = \frac{GM_{\rm BH}}{\sigma^2}, \tag{8.36}$$

which is approximately $10^5-10^6\,r_s$, roughly equal to the Bondi radius of the accretion flow. For their standard model, the radial simulation range was $0.3-40\,R_{\rm A}$. They found that although the accretion rate decreased inward, trajectory analysis revealed almost no real outflow. Further analysis indicated that irrespective of the magnetic field, the region beyond $R_{\rm A}$ is convectively unstable. These results suggest that the inward decrease in accretion rate here is caused by convection, which is consistent with the CDAF model.

When they moved the simulation area closer to the black hole, changing the radial simulation range to $0.02-4\,R_{\rm A}$ while keeping other parameters unchanged, they found that the accretion flow within $0.4\,R_{\rm A}$ is convectively stable and exhibits very strong wind, comparable to that described by Eq. 8.18. This indicates that within a certain radius, strong wind generation persists, but beyond that radius, the dynamics are dominated by convection rather than continuous outflow generation.

The difference between the two simulations lies solely in the radial range considered, which alters the gravitational potential affecting the accretion flow. In the simulation closer to the black hole, the black hole's gravitational potential dominates over that of the star cluster, leading to the generation of new local wind. In the simulation farther from the black hole, the star cluster's gravity gradually surpasses that of the black hole and becomes dominant. Beyond $r > R_A$, the accretion flow becomes convectively unstable, and no new local wind is produced. This instability may result from changes in the gravitational potential, which in turn alter turbulence properties and thus affect wind generation. Since R_A is close to the Bondi radius, no new wind is generated locally outside the Bondi radius. Therefore, Eqs. 8.18–8.21 are valid only within the Bondi radius.

Due to technical limitations, simulations typically extend only to about $10^3 - 10^4\, r_g$. However, how far can the wind propagate outward once it escapes this boundary? To explore this question, Cui et al. [19, 20] investigated the large-scale behavior of winds generated by accretion flows.

They initially studied wind propagation on a large scale using HD simulations, which included analytic and 2D numerical simulations. Their approach considered not only the gravitational potential of the black hole but also that of the host galaxy. They employed a small-scale hot accretion flow [115] as the inner boundary conditions, where they found:

$$v_r \approx v_{r_0}, \tag{8.37}$$

with v_{r_0} representing the wind velocity at the boundary ($r_0 \equiv 10^3\, r_g$). Equation 8.37 illustrates that the wind velocity remains nearly constant as it propagates outward, as the increased gravitational potential is almost offset by enthalpy and rotational energy. They also found that the gravitational potential of the galaxy did not significantly alter the radial velocity of the wind, likely due to the high initial velocity of the wind from the small-scale source. They concluded that the wind can extend to the scale of the galaxy, though this was considered an upper limit without accounting for the interstellar medium (ISM).

They then examined the role of magnetic fields, employing 1D MHD simulations for simplicity. Despite this simplification, they maintained the wind properties from Yuan et al. [115] at the inner boundary. Their findings indicated that $v_p \propto$ const., meaning the poloidal velocity of the wind remains largely unchanged as it propagates outward. Results from the MHD simulations mirrored those from HD simulations: in cases of weak magnetization, thermal pressure, and magnetic forces jointly accelerated the wind, while in highly magnetized scenarios, magnetic forces dominated and accelerated the wind primarily. Higher magnetization led to increased terminal velocities, with terminal wind velocities reaching approximately $0.016\,c$ for the fiducial model parameters.

It is important to note that their MHD simulations did not incorporate the gravitational potential of the galaxy, despite HD simulations showing minimal impact on wind velocities. Overall, their work demonstrated that small-scale winds originating from accretion flows can propagate far beyond the Bondi radius under favorable conditions.

8.3 Observations of Wind from Hot Accretion Flows

Theoretical research suggests the presence of outflows in the ADAF model with low mass accretion rates. However, the question remains: are there any corresponding observational discoveries? Due to the high temperature of the wind within the hot accretion flow, where gas is nearly entirely ionized, observation becomes challenging. Nonetheless, recent studies have unveiled the presence of winds in several low-luminosity AGNs, including Sagittarius A* (Sgr A*) [53, 100], M81 [85, 87], NGC 7213 [86, 87], M87 [73], and M32 [75], among others [17, 94]. Additionally, winds have been observed in low-mass X-ray binaries (LMXBs) during their hard state [39].

Sgr A* stands as the initial radio source detected [8] at the center of our Galaxy. Observations have unveiled its mass, estimated to be $(4.1 \pm 0.4) \times 10^6 M_\odot$ [33], confined within tens of light-hours [32], confirming its status as a SMBH. Typically, the outer boundary of an accretion disk surrounding a central black hole is described by the Bondi radius [13]. This radius denotes the equilibrium between the thermal expansion energy of the gas and the gravitational potential energy exerted by the black hole. For Sgr A*, $R_B \sim 10^5 R_s \approx 0.04\ pc$. The *Chandra X-ray Observatory* furnishes insights into the gas density and temperature near the Bondi radius, enabling estimation of its mass accretion rate at this boundary, $\dot{M}_B \sim 10^{-5} M_\odot\,\mathrm{yr}^{-1}$ [6]. Analysis of Sgr A*'s spectral energy distribution reveals its remarkably low bolometric luminosity, $L_{bol} \sim 10^{36} \mathrm{erg\,s}^{-1} \sim 2 \times 10^{-9} L_{\mathrm{Edd}}$. Should the matter's accretion rate into the black hole equate to its Bondi radius accretion rate, its radiation efficiency would be markedly low, at only 10^{-6}, far below the typical value for a traditional standard thin disk (approximately 10%). This discrepancy strongly suggests the presence of a hot accretion flow. Furthermore, its spectral energy distribution starkly differs from that of a standard thin disk, displaying a multi-temperature blackbody spectrum.

Polarization observations of Sgr A* at 227 and 343 GHz using the Submillimeter Array polarimeter revealed a mean rotation measure of $(-5.6 \pm 0.7) \times 10^5 \mathrm{rad\,m}^{-2}$. This rotation measure sets constraints on the accretion rate within the innermost region ($r \lesssim 100 R_{\mathrm{s}}$). For magnetic fields near equipartition, predominantly ordered and radially oriented, the accretion rate is constrained to be less than $2 \times 10^{-7} M_\odot\,\mathrm{yr}^{-1}$. Even in scenarios with nearly subequipartition, disordered, or toroidal fields, a lower limit of $2 \times 10^{-9} M_\odot\,\mathrm{yr}^{-1}$ is established [54]. Recent work by the Event Horizon Telescope Collaboration [26] indicates an accretion rate near the event horizon of Sgr A* ranging from $(5.2 - 9.5) \times 10^{-9} M_\odot\,\mathrm{yr}^{-1}$. Notably, this accretion rate near the black hole is considerably smaller than the accretion rate at the

Bondi radius, suggesting a decrease in the accretion rate inward. This result aligns with numerical simulations, hinting at the possibility of outflow.

In 2013, Wang et al. [100] analyzed high-resolution X-ray emission data from 3 mega-seconds of Chandra observations, revealing a significant outflow (wind) emanating from the center of the Milky Way. The quiescent spectrum of Sgr A* exhibits notable emission lines. In their investigation into the origins of these emissions, they initially discounted the stellar coronal hypothesis due to its inability to account for the predicted Fe Kα emission. The presence of a very faint H Fe-like Kα line suggested either a radiatively inefficient accretion flow (RIAF) characterized by a very low plasma mass fraction at $kT \gtrsim 9$ keV or the existence of a potent outflow at a radius $r \gtrsim 10^4 R_s$. Employing quantitative analyses, they dismissed the non-outflow scenario characterized by a radial plasma density profile of $n \propto r^{-3/2+s}$, with $s = 0$, as it would overestimate the H-like Fe Kα line. Instead, they found that a RIAF with a flat-density profile ($s \sim 1$) could aptly fit both continuum and line data. This suggests a decrease in mass accretion rate inward, accompanied by the presence of outflow.

Tombesi et al. [94] conducted an analysis of Fe K absorption lines observed in 26 radio-loud AGN spectra from *XMM-Newton* and *Suzaku*. They identified ultra-fast outflows (UFOs) with a broad range of velocity distributions, spanning from $1000\,\mathrm{km\,s^{-1}}$ to $0.4\,c$. These outflows exhibited characteristics of high ionization and significant column densities. The inclination angles of the outflows ranged from approximately 10° to 70°. These findings suggest a plausible origin from a hot accretion flow, particularly given recent observations indicating that low-luminosity AGNs tend to be radio-loud [30, 38, 62], implying a potential association between radio-loudness and hot accretion flows. Furthermore, observational evidence suggests that cool thin disks typically exhibit weak or absent jets [114].

Cheung et al. [17] observed a bisymmetric ionized gas emission feature spanning the entire galaxy in low-luminosity AGN. This feature was found to be aligned with gas velocity gradients, indicating the presence of multi-component gas with varying temperatures and velocities. Their model of centrally driven winds successfully reproduced these observed features. Their results not only supported the existence of winds but also implied that low-luminosity active galactic nuclei possess sufficient energy to generate winds, and heat surrounding cold gas, and potentially suppress star formation.

However, the aforementioned outflow observations were primarily obtained indirectly. In 2021, Shi et al. [85] provided direct evidence for the existence of wind from M81*, one of the nearest SMBHs with a mass of approximately $M_{\rm BH} \approx 10^7 M_\odot$ [22]. Despite its very low bolometric luminosity, roughly $L_{\rm bol} = 3 \times 10^{-5} L_{\rm Edd}$ [67], M81* exhibits a pronounced radio jet [9] and possibly lacks the classic thin accretion disk [37, 110], indicating that it may be powered by a radiatively inefficient, hot accretion flow.

Shi et al. [85] analyzed the X-ray spectrum of M81* using high-resolution data from the *Chandra* observatory [85]. They detected emission lines but no absorption lines, with four prominent emission lines located at energies of 6.40, 6.69, 6.90, and

7.05 keV. These emission lines were interpreted as arising from a pair of collisionally-ionized, optically-thin plasmas-one red-shifted and the other blue-shifted. Treating the 6.90/7.05 keV lines as symmetrically redshifted/blueshifted Fe XXVI lines, they derived a best-fitting absolute line-of-sight velocity of $(2.8 \pm 0.2) \times 10^3 \mathrm{km\,s^{-1}}$. Additionally, they determined a best-fitting temperature of 1.3×10^8 K based on the relative intensity between Fe XXVI Lyα and Fe XXV Kα. Such high velocities and temperatures are inconsistent with stellar activities or accretion inflow, suggesting that only winds originating from hot accretion flows could produce such features.

By conducting MHD simulations of the wind launched from the hot accretion flow in M81*, Shi et al. obtained spatial distributions of density, velocity, temperature, and magnetic field. They utilized this information to generate synthetic X-ray spectra, which exhibited two main features consistent with the observed results: a double-peak Fe XXVI Lyα profile and a high Fe XXVI-to-Fe XXV flux ratio.

Conclusively, the observed high-velocity and high-temperature plasma represents the wind originating from the hot accretion flow, marking the first direct observation of such a phenomenon. Shi et al. [87] subsequently identified novel blue-shifted emission lines, hypothesizing that these emanated from circumnuclear gas shock-heated by the hot wind. Their findings provide compelling evidence for energy and momentum feedback from the AGN's hot wind.

8.3.1 Application of the Wind from Hot Accretion Flows

An intriguing aspect of Sgr A* is that, despite its current low mass accretion rate of only $\dot{M}_{\rm BH} \sim 10^{-6} \dot{M}_{\rm Edd}$, numerous observations indicate that it was significantly more active in the past few million years, with an accretion rate approximately four orders of magnitude higher than it is now [95]. A clear piece of observational evidence supporting this increased activity is the recently discovered *Fermi* Bubbles. Su et al. [91] identified two giant gamma-ray bubbles above and below the Galactic plane using *Fermi*-LAT. In galactic coordinates, each bubble has a height of approximately 40° and a width of 50°. The surface brightness of these bubbles is roughly uniform with sharp edges. In the $1 - 100$ eV band, the total luminosity of the bubbles is $4 \times 10^{37}\,\mathrm{erg\,s^{-1}}$, and the total energy of the two bubbles is about $10^{55} - 10^{56}$ erg.

To explain the observed *Fermi* Bubbles, several models have been proposed, such as the "jet" model [34, 35] and the "quasar outflow" model [118]. These models typically explain the bubbles' morphology through interactions between a jet or wind (from super-Eddington accretion flow) and the ISM. However, wind from hot accretion flows could also produce such bubbles. Research by Mou et al. [59, 60] demonstrates that wind from a hot accretion flow can adequately explain the observed bubbles' morphology and gamma-ray spectral energy distribution. Although the accretion rate of Sgr A* was $\sim 10^{-2}\,\dot{M}_{\rm Edd}$ at that time, it remained in a hot accretion mode. At this accretion rate, the accretion disk typically consists of a hot accretion flow near the black hole and a thin disk further away, with the transition radius $R_{\rm tr}$ being a function of the accretion rate. For $\sim 10^{-2}\,\dot{M}_{\rm Edd}$, $R_{\rm tr}$ is approximately $200\,r_{\rm s}$ [114].

The properties of the wind are not arbitrary but are based on numerical simulation results of the hot accretion flow. The wind parameters, including mass flux and velocity, were derived from the results of Yuan et al. [115], such as Eqs. 8.9 and 8.11. The wind used in their simulations has a mass flux of about $18\%\dot{M}_{\rm Edd}$ and a velocity of approximately $v_{\rm k}(R_{\rm tr}) \sim 0.046\,c$, resulting in kinetic energy of the thermal gas in the wind of 10^{42} erg s^{-1}. They conducted a series of 3D MHD simulations based on these wind parameters. By using the boundary between the shocked ISM and the shocked wind as the edge of the bubbles, they found that the morphology of the bubbles closely matched the observations.

To calculate the gamma-ray spectra, cosmic rays (CRs), mainly CR protons, were injected into the wind at the beginning of the simulation, with about 50% of the wind's power. The thermal gas and CR protons were simulated simultaneously. The gamma-ray spectrum, obtained by combining the CR results with the "hadronic" model, was highly consistent with observations. Additionally, their results showed a slight limb-brightened surface brightness, consistent with the observed blurred limb-brightened feature reported by [4].

In their work, they ejected the wind isotropically without considering the jet. They posited that the jet would pass through a narrow low-density channel, allowing it to move freely without generating a bubble-like structure [97]. Additionally, they calculated the energy transformation efficiency, finding that approximately 60% of the wind's energy is transferred to the ISM, likely due to the wind's large opening angle. These results suggest that we should seriously consider the possibility that cavities and bubbles observed in other environments, such as galaxy clusters, may be formed by winds rather than jets.

8.4 Wind from Super-Eddington Accretion Flows

In this section, we primarily investigate wind originating from super-Eddington accretion flow. Observational data suggest that certain quasars experience super-Eddington accretion phases during their growth [49], particularly in their early developmental stages [45]. The feedback mechanisms of AGN profoundly influence the formation and evolution of their host galaxies [28, 52, 61]. Although AGNs may only exhibit super-Eddington accretion sporadically, the resulting outflows and radiation remain substantial, potentially exerting significant feedback effects. The outflows from super-Eddington accretion flow impact the observed spectra and have been invoked to explain certain TDEs [21, 93].

Super-Eddington accretion flows, characterized by higher accretion rates, exhibit distinct properties in both their accretion flow and outflow compared to the thin disk model. Consequently, the slim disk model was proposed [1]. However, due to its one-dimensional nature, this model failed to provide comprehensive insights into the properties of the outflow. Subsequent analytical studies [16, 50] demonstrated the existence of winds. For instance, they found that when the disk accretes at its outer edge at moderate Eddington scale rates (up to $\sim$100), a significant portion of the inflow is expelled as outflow, with only a fraction accreted by the black hole. The outflow typically exhibits slightly subrelativistic speeds, reaching velocities of $\sim$0.1$-$0.2 c for Eddington accretion factors $\dot{m}_{\rm acc} \sim 10 - 100$, and around $\sim$1500 kms^{-1} for $\dot{m}_{\rm acc} \sim 10^4$. However, significant advancements emerged from more realistic numerical simulations [5, 24, 40, 46, 58, 68, 81, 105]. These simulations have provided substantial progress in understanding the dynamics of super-Eddington accretion flow and their associated outflows.

The pioneering numerical simulations on radiation hydrodynamics (RHD) of super-Eddington accretion flows were conducted by Eggum et al. [24]. They observed the emergence of high-velocity outflows near the axis of rotation. Subsequently, Ohsuga et al. [68] extended this seminal work by scaling up the simulation domain and prolonging the runtime to attain a quasi-steady state. In their investigation, they confirmed the presence of radiation-driven wind. Subsequent advancements included the incorporation of the magnetic field in simulations [69, 70], revealing that radiation pressure exerted at the disk's surface significantly exceeded gravity, thereby driving outflows.

However, it's noteworthy that these simulations relied on the flux-limited diffusion (FLD) method to handle radiation transport. While effective for optically thick accretion disks, the FLD approximation assumes isotropic emission relative to the fluid frame. This approximation is adequate for optically thick regions but may lead to inaccuracies in optically thin coronal regions where radiation tends to be anisotropic. Thus, despite their larger contributions, these simulations may provide only a rough approximation due to the limitations of the FLD method.

Subsequently, Jiang et al. [46] conducted a comprehensive global three-dimensional radiation magneto-hydrodynamical simulation. By directly solving the time-dependent radiative transfer equations within the Newtonian limit, they identified prominent outflow regions within approximately 45° from the rotation axis. These outflows, primarily propelled by radiation forces, were observed to be collimated by magnetic pressure [70]. Notably, the wind velocities ranged from about 0.1 to 0.4 times the speed of light. Their findings indicated that for a stellar-mass black hole accreting at an Eddington rate of $\sim$220 $L_{\rm Edd}/c^2$, radiation-driven outflows predominantly formed near the rotation axis, with the total mass flux of outflow constituting approximately 29% of the net accretion rate.

While the pseudo-Newtonian potential can adequately represent black holes without spin to some extent, the influence of general relativity (GR) effects on the spectrum cannot be neglected, particularly considering that a significant portion of radiation emanates from the inner regions. Additionally, GR effects must be considered in discussions concerning jet formation and the impact of black hole spin.

To address these complexities, a series of General Relativistic Radiation Magnetohydrodynamics (GRRMHD) codes, such as KORAL [80] and HARMRAD [57], were developed. These codes employed the M1 closure to effectively handle radiative momentum equations, offering a suitable approximation for both optically thick and optically thin regions under GR. Sadowski et al. [82] introduced a novel mean-field magnetic dynamo into KORAL, enabling 2D axisymmetric simulations to achieve a larger steady-state radius. They observed that for non-spinning stellar-mass black holes ($10\,M_\odot$), the relative strength of the wind mass flux remained relatively constant across varying accretion rates ($2 < \dot{M}/\dot{M}_{\rm Edd} < 500$), with the outflow rate approximately matching the net accretion rate at $40\,r_{\rm g}$.

Moreover, Utsumi et al. [96] investigated the impact of black hole spin on the outflow of super-Eddington accretion flows through a 2D axisymmetric GR-RMHD simulation. They observed the ejection of disk material from the disk surface, primarily driven by radiative forces. However, at very high spins, the magnetic force also became significant. Notably, even in cases where the accretion flow exhibited a SANE rather than MAD configuration, the magnetic field's influence became pronounced at high spin values.

Given the pivotal role of radiation, the ratio of radiation pressure to gas pressure differs significantly between the accretion flows of stellar-mass black holes and SMBHs, even at the same Eddington accretion rate. To investigate this, Jiang et al. [47] conducted 3D radiation magnetohydrodynamics (RMHD) simulations of super-Eddington accretion flows around SMBHs. Their findings revealed that the outflow mass flux ranged from 15% to 50% of the net accretion rate, with velocities in the funnel region reaching $0.3-0.4c$. These results underscored that, for equivalent accretion rates, AGNs exhibit larger outflow mass fluxes compared to stellar-mass black holes.

Additionally, Sadowski et al. [83] investigated mildly SMBHs ($3 \times 10^5\,M_\odot$) through 2D GRRMHD simulations at $45\,\dot{M}_{\rm Edd}$. They identified high-speed outflows, with velocities increasing gradually from the disk to the axis, reaching speeds of up to $0.5c$. The density-weighted velocity averaged approximately $0.3c$. These high-speed outflows in the funnel regions were primarily radiation-driven, attributed to the absence of black hole spin and the presence of minimal magnetic fields. For higher accretion rates ($310\,\dot{M}_{\rm Edd}$ and $4800\,\dot{M}_{\rm Edd}$), the termination velocity decreased to $0.3c$, possibly due to increased mass loading in the funnel region.

While numerous simulations have hinted at the existence of winds, few studies have comprehensively examined and compared them to jets. Some research has suggested that the inflow rate of the accretion flow follows a power-law distribution $\dot{M}_{\rm in} \propto r^s$, with an index s typically ranging from $\sim$0.4 to 1.0 [40, 105, 107]. This implies that only a fraction of matter is accreted into the black hole. However, accurately determining the true outflow is challenging due to the turbulent nature of the accretion flow. Conventionally, the Be parameter has been employed to identify outflows, albeit with limitations. It tends to underestimate the mass flux of the outflow and can only provide a lower limit. Studies on both sub-Eddington and super-Eddington accretion flows have revealed that the Be parameter is not a robust criterion for distinguishing outflows [109, 115].

Yang et al. [105] employed the "virtual particle trajectory" method to analyze super-Eddington accretion flow at the MAD state in 3D GRRMHD. This method, compared to traditional streamline approaches, offers improved discernment between real outflows (wind and jet) and turbulent flows. Moreover, the definition of jets and winds has been a subject of ambiguity. Jets are typically delineated by magnetization parameters ($\sigma > 1$) [56] and Bernoulli constant ($Be > 0.05$) [57, 83]. In Yang et al.'s study [104], the outermost magnetic field line connected to the black hole event horizon (away from the rotation axis) was deemed the boundary of the BZ jet. Subsequent research corroborated that the observed jet indeed corresponded to a BZ jet [106].

Furthermore, the results of Yang et al. [105] revealed that the wind mass flux follows a power-law distribution

$$\dot{M}_{\rm wind} = \left(\frac{r}{45r_{\rm s}}\right)^{0.83} \dot{M}_{\rm BH}, \tag{8.38}$$

and the mass-flux-weighted poloidal velocity of approximately $0.15\,c$. The momentum and kinetic energy flux of the wind at $r = 200\,r_{\rm g}$ are

$$\dot{P}_{\rm wind} \approx 0.15\,\dot{M}_{\rm BH}c, \tag{8.39}$$

$$\dot{E}_{\rm wind} \approx 0.01\,\dot{M}_{\rm BH}c^2, \tag{8.40}$$

respectively. Additionally, a comparison of wind and jet properties indicated that the momentum flux of wind exceeds that of jets ($\dot{P}_{\rm wind} \approx 1.2\,\dot{P}_{\rm jet}$), while the total energy flux of jets is only about three times greater than that of wind ($\dot{E}_{\rm jet} \approx 2.5\,\dot{E}_{\rm wind}$). In MAD disks, the primary acceleration force for both wind and jets is the Lorentz force, with radiation and centrifugal forces serving as secondary accelerators.

Recently, several GRRMHD codes have been developed, incorporating magnetic fields, GR effects, and more accurate radiation treatments [5, 42, 79, 103]. Asahina et al. [5] conducted a comparative study between methods directly solving radiative transfer equations to obtain the Eddington tensor and employing the M1 closure in GRRMHD simulations. Their findings revealed the emergence of accretion disks and disk winds in both scenarios, with no significant disparities observed in accretion rates, outflow rates, or luminosity. However, discrepancies were noted in the radiation field near the rotation axis, with the M1 method amplifying the radial component of the radiation flux.

Observations of winds from Super-Eddington AGNs remain limited. While investigating the X-ray spectrum of the narrow-line Seyfert 1 galaxy I Zwicky 1, Ding et al. [23] discovered multiple variable blue-shifted absorption features attributed to outflowing material along the line of sight. These features are believed to originate from ionization winds and ultrafast outflows. Notably, the disk ionization parameter is correlated with the outflow velocity, following a relationship of $\xi \propto v_{\rm wind}^{3.24}$, indicative of super-Eddington winds. The wind velocity ranges from approximately 0.1 to $0.3\,c$.

In a separate study, Vietri et al. [98] identified three Broad Absorption Line (BAL) UFOs ($\sim$0.13–0.18 c) during observations of hyper-luminous quasars. These UFOs exhibited variations in intensity and shape, along with instances of disappearance and reappearance at different temporal intervals.

Acknowledgements This work was supported in part by the Natural Science Foundation of China (grants 12133008, 12192220, and 12192223). H.Y. acknowledges the support from the National Key R&D Program of China (No. 2023YFE0101200), the National Natural Science Foundation of China (Grant No. 12273022), and the Shanghai Municipality orientation program of Basic Research for International Scientists (grant no. 22JC1410600).

References

1. M.A. Abramowicz, B. Czerny, J.P. Lasota et al., APJ **332**, 646 (1988). https://doi.org/10.1086/166683
2. M.A. Abramowicz, W. Kluźniak, J.-P. Lasota, A & A **396**, L31 (2002). https://doi.org/10.1051/0004-6361:20021645
3. M.A. Abramowicz, P.C. Fragile, Living Rev. Relativ. **16**, 1 (2013). https://doi.org/10.12942/lrr-2013-1
4. M. Ackermann, A. Albert, W.B. Atwood et al., APJ **793**, 64 (2014). https://doi.org/10.1088/0004-637X/793/1/64
5. Y. Asahina, K. Ohsuga, APJ **929**, 93 (2022). https://doi.org/10.3847/1538-4357/ac5d37
6. F.K. Baganoff, Y. Maeda, M. Morris et al., APJ **591**, 891 (2003). https://doi.org/10.1086/375145
7. M.C. Begelman, MNRAS **420**, 2912 (2012). https://doi.org/10.1111/j.1365-2966.2011.20071.x
8. B. Balick, R.L. Brown, APJ **194**, 265 (1974). https://doi.org/10.1086/153242
9. M.F. Bietenholz, N. Bartel, M.P. Rupen, APJ **532**, 895 (2000). https://doi.org/10.1086/308623
10. R.D. Blandford, R.L. Znajek, MNRAS **179**, 433 (1977). https://doi.org/10.1093/mnras/179.3.433
11. R.D. Blandford, M.C. Begelman, MNRAS **303**, L1 (1999). https://doi.org/10.1046/j.1365-8711.1999.02358.x
12. R.D. Blandford, M.C. Begelman, MNRAS **349**, 68 (2004). https://doi.org/10.1111/j.1365-2966.2004.07425.x
13. H. Bondi, MNRAS **112**, 195 (1952). https://doi.org/10.1093/mnras/112.2.195
14. D.-F. Bu, F. Yuan, Z.-M. Gan et al., APJ **818**, 83 (2016). https://doi.org/10.3847/0004-637X/818/1/83
15. D.-F. Bu, F. Yuan, Z.-M. Gan et al., APJ **823**, 90 (2016). https://doi.org/10.3847/0004-637X/823/2/90
16. X. Cao, W.-M. Gu, APJ **936**, 141 (2022). https://doi.org/10.3847/1538-4357/ac8980
17. E. Cheung, K. Bundy, M. Cappellari et al., Nature **533**, 504 (2016). https://doi.org/10.1038/nature18006
18. L. Ciotti, J.P. Ostriker, D. Proga, APJ **717**, 708 (2010). https://doi.org/10.1088/0004-637X/717/2/708
19. C. Cui, F. Yuan, B. Li, APJ **890**, 80 (2020). https://doi.org/10.3847/1538-4357/ab6e6e
20. C. Cui, F. Yuan, APJ **890**, 81 (2020). https://doi.org/10.3847/1538-4357/ab6e6f
21. L. Dai, J.C. McKinney, N. Roth et al., APJL **859**, L20 (2018). https://doi.org/10.3847/2041-8213/aab429
22. N. Devereux, H. Ford, Z. Tsvetanov et al., Astron. J. **125**, 1226 (2003). https://doi.org/10.1086/367595
23. Y. Ding, R. Li, L.C. Ho et al., APJ **931**, 77 (2022). https://doi.org/10.3847/1538-4357/ac6955
24. G.E. Eggum, F.V. Coroniti, J.I. Katz, APJ **330**, 142 (1988). https://doi.org/10.1086/166462
25. K. Akiyama, J.C. Algaba et al., Event horizon telescope collaboration. APJL **910**, L13 (2021). https://doi.org/10.3847/2041-8213/abe4de
26. K. Akiyama, A. Alberdi et al., Event horizon telescope collaboration. APJ **930**, L16 (2022). https://doi.org/10.3847/2041-8213/ac6672

27. K. Akiyama, A. Alberdi et al., Event horizon telescope collaboration. APJL **964**, L26 (2024). https://doi.org/10.3847/2041-8213/ad2df1
28. A.C. Fabian, ARA & A **50**, 455 (2012). https://doi.org/10.1146/annurev-astro-081811-125521
29. S. Fabrika, ASPRv **12**, 1 (2004). https://doi.org/10.48550/arXiv.astro-ph/0603390
30. H. Falcke, S. Markoff, Astron. Astrophys. **362**, 113 (2000). https://doi.org/10.48550/arXiv.astro-ph/0102186
31. M. Gaspari, F. Brighenti, P. Temi, MNRAS **424**, 190 (2012). https://doi.org/10.1111/j.1365-2966.2012.21183.x
32. R. Genzel, F. Eisenhauer, S. Gillessen, Rev. Mod. Phys. **82**, 3121 (2010). https://doi.org/10.1103/RevModPhys.82.3121
33. A.M. Ghez, G. Duchêne, K. Matthews et al., APJ **586**, L127 (2003). https://doi.org/10.1086/374804
34. F. Guo, W.G. Mathews, APJ **756**, 181 (2012). https://doi.org/10.1088/0004-637X/756/2/181
35. F. Guo, W.G. Mathews, G. Dobler et al., APJ **756**, 182 (2012). https://doi.org/10.1088/0004-637X/756/2/182
36. J.F. Hawley, S.A. Balbus, J.M. Stone, APJL **554**, L49 (2001). https://doi.org/10.1086/320931
37. L.C. Ho, A.V. Filippenko, W.L.W. Sargent, APJ **462**, 183 (1996). https://doi.org/10.1086/177140
38. L.C. Ho, APJ **564**, 120 (2002). https://doi.org/10.1086/324399
39. J. Homan, J. Neilsen, J.L. Allen et al., APJL **830**, L5 (2016). https://doi.org/10.3847/2041-8205/830/1/L5
40. H. Hu, K. Inayoshi, Z. Haiman et al., APJ **934**, 132 (2022). https://doi.org/10.3847/1538-4357/ac75d8
41. J. Li, J. Ostriker, R. Sunyaev, APJ **767**, 105 (2013). https://doi.org/10.1088/0004-637X/767/2/105
42. M.T.P. Liska, N. Kaaz, G. Musoke et al., APJL **944**, L48 (2023). https://doi.org/10.3847/2041-8213/acb6f4
43. I.V. Igumenshchev, APJL **577**, L31 (2002). https://doi.org/10.1086/344148
44. I.V. Igumenshchev, R. Narayan, M.A. Abramowicz, APJ **592**, 1042 (2003). https://doi.org/10.1086/375769
45. K. Inayoshi, E. Visbal, Z. Haiman, ARA & A **58**, 27 (2020). https://doi.org/10.1146/annurev-astro-120419-014455
46. Y.-F. Jiang, J.M. Stone, S.W. Davis, APJ **796**, 106 (2014). https://doi.org/10.1088/0004-637X/796/2/106
47. Y.-F. Jiang, J.M. Stone, S.W. Davis, APJ **880**, 67 (2019). https://doi.org/10.3847/1538-4357/ab29ff
48. Y. Kato, S. Mineshige, K. Shibata, APJ **605**, 307 (2004). https://doi.org/10.1086/381234
49. B.C. Kelly, Y. Shen, APJ **764**, 45 (2013). https://doi.org/10.1088/0004-637X/764/1/45
50. A. King, S.I. Muldrew, MNRAS **455**, 1211 (2016). https://doi.org/10.1093/mnras/stv2347
51. A. Koratkar, O. Blaes, PASP **111**, 1 (1999). https://doi.org/10.1086/316294
52. J. Kormendy, L.C. Ho, ARA & A **51**, 511 (2013). https://doi.org/10.1146/annurev-astro-082708-101811
53. R.-Y. Ma, S.R. Roberts, Y.-P. Li et al., MNRAS **483**, 5614 (2019). https://doi.org/10.1093/mnras/sty3039
54. D.P. Marrone, J.M. Moran, J.-H. Zhao et al., APJ **654**, L57 (2007). https://doi.org/10.1086/510850
55. J.E. McClintock, R. Narayan, J.F. Steiner, SSRv **183**, 295 (2014). https://doi.org/10.1007/s11214-013-0003-9
56. J.C. McKinney, A. Tchekhovskoy, R.D. Blandford, MNRAS **423**, 3083 (2012). https://doi.org/10.1111/j.1365-2966.2012.21074.x
57. J.C. McKinney, A. Tchekhovskoy, A. Sadowski et al., MNRAS **441**, 3177 (2014). https://doi.org/10.1093/mnras/stu762

58. J.C. McKinney, L. Dai, M.J. Avara, MNRAS **454**, L6 (2015). https://doi.org/10.1093/mnrasl/slv115
59. G. Mou, F. Yuan, D. Bu et al., APJ **790**, 109 (2014). https://doi.org/10.1088/0004-637X/790/2/109
60. G. Mou, F. Yuan, Z. Gan et al., APJ **811**, 37 (2015). https://doi.org/10.1088/0004-637X/811/1/37
61. T. Naab, J.P. Ostriker, ARA & A **55**, 59 (2017). https://doi.org/10.1146/annurev-astro-081913-040019
62. N.M. Nagar, H. Falcke, A.S. Wilson et al., APJ **542**, 186 (2000). https://doi.org/10.1086/309524
63. R. Narayan, I. Yi, APJL **428**, L13 (1994). https://doi.org/10.1086/187381
64. R. Narayan, I.V. Igumenshchev, M.A. Abramowicz, APJ **539**, 798 (2000). https://doi.org/10.1086/309268
65. R. Narayan, I.V. Igumenshchev, M.A. Abramowicz, PASJ **55**, L69 (2003). https://doi.org/10.1093/pasj/55.6.L69
66. R. Narayan, A. SÄ dowski, R.F. Penna et al., MNRAS **426**, 3241 (2012). https://doi.org/10.1111/j.1365-2966.2012.22002.x
67. R.S. Nemmen, T. Storchi-Bergmann, M. Eracleous, MNRAS **438**, 2804 (2014). https://doi.org/10.1093/mnras/stt2388
68. K. Ohsuga, M. Mori, T. Nakamoto et al., APJ **628**, 368 (2005). https://doi.org/10.1086/430728
69. K. Ohsuga, S. Mineshige, M. Mori et al., PASJ **61**, L7 (2009). https://doi.org/10.1093/pasj/61.3.L7
70. K. Ohsuga, S. Mineshige, APJ **736**, 2 (2011). https://doi.org/10.1088/0004-637X/736/1/2
71. J.P. Ostriker, E. Choi, L. Ciotti et al., APJ **722**, 642 (2010). https://doi.org/10.1088/0004-637X/722/1/642
72. B. Pang, U.-L. Pen, C.D. Matzner et al., MNRAS **415**, 1228 (2011). https://doi.org/10.1111/j.1365-2966.2011.18748.x
73. J. Park, K. Hada, M. Kino et al., APJ **871**, 257 (2019). https://doi.org/10.3847/1538-4357/aaf9a9
74. U.-L. Pen, C.D. Matzner, S. Wong, APJL **596**, L207 (2003). https://doi.org/10.1086/379339
75. S. Peng, Z. Li, L.O. Sjouwerman et al., APJ **894**, 61 (2020). https://doi.org/10.3847/1538-4357/ab855d
76. R.F. Penna, A. Kulkarni, R. Narayan, A & A **559**, A116 (2013). https://doi.org/10.1051/0004-6361/201219666
77. J.E. Pringle, ARA & A **19**, 137 (1981). https://doi.org/10.1146/annurev.aa.19.090181.001033
78. E. Quataert, A. Gruzinov, APJ **539**, 809 (2000). https://doi.org/10.1086/309267
79. B.R. Ryan, J.C. Dolence, C.F. Gammie, APJ **807**, 31 (2015). https://doi.org/10.1088/0004-637X/807/1/31
80. A. Sadowski, R. Narayan, A. Tchekhovskoy et al., MNRAS **429**, 3533 (2013). https://doi.org/10.1093/mnras/sts632
81. A. Sadowski, R. Narayan, J.C. McKinney et al., MNRAS **439**, 503 (2014). https://doi.org/10.1093/mnras/stt2479
82. A. Sadowski, R. Narayan, A. Tchekhovskoy et al., MNRAS **447**, 49 (2015). https://doi.org/10.1093/mnras/stu2387
83. A. Sadowski, R. Narayan, MNRAS **453**, 3213 (2015). https://doi.org/10.1093/mnras/stv1802
84. N.I. Shakura, R.A. Sunyaev, A & A **24**, 337 (1973)
85. F. Shi, Z. Li, F. Yuan et al., Nature Astronomy **5**, 928 (2021). https://doi.org/10.1038/s41550-021-01394-0
86. F. Shi, B. Zhu, Z. Li et al., APJ **926**, 209 (2022). https://doi.org/10.3847/1538-4357/ac4789
87. F. Shi, F. Yuan, Z. Li et al., arXiv:2403.00479 (2024). https://doi.org/10.48550/arXiv.2403.00479
88. J.M. Stone, J.E. Pringle, M.C. Begelman, MNRAS **310**, 1002 (1999). https://doi.org/10.1046/j.1365-8711.1999.03024.x

89. J.M. Stone, J.E. Pringle, MNRAS **322**, 461 (2001). https://doi.org/10.1046/j.1365-8711.2001.04138.x
90. J.M. Stone, K. Tomida, C.J. White et al., APJS **249**, 4 (2020). https://doi.org/10.3847/1538-4365/ab929b
91. M. Su, T.R. Slatyer, D.P. Finkbeiner, APJ **724**, 1044 (2010). https://doi.org/10.1088/0004-637X/724/2/1044
92. A. Tchekhovskoy, R. Narayan, J.C. McKinney, MNRAS **418**, L79 (2011). https://doi.org/10.1111/j.1745-3933.2011.01147.x
93. L.L. Thomsen, T.M. Kwan, L. Dai et al., APJL **937**, L28 (2022). https://doi.org/10.3847/2041-8213/ac911f
94. F. Tombesi, F. Tazaki, R.F. Mushotzky et al., MNRAS **443**, 2154 (2014). https://doi.org/10.1093/mnras/stu1297
95. T. Totani, PASJ **58**, 965 (2006). https://doi.org/10.1093/pasj/58.6.965
96. A. Utsumi, K. Ohsuga, H.R. Takahashi et al., APJ **935**, 26 (2022). https://doi.org/10.3847/1538-4357/ac7eb8
97. J.C. Vernaleo, C.S. Reynolds, APJ **645**, 83 (2006). https://doi.org/10.1086/504029
98. G. Vietri, T. Misawa, E. Piconcelli et al., A & A **668**, A87 (2022). https://doi.org/10.1051/0004-6361/202243285
99. K.-Y. Watarai, T. Mizuno, S. Mineshige, APJL **549**, L77 (2001). https://doi.org/10.1086/319125
100. Q.D. Wang, M.A. Nowak, S.B. Markoff et al., Science **341**, 981 (2013). https://doi.org/10.1126/science.1240755
101. R. Weinberger, V. Springel, L. Hernquist et al., MNRAS **465**, 3291 (2017). https://doi.org/10.1093/mnras/stw2944
102. C.J. White, J.M. Stone, C.F. Gammie, APJS **225**, 22 (2016). https://doi.org/10.3847/0067-0049/225/2/22
103. C.J. White, P.D. Mullen, Y.-F. Jiang et al., APJ. B.R. Ryan, J.C. Dolence, C.F. Gammie, APJ **807**, 31 (2015). https://doi.org/10.1088/0004-637X/807/1/31
104. H. Yang, F. Yuan, Y.-F. Yuan et al., APJ **914**, 131 (2021). https://doi.org/10.3847/1538-4357/abfe63
105. H. Yang, F. Yuan, T. Kwan et al., MNRAS **523**, 208 (2023). https://doi.org/10.1093/mnras/stad1444
106. H. Yang, F. Yuan, H. Li et al., Science Advances, **10**, eadn3544 (2024). https://doi.org/10.1126/sciadv.adn3544
107. X.-H. Yang, F. Yuan, K. Ohsuga et al., APJ **780**, 79 (2014). https://doi.org/10.1088/0004-637X/780/1/79
108. D. Yoon, F. Yuan, J.P. Ostriker et al., APJ **885**, 16 (2019). https://doi.org/10.3847/1538-4357/ab45e8
109. S. Yoshioka, S. Mineshige, K. Ohsuga et al., PASJ **74**, 1378 (2022). https://doi.org/10.1093/pasj/psac076
110. A.J. Young, I. McHardy, D. Emmanoulopoulos et al., MNRAS **476**, 5698 (2018). https://doi.org/10.1093/mnras/sty509
111. F. Yuan, D.-F. Bu, MNRAS **408**, 1051 (2010). https://doi.org/10.1111/j.1365-2966.2010.17175.x
112. F. Yuan, M. Wu, D. Bu, APJ **761**, 129 (2012). https://doi.org/10.1088/0004-637X/761/2/129
113. F. Yuan, D. Bu, M. Wu, APJ **761**, 130 (2012). https://doi.org/10.1088/0004-637X/761/2/130
114. F. Yuan, R. Narayan, Rev. Astron. Astrophys. **52**, 529 (2014). https://doi.org/10.1146/annurev-astro-082812-141003
115. F. Yuan, Z. Gan, R. Narayan et al., APJ **804**, 101 (2015). https://doi.org/10.1088/0004-637X/804/2/101
116. F. Yuan, D. Yoon, Y.-P. Li et al., APJ **857**, 121 (2018). https://doi.org/10.3847/1538-4357/aab8f8
117. F. Yuan, H. Wang, H. Yang, APJ **924**, 124 (2022). https://doi.org/10.3847/1538-4357/ac4714
118. K. Zubovas, A.R. King, S. Nayakshin, MNRAS **415**, L21 (2011). https://doi.org/10.1111/j.1745-3933.2011.01070.x

Chapter 9
Wind from Cold Accretion Flows

De-Fu Bu and Feng Yuan

Abstract We describe the physical mechanisms of launching and acceleration of wind from an active galactic nucleus (AGN) accretion disk. We focus on the radiation line force driven and magnetic driven wind models, which operate on the accretion disk scale. We review the investigation histories of the two mechanisms and the most new results obtained recently. The ultra-fast outflows (UFOs) found in the hard X-ray bands are believed to directly originate from AGN accretion disks. We review the theoretical works applying the two mechanisms of wind to explain the UFOs. We briefly introduce the propagation of winds on a large scale which is important for AGN wind feedback study. Finally, the roles of wind in AGN feedback are briefly reviewed.

9.1 Introduction

The standard thin accretion disk (SSD, [84]) has now been widely accepted as the central engine for luminous AGNs and the high-soft state of black hole X-ray binaries. For the SSD, it is radiatively very efficient. The viscous heating rate is balanced with the radiative cooling rate. Therefore, the gas temperature is very low. The gas pressure gradient force is negligibly small compared to black hole gravity. Due to the weak vertical pressure gradient force, the disk is geometrically thin. The SSD rotates with almost Keplerian velocity and its inflow velocity is quite small. The SSD is optically thick and its spectrum is a multi-color black body.

Observations have shown that for luminous AGNs, winds with different ionization states are quite common phenomena. There is also a long history of wind study from

D.-F. Bu (✉)
Shanghai Astronomical Observatory, 80 Nandan Road, Shanghai 200030, China
e-mail: dfbu@shao.ac.cn

F. Yuan
Center for Astronomy and Astrophysics and Department of Physics, Fudan University, 2005 Songhu Road, Shanghai 200438, China
e-mail: fyuan@fudan.edu.cn

C. Bambi et al. (eds.), *New Frontiers in GRMHD Simulations*, Springer Series in Astrophysics and Cosmology, https://doi.org/10.1007/978-981-97-8522-3_9

the theoretical point of view. It is now widely accepted that the accretion process is accompanied by wind. There are three main mechanisms driving wind from an accretion disk: radiation pressure, magnetic force and thermal pressure.

For highly ionized gas, radiation pressure plays a role through Compton scattering. If an accretion system radiates with luminosity close to or higher than the Eddington luminosity, the radiation pressure can exceed black hole gravity, and winds can be launched from the accretion flow. The winds driven by radiation pressure due to Compton scattering have been intensively studied by numerical simulations ([19, 21, 41, 63, 82, 83, 96, 101, 104]). The radiation pressure (through Compton scattering) driven wind process can occur in accretion flows both in AGNs and black hole X-ray binaries.

If the gas in an accretion disk is partially ionized, the UV photons emitted from the accretion disk can be absorbed, and the radiation pressure on resonance lines (radiation line force) can launch wind. Under proper conditions, the cross section of line resonance can be 1000 times that of Compton scattering. Therefore, radiation line force can launch wind even from sub-Eddington luminosity systems. It should be noted that radiation line force can only operate in accretion system around super-massive black holes in AGNs. For accretion disks in black hole X-ray binaries, the gas temperature is too high that radiation line force does not exist. Radiation line force driven wind from an AGN accretion disk has been studied both in analytical work [59] and numerical simulations ([60–62, 71–74, 102, 103]).

The magnetic driven wind models include magneto-centrifugal driven and magnetic pressure gradient force driven wind (see Sect. 1.2.2 for details). Because magnetic field must be present in black hole accretion systems which is responsible for the angular momentum transfer of the accretion disk, the magnetic driven wind models are believed to play a role in driving the wind from accretion disk in both AGNs and black hole X-ray binaries.

The wind can also be thermally driven ([4, 100]). The X-rays generated very close to the vicinity of the black hole can heat gas on the accretion disk. If the gas can be heated up to the X-ray Compton temperature ($T_{\rm compton}$) and if $T_{\rm compton}$ is higher than the local Virial temperature, gas can be launched from the accretion disk and form wind. The value $T_{\rm compton} \sim 10^7$K, therefore, wind can roughly be launched at location $>10^5 R_{\rm s}$ (with $R_{\rm s}$ being Schwarzschild radius), which may be beyond an accretion disk scale.

In addition to the extensive theoretical development of wind models, there are lots of observational signatures of winds from black hole accretion systems. In this Chapter, we focus on winds from luminous AGNs which are powered by cold thin accretion disks.

For AGNs, there are abundant observations of wind through blueshifted absorption lines in both UV and X-ray bands. The AGN UV absorption lines are usually classified into three groups depending on their widths, (1) the broad absorption lines (BALs), (2) the narrow absorption lines (NALs) and (3) the intermediate 'mini-BALs'. The BALs are blueshifted with a velocity $v_{\rm w} \sim 10000 - 60000$ km s^{-1}, with a representative width being 10000 km s^{-1}. The NALs wind has velocity $v_{\rm w} \sim 100 - 1000$ km s^{-1}. The properties of 'mini-BALs' lie between those of BLAs and NALs.

Nearly ~15–20% of quasars exhibit signatures of blueshifted BALs in UV spectra ([29, 99]), at transitions from ions such as Si IV λ1400, C IV λ1549, Al III λ 1857 and Mg II λ2799 with speed ~0.1–0.2 c (c is the speed of light; [34]). The location of BALs (or the distance from the black hole) is quite difficult to be measured. In the case of BALs, one usually uses the absorption line ratios to obtain the density of BALs. Combined with the ionization parameter ξ (which is defined as $\xi = L_{\rm ion}/n_{\rm e}r^2$, with $L_{\rm ion}$ being the ionizing luminosity integrated over 13.6 eV-100 keV, $n_{\rm e}$ being the electron number density and r being the distance from the ionizing source) and the ionizing luminosity $L_{\rm ion}$, one can obtain the distance of BALs. Recent studies of BALs found that 50% of the BALs winds are at a distance larger than 100 pc and at least 12% are at distances >1 kpc ([1]). Very recently, He et al. ([38]) found that the variability of the BALs wind is due to the variation of the ionizing continuum. By using the statistical properties of the variability of BALs in a sample of 915 quasars, it is found that the typical BALs wind location is tens of parsecs ([39]). The ionization parameter of BALs wind is $\log\xi = 0.5 - 2.5$ erg cm s^{-1}, and the column density is $N_{\rm H} \sim 10^{20} - 10^{23}$ cm^{-2} ([50]). Theoretical models have shown that the BALs winds can have important feedback effects on their host galaxies ([22, 40]). The NALs wind has a much lower velocity. The location of the NALs is ~1 pc - 1 kpc from the central black hole ([50, 53]). The ionization parameter and column density of NALs wind are $\log\xi = 0 - 1.5$ erg cm s^{-1} and $N_{\rm H} \sim 10^{18} - 10^{20}$ cm^{-2}, respectively. Due to the low velocity, the kinetic power of NALs winds is small. Therefore, the NALs winds are thought to be insignificant in feedback.

In the soft X-ray band, warm absorbers (WAs) with velocity $v_{\rm w} \sim 100 - 2000$ km s^{-1} are frequently detected ([6, 8, 28, 33, 48, 49, 54, 81]). WAs are mildly ionized with $\log\xi = -1 - 3$ erg cm s^{-1}. The column density of WAs is in the range $N_{\rm H} \sim 10^{20} - 10^{22.5}$. The observed location of WAs is 1-1000 pc ([6]). Due to the low kinetic power, the WAs wind may not be sufficient for feedback. The WAs has been suggested to be a magneto-centrifugal wind ([26]) or thermally driven wind ([20, 46, 47, 56]). Very recently, it is found that the WAs may be partially due to the collision of UFOs with the interstellar medium gas ([7]).

In the hard X-ray bands, the ultra-fast outflows (UFOs) with velocity >10000 km s^{-1} are observed very recently ([89]). The UFOs are highly ionized and having highest column density compared to BALs, NALs and WAs (see the detailed introduction of UFOs in Sect. 1.3). The UFOs are believed to be launched directly from the black hole accretion disk. The kinetic power of UFOs is in the range of 0.1–10% of the Eddington luminosity $L_{\rm Edd}$ ([31]). Therefore, UFOs may play an important role in AGN feedback.

The current observations of BALs, NALs and WAs winds indicate that they may be launched at a scale much larger than the size of an accretion disk. Only the UFOs are launched directly from accretion disk. In the following sections, we focus on the discussion of UFOs.

9.2 Theoretical Models of Wind Launching from Cold Accretion Flows

Both the radiation pressure and magnetic driven wind mechanisms can take place on accretion disk scale. However, the thermal mechanism takes place at a location $>10^5 R_s$, which seems to be beyond the accretion disk scale of an AGN. In this Chapter, we focus on the radiation pressure and magnetic driven wind models.

9.2.1 *Radiation Line Force Driven Wind*

A cold accretion disk is just partially ionized. In AGNs, the UV photons emitted by the disk can be absorbed by the gas on the disk, the radiation pressure on resonance lines can very effectively lift up gas from the disk and accelerate gas to form high velocity wind. The line force per gram in wind exposed to a UV flux $F_{\rm UV}$ is

$$g_{\rm line} = (\sigma_{\rm e} F_{\rm UV}/c) M(t) \tag{9.1}$$

where the first factor is the radiation pressure due to electron scattering and $\sigma_{\rm e}$ is the mass-scattering coefficient for free electrons. The effect of all the lines is accounted for by the force multiplier $M(t)$, which is a function of the optical depth parameter t,

$$t = \frac{\sigma_e \rho v_{\rm th}}{|dv_l/dl|} \tag{9.2}$$

where ρ is gas density, $v_{\rm th}$ is the thermal velocity, and dv_l/dl is the velocity gradient along the line of sight. The force multiplier is as follows ([12, 66]),

$$M(t) = kt^{-\alpha} \frac{(1 + \tau_{\rm max})^{(1-\alpha)} - 1}{\tau_{\rm max}^{(1-\alpha)}} \tag{9.3}$$

where k is proportional to the total number of lines, α is the ratio of optically thick to optically thin lines, $\tau_{\rm max} = t\eta_{\rm max}$, and $\eta_{\rm max}$ is a parameter determining the maximum value $M_{\rm max}$. k is a function of both ionization parameter ξ and gas temperature. $\eta_{\rm max}$ is a function of ionization parameter.

The value of $M_{\rm max}$ is a function of both ξ and gas temperature T. It is found that $M_{\rm max}$ goes gradually from $\sim$2000 to $\sim$5000 as ξ increases from 0 to $\sim$3 and then decreases to $\sim$1 at $\xi = 100$. Also, the line force becomes negligible if $T > 10^5$K for any ξ ([76]). Under proper conditions of ξ and T, the line force can be $\sim$1000 times higher than the radiation pressure due to free electrons scattering. Therefore, line force can very effectively launch and accelerate wind even in AGNs which have luminosity much lower than the Eddington value.

Murray et al. ([59]) analytically calculated the line force driven wind from an accretion disk around a $10^8 M_\odot$ (with $M_\odot$ being solar mass) black hole. The luminosity of the disk is assumed to be half of the Eddington value. The innermost radius from which wind is launched is $\sim 300 R_s$. It is assumed that there is entrained (hitchhiking) dense gas located on the inner edge of the wind, which blocks soft X-rays from the center but transmits UV photons. The hitchhiking gas keeps the wind from being strongly ionized. Wind gas is first lifted vertically up from the disk by the line resonance pressure due to UV photons locally. Then the UV photons from the central region exert line force on the wind almost radially. The wind can be accelerated to velocity $v_w \sim 0.1c$. The wind is nearly along the plane of the disk, with an opening angle of $\sim 5°$. The line force driven wind is used to explain the origin of the broad absorption lines seen in some quasars. It is found that the covering fraction of the wind is $\sim 10\%$, which naturally explains the fraction of quasars showing broad absorption lines. In addition, the wind generates broad emission lines, therefore, the emitting gas in broad emission line regions may be line force driven wind.

Two dimensional axial-symmetric numerical simulations of line force driven wind were first performed in the late 1990s ([68, 71, 72]). We note that these simulations are designed to study the wind from cataclysmic variables (CVs). Therefore, in their simulations, the central object is a star with mass and radius appropriate to a white dwarf. Both the central star and the surrounding cold thin accretion disk can emit photons which exert line force on wind. We also mention that gas in these simulations was assumed to be globally isothermal. A dense, nearly equatorial, low-velocity wind is found which is bounded by a low density, high velocity wind in a channel at larger angles.

Numerical simulations of line force driven wind from an AGN accretion disk were carried out by Proga et al. ([73]). In these simulations, the gas temperature is self-consistently calculated by taking into account radiative cooling/heating processes. The radiative heating processes include Compton heating and photoionization heating. The radiative cooling processes include Compton cooling, recombination cooling, bremsstrahlung and line coolings. The inner radial boundary of these simulations is located at 300 R_s. They found that the accretion disk atmosphere can 'shield' itself from external X-rays so that the gas on the accretion disk can not be over-ionized. The local disk radiation can launch gas off the disk photosphere by radiation line force. It is also found that for a $10^8 M_\odot$ with luminosity of 0.5 $L_{\rm Edd}$, an equatorial wind can be launched from $\sim 10^{16}$cm from the black hole with a covering factor being $\sim$0.2. The highest velocity of wind is 0.05c and the mass loss rate of wind is $0.5 M_\odot$ yr^{-1}. We note that the innermost radius at which wind is launched found by Proga et al. ([73]) roughly equals to the inner radial boundary of the simulation. Therefore, the strength of wind may be underestimated, since the wind from radius smaller than $300 R_s$ has not been simulated in the simulations. In order to capture as much as possible of the line force driven wind, Proga and Kallman ([75]) decrease the inner radial boundary of the simulations to 30 R_s, the new simulations confirm the main results from Proga et al. ([73]).

Very recently, Nomura et al. ([60, 61]) re-perform simulations of the line force driven accretion disk wind. They study the dependence of the properties of line force driven disk wind on black hole mass ($M_{\rm BH}$) and the Eddington ratio of radiation (ϵ). They found that line force driven winds can be successfully launched for the range of $M_{\rm BH} = 10^{6-9} M_\odot$ and $\epsilon = 0.1 - 0.5$. They also found that no winds can be line force driven for AGNs with $\epsilon \lesssim 0.01$. They also apply the line force driven wind model to explain the ultra-fast outflows found in hard X-ray bands (see Sect. 1.3.1.1 for details). The above simulations all assume that the mass accretion rate in the disk is fixed and does not decrease even in the presence of wind. Actually, in the presence of wind, the disk accretion rate should be decreased and the disk should be less UV-luminous, which may attenuate the strength of line force and affect the properties of wind. In 2020, Nomura et al. ([62]) performed two-dimensional numerical simulations to study the line force driven accretion disk wind by taking into account the reduction of the disk mass accretion rate in response to the wind mass-loss. They found that even though the disk is less luminous compared to the case in which a constant accretion rate is assumed, the line force driven wind is still powerful. Roughly half the supplied accretion rate is ejected out in the wind for black holes with mass $10^8 - 10^{10} M_\odot$, accreting with luminosity at $L/L_{\rm Edd} = 0.5 - 0.9$.

There are also simulations studying the effects of line force on the mass supply to an AGN ([52, 58, 76, 77]). In these simulations, the authors focus on much larger scale, with the inner radial boundary located at $\sim 1000 R_{\rm s}$. The outer radial boundary $\sim 10^6 R_{\rm s}$. The AGN corona emitting X-ray photons and the accretion disk emitting UV photons are both not resolved and treated as point sources in the simulations. Gas is injected into the computational domain from the outer radial boundary to mimic the gas supply to AGNs from interstellar medium (ISM). They found that strong wind can be driven by line force. The presence of wind significantly reduces the rate at which the central black hole is fed compared to the mass supply rate from ISM.

9.2.2 *Magnetically Driven Wind*

There are two magnetically driven wind model, namely magnetocentrifugal force driven wind model and magnetic pressure driven wind model. The magnetocentrifugal force driven wind model (hereafter BP model) was established by Blandford and Payne ([5]). In this model, a large scale, sufficiently strong, open magnetic field is anchored in a thin, cold and Keplerian rotation accretion disk. In the BP model, the poloidal component of the magnetic field is at least comparable to the toroidal magnetic field, $|B_\phi/B_p| \lesssim 1$. The foot points of the magnetic field lines co-rotate with the accretion disk. For a steady state case, the angular velocity at any point of a magnetic field line equals that at the foot point. When a fluid element is loaded to a magnetic field line at a position higher than the foot point of the magnetic field line, the element will rotate with a super-Keplerian speed. When the field line has an angle $>30^\circ$ relative to the rotational axis of the accretion disk, the component of the centrifugal force along the field line exceeds the component of gravity along the field line. The fluid element is accelerated out along the field line. An important feature of the BP model winds is that they require some assistance to launch the gas steadily from the surface of the accretion disk to the magnetic field lines. We note that numerical studies on BP winds usually do not resolve the vertical structure of the accretion disk but treat it as a boundary surface through which mass is loaded on to the magnetic field lines at a specified rate.

In the magnetic pressure driven wind model, it is believed that the toroidal magnetic field can be very quickly generated due to the differential rotation of accretion disk and built up with $|B_\phi/B_p| \gg 1$. The magnetic pressure of the toroidal magnetic field can launch a self-starting wind ([78, 85, 95]). The wind is launched by the magnetic pressure gradient force. We note that in order to maintain a steady wind driven by the magnetic pressure, a steady supply of advected toroidal magnetic flux at the wind base is required. Otherwise, the winds are much likely to be a transient phenomenon. However, it is still not clear whether a steady supply of the toroidal magnetic flux can be maintained by the accretion disk differential rotation to compensate the escape of the magnetic flux in the wind.

9.2.2.1 Analytical Works

The effects of magnetic field inclination angle with respect to the accretion disk on the BP model wind have been studied analytically ([9, 10]). It is found that the lower the inclination angle, the easier for wind to be accelerated. Cao ([11]) analytically includes the effects of radiation pressure due to electron scattering into the BP model. It is found that with the help of the radiation force, the mass loss rate in the wind is high, which leads to a slow wind.

The MHD equations describing the winds are coupled, partial differential equations, which are quite hard to be solved. In order to simplify the MHD equations, radial self-similarity assumption is also usually employed. For the self-similar solution, the physical variables (e.g., gas density, velocity, internal energy, magnetic field) are all assumed to have a power-law form of radius. Under these assumptions, the two-dimensional structure of wind can be solved analytically. The self-similarity BP wind model has been used to explain the observations of warm absorber in soft X-ray bands ([24]) and ultra-fast outflows in hard X-ray band ([25, 27]).

9.2.2.2 Numerical Simulations

The cold thin disk has an aspect ratio $z/R \ll 1$, with z and R being disk scale height and radius in cylindrical coordinates, respectively. It is numerically challenging to simulate a global thin disk with the disk inside well resolved. In order to well resolve a thin disk and include the magnetorotational instability (MRI; [3]) which is responsible for angular momentum transfer, one usually adopts the shearing box simulation. In the shearing box simulation, only a small patch of a thin disk is simulated with $\Delta R \ll R$ and $z \ll R$. The box is rotating with respect to the black hole with the keplerian speed Ω at the box center. The effective gravity is the tidal expansion of the effective potential (centrifugal + gravitational). The effective gravity in the radial and vertical direction are $3\Omega^2 x$ (with $x = R - R_0$) and $-\Omega^2 z$, respectively. It is clear that the gravity g in radial direction is symmetric with respect to the box center $g_x = g_{-x}$. In this case, actually, one can not know which side (relative to the box center) the black hole resides. For the vertical component of gravity, it keeps increasing with height, which can only be used to the region with $z \ll R_0$. Actually, in a global case, in the region $z \gtrsim R_0$, the vertical component of gravity decreases with height.

Three-dimensional shearing box simulations have been performed to study the wind from a thin disk. In these simulations, the MRI process can be well captured. Therefore, the turbulence inside of the accretion disk is well included in the simulations. It is found that the wind is magneto-centrifugally driven ([2, 23]), consistent with the BP model. However, the properties of wind provided by shearing box simulations should be quite different from those of wind launched from a real accretion disk. For example, due to the symmetry of the radial component of effective gravity about the box center, the winds below and above the midplane have opposite radial flowing direction. The time-averaged net wind mass flux in radial direction is zero and the direction of the wind constantly swap between radially inward and outward due to the constant flipping of the mean azimuthal field lines ([2, 51, 86]). As introduced above, in the shearing box assumptions, the vertical component of the gravity becomes stronger and stronger as the vertical box size increases. It means that larger and larger energy for gas is required to climb out of that potential, while in reality the gravitational potential has a finite depth. This may result in the wind mass loss rate dropping dramatically when a taller box is used ([23]). Although the MRI process can be well captured in shearing box simulations, the numerical setups of these simulations prevent a reliable estimate of the properties of wind from being made. Therefore, global simulations with the MRI process inside the accretion disk well captured are needed.

Global simulations of disk wind can be traced back to the 1990s. Stone and Norman ([85]) performed two-dimensional axis-symmetric global simulations of disk wind driven from a thin disk threaded by uniform vertical magnetic field lines. In their simulations, due to the coarse resolution, the MRI in the disk interior is not captured. They found that wind can be launched from the disk by the magnetic pressure gradient force. The mass flux of wind at the outer radial boundary can even exceed the mass inflow rate there. Kato et al. ([42]) performed similar global simulations. They found that the driving mechanism of wind depends on the strength of the initial magnetic field. When the initial magnetic field is weak, the acceleration force is the magnetic pressure gradient. However, when the initial magnetic field is strong, the centrifugal force is responsible for wind acceleration which is consistent with the BP model. There are also 2D axsi-symmetric simulations in which only the wind region has been studied. In these simulations, the accretion disk is put outside the computational domain ([44, 45, 65, 69]). The accretion disk is put at the midplane boundary, which just supplies gas into the computational domain. Krasnopolsky et al. ([44, 45]) found that the wind can be centrifugally accelerated through the Alfven and fast magnetosonic surfaces and collimated into cylinders parallel to the disk's axis, which is consistent with the BP model. Porth and Fendt ([69]) found that the wind arises around 100 R_s and reaches only mildly relativistic speeds with Lorentz factor $\Gamma < 1.5$.

Magnetohydrodynamic turbulence induced by MRI is believed to be responsible for angular momentum transfer in accretion disks, which allows the mass be accreted to the central black hole. Wind is launched from the turbulent accretion disk. It is pointed out that the MRI turbulence in accretion disks could play a role in driving disk winds, particularly in mass loading to the surface region ([86, 87]). Therefore, ideally, simulations of wind should also capture the MRI process in accretion disk, which should leave an impression on the properties of wind. There have been several attempts. Suzuki and Inutsuka ([88]) performed three-dimensional global simulations which resolved the MRI process inside the disk. The disk is assumed to be locally isothermal. Initially, a weak vertical magnetic field is threading the disk. They found that after a quasi-steady state is achieved, winds can be produced intermittently. Wind is mainly accelerated by the magnetic pressure gradient forces. However, we note that the simulations have very limited radial and angular range with $\theta \in [\pi/2 - 0.5, \pi/2 + 0.5]$. It is not clear that whether wind has been sufficiently developed in the limited simulation box. It has been pointed that wind can be properly studied when the simulation domain is larger than the Alfven surface and fast magnetosonic surface ([5]). Efforts have been made by Zhu and Stone ([106]), who performed simulations with a large radial dynamical range with the outer radial boundary being 3 orders of magnitude larger than the inner radial boundary. The whole angular region with $\theta \in [0, \pi]$ and $\phi \in [0, 2\pi]$ is simulated. By a artificial setting of disk temperature, the disk is set to have an aspect ratio $H/R = 0.1$. A weak vertical magnetic field is threading the disk initially. It is found that the disk accretion mainly occurs at the coronal region. At the midplane, little inflow or even outflow is found. A BP style wind is fully developed. The wind passes though both the Alfven surface and fast magnetosonic surface. However, the mass flux of wind is only 0.4% of the disk accretion rate. Both the Maxwell stress from MRI turbulence and wind torque contribute to disk angular momentum transfer. The contribution of wind torque to disk accretion is low, only 5% of the disk accretion is due to the wind torque.

In all of the above mentioned simulations (including shearing-box and global simulations), in order to simplify the equations, the gas is usually assumed to evolve isothermally or adiabatically. However, in reality, the gas temperature should be determined by the exact radiative cooling and heating processes. It is not known whether adiabatic or isothermal assumption can be a good approximation when one studies the dynamics of wind. Wang et al. ([97]) compares the properties of wind under different assumptions of gas thermal dynamics by performing 2D simulations. In their simulations, a not resolved thin disk is put on the midplane. A Large scale open magnetic field is initially anchored on the accretion disk. They have four models, the first one assumes adiabatic evolution of gas internal energy; the second and third models assume globally isothermal gas; and the fourth one self-consistently calculates gas temperature with proper heating and cooling functions for AGNs. In the fourth model, the heating processes include the X-ray Compton scattering and photoionization heatings. The X-ray is from the AGN corona. The cooling processes include bremsstrahlung, line and recombination coolings. It is found that compared to the fourth model solving the energy equation with radiative cooling and heating, both the isothermal model and adiabatic model overestimate the temperature, underestimate the wind power, and can not predict the local structure of the winds (see Fig. 9.1). Therefore, it is recommended that simulations studying the properties of wind should self-consistently calculating gas temperature with proper radiative cooling and heating processes.

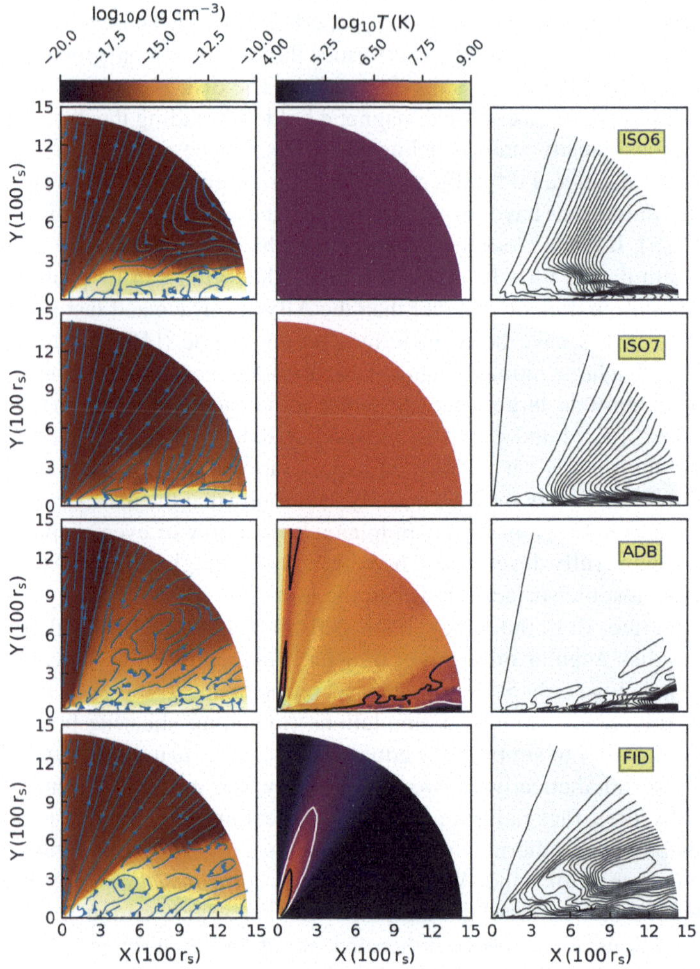

Fig. 9.1 From left to right columns show logarithmic time-averaged density (colormap) overlaid by streamlines (blue lines with arrow), temperature (colormap) overlaid by contours with $T = 10^{6,7}$K (white, black) and magnetic field lines (black contours), respectively. From top to bottom are quantities for isothermal model with $T = 10^6$K, isothermal model with $T = 10^7$K, adiabatic model and mole with radiative cooling/heating, respectively. It is clear that the properties of wind differ significantly from model to model. Figure from Wang et al. 2022, MNRAS, 513, 5818

9.2.3 Combining Radiation Line Force and Magnetic Field

The Magnetorotational instability has been shown to be a very robust and universal mechanism to produce turbulence and the transport of angular momentum in accretion disks at all radii. Therefore, magnetic fields are very likely crucial for all accretion disks. It is very likely that magnetic field plays an important role both in controlling mass accretion inside the disk and in launching winds from the disk. As introduced above, under the most favorable conditions (i.e., high UV flux, low ionization gas), the radiation line force can exceed the radiation force due to electron scattering by a factor as high as $\sim$2000. Thus, theoretically, even in accretion disks with luminosity as low as $\sim 10^{-3} L_{\rm Edd}$, can produce strong wind. It is likely that in an AGN accretion disk both magnetic field and line force are important in driving winds. It is very important to study the two mechanisms simultaneously.

Proga ([74]) studied the hybrid model including line force driven wind and magnetic driven wind. This model is trying to study wind from nonmagnetic cataclysmic variables (nMCVs). In the model, the mass of the central object is $0.6 M_{\odot}$. The thin disk is not put inside the computational domain (but instead in the boundary zoos). The first active zone above the midplane represents the photosphere of the thin disk. A uniform vertical magnetic field is threading the photosphere initially. The gas is assumed to be globally isothermal with a sound speed $c_{\rm s} = 14$ km/s, which corresponds to gas temperature of 1.4×10^4 K. They found that the magnetic field deviates very quickly from the purely vertical. The toroidal component of the magnetic field grows quickly due to the differential rotation of the disk. It is also found that the toroidal field dominates over the poloidal field above the disk. Whether wind is magnetic driven or line force driven depends on the plasma beta $\beta = p_{\rm gas}/p_{\rm mag}$. When $\beta \gtrsim 1$, the wind is radiation line force driven. In this case, the wind consists of a dense, slow part that is bounded on the polar side by a high-velocity stream. The mass-loss rate is mainly due to the fast stream. As the increase of the magnetic field strength (i.e., $\beta \lesssim 1$), the dense, slow part is first affected. The slow part becomes denser and faster than its pure line force driven counterpart by a factor of $\gtrsim 100$ and $\gtrsim 3$, respectively. The dense wind begins to dominate the mass-loss rate. When the flow is strongly magnetized (i.e., $\beta \lesssim 10^{-3}$), the high-velocity stream disappears, and the wind only consists of a dense, slow outflow. The magnetic driven dense, slow wind is accelerated by the gradient of the magnetic pressure due to the toroidal field. A magnetocentrifugal wind is not found.

The hybrid model in the context of AGNs is studied recently by Yang et al. ([102, 103]). As done by Proga ([74]), the accretion disk itself is not simulated. The first active zoo above the midplane represents the photosphere of the thin disk. Initially, a large scale open magnetic field is threading the photosphere. There are several key differences between the work down by Proga ([74]) and Yang's works. First, in Yang's works, the central accreting object is a supermassive black hole. Second, in Yang's simulations, isothermal equation of state for gas is abandoned and the gas temperature is self-consistently calculated by taking into account of radiative heating/cooling processes. The gas is irradiated by X-rays from the AGN corona, which can heat gas. Also, gas can cool down due to radiative cooling. The magnitude of the line force sensitively depends on the ionization parameter and temperature of gas. It is very necessary to treat the gas thermal properties properly. It is found that when the magnetic field is weak, the winds are line force driven and move outward mainly in the region $\theta \gtrsim 40°$, which is consistent with that found by Proga et al. ([73]). When a relative strong magnetic field is taking into account, the vertical component of the Lorentz force can push gas into the region close to the pole, a high velocity wind can form in this region. The authors argued that the high velocity wind in the polar region can explain the UFOs in some radio-loud AGNs, which have an inclination angle of $\sim 10° - 70°$ away from jets ([93]). The maximum velocity of wind found in Yang's works is around $\sim$0.1c. However, we note that highest velocity of the UFOs can even exceed 0.3c. The reason for a relative lower velocity found in Yang's work may be that the magnetic field is still not strong enough. Another possible reason is that the simulations are just run for 0.18 orbital times measured at the outer radial boundary, which may be too short for a higher velocity wind to develop.

9.3 Comparison with Observations

9.3.1 UFOs in Hard X-Rays Band

Recent surveys with XMM-Newton and Suzaku have shown that UFOs found through blueshifted Fe xxv and Fe xxvi K-shell absorption lines in the hard X-ray bands are observed in a significant fraction ($\sim$40–50%; [31]) of active galaxies ([30, 31, 89, 90]). The outflowing velocity of the absorbers is in the range $v_{\rm w} \sim 0.03 - 0.3$c ([13, 30, 31, 70, 79, 89, 90]). The UFOs in hard X-ray bands have been observed in both radio-quiet ([30, 89]) and radio-loud ([30, 93]) AGNs. The column densities of UFOs given by photoionization modeling are on the order of $N_{\rm H} \sim 10^{22} - 10^{24} {\rm cm}^{-2}$. The ionization parameter of UFOs spans $3 \lesssim \log \xi \lesssim 6$ ([91]). Given the high velocity of UFOs, it is believed that they originate from the accretion disk. On average, the UFOs are located at $r \sim 10^{15-18}$ cm (typically $\sim 10^{2-4} R_{\rm s}$) from their black hole ([31]). The mass outflow rates are of the order $\dot{M}_{\rm w} \sim 0.01 - 1 M_{\odot} {\rm yr}^{-1}$ or $\sim (0.01 - 1)\dot{M}_{\rm Edd}$, kinetic power is in the range $L_{\rm w} \sim 10^{43-45} {\rm erg\ s}^{-1}$, equivalent to $(0.1 - 10)\% \ L_{\rm Edd}$ ([31]). Positive correlation between velocity of UFOs and the bolometric luminosity of its host AGNs have been found $v_{\rm w} \propto L_{\rm bol}^{\beta}$, with $\beta = 0.4^{+0.3}_{-0.2}$ ([31]). Positive relationship between the mass outflow rate of UFOs and $L_{\rm bol}$ is also found $\dot{M}_{\rm w} \propto L_{\rm bol}^{\beta}$, with $\beta = 0.9^{+0.8}_{-0.6}$, such that the UFOs in more luminous AGN are correspondingly more massive. Similar is also true for $L_{\rm w} - L_{\rm bol}$ and $\dot{p}_{\rm w} - L_{\rm bol}$ (with $\dot{p}_{\rm w}$ being the momentum flux carried by UFOs), which share strong positive slopes of $\beta = 1.5^{+1.0}_{-0.8}$ and $\beta = 1.2^{+0.8}_{-0.7}$, respectively ([31]). Because,

$L_w \propto v_w^3$, the observed slopes $L_w \propto L_{bol}^{\sim 1.5}$ and $\dot{p}_w \propto L_{bol}^{\sim 1.0}$ likely stem from their mutual dependence on v_w. Gofford et al. ([31]) argued that the velocity may be the driving factor in the observed relationships.

9.3.1.1 Acceleration Mechanism: Radiation Pressure

The UFOs are highly ionized. Therefore, one may first recall the radiation pressure on highly ionized gas through Compton scattering.

Tombesi et al. ([92]) found that the momentum flux carried by UFOs roughly equals the momentum flux of the radiation field ($\dot{p}_{rad} = L_{bol}/c$) with $\dot{p}_w \simeq \dot{p}_{rad}$. The average value of the ratio is $< \dot{p}_w/\dot{p}_{rad} >= 1.6 \pm 1.1$. If UFOs are accelerated by radiation pressure due to Compton scattering, there should be a direct proportionality, $\dot{p}_w \simeq C_f \tau_e \dot{p}_{rad}$, where C_f is the covering factor of UFOs, τ_e is the electron optical depth to Compton scattering. If this is the dominating acceleration mechanism, the product $C_f \tau_e$ should be of the order of unity. The fraction of sources with detected UFOs in the sample studied by Tombesi et al. ([89]) is ~0.4–0.6. The average covering factor is $C_f \simeq 0.5$. With the average column density of $N_H \simeq 10^{23} cm^{-2}$, one can obtain $\tau_e \simeq \sigma_T N_H \simeq 0.05$ ([92]). The product $C_f \tau_e$ is much lower than unity expected from the relation $\dot{p}_w \simeq \dot{p}_{rad}$. It seems that radiation pressure due to Compton scattering is not the acceleration mechanism of UFOs. There may be one possibility that the observed UFOs are not in their acceleration phase. Therefore, there is one possibility that the observed column density is much lower than that during the acceleration phase. Through extrapolating the relation between the column density and the distance, Tombesi et al. ([92]) found that at the innermost radii of $\log(r/R_s) \sim 1$, the column density can be $N_H = 10^{24} cm^{-2}$. In addition, there may be presence of some ionized material that is not visible through X-ray spectroscopy. Taking all of this into account, the material may be Compton-thick with $\tau_e \sim 1$. The product $C_f \tau_e$ is closer to unity.

There are several numerical simulation works, which study the origin of UFOs in the context of the line force driven wind model ([57, 60, 61, 102, 103]). The UFOs are highly ionized with $3 \lesssim \log \xi \lesssim 6$. However, as mentioned above, when $\xi > 100$, the line force can be neglected. Therefore, at first glance, it seems that the UFOs can not be line force driven. For example, Higginbottom et al. ([37]) think UFOs may not be driven by radiation line force due to the high ionization and less of opacity in UV or X-ray lines. However, the situation is not such simple. At small radii, line force can lift gas up from the accretion disk photosphere. The lifted gas is illuminated by the central X-ray source. The X-ray illumination will ionize the gas, stopping the acceleration before the wind reaches the escape velocity so that the gas falls back as a failed wind. The failed wind shields outer gas from the central X-ray radiation. Therefore, at large radii, the gas can not be highly ionized. The line force can accelerate the weakly ionized gas at large radii to escape velocity before it is directly illuminated and overionized. The weakly ionized high velocity wind can not be recognized as UFOs. When the high velocity winds are exposed to X-rays and get highly ionized, they will be recognized as UFOs.

Nomura et al. ([60]) performed hydrodynamic simulations (in spherical coordinates (r, θ, ϕ)) of line force driven wind and applied it to explain the origin of UFOs. In the simulations, the thin accretion disk is not resolved. The midplane ($\theta = 90°$) corresponds to the surface of the accretion disk. It is assumed that the density of the disk surface (ρ_0) is a constant with radius. In most of the models, they set $\rho_0 = 10^{-9}$g cm^{-3}. The temperature of the disk surface is set to decrease with radius $T(r) = T_{\rm in}\,(r/r_{\rm in})^{-3/4}$, with $r_{\rm in} = 3R_{\rm s}$. The value $T_{\rm in}$ is set to meet the condition of $L_{\rm D} = \epsilon L_{\rm Edd} = \int_{r_{\rm in}}^{r_{\rm out}} 2\pi r \sigma T^4 dr$, with ϵ being the Eddington ratio. $r_{\rm out}$ is the outer radius of the disk where the temperature is 3×10^3K. In the region $r > r_{\rm out}$, the gas temperature is too low to emit photons contributing to the line force effectively. Above the disk surface, the gas is assumed to be in hydrostatic equilibrium state. In addition to disk radiation, there is a spherical X-ray radiation source which contributes to the ionization of metals. The X-ray flux is assumed to be 10% of the disk radiation. The UFOs are recognized following two conditions: (A) The outward velocity of matter with $2.5 \le \log\xi \le 5.5$ exceeds 10000 km s^{-1}. (B) The column density of matter with $2.5 \le \log\xi \le 5.5$ is larger than 10^{22} cm^{-2}. In their fiducial model with $M_{\rm BH} = 10^8 M_\odot$ and $\epsilon = 0.5$, the UFOs are found to moving outward in the region $75° \lesssim \theta \lesssim 86°$. The detection probability calculated as $\Omega/4\pi$ (with Ω being the solid angle occupied by UFOs) is 20%. The Eddington ratio dependence of the UFO probability is calculated by performing simulations with different Eddington ratios but keeping $M_{\rm BH} = 10^8 M_\odot$. It is found that the probabilities are 17%, 19%, 20%, 27% for $\epsilon = 0.1$, 0.3, 0.5, and 0.7, respectively. The detection probability is not sensitive to the Eddington ratio given that $\epsilon > 0.1$. It is also found that in the case $\epsilon = 0.01$, the radiation line force is too small to accelerate disk wind. Thus, the UFO probability is zero. The black hole mass dependence of the UFO probability is calculated by simulations with different black hole masses but keeping $\epsilon = 0.5$. The UFO probability just slightly decreases with increasing black hole mass: 28%, 22%, 20% and 13% for $M_{\rm BH} = 10^6 M_\odot$, $10^7 M_\odot$, $10^8 M_\odot$, and $10^9 M_\odot$, respectively. Tombesi et al. ([91]) have reported that the UFO probability to be ~40% based on the frequency of detecting UFOs in the Seyfert galaxies. Therefore, the UFOs detection probability given by the line force disk wind model is much smaller than that given by observations.

In 2017, Nomura and Ohsuga ([61]) continued to study the line force driven wind. The only difference in numerical settings between this work and Nomura et al. ([60]) lies in the density profile of the disk surface at the midplane. In 2016, Nomural et al. ([60]) set the density for the disk surface to be a constant with radius. However, in 2017, Nomura and Ohsuga ([61]) set the density at the disk surface at the midplane $\theta = 90°$ according to the standard accretion disk model. In this paper, the authors study the dependence of the mass outflow rate, momentum flux, and kinetic power of UFOs on the disk luminosity. It is found that the observed scaling laws $\dot{M}_{\rm w} \propto L_{\rm bol}^{0.9}$, $L_{\rm w} \propto L_{\rm bol}^{1.5}$, and $\dot{p}_{\rm w} \propto L_{\rm bol}^{1.2}$ ([31]) can roughly being explained by the line force driven wind model. However, the velocity dependence on $L_{\rm bol}$ found by the simulations is $v_{\rm w} \propto L_{\rm bol}^{1/8}$, which is different from the observed correlation of $v_{\rm w} \propto L_{\rm bol}^{1/2}$ ([31]). It is also found that if $\epsilon \lesssim 0.01$, the line force driven wind does not appear, implying that the UFOs can not be observed in the AGNs with small Eddington ratio. Gofford et al. ([31]) found that a large sample of UFOs is detected in AGNs with $\epsilon \sim 0.1$, and UFOs are not detected in the AGNs with $\epsilon \lesssim 0.01$. Therefore, the simulation result is consistent with observations. From Fig. 5 of the work down by Nomura and Ohsuga ([61]), we can see that the maximum velocity of line force driven wind is smaller than $0.2c$. However, observations show that the maximum velocity of UFOs can reach or even exceed $0.3c$. For example, Reeves et al. ([80]) found an even faster Fe K UFO with speed $v_{\rm w}/c \sim 0.42$ in their 2017 *XMM-Newton/NuSTAR* observation.

The simulations performed by Nomura et al. ([60, 61]) assume that the mass accretion rate is a constant with radius, although the wind mass-loss rate is a large fraction of the mass accretion rate through the disk. In 2020, Nomura et al. ([62]) improved the calculation of the simulations so as to consider the reduction of mass accretion rate due to the wind mass loss. In the presence of wind, the mass accretion rate of the disk decreases. Therefore, the UV flux from the inner disk is reduced. However, it is found that powerful UV winds are still produced which carry enough energy and momentum to affect its host galaxy.

Based on the simulation result of Nomura et al. [62], Monte Carlo radiation transfer calculations have been done by Mizumoto et al. ([57]) to test whether the observed UFOs can be explained by the UV line-driven disk wind model. Spectra along different lines of sight are calculated, which shows that the line driven disk wind includes sightlines where the wind is both highly ionized and fast, as required by UFOs. The PCygni-like feature in the PG1211+143 X-ray spectrum is successfully explained. This is the first time that the X-ray data from a UFO has been convincingly fit by the direct results from a UV line-driven disk wind simulation.

9.3.1.2 Acceleration Mechanism: Magnetocentrifugal Acceleration

Although strong correlation between the bolometric luminosity of AGNs and the kinetic power of UFOs has been detected, indicating the faster winds are more likely to be driven in higher luminosity AGNs, however, UFOs may not necessarily be radiation pressure driven ([31]). There are some efforts trying to explain the origin of UFOs in the context of magnetocentrifugal driven wind model ([25–27]).

The first attempt is the application of the BP model to the UFOs found in PG 1211+143 ([25]). In this work, the authors first solve the steady-state axisymmetric MHD equations by assuming self-similarity. The fundamental quantity of the axisymmetric MHD equations is the magnetic stream function ψ, assumed to have a self-similar form $\psi(r,\theta) = (R/R_0)^q \widetilde{\psi}(\theta)\psi_o$, with ψ_o as the poloidal magnetic flux through the innermost disk radius at R_o. $\widetilde{\psi}(\theta)$ is the angular dependence of the magnetic flux which needs to be solved, and q is a free parameter which determines the radial dependence of the poloidal current. The self-similar forms of magnetic field, wind velocity, and wind number density are as follows,

$$\mathbf{B}(r,\theta) = (R/R_0)^{q-2}\widetilde{\mathbf{B}}(\theta)B_o \tag{9.4}$$

$$\mathbf{v}(r,\theta) = (R/R_0)^{-1/2}\widetilde{\mathbf{v}}(\theta)v_o \tag{9.5}$$

$$n(r,\theta) = (R/R_0)^{2q-3}\widetilde{n}(\theta)B_o^2 v_o^{-2} m_{\mathrm{p}}^{-1} \tag{9.6}$$

where, m_{p} is the proton mass. The angular dependence of the variables denoted by a tilde ($\sim$) can be obtained by solving the conservation equation and the momentum equations, with the boundary values on the disk (denoted by the subscript 'o') at ($R = R_o$, $\theta = 90°$). With the wind mass flux, the value of q, the inner and outer launching radii fixed, the wind solutions can be obtained. Second, the authors calculate the ionization structure of the wind assuming the presence of a point-like ionization spectrum at the origin. The ionization spectrum includes a multicolor disk component and a X-ray power-law component. The ionizing luminosity (X-ray plus EUV) is $\sim 1.25 \times 10^{44}$erg s^{-1}. The photoionization code XSTAR is used, which simultaneously solves the ionization balance and thermal equilibrium of the locally illuminated wind. It is found that the BP wind with $n \propto r^{\sim -1}$ can explain the observed UFOs in PG 1211+143. The observed UFOs of PG 1211+143, as manifested by the Fe XXV/Fe XXVI transition properties, are determined mainly by the wind mass flux, the temperature of the ionizing spectrum and the observer's viewing angle θ. It is found that under the BP model, the UFOs are mainly at a location of $\sim 1200 R_{\mathrm{s}}$, outflowing with a characteristic velocity of $v_{\mathrm{w}}/c \sim 0.1 - 0.2$.

The properties of UFOs vary with AGN continuum flux. For example, for PDS 456, the centroid energy of the blueshifted Fe K absorption profile increases with AGN luminosity and wind velocity v/c $\propto L_{7-30kev}^{0.22\pm0.04}$, which indicates the wind is predominantly radiatively driven due to the high Eddington ratio of the source ([55]). Such occurences have also been found in other sources accreting at high Eddington ratio, indicating that the UFOs may be driven by intense radiation pressure in sources accreting at or close to the Eddington rate, such as in APM 08279+5255 ([13]), PG 1211+143 ([70]), IRAS F11119+3257 ([94]), 1 H 0707-495 ([32]), and IRAS 13224-3809 ([67]).

However, we note that one can not rule out the magnetic driven origin of UFOs based on the phenomenon that the properties of UFOs vary with the host AGN continuum flux. For example, the variations of the properties of UFOs found in the radio-quiet quasar PDS 456 were modeled in the context of the BP wind model ([26]). In this work, the self-similar BP wind solution is applied. Assuming plausible dependence of X-ray luminosity on the accretion rate applicable to near-Eddington state, it is found that the photoionization calculations of the BP wind model can reproduce the observed correlations of the UFO velocity and the anti-correlation of the UFO absorber's equivalent width with the X-ray strength of PDS 456. This result indicates that even without radiation pressure, UFOs can be magnetically driven while also producing the observed spectrum and correlations with the AGN host spectrum.

9.4 Propagation of Wind on Large-Scale

All the above mentioned theoretical works (including both analytical works and numerical simulations) have focused on the accretion disk scale. For example, the simulations studying wind usually have an outer radial boundary located at several hundreds of R_s. Thus, the above works can only study the launching and propagation of winds on the accretion disk scale. Once the winds move out of the outer radial boundary of the simulations, one can not know the evolution of wind on scales much larger than the accretion disk. However, the evolution of winds on much larger scales is important for AGNs feedback study. Recent works have shown that winds play a significant role in the interaction between the AGNs and their host galaxies ([14–16, 35, 64, 98, 105]). Therefore, we must know whether the accretion disk winds can move to galaxy or ever larger scales. After arriving at the larger scales, what will the properties of wind be?

Cui et al. ([17, 18]) study the dynamics of wind on large scale. In these works, it is assumed that when the wind moves at large scales, the radiation pressure exerted by the radiation from the accretion disk on wind is neglected. They also assume that the wind evolves adiabatically. They found that for wind from cold thin disk, when the Bernoulli parameter is small, it cannot propagate far, but stops at a certain radius where the Bernoulli parameter is equal to the gravitational potential energy.

The assumption that radiation pressure is neglected when wind propagates at large scale may be problematic. Radiation line force plays an important role in driving winds process. When winds move at large-scale, line force may continuously accelerate winds. For example, Proga ([76]) found that even at parsec scale, the radiation line force can be important. Zhu et al. ([107]) restudied the propagation of cold line force driven wind at large-scale by numerical simulations. In these simulations, radiation line force on wind is included when wind moves at large-scale. It is found that when the disk luminosity exceeds $0.6L_{\rm Edd}$, independent of $M_{\rm BH}$, the line force driven wind has kinetic energy flux exceeding $1\%L_{\rm Edd}$, and can escape from the black hole potential (see Fig. 9.2 for an example for $M_{\rm BH} = 10^8 M_\odot$

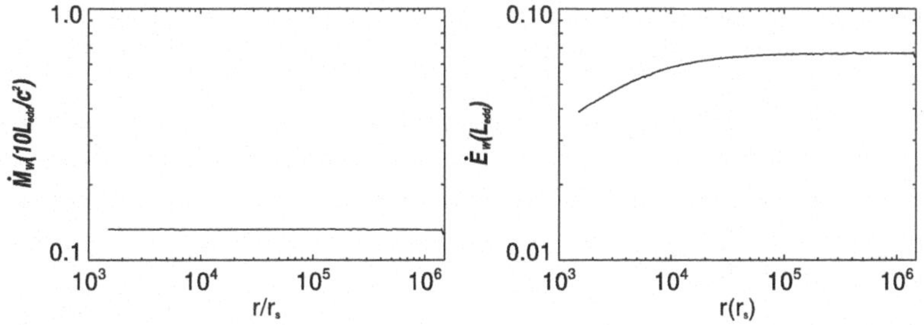

Fig. 9.2 Time-averaged radial profiles of mass flux (left panel) and kinetic power (right panel) of winds for model with $M_{\rm BH} = 10^8 M_\odot$ and $\epsilon = 0.6$. Figrure from Zhu et al. 2022, MNRAS, 513, 1141

and $\epsilon = 0.6$. In this figure, the radial profiles of the mass flux and kinetic power of wind are plotted). When the disk luminosity $\sim 0.3 L_{\rm Edd}$, in the case of $M_{\rm BH} \geq 10^7 M_\odot$, winds can escape from black hole potential; however, in the case $M_{\rm BH} \leq 10^6 M_\odot$, the line force driven wind can not escape from the black hole potential.

9.5 Role of Cold Wind in AGN Feedback

Supermassive black holes (SMBHs), with masses of $10^{6-10} M_\odot$, are universal in the center of galaxies with massive bulges. Observations show tight correlations between the black hole mass and various bulge properties, which indicate that the SMBHs co-evolves with their host galaxies ([36, 43]). The SMBH at galaxy center accretes gas from its host galaxy. In the accretion process, both radiation and outflow (including winds and jet) can be produced, which can interact with the ISM in the galaxy. The radiation can directly heat/cool the ISM. Also, radiation from the center can exert radiation pressure on ISM. Winds can directly blow away ISM. The interactions can significantly affect the properties of ISM, changing the star formation rate. In turn, the changing of properties of ISM will change the accretion circumstance of SMBHs, thus, changing the SMBH accretion rate and its growth. In this Chapter, we focus on the wind feedback.

There have been analyses about the effects of wind feedback on star formation and growth of SMBH. It has been suggested that it may be hard for the wind from the black hole accretion disk to directly couple to cold molecular gas, especially at kpc scales, the dominant reservoir for star formation ([40]). However, the wind can drive a wind in the hot diffuse ISM. When the hot ISM wind passes over a cold cloud, the cold gas can expand in the direction perpendicular to the incident outflow. The expansion of the cold gas may alone be enough to substantially suppress star formation in the host. Moreover, the expansion of cold gas can make itself susceptible to both ionization and momentum coupling from absorption of the incident quasar radiation field. This will drive the cold cloud at large radii to be fully ionized and driven into an outflow by radiation pressure, dramatically altering the gas properties and star formation rate in the host galaxy ([40]). The effects of wind feedback in an isolated elliptical galaxy have been studied ([15]). It is found that wind from AGN can effectively affect the inner parts ($\sim$100 pc) of elliptical galaxies, significantly reducing the gas infall to the central black hole. Therefore, in addition to star formation, the wind feedback can also significantly influence the black hole growth. For example, Ostriker et al. ([64]) show that the omission of wind feedback can lead to a hundred-fold increase in the mass of the SMBH to over $10^{10} M_\odot$. When the wind feedback is taken into account, the final SMBH mass is much lower.

Very recently, the roles of AGN feedback on the late-stage evolution of a massive elliptical galaxies are investigated by high resolution numerical simulations ([108]). In this work, the state-of-the-art AGN physics are adopted. The AGN feedback is found to be crucial in keeping the low black hole accretion rate and suppressing the star formation. AGN wind can compensate for the radiative cooling of the gas in the galaxy. Although AGN spends most of its time in hot (radio) mode, the cumulative energy output is dominated by the outburst of the code (quasar) mode. The wind from cold mode can sweep up the gas from stellar mass-loss. The cold mode wind plays a dominant role in the late-stage evolution of massive elliptical galaxies.

Acknowledgements The authors are supported by the Natural Science Foundation of China (grants 12173065, 12133008, 12192220, 12192223).

References

1. N. Arav et al., APJ **857**, Article id. 60 (2018)
2. X. Bai, J.M. Stone, APJ **767**, Article id. 30 (2013)
3. S.A. Balbus, J.F. Hawley, APJ **376**, 214 (1991)
4. M.C. Begelman, C.F. McKee, G.A. Shields, APJ **271**, 70–88 (1983)
5. R.D. Blandford, D.G. Payne, MNRAS **199**, 883–903 (1982)
6. A.J. Blustin et al., A & A **431**, 111–125 (2005)
7. D. Bu, X. Yang, APJ **921**, Id. 100 (2021)
8. C.R. Canizares, J. Kruper, APJ **278**, L99–L102 (1984)
9. X. Cao, H.C. Spruit, A & A **287**, 80–86 (1994)
10. X. Cao, AP & SS **226**, 165–168 (1995)
11. X. Cao, APJ **783**, Article id. 51 (2014)
12. J.I. Castor et al., APJ **195**, 157–174 (1975)
13. G. Chartas, W.N. Brandt, S.C. Gallagher, APJ **595**, 85–93 (2003)
14. Z. Chen et al., Nat. Astron. **6**, 339–343 (2022)
15. L. Ciotti, J.P. Ostriker, D. Proga, APJ **717**, 708–723 (2010)
16. L. Ciotti, et al., APJ **835**, Article id. 15 (2017)
17. C. Cui, F. Yuan, B. Li, APJ **890**, Id. 80 (2020)
18. C. Cui, F. Yuan, APJ **890**, Id. 81 (2020)
19. L. Dai et al., APJ **859**, Article id. L20 (2018)
20. A. Dorodnitsyn, T. Kallman, D. Proga, APJ **687**, 97–110 (2008)
21. G.E. Eggum, F.V. Coroniti, J.I. Katz, APJ **330**, 142 (1988)
22. C. Faucher-Giguere, E. Quataert, MNRAS **425**, 605–622 (2012)
23. S. Fromang et al., A & A. **552**, Id. A71 (2013)
24. K. Fukumura et al., APJ **715**, 636–650 (2010)
25. K. Fukumura et al., APJ **805**, Article id. 17 (2015)
26. K. Fukumura et al., APJL **864**, Article id. L27 (2018)
27. K. Fukumura et al., APJ **940**, Id. 6 (2022)
28. I.M. George et al., APJS **114**, 73–120 (1998)
29. R.R. Gibson et al., APJ **692**, 758–777 (2009)
30. J. Gofford et al., MNRAS **430**, 60–80 (2013)
31. J. Gofford et al., MNRAS **451**, 4169–4182 (2015)
32. K. Hagino, MNRAS **461**, 3954–3963 (2016)
33. J.P. Halpern, APJ **281**, 90–94 (1984)
34. F. Hamann et al., MNRAS **435**, 133–148 (2013)
35. Z. He et al., Sci. Adv. **8**, Id. eabk3291 (2022)
36. T.M. Heckman, P.N. Best, ARA & A **52**, 589–660 (2014)
37. N. Higginbottom et al., APJ **789**, Article id. 19 (2014)
38. Z. He et al., APJS **229**, Article id. 22 (2017)
39. Z. He et al., NatAs **3**, 265–271 (2019)
40. P.F. Hopkins, M. Elvis, MNRAS **401**, 7–14 (2010)

41. H. Hu et al., APJ **934**, Id. 132 (2022)
42. S.X. Kato, T. Kudoh, K. Shibata, APJ **565**, 1035–1049 (2002)
43. J. Kormendy, L.C. Ho, ARA & A **51**, 511–653 (2013)
44. R. Krasnopolsky, Z. Li, R. Blandford, APJ **526**, 631–642 (1999)
45. R. Krasnopolsky, Z. Li, R. Blandford, APJ **595**, 631–642 (2003)
46. J.H. Krolik, G.A. Kriss, APJ **447**, 512 (1995)
47. J.H. Krolik, G.A. Kriss, APJ **561**, 684–690 (2001)
48. S. Laha et al., MNRAS **441**, 2613–2643 (2014)
49. S. Laha et al., MNRAS **457**, 3896–3911 (2016)
50. S. Laha et al., NatAs **5**, 13–24 (2021)
51. G. Lesur, M. W. Kunz, S. Fromang, A & A **566**, Id. A56 (2014)
52. C. Liu et al., MNRAS **434**, 1721–1735 (2013)
53. G. Liu, N. Arav, D.S.N. Rupke, APJS **221**, Article id. 9 (2015)
54. G.M. Madejski et al., APJ **370**, 198 (1991)
55. G.A. Matzeu et al., MNRAS **472**, L15–L19 (2017)
56. M. Mizumoto et al., MNRAS **489**, 1152–1160 (2019)
57. M. Mizumoto et al., MNRAS **503**, 1442–1458 (2021)
58. A. Mosallanezhad et al., MNRAS **490**, 2567–2578 (2019)
59. N. Murray et al., APJ **451**, 498 (1995)
60. M. Nomura et al., PASJ **68**, Id. 16 (2016)
61. M. Nomura, K. Ohsuga, MNRAS **465**, 2873–2879 (2017)
62. M. Nomura, K. Ohsuga, C. Done, MNRAS **494**, 3616–3626 (2020)
63. K. Ohsuga, M. Mori, T. Nakamoto, S. Mineshige, APJ **628**, 368–381 (2005)
64. J.P. Ostriker et al., APJ **722**, 642–652 (2010)
65. R. Ouyed, R.E. Pudritz, APJ **482**, 712–732 (1997)
66. S.P. Owocki, S.R. Cranmer, K.G. Gayley, AP & SS **260**, 149–159 (1998)
67. M.L. Parker et al., MNRAS **469**, 1553–1558 (2017)
68. N.A. Pereyra, T.R. Kallman, J.M. Blondin, APJ **477**, 368–378 (1997)
69. O. Porth, C. Fendt, APJ **709**, 1100–1118 (2010)
70. K.A. Pounds et al., MNRAS **345**, 705–713 (2003)
71. D. Proga, J.M. Stone, J.E. Drew, MNRAS **295**, 595–617 (1998)
72. D. Proga, J.M. Stone, J.E. Drew, MNRAS **310**, 476–482 (1999)
73. D. Proga, J.M. Stone, T.R. Kallman, APJ **543**, 686–696 (2000)
74. D. Proga, APJ **585**, 406–417 (2003)
75. D. Proga, T.R. Kallman, APJ **616**, 688–695 (2004)
76. D. Proga, APJ **661**, 693–702 (2007)
77. D. Proga, J.P. Ostriker, R. Kurosawa, APJ **676**, 101–112 (2008)
78. R.E. Pudritz, C.A. Norman, APJ **301**, 571 (1986)
79. J.N. Reeves, P.T. O'Brien, M.J. Ward, APJ **593**, L65–L68 (2003)
80. J.N. Reeves et al., APJL **854**, Article id. L8 (2018)
81. C.S. Reynolds, MNRAS **286**, 513–537 (1997)
82. A. Sadowski, R. Narayan, MNRAS **453**, 3213–3221 (2015)
83. A. Sadowski, J. Lasota, M.A. Abramowicz, R. Narayan, MNRAS **456**, 3915–3928 (2016)
84. N.I. Shakura, R.A. Sunyaev, A & A **24**, 337-355
85. J.M. Stone, M.L. Norman, APJ **433**, 746 (1994)
86. T.K. Suzuki, S. Inutsuka, APJL **691**, L49–L54 (2009)
87. T.K. Suzuki, T. Muto, S. Inutsuka, APJL **718**, 1289–1304 (2010)
88. T. K. Suzuki, S. Inutsuka, APJ **784**, Article id. 121 (2014)
89. F. Tombesi et al., A & A **521**, Id. A57 (2010)
90. F. Tombesi et al., APJ **719**, 700–715 (2010)
91. F. Tombesi et al., APJ **742**, Article id. 44 (2011)
92. F. Tombesi et al., MNRAS **430**, 1102–1117 (2013)
93. F. Tombesi et al., MNRAS **443**, 2154–2182 (2014)
94. F. Tombesi et al., Nature **519**, 436–438 (2015)

95. Y. Uchida, K. Shibata, PASJ **37**, 515–535 (1985)
96. A. Utsumi et al., APJ **935**, Id. 26 (2022)
97. W. Wang, D. Bu, F. Yuan, MNRAS **513**, 5818–5828 (2022)
98. R. Weinberger et al., MNRAS **479**, 4056–4072 (2018)
99. R.J. Weymann, R.F. Carswell, M.G. Smith, ARA & A **19**, 41–76 (1981)
100. D.T. Woods et al., APJ **461**, 767 (1996)
101. X. Yang, F. Yuan, K. Ohsuga, D. Bu, APJ **780**, Article id. 79 (2014)
102. X. Yang, K. Ablimit, Q. Li, APJ **914**, Id. 31 (2021)
103. X. Yang, APJ **922**, Id. 262 (2021)
104. S. Yoshioka, S. Mineshige, K. Ohsuga, PASJ **74**, 1378–1395 (2022)
105. F. Yuan et al., APJ **857**, Article id. 121 (2018)
106. Z. Zhu, J.M. Stone, APJ **857**, Article id. 34 (2018)
107. Y. Zhu et al., MNRAS **513**, 1141–1153 (2022)
108. B. Zhu et al., MNRAS **525**, 4840–4853 (2023)

Chapter 10
Thin Accretion Disks in GR-MHD Simulations

Indu K. Dihingia and Christian Fendt

Abstract We review some recent results of general relativistic magnetohydrodynamic (GR-MHD) simulations considering the evolution of geometrically thin disks around a central black hole. Thin disk GR-MHD simulations complement the widely used MAD (Magnetically Arrested Disk) or SANE (Standard And Normal Evolution) approaches of evolving from an initial disk torus. In particular, we discuss the dynamical evolution of the disk, its role in the formation of disk winds or jets, the impact of disk resistivity, and its potential role in generating magnetic flux by an internal disk dynamo. The main characteristics of a thin disk in our approach are the Keplerian rotation of the disk material, which allows to launch disk outflows by the Blandford-Payne magneto-centrifugal effect, in addition to the Blandford-Znajek-driven spine jet from the black hole ergosphere. Thus, for this approach, we neglect disk thermodynamics and radiative effects, concentrating predominantly on the dynamical evolution of the system. Resistive MHD further allows the investigation of physical reconnection and also dynamo action. Magnetic reconnection may generate magnetic islands of plasmoids that are ejected from the disk along with the outflow. We also discussed potential applications of thin disk in explaining the decaying phase of an outburst in black hole X-ray binaries (BH-XRBs). Post-processing of radiation using the simulated dynamical data allows to derive spectra or fluxes, e.g., in the X-ray band, and to derive potential variability characteristics.

I. K. Dihingia (✉)
Tsung-Dao Lee Institute, Shanghai Jiao-Tong University, Shanghai, 520 Shengrong Road, Shanghai 201210, People's Republic of China
e-mail: ikd4638@gmail.com; ikd4638@sjtu.edu.cn

C. Fendt
Max Planck Institute for Astronomy, Königstuhl 17, Heidelberg 69117, Germany
e-mail: fendt@mpia.de

C. Bambi et al. (eds.), *New Frontiers in GRMHD Simulations*, Springer Series in Astrophysics and Cosmology, https://doi.org/10.1007/978-981-97-8522-3_10

10.1 Introduction

Thin accretion disks have been instrumental for several decades in explaining a wide range of observed astrophysical sources, including X-ray binaries (XRBs) and active galactic nuclei (AGNs). This is evident from the enormous citation metric for the paper by Shakura and Sunyaev from 1973 [108] (SS73, see Fig. 10.1), which is the milestone article for thin disks. SS73 deals the accretion flow in Newtonian approach, which was subsequently developed to its relativistic version by Novikov and Thorne [87]. Thin disk models are most relevant if the astrophysical source is emitting multi-colour black body radiation (e.g., in the soft state of XRBs).

Along with the thin disk model, many other accretion disk models have been proposed with the increase in varieties of astrophysical sources. For example, spherically symmetric Bondi flows [15], advection dominated accretion flows (ADAF) [79], trans-sonic accretion disks [20, 48], ultra-luminous optically thick geometrically thick disks [1], two-component accretion flows (TCAF) [22], etc. However, due to the complex theoretical nature of these models, they are not widely used as the thin disk model by astronomers. With the development of cutting edge general relativistic magneto-hydrodynamic (GR-MHD) simulations, further families of accretion disks have evolved.

Typically, GR-MHD simulations of accretion start with the initial condition of a hydrodynamically stable torus developed by Fishbone and Moncrief [46]. Threaded by a weak magnetic field, the initial stable torus configuration becomes unstable and evolves into a disky structure. Depending on the initial magnetic flux, the torus evolves into a SANE disk (standard and normal evolution) or a MAD disk (magnetically arrested disk) [e.g., 97, 114]. These two solutions are seen as standard solutions by many in the scientific community. These models (SANE and MAD) have been used to understand near horizon physics around supermassive black holes Sgr A* and M 87*. Undoubtedly, these models adequately constrain the parameters of these two

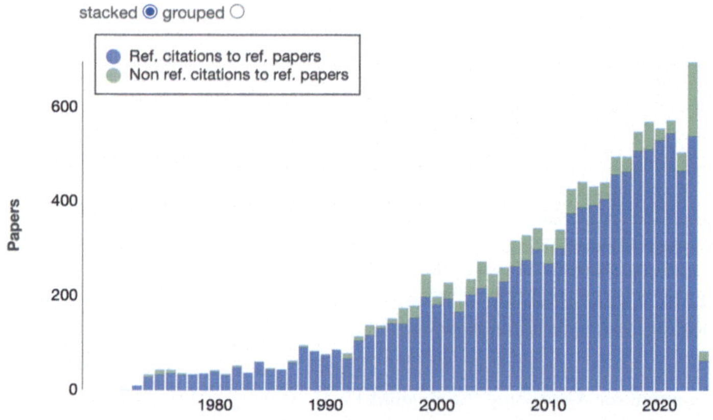

Fig. 10.1 Citation metric for Shakura and Sunyaev [108], accumulating 11,850 citations by March 2024

supermassive black holes, such as mass, accretion rate, inclination, and black hole spin [e.g., 38–41]. However, many observations of the "soft excess" feature found in many Seyfert galaxies (e.g., [14, 120, 121]) can be understood with the high-density reflection model (e.g., [49]). Note that the "warm corona" model is one of the candidates to explain "soft excess" in Seyfert galaxies (e.g., [8, 70]). This is also supported by the soft X-ray reverberation signature (e.g., [25, 53]). These models essentially rely on accretion through a thin disk geometry. X-ray reflection models have indicated that the inner disk radius of the accretion disk is very close to the innermost stable circular orbit (ISCO) for several XRBs even in luminous hard-state [see for detail 63]. The caveat is that this model approach uses a "lamp-post" corona model, where a hot corona is considered at some vertical height from the black hole (e.g., [64–66]). In summary, it is clear that at least some astrophysical sources (both AGNs and BH-XRBs), if not all, require a thin disk geometry around black holes in order to explain their observations.

Thus, the model of a thin disk evolution in GR-MHD is still appealing. We know from non-relativistic simulations that such Keplerian thin disks can easily launch a strong disk wind that can evolve into a disk jet. There is no reason to doubt that such a thin disk can also be present around (supermassive) black holes. With such a disk, a two-component jet structure for rotating black holes is feasible: a Blandford-Znajek (hereafter BZ; [12]) driven Poynting flux driven high-speed jet launched from the spinning black hole and a mass loaded disk jet (driven by the Blandford-Payne magneto-centrifugal mechanism; hereafter BP; [13]) from the surrounding accretion disk. Recent GR-MHD simulations were indeed able to compare the energy output of both jet components [100, 101, 119].

A number of studies have been conducted considering different accretion flow models, including thin disks. In this book chapter, we will discuss recent developments in GR-MHD simulations of jet launching and other aspects considering thin disks. Other chapters will provide a comprehensive understanding of the current research on various other accretion flow models and their implications for astrophysical phenomena. Additionally, we will explore the current limitations and future directions of GR-MHD simulations in understanding the dynamics of thin disk accretion flows.

10.2 Basic Equations

In order to derive the magnetized flow properties, a set of GR-MHD equations need to be solved. The basic resistive GR-MHD equations are as follows:

$$
\begin{aligned}
\nabla_\mu \left(\rho u^\mu\right) &= 0, \\
\nabla_\mu T^{\mu\nu} &= 0, \\
\nabla_\mu F^{\mu\nu} &= \mathcal{J}^\mu, \\
\nabla_\mu {}^*F^{\mu\nu} &= 0,
\end{aligned}
\tag{10.1}
$$

where different symbols have their usual meaning, such as $\rho, u^\mu, T^{\mu\nu}, F^{\mu\nu}, {}^*F^{\mu\nu}$, and $\mathscr{J}^\mu$ represent the rest-mass density, four velocity, energy-momentum tensor, Faraday tensor, dual of the Faraday tensor, and electric 4-current, respectively. The details of these terms, as well as the explicit procedures for solving them, are given in [88, 98, 102, 105, 119].

These equations are solved using a Modified Kerr-Schild (MKS) geometry. By selecting an appropriate MKS stretching parameter, we ensure that the concentration of maximum resolution is near the equatorial plane of the simulation domain [67]. In this chapter, we adopt a generalized unit system with $G = M = c = 1$ to express any quantities. Here, M represents the mass of the black hole, G represents the universal gravitational constant, and c represents the speed of light. The unit system employed here represents mass, length, and time using M, $r_g = GM/c^2$, and $t_g = GM/c^3$, respectively. Also, we will be following the sign convention for the metric as $(-, +, +, +)$, with four velocities satisfying $u_\mu u^\mu = -1$.

In this chapter, we will only concentrate on thin disks around Kerr black holes. Accordingly, the Kerr metric in Kerr-Schild (KS) coordinates (t, r, θ, ϕ) is given by,

$$\begin{aligned} ds^2 = &-\left(1 - \frac{2r}{\Sigma}\right) dt^2 + \left(\frac{4r}{\Sigma}\right) dr\, dt + \left(1 + \frac{2r}{\Sigma}\right) dr^2 + \Sigma d\theta^2 \\ &+ \sin^2\theta \left[\Sigma + a^2 \left(1 + \frac{2r}{\Sigma}\right) \sin^2\theta \right] d\phi^2 - \left(\frac{4ar\sin^2\theta}{\Sigma}\right) d\phi\, dt \\ &- 2a\left(1 + \frac{2r}{\Sigma}\right) \sin^2\theta d\phi\, dr, \end{aligned} \tag{10.2}$$

McKinney and Gammie [67], where $\Sigma \equiv r^2 + a^2\cos^2\theta$ and a is the rotation parameter for the Kerr black hole. The determinant of the metric will also be often used in some calculations, which is given by $g = -\Sigma^2 \sin^2\theta$. Finally, the coordinate transformations from KS (t, r, θ, ϕ) to MKS (x_0, x_1, x_2, x_3) are related by,

$$t = x_0\ ;\ r = R_0 + e_1^x\ ;\ \theta = \pi x_2 + \frac{1}{2} h \sin(2\pi x_2)\ ;\ \text{and}\ \phi = x_3. \tag{10.3}$$

The parameter h can be adjusted to concentrate grid zones towards the equator as it is increased from 0 to 1. In practice, the initial conditions are supplied to the GR-MHD codes in Boyer-Lindquist (BL) coordinates, and after proper transformations, codes solve them in MKS coordinates. Below, we show all the initial conditions in BL coordinates.

Accretion disks may also exist around other exotic objects or on non-Kerr backgrounds proposed in the literature [e.g., 4, 9, 43, 116, and many more]. However, there are very few GR-MHD thin disk simulations performed on such backgrounds (e.g., [76]); only semi-analytical accretion solutions are available [e.g., 10, 23, 27, 42, 93, 117]. It would be interesting to explore these alternative scenarios in future studies to understand the behaviour of accretion disks in diverse gravitational environments. Such investigations could provide valuable insights into the dynamics and observational signatures of accretion processes beyond Kerr black holes.

10.3 The Thin Disk Prescription

GR-MHD simulations rely on the initial conditions, such as initial density, velocities, magnetic field components, resistivity profiles, etc. In this section, we give the basic profiles of these initial conditions. Generally, for density and velocities, there are two broad ways of setting them: a (i) Shakura-Sunyaev-like disk model [108], or a (ii) Novikov-Thorne relativistic disk model [87]. In the following, we briefly discuss them explicitly, along with other initial conditions.

10.3.1 The Shakura-Sunyaev Disk (SSD)

One option to prescribe an initially thin disk is to apply a density and pressure distribution typically used in non-relativistic MHD simulations of jet launching [19, 110, 112, 125]. This disk setup has been proven to be stable in the hydrodynamical limit and evolves smoothly for the MHD case, in which strong disk winds can be launched that may turn into collimated jets. The idea for applying this setup to the GR-MHD approach is that relativistic deviations are minor for most parts of the disk, assuming an inner disk radius outside the ISCO. After all, there is a dynamical evolution of the disk-outflow system that will settle into a new dynamical equilibrium considering the effects of accretion, outflow, and relativity.

In the simulations described below, an initial disk density distribution is applied as described by a non-relativistic vertical equilibrium profile, such as applied in [100, 110, 119, 125], slightly modified to fit into the GR-MHD code,

$$\rho(r,\theta) = \left[\frac{\Gamma-1}{\Gamma}\frac{r_{\rm in}}{r}\frac{1}{\epsilon^2}\left(\sin\theta + \epsilon^2\frac{\Gamma}{\Gamma-1}\right)\right]^{1/(\Gamma-1)}, \tag{10.4}$$

Here, $r_{\rm in}$ is the initial inner disk, and $\epsilon = H/r$ is the initial disk aspect ratio as defined by the vertical equilibrium of a disk with a local pressure scale height $H(r)$. The pressure and internal energy are given by the polytropic equation of state $p = K\rho^\Gamma$, where K is the polytropic constant, and the internal energy $u = p/(\Gamma-1)$.

Conveniently, for the initial orbital velocity a profile following [90] can be applied,

$$\tilde{u}^\phi = r^{-3/2}\left(\frac{r}{r - R_{\rm PW}}\right), \tag{10.5}$$

where the smoothing length scale for the gravity $R_{\rm PW}$ is a constant of choice, for example chosen to be equal to the gravitational radius. This approximation is applied in the ϕ-component of the fluid velocity u^ϕ.

The initial velocity profile of the disk will not be in equilibrium with the underlying general relativistic gravity around the black hole. However, with temporal evolution, the disk rotational velocity will soon approach a new dynamical equilibrium, which depends on governing GR-MHD variables such as gas pressure and Lorentz forces.

10.3.2 The Novikov-Thorne Disk (NTD)

In thin disk approximation, most of the matter occupies the equatorial plane of the coordinate system ($\theta = \pi/2$) and the θ component of the four-velocity u^μ can be neglected (*i.e.*, $u^\theta = 0$). With these approximations, the accretion rate ($\dot{M}_0$), time average radiation flux (F), and time average torque (W^r_ϕ) are obtained as [87, 91],

$$\begin{aligned}
\dot{M}_0 &= -2\pi r \Sigma u^r, \\
F &= \left(\dot{M}_0/4\pi r\right) f, \\
W^r_\phi &= \left(\dot{M}_0/2\pi r\right)\left[(E^\dagger - \Omega L^\dagger)/(-\Omega_{,r})\right] f,
\end{aligned} \tag{10.6}$$

where $E^\dagger = -u_{t,e}$, $L^\dagger = u_{\phi,e}$, and $\Omega = u^\phi_e/u^t_e$. The subscript e represents quantities defined on the equatorial plane. Considering the flow motion to be Keplerian ($u^r \sim 0$), i.e., rotation supported accretion disk, the explicit expression of f is obtained as

$$f = -\Omega_{,r}(E^\dagger - \Omega L^\dagger)^{-2}\int_{r_{ms}}^{r}(E^\dagger - \Omega L^\dagger)L^\dagger_{,r}dr. \tag{10.7}$$

Here, r_{ms} is the marginally stable radius or the ISCO radius: (for Kerr black hole it is the radius at which $\frac{dE^\dagger}{dr} = \frac{dL^\dagger}{dr} = 0$). The explicit expression of f is given by

$$\begin{aligned}
f =& \frac{3}{2}\frac{1}{x^2\left(2a + x^3 - 3x\right)}\left[x - x_0 - \frac{3}{2}\ln\left(\frac{x}{x_0}\right)\right. \\
&- \frac{3\,(s_1 - a)^2}{s_1(s_1 - s_2)(s_1 - s_3)}\ln\left(\frac{x - s_1}{x_0 - s_1}\right) \\
&- \frac{3\,(s_2 - a)^2}{s_2(s_2 - s_1)(s_2 - s_3)}\ln\left(\frac{x - s_2}{x_0 - s_2}\right) \\
&\left.- \frac{3\,(s_3 - a)^2}{s_3(s_3 - s_1)(s_3 - s_2)}\ln\left(\frac{x - s_3}{x_0 - s_3}\right)\right],
\end{aligned} \tag{10.8}$$

where we use $x = \sqrt{r}$, implying $x_0 = r_{ms}^{1/2}$, and s_1, s_2 and s_3 are the three real roots of $s^3 - 3s + 2a = 0$, which are given bellow,

$$\begin{aligned} s_1 =& 2\cos\left(\frac{1}{3}\cos^{-1}(a) - \frac{\pi}{3}\right), \\ s_2 =& 2\cos\left(\frac{1}{3}\cos^{-1}(a) + \frac{\pi}{3}\right), \\ s_3 =& -2\cos\left(\frac{1}{3}\cos^{-1}(a)\right). \end{aligned} \tag{10.9}$$

By considering the flux to be emitted as a multi-color black body, we can calculate the temperature of the thin disk. The explicit form of the temperature is given by:

$$\frac{p_e}{\rho_e} \propto T_{\rm bb}(x) \propto F^{1/4} = \Theta_0 \left(\frac{f(x)}{x^2}\right)^{1/4}, \tag{10.10}$$

where p_e and ρ_e are the pressure and density at the equatorial plane. Θ_0 is a constant related to the initial temperature of the disk. With the polytropic equation of state $p = \mathscr{K}\rho^{\Gamma}$, the density of the fluid at the equatorial plane is expressed as,

$$\rho_e = \left(\frac{\Theta_0}{\mathscr{K}}\right)^{1/(\Gamma-1)} \left(\frac{f(x)}{x^2}\right)^{1/(4(\Gamma-1))}, \tag{10.11}$$

where $\mathscr{K}$ is a constant associated with the flow's entropy. The density profile at the equatorial plane ($\theta = \pi/2$) is provided by Eq. (10.11). The general density profile for the off-equatorial plane can be formulated as follows:

$$\rho(r,\theta) = \rho_e \exp\left(-\frac{\alpha^2 z^2}{H^2}\right); \quad z = r\cos(\theta). \tag{10.12}$$

Equation (10.12) is written considering density drops along the vertical direction that follow a normal distribution. In order to ensure a thin disk geometry, we choose $\alpha = 2$. Here, H is the scale height of the initial accretion disk. We follow Riffert and Herold [103] and Peitz and Appl [94] to obtain H,

$$H^2 = \frac{p_e r^3}{\rho_e \mathscr{F}}, \tag{10.13}$$

where

$$\mathscr{F} = \gamma_\phi^2 \frac{\left(a^2 + r^2\right)^2 + 2a^2\Delta}{\left(a^2 + r^2\right)^2 - 2a^2\Delta},$$

with $\gamma_\phi^2 = (1 - \Omega\lambda)^{-1}$ and $\Delta = r^2 - 2r + a^2$ where $\lambda = -u_{\phi,e}/u_{t,e}$.

Along with the density profile, we need to supply the initial four velocities (u^μ). To calculate u^ϕ, we consider that the fluid element follows the geodesic equation,

$$u^t u^t \Gamma^r_{tt} + 2u^t u^\phi \Gamma^r_{t\phi} + u^\phi u^\phi \Gamma^r_{\phi\phi} = 0. \tag{10.14}$$

Also, the four velocities must satisfy

$$g_{\mu\nu} u^\mu u^\nu = g_{\phi\phi} u^\phi u^\phi + g_{tt} u^t u^t + 2g_{t\phi} u^t u^\phi = -1. \tag{10.15}$$

Here, $g_{\mu\nu}$ is the metric tensor, and $\Gamma^\alpha_{\beta\gamma}$ are the Christoffel symbols. Solving Eqs. (10.14) and (10.15), we obtain the expressions of u^ϕ and u^t. The explicit expression of u^ϕ is given by

$$u^\phi(r,\theta) = \left(\frac{\mathscr{A}}{\mathscr{B} + 2\mathscr{C}^{1/2}}\right)^{1/2}, \tag{10.16}$$

where

$$\begin{aligned}
\mathscr{A} &= \left(\Gamma^r_{tt}\right)^2,\\
\mathscr{B} &= g_{tt}\left(\Gamma^r_{tt}\Gamma^r_{\phi\phi} - 2\Gamma^{r\,2}_{t\phi}\right) + 2g_{t\phi}\Gamma^r_{tt}\Gamma^r_{t\phi} - g_{\phi\phi}\Gamma^{r2}_{tt},\\
\mathscr{C} &= \left(\Gamma^{r\,2}_{t\phi} - \Gamma^r_{tt}\Gamma^r_{\phi\phi}\right)\left(g_{t\phi}\Gamma^r_{tt} - g_{tt}\Gamma^r_{t\phi}\right)^2.
\end{aligned}$$

The explicit expression of $u^t(r,\theta)$ can be calculated by substituting $u^\phi(r,\theta)$ to any of the equations, either Eqs. (10.14) or (10.15). Thus, the expressions of u^ϕ and u^t on the equatorial plane ($\theta = \pi/2$) are given by

$$\begin{aligned}
u^\phi_e &= \frac{1}{\sqrt{x^3\left(2a + x^3 - 3x\right)}},\\
u^t_e &= \frac{a + x^3}{\sqrt{x^3\left(2a + x^3 - 3x\right)}}.
\end{aligned} \tag{10.17}$$

After all, $\rho(r,\theta)$ and $u^\phi(r,\theta)$ obtained in Eqs. (10.12) and (10.16) serve as initial conditions for the thin disk for simulation models.

10.3.3 The Initial Magnetic Field

Along with the velocities and density of a rotation-supported thin disk, one needs to supply initial magnetic fields that trigger magneto-rational instabilities (MRI). MRI transports angular momentum [7], thereby setting accretion in the system. Moreover, MRI aids in the development of large-scale magnetic fields, which are responsible for launching relativistic jets and winds. Typically, in GR-MHD codes, a vector potential for inclined large-scale magnetic fields threading the disk is supplied. One such vector potential, involving non-zero ϕ components, is described by [119, 125].

$$A_\phi \propto (r \sin\theta)^{3/4} \frac{m^{5/4}}{\left(m^2 + (\tan\theta)^{-2}\right)^{5/8}}. \tag{10.18}$$

The parameter m is related to the initial inclination of the field lines and determines the system's magnetic flux. The parameter m is crucial for launching BP type wind from the accretion disk [13, 32]. The strength of the poloidal magnetic field depends on the initial supplied minimum plasma-β parameter $\beta_{\rm min}$.

10.3.4 Resistivity and Magnetic Diffusivity

The major breakthrough of the thin disk model by Shakura and Sunyaev came with their introduction of a *turbulent viscosity* ν as the major agent for angular momentum transport in accretion disks. This was accompanied by viscous heating that can result in the observed luminosities in stellar compact binaries or quasars.

This turbulent viscosity is understood as an *anomalous* viscosity, thus much stronger than the microscopic molecular viscosity. At that time, the cause of the turbulence was yet unknown.[1] Fundamentally, [108] introduced the terminology of an α-viscosity, which can be approximated in strength by the typical sound speed c_S in the disk and a typical length scale H (the disk scale height), resulting in values of $\alpha < 1$ that determine the strength of the viscous stress and the disk angular momentum transport, involving the turbulent viscosity $\nu = \alpha c_s H$, and involving the tangential stress between two disk layers, $w_{r,\phi} = -\alpha \rho c_s^2$.

Similar to the turbulent viscosity, one may infer a turbulent resistivity or, respectively, a turbulent magnetic diffusivity η. The interpretation is that turbulent motions in the disk allow the material to diffuse across the magnetic field. In particular, this allows for mass loading onto the disk wind from the accreting material. Connected to turbulent diffusivity and resistivity, there are two major physical effects that play roles in disk-jet evolution. One is the Ohmic heating of the disk gas, and the other is the reconnection of the magnetic field.

In non-relativistic MHD, the resistive term enters the system of equations via the induction equation for the field evolution and the energy equation for the Ohmic heating. In general relativity, the situation is more complex as resistivity enters via Ohm's law and the prescription of the electric field,

$$e^\mu = \eta j^\mu \tag{10.19}$$

[1] It is now accepted that it is caused by the MRI [6].

with the electric four vector e^μ and the electric current density in the comoving observer's frame (in contrast to $\mathscr{J}^\mu$ as seen by the normal observer). The electric field observed by a normal observer, split into the usual 3+1 form is $\mathscr{E}^\mu$. Bucciantini and Del Zanna [17] have worked out a closure relation for general relativistic resistive MHD that treats the time evolution of the electric field four vector $\mathcal{E}^\mu$ as

$$\gamma^{-1/2}\partial_t\left(\gamma^{1/2}\boldsymbol{\mathcal{E}}\right) - \nabla \times (\alpha\boldsymbol{\mathcal{B}} - \boldsymbol{\beta} \times \boldsymbol{\mathcal{E}}) + (\alpha\mathbf{v} - \boldsymbol{\beta}) = -\frac{\alpha\Gamma}{\eta}\left[\boldsymbol{\mathcal{E}} + \mathbf{v} \times \boldsymbol{\mathcal{B}} - (\boldsymbol{\mathcal{E}} \cdot \mathbf{v})\,\mathbf{v}\right] \tag{10.20}$$

with a Lorentz factor Γ, the determinant of its spatial 3-metric $\gamma = \sqrt{-g}/\alpha$, while $\boldsymbol{\beta} = \beta^i$ is the spatial shift vector in the 3+1 formalism. We refer to [17, 100, 119] for a detailed derivation. Note the η in the denominator on the r.h.s which makes the resistive treatment stiff for low diffusivity.

For alternative implementations of resistivity in (general) relativistic MHD codes and further applications, do refer to [18, 34–36, 55, 71, 92, 104, 105].

10.3.5 A Mean-Field Disk Dynamo

An interesting property of the closure relations for relativistic resistivity discussed above is that, in addition to the diffusive (loss) term for the magnetic field, a source term can also be added in a similar fashion. That provides the option to implement a mean-field dynamo term [17, 118], similar to previous non-relativistic studies [113, 126], that apply the mean-field dynamo theory [56, 99, 111].

Here, the mean-field usual α-dynamo parameter is replaced by $\xi \equiv -\alpha$ to avoid confusion with the gravitational lapse. This parameter measures the strength of the dynamo and is thought to result from a mean effect of disk turbulence. Consequently, when turbulence decays, the dynamo effect becomes weaker. This is also understood as dynamo quenching.

In covariant form and in the fluid frame, Ohm's law is written as

$$e^\mu = \eta j^\mu + \xi b^\mu, \tag{10.21}$$

where j^μ is again the four vector of the electric current density. Essentially, now the electric field can no longer be calculated by the cross-product of fluid velocity and magnetic field, and new equations need to be formulated for the evolution of the electric field,

$$\begin{aligned}
&\gamma^{-1/2}\partial_t\left(\gamma^{1/2}\mathcal{E}^i\right) - \epsilon^{ijk}\partial_j\left(\alpha\mathcal{B}_k - \epsilon_{knm}\mathcal{E}^m\beta^n\right) \\
&= -\alpha\left[\mathcal{E}^i + \epsilon^{ijk}v_j\mathcal{B}_k - (\mathcal{E}^k v_k)v^i\right]/\eta \\
&+\alpha\xi\left[\mathcal{B}^i - \epsilon^{ijk}v_j\mathcal{E}_k - (\mathcal{B}^k v_k)v^i\right]/\eta \\
&-q(\alpha v^i - \beta^i).
\end{aligned} \tag{10.22}$$

(Note the difference to Eq. 10.20). For a more detailed derivation and tests of the dynamo rHARM3D code we refer to [17, 118].

10.4 Characteristics of the Black Hole-Disk-Jet Structure

In Fig. 10.2, we present a schematic diagram of the disk-jet structure surrounding a central black hole as based on our simulation results. We observe such a structure in our reference model developing after several tens of inner disk revolutions and see it remaining like that up to the end of the simulation, corresponding to several hundreds of inner disk revolutions.

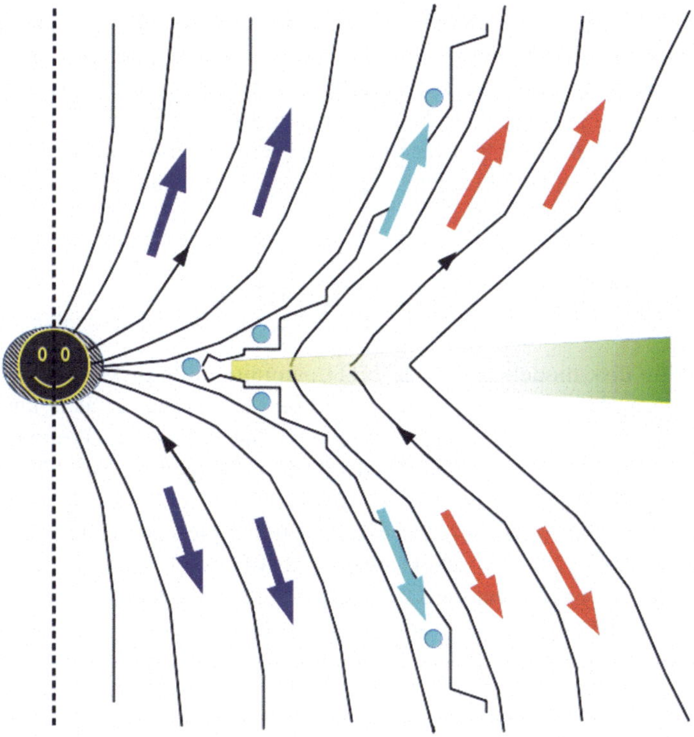

Fig. 10.2 Schematic diagram of the disk-jet-wind configuration. A spinning black hole (black circle, rotational axis dashed) is surrounded by a thin disk in Keplerian rotation (greenish block). Magnetic field lines (black lines) that are rooted in the ergosphere (hatched ellipse) launch a BZ jet (purple arrows). Magnetic field lines rooted in the disk launch a BP disk wind (red arrows). The anti-aligned magnetic field in the plunging region leads to reconnection, magnetic islands and plasmoids (turquoise circles) that are ejected along a boundary layer (jittered field line) between disk wind and spine jet. Note the upward magnetic field direction (black arrows)

The high-density thin accretion disk occupies the area near the disk equatorial plane. The region near the rotation axis is occupied by very low density, relativistic, and highly magnetized BZ-jets, where the ratio $B_{\rm tor}/B_{\rm p} \ll 1$. The shape of the BZ-jet is supported by a region with $B_{\rm tor}/B_{\rm p} \gg 1$. We called this region as a $B_{\rm tor}$ dominated disk wind, and the driving mechanism for this disk wind is the excess toroidal magnetic pressure. $B_{\rm tor}$ dominated disk wind is turbulent due to the different instabilities present in the thin accretion disk, and we observe plasmoids in this region. Further, such a disk wind could play a very important role in the collimation or the de-collimation of the BZ-jet due to external support.

The $B_{\rm tor}$ dominated disk wind region is a potential site for turbulent mass loading from the accretion disk [16]. The area between the $B_{\rm tor}$ dominated wind and the accretion disk is occupied by BP disk wind. In this region, the flow around the disk surface has $B_{\rm tor}/B_{\rm p} \lesssim 1$, the driving mechanism of the BP disk wind is the magneto-centrifugal acceleration as proposed by [13]. The flow leaves the surface of the accretion disk with sub-Alfvénic poloidal velocity ($M_{\rm A,p} < 1$). As the BP disk wind moves far from the equatorial plane, the ratio $B_{\rm tor}/B_{\rm p}$ increases, eventually the ratio becomes $B_{\rm tor}/B_{\rm p} \gg 1$ and the flow attains a super-Alfvénic poloidal velocity ($M_{\rm A,p} > 1$). Thus, both $B_{\rm tor}$ dominated disk wind and BP disk wind are super-Alfvénic and $B_{\rm tor}/B_{\rm p} \gg 1$ far from the equatorial plane, in such scenario, the flow self-collimates and evolves into a large-scale jet [44, and references therein].

10.5 Transition from Thick to Thin Disks

In the two thin disk models (SSD and NTD), an initial thin disk is considered, and its evolution can be studied [e.g., 32, 62, 100, 119]. However, it is known from the basic understanding of the accretion process that for cooling dominated flow, a geometrically thick flow should be transformed into a geometrically thin flow with time. Many studies of thin disks follow this idea and supply a rotation supported initial torus (e.g., FM torus [46]) and evolve it in the presence of radiation/cooling etc. [76, 84–86, 107, 109]. To radiate the majority of heat, these models use a cooling function to keep the gas temperature at a fraction of the virial temperature. This allows for control over the disk's aspect ratio (H/r), which is a crucial parameter in these models. By considering the emissivity of a thin disk, the target temperature (T_*) is expressed as,

$$T_* = \frac{\pi}{2}\frac{R_z(r)}{r}\left[\frac{H(r)}{r}\right]^2. \tag{10.23}$$

where $R_z(r)$ is the correction due to the vertical component of gravity [3, 57]. In this approach, the thickness of the disk is maintained by introducing cooling bound portions where the local temperature is greater than that of the target temperature $T_*(r)$. The bound matter satisfies $-hu_t < 1$ and $p/\rho > T_*$, where h is the specific enthalpy of the flow. With this, a cooling function is introduced in the stress-energy conservation equation, i.e., $\nabla_\mu T^{\mu\nu} = -\mathscr{L} u^\nu$ (e.g., Eq. (17) of Noble et al. [84]).

This model has been used to establish the NTD model by making one-to-one comparisons between simulations and analytical expectations. Early simulations showed differences from the NTD model at the ISCO of about ~10% [50, 84]. Other studies [95] found smaller differences of only < 3%. Penna et al. [95, 96] analyzed potential causes for the differences and proposed that the findings are actually more aligned than they initially seemed. Figure 10.3 (Fig. 1 of [96]) shows time-average density distributions for four different GR-MHD runs for different disk heights and with $a = 0$. These panels suggest the formation of NTD for aspect ratio $h_s \ll \alpha$ (viscosity parameter). They concluded that the NTD model aligns closely with numerical results quite well [95, 96]. These findings give leverage to directly use NTD or SSD

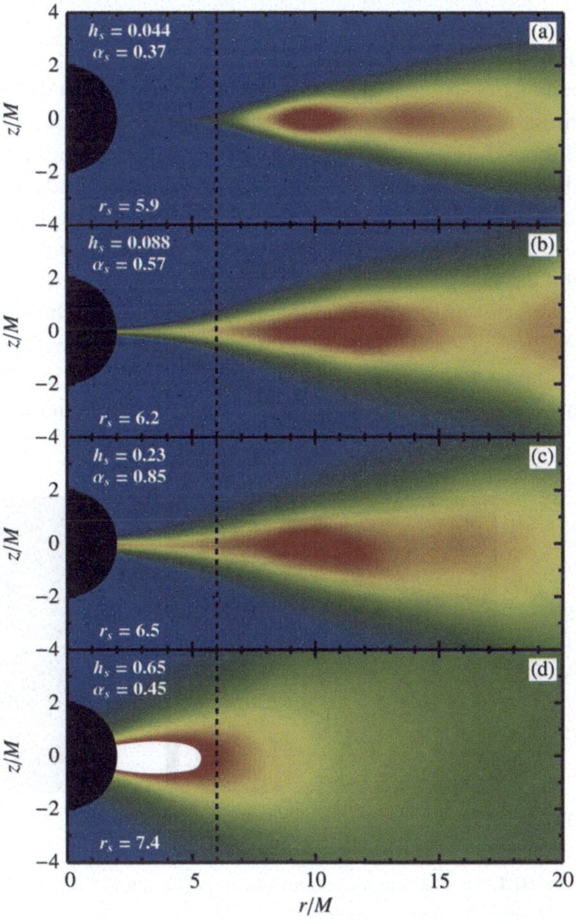

Fig. 10.3 Time-averaged density distribution in the $r - z$ plane for four GR-MHD simulations with various disk thicknesses around a non-rotating black hole ($a = 0$). The dashed vertical line shows the radial location of the ISCO. The disk aspect ratio, $h_s = H/r$, and effective Shakura-Sunyaev viscosity, α, are marked on each panel. This figure is reproduced from [96], with kind permission of Robert F. Penna

as the initial condition for performing GR-MHD simulations [2]. Rather than performing elaborate simulations, researchers can save time and resources by utilizing the NTD or SSD models as initial conditions. This allows for more efficient and accurate simulations of astrophysical phenomena.

10.6 Temporal Evolution of the Disk

10.6.1 Hydrodynamic Evolution

The initial setup (irrespective of SSD or NTD) is expected to be in rotation supported hydrodynamic equilibrium, provided the aspect ratio of the accretion disk is $H/r \ll 1$. To test this, the evolution of a hydrodynamical NTD asymmetric setup with $A_{r,\theta,\phi} = 0$ is shown (for detailed parameters, see [32]) in Fig. 10.4. The figure shows the density distribution of the setup at simulation times $t = 0$ and $t = 10000$ in panels Fig. 10.4a and b, respectively. We note that the thin disk structure experiences minimal alterations over time, particularly in the outer region of the disk. It should be emphasized that the hydrodynamical model examined does not include any explicit effects of viscous heating or radiative cooling. Indeed, radiative cooling is essential for preserving the thin disk structure of the accretion disk. Nevertheless, the local heat balance condition of the thin disk ensures that the combined impact of cooling and explicit viscous heating is insignificant (refer to [89, 108], etc.). Consequently, it does not influence the dynamic progression of the NTD setup. The setup or similar setups (e.g., SSD) are well-suited for investigating the impact of the magnetic field on the dynamic characteristics of the accretion disk, as well as the launch of the jet and the driving of the disk wind, even under the ideal GR-MHD conditions.

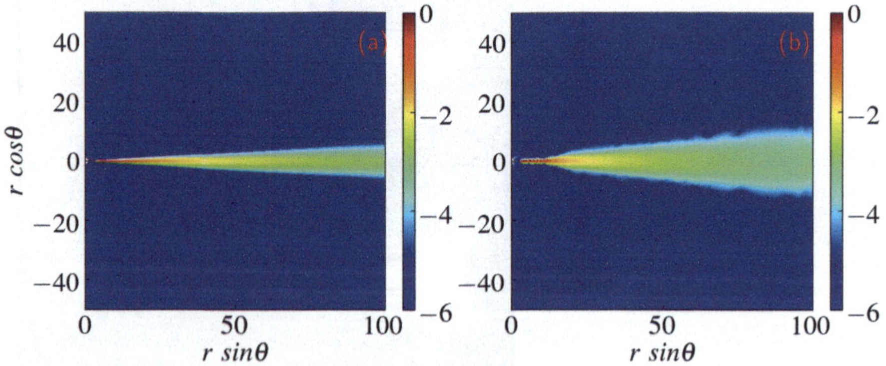

Fig. 10.4 Hydrodynamic evolution of NTD setup from [32]. Panel (**a**) and (**b**) show density distribution at time $t = 0$ and $t = 10000\, t_g$, respectively

10.6.2 Magnetohydrodynamic Evolution

Recently, Dihingia et al. and Dihingia and Vaidya [31, 32] have done an extensive parameter-space survey of axisymmetric thin accretion disks, considering different values of magnetic field strengths, inclination of field lines, and disk aspect ratio (H/r).

In Fig. 10.5, the time evolution of a thin disk in the presence of an inclined magnetic field (Eq. 10.18) from [32] is depicted. Following the evolution, the accretion process transitions to accretion-ejection and a disk-jet, where three major components can be identified.

1. Low density funnel near the black hole rotation axis.
2. The off-equatorial disk wind part has a lower density than the accretion disk part.
3. Thin, high-density accretion disk near the equatorial plane.

Note that relativistic jets launched due to the BZ mechanism can be seen in the low density funnel region. This is commonly seen in magnetized accretion flow, irrespective of the geometrically thin or thick disk [24, 72]. Other chapters from the books will describe its properties in detail; we avoid doing it here to prevent repetitions. Next, we discuss phenomena that are more specific to a thin disk geometry.

Dihingia et al. [32] have also discussed the launching of disk winds from the a thin accretion disk in both channels, i.e., $B_{\rm tor}$ dominated and BP driven. Figure 10.6 shows a collection of some representative results. Panel Fig. 10.6a shows distribution of the Alfvénic poloidal Mach number ($M_{\rm A,P}$), where black and blue contours correspond to $\sigma = b^2/\rho = 1$ and $M_{\rm A,P} = 1$, respectively, for a typical simulation model. Panel Fig. 10.6b shows the distribution of the ratio of toroidal to poloidal magnetic field strength ($B^2_{\rm tor}/B^2_{\rm p}$), where the solid black, grey, and blue lines correspond to contours of $B_{\rm tor}/B_{\rm p} = 1, 5$, and $M_{\rm A,P} = 1$, respectively.

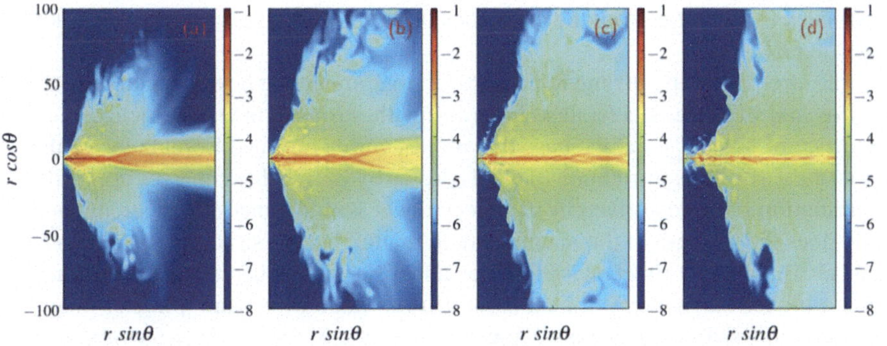

Fig. 10.5 Magneto-hydrodynamic evolution of NTD setup from [32]. Panel (**a**), (**b**), (**c**), and (**d**) show density distribution at time $t = 500\,t_g$, $1000\,t_g$, $2000\,t_g$, and $t = 4000\,t_g$, respectively

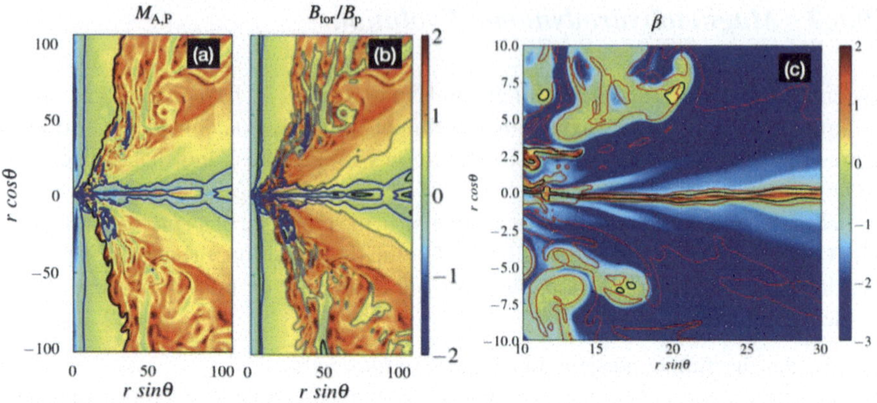

Fig. 10.6 Distribution of **a** poloidal Alfvénic Mach number ($M_{A,P}$), **b** B_{tor}/B_p, and **c** plasma-β for a magnetised thin accretion disk evolution from [32]. In panel (**a**), the solid black and blue lines correspond to contours of magnetisation $\sigma = 1$ and $M_{A,P} = 1$, respectively. In panel (**b**), the solid black, grey, and blue lines correspond to contours of $B_{tor}/B_p = 1, 5$, and $M_{A,P} = 1$, respectively. In panel (c), the black and red lines correspond to the contours $\beta = 1$, and $M_{A,P} = 1$, respectively

These two figures suggest that in this model for radii $r \gtrsim 10$, the disk wind leaves the disk surface with sub-Alfvénic velocity and becomes super-Alfvénic after travelling a certain distance along the vertical direction. Within the high-density region, we observe that the flow is super-Alfvénic $M_{A,P} > 1$ and the ratio $B_{tor}/B_p \sim 1$. Also, noticed that in the region with sub-Alfvénic ($M_{A,P} < 1$) poloidal velocity, the ratio B_{tor}/B_p is of the order of unity ($\gtrsim 5$). Finally, by looking at the plasma-β distribution in Fig. 10.6c, we observe that the flow on the surface of the disk is magnetically dominated ($\beta < 1$, see the black contour), where the black and red lines correspond to the contours $\beta = 1$, and $M_{A,P} = 1$, respectively. These observations strongly suggest that the matter is magnetically driven in the form of a wind. This process shows similarities with the properties of disk winds, as suggested by Blandford and Payne [13].

The strength of the magnetic field has a large impact on how jets and disk winds evolve and become accelerated. For runs with stronger magnetic fields, there is evidence of a wider jet at the base that moves with a terminal Lorentz factor that is about $\gamma \sim 10$. The contribution of winds driven by BP mechanism goes up, while the contribution of winds driven by B_{tor} goes down with magnetic field strengths. Not only does the strength of the magnetic field matter, but the structure of the magnetic field also has a large effect on the disk, jet, and disk-wind structures. It seems that runs with a higher magnetic field inclination (lower value of m) and a higher magnetic flux around the disk surface have a wider jet with a higher Lorentz factor than runs with more vertical magnetic field lines at the disk surface. For low inclined magnetic fields, the wind is most likely be driven by BP mechanism. In cases with vertical field lines, extra toroidal magnetic pressure is more likely to be the main factor that launches the disk wind, i.e., B_{tor} driven disk wind.

In the reference run of Dihingia et al. [32] with $m = 0.1$ and $\beta_{\max} \sim 188$, featuring a MAD, the appearance of plasmoids due to reconnection close to the black hole is observed. Toroidal magnetic fields carry these plasmoids away from the disk, resulting in stronger magnetic tension from reconnected poloidal field lines in the inner region of the accretion disk. Consequently, the process of accretion is slowed down, leading to the truncation of the disk from the black hole, producing a truncated disk. Such scenarios are discussed in a later section of this chapter.

10.6.3 Resistive Accretion Disks

Here we discuss exemplary results of resistive GR-MHD mass loading from disk accretion into a corresponding disk wind [119]. Figure 10.7 shows a set of physical variables that characterize the transition from accretion to ejection.

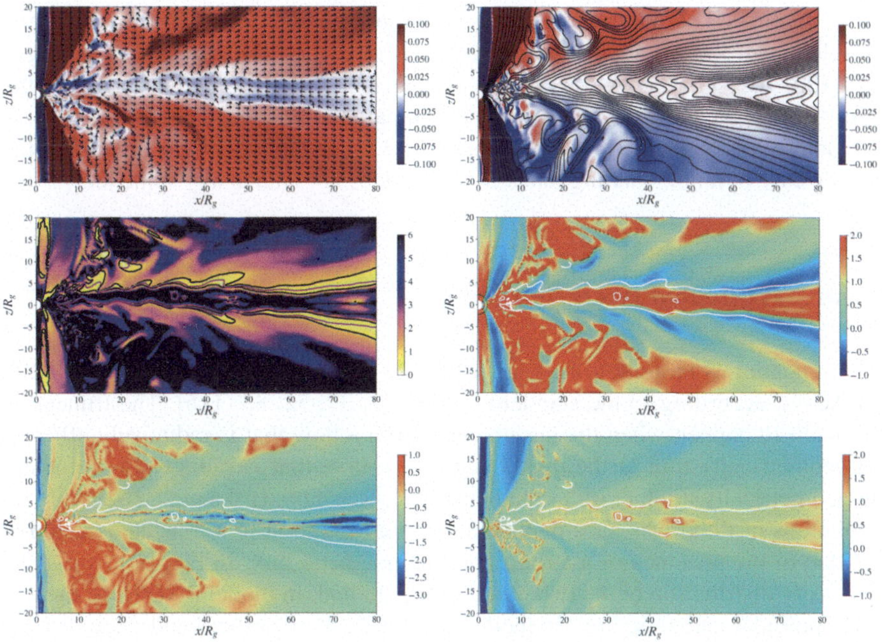

Fig. 10.7 Snapshots at time $t = 4000\, t_g$ of the disk wind launching area for different physical variables. From top left to bottom right we show the radial velocity (in color) with normalized poloidal velocity vectors (black arrows), the vertical velocity (in color) with poloidal magnetic field lines (black lines), the poloidal Alfvén Mach number (in color) superimposed with the Alfvén surface (black lines), the plasma-β (in color), the ratio between the toroidal and poloidal magnetic field components (in color, log scale), and the ratio between toroidal and poloidal velocity (color, log scale). The white line (last three panels) indicate the reversal in u_r, thus where $u_r = 0$, and where the flow changes from inflow to outflow. Panels taken from [119]

We first look at the velocity distribution, in particular the radial component u_r. This is essentially negative in the disk (accretion) and changes to positive values above the disk (outflow). Correspondingly, the poloidal velocity vectors (normalized in the figure) turn from inward direction to outward direction (the last three panels indicate this transition with a white line). This can be considered the launching surface of the disk wind.

A similar behavior is seen in the vertical speed u_z (i.e. vertical to the disk midplane), where the disk wind shows up as upward (red) and downward (blue) motion of the material above (or below, respectively) the disk surfaces. The color coding is chosen for the low speed disk wind, but note (i) the high speed BZ jet ejected from the ergosphere and (ii) the infall of material along the axis (as B_ϕ along the axis, no Poynting flux driven mass flow is possible here).

The disk wind becomes trans-Alfvénic about 5 scale heights above the midplane. Naturally, this flow is launched magnetically (see the distribution of the plasma-β, or magnetization); and, as an MHD wind, it becomes matter-dominated at some point.

The energetics of the disk outflow and respective disk jet can now be compared to the BZ-driven spine jet from the black hole. The disk wind is clearly matter-dominated, while the spine jet is Poynting flux dominated. Correspondingly, the velocity of the spine jet is highly relativistic,[2] while the disk jet reaches 0.1–0.2c [100, 101, 119]. However, due to the high mass flux of the disk outflow, the energy flux of the spine jet and disk jet is comparable. In some cases, the disk jet is even dominating energetically the spine jet [101]. The latter is obviously true in the limiting case of a non-rotating black hole that does not launch any spine jet.

10.6.4 Truncated Accretion Disks

Disk truncation is thought to be responsible for the different features of BH-XRBs and AGNs, according to the standard model [33, 37, 52, 83, 115]. The truncation is thought to happen when the gas changes from a flow that is radiatively efficient and geometrically thin at the disk's edges to a flow that is radiatively inefficient and geometrically thick close to the black hole. Recently, Liska et al. [62] with 3D simulations of thin disks have demonstrated the formation of a truncated accretion disk due to the presence of a strong poloidal magnetic flux. In Fig. 10.8, we show the density distribution of the two models they studied. In model RADTOR, there is no net poloidal magnetic flux. In model RADPOL, on the other hand, there is only poloidal magnetic flux, and it reaches a MAD state after time evolution. They show that when the magnetic flux in the model RADPOL's inner disk reaches its maximum, there is a sharp change between an outer disk supported by radiation pressure and an inner corona supported by magnetic pressure. For their model, the truncation radius

[2] As for all GR-MHD simulations, the kinematics, thus the mass flux and also the speed of the BZ jet depend on the choice of the floor-density for the simulation.

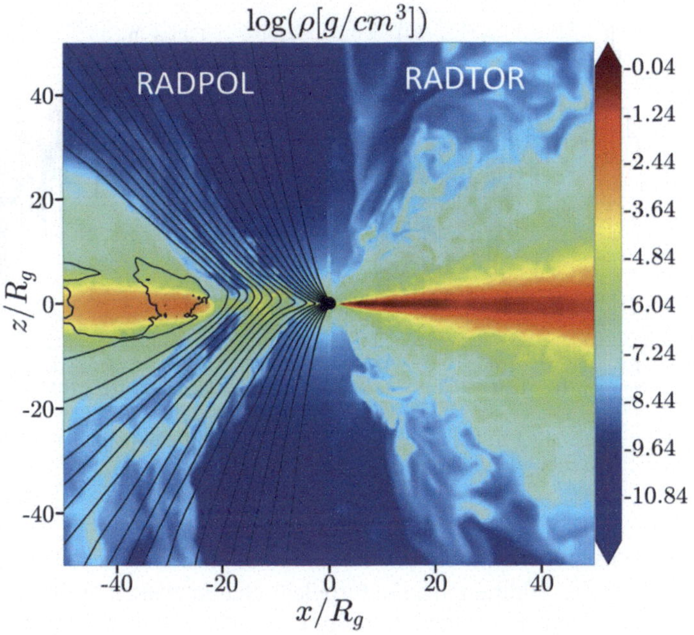

Fig. 10.8 Density distribution and the poloidal magnetic field lines for model RADPLO and RADTOR from [62]

is found to be $r_{\mathrm{tr}} = 20r_g$. On the other hand, we do not observe such truncation in the RADTOR model (see Fig. 10.8 for reference).

More interesting features of the truncated accretion disk have been reported in Dihingia et al. [33]. According to them, the reconnection events and the plasmoids that form because of active tearing mode instabilities are very important to the accretion flow dynamics of the truncated disk. They make poloidal magnetic fields vertical to the equatorial plane, and that creates a strong magnetic tension force in the mid-plane of the disk. The accretion flow is temporarily stopped by this force's resistance. Over time, the ram pressure of the flow in the inner edge becomes higher against this force because of the transport of angular momentum outwards (because of MRI and disk winds), which causes accretion to happen again. As the simulation goes on, this process keeps going as the inner accretion flow oscillates. They show that the flow inside the truncation radius oscillates or the inner radius oscillates in a quasi-periodic manner. The frequency of oscillations depends on the magnetic field strengths, truncation radius, and resistivity of the magnetized flow. We reproduced the power density spectra (PDS) of accretion rate at the horizon from their paper in Fig. 10.9.

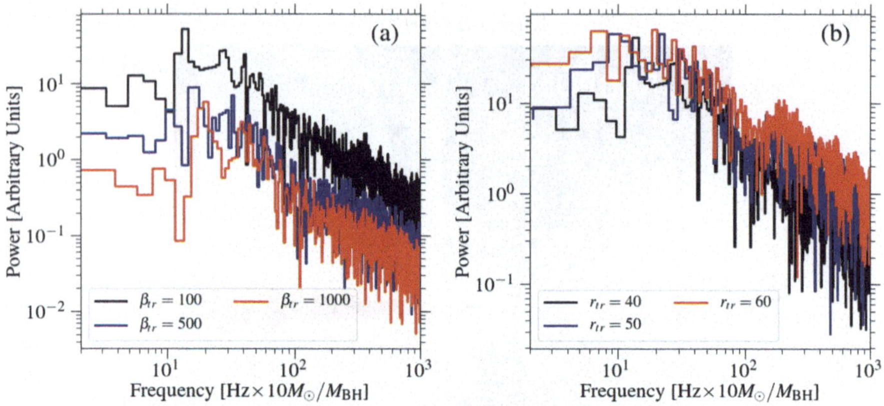

Fig. 10.9 PDS corresponding to the accretion rate profile for different simulation models with different magnetic field strengths and truncation radius from [33]

In contrast to ideal MHD models, no formation of plasmoids is observed with the increase in resistivity. Additionally, the turbulent characteristics of the inner accretion flow are subdued as magnetic resistivity rises. The Kelvin-Helmholtz instability (KHI) across the inflow and outflow boundaries experiences suppression due to increased magnetic resistivity. Moreover, the frequency of oscillation rises with the increase in magnetic resistivity. Furthermore, the radius at which the balancing magnetic tension force and the ram pressure occur decreases proportionally with the escalation of magnetic resistivity (see [33]). Further evidence of transient shocks during the highly accreting stage is uncovered during oscillation. These transient shocks manifest as an extended hot post-shock corona surrounding the black hole, exerting an influence on its radiative properties. Radiative transfer calculations reveal diskernible signatures of these oscillations through modulation in the edge-brightened structure of the accretion disk. These modulations strongly relate to the observed quasi-periodic oscillations (QPOs) in BH-XRBs during their outburst [11].

10.6.5 High-Soft to Low-Hard Transition

It has been well known for many decades that a geometrically thin disk can suitably describe the high-soft state of an outburst. Similarly, geometrically thick flow close to the black hole can describe low-hard state of an outburst [21, 22, 37, 58, 59, 78, 79]. Recent, GR-MHD simulations also suggest that geometrically thick hot magnetised flow could be a reasonable choice for the hard state of BH-XRBs [e.g., 26]. Although there is doubt if the flow in the hard state is MAD or not [e.g., 47, 124]. Still, no one knows how or why the different states change or how the different states are connected together. Accrution flow simulations that look like the real thing might

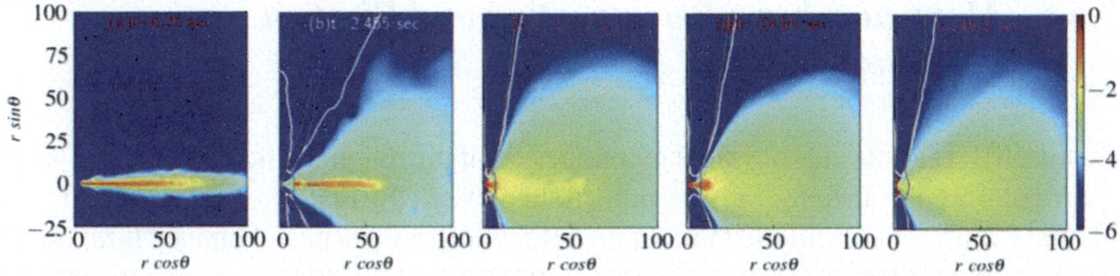

Fig. 10.10 Logarithmic time-average density distribution at different times marked on the top of the figures. The time averaging is performed within 0.1 sec around the marked time. The white and green solid lines show $-hu_t = 1$ and $\sigma = 1$ contours, respectively. The panels are taken from [30]

help us understand the changes in states. Still, simulating an outburst over a realistic time range (hours to tens of days) is hard, especially when using GR-MHD.

Recently, Dihingia et al. [30] investigated the decaying phase of an outburst in BH-XRBs. For this setup, an initial thin disk is considered near the black hole, and low-angular matter is injected far from the black hole. The density evolution of their simulation is reproduced in Fig. 10.10. These simulations include radiative cooling following [28, 29]. Initially, the injected matter interacts with the thin disk and forms a TCAF. The thin disk at the equatorial plane is where the majority of the radiation in such a flow originates. However, some hard X-rays are expected to be observed by the inverse Compton process because of the hot corona surrounding the thin disk. The soft-intermediate phases of an outburst could therefore be explained by such a TCAF [21, 22, and references therein]. Over time, more matter is injected, leading to the transformation of the accretion flow into a geometrically thick and hot torus. The size of the torus diminishes, leaving behind very low-density hot matter surrounding the BH. Such a flow structure resembles a radiatively inefficient quiescent state (RIAF) e.g.,[45, 77, 80]. Thus, the simulation shows a progression from a soft state to a soft-intermediate state, then to a hard-intermediate state, followed by a hard state, and finally to a quiescent state. The detection of jet properties aligns with expectations from many observations, indicating the presence of jets only in intermediate states. A distinct variation in QPO frequency is observed throughout the evolution. On an equal decaying time scale for mass and angular momentum, the accretion rate at different radii decays with time. This feature is expected in the decaying phase of commonly known outbursts in BH-XRBs [37, 45, 58]. The high-resolution run shows similar results for accretion rates and QPO frequencies, but a higher accumulation of magnetic flux around the event horizon results in stronger jets.

10.6.6 Magnetospheric Reconnection and Ejection of Plasmoids

Resistive MHD allows for physical reconnection of the magnetic field. This process is of particular importance for the magnetic field close to the black hole, where we typically have a strong anti-aligned magnetic field along the equatorial plane. Here, magnetic reconnection works very efficiently, effectively disturbing the smooth inflow-outflow close to the inner disk radius. As a result, magnetic islands and plasmoids are ejected along the transition region between the inner BZ jet and the surrounding disk wind.

Figure 10.11 visualizes the processes involved. The high speed BP jet is visible predominantly in the Lorentz factor. This jet starts sub-Alfvénic and becomes super-Alfvénic further downstream. We can see the turbulent flow region that the reconnection along the mid-plane in the plunge region towards the black hole has perturbed right outside this jet. This region is not magnetically dominated, with rather high plasma-β, respectively, low magnetization.

In our simulations, these plasmoids are rather long-lived, as resistivity decreases with altitude. When these magnetic islands reach the ideal MHD regime at sufficiently large distances from the black hole and the disk, they may reconnect only by numerical resistivity. In our case, they are just ejected out of the simulation domain. However, on their way downstream, they also perturb the (light) BZ jet (see Fig. 10.12). Plasmoid generation and ejection have been investigated in more detail and have been confirmed by [81, 82].

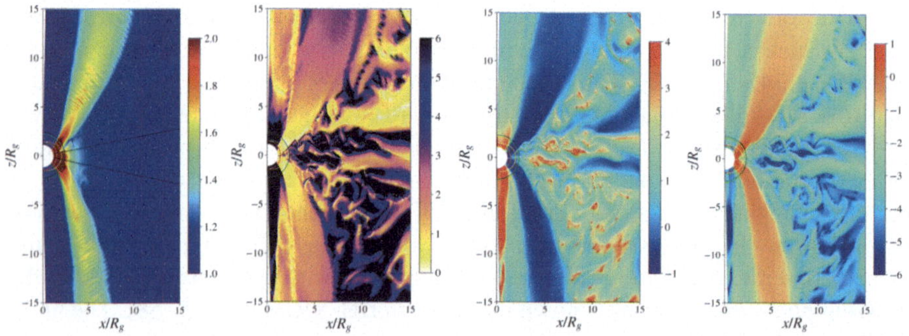

Fig. 10.11 Snapshots at time $t = 4000\,t_g$ of magnetohydrodynamic accretion-ejection structure close to the black hole, $r < 15$. Shown is the Lorentz factor (left), the poloidal Alfvén Mach number (log scale; second from left), the plasma-β (log scale; second from right), and the magnetization (log scale; right). The high Alfvén Mach number indicates flows that are dominated by kinetic energy, regardless of the highly magnetized area. The red semicircle marks the horizon, the yellow line marks the ergosphere, and the green line indicates the radius of the ISCO. Panels taken from [119]

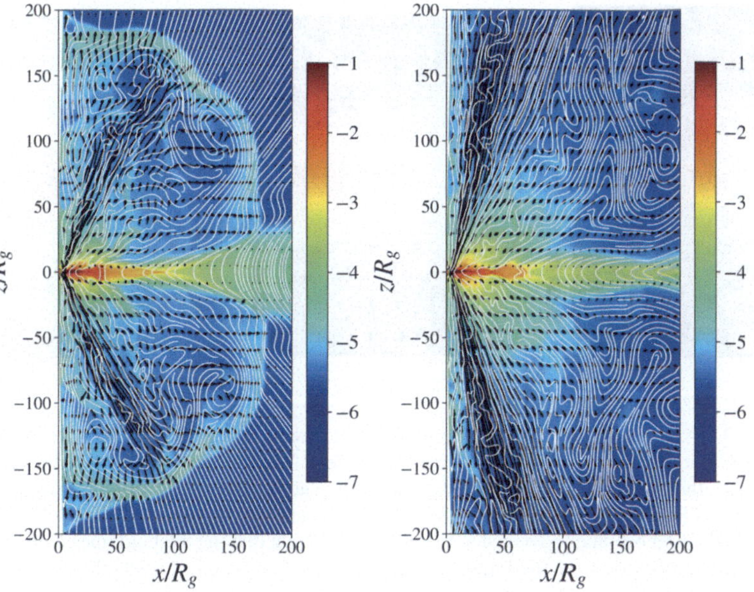

Fig. 10.12 Snapshots at time $t = 2000, 4000\,t_g$ (from left to right) of magnetohydrodynamic accretion-ejection structure enclosing a BZ jet at larger distanced from th elaunching reqion. Shown is the density distribution (in log scale) and the poloidal magnetic field lines (white lines), while the poloidal velocity field is represented by the black arrows. Panels taken from [119]

10.6.7 A GR-MHD Disk Dynamo-Generated Disk Jet

We subsequently discuss results considering the action of a mean-field GR-MHD disk dynamo. As found in previous studies for non-relativistic disks, the GR-MHD disk dynamo is also able to generate a strong poloidal magnetic flux anchored in the disk. This magnetic field can (i) drive a disk wind that potentially collimates into a disk jet and can also (ii) be advected to the central spinning black hole, thereby generating a strong BZ jet.

Thus, our approach self-consistently generates a two-component jet structure from a disk seed field. In comparison to the non-relativistic studies, which result in a smooth disk-jet structure after about 100, 000 inner disk revolutions, the magnetic field of the GR-MHD dynamo is much more disturbed due to the strong reconnection going on in the plunge region and the very inner disk. This is shown in Fig. 10.13 with the time evolution of the global magnetic field structure. The simulation starts with a weak, purely radial seed field that is confined to the disk. At the final state, a global magnetic field has evolved, in particular a strong field that has been advected to the black hole ergosphere and drives a strong BZ jet, in addition to a disk wind.

The increase in magnetic field strength is shown in Figs. 10.14 and 10.15. Figure 10.14 shows the evolution of the magnetic energy in the disk for selected

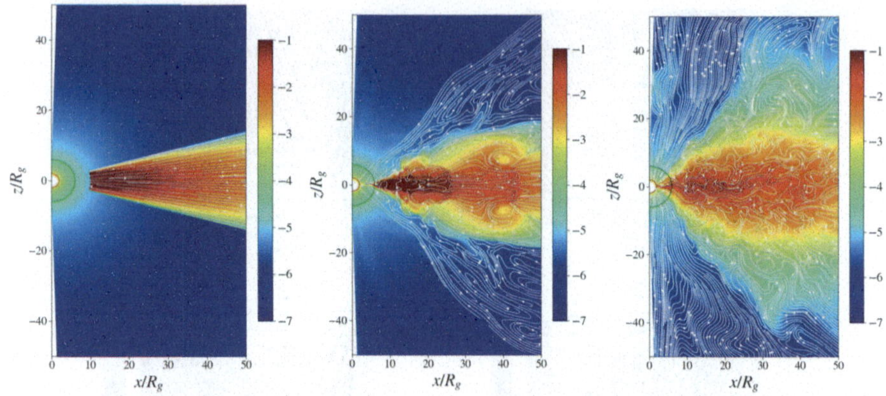

Fig. 10.13 Evolution of density and the poloidal magnetic field (white lines) over time. Left is the initial setup with a purely radial (weak) seed field. The middle panel show the situation at $t = 3000\,t_g$, the right panel at $t = 12000\,t_g$. The red line around the (0, 0) point is the event horizon, and the green line denotes the radius of the ISCO. It is clearly seen how the field evolves, expands to larger radii, and becomes advected to the black hole. Note that the the field lines plotted indicate the field structure, but not the field strength. The latter increase by a factor 10,000 over the run time of this simulation. Panels taken from [118]

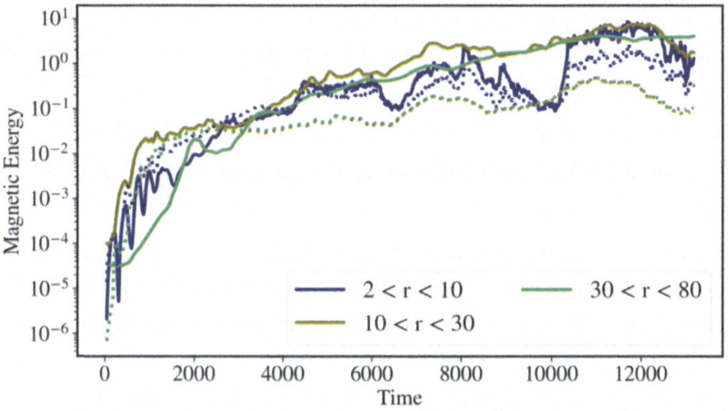

Fig. 10.14 Evolution magnetic field strength during the dynamo process. Shown is the averaged disk magnetic energy over time. Different disk areas are distinguished, such as $2 < r < 10$, $10 < r < 30$, and $30 < r < 80$ (radii r normalized to R_g). Panels taken from [118]

disk areas. We clearly see an exponential increase with a flattening due to dynamo quenching, resulting in a 10^6 increase of the disk magnetic energy (for this choice of dynamo strength ξ and dynamo quenching). We want to stress the point that jet launching, in particular by the BP process, requires a large global magnetic flux. In this respect, plotting the magnetic energy or the absolute field strength could be misleading, as for the global magnetic flux, the small-scale field components may actually cancel out.

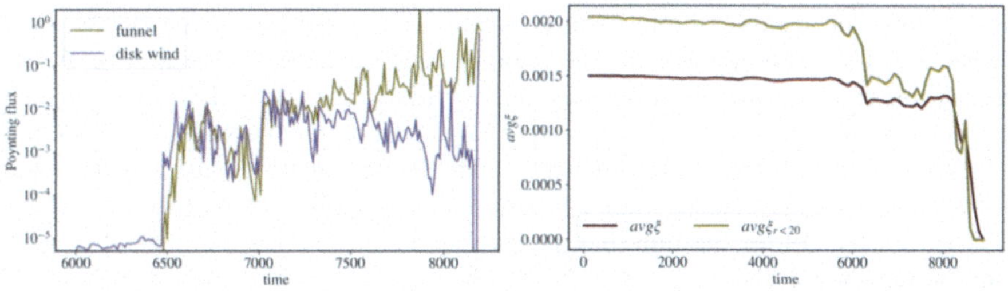

Fig. 10.15 Evolution magnetic field strength during the dynamo process. Shown is the Poynting flux for the funnel BZ jet, and the disk wind (left). The flux is integrated along a sphere with radius $r = 35\,R_g$, while the BZ considered for the segment $0\,\text{deg} < \theta < 25\,\text{deg}$ and the disk wind for the segment $25\,\text{deg} < \theta < 65\,\text{deg}$. The values are averaged over the corresponding segments at the lower hemisphere. The corresponding evolution of the absolute dynamo parameter ξ is shown at the bottom, averaged in space for the whole disk, and for the inner disk area, $r < 20\,R_g$. Panels taken from [118]

It is therefore interesting to look at the Poynting flux of the two-component jet structure. This is shown in Fig. 10.15, where we see how the Poynting flux evolves in time. The flux is measured along a sphere around the black hole with a radius of $35\,R_g$. Note that it takes some time ($t = 6000\,t_g$) until the flux generated in the disk has reached that distance. After that, the flux increases with similar strength for both components: the inner funnel flow, i.e., the BZ jet, and the outer disk wind. After a further period of time ($t \simeq 8000\,t_g$) the disk dynamo is quenched as the disk magnetic field has reached a (chosen) equipartition value. Thus, the disk Poynting flux levels off, even slightly decreasing. On the other hand, the advecting of magnetic flux to the black hole is still going on, and the Poynting flux of the funnel jet still increases. That increase is actually on the costs of the disk flux, as it is advected away from the disk area towards the black hole.

Corresponding to the magnetic flux evolution, the lower panel of Fig. 10.15 shows the evolution of the dynamo parameter ξ. The onset of dynamo quenching is simply seen as a decrease of ξ.

10.7 Summary and Outlook

Over the past few decades, GR-MHD simulations have been performed for different astrophysical scenarios, including thin disk configurations. In this chapter, we have discussed some of the recent results considering highly resolved GR-MHD simulations of thin disks. In particular, we have discussed non-ideal MHD effects such as resisitivity, reconnection, and a mean-field dynamo.

Future work should be devoted to high-resolution 3D simulations of the inner disk area in order to investigate variable accretion events and the time scales of the same. Also, as thin disks are thought to be cold and radiative, radiation modeling of the disk material seems essential. This is because of two reasons (at least): one is the connection and the comparison to observations, in particular the opportunity of fitting distinct sources. The other is the scaling of mass fluxes, as in pure MHD modeling the disk mass is unconstrained (as in the simple Keplerian problem). Radiation modeling in fact requires applying physical densities, which can constrain mass accretion and ejection rates. Essential steps in this direction have already been performed [61, 62, 69, 84], mostly not for the thin disk ansatz, but also a few for thin disks [74, 86], none for resistive disks.

In this book chapter, we show that an initial thin disk could explain the decaying stage of an outburst with GR-MHD simulations. It should now be a priority to simulate the full cycle of an outburst (including the rising phase). Note that a realistic outburst in BH-XRBs could be a few hours to months long and rarely even years [for e.g., 122], which is impossible to perform under the current framework. Therefore, more quantitative and parameter space surveys are required to correlate the realistic decay time with the physical properties of injected matter, the initial thin accretion disk, or the properties of the central BH (a, $M_{\rm BH}$, etc.).

Most of the studies do not include self-consistent radiative cooling processes that may be present in the flow. The radiative cooling may not be important for lower accretion rates ($\dot{M} < 10^{-7}$ Eddington units, [29, 123]). Also, the radiative calculations require the electron temperature. Throughout the different studies we discussed here, a single-temperature approximation has been employed. Such studies usually rely on temperature scaling relations, e.g., $R - \beta$ model [75]. Such relations may result in an overestimation of electron temperatures and are hence not recommended [e.g., 29, 73, 106]. Hence, the adoption of a robust two-temperature approach is recommended for achieving self-consistent results.

After all the efforts in the field, we still don't know the mass composition of relativistic jets or the very physical mechanism that leads to the ejection of the observed radio knots. In interpreting the highly relativistic jet region from GR-MHD simulations, one always needs to take prior caution. This region is filled with floor quantities and is therefore not perfectly reliable [e.g., 68, 98]. As a result, GR-MHD alone cannot explain the fundamental physical mechanisms behind the launch of highly relativistic jets. General relativistic perticle-in-cell (GR-PIC) simulations may be needed, which are currently under development [e.g., 5, 51, 54, 60]. These simulations may provide a more accurate representation of the complex processes involved in the formation of relativistic jets. By combining GR-MHD and GR-PIC simulations, scientists may be able to gain a more comprehensive understanding of the physics driving these phenomena.

Acknowledgements C.F. is grateful to Scott Noble for permission to use the original HARM3D code for further development (resistivity, dynamo). C.F. acknowledges fundamental contributions during code development by PhD students Qian Qian and Christos Vourellis, as well as critical help during the implementation of resistivity in HARM3D by Matteo Bugli and Scott Noble. C.F. acknowledges support by the German Research Foundation DFG via the research unit FORS 5195. Simulations co-authored by C.F. were performed on the ISAAC and VERA clusters of the Max Planck Institute for Astronomy. I.K.D. acknowledges the support from the National Key R&D Program of China (No. 2023YFE0101200), the National Natural Science Foundation of China (Grant

No. 12273022), and the Shanghai Municipality orientation program of Basic Research for International Scientists (grant no. 22JC1410600), and Max Planck partner group award at Indian Institute Technology of Indore, India. Simulations co-authored by I.K.D. were performed on the TDLI-Astro cluster in Tsung-Dao Lee Institute, Pi2.0, Siyuan Mark-I clusters in the High-Performance Computing Center at Shanghai Jiao Tong University, and Max Planck Gesellschaft (MPG) super-computing resources.

References

1. M.A. Abramowicz, B. Czerny, J.P. Lasota, E. Szuszkiewicz, Slim accretion disks. Astrophys. J. **332**, 646 (1988). https://doi.org/10.1086/166683
2. M.A. Abramowicz, P.C. Fragile, Foundations of black hole accretion disk theory. Living Rev. Rel. **16**, 1 (2013). https://doi.org/10.12942/lrr-2013-1 [arXiv:1104.5499 [astro-ph.HE]]
3. M.A. Abramowicz, A. Lanza, M.J. Percival, Accretion disks around Kerr black holes: vertical equilibrium revisited. Astrophys. J. **479**(1), 179–183 (1997). https://doi.org/10.1086/303869
4. L. Amarilla, E.F. Eiroa, G. Giribet, Null geodesics and shadow of a rotating black hole in extended Chern-Simons modified gravity. Phys. Rev. D **81**, 124045 (2010). https://doi.org/10.1103/PhysRevD.81.124045 [arXiv:1005.0607 [gr-qc]]
5. F. Bacchini, B. Ripperda, O. Porth, L. Sironi, Generalized, energy-conserving numerical simulations of particles in general relativity. II. Test particles in electromagnetic fields and GRMHD. Astrophys. J. Suppl. **240**(2), 40 (2019). https://doi.org/10.3847/1538-4365/aafcb3[arXiv:1810.00842 [astro-ph.HE]]
6. S.A. Balbus, J.F. Hawley, A powerful local shear instability in weakly magnetized disks. 1. Linear analysis. 2. Nonlinear evolution. Astrophys. J. **376**, 214–233 (1991). https://doi.org/10.1086/170270
7. S.A. Balbus, J.F. Hawley, S.A. Balbus, J.F. Hawley, Instability, turbulence, and enhanced transport in accretion disks. Rev. Mod. Phys. **70**, 1–53 (1998). https://doi.org/10.1103/RevModPhys.70.1
8. D.R. Ballantyne, Examining the physical conditions of a warm corona in active galactic nuclei accretion discs. Mon. Not. R. Astron. Soc. **491**(3), 3553–3561 (2020). https://doi.org/10.1093/mnras/stz3294 [arXiv:1911.10029 [astro-ph.HE]]
9. C. Bambi, N. Yoshida, Shape and position of the shadow in the $\delta = 2$ Tomimatsu-Sato space-time. Class. Quant. Grav. **27**, 205006 (2010). https://doi.org/10.1088/0264-9381/27/20/205006 [arXiv:1004.3149 [gr-qc]]
10. C. Bambi, E. Barausse, Constraining the quadrupole moment of stellar-mass black-hole candidates with the continuum fitting method. Astrophys. J. **731**, 121 (2011). https://doi.org/10.1088/0004-637X/731/2/121 [arXiv:1012.2007 [gr-qc]]
11. T.M. Belloni, S.E. Motta, T. Muñoz-Darias, Black hole transients. Bull. Astron. Soc. India **39**, 409–428 (2011)
12. R.D. Blandford, R.L. Znajek, Electromagnetic extractions of energy from Kerr black holes. Mon. Not. R. Astron. Soc. **179**, 433–456 (1977). https://doi.org/10.1093/mnras/179.3.433
13. R.D. Blandford, D.G. Payne, Hydromagnetic flows from accretion discs and the production of radio jets. Mon. Not. R. Astron. Soc. **199**(4), 883–903 (1982). https://doi.org/10.1093/mnras/199.4.883
14. R. Boissay, C. Ricci, S. Paltani, A hard X-ray view of the soft excess in AGN. Astron. Astrophys. **588**, A70 (2016). https://doi.org/10.1051/0004-6361/201526982 [arXiv:1511.08168 [astro-ph.HE]]
15. H. Bondi, Mon. Not. R. Astron. Soc. **112**, 195 (1952). https://doi.org/10.1093/mnras/112.2.195
16. S. Britzen, C. Fendt, A. Eckart et al., A new view on the M 87 jet origin: turbulent loading leading to large-scale episodic wiggling. Astron. Astrophys. **601**, A52 (2017). https://doi.org/10.1051/0004-6361/201629469

17. N. Bucciantini, L. Del Zanna, A fully covariant mean-field dynamo closure for numerical 3+1 resistive GRMHD. Mon. Not. R. Astron. Soc. **428**, 71 (2013). https://doi.org/10.1093/mnras/sts005 [arXiv:1205.2951 [astro-ph.HE]]
18. M. Bugli, L. Del Zanna, N. Bucciantini, Dynamo action in thick discs around Kerr black holes: high-order resistive GRMHD simulations. Mon. Not. R. Astron. Soc. **440**, L41–L45 (2014)
19. F. Casse, R. Keppens. F. Casse, R. Keppens, Magnetized accretion-ejection structures: 2.5D MHD simulations of continuous ideal jet launching from resistive accretion disks. Astrophys. J. **581**, 988–1001 (2002). https://doi.org/10.1086/344340 [arXiv:astro-ph/0208459 [astro-ph]]
20. S. Chakrabarti, Standing Rankine-Hugoniot shocks in the hybrid model flows of the black hole accretion and winds. Astrophys. J. **347**, 365 (1989)
21. S. Chakrabarti, L.G. Titarchuk, Spectral properties of accretion disks around galactic and extragalactic black holes. Astrophys. J. **455**, 623 (1995)
22. S.K. Chakrabarti, Study of accretion processes around black holes becomes 'Science': tell tale observational signatures of two component advective flows. https://doi.org/10.1142/9789813226609_0020 [arXiv:1604.05955 [astro-ph.HE]]
23. S. Chen, J. Jing, Properties of a thin accretion disk around a rotating non-Kerr black hole. Phys. Lett. B **711**, 81–87 (2012). https://doi.org/10.1016/j.physletb.2012.03.047 [arXiv:1110.3462 [gr-qc]]
24. S.W. Davis, A. Tchekhovskoy, Magnetohydrodynamics simulations of active galactic nucleus disks and jets. Ann. Rev. Astron. Astrophys. **58**, 407–439 (2020). https://doi.org/10.1146/annurev-astro-081817-051905 [arXiv:2101.08839 [astro-ph.HE]]
25. B. De Marco, G. Ponti, M. Cappi, M. Dadina, P. Uttley, E.M. Cackett, A.C. Fabian, G. Miniutti, Discovery of a relation between black hole mass and soft X-ray time lags in active galactic nuclei. Mon. Not. R. Astron. Soc. **431**, 2441 (2013). https://doi.org/10.1093/mnras/stt339 [arXiv:1201.0196 [astro-ph.HE]]
26. J. Dexter, N. Scepi, M.C. Begelman, Radiation GRMHD simulations of the hard state of black hole x-ray binaries and the collapse of a hot accretion flow. Astrophys. J. Lett. **919**(2), L20 (2021). https://doi.org/10.3847/2041-8213/ac2608 [arXiv:2109.06239 [astro-ph.HE]]
27. I.K. Dihingia, S. Das, G. Prabhakar, S. Mandal, Properties of two-temperature magnetized advective accretion flow around rotating black hole. Mon. Not. R. Astron. Soc. **496**(3), 3043–3059 (2020). https://doi.org/10.1093/mnras/staa1687 [arXiv:1911.02757 [astro-ph.HE]]
28. I.K. Dihingia, Y. Mizuno, C.M. Fromm, Z. Younsi, Impact of radiative cooling on the magnetised geometrically thin accretion disk around Kerr black hole. [arXiv:2305.09698 [astro-ph.HE]]
29. I.K. Dihingia, Y. Mizuno, C.M. Fromm, L. Rezzolla, Temperature properties in magnetized and radiatively cooled two-temperature accretion flows on to a black hole. Mon. Not. R. Astron. Soc. **518**(1), 405–417 (2022). https://doi.org/10.1093/mnras/stac3165 [arXiv:2206.13184 [astro-ph.HE]]
30. I.K. Dihingia, Y. Mizuno, P. Sharma, High-soft to low-hard state transition in black hole X-ray binaries with GRMHD simulations. Astrophys. J. **958**(2), 105 (2023). https://doi.org/10.3847/1538-4357/ad0049 [arXiv:2309.16092 [astro-ph.HE]]
31. I.K. Dihingia, B. Vaidya, Properties of the accretion disc, jet and disc-wind around Kerr black hole. J. Astrophys. Astron. **43**(1), 23 (2022). https://doi.org/10.1007/s12036-022-09804-z
32. I.K. Dihingia, B. Vaidya, C. Fendt, Jets, disc-winds, and oscillations in general relativistic, magnetically driven flows around black hole. Mon. Not. R. Astron. Soc. **505**(3), 3596–3615 (2021). https://doi.org/10.1093/mnras/stab1512 [arXiv:2105.11468 [astro-ph.HE]]
33. I.K. Dihingia, B. Vaidya, C. Fendt, Truncated accretion discs in black hole X-ray binaries: dynamics and variability signatures. Mon. Not. R. Astron. Soc. **517**(4), 5032–5049 (2022). https://doi.org/10.1093/mnras/stac3021 [arXiv:2208.10726 [astro-ph.HE]]
34. K. Dionysopoulou, D. Alic, C. Palenzuela, L. Rezzolla, B. Giacomazzo, General-relativistic resistive magnetohydrodynamics in three dimensions: formulation and tests. Phys. Rev. D **88**, 044020 (2013). https://doi.org/10.1103/PhysRevD.88.044020. [arXiv:1208.3487 [gr-qc]]

35. K. Dionysopoulou, D. Alic, L. Rezzolla, General-relativistic resistive-magnetohydrodynamic simulations of binary neutron stars. Phys. Rev. D **92**(8), 084064 (2015). https://doi.org/10.1103/PhysRevD.92.084064 [arXiv:1502.02021 [gr-qc]]
36. M. Dumbser, O. Zanotti, Very high order P(N) P(M) schemes on unstructured meshes for the resistive relativistic MHD equations. J. Comput. Phys. **228**, 6991–7006 (2009). https://doi.org/10.1016/j.jcp.2009.06.009 [arXiv:0903.4832 [gr-qc]]
37. A.A. Esin, J.E. McClintock, R. Narayan, Advection-dominated accretion and the spectral states of black hole X-ray binaries: application to Nova Muscae 1991. Astrophys. J. **489**(2), 865–889 (1997)
38. K. Akiyama et al., [Event Horizon Telescope], First M87 event horizon telescope results. I. The shadow of the supermassive black hole. Astrophys. J. Lett. **875**, L1 (2019). https://doi.org/10.3847/2041-8213/ab0ec7 [arXiv:1906.11238 [astro-ph.GA]]
39. K. Akiyama et al. [Event Horizon Telescope], First M87 event horizon telescope results. VII. Polarization of the ring. Astrophys. J. Lett. **910**(1), L12 (2021). https://doi.org/10.3847/2041-8213/abe71d [arXiv:2105.01169 [astro-ph.HE]]
40. K. Akiyama et al., [Event Horizon Telescope], First Sagittarius A* Event Horizon Telescope Results. V. Testing Astrophysical Models of the Galactic Center Black Hole. Astrophys. J. Lett. **930**(2), L16 (2022). https://doi.org/10.3847/2041-8213/ac6672 [arXiv:2311.09478 [astro-ph.HE]]
41. K. Akiyama et al., [Event Horizon Telescope], The persistent shadow of the supermassive black hole of M 87. I. Observations, calibration, imaging, and analysis. Astron. Astrophys. **681**, A79 (2024). https://doi.org/10.1051/0004-6361/202347932.
42. S. Faraji, E. Hackmann, Thin accretion disk around the distorted Schwarzschild black hole. Phys. Rev. D **101**(2), 023002 (2020). https://doi.org/10.1103/PhysRevD.101.023002 [arXiv:2010.02786 [astro-ph.HE]]
43. M. Fathi, M. Olivares, J.R. Villanueva, Analytical study of light ray trajectories in Kerr spacetime in the presence of an inhomogeneous anisotropic plasma. Eur. Phys. J. C **81**(11), 987 (2021). https://doi.org/10.1140/epjc/s10052-021-09787-1 [arXiv:2104.07721 [gr-qc]]
44. C. Fendt, Collimation of astrophysical jets: the role of the accretion disk magnetic field distribution. Astrophys. J. **651**, 272–287 (2006). https://doi.org/10.1086/507976 [arXiv:astro-ph/0511611 [astro-ph]]
45. J. Ferreira, P.O. Petrucci, G. Henri, L. Sauge, G. Pelletier, A unified accretion-ejection paradigm for black hole x-ray binaries. Astron. Astrophys. **447**, 813 (2006). https://doi.org/10.1051/0004-6361:20052689 [arXiv:astro-ph/0511123 [astro-ph]]
46. L.G. Fishbone, V. Moncrief, Relativistic fluid disks in orbit around Kerr black holes. Astrophys. J. **207**, 962–976 (1976)
47. P.C. Fragile, K. Chatterjee, A. Ingram, M. Middleton, The luminous, hard state can't be MAD. Mon. Not. R. Astron. Soc. **525**(1), L82–L86 (2023). https://doi.org/10.1093/mnrasl/slad099 [arXiv:2307.08820 [astro-ph.HE]]
48. J. Fukue, Transonic disk accretion revisited. Publ. Astron. Soc. Jpn. **39**, 2 (1987)
49. J.A. Garcia, E. Kara, D. Walton, T. Beuchert, T. Dauser, E. Gatuzz, M. Balokovic, J.F. Steiner, F. Tombesi, R.M.T. Connors et al., Implications of the warm corona and relativistic reflection models for the soft excess in mrk 509. Astrophys. J. **871**(1), 88 (2019). https://doi.org/10.3847/1538-4357/aaf739 [arXiv:1812.03194 [astro-ph.HE]]
50. J.F. Hawley, J.H. Krolik, High resolution simulations of the plunging region in a pseudo-Newtonian potential: dependence on numerical resolution and field topology. Astrophys. J. **566**, 164 (2002). https://doi.org/10.1086/338059 [arXiv:astro-ph/0110118 [astro-ph]]
51. K. Hirotani, H. Shang, R. Krasnopolsky, K.I. Nishikawa, Two-dimensional particle-in-cell simulations of axisymmetric black hole magnetospheres: angular dependence of the Blandford-Znajek Flux. Astrophys. J. **943**(2), 164 (2023). https://doi.org/10.3847/1538-4357/aca8b0 [arXiv:2209.13140 [astro-ph.HE]]
52. J.D. Hogg, C.S. Reynolds, The dynamics of truncated black hole accretion disks II: magnetohydrodynamic case. Astrophys. J. **854**(1), 6 (2018). https://doi.org/10.3847/1538-4357/aaa6c6 [arXiv:1801.05836 [astro-ph.HE]]

53. E. Kara, A.C. Fabian, E.M. Cackett, J.F. Steiner, P. Uttley, D.R. Wilkins, A. Zoghbi, The Closest look at 1H0707-495: X-ray reverberation lags with 1.3 Ms of data. Mon. Not. R. Astron. Soc. **428**, 2795 (2013). https://doi.org/10.1093/mnras/sts155 [arXiv:1210.1465 [astro-ph.HE]]
54. S. Kisaka, A. Levinson, K. Toma, Comprehensive analysis of magnetospheric gaps around Kerr black holes using 1D GRPIC simulations. Astrophys. J. **902**(1), 80 (2020). https://doi.org/10.3847/1538-4357/abb46c [arXiv:2007.02838 [astro-ph.HE]]
55. S.S. Komissarov, Multi-dimensional numerical scheme for resistive relativistic MHD. Mon. Not. R. Astron. Soc. **382**, 995 (2007). https://doi.org/10.1111/j.1365-2966.2007.12448.x [arXiv:0708.0323 [astro-ph]]
56. F. Krause, K.H. Raedler, *Mean-Field Magnetohydrodynamics and Dynamo Theory* (1980)
57. J.H. Krolik, *Active Galactic Nuclei: from the Central Black Hole to the Galactic Environment* (Princeton University Press, Princeton, NJ, 1999)
58. T.M. Belloni, S.E. Motta, Transient black hole binaries. https://doi.org/10.1007/978-3-662-52859-4_2 [arXiv:1603.07872 [astro-ph.HE]]
59. N. Kylafis, P. Reig, Correlation of time lag and photon index in GX 339–4. Astron. Astrophys. **614**, L5 (2018). https://doi.org/10.1051/0004-6361/201833339 [arXiv:1806.03052 [astro-ph.HE]]
60. A. Levinson, B. Cerutti, Particle-in-cell simulations of pair discharges in a starved magnetosphere of a Kerr black hole. Astron. Astrophys. **616**, A184 (2018). https://doi.org/10.1051/0004-6361/201832915 [arXiv:1803.04427 [astro-ph.HE]]
61. M.T.P. Liska, N. Kaaz, G. Musoke, A. Tchekhovskoy, O. Porth, Radiation transport two-temperature GRMHD simulations of warped accretion disks. Astrophys. J. Lett. **944**(2), L48 (2023). https://doi.org/10.3847/2041-8213/acb6f4 [arXiv:2210.10198 [astro-ph.HE]]
62. M.T.P. Liska, G. Musoke, A. Tchekhovskoy, O. Porth, A.M. Beloborodov, Formation of magnetically truncated accretion disks in 3D radiation-transport two-temperature GRMHD simulations. Astrophys. J. Lett. **935**(1), L1 (2022). https://doi.org/10.3847/2041-8213/ac84db [arXiv:2201.03526 [astro-ph.HE]]
63. H. Liu, J. Jiang, Z. Zhang, C. Bambi, A.C. Fabian, J.A. Garcia, A. Ingram, E. Kara, J.F. Steiner, J.A. Tomsick et al., High-density reflection spectroscopy of black hole X-ray binaries in the hard state. Astrophys. J. **951**(2), 145 (2023). https://doi.org/10.3847/1538-4357/acd8b9 [arXiv:2303.10593 [astro-ph.HE]]
64. S. Markoff, M.A. Nowak, J. Wilms, Going with the flow: can the base of jets subsume the role of compact accretion disk coronae? Astrophys. J. **635**, 1203–1216 (2005). https://doi.org/10.1086/497628 [arXiv:astro-ph/0509028 [astro-ph]]
65. A. Martocchia, G. Matt, Iron Kalpha line intensity from accretion discs around rotating black holes. Mon. Not. R. Astron. Soc. **282**(4), L53–L57 (1996)
66. G. Matt, G.C. Perola, L. Piro, The iron line and high energy bump as X-ray signatures of cold matter in Seyfert 1 galaxies. Astron. Astrophys. **247**, 25 (1991)
67. J.C. McKinney, C.F. Gammie, A Measurement of the electromagnetic luminosity of a Kerr black hole. Astrophys. J. **611**, 977–995 (2004). https://doi.org/10.1086/422244 [arXiv:astro-ph/0404512 [astro-ph]]
68. J.C. McKinney, A. Tchekhovskoy, R.D. Blandford, General relativistic magnetohydrodynamic simulations of magnetically choked accretion flows around black holes. Mon. Not. R. Astron. Soc. **423**, 3083 (2012). https://doi.org/10.1111/j.1365-2966.2012.21074.x [arXiv:1201.4163 [astro-ph.HE]]
69. J.C. McKinney, A. Tchekhovskoy, A. Sadowski, R. Narayan, Three-dimensional general relativistic radiation magnetohydrodynamical simulation of super-Eddington accretion, using a new code HARMRAD with M1 closure. Mon. Not. R. Astron. Soc. **441**(4), 3177–3208 (2014). https://doi.org/10.1093/mnras/stu762 [arXiv:1312.6127 [astro-ph.CO]]
70. M. Mehdipour, G.A. Kriss, J.S. Kaastra, E. Costantini, J. Mao, Dimming of continuum captured in Mrk 841: new clues on the nature of the soft X-ray excess. Astrophys. J. Lett. **952**(1), L5 (2023). https://doi.org/10.3847/2041-8213/ace053. [arXiv:2306.11960 [astro-ph.HE]]

71. A. Mignone, G. Mattia, G. Bodo, L. Del Zanna, A constrained transport method for the solution of the resistive relativistic MHD equations. Mon. Not. R. Astron. Soc. **486**(3), 4252–4274 (2019). https://doi.org/10.1093/mnras/stz1015 [arXiv:1904.01530 [physics.comp-ph]]
72. Y. Mizuno, GRMHD simulations and modeling for jet formation and acceleration region in AGNs. Universe **8**(2), 85 (2022). https://doi.org/10.3390/universe8020085 [arXiv:2201.12608 [astro-ph.HE]]
73. Y. Mizuno, C.M. Fromm, Z. Younsi, O. Porth, H. Olivares, L. Rezzolla, Comparison of the ion-to-electron temperature ratio prescription: GRMHD simulations with electron thermodynamics. Mon. Not. R. Astron. Soc. **506**(1), 741–758 (2021). https://doi.org/10.1093/mnras/stab1753 [arXiv:2106.09272 [astro-ph.HE]]
74. D.M. Teixeira, M.J. Avara, J.C. McKinney, General relativistic radiation magnetohydrodynamic simulations of thin magnetically arrested disks. Mon. Not. R. Astron. Soc. **480**(3), 3547–3561 (2018). https://doi.org/10.1093/mnras/sty2044 [arXiv:1706.08982 [astro-ph.HE]]
75. M. Moscibrodzka, H. Falcke, H. Shiokawa, General relativistic magnetohydrodynamical simulations of the jet in M 87. Astron. Astrophys. **586**, A38 (2016). https://doi.org/10.1051/0004-6361/201526630 [arXiv:1510.07243 [astro-ph.HE]]
76. S. Nampalliwar, A.I. Yfantis, K.D. Kokkotas, Extending GRMHD for thin disks to non-Kerr spacetimes. Phys. Rev. D **106**(6), 063009 (2022). https://doi.org/10.1103/PhysRevD.106.063009
77. R. Narayan, D. Barret, J.E. McClintock, Advection-dominated accretion model of the black hole V404 Cygni in quiescence. Astrophys. J. **82**(1), 448–464 (1997)
78. R. Narayan, J.E. McClintock, Observational evidence for a correlation between jet power and black hole spin. Mon. Not. R. Astron. Soc. **419**, L69–L73 (2012). https://doi.org/10.1111/j.1745-3933.2011.01181.x [arXiv:1112.0569 [astro-ph.HE]]
79. R. Narayan, I. Yi, Astrophys. J. **452**, 710 (1995). https://doi.org/10.1086/176343. [arXiv:astro-ph/9411059 [astro-ph]]
80. R. Narayan, J.E. McClintock, I. Yi, A New model for black hole soft x-ray transients in quiescence. Astrophys. J. **457**, 821 (1996). https://doi.org/10.1086/176777 [arXiv:astro-ph/9508014 [astro-ph]]
81. A. Nathanail, C.M. Fromm, O. Porth, H. Olivares, Z. Younsi, Y. Mizuno, L. Rezzolla, Plasmoid formation in global GRMHD simulations and AGN flares. Mon. Not. R. Astron. Soc. **495**(2), 1549–1565 (2020). https://doi.org/10.1093/mnras/staa1165 [arXiv:2002.01777 [astro-ph.HE]]
82. A. Nathanail, V. Mpisketzis, O. Porth, C.M. Fromm, L. Rezzolla, Magnetic reconnection and plasmoid formation in three-dimensional accretion flows around black holes. Mon. Not. R. Astron. Soc. **513**(3), 4267–4277 (2022). https://doi.org/10.1093/mnras/stac1118 [arXiv:2111.03689 [astro-ph.HE]]
83. R.S. Nemmen, T. Storchi-Bergmann, M. Eracleous, Spectral models for low-luminosity active galactic nuclei in liners: the role of advection-dominated accretion and jets. Mon. Not. R. Astron. Soc. **438**(4), 2804–2827 (2014). https://doi.org/10.1093/mnras/stt2388 [arXiv:1312.1982 [astro-ph.HE]]
84. S.C. Noble, J.H. Krolik, J.F. Hawley, Dependence of inner accretion disk stress on parameters: the Schwarzschild case. Astrophys. J. **711**, 959–973 (2010). https://doi.org/10.1088/0004-637X/711/2/959 [arXiv:1001.4809 [astro-ph.HE]]
85. S.C. Noble, J.H. Krolik, J.F. Hawley, Dependence of inner accretion disk stress on parameters: the Schwarzschild case. Astrophys. J. **711**, 959–973 (2010)
86. S.C. Noble, J.H. Krolik, J.D. Schnittman, J.F. Hawley, Radiative efficiency and thermal spectrum of accretion onto Schwarzschild black holes. Astrophys. J. **743**, 115 (2011). https://doi.org/10.1088/0004-637X/743/2/115 [arXiv:1105.2825 [astro-ph.HE]]
87. I.D. Novikov, K.S. Thorne, Astrophysics of black holes. Black Holes (Les Astres Occlus), pp. 343–450 (1973)
88. H. Olivares, O. Porth, J. Davelaar, E.R. Most, C.M. Fromm, Y. Mizuno, Z. Younsi, L. Rezzolla, Constrained transport and adaptive mesh refinement in the Black Hole Accretion

Code. Astron. Astrophys. **629**, A61 (2019). https://doi.org/10.1051/0004-6361/201935559 [arXiv:1906.10795 [astro-ph.HE]]
89. B. Paczynski, G. Bisnovatyi-Kogan, A model of a thin accretion disk around a black hole. Acta Astron. **31**, 283 (1981)
90. B. Paczynski, P.J. Wiita, Thick accretion disks and supercritical luminosities. Astron. Astrophys. **88**, 23–31 (1980). EFI-78-68-CHICAGO
91. D.N. Page, K.S. Thorne, Disk-accretion onto a black hole. Time-averaged structure of accretion disk. Astrophys. J. **191**, 499–506 (1974). https://doi.org/10.1086/152990
92. C. Palenzuela, L. Lehner, O. Reula, L. Rezzolla, Beyond ideal MHD: towards a more realistic modeling of relativistic astrophysical plasmas. Mon. Not. R. Astron. Soc. **394**, 1727–1740 (2009). https://doi.org/10.1111/j.1365-2966.2009.14454.x [arXiv:0810.1838 [astro-ph]]
93. S. Patra, B.R. Majhi, S. Das, Accretion flow in deformed Kerr spacetime: spectral energy distributions from free-free emission. JCAP **01**, 060 (2024). https://doi.org/10.1088/1475-7516/2024/01/060 [arXiv:2308.12839 [astro-ph.HE]]
94. J. Peitz, S. Appl, Viscous accretion discs around rotating black holes. Mon. Not. R. Astron. Soc. **286**, 681–695 (1997)
95. R.F. Penna, J.C. McKinney, R. Narayan et al., Simulations of magnetized discs around black holes: effects of black hole spin, disc thickness and magnetic field geometry. Mon. Not. R. Astron. Soc. **408**(2), 752–782 (2010)
96. R.F. Penna, A. Sádowski, J.C. McKinney, Thin-disc theory with a non-zero-torque boundary condition and comparisons with simulations. Mon. Not. R. Astron. Soc. **420**(1), 684–698 (2012)
97. O. Porth, Y. Mizuno, Z. Younsi, C.M. Fromm, Flares in the galactic centre–I. Orbiting flux tubes in magnetically arrested black hole accretion discs. Mon. Not. R. Astron. Soc. **502**(2), 2023–2032 (2021). https://doi.org/10.1093/mnras/stab163 [arXiv:2006.03658 [astro-ph.HE]]
98. O. Porth, H. Olivares, Y. Mizuno, Z. Younsi, L. Rezzolla, M. Moscibrodzka, H. Falcke, M. Kramer, The black hole accretion code. Comput. Astrophys. Cosmol. **4**(1), 1 (2017). https://doi.org/10.1186/s40668-017-0020-2 [arXiv:1611.09720 [gr-qc]]
99. R.E. Pudritz, Dynamo action in turbulent accretion discs around black holes. I—The fluctuations. II—The mean magnetic field. Mon. Not. R. Astron. Soc. **195**, 881–914 (1981)
100. Q. Qian, C. Fendt, S. Noble, M. Bugli, rHARM: accretion and ejection in resistive GR-MHD. Astrophys. J. **834**(1), 29 (2017). https://doi.org/10.3847/1538-4357/834/1/29 [arXiv:1610.04445 [astro-ph.HE]]
101. Q. Qian, C. Fendt, C. Vourellis, Jet launching in resistive GR-MHD black hole-accretion disk systems. Astrophys. J. **859**(1), 28 (2018). https://doi.org/10.3847/1538-4357/aabd36 [arXiv:1804.09652 [astro-ph.HE]]
102. L. Rezzolla, O. Zanotti, *Relativistic Hydrodynamics* (2013)
103. H. Riffert, H. Herold, Relativistic accretion disk structure revisited. Astrophys. J. **450**, 508 (1995)
104. B. Ripperda, F. Bacchini, A. Philippov, Magnetic reconnection and hot spot formation in black hole accretion disks. Astrophys. J. **900**(2), 100 (2020). https://doi.org/10.3847/1538-4357/ababab [arXiv:2003.04330 [astro-ph.HE]]
105. B. Ripperda, F. Bacchini, O. Porth, E.R. Most, H. Olivares, A. Nathanail, L. Rezzolla, J. Teunissen, R. Keppens, General relativistic resistive magnetohydrodynamics with robust primitive variable recovery for accretion disk simulations. Astrophys. J. Suppl. **244**(1), 10 (2019). https://doi.org/10.3847/1538-4365/ab3922 [arXiv:1907.07197 [physics.comp-ph]]
106. A. Sádowski, M. Wielgus, R. Narayan, D. Abarca, J.C. McKinney, A. Chael, Radiative, two-temperature simulations of low luminosity black hole accretion flows in general relativity. Mon. Not. R. Astron. Soc. **466**(1), 705–725 (2017). https://doi.org/10.1093/mnras/stw3116 [arXiv:1605.03184 [astro-ph.HE]]
107. J.D. Schnittman, J.H. Krolik, S.C. Noble, Disk emission from magnetohydrodynamic simulations of spinning black holes. Astrophys. J. **819**(1), 48 (2016). https://doi.org/10.3847/0004-637X/819/1/48 [arXiv:1512.00729 [astro-ph.HE]]

108. N.I. Shakura, R.A. Sunyaev, Black holes in binary systems. Observational appearance. Astron. Astrophys. **24**, 337–355 (1973)
109. S. Shashank, S. Riaz, A.B. Abdikamalov, C. Bambi, Testing relativistic reflection models with GRMHD simulations of accreting black holes. Astrophys. J. **938**(1), 53 (2022). https://doi.org/10.3847/1538-4357/ac9128 [arXiv:2207.11526 [astro-ph.HE]]
110. S. Sheikhnezami, C. Fendt, O. Porth, B. Vaidya, J. Ghanbari, Bipolar jets launched from magnetically diffusive accretion disks. I. Ejection efficiency vs field strength and diffusivity. Astrophys. J. **757**, 65 (2012). https://doi.org/10.1088/0004-637X/757/1/65 [arXiv:1207.6086 [astro-ph.HE]]
111. M. Steenbeck, F. Krause, On the dynamo theory of stellar and planetary magnetic fields. I. AC dynamos of solar type. Astronomische Nachrichten **291**, 49–84 (1969)
112. D. Stepanovs, C. Fendt, An extensive numerical survey of the correlation between outflow dynamics and accretion disk magnetization. Astrophys. J. **825**(1), 14 (2016). https://doi.org/10.3847/0004-637X/825/1/14 [arXiv:1604.07313 [astro-ph.GA]]
113. D. Stepanovs, C. Fendt, S. Sheikhnezami, Modelling MHD accretion-ejection–episodic ejections of jets triggered by a mean-field disk dynamo. Astrophys. J. **796**(1), 29 (2014). https://doi.org/10.1088/0004-637X/796/1/29 [arXiv:1409.7003 [astro-ph.HE]]
114. A. Tchekhovskoy, R. Narayan, J.C. McKinney, Black hole spin and the radio loud/quiet dichotomy of active galactic nuclei. Astrophys. J. **711**, 50–63 (2010). https://doi.org/10.1088/0004-637X/711/1/50 [arXiv:0911.2228 [astro-ph.HE]]
115. J.R. Trump, C.D. Impey, B.C. Kelly, F. Civano, J.M. Gabor, A.M. Diamond-Stanic, A. Merloni, C.M. Urry, H. Hao, K. Jahnke et al., Accretion rate and the physical nature of unobscured active galaxies. Astrophys. J. **733**, 60 (2011). https://doi.org/10.1088/0004-637X/733/1/60 [arXiv:1103.0276 [astro-ph.CO]]
116. N. Tsukamoto, T. Kitamura, K. Nakajima, H. Asada, Gravitational lensing in Tangherlini spacetime in the weak gravitational field and the strong gravitational field. Phys. Rev. D **90**(6), 064043 (2014). https://doi.org/10.1103/PhysRevD.90.064043 [arXiv:1402.6823 [gr-qc]]
117. A. Uniyal, S. Chakrabarti, S. Das, Study of relativistic accretion flow in the f(R) theory of gravity. Phys. Dark Univ. **44**, 101429 (2024). https://doi.org/10.1016/j.dark.2024.101429 [arXiv:2306.14434 [astro-ph.HE]]
118. C. Vourellis, C. Fendt, Relativistic outflows from a GRMHD mean-field disk dynamo. Astrophys. J. **911**(2), 85 (2021). https://doi.org/10.3847/1538-4357/abe93b [arXiv:2102.12482 [astro-ph.HE]]
119. C. Vourellis, C. Fendt, Q. Qian, S. Noble, GR-MHD disk winds and jets from black holes and resistive accretion disks. Astrophys. J. **882**, 2 (2019). https://doi.org/10.3847/1538-4357/ab32e2 [arXiv:1907.10622 [astro-ph.HE]]
120. K.A. Weaver, R.F. Mushotzky, P.J. Serlemitsos et al., The origin of the soft X-ray excess in the Seyfert 2 galaxy NGC 2110. Astrophys. J. **442**, 597 (1995)
121. X. Xu, N. Ding, Q. Gu, X. Guo, E. Contini, The origin of the soft X-ray excess in the narrow-line Seyfert 1 galaxy SBS 1353 + 564. Mon. Not. R. Astron. Soc. **507**(3), 3572–3582 (2021). https://doi.org/10.1093/mnras/stab2278 [arXiv:2108.05265 [astro-ph.HE]]
122. Z. Yan, W. Yu, X-ray outbursts of low-mass X-ray binary transients observed in the RXTE era. Astrophys. J. **805**(2), 87 (2015). https://doi.org/10.1088/0004-637X/805/2/87 [arXiv:1408.5146 [astro-ph.HE]]
123. D. Yoon, K. Chatterjee, S. Markoff, D. van Eijnatten, Z. Younsi, M. Liska, A. Tchekhovskoy, Spectral and imaging properties of Sgr A* from high-resolution 3D GRMHD simulations with radiative cooling. Mon. Not. R. Astron. Soc. **499**(3), 3178–3192 (2020). https://doi.org/10.1093/mnras/staa3031 [arXiv:2009.14227 [astro-ph.HE]]
124. B. You, X. Cao, Z. Yan, J.M. Hameury, B. Czerny, Y. Wu, T. Xia, M. Sikora, S.N. Zhang, P. Du et al., Observations of a black hole x-ray binary indicate formation of a magnetically arrested disk. Science **381**(6661), abo4504 (2023). https://doi.org/10.1126/science.abo4504 [arXiv:2309.00200 [astro-ph.HE]]

125. C. Zanni, A. Ferrari, R. Rosner, G. Bodo, S. Massaglia, MHD simulations of jet acceleration from Keplerian accretion disks: the effects of disk resistivity. Astron. Astrophys. **469**, 811 (2007). https://doi.org/10.1051/0004-6361:20066400 [arXiv:astro-ph/0703064 [astro-ph]]
126. H. Zhou, Helical and non-helical large-scale dynamos in thin accretion discs. Mon. Not. R. Astron. Soc. **527**(2), 3018–3028 (2023). https://doi.org/10.1093/mnras/stad3406 [arXiv:2309.15565 [astro-ph.HE]]

Chapter 11
Tilted Accretion Disks

P. Chris Fragile and Matthew Liska

Abstract In this chapter, we review some of the interesting consequences that tilt between the spin axis of the black hole and angular momentum axis of the accretion disk can have on the dynamics, thermodynamics, and observational appearance of accreting systems, from precessing coronae and jets to standing nozzle shocks and quasi-periodic oscillations. We begin the chapter by examining some of the reasons tilted disks are interesting as well as present arguments for how ubiquitous they may be. We then review the existing simulation results in the literature, broadly dividing them into sections on thick disks, thin disks, and magnetically arrested disks (MADs). We finish by highlighting some of the phenomenology that is unique to tilted disk simulations and discuss how this may apply to observations.

11.1 General Overview

11.1.1 Why Is Tilt Interesting?

If an accretion disk is misaligned, or "tilted," with respect to a compact rotating object, it will be subject to Lense-Thirring (LT) precession [46]. For an ideal test particle in a slightly tilted orbit at a radius r around a black hole of mass M and specific angular momentum $a_* = J/M^2 = a/M$, this precession occurs at an angular frequency $\Omega_{\rm LT} \approx 2a_* M/r^3$ (in units where $G = c = 1$). Close to the black hole, this is comparable to the orbital angular frequency $\Omega = (M/r^3)^{1/2}[1 + a_*(M/r^3)^{1/2}]$. However, because of its strong radial dependence, LT precession becomes much weaker far from the hole. Therefore, a tilted disk made of non-interacting rings would experience differential precession that would twist and warp it over time, or even tear

P. C. Fragile (✉)
College of Charleston, 66 George Street, Charleston, SC, USA
e-mail: fragilep@cofc.edu

M. Liska
Georgia Institute of Technology, Howey Physics Bldg, 837 State St NW, Atlanta, GA, USA
e-mail: mliska3@gatech.edu

C. Bambi et al. (eds.), *New Frontiers in GRMHD Simulations*, Springer Series in Astrophysics and Cosmology,https://doi.org/10.1007/978-981-97-8522-3_11

it if the misalignment was large enough [76]. However, in a real disk, there can be strong interactions between successive radial rings, such that any warping disturbance will be communicated over some range of the disk. This communication is generally thought of as happening in either a diffusive or wave-like manner, depending on the properties of the disk. In the diffusive case, the warping is limited by secular (i.e., "viscous") responses within the disk. In such a case, the misaligned angular momentum is transported outward by viscous torques until the timescale for outward transport roughly equals the precession timescale [44, 75], leading to a unique, nearly constant transition radius r_{BP} [2, 44], inside of which the disk is expected to be flat and aligned with the black-hole symmetry plane, and outside of which the disk is also expected to be flat but in a plane determined by the angular momentum vector of the larger gas supply. This is often termed the "Bardeen-Petterson" (BP) effect and is illustrated in Fig. 11.1.

The BP effect is expected to apply whenever the dimensionless stress parameter α within the disk is larger than the scale height $\delta = H(r)/r$, where $H(r)$ is the thermal semi-thickness of the disk. Given that α is usually thought to be significantly less than one, this implies geometrically thin disks. However, neglecting relativistic correction factors, the scale height in the innermost, radiation-pressure- and electron-scattering-dominated portion of thin disks is roughly given by

$$\delta \sim \eta^{-1} \left(\frac{L}{L_{\mathrm{Edd}}} \right) \left(\frac{r}{r_G} \right)^{-1} ,$$

where $\eta \sim 0.1$ is the radiative efficiency, L/L_{Edd} is the luminosity in units of Eddington, and $r_G = GM/c^2$ is the gravitational radius. This means that the BP effect will only be relevant for the inner regions of thin disks at very small Eddington ratios, $L/L_{\mathrm{Edd}} \leq \alpha\eta \ll 1$, but at those luminosities, real black hole accretion systems are

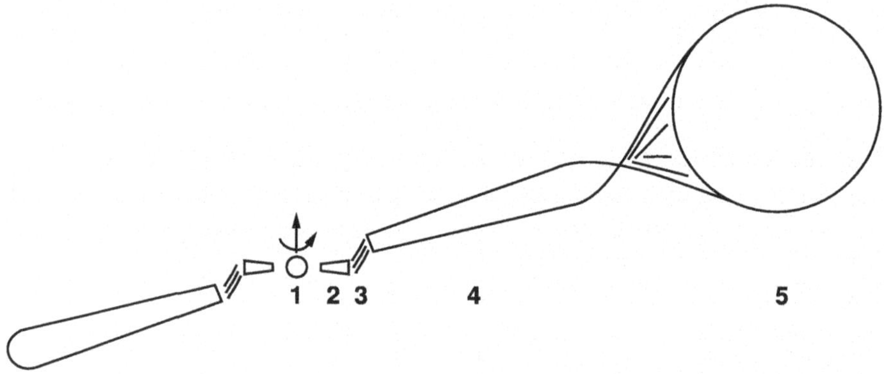

Fig. 11.1 Schematic diagram of the Bardeen-Petterson effect in a black hole X-ray binary showing (1) the central rotating black hole, (2) the inner, aligned accretion disk, (3) the transition region, (4) the outer, tilted accretion disk, and (5) the companion star. Image reproduced by permission from [25], copyright by AAS

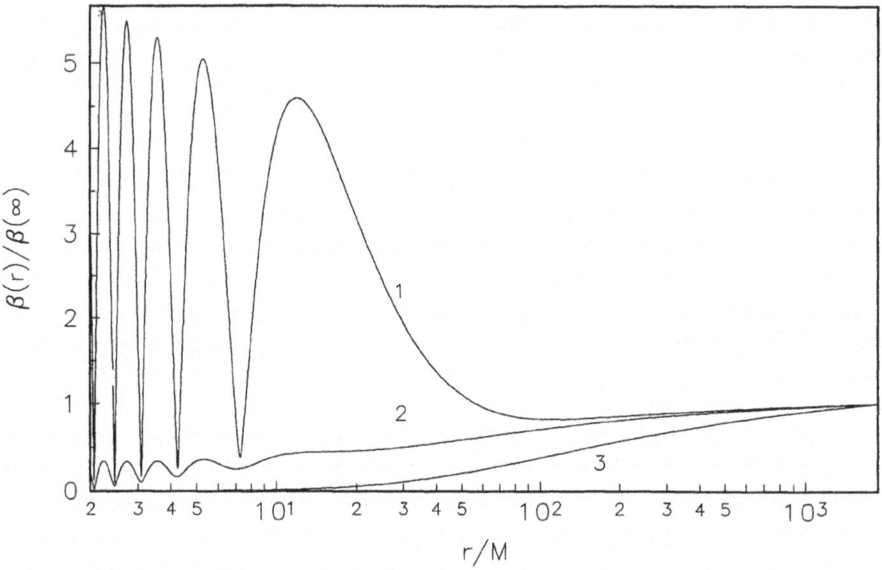

Fig. 11.2 Radial dependence of the tilt angle $\beta(r)/\beta(\infty)$ for values of δ/α of 10 (curve 1), 1 (curve 2), and 0.1 (curve 3). Curve 1 would be in the wave-like limit, whereas curve 3 would be in the Bardeen-Petterson limit. Image reproduced by permission from [37], copyright by RAS

generally found in the hard state, so the thin-disk solution may not be applicable. Furthermore, other accretion solutions, such as radiatively inefficient, geometrically slim and thick flows, will clearly not be in the Bardeen-Petterson regime. Therefore, the prevalence of this configuration may be limited. Nevertheless, we review numerical results applicable to the Bardeen-Petterson limit in Sect. 11.3.1.

In the thick-disk, $\delta > \alpha$ regime, LT precession is expected to produce warps that propagate in a wave-like manner [79], ultimately developing a standing, bending-wave pattern [37], where the tilt varies as a function of radius and can even take on values greater than in the outer disk, as illustrated in Fig. 11.2 (curve 1). Additionally, in the thick-disk limit, when the sound-crossing time is short compared to the precession timescale, a finite, isolated disk can precess as a solid body [20, 22, 70], which can have numerous interesting observational consequences. We discuss simulations in this regime and some of the corresponding phenomenology in Sect. 11.2.1.

11.1.2 How Ubiquitous Is Tilt?

Although tilt introduces new physical solutions and interesting corresponding phenomenology, there still seems to be some lingering doubt within the community as to how applicable it is to real black hole accretion systems. Hence, before proceeding with our review of tilted disk simulations, we lay out some of the arguments that support such disks being present, if not commonplace, in nature. Let's start with the simple cases: tidal disruption events (TDEs) and active galactic nuclei (AGN) after major merger events. In both cases, there is no way for the accretion flow at large scales to know about or be affected by the orientation of the spin axis of the black hole; therefore, there is absolutely no reason to expect things to be aligned at the start of such events. In fact, in such cases, retrograde accretion, where the angular momentum vector of the accreting gas points opposite the spin axis of the black hole, may be just as common as prograde [40].

Even in stellar-mass accretion systems, misalignment may be commonplace [25, 57]. Consider the formation avenues for black hole X-ray binaries (BHBs). Here, the orientation of the outer disk is fixed by the binary orbit, whereas the orientation of the black hole is set at the time it becomes part of the system. If the black hole joined the binary through multi-body interactions, such as binary capture or replacement, which may be common for X-ray binaries in globular clusters, then there would have been no preexisting symmetry, so the resulting system would nearly always harbor a tilted black hole. Even if the black hole formed from a member of a preexisting binary via a supernova, as is expected among field binaries, the black hole could still end up tilted if the supernova explosion were asymmetric [26], as it appears many are.

Following their formation, there is a tendency for tilted accretion disks to eventually align with the black hole due to their mutual tidal interactions [86]. However, for BHBs and TDEs, the alignment timescale is comparable to or longer than their observational lifetimes [91], and for AGN, each merger event has the possibility of reorienting the central black hole or its fuel supply, potentially resulting in repeated tilted configurations [42]. Thus, while we may not know the exact prevalence of tilted accretion disks, there are ample theoretical arguments to suggest they may be quite common.

Additionally, there is now mounting observational evidence to support the same conclusion. Several BHBs, e.g. GRO J1655-40 [77], XTE J1550-564 [29, 78], V4641 Sgr [64], GX 339-4 [65], and MAXI J1820+070 [83], and AGN, e.g. NGC 3079 [43], NGC 1068 [8], and NGC 4258 [7], have accretion disks that appear to be tilted with respect to the symmetry plane of their central black hole spacetimes. At present, most of these claims are circumstantial, as they rely on observations of relativistic bipolar jets (assumed to be aligned with the spin axis of the black hole) that are not perpendicular to the plane of the accretion disk observed at large scales [42, 92] or because the disk is warped in a way that may be consistent with the expectations of the BP effect [63]. In the future, more direct evidence for tilted disks may come from polarization measurements, which can be sensitive to the inclination and viewing angle of the inner accretion disk [11].

11.1.3 Numerical Considerations

Before proceeding with our review of the simulations themselves, we feel it may be worthwhile to mention a few of the special challenges associated with tilted disks. Tilted disk simulations are a level up in complexity in many ways when compared to their untilted counterparts. First off, tilted disks break any symmetries that may be exploited in simulations of untilted disks. This has two significant implications: (1) the simulations must always be done in full 3D, as there is no way to apply reflection or axial symmetry to reduce the computational cost; and (2) if one chooses to tilt the black hole with respect to the grid coordinates, then all the components of the resulting spacetime metric for gravity will be non-zero, which can potentially add computational cost relative to sparser, flat-space or symmetric metrics. A second challenge is choosing an appropriate grid and coordinate combination. Spherical-polar coordinates are commonly used in untilted black hole accretion simulations, which makes sense as such coordinates can be aligned with the predominant (azimuthal) direction of the flow to reduce numerical errors associated with advection oblique to cell faces. However, again, such symmetries are not as prevalent in tilted disks. Furthermore, the magnetohydrodynamic (MHD) equations are especially challenging to solve around the polar singularity. To minimize numerical artifacts when parts of the accretion disk pass through this artifact of the grid, transmissive polar boundary conditions can be used to minimize disturbances to the fluid [47, 51, 60]. These allow the gas to pass through the polar singularity more naturally than the more commonly used reflecting boundary conditions, but they still induce excess dissipation. Thus, in cases of extreme tilt, especially when the disk passes through the pole, it may make sense to consider alternative coordinate and grid choices, such as the cubed sphere [24] or even a Cartesian grid (see Fig. 11.3 for a comparison of the cubed-sphere and spherical-polar grids).

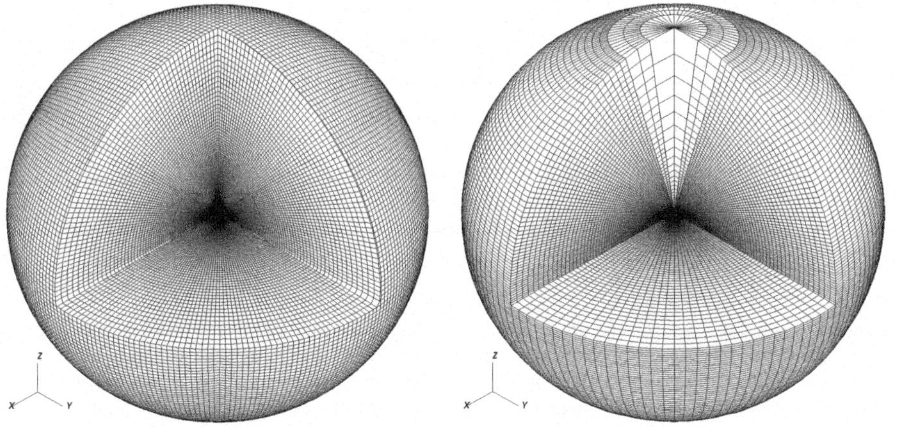

Fig. 11.3 Left: plot of the cubed-sphere grid geometry. Right: plot of a comparable spherical-polar grid with multiple resolution layers. Image reproduced by permission from [24], copyright by AAS

11.2 Simulations of Tilted Thick Disks

11.2.1 Discovery of Global LT Precession

We begin our review of tilted disk simulations by considering those in the wave-like (thick-disk) limit, as this is numerically the easiest case and was, hence, the first one considered in general relativistic simulations. The early simulations all started with finite, rather small (tens of r_G), reservoirs of gas, usually in the form of a torus orbiting the black hole. These were largely numerical experiments, as there were very few published predictions for what should come out of such simulations.

One of the biggest surprises was the discovery of global precession of the torus following a brief period of differential twisting [20]. A nice illustration of this behavior is provided in Fig. 11.4, where the disk twist profile shows both differential precession (mostly at small radii) and steady growth of the total precession angle with time

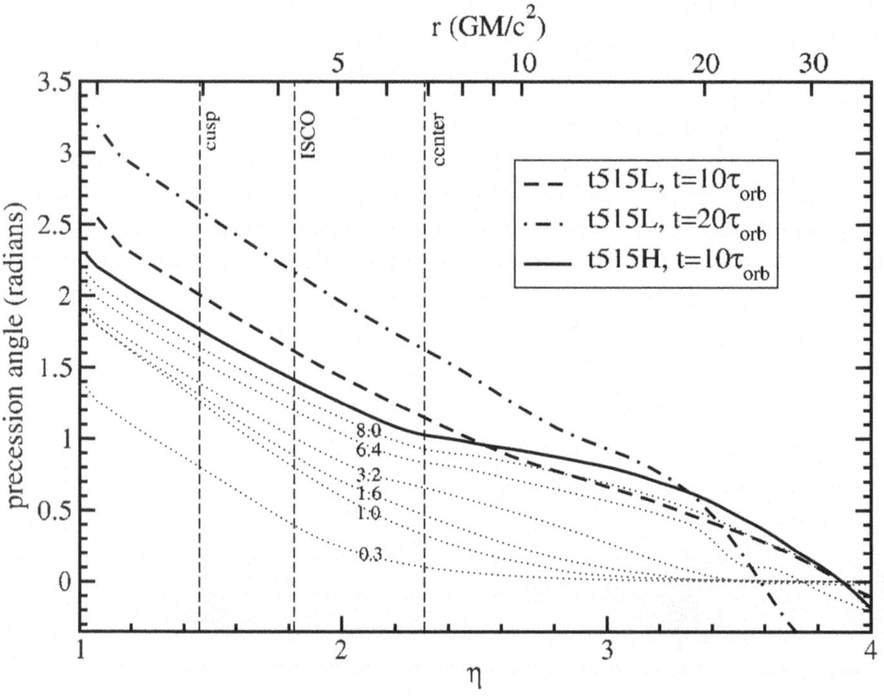

Fig. 11.4 Cumulative precession angle γ in radians as functions of radius for times spanning from 0.3 to 20 orbital timescales for a tilted disk simulation with $\beta = 15°$ and $a_* = 0.5$. Initially, $\gamma = 0$ throughout the disk. The bend in the profile around $6r_G$ reflects the initial period of relatively rapid differential precession. However, once that profile is established, it remains largely unchanged as the disk transitions to more global precession. Image reproduced by permission from [20], copyright by AAS

at all radii. Similar results have now been confirmed in multiple general relativistic magnetohydrodynamic (GRMHD) studies [21, 22, 47, 97].

In retrospect, it seems obvious that global precession must occur whenever the sound-crossing time in the accretion flow is shorter than the precession timescale, which it almost always is for compact tori and disks. In fact, earlier Newtonian simulations had already pointed to this behavior [45, 75]. We can even write down an expression for the rate of disk precession. Generically, the precession period for a solid-body rotator with angular momentum **J** subject to a torque τ is simply $T_{\rm prec} = 2\pi(\sin\beta)(J/\tau)$ [54], where β is the tilt angle between **J** and the black hole spin axis. The important point about this is that the LT torque gets integrated over the entire disk, so it has a cumulative effect. To illustrate, if we assume a power-law radial dependence to the surface density of the form $\Sigma \propto r^{-\zeta}$ and ignore higher order general relativistic corrections, we can write that

$$T_{\rm prec} = \frac{\pi(1+2\zeta)}{(5-2\zeta)} \frac{r_o^{5/2-\zeta} r_i^{1/2+\zeta}\left[1-(r_i/r_o)^{5/2-\zeta}\right]}{aM\left[1-(r_i/r_o)^{1/2+\zeta}\right]} \tag{11.1}$$

(in units of $G = c = 1$), where r_i and r_o are the inner and outer radii of the disk, respectively. This quantitative prediction was first confirmed in the simulations of [22]. Interestingly, it was pointed out that for small disks, or isolated disk components, with $r_o \approx 10 r_G$, this precession period falls in the proper range to match what is known as the type-C class of quasi-periodic oscillations (QPOs) seen in the X-ray light curves of BHBs [34]. We elaborate further on the potential association of QPOs with tilted disks in Sect. 11.6.

11.2.2 Unique Structural Features of Tilted Thick Disks

Early tilted thick-disk simulations also revealed interesting non-axisymmetric structures, particularly in the innermost regions of the disk, that had not been seen in comparable simulations of untilted disks. One example is the pair of over-dense bridges of matter appearing to connect the main body of the disk to the black hole, as seen in Fig. 11.5. These features were originally referred to as "plunging streams" [22], although it was later shown that they are better described as post-shock, over-density features, since the actual fluid trajectories are roughly normal to these structures, as opposed to along them [27]. Regardless, it is remarkable that these over-dense bridges (and the intervening low-density pockets) are *persistent* features, lasting the entire duration of the simulations and even precessing along with the rest of the inner disk [21].

Since the persistent, over-dense bridges are actually post-shock enhancements, this means that the shocks that create them are *standing* shocks, having fixed orientations with respect to the line-of-nodes between the disk and black hole midplanes

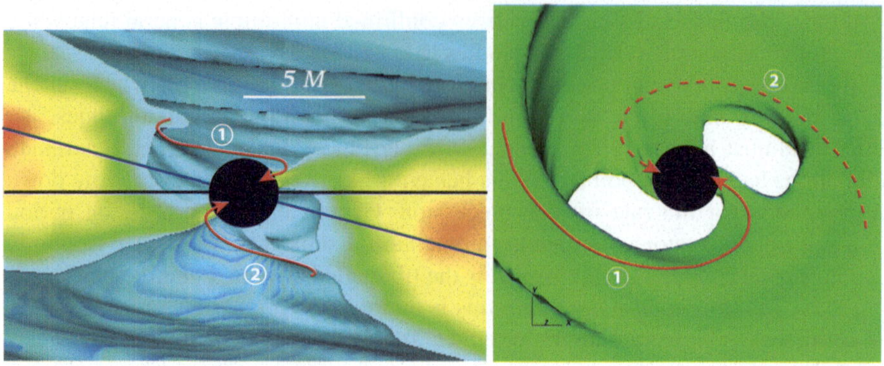

Fig. 11.5 Left: Cut-away view of the inner $10r_G$ of a tilted accretion flow revealing two opposing, high-latitude features connecting the disk to the horizon (red arrows). The figure is oriented with the black hole spin ($a_* = 0.9$) axis vertical. The black hole symmetry plane (black line) and initial disk midplane (blue line) are marked for reference. Right: Isodensity contour viewed down the spin axis of the black hole, highlighting the same two features as the left panel. Note that feature (1) remains entirely above both the black hole symmetry plane and disk midplane, while feature (2) remains below them. Images reproduced by permission from [22], copyright by AAS

and also persisting for the duration of current simulations. Standing shocks could profoundly affect the behavior of tilted disks by enhancing the angular momentum transport, increasing the mass accretion rate, increasing the energy dissipation, changing the effective innermost radius of the disk [19], and accelerating particles [94].

Further study revealed that the non-axisymmetric features owe their existence to the fact that orbital trajectories are more eccentric in tilted disks than untilted ones, and the eccentricity rises strongly with decreasing radius, especially inside $15r_G$ (see Fig. 16 of [13], for example). This rise in eccentricity with decreasing radius inevitably leads to a crowding of orbital trajectories near their respective apocenters [37]. Such crowding of the orbital trajectories provides the necessary conditions for standing shocks to arise. The reason there are *two* standing shocks, located on opposite sides of the black hole with one above the midplane and one below, is because the eccentricity of the orbital trajectories is 180° out of phase across the disk midplane. In other words, the apocenters of the orbits above the midplane come together on the opposite side of the black hole from where they come together below the midplane. Another way to picture this is in terms of the radial epicyclic motion of the gas in the disk. Figure 11.6 shows material moving radially outward on one side of the black hole and radially inward on the other side (indicative of eccentric trajectories), with the pattern being reversed across the disk midplane (left panel). Similar behavior is not seen in an equivalent untilted simulation (right panel).

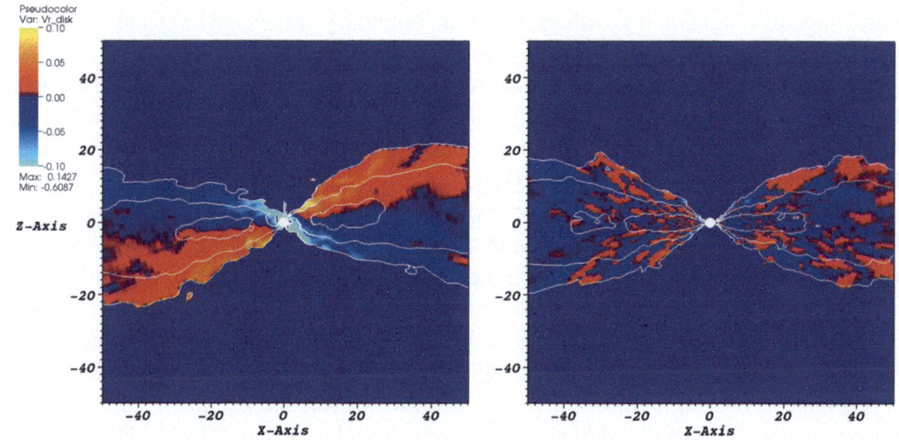

Fig. 11.6 Meridional ($\phi = 0$) plots for tilted (*left*) and untilted (*right*) simulations showing a pseudocolor representation of V^r for outflowing ($V^r > 0$) and inflowing ($V^r < 0$) gas as *hot* and *cold* colors, respectively. The velocity scale is normalized to the speed of light c. The plots are overlaid with isocontours of density at $\rho = 0.4$, 0.04, and $0.004\rho_{\text{max},0}$. The figure in the *left* panel is oriented such that the black hole is tilted 15° to the left. Note the pattern of radial epicyclic motion in the image on the *left* that is clearly absent from the image on the *right*. Image reproduced by permission from [21], copyright by AAS

11.2.3 SANE Versus MAD

Before discussing results for thin disks and jets, we first need to distinguish between two general types of black hole accretion simulations: standard and normal evolution (SANE) [73] and magnetically arrested disks (MAD) [72]. Most of the early work, for both untilted and tilted disks, focused on the weakly magnetized (SANE) case, with MADs becoming more popular in recent years. This has also begun to translate into the realm of tilted disks as well. The thick disks that we have discussed so far that precess globally [22, 47, 97] were all SANE disks.

However, for MADs, the case can be very different. MADs are formed when the accretion disk is saturated by vertical magnetic flux, and the corresponding magnetic pressure comes within a factor of a few of equipartition with the gas pressure. Whenever this happens, the disk attempts to get rid of the excess magnetic flux by rapidly ejecting flux bundles toward larger radii, allowing the magnetic pressure to drop. The radius out to which the disk is saturated, which we will call r_{MAD}, is set by, among other things, the amount of vertical magnetic flux in the system. In some MADs, the disk will only be saturated very close to the black hole, while in other cases, excess magnetic flux might be present out to hundreds, if not thousands, of r_G. The plasma within r_{MAD} typically has a much larger radial infall speed, and, because of mass continuity, a correspondingly lower density than the plasma beyond this radius. While many numerical studies include MAD and SANE models, it is important to note that not all MAD or SANE models are the same. Each can experience a wide range of magnetic flux values. Even the boundary between MAD and SANE is merely determined by the saturation value of the magnetic flux at the event horizon. There can be remarkable differences between SANE states that are, for example, just below the saturation level and SANE states that contain no vertical magnetic flux at all. Likewise, MADs will depend on the total amount of excess magnetic flux in the disk, in other words on the value of $r_{\rm MAD}$.

Going back to the question of precession, simulation results obtained so far suggest that the disk *does not* precess when the accretion is MAD [10, 60, 89]. This seems to be because, in the MAD state, the black hole ergosphere and inner accretion disk are saturated with magnetic flux, which, as it is twisted by the rotating black hole, is forced to align with the black hole spin axis. Because the dynamics of the inner disk are dominated by the magnetic field in the MAD state, the inner disk aligns in response to the alignment of the field. In other words, the tilt of the inner disk is erased by the effects of the MAD field. Even for initial tilts up to 60°, MADs are seen to be forced to align with the symmetry plane of the black hole out to $r \gtrsim 10 r_G$ [10]. This forced alignment obviously prevents LT precession of the disk. However, we caution that all numerical work considering tilted MADs so far has used rather large values of $r_{\rm MAD}$, meaning more work needs to be done on tilted MADs.

11.3 Simulations of Tilted Thin Disks

11.3.1 Status of the Bardeen-Petterson Effect

The BP effect is often quoted as one of the reasons why tilt might actually not be an important parameter for accretion disks. The logic is that if the BP effect aligns the disk with the black hole out to a relatively large radius (100–1000 r_G), then the disk can, for all intents and purposes, be treated as an untilted disk. The trouble with this thinking is that the validity of the BP effect, at least based on numerical studies, is somewhat ambiguous to say the least. On the one hand, smoothed-particle hydrodynamic (SPH) simulations [55, 56, 75] have largely confirmed the analytic predictions of [80], but were subject to the same assumptions, namely that the physics is adequately described by Newtonian gravity plus a torque term, that the "viscosity" is isotropic, and that the Shakura & Sunyaev [93] α-parameter is constant. However, each of those assumptions is known to be incorrect [68, 95].

On the other hand, it has taken a while for full GRMHD simulations to weigh in on the BP effect. With grid-based codes, tilted thin-disks are very challenging because of the difficulty in resolving all three dimensions. As an example, if you want 20 zones to cover the disk vertical scale height δ, then to have comparable resolution in the radial and azimuthal directions would require something like $80/\delta$ and $40\pi/\delta$ zones, respectively. For $\delta \approx 0.1$, that is approaching a 1000^3 computational domain, which is still challenging for all but the largest supercomputers. Nevertheless, [68] probed the regime where $\alpha \approx \delta \lesssim 0.1$, i.e., when the BP effect should begin to apply. They

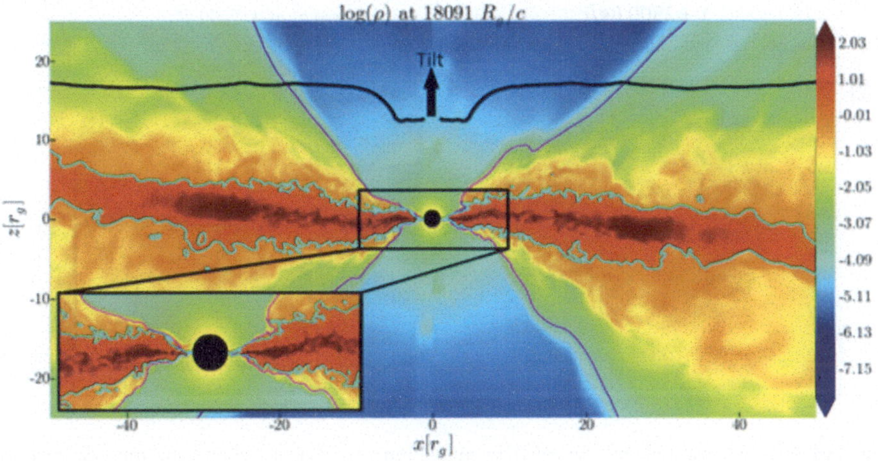

Fig. 11.7 Pseudocolor plot of density for a thin ($\delta = 0.03$), tilted disk, showing that only the innermost part of the disk ($r \lesssim 5r_G$) has aligned with the black hole symmetry plane (horizontal in this image). Image reproduced by permission from [50], copyright by RAS

found that their prograde tilted disks showed *no sign* of BP alignment even in the innermost regions. Only their retrograde case showed even modest evolution toward anti-alignment inside $10r_G$. [50] pushed to even thinner disks ($\delta \approx 0.015 - 0.03$) and saw modest BP alignment in prograde disks out to $\sim 5 - 10r_G$ (an illustration is provided in Fig. 11.7). The implication of these results is that the parameter space associated with the BP effect may be rather small, only including very thin disks, and even so, the alignment appears to be much more modest than what was originally predicted in [2].

11.3.2 Disk Tearing

In Sect. 11.2.1, we discussed precession in the context of relatively small, isolated, thick accretion disks (few 10 s of r_G). The precession period thus found was in a range consistent with the frequency of QPOs in astrophysical BHBs. However, accretion disks in real, physical systems are orders of magnitude larger than those used in early simulations. Since the Lense-Thirring precession rate decreases proportional to r^{-3}, we would thus expect such large accretion disks to precess at a much lower frequency due to their immense size (and much larger total angular momentum).

One way to circumvent this problem, and recover a reasonable precession period, would be to tear off a smaller, precessing sub-disk from the larger, non-precessing parent disk. This process, called disk tearing, was first demonstrated in SPH simulations [74, 76] in which disks governed by an isotropic α-viscosity tore apart into multiple independently precessing rings. Later, GRMHD simulations confirmed that

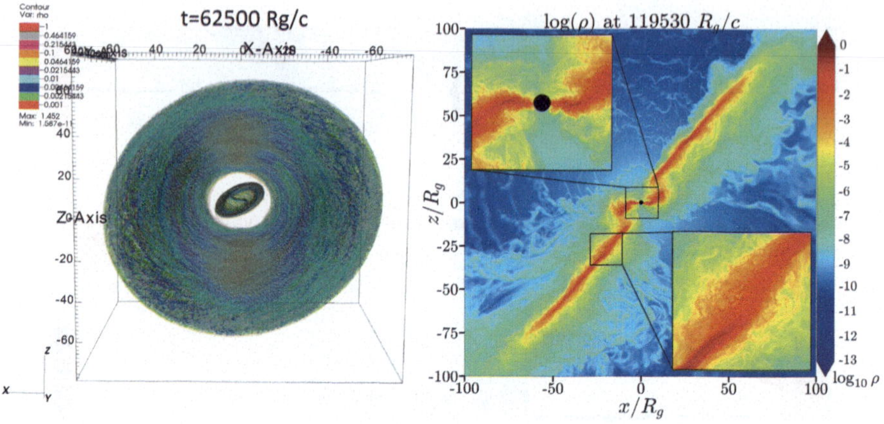

Fig. 11.8 An extreme resolution GRMHD simulation [51, 71] of a thin accretion disk (aspect ratio $\delta = 0.02$), which tears apart into an outer, non-precessing disk and inner, precessing one. The initial disk tilt is $\beta = 65°$ relative to the (horizontal) equatorial plane of the black hole. The right panel shows a vertical slice through the density with the black hole spin pointing vertically. Image reproduced by permission from [51], copyright by AAS

disks can tear [49, 51], though they have so far found the disk to only tear off into a single, radially extended, precessing sub-disk rather than a multitude of individual rings (see Fig. 11.8). The reason for this discrepancy is still an area of investigation, but magnetic torques [49], which are not included in SPH simulations, may be a key difference.

The exact location of the tearing radius is difficult to predict. Naively, one might expect a disk to tear if the viscous torques within the disk are not strong enough to counter the differential Lense-Thirring precession of the surrounding spacetime. In other words, if we write the Lense-Thirring torque as τ_{LT} and the viscous torque as τ_ν, then tearing may be expected whenever $\tau_{LT} \gtrsim \tau_\nu$. The radius where this criterion is met, or the so-called tearing radius $r_{\rm tear}$, can be written [76]

$$r_{\rm tear} \lesssim \frac{4}{3} |\sin\beta| \frac{a}{\alpha\delta} . \tag{11.2}$$

However, it turns out that this expression does not agree well with the results of GRMHD simulations [49]. This could be because the viscosity in these simulations is not isotropic, as it is assumed to be in analytic work and SPH simulations, or because the viscosity coefficient is not constant throughout the disk. The behavior of α in regions of large warp amplitude is also poorly understood. What is known is that when the LT torque is increased, the disk needs to compensate by increasing the warp amplitude to remain a rigid body. Typically, such an increase in warp amplitude would also lead to an increase in the viscous torques responsible for transporting misaligned angular momentum. However, some semi-analytical work suggests that, at a critical warp amplitude, the viscous torque becomes weaker instead of stronger, allowing the disk to tear apart [15].

11.3.3 Nozzle Shocks

As described in Sect. 11.2.2, the increasing eccentricity of particle orbits towards smaller radii in warped accretion disks can lead to a crowding that produces standing shocks. In thick disks with modest tilts, these standing shocks have only a modest effect on the accretion rate [22]. However, in thin disks with large tilts, the disk undergoes vertical compression twice per orbit (see Fig. 11.9) that is strong enough to produce nozzle shocks, which increase the radial infall speed of the gas by 2–3 orders of magnitude [39]. Whenever a particle passes through one of the nozzle shocks, it can lose several percent of its orbital kinetic energy, leading to an effective α of up to $10^2 - 10^3$, possibly providing an explanation for the rapid variability seen in certain changing-look AGN.

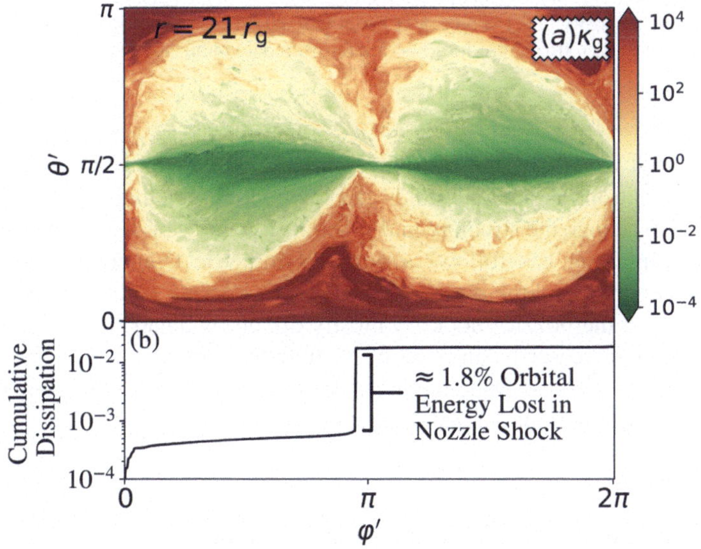

Fig. 11.9 Vertical compression in simulations of warped, thin disks leads to nozzle shocks at two azimuths. Panel (a): Fluid-frame entropy (κ_g), at radius $r = 21r_G$ in the $\theta' - \varphi'$ plane showing spikes at $\varphi' \approx 0$ (or 2π) and π, where the disk is most compressed. Panel (b): The cumulative fraction of dissipated energy along the annulus, normalized to the orbital energy. Across the $\varphi' = \pi$ nozzle shock, $\approx$1.8% of the orbital energy is dissipated. Image reproduced by permission from [39], copyright by AAS

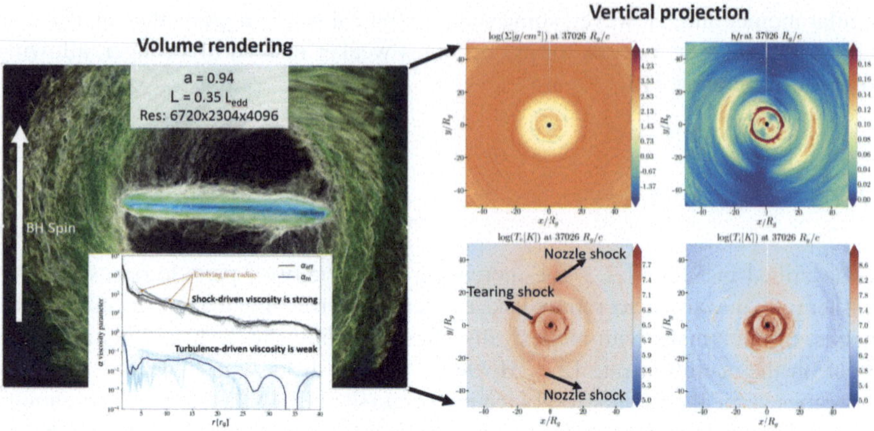

Fig. 11.10 Volume rendering of a radiative GRMHD simulation of a misaligned accretion disk in the "high-soft" state (left panel) alongside the vertically integrated density Σ, scale height h/r, electron temperature T_e, and ion temperature T_i (right panel). In this accretion disk the warping of space-time causes the disk to tear apart into two sub-disks: a precessing, inner disk and a non-precessing, outer one. It demonstrates that shock dissipation occurs at two locations, where it leads to a substantial increase in the electron and ion temperatures: 1) Perpendicular to the line of nodes along the z-axis; and 2) at the tearing radius. The inset in the left panel demonstrates that the shock-induced effective viscosity (α_{eff}) is orders of magnitude larger than the turbulent viscosity seeded by MRI turbulence (α_{m}). Image produced by author from data published in [39, 52], copyright by AAS

More recent two-temperature, radiative GRMHD simulations [52], which treated the ions and electrons as separate fluids and solved for the radiation, found that nozzle shocks also change the structure and spectral signatures of disks (see Fig. 11.10). Namely, the infalling gas follows a non-axisymmetric temperature profile with hotspots centered around the nozzle shocks and the tearing radius. While the heating of the gas in the nozzle shocks is mostly driven by adiabatic compression, as the scale-height decreases by an order of magnitude within the nozzle, gas crossing the tearing radius heats up dramatically as it undergoes a rapid orbital plane change crossing from the outer, parent disk to the inner, sub-disk.

11.4 Simulations of Tilted Truncated Disks

In many systems we are not dealing simply with either a thin disk or a thick one, but with a hot, thick inner flow coupled to a cold, thin outer disk at a "truncation radius" [14, 16, 18]. Such a truncated disk picture is typically used to explain the hard state of BHBs whenever $L \gtrsim 10^{-3} L_{\text{Edd}}$. Truncated disks naturally produce a spectrum that is dominated by Compton-scattered X-ray emission on top of a weaker, thermal blackbody component. As of the time of this writing, the physics behind disk truncation is still poorly understood. The best known models invoke thermal conduction from the corona to the disk leading to evaporation [12, 54, 61, 85]. However, recent shearing-box MHD simulations [1] suggest that the plasma may cool too quickly to be able to efficiently conduct heat, thus raising doubt about the feasibility of this picture.

As an alternative, recent radiative, two-temperature GRMHD simulations [53] have suggested that truncation might instead be set by $r_{\rm MAD}$. In these simulations, the saturation of magnetic flux in the inner disk can cause it to become magnetic-pressure-dominated, allowing reconnection layers to effectively heat up the gas to extreme temperatures, consistent with a hot corona. Interestingly, within this hot corona were found cold clumps of much denser gas (see Fig. 11.11). These cold clumps might explain the presence of broadened iron lines seen in the hard state [87], which are difficult to explain in the standard truncated disk picture.

Sidestepping the question of how truncated disks might form, recent GRMHD simulations [6] investigated the question of precession in tilted, truncated disks by simply prescribed a radially dependent scale-height for the disk using a cooling function. They confirmed that the corona region inside the truncation radius can

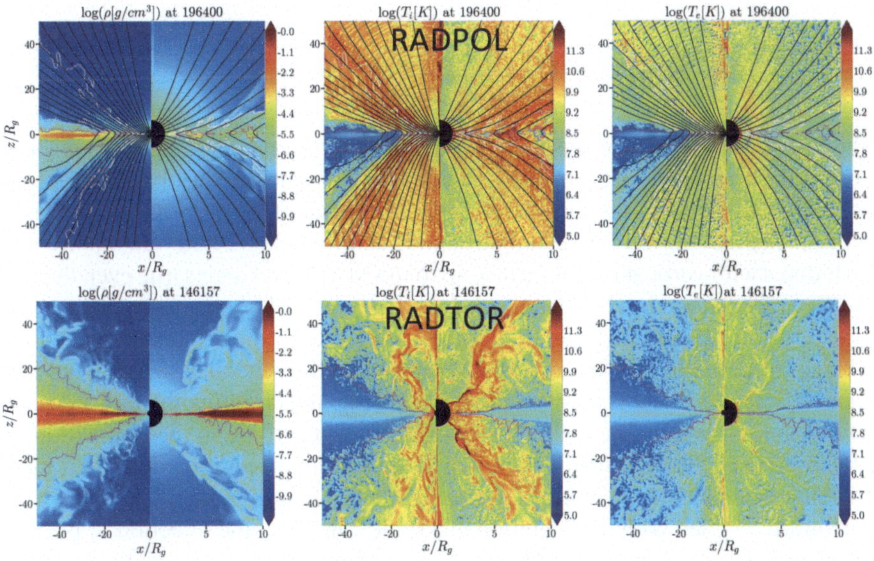

Fig. 11.11 The presence of large-scale poloidal magnetic flux leads to the development of a two-phase medium: a low-density, thick, hot corona-like accretion flow with clumps of cold gas floating through it (top row). In contrast, in the absence of the poloidal magnetic flux, the cold, thin flow extends down to the black hole (bottom row). The three columns, from left to right, show vertical slices through the density, ion temperature, and electron temperature; the right hemispheres show zoom-ins on the left hemispheres. Magnetic field lines are shown in black, jet boundary ($p_{mag} = \rho c^2$) in white, and the last scattering surface ($\tau_{es} = 1$, integrated in the vertical direction) in pink. Image reproduced by permission from [53], copyright by AAS

precess as a rigid body independent of the outer disk, though possibly at a lower frequency than an isolated corona. This is in line with semi-analytical work [58] suggesting there might exist some correlation between the frequency of precession-induced type-C QPOs and the disk truncation radius.

11.5 Jets from Tilted Disks

Accretion disks throughout nature are very often associated with "jets"—narrow streams of fast-moving plasma usually ejected perpendicular to the disk. Studying these jets has been a popular goal of black hole accretion simulations [9, 59, 96]. In this section, we look particularly at the case of jets from tilted disk simulations. Naively, one might expect that jets powered by the Blandford-Znajek process [4], which tap into the spin of the black hole, should align with the black hole spin axis, while jets powered by the Blandford-Payne process [3], which tap into the spin of the disk, should precess along with the disk. Current simulation results, though, present a more nuanced picture.

11.5.1 SANE Versus MAD (Revisited)

In the SANE case, where as we already mentioned in Sect. 11.2.1, the disks precess globally, the corresponding jets also precess *in phase* with the disk [47, 48, 89]. In other words, the jet aligns with, *and follows*, the angular momentum axis of the disk, not the black hole spin axis. An example of this is shown in Fig. 11.12. This applies for SANE disks with vertical magnetic flux up to the MAD level and is true even though these jets are powered by the Blandford-Znajek process. This apparent paradox, that the jet seems to be aligned with the accretion disk even though it is powered by the black hole, can be explained by the torque exerted by the disk-driven winds on the jets. These winds are responsible for collimating and orienting the jets. Because the winds are driven by the disk, they orient the jet according to the orientation of the accretion disk. In simulations at least, it appears that accretion disks that have sufficient magnetic flux to launch meaningful jets also launch powerful winds, which are naturally aligned with the accretion disks and can affect the direction of the jet.

However, as we mentioned in Sect. 11.2.3, published MAD simulations *do not* precess. This appears to be because the magnetic flux aligns with the black hole spin axis, which then forces the inner disk to align. Figure 11.13 provides one example of this behavior (see also Table 1 of [60] and Fig. 14 of [89] for further evidence). This forced alignment prevents LT precession of the disk, which, in turn, precludes any LT-driven precession of the jet. (Because the tilt is effectively 0° for much of the jet, it is not even possible to define a reasonable precession angle.) Likewise, in cases where thin disks undergo BP alignment (Sect. 11.3.1), there is also no precession of the jet, at least at small scales [50].

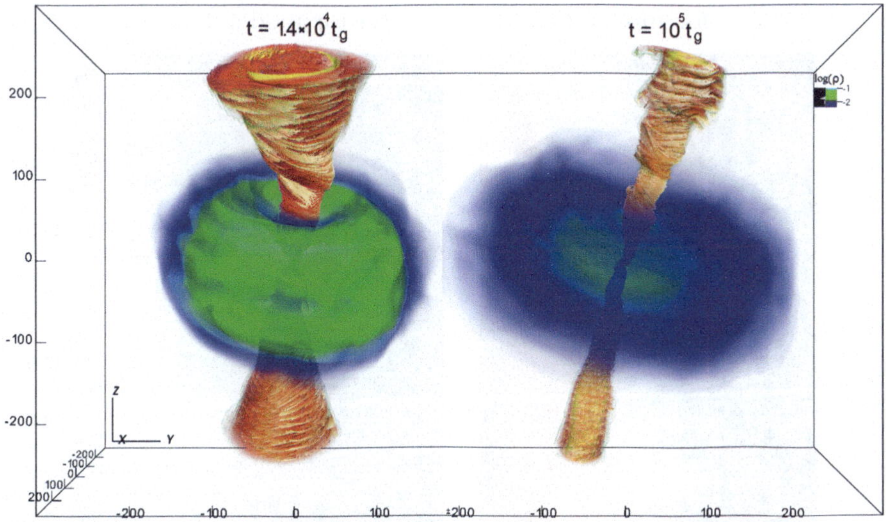

Fig. 11.12 Volume rendering of density [log(ρ) in blue and green, indicating the disk] and magnetic energy density [$p_B r/\rho c^2 > 0.5$ in red and yellow, indicating the jets] at early (left) and late times (right). The disk-jet system precesses as a whole around the black hole spin vector, which is vertical in the figure. Image reproduced by permission from [47], copyright by RAS

Finally, we note that even in cases where the inner disk precesses (Sects. 11.3.2 and 11.4), some GRMHD simulations show that the jet at larger scales can be torqued into partial alignment with the outer, non-precessing disk [49]. This suggests that it may be difficult to detect LT jet precession on scales beyond $r \gtrsim 10^2 r_G$. However, those particular simulations had relatively small accretion disks that were not continuously fed from larger scales. Future simulations that model larger accretion disks will be necessary to test whether or not jet precession can persist on scales order(s) of magnitude larger than the size of the inner precessing disk.

11.6 Variability and Quasi-Periodic Oscillations

One of the earliest interests in tilted accretion disks stemmed from the possibility that the additional timescales associated with LT precession or the BP effect might correspond with one or more of the unexplained QPOs seen in the light curves of many accreting black hole systems [25]. Although there is still much work to do to connect all the dots, the exciting news is that numerical simulations of tilted accretion disks have indeed demonstrated a richer, stronger, and more persistent set of timing signals than untilted disks, as we describe in this section.

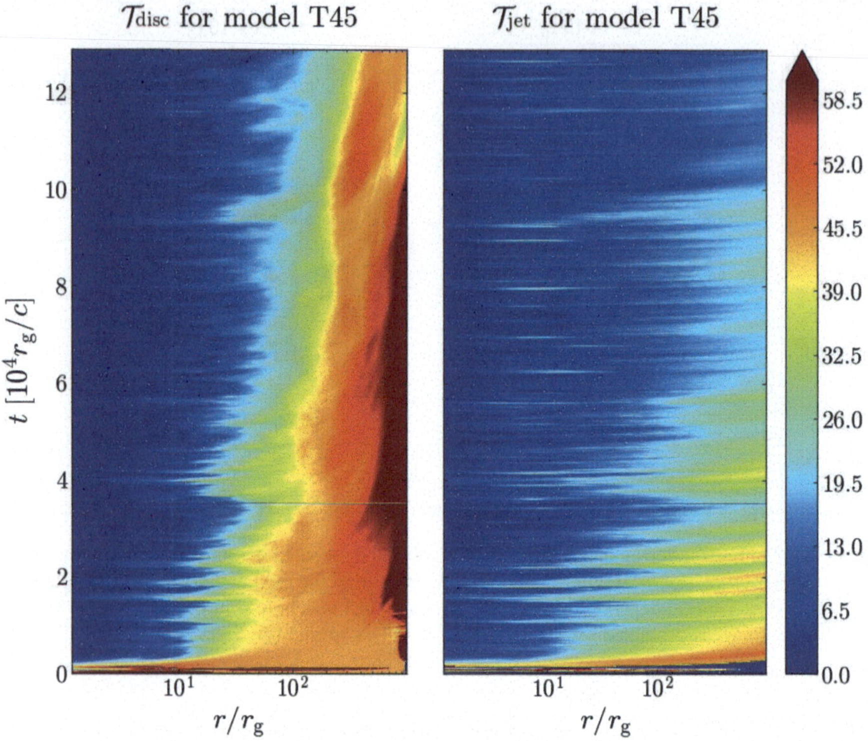

Fig. 11.13 Spacetime plots of the disk (left panel) and jet (right panel) tilt angles (in degrees) for model T45 of [10]. The disk quickly aligns (tilt goes to ≈0) to beyond a radius of $10r_G$, while the jet is also closely aligned with the black hole spin axis to beyond $10^2 r_G$. Image reproduced by permission from [23], copyright by RAS

11.6.1 Low-Frequency QPOs

The first "timing signals" that were observed in tilted disk simulations were those of LT precession itself, as already discussed in Sect. 11.2.1. It was recognized fairly early on that variations in the light curve from this type of precession would likely provide an excellent match to a particular type of low-frequency QPO called the type-C [33, 34].[1] The strengths of this association are that: (1) it produces the correct range of QPO frequencies for reasonable black hole and disk parameters [34, 70]; (2) it naturally yields a rise in the QPO frequency at the start of each outburst cycle when coupled with the truncated disk model [33]; (3) it easily reproduces the correlation between system inclination and QPO strength [30, 69]; and (4) it also explains why

[1] For a thorough review of QPO phenomenology in BHBs, see [36].

the centroid energies of reflection features seen in phase-resolved spectroscopy vary with QPO phase [35]. Together, these strengths make LT precession one of the most successful QPO models out there.

11.6.2 High-Frequency QPOs

Tilted disk simulations have also yielded some promising candidates for "high-frequency" QPOs ($\gtrsim$60 Hz). For instance, [32] found evidence for excess power around $118(M/10M_{\odot})^{-1}$ Hz over an extended radial range in one early tilted-disk simulation. Later analysis revealed this to be a transient feature related to the orbit of an over-dense clump passing through the standing shocks [31]; however, this effort was still important for opening the doors for later discoveries.

It now appears that it may be necessary to have both tilt *and* a distinct transition radius in the disk to trigger sustained QPO activity [5]. For example, [71] found that, in a tearing disk simulation (tearing disks were discussed in Sect. 11.3.2), the inner accretion disk oscillates at the *radial* epicyclic frequency associated with the tearing radius (see Fig. 11.14). Oscillations at *local* mode frequencies have been seen in many accretion disk simulations [67, 90], but because each radius excites a different frequency, these usually do not produce a QPO-like peak. The key in the tearing disk results is that the tearing radius provides a preferred frequency that is amplified with respect to the others.

However, it may not be enough just to have a transition radius; tilt also appears to be important. For instance, [5] found excitation of *vertical* epicyclic motion in their truncated disk simulations, but only ones that included tilt. Figure 11.15 provides visual evidence for both low- and high-frequency variations, specifically in the midplane θ value, for one such tilted, truncated disk simulation.

11.7 Future Directions

In this chapter, we have reviewed much of the interesting phenomenology associated with simulations of tilted accretion disks, including precession (Sect. 11.2.1), the BP effect (Sect. 11.3.1), and disk tearing (Sect. 11.3.2). While we gave some basic arguments for when the disk might undergo each of these processes (see Sects. 11.1 and 11.3.2), there are still large uncertainties within the available parameter space and many more simulations need to be done to fill in the gaps. In this closing section, we outline what we see as some of the most promising directions to explore in the near future.

One very important issue to settle is when do the jets associated with tilted disks precess and when do they align? As we mentioned in Sect. 11.5, jets in some simulations appear to precess with the same frequency as the inner disk, while in others, specifically tilted, MAD simulations, the whole inner disk is forced to align with the

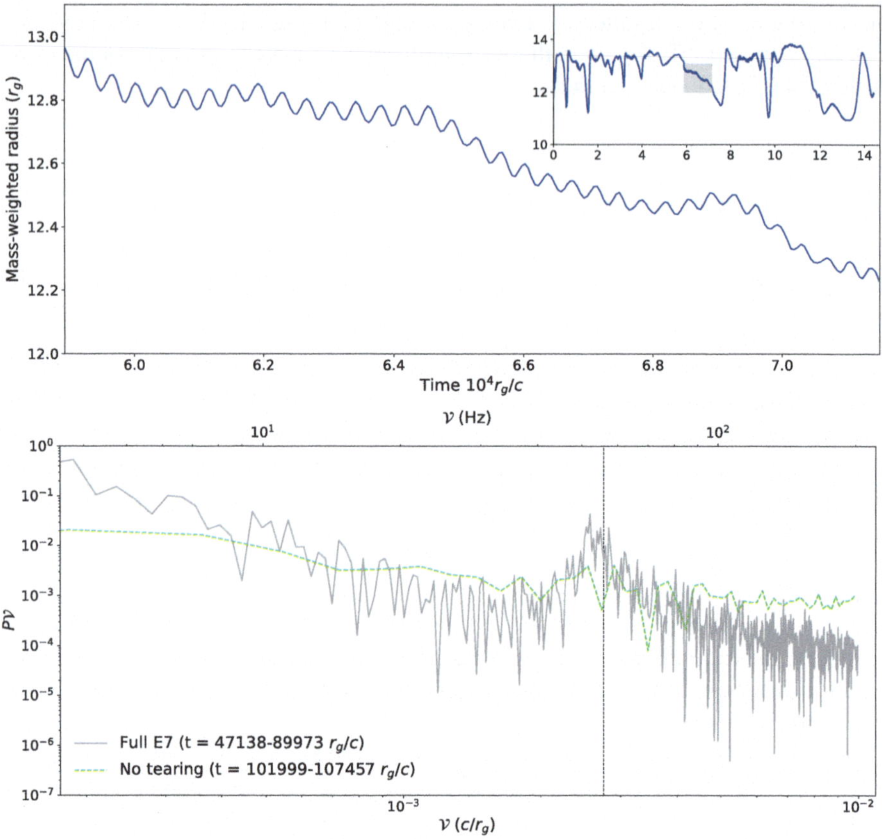

Fig. 11.14 Top: Radial oscillations of the inner sub-disk in a tearing-disk simulation. The inset panel shows the mass-weighted radius for the entire simulation duration, while the main panel shows a zoom during a period of coherent oscillations. Bottom: Power spectra of the mass-weighted radius computed over one tearing cycle (grey curve) and over a period of time in which the disk does not tear (dashed green curve). The vertical dashed black line shows the radial epicyclic frequency at $13r_G$ (55 Hz for a $10M_\odot$ black hole). Image reproduced by permission from [71], copyright by RAS

black hole and the jet does not precess. However, these conclusions are currently based on a very small number of simulations. More work needs to be done to flesh out the rough conclusion that SANE disks and jets precess, while MAD ones do not. This is pertinent to, among other things, the question of whether the hard state of BHBs can be MAD [23]. There is also a question as to what is the distance scale over which jet precession can be maintained, and therefore, observed? This is critically important as it may be through X-ray or radio jets that LT precession is first confirmed in nature (see e.g. [66]).

Another issue is that current GRMHD simulations [49, 51] suggest that, for tearing to occur, the disk needs be extremely thin ($\delta \lesssim 0.02$) and highly misaligned ($\beta \gtrsim$

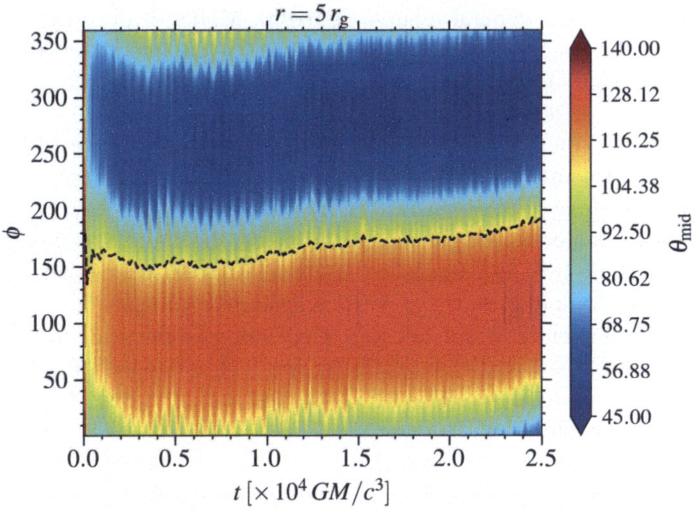

Fig. 11.15 A time-space plot of the mid-plane angle, $\theta_{\rm mid}$, near the inner edge of a tilted, truncated disk simulation. A clear $m = 1$ pattern is seen, which precesses along with the rest of the thick-disk region (black, dashed line) and also oscillates with a higher-frequency period close to $500GM/c^3$. Image reproduced by permission from [5], copyright by RAS

45°). This would imply that tearing can only occur at luminosities $L \gtrsim 0.01 L_{\rm Edd}$ and for highly misaligned sources, which would seem to make it inapplicable to most BHBs. However, tearing from a thin disk might not be the only avenue to form a precessing disk. If some other process (e.g., thermal conduction and evaporation) can separate the disk into distinct components, then it may be the inner one can precess semi-independent of the outer. Or, super-Eddington disks might precess inside their spherization radius [62]. Finally, GRMHD simulations have yet to consider even thinner disks ($\delta \lesssim 0.02$) and whether they can tear at smaller misalignment angles.

Speaking of super-Eddington accretion, most ultra-luminous X-ray sources (ULXs) are now believed to be powered by stellar-mass compact objects accreting at super-Eddington rates.[2] What effect would tilt have in those cases? How does the addition of dynamically important radiative forces within the disk interplay with the LT torques to change the resulting disk structure? How apparent would those changes be out at the effective photosphere of the disk, i.e., is there an observational signature of tilted ULXs? Here, LT warping may also compete with radiative warping [84], such that the final disk structure could be quite complex. Currently, there is very little published work in this regime.

What about LT precession in TDEs? As we argued in Sect. 11.1.2, TDE disks should almost always be tilted. Is there some way to exploit TDE light curves to look for evidence of LT precession or BP alignment? Before we can answer that question, we probably need more simulations of TDEs that properly account for the effects

[2] See [38, 41] for reviews of ULXs.

of black hole spin (i.e., general relativistic hydro, MHD, or even radiation MHD simulations) and consider a range of inclinations for the encounter. Interestingly, some TDEs show QPO signals that may be consistent with LT precession [81, 82], so there is motivation to do this work.

Finally, Sgr A*, the $4 \times 10^6 M_\odot$ black hole at the center of the Milky Way, is another important source to compare simulations with. In this case, the evidence points to the black hole primarily being fed by winds from Wolf-Rayet stars orbiting near the Galactic center, making it possible to model the entire accretion flow from source to black hole [88, 89]. Because these stars orbit on scales that are unlikely to be affected by the spin axis of Sgr A*, this again likely means the accretion flow is tilted and might even be warped or torn. Depending on the scale of the warping, instruments such as the Event Horizon Telescope [17] or GRAVITY [28] may even give us a chance to directly image a tilted, warped accretion disk. On that hopeful note, we end our chapter.

Acknowledgements PCF gratefully acknowledges the support of NASA through award No 23-ATP23-0100.

References

1. C.J. Bambic, E. Quataert, M.W. Kunz, Y.-F. Jiang, Local models of two-temperature accretion disc coronae. II. Ion thermal conduction and the absence of disc evaporation. arXiv e-prints. arXiv:2401.05482 (2024)
2. James M. Bardeen, Jacobus A. Petterson, The Lense-Thirring effect and accretion disks around Kerr Black holes. Astrophys. J. **195**, L65 (1975)
3. R.D. Blandford, D.G. Payne, Hydromagnetic flows from accretion discs and the production of radio jets. Mon. Not. R. Astron. Soc. **199**, 883–903 (1982)
4. R.D. Blandford, R.L. Znajek, Electromagnetic extraction of energy from Kerr black holes. Mon. Not. R. Astron. Soc. **179**, 433–456 (1977)
5. D.A. Bollimpalli, P.C. Fragile, J.W. Dewberry, W. Kluźniak, Truncated, tilted discs as a possible source of quasi-periodic oscillations. Mon. Not. R. Astron. Soc. **528**(2), 1142–1157 (2024)
6. D.A. Bollimpalli, P.C. Fragile, W. Kluźniak, Effect of geometrically thin discs on precessing, thick flows: relevance to type-C QPOs. Mon. Not. R. Astron. Soc. **520**(1), L79–L84 (2023)
7. A. Caproni, Z. Abraham, M. Livio, H.J. Mosquera Cuesta, Is the Bardeen-Petterson effect responsible for the warping and precession in NGC4258? Mon. Not. R. Astron. Soc. **379**(1), 135–142 (2007)
8. A. Caproni, Z. Abraham, H.J. Mosquera Cuesta, Bardeen-Petterson effect and the disk structure of the Seyfert galaxy NGC 1068. Astrophys. J. **638**(1), 120–124 (2006)
9. K. Chatterjee, M. Liska, A. Tchekhovskoy, S.B. Markoff, Accelerating AGN jets to parsec scales using general relativistic MHD simulations. Mon. Not. R. Astron. Soc. **490**(2), 2200–2218 (2019)
10. K. Chatterjee, M. Liska, A. Tchekhovskoy, S. Markoff, Misaligned magnetized accretion flows onto spinning black holes: magneto-spin alignment, outflow power and intermittent jets. arXiv e-prints. arXiv:2311.00432 (2023)
11. Yifan Cheng, Dan Liu, Sourabh Nampalliwar, Cosimo Bambi, X-ray spectropolarimetric signature of a warped disk around a stellar-mass black hole. Classical Quant. Gravity **33**(12), 125015 (2016)

12. Hyerin Cho, Ramesh Narayan, Analytical model of disk evaporation and state transitions in accreting black holes. Astrophys. J. **932**(2), 97 (2022)
13. J. Dexter, P.C. Fragile, Observational signatures of tilted black hole accretion disks from simulations. Astrophys. J. **730**(1), 36 (2011)
14. C. Done, M. Gierliński, A. Kubota, Modelling the behaviour of accretion flows in X-ray binaries. Everything you always wanted to know about accretion but were afraid to ask. Astron. Astrophys. Rev. **15**, 1–66 (2007)
15. S. Doğan, C.J. Nixon, A.R. King, J.E. Pringle, Instability of warped discs. Mon. Not. R. Astron. Soc. **476**, 1519–1531 (2018)
16. Ann A. Esin, Jeffrey E. McClintock, Ramesh Narayan, Advection-dominated accretion and the spectral states of black hole X-ray binaries: application to Nova Muscae 1991. Astrophys. J. **489**(2), 865–889 (1997)
17. Event Horizon Telescope Collaboration, K. Akiyama, A. Alberdi, W. Alef, J.C. Algaba, R. Anantua, K. Asada, R. Azulay, U. Bach, A.-K. Baczko, D. Ball, M. Baloković, J. Barrett, M. Bauböck, B.A. Benson, D. Bintley, L. Blackburn, R. Blundell, K.L. Bouman, G.C. Bower, H. Boyce, M. Bremer, C.D. Brinkerink, R. Brissenden, S. Britzen, A.E. Broderick, D. Broguiere, T. Bronzwaer, S. Bustamante, D.-Y. Byun, J.E. Carlstrom, C. Ceccobello, A. Chael, C.-K. Chan, K. Chatterjee, S. Chatterjee, M.-T. Chen, Y. Chen, X. Cheng, I. Cho, P. Christian, N.S. Conroy, J.E. Conway, J.M. Cordes, T.M. Crawford, G.B. Crew, A. Cruz-Osorio, Y. Cui, J. Davelaar, M. De Laurentis, R. Deane, J. Dempsey, G. Desvignes, J. Dexter, V. Dhruv, S.S. Doeleman, S. Dougal, S.A. Dzib, R.P. Eatough, R. Emami, H. Falcke, J. Farah, V.L. Fish, E. Fomalont, H.A. Ford, R. Fraga-Encinas, W.T. Freeman, P. Friberg, C.M. Fromm, A. Fuentes, P. Galison, C.F. Gammie, R. García, O. Gentaz, B. Georgiev, C. Goddi, R. Gold, A.I. Gómez-Ruiz, J.L. Gómez, M. Gu, M. Gurwell, K. Hada, D. Haggard, K. Haworth, M.H. Hecht, R. Hesper, D. Heumann, L.C. Ho, P. Ho, M. Honma, C.-W.L. Huang, L. Huang, D.H. Hughes, S. Ikeda, C.M.V. Impellizzeri, M. Inoue, S. Issaoun, D.J. James, B.T. Jannuzi, M. Janssen, B. Jeter, W. Jiang, A. Jiménez-Rosales, M.D. Johnson, S. Jorstad, A.V. Joshi, T. Jung, M. Karami, R. Karuppusamy, T. Kawashima, G.K. Keating, M. Kettenis, D.-J. Kim, J.-Y. Kim, J. Kim, J. Kim, M. Kino, J.Y. Koay, P. Kocherlakota, Y. Kofuji, P.M. Koch, S. Koyama, C. Kramer, M. Kramer, T.P. Krichbaum, C.-Y. Kuo, N. La Bella, T.R. Lauer, D. Lee, S.-S. Lee, P.K. Leung, A. Levis, Z. Li, R. Lico, G. Lindahl, M. Lindqvist, M. Lisakov, J. Liu, K. Liu, E. Liuzzo, W.-P. Lo, A.P. Lobanov, L. Loinard, C.J. Lonsdale, R.-S. Lu, J. Mao, N. Marchili, S. Markoff, D.P. Marrone, A.P. Marscher, I. Martí-Vidal, S. Matsushita, L.D. Matthews, L. Medeiros, K.M. Menten, D. Michalik, I. Mizuno, Y. Mizuno, J.M. Moran, K. Moriyama, M. Moscibrodzka, C. Müller, A. Mus, G. Musoke, I. Myserlis, A. Nadolski, H. Nagai, N.M. Nagar, M. Nakamura, R. Narayan, G. Narayanan, I. Natarajan, A. Nathanail, S.N. Fuentes, J. Neilsen, R. Neri, C. Ni, A. Noutsos, M.A. Nowak, J. Oh, H. Okino, H. Olivares, G.N. Ortiz-León, T. Oyama, F. Özel, D.C.M. Palumbo, G.F. Paraschos, J. Park, H. Parsons, N. Patel, U.-L. Pen, D.W. Pesce, V. Piétu, R. Plambeck, A. PopStefanija, O. Porth, F.M. Pötzl, C. Prather, J.A. Preciado-López, D. Psaltis, H.-Y. Pu, V. Ramakrishnan, R. Rao, M.G. Rawlings, A.W. Raymond, L. Rezzolla, A. Ricarte, B. Ripperda, F. Roelofs, A. Rogers, E. Ros, C. Romero-Cañizales, A. Roshanineshat, H. Rottmann, A.L. Roy, I. Ruiz, C. Ruszczyk, K.L.J. Rygl, S. Sánchez, D. Sánchez-Argüelles, M. Sánchez-Portal, M. Sasada, K. Satapathy, T. Savolainen, F.P. Schloerb, J. Schonfeld, K.-F. Schuster, L. Shao, Z. Shen, D. Small, B.W. Sohn, J. SooHoo, K. Souccar, H. Sun, F. Tazaki, A.J. Tetarenko, P. Tiede, R.P.J. Tilanus, M. Titus, P. Torne, E. Traianou, T. Trent, S. Trippe, M. Turk, I. van Bemmel, H.J. van Langevelde, D.R. van Rossum, J. Vos, J. Wagner, D. Ward-Thompson, J. Wardle, J. Weintroub, N. Wex, R. Wharton, M. Wielgus, K. Wiik, G. Witzel, M.F. Wondrak, G.N. Wong, Q. Wu, P. Yamaguchi, D. Yoon, A. Young, K. Young, Z. Younsi, F. Yuan, Y.-F. Yuan, J.A. Zensus, S. Zhang, G.-Y. Zhao, S.-S. Zhao, C. Agurto, A. Allardi, R. Amestica, J.P. Araneda, O. Arriagada, J.L. Berghuis, A. Bertarini, R. Berthold, J. Blanchard, K. Brown, M. Cárdenas, M. Cantzler, P. Caro, E. Castillo-Domínguez, T.L. Chan, C.-C. Chang, D.O. Chang, S.-H. Chang, S.-C. Chang, C.-C. Chen, R. Chilson, T.C. Chuter, M. Ciechanowicz, E. Colin-Beltran, I.M. Coulson, J. Crowley, N. Degenaar, S. Dornbusch, C.A. Durán, W.B. Everett, A. Faber, K. Forster, M.M. Fuchs, D.M. Gale, G. Geertsema, E. González, D. Graham, F. Gueth,

N.W. Halverson, C.-C. Han, K.-C. Han, Y. Hasegawa, J.L. Hernández-Rebollar, C. Herrera, R. Herrero-Illana, S. Heyminck, A. Hirota, J. Hoge, S.R. Hostler Schimpf, R.E. Howie, Y.-D. Huang, H. Jiang, H. Jinchi, D. John, K. Kimura, T. Klein, D. Kubo, J. Kuroda, C. Kwon, R. Lacasse, R. Laing, E.M. Leitch, C.-T. Li, C.-T. Liu, K.-Y. Liu, L.C.C. Lin, L.-M. Lu, F. MacAuliffe, P. Martin-Cocher, C. Matulonis, J.K. Maute, H. Messias, Z. Meyer-Zhao, A. Montaña, F. Montenegro-Montes, W. Montgomerie, M.E. Moreno Nolasco, D. Muders, H. Nishioka, T.J. Norton, G. Nystrom, H. Ogawa, R. Olivares, P. Oshiro, J.P. Pérez-Beaupuits, R. Parra, N.M. Phillips, M. Poirier, N. Pradel, R. Qiu, P.A. Raffin, A.S. Rahlin, J. Ramírez, S. Ressler, M. Reynolds, I. Rodríguez-Montoya, A.F. Saez-Madain, J. Santana, P. Shaw, L.E. Shirkey, K.M. Silva, W. Snow, D. Sousa, T.K. Sridharan, W. Stahm, A.A. Stark, J. Test, K. Torstensson, P. Venegas, C. Walther, T.-S. Wei, C. White, G. Wieching, R. Wijnands, J.G.A. Wouterloot, C.-Y. Yu, W. Yu, M. Zeballos, First Sagittarius A* event horizon telescope results. I. The shadow of the supermassive black hole in the center of the milky way. Astrophys. J. **930**(2), L12 (2022)
18. J. Ferreira, P.O. Petrucci, G. Henri, L. Saugé, G. Pelletier, A unified accretion-ejection paradigm for black hole X-ray binaries. I. The Dyn. Constituents. Astron. Astrophys. **447**, 813–825 (2006)
19. P.C. Fragile, Effective inner radius of tilted black hole accretion disks. Astrophys. J. **706**(2), L246–L250 (2009)
20. P.C. Fragile, P. Anninos, Hydrodynamic simulations of tilted thick-disk accretion onto a Kerr Black hole. Astrophys. J.**623**(1), 347–361 (2005)
21. P.C. Fragile, O.M. Blaes, Epicyclic motions and standing shocks in numerically simulated tilted black hole accretion disks. Astrophys. J.**687**(2), 757–766 (2008)
22. P.C. Fragile, O.M. Blaes, P. Anninos, J.D. Salmonson, Global general relativistic magneto-hydrodynamic simulation of a tilted black hole accretion disk. Astrophys. J. **668**(1), 417–429 (2007)
23. P.C. Fragile, K. Chatterjee, A. Ingram, M. Middleton, The luminous, hard state can't be MAD. Mon. Not. R. Astron. Soc. **525**(1), L82–L86 (2023)
24. P.C. Fragile, C.C. Lindner, P. Anninos, J.D. Salmonson, Application of the cubed-sphere grid to tilted black hole accretion disks. Astrophys. J. **691**(1), 482–494 (2009)
25. P.C. Fragile, G.J. Mathews, J.R. Wilson, Bardeen-Petterson effect and quasi-periodic oscillations in X-ray binaries. Astrophys. J. **553**(2), 955–959 (2001)
26. T. Fragos, M. Tremmel, E. Rantsiou, K. Belczynski, Black hole spin-orbit misalignment in galactic X-ray binaries. Astrophys. J. **719**(1), L79–L83 (2010)
27. A. Generozov, O. Blaes, P.C. Fragile, K.B. Henisey, Physical properties of the inner shocks in hot, tilted black hole accretion flows. Astrophys. J. **780**(1), 81 (2014)
28. Gravity Collaboration, R. Abuter, A. Amorim, M. Bauböck, J.P. Berger, H. Bonnet, W. Brandner, Y. Clénet, V. Coudé D. Foresto, P.T. de Zeeuw, C. Deen, J. Dexter, G. Duvert, A. Eckart, F. Eisenhauer, N.M. Förster Schreiber, P. Garcia, F. Gao, E. Gendron, R. Genzel, S. Gillessen, P. Guajardo, M. Habibi, X. Haubois, T. Henning, S. Hippler, M. Horrobin, A. Huber, A. Jiménez-Rosales, L. Jocou, P. Kervella, S. Lacour, V. Lapeyrère, B. Lazareff, J.B. Le Bouquin, P. Léna, M. Lippa, T. Ott, J. Panduro, T. Paumard, K. Perraut, G. Perrin, O. Pfuhl, P.M. Plewa, S. Rabien, G. Rodríguez-Coira, G. Rousset, A. Sternberg, O. Straub, C. Straubmeier, E. Sturm, L.J. Tacconi, F. Vincent, S. von Fellenberg, I. Waisberg, F. Widmann, E. Wieprecht, E. Wiezorrek, J. Woillez, S. Yazici, Detection of orbital motions near the last stable circular orbit of the massive black hole SgrA*. Astron. Astrophys. **618**, L10 (2018)
29. Diana Hannikainen, Duncan Campbell-Wilson, Richard Hunstead, Vince McIntyre, Jim Lovell, John Reynolds, Tasso Tzioumis, Wu. Kinwah, XTE J1550–564: a superluminal ejection during the September 1998 outburst. Astrophys. Space Sci. Supplement **276**, 45–48 (2001)
30. L.M. Heil, P. Uttley, M. Klein-Wolt, Inclination-dependent spectral and timing properties in transient black hole X-ray binaries. Mon. Not. R. Astron. Soc. **448**(4), 3348–3353 (2015)
31. K.B. Henisey, O.M. Blaes, P.C. Fragile, Variability from non-axisymmetric fluctuations interacting with standing shocks in tilted black hole accretion disks. Astrophys. J. **761**(1), 18 (2012)
32. K.B. Henisey, O.M. Blaes, P.C. Fragile, B.T. Ferreira, Excitation of trapped waves in simulations of tilted black hole accretion disks with magnetorotational turbulence. Astrophys. J. **706**(1), 705–711 (2009)

33. Adam Ingram, Chris Done, A physical model for the continuum variability and quasi-periodic oscillation in accreting black holes. Mon. Not. R. Astron. Soc. **415**(3), 2323–2335 (2011)
34. A. Ingram, C. Done, P.C. Fragile, Low-frequency quasi-periodic oscillations spectra and Lense-Thirring precession. Mon. Not. R. Astron. Soc. **397**(1), L101–L105 (2009)
35. Adam Ingram, Michiel van der Klis, Matthew Middleton, Chris Done, Diego Altamirano, Lucy Heil, Phil Uttley, Magnus Axelsson, A quasi-periodic modulation of the iron line centroid energy in the black hole binary H1743–322. Mon. Not. R. Astron. Soc. **461**(2), 1967–1980 (2016)
36. Adam R. Ingram, Sara E. Motta, A review of quasi-periodic oscillations from black hole X-ray binaries: observation and theory. New Astron. Rev. **85**, 101524 (2019)
37. P.B. Ivanov, A.F. Illarionov, The oscillatory shape of the stationary twisted disc around a Kerr black hole. Mon. Not. R. Astron. Soc. **285**(2), 394–402 (1997)
38. Philip Kaaret, Hua Feng, Timothy P. Roberts, Ultraluminous X-ray sources. Annu. Rev. Astron. Astrophys. **55**(1), 303–341 (2017)
39. N. Kaaz, M.T.P. Liska, J. Jacquemin-Ide, Z.L. Andalman, G. Musoke, A. Tchekhovskoy, O. Porth, Nozzle shocks, disk tearing and streamers drive rapid accretion in 3D GRMHD simulations of warped thin disks. arXiv e-prints. arXiv:2210.10053 (2022)
40. A.R. King, J.E. Pringle, J.A. Hofmann, The evolution of black hole mass and spin in active galactic nuclei. Mon. Not. R. Astron. Soc. **385**(3), 1621–1627 (2008)
41. Andrew King, Jean-Pierre. Lasota, Matthew Middleton, Ultraluminous X-ray sources. New Astron. Rev. **96**, 101672 (2023)
42. A.L. Kinney, H.R. Schmitt, C.J. Clarke, J.E. Pringle, J.S. Ulvestad, R.R.J. Antonucci, Jet directions in Seyfert galaxies. Astrophys. J. **537**(1), 152–177 (2000)
43. Paul T. Kondratko, Lincoln J. Greenhill, James M. Moran, Evidence for a geometrically thick self-gravitating accretion disk in NGC 3079. Astrophys. J. **618**(2), 618–634 (2005)
44. S. Kumar, J.E. Pringle, Twisted accretion discs—the Bardeen-Petterson effect. Mon. Not. R. Astron. Soc. **213**, 435–442 (1985)
45. J.D. Larwood, R.P. Nelson, J.C.B. Papaloizou, C. Terquem, The tidally induced warping, precession and truncation of accretion discs in binary systems: three-dimensional simulations. Mon. Not. R. Astron. Soc. **282**(2), 597–613 (1996)
46. Josef Lense, Hans Thirring, Über den Einfluß der Eigenrotation der Zentralkörper auf die Bewegung der Planeten und Monde nach der Einsteinschen Gravitationstheorie. Physikalische Zeitschrift **19**, 156 (1918)
47. M. Liska, C. Hesp, A. Tchekhovskoy, A. Ingram, M. van der Klis, S. Markoff, Formation of precessing jets by tilted black hole discs in 3D general relativistic MHD simulations. Mon. Not. R. Astron. Soc. **474**(1), L81–L85 (2018)
48. M. Liska, C. Hesp, A. Tchekhovskoy, A. Ingram, M. van der Klis, S.B. Markoff, A phase lag between disc and corona in GRMHD simulations of precessing tilted accretion discs. New Astron. **101**, 102012 (2023)
49. M. Liska, C. Hesp, A. Tchekhovskoy, A. Ingram, M. van der Klis, S.B. Markoff, M. Van Moer, Disc tearing and Bardeen-Petterson alignment in GRMHD simulations of highly tilted thin accretion discs. arXiv e-prints. arXiv:1904.08428 (2019)
50. M. Liska, A. Tchekhovskoy, A. Ingram, M. van der Klis, Bardeen-Petterson alignment, jets, and magnetic truncation in GRMHD simulations of tilted thin accretion discs. Mon. Not. R. Astron. Soc. **487**(1), 550–561 (2019)
51. M.T.P. Liska, K. Chatterjee, D. Issa, D. Yoon, N. Kaaz, A. Tchekhovskoy, D. van Eijnatten, G. Musoke, C. Hesp, V. Rohoza, S. Markoff, A. Ingram, M. van der Klis, H-AMR: a new GPU-accelerated GRMHD code for exascale computing with 3D adaptive mesh refinement and local adaptive time stepping. Astrophys. J. Suppl. **263**(2), 26 (2022)
52. M.T.P. Liska, N. Kaaz, G. Musoke, A. Tchekhovskoy, O. Porth, Radiation transport two-temperature GRMHD simulations of warped accretion disks. Astrophys. J. **944**(2), L48 (2023)
53. M.T.P. Liska, G. Musoke, A. Tchekhovskoy, O. Porth, A.M. Beloborodov, Formation of magnetically truncated accretion disks in 3D radiation-transport two-temperature GRMHD simulations. Astrophys. J. **935**(1), L1 (2022)

54. Siming Liu, Fulvio Melia, Spin-induced disk precession in the supermassive black hole at the galactic center. Astrophys. J. **573**(1), L23–L26 (2002)
55. Giuseppe Lodato, Daniel J. Price, On the diffusive propagation of warps in thin accretion discs. Mon. Not. R. Astron. Soc. **405**(2), 1212–1226 (2010)
56. G. Lodato, J.E. Pringle, Warp diffusion in accretion discs: a numerical investigation. Mon. Not. R. Astron. Soc. **381**(3), 1287–1300 (2007)
57. Thomas J. Maccarone, On the misalignment of jets in microquasars. Mon. Not. R. Astron. Soc. **336**(4), 1371–1376 (2002)
58. G. Marcel, F. Cangemi, J. Rodriguez, J. Neilsen, J. Ferreira, P.O. Petrucci, J. Malzac, S. Barnier, M. Clavel, A unified accretion-ejection paradigm for black hole X-ray binaries. V. Low-frequency quasi-periodic oscillations. Astron. Astrophys. **640**, A18 (2020)
59. Jonathan C. McKinney, General relativistic magnetohydrodynamic simulations of the jet formation and large-scale propagation from black hole accretion systems. Mon. Not. R. Astron. Soc. **368**(4), 1561–1582 (2006)
60. Jonathan C. McKinney, Alexander Tchekhovskoy, Roger D. Blandford, Alignment of magnetized accretion disks and relativistic jets with spinning black holes. Science **339**(6115), 49 (2013)
61. F. Meyer, E. Meyer-Hofmeister, Accretion disk evaporation by a coronal siphon flow. Astron. Astrophys. **288**, 175–182 (1994)
62. M.J. Middleton, P.C. Fragile, M. Bachetti, M. Brightman, Y.F. Jiang, W.C.G. Ho, T.P. Roberts, A.R. Ingram, T. Dauser, C. Pinto, D.J. Walton, F. Fuerst, A.C. Fabian, N. Gehrels, Lense-Thirring precession in ULXs as a possible means to constrain the neutron star equation of state. Mon. Not. R. Astron. Soc. **475**(1), 154–166 (2018)
63. Matthew J. Middleton, Michael L. Parker, Christopher S. Reynolds, Andrew C. Fabian, Anne M. Lohfink, The view of AGN-host alignment via reflection spectroscopy. Mon. Not. R. Astron. Soc. **457**(2), 1568–1576 (2016)
64. J.M. Miller, A.C. Fabian, J.J.M. in't Zand, C.S. Reynolds, R. Wijnands, M.A. Nowak, W.H.G. Lewin, A relativistic Fe Kα emission line in the intermediate-luminosity BeppoSAX spectrum of the galactic microquasar V4641 Sgr. Astrophys. J. **577**(1), L15–L18 (2002)
65. J.M. Miller, C.S. Reynolds, A.C. Fabian, G. Miniutti, L.C. Gallo, Stellar-mass black hole spin constraints from disk reflection and continuum modeling. Astrophys. J. **697**(1), 900–912 (2009)
66. J.C.A. Miller-Jones, A.J. Tetarenko, G.R. Sivakoff, M.J. Middleton, D. Altamirano, G.E. Anderson, T.M. Belloni, R.P. Fender, P.G. Jonker, E.G. Körding, H.A. Krimm, D. Maitra, S. Markoff, S. Migliari, K.P. Mooley, M.P. Rupen, D.M. Russell, T.D. Russell, C.L. Sarazin, R. Soria, V. Tudose, A rapidly changing jet orientation in the stellar-mass black-hole system V404 Cygni. Nature **9**(7756), 374–377 (2019)
67. B. Mishra, W. Kluźniak, P.C. Fragile, Breathing oscillations in a global simulation of a thin accretion disk. Mon. Not. R. Astron. Soc. **483**(4), 4811–4819 (2019)
68. D. Morales Teixeira, P.C. Fragile, V.V. Zhuravlev, P.B. Ivanov, Conservative GRMHD simulations of moderately thin, tilted accretion disks. Astrophys. J. **796**(2), 103 (2014)
69. S.E. Motta, P. Casella, M. Henze, T. Muñoz-Darias, A. Sanna, R. Fender, T. Belloni, Geometrical constraints on the origin of timing signals from black holes. Mon. Not. R. Astron. Soc. **447**(2), 2059–2072 (2015)
70. S.E. Motta, A. Franchini, G. Lodato, G. Mastroserio, On the different flavours of Lense-Thirring precession around accreting stellar mass black holes. Mon. Not. R. Astron. Soc. **473**(1), 431–439 (2018)
71. G. Musoke, M. Liska, O. Porth, M. van der Klis, A. Ingram, Disc tearing leads to low and high frequency quasi-periodic oscillations in a GRMHD simulation of a thin accretion disc. Mon. Not. R. Astron. Soc. **518**(2), 1656–1671 (2023)
72. Ramesh Narayan, Igor V. Igumenshchev, Marek A. Abramowicz, Magnetically arrested disk: an energetically efficient accretion flow. Publ. Astron. Soc. Japan **55**, L69–L72 (2003)
73. R. Narayan, A. SÄ dowski, R.F. Penna, A.K. Kulkarni, GRMHD simulations of magnetized advection-dominated accretion on a non-spinning black hole: role of outflows. Mon. Not. R. Astron. Soc. **426**(4), 3241–3259 (2012)

74. R. Nealon, D.J. Price, C.J. Nixon, On the Bardeen-Petterson effect in black hole accretion discs. Mon. Not. R. Astron. Soc. **448**, 1526–1540 (2015)
75. R.P. Nelson, J.C.B. Papaloizou, Hydrodynamic simulations of the Bardeen-Petterson effect. Mon. Not. R. Astron. Soc. **315**(3), 570–586 (2000)
76. Chris Nixon, Andrew King, Daniel Price, Juhan Frank, Tearing up the disk: how black holes accrete. Astrophys. J. **757**(2), L24 (2012)
77. J.A. Orosz, C.D. Bailyn, Optical observations of GRO J1655-40 in quiescence. I. A precise mass for the black hole primary. Astrophys. J. **477**(2), 876–896 (1997)
78. Jerome A. Orosz, Paul J. Groot, Michiel van der Klis, Jeffrey E. McClintock, Michael R. Garcia, Ping Zhao, Raj K. Jain, Charles D. Bailyn, Ronald A. Remillard, Dynamical evidence for a black hole in the Microquasar XTE J1550–564. Astrophys. J. **568**(2), 845–861 (2002)
79. J.C.B. Papaloizou, D.N.C. Lin, On the dynamics of warped accretion disks. Astrophys. J. **438**, 841 (1995)
80. J.C.B. Papaloizou, J.E. Pringle, The time-dependence of non-planar accretion discs. Mon. Not. R. Astron. Soc. **202**, 1181–1194 (1983)
81. D.R. Pasham, R.A. Remillard, P.C. Fragile, A. Franchini, N.C. Stone, G. Lodato, J. Homan, D. Chakrabarty, F.K. Baganoff, J.F. Steiner, E.R. Coughlin, N.R. Pasham, A loud quasi-periodic oscillation after a star is disrupted by a massive black hole. Science **363**(6426), 531–534 (2019)
82. D.R. Pasham, M. Zajacek, C.J. Nixon, E.R. Coughlin, M. Sniegowska, A. Janiuk, B. Czerny, T. Wevers, M. Guolo, Y. Ajay, M. Loewenstein, Lense-Thirring precession after a supermassive black hole disrupts a star. arXiv e-prints. arXiv:2402.09689 (2024)
83. J. Poutanen, A. Veledina, A.V. Berdyugin, S.V. Berdyugina, H. Jermak, P.G. Jonker, J.J.E. Kajava, I.A. Kosenkov, V. Kravtsov, V. Piirola, M. Shrestha, M.A. Perez Torres, S.S. Tsygankov, Black hole spin–orbit misalignment in the x-ray binary MAXI J1820+070. Science **375**(6583), 874–876 (2022)
84. J.E. Pringle, Self-induced warping of accretion discs. Mon. Not. R. Astron. Soc. **281**(1), 357–361 (1996)
85. L. Qian, B.F. Liu, X.-B. Wu, Disk evaporation-fed corona: structure and evaporation features with magnetic field. The Astrophys. J. **668**(2), 1145 (2007)
86. M.J. Rees, Relativistic jets and beams in radio galaxies. Nature **275**(5680), 516–517 (1978)
87. R.C. Reis, A.C. Fabian, J.M. Miller, Black hole accretion discs in the canonical low-hard state. Mon. Not. R. Astron. Soc. **402**(2), 836–854 (2010)
88. S.M. Ressler, E. Quataert, J.M. Stone, Hydrodynamic simulations of the inner accretion flow of Sagittarius A* fuelled by stellar winds. Mon. Not. R. Astron. Soc. **478**(3), 3544–3563 (2018)
89. S.M. Ressler, C.J. White, E. Quataert, Wind-fed GRMHD simulations of Sagittarius A*: tilt and alignment of jets and accretion discs, electron thermodynamics, and multiscale modelling of the rotation measure. Mon. Not. R. Astron. Soc. **521**(3), 4277–4298 (2023)
90. C.S. Reynolds, M.C. Miller, The time variability of geometrically thin black hole accretion disks. I. The search for modes in simulated disks. Astrophys. J. **692**(1), 869–886 (2009)
91. P.A.G. Scheuer, R. Feiler, The realignment of a black hole misaligned with its accretion disc. Mon. Not. R. Astron. Soc. **282**, 291 (1996)
92. H.R. Schmitt, J.E. Pringle, C.J. Clarke, A.L. Kinney, The orientation of jets relative to dust disks in radio galaxies. Astrophys. J. **575**(1), 150–155 (2002)
93. N.I. Shakura, R.A. Sunyaev, Black holes in binary systems. Observational appearance. Astron. Astrophys. **24**, 337–355 (1973)
94. L. Sironi, A. Tran, Electron heating in the trans-relativistic perpendicular shocks of tilted accretion flows. arXiv e-prints. arXiv:2402.13317 (2024)
95. Kareem A. Sorathia, Julian H. Krolik, John F. Hawley, Magnetohydrodynamic simulation of a disk subjected to Lense-Thirring Precession. Astrophys. J. **777**(1), 21 (2013)
96. Alexander Tchekhovskoy, Ramesh Narayan, Jonathan C. McKinney, Efficient generation of jets from magnetically arrested accretion on a rapidly spinning black hole. Mon. Not. R. Astron. Soc. **418**(1), L79–L83 (2011)
97. Christopher J. White, Eliot Quataert, Omer Blaes, Tilted disks around black holes: a numerical parameter survey for spin and inclination angle. Astrophys. J. **878**(1), 51 (2019)

Chapter 12
Numerical Simulations of Super-Eddington Accretion Flows

Yan-Fei Jiang and Lixin Dai

Abstract In this chapter, we summarize recent progress on the properties of accretion disks when the accretion rate exceeds the so-called Eddington limit based on multi-dimensional radiation magnetohydrodynamic simulations. We first summarize the classical models that are used to describe the accretion disks in the super-Eddington regime with an emphasis on the key uncertainties in these models. Then we show that radiation-driven outflows are ubiquitously found by numerical simulations of super-Eddington accretion disks. Some key physical processes on energy transport inside the disk are also identified by numerical simulations. Radiative and mechanical output as a function of mass accretion rates, black hole mass, spin, and magnetic field topology are summarized. Applications of super-Eddington accretion disks to different astrophysical systems, particularly tidal disruption events, are also discussed.

12.1 Why Is Super-Eddington Accretion Not only Interesting but Also Relevant for Astrophysics?

Eddington luminosity $L_{\rm Edd}$ [25] is defined as the luminosity when the radiation force balances the gravity for spherical symmetric system $L_{\rm Edd} \equiv 4\pi GMc/\kappa$, where G is the gravitational constant, c is the speed of light, M is the mass of the central object while κ is the opacity. This is a well defined quantity for a given opacity κ only when radiation force (via radiation flux) and gravity both change with radius as r^{-2}. If the luminosity is generated via the accretion process, the corresponding mass accretion rate is the Eddington accretion rate $\dot{M}_{\rm Edd} \equiv L_{\rm Edd}/(\eta c^2)$, where η is the radiative efficiency and c is the speed of light. Typically, electron scattering is

Y.-F. Jiang (✉)
Center for Computational Astrophysics, Flatiron Institute, New York, NY 10010, USA
e-mail: yjiang@flatironinstitute.org

L. Dai (✉)
Department of Physics, The University of Hong Kong, Pokfulam Road, Hong Kong, China
e-mail: lixindai@hku.hk

C. Bambi et al. (eds.), *New Frontiers in GRMHD Simulations*, Springer Series in Astrophysics and Cosmology, https://doi.org/10.1007/978-981-97-8522-3_12

used to estimate $L_{\rm Edd}$ for the fully ionized gas, which gives a value $L_{\rm Edd} = 1.47 \times 10^{39}\,(M/10M_\odot)\left(0.34\ {\rm cm^2\ g^{-1}}/\kappa\right)$ erg/s, where $M_\odot$ is the solar mass. For spherical symmetric systems, as in the case of stars where the concept of Eddington limit was originally discussed, it is generally expected that the large radiation force will significantly modify or even stop the accretion flow. Some models have proposed that with sufficiently high accretion rates (e.g., $> 5000L_{\rm Edd}/c^2$), photons are fully coupled to the gas and radiation pressure will behave like the thermal pressure with a different adiabatic index. Therefore, accretion can proceed in a similar way as described by the Bondi solution [36, 39]. However, the issue is what will happen beyond the so-called photon trapping radius [9], where photon diffusion speed becomes larger than the inflow speed as density and optical depth will drop with increasing radius. Variations of opacity with temperature will also make the situation more complicated. If the radiation field is fully thermalized, the effective temperature at 1000 gravitational radii r_g for a given Eddington luminosity will vary with black hole mass as $T = 1.52 \times 10^3\left(M/10^8M_\odot\right)^{-1/4}$ K. Therefore, the gas may actually not be fully ionized for supermassive black holes with an accretion rate significantly above the Eddington limit around the photon trapping radius. A more realistic and common situation is that accretion will happen via the disk configuration due to finite angular momentum. Once the spherical symmetry is broken, it is possible that the mass inflow and escape photons can happen in different directions [61, 79].

Super-Eddington accretions are believed to happen in many astrophysical systems. The one with the most extensive evidence is ultraluminous X-ray (ULX) sources. These are non-nuclear, point-like objects with an apparent isotropic X-ray luminosity at least once observed with a value larger than 10^{39} erg s^{-1}, which is the Eddington luminosity of typical stellar mass black holes [30, 53]. Coherent pulsations are detected in some ULXs [5, 31, 41, 57], which indicate that at least some ULXs are accreting neutron stars. Although sub-Eddington accretion onto intermediate mass black holes is still possible for some ULXs (such as HLX-1) [29], spectra of most ULXs are inconsistent with sub-Eddington accretion disks as found for many X-ray binaries. Outflows are commonly observed for ULXs. The inferred kinetic energy flux based on the properties of shock-ionized ULX nebulae is typically comparable or even larger than the radiation flux [30]. Photoionized nebulae are also found for at least a few cases [51–53, 63, 82], which suggests that the observed super-Eddington luminosity in this case is comparable to the intrinsic luminosity of the system and it cannot be just caused by strong relativistic beaming as suggested by some models [56, 59, 62]. All these observational properties of ULXs put significant constraints on models of super-Eddington accretions onto black hole and neutron star systems.

Super-Eddington accretion has also been commonly used to explain the rapid growth of black holes in the early Universe [40, 109]. More than 200 quasars with black hole mass $\gtrsim 10^9 M_\odot$ have been found at redshift $z \geq 6$ when the Universe is less than ~ 1 Gyr old [8, 28, 46, 113]. It is necessary for the black holes to accrete with a super-Eddington rate at least for a fraction of the lifetime in order to grow from the stellar mass black hole range if the mass of the black hole seeds is $\lesssim 10^4 M_\odot$ [72, 110], although this can be avoided in some models if black holes can grow via mergers [104]. The biggest uncertainty for the growth of black holes via super-Eddington accretion is how the radiative and mechanical output from the accretion disk will interact with the gas supply at large distances, which may limit the duty cycle of the super-Eddington accretion phase [71, 72, 83]. It has also been suggested that supermassive black holes in narrow-line Seyfert-1 galaxies or some X-ray weak quasars are also going through super-Eddington accretion phases [15, 50, 60, 67, 69, 70]. This topic has been recently reviewed by [72] and we will not repeat it here.

Moreover, super-Eddington accretion happens in various transient accretion phenomena around black holes. For example, stars venturing too close to massive black holes (MBHs) can get destroyed by the tidal force [86]. In such a tidal disruption event (TDE), roughly half of the stellar debris stays bound to the black hole and is then supplied to the black hole in the timescale of years. The peak debris mass "fall-back rate" typically exceeds the black hole Eddington accretion rate by 1–2 orders of magnitude [27, 33]. Therefore, TDEs are often considered ideal local laboratories for studying super-Eddington accretion and outflow physics around MBHs [21].

12.2 Classical Models for Super-Eddington Accretion Disks

Properties of accretion disks are mainly controlled by the angular momentum and energy transport within the disk. In the standard α disk model [68, 95], angular momentum transport is assumed to be caused by an effective viscosity, which is proportional to the total thermal pressure with a constant α. Dissipation in the disk, which converts the gravitational potential energy of the accreted gas to the thermal energy, is entirely balanced by radiative cooling locally. The typical inflow speed of such a disk is [24, 85]

$$\begin{aligned} v_{\rm in} &= \alpha \Omega r \left(\frac{h}{r}\right)^2 \\ &= 5.4 \times 10^4 \left(\frac{\alpha}{0.1}\right) \left(\frac{h}{0.01 r}\right)^2 \left(\frac{30 r_g}{r}\right)^{1/2} \ \mathrm{cm/s}, \end{aligned} \tag{12.1}$$

where Ω and h are the angular velocity and the disk scale height at r, while r_g is the gravitational radius for a $10^8 M_\odot$ black hole. The averaged vertical photon diffusion speed is c/τ, where τ is the vertical optical depth from the disk midplane to the photosphere. The assumptions made in the standard thin disk model is justified when $\tau \lesssim 10^5$ as the photon diffusion speed is faster than the inflow speed.

When the accretion rate exceeds the Eddington limit, h/r is expected to increase and the inflow speed can be comparable or even larger than the vertical photon diffusion speed as τ is also likely to be larger. Therefore, some of the dissipated energy will be advected towards the black hole with the accretion flow before the photons have time to diffuse out of the system. This is the most important modification introduced by the slim disk model in order to describe the structures of super-Eddington accretion disks. A stationary, vertically integrated slim accretion disk model was constructed by [1]. It represents a natural extension of the standard thin disk model by including the pressure gradient and advective energy transport along the radial direction. When these terms become significant for $\dot{M} > \dot{M}_{\rm Edd}$, the surface density Σ starts to increase with increasing mass accretion rate in contrast to the $\dot{M} - \Sigma$ relation in the radiation pressure dominated thin disk solution. This suggests that the super-Eddington disk model will not subject to the viscous instability as in the thin disk case [66]. The additional energy transport mechanism due to radial advection also helps avoid the thermal instability [43, 84, 96]. When the accretion rate exceeds $\dot{M}_{\rm Edd}$, the total radiative luminosity will also become larger than $L_{\rm Edd}$. But then it will only increase with mass accretion rate logarithmically instead of linearly for $\dot{M} > \dot{M}_{\rm Edd}$. The ratio between disk scale height and radius h/r can increase to ~ 0.7 when $\dot{M} = 10\dot{M}_{\rm Edd}$.

Various modifications to the original slim disk model have been proposed to account for different physical processes that may be important for accretion disks in the super-Eddington regime. It has been suggested that large density inhomogeneities on scales much smaller than the disk scale height can be produced in the radiation pressure dominated accretion disks due to photon bubble instability [10]. This may allow photons to leave the disk much quicker compared with the photon diffusion rate as estimated based on the vertically integrated optical depth because photons will preferentially go through low density regions as suggested in the envelope of massive stars [97]. Therefore super-Eddington luminosity may be achieved without blowing away most of the mass. The disk scale height will also be reduced due to the enhanced cooling. Outflows have also been expected for accretion disks with super-Eddington accretion rate since the thin disk model was originally proposed [95]. King and Pounds [55] suggested that the outflow will be optically thick and cause the photospheric radius as well as outflow speed to vary with the outflow rate as $\dot{M}_o^2$ and $\dot{M}_o^{-1}$ respectively, although the actual outflow rate $\dot{M}_o$ for a given mass accretion rate cannot be determined self-consistently in the model.

12.3 Results of Recent Numerical Simulations

12.3.1 Radiation Hydrodynamic Simulations of Super-Eddington Accretion Disks and the Missing Physics

To self-consistently determine the structures of super-Eddington accretion disks, including the outflow rate and radiative efficiency, multi-dimensional numerical simulations are necessary. Early two dimensional radiation hydrodynamic simulations

[26, 79, 80] model the angular momentum transport via the α prescription as done in the standard thin disk model. The energy and momentum exchanges between photons and gas, as well as the transport of photons, are modeled using the flux-limited diffusion approach. These calculations confirm that super-Eddington accretion is possible when mass accretion happens near the midplane while photons leave the disk through the polar region. Temperature gradient along the radial direction is small and the optical depth near the midplane is large. Therefore, radiation force is not significant to stop the accretion flow near the midplane. For an accretion rate of $100L_{\rm Edd}/c^2$, the viscous heating rate is $10L_{\rm Edd}$ while the radiative luminosity is only $3L_{\rm Edd}$ with the rest of the energy gets advected inward with the accretion flow. This is the "photon trapping" effect discussed in the literature. Significant optical thick outflow is launched along the polar direction, which will impact the observational properties of these systems (see Sect. 12.4). These early simulations typically didn't run long enough to determine a reliable outflow rate.

Since magneto-rotational instability (MRI) is now believed to be the physical mechanism that generates the turbulence responsible for the angular momentum transport in accretion disks [6, 7], it is crucial to include magnetic fields in simulations of super-Eddington accretion disks. Simulations in 3D are also essential to capture the properties of turbulence correctly. In addition, magnetic fields are necessary to produce the photon bubbles, which can enhance the radiative efficiency as proposed in some models [10].

Another important physical process that needs to be modeled accurately is radiation transport, as radiation pressure is much larger than gas pressure for these disks, particularly for supermassive black holes. The anisotropic transport of the photons is also crucial for the launch of outflow in the super-Eddington accretion disks. Flux limited diffusion (FLD, [64, 106]) is typically adopted for its simplicity, where transport of photons is approximately treated as a diffusion process. It is a valid approximation near the midplane of the disk, where it is very optically thick. However, near the photosphere, particularly near the funnel region where the outflow is launched, the diffusion approximation may not be valid. One extension is to evolve both the zeroth and first moment of specific intensities, which are radiation energy density E_r and radiation flux $\boldsymbol{F}_r$ respectively, in contrast to FLD where only E_r is evolved. A closure relation is needed in this case to relate the radiation pressure tensor $\boldsymbol{P}_r$ with E_r and $\boldsymbol{F}_r$ in order to close the evolution equations. A commonly adopted approach is the *M1* closure, where the ratio $\boldsymbol{P}_r/E_r$ is assumed to be a function of $\boldsymbol{F}_r/|\boldsymbol{F}_r|$ and $|\boldsymbol{F}_r|/E_r$ only to connect the free streaming limit and diffusion limit [32, 65, 88, 94, 100]. This closure relation is local and treats photons as a fluid with an effective equation of state, which has its own limitations when the photon mean free path becomes large. A well known example is that when two streams of photons collide in the vacuum, they will merge into one beam along the direction of the net radiation flux [88]. The radiation transport equation is essentially not local in this regime as photons at a particular spatial location will depend on the distribution of the sources over a large volume. The only way to get the solutions to the transport equation accurately for all parameter regimes is to calculate the solutions for specific intensities. This approach is typically more expensive as the number of variables that need to be evolved is much larger. Specific intensities can be used to calculate the Eddington tensor, which is needed to close the moment equations of radiation transport [3, 22, 35, 42, 99, 112]. In this scheme, which is called Variable Eddington Tensor (VET) method in the literature, the radiation field is represented by two independent sets of variables: the moments (E_r and $\boldsymbol{F_r}$) and specific intensities. They are only coupled via the Eddington tensor and the moments can be inconsistent with the direct angular quadrature of the intensities. Time dependent evolution equations for specific intensities can also be solved and directly coupled to the gas without evolving separate moment equations. This approach has been successfully applied for flows in both the Newtonian limit [45, 48, 49] and general relativity (GR) [111].

12.3.2 Radiation MHD Simulations with GR and M1 Radiation Transport for Standard and Normal Evolution Disks

General properties of super-Eddington accretion disks are studied based on two or three dimensional radiation magneto-hydrodynamic (MHD) simulations using the M1 closure in GR [73, 74, 89, 91, 92, 108]. The numerical artifact of M1 can cause shock-like structures in the radiation field near the rotation axis of the disk, because this scheme will merge photons emitted from different azimuthal regions of the disk to one direction near the polar axis. An additional viscosity term is added to the radiation moment equations to smooth out these artificial shock structures [90, 92]. These simulations typically start with a rotating torus located at $15 - 30r_g$ initially, where r_g is the gravitational radius of the black hole. Simulation parameters are chosen to be appropriate for stellar mass black holes, up to $10^6 M_\odot$ black holes with the opacity set to be electron scattering plus free-free absorption. Radiation MHD simulations, unlike the MHD calculations, are not scale free and therefore one simulation cannot be used to describe the accretion flow around black holes with different masses. A single poloidal loop of magnetic field is typically used in the torus to provide the seed magnetic field, which is needed for MRI. With this setup, if a significant fraction of the torus is allowed to flow towards the black hole, enough magnetic flux will be accreted near the black hole horizon to cause the disk to enter the magnetic arrested state (the MAD state). It is found that MAD super-Eddington accretion flows around fastly-spinning black holes can have unique properties, which we will describe in Sect. 12.3.4. Here in this section we focus on describing the general properties of super-Eddington accretion flows which are SANE ("standard and normal evolution") or MAD but around slowly spinning black holes. These simulations usually can only study the very inner region of the disk due to a limited run time. Some analyses of these simulations only focus on the initial period of $t \lesssim 1.5 \times 10^4 r_g/c$, which only corresponds to one orbital period at $178.6r_g$. Therefore, these simulations can only reach a steady state for the inner $10 \sim 20r_g$.

Consistent with previous 2D hydrodynamic simulations, these simulations show that gas can flow towards the black hole with a super-Eddington accretion rate near the midplane while strong outflow driven by the photons emitted by the disk is formed near the rotation axis. Outgoing mass flux only starts to show up near $\sim 10r_g$ and can become comparable to the net mass accretion rate beyond 20–50r_g. The mass flux is typically calculated using the time and shell averaged radial momentum for each radius. It is important to point out that not all the outflowing mass flux should be considered as truly unbound gas and different ways to determine the outflow may result in different results. For example, if outflowing gas is firstly determined for each snapshot and then the outflow is averaged over time, the total amount of inferred outflow gas can be different from the value as determined using the time averaged total radial momentum [115]. The radiative luminosity integrated over the funnel region saturates to $\approx L_{\rm Edd}$ while the mass accretion rate is $\approx 100L_{\rm Edd}/\left(0.057c^2\right)$ for the zero spin case shown in [89]. When the spin value is increased to 0.9, the radiative luminosity is $\approx L_{\rm Edd}$ at $20r_g$ but it continues to increase with radius to $10L_{\rm Edd}$ at $200r_g$. The outflow rate is also larger with a larger spin value. The outflow generated by the super-Eddington accretion disks is also referred as "radiative jet" by [91], although the mass weighted outflow speed is only 0.2–0.3c for a wide range of mass accretion rates. This is very different from the ultra-relativistic jet generated by the Blandford-Znajek mechanism with magnetic fields [24]. Similar outflow properties are found by simulations of super-Eddington accretion disks performed by different groups and the terminal speed is limited by strong radiation drag [44, 101]. The terminal speed of the outflow decreases to a few percent of the speed of light when the viewing angle is larger than $\sim 45°$ from the rotational axis. The total kinetic power of the outflow is $\approx L_{\rm Edd}$ when the mass accretion rate is $10^3 L_{\rm Edd}/c^2$ for the simulation done by [101].

The isotropic equivalent luminosity as a function of polar angle and mass accretion rate for a non-spinning black hole is found by [91] to be

$$L_{\rm iso} = 4 \times 10^{47} \exp\left(\theta/\theta_0\right) \frac{\dot{M}}{10^3 \dot{M}_{\rm Edd}} \frac{M_{\rm BH}}{10^6 M_\odot} {\rm erg/s}. \tag{12.2}$$

The total luminosity integrated over all the angles is only slightly larger than $L_{\rm Edd}$. It can increase with larger black hole spins. Although the characteristic opening angle is fitted to be $\theta_0 = 0.2$, escaping photons are typically more strongly beamed towards the rotation axis with larger accretion rates (Fig. 6 of [91]). The dependence on black hole mass also assumes that the properties of these super-Eddington disks are scale free, which cannot be tested by these simulations.

12.3.3 Simulations with Full Radiation Transport

Because radiation pressure plays an important role in determining the properties of super-Eddington accretion disks, other simulations that solve the full angular distributions of specific intensities are also available [3, 37, 44, 47]. One approach is to solve the full time dependent evolution of specific intensities with a Newtonian flow. The GR effects around a Schwarzschild black hole are typically captured using a pseudo-Newtonian potential [81]. Using this method, [44] studied a super-Eddington accretion disk onto a $6.62M_{\odot}$ black hole with the accretion rate $22\dot{M}_{\rm Edd}$. This simulation was done in cylindrical coordinates and initialized with a torus centered at $50r_g$. Typical structures for the density and radiation energy density from the disk are shown in Fig. 12.1. The overall properties of the disk are similar to results obtained from the simulations performed with GR and M1 closure for radiation transport qualitatively but not quantitatively. The gas is accreted near the disk midplane region while strong radiation driven outflow is launched near the rotation axis. Total radiative luminosity emitted by the inner $40r_g$ of the disk, where a steady state is reached, is $\sim 10L_{\rm Edd}$. Mass flux carried by the outflow is 29% of the net mass accretion rate.

One important physical process that can enhance the cooling rate of super-Eddington accretion disks as pointed out by [44] is magnetic buoyancy. Density fluctuations generated by MRI turbulence in the main body of the disk are generically anti-correlated with fluctuations of magnetic pressure. Therefore, low density

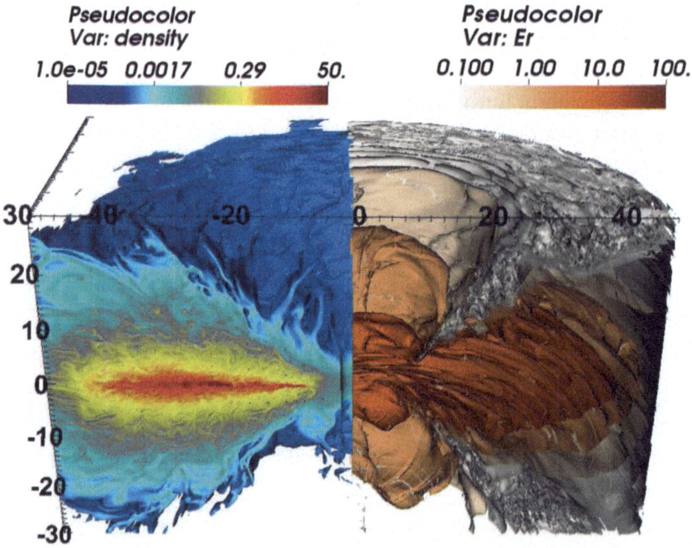

Fig. 12.1 Typical structures of Super-Eddington accretion disks onto a stellar mass black hole. The left panel shows the density in unit of 10^{-2} g/cm^3 while the right panel shows the radiation energy density in unit of $a_r T_0^4$ with $T_0 = 10^7$K. Length scale labeled in this Figure is in unit of the Schwarzschild radius for $6.62M_{\odot}$ black hole. This Figure is adopted from [44]

regions of the disk will buoyantly rise towards the disk surface, which also carries photons with them as the disk is still very optically thick. This effectively reduces the optical depth of the photons as they do not need to diffuse all the way from the midplane, but a region closer to the surface. Total radiative luminosity emitted by the inner disk that has reached a steady state is found to be $10L_{\rm Edd}$ for the simulation shown in [44]. Similar properties are also found for super-Eddington accretion disks with different accretion rates and different black hole masses [37, 47], even though these simulations are done in spherical polar coordinates not in cylindrical coordinates. Furthermore, [47] shows that the opening angle of the funnel region gets smaller when the accretion rate is increased. The radiative efficiency is reduced to $\sim 1\%$ when $\dot{M}$ reaches $150\dot{M}_{\rm Edd}$.

One important difference between super-Eddington accretion disks around supermassive black holes and stellar mass black holes is that the ratio between radiation pressure P_r and gas pressure P_g is much larger for the disks around supermassive black holes. For the two super-Eddington accretion disks onto a $5 \times 10^8 M_\odot$ black hole shown in [47], the ratio P_r/P_g can vary between $\sim 10^3$ and $\sim 10^5$ for the inner region $r < 40r_g$, while the ratio is typically 10–10^3 for the stellar mass black hole case. This difference causes the accretion flow around supermassive black holes to be much more compressible. Spiral shock structures are clearly identified in these accretion flows as shown in Fig. 12.2. Reynolds stress caused by these spiral structures becomes comparable to or even larger than the Maxwell stress as produced by the MRI turbulence. No significant spiral shocks are found for the stellar mass black hole case. Other properties of the accretion flow, such as radiative efficiency, mass flux and energy flux carried by the outflow for comparable accretion rates in the Eddington unit, are actually insensitive to the black hole masses.

The numerical scheme to directly solve the equation for specific intensities with GR has been developed [23, 111] and it is currently used to model the super-Eddington accretion flows. Another approach is to use the specific intensities to

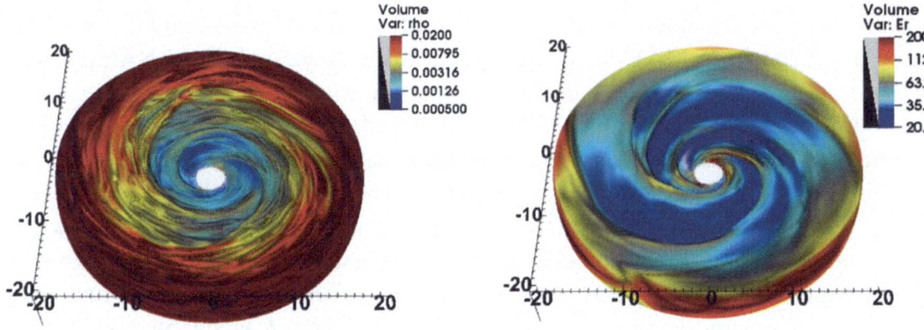

Fig. 12.2 Spiral structures found in Super-Eddington accretion disks onto a $5 \times 10^8 M_\odot$ black holes. The left panel shows the density snapshot (in unit of 10^{-8} g/cm^3) within $40r_g$ for an accretion disk onto a $5 \times 10^8 M_\odot$ black hole with mass accretion rate $\dot{M} = 25\dot{M}_{\rm Edd}$. The right panel is the corresponding radiation energy density (in unit of $a_r T_0^4$ with $T_0 = 2 \times 10^5$ K). Length scale labeled in this Figure is $2r_g$. This Figure is adopted from [47]

calculate the Eddington tensor, which is then used to close the moment equations for radiation energy density and radiation flux. Only solutions to the radiation moment equations are used to update gas energy and momentum [3]. This can be done in the curved spacetime and it is more accurate than the M1 approach as no assumption about the closure relation is made. However, the radiation field is described by two independent sets of variables in this scheme, as the moments are not derived from the specific intensities that are used to calculate the Eddington tensor. Special care is needed to make sure they are consistent. Asahina et al. [3] also compared the radiation field they obtained using this scheme and the M1 closure, based on the flow properties as calculated by an axisymmetric general relativistic magneto-hydrodynamic (GRMHD) simulation of super-Eddington accretion disk onto a non-magnetized neutron Star as done by [102]. The solutions are very similar near the optically thick disk midplane. However, significant differences show up near the disk surface and the funnel region. In particular, the radial component of the radiation flux near the rotation axis is unphysically enhanced with the M1 closure scheme [4].

12.3.4 Simulations of MAD Super-Eddington Accretion Flows

Strongly magnetized accretion flows show behaviors different from weakly magnetized flows, such as exhibiting non-axisymmetry in the flow structure and fast variability in the accretion rate. Also, MAD disks can potentially produce relativistic jets through magnetic processes. Several numerical works have focused on studying the properties of supper-Eddington accretion flows in the MAD limit [16, 20, 74, 78, 87, 105, 114]. These works are all based on 2D or 3D general relativistic radiation magneto-hydrodynamic (GRRMHD) simulations using the M1 closure.

Winds are launched from MAD super-Eddington disks not only because of radiation pressure but also magnetic pressure. Therefore, more massive and faster winds can be produced from MAD super-Eddington disks compared to the SANE disks. We illustrate the general accretion flow structure in Fig. 12.3 by plotting the accretion flow in the final quasi-equilibrium state simulated by [20]. One can see that both the accretion disk and the wind are geometrically thick. Furthermore, the wind has a highly anisotropic structure in both density and velocity. "Ultra-fast outflows" (UFOs) with speeds faster than $0.1c$ are seen surrounding the polar region, while slower and denser winds reside between the UFOs and the disk. Lastly, as this specific simulation involves a MAD flow around a fastly spinning black hole, a relativistic jet is launched mainly magnetically by the Blandford-Znajek mechanism [11].

The magnetized, optically thick wind launched from MAD disks helps carry radiation to an outer radius, where the wind becomes optically thin and therefore releases the radiation. When a relativistic jet is produced, the jet pushes away the wind and drills out an optically thin funnel, which allows more photons to leak out.

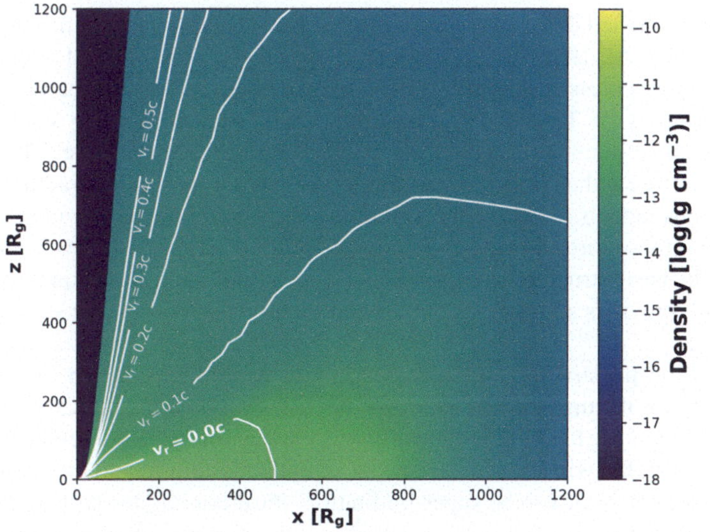

Fig. 12.3 The time and azimuthally averaged vertical profile of a MAD super-Eddington accretion flow simulated by [20]. The simulation has $M_{\rm BH} = 5 \times 10^6 M_\odot$, $a = 0.8$, and $\dot{M}_{\rm acc} \approx 15\dot{M}_{\rm Edd}$. The gas density is shown using the background color. The white lines are contours of constant radial velocity in the wind region. A geometrically thick accretion inflow region exists around the midplane. A strong and geometrically thick wind is launched, which has an anisotropic structure. The jet region (where magnetic pressure dominates the gas pressure) is marked using a dark blue colour. This figure is adapted from [21]

Putting together all these factors, MAD disks around fastly-spinning black holes can reach maximum radiative efficiency and such disks can even be as radiatively efficient as thin disks.

MAD super-Eddington disks around fastly-spinning black holes also have the unique property that the disk energy efficiency generally increases with the Eddington ratio, which is opposite to the behavior of SANE disks or MAD disks around black holes of low spins. Ricarte et al. [87] further investigated the jets produced from MAD super-Eddington flows. They showed that the jets produced from such super-Eddington disks are as powerful as the jets from nonradiative MAD disks. They also gave fitting functions for how the jet power and efficiency vary with black hole spin and Eddington ratio for such disks.

12.3.5 Simulations of Super-Eddington Accretion Flows in Tidal Disruption Events

A series of numerical simulations have been carried out to investigate super-Eddington accretion flows formed in TDEs [2, 13, 14, 16–18, 20, 93, 98, 105]. One typical approach is to set up the initial disk conditions by assuming that circularized, aligned disks have already formed in TDEs, as in the simulations conducted by [16, 20, 105]. These simulations place the disks around supermassive black holes with masses around $10^6 M_\odot$ which are optimal for disrupting stars. One distinct property of TDE disks is that they are supposed to carry small angular momenta, since the stars are disrupted very close to the black hole. These simulations therefore either truncate the initial torus at an appropriate radius or adopt specific torus models to ensure the initial disk angular momenta is within the limit. In addition, these three works all employ 3D GRRMHD simulations with the M1 closure. The basic structure of the simulated super-Eddington accretion flows is consistent with previous simulations of super-Eddington accretion flows starting with a torus set up.

The anisotropic structure of the accretion flow naturally leads to a viewing angle dependence for the disk emission. Along the disk direction, the optical depth of the accretion flow is high so the radiative flux is capped around Eddington. Along the polar direction, the wind density is low so an optically thin funnel can form, which allows the radiative flux to be super-Eddington. Furthermore, for typical Eddington ratios in TDEs ($\dot{M}_{\rm acc} \approx$ few $\times$ $(1 - 10)\dot{M}_{\rm Edd}$), the electron scattering photosphere has a size of few$\times(100 - 1000)R_g$ in the wind region and truncates near the pole due to the lower wind density there, as shown in Fig. 12.4. The radiation temperature of the disk and wind ranges from 10^6 K near the black hole to 10^4 K in the outer wind region. The gas temperature is usually much higher than the radiation temperature,

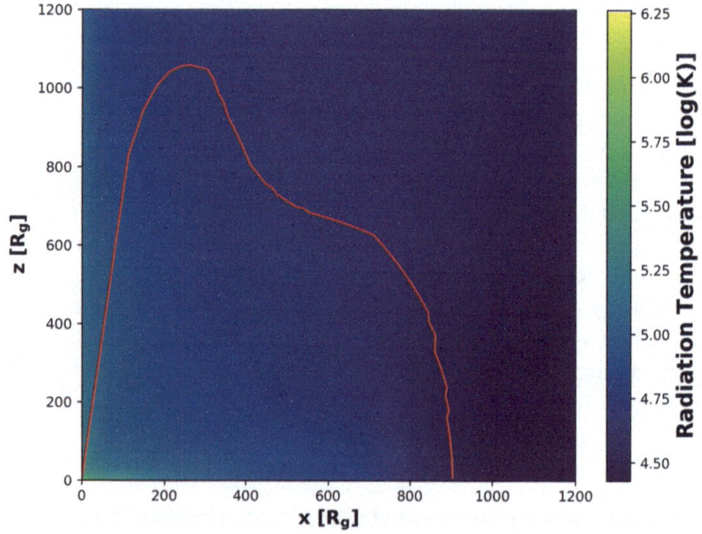

Fig. 12.4 The time and azimuthally averaged vertical profile of the same TDE super-Eddington accretion flow as in Fig. 12.3. The radiation temperature is shown using the background color. The radial electron scattering photosphere with $\tau_{\rm ES} = 1$ is plotted using the red line

except in the disk and slow winds where the gas has a very high optical depth. More details of the emission properties of these accretion flows will be discussed later.

The detailed properties of super-Eddington accretion flows for TDEs are demonstrated to be affected by the black hole spin and accretion rate as well as the strength of the magnetic field. The dependence on the black hole spin and the magnetic field strength has been checked by [16], which are consistent with previous results [92]. For example, they show that only when the black hole spin is large and the accretion flow is MAD, can the accretion flow have a larger efficiency in converting the gas rest-mass energy into energy, and under this scenario, most of the energy escapes in the form of a relativistic jet. The dependence on the black hole accretion rate has been investigated by [18, 105], both under the MAD limit. The former work [105] conducts a few simulations that achieve different accretion rates ranging from a few to a few tens of $\dot{M}_{\rm Edd}$ in the quasi-equilibrium state. They show that the wind structure is insensitive to the change in the accretion rate within this range. However, the wind density drops fast when the accretion rate decreases, which means that the optically thin funnel effectively opens up at a later time in TDEs. As a consequence, the inner disk emissions can more easily escape. On the other hand, the latter work [18] conducts simulations with the accretion rate decreasing from a super-Eddington accretion rate of a few $\dot{M}_{\rm Edd}$ to a sub-Eddington accretion rate, so the disks are dynamically evolving throughout the simulations. They show that the disk energy efficiency can increase or decrease with the accretion rate, depending on the black hole spin. For a black hole with $a = 0$, the disk becomes more radiatively efficient as the accretion rate drops. However, for the case of $a = 0.9$, both the radiative efficiency and the jet efficiency generally drop with the accretion rate. These results can be applied to studying TDEs, especially the ones producing relativistic jets. Also, interestingly, [18] showed that only their $a = 0.9$ black hole model could produce coherent variability in the light curve, which can possibly explain the quasi-periodic oscillations (QPOs) observed in a few TDEs.

Alternatively, some TDE disk simulations do not start with an already-assembled disk or torus but instead have gas injected in certain ways mimicking the TDE disk formation process [2, 13, 14, 17, 38, 93, 98], such as injecting gas at the TDE debris stream self-crossing points with high eccentricities [19]. Since simulations with such initial setups are more time-consuming, they usually do not include radiation, magnetic field, or GR, and some simulations are axisymmetric. We refer the readers to the review by [12] for the disk formation aspects and here focus on the properties of the accretion flow after the disk is more or less assembled in these simulations. These simulations show that a significant fraction of the injected gas becomes outflows, both due to the shocks and disk radiation pressure. The structure of the outflow somewhat resembles the super-Eddington winds launched from a circularized disk such as having faster speeds with $v > 0.1c$ at smaller inclinations. The outer disk usually retains an elliptical configuration at least for some period after the disk assembly, yet the inner disk eventually has an angular momentum closer to Keplerian. The disks can have a super-Eddington accretion rates under such a configuration, which makes the inner disk advection-dominated and radiatively inefficient. Furthermore, the simulation by [93] shows that the angular momentum transfer in the newborn disks is dominated by hydrodynamical turbulence instead of magnetic fields. We note that these simulations are usually run for only a few days since the disk assembly and some of them only represent non-typical TDE parameter spaces. The long-term evolution of elliptical super-Eddington TDE disks and their interaction with the still in-falling stellar debris require further studies.

12.4 Observational Properties of Super-Eddington Flows

One generic outcome of all the simulations of super-Eddington accretion disks is that optically thick outflow will be launched from the inner part of the accretion disk with typical speed varying from $\sim 0.1c - 0.4c$. These outflows will push the photosphere away from the black hole and reduce the effective temperature of the emitted photons. Particularly for observers that do not directly see the funnel region, X-ray emission will be much weaker compared with the standard thin accretion disks. Since the opening angle of the funnel generally decreases and the outflow rate increases with increasing mass accretion rate, the disk spectrum will be more likely dominated by the emission from the outflow for systems with high accretion rates, except for face on observers. Narayan et al. [78] performed a series of 2D and 3D simulations of super-Eddington accretion disks onto a $10M_\odot$ black hole. They cover accretion rates from $\sim\dot{M}_{\rm Edd}$ to $\sim 40\dot{M}_{\rm Edd}$ with either zero spin or dimensionless spin parameter 0.9. In order to predict the spectra for ULXs, these simulations are post-processed using a ray-tracing scheme by extrapolating the disk profiles to a much larger radius than the radial range where a steady state is achieved for these simulations. Spectra from the SANE models are not sensitive to the accretion rate but strongly depend on the viewing angle. For viewing angles near edge-on, the spectra are very soft with peaks located below < 1 keV and the apparent luminosity is much smaller than the observed luminosity for ULXs. The spectra are much harder for face-on observers and the observed luminosity can be boosted because of beaming. These results are actually difficult to explain the super-Eddington luminosity for the supersoft ULXs [107] and isotropic illumination favored by observations of ULX bubbles. The spectra from MAD disks are less sensitive to viewing angles. For the zero spin case, the disk shows a soft spectrum peaked around 1 keV for $\dot{M} = 1.3\dot{M}_{\rm Edd}$ and the spectrum becomes harder with a much broader peak between 1 and 10 keV when $\dot{M}$ is increased to $23\dot{M}_{\rm Edd}$. When the dimensionless spin value is 0.9, the spectra are very hard and broadly extend from 1 to 100 keV for a variety of accretion rates and viewing angles.

Takeuchi et al. [103] performed 2D radiation hydro and MHD simulations of super-Eddington accretion disks onto black holes covering the radial range up to $1024r_g$. For the disk with a net accretion rate reaching $10\dot{M}_{\rm Edd}$ and an outflow with mass flux $\approx\dot{M}_{\rm Edd}$, they find that the outflow becomes clumpy at the scale $> 400r_g$. Electron scattering optical depth across each filament is typically ~ 1. The spatial covering factor of these clumps is ~ 0.3. They attribute the formation of these clumps to Rayleigh-Taylor instability. Similar clumpy structures are also confirmed in 3D simulations [58] using the M1 closure scheme for radiation transport. If these results are robust, they may explain the rapid variability observed for some ULXs [75]. These results also demonstrate that extrapolating the simulation result beyond the radial range they have simulated can be very unreliable.

Mills et al. [76] used Monte Carlo scheme to post process simulations of super-Eddington accretion disks onto a $6.62\dot{M}_{\rm Edd}$ black hole as described by [37]. These calculations focus on the spectra from the funnel region (so the face-on case) as only the inner steady state region can be trusted without the extrapolation as used by [78]. The Monte Carlo calculations rely on gas properties (particularly temperature) given by the simulations, which can be unreliable for the funnel region. This is because Compton scattering is important to determine the gas temperature there and frequency integrated radiation transport, which is the case for most radiation MHD simulations, cannot treat Compton scattering accurately. Mills et al. [76] restarted the simulations using the multi-group approach [49], which solves the Kompaneets equation for Compton scattering using 20 frequency groups. Gas temperature in the funnel region typically gets hotter compared with the original solution obtained by the frequency integrated simulations. Using the improved gas properties, [76] obtained spectra that can fit the observed hard X-rays (photon frequency $\gtrsim$ 3 keV) of the ULX NGC 1313 X-1 very well for both total luminosity and spectrum shape.

For the case of SMBHs/TDEs, [20] and [105] post-processed their 3D simulations of TDE super-Eddington accretion disks as described in Sect. 12.3.5 using the Monte Carlo radiative transfer code Sedona [54]. Sedona evolves non-local thermodynamic equilibrium equations and can capture the electron scattering, thermal Comptonization, and the free-free, bound-free and bound-bound processes. The latter two processes are shown to be important for accurately computing the emissions from super-Eddington accretion flows around supermassive black holes, especially in the regions where the gas temperature is in the range of 10^4–10^5 K. Dai et al. [20] and Thomsen et al. [105] grouped the simulations into different inclination bins and then conducted 1D radiative transfer calculations for each bin. They showed that in the low-density fast wind regions near the pole, photons go through multiple scatterings in the expanding outflow and lose energy through adiabatic reprocessing. In the high-density slow wind or disk region, the X-ray photons are mainly absorbed by photoionization and then re-emitted as UV/optical photons due to recombination, free-free emission, and line emission. Notably, a viewing-angle dependent unification model for TDEs is proposed based on the simulation results, which states that X-ray TDEs are observed at low inclination angles and optical TDEs are viewed at high inclination angles upon super-Eddington accretion flows formed in TDEs.

Curd and Narayan [16] also post-processed their simulated TDE super-Eddington accretion flows, using the code HEROIC [77, 116]. Compared to the Sedona simulations mentioned above, HEROIC is a 3D post-processor but it uses a simpler model for the frequency-dependent opacity of atomic processes. In particular, HEROIC does not fully capture bound-free processes and does not include bound-bound processes. As a result of all these factors, the spectra obtained by [16] carry less viewing-angle dependence compared to [20] and [105]. Moreover, [16] also computed the spectra from both TDE disks and jets, and showed that the hard X-ray emissions produced by a relativistic jet can be seen from all viewing angles.

12.5 Future Directions for Simulating Super-Eddington Accretion Disks

One limitation of current multi-dimensional radiation MHD simulations is that they can only cover a small portion of the inner accretion disks. This is fundamentally limited by a large scale separation in the system, as the time step of these simulations is limited by the light crossing time of each resolution element at the inner boundary while the simulation needs to run at least a few thermal timescales at the radius of interest to get some reliable results. Since low energy photons will be emitted from large radii, this means the long wavelength part of the spectra from super-Eddington accretion disks cannot be easily predicted from these simulations. Simple extrapolation of the numerical solutions to larger radii [78] may not be reliable. With the increasing computational power, particularly the recent development of simulation tools that can take full advantage of GPUs, it will be very interesting to extend these simulations to achieve a steady state over a much larger radial range.

Another important uncertainty of current simulations for super-Eddington accretion disks is how the disk is fed from large scales, which will vary for different astrophysical systems. For simplicity, a majority of current simulations start from a rotating torus that is already close to the black hole, which was the same initial condition as adopted for the first global MHD simulations of accretion disks more than 20 years ago [34]. The choice of initial magnetic fields can result in very different properties of the disks (such as SANE versus MAD disks). For accretion onto stellar mass black holes, the condition will depend on the properties of the companions if the disk is fed by stellar winds or Roche lobe overflow. In the case of supermassive black holes, input from studies of interstellar medium and galactic properties will be necessary as summarized by [72]. When the accretion rate exceeds $100\dot{M}_{\rm Edd}$, it is often assumed that spherical symmetric, adiabatic flow can be a good approximation to describe super-Eddington flows onto black hole seeds in the early Universe [36]. This can only be true up to a certain radius. The gas temperature will drop below 10^4 K beyond $\sim 10^3 r_g$ for $10^8 M_\odot$ with Eddington luminosity. Opacity will drop by several orders of magnitude compared with the electron scattering value. There is just not enough opacity to keep the adiabatic condition. Therefore, understanding the dynamics at the large scale where the gas is fed is crucial for the long-term evolution of super-Eddington accretion.

For TDEs, the current numerical studies can be extended to address two important, unique aspects of the disks formed in these transient events. First, it is generally expected that a TDE disk, when just formed, should retain an elliptical configuration. The long-term evolution of elliptical super-Eddington disks such as whether and how quickly they can get circularized and their emission properties throughout this course are worthy of further studies. Second, the disrupted star likely has angular momentum not along the black hole spin axis, which will lead to the formation of a misaligned super-Eddington disk. However, up to now only aligned super-Eddington accretion disks have been studied using radiation MHD simulations. It is generally expected that thick, misaligned disks should precess like solid bodies (see Chap. 11). If this is indeed the case for misaligned super-Eddington accretion disks, then we can expect that the winds launched from such precessing disks should eventually achieve a quasi-spherical geometry, and the jets, if produced, may or may not be able to drill out of the wind materials.

Acknowledgements LD acknowledges the support from the National Natural Science Foundation of China and the Hong Kong Research Grants Council (HKU12122309, N_HKU782/23, 17305920, 27305119, 17305523).

References

1. M.A. Abramowicz, B. Czerny, J.P. Lasota, E. Szuszkiewicz, Astrophys. J. **332**, 646 (1988). https://doi.org/10.1086/166683
2. Z.L. Andalman, M.T.P. Liska, A. Tchekhovskoy, E.R. Coughlin, N. Stone, Mon. Not. R. Astron. Soc. **510**(2), 1627–1648 (2022).https://doi.org/10.1093/mnras/stab3444[arXiv:2008.04922 [astro-ph.HE]]
3. Y. Asahina, H.R. Takahashi, K. Ohsuga, Astrophys. J. **901**, 96A (2020). https://doi.org/10.3847/1538-4357/abaf51
4. Y. Asahina, K. Ohsuga, Astrophys. J. **929**, 93 (2022). https://doi.org/10.3847/1538-4357/ac5d37
5. M. Bachetti, F.A. Harrison, D.J. Walton, B.W. Grefenstette, D. Chakrabarty, F. Fürst, D. Barret, A. Beloborodov, S.E. Boggs, F.E. Christensen et al., Nature **514**, 202–204 (2014). https://doi.org/10.1038/nature13791[arXiv:1410.3590 [astro-ph.HE]]
6. S.A. Balbus, J.F. Hawley, Astrophys. J. **376**, 214–233 (1991). https://doi.org/10.1086/170270
7. S.A. Balbus, J.F. Hawley, Rev. Mod. Phys. **70**, 1–53 (1998). https://doi.org/10.1103/RevModPhys.70.1
8. E. Banados, B.P. Venemans, C. Mazzucchelli, E.P. Farina, F. Walter, F. Wang, R. Decarli, D. Stern, X. Fan, F. Davies et al., Nature **553**(7689), 473–476 (2018). https://doi.org/10.1038/nature25180[arXiv:1712.01860 [astro-ph.GA]]
9. M.C. Begelman, Mon. Not. R. Astron. Soc. **187**(2), 237–251 (1979)
10. M.C. Begelman, Astrophys. J. Lett. **568**, L97–L100 (2002).https://doi.org/10.1086/340457[arXiv:astro-ph/0203030 [astro-ph]]
11. R.D. Blandford, R.L. Znajek, Mon. Not. R. Astron. Soc. **179**, 433–456 (1977). https://doi.org/10.1093/mnras/179.3.433
12. C. Bonnerot, N. Stone, Space Sci. Rev. **217**(1), 16 (2021). https://doi.org/10.1007/s11214-020-00789-1[arXiv:2008.11731 [astro-ph.HE]]
13. D.F. Bu, E. Qiao, X.H. Yang, J. Liu, [arXiv:2208.04118 [astro-ph.HE]]
14. D.F. Bu, E. Qiao, X.H. Yang, Mon. Not. R. Astron. Soc. **523**(3), 4136–4145 (2023). https://doi.org/10.1093/mnras/stad1696[arXiv:2306.04313 [astro-ph.HE]]
15. S. Collin, T. Kawaguchi, Astron. Astrophys. **426**, 797–808 (2004).https://doi.org/10.1051/0004-6361:20040528[arXiv:astro-ph/0407181 [astro-ph]]
16. B. Curd, R. Narayan, Mon. Not. R. Astron. Soc. **483**(1), 565–592 (2019).https://doi.org/10.1093/mnras/sty3134[arXiv:1811.06971 [astro-ph.HE]]

17. B. Curd, Mon. Not. R. Astron. Soc. **507**(3), 3207–3227 (2021).https://doi.org/10.1093/mnras/stab2172[arXiv:2105.09904 [astro-ph.HE]]
18. B. Curd, R. Narayan, Mon. Not. R. Astron. Soc. **518**(3), 3441–3461 (2022). https://doi.org/10.1093/mnras/stac3330[arXiv:2209.12081 [astro-ph.HE]]
19. L. Dai, J.C. McKinney, M.C. Miller, Astrophys. J. Lett. **812**(2), L39 (2015). https://doi.org/10.1088/2041-8205/812/2/L39[arXiv:1507.04333 [astro-ph.HE]]
20. L. Dai, J.C. McKinney, N. Roth, E. Ramirez-Ruiz, M.C. Miller, Astrophys. J. Lett. **859**(2), L20 (2018). https://doi.org/10.3847/2041-8213/aab429[arXiv:1803.03265 [astro-ph.HE]]
21. J.L. Dai, G. Lodato, R.M. Cheng, Space Sci. Rev. **217**(1), 12 (2021).https://doi.org/10.1007/s11214-020-00747-x[arXiv:2101.05195 [astro-ph.HE]]
22. S.W. Davis, J.M. Stone, Y.F. Jiang, Astrophys. J. Suppl. **199**, 9 (2012). https://doi.org/10.1088/0067-0049/199/1/9. ([arXiv:1201.2222 [astro-ph.IM]].)
23. S.W. Davis, C.F. Gammie, Astrophys. J. **888**, 94 (2020). https://doi.org/10.3847/1538-4357/ab5950. ([arXiv:1911.07950 [astro-ph.HE]].)
24. S.W. Davis, A. Tchekhovskoy, Ann. Rev. Astron. Astrophys. **58**, 407–439 (2020). https://doi.org/10.1146/annurev-astro-081817-051905[arXiv:2101.08839 [astro-ph.HE]]
25. A.S. Eddington, *The Internal Constitution of the Stars* (Cambridge University Press, Cambridge, 1926)
26. G.E. Eggum, F.V. Coroniti, J.I. Katz, Astrophys. J. **330**, 142–167 (1988). https://doi.org/10.1086/166462
27. C.R. Evans, C.S. Kochanek, Astrophys. J. Lett. **346**, L13–L16 (1989). https://doi.org/10.1086/168002
28. X. Fan et al., [SDSS], Astron. J. **122**, 2833 (2001). https://doi.org/10.1086/324111[arXiv:astro-ph/0108063 [astro-ph]]
29. S. Farrell, N. Webb, D. Barret, O. Godet, J. Rodrigues, Nature **460**, 73–75 (2009). https://doi.org/10.1038/nature08083[arXiv:1001.0567 [astro-ph.HE]]
30. H. Feng, R. Soria, New Astron. Rev. **55**, 166–183 (2011). https://doi.org/10.1016/j.newar.2011.08.002[arXiv:1109.1610 [astro-ph.HE]]
31. F. Fürst, D.J. Walton, F.A. Harrison, D. Stern, D. Barret, M. Brightman, A.C. Fabian, B. Grefenstette, K.K. Madsen, M.J. Middleton et al., Astrophys. J. Lett. **831**(2), L14 (2016). https://doi.org/10.3847/2041-8205/831/2/L14[arXiv:1609.07129 [astro-ph.HE]]
32. M. González, E. Audit, P. Huynh, Astro. & Astrophs. **464**, 429–435 (2007). https://doi.org/10.1051/0004-6361:20065486
33. J. Guillochon, E. Ramirez-Ruiz, Astrophys. J. **767**, 25 (2013) [erratum: Astrophys. J. **798**(1), 64 (2015)].https://doi.org/10.1088/0004-637X/767/1/25[arXiv:1206.2350 [astro-ph.HE]]
34. J.F. Hawley, Astrophys. J. **528**, 462 (2000). https://doi.org/10.1086/308180. ([arXiv:astro-ph/9907385 [astro-ph]].)
35. J.C. Hayes, M.L. Norman, Astrophys. J. Suppl. **147**, 197 (2003). https://doi.org/10.1086/374658. ([arXiv:astro-ph/0207260 [astro-ph]].)
36. H. Hu, K. Inayoshi, Z. Haiman, E. Quataert, R. Kuiper, Astrophys. J. **934**(2), 132 (2022). https://doi.org/10.3847/1538-4357/ac75d8[arXiv:2203.14994 [astro-ph.HE]]
37. J. Huang, Y.F. Jiang, H. Feng, S.W. Davis, J.M. Stone, M.J. Middleton, Astrophys. J. **945**(1), 57 (2023). https://doi.org/10.3847/1538-4357/acb6fc[arXiv:2301.12679 [astro-ph.HE]]
38. X. Huang, S.W. Davis, Y.F. Jiang, [arXiv:2404.18446 [astro-ph.HE]]
39. K. Inayoshi, Z. Haiman, J.P. Ostriker, Mon. Not. R. Astron. Soc. **459**(4), 3738-3755 (2016). https://doi.org/10.1093/mnras/stw836[arXiv:1511.02116 [astro-ph.HE]]
40. K. Inayoshi, E. Visbal, Z. Haiman, Ann. Rev. Astron. Astrophys. **58**, 27–97 (2020). https://doi.org/10.1146/annurev-astro-120419-014455[arXiv:1911.05791 [astro-ph.GA]]
41. G.L. Israel, A. Papitto, P. Esposito, L. Stella, L. Zampieri, A. Belfiore, G.A.R. Castillo, A. De Luca, A. Tiengo, F. Haberl et al., Mon. Not. R. Astron. Soc. **466**, L48–L52 (2017). https://doi.org/10.1093/mnrasl/slw218[arXiv:1609.06538 [astro-ph.HE]]
42. Y.F. Jiang, J.M. Stone, S.W. Davis, Astrophys. J. Suppl. **199**, 14 (2012). https://doi.org/10.1088/0067-0049/199/1/14. ([arXiv:1201.2223 [astro-ph.HE]].)

43. Y.F. Jiang, J. Stone, S. Davis, Astrophys. J. **778**, 65 (2013). https://doi.org/10.1088/0004-637X/778/1/65. ([arXiv:1309.5646 [astro-ph.HE]].)
44. Y.F. Jiang, J.M. Stone, S.W. Davis, Astrophys. J. **796**, 106 (2014). https://doi.org/10.1088/0004-637X/796/2/106. ([arXiv:1410.0678 [astro-ph.HE]].)
45. Y.F. Jiang, J.M. Stone, S.W. Davis, Astrophys. J. Suppl. **213**, 7 (2014). https://doi.org/10.1088/0067-0049/213/1/7
46. L. Jiang, I.D. McGreer, X. Fan, M.A. Strauss, E. Banados, R.H. Becker, F. Bian, K. Farnsworth, Y. Shen, F. Wang et al., Astrophys. J. **833**(2), 222 (2016).https://doi.org/10.3847/1538-4357/833/2/222[arXiv:1610.05369 [astro-ph.GA]]
47. Y.F. Jiang, J. Stone, S.W. Davis, Astrophys. J. **880**, 67 (2019). https://doi.org/10.3847/1538-4357/ab29ff. ([arXiv:1709.02845 [astro-ph.HE]].)
48. Y.F. Jiang, Astrophys. J. Supp. **253**, 49 (2021). https://doi.org/10.3847/1538-4365/abe303
49. Y.F. Jiang, Astrophys. J. Supp. **263**(1), 4 (2022). https://doi.org/10.3847/1538-4365/ac9231[arXiv:2209.06240 [astro-ph.IM]]
50. C. Jin, C. Done, M. Ward, E. Gardner, Mon. Not. R. Astron. Soc. **471**(1), 706–721 (2017). https://doi.org/10.1093/mnras/stx1634[arXiv:1706.08125 [astro-ph.HE]]
51. P. Kaaret, M.J. Ward, A. Zezas, Mon. Not. R. Astron. Soc. **351**, L83 (2004). https://doi.org/10.1111/j.1365-2966.2004.08020.x. ([arXiv:astro-ph/0407031 [astro-ph]].)
52. P. Kaaret, S. Corbel, Astrophys. J. **697**, 950–956 (2009). https://doi.org/10.1088/0004-637X/697/1/950[arXiv:0903.5329 [astro-ph.HE]]
53. P. Kaaret, H. Feng, T.P. Roberts, Ann. Rev. Astron. Astrophys. **55**, 303–341 (2017). https://doi.org/10.1146/annurev-astro-091916-055259[arXiv:1703.10728 [astro-ph.HE]]
54. D. Kasen, R.C. Thomas, P. Nugent, Astrophys. J. **651**, 366–380 (2006).https://doi.org/10.1086/506190[arXiv:astro-ph/0606111 [astro-ph]]
55. A.R. King, K.A. Pounds, Mon. Not. R. Astron. Soc. **345**, 657 (2003). https://doi.org/10.1046/j.1365-8711.2003.06980.x. ([arXiv:astro-ph/0305541 [astro-ph]].)
56. A.R. King, Mon. Not. R. Astron. Soc. **393**, 41 (2009). https://doi.org/10.1111/j.1745-3933.2008.00594.x. ([arXiv:0811.1473 [astro-ph]].)
57. A. King, J.P. Lasota, M. Middleton, New Astron. Rev. **96**, 101672 (2023). https://doi.org/10.1016/j.newar.2022.101672. ([arXiv:2302.10605 [astro-ph.HE]].)
58. H. Kobayashi, K. Ohsuga, H.R. Takahashi, T. Kawashima, Y. Asahina, S. Takeuchi, S. Mineshige, Publ. Astron. Soc. Jap. **70**(2), 22 (2018).https://doi.org/10.1093/pasj/psx157[arXiv:1802.00567 [astro-ph.HE]]
59. E. Koerding, H. Falcke, S. Markoff, Astron. Astrophys. **382**, L13 (2002). https://doi.org/10.1051/0004-6361:20011776. ([arXiv:astro-ph/0112385 [astro-ph]].)
60. S. Komossa, W. Voges, D.W. Xu, S. Mathur, H.M. Adorf, G. Lemson, W.J. Duschl, D. Grupe, Astron. J. **132**, 531–545 (2006).https://doi.org/10.1086/505043[arXiv:astro-ph/0603680 [astro-ph]]
61. M.R. Krumholz, R.I. Klein, C.F. McKee, S.S.R. Offner, A.J. Cunningham, Science **323**, 754–757 (2009). https://doi.org/10.1126/science.1165857[arXiv:0901.3157 [astro-ph.SR]]
62. J.P. Lasota, A. King, Mon. Not. R. Astron. Soc. **526**(2), 2506–2509 (2023). https://doi.org/10.1093/mnras/stad2926[arXiv:2309.00034 [astro-ph.HE]]
63. I. Lehmann, T. Becker, S. Fabrika, M. Roth, T. Miyaji, V. Afanasiev, O. Sholukhova, S.F. Sanchez, J. Greiner, G. Hasinger et al., Astron. Astrophys. **431**, 847 (2005). https://doi.org/10.1051/0004-6361:20035827. ([arXiv:astro-ph/0410458 [astro-ph]].)
64. C.D. Levermore, G.C. Pomraning, Astrophys. J. **248**, 321–334 (1981). https://doi.org/10.1086/159157
65. C.D. Levermore, J. Quant. Spectrosc. Radiat. Transfer **31**, 149–160 (1984). https://doi.org/10.1016/0022-4073(84)90112-2
66. A.P. Lightman, D.M. Eardley, Astrophys. J. Lett. **187**, L1 (1974). https://doi.org/10.1086/181377
67. B. Luo, W.N. Brandt, P.B. Hall, J. Wu, S.F. Anderson, G.P. Garmire, R.R. Gibson, R.M. Plotkin, G.T. Richards, D.P. Schneider et al., Astrophys. J. **805**(2), 122 (2015). https://doi.org/10.1088/0004-637X/805/2/122[arXiv:1503.02085 [astro-ph.GA]]

68. D. Lynden-Bell, J.E. Pringle, Mon. Not. R. Astron. Soc. **168**, 603–637 (1974)
69. Y. Ma, A. Goulding, J. Greene, N. Zakamska, D. Wylezalek, Y.F. Jiang, arXiv:2401.04177 (2024)
70. S. Martocchia, E. Piconcelli, L. Zappacosta, F. Duras, G. Vietri, C. Vignali, S. Bianchi, M. Bischetti, A. Bongiorno, M. Brusa et al., Astron. Astrophys. **608**, A51 (2017). https://doi.org/10.1051/0004-6361/201731314. ([arXiv:1708.00452 [astro-ph.GA]].)
71. W. Massonneau, M. Volonteri, Y. Dubois, R.S. Beckmann, Astron. Astrophys. **670**, A180 (2023). https://doi.org/10.1051/0004-6361/202243170. ([arXiv:2201.08766 [astro-ph.GA]].)
72. L. Mayer, https://doi.org/10.1142/9789813227958_0011[arXiv:1807.06243 [astro-ph.HE]]
73. J.C. McKinney, A. Tchekhovskoy, A. Sadowski, R. Narayan, Mon. Not. R. Astron. Soc. **441**(4), 3177–3208 (2014). https://doi.org/10.1093/mnras/stu762[arXiv:1312.6127 [astro-ph.CO]]
74. J.C. McKinney, L. Dai, M. Avara, Mon. Not. R. Astron. Soc. **454**(1), L6–L10 (2015). https://doi.org/10.1093/mnrasl/slv115[arXiv:1508.02433 [astro-ph.HE]]
75. M.J. Middleton, T.P. Roberts, C. Done, F.E. Jackson, Mon. Not. R. Astron. Soc. **411**, 644 (2011). https://doi.org/10.1111/j.1365-2966.2010.17712.x. ([arXiv:1009.2686 [astro-ph.CO]].)
76. B.S. Mills, S.W. Davis, Y.F. Jiang, M.J. Middleton, [arXiv:2304.07977 [astro-ph.HE]]
77. R. Narayan, Y. Zhu, D. Psaltis, A. Sadowski, Mon. Not. R. Astron. Soc. **457**(1), 608–628 (2016). https://doi.org/10.1093/mnras/stv2979[arXiv:1510.04208 [astro-ph.HE]]
78. R. Narayan, A. Sadowski, R. Soria, Mon. Not. R. Astron. Soc. **469**(3), 2997–3014 (2017). https://doi.org/10.1093/mnras/stx1027[arXiv:1702.01158 [astro-ph.HE]]
79. K. Ohsuga, M. Mori, T. Nakamoto, S. Mineshige, Astrophys. J. **628**, 368–381 (2005). https://doi.org/10.1086/430728[arXiv:astro-ph/0504168 [astro-ph]]
80. T. Okuda, Publ. Astron. Soc. Jap. **54**, 253 (2002). https://doi.org/10.1093/pasj/54.2.253. ([arXiv:astro-ph/0202446 [astro-ph]].)
81. B. Paczynski, P.J. Wiita, Astron. Astrophys. **88**, 23–31 (1980) EFI-78-68-CHICAGO
82. M.W. Pakull, L. Mirioni, [arXiv:astro-ph/0202488 [astro-ph]]
83. E. Pezzulli, M. Volonteri, R. Schneider, R. Valiante, Mon. Not. R. Astron. Soc. **471**(1), 589–595 (2017). https://doi.org/10.1093/mnras/stx1640[arXiv:1706.06592 [astro-ph.GA]]
84. T. Piran, Astrophys. J. **221**, 652–660 (1978). https://doi.org/10.1086/156069
85. J.E. Pringle, Ann. Rev. Astron. Astrophys. **19**, 137–160 (1981). https://doi.org/10.1146/annurev.aa.19.090181.001033
86. M.J. Rees, Nature **333**, 523–528 (1988). https://doi.org/10.1038/333523a0
87. A. Ricarte, R. Narayan, B. Curd, Astrophys. J. Lett. **954**(1), L22 (2023). https://doi.org/10.3847/2041-8213/aceda5[arXiv:2307.04621 [astro-ph.HE]]
88. J. Rosdahl, R. Teyssier, Mon. Not. R. Astron. Soc. **449**(4), 4380–4403 (2015). https://doi.org/10.1093/mnras/stv567[arXiv:1411.6440 [astro-ph.IM]]
89. A. Sadowski, R. Narayan, J.C. McKinney, A. Tchekhovskoy, Mon. Not. R. Astron. Soc. **439**(1), 503–520 (2014). https://doi.org/10.1093/mnras/stt2479[arXiv:1311.5900 [astro-ph.HE]]
90. A. Sadowski, R. Narayan, A. Tchekhovskoy, D. Abarca, Y. Zhu, J.C. McKinney, Mon. Not. R. Astron. Soc. **447**(1), 49–71 (2015).https://doi.org/10.1093/mnras/stu2387[arXiv:1407.4421 [astro-ph.HE]]
91. A. Sadowski, R. Narayan, Mon. Not. R. Astron. Soc. **453**(3), 3213–3221 (2015). https://doi.org/10.1093/mnras/stv1802[arXiv:1503.00654 [astro-ph.HE]]
92. A. Sadowski, R. Narayan, Mon. Not. R. Astron. Soc. **456**(4), 3929–3947 (2016). https://doi.org/10.1093/mnras/stv2941[arXiv:1509.03168 [astro-ph.HE]]
93. A. Sadowski, E. Tejeda, E. Gafton, S. Rosswog, D. Abarca, Mon. Not. R. Astron. Soc. **458**(4), 4250–4268 (2016). https://doi.org/10.1093/mnras/stw589[arXiv:1512.04865 [astro-ph.HE]]
94. A. Sdowski, R. Narayan, A. Tchekhovskoy, Y. Zhu, Mon. Not. R. Astron. Soc. **429**, 3533 (2013). https://doi.org/10.1093/mnras/sts632. ([arXiv:1212.5050 [astro-ph.HE]].)
95. N.I. Shakura, R.A. Sunyaev, Astron. Astrophys. **24**, 337–355 (1973)

96. N.I. Shakura, R.A. Sunyaev, Mon. Not. R. Astron. Soc. **175**, 613–632 (1976)
97. N.J. Shaviv, Astrophys. J. Lett. **494**, L193–L197 (1998). https://doi.org/10.1086/311182
98. H. Shiokawa, J.H. Krolik, R.M. Cheng, T. Piran, S.C. Noble, Astrophys. J. **804**(2), 85 (2015). https://doi.org/10.1088/0004-637X/804/2/85[arXiv:1501.04365 [astro-ph.HE]]
99. J.M. Stone, D. Mihalas, M.L. Norman, Astrophys. J. Suppl. **80**, 819 (1992). https://doi.org/10.1086/191682
100. H.R. Takahashi, K. Ohsuga, Astrophys. J. **772**, 127 (2013). https://doi.org/10.1088/0004-637X/772/2/127. ([arXiv:1306.0049 [astro-ph.HE]].)
101. H.R. Takahashi, K. Ohsuga, Publ. Astron. Soc. Jap. **67**(4), 60 (2015).https://doi.org/10.1093/pasj/psu145[arXiv:1411.1484 [astro-ph.HE]]
102. H.R. Takahashi, S. Mineshige, K. Ohsuga, Astrophys. J. **853**(1), 45 (2018). https://doi.org/10.3847/1538-4357/aaa082[arXiv:1712.02068 [astro-ph.HE]]
103. S. Takeuchi, K. Ohsuga, S. Mineshige, Publ. Astron. Soc. Jap. **65**, 88 (2013). https://doi.org/10.1093/pasj/65.4.88. ([arXiv:1305.1023 [astro-ph.HE]].)
104. T. Tanaka, Z. Haiman, Astrophys. J. **696**, 1798–1822 (2009). https://doi.org/10.1088/0004-637X/696/2/1798[arXiv:0807.4702 [astro-ph]]
105. L.L. Thomsen, T.M. Kwan, L. Dai, S.C. Wu, E. Ramirez-Ruiz, Astrophys. J. Lett. **937**(2), L28 (2022). https://doi.org/10.3847/2041-8213/ac911f[arXiv:2206.02804 [astro-ph.HE]]
106. N.J. Turner, J.M. Stone, Astrophys. J. Suppl. **135**, 95–108 (2001).https://doi.org/10.1086/321779[arXiv:astro-ph/0102145 [astro-ph]]
107. R. Urquhart, R. Soria, Mon. Not. R. Astron. Soc. **456**(2), 1859–1880 (2016). https://doi.org/10.1093/mnras/stv2293[arXiv:1511.05275 [astro-ph.HE]]
108. A. Utsumi, K. Ohsuga, H.R. Takahashi, Y. Asahina, Astrophys. J. **935**(1), 26 (2022). https://doi.org/10.3847/1538-4357/ac7eb8[arXiv:2207.02560 [astro-ph.HE]]
109. M. Volonteri, M.J. Rees, Astrophys. J. **633**, 624–629 (2005). https://doi.org/10.1086/466521[arXiv:astro-ph/0506040 [astro-ph]]
110. M. Volonteri, M. Habouzit, M. Colpi, Nature Rev. Phys. **3**(11), 732–743 (2021). https://doi.org/10.1038/s42254-021-00364-9[arXiv:2110.10175 [astro-ph.GA]]
111. C.J. White, P.D. Mullen, Y.F. Jiang, S.W. Davis, J.M. Stone, V. Morozova, L. Zhang, Astrophys. J. **949**(2), 103 (2023).https://doi.org/10.3847/1538-4357/acc8cf[arXiv:2302.04283 [astro-ph.HE]]
112. B.D. Wibking, M.R. Krumholz, Mon. Not. R. Astron. Soc. **512**, 1430 (2022). https://doi.org/10.1093/mnras/stac439
113. X.B. Wu, F. Wang, X. Fan, W. Yi, W. Zuo, F. Bian, L. Jiang, I.D. McGreer, R. Wang, J. Yang et al., Nature **518**(7540), 512–515 (2015). https://doi.org/10.1038/nature14241
114. H. Yang, F. Yuan, T. Kwan, L. Dai, Mon. Not. R. Astron. Soc. **523**(1), 208–220 (2023). https://doi.org/10.1093/mnras/stad1444[arXiv:2211.10710 [astro-ph.HE]]
115. F. Yuan, D. Bu, M. Wu, Astrophys. J. **761**, 130 (2012). https://doi.org/10.1088/0004-637X/761/2/130. ([arXiv:1206.4173 [astro-ph.HE]].)
116. Y. Zhu, R. Narayan, A. Sadowski, D. Psaltis, Mon. Not. R. Astron. Soc. **451**(2), 1661–1681 (2015). https://doi.org/10.1093/mnras/stv1046[arXiv:1505.04838 [astro-ph.HE]]

Chapter 13
Modeling and Comparison with Observations by the EHT and the GMVA

Christian M. Fromm

Abstract Active Galactic Nuclei (AGN) are one of the most enigmatic and powerful objects in the universe. They are powered by supermassive black holes and launch highly collimated relativistic jets while accelerating particle to the highest energies. As such, they represent the ideal laboratory to test our current understanding of accretion on to supermassive black holes and jet formation using general relativistic magneto hydrodynamics (GRMHD) simulations. Due to the progress in observational astronomy using the technique of Very Long Baseline Interferometry (VLBI) we can nowadays image and study relativistic jets from their outskirts down to their launching site in the direct vicinity of supermassive black holes. Making use of the VLBI technique at 1.3 *mm* wavelength the recent observations of the Event Horizon Telescope (EHT) can resolve and image the accretion flow around the supermassive black hole in the giant elliptical galaxy Messier 87 (M 87) and in the galactic center Sagittarius A* (SgrA*). Using the Global Millimetre VLBI Array (GMVA) at 3.5 *mm* the propagation of relativistic jets can be observed and monitored with great details. There is vast and growing literature on methods and tools for comparing GRMHD simulations to various observations and sources. In order to provide an pedagogical introduction to the field we will focus in this Chapter on how to compare GRMHD simulations with recent M 87 observations of the EHT and GMVA.

13.1 Introduction and Overview

In April 2019 the Event Horizon Telescope (EHT[1]) published the first image of the horizon scale structure in M 87 [12] This iconic image represents the missing link in a century long quest for imaging the central engine of AGNs and the launching site of relativistic jets. This achievement was only possible due to the technique of Very

[1] https://eventhorizontelescope.org/.

C. M. Fromm (✉)
Institut für Theoretische Physik und Astrophysik, Universität Würzburg, Emil-Fischer-Strasse 31, 97074 Würzburg, Germany
e-mail: christian.fromm@uni-wuerzburg.de

C. Bambi et al. (eds.), *New Frontiers in GRMHD Simulations*, Springer Series in Astrophysics and Cosmology,https://doi.org/10.1007/978-981-97-8522-3_13

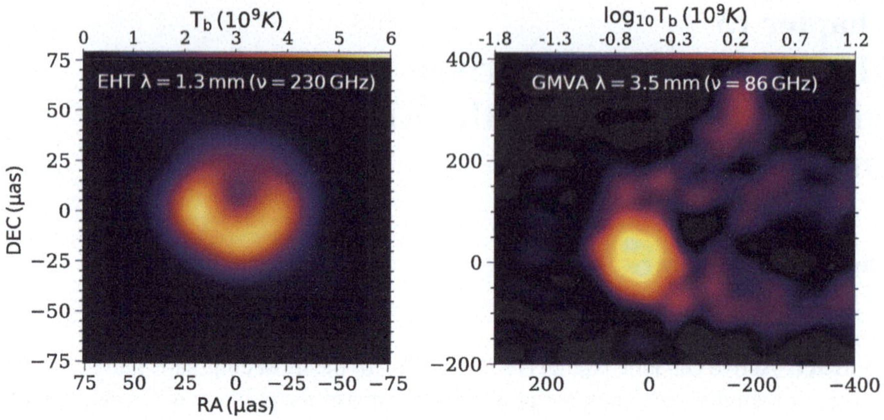

Fig. 13.1 Reconstructed M 87 images from the EHT observations (left) and the GMVA observations (right). The images show the distribution of the brightness temperature T_b blurred by a circular gaussian with a FWHM of 20 μas for the EHT and 37 μas for the GMVA. Notice the difference in the image dimensions and in the scaling: linear for the EHT image and logarithmic for the GMVA one. The images are reproduced from [12] and from [24]

Long Baseline Interferometry (VBLI) which use radio telescopes scattered across the globe to form a synthetic earth size telescope combined with its short observing wavelength of 1.3 mm. In addition to the EHT there are complementary VLBI arrays operating at longer wavelength which provide high-resolution images of the M 87 jet at various scales. Until the commissioning of the EHT, the Global Millimetre VLBI Array (GMVA[2]) observing at a wavelength of 3.5 mm was providing the highest angular resolution images of M 87 [23]. The most recent GMVA observations of M 87 including the phased Atacama Large Millimetre/ submillimetre Array (ALMA) and the Greenland Telescope (GLT) detect for the first time both the ring-like horizon scale structure and the edge-brightened jet structure with an additional faint inner jet ridge [24]. In Fig. 13.1 we show reconstructed images of M 87 from the EHT (left) and the GMVA observations (right). The images display the brightness temperature $T_b = \lambda^2 S/(2k_b\Omega)$ where λ is the wavelength, S is the flux density in Jansky, k_b is the Boltzmann constant and Ω is the solid angle.

The EHT image shows a ring-like structure with a diameter of $d \sim 43 \pm 3\,\mu$as together with a central flux depression. The brightness distribution along the ring is asymmetric and shows a brighter southern part. Due to the currently limited dynamical range of the EHT array no large-scale jet structure is visible. On the contrary the GMVA observation of M 87 shows a bright central region together with a faint jet. The jet shows a large apparent opening angle $\phi_{\mathrm{app}} = 127^\circ \pm 22^\circ$ [23] and three distinguishable ridges extending up to 1000 μ as [24]. These properties are among the key characteristics which need to be recovered by a physical

[2] https://www3.mpifr-bonn.mpg.de/div/vlbi/globalmm/.

model of black hole accretion and jet launching. In Sect. 13.3 we will provide a basic introduction to the VLBI technique and image reconstruction methods.

Among the most promising models utilised to interpret VLBI observations are three-dimensional (3d) general relativistic magneto hydrodynamics (GRMHD) simulations. The initial conditions of these simulations consist of a torus in hydro-static equilibrium which is seeded with a poloidal magnetic field [18]. This configuration is perturbed by a small pressure variation which triggers the magneto-rotational stability (MRI) [22]. The MRI reduces the angular momentum and the accretion process starts. The mass accretion and the magnetic flux are rising and approximate a steady state during the cause of the simulation [32]. Depending on the strength and geometry of the initial magnetic field two main configurations are obtained, i.e. a weakly magnetised one typically referred as Standard And Normal Evolution (SANE) and a highly magnetised one leading to a Magnetically Arrested Disk (MAD) [38]. In contrast to the ideal GRMHD simulations which evolve a single fluid (i.e., evolving only protons) more advanced simulations consider a two fluid model (i.e., evolving both protons and electrons) while including heating processes, e.g., reconnection heating [26] and radiative cooling processes e.g., synchrotron cooling [5, 9]. Finally the 3D GRMHD simulations provide us with the spatial and temporal evolution of the plasma properties, i.e., the velocity, pressure, density and magnetic field. However, the EHT and GMVA observations report the projected brightness distribution of the source in the sky. As such a direct comparison between the GRMHD simulations and the observations is not possible and an additional post-processing step is required to translate the plasma properties to the observed radiation.

During the general relativistic radiative transfer (GRRT) calculations the radiative transfer is solved along geodesics taking into account the distribution of the radiating electrons and their radiation microphysics as well as the curved space time in the vicinity of the black hole. Figure 13.2 displays on the left a volume rendering of a 3D GRMHD simulation for black hole with spin $a^{\star} = 0.94$ and a MAD torus. The spin axis of the black hole is pointing away from us and the system is seen under a viewing angle $\vartheta \sim 160°$ and a position angle[3] PA of 288° [40]. The red-orange colour corresponds to the dense torus region and the blue-white colour indicates the low density jet region. The right side reports the emission from the GRMHD simulation after performing the radiative transfer via GRRT calculations. The image shows the logarithm of the emission with a bright central ring and a faint large scale jet with large opening angle. The images are reproduced from the simulations used in [19]. In Sect. 13.2 we will provide details on both the GRMHD simulations and the GRRT calculations which were used to generate Fig. 13.2.

After the emission calculations are done the numerical simulations can be compared to the observations. This comparison can be performed either in the image plane or in the Fourier plane. Given the nature of VLBI observations (see more details in Sect. 13.3) carrying out the comparison in the Fourier plane is the more natural/direct approach. However, some insights can already be gained by comparing the GRRT results with the VLBI observations in image plane. Therefore, we need to adjust the

[3] Measured from North through South.

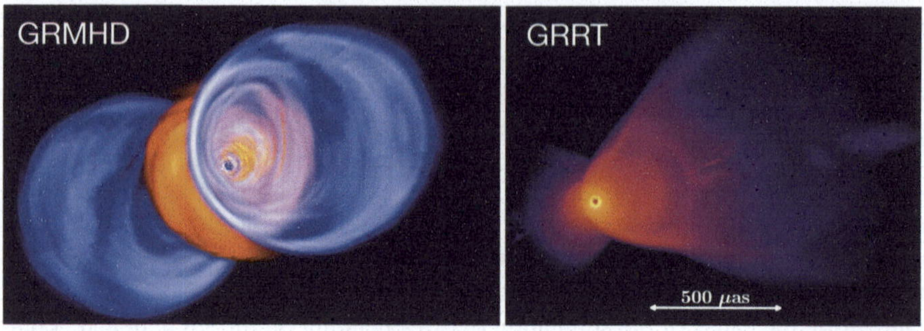

Fig. 13.2 Left: 3D rendering of GRMHD simulation for a black hole with spin $a^{\star}$=0.94. The red colors correspond to the dense torus and blue colors to the dilute jet. Right: GRRT calculation showing the emission from the jet-torus system at frequency of 230 GHz. The images are generated from the simulations presented in [19]

resolution of the GRRT images to the observed images. This can be done by convolving the GRRT images with the nominal resolution of the VLBI observations given by the beam of the array. The beam can be regarded as a 2D point spread function with a certain full width half maximum (FWHM). Once both images have the same resolution, image dimensions and number of pixels different image metrics i.e., root mean square error, can be applied to quantify the similarity or dis-similarity of the images and thus the validity of the underlying GRMHD model [27]. An example of a convolved GRRT horizon scale image using the resolution of the EHT array ($\sim$20 μ as) is give in the top panel of Fig. 13.3.

For the more detailed comparison in the Fourier plane it is necessary to generated from the GRRT image synthetic data which is as similar as possible to the VLBI data. This implies the Fourier transformation of the image and its sampling with the projected distances between the telescopes of the VLBI array the so-called baselines. The results of this step will be a complex number typically referred as complex visibilities. The absolute value of the visibility is called visibility amplitude and the angular part as visibility phase. Once the GRRT image is Fourier-transformed and sampled the χ^2 (reduced χ^2) between the data and the model can be computed. In addition to the calculation of the χ^2 the GRRT image can be fitted to some extend to the EHT data by scaling the brightness, rotating and stretching the image from which an estimate for the black hole mass can be extracted assuming a distance to the source [15, 16]. The bottom panel of Fig. 13.3 shows the comparison of the visibility amplitude from a GRRT image (top panel) with the visibility amplitude of the EHT observation. The data points are given in blue and the Fourier-transformed GRRT image is represented by green points.

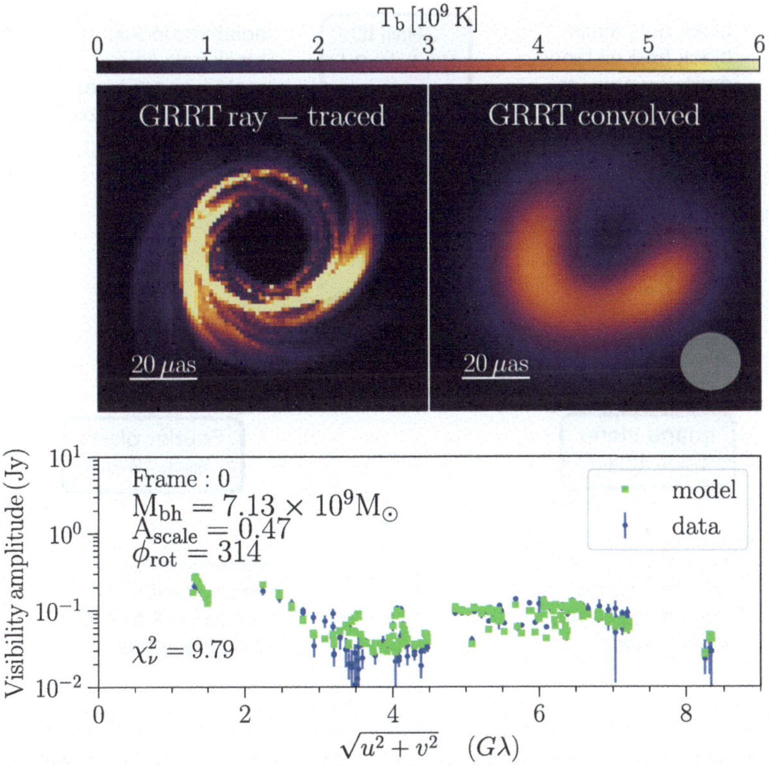

Fig. 13.3 Results of fitting a GRMHD/GRRT snapshot to the EHT observations of M87. Top: GRRT image ray-traced (left) and image convolved with a circular gaussian with FHWM=20 μas. Bottom: Matching of the visibility amplitudes between GRRT image and observation by scaling the black hole mass, flux amplitude and the rotation of the GRRT image

In this chapter we will cover three major topics: Numerical modelling including GRMHD simulations and GRRT calculations, the basics of VLBI observations and image reconstruction and the comparison between numerical models and EHT and GMVA observations. Given that each topic can easily fill a text book we will only provide some basic introduction to the individual topics and will refer to individual Chapters of this book and/or to external references for details. The typical working steps involved in the GRMHD-data comparison are summarised in a work-flow diagram in Fig. 13.4 and we will apply them to the modelling of the total intensity of M87 using EHT and GMVA observations.

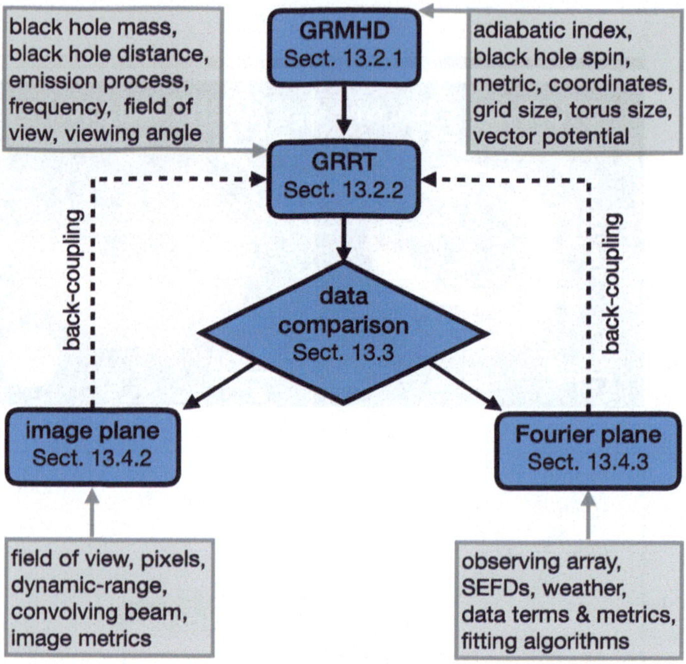

Fig. 13.4 Work-flow diagram illustrating the different working packages (blue boxes including the Sect. numbers within this Chapter) required for the comparison between GRMHD simulations and VLBI observations together with the most important input parameters (grey boxes) which will be explained in the following sections

13.2 Numerical Simulations

In this Section we will setup a 3D aligned i.e., the spin axis the and rotation axis of the torus are aligned, GRMHD simulation for a fast rotating Kerr Black hole with spin $a^{\star} = 0.95$ surrounded by a MAD torus. For tilted black hole torus systems see Chap. 11 of this book and for accretion onto non-Kerr objects see Chap. 15. The simulation will be carried out with the publicly available Black Hole Accretion Code `BHAC` [30, 34]. After a steady-state for the mass accretion rate $\dot{m}$ and the accreted magnetic flux $\phi_{\rm mag}$ is obtained we will perform GRRT calculations. For the radiative transfer calculations we will use the Black Hole Observations in Stationary Space times `BHOSS` code [42, 43]. Furthermore, we will scale the GRRT calculations to M 87 i.e., assuming a black hole mass of $m_{bh,M87} = 6.5 \times 10^9\,M_{\odot}$ and a distance of 16.9 Mpc [15].

13.2.1 GRMHD Simulations

Within this Chapter Greek indices run from [0, 1, 2, 3] while Roman indices run over [1, 2, 3]. BHAC solves the conservation of mass, the local conservation of the energy-momentum and the covariant Maxwell equations for Kerr space time using spherical coordinates and using geometric units ($GM = c = 1$ and $1/\sqrt{4\pi}$ is absorbed in the magnetic field). For the detailed derivation of the equations see Chap. 1 of this book. However for reasons of completeness we provide the equations and definitions below.

$$\nabla_\mu(\rho u^\mu) = 0\,, \quad \nabla_\mu T^{\mu\nu} = 0\,, \quad \nabla_\mu {}^*F^{\mu\nu} = 0. \tag{13.1}$$

where ρ is the rest-mass density and u^μ is the four-velocity. $T^{\mu\nu}$ is the stress-energy tensor for ideal MHD which is given below

$$T^{\mu\nu} = (\rho h_{\rm tot}) u^\mu u^\nu + \left(p + \frac{1}{2}b^2\right) g^{\mu\nu} - b^\mu b^\nu, \tag{13.2}$$

. In the equation above $h_{\rm tot} = h + b^2/\rho$ is the total specific enthalpy and p the fluid pressure. Furthermore, the magnetic four-vector b^μ can be written as:

$$b^t = B^i u^\mu g_{i\mu} \quad b^i = (B^i + b^t u^i)/u^t, \tag{13.3}$$

where B^i is the magnetic field measured by an Eulerian observer. We assume an ideal equation of state using a fixed adiabatic index $\hat{\gamma} = 4/3$. The equations above are solved within BHAC on a spherical polar grid (r, θ, ϕ) and to keep the required numerical resources with in a meaningful limit we use a logarithmic spacing in the radial coordinate and employ three levels of adaptive mesh refinement (AMR). In Table 13.1 we list the properties of the used grid. In addition to the dimension and type of the grid we need to specify the boundary conditions i.e., how should the physical parameters behave once the reach the edge of the grid. In the radial direction we use outflow conditions, while in the polar direction we set a reflective wall and in the azimuthal direction we employ periodic boundary conditions (for details on the boundary conditions see [34]). Since the numerical methods used to solve and

Table 13.1 Properties of the used GRMHD grid

r [M]	θ [rad]	ϕ [rad]	$(N_r, N_\theta, N_\phi)^a$
$1.18 r_{\rm eh} - 2500$	$0 - \pi$	$0 - 2\pi$	96,48,48 (384,192,192)

[a] Values in brackets correspond to the effective resolution at highest AMR level

[b] Radius of the event horizon $r_{\rm eh} = 1 + \sqrt{1 - a^{\star 2}}$

evolve the GRMHD equations lead to numerical issues in the low density regions (jet funnel) we fix this issue by applying a geometric floor model to the rest-mass density and gas pressure once their values drop below their floor values:

$$\rho_{\text{fl}} = 10^{-4} r^{-2} \quad p_{\text{fl}} = (10^{-6}/3) r^{-2\hat{\gamma}} \tag{13.4}$$

After the numerical grid and the floor model is set we need to define the initial conditions for the torus surrounding the black hole. As already mentioned in the introduction we use a torus in hydrostatic equilibrium following [18, 35]. In our case we place the inner edge of the torus at $r_{\text{in}} = 20\,M$ and location of the pressure maximum at $r_{\text{pm}} = 40\,M$. This torus is filled with a weak poloidal field which we initialise by a vector potential of the form:

$$A_\phi \propto \max(q - 0.2, 0), \tag{13.5}$$

where

$$q = \frac{\rho}{\rho_{\max}} \left(\frac{r}{r_{\text{in}}}\right)^3 \sin^3\theta \exp\left(\frac{-r}{400}\right) \tag{13.6}$$

This special form of the vector potential above will lead to the MAD state. To initiate the accretion onto the black hole we trigger the magneto-rotational instability (MRI), by perturbing the equilibrium torus by a small pressure variation so that its resulting value is $p = p(1 + X_p)$ with a random number $|X_p| < 0.04$.

This setup can be found in *tests/rmhd/FMtorus* within the BHAC repository[4]

In order to keep accretion process activated throughout the course of the simulation it is necessary to resolve at least the fastest growing mode of the MRI with sufficient numerical resolution in the disk region. This necessary condition can be quantified by computing MRI quality factors in the different spatial directions, $Q^{r,\theta,\phi}$. The used definition of $Q^{r,\theta,\phi}$ can be found in [32]. The values for the required Q-factors is a matter of discussion and typical values are between 6 and 10 (see [32] and references therein). In Fig. 13.5 we show the azimuthal averaged Q-factors computed from a MAD GRMHD simulation of a fast spinning black hole ($a^\star = 0.94$) at t=2944 M. The disk area is marked by black lines at the lowest Q-factors are obtained in θ direction. However, the values are still in agreement with minimal Q-factors $\sim$6.

During the run time of the simulation we monitor the mass accretion rate, $\dot{M}$ and the magnetic flux, Φ, across the horizon which are defined as:

$$\dot{M} = \int_0^{2\pi} \int_0^{\pi} \rho u^r \sqrt{-g} d\theta d\phi \tag{13.7}$$

$$\phi = \int_0^{2\pi} \int_0^{\pi} \left|{}^\star F^{rt}\right| \sqrt{-g} d\theta d\phi \tag{13.8}$$

[4] https://bhac.science/.

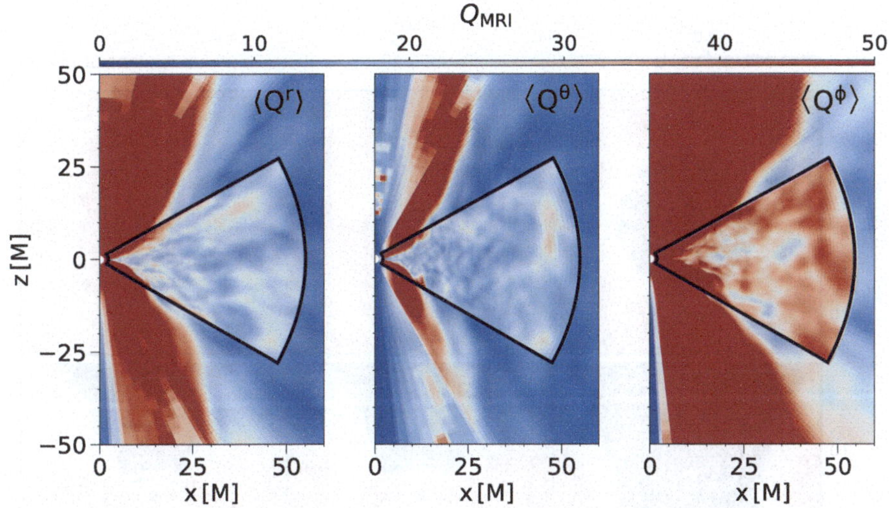

Fig. 13.5 Azimuthal averaged Q-factors in r, θ and ϕ direction computed from MAD GRMHD simulation of a fast spinning black hole ($a^{\star}$ = 0.94) at t = 2944 M. The disk region is highlighted by black lines

with ${}^{\star}F^{\mu\nu}$ the dual of the Faraday tensor and $\sqrt{-g}$ the determinant of the metric tensor.

In Fig. 13.6 we show snapshots for the density distribution (top panels) as well as the evolution of the mass accretion rate and the normalised magnetic flux, $\Phi = \phi/\sqrt{\dot{M}}$ from our 3D MAD simulation. The top panels show the density distribution in the equatorial plane and below in the meridional plane for three different times. The first snapshot at $t = 1000\,M$, shortly after the start of the simulation, shows the inner edge of the initial torus which is still disconnected from the black hole. The accretion process is not yet developed which can also be seen in the plots for the mass accretion rate and the magnetic flux (first vertical blue line in the bottom panels). The second time frame is located at $t = 18760\,M$. At this time the accretion flow is fully developed (see second vertical blue line in the bottom panels). A deeper look at the density distribution close to the black holes reveals that some part of the accretion flow is disconnected from the black hole visible by the lower density. At this time and location there are local ejections of so-called magnetic flux bundles which are common features connected to the MAD state (see [33] for more details). The last time frame is taken at the end of our simulation and shows a well established accretion flow around the black hole (best seen in the equatorial plane) together with a dilute jet formed above the poles of the black hole (best seen in the meridional plane). From the temporal evolution of the mass accretion rate and the magnetic flux we can conclude that a quasi steady-state is obtained after 15000 M. Given an average normalised magnetic flux of $\langle\Phi\rangle \sim 15$ our simulation has entered the MAD state according to the definition of [38].

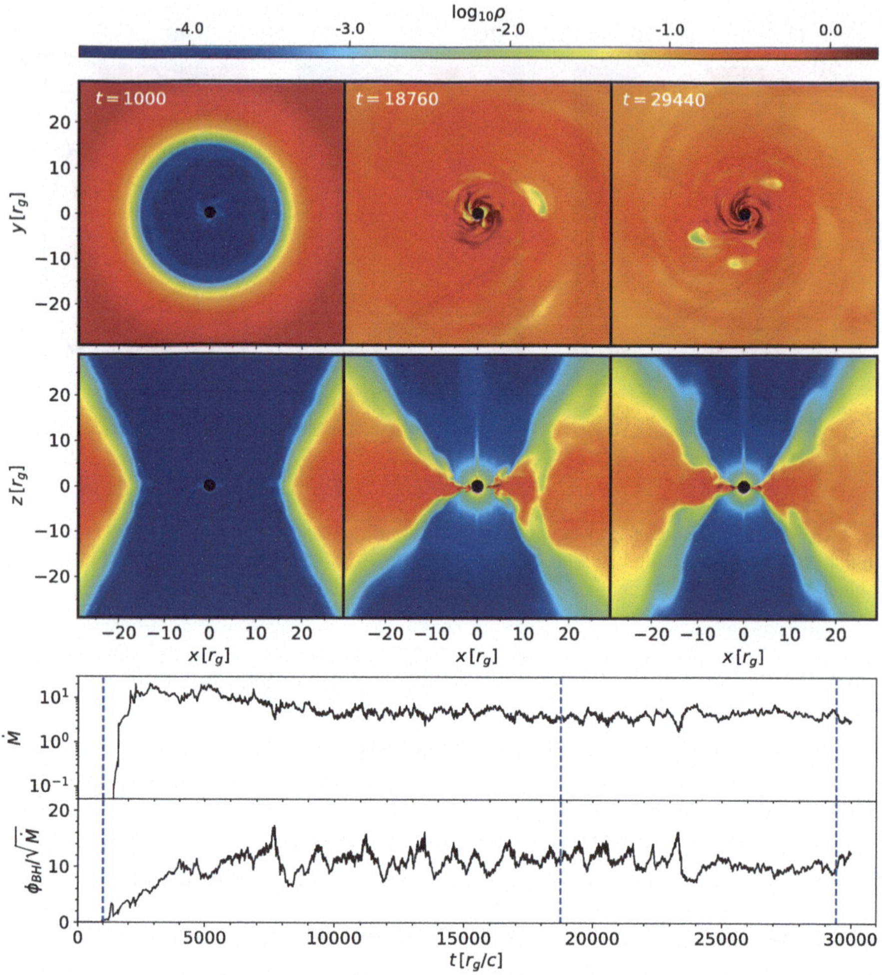

Fig. 13.6 Visualisation of a MAD GRMHD simulation with black hole spin $a^{\star} = 0.94$. The top panels show the density in equatorial plane and the middle ones the meridional plane for three different times. The bottom panels show the temporal evolution of the mass accretion rate, $\dot{M}$, and the MAD parameter Φ. The blue vertical dashed lines correspond to the time frames of the density plots

In addition to the horizon penetrating fluxes, $\dot{M}$ and ϕ it is useful to study disk or jet-averaged quantities. We use the definition of the average quantities from [32]:

$$\langle q(r) \rangle = \frac{1}{\Delta t} \int_{t_{\min}}^{t_{\max}} \frac{\int_{-\pi}^{\pi} \int_{0}^{2\pi} q(t, r, \theta, \phi) \sqrt{-g(r, \theta)} d\phi d\theta}{\int_{-\pi}^{\pi} \int_{0}^{2\pi} \sqrt{-g(r, \theta)} d\phi d\theta} dt, \tag{13.9}$$

where $q(t, r, \theta, \phi)$ is the quantity under investigation and $\sqrt{-g(r, \theta)}$ is the determinant of the metric. We compute the averaged values for the density, ρ, magnetic field B, and the electron temperature Θ_e (see Eq. 13.19) for a time range of $\Delta t = 2000\,\mathrm{M}$ from $28000\,M < t < 30000\,M$. The radial profiles can be compared to analytical models of relativistic jets [3, 17]. In analytical jet models the density decreases as $\rho \propto r^{-2}$ and the magnetic field declines as $B \propto r^{-1}$ indicating a dominant toroidal magnetic field. In addition, the equipartition temperature in jet should be constant where as the temperature in the disk should behave as $\Theta_e \propto r^{-1}$. In Moreover, the obtained profiles elucidate on the temporal and spatial structures in the GRMHD simulations which are difficult to process and understand in full 3D. The radial profiles from our MAD $a^\star = 0.94$ GRMHD simulation are presented in Fig. 13.7. As can be seen the profiles follow the behaviours from analytical models indicated by the dashed black lines. The electron temperature in the jet is constant until $r \sim 300\,M$ and increases for larger distances. A similar behaviour is reported by [7] and could be triggered by the de-confinement of the jet.

To summarise: The free parameters of our ideal GRMHD simulation used in this chapter are: the spin of the black hole, $a^\star$, the adiabatic index, $\hat{\gamma}$, the inner edge of the torus, $r_{\rm in}$, the location of the pressure maximum of the torus, $r_{\rm pm}$, and the vector potential A_ϕ of the torus. In addition, the numerical resolutions needs to be sufficient to ensure Q-factors above 6 throughout the course of the simulation.

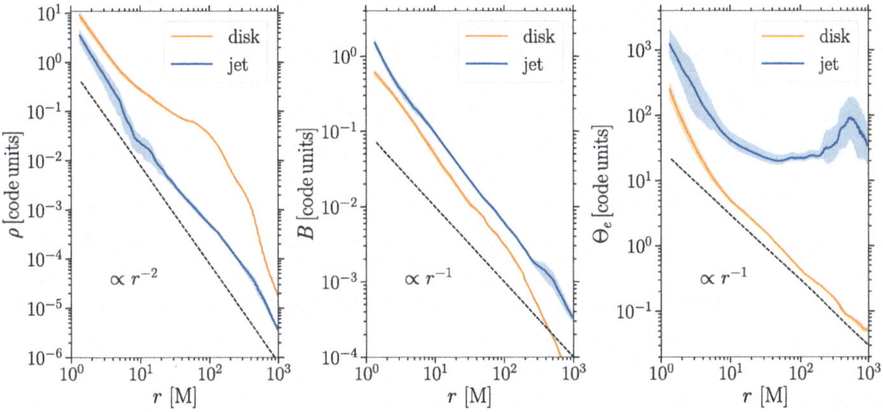

Fig. 13.7 Temporal averaged radial profiles for jet and disk together with analytical expectations (dashed lines) obtained from a MAD $a^\star = 0.94$ GRMHD simulations. The panels show the density, ρ, (left), magnetic field, B, (middle) and electron temperature, Θ_e (right)

13.2.2 GRRT Calculations

Once our GRMHD simulations have reached a steady-state in the mass accretion rate and in the magnetic flux ($t > 15000\,\mathrm{M}$ see bottom panels in Fig. 13.6). We can use the GRMHD data to generate light-curves, images and spectra by solving the radiative transport equations. The first step in the calculation of radiative signature of the GRMHD simulations is the calculation of light rays, i.e., the photon trajectories. This is achieved by solving the geodesic equation which is set of eight coupled first order differential equations defined by:

$$\frac{dx^\alpha}{d\lambda} = k^\alpha \tag{13.10}$$

$$\frac{dk^\alpha}{d\lambda} = -\Gamma^\alpha_{\mu\nu} k^\mu k^\nu \tag{13.11}$$

In the equations above x^α is the photon position and k^α is the photon 4-momentum and λ is the affine parameter. The Christoffel symbol $\Gamma^\alpha_{\mu\nu}$ is computed from the first derivatives of the metric g:

$$\Gamma^\alpha_{\mu\nu} = \frac{1}{2} g^{\alpha\beta} \left[\partial_\mu g_\nu \beta + \partial_\nu g_{\mu\beta} - \partial_\beta g_{\mu\nu} \right] \tag{13.12}$$

The formulation above allows for the calculation of the photon path in any space-time by specifying the metric $g_{\mu\nu}$. For non-Kerr space times see Chap. 15 of this book. In order to solve the geodesic equations we need to provide appropriate boundary conditions which can be obtained from a pre-defined observers camera positions. Once the boundary conditions are set the differential equations can be solved numerically for example by using Runge-Kutta Methods (more details can be found in [8, 29, 43]). As an example for the ray-tracing we present in Fig. 13.8 the geodesics computed in the equatorial plane for a Schwarzschild black hole, $a^\star = 0$, (left) and a Kerr black hole with $a^\star = 0.94$ (right). The black hole horizon is marked by the black sphere with radius $r_{\mathrm{eh}} = 1 + \sqrt{1 - a^{\star 2}}$ and we colour-coded the geodesics according to their winding parameter $n = \phi/(2\pi)$ where ϕ is the change in the azimuthal angle [21]. The colours correspond to $n < 0.75$ (black), $0.75 \leq n \leq 1.25$ (gold), $n > 1.25$ (red). In addition, we use black colour for photons terminating at the black holes horizon. Notice, that in the Schwarzschild case the image is symmetric across $y = 0$ and effect of frame-dragging (change in azimuthal direction of rays) in the case of the rotating black hole. In addition it is clearly visible that the area formed from rays captured by the black hole (black rays) is larger than the black holes event horizon (black circle).

As mentioned before we use ideal non-radiative GRMHD simulations using geometric units. As such the GRMHD simulations provide the absolute velocity field but are scale-free for the density, magnetic field and pressure. More precisely, the simulations are invariant for the following scalings:

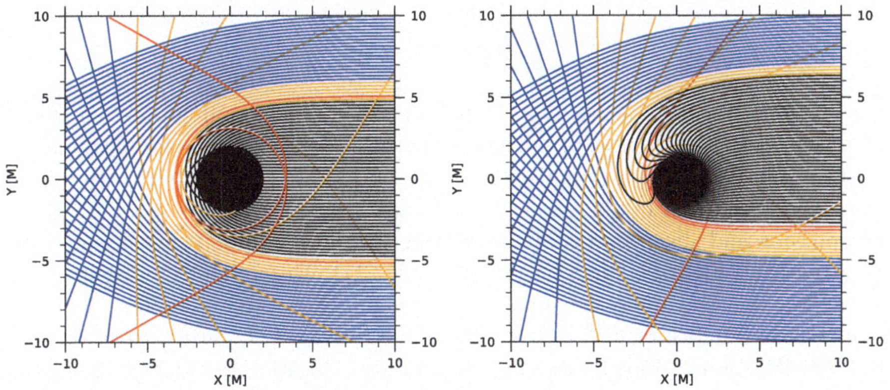

Fig. 13.8 Geodesics in the equatorial plane around a Schwarzschild black hole with $a^{\star} = 0$, (left) and a Kerr black hole with $a^{\star} = 0.94$ (right). The colour of the rays indicate their winding parameter $n = \phi/(2\pi)$: $n < 0.75$ (black), $0.75 \leq n \leq 1.25$ (gold), $n > 1.25$ (red)

$$\rho \to \mathcal{M}\rho \quad B \to \mathcal{M}^{1/2}B \quad p \to \mathcal{M}p, \tag{13.13}$$

where $\mathcal{M}$ is the mass unit which sets the black hole mass accretion rate $\dot{M}$. Later we will use $\mathcal{M}$ to adjust our GRRT calculations and thus our GRMHD simulations to the observed total flux of M 87 at 230 GHz. In addition to the scaling of the GRMHD parameters we need to define a length scale of our GRRT images. This can be done by providing a black hole mass and distance. In this Chapter we will use the black hole mass and distance estimates for M 87 which are: $m_{bh} = 6.5 \times 10^9\, M_{\odot}$ and $d_{bh} = 16.9\, Mpc$ [16]. Using this number one gravitational radius corresponds to 3.8 μas . Thus, once we scaled our GRMHD simulations to M87 we can compute the radiative transfer along the geodesics. The covariant radiative transfer equation can be written as:

$$\frac{d\mathcal{I}}{d\tau_{\nu}} = -\mathcal{I} + \frac{\eta}{\chi}, \tag{13.14}$$

where $\mathcal{I}$ is the Lorentz invariant specific intensity and is connected to the specific intensity, I_{ν}, at frequency, ν by:

$$\mathcal{I} = \frac{I_{\nu}}{\nu^3}. \tag{13.15}$$

and Lorentz invariant emissivity, η, and absorptivity, χ, are related to the emission, j_{ν} and absorption coefficients, α_{ν} evaluated at frequency ν via:

$$\eta = j_{0,\nu}/\nu^2 \qquad \chi = \alpha_{0,\nu}\nu, \tag{13.16}$$

where the subscript "0" indicates quantities measured in the local-rest frame of the plasma. Using the definition of the optical depth $d\tau_{\nu} = \int \alpha_{\nu} ds$, the equation for the covariant radiative transport can be decoupled into two differential equations:

$$\frac{d\tau_\nu}{d\lambda} = \gamma^{-1}\alpha_{0,\nu}\,, \qquad \frac{d\mathcal{I}}{d\lambda} = \gamma^{-1}\left(\frac{j_{0,\nu}}{\nu^3}\right)\exp\left(-\tau_\nu\right)\,, \tag{13.17}$$

with affine parameter λ and energy shift between oberver's and co-moving frame $\gamma^{-1} = \nu_0/\nu = -k_\alpha u^\alpha|_\lambda / k_\beta u^\beta|_\infty$. The coupled differential equation for the radiative transfer can be solved with a simple forward Euler step since the step size in the affine parameter is provided by the geodesic integrator (for more details see, e.g., [43]). The emission and absorption coefficients depend on the assumed emission process and electron distribution function see, e.g., [25, 31]. Throughout this Chapter we will assume that the emission is mainly generated by electrons. Therefore, we need to know the (dimensionless) electron temperature, $\Theta_e = k_b T_e / m_e c^2$. However, our ideal GRMHD simulation provides the temperature of the ions. In order to reconstruct the electron temperature from the ion temperature we assume the following ration [28]:

$$R = \frac{T_p}{T_e} = \frac{R_{\rm low} + R_{\rm high}\beta_p^2}{1+\beta_p^2}, \tag{13.18}$$

where $R_{\rm low,high}$ are free parameters. Thus the electron temperature can be written as:

$$\Theta_e = \frac{p m_p}{\rho m_e \left(R+1\right)}, \tag{13.19}$$

where p is the pressure, ρ is the density (both are provided from the GRMHD simulation) and $m_{e,p}$ is the electron and proton mass. In Fig. 13.9 we show the distribution of the plasma beta, β_p, (left column) and the electron temperature, Θ_e for two different values of $R_{\rm high}$ while keeping $R_{\rm low} = 1$.

The underlying GRMHD snapshot corresponds to black hole with spin $a^\star = 0.94$ at a time of 29440 M (see right column in Fig. 13.2). The inverse temperature ratio R^{-1} approaches $1/R_{\rm low}$ in low β_p regions while for high β_p regions the asymptotic values is $1/R_{\rm high}$. This behaviour can be translated to high electron temperature in the jet funnel and low electron temperature in the disk region. Notice the changes in the disk electron temperature with increased $R_{\rm high}$ by comparing the middle ($R_{\rm high} = 1$) and the right columns ($R_{\rm high} = 160$) in Fig. 13.9.

After we set the procedure to compute the electron temperature we need to decide on the distribution function of the emitting electrons (eDF). There are in principle two possibilities: a relativistic Maxwell-Jüttner distribution, commonly refereed as thermal eDF, and non-thermal eDF. For both eDFs we provide approximations to the emissivities and absorptivities in Appendix 13.5. The most common non-thermal eDFs are power-law and kappa-eDF. In this Chapter we will use a relativistic Maxwell-Jüttner eDF and a kappa eDF. This choice is motivated from the modelling of the horizon scale structure of M87 using the EHT observations [15] and the observed spectral indices of large scale jet in M87 at longer wavelengths [11, 37]. The thermal eDF is given by:

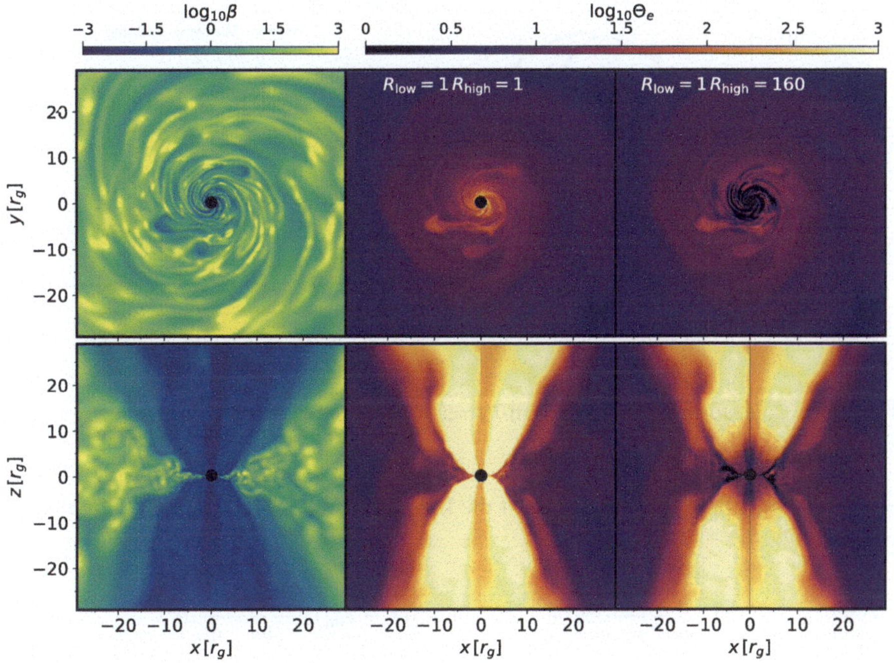

Fig. 13.9 Visualisation of the $R-\beta$ electron temperature model for a GRMHD model with black hole spin $a^\star = 0.94$ at time $t = 29440\,M$. The top panels show the equatorial plane and the bottom ones the meridional plane. The first column displays the plasma-beta parameter β_p and the middle and right column the electron temperature, Θ_e, for $R_{\rm high} = 1$ and $R_{\rm high} = 160$ while keeping $R_{\rm low} = 1$ fixed

$$\frac{dn_{\rm e}}{d\gamma_{\rm e}} = \frac{n_{\rm e}}{4\pi\,\Theta_{\rm e}}\frac{\gamma_{\rm e}\sqrt{\gamma_{\rm e}^2-1}}{K_2\left(1/\Theta_{\rm e}\right)}\exp\left(-\frac{\gamma_{\rm e}}{\Theta_{\rm e}}\right), \tag{13.20}$$

with electron number density $n_{\rm e}$ electron Lorentz factor γ_e and K_2 corresponds to the Bessel functions of second kind. The kappa eDF can be written as:

$$\frac{dn_{\rm e}}{d\gamma_{\rm e}} = \frac{N}{4\pi}\gamma_{\rm e}\sqrt{\gamma_{\rm e}^2-1}\left(1+\frac{\gamma_{\rm e}-1}{\kappa w}\right)^{-(\kappa+1)}, \tag{13.21}$$

where N is a normalisation factor, kappa parameter κ and width of the distribution is given by w. In Fig. 13.10 we compare the Maxwell-Jüttner to the kappa-eDF for different values of the kappa parameter. Notice, that the kappa-eDF approaches the Maxwell-Jüttner eDF for large values of κ.

The kappa parameter, κ , is related to the exponent s of a power-law eDF $dn_{\rm e}/d\gamma_{\rm e} \propto \gamma_{\rm e}^{-s}$ by $s = \kappa - 1$.We follow the works of [6, 7, 19] and connect the width of the kappa eDF, w, to the thermal energy of the electrons and a fraction ε of the magnetic energy. Furthermore, we assume that the fraction of the magnetic

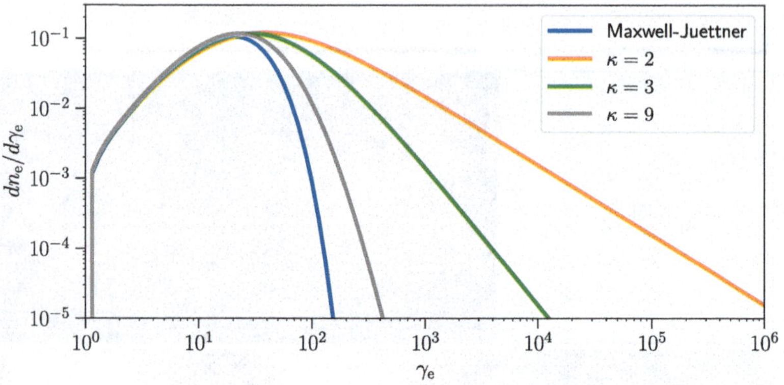

Fig. 13.10 Comparison between Maxwell-Jüttner and kappa eDF

energy is contributing to the width of the kappa eDF after a distance $r_{\rm inj}$. Finally, w is given by:

$$w := \frac{\kappa - 3}{\kappa}\Theta_{\rm e} + \frac{\varepsilon}{2}\left[1 + \tanh(r - r_{\rm inj})\right]\frac{\kappa - 3}{6\kappa}\frac{m_{\rm p}}{m_{\rm e}}\sigma \tag{13.22}$$

where $\sigma = b^2/\rho$ is the magnetisation. For the value of the κ parameter we use a parametrisation based on particle-in-cell simulations of magnetic reconnection using the work of [2]:

$$\kappa := 2.8 + 0.7\sigma^{-1/2} + 3.7\sigma^{-0.19}\tanh\left(23.4\sigma^{0.26}\beta\right) \tag{13.23}$$

Given the width, w, of the kappa distribution (see Eq. 13.22) together with the validity of the numerical approximations for the emission, j_ν, and absorption coefficients, α_ν, [31] and Appendix 1 we restrict the values for κ to the interval $3 < \kappa \leq 8$ and switch to a thermal eDF otherwise. In Fig. 13.11 we display the distribution of w and κ for a GRMHD snapshot at $t = 29440\,M$ using a $R_{\rm low,high} = 1, 160$ for the calculation of the electron temperature Θ_e (see right column in Fig. 13.9). Notice, with this parameterisation of the κ parameter (Eq. 13.23) the non-thermal electrons are mainly located in the jet region (small κ values) and not in the disk (large κ values). This behaviour is best seen in the middle column of Fig. 13.11.

As mentioned in Sect 13.2.1 we include a floor model in the jet funnel. In order not to include this region in our GRRT images we introduce a cut-off on the magnetisation, $\sigma_{\rm cut}$. This magnetisation cut-off, $\sigma_{\rm cut} > 1$ is set during the radiative transfer calculation and is a matter of debate in the literature (see, e.g., [5]).

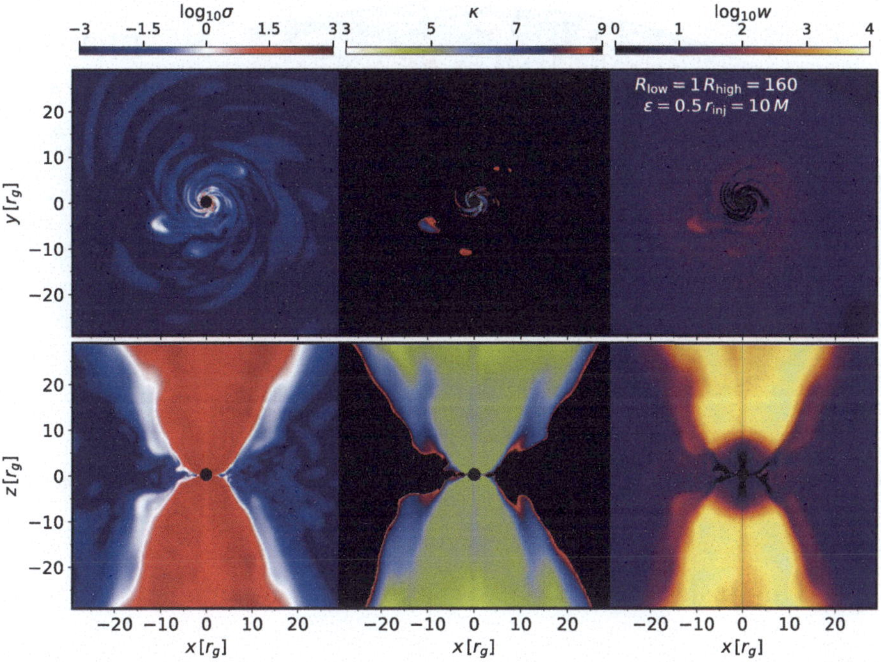

Fig. 13.11 Visualisation of the κ parameter and the width of the kappa eDF for a GRMHD model with black hole spin $a^{\star} = 0.94$ at time $t = 29440\,M$. The top panels show the equatorial plane and the bottom ones the meridional plane. The first column displays the magnetisation σ, middle column the κ parameter and right column the width of the kappa eDF, w, for $R_{\rm high} = 1$, $R_{\rm high} = 160$, $\varepsilon = 0.5$ and injection radius, $r_{\rm inj} = 10\,M$

To summarise: The free parameters of our GRRT calculations used in this chapter are: the mass unit $\mathcal{M}$ (mass accretion rate) as well as the black hole mass, $m_{\rm bh}$, and distance, $d_{\rm bh}$, of the source of interest. In addition, the calculation of the electron temperature from the ideal GRMHD simulations via the $R - \beta$ parametrisation introduces two free parameters, namely $R_{\rm low,high}$. Furthermore, the choice of the eDF introduces additional free parameters, e.g., ε and $r_{\rm inj}$ in case of kappa eDF. Lastly, the exclusion of the floor-regions in the jet funnel is implemented via $\sigma_{\rm cut}$ parameter.

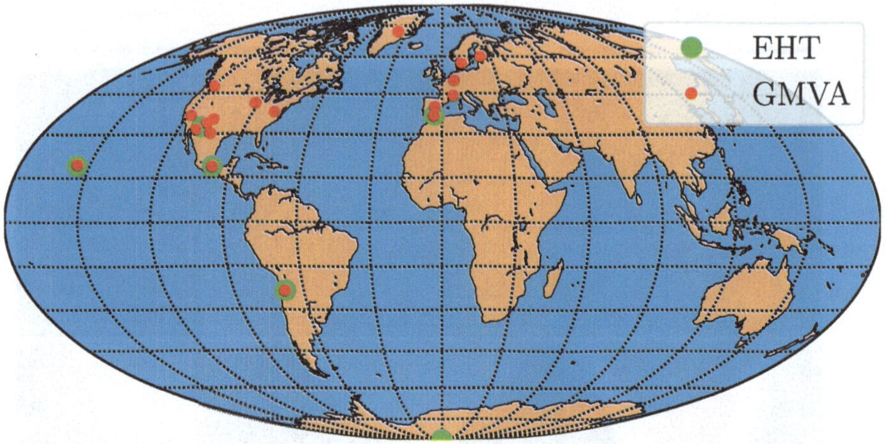

Fig. 13.12 Location of the individual radio telescopes of the EHT (2017 configuration) and the GMVA

13.3 The Basics of VLBI Observations and Imaging

The EHT and the GMVA consist of several radio telescopes scattered across the globe and in Fig. 13.12 we show the locations of the individual telescopes. In Appendix 2 we provide the locations of the telescopes in earth-centred coordinates.

These individual telescopes are connected via a technique called Very Long Baseline interferometry (VLBI) (see, e.g., [39]). During an observational campaign each telescope records the electric field across its aperture together with a precise clock signal and the frequency. For each pair of telescopes i, j we can compute from the measurements the cross correlation function with is referred as complex visibilities V_{ij}. This complex visibility is related to the brightness distribution of the source in the sky, $I(x, y)$, via the van-Cittert-Zernike theorem (for a detailed derivation see [39]):

$$V_{ij} = \int\int I(x, y)e^{-2\pi\rangle(ux+vy)}dxdy, \tag{13.24}$$

where x, y are the location in the sky (angular coordinates). The coordinates u, v correspond to the projection of the baseline (i.e. distance) between the telescopes i and j in units of the observing wavelength. The measured visibilities are affected by several corrupting factors. These factors include, e.g., thermal noise from the telescope receives and atmospheric turbulence. In this introductory text we focus only on the thermal receiver noise and refer to [4] for details. The thermal noise can be modelled as an additive complex Gaussian random variable with a zero mean and standard deviation σ_{ij}:

$$\sigma_{ij} = \frac{1}{\eta}\sqrt{\frac{\mathrm{SEFD_i SEFD_j}}{2\Delta\nu\Delta t}}, \tag{13.25}$$

where $\Delta\nu$ is the bandwidth of the observation and Δt is the integration time. In the equation above SEFD_i corresponds to the system equivalent flux density of the i-th telescope and is given by:

$$\mathrm{SEFD_i} = \frac{2k_b T_{\mathrm{sys}}}{A_{\mathrm{eff}}}, \tag{13.26}$$

with system temperature T_{sys} and effective area A_{eff}. In Appendix 2 we list the SEFDs for telescopes of the EHT and GMVA. Given that the visibilities are complex numbers we can compute a visibility amplitude, VA, and a visibility phase, VP. The number of baselines depends on the number of telescopes, N_t, within the array and is given by:

$$N_{bl} = N_t\,(N_t - 1)\,/2 \tag{13.27}$$

In order to densely sample the Fourier plane (u-v plane) during a short time instance a large number of telescopes at various locations would be required. However, the EHT and the GMVA consist only of a small number of telescopes. Nonetheless it is still possible to sample the Fourier plane using the effect of Earth synthesis i.e. the rotation of the Earth during the observation. While the Earth is rotating the projected distances between the telescopes are changing and the Fourier plane is traced-out by elliptical arcs leading to new measurements i.e. visibilities.

In Fig. 13.13 we illustrate the effect of Earth synthesis for the EHT observing M87 using three selected observing times. The first column shows the EHT telescopes and we colour-coded the different baselines. In the second column we plot the u-v tracks of the different baselines (same colour-code as in the first column) on top of the 2D Fourier-transform of a M 87 model (Eq. 13.24). The right column shows the visibility amplitude, VA, sampled by the u-v tracks. Notice that despite the Earth synthesis, the u-v plane is still only sparsely sampled which can be seen by the gaps of missing data in the right column of Fig. 13.13.

As can be seen from Eq. 13.24 and Fig. 13.13 the data product of an VLBI array is not an image. If we want an image we need to reconstruct it from the visibilities via the inverse Fourier transformation. However, the u-v plane is only sparsely sampled which makes the image reconstructed a highly under-determined procedure, i.e., we have more degrees of freedom than data points. There exist several algorithms to reconstruct an image from visibility (see, e.g., [39]). In the last years so-called maximum-entropy method (MEM) have been widely used in the literature to reconstruct VLBI observations [14]. The `ehtim` package[5] provides an excellent and well documented framework for the generation of synthetic VLBI data and for the image reconstruction from VLBI observations [4]. Throughout this Chapter we will use `ehtim` for generating synthetic data and for reconstructing the images. For the details of the different implementations of MEMs we refer to [4] and provide below only some basics.

[5] https://achael.github.io/eht-imaging/.

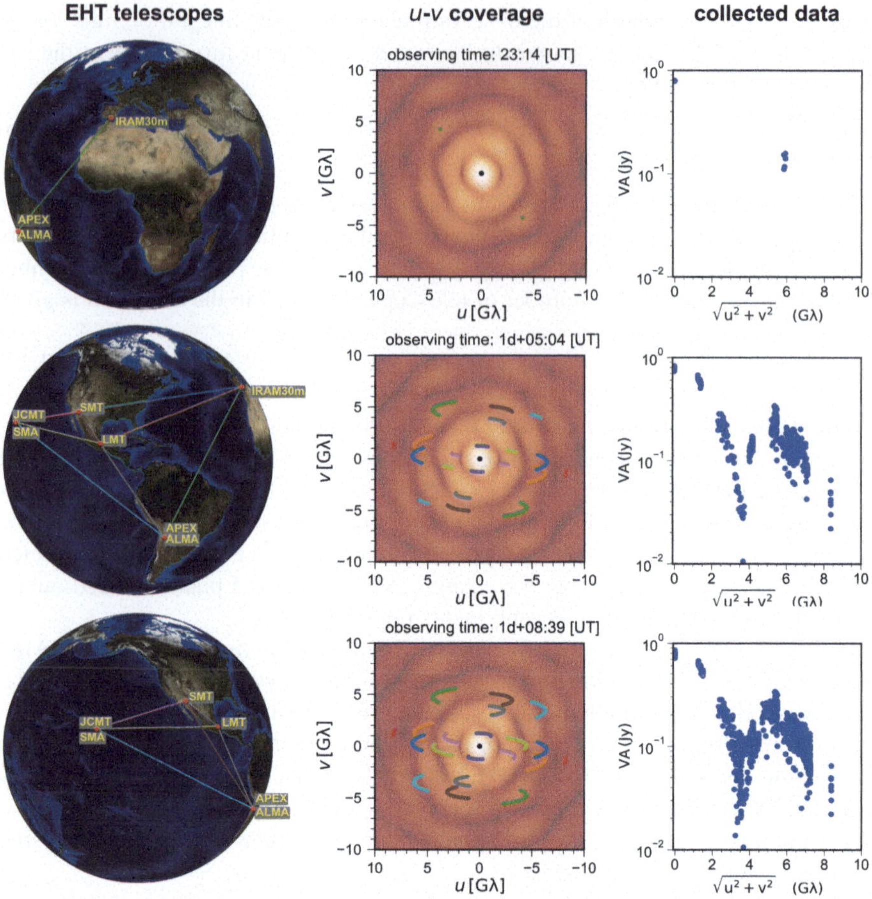

Fig. 13.13 Illustration of Earth synthesis for an EHT observation of M 87. The left column shows the baselines of the EHT telescopes. The middle panel displays the sampling of the u-v plane by the different baselines on top of 2D Fourier transform of a M87 model. The right column displays the collected visibility amplitude

During the MEM image reconstruction we want to obtain an image,**I**, that minimises an objective function J(**I**). The objective function is the sum of data terms and image properties (i.e. entropy) and can be written as:

$$J(\mathbf{I}) = \sum_{\text{data terms}} \alpha_D \chi^2_D (\mathbf{I}, \mathbf{d}) - \sum_{\text{regularizers}} \beta_R S_R (\mathbf{I}) \,, \tag{13.28}$$

where χ^2 evaluates the goodness of the fit to the data term **d** and S_R is a regularising function, e.g. the image entropy. α_D and β_R are hyper-parameters which need to be set/adjusted during the initialisation of the imaging process.

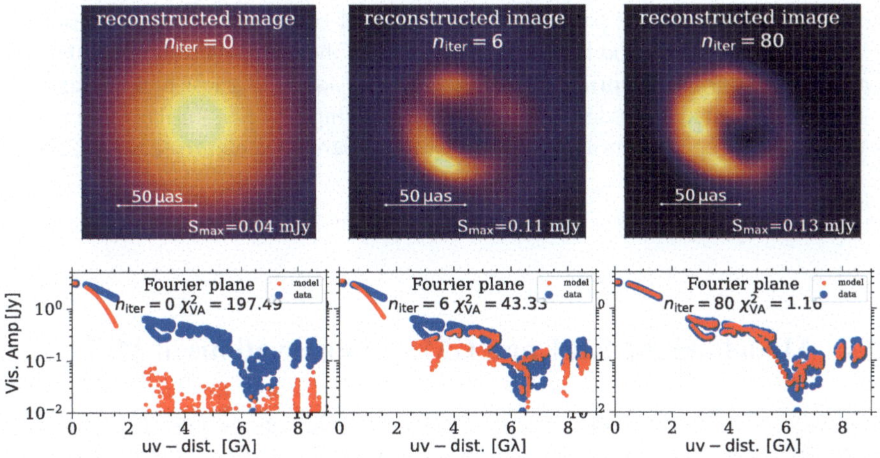

Fig. 13.14 Image reconstruction using MEM. The top panels show the reconstruction in the image plane for different numbers of iterations while the bottom panels display the image-reconstruction in the Fourier plane for different iteration steps

In Fig. 13.14 we show an example for the image reconstruction using MEM. In this case we use a simple image entropy for the regularizer term and complex visibilities for the data term. Furthermore, we set the both hyper-parameters to one. The MEM process is initialised on the $M \times M$ image grid with a large circular gaussian as initial guess. During the optimisation process the individual pixels are adjusted in way that the objective function is minimised. The top panels show the reconstructed images after 0, 6, and 80 iterations and the bottom panels display the image reconstruction in the Fourier plane or more precise on the visibility amplitude. Notice the improved matching between the data points (blue) and the model (red) with increased number of iterations together with the developing of the image structure.

An example script for the image reconstruction can be found at *examples/example.py* within the `ehtim` repository[6]

To summarise: In contrast to optical observations VLBI observations provide the Fourier-transformed brightness distribution of a source in the sky, the so-called complex visibilities in the Fourier plane and not an image. The sampling of the Fourier plane depends on the number of telescopes and their distances i.e. baselines. For the generation of images we need to fill the missing gaps in the sampled Fourier plane while fitting the observed visibilities. This can be accomplished by means of maximum entropy methods (MEM). During an iterative process of Fourier- and inverse Fourier transformation an image is

[6] https://github.com/achael/eht-imaging.

reconstructed. The free parameters for the MEM-based image reconstruction are: the field of view and the number of pixels in the image as well as the data terms used during the fitting process. In addition an initial guess of the image, i.e. a prior image is needed to initialise the imaging process. Furthermore the choice of the regularisers and the values of the hyper-parameters (see Eq. 13.28) need to be provided.

13.4 Modelling the EHT and GMVA Observations of M87

After we introduced the basic concepts of GRMHD simulations, GRRT calculations and VLBI observations we can now turn towards the modelling of M 87 at 86 GHz and 230 GHz. For reasons of clarity we restrict ourselves to one ideal GRMHD simulation of a black hole spin $a^\star = 0.94$ and an initial torus seeded with a MAD vector potential and an adiabatic index of $\hat{\gamma} = 4/3$ (see Sect. 13.2.1).

13.4.1 Numerical Modelling

As mentioned in Sect. 13.2.2 there are several free parameters for GRRT calculations and we list the common ones used Table 13.2. In addition to the parameters listed in Table 13.2 we explore the impact of a pure thermal eDF and kappa eDF with variable κ parameter (see Eq. 13.23). The first step in our modelling is to scale the GRMHD to the properties of M 87 by adjusting the mass unit, $\mathcal{M}$. For this procedure we assume a total flux density of 1 Jy at an observing frequency of 230 GHz [1, 10] and iterate the mass unit, $\mathcal{M}$ until the target flux is obtained (see, e.g., [19]). In Table 13.3 we provide the obtained the mass accretion rates for the different models under investigation.

Table 13.2 Common parameters used during the GRRT

$a^\star$[1]	$m_{\rm bh}[10^9 M_\odot]$	$d_{\rm bh}$ [Mpc]	θ [°]	FOV [μas]	npix[1]	$\nu_{\rm min,max}$ [Hz]	N_ν [1]
0.94	6.5	16.8	160	1000	4096	10^9, 10^{16}	50

Table 13.3 Parameters for the eDFs used during the GRRT

$R_{\rm low,high}$[1]	$\sigma_{\rm cut}$[1]	$r_{\rm inj}[M]$	ε[1]	$\dot{M}10^{-5} M_\odot/yr$
1,160	3	Thermal only model		5.2
1,160	3	10	0, 0.2, 0.3, 0.4, 0.5	5.3, 4.5, 4.4, 4.3, 4.2

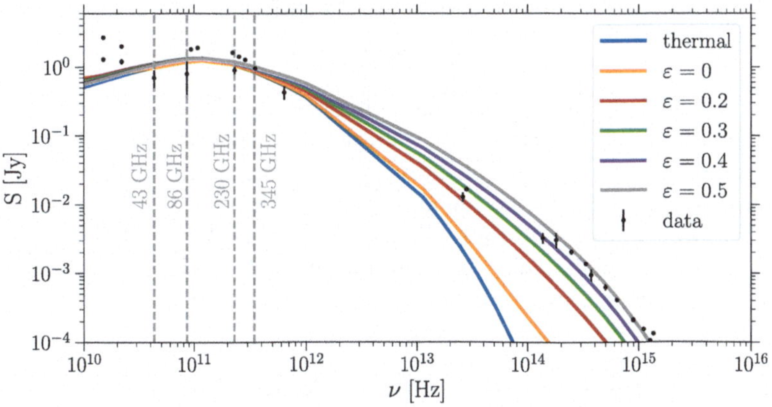

Fig. 13.15 Broad band spectrum of M 87 for different models. The data is taken from [19]

Notice, that this procedure needs to be repeated every time the initial parameters of the GRRT model are changed.

To judge the impact of the different parameters on our M 87 models, it is useful to investigate first the changes in the broad band spectrum between $10^9 < \nu < 10^{16}$ Hz. In Fig. 13.15 we show the broad band spectrum of M 87 together with different models. First we focus on the low frequency part of the spectrum $\nu < 345$ GHz. All of our models show a flat spectrum with an spectral peak around $S_{\rm peak} \sim 1$ Jy at a frequency $\nu_{\rm peak} \sim 100$ GHz. Given that thermal and kappa eDF show a common shape at low electron Lorentz factors, $\gamma_{\rm e}$ (see Fig. 13.10), we would expect such a behaviour. As such it is difficult to quantify the models under investigation based on this regime in the spectrum. However, the eDFs differ strongly for high electron Lorentz factors and therefore we anticipate differences a high frequencies (see Fig. 13.10). Indeed, the high frequency ($\nu > 10^{13}$ Hz) observations of M 87 are better matched by our kappa models. Notice also the difference in the slope of the different spectra: The thermal model displays an exponential decay where as the kappa models show a power-law behaviour. This can be explained by the functional form of the emission coefficients (see Appendix 1). At frequencies higher than the peak frequency, $\nu_{\rm peak}$, the flux density can be approximated by $S_\nu \propto j_\nu$. Given that the thermal emissivities at high frequencies can be approximated by $j_{\nu,\rm thermal} \propto \exp^{-\nu^{1/3}}$ where as the kappa emissivities can be written as $j_{\nu,\rm kappa} \propto \nu^{-(\kappa-2)/2}$ which explains the observed spectral behaviour.

After the comparison with the broad band spectrum we turn now to the comparison in the image plane using the best fitting model with $\varepsilon = 0.5$ together with the thermal model. Therefore, we generate from the GRRT calculations images at different frequencies and we compute the spectral index, α, between them. The spectral index, α, is defined as:

$$\alpha_{\nu_1,\nu_2} = \frac{\log\left[S(\nu_1)/S(\nu_2)\right]}{\log\left(\nu_1/\nu_2\right)}, \tag{13.29}$$

where $S(\nu_1)$ and $S(\nu_2)$ are flux densities at frequencies ν_1 and ν_2.

Regions with $\alpha < 0$ are referred as optically thin whereas regions with $\alpha > 0$ a labeled as optically thick. A detailed discussion on the spectral indices on horizon scale for different eDFs can be found in [36]. In Fig. 13.16 we show the images for the Maxwell-Jüttner eDF (left column) and for the kappa-eDF with $\varepsilon = 0.5$ (right column). For both eDFs we obtain images with a bright inner region ($r < 100\,\mu$as) and a faint jet with wide opening angle. Therefore, the two main features

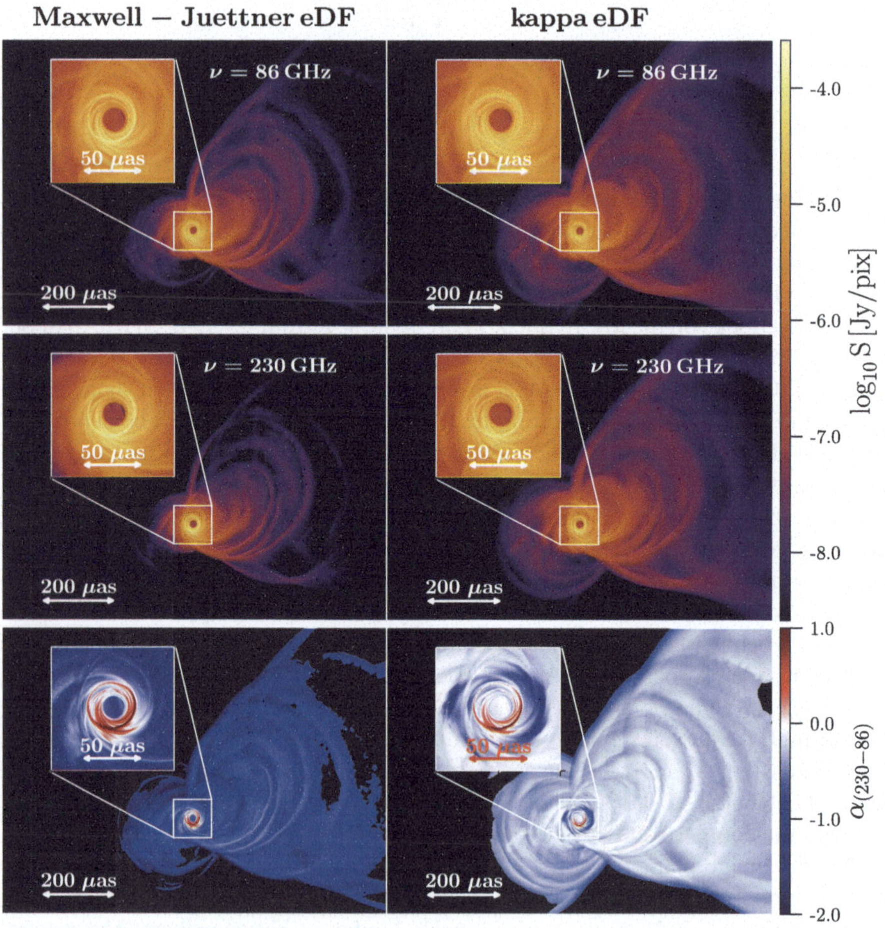

Fig. 13.16 High resolution GRRT images for M 87 using a Maxwell-Jüttner (left) and a kappa eDF (right). The panels show from top to bottom the images at 86 GHz and 230 GHz as well as the spectral index between the aforementioned frequencies. In each panel a zoom into the horizon structure is provided

of the observations are in principle recovered by the numerical modelling. However, there are some details in which both models differ: kappa eDF models typically generate brighter and longer jets as compared to their thermal counter parts (best seen at $\nu = 230$ GHz). In addition, the kappa eDF models produce flatter spectral indices than thermal eDFs.This behaviour can be explained in the same way as the spectral slopes in the broad band spectrum. Currently, there are no spectral index maps between 86 GHz and 230 GHz using GMVA and the EHT observations. However, at lower frequencies there are spectral index maps computed from 43 GHz and 86 GHz observations which show a $-1 < \alpha < 0$ for the innermost regions [11] which are in the range of our models.

13.4.2 Image Plane Comparison

A more quantitative comparison between theoretical models and observations in the image plan can be performed by applying image metrics [20, 27]. Before we can compute the metrics we need to re-grid the GRRT images to same number of pixels and to the same field-of-view as the GMVA and EHT observation. In addition to mimic the finite resolution of the observations we convolve the GRRT images with the observing beam of the EHT (20 μas) and the GMVA (37 μas). In Fig. 13.17 we

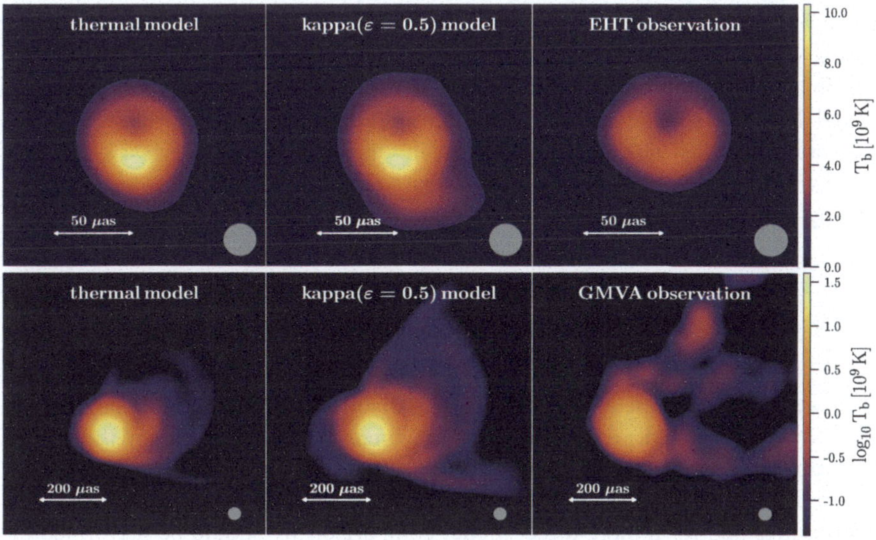

Fig. 13.17 Convolved images for M87. Top row: Model images at 230 GHz (left and middle) convolved with a beam of 20 μas and the EHT image of M87 [12] . Bottom row: Model images at 86 GHz (left and middle) convolved with a beam of 37 μas and the GMVA image of M87 [24]. The size of the beam is indicated by the grey circle in bottom right corner of each panel

Table 13.4 Results for the image plane comparison. Best matched models in are in bold-face

Model	NRMSE (EHT)	ISSM (EHT)	NRMSE (GMVA)	ISSM (GMVA)
	[1]	[1]	[1]	[1]
Thermal	0.773	0.895	0.858	0.561
kappa ($\varepsilon = 0$)	**0.768**	**0.901**	0.866	0.556
kappa ($\varepsilon = 0.2$)	0.977	0.787	0.811	0.578
kappa ($\varepsilon = 0.3$)	0.985	0.753	0.803	0.580
kappa ($\varepsilon = 0.4$)	0.988	0.729	0.797	0.581
kappa ($\varepsilon = 0.5$)	0.991	0.712	**0.795**	**0.581**

display the convolved GRRT images for a thermal and a kappa model ($\varepsilon = 0.5$) next to the observations of the EHT [12] and GMVA [24] using a dynamic range of 10 (EHT) and 10^3 (GMVA).

In the next step we compute the normalised root mean square error (NRMSE) and image structural similarity measure (ISSM) for details see [41] between the convolved model images and the EHT and GMVA images of M 87. The NRMSE tends towards zero for perfectly matched images while the ISSM approaches one. In Table 13.4 we list the results of these metrics.

From the images in Fig. 13.17 and the obtained image metrics in Table 13.4 we can conclude that the best matching model for the comparison with the EHT image of M 87 is a kappa model with $\varepsilon = 0$. This result is not surprising since the kappa models with larger ε show a bright southern flux arc which is not visible in the EHT image. However, this missing arc in the EHT observations could be due to the limited dynamic range of the EHT observations. For the comparison with the GMVA observations of M87 the best model according to the image metrics is a kappa eDF model with $\varepsilon = 0.5$. This model is characterised by long and bright jet with large jet opening angle (see middle panel in the bottom row of Fig. 13.17.

13.4.3 Fourier Plane Comparison

So far we compared convolved GRRT images to reconstructed images from observations. Notice, that for the reconstruction of the images from the EHT and the GMVA observations we need to set hyper-parameter (see Sect. 13.3) which influence to some extended the structure seen in the final images. To avoid this additional source of uncertainties and for a more direct model comparison we should be performed the comparison in the Fourier plane (see Sect. 13.3). For the modelling in the Fourier plane several methods exist (see [16] for details) and we use the `GENA` code [16, 20] for the fitting of our models to the EHT observations. In this Chapter we restrict the modelling in the Fourier plane to the EHT observations of M 87. During the fitting

Table 13.5 Parameters for the best fitting model obtained from the fitting in the Fourier plane

Model	$\phi_{\rm rot}$ [deg]	$A_{\rm scale}$ [1]	$m^a_{\rm bh}$ [$10^9\,M_\odot$]
Thermal	−9	0.33	6.54
kappa ($\varepsilon = 0$)	−9	0.33	6.59
kappa ($\varepsilon = 0.2$)	−13	0.44	6.67
kappa ($\varepsilon = 0.3$)	−12	0.48	6.70
kappa ($\varepsilon = 0.4$)	−12	0.50	6.71
kappa ($\varepsilon = 0.5$)	−11	0.52	6.72

Assuming a distance of 16.9 Mpc

of the EHT data we scale, rotate and stretch the model images in the Fourier plane and compute the χ^2 between the data terms, here the visibility amplitude and the closure phase (i.e. the phase across a closed triangle of baselines). The results of the fitting can be found in Table 13.5. In Fig. 13.18 we show as an example the fitting of a model with a kappa eDF using $\varepsilon = 0.5$ to the EHT observation. The top panels report the scaled, rotated and stretched model images at 230 GHz using the best fitting parameters (reported in the bottom panel). The image on the left is plotted using the resolution of the ray-tracing while the one on the right is displayed with the resolution of the EHT in 2017 i.e., convolved by a gaussianwith a FWHM of 20 μas. The bottom panel displays the matching between the visibility amplitude of the EHT observation (blue) and the model (green) together with the best fitting parameters for the black hole mass, flux scaling and rotation in the plane of the sky.

The results of the fitting show that on horizon scale we cannot distinguish between our models using the EHT 2017 observations of M 87. However, we obtain a good agreement between our models and the observations. We found an average black hole mass of $\langle m_{\rm bh}\rangle = 6.65 \pm 0.07$ which matches nicely the value reported by the EHT [16]. The rotation angle is least constrained parameter in our analysis as can be seen in top left panel of Fig. 13.18. Given that the flux scaling parameter, $A_{\rm scale}$ is always less than one indicates that our models are to bright on horizon scales as compared to the EHT observations. In addition, there is an interesting trend in the flux scaling: The value increases with larger values of ε. Remember, that ε sets the fraction of magnetic energy contributing to the acceleration of non-thermal particles (see Eq. 13.22). With increasing ε more emission is pushed in the extend jet while the horizon scale structure is getting dimmer. During the comparison with the EHT observations we restrict the fitting to $r < 80\mu as$, i.e., the horizon structure in M 87. Therefore, the thermal model exhibits a higher flux within this region than the non-thermal ones. As a consequence the flux scaling is stronger for the thermal model than for the non-thermal ones.

The posterior distribution for the fitted parameters, $\phi_{\rm rot}$, $A_{\rm scale}$ and m_{bh} (assuming a distance of 16.9 Mpc) for a thermal (blue) and non-thermal model with $\varepsilon = 0.5$ (orange) are presented in Fig. 13.19. We initialised the fitting with flat priors in order

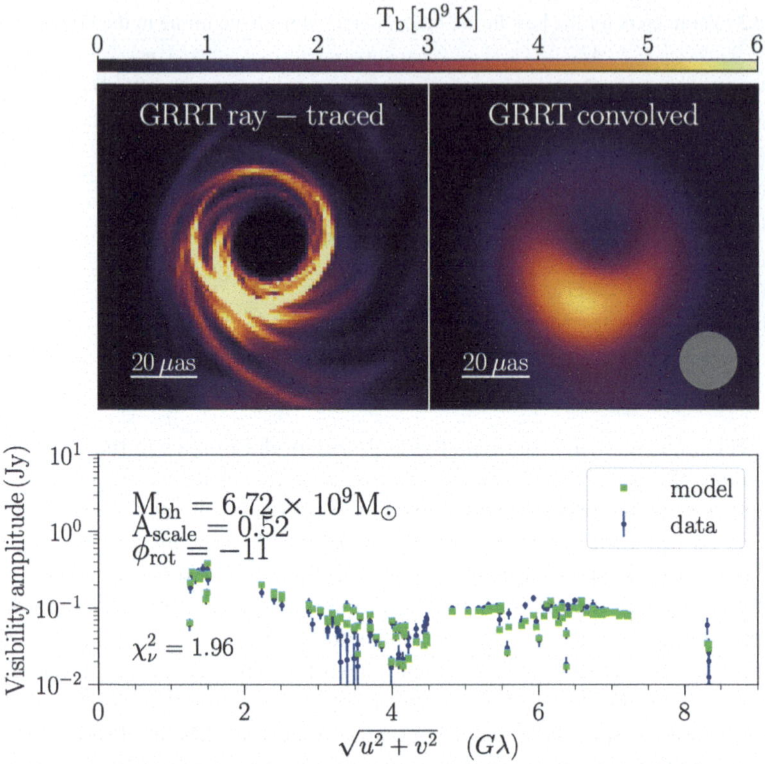

Fig. 13.18 Illustration of the model fitting in the Fourier plane for a kappa eDF model with $\varepsilon = 0.5$. Top panels display the rotated, scaled and stretched images at infinite (left) and EHT resolution (right). Bottom panel: Matching between the visibility amplitude of the model (green) and EHT observation (blue) together with best fitting values for the stretching, flux scaling and rotation of the model

not to bias the numerical method and we removed the first 3000 iterations from the plotting.[7] The fitting method converged to small regions in the parameter space which display a gaussian shape. The distributions exhibit small standard deviations except for the rotation angle which we could not constrain well enough. Notice, that we only used a single GRMHD snapshot for our analysis and in order to provide a statistically solid analysis we would need to repeat the above described procedure for several snapshots (see [16] for more details).

Finally, we apply the found scaling parameters (see Table 13.5) to our models and generate synthetic data using the EHT configuration from 2017 [13] and the GMVA

[7] Typically the first iterations explore the parameter space globally before settling into an interesting region of the parameter space.

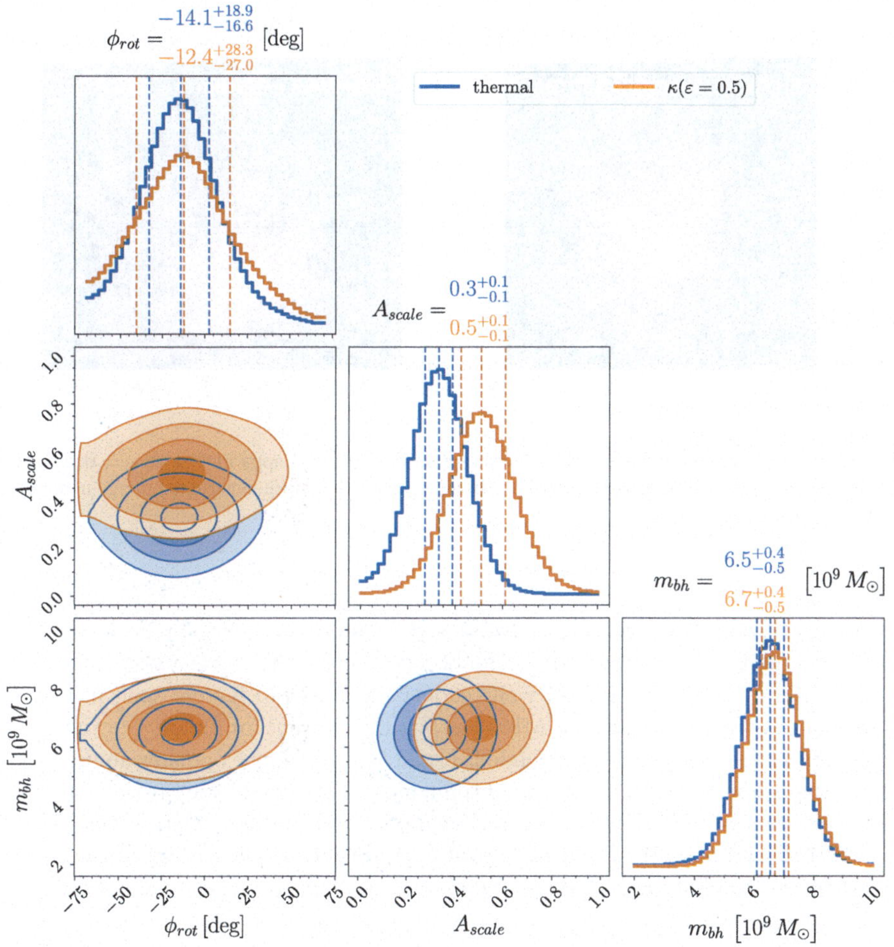

Fig. 13.19 Posterior distribution of the fitted parameters, $\phi_{\rm rot}$, $A_{\rm scale}$ and $m_{\rm bh}$ (assuming a fixed distance of 16.9 Mpc) for a thermal (blue) and kappa (orange) model. The mean values of the parameters together with their 1σ derivations are indicated by the dashed lines and the exact values are provided on top of each panel

array.[8] As mentioned in Sect. 13.3 we use the `ehtim`[9] package [5] to generate the synthetic data and to reconstruct the images. The reconstructed images are presented in Fig. 13.20 and should be compared to the observations shown in Fig. 13.1.

[8] https://www3.mpifr-bonn.mpg.de/div/vlbi/globalmm/.

[9] https://achael.github.io/eht-imaging/.

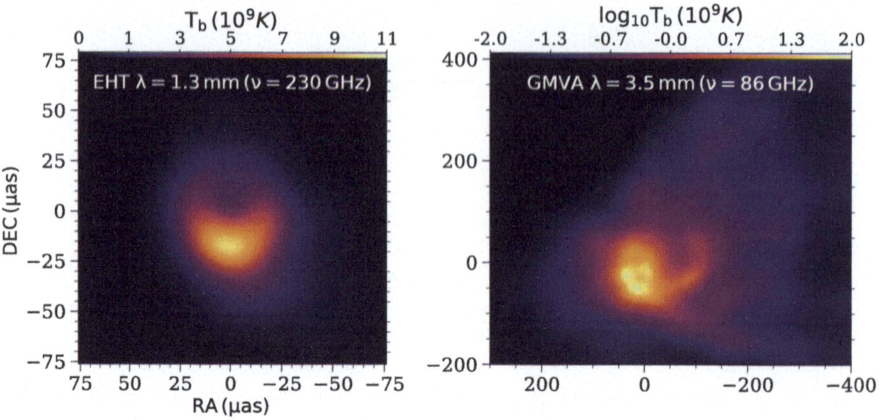

Fig. 13.20 Reconstructed M87 images from our model using a kappa eDF with $\varepsilon = 0.5$. Left: reconstructed image using the EHT 2017 as observing array. Right: reconstructed image using the GMVA as observing array

As can be seen from the images above and from the modelling in the Fourier plane our model describes the EHT observations of M87 very well. However, the used models cannot well recover the ring-like feature seen in the recent GMVA observation of M87. Despite this current limitation, our models are able to produce an extended, bright jet with large opening angle as well as some indications of edge-brightening. This result demonstrates the capabilities of GRMHD simulations to model and interpret high-resolution VLBI observations of relativistic jets and accreting black holes. More advanced GRMHD simulations including heating and cooling processes together with future improvements in the electron thermodynamics will allow us to model relativistic jets in great details and will improve our understanding on the formation and acceleration of relativistic jets.

13.5 Summary

In this chapter we used GRMHD simulations and GRRT calculations to model the M87 observations of the EHT and GMVA. We presented the basic setup of a MAD GRMHD simulation together with the required initial parameters. Furthermore, we introduced some concepts for checking the quality of the GRMHD simulations such as MRI quality factors and radial profiles (Sect. 13.2.1). The second step towards the comparison of the numerical model with observations dealt with the generation of emission maps and spectra. During the GRRT we computed geodesics and solved the radiative transfer along them. We explained in detail how to scale the scale-free GRMHD simulation to observations and provided insights into the different electron distribution functions (eDF). In addition, we illustrated the coupling between GRMHD and GRRT for different eDFs and explained the computation of the electron temperature (Sect. 13.2.2). In Sect. 13.3 we provided a basic introduction to the VLBI technique and demonstrated in an example the generation of an image from observed visibilities via Maximum Entropy Methods. After the introduction to GRMHD, GRRT and VLBI observations we applied these concepts to the modelling of the EHT and GMVA observations of M87 (Sect. 13.4). We computed images and spectra for one thermal and several non-thermal eDFs and showed their common as well as their distinguishing features in images and in the broad-band spectra. Finally, we compared the models to the observations in the image plane by using image metrics and fitted the EHT observations in the Fourier plane deriving estimates, e.g., for the black hole mass.

In summary, we provided in this Chapter the numerical methods and concepts for the direct comparison of GRMHD simulations to VLBI observations of accreting black holes and relativistic jets.

Acknowledgements CMF is supported by the DFG research unit FOR195 "relativistic jets in AGNs" and "jet physics on horizon scale and beyond" (Grant Nr. 443220636). The simulations and calculations were performed on `iboga` at the Institute for theoretical physics at the University of Frankfurt and on `MISTRAL` at the Institute for theoretical physics and astrophysics at the University of Wuerzburg. CMF would like thank Luciano Rezzolla, Yosuke Mizuno, Ziri Younsi, Oliver Porth, Mariafelica DeLaurentis, Hector Olivares, Alejandro Cruz-Osorio, Karl Mannheim, Matthias Kadler, Eduardo Ros, Anton Zensus and Manel Perucho for fruitful discussion and continuous collaboration on GRMHD simulations and GRRT calculations over the past years.

Appendix 1

For reasons of completeness we provide here the Stokes I emission and absorption coefficients for a Maxwell-Jüttner and kappa eDF following [31]. The synchrotron emission, j_ν, and absorption coefficients, α_ν, independent of the eDF can be written as:

$$j_\nu = \frac{n_e e^2 \nu_c}{c} J_S\left(\frac{\nu}{\nu_c}, \theta\right) \tag{13.30}$$

$$\alpha_\nu = \frac{n_e e^2}{\nu m_e c} A_S\left(\frac{\nu}{\nu_c}, \theta\right) \tag{13.31}$$

with electron charge e, speed of light c, electron cyclotron frequency $\nu_c = eB/(2\pi m_e c)$ and pitch angle θ. In the equations above J_S and A_S are functions wich are dependent on the eDF and are typically referred as dimensionless emissivity and absorptivity.

Maxwell-Jüttner (Thermal) eDF

For a Maxwell-Jüttner (thermal) eDF the dimensionless emissivity can be expressed by:

$$J_S = e^{-X^{1/3}} \frac{\sqrt{2}\pi}{27} \sin(\theta)(X^{1/2} + 2^{11/12} X^{1/6})^2, \tag{13.32}$$

where $X = \nu/\nu_s$ and $\nu_s = 2/9\nu_c\Theta_e^2 \sin\theta$. The dimensionless absorptivity can be computed using Kirchoffs law

$$A_S = \frac{J_S}{B_\nu} = J_S \frac{m_e c^2 \nu_c}{2h\nu^2} \left(e^{h\nu/(k_b T_e)} - 1\right) \tag{13.33}$$

where h is the Plancks constant and k_b is the Boltzmann constant.

Kappa (Non-thermal) eDF

In order to provide approximations for the emissivities and absorptivities for the kappa eDF their functional form is divided into a low and high-frequency fit which are bridged by an interpolation formulae (for more details see [31]). The equations below are valid for $3 \leq \kappa \leq 7$. Similar to the thermal eDF we define a characteristic frequency $\nu_\kappa = \nu_c(w\kappa)^2 \sin\theta$, and $X_\kappa \equiv \nu/\nu_\kappa$ where w is the width of the kappa eDF (see Eq. 13.22). The low and high frequency limit of the kappa eDF can be approximated by:

$$J_{S,lo} = X_\kappa^{1/3} \sin(\theta) \frac{4\pi\Gamma(\kappa - 4/3)}{3^{7/3}\Gamma(\kappa - 2)} \tag{13.34}$$

$$J_{S,hi} = X_\kappa^{-(\kappa-2)/2} \sin(\theta)\, 3^{(\kappa-1)/2} \frac{(\kappa - 2)(\kappa - 1)}{4} \Gamma(\frac{\kappa}{4} - \frac{1}{3})\Gamma(\frac{\kappa}{4} + \frac{4}{3}), \tag{13.35}$$

where $\Gamma(x)$ is the incomplete gamma function. Finally, both approximations can be combined via the bridging function below:

$$J_S = \left(J_{S,lo}^{-3\kappa^{-3/2}} + J_{S,hi}^{-3\kappa^{-3/2}}\right)^{-1/3\kappa^{-3/2}}. \tag{13.36}$$

Similar to the emissivities the absorptivities are split into a low and high frequency fit

$$A_{S,lo} = X_\kappa^{-2/3} 3^{1/6} \frac{10}{41} \frac{(2\pi)}{(w\kappa)^{10/3-\kappa}} \frac{(\kappa-2)(\kappa-1)\kappa}{3\kappa-1} \Gamma(\frac{5}{3}) \\ {}_2F_1(\kappa - \frac{1}{3}, \kappa+1, \kappa+\frac{2}{3}, -\kappa w) \tag{13.37}$$

$$A_{S,hi} = X_\kappa^{-(1+\kappa)/2} \frac{\pi^{3/2}}{3} \frac{(\kappa-2)(\kappa-1)\kappa}{(w\kappa)^3} \left(\frac{2\Gamma(2+\kappa/2)}{2+\kappa} - 1\right) \\ \left(\left(\frac{3}{\kappa}\right)^{19/4} + \frac{3}{5}\right) \tag{13.38}$$

and are matched by the following bridging function

$$A_S = \left(A_{S,lo}^{-(-\frac{7}{4}+\frac{8}{5}\kappa)^{-43/50}} + A_{S,hi}^{-(-\frac{7}{4}+\frac{8}{5}\kappa)^{-43/50}}\right)^{-1/(-\frac{7}{4}+\frac{8}{5}\kappa)^{-43/50}}. \tag{13.39}$$

Appendix 2

Here we provide the array configuration files for the EHT in its 2017 configuration and the GVMA (Tables 13.6 and 13.7).

Table 13.6 Antenna locations and SEFDs for the GMVA

ID	x [m]	y [m]	z [m]	SEFD [Jy]
FD	−1324008.669	−5332181.649	3231962.333	4100
LA	−1449752.366	−4975298.112	3709123.806	4100
PT	−1640953.555	−5014817.328	3575410.707	4100
KP	−1995678.595	−5037317.661	3357328.236	4100
OV	−2409149.405	−4478573.152	3838617.400	4100
BR	−2112064.710	−3705356.001	4726813.642	4100
MK	−5464074.943	−2495249.078	2148296.797	4100
NL	−130871.917	−4762316.904	4226850.876	4100
EF	4033903.994	486927.367	4900346.030	1000
GBT	882600.586	−4924875.801	3943723.181	133
NOEMA	4523973.676	468138.972	4460327.201	163
PV	5088967.711	−301691.460	3825015.742	654
YB	4848714.404	−261640.038	4123130.688	990
ON	3370947.110	711462.434	5349630.780	2650
MET	2892550.449	1311747.770	5512673.555	15000
LMT	−768716.239	−5988504.353	2063361.540	513
ALMA	2225371.536	−5440399.583	−2480671.175	69
GLT	1500692.0	−1191735.0	6066409.0	5312

Table 13.7 Antenna locations and SEFDs for the EHT in its 2017 configuration

ID	x [m]	y [m]	z [m]	SEFD [Jy]
ALMA	2225065	−5440057	−2481679	90
APEX	2225043	−5441195	−2479305	3500
JCMT	−5464585	−2493004	2150650	6000
LMT	−768712	−5988542	2063276	600
PV	5088966	−301681	3825019	1400
SMA	−5464521	−2493148	2150615	4900
SMT	−1828792	−5054409	3427864	5000
SPT	0	0	−6359610	5000

References

1. K. Akiyama, R.-S. Lu et al., 230 GHz VLBI observations of M87: event-horizon-scale structure during an enhanced very-high-energy γ-ray state in 2012. Astrophys. J. **807**, 150 (2015)
2. D. Ball, L. Sironi, F. Özel, Electron and proton acceleration in trans-relativistic magnetic reconnection: dependence on plasma Bbeta and magnetization. Astrophys. J. **862**(1), 80 (2018)
3. R.D. Blandford, A. Königl, Relativistic jets as compact radio sources. Astrophys. J. **232**, 34–48 (1979)

4. A. Chael, M.D. Johnson et al., Interferometric imaging directly with closure phases and closure amplitudes. Astrophys. J. **857**(1), 23 (2018)
5. A. Chael, R. Narayan, M.D. Johnson, Two-temperature, magnetically arrested disc simulations of the jet from the supermassive black hole in M87. Mon. Not. R. Astron. Soc. **486**(2), 2873–2895 (2019)
6. A. Cruz-Osorio, C.M. Fromm et al., State-of-the-art energetic and morphological modelling of the launching site of the M87 jet. Nature Astron. (2021)
7. J. Davelaar, H. Olivares et al., Modeling non-thermal emission from the jet-launching region of M 87 with adaptive mesh refinement. Astron. Astrophys. **632**, A2 (2019)
8. J. Dexter, A public code for general relativistic, polarised radiative transfer around spinning black holes. Mon. Not. R. Astron. Soc. **462**(1), 115–136 (2016)
9. I.K. Dihingia, Y. Mizuno et al., Temperature properties in magnetized and radiatively cooled two-temperature accretion flows on to a black hole. Mon. Not. R. Astron. Soc. **518**(1), 405–417 (2023)
10. S.S. Doeleman, V.L. Fish et al., Jet-launching structure resolved near the supermassive black hole in M87. Science **338**(6105), 355 (2012)
11. EHT Mwl Science Working Group, Broadband multi-wavelength properties of M87 during the 2017 event horizon telescope campaign. Astrophys. J. Let. **911**(1), L11 (2021)
12. Event Horizon Telescope Collaboration, First M87 event horizon telescope results. I. The shadow of the supermassive black hole. Astrophys. J. Lett. **875**, L1 (2019)
13. Event Horizon Telescope. Collaboration, First M87 event horizon telescope results. II. Array and instrumentation. Astrophys. J. **875**(1), L2 (2019)
14. Event Horizon Telescope Collaboration, First M87 event horizon telescope results. IV. Imaging the central supermassive black hole. Astrophys. J. Lett. **875**(1), L4 (2019)
15. Event Horizon Telescope Collaboration, First M87 event horizon telescope results. V. Physical origin of the asymmetric ring. Astrophys. J. Lett. **875**(1), L5 (2019)
16. Event Horizon Telescope Collaboration, First M87 event horizon telescope results. VI. The shadow and mass of the central black hole. Astrophys. J. Lett. **875**(1), L6 (2019)
17. H. Falcke, P.L. Biermann, The jet-disk symbiosis. I. Radio to X-ray emission models for quasars. Astron. Astrophys. **293**, 665–682 (1995)
18. L.G. Fishbone, V. Moncrief, Relativistic fluid disks in orbit around Kerr black holes. Astrophys. J. **207**, 962–976 (1976)
19. C.M. Fromm, A. Cruz-Osorio et al., Impact of non-thermal particles on the spectral and structural properties of M87. Astron. Astrophys. **660**, A107 (2022)
20. C.M. Fromm, Y. Mizuno et al., Using space-VLBI to probe gravity around Sgr A*. Astron. Astrophys. **649**, A116 (2021)
21. S.E. Gralla, D.E. Holz, R.M. Wald, Black hole shadows, photon rings, and lensing rings. Phys. Rev. D **100**(2), 024018 (2019)
22. John F. Hawley, Three-dimensional simulations of black hole tori. Astrophys. J. **381**, 496 (1991)
23. J.Y. Kim, T.P. Krichbaum et al., The limb-brightened jet of M87 down to the 7 Schwarzschild radii scale. Astron. Astrophys. **616**, A188 (2018)
24. R. Lu, K. Asada et al., A ring-like accretion structure in M87 connecting its black hole and jet. Nature **616**(7958), 686–690 (2023)
25. A. Marszewski, C. Prather et al., Updated transfer coefficients for magnetized plasmas. Astrophys. J. **921**(1), 17 (2021)
26. Y. Mizuno, C.M. Fromm et al., Comparison of the ion-to-electron temperature ratio prescription: GRMHD simulations with electron thermodynamics. Mon. Not. R. Astron. Soc. **506**(1), 741–758 (2021)
27. Y. Mizuno, Z. Younsi et al., The current ability to test theories of gravity with black hole shadows. Nature Astron. **2**, 585–590 (2018)
28. M. Mościbrodzka, H. Falcke, H. Shiokawa, General relativistic magnetohydrodynamical simulations of the jet in M 87. Astron. Astrophys. **586**, A38 (2016)
29. M. Mościbrodzka, C.F. Gammie, IPOLE—semi-analytic scheme for relativistic polarized radiative transport. Mon. Not. R. Astron. Soc. **475**(1), 43–54 (2018)

30. H. Olivares, O. Porth et al., Constrained transport and adaptive mesh refinement in the black hole accretion code. Astron. Astrophys. **629**, A61 (2019)
31. A. Pandya, Z. Zhang et al., Polarized synchrotron emissivities and absorptivities for relativistic thermal, power-law, and Kappa distribution Functions. Astrophys. J. **822**(1), 34 (2016)
32. O. Porth, K. Chatterjee et al., The event horizon general relativistic magnetohydrodynamic code comparison project. Astrophys. J. Suppl. **243**(2), 26 (2019)
33. O. Porth, Y. Mizuno et al., Flares in the galactic centre—I. Orbiting flux tubes in magnetically arrested black hole accretion discs. Mon. Not. R. Astron. Soc. **502**(2), 2023–2032 (2021)
34. O. Porth, H. Olivares et al., The black hole accretion code. Comput. Astrophys. Cosmol. **4**(1), 1 (2017)
35. L. Rezzolla, O. Zanotti, *Relativistic Hydrodynamics* (Oxford University Press, Oxford, UK, 2013)
36. A. Ricarte, C. Gammie et al., Probing plasma physics with spectral index maps of accreting black holes on event horizon scales. Mon. Not. R. Astron. Soc. **519**(3), 4203–4220 (2023)
37. H. Ro, M. Kino et al., Spectral analysis of a parsec-scale jet in M 87: pbservational constraint on the magnetic field strengths in the jet. Astron. Astrophys. **673**, A159 (2023)
38. A. Tchekhovskoy, R. Narayan, J.C. McKinney, Efficient generation of jets from magnetically arrested accretion on a rapidly spinning black hole. Mon. Not. R. Astron. Soc. **418**, L79–L83 (2011)
39. A.R. Thompson, J.M. Moran, G. W. Jr. Swenson, *Interferometry and Synthesis in Radio Astronomy*, 3rd ed. (2017)
40. R.C. Walker, P.E. Hardee et al., The structure and dynamics of the Subparsec jet in M87 based on 50 VLBA observations over 17 years at 43 GHz. Astrophys. J. **855**(2), 128 (2018)
41. Z. Wang, A.C. Bovik et al., Image quality assessment: from error visibility to structural similarity. IEEE Trans. Image Process. **13**(4), 600–612 (2004)
42. Z. Younsi, O. Porth et al., Modelling the polarised emission from black holes on event horizon-scales, in *Perseus in Sicily: From Black Hole to Cluster Outskirts*, ed. by K. Asada, E. de Gouveia Dal Pino, M. Giroletti, H. Nagai, R. Nemmen, vol. 342 (2020), pp. 9–12
43. Z. Younsi, K. Wu, S.V. Fuerst, General relativistic radiative transfer: formulation and emission from structured tori around black holes. Astron. Astrophys. **545**, A13 (2012)

Chapter 14
Accretion Onto Supermassive Black Hole Binaries

Eduardo M. Gutiérrez, Luciano Combi, and Geoffrey Ryan

Abstract In this chapter, we give an overview of our current understanding of the physics of accreting massive black hole binaries (MBHBs), with a special focus on the latest developments in numerical simulations and General-Relativistic Magneto-hydrodynamics (GRMHD) simulations in particular. We give a self-contained global picture of how to model accretion onto MBHBs, analyzing different aspects of the system such as the dynamics of the circumbinary disk, mini-disks, outflows, the role of magnetic fields, and electromagnetic signatures. We discuss important questions and open problems related to these systems, what are the advantages and disadvantages of the different numerical approaches, and what robust knowledge we have built from simulations.

14.1 Introduction

The numerous detections of gravitational wave (GW) signals made in the last few years by the LIGO[1], Virgo[2] and KAGRA[3] collaborations [1, 4] have opened a new and

[1] Laser Interferometer Gravitational-Wave Observatory; https://www.ligo.org/.

[2] https://www.virgo-gw.eu/.

[3] Kamioka Gravitational Wave Detector; https://gwcenter.icrr.u-tokyo.ac.jp/en/.

E. M. Gutiérrez (✉)
Institute for Gravitation and the Cosmos, The Pennsylvania State University, University Park, PA 16802, USA

Department of Physics, The Pennsylvania State University, University Park, PA 16802, USA
e-mail: emgutierrez@psu.edu

L. Combi · G. Ryan
Perimeter Institute for Theoretical Physics, Waterloo, ON N2L 2Y5, Canada
e-mail: lcombi@perimeterinstitute.ca

G. Ryan
e-mail: gryan@perimeterinstitute.ca

L. Combi
Department of Physics, University of Guelph, Guelph, ON N1G 2W1, Canada

C. Bambi et al. (eds.), *New Frontiers in GRMHD Simulations*, Springer Series in Astrophysics and Cosmology, https://doi.org/10.1007/978-981-97-8522-3_14

unique observational window to the Universe, giving start to a new era of multimessenger astrophysics. The enormous potential of these observations was demonstrated with the event GW170817 [2], where the GW signal detected from the merger of two neutron stars was followed by the detection of multi-wavelength electromagnetic (EM) radiation on timescales that ranged from seconds to years after the merger [3]. This sole breakthrough marked an enormous step forward in our understanding of the most violent events in our Universe.

The vast majority ($>90\%$) of the detected GW signals to date involved only black holes, and none of them was accompanied by the detection of a clear EM counterpart [4]. This is not surprising: black holes only shine provided they are accreting matter from their surroundings [119]. In the case of galactic stellar-mass black holes, this usually happens when they are fed by a nearby companion star, either through winds or through the overflow of its Roche lobe. The merger of two stellar black holes, however, generally occurs at a later evolutionary stage of a binary system, when the material to feed either black hole has already been exhausted. It is reasonable then to expect mergers of stellar-mass black holes to be "dry", that is, without enough matter around them that makes them accrete and shine. The chance of stellar-mass black hole binaries to be EM bright during the merger may significantly increase if they are located in a particularly dense medium, for example, within the accretion disk of an active galactic nucleus (AGN) [17, 109, 123, 169, 189, 206].

The situation is much more promising for *mergers of supermassive black holes*. Massive black hole binaries (MBHBs) were first proposed to exist in the 1980 s with the pioneering works by Begelman et al. [20] and Roos [173], who suggested that as a by-product of the coalescence of two galaxies, the supermassive black holes (SMBHs) at their centers could end up sinking towards the common gravitational potential of the merged galaxy and form a bound pair. Dynamic processes in the galactic center, mediated by stars or gas, may extract angular momentum from the binary and reduce the orbital separation to $\sim$ milli-parsec scales [125]. The orbital movement of such massive objects during the inspiral releases enormous amounts of energy as GWs; the merger itself is among the most luminous events in the Universe [36].

The prospect of detecting EM counterparts to GWs from MBHBs increases (compared to stellar black holes) mainly due to two reasons. First, accretion is stronger the larger the central mass: for fixed conditions of the ambient medium (rest-mass density, temperature, etc.), the accretion rate towards a central gravitating object of mass M scales as the area of the Bondi sphere[4] $\sim M^2$. Second, these sources would be located in the center of galaxies that underwent a merger sometime in the past. Such an environment may present abundant amounts of gas in their nuclei to feed the black hole pair. Hence, in principle, an MBHB could be surrounded by a bright accretion flow in a similar fashion to normal AGNs, which makes these sources highly promising multimessenger sources of GWs and electromagnetic waves.

[4] The Bondi sphere determines the outer boundary of the region where the gravitational influence of an object dominates the dynamics of matter. It is given by $\sim 4\pi r_{\mathrm{B}}^2$, where $r_{\mathrm{B}} \approx 0.1(kT/\mathrm{keV})^{-1}(M/10^6 M_{\odot})$ pc is the Bondi radius [28].

Without direct observational evidence, advancements in the field have been mainly driven by theoretical studies. The study of binary black hole accretion adds several difficulties to the already complex field of accretion onto compact objects. Binary accretion covers a particularly large parameter space, including the accretion rate $\dot{M}$, total mass M, disk scale height h/r and equation of state, viscosity ν or magnetic field B, as well as the binary separation a, mass ratio q, orbital eccentricity e, disk inclination i, and black hole spins $\mathbf{a}_1$, $\mathbf{a}_2$, all of these potentially playing a role in the accretion dynamics and resulting EM emission. *The highly nonlinear and dynamical nature of the problem forces us to depend much more on numerical simulations, which in turn are more expensive and difficult to perform than in single black hole accretion.* The large majority of work on binary black hole accretion has relied on 2-D (vertically integrated) hydrodynamical simulations with Newtonian physics (see [70] and references therein). A much fewer number of works used three-dimensional General Relativistic Magnetohydrodynamical (GRMHD) simulations, and in general they have explored a subrange of the parameter space or are limited in total duration due to the computational cost [15, 32–34, 40, 41, 48, 75, 86, 87, 145, 147, 153, 175].

Several reviews have come out in recent years overviewing MBHBs, each of them focusing on different aspects of the problem: formation and evolution of MBHBs [45, 68], Newtonian [179] and relativistic [39, 85] aspects of the accretion process, circumbinary disk physics [114], and EM counterparts [27, 56, 63]. This chapter aims to be complementary to these and tackle a subset of the problems that accretion onto MBHB rises. First, we discuss the main modeling approaches taken in the literature, for which we present the differences, advantages, and disadvantages among them. Second, we choose a few specific questions or open problems related to accretion onto MBHBs on the way to the merger. In particular, we focus on relatively short separations $a < 100\,M$, where a is the major semi-axis of the orbit, and where general relativity (GR) becomes important to describe some aspects of the accretion flow.

The remainder of this chapter is structured as follows. In Sect. 14.2, we give a brief overview of our current understanding of how MBHBs may form and their GW emission, and in Sect. 14.3, we summarize the main known phenomenology of accretion onto these systems. In Sect. 14.4, we discuss the various numerical approaches that have been taken in the field to investigate MBHBs, what are the main difficulties in doing so, what are the key differences with the modeling of single black hole accretion disks, and what are the main approximations, advantages, and disadvantages of the different approaches, emphasizing what we can learn from each of them. Then, in Sect. 14.5, we analyze various of the most compelling aspects related to the physics of the circumbinary disk and the mini-disks, and in Sect. 14.6, we discuss the potential outflows. In Sect. 14.7, we discuss what are the main EM signatures that may be associated with each of these components. Finally, in Sect. 14.8, we analyze what are the main open questions and what needs to be accomplished in the theoretical modelling of accreting MBHBs to answer them.

14.2 Formation of Massive Black Hole Binaries and Gravitational Wave Emission

Massive galaxies form by successive mergers of smaller galaxies in a hierarchical process [26]. The relationship between the merger of the galaxy and that of the SMBHs they host is however difficult to establish through observations. Both processes are temporally separated by millions or billions of years and, in turn, the coalescence of SMBHs occurs in an extremely short period compared to the overall evolution of the merged galaxy.

Several aspects of the evolution of SMBH pairs in a merging galaxy are still under debate. We give here a brief overview of the various stages that such a pair needs to follow before merging and the main physical processes driving the evolution. For a more in-depth discussion on this, see Ref. [45].

1. **Galaxy merger** (*t* $\sim$ **a few Gyr**): Two colliding galaxies merge into a single galaxy within a few orbital periods, typically $t_{\rm orb} \sim 2\pi\sqrt{a^3/GM} = 1.5 \times (M/10^{11}M_{\odot})^{-1/2}(a/30\,\text{kpc})^{3/2}$ Gyr. Initially, the bulges of these galaxies, which contain the SMBHs, behave dynamically as individual objects. When the bulges of the two galaxies get close enough, they merge and their nuclei begin to dynamically interact with each other. At this point, the two black holes orbit separately in the gravitational potential of the merged galaxy.
2. **Dynamical friction** ($t \sim 10^7 - 10^9$ yr): The interaction of the black holes with the stars and gas exerts an effect of "dynamical friction" on them that slows them down and brings them closer together [128]. Assuming a spherically symmetric distribution of stars in the nucleus, this process is capable of reducing the distance between the two black holes from a few hundred pc to $\sim$ a pc. This occurs over a timescale of $t_{\rm df} \sim 2 \times 10^8 \ln^{-1} N \times (M/10^6 M_{\odot})^{-1}(a/100\,\text{pc})^2(\sigma_*/100\,\text{km s}^{-1})$ yr, where N is the number of background stars, and σ_* is their dispersion velocity.
   Once the SMBHs are sufficiently close to each other, at a separation $a \sim 0.1q/(1+q)^2(M/10^6M_{\odot})(\sigma_*/100\text{km s}^{-1})^{-2}$ pc, the mass of the gas and stars contained in their orbit is less than the black hole masses and dynamical friction becomes inefficient. At this stage, the pair forms a *gravitationally bound system* but its orbit can stall and stop shrinking. The dilemma of whether the orbit continues shrinking further down to a merger or not is known as the *"last parsec problem"* [20, 129]
3. **Evolution in the last parsec** (*t* $\sim$ uncertain): There is great uncertainty about the dominant orbital evolution mechanism when the separation of the SMBH pair reaches a parsec or less. The current understanding of this problem considers that it may be solved by relaxing some of the assumptions in the simple spherical dynamical friction model, such as considering a non-spherical gravitational potential [21, 108, 157, 200] or rotation of the galactic nucleus [10, 100, 134, 164, 180]. Alternatively, additional interactions may play a role, such as the presence of more than two SMBHs [11, 29, 99, 177] or interaction with disk gas [50, 118, 170].

4. **Relativistic regime** ($t \sim$ variable): If the black hole pair reaches separations of $\sim 10^{-2}$–10^{-3} pc, the orbital evolution becomes dominated by the emission of GWs, and the merger typically occurs in less than a Hubble time [130].

From the point of view of the GW emission, the relativistic regime is divided into three periods: the *inspiral*, where the SMBHs slowly approach in a quasi-stationary way, the *merger* itself, highly nonlinear, and the *ring-down*, where the remnant compact object emits in the form of GWs the energy stored in the asymmetries of quadrupole order or greater while it relaxes to equilibrium. In the fraction of time around the merger, the GW luminosity can be so high that it exceeds the EM luminosity of the entire Universe. The frequency of these GWs can cover a very broad range, from $\sim$ nHz frequencies for masses of $\sim 10^{9-10} M_{\odot}$ during the inspiral, up to 100 mHz for masses of $\sim 5 \times 10^4 M_{\odot}$ during the merger itself.

The lowest frequency GW from massive binaries may be detected using a *Pulsar Timing* array (PTA) [58, 178]: an array of many milli-second pulsars which are reliable rotators acting as precise clocks in the sky. The passing of a GW through the galaxy can be detected by measuring carefully the time of arrival of each pulse. Pulsar timing array collaborations including the NANOGrav collaboration have recently found evidence of an ultra-low frequency "stochastic background" of GWs [5]. The leading hypothesis for the origin of such a stochastic background is that it is the result of the collective effect of all the MBHBs in the Universe [6, 36, 163, 181]. If this is the case, this discovery marks the first direct evidence of the presence of MBHBs in the inspiralling regime. In the next decade, PTAs will continue collecting data, increasing the statistical significance of their results. This will hopefully enable the detection of the continuous GW signal of individual MBHBs.

The higher frequency GW from massive binaries could be detected by long-baseline space interferometers such as the Laser Interferometer Space Antenna[5] (LISA) [9, 51, 110], from the European Space Agency[6] (ESA), which has recently been adopted and is expected to be launched in the next decade [46]. LISA will be sensitive to GWs with frequencies in the range 100μHz–100mHz, which corresponds to MBHBs with total masses between 10^4 and 10^7 $M_{\odot}$ in the last stages of their evolution. LISA is expected to detect black hole mergers up to a *redshift* of $z \sim 20$, corresponding to cosmic epochs where the first seeds of SMBHs were formed [8].

14.3 The Standard Picture of Accretion Onto Massive Black Hole Binaries

MBHBs are located in the center of merged galaxies. These environments may have sufficient amounts of gas for the MBHB to develop similar features to normal AGNs; in particular, the formation of an accretion flow and (potentially) outflows.

[5] https://www.elisascience.org/.

[6] https://www.esa.int/.

At sufficiently large distances from the center of mass of the system, $r \gg a$, matter from the ambient medium feels the gravitational influence of the binary system as that of a single object with a total mass $M = m_1 + m_2$. The orbiting gas would gradually fall inward by transferring angular momentum outwards while generating heat [81]. This process would be primarily driven by viscous dissipation [158] triggered by the magneto-rotational instability (MRI) and the resulting MHD turbulence [16].

The evolution of the flow may be drastically different depending mainly on the accretion rate and angular momentum supplied at the Bondi radius. For low accretion rates, i.e., much smaller than the Eddington rate, $\dot{M}_{\rm edd} = 1.38 \times 10^{18}(M/M_\odot)\,{\rm g\,s^{-1}}$, the densities will be low and the gas will form a radiatively inefficient accretion flow (RIAF) [214]; here, most of the thermal energy is advected to the central object with large radial velocities and sub-Keplerian orbits. A similar flow structure arises when the accretion rate is super-Eddington and the photon energy gets trapped in the gas, radiating very little. Given that the sound speed is comparable with the Keplerian orbital velocity in these cases, the accretion flow tends to be geometrically thick. We will focus on radiatively efficient flows, which have accretion rates close to the Eddington limit and are more promising for multimessenger astrophysics. For the reader interested in the RIAF-like regime in binaries, we refer to Ref. [39] and references therein.

Far from the binary, a luminous accretion flow will resemble a standard thin disk around a point mass in which the disk is cooled by radiating the dissipated energy of accretion into photons [182]. As the gas gets close enough to the binary, it will start to feel the effects of the *dynamic and non-axisymmetric gravitational field*, which will significantly modify the flow structure. The main influence of the binary on the accreting gas is through the positive torque that the rotating potential imprints on the matter [88, 120]. Indeed, the time-dependent quadrupolar potential of the binary does not conserve the energy or the angular momentum of the plasma. It was first noticed by Artymowicz & Lubow [14] that if this effect dominates the flow's angular momentum evolution, the torque will push the gas away from the binary *carving a low-density cavity in the inner region where torques and viscous stresses balance each other.* For close to equal-mass binaries, the accretion disk is then truncated at a distance $r_{\rm edge} \sim (2-3)a$ from the center of mass, as shown by many different simulations [e.g., 67, 73, 147].

The truncated disk orbiting the binary is usually referred to as the *circumbinary disk* (CBD), and it was first suggested to exist around MBHBs by Begelman et al. [20]. The first models on CBDs [159] assumed that no accretion would occur within the cavity. If this were the case, we may conclude that since the luminosity of an accretion disk depends on how close its inner edge is from the gravitating source ($L_{\rm disk} \propto R_{\rm in}^{-1}$; see, e.g., [81]), accreting MBHBs would be much dimmer than normal AGNs. This would make it much harder to detect EM counterparts of SMBH mergers, decreasing the multimessenger prospects for these sources.

Fortunately, in more recent works a different and more promising picture has consistently emerged from results from various types of simulations using Newtonian hydrodynamic [57, 59, 66, 67, 71, 73, 121, 136, 138, 139, 141, 143, 196, 217] with and without viscosity, Newtonian MHD [184], GRMHD with approximate spacetimes [12, 147], and full NR + MHD simulations [78, 86, 153]. It is now accepted that a significant fraction of the matter in the inner edge of the CBD falls towards the binary system, mainly through thin streams. Moreover, it was shown that, generally, these streams of matter have enough angular momentum to start orbiting the individual holes forming a pair of circum-single disks, also known as *mini-disks*.

A unique characteristic of CBD accretion of binaries with mass-ratios $q \lesssim 1$ is the presence of a 'lump'; this is a *coherent* overdense region ($m = 1$ mode) that orbits the binary close to the inner edge of the CBD. The formation of a lump has been consistently demonstrated through simulations of CBD accretion with different approaches: 2D hydro [60, 66, 67, 73, 121, 131, 136, 139], 3D Newtonian MHD [183, 184] and 3D-GRMHD [12, 86, 145, 147]. As we will discuss later, the lump is a very important feature of a CBD because it modulates how the mass flows onto the cavity.

To summarize, we can draw a schematic picture of 'standard' thin-disk accretion onto MBHBs as follows: (a) the matter flows inwards from large distances transporting angular momentum outward through viscous stresses due to MHD turbulence; (b) binary torques carve an eccentric cavity with a mean radius of $r_{\rm cav} \sim (2-3)a$, which also determines the inner boundary of the truncated CBD; (c) part of the matter reaching this inner boundary is accreted onto the binary through free-fall thin streams connecting it with the black holes and, potentially, forming mini-disks around each binary component. In Fig. 14.1, we show the various components of the accretion flow onto an MBHB on the orbital plane.

We should remark that this picture is approximately valid for near-equal-mass binaries with an aligned cooled thin disk. Accretion onto either eccentric binaries, misaligned disk, or small mass-ratios $m_1/m_2 \ll 1$, can drastically change the system. If we focus on relatively short binary separations, $a \lesssim 100\,M$, i.e. where the binary is efficiently emitting GWs, quasi-circular, and aligned equal-mass binaries may be a natural outcome of long-term evolution. As the binary shrinks, its initial eccentricity would also be radiated away by GWs [155]; on the other hand, for unequal mass binaries, the less massive component will capture most of the mass and grow through accretion, driving the system to equal-mass; finally, if the flow is initially misaligned with the orbit, the binary torque may align the gas close to the center of mass. Of course, this all assumes that the gas supply at the Bondi radius remains roughly in a steady state for a sufficiently long time.

Accreting MBHBs present several hydrodynamical quasi-periodicities that might help distinguish them from single AGN [30, 64, 121], as shown in the quasi-periodic accretion rate onto the BHs, see for instance Fig. 14.2. The light curves of AGNs have large variabilities dominated by red noise, which makes the identification of periodicities extremely challenging [63]. Accurate models of binary accretion are thus essential to finding a real 'smoking gun' in the EM signatures which may lead to detection.

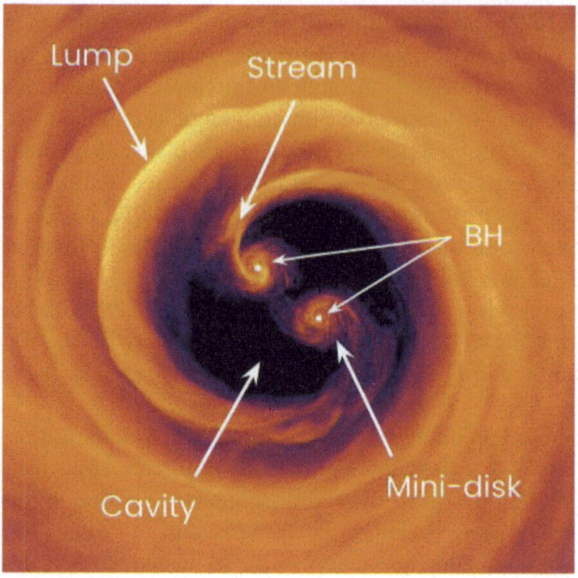

Fig. 14.1 Gas density in the equatorial plane around a quasi-circular equal-mass binary at a separation of $a \sim 20\,M$ from a 3D GRMHD simulations [48]. The different components of the accretion are indicated through arrows

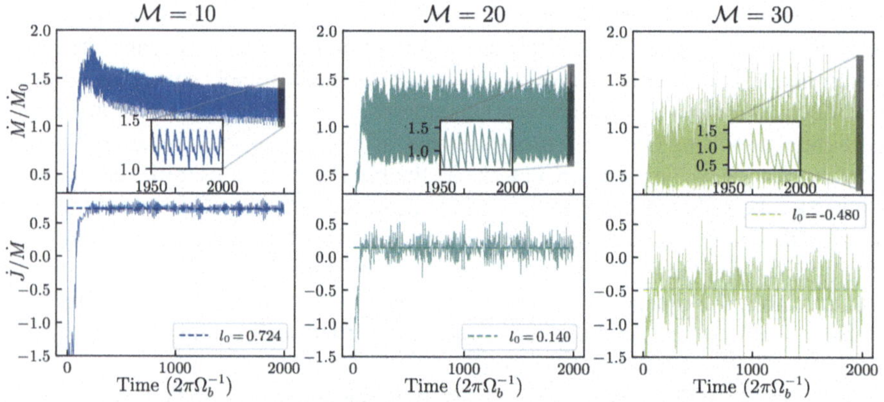

Fig. 14.2 Accretion onto the cavity for different temperatures (Mach numbers) in an equal-mass binary, showing strong modulations, from a 2D VH simulation. Figure from Ref. [60]

14.4 Theoretical Models of Binary Accretion

The accretion flow onto a binary will differ from that with a single black hole only close to the source, where the effect of the binary quadrupole ($\sim r^{-3}$) becomes important relative to the gravitational monopole ($\sim 1/r^2$). As we go closer to the black holes ($r < 2a$) and binary separations are small ($a < 100M$), GR becomes important to describe the interaction between the plasma and spacetime. The gravitational field is non-axisymmetric and non-stationary, providing an additional source of torque and energy, changing how angular momentum is transported in the system.

The physical ingredients for describing binary accretion do not differ greatly from single black hole accretion except that now we must account for the dynamical spacetime, or in Newtonian language, the time-dependent quadrupolar gravitational field. We must then solve Einstein's equations, coupled with the fluid conservation equations and, ideally, radiation transport. On this last point: many simulations to date treat radiation at best with a crude effective cooling function [86, 146, 147, 208], whereas most Newtonian simulations neglect radiation entirely and instead evolve an isothermal gas [70].

Because the accretion flows coupled with magnetic fields are essentially turbulent and the gravitational potential is time-dependent and non-axisymmetric, analytical models can only treat a few qualitative aspects of binary accretion [22]. Even to get qualitative predictions, numerical solutions are needed. A big challenge for these simulations is resolving the large dynamic range of time and length scales that govern the system's evolution. Global simulations must simultaneously include the black hole horizon $\sim R_g$ and the CBD of size $\gtrsim 3a \gg R_g$, as well as resolving time-scales of the order of the orbital period $\sim P_{bin}$ while evolving to much longer viscous timescales for the disk to reach an inflow-outflow equilibrium [114]. The viscous timescale $t_{visc} \sim r^2/\nu \sim \alpha^{-1}(h/r)^{-2}(r/a)^{3/2} P_{bin}$, can be easily longer than $100 P_{bin}$ at moderate radii. This implies that simulations must be evolved for hundreds of orbits to reach a steady state independent in some measure from initial conditions. Because of the computational cost, long-term simulations can be unfeasible unless approximations are made. These will depend on the specific questions one wants to address. On this aspect, the literature is roughly divided into Newtonian 2D viscous-hydrodynamics (VH) and 3D GR-magnetohydrodynamics, with a few simulations in between studying binaries in 3D Newtonian MHD [183, 184], 3D Newtonian VH [60], and a few 2D GR-hydrodynamic (GRHD) simulations [126]. Although 3D GRMHD can capture the correct angular momentum transport and resolve the physics around the black hole horizon, the current computational cost of each simulation is extremely expensive to run for $\gtrsim$1000 orbits. Two-dimensional VH simulations, instead, are much cheaper (by a factor of > 10) and can be used to model long-term evolution and explore the parameter space. However, because VH simulations do not model magnetic fields and cannot resolve the physics close to the black holes, they cannot accurately model the EM emission that comes out from the innermost regions.

In the next subsections, we briefly present the main aspects of each of these two approaches.

14.4.1 Newtonian 2D Viscous Hydrodynamics

For luminous AGNs, those that we can observe up to very large distances, the accretion disk must be efficiently converting gravitational potential energy in radiation and thus maintaining (relatively) low temperatures [81]. If the thermal pressure remains low, the disk must be geometrically thin $H/r \ll 1$. The vertically integrated 2D hydrodynamical equations on the azimuthal plane (r, ϕ) are thus a reasonable zeroth order representation of the system. In addition, if one is mostly interested in the orbital secular evolution due to accretion, we may assume that the binary separation is large and neglects its evolution through GW emission. As long as we focus on scales far from the black holes, we can also assume that the gravitational field of the binary is well described by a Newtonian potential of two-point masses of the form

$$\Phi(t, \mathbf{r}) = -\frac{m_1}{|\mathbf{r} - \mathbf{r}_1(t) + \epsilon_g|} - \frac{m_2}{|\mathbf{r} - \mathbf{r}_2(t) + \epsilon_g|}, \tag{14.1}$$

where $r_i(t)$ follows a Keplerian orbit and ϵ_g is the softening gravitational length to avoid numerical singularities in the simulation [60]. Because we are not resolving the black hole horizon explicitly, a subgrid model of accretion is needed to account for the amount of mass and angular momentum that exits the system. For this purpose, sink terms for mass and momentum are included in the equations to extract the gas that is accreted. Several methods have been used in the literature, including special techniques to accrete without introducing spurious torques [59]. An alternative solution pursued in early work was instead to excise the inner region entirely, placing an inflow-only boundary condition inside the cavity at a radius $r \sim a$ [12, 121, 147, 184]. While this alleviates the need for sub-grid source terms, this approach precludes the possibility of gas being recycled from the mini-disks back into the cavity, does not allow simulation of the mini-disks at all, and cannot measure the accretion rates onto the individual black holes.

In the vertically-integrated MHD equations, magnetic stresses due to MRI cannot be self-consistently generated nor sustained due to the necessity of a vertical field [84]. Hence, the source of viscous stresses needs to be approximated with a prescription that relates this to the local properties of the gas. In the equatorial plane, the relevant stresses operate in the $r\phi$ direction and can be approximated by a kinematical viscosity, ν, which might be fixed directly or indirectly through an α-prescription, where $\nu = \alpha c_s H \propto \alpha \mathcal{M}^{-2}$. The 2D Newtonian hydrodynamical vertically integrated equations can be found explicitly in Eqs. (1–3) in [59]. The equation of state is usually an isothermal one, $\Pi = c_s^2 \Sigma$, where $\Sigma = \int \rho dz$ is the surface density, $\Pi = \int P dz$ is the vertically integrated pressure, and the sound speed c_s is chosen proportional to the gravitational potential and Mach number $\mathcal{M}$, $c_s^2 = -\Phi \mathcal{M}^{-2}$. Other works have used a gamma-law gas equation of state, $\Pi = (\Gamma - 1)\Sigma\epsilon$, and the so-called $\beta-$cooling, which adds a cooling function to the equations [74, 194, 209] and can be used as a proxy for radiated energy.

Several finite-volume and smooth-particle hydrodynamical codes have been used to evolve the 2D Newtonian equations of the state including the moving-mesh `DISCO`, the multi-purpose Cartesian/spherical `Athena++`, `Pluto`, and `AREPO`, and smooth particle hydrodynamical (SPH) codes like `Phantom`. A code comparison project has recently shown good agreement among these codes [70]. These simulations are sufficiently cheap to run with relatively high resolution for $\sim$1000 orbits which is needed to reach a steady state on longer viscous timescales. Many crucial insights in binary accretion have been solved using 2D hydro, especially related to binary evolution [114].

Two-dimensional VH Newtonian simulations can capture several sources of angular momentum flux, including most importantly, the gravitational torques. These simulations, however, use a very simplified representation of the internal magnetic stresses present in real systems. Moreover, most of these use fairly unrealistic thermodynamics given by isothermal EOS and neglect all 3D effects. As pointed out in [145], this means that *VH simulations are never turbulent and show smooth and steady flow*, which are quite different from real AGNs.

14.4.2 General Relativistic 3D Magneto-Hydrodynamics

Because magnetic stresses are the main mechanism of angular momentum transport in accretion disks, 3D MHD simulations are needed to accurately model this aspect of accreting MBHBs. Close to the black holes, where most of the 'viscous' dissipation and emission occurs, a general relativistic description of the gravitational field becomes essential too. Moreover, if the black holes are spinning, accretion can drive magnetic flux to the ergosphere and launch powerful double jets, which is a purely GR effect. If we want our emission models from binary black hole accretion to be accurate and realistic, we need to incorporate both MHD and GR. This is challenging because the flow around the black holes is still highly coupled with the outer circumbinary accretion disk, which evolves much more slowly.

The equations of GRMHD are obtained from (a) the conservation of the energy-momentum tensor, (b) the conservation of rest-mass density, and (c) the Maxwell's equations in the ideal MHD limit of infinite conductivity. These equations must be supplemented with an equation of state, e.g. relating pressure and internal energy. In addition, the spacetime evolution is coupled to matter through the covariant derivatives in the equations. The (dynamical) metric of the binary can be obtained by solving Einstein's equations numerically or using some approximation (see below).

To solve the equations numerically in a computer, these can be cast in a conservative form using an explicit 3+1 decomposition as in the Valencia formulation used by the numerical relativity community [79], or in a coordinate split as used in the `Harm3D` code [147]. The coupled equations of motions, following [147], can be written as

$$\partial_t \mathbf{U}(\mathbf{P}) = -\partial_i \mathbf{F}^i(\mathbf{P}) + \mathbf{S}(\mathbf{P}), \tag{14.2}$$

where **P** are the *primitive* variables, **U** are the *conserved* variables, $\mathbf{F}^i$ are the *fluxes*, and **S** are the *source* terms. These are given explicitly as

$$\mathbf{P} := [\rho, u, \tilde{u}^j, B^j], \tag{14.3}$$

$$\mathbf{U}(\mathbf{P}) := \sqrt{-g}\left[\rho u^t,\ T^t{}_t + \rho u^t,\ T^t{}_j,\ B^j\right], \tag{14.4}$$

$$\mathbf{F}^i(\mathbf{P}) := \sqrt{-g}\left[\rho u^i,\ T^i{}_t + \rho u^i,\ T^i{}_j,\ \left(b^i u^j - b^j u^i\right)\right], \tag{14.5}$$

$$\mathbf{S}(\mathbf{P}) := \sqrt{-g}\left[0,\ T^\kappa{}_\lambda \Gamma^\lambda{}_{t\kappa} - \mathcal{F}_t,\ T^\kappa{}_\lambda \Gamma^\lambda{}_{j\kappa} - \mathcal{F}_j,\ 0\right], \tag{14.6}$$

where ρ is the rest-mass density, T^a_b are the components of the energy-momentum tensor, g is the determinant of the metric, $\Gamma^\lambda{}_{\alpha\beta}$ are the Christoffel symbols, $u := \rho\epsilon$ is the internal energy, $\tilde{u}^j := u^j - g^{tj}/g^{tt}$ is the velocity relative to the normal spacelike hypersurface, and $B^j := {}^*F^{it}$ is the magnetic field, which is both a conserved and a primitive variable. Finally, $\mathcal{F}_j$ is a contribution to the source term that can account for radiative cooling [146].

Although relatively less explored than 2D VH simulations, in the last decade several 3D GRMHD simulations have investigated binary black hole accretion [85]. These simulations have focused on the late inspiral stage before the merger, starting from separations of $\sim 10-20\ M$ and evolving for a few tens of orbits when the black holes are included in the domain. 3D GRMHD simulations have to deal with two main issues: a) long evolutions to reach a true reliable steady state, which makes them very expensive, and b) accurate modeling of the time-dependent binary spacetime.

A 3D GRMHD simulation of binary black hole accretion must include a dynamic representation of the spacetime. Depending on the regime of binary evolution, different approaches have been taken in the literature, from PN approximations to full numerical relativity calculations, i.e., numerically solving Einstein's equations coupled with the MHD equations. We summarize the various approaches below:

- **Weak-field approximation**: For modeling the circumbinary gas flow far away from the black hole, one can use a weak-field PN approximation for the metric [23]. This metric is an expansion in powers of v/c and is usually derived in harmonic coordinates. It represents two point particles following PN trajectories and it is valid for binary separations of $a > 10\ M$. It has been used in several simulations that focused on the CBD while excising the inner region of the binary [145, 146, 216].
- **Strong-field approximation**: Well before the merger, tidal interactions between the black holes are small and GW effects are mostly relevant for the trajectories of the black holes. A strong field representation that includes black hole horizons explicitly can be built in a variety of ways without numerical relativity (i) using a matching metric, where different approximations are sewed together [142], (ii) using a conformal thin sandwich approximation for binaries on quasi-circular

orbits [78], and (iii) using a linear superposition of black holes e.g. in Kerr-Schild coordinates [48]. Simulations using approximations like these are much cheaper than those using full numerical relativity and can be used to evolve the flow near the black holes for separations of $a > 10M$ (however, see also the approximation of [49] which is valid even at merger).

- **Numerical relativity**: For evolving the binary from small separation to merger, several simulations have performed full GRMHD calculations where Einstein's equations are solved along with the fluid equations [34, 75, 86, 87, 153, 175]. In this approach, the energy-momentum of matter is usually neglected and the vacuum equations are evolved using a 3+1 decomposition in the Baumgarte–Shapiro–Shibata–Nakamura (BSSN) formalism [18]. The first GRMHD simulation of a binary black hole with numerical relativity including a CBD was conducted by Farris et al. [75], who considered equal-mass black holes and explored disks with and without cooling for $\sim$45 orbits starting at 10 M

Most 3D GRMHD simulations have used an ideal gas equation of state, with adiabatic index given by $\Gamma = 4/3$ or $\Gamma = 5/3$. There are no simulations, as far as we know, that include proper photon transport in this context. An effective isotropic cooling function was included in several works to cool the disk to a target entropy on timescales related to the local Keplerian time [31, 33, 48, 75, 86, 87, 94, 145, 147]; various prescriptions for this cooling time were proposed and tested in Ref. [86] using ideas from Ref. [146]. This cooling function is physically ad-hoc but maintains the disk geometry to a fixed aspect-height ratio, similar to the standard isothermal approach of 2D VH but allowing for dynamic cooling and heating. The heat extracted from the system via the cooling function is used then to calculate the radiation output of accretion, see Sect. 14.7.

GRMHD simulations of single black hole accretion usually start with a hydrostatic axisymmetric disk solution which can be constructed around a Kerr black hole using different prescriptions, e.g. with constant specific angular momentum as proposed by Fishbon and Moncrief [80] or with a power-law relationship between the specific angular momentum and angular velocity [43, 201]. This equilibrium is then perturbed by the weak magnetic field threading the disk (plus some pressure perturbations) and a turbulent state is developed. In most GRMHD simulations of binary accretion, the circumbinary disk was initialized following Ref. [43] and setting up the inner edge of the disk at $\sim 3a$, where the cavity would be approximately located, and the radius of maximum pressure at $\sim 5a$. In [76] and subsequent work by the same collaboration, a hydrostatic torus solution is initialized on a single black hole spacetime. Differently, in [147] and subsequent work by the same collaboration, the hydrostatic solution was set up in an axisymmetric metric constructed as the azimuthal average of the binary metric.

To obtain inflow equilibrium on the circumbinary disk, where the system evolves from an artificial equilibrium state to a fully turbulent steady state, 3D MHD simulations must evolve for more than $\sim$100 orbits, which is challenging when the black holes are included in the simulation grid. For this purpose, Refs. [15, 33, 48] used a hybrid approach where the circumbinary disk is evolved first, excising a sphere around the binary, until a steady state is achieved at the inner edge, and then starting the simulation with the black holes on the grid. For quasi-circular orbits, this approach is likely justified because the inner part of the binary has an effective "accretion horizon" where the matter is trapped and accreted [197].

The magnetic field in all these simulations is set as a single poloidal loop with a maximum strength set by a ratio of fluid pressure and magnetic pressure of ~ 100. This leads to a regime of standard accretion, in contrast to a magnetically-arrested disk (MAD)[195]. We notice that the MAD regime has not been investigated for equal-mass binaries that develop a cavity (see however [145] for a discussion on magnetic field effects on the CBD). On the other hand, Ref. [165] investigated the case of a MAD, hot accretion flow around a small mass-ratio inclined binary, finding the precession of the jet due to spin-orbit coupling and periodicities in the mass outflows.

In Table 14.1, we collect all the Refs. we are aware of where results from 3D-GRMHD simulations of accreting MBHBs were presented. We include in the table the numerical approach used to model the dynamical spacetime, the initial data, the grid, and the code used in each paper. We also discriminate those simulations that included the black holes on the grid (BHOG) from those that excised the inner region, as well as those that included cooling from those that did not include it.

14.5 Aspects of Supermassive Binary Black Hole Accretion on the Way to Merger

There are still many open questions and ongoing work on binary black hole accretion. Some aspects related to circumbinary accretion are shared among other binary accreting systems such as in protoplanetary disks [114]; other aspects of the system where GR effects are important are unique to MBHB accretion. Even within the simple 2D VH approximation, the literature has debated fundamental questions such as whether the net angular momentum flux from the gas to the binary is positive or negative. There is also a huge range of parameter space to explore both for the gas properties and the binary. Moreover, recent full 3D MHD calculations of CBD accretion that are evolved for ~ 200 orbits [12, 145] show some qualitative differences with respect to 2D VH simulation.

As we stated before, we are mainly interested in the MBHB problem when the separations are small and the prospects of multimessenger observations are higher. Within this small subset of possible models, we will discuss the following questions:

- What are the mechanisms for transporting angular momentum in the system? What drives accretion from the CBD to the black holes?
- What are the properties of the circumbinary cavity?
- How does the *lump* form, and what are their properties? What influence do magnetic fields play on this matter?
- Does the binary decouple from the CBD?

Table 14.1 List of references presenting results from GRMHD simulations of accreting MBHBs. The abbreviations shown in the table denote the following. NR: Numerical Relativity; SKS: Superposed Kerr–Schild; SKS-PN: Superposed Kerr–Schild + Post-Newtonian; SHPN: Superposed Harmonic + Post-Newtonian; BHOG: Black holes on the gird. *Cart.* refers to simulations using Cartesian mesh refinement with box-in-box grids following the BHs. *Warped* refers to a topological spherical grid with more resolution on the BHs [216]

Refs.	Spacetime	BHOG	Cooling	Initial data	Grid	Code
Farris et al. [75]	NR	✓	✓	Hydrostatic torus	Cart.	`IllinoisGRMHD`
Noble et al. [147, 216], Zilhão et al. [145]	PN	✗	✓	Hydrostatic torus	Sph.	`Harm3D`
Gold et al. [86, 87]	NR	✓	✓	Hydrostatic torus	Cart.	`IllinoisGRMHD`
Bowen et al. [32, 33]	Matching [142]	✓	✓	Relaxed CBD [147]	Warped	`Harm3D`
Armengol et al. [12]	SKS	✗	✓	Hydrostatic torus	Sph.	`Harm3D`
Combi et al. [48]	SHPN [47]	✓	✓	Relaxed CBD	Warped	`Harm3D`
Avara et al. [15]	Matching	✓	✓	Relaxed CBD	Multipatch	`PatchworkMHD+Harm3D`
Cattorini et al. [40, 41]	NR	✓	✗	Uniform gas	Cart.	`IllinoisGRMHD`
Ruiz et al. [175]	NR	✓	✗	Hydrostatic torus	Cart.	`IllinoisGRMHD`
Bright and Paschalidis [153], Paschalidis et al. [34]	NR	✓	✗	Hydrostatic torus	Cart.	`IllinoisGRMHD`
Giacomazzo et al. [82]	NR	✓	✗	Uniform gas	Cart.	`Whisky`
Kelly et al. [106]	NR	✓	✗	Uniform gas	Cart.	`IllinoisGRMHD`
Ressler et al. [165]	SKS-PN [49]	✓	✗	Relaxed MAD disk	Cart.	`Athena++`

- What are the differences between single black hole disks and mini-disks? What drives accretion in the mini-disks?

We purposely neglect other important topics, for instance, related to the long-term evolution of the binary and the influence of the gas, misaligned disks, and large eccentricities. These are crucial aspects of binary accretion but in service of focusing the review on results where GRMHD is more relevant, we recommend other reviews such as [114] for the interested reader.

14.5.1 Angular Momentum Transport in Circumbinary Disks

In steady accretion disks around a point-mass potential, angular momentum transport is driven by internal magnetic stresses generated through the MRI [16, 96]. In the case of binary black hole accretion, this is likely the main mechanism for angular momentum transport far from the binary, but the inner regions will be heavily influenced by gravitational torques and the non-axisymmetric, time-dependent gravitational potential. In addition to MHD turbulence and gravitational torques, the binary potential can induce spiral waves in the disk, which can also transport angular momentum as they shock and become non-linear [183]. In 2D VH, the internal (viscous) stresses are modeled by an α prescription or a fixed kinematic viscosity. However, other mechanisms of angular transport such as the gravitational torques are captured in 2D VH with some limitations related to isothermal equations of state [132] and the lack of 3D effects.

Before describing in more detail how these mechanisms operate, we should say a few things about how to measure the different contributions to the angular momentum flux. Because the gravitational potential is time-dependent and non-axisymmetric, angular momentum and energy are not conserved. We can however define a preferred coordinate system on the center of mass where a total angular momentum can be defined even if it is time-dependent (in this frame, the total angular momentum at infinity will be, however, conserved). The procedures to obtain this quantity are different in GRMHD and Newtonian physics. In 2D VH [131], we start defining the Newtonian total angular momentum per unit r as

$$\frac{dJ_{\rm Newt}}{dr} = \oint r\Sigma(r^2\Omega)d\phi, \tag{14.7}$$

where Ω is the angular velocity.

Using the equations of motion, it is possible to derive a balance law as

$$\frac{d\dot{J}_{\rm Newt}}{dr} = \frac{d\dot{J}_{\rm adv}}{dr} - \frac{d\dot{J}_{\rm visc}}{dr} + \frac{dT_{\rm Newt}}{dr}. \tag{14.8}$$

which can be integrated in r to yield the total angular momentum flux

$$\dot{J}_{\rm Newt} = \dot{J}_{\rm adv} - \dot{J}_{\rm visc} + T_{\rm Newt} \tag{14.9}$$

where $\dot{J}_{\rm adv}$ is the advected angular momentum flux, $-\dot{J}_{\rm visc}$ the angular momentum flux due to internal viscosity and $T_{\rm Newt}$ is the binary torque, which is expressed more naturally as a torque density:

$$\frac{dT_{\rm Newt}}{dr} = \int r\Sigma\partial_\phi\Phi d\phi \tag{14.10}$$

In GR, energy and angular momentum cannot be locally localized and gauge ambiguities may arise when we try to find balance laws for fluxes. One could use for instance pseudo-tensors such as Landau-Lifshitz's or quasi-local surface tensors such as Brown-Yorks's, both of which depend on specific gauge (or frame) choices [190]. Despite these gauge dependencies, being able to formulate balance laws can give us crucial information about how the system transfers angular momentum.

To derive a balance law for angular momentum transport we can start from a current that we can define as $j^a = T^{ab}\phi_b$, where ϕ is the azimuthal direction. If ϕ_a were a Killing vector, then angular momentum would be conserved. In the general case, we end up instead with $\nabla_a j^a = T^{ab}\nabla_a\phi_b$. As shown in Refs. [77] and [147], this quantity can be integrated to obtain the following balance law:

$$\frac{d^2J}{dtdr} = \frac{dT}{dr} - \{\mathcal{F}_\phi\} - \partial_r\{M^r_\phi\} - \partial_r\{R^r_\phi\} - \partial_r\{A^r_\phi\}, \tag{14.11}$$

where braces indicate surface integrals, $M^r_\phi = 2p_{\rm m}u^r u_\phi - b^r b_\phi$ denotes the Maxwell stresses, and R^r_ϕ and A^r_ϕ denote the Reynolds and advection stresses, respectively, which sum to the hydrodynamic part of the stress-energy tensor: $(T_{\rm H})^r_\phi \equiv R^r_\phi + A^r_\phi = \rho h u^r u_\phi$. The spacetime torques are described in a relativistic framework as

$$\frac{dT}{dr} = \int T^{ab}\partial_\phi g_{ab}\sqrt{-g}d\theta d\phi. \tag{14.12}$$

If ϕ is chosen as the orbital azimuthal direction, we recover the Newtonian limit at large distances [147]. We can integrate over r these quantities to obtain the total angular momentum flux as before, although most of the existing literature in GRMHD works directly with $d^2J/drdt$ which is in a sense a more local quantity (a torque density) [12, 145, 147].

These quantities are usually averaged in time to measure the net transport. Once the disk reaches an inflow equilibrium, the total angular momentum flux does not depend on r. This is important to understand whether the gas delivers or extracts angular momentum from the binary. Several 2D VH simulations (and 3D VH simulation [136]) have recently found that for equal-mass binaries, $\alpha = 0.1$, and $h/r = 0.1$, the angular momentum delivered to the binary is positive $\ell_0 = \langle \dot{J}_{\rm bin}\rangle / \langle \dot{M}_{\rm bin}\rangle = 0.68a\Omega_{\rm bin}$, where $\langle \dot{J}_{\rm bin}\rangle \equiv \langle \dot{J}_{\rm Newt}\rangle$ is the angular momentum delivered to the binary. A sufficiently large ℓ_0 (e.g., greater than 3/8 for circular equal-mass binaries) determines that the orbital separation will increase as the system accretes. These results contradict the early 3D MHD Newtonian calculations of [184], which might be attributed to short evolution times $100P_{\rm bin}$ in this simulations; it is an open question what happens with longer 3D MHD evolutions. On the other hand, a disk with higher Mach numbers (lower h/r) seems to extract angular momentum from the system [60]. The torque delivered to the binary is also tied to the eccentricity evolution of the orbit; recent 2D VH simulations found that equal-mass binaries tend to evolve towards an eccentricity of $\sim$0.4 [72, 185, 217].

Hydrodynamic simulations cannot capture the true mechanism for the viscous stresses: MHD turbulence. Therefore, they need to include it with an ad-hoc prescription such as the α-prescription, $\nu = \alpha c_s^2 \Omega_K^{-1}$ [182]), or by considering a constant viscosity coefficient ν. Some studies have explored the impact of viscosity on the properties of the accretion disk [60]. Spiral density waves generated by the binary, usually exerted as a single-arm $m = 1$ density mode propagates through the disk but they seem to carry little angular momentum ([183]; notice, however, the dynamics of these spiral waves are very sensitive to thermodynamics [133]).

There are very few long-term 3D MHD simulations of CBD accretion [12, 145, 147, 183, 184]. Shi et al. [184] conducted the first 3D-MHD simulation of circumbinary accretion onto MBHBs using the `ZEUS` code [97, 187, 188]. Their study focused on equal-mass black holes at a fixed separation a. They found that the Reynolds stresses are larger than Maxwell stresses near the binary. This finding is consistent with previous research on protoplanetary disk simulations (e.g., [211]). While Shi et al. [184] focused on large binary separations, where orbital evolution is negligible, Noble et al. [147] explored the inspiralling regime by including a PN metric and taking into account the shrinkage of the orbit. These simulations were conducted using the code `Harm3D` with a PN approximation for the spacetime metric. They traced the evolution of an equal-mass binary from an initial separation of $20M$ to $\sim 8M$, where the PN approximation becomes too inaccurate. In Fig. 14.3, we show the time-averaged contributions to the angular momentum change in the simulation named `RunSE` in Ref. [147]. The figure reveals that binary torques dominate in the gap region $a \lesssim r \lesssim 2a$, and the angular momentum is delivered to the fluid, which shows large and positive Reynolds stresses at $r \sim 2a$. On the other hand, Maxwell stresses dominate the internal stresses in outside regions, as expected. A similar analysis has been done in later works [12, 145].

Additionally, Noble et al. [145] investigated the internal stress dependence on the mass ratio q and demonstrated a weak correlation. The primary observable effect is the inward shift of the gravitational torque for lower mass ratios. Using 2D VH,

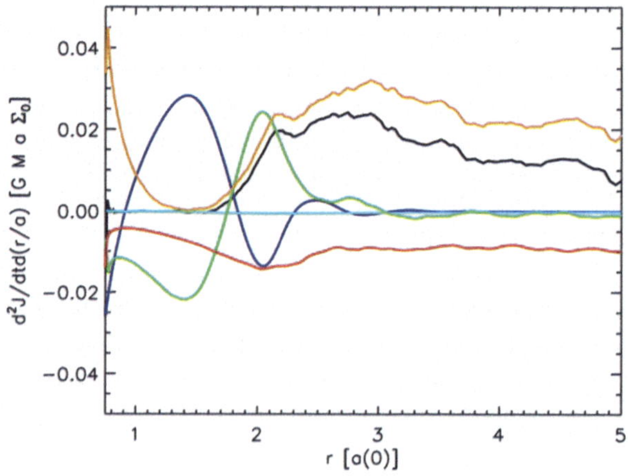

Fig. 14.3 Time-averaged torque density as a function of radius for a CBD simulation, caused by Maxwell stresses (red), Reynolds stresses (green), and advected angular momentum (gold). The gravitational torque density is shown in blue, while the radiation losses are shown in cyan. The black line shows the total angular momentum flux. Figure from Ref. [147] with permission

Dittman & Ryan [59] showed that the gravitational torque becomes very weak at low mass ratios with $q \sim 0.02$, promoting the outspiral of the binary orbit.

14.5.2 Inner Parts of a Circumbinary Disk: The Cavity and the Lump

The balance of gravitational torques and viscous stresses gives rise to an inner truncation of the CBD and the formation of a low-density cavity surrounding the black holes. For a disk weakly perturbed by a binary potential, this inner edge is determined by the outer Lindblad resonance [89] which is located at $r \sim 1.3a$. All recent numerical simulations of binary accretion have shown that the cavity is usually at a larger radius of $r_{\rm trunc} \sim (2-3)a$ [130]. The specific shape of the cavity depends on viscous stresses and thermodynamics of the disk. For instance, Ref. [60] showed that low viscosities and high Mach numbers lead to an increase in the size of the cavity. In particular, high Mach numbers show large surface density in the inner parts of the disk [196].

Simulations show that the binary torque can induce eccentricity in the cavity. We can define the disk's inner eccentricity (sometimes called lopsidedness) as:

$$\epsilon \equiv \left\langle \frac{\left|\langle \rho u^r e^{i\phi} \rangle_\phi\right|}{\langle \rho r u^\phi \rangle_\phi} \right\rangle_{(r,t)}, \tag{14.13}$$

where ρ is the rest-mass density, u^μ are the components of the fluid four-velocity, and the brackets indicate averages in the subscripted variable. 2D hydrodynamical studies have found that the CBD's eccentricity increases with the mass ratio of the binary from $q \gtrsim 0.05$ up to $q \sim 0.4$, above which it remains roughly constant at $\epsilon \approx 0.1$ [73]. Simulations in Ref. [12] have found that the cavity is less eccentric, stabilizing at $\epsilon \gtrsim 0.05$ for equal-mass binaries, though Ref. [145] found values about 50% smaller. Given the large turbulence in MHD disks, the lower value can result from the damping of coherent motions and resonances. However, it can also be that these simulations did not reach a full steady state in terms of this variable [145].

So far, we discussed average properties of angular momentum transport but we have not discussed in what way matter actually leaves the inner edge of the CBD and is accreted to the binary. There must be some time-dependent process that allows matter in the inner edge to lose enough angular momentum to be accreted. We know that once a binary component gets closer to the CBD, a thin ballistic stream is pulled from the CBD to the black hole. As shown in Ref. [183] using 3D MHD with an excised binary, this stream still has too much specific angular momentum to be accreted and is instead flung back to the CBD; the authors argue that the stream will shock against the CBD wall and lose enough angular momentum to be accreted. In a recent paper, Tiede et al. [198] used tracer particles in a 2D VH simulation and argued, on the contrary, that this process is mainly gravitational, driven by tidal stressing (although shocks can still play a role).

For equal-mass binaries on low eccentricity orbits and with cool accretion flows, the inner edge of the cavity develops a strong coherent density structure known as the *lump* [131, 147, 184]. This structure orbits the inner edge of the disk at a Keplerian orbit, given by $P_{\rm lump} \sim 4 - 5P_{\rm bin}$ for cases with $h/r \sim 0.1$. The lump is a key feature of the system because it modulates the accretion onto the binary [12, 59, 66, 67, 71, 74, 121, 131, 136, 139, 143, 147, 184, 216]. The modulated accretion depends on the cavity eccentricity as well, which appears to be different in 3D MHD and 2D VH [12]. However, most simulations show that the lump periodicity is the strongest periodicity in the CBD accretion. Accretion in this case happens when one of the binary components gets very close to the lump and a large portion of the fluid mass is pulled to the binary.

The formation of a lump requires a mechanism to pile up matter on the inner edge of the CBD in a coherent way. As the gas falls from the CBD's inner edge toward the cavity, it is thrown back to the disk by the positive gravitational torque exerted by the binary [67, 184]; this stream of gas hits the disk's inner edge and creates a shockwave that compresses the fluid and produces an overdensity moving with the disk. This occurs coherently since the orbital frequency at the CBD's inner edge ($r_{\rm edge} \sim 2.5a$) is a multiple of the orbital frequency ($\Omega(r_{\rm edge}) \sim \Omega_{\rm bin}/4$). Ref. [140] showed that due to the strong density gradient a long-lived $m = 1$ mode can exist freely on the inner edge of the CBD[7]. In its asymptotic state, the lump spans an angular extension of $\pi/3 \lesssim \delta\phi_{\rm lump} \lesssim \pi$ and a radial extension $\Delta r_{\rm lump} \lesssim a$.

Other processes that may contribute to sustaining the lump are the so-called *stream reabsorption* and *lump feeding*. The first occurs because the binary peels off more matter when the lump is close to the pericenter; then, this stream is thrown back with strong torque and hits the CBD still within the lump since $\delta\phi_{\rm lump}$ can be quite large. The latter process occurs when the lump approaches the apocenter and its velocity reaches a minimum, while reverse streams are hitting the disk at other phases of the orbit with larger orbital velocities, forming mini-lumps. These new smaller lumps can catch up with the main lump and join it [184].

The lump seems to form more easily in 2D hydrodynamical simulations because α-viscosity naturally implies little internal stresses in a coherently orbiting structure such as the lump, making it less likely to be pulled apart. In support of this concept, Ragusa [161] found that the vorticity component perpendicular to the disk plane shows a significant local minimum inside the lump. This results in the α-viscosity effectively emulating the role of MHD physics in promoting the growth and persistence of the lump in CBDs. We should note, however, that the lump seems to even appear in 2D inviscid simulations; [126] attribute the formation of the lump in this case to a Rossby-wave type instability related to vorticity. In other words, there might be different paths to creating the lump in a binary.

---

[7] Because MHD turbulence usually puts more power on longer wavelengths, small azimuthal densities are naturally preferred. However, this strong breaking of symmetry should be further investigated.

Finally, colder and less viscous disks give rise to larger cavities. This produces a longer period for the lump orbit and thus for the accretion rate modulation into the cavity. Ref. [162] found $t_{\rm lump} \sim 5P_{\rm bin}$ ($t_{\rm lump} \sim 4P_{\rm bin}$) for $h/r = 0.04$ ($h/r = 0.1$), and Ref. [160] reported modulations at $(7-8)P_{\rm bin}$ from a cavity of radius $r \approx 3.5a$, while Ref. [60] found that the cavity radius shrink from $3a$ to $2a$ as the viscosity ν increased from 0.0005 to 0.008.

What is the influence of magnetic fields on the lump?
When binary torques throw matter back to the CBD, this also carries with it magnetic field flux. This flux may increase the magnetic field strength in the region where it shocks the CBD, namely the lump. To form a lump, the implied magnetic field strength growth rate in the lump must be lower than the density growth rate. Noble et al. [145] conducted a thorough analysis of this process and found that the magnetic field carried by the torqued streams tends to have opposite polarity to that in the CBD where the stream hits. This results in large-scale magnetic reconnection, which dissipates the field into heat. The timescale for this dissipation can be estimated as the period between successive lump-black hole interactions, which is known as the beat period:

$$t_{\rm diss} \sim \frac{2\pi}{\Omega_{\rm beat}} \approx \frac{2\pi}{2(\Omega_{\rm bin} - \Omega_{\rm lump})} \sim 0.7P_{\rm bin}. \tag{14.14}$$

On the other hand, the magnetic field is replenished in the lump from the far CBD with the advection timescale, defined as

$$t_{\rm adv} \sim \frac{\Delta r_{\rm lump}}{v_r} \approx \frac{\Delta r_{\rm lump} r_{\rm edge} \Omega_{\rm K}(r_{\rm edge})}{W^r_\phi} \tag{14.15}$$

$$\sim 5P_{\rm bin}\left(\frac{\Delta r_{\rm lump}}{0.1a}\right)\left(\frac{r_{\rm edge}}{2.5a}\right)^{1/2}\left(\frac{a}{20M}\right)^{-1}\left(\frac{W^r_\phi}{10^{-4}}\right)^{-1}, \tag{14.16}$$

where we have used the fact that the radial velocity is $v_r \approx W^r_\phi/[r\Omega_{\rm K}(r)]$, with W^r_ϕ the $(r\phi)$- component of the Maxwell stress tensor, which is the dominant contribution to the accretion in this region (see Sect. 14.5). A lump develops provided $t_{\rm diss} < t_{\rm adv}$, which is indeed true case for typical parameters of binaries in the inspiral regime. MRI can also operate to increase the magnetic field but this occurs on an even longer timescale: $t_{\rm MRI} \sim P_{\rm lump} \gg t_{\rm diss}$.

The formation of a lump in a binary system is significantly influenced by the mass ratio between the two black holes. A lower mass ratio results in a lower quadrupole moment, which reduces the strength of the gravitational torques. This, in turn, reduces the efficiency of the throwing back of matter that feeds the lump. Additionally, the magnetic field dissipation at the lump will decrease because less magnetic flux will return to the disk.

In their study, Ref. [145] surveyed the impact of mass ratio (1:1, 1:2, 1:5, and 1:10), and of varying amounts of mass and magnetic flux on the accretion disk. Their findings revealed that a significant lump formation requires a sufficiently large mass

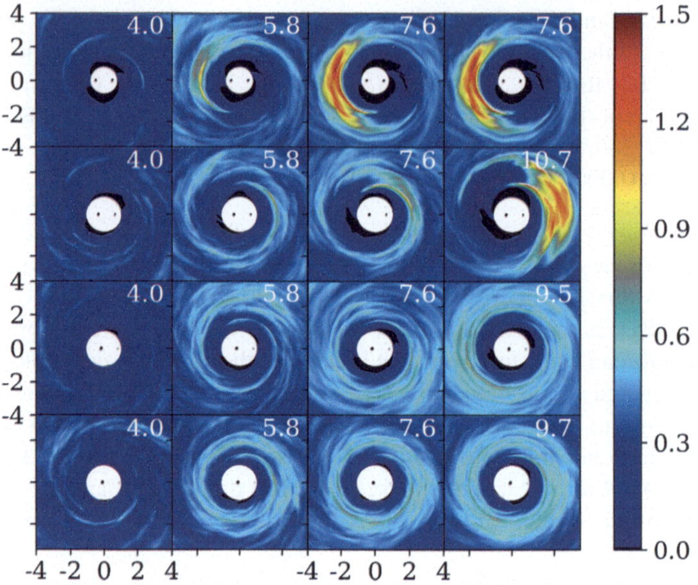

Fig. 14.4 Colormaps of surface density at four different times indicated by the number in the upper-right corner of each image (in units of $10^4 M$). The surface density is shown in units of the maximum surface density in the initial state of the disk for the CBD run. From top to bottom, runs with decreasing mass ratio are shown: $q = 1, 1/2, 1/5, 1/10$. Figure from Ref. [145] with permission

ratio, as can be seen in Fig. 14.4. Specifically, the relative amplitude of the overdensity decreases as the mass ratio diminishes, ultimately vanishing somewhere between 1:2 and 1:5. Regarding the magnetic flux series, they concluded that, whereas the amount of mass in the disk has little influence on the lump properties, the amount of magnetic flux plays a significant role in perturbing the CBD: even a modest, ordered magnetic field was enough to significantly delay the lump formation. In addition, a greater magnetization weakens any modulation of the light output.

14.5.3 Decoupling of the Circumbinary Disk from the Binary

Once the binary system reaches a sufficiently short distance ($a \lesssim 10^{-3}$ pc), its evolution starts to be dominated by the loss of energy through the emission of GWs. The timescale for such evolution can be estimated using the post-Newtonian (PN) approximation [155]

$$t_{\rm GW} := \left(\frac{\dot{a}}{a}\right)^{-1} = \frac{5}{256}\frac{c^5}{G^3}\frac{a^4}{M^2\mu} \sim 140\frac{(1+q)^2}{4q}\left(\frac{a}{30R_{\rm g}}\right)^4 M_7 \text{ days}, \quad (14.17)$$

where $R_g := GM/c^2$ is the gravitational radius associated with the total mass of the system M, $\mu := qM/(1+q)^2$ is the reduced mass of the system, and $M_7 = M/(10^7\ M_\odot)$. On the other hand, the accretion timescale for matter in the inner edge of the CBD, assuming it evolves due to viscous stresses, is given by

$$t_{\rm vis} := \left(\left|\frac{v_r}{r}\right|_{r_{\rm in}}\right)^{-1} \sim \frac{2r_{\rm in}^2}{3\nu} \sim 30\left(\frac{r_{\rm in}}{2.5a}\right)^{3/2}\left(\frac{a}{30R_{\rm g}}\right)^{3/2}\left(\frac{h/r}{0.3}\right)^{-2}\left(\frac{\alpha}{0.15}\right)^{-1} M_7 \text{ days} \tag{14.18}$$

where we have used the α prescription for the kinematic viscosity, $\nu = \alpha H c_{\rm s} \approx \alpha(h/r)^2 l_{\rm K}$, and $l_{\rm K} := r^2 \Omega_{\rm K}$ is the Keplerian angular momentum.

One could predict that when the binary reaches a separation such that $t_{\rm GW} \lesssim t_{\rm vis}$, the shrinking of the orbit becomes fast enough so that the CBD will *decouple* from the binary evolution, and the accretion rate would largely decrease after that. These two timescales become equal at a separation

$$a_{\rm d}^{(1)} \sim 16R_{\rm g}\left(\frac{r_{\rm in}}{2.5a}\right)^{3/5}\left(\frac{h/r}{0.3}\right)^{-4/5}\left(\frac{\alpha}{0.15}\right)^{-2/5}\left(\frac{4q}{(1+q)^2}\right)^{2/5}. \tag{14.19}$$

Nevertheless, this criterion was found to overestimate the decoupling separation and simulations show a better agreement [61, 98, 111, 117, 121, 130, 191, 215] with a velocity-based decoupling criterion [13], namely, one obtained by equating $\dot{a}_{\rm GW}$ with $v_r|_{r_{\rm edge}}$. This gives $a_{\rm d}^{(2)} \sim (r_{\rm in}/a)^{-2/5} a_{\rm d}^{(1)} \approx 0.7 a_{\rm d}^{(1)}$, where the last equality holds for $r_{\rm in} \approx 2.5$. Figure 14.5 shows the two timescales for three different mass-ratios and using the fiducial values in Eq. 14.19.

The expressions above show that the decoupling occurs at closer separations for high CBD's viscosity coefficient and height scale of the disk and for lower mass ratios of the binary. The dependence of the decoupling separation with α motivates the self-consistent treatment of accretion through MHD simulations, where viscous stresses naturally arise due to turbulence generated by the MRI [16]. Noble et al. [147] performed the first GRMHD simulations in the inspiralling regime, reaching separations when decoupling is predicted to occur: $\sim(10-20)R_{\rm g}$. They found an average α parameter of ≈ 0.2.

This simple picture described above is more complicated in a realistic scenario due to several factors, such as the non-steady behavior or the presence of the lump, but overall simulations show an agreement with the analytical estimations. Ref. [147] found that after the theoretically predicted decoupling, a significant amount of mass still follows the binary's inspiral within the gap region, and the accretion rate is still $\sim$70% the one in the steady state. Ref. [15] also found that the binary sustains large accretion rates throughout the inspiral and this is only modestly reduced after decoupling.

The decoupling between the CBD and the binary triggers the following important question: *can MBHBs shine during the merger?* Since most of the radiation from accretion flows comes out very close to the gravitating source, a CBD decoupling from the binary early on may deplete the inner cavity of gas leading to an EM-dim merger. With the same argument, the falling back of the CBD matter after the merger

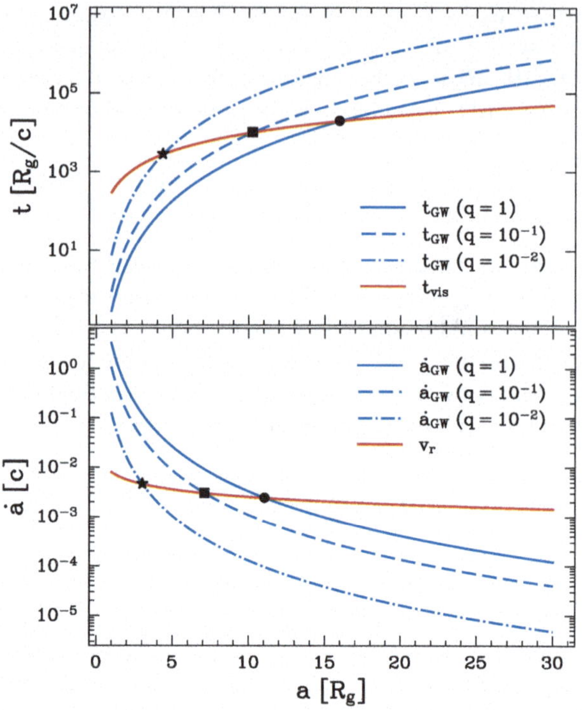

Fig. 14.5 Timescale and rates for binary orbit evolution due to GW emission and viscosity-driven accretion in the CBD. *Top panel*: Timescale-based criterion. *Bottom panel*: Velocity-based criterion. Black markers indicate the decoupling separation for each mass-ratio

can give rise to a rebrightening and a delayed post-merger EM counterpart, with a delay

$$t_{\rm rb} \sim t_{\rm vis}(a_{\rm d}) \sim 6 \left(\frac{r_{\rm in}}{2.5a}\right)^{3/2} \left(\frac{h/r}{0.3}\right)^{-2} \left(\frac{\alpha}{0.15}\right)^{-1} M_7 \text{ days.} \tag{14.20}$$

On the other hand, if the mini-disks are present and they remain healthy after decoupling, there might still be a dense enough environment during the merger to produce EM emission. GRMHD simulations of mini-disk accretion show an upper bound of mass in the cavity of $M_{\rm cav} \lesssim 2 \times 10^{28} M_7^2 \left(\dot{M}/\dot{M}_{\rm Edd}\right)$ g [94]. Nevertheless, this matter would be accreted before the merger if, with an equivalent argument as before, the accretion timescale for the outer boundary of the minidisks is lower than the GW shrinking timescale after decoupling. In such a scenario, the merger itself could be rather empty of gas and proceed without an EM counterpart, for instance, showing a decrease in X-rays [122].

14.5.4 Mini-Disks

Mini-disks (sometimes referred to as circumsingle disks) are rotationally supported accretion flow structures formed around each individual black hole. In a binary system with an external gas supply, the properties of mini-disks are determined strongly by tidal effects and thus by the mass ratio and, once we include GR, orbital separation, and spin.

A key quantity to characterize a mini-disk is the Hill sphere (or truncation) radius which determines at what point the gravity of the companion disrupts stable orbits. As with the CBD, we can distinguish two regimes of small $q = M_1/M_2 \ll 1$ and near unity $q \sim 1$ mass-ratios. When $M_2 \ll M_1$, a simple Newtonian analysis shows that the truncation radius, defined as the radius where centrifugal force and gravitational force of the companion balance each other, is given by

$$r_{\rm H} = (q/3)^{-1/3} a \tag{14.21}$$

For small mass-ratio systems and low inclination angles with respect to the disk, the secondary would be embedded in the disk of the primary, e.g. a stellar-mass black hole in an AGN disk. The bounded flow will then be determined by the minimum between the Bondi radius, and the Hill sphere.

For $q \sim 1$, Paczynski [151] showed with a simple numerical experiment using a pressureless gas that the expression

$$r_{\rm H} \sim 0.27 q^{-0.3} a, \tag{14.22}$$

can fit the truncation radius fairly well. This has been verified in both 3D MHD and 2D VH simulations [30, 33]. Notice that for large separations, the Newtonian approximation allows us to scale everything with the separation a; this scaling is broken when the size of the black holes becomes important. At small separations, the radius of the innermost stable circular orbit (ISCO) is important to determine the size of the mini-disk. The mini-disk will extend from $r_{\rm H}$ to $r_{\rm ISCO} \sim 6\,M$ and will exist as a proper disk as long as $r_{\rm H}/r_{\rm ISCO} > 1$, which for non-spinning black holes requires $a > 20M$ (see below).

14.5.4.1 Mass Supply and Angular Momentum Flux in Mini-Disks

Due to tidal forces and periodic mass supply, mini-disks can behave quite differently than single, steady-state, black hole disks. How mini-disks are formed, in what regime, and how similar they are to single black hole disks are crucial questions to understand what sort of EM signatures we obtain from these systems. Before hydrodynamical simulations, it was unclear whether steady mini-disks could form at all [14]. For instance, [192] argued that the gas from the elongated stream falling to the black hole will be shock-heated, enhancing stresses in the incipient mini-disk. The large viscosity induced by this would then deplete the mini-disk in short timescales and produce quasi-periodic flaring emission similar to a tidal disruption event. However, as pointed out by Refs. [127, 171], the main effect of angular momentum transport in a small disk structure is to viscously spread the gas until the mini-disk reaches the Hill sphere; only then does the gas proceed to be viscously accreted to the black hole. This picture seems confirmed by simulations. Quasi-periodic emission could arise, however, at small binary separations, as we discuss below.

In a mini-disk, contrary to single black hole disks, there may be different sources of angular momentum flux: spiral shocks due to circumbinary stream, tidal-induced shocks by the companion, MRI-driven stresses, and large-scale motions of matter advected onto the black holes [15, 176]. The relevance of each contribution is a function of radius (from the black hole) and determines the inflow time of the mini-disk. The inflow time can be defined as $t_{\rm in} \sim \int dr/v^r$, where v^r is the radial velocity of the gas in the frame of the black hole and the integral is computed from $r_{\rm H}$ to $r_{\rm ISCO}$. The binary separation, and more weakly the spin of the black hole, determine the size of the mini-disk and the inflow time. Mini-disks in large-separation binaries are likely to be healthy, similar to single black hole accretion disks near the black hole, while in small separation binaries, the inflow time becomes shorter than the accretion period, driving the system out of a steady state. We discuss this distinction in the next section.

14.5.4.2 Mini-Disks in and Out of Equilibrium

For large separations, the mini-disk size is big and the inflow time is longer than the accretion periodicity from the CBD. In that case, the mini-disks can reach a steady state, where internal magnetic stresses dominate angular momentum transport. We should notice, however, that there are no self-consistent calculations in 3D GRMHD to quantify this in the regime where separations are $a > 20\,M$. Two-dimensional Newtonian VH simulations are scale-free and can model this large separation regime (as long as the sink that determines the accretion onto the point masses acts on a small region). These simulations have shown that the inflow time is sufficiently long to form coherent structures: the mini-disks never fully deplete and variations in the mass are small [73, 210]. In an interesting numerical experiment, Westernacher-Schneider et al. [210] showed that if the CBD is removed abruptly from an equal-mass binary, the mini-disks take $\sim$20 binary orbits to deplete half their mass. They also show that the mini-disks can grow in eccentricity due to resonant instabilities. On the other hand, in analogy to cataclysmic variables (e.g. Refs. [104, 176]) analyzed spiral shocks in 2D GRHD simulation of a single mini-disk in the corotating frame, showing that even in the absence of viscosity, tidal shocks induce accretion onto the black holes.

For small binary separations, tidal effects are important, mini-disks are smaller, and the inflow time becomes comparable to accretion timescales from the CBD. In that case, the Hill sphere of the mini-disk approaches the ISCO, and the spin of the black holes becomes important to determine the size of the mini-disk. To get an estimate of when the mini-disks start being out of equilibrium, we can calculate the instantaneous inflow time of the mini-disk as $t_{\rm in} \sim r/v^r \sim (2/3)(1 - \sqrt{r_{\rm ISCO}/r})\, r^2/\nu$, assuming a vertically-integrated 2D stationary disk where the stresses go to zero at the ISCO. Taking $r = r_{\rm H}$ and considering $\nu = \alpha c_{\rm s}^2 \Omega^{-1} \sim \alpha (h/r_{\rm H})^2 \sqrt{r_{\rm H} M/2}$, we have:

$$t_{\rm in} \sim \frac{2\sqrt{2}}{3} \frac{(r_{\rm H}/M)^{3/2}}{\alpha (h/r_{\rm H})^2} \left(1 - \sqrt{\frac{r_{\rm ISCO}}{r_{\rm H}}}\right) M. \tag{14.23}$$

For an equal-mass binary, setting $\alpha = 0.1$, $(h/r_{\rm H}) = 0.1$, and $r_{\rm H} \sim 0.3a$ we obtain $t_{\rm in} \sim 150 \times (a/M)^{3/2}(1 - \sqrt{r_{\rm ISCO}/r_{\rm H}})M$, which is a rough upper limit to the inflow time. This crude estimate, which assumes that accretion occurs due to viscous stresses, must be compared with the timescales associated with CBD accretion. The largest influx of mass from the CBD is related to the motion of the lump, which has a Keplerian frequency at the inner edge. Accretion from the lump to the black holes occurs then with a timescale given by $t_{\rm acc} = 5 P_{\rm bin} = 5 \times 2\pi/\Omega_{\rm bin} = 10\pi (a/M)^{3/2} M$. We see here that for $a < 20\, M$, the timescales start to be comparable, $t_{\rm in}/t_{\rm acc} \sim 1$, in agreement with 3D GRMHD simulations [48].

Starting from very small separations, $a = 10\, M$, Refs. [78, 87] showed using full numerical relativity GRMHD simulations that mini-disks do not form in the system; instead, accretion occurs directly via streams from the inner edge of the CBD to the black holes. These results hold for different equations of states and with/without cooling. We should notice, however, that the accretion picture might change if decoupling occurs before those separations are reached.

For slightly larger initial separations, $a = 20\, M$, Bowen et al. [31] first found using a cooled, relaxed circumbinary accretion disk (taking as initial data the final state of the simulation in Ref. [147]), that mini-disks are in fact present, but their mass and accretion rate are highly variable. This is due to the inflow time being comparable to the accretion timescales, as we showed above, and as discussed in Ref. [86]. When the mini-disk enters this starvation mode (or a filling and depleting cycle), the specific angular momentum of the gas falling into the black holes determines whether the material will form a mini-disk or a transient structure.

In general, the gas falling into the cavity from the CBD needs to have low specific angular momentum $\ell_{\rm in}$ to pass the centrifugal barrier of the binary. Refs. [183] and [197] analyzed the distribution of specific angular momentum of the gas that falls into the cavity using Newtonian 3D MHD and 2D VH respectively. In particular, Tiede et al. [197] showed that the specific angular momentum distribution of the gas that is accreted by the binary is $\ell_{\rm in} \sim 2 - 0.5\sqrt{a/M} M$. Once the infalling matter is captured by one of the black holes, the fate of the stream is decided by whether $l_{\rm in}$ is high enough to sustain orbital motion around the black hole. We thus need to compare $l_{\rm in}$, transformed to the frame of the black hole, with the specific angular momentum at the ISCO, $l_{\rm ISCO} \sim 2\sqrt{3}(M/2)$.

The angular momentum budget of mini-disk for small separation binaries was investigated using GRMHD simulations and an approximated strong-field metric in Ref. [48] for spinning and non-spinning black holes, and in Ref. [15] for non-spinning black holes. These two papers found that mini-disks at these separations have two states: a disk-dominated phase, where the material orbits a number of times the black hole, and a stream-dominated phase, where most of the mass is advected directly to the black hole. For spinning black holes, the mini-disk is larger, which allows for more thermal radiation due to turbulent dissipation [94]. Moreover, Avara et al. [15] showed that at these separations hydrodynamic stresses dominate the angular momentum transport.

Paschalidis et al. [153] were the first to demonstrate, using equal-mass 3D GRMHD simulations of hot flows, that mini-disk mass increases with higher spin black holes.

14.6 Magnetic Fields and Outflows

Magnetic fields are a crucial ingredient in the engine of an AGN. On a (Keplerian) rotationally supported flow, magnetic fields can trigger the development of turbulent stresses through the MRI driving accretion, and they are also responsible for launching powerful outflows that take energy and angular momentum out from the system. The rotating spacetime of a single black hole can interact with the large-scale fields present in the flow [19, 115] and produce relativistic jets through the Blandford–Znajeck (BZ) mechanism [24]. Many GRMHD simulations have been performed in the last 20 years showing how jets are launched and how they interact with the surrounding medium [55].

The BZ mechanism is an example of how curved spacetime can drastically affect the dynamics of the magnetic field [92]. In a binary black hole spacetime, the acceleration of the black holes, the quadrupole gravitational field, and the spin of the black holes can all contribute to generating a Poynting flux that propagates energy to infinity. This has been demonstrated in the context of force-free plasma analytically [212] and in various numerical relativistic simulations [7, 135, 152]. These simulations evolved a binary on a uniform force-free plasma with a vertical magnetic field, showing that a double jet structure emerges during inspiral, and the Poynting flux increases as they approach the merger. The EM luminosity in this binary merger scenario has two contributions due to spin (BZ) and orbital motion, which is a strong function of velocity :

$$L_{\rm EM} \sim L_{\rm bin} + L_{\rm spin}, \quad L_{\rm bin} \propto v^2 \sim a^2 \Omega_{\rm K}^2, \quad L_{\rm spin} \sim a_\bullet^2, \tag{14.24}$$

where $a_\bullet$ is the black hole spin. This also happens in single-boosted black holes, which acquire an effective ergosphere and generate a Poynting flux as well [42, 144, 154]. Ref. [135] found that the Poynting flux produced by the binary can be divided into a collimated and uncollimated part, which arises from the spin and quadrupole field respectively, see the analysis done in Ref. [212].

A similar scenario was explored in MHD simulations for a uniform plasma with an initial $\beta \sim P/b^2 \sim 10^2$ [83, 106] including work with different spin inclinations [40,

41]. These simulations show a similar phenomenology to force-free: a magnetically-dominated funnel producing a Poytning flux above and below the black hole spin axis, increasing during the merger. Even though a magnetized region is formed, these simulations do not seem to produce outflows but rather inflows, most likely because they start at very small separations, $a \sim 10M$ and there is no time to produce a proper jet before the merger. We notice that simulations of single black holes embedded in uniform magnetized gas have indeed shown jet production, e.g., [166].

The non-linear interaction between magnetic fields and a binary spacetime in a force-free environment ($b^2/\rho \gg 1$) will only occur if the cavity is strongly depleted and the disk still provides sufficient magnetic flux, e.g. at late stages of decoupling (although it is still unclear whether the density can drop to reach a force-free state). On the other hand, the mildly magnetized ($b^2/\rho \sim 1$) uniform gas-cloud scenario could well be realized in underfed AGNs or the interior regions of super-Eddington radiatively inefficient flows; however little radiation will emerge from the accretion in these flows and thus will be difficult to detect in distant galaxies.

Unlike the uniform plasma scenario, most luminous AGNs with a binary engine may have an accretion structure described in earlier sections: a cooled CBD that feeds the black holes and forms mini-disks. To understand what sort of outflows these systems produce, we should ask how large-scale magnetic fields are generated and advected onto the binary.

In this scenario, the magnetic field must be accreted from the CBD. Contrary to single black hole disks where viscous stresses drive the flow near the black hole ISCO, the gas falls into the cavity through thin streams, which compress and amplify the field. Because the lump orbiting the edge of the CBD is a low-magnetized region, the field topology remains an open question. In this direction, Avara et al. [15] showed that the magnetic flux seems to change polarity at the black hole for non-spinning black holes. Noble et al. [145] showed that greater magnetization weakens the lump; this has been shown using an excised cavity (i.e. simulating only the outer disk) and manually injecting flux.

Three-dimensional GRMHD simulations of spinning binary black holes starting with a CBD have been performed in Refs. [48, 153]. Paschalidis et al. [153] evolved a hot accretion torus starting with an initial magnetic seed loop in the CBD for ~ 20 orbits; see Fig. 14.6. For spinning black holes with $a_\bullet = \pm 0.7$, aligned and anti-aligned, they show the emergence of double Poynting-dominated jets, with efficiencies of $\eta \sim 10\%$ as in single black hole jets; they find that the maximum jet efficiency is reached for the aligned case, as expected. Combi et al. [48] evolved a cooled, quasi-steady-state CBD starting also with poloidal magnetic field seeds. In this work, for spinning black holes with $a = 0.6$, it was also shown that the efficiencies are of the same order $\eta \sim 10\%$. The same work demonstrated that the Poynting luminosity acquires the same variability as the accretion rate due to the filling and depleting state in the mini-disks, which might translate to an observable in jet emission.

In Ref. [34], it was remarked and demonstrated that the strength of the variability in the Poynting flux depends on the spin of the black holes: higher spins lead to more steady disks and steady luminosity. This was expected from previous arguments since

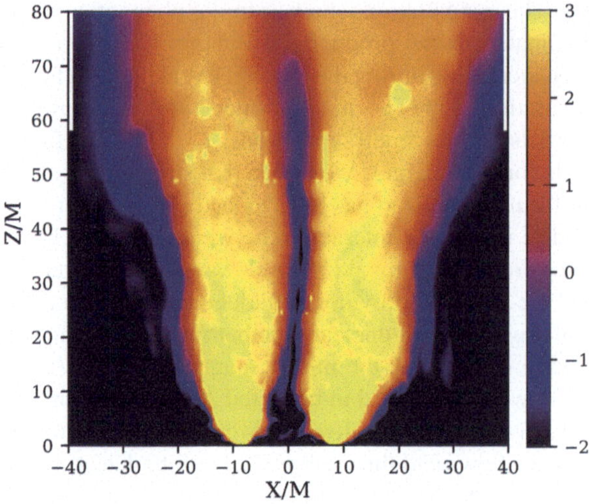

Fig. 14.6 Magnetization ($\sigma = b^2/2\rho$) for orbiting black holes in a full 3D GRMHD simulation of equal-mass binaries with aligned spines of $a = 0.7$. Figure from Ref. [34]

the variability depends on the accretion periodicity of the CBD and the inflow time [31], which in turn, depends on the size of the mini-disks, and thus the radius of the ISCO. To reduce the variability of the accretion rate and the Poynting flux, the inflow time must be much larger than the accretion timescales, see Sect. 1.5.6. One must be careful here because the accretion timescale onto the cavity is very sensitive to the initial data and thermodynamics of the disk. For instance, the larger accretion timescale is associated with the lump $t_{\rm acc} \sim 5P_{\rm bin}$ in a cooled accretion disk; a non-cooled accretion disk however does not show this variability. More simulations with realistic (steady-state) initial conditions are needed to address these questions. In addition, for short orbital separations, [15] showed that the magnetic flux is less variable than the accretion rate in the mini-disk.

14.7 Electromagnetic Signatures and Variability

As we have discussed throughout this chapter, MBHBs surrounded by accretion flows are promising EM emitters. Nevertheless, large uncertainties still exist on the properties of the EM emission and signatures may serve to distinguish them from conventional AGNs with single black holes. In what follows, we summarize what numerical simulations have taught us about the EM emission from MBHBs and how this depends on the binary and accretion disk properties. For more extensive reviews on the EM search of MBHBs, both from a theoretical and an observational point of view, the reader may see Refs. [8, 27, 35, 63, 205]

Several techniques are used to search for accreting MBHBs during the various stages of their evolution. At earlier phases in the merger of the hosting galaxies, the pair of black holes are at large separations and they can appear in the sky as two independent AGNs but in the same galactic nucleus. The best candidate for such a *dual AGN* is found in the radio galaxy 0402+379 [137, 167, 168], but tens of these sources have been identified so far (see Ref. [27] and references therein).

MBHBs at later stages of the evolution are much more difficult to distinguish from single AGNs, and we must rely on periodic modulations in the emission or specific spectral features. Currently, though there are several candidates (e.g., see [65, 91, 101, 149, 199]), none of them have been unambiguously confirmed as a close MBHB (see Ref. [63], Sect. 5, for an up-to-date list of candidates and the mechanism by which they were detected). Theoretical studies predict that a fraction $< 10^{-2}$ of AGNs at low redshift ($z < 0.6$) may host MBHBs [105] (this fraction increases at higher redshifts [204]). However, only a fraction of these binaries could be effectively identified as such through its EM emission [62, 113, 203].

The main distinctive characteristic we expect for the EM emission from close MBHBs is a periodic modulation related to the orbital motion of the SMBHs. This can be caused both by *i) relativistic effects* such as Doppler *beaming* of the emission due to the motion of the mini-disks attached to the black holes [44, 65, 90] or self-lensing flares for edge-on systems where one of the black holes passes in front of the other and largely amplifies its emission [53, 54, 64, 101, 102, 107, 156]; or by *ii) intrinsic variability* in the emitting plasma (accretion disk or jets) [52, 74, 94, 193, 209].

Finally, when the MBHB merges due to the emission of GWs [37], other characteristic EM signals may appear, although these predictions are much less robust than those listed above. Possibilities include disruption of emission from the jet [116], immediate thermal radiation limited by Eddington [112], and a variety of signals due to black hole recoil resulting from the merger [25, 38, 150, 174, 202]. See Ref. [27] for a more extensive discussion on EM signatures at the time of the merger.

14.7.1 Quasi-Periodicities in the Emission

Numerical simulations have shown that the accretion rate onto an MBHB can be strongly modulated, see Sect. 14.5.4 and Fig. 14.2. This is a natural consequence of the influence that the binary motion (which is periodic) exerts on the accretion flow. The amplitude and period of these modulations may depend on the binary properties, such as mass-ratio and eccentricity or black hole spin, and intrinsic physical properties of the accretion disk, such as the magnetization or the cooling.

D'Orazio et al. [66, 67] performed 2D hydrodynamical simulations to investigate the influence of the mass ratio on the EM modulation. The main difference is mediated by the presence (or not) of a lump in the CBD, and the influence it exerts on how fast matter flows towards the cavity and the mini-disks. The lump typically forms for $q \gtrsim 0.2$. Since the cavity is predicted to have some degree of eccentricity, every time the lump approaches the pericenter of its orbit ($P \sim 5P_{\rm bin}$), it overfeeds the binary, which in turn increases the accretion rate and the luminosity. This has been confirmed in numerous 2D VH studies and it has also been found in GRMHD simulations [12,

48, 146]. On the other hand, when the CBD lacks a lump ($q \lesssim 0.2$), this long-period modulation disappears, though modulations at the orbital period and its harmonics persist. These are caused by the interaction of the periodic thrown-back of matter towards the CBD through the streams.

Another property that influences the variability is the eccentricity of the orbit, but for the same fact: for orbits with high eccentricity the lump does not form and the modulations are similar to the circular $q < 0.2$ case, namely at the orbital period and their harmonics. Nevertheless, the variabilities become more bursty for very high eccentricities [185, 217]. Ref. [207] predict modulations on the orbital period that are in phase with the orbit.

Other properties that may alter how emission from accretion MBHB varies are the misalignment between the CBD and binary angular momenta [69, 136, 186] or variable Fe Kα lines due to relativistic effects [103, 124, 213].

Finally, jet emission may dominate over that of the accretion disk during some stages of the evolution or at specific wavelengths. The precession of the jets due to the influence of the binary has been proposed as an explanation of sinusoidal modulation observed in some sources [149]. Finally, for close binaries with mass ratios $q \approx 1$, Gutiérrez et al. [95] proposed that the two jets may interact between them giving rise to flares of nonthermal emission.

14.7.2 Spectral Signatures

Another possibility for the identification of MBHB is by looking at differences in their spectral energy distribution (SED) with respect to standard AGNs, though this may be much more difficult to observe. One predicted signature is the presence of a "notch" in the thermal spectrum [93, 172]: the truncation of the CBD and the presence of a low-density cavity would predict a lack of thermal emission at high energies; if minidisks are present in the cavity and are bright, their emission could cover the highest energy part of the spectrum, leaving a "notch" between the thermal peaks of the CBD and the minidisks. This prediction relies on the assumption that the minidisks and the CBD radiate efficiently, while the *streams* that link both components do not. If these conditions are not fulfilled, the spectrum may lack a clear notch [94].

The presence of shocks both in the CBD and in the mini-disks may change the disk temperature profile and produce a thermal spectrum peaking at higher energies, typically in X-rays [172]. Viscous hydrodynamical simulations with optically thick cooling do predict a high-energy excess but this may be due to the assumption of optically thick emission in the cavity, which may not be true [74, 193, 209].

Most simulations link the luminosity of the disk to the accretion rate. Though this is the case for a standard steady-state disk, it is not clear that the observed luminosity in accreting MBHBs will directly track variations in the simulation-measured accretion. If the accretion timescale is longer than the modulation period for the accretion rate, then the mini-disks may buffer the accretion rate variability and decrease the amplitude of the variability. GRMHD simulations of mini-disk accretion [34, 48] have shown that higher spins produce more massive and larger minidisks, and the viscous timescale becomes longer than the re-filling period. Though these mini-disks are brighter than for non-spinning black holes, the variability is damped. The damping depends, however, on the accretion timescales and so it is not clear at what point the variability disappears. Shocks can also induce periodic modulations which depend on the thermodynamics of the gas.

d'Ascoli et al. [52] used the ray tracing code `Bothros` to predict spectra and light curves of GRMHD simulations performed with `Harm3D` [146]. Gutiérrez et al. [94] extended this approach and studied in detail the spectral and timing properties in the luminosity of accretion flows onto close binaries for both spinning and nonspinning black holes [31, 33, 48]. In these simulations, the cooling function derived from the GRMHD simulations is transformed into a two-component spectrum composed of a blackbody component, associated to the optically-thick emission from the disk, and a power-law component, which represents the inverse Compton emission usually observed in AGNs. The left panel of Fig. 14.7 shows the averaged SED from the simulation with spinning black holes in Ref. [94]. The CBD dominates the emission at lower frequencies whereas the minidisks dominate at higher energies. The emission from the streams falls in the middle of these two regimes and is comparable in brightness to that of the mini-disks. The right panel shows the total SED (CBD+streams+mini-disks) compared to the thermal spectrum from an "equivalent" (same mass and accretion rate) standard Novikov–Thorne (NT) disk [148] around a single black hole. In the binary case, the SED peaks at a lower frequency and the spectrum at frequencies above the peak is a broken power law instead of a single decaying exponential. The overall luminosity is lower in the binary scenario mainly due to the smaller radiative efficiency of mini-disks with respect to standard

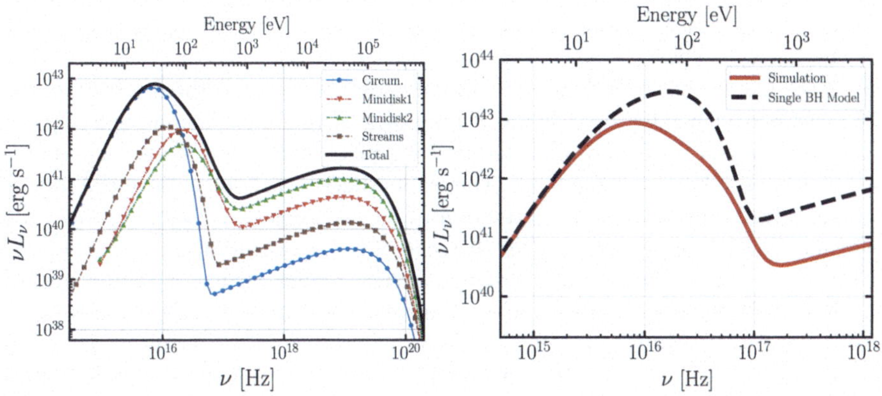

Fig. 14.7 SED from the accretion flow around a binary with equal-mass black holes with spin $a = 0.6$. *Left panel*: Average spectrum for the different components contributing to the spectrum. *Right panel*: Total SED of the MBHB compared with an NT model around a single black hole for the same accretion rate. Figure from Ref. [94]

NT disks. This is because part of the material falling towards the mini-disks has very little angular momentum and thus it can fall into the black hole without suffering very much dissipation.

14.8 The Future: Open Issues

The coalescence of MBHBs is among the most promising events of multimessenger emission for the upcoming years. Hence, it is essential to build a clear understanding of how accretion develops in these systems and what is their EM emission. At the moment, it is still very hard to unambiguously identify such systems and differentiate them from single AGNs.

Accretion flows in MBHBs are more complex than those around single black holes, mainly due to the non-axisymmetric and non-stationary nature of the space-time, which significantly affects the accretion in the innermost regions where most of the EM emission comes out. This also implies that analytical modeling is largely limited, and we need to rely much more on numerical simulations. Moreover, simulating accretion onto MBHBs is itself more complicated and/or more expensive than for single black hole accretion; for this reason, a large majority of the simulations carried out have been 2D Newtonian viscous hydrodynamic simulations. GRMHD simulations have started to be performed in the last years, though with significant limitations in the duration of the simulation or the range of the parameter space that can be covered. GRMHD simulations are essential to understand the interplay of accretion with magnetic fields, to study the production of jets, and to accurately predict the EM emission from the innermost regions of the disks or during the latest stages of the binary evolution, including the merger and the post-merger.

Despite these limitations, we have built robust knowledge about some of the properties of accretion flows around MBHBs. We know that the CBD is typically truncated at a distance $r \sim (2-3)a$, giving rise to a cavity that may be eccentric. Part of the matter that falls from the CBD towards the cavity is thrown back to the CBD and part of it falls towards the black holes. If the latter has enough angular momentum, it can form mini-disks around the black holes, and if the black holes have large spins, they can launch jets. The matter that comes back to the CBD shocks its inner edge and produces the formation of an overdensity, called the lump, that orbits the cavity at approximately the distance of CBD truncation.

The lump plays a large role in mediating the amount of matter falling toward the cavity for a near-equal mass binary. For instance, when one of the black holes passes close to the lump, or when this is at the pericenter of its orbit, accretion towards the cavity increases. These phenomena give rise to a rich set of periodicities in the accretion rate and mass of the mini-disks, which may then translate to the EM emission as a modulated signal.

In the next years, we will need to improve the capabilities of GRMHD simulations to obtain more accurate and realistic predictions and to expand our coverage of the parameter space. For instance, full 3D GRMHD simulations [165] using a semi-analytical metric are now starting to investigate inclined binaries and related effects such as spin-orbit coupling and jet precession. Some of the topics that need to be explored more in-depth are

- How do the initial conditions for the simulations (disk profile and magnetic field topology and amplitude) influence the disk and jet evolution and the EM outcome?
- How long does the disk take to reach a clear steady state? Are there differences with respect to results obtained from hydrodynamic simulations?
- How does the global picture of accreting MBHB change if the disk is misaligned with respect to the orbit?
- Do MBHBs produce jets? How frequent is jet production compared to a single-black hole scenario? If there are two jets, do they interact?
- How does a realistic thermodynamic and radiation transport treatment influence the properties of the CBD and the mini-disks?
- Are there unique signatures in the SED of accreting MBHBs?

Acknowledgements We thank Manuela Campanelli, Julian Krolik, Scott Noble, Huan Yang, Sean Ressler, and Bart Ripperda for fruitful discussions. EG acknowledges funding from the National Science Foundation under grant No AST-2108467 and from an Institute for Gravitation and the Cosmos fellowship. This research was enabled in part by support provided by SciNet (www.scinethpc.ca) and Compute Canada (www.computecanada.ca). LC is a CITA National fellow and acknowledges the support of the Natural Sciences and Engineering Research Council of Canada (NSERC), funding reference DIS-2022-568580. Perimeter Institute for Theoretical Physics. Research at Perimeter Institute is supported in part by the Government of Canada through the Department of Innovation, Science and Economic Development Canada and by the Province of Ontario through the Ministry of Colleges and Universities.

References

1. B.P. Abbott, R. Abbott, T.D. Abbott, M.R. Abernathy, F. Acernese, K. Ackley, C. Adams, T. Adams, P. Addesso, R.X. Adhikari et al., Observation of gravitational waves from a binary black hole merger. Phys. Rev. Lett. **116**(6), 061102 (2016)
2. B.P. Abbott, R. Abbott, T.D. Abbott, F. Acernese, K. Ackley, C. Adams, T. Adams, P. Addesso, R.X. Adhikari, V.B. Adya et al., GW170817: observation of gravitational waves from a binary neutron star inspiral. Phys. Rev. Lett. **119**(16), 161101 (2017)
3. B.P. Abbott, R. Abbott, T.D. Abbott, F. Acernese, K. Ackley, C. Adams, T. Adams, P. Addesso, R.X. Adhikari, V.B. Adya et al., Multi-messenger observations of a binary neutron star merger. Astrophys. J. **848**(2), L12 (2017)
4. R. Abbott, T.D. Abbott, F. Acernese, K. Ackley, C. Adams, N. Adhikari, R.X. Adhikari, V.B. Adya, C. Affeldt, D. Agarwal et al., GWTC-3: compact binary coalescences observed by LIGO and Virgo during the second part of the third observing run. Phys. Rev. X **13**(4), 041039 (2023)
5. G. Agazie, A. Anumarlapudi, A.M. Archibald, Z. Arzoumanian, P.T. Baker, B. Bécsy, L. Blecha, A. Brazier, P.R. Brook, S. Burke-Spolaor, R. Burnette, R. Case, M. Charisi, S. Chatterjee, K. Chatziioannou, B.D. Cheeseboro, S. Chen, T. Cohen, J.M. Cordes, N.J. Cornish, F. Crawford, H.T. Cromartie, K. Crowter, C.J. Cutler, M.E. Decesar, D. Degan, P.B. Demorest, H. Deng, T. Dolch, B. Drachler, J.A. Ellis, E.C. Ferrara, W. Fiore, E. Fonseca, G.E. Freedman, N. Garver-Daniels, P.A. Gentile, K.A. Gersbach, J. Glaser, D.C. Good, K. Gültekin, J.S.

Hazboun, S. Hourihane, K. Islo, R.J. Jennings, A.D. Johnson, M.L. Jones, A.R. Kaiser, D.L. Kaplan, L.Z. Kelley, M. Kerr, J.S. Key, T.C. Klein, N. Laal, M.T. Lam, W.G. Lamb, T.J.W. Lazio, N. Lewandowska, T.B. Littenberg, T. Liu, A. Lommen, D.R. Lorimer, J. Luo, R.S. Lynch, C.-P. Ma, D.R. Madison, M.A. Mattson, A. McEwen, J.W. McKee, M.A. McLaughlin, N. McMann, B.W. Meyers, P.M. Meyers, C.M.F. Mingarelli, A. Mitridate, P. Natarajan, C. Ng, D.J. Nice, S.K. Ocker, K.D. Olum, T.T. Pennucci, B.B.P. Perera, P. Petrov, N.S. Pol, H.A. Radovan, S.M. Ransom, P.S. Ray, J.D. Romano, S.C. Sardesai, A. Schmiedekamp, C. Schmiedekamp, K. Schmitz, L. Schult, B.J. Shapiro-Albert, X. Siemens, J. Simon, M.S. Siwek, I.H. Stairs, D.R. Stinebring, K. Stovall, J.P. Sun, A. Susobhanan, J.K. Swiggum, J. Taylor, S.R. Taylor, J.E. Turner, C. Unal, M. Vallisneri, R. van Haasteren, S.J. Vigeland, H.M. Wahl, Q. Wang, C.A. Witt, O. Young, Nanograv Collaboration, The NANOGrav 15 yr data set: evidence for a gravitational-wave background. Astrophys. J. **951**(1), L8 (2023)

6. G. Agazie, A. Anumarlapudi, A.M. Archibald, P.T. Baker, B. Bécsy, L. Blecha, A. Bonilla, A. Brazier, P.R. Brook, S. Burke-Spolaor, R. Burnette, R. Case, J.A. Casey-Clyde, M. Charisi, S. Chatterjee, K. Chatziioannou, B.D. Cheeseboro, S. Chen, T. Cohen, J.M. Cordes, N.J. Cornish, F. Crawford, H.T. Cromartie, K. Crowter, C.J. Cutler, D.J. D'Orazio, M.E. Decesar, D. Degan, P.B. Demorest, H. Deng, T. Dolch, B. Drachler, E.C. Ferrara, W. Fiore, E. Fonseca, G.E. Freedman, E. Gardiner, N. Garver-Daniels, P.A. Gentile, K.A. Gersbach, J. Glaser, D.C. Good, K. Gültekin, J.S. Hazboun, S. Hourihane, K. Islo, R.J. Jennings, A. Johnson, M.L. Jones, A.R. Kaiser, D.L. Kaplan, L.Z. Kelley, M. Kerr, J.S. Key, N. Laal, M.T. Lam, W.G. Lamb, T.J.W. Lazio, N. Lewandowska, T.B. Littenberg, T. Liu, J. Luo, R.S. Lynch, C.-P. Ma, D.R. Madison, A. McEwen, J.W. McKee, M.A. McLaughlin, N. McMann, B.W. Meyers, P.M. Meyers, C.M.F. Mingarelli, A. Mitridate, P. Natarajan, C. Ng, D.J. Nice, S.K. Ocker, K.D. Olum, T.T. Pennucci, B.B.P. Perera, P. Petrov, N.S. Pol, H.A. Radovan, S.M. Ransom, P.S. Ray, J.D. Romano, J.C. Runnoe, S.C. Sardesai, A. Schmiedekamp, C. Schmiedekamp, K. Schmitz, L. Schult, B.J. Shapiro-Albert, X. Siemens, J. Simon, M.S. Siwek, I.H. Stairs, D.R. Stinebring, K. Stovall, J.P. Sun, A. Susobhanan, J.K. Swiggum, J. Taylor, S.R. Taylor, J.E. Turner, C. Unal, M. Vallisneri, S.J. Vigeland, J.M. Wachter, H.M. Wahl, Q. Wang, C.A. Witt, D. Wright, O. Young, Nanograv Collaboration, The NANOGrav 15 yr data set: constraints on supermassive black hole binaries from the gravitational-wave background. Astrophys. J. **952**(2), L37 (2023)
7. D. Alic, P. Moesta, L. Rezzolla, O. Zanotti, J.L. Jaramillo, Accurate simulations of binary black hole mergers in force-free electrodynamics. The Astrophys. J. **754**(1), 36 (2012)
8. P. Amaro-Seoane, J. Andrews, M. Arca Sedda, A. Askar, Q. Baghi, R. Balasov, I. Bartos, S.S. Bavera, J. Bellovary, C.P.L. Berry, E. Berti, S. Bianchi, L. Blecha, S. Blondin, T. Bogdanović, S. Boissier, M. Bonetti, S. Bonoli, E. Bortolas, K. Breivik, P.R. Capelo, L. Caramete, F. Cattorini, M. Charisi, S. Chaty, X. Chen, M. Chruślińska, A.J.K. Chua, R. Church, M. Colpi, D. D'Orazio, C. Danielski, M.B. Davies, P. Dayal, A. De Rosa, A. Derdzinski, K. Destounis, M. Dotti, I. Dutan, I. Dvorkin, G. Fabj, T. Foglizzo, S. Ford, J.-B. Fouvry, A. Franchini, T. Fragos, C. Fryer, M. Gaspari, D. Gerosa, L. Graziani, P. Groot, M. Habouzit, D. Haggard, Z. Haiman, W.-B. Han, A. Istrate, P.H. Johansson, F.M. Khan, T. Kimpson, K. Kokkotas, A. Kong, V. Korol, K. Kremer, T. Kupfer, A. Lamberts, S. Larson, M. Lau, D. Liu, N. Lloyd-Ronning, G. Lodato, A. Lupi, C.-P. Ma, T. Maccarone, I. Mandel, A. Mangiagli, M. Mapelli, S. Mathis, L. Mayer, S. McGee, B. McKernan, M.C. Miller, D.F. Mota, M. Mumpower, S.S. Nasim, G. Nelemans, S. Noble, F. Pacucci, F. Panessa, V. Paschalidis, H. Pfister, D. Porquet, J. Quenby, A. Ricarte, F.K. Röpke, J. Regan, S. Rosswog, A. Ruiter, M. Ruiz, J. Runnoe, R. Schneider, J. Schnittman, A. Secunda, A. Sesana, N. Seto, L. Shao, S. Shapiro, C. Sopuerta, N.C. Stone, A. Suvorov, N. Tamanini, T. Tamfal, T. Tauris, K. Temmink, J. Tomsick, S. Toonen, A. Torres-Orjuela, M. Toscani, A. Tsokaros, C. Unal, V. Vázquez-Aceves, R. Valiante, M. van Putten, J. van Roestel, C. Vignali, M. Volonteri, K. Wu, Z. Younsi, S. Yu, S. Zane, L. Zwick, F. Antonini, V. Baibhav, E. Barausse, A. Bonilla Rivera, M. Branchesi, G. Branduardi-Raymont, K. Burdge, S. Chakraborty, J. Cuadra, K. Dage, B. Davis, S.E. de Mink, R. Decarli, D. Doneva, S. Escoffier, P. Gandhi, F. Haardt, C.O. Lousto, S. Nissanke, J. Nordhaus, R. O'Shaughnessy, S. Portegies Zwart, A. Pound, F. Schussler, O. Sergijenko,

A. Spallicci, D. Vernieri, A. Vigna-Gómez, Astrophysics with the laser interferometer space antenna. Living Rev. Relativity **26**(1), 2 (2023)

9. P. Amaro-Seoane, H. Audley, S. Babak, J. Baker, E. Barausse, P. Bender, E. Berti, P. Binetruy, M. Born, D. Bortoluzzi, J. Camp, C. Caprini, V. Cardoso, M. Colpi, J. Conklin, N. Cornish, C. Cutler, K. Danzmann, R. Dolesi, L. Ferraioli, V. Ferroni, E. Fitzsimons, J. Gair, L. Gesa Bote, D. Giardini, F. Gibert, C. Grimani, H. Halloin, G. Heinzel, T. Hertog, M. Hewitson, K. Holley-Bockelmann, D. Hollington, M. Hueller, H. Inchauspe, P. Jetzer, N. Karnesis, C. Killow, A. Klein, B. Klipstein, N. Korsakova, S.L. Larson, J. Livas, I. Lloro, N. Man, D. Mance, J. Martino, I. Mateos, K. McKenzie, S. T. McWilliams, C. Miller, G. Mueller, G. Nardini, G. Nelemans, M. Nofrarias, A. Petiteau, P. Pivato, E. Plagnol, E. Porter, J. Reiche, D. Robertson, N. Robertson, E. Rossi, G. Russano, B. Schutz, A. Sesana, D. Shoemaker, J. Slutsky, C.F. Sopuerta, T. Sumner, N. Tamanini, I. Thorpe, M. Troebs, M. Vallisneri, A. Vecchio, D. Vetrugno, S. Vitale, M. Volonteri, G. Wanner, H. Ward, P. Wass, W. Weber, J. Ziemer, P. Zweifel, Laser interferometer space antenna. arXiv e-prints. arXiv:1702.00786 (2017)
10. P. Amaro-Seoane, C. Eichhorn, E.K. Porter, R. Spurzem, Binaries of massive black holes in rotating clusters: dynamics, gravitational waves, detection and the role of eccentricity. Mon. Not. R. Astron. Soc. **401**(4), 2268–2284 (2010)
11. P. Amaro-Seoane, A. Sesana, L. Hoffman, M. Benacquista, C. Eichhorn, J. Makino, R. Spurzem, Triplets of supermassive black holes: astrophysics, gravitational waves and detection. Mon. Not. R. Astron. Soc. **402**(4), 2308–2320 (2010)
12. F.G.L. Armengol, L. Combi, M. Campanelli, S.C. Noble, J.H. Krolik, D.B. Bowen, M.J. Avara, V. Mewes, H. Nakano, Circumbinary disk accretion into spinning black hole binaries. The Astrophys. J. **913**(1), 16 (2021)
13. P.J. Armitage, P. Natarajan, Accretion during the merger of supermassive black holes. Astrophys. J. **567**(1), L9–L12 (2002)
14. P. Artymowicz, S.H. Lubow, Dynamics of binary-disk interaction. I. Resonances and disk gap sizes. Astrophys. J. **421**, 651 (1994)
15. M.J. Avara, J.H. Krolik, M. Campanelli, S.C. Noble, D. Bowen, T. Ryu, Accretion onto a supermassive black hole binary before merger. arXiv preprint. arXiv:2305.18538 (2023)
16. S.A. Balbus, J.F. Hawley, A powerful local shear instability in weakly magnetized disks. I. Linear analysis. Astrophys. J. **376**, 214 (1991)
17. I. Bartos, B. Kocsis, Z. Haiman, S. Márka, Rapid and bright stellar-mass binary black hole mergers in active galactic muclei. Astrophys. J. **835**(2), 165 (2017)
18. T.W. Baumgarte, S.L. Shapiro, *Numerical Relativity: Solving Einstein's Equations on the Computer* (Cambridge University Press, 2010)
19. K. Beckwith, J.F. Hawley, J.H. Krolik, Transport of large-scale poloidal flux in black hole accretion. The Astrophys. J. **707**(1), 428 (2009)
20. M.C. Begelman, R.D. Blandford, M.J. Rees, Massive black hole binaries in active galactic nuclei. Nature **287**(5780), 307–309 (1980)
21. P. Berczik, D. Merritt, R. Spurzem, H.-P. Bischof, Efficient merger of binary supermassive black moles in nonaxisymmetric galaxies. Astrophys. J. **642**(1), L21–L24 (2006)
22. J. Binney, S. Tremaine, *Galactic Dynamics* (1987)
23. L. Blanchet, Gravitational radiation from post-Newtonian sources and inspiralling compact binaries. Living Rev. Relativity **17** (2014)
24. R.D. Blandford, R.L. Znajek, Electromagnetic extraction of energy from Kerr black holes. Mon. Not. R. Astron. Soc. **179**(3), 433–456 (1977)
25. L. Blecha, D. Sijacki, L.Z. Kelley, P. Torrey, M. Vogelsberger, D. Nelson, V. Springel, G. Snyder, L. Hernquist, Recoiling black holes: prospects for detection and implications of spin alignment. Mon. Not. R. Astron. Soc. **456**(1), 961–989 (2016)
26. G.R. Blumenthal, S.M. Faber, J.R. Primack, M.J. Rees, Formation of galaxies and large-scale structure with cold dark matter. Nature **311**, 517–525 (1984)
27. T. Bogdanović, M.C. Miller, L. Blecha, Electromagnetic counterparts to massive black-hole mergers. Living Rev. Relativity **25**(1), 3 (2022)
28. H. Bondi, On spherically symmetrical accretion. Mon. Not. R. Astron. Soc. **112**, 195 (1952)

29. M. Bonetti, A. Sesana, F. Haardt, E. Barausse, M. Colpi, Post-Newtonian evolution of massive black hole triplets in galactic nuclei—IV. Implications for LISA. Mon. Not. R. Astron. Soc. **486**(3), 4044–4060 (2019)
30. D.B. Bowen, M. Campanelli, J.H. Krolik, V. Mewes, S.C. Noble, Relativistic dynamics and mass exchange in binary black hole mini-disks. Astrophys. J. **838**(1), 42 (2017)
31. D.B. Bowen, M. Campanelli, J.H. Krolik, V. Mewes, S.C. Noble, Relativistic dynamics and mass exchange in binary black hole mini-disks. Astrophys. J. **838**, 42 (2017)
32. D.B. Bowen, V. Mewes, M. Campanelli, S.C. Noble, J.H. Krolik, M. Zilhão, Quasi-periodic behavior of mini-disks in binary black holes approaching merger. Astrophys. J. **853**(1), L17 (2018)
33. D.B. Bowen, V. Mewes, S.C. Noble, M. Avara, M. Campanelli, J.H. Krolik, Quasi-periodicity of supermassive binary black hole accretion approaching merger. Astrophys. J. **879**(2), 76 (2019)
34. J.C. Bright, V. Paschalidis, Minidisc influence on flow variability in accreting spinning black hole binaries: simulations in full general relativity. Mon. Not. R. Astron. Soc. **520**(1), 392–401 (2023)
35. S. Burke-Spolaor, L. Blecha, T. Bogdanović, J.M. Comerford, J. Lazio, X. Liu, T.J. Maccarone, D. Pesce, Y. Shen, G. Taylor, Supermassive black hole pairs and binaries, in *Science with a Next Generation Very Large Array, volume 517 of Astronomical Society of the Pacific Conference Series*, ed. by E. Murphy (2018), p. 677
36. S. Burke-Spolaor, S.R. Taylor, M. Charisi, T. Dolch, J.S. Hazboun, A.M. Holgado, L.Z. Kelley, T.J.W. Lazio, D.R. Madison, N. McMann, C.M.F. Mingarelli, A. Rasskazov, X. Siemens, J.J. Simon, T.L. Smith, The astrophysics of nanohertz gravitational waves. Astron. Astrophys. Rev. **27**(1), 5 (2019)
37. M. Campanelli, C.O. Lousto, Y. Zlochower, Spin-orbit interactions in black-hole binaries. Phys. Rev. D **74**, 084023 (2006)
38. M. Campanelli, C.O. Lousto, Y. Zlochower, D. Merritt, Large merger recoils and spin flips from generic black-hole binaries. Astrophys. J. **659**, L5–L8 (2007)
39. F. Cattorini, B. Giacomazzo, GRMHD study of accreting massive black hole binaries in astrophysical environment: a review. Astroparticle Phys. **154**, 102892 (2024)
40. F. Cattorini, B. Giacomazzo, F. Haardt, M. Colpi, Fully general relativistic magnetohydrodynamic simulations of accretion flows onto spinning massive black hole binary mergers. Phys. Rev. D **103**(10), 103022 (2021)
41. F. Cattorini, S. Maggioni, B. Giacomazzo, F. Haardt, M. Colpi, S. Covino, Misaligned spinning binary black hole mergers in hot magnetized plasma. The Astrophys. J. Lett. **930**(1), L1 (2022)
42. R. Cayuso, F. Carrasco, B. Sbarato, O. Reula, Astrophysical jets from boosted compact objects. Phys. Rev. D **100**(6), 063009 (2019)
43. S.K. Chakrabarti, The natural angular momentum distribution in the study of thick disks around black holes. Astrophys. J., Part 1 **288**, 1–6 (1985). ISSN 0004-637X
44. M. Charisi, Z. Haiman, D. Schiminovich, D.J. D'Orazio, Testing the relativistic doppler boost hypothesis for supermassive black hole binary candidates. Mon. Not. R. Astron. Soc. **476**(4), 4617–4628 (2018)
45. M. Colpi, Massive binary black holes in galactic nuclei and their path to coalescence. Space Sci. Rev. **183**(1–4), 189–221 (2014)
46. M. Colpi, K. Danzmann, M. Hewitson, K. Holley-Bockelmann, P. Jetzer, G. Nelemans, A. Petiteau, D. Shoemaker, C. Sopuerta, R. Stebbins, N. Tanvir, H. Ward, W.J. Weber, I. Thorpe, A. Daurskikh, A. Deep, I. Fernández Núñez, C. García Marirrodriga, M. Gehler, J.-P. Halain, O. Jennrich, U. Lammers, J. Larrañaga, M. Lieser, N. Lützgendorf, W. Martens, L. Mondin, A. Piris Niño, P. Amaro-Seoane, M. Arca Sedda, P. Auclair, S. Babak, Q. Baghi, V. Baibhav, T. Baker, J.-B. Bayle, C. Berry, E. Berti, G. Boileau, M. Bonetti, R. Brito, R. Buscicchio, G. Calcagni, P.R. Capelo, C. Caprini, A. Caputo, E. Castelli, H.-Y. Chen, X. Chen, A. Chua, G. Davies, A. Derdzinski, V.F. Domcke, D. Doneva, I. Dvorkin, J. María Ezquiaga, J. Gair, Z. Haiman, I. Harry, O. Hartwig, A. Hees, A. Heffernan, S. Husa, D. Izquierdo, N. Karnesis, A. Klein, V. Korol, N. Korsakova, T. Kupfer, D. Laghi, A. Lamberts, S. Larson,

M. Le Jeune, M. Lewicki, T. Littenberg, E. Madge, A. Mangiagli, S. Marsat, I.M. Vilchez, A. Maselli, J. Mathews, M. van de Meent, M. Muratore, G. Nardini, P. Pani, M. Peloso, M. Pieroni, A. Pound, H. Quelquejay-Leclere, A. Ricciardone, E.M. Rossi, A. Sartirana, E. Savalle, L. Sberna, A. Sesana, D. Shoemaker, J. Slutsky, T. Sotiriou, L. Speri, M. Staab, D. Steer, N. Tamanini, G. Tasinato, J. Torrado, A. Torres-Orjuela, A. Toubiana, M. Vallisneri, A. Vecchio, M. Volonteri, K. Yagi, L. Zwick, LISA definition study report. arXiv e-prints. arXiv:2402.07571 (2024)
47. L. Combi, F.G.L. Armengol, M. Campanelli, B. Ireland, S.C. Noble, H. Nakano, D. Bowen, Superposed metric for spinning black hole binaries approaching merger. Phys. Rev. D **104**(4), 044041 (2021)
48. L. Combi, F.G.L. Armengol, M. Campanelli, S.C. Noble, M. Avara, J.H. Krolik, D. Bowen, Minidisk accretion onto spinning black hole binaries: quasi-periodicities and outflows. The Astrophys. J. **928**(2), 187 (2022)
49. L. Combi, S.M. Ressler, A binary black hole metric approximation from inspiral to merger. arXiv e-prints. arXiv:2403.13308 (2024)
50. J. Cuadra, P.J. Armitage, R.D. Alexander, M.C. Begelman, Massive black hole binary mergers within subparsec scale gas discs. Mon. Not. R. Astron. Soc. **393**(4), 1423–1432 (2009)
51. K. Danzmann, LISA Study Team, LISA: laser interferometer space antenna for gravitational wave measurements. Classical Quant. Gravity **13**(11A), A247–A250 (1996)
52. S. d'Ascoli, S.C. Noble, D.B. Bowen, M. Campanelli, J.H. Krolik, V. Mewes, Electromagnetic emission from supermassive binary black holes approaching merger. Astrophys. J. **865**(2), 140 (2018)
53. J. Davelaar, Z. Haiman, Self-lensing flares from black hole binaries: general-relativistic ray tracing of black hole binaries. Phys. Rev. D **105**(10), 103010 (2022)
54. J. Davelaar, Z. Haiman, Self-lensing flares from black hole binaries: observing black hole shadows via light curve tomography. Phys. Rev. Lett. **128**(19), 191101 (2022)
55. S.W. Davis, A. Tchekhovskoy, Magnetohydrodynamics simulations of active galactic nucleus disks and jets. Ann. Rev. Astron. Astrophys. **58**, 407–439 (2020)
56. A. De Rosa, C. Vignali, T. Bogdanović, P.R. Capelo, M. Charisi, M. Dotti, B. Husemann, E. Lusso, L. Mayer, Z. Paragi, J. Runnoe, A. Sesana, L. Steinborn, S. Bianchi, M. Colpi, L. del Valle, S. Frey, K. É. Gabányi, M. Giustini, M. Guainazzi, Z. Haiman, N. Herrera Ruiz, R. Herrero-Illana, K. Iwasawa, S. Komossa, D. Lena, N. Loiseau, M. Perez-Torres, E. Piconcelli, M. Volonteri, The quest for dual and binary supermassive black holes: a multi-messenger view. New Astron. Rev. **86**, 101525 (2019)
57. A. Derdzinski, D. D'Orazio, P. Duffell, Z. Haiman, A. MacFadyen, Evolution of gas disc-embedded intermediate mass ratio inspirals in the LISA band. Mon. Not. R. Astron. Soc. **501**(3), 3540–3557 (2021)
58. S. Detweiler, Pulsar timing measurements and the search for gravitational waves. Astrophys. J. **234**, 1100–1104 (1979)
59. A. Dittmann, G. Ryan, Preventing anomalous torques in circumbinary accretion simulations. arXiv e-prints. arXiv:2102.05684 (2021)
60. A.J. Dittmann, G. Ryan, A survey of disc thickness and viscosity in circumbinary accretion: binary evolution, variability, and disc morphology. Mon. Not. R. Astron. Soc. **513**(4), 6158–6176 (2022)
61. A.J. Dittmann, G. Ryan, M.C. Miller, The decoupling of binaries from their circumbinary disks. Astrophys. J. **949**(2), L30 (2023)
62. C.A. Dong-Páez, M. Volonteri, R.S. Beckmann, Y. Dubois, A. Mangiagli, M. Trebitsch, S.D. Vergani, N.A. Webb, Multi-messenger study of merging massive black holes in the OBELISK simulation: gravitational waves, electromagnetic counterparts, and their link to galaxy and black-hole populations. Astron. Astrophys. **676**, A2 (2023)
63. D.J. D'Orazio, M. Charisi, Observational signatures of supermassive black hole binaries. arXiv e-prints, arXiv:2310.16896 (2023)
64. D.J. D'Orazio, R. Di Stefano, Periodic self-lensing from accreting massive black hole binaries. Mon. Not. R. Astron. Soc. **474**(3), 2975–2986 (2018)

65. D.J. D'Orazio, Z. Haiman, P. Duffell, B.D. Farris, A.I. MacFadyen, A reduced orbital period for the supermassive black hole binary candidate in the quasar PG 1302–102? Mon. Not. R. Astron. Soc. **452**, 2540–2545 (2015)
66. D.J. D'Orazio, Z. Haiman, P. Duffell, A. MacFadyen, B. Farris, A transition in circumbinary accretion discs at a binary mass ratio of 1:25. Mon. Not. R. Astron. Soc. **459**, 2379–2393 (2016)
67. D.J. D'Orazio, Z. Haiman, A. MacFadyen, Accretion into the central cavity of a circumbinary disc. Mon. Not. R. Astron. Soc. **436**, 2997–3020 (2013)
68. M. Dotti, A. Sesana, R. Decarli, Massive black hole binaries: dynamical evolution and observational signatures. Adv. Astron. **2012**, 940568 (2012)
69. S. Doğan, C. Nixon, A. King, D.J. Price, Tearing up a misaligned accretion disc with a binary companion. Mon. Not. R. Astron. Soc. **449**(2), 1251–1258 (2015)
70. P.C. Duffell, A.J. Dittmann, D.J. D'Orazio, A. Franchini, K.M. Kratter, A.B. Penzlin, E. Ragusa, M. Siwek, C. Tiede, H. Wang et al., The santa barabara binary-disk code comparison. arXiv preprint. arXiv:2402.13039 (2024)
71. P.C. Duffell, D. D'Orazio, A. Derdzinski, Z. Haiman, A. MacFadyen, A.L. Rosen, J. Zrake, Circumbinary disks: accretion and torque as a function of mass ratio and disk viscosity. Astrophys. J. **901**(1), 25 (2020)
72. D.J. D'Orazio, P.C. Duffell, Orbital evolution of equal-mass eccentric binaries due to a gas disk: eccentric inspirals and circular outspirals. The Astrophys. J. Lett. **914**(1), L21 (2021)
73. B.D. Farris, P. Duffell, A.I. MacFadyen, Z. Haiman, Binary black hole accretion from a circumbinary disk: gas dynamics inside the central cavity. Astrophys. J. **783**, 134 (2014)
74. B.D. Farris, P. Duffell, A.I. MacFadyen, Z. Haiman, Binary black hole accretion during inspiral and merger. Mon. Not. R. Astron. Soc. **447**(1), L80–L84 (2015)
75. B.D. Farris, R. Gold, V. Paschalidis, Z.B. Etienne, S.L. Shapiro, Binary black hole mergers in magnetized disks: simulations in full general relativity. Phys. Rev. Lett. **109**, 221102 (2012)
76. B.D. Farris, R. Gold, V. Paschalidis, Z.B. Etienne, S.L. Shapiro, Binary black-hole mergers in magnetized disks: simulations in full general relativity. Phys. Rev. Lett. **109**, 221102 (2012)
77. B.D. Farris, Y.T. Liu, S.L. Shapiro, Binary black hole mergers in gaseous environments: "Binary BondiâŁž and "binary Bondi-Hoyle-LyttletonâŁž accretion. Phys. Rev. D **81**(8), 084008 (2010)
78. B.D. Farris, Y.T. Liu, S.L. Shapiro, Binary black hole mergers in gaseous disks: simulations in general relativity. Phys. Rev. D **84**(2), 024024 (2011)
79. B.D. Farris, Y.T. Liu, S.L. Shapiro, Binary black hole mergers in gaseous disks: simulations in general relativity. Phys. Rev. D **84**(2), 024024 (2011)
80. L.G. Fishbone, V. Moncrief, Relativistic fluid disks in orbit around Kerr black holes. Astrophys. J. **207**, 962–976 (1976)
81. J. Frank, A. King, D.J. Raine, *Accretion Power in Astrophysics*, 3rd ed. (2002)
82. B. Giacomazzo, J.G. Baker, M.C. Miller, C.S. Reynolds, J.R. van Meter, General relativistic simulations of magnetized Masmas around merging supermassive black holes. Astrophys. J. **752**, L15 (2012)
83. B. Giacomazzo, J.G. Baker, M.C. Miller, C.S. Reynolds, J.R. van Meter, General relativistic simulations of magnetized plasmas around merging supermassive black holes. The Astrophys. J. Lett. **752**(1), L15 (2012)
84. A. Gilbert, S. Friedlander, D. Serre, *Handbook of Mathematical Fluid Dynamics* (2003)
85. R. Gold, Relativistic aspects of accreting supermassive black hole binaries in their natural habitat: a review. Galaxies **7**(2), 63 (2019)
86. R. Gold, V. Paschalidis, Z.B. Etienne, S.L. Shapiro, H.P. Pfeiffer, Accretion disks around binary black holes of unequal mass: General relativistic magnetohydrodynamic simulations near decoupling. Phys. Rev. D **89**(6) (2014)
87. R. Gold, V. Paschalidis, M. Ruiz, S.L. Shapiro, Z.B. Etienne, H.P. Pfeiffer, Accretion disks around binary black holes of unequal mass: general relativistic MHD simulations of postdecoupling and merger. Phys. Rev. D **90**(10) (2014)

88. P. Goldreich, R. Sari, Eccentricity evolution for planets in gaseous disks. Astrophys. J. **585**(2), 1024–1037 (2003)
89. P. Goldreich, S. Tremaine, The excitation of density waves at the Lindblad and corotation resonances by an external potential. Astrophys. J. **233**, 857–871 (1979)
90. M.J. Graham, S.G. Djorgovski, D. Stern, A.J. Drake, A.A. Mahabal, C. Donalek, E. Glikman, S. Larson, E. Christensen, A systematic search for close supermassive black hole binaries in the Catalina real-time transient survey. Mon. Not. R. Astron. Soc. **453**(2), 1562–1576 (2015)
91. M.J. Graham, S.G. Djorgovski, D. Stern, E. Glikman, A.J. Drake, A.A. Mahabal, C. Donalek, S. Larson, E. Christensen, A possible close supermassive black-hole binary in a quasar with optical periodicity. Nature **518**(7537), 74–76 (2015)
92. S.E. Gralla, T. Jacobson, Spacetime approach to force-free magnetospheres. Mon. Not. R. Astron. Soc. **445**(3), 2500–2534 (2014)
93. K. Gültekin, J.M. Miller, Observable consequences of merger-driven gaps and holes in black hole accretion disks. Astrophys. J. **761**(2), 90 (2012)
94. E.M. Gutiérrez, L. Combi, S.C. Noble, M. Campanelli, J.H. Krolik, F. López Armengol, F. García, Electromagnetic signatures from supermassive binary black holes approaching merger. Astrophys. J. **928**(2), 137 (2022)
95. E.M. Gutiérrez, L. Combi, G.E. Romero, M. Campanelli, Flares from dual jets in supermassive black hole binaries. arXiv e-prints. arXiv:2301.04280 (2023)
96. J.F. Hawley, S.A. Balbus, A powerful local shear instability in weakly magnetized disks. II. Nonlinear evolution. Astrophys. J. **376**, 223 (1991)
97. J.F. Hawley, J.M. Stone, MOCCT: a numerical technique for astrophysical MHD. Comput. Phys. Commun. **89**(1–3), 127–148 (1995)
98. K. Hayasaki, S. Mineshige, H. Sudou, Binary black hole accretion flows in merged galactic nuclei. Publ. Astron. Soc. Japan **59**, 427–441 (2007)
99. L. Hoffman, A. Loeb, Dynamics of triple black hole systems in hierarchically merging massive galaxies. Mon. Not. R. Astron. Soc. **377**(3), 957–976 (2007)
100. K. Holley-Bockelmann, F.M. Khan, Galaxy rotation and rapid supermassive binary coalescence. Astrophys. J. **810**(2), 139 (2015)
101. B.X. Hu, D.J. D'Orazio, Z. Haiman, K.L. Smith, B. Snios, M. Charisi, R. Di Stefano, Spikey: self-lensing flares from eccentric SMBH binaries. Mon. Not. R. Astron. Soc. **495**(4), 4061–4070 (2020)
102. A. Ingram, S.E. Motta, S. Aigrain, A. Karastergiou, A self-lensing binary massive black hole interpretation of quasi-periodic eruptions. Mon. Not. R. Astron. Soc. **503**(2), 1703–1716 (2021)
103. P. Jovanović, V. Borka Jovanović, D. Borka, T. Bogdanović, Composite profile of the Fe Kα spectral line emitted from a binary system of supermassive black holes. Adv. Space Res. **54**(7), 1448–1457 (2014)
104. W. Ju, J.M. Stone, Z. Zhu, Global MHD simulations of accretion disks in cataclysmic variables. I. The importance of spiral shocks. Astrophys. J. **823**, 81 (2016)
105. L. Kelley, M. Charisi, S. Burke-Spolaor, J. Simon, L. Blecha, T. Bogdanovic, M. Colpi, J. Comerford, D. D'Orazio, M. Dotti, M. Eracleous, M. Graham, J. Greene, Z. Haiman, K. Holley-Bockelmann, E. Kara, B. Kelly, S. Komossa, S. Larson, X. Liu, C.P. Ma, S. Noble, V. Paschalidis, R. Rafikov, V. Ravi, J. Runnoe, A. Sesana, D. Stern, M.A. Strauss, V.U, M. Volonteri, Nanograv Collaboration, Multi-messenger astrophysics with pulsar timing arrays. Bull. Am. Astron. Soc. **51**(3), 490 (2019)
106. B.J. Kelly, J.G. Baker, Z.B. Etienne, B. Giacomazzo, J. Schnittman, Prompt electromagnetic transients from binary black hole mergers. Phys. Rev. D **96**, 123003 (2017)
107. B.J. Kelly, Z.B. Etienne, J. Golomb, J.D. Schnittman, J.G. Baker, S.C. Noble, G. Ryan, Electromagnetic emission from a binary black hole merger remnant in plasma: field alignment and plasma temperature. Phys. Rev. D **103**(6), 063039 (2021)
108. F.M. Khan, A. Just, D. Merritt, Efficient merger of binary supermassive black holes in merging galaxies. Astrophys. J. **732**, 89 (2011)

109. S.S. Kimura, K. Murase, I. Bartos, Outflow bubbles from compact binary mergers embedded in active galactic nuclei: cavity formation and the impact on electromagnetic counterparts. Astrophys. J. **916**(2), 111 (2021)
110. A. Klein, E. Barausse, A. Sesana, A. Petiteau, E. Berti, S. Babak, J. Gair, S. Aoudia, I. Hinder, F. Ohme, B. Wardell, Science with the space-based interferometer eLISA: supermassive black hole binaries. Phys. Rev. D **93**(2), 024003 (2016)
111. B. Kocsis, N. Yunes, A. Loeb, Observable signatures of EMRI black hole binaries embedded in thin accretion disks. Phys. Rev. D **84**, 024032 (2011)
112. J.H. Krolik, Estimating the prompt electromagnetic luminosity of a black hole merger. The Astrophys. J. **709**(2), 774–779 (2010)
113. J.H. Krolik, M. Volonteri, Y. Dubois, J. Devriendt, Population estimates for electromagnetically distinguishable supermassive binary black holes. The Astrophys. J. **879**(2), 110 (2019)
114. D. Lai, D.J. Muñoz, Circumbinary accretion: from binary stars to massive binary black holes. Annu. Rev. Astron. Astrophys. **61**, 517–560 (2023)
115. M. Liska, A. Tchekhovskoy, E. Quataert, Large-scale poloidal magnetic field dynamo leads to powerful jets in grmhd simulations of black hole accretion with toroidal field. Mon. Not. R. Astron. Soc. **494**(3), 3656–3662 (2020)
116. F.K. Liu, X.-B. Wu, S.L. Cao, Double?double radio galaxies: remnants of merged supermassive binary black holes. Mon. Not. R. Astron. Soc. **340**(2), 411–416 (2003)
117. Y.T. Liu, S.L. Shapiro, Accretion disks around binary black holes: a quasistationary model. Phys. Rev. D **82**(12), 123011 (2010)
118. G. Lodato, S. Nayakshin, A.R. King, J.E. Pringle, Black hole mergers: can gas discs solve the 'final parsec' problem? Mon. Not. R. Astron. Soc. **398**, 1392–1402 (2009)
119. D. Lynden-Bell, Galactic nuclei as collapsed old qasars. Nature **223**(5207), 690–694 (1969)
120. D. Lynden-Bell, A.J. Kalnajs, On the generating mechanism of spiral structure. Mon. Not. R. Astron. Soc. **157**, 1 (1972)
121. A.I. MacFadyen, M. Milosavljević, An eccentric circumbinary accretion disk and the detection of binary massive black holes. Astrophys. J. **672**(1), 83–93 (2008)
122. L. Major Krauth, J. Davelaar, Z. Haiman, J.R. Westernacher-Schneider, J. Zrake, A. MacFadyen, Self-lensing flares from black hole binaries III: general-relativistic ray tracing of circumbinary accretion simulations. arXiv e-prints. arXiv:2310.19766 (2023)
123. B. McKernan, K.E.S. Ford, I. Bartos, M.J. Graham, W. Lyra, S. Marka, Z. Marka, N.P. Ross, D. Stern, Y. Yang, Ram-pressure stripping of a kicked hill sphere: prompt electromagnetic emission from the merger of stellar mass black holes in an AGN accretion disk. Astrophys. J. **884**(2), L50 (2019)
124. B. McKernan, K.E.S. Ford, B. Kocsis, Z. Haiman, Ripple effects and oscillations in the broad Fe Kα line as a probe of massive black hole mergers. Mon. Not. R. Astron. Soc. **432**(2), 1468–1482 (2013)
125. D. Merritt, M. Milosavljević, Massive black hole binary evolution. Living Rev. Relativity **8**, 8 (2005)
126. R. Mignon-Risse, P. Varniere, F. Casse, On the origin of the lump in circumbinary discs. Mon. Not. R. Astron. Soc. **520**(1), 1285–1295 (2023)
127. M.C. Miller, J.H. Krolik, Alignment of supermassive black hole binary orbits and spins. Astrophys. J. **774**, 43 (2013)
128. M. Milosavljević, D. Merritt, Formation of galactic nuclei. Astrophys. J. **563**(1), 34–62 (2001)
129. M. Milosavljević, D. Merritt, Long-term evolution of massive black hole binaries. Astrophys. J. **596**(2), 860–878 (2003)
130. M. Milosavljević, E.S. Phinney, The afterglow of massive black hole coalescence. Astrophys. J. **622**(2), L93–L96 (2005)
131. R. Miranda, D.J. Muñoz, D. Lai, Viscous hydrodynamics simulations of circumbinary accretion discs: variability, quasi-steady state and angular momentum transfer. Mon. Not. R. Astron. Soc. **466**(1), 1170–1191 (2017)
132. R. Miranda, R.R. Rafikov, Multiple spiral arms in protoplanetary disks: linear theory. The Astrophys. J. **875**(1), 37 (2019)

133. R. Miranda, R.R. Rafikov, Planet-disk interaction in disks with cooling: basic theory. The Astrophys. J. **892**(1), 65 (2020)
134. M.A. Mirza, A. Tahir, F.M. Khan, H. Holley-Bockelmann, A.M. Baig, P. Berczik, F. Chishtie, Galaxy rotation and supermassive black hole binary evolution. Mon. Not. R. Astron. Soc. **470**(1), 940–947 (2017)
135. P. Moesta, D. Alic, L. Rezzolla, O. Zanotti, C. Palenzuela, On the detectability of dual jets from binary black holes. The Astrophys. J. Lett. **749**(2), L32 (2012)
136. M.S.L. Moody, J.-M. Shi, J.M. Stone, Hydrodynamic torques in circumbinary accretion disks. Astrophys. J. **875**(1), 66 (2019)
137. R. Morganti, B. Emonts, T. Oosterloo, Broad H I absorption in the candidate binary black hole 4C37.11 (B2 0402+379). Astron. Astrophys. **496**(2), L9–L12 (2009)
138. P. Mösta, R.E. Taam, P.C. Duffell, Gas flows within cavities of circumbinary disks in eccentric binary protostellar systems. Astrophys. J. **875**(2), L21 (2019)
139. D.J. Muñoz, D. Lai, Pulsed accretion onto eccentric and circular binaries. Astrophys. J. **827**, 43 (2016)
140. D.J. Muñoz, Y. Lithwick, Long-lived eccentric modes in circumbinary disks. Astrophys. J. **905**(2), 106 (2020)
141. D.J. Muñoz, R. Miranda, D. Lai, Hydrodynamics of circumbinary accretion: angular momentum transfer and binary orbital evolution. Astrophys. J. **871**, 84 (2019)
142. B.C. Mundim, H. Nakano, N. Yunes, M. Campanelli, S.C. Noble, Y. Zlochower, Approximate black hole binary spacetime via asymptotic matching. Phys. Rev. D **89**, 084008 (2014)
143. D.J. Muñoz, D. Lai, K. Kratter, R. Miranda, Circumbinary accretion from finite and infinite disks. The Astrophys. J. **889**(2), 114 (2020)
144. D. Neilsen, L. Lehner, C. Palenzuela, E.W. Hirschmann, S.L. Liebling, P.M. Motl, T. Garrett, Boosting jet power in black hole spacetimes. Proc. Natl. Acad. Sci. **108**(31), 12641–12646 (2011)
145. S.C. Noble, J.H. Krolik, M. Campanelli, Y. Zlochower, B.C. Mundim, H. Nakano, M. Zilhão, Mass-ratio and magnetic flux dependence of modulated accretion from circumbinary disks. Astrophys. J. **922**(2), 175 (2021)
146. S.C. Noble, J.H. Krolik, J.F. Hawley, Direct calculation of the radiative efficiency of an accretion disk around a black hole. Astrophys. J. **692**, 411–421 (2009)
147. S.C. Noble, B.C. Mundim, H. Nakano, J.H. Krolik, M. Campanelli, Y. Zlochower, N. Yunes, Circumbinary magnetohydrodynamic accretion into inspiraling binary black holes. The Astrophys. J. **755**(1), 51 (2012)
148. I.D. Novikov, K.S. Thorne, Astrophysics of black holes, in *Black Holes (Les Astres Occlus)*, ed. by C. Dewitt, B.S. Dewitt (1973), pp. 343–450
149. S. O'Neill, S. Kiehlmann, A.C.S. Readhead, M.F. Aller, R.D. Blandford, I. Liodakis, M.L. Lister, P. Mróz, C.P. O'Dea, T.J. Pearson, V. Ravi, M. Vallisneri, K.A. Cleary, M.J. Graham, K.J.B. Grainge, M.W. Hodges, T. Hovatta, A. Lähteenmäki, J.W. Lamb, T.J.W. Lazio, W. Max-Moerbeck, V. Pavlidou, T.A. Prince, R.A. Reeves, M. Tornikoski, P. Vergara de la Parra, J.A. Zensus, The unanticipated phenomenology of the blazar PKS 2131–021: a unique supermassive black hole binary candidate. Astrophys. J. **926**(2), L35 (2022)
150. S.M. O'Neill, M.C. Miller, T. Bogdanović, C.S. Reynolds, J.D. Schnittman, Reaction of accretion disks to abrupt mass loss during binary black hole merger. Astrophys. J. **700**, 859–871 (2009)
151. B. Paczynski, A model of accretion disks in close binaries. Astrophys. J. **216**, 822–826 (1977)
152. C. Palenzuela, L. Lehner, S.L. Liebling, Dual jets from binary black holes. Science **329**(5994), 927–930 (2010)
153. V. Paschalidis, J. Bright, M. Ruiz, R. Gold, Minidisk dynamics in accreting, spinning black hole binaries: simulations in full general relativity. Astrophys. J. **910**(2), L26 (2021)
154. R.F. Penna, Energy extraction from boosted black holes: penrose process, jets, and the membrane at infinity. Phys. Rev. D **91**(8), 084044 (2015)
155. P.C. Peters, Gravitational radiation and the motion of two point masses. Phys. Rev. **136**, B1224–B1232 (1964)

156. K. Porter, S. Noble, E. Gutiérrez, J. Pelle, M. Campanelli, in preparation (2024)
157. M. Preto, I. Berentzen, P. Berczik, R. Spurzem, Fast coalescence of massive black hole binaries from mergers of galactic nuclei: implications for low-frequency gravitational-wave astrophysics. Astrophys. J. **732**(2), L26 (2011)
158. J.E. Pringle, Accretion discs in astrophysics. Annu. Rev. Astron. Astrophys. **19**, 137–162 (1981)
159. J.E. Pringle, The properties of external accretion discs. Mon. Not. R. Astron. Soc. **248**(4), 754–759 (1991)
160. E. Ragusa, R. Alexander, J. Calcino, K. Hirsh, D.J. Price, The evolution of large cavities and disc eccentricity in circumbinary discs. Mon. Not. R. Astron. Soc. **499**(3), 3362–3380 (2020)
161. E. Ragusa, G. Dipierro, G. Lodato, G. Laibe, D.J. Price, On the origin of horseshoes in transitional discs. Mon. Not. R. Astron. Soc. **464**(2), 1449–1455 (2017)
162. E. Ragusa, G. Lodato, D.J. Price, Suppression of the accretion rate in thin discs around binary black holes. Mon. Not. R. Astron. Soc. **460**(2), 1243–1253 (2016)
163. M. Rajagopal, R.W. Romani, Ultra-low-frequency gravitational radiation from massive black hole binaries. Astrophys. J. **446**, 543 (1995)
164. A. Rasskazov, D. Merritt, Evolution of binary supermassive black holes in rotating nuclei. Astrophys. J. **837**(2), 135 (2017)
165. S.M. Ressler, L. Combi, X. Li, B. Ripperda, H. Yang, Black hole-disk interactions in magnetically arrested active galactic nuclei: general relativistic magnetohydrodynamic simulations using a time-dependent, binary metric. arXiv preprint. arXiv:2404.02193 (2024)
166. S.M. Ressler, E. Quataert, C.J. White, O. Blaes, Magnetically modified spherical accretion in grmhd: reconnection-driven convection and jet propagation. Mon. Not. R. Astron. Soc. **504**(4), 6076–6095 (2021)
167. C. Rodriguez, G.B. Taylor, R.T. Zavala, A.B. Peck, L.K. Pollack, R.W. Romani, A compact supermassive binary black hole system. The Astrophys. J. **646**(1), 49–60 (2006)
168. C. Rodriguez, G.B. Taylor, R.T. Zavala, Y.M. Pihlström, A.B. Peck, H I observations of the supermassive binary black hole system in 0402+379. Astrophys. J. **697**(1), 37–44 (2009)
169. J.C. Rodríguez-Ramírez, C.R. Bom, B. Fraga, R. Nemmen, Optical emission model for binary black hole merger remnants travelling through discs of active galactic nuclei. Mon. Not. R. Astron. Soc. **527**(3), 6076–6089 (2024)
170. C. Roedig, M. Dotti, A. Sesana, J. Cuadra, M. Colpi, Limiting eccentricity of subparsec massive black hole binaries surrounded by self-gravitating gas discs. Mon. Not. R. Astron. Soc. **415**(4), 3033–3041 (2011)
171. C. Roedig, J.H. Krolik, M.C. Miller, Observational signatures of binary supermassive black holes. The Astrophys. J. **785**(2), 115 (2014)
172. C. Roedig, J.H. Krolik, M.C. Miller, Observational signatures of binary supermassive black holes. Astrophys. J. **785**, 115 (2014)
173. N. Roos, Galaxy mergers and active galactic nuclei. Astron. Astrophys. **104**, 218–228 (1981)
174. E.M. Rossi, G. Lodato, P.J. Armitage, J.E. Pringle, A.R. King, Black hole mergers: the first light. ArXiv e-prints (2009)
175. M. Ruiz, A. Tsokaros, S.L. Shapiro, General relativistic magnetohydrodynamic simulations of accretion disks around tilted binary black holes of unequal mass. Phys. Rev. D **108**(12), 124043 (2023)
176. G. Ryan, A. MacFadyen, Minidisks in binary black hole accretion. Astrophys. J. **835**, 199 (2017)
177. T. Ryu, R. Perna, Z. Haiman, J.P. Ostriker, N.C. Stone, Interactions between multiple supermassive black holes in galactic nuclei: a solution to the final parsec problem. Mon. Not. R. Astron. Soc. **473**(3), 3410–3433 (2018)
178. M.V. Sazhin, Opportunities for detecting ultralong gravitational waves. Soviet Astron. **22**, 36–38 (1978)
179. J.D. Schnittman, Astrophysics of super-massive black hole mergers. Classical Quant. Gravity **30**(24), 244007 (2013)

180. A. Sesana, A. Gualandris, M. Dotti, Massive black hole binary eccentricity in rotating stellar systems. Mon. Not. R. Astron. Soc. **415**(1), L35–L39 (2011)
181. A. Sesana, A. Vecchio, C.N. Colacino, The stochastic gravitational-wave background from massive black hole binary systems: implications for observations with Pulsar Timing Arrays. Mon. Not. R. Astron. Soc. **390**(1), 192–209 (2008)
182. N.I. Shakura, R.A. Sunyaev, Black holes in binary systems. Observational appearance. Astron. Astrophys. **24**, 337–355 (1973)
183. J.-M. Shi, J.H. Krolik, Three-dimensional MHD simulation of circumbinary accretion disks. II. Net accretion rate. Astrophys. J. **807**, 131 (2015)
184. J.-M. Shi, J.H. Krolik, S.H. Lubow, J.F. Hawley, Three-dimensional magnetohydrodynamic simulations of circumbinary accretion disks: disk structures and angular momentum transport. Astrophys. J. **749**, 118 (2012)
185. M. Siwek, R. Weinberger, D.J. Muñoz, L. Hernquist, Preferential accretion and circumbinary disc precession in eccentric binary systems. Mon. Not. R. Astron. Soc. **518**(4), 5059–5071 (2023)
186. J.L. Smallwood, S.H. Lubow, R.G. Martin, Accretion on to a binary from a polar circumbinary disc. Mon. Not. R. Astron. Soc. **514**(1), 1249–1257 (2022)
187. J.M. Stone, M.L. Norman, ZEUS-2D: a radiation magnetohydrodynamics code for astrophysical flows in two space dimensions. I. The hydrodynamic algorithms and tests. Astrophys. J. Suppl. **80**, 753 (1992)
188. J.M. Stone, M.L. Norman, ZEUS-2D: a radiation magnetohydrodynamics code for astrophysical flows in two space dimensions. II. The magnetohydrodynamic algorithms and tests. Astrophys. J. Suppl. **80**, 791 (1992)
189. N.C. Stone, B.D. Metzger, Z. Haiman, Assisted inspirals of stellar mass black holes embedded in AGN discs: solving the 'final au problem'. Mon. Not. R. Astron. Soc. **464**(1), 946–954 (2017)
190. L.B. Szabados, Quasi-local energy-momentum and angular momentum in gr: a review article. Living Rev. Relativity **7**, 1–140 (2004)
191. T. Tanaka, K. Menou, Time-dependent models for the afterglows of massive black hole mergers. Astrophys. J. **714**, 404–422 (2010)
192. T.L. Tanaka, Recurring flares from supermassive black hole binaries: implications for tidal disruption candidates and OJ 287. Mon. Not. R. Astron. Soc. **434**(3), 2275–2288 (2013)
193. Y. Tang, Z. Haiman, A. MacFadyen, The late inspiral of supermassive black hole binaries with circumbinary gas discs in the LISA band. Mon. Not. R. Astron. Soc. **476**(2), 2249–2257 (2018)
194. Y. Tang, A. MacFadyen, Z. Haiman, On the orbital evolution of supermassive black hole binaries with circumbinary accretion discs. Mon. Not. R. Astron. Soc. **469**(4), 4258–4267 (2017)
195. A. Tchekhovskoy, R. Narayan, J.C. McKinney, Efficient generation of jets from magnetically arrested accretion on a rapidly spinning black hole. Mon. Not. R. Astron. Soc.: Lett. **418**(1), L79–L83 (2011)
196. C. Tiede, J. Zrake, A. MacFadyen, Z. Haiman, Gas-driven inspiral of binaries in thin accretion disks. Astrophys. J. **900**(1), 43 (2020)
197. C. Tiede, J. Zrake, A. MacFadyen, Z. Haiman, How binaries accrete: hydrodynamics simulations with passive tracer particles. arXiv e-prints. arXiv:2111.04721 (2021)
198. C. Tiede, J. Zrake, A. MacFadyen, Z. Haiman, How binaries accrete: hydrodynamic simulations with passive tracer particles. Astrophys. J. **932**(1), 24 (2022)
199. M.J. Valtonen, H.J. Lehto, K. Nilsson, J. Heidt, L.O. Takalo, A. Sillanpää, C. Villforth, M. Kidger, G. Poyner, T. Pursimo, S. Zola, J.H. Wu, X. Zhou, K. Sadakane, M. Drozdz, D. Koziel, D. Marchev, W. Ogloza, C. Porowski, M. Siwak, G. Stachowski, M. Winiarski, V.P. Hentunen, M. Nissinen, A. Liakos, S. Dogru, A massive binary black-hole system in OJ287 and a test of general relativity. Nature **452**(7189), 851–853 (2008)
200. E. Vasiliev, F. Antonini, D. Merritt, The final-parsec problem in the collisionless limit. Astrophys. J. **810**(1), 49 (2015)

201. J.-P. D. Villiers, J.F. Hawley, J.H. Krolik, Magnetically driven accretion flows in the kerr metric. i. models and overall structure. The Astrophys. J. **599**(2), 1238–1253 (2003)
202. M. Volonteri, P. Madau, Off-nuclear AGNS as a signature of recoiling massive black holes. The Astrophys. J. Lett. **687**(2), L57 (2008)
203. M. Volonteri, J.M. Miller, M. Dotti, Sub-parsec supermassive binary quasars: expectations at $z < 1$. Astrophys. J. **703**(1), L86–L89 (2009)
204. M. Volonteri, H. Pfister, R. Beckmann, M. Dotti, Y. Dubois, W. Massonneau, G. Musoke, M. Tremmel, Dual AGN in the Horizon-AGN simulation and their link to galaxy and massive black hole mergers, with an excursus on multiple AGN. Mon. Not. R. Astron. Soc. **514**(1), 640–656 (2022)
205. J.-M. Wang, Y.-R. Li, Observational signatures of close binaries of supermassive black holes in active galactic nuclei. Res. Astron. Astrophys. **20**(10), 160 (2020)
206. J.-M. Wang, J.-R. Liu, L.C. Ho, Y.-R. Li, P. Du, Accretion-modified stars in accretion disks of active galactic nuclei: gravitational-wave bursts and electromagnetic counterparts from merging stellar black hole binaries. Astrophys. J. **916**(2), L17 (2021)
207. J.R. Westernacher-Schneider, J. Zrake, A. MacFadyen, Z. Haiman, Multi-band light curves from eccentric accreting supermassive black hole binaries. arXiv e-prints. arXiv:2111.06882 (2021)
208. J.R. Westernacher-Schneider, J. Zrake, A. MacFadyen, Z. Haiman, Multiband light curves from eccentric accreting supermassive black hole binaries. Phys. Rev. D **106**(10), 103010 (2022)
209. J.R. Westernacher-Schneider, J. Zrake, A. MacFadyen, Z. Haiman, Multiband light curves from eccentric accreting supermassive black hole binaries. Phys. Rev. D **106**(10), 103010 (2022)
210. J.R. Westernacher-Schneider, J. Zrake, A. MacFadyen, Z. Haiman, Eccentric minidisks in accreting binaries. The Astrophys. J. **962**(1), 76 (2024)
211. W.F. Winters, S.A. Balbus, J.F. Hawley, Gap formation by planets in turbulent protostellar disks. Astrophys. J. **589**(1), 543–555 (2003)
212. H. Yang, F. Zhang, Plasma-wave generation in a dynamic spacetime. The Astrophys. J. **817**(2), 183 (2016)
213. Q. Yu, Y. Lu, Fe Kα line: a tool to probe massive binary black holes in active galactic nuclei? Astron. Astrophys. **377**, 17–22 (2001)
214. F. Yuan, R. Narayan, Hot accretion flows around black holes. Annu. Rev. Astron. Astrophys. **52**, 529–588 (2014)
215. N. Yunes, B. Kocsis, A. Loeb, Z. Haiman, Imprint of accretion disk-induced migration on gravitational waves from extreme mass ratio inspirals. Phys. Rev. Lett. **107**, 171103 (2011)
216. M. Zilhão, S.C. Noble, M. Campanelli, Y. Zlochower, Resolving the relative influence of strong field spacetime dynamics and MHD on circumbinary disk physics. Phys. Rev. D **91**(2), 024034 (2015)
217. J. Zrake, C. Tiede, A. MacFadyen, Z. Haiman, Equilibrium eccentricity of accreting binaries. Astrophys. J. **909**(1), L13 (2021)

Chapter 15
GRMHD Simulations of Accretion Onto Exotic Compact Objects

Héctor R. Olivares Sánchez, Prashant Kocherlakota, and Carlos A. R. Herdeiro

Abstract Some of the extensions to general relativity and to the Standard Model of particle physics predict families of hypothetical compact objects, collectively known as exotic compact objects (ECOs). This category can be defined to encompass non-Kerr black holes both within and beyond general relativity, as well as horizonless compact objects such as boson stars. In order to model observational signatures and identify possible detections, it is crucial to understand the interaction between these objects and their surrounding medium, usually plasmas described by the equations of general relativistic magnetohydrodynamics (GRMHD). To this end, we review the existent literature on GRMHD simulations of accretion onto these objects. These cover a variety of objects and accretion patterns. We conclude by listing possible directions to continue exploring this relatively young field.

15.1 Introduction

Many of the effects from extensions to general relativity (GR) and the Standard Model of particle physics are expected to manifest in strong gravity environments, and are directly linked to relativistically compact objects. In fact, some of them directly predict new families of hypothetical compact objects, collectively known as exotic compact objects (ECOs). The advances in observing technologies of the

H. R. Olivares Sánchez (✉)
Departamento de Matemática da Universidade de Aveiro and Centre for Research and Development in Mathematics and Applications (CIDMA), Campus de Santiago, 3810-193 Aveiro, Portugal
e-mail: h.sanchez@ua.pt

P. Kocherlakota
Black Hole Initiative at Harvard University, 20 Garden St., Cambridge, MA 02138, USA
e-mail: pkocherlakota@fas.harvard.edu

Center for Astrophysics, Harvard & Smithsonian, 60 Garden St., Cambridge, MA 02138, USA

C. A. R. Herdeiro
Departamento de Matemática da Universidade de Aveiro and Centre for Research and Development in Mathematics and Applications (CIDMA), Campus de Santiago, 3810-193 Aveiro, Portugal
e-mail: herdeiro@ua.pt

C. Bambi et al. (eds.), *New Frontiers in GRMHD Simulations*, Springer Series in Astrophysics and Cosmology, https://doi.org/10.1007/978-981-97-8522-3_15

last decades, such as Very Long Baseline Interferometry (VLBI), gravitational wave (GW) detectors, and the possibility of multi-messenger astronomy, allow us to zoom into the environments of compact objects, with a great potential for new discoveries in astro- and fundamental physics. However, without a proper understanding of the interaction between these objects and their environments, we risk to misinterpret environmental effects as new fundamental physics, or to miss the opportunity to make important discoveries.

Most of the baryonic matter in the Universe is in the state of plasma, and in general can be well described by the equations of general relativistic magnetohydrodynamics (GRMHD). Understanding the observational properties of ECOs in realistic astrophysical environments is one of the motivations to perform GRMHD simulations of accretion onto these objects.

Although the field is still quite young, in this review we aim to encompass the existent literature featuring such simulations. We will include not only GRMHD simulations, but also simulations performed in pure general relativistic hydrodynamics (GRHD).

This chapter is organized as follows. In Sect. 15.2, we give an overview of some of the most interesting families of ECOs, as well as some pre-requisites for their consideration as viable models for compact objects in our Universe. Later, in Sect. 15.3, we discuss some important issues that is necessary to consider from the point of view of modeling before embarking in performing GRMHD simulations. An actual review of the results currently available in the literature is presented in Sect. 15.4. Finally, in Sect. 15.5 we conclude by giving some future perspectives on the field.

15.2 A View on the Landscape of Exotic Compact Objects

The most compact objects in the Universe are now widely believed to be black holes (BHs). Under the *Kerr hypothesis*, moreover, they are believed to be well modelled by the vacuum BHs of GR [86, 127]. For a large part of their century-old history, however, BHs were speculative, or exotic, solutions of GR. Observational evidence [105] together with the (unexpected) theoretical robustness of these solutions [116, 142, 147], slowly established their case as physical objects, abundantly realized in the Cosmos. Yet, their subtle phenomenology together with the theoretical puzzles they introduce [7, 65, 112] invite us to ask if the astrophysical BH *candidates* are - all of them [70]—well modelled by the Kerr solution.

There are two qualitatively different departures from the Kerr hypothesis. The first way is that the astrophysical BH candidates are in fact BHs, in the sense of GR, but not Kerr BHs. It is important to observe they could be either non-Kerr BHs in GR or beyond GR, as will be expanded below. The second way is that the astrophysical BH candidates are not BHs; they are compact objects without a horizon. We shall call all such speculative, alternative, dark compact objects ECOs, encompassing both *non-Kerr BHs* and *horizonless ECOs*.

15.2.1 Theoretical Robustness

Precise strong gravity observables, such as gravitational waves and BH imaging, are largely theory-driven: one needs theoretical templates in order to interpret the results. By the same token, one needs such templates for ECOs, if these are to be considered as contenders for the data and to allow a hypothetical detection.

Still, before embarking on the considerable task of obtaining precision phenomenological templates for a generic ECO model, the model should have surpassed several important theoretical criteria, making it worthy of the phenomenological endeavour.

The first criterion is that the ECO should arise in a consistent theoretical model. This does not requires the model to be complete. GR, for instance, is not UV complete; but within its regime of validity it is self-consistent, in particular offering a well-posed Cauchy problem (the hyperbolicity issue) [29]. This is by no means guaranteed for an ECO model. It is elucidative to consider two examples of its failure. The first one is in GR with "matter". Consider the "matter" to be a Proca field, which is appealing because it leads to a popular model of horizonless ECOs called Proca stars [23]. Including apparently innocuous self-interactions for the Proca field leads to a breakdown of hyperbolicity [30–32, 103]. The second one is in modified gravity. Consider Einstein-scalar-Gauss-Bonnet models, which are appealing within higher curvature models because the equations of motion are still second order. Additionally, it has known examples of interesting non-Kerr BHs [9, 17, 46, 50, 74, 83, 87, 131]. However, it was observed that there is a breakdown of hyperbolicity in dynamical evolutions, at least for some classes of scalar-curvature couplings [117].

The second criterion is that the ECO should be sufficiently stable, so that it could play a role in astrophysical processes. Establishing mode stability of a given ECO model may be rather challenging, especially beyond spherical symmetry. Moreover there may be even more subtle instabilities, possibly unseen in a mode stability analysis. One such pathology of recent interest in the context of horizonless ECOs is the *light ring instability*, or *trapping instability*. Light rings are circular, bound, photon orbits that generically exist around BHs [42]. They are closely connected to strong gravity phenomenology, such as the ringdown of a perturbed BH/ECO [25] and the BH shadow [44]. For horizonless ECOs that can emerge from an incomplete gravitational collapse in GR, when they become sufficiently compact to have light rings, these come in pairs [39]. Together with a light ring similar to that of Kerr BHs, that is associated to the aforementioned strong gravity phenomenology, there is a second light ring which, unless, the null energy condition is violated, is a stable, bound photon orbit. These stable orbits are potentially problematic; they form a potential well that can trap modes of null fields (such as gravity), leading to a strong backreaction and thus a spacetime instability. It was argued that this instability is non-linear [85]; moreover, the instability was confirmed to have short time scales in concrete ECO models [41]. As such, this may be a generic issue with horizonless ECO models that have light rings and have a clear formation mechanism. Another possible cause of instability of rotating horizonless ECOs is the existence of an ergoregion [35]. It has been shown, however, that generically the existence of an ergoregion in horizonless ECOs is accompanied by light rings [58]; in other words, light rings arise first when making a spinning object compact, making their associated instability a more fundamental one.

A third and final criterion is that the ECO should have a formation mechanism. Non-Kerr BHs in GR or modified gravity may, in general, form by gravitational collapse, in appropriate circumstances. But the formation of horizonless ECOs may be more delicate. Popular models, such as gravastars [94], fuzzballs [93] or wormholes [102], have no clear, under control formation mechanism, that has been dynamically tested, even though some general principles may exist. By contrast, a model of horizonless ECOs with a tested formation mechanism is the family of bosonic stars, the mechanism being known as gravitational cooling [128]. Together with the established stability of some of these stars (in regions of their parameter space), this has made bosonic stars a popular model of horizonless ECOs.

15.2.2 Horizonless ECOs

Many models of horizonless ECOs have been proposed throughout the years—see the review [26]. They have had different motivations, not always to replace BHs and solve their aforementioned theoretical puzzles; some, in fact, co-exist with BHs. The latter have other motivations, that could be theoretical—such as geons [141], to model classical particles in field theory—or phenomenological—such as bosonic stars (see the review [92])—that could relate to fuzzy dark matter [76]. Most of these, however, do not fulfil one or more of the above criteria. Whereas this does not prevent some phenomenological considerations, e.g. computing geodesics or imaging properties under some academic models, or even full GRMHD simulations on a fixed background, their linear and (especially) non-linear dynamics is often still out of reach.

One exception in which dynamics is under control is the broad family of bosonic stars. These are lumps of a complex, massive scalar or vector field. The scalar stars, known as boson stars, have been known for over half a century [84, 123], although their rotating versions were only found much later [126, 143]. The vector stars, known as Proca stars, have only been constructed and studied over the last decade [23]—see also [68]. To be in the astrophysical BHs mass range, these stars—at least in the simplest models—must be composed by ultralight bosonic particles, in fact, as a Bose-Einstein condensate of very many such ultralight particles in the same quantum state. Their collapse into a BH is prevented from Heisenberg's uncertainty principle, up to a (model dependent) maximal mass. They can form via gravitational cooling [128] and many models (static and spinning) have dynamically stable regions in the parameter space [92], even though some unexpected instabilities have been found as well [73, 124].

It is not surprising, therefore, that there is a considerable literature on strong gravity phenomenology of bosonic stars (by which we mean both the scalar and vector cases, collectively). In particular GRMHD simulations on the background of bosonic stars have been performed, as described below [109]. Since some of the models in [109] are not dynamically stable, this has led to exploring the imaging of other bosonic stars, within the program of the *imitation game*: assessing possible degeneracies in imaging properties between BHs and horizonless ECOs. In [71] it was observed that spherical Proca stars could produce an effective shadow, due to a special property of timelike circular orbits, that could quench the magneto-rotational instability (MRI) driving matter inwards in an accretion disk, effectively creating a void of emitting matter near the centre of the star. It was even possible to find a Proca star for which this effective shadow is similar in size to that of a Schwarzschild BH with the same mass. Subsequently, however, it was found that such Proca stars have an unexpected instability against non-spherical dynamics [73], undermining this model of a BH foil. Still, as a proof of principle, it would be interesting to understand if such void creating mechanism actually holds in a more complete GRMHD analysis.

Spinning Proca stars, on the other hand, are dynamically robust in their fundamental state [124]. It was observed that these stars have a peculiar structure of their timelike, equatorial, counter-rotating, circular orbits, with disconnected regions of stable orbits [49]. This has led to the suggetions that matter accumulated in the regions of such stable orbits could mimic some imaging features of BHs [130]: not only a shadow, but also fine features attributed to the existence of light rings, but in a spacetime without light rings—Fig. 15.1. Again, it would be interesting to test if such features are kept or washed way in a more realistic GRMHD analysis. For other imaging studies of bosonic stars and their comparison with BHs see e.g. [40, 47, 119–122, 138].

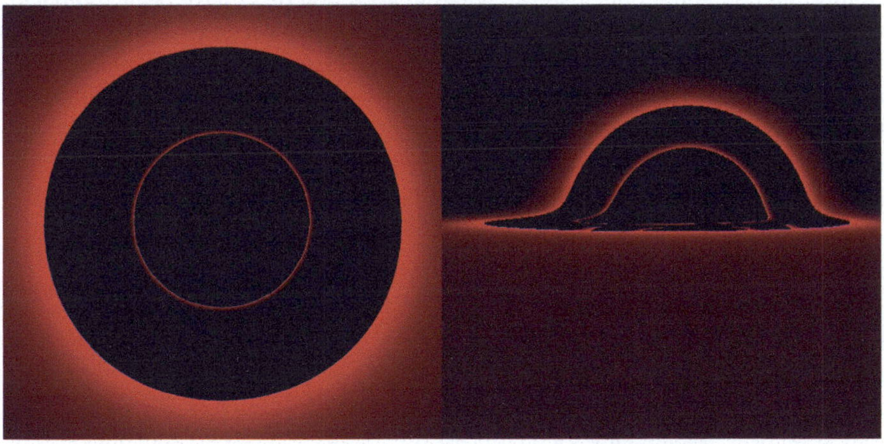

(a) $\theta = 5^o$ (b) $\theta = 88^o$

Fig. 15.1 Lensing images for a spinning Proca star without light rings but with discontinuous regions of equatorial, timelike circular orbits, wherein emitting matter is placed. The images are for two different co-latitude θ observations. From [130]

15.2.3 Non-kerr BHs

The spectrum of non-Kerr BHs both in GR and beyond GR is immense. Imaging properties of such ECOs are now a very popular subject with hundreds of models analysed in more or less academic setups. Again, however, restricting to ECOs abiding the above criteria narrows down significantly the space of models. But since most of imaging analyses rely simply on computing geodesics, this program can be pursued even if the model lacks a strong theoretical support. A broad conclusion is that for most models current observations of Sgr A* and M87* can only constraint part (if any at all) of the parameter space.

GRMHD simulations, on the other hand, have been considered only for a handful of models. One case is dilatonic BHs [101, 118], which is a GR model, since the dilatonic (and axionic) field are minimally coupled to gravity, while non-minimally coupled to the electromagnetic field. These explorations support the expectation that in real-world observations imaging approaches will not be able to clearly distinguish such alternative models from Kerr using the current finite resolution of the EHT.

A similar conclusion can be found for many beyond GR models: the non-Kerrness is small and differences hard to tell from imaging e.g. [43, 46, 52]. But examples where dramatic differences from Kerr occur are also known. One such family is that of BHs with synchronised hair, which interpolates between Kerr BHs and spinning bosonic stars [69, 72]. Some dramatic imaging features can occur, such as chaotic lensing and multiple shadows of a single object [47]. In part of the parameter space such solutions have a natural formation mechanism via the phenomenon of superradiance [24], but in this region deviations from Kerr are smaller [45]. It would be quite interesting to perform GRMHD simulations on these backgrounds. In this respect, equilibrium configurations of self-gravitating accretion disks have been computed, which could be taken as initial data for such simulations [61, 62].

Let us comment that lensing has also been considered for even more exotic (and problematic) spacetimes, such as naked singularities [82, 110, 140]. Pushing GRMHD simulations on these background requires to face issues related to the arbitrariness of boundary conditions near the singularity, even though this may be mitigated for repulsive singularities. An interesting work performing GRHD simulations on a spacetime containing naked singularities can be found in Reference [16].

15.3 Practical Considerations

In broad terms, the aim of a GRMHD simulation can be described as to increase the level of realism of a model for a relativistically compact object by modeling also its interactions with a surrounding medium. In most of the works based on GRMHD simulations of ECOs, the simulation data is further postprocessed to connect this modeling with observables.

During this process, physical models are transformed into numerical models, and it is crucial to ensure that the essential features of physical models are preserved. This translates into making some choices for the numerical model, that are generally related to accommodating the physical model to the limitations of a numerical model. Here we briefly discuss a selection of these practical considerations, and some of the choices that require special attention when simulating ECOs surrounded by the interstellar medium.

General assumptions. All of the works reviewed in this chapter make the assumption that the fluid that makes the interstellar medium is non self-gravitating and its evolution does not back-react on the spacetime metric, so that the only source of gravity is the ECO. This is known as the *test fluid approximation*, and is justified for systems such X-ray binaries and supermassive black hole (SMBH) candidates, where the central object is several orders of magnitude more massive than the surrounding medium. In addition to this, the spacetime metric is not expected to evolve, so simulations are performed on *stationary spacetimes*. Other important assumptions are that the fluid is described by a gas with a simple equation of state (be it ideal or polytropic), and (for those involving magnetic fields), that it is a very good electric conductor. The latter is known as the *ideal magnetohydrodynamics* assumption, and it is generally an accurate description of interstellar plasmas. Finally, even though some of the ECOs that are simulated involve the presence of additional fields (general scalar, axion or dilaton fields), their couplings to the matter in the fluid or to the simulated electromagnetic field are ignored. Further comments on this assumption can be encountered in parts of Sect. 15.4 dealing with specific works in the literature.

Flexibility to adopt custom metrics. To simulate most of the ECO models described in Sect. 15.2, it is essential to endow GRMHD codes with the capability of performing covariant simulations on non-Kerr spacetimes. For analytic metrics, this just involves replacing the expressions for metric functions in the code by the appropriate ones. The applicability of this procedure can be significantly enlarged by the use of parameterized expansions, that can represent several theories by different values of the parameters [27, 28, 101, 104]. For cases where the metric is numerical (an important example are bosonic stars and BHs with synchronized hair), simulations may resort to tabulated metrics [108, 136] or to evaluations of spectral representations of the metric [96, 97].

Initial conditions. Due to the finiteness in time of numerical models, it is necessary to specify the configuration of the fluid at the start of the simulation. Naturally, this is problem-dependent, but we can mention here some generalities. Most of the simulations of magnetized accretion onto BHs and other compact objects start from tori in hydrodynamic equilibrium to which magnetic loops are added. Even though these are the most widely used initial condition for this type of simulations, their applicability is limited to initially small magnetic fields, as their inclusion breaks this equilibrium [36]. Solutions for magnetized tori in true equilibrium have been obtained for boson stars and BHs with scalar hair in [61, 62], although they have so far not been utilized in simulations. In fact, the sub-field of equilibrium solutions in this kind of spacetimes features quite interesting results, that differ in significant ways from those that are possible around Kerr BHs (see e.g. [134, 135]).

Boundary conditions. We also need to specify the behavior of the fluid when interacting with the spatial boundaries of the simulation. Simulations of ECOs can present very different challenges to those of accretion onto Kerr BHs. For objects that possess an event horizon, it is customary to extend the simulation domain slightly inside it in order to prevent signals associated to the boundary from propagating into the simulation and producing spurious effects (see e.g. [37]). For this reason, it is important to use horizon penetrating expressions for the metric. These have been obtained for some general classes of metrics, and due to the importance of the subject, we have dedicated Sect. 15.3.1 to the procedure for obtaining them.

For the case of horizonless compact objects, there are two possibilities: either the simulation domain has a boundary at a physical surface, or it continues down to the center of the object. ECOs with surfaces have not been considered so far for GRMHD simulations, however there are works in the literature that simulate accretion onto objects with physical surfaces such as neutron stars (see Sect. 15.4.1). For cases where the center of the object is part of the computational domain, care must be taken to avoid problems related to the coordinate singularity at the origin of spherical coordinates. This can be done by selecting boundary conditions with the appropriate symmetry properties or by ensuring communication between neighboring cells at each side of the singularity [97, 109]. Another approach is to remove the coordinate singularity by resorting to Cartesian coordinate systems (see e.g. [96, 136]). The case of GRHD simulations in spacetimes with true curvature singularities was explored in [16]. Although the proper way to treat numerically domains that contain such singularities is still an open problem; one can consider (as argued by the authors of [16]) that a naked singularity is a pathology of classical GR which should be replaced by something else in a yet unknown full theory. This motivates setting, for instance, absorbing boundary conditions around the region containing the unknown physics, and excising it from the domain, as done in the cited work.

15.3.1 Horizon-Penetrating Coordinates

Most current codes used to perform GMRHD simulations require, reasonably, the spacetime metric in a horizon-penetrating form, to avoid numerical issues due to the presence of the coordinate singularity at the horizon. For the Kerr metric, specific coordinates, suitable for simulations, were introduced in Refs. [53, 95]. A large class of geodesically-integrable, asymptotically-flat BH spacetimes, characterized by four free functions, were discovered in Ref. [77]. The horizon-penetrating Kerr-Schild form of such Johannsen(-Psaltis) spacetimes was also obtained there (Sect. V) and has recently been used to simulate accretion in such spacetimes [28, 104].

The horizon-penetrating form for yet another broad class of stationary and axisymmetric spacetimes was reported in Ref. [88]. The Azreg-Aïnou (AA) metric, used there, was discovered in Refs. [11–13] in an attempt to obtain stationary and axisymmetric solutions, through an algorithmic procedure similar to the one employed by Newman and Janis [107], from a static and spherically-symmetric "seed" metric. The

AA metric has the value of describing null geodesically-integrable spacetimes, rigidly rotating BHs, and of capturing several well-known solutions to GR and alternative theories, some of which we discuss in Sect. 15.4.2 below.

We now briefly review the steps involved in arriving at coordinates that are suitable for GRMHD simulations, using the AA metric as a simple case study. Its line element in Boyer-Lindquist (BL; [22]) coordinates, $x^\mu = (t, r, \vartheta, \varphi)$, is given as [12, 88]

$$\begin{aligned} \mathrm{d}s^2 = \frac{X}{\Sigma}\Bigg[&-\left(1-\frac{2F}{\Sigma}\right)\mathrm{d}t^2 - 2\left(\frac{2F}{\Sigma}\right)a\sin^2\vartheta\,\mathrm{d}t\mathrm{d}\varphi + \frac{\Pi}{\Sigma}\sin^2\vartheta\,\mathrm{d}\varphi^2 \\ &+\frac{\Sigma}{\Delta}\mathrm{d}r^2 + \Sigma\mathrm{d}\vartheta^2\Bigg], \end{aligned} \tag{15.1}$$

where a denotes the spin parameter and the metric functions are not all independent. In addition to the conformal factor X, there are only two other free functions, $A = A(r)$ and $B = B(r)$. The latter pair determines the remaining metric functions via $2F = A - B$, $\Delta = B + a^2$, $\Sigma = A + a^2\cos^2\vartheta$, and $\Pi = (A + a^2)^2 - \Delta a^2 \sin^2\vartheta$.

To obtain the Kerr metric, one sets $X = \Sigma$, $A = r^2$, and $B = r^2 - 2Mr$. For the Kerr-Newman metric [106], which is the stationary electrovacuum BH solution of the Einstein-Maxwell equations, B is modified to $B = r^2 - 2Mr + Q^2$. In the above, M and Q are the Arnowitt-Deser-Misner (ADM; [10]) mass and charge of the spacetime respectively.

When used to describe BH spacetimes, horizons are permitted at the roots of Δ, where the rr−component of the metric diverges. This coordinate singularity can be eliminated by moving to coordinates, $x^{\bar{\mu}} = (\tau, r, \vartheta, \phi)$, in which the one-form field associated with the ingoing principal null congruence (PNC) of the spacetime, $(\ell_-)_\mu = \left[-\Delta, -\Sigma, 0, \Delta a \sin^2\vartheta\right]$, becomes $(\ell_-)_{\bar{\mu}} = [-1, -1, 0, a\sin^2\vartheta]$. The Jacobian, $\Lambda^{\bar{\mu}}{}_{\mu} = \partial_\mu x^{\bar{\mu}}$, for the desired transformation from BL to these so-called "spherical ingoing Kerr-Schild" (siKS) coordinates is given as

$$\Lambda^{\bar{\mu}}{}_{\mu} = \begin{bmatrix} 1 & 2F/\Delta & 0 & 0 \\ 0 & 1 & 0 & 0 \\ 0 & 0 & 1 & 0 \\ 0 & a/\Delta & 0 & 1 \end{bmatrix}. \tag{15.2}$$

The horizon-penetrating form of the metric (15.1) is then [88]

$$\begin{aligned} \mathrm{d}s^2 = \frac{X}{\Sigma}\Bigg[&-\left(1-\frac{2F}{\Sigma}\right)\mathrm{d}\tau^2 + \left(1+\frac{2F}{\Sigma}\right)\mathrm{d}r^2 + \Sigma\mathrm{d}\vartheta^2 + \frac{\Pi}{\Sigma}\sin^2\vartheta\,\mathrm{d}\phi^2 \\ &-2\left(\frac{2F}{\Sigma}\right)a\sin^2\vartheta\,\mathrm{d}\tau\mathrm{d}\phi + 2\left(\frac{2F}{\Sigma}\right)\mathrm{d}\tau\mathrm{d}r - 2\left(1+\frac{2F}{\Sigma}\right)a\sin^2\vartheta\,\mathrm{d}r\mathrm{d}\phi\Bigg], \end{aligned} \tag{15.3}$$

from which we see that there are no divergences at the horizons ($\Delta = 0$).

The two key ingredients therefore are finding both the 1-form field associated with the ingoing PNC in BL coordinates and the coordinate transformation that transforms it to an accepted standard form, $(\ell_-)_{\bar{\mu}} = [-1, -1, 0, a\sin^2\vartheta]$. Thus, the siKS coordinates are adapted to the ingoing PNC in a specific way, and, thus, penetrate the future horizon. The metric above has recently been used to perform GRMHD simulations in "dilaton-axion" BH spacetimes, as discussed below in Sect. 15.4.2.3. Furthermore, the Jacobian (15.2) can be used to conveniently transform GRMHD data back to BL coordinates, if of interest.

The 3+1 split of the metric (15.3) has also been discussed in Ref. [88], from which it becomes clear that the fiducial choice for the congruence of hypersurface-forming Eulerian observers corresponds to a timelike one that has the same angular velocity as that of the zero angular momentum observer congruence.

It is worth noting that the closely-related set of spherical ingoing *null* Kerr-Schild coordinates, $x^{\bar{\mu}} = (v, r, \vartheta, \phi)$, more commonly used in the GR literature, are obtained by transforming the ingoing PNC generator to the form $(\ell_-)_{\bar{\bar{\mu}}} = [-1, 0, 0, a\sin^2\vartheta]$. From $\ell_-^{\bar{\bar{\mu}}} = -\delta_r^{\bar{\bar{\mu}}}$, we see that these are truly adapted to the ingoing PNC. For the Kerr metric, in particular, this form can be seen, e.g., in Sect. 5.3.6 of Ref. [113]. The Jacobian for the coordinate transformation from BL coordinates is similar to the one in Eq. 15.2 with the term $2F/\Delta$ being modified to $2F/\Delta + 1$.

15.4 GRMHD Simulations of Accretion Onto Exotic Compact Objects

15.4.1 Horizonless Compact Objects

Accretion onto compact objects with no event horizon is of particular interest, as it is expected to produce a qualitatively different phenomenology from that onto BHs, as matter does not disappear behind the event horizon. For compact objects with a surface, such as neutron stars, one would expect an interaction between this surface and the accreting matter, which should result in heating the surface. However, in practice, GRMHD simulations of accretion onto neutron stars employ absorbing boundary conditions [111] or simply remove the matter that falls onto the surface from the simulation [48]. Instead, the physics of the surface is usually accounted for in post-processing, at the stage of calculating images and spectra. For instance, [33] considers the cases of thermally emitting and reflecting surfaces. The results of such analysis apply also to ECOs with surfaces for which the exterior metric is very close to that of Schwarzschild and Kerr BHs, such as gravastars.

More interesting from the point of view of GRMHD simulations are horizonless compact objects that also lack surfaces. Most of the literature on accretion onto this kind of objects refers to boson stars, although there exist at least one work simulating accretion onto Kerr superspinars (Sect. 15.4.1.4) and another one considering the case of wormholes (Sect. 15.4.1.5). Boson stars can be considered as surfaceless objects when the scalar field that constitutes them has very weak couplings with the particles of the standard model. In such case, the exotic object interacts with the accretion flow only through gravity, behaving as a clump of dark matter. A similar reasoning can be applied to other ECOs such as Proca stars; however, the available literature is limited to the simulation of accretion onto boson stars, likely due to the simplicity of their construction.

15.4.1.1 Accretion-Ejection in Boson Stars

The earliest simulations are those of [97], which were performed in pure hydrodynamics using the code GR-AMRVAC. In such work, Meliani and collaborators explore different accretion scenarios onto boson star metrics generated by the spectral solver KADATH [64] for the case of a free scalar field (no self-interaction), that is, subject to the potential

$$U(|\Phi|) = m^2|\Phi|^2\,, \tag{15.4}$$

where m is the mass of the scalar field. Part of the aim of such work is to introduce tests to evaluate the reliability of GRHD codes in boson star spacetimes.

They start by considering spherical accretion onto boson stars, the analogue of Bondi-Michel accretion onto BHs [20, 98], which is widely used as a test for GRHD codes. These simulations are one-dimensional, in spherical symmetry, and exclude the origin, placing the inner boundary at $r = 1.8\ M$ or $r = 3.1\ M$, where from now on M is the ADM-mass of the boson star. This avoids the fluid shocking with itself and accumulating at the center of the boson star, preserving the stationary character of the Bondi-Michel solution. The authors show that at small distances from the center of the boson star, the behavior of matter differs considerably from the case of BHs.

The second test they present is a stationary torus in hydrodynamic equilibrium around an axially symmetric boson star ($k = 1$, $\omega = 0.81\ m/\hbar$). The torus has constant specific angular momentum $\ell = 4.206$ and is constructed in the way of [1]. The code evolves this stationary solution up to $t = 10^3\ M$ showing still an excellent agreement with the initial condition.

Finally, they consider a non-spherical accretion scenario, where matter is injected in a 20° wedge around the equatorial plane. This wedge is initially a Bondi-Michel flow truncated at $r = 2\ M$. The boson star used for this simulation is spherically symmetric ($k = 0, \omega = 0.771\ m/\hbar$), and the simulation is performed in axisymmetry (2.5D). As the system evolves, at first the inflow is decelerated by the increasing thermal pressure. Soon, the the flow reaches the center of the boson star and produces a shock that propagates outwards until it is halted by the supersonic inflow. Later, a spherical structure sustained by pressure gradients appear and propagates out of the boson star. The only direction in which matter can escape is along the poles. Figure 15.2 shows the accretion-ejection region near the center of the boson star. Part of the flow is redirected toward the accretion flow, while another fraction escapes

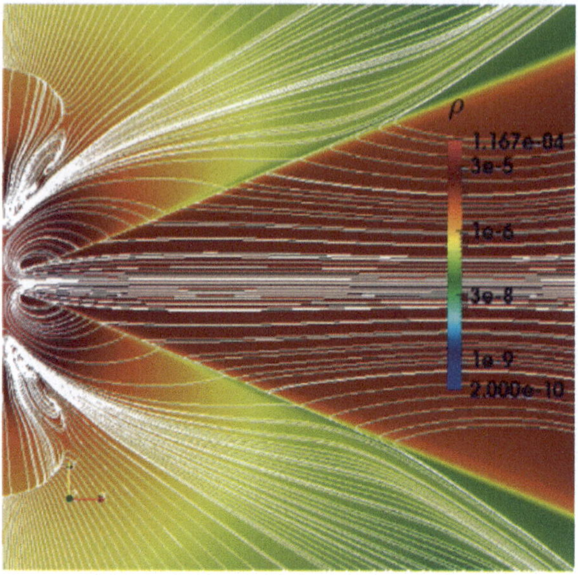

Fig. 15.2 Accretion-ejection region near the center of the boson star in the last simulation by [97]. The color represents logarithmic density, and the white curves are velocity streamlines. From [97]

as a thermal wind. The strong radial pressure gradient causes the acceleration of the outflow, which becomes supersonic for $r > 4\ M$. Finally, interactions between inflow and outflow render the structure unstable, and the shock front changes shape from spherical to triangular. At later stages, the system reaches a quasi-steady state with some variability in the wind intensity. The authors mention that this model does not indicate the presence of Type I (i.e. thermonuclear) X-ray bursts, which are expected to be produced in some models of accretion onto ECOs, and whose absence is considered to strengthen the case of BH candidates in X-ray binaries as true BHs [146]. Overall, this simulation presents a scenario with a phenomenology that is radically different from the case of accretion onto BHs, and invites to consider more diverse scenarios for accretion-ejection flows in compact objects.

15.4.1.2 Tidal Disruption of Clouds by Boson Stars

GRHD simulations have also been employed to study the tidal disruption of gas clouds by boson stars. These studies were in part motivated by the pericenter passage in 2014 of the object G2, likely a dusty, ionized gas cloud, in its eccentric orbit around Sgr A* [60]. simulations of this event were performed by [8, 125], among others. However, for horizonless and surfaceless compact objects such as boson stars, tidal disruption would produce a qualitatively different behavior as for BHs. For example, boson stars allow closed orbits with very low angular momentum (including zero). In contrast, for BHs, a cloud approaching with an angular momentum below some threshold would lose a large fraction of its matter behind the event horizon. In addition, matter that falls inside a boson star can in principle form structures inside and continue to emit electromagnetic radiation that may reach a distant observer.

The first GRHD simulations of tidal disruption of clouds by boson stars appeared in [96], also using the code GR-AMRVAC. They consider three cases of falling clouds and two cases of boson stars. Both of the boson stars are made from a free field and are rotating ($k = 1$, $\omega = 0.8\, m/\hbar$ and $k = 4$, $\omega = 0.8\, m/\hbar$). The gas cloud is initialized at rest at different distances from the boson star (23 and 3 M). It has a Gaussian density profile and is in pressure equilibrium with the ambient medium. Self-gravity is ignored, which is a reasonable approximation, as the gravitational force of the compact object is expected to be much larger due to its proximity. Simulations were two-dimensional and performed in Cartesian quasi-isotropic coordinates on the equatorial plane of the boson stars. This choice avoids spurious effects due to the singularity of the origin of spherical coordinates.

The trajectory of the center of mass of the clouds approximately follows a geodesic. However, close to the boson star, pressure gradients caused by deformations and heating of the cloud produce significant deviations from geodesic motion. At this stage, the cloud acquires a complex structure, which can include shocks propagating against the rotation of the star.

For the case of one of the clouds initialized at a large distance, the combination of tidal stretching, frame dragging and deviations produced by hydrodynamic effects cause the center of mass of the cloud to scatter in the *opposite* direction with respect to the star rotation. In general, the trajectories of the clouds are petal-shaped, with those initialized at large distance showing more elongated petals (left panel of Fig. 15.3).

Following close encounters between the cloud and the boson star, a fraction of the gas is teared off from the cloud by tidal forces. The simulations by [96] show that most of this gas is recaptured by the boson star and forms a disk at its interior, between the last stable orbit and the center of the scalar field torus.

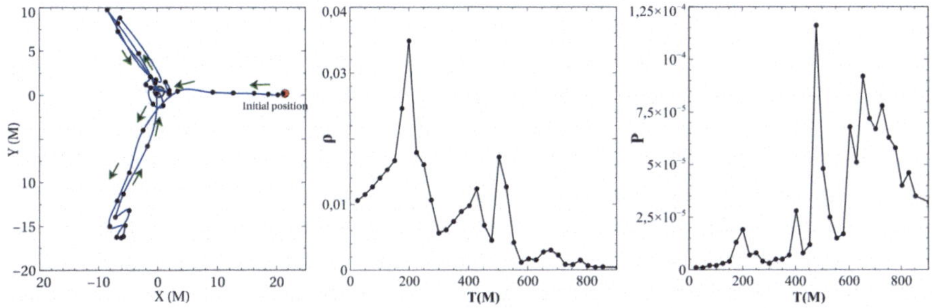

Fig. 15.3 *Left panel:* Trajectory of the center of mass of a cloud initialized at 23 M away from a boson star with $k = 1$ and $\omega = 0.8\ m/\hbar$. *Center and right panels:* Evolution of the density and pressure maxima for the same simulation, showing spikes that can be associated to flares. From [96]

As it could be expected, the passage through the boson star is more violent for the clouds that are initialized at a larger distance. For those cases, the evolution of the pressure and density maximum exhibits several spikes that could be observed as flares (center and right panels of Fig. 15.3). The time between flares ΔT decreases significantly following the decreasing orbit size of the cloud ($\Delta T \sim 150, 250, 100, 60\,M$). In contrast, the cloud that is initialized close to the boson star requires many passages to turn the gas expelled from the cloud into a disk. The spikes in this case appear in larger number and at a more regular separation in time ($\Delta T \sim 50M$). In fact, the orbit of the center of mass is almost completely well described by an ellipse affected by strong precession with an important contribution from frame dragging. The authors suggest that this simulation can be used to study the dynamics of a hot spot orbiting a compact object.

Further simulations of tidal disruption by boson stars were presented by [136]. In such work, the authors perform several simulations for the same boson star model. This is no longer a mini-boson star, but a solitonic boson star, subject to a sextic self-interaction potential

$$U(|\Phi|) = |\Phi|^2(m^2 + a|\Phi|^2 + b|\Phi|^4)\,. \tag{15.5}$$

The parameters chosen for the scalar field are $m = 1, a = -2$, and $b = 1$. The boson star considered has a frequency $\omega = 0.27\,m/\hbar$ and is nonrotating. This object possess a compactness[1] $C = 0.4$, very close to the value $C = 0.5$ characteristic of BHs. The boson star considered in this work is nonrotating. These simulations use the GRMHD code BHAC [108, 114] and similarly to those of [96, 97] are performed in a Cartesian coordinate system. Also in this case, the cloud is not magnetized.

The authors explore four different scenarios for the encounter between a spherical cloud and the boson star, namely, a zero angular momentum orbit, an elliptic orbit, and two circular orbits for different sizes and initial positions of the cloud. For the first three simulations, the cloud is very compact, with a size of $0.03\ M$. Its initial position is at $r = 4\ M$, that is, comparable with one initialized at $r = 3\ M$ by [96]. However, the observed behavior differs in important ways from that reported there. This is due to the fact that the boson star is more compact and no longer rotating.

For the the zero-angular momentum case, the motion is now dominated by local tidal forces, instead of the effect of frame dragging. This results in a much faster deformation of the cloud than for the rotating case, due to effects such as compression/expansion cycles and collisions between the cloud and its tail or other debris. For this reason, the cloud survives for a longer time for the case of the elliptic orbit, and even longer for the case of a circular orbit, where interactions between the cloud and its tail are more difficult.

Similarly to the $r = 3\ M$ case reported by [96], the motion of the center of mass for the case with an elliptic orbit can be well described by a geodesic. This is still

[1] The compactness of an object is defined as $C = R/M$, where R is the radius of the object and M is its mass. Since bosonic stars are extended objects with no clear surface, the authors of [136] define R as the average radial coordinate weighted by the number density of the bosonic particles.

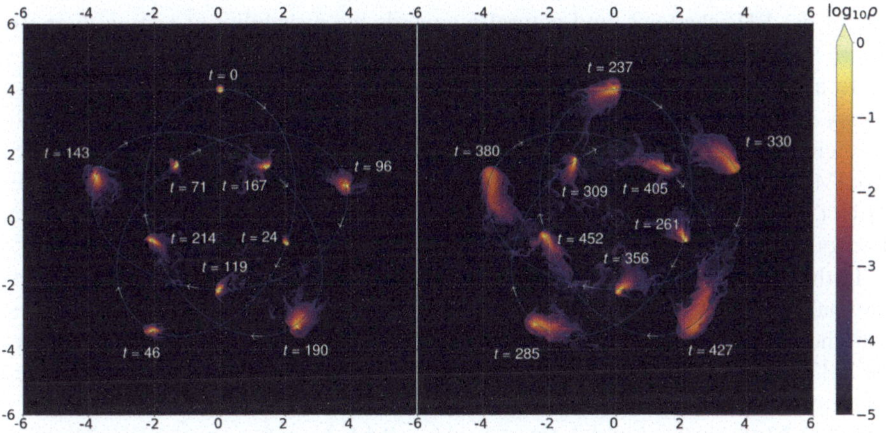

Fig. 15.4 Composite image of a gas cloud closely following a five-petals geodesic around a solitonic star. The cloud experiences several compression-expansion cycles. From [136]

a precessing ellipse (Fig. 15.4), although frame dragging does not contribute to this precession.

For the last boson star-cloud simulation, they place the cloud at a larger distance of $r = 10\ M$ and use a more extended cloud of size $0.3\ M$. The initial velocity of the cloud is that corresponding to a local circular geodesic. They complement this simulation setup with an identical one where the boson star is replaced by a Schwarzschild BH in order to perform a comparison. For this test, they calculate relevant diagnostics such as the mass accretion rate, a proxy from Bremsstrahlung luminosity to produce a synthetic lightcurve, and a shock detector.

They report that after the tearing of the cloud by tidal forces, the debris starts to accumulate and form a ringlike structure around the boson star. The structure also exhibits an extended spiral shock. The persistence of this structure is in stark contrast with the BH case, where matter disappears behind the horizon, and could result in observable differences. In fact, the synthetic lightcurve produced by the authors not only raises orders of magnitude above that of the BH simulation, but also persists with a nearly constant luminosity for the rest of the simulation at $t = 1000\ M$. In contrast the lightcurve for the BH case drops to zero once the entire cloud disappears behind the horizon, at $t = 500\ M$.

15.4.1.3 Magnetized Accretion Onto Boson Stars

Another set of observations that motivated studies on accretion onto boson stars were the very long baseline interferometry (VLBI) campaigns to observe Sgr A*and M 87* by the Event Horizon Telescope (EHT) Collaboration. There were great expectations towards the first images of SMBH candidates at a resolution comparable to the angular size of the BHs shadows, and naturally towards the possibility of testing their BH nature (see e.g. Ref. [139]).

To aid with the interpretation of observations, the EHT Collaboration created a library of GRMHD simulations of Kerr BHs with several spins and accreting in different modes (standard and normal evolution (SANE) and magnetically arrested disk (MAD)). These were later postprocessed by general relativistic radiative transfer (GRRT) codes to produce another library, this time of synthetic images and spectra, that could be compared with the real data [2, 33].

In this context, the work of [109] applied the same pipeline to study the observational properties of a boson star in the same accretion regime as the EHT targets, with the goal of assessing the possibility of distinguishing it from a BH by means of the EHT observations. To this end, the authors performed simulations of magnetized accretion onto two members of the family of nonrotating mini boson stars, namely $\omega M = 0.32$ (hereafter 'model A') and $\omega M = 0.54$ (hereafter 'model B'). Model A belongs to the unstable branch of the mini boson star family; however, the choice of these models was not based on their stability properties, but had the purpose of exploring two qualitatively different behaviors that can take place also in more general surfaceless and horizonless compact objects, and that may play an important role in their observational properties, as we will discuss later.

The GRMHD simulation setup was similar to that used for the simulations in the EHT library. That is, a torus in hydrodynamic equilibrium around a BH was seeded with a poloidal magnetic field loop, and was slightly perturbed by white noise added to the pressure. This triggered the MRI, which amplified the magnetic field, and started transporting angular momentum outwards, causing the torus material to accrete onto the central object. Due to the prescription chosen for the magnetic field, the simulations were in the SANE regime, that is, Maxwell stresses played a secondary role, and most of the angular momentum transport was mediated by the MRI.

The first and most important difference between the two boson star models originates precisely at this stage. In order for the MRI to be active, the angular velocity profile of the accretion disk $\Omega(r) := u^r/u^t$ must decrease with the distance from the center of rotation, that is, $d\Omega/dr < 0$, where u^μ is the 4-velocity of the fluid elements. As the angular momentum is redistributed by the MRI, $\Omega(r)$ acquires a profile close to that of circular timelike geodesics $\Omega_K(r)$, as it is common for SANE simulations (even though not as close as, for example, in thin disk theory). In the Kerr spacetime, $d\Omega_K/dr < 0$ for all radii, and the instability is always active. However, while this is also the case for boson star B, for boson star A there is a radius $r_{\rm turn} > 0$ such that $d\Omega_K/dr > 0$ for $r < r_{\rm turn}$. As $\Omega(r)$ gets closer to $\Omega_K(r)$, it acquires a maximum where $d\Omega(r)/dr \approx 0$, and the instability is suppressed. With no other means of efficient angular momentum transport, the accretion flow becomes stalled and forms a torus inside the boson star, with its inner edge close to $r = r_{\rm turn}$. This structure appears to be stable and lasts until the end of the simulation.

When ray-tracing these simulations with the GRRT code BHOSS [144, 145], the images produced for each of the boson star models are clearly different (left column of Fig. 15.5). Boson star model B has the appearance of a spherical star-like object,

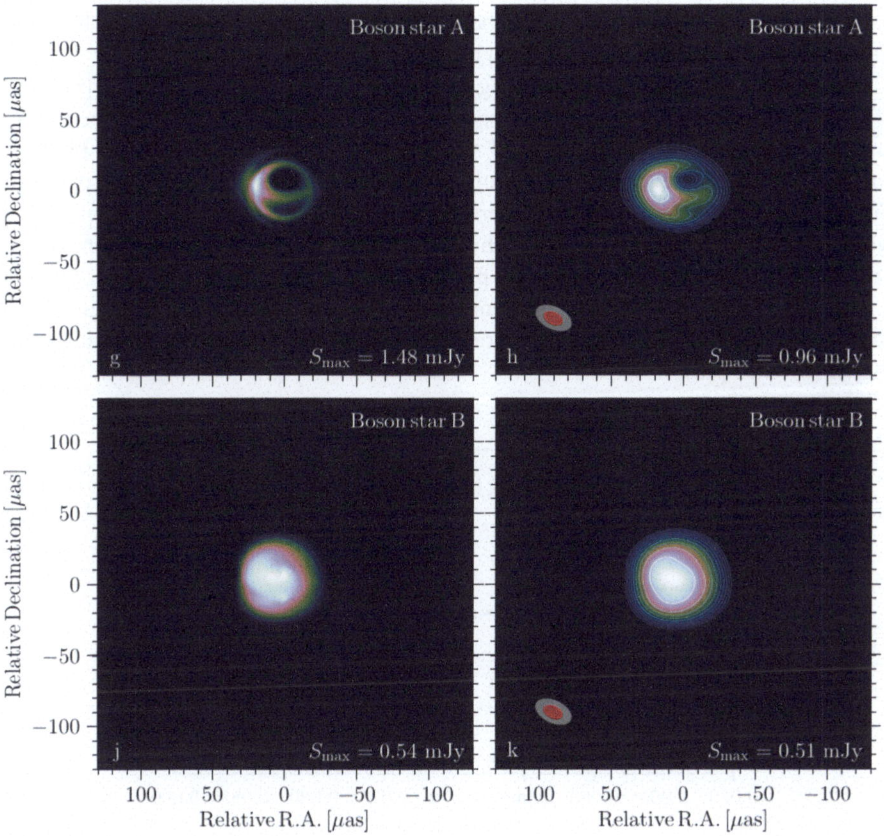

Fig. 15.5 Ray-traced (*left column*) and synthetic (*right column*) images at 230 GHz and inclination angle 60° of boson star models A (*top row*) and B (*bottom row*). From [109]

while boson star model A exhibits a central brightness depression, resembling the typical image of a BH shadow. This brightness depression corresponds, however, to the evacuated region at the center of the torus enhanced by gravitational lensing. The differences are still clearly noticeable when convolving the GRRT image with 50% of the EHT beam in order to simulate a resolution similar to that achievable by the instrument (right column of Fig. 15.5). In particular, the central brightness depression survives.

However, one needs to be cautious before concluding that mini boson stars can mimic supermassive BHs in VLBI observations. One needs to consider, first, that the size of the brightness depression for this boson star is considerably smaller than the one expected for a Kerr BH of the same mass (by 40%), and second, that model A belongs to the unstable branch of the family. As the instability can cause the star to collapse to a BH in a timescale of the order of less than an hour for Sgr A*and less than a month for M 87*, this particular model is clearly not viable. It is therefore natural to ask whether there is a member of the family that is stable and at the same time produces a brightness depression of the appropriate size.

This can be estimated by calculating the radius $r_{\rm turn}$ and the size of the depression after the enhancement by lensing, $b(r_{\rm turn})$, both of which depend only on the metric and are easy to compute in spherical symmetry, even when the metric is numerical. The results for several members of the mini boson star family are presented in Fig. 15.6. It can be seen that no members of the family produce brightness depression with the required size to be mistaken by Kerr BHs given the EHT resolution (of the order of 10 μas), and that these are all produced on the unstable branch. This numerical experiment therefore rules out mini boson stars as viable models for Sgr A*and M 87* under the accretion scenario considered by [109] and under the assumption that $\Omega(r) \approx \Omega_K(r)$.

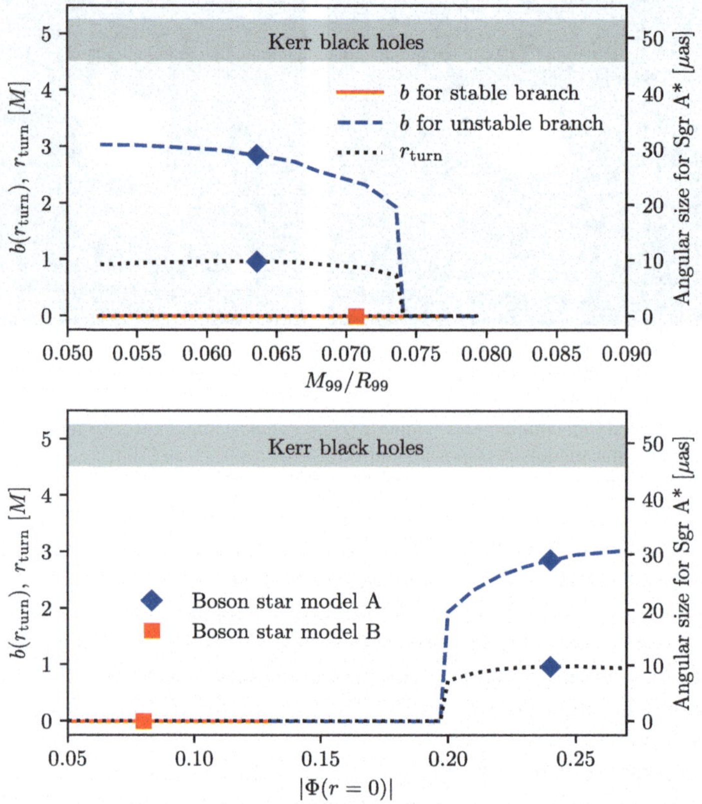

Fig. 15.6 Radial position $r_{\rm turn}$ at which the MRI is suppressed and apparent size of the central brightness depression after lensing $b(r_{\rm turn})$ for the family of mini-boson stars parametrized by compactness (*top panel*) and central amplitude of the scalar field (*bottom panel*). The expected sizes of Kerr BH shadows of the same mass are shown as a shaded region for comparison. From [109]

It is important to remark that this result does not rule out boson stars in general, and consequently neither other horizonless compact objects, as viable models to SMBH canditates. Rather, it presents a scenario where they could act as BH mimickers by a previously unknown mechanism. In fact, it has been shown that other horizonless compact objects, namely Proca stars, could in principle produce brightness depressions comparable to those of Kerr BHs by this mechanism [71]. On the other hand, it is conceivable that other kinds of ECOs produce central brightness depressions by different mechanisms, such as the nonexistence of timelike circular geodesics below some finite radius, a property present in some rotating boson and Proca stars [130].

It is also important to remark that there are important differences that may help distinguishing it from a BH shadow. Most importantly, it would not be achromatic phenomenon as it is the case for the latter. For example, its size may vary with the observing frequency, as it is dependent on the optical depth of the fluid. Although at present the images released by the EHT Collaboration are at the single frequency of 230 GHz, there are plans to provide the instrument with multi-frequency capabilities. This would enable us to test the achromaticity of the BH shadow and potentially rule out scenarios as the one presented here.

Besides this, [109] report other differences in simulations of magnetized accretion onto boson stars with respect to what is observed in BH simulations, such as the absence of a jet, which can be explained in part by the lack of rotation, and therefore of the absence of the Blandford-Znajek mechanism; and the presence of quasi-periodic oscillations in the lightcurve of model B, which correspond to the motion of the plasma density maximum around the center of the boson star.

Before concluding this section, it is worth mentioning that the results presented in [109] have the limitation that they only consider the SANE accretion scenario. However, increasing evidence favors instead the MAD scenario for both Sgr A*and M87* [2, 33]. It is uncertain whether the formation of the central toroidal structure would also occur for the latter, due to the larger deviations from geodesic motion, the secondary role played by the MRI in angular momentum transport, and the dynamical importance of coherent magnetic fields, which may introduce new unexpected effects. This, however, highlights the importance of performing GRMHD for different models of ECOs under a variety of accretion scenarios.

15.4.1.4 Spherical Accretion Onto Naked Singularities

In Reference [16], the authors performed GRHD simulations of quasi-spherical accretion onto a spacetime that contains a naked singularity, namely Kerr spacetimes with dimensionless spin parameter $a \geq 1$, also known as *superspinars* [63].

Their simulations adopt as initial and boundary condition a fluid with constant density and pressure, and excise the singularity from the domain by placing an outflow boundary condition at $r = 0.5\ M$ in Boyer-Lindquist coordinates. This is justified by assuming that the singularity is a pathology of GR that should be replaced by something else in the correct underlying theory, but that the detailed description of its unknown properties will not affect significantly the simulation results. Simulations are 2.5D and are performed using the code PLUTO [99].

The authors found accretion patterns that differ qualitatively from the analogous case for BHs, namely Bondi-Michel accretion. The properties of these accretion patterns appear to depend on spin. Below a critical value of $a \approx 1.4$, matter is unable to reach the central object due to short-distance repulsion, and instead forms a cloud that continues growing without reaching stationarity. In contrast, simulations above that critical value reach a stationary state in which the central object accretes through a thin equatorial stream. The authors discuss some of the astrophysical implications of both scenarios. In particular, a harder and more intense radiation than for the case of accreting BHs is expected, which they link to the abundance of γ-rays observed in some blazars. They also discuss the possibility of violent events resulting from the collapse of the accreting system onto a black hole.

15.4.1.5 Magnetized Accretion Onto Traversable Wormholes

Another interesting example of accretion onto an horizonless, regular, exotic compact object is presented in Reference [34]. In that work, the authors simulate magnetized accretion onto traversable wormholes. Specifically, the spacetimes they consider are described by the Simpson-Visser metric, a two-way traversable wormhole with no rotation. This solution is similar to Schwarzschild from the outside, however it is not in vacuum and requires exotic matter to be sustained. To perform the simulation, the authors assume that this exotic field does not interact with the accreting plasma, and that the latter does not back-react on the field. The simulation set-up is similar to the one described for the boson star case in Sect. 15.4.1.3 and in typical BH-torus simulations with MRI-driven accretion. However, the disk is initialized only at one side of the wormhole.

The authors employ the code Harm3D and spherical Boyer-Lindquist type coordinates. More specifically, the wormhole metric is given by

$$ds^2 = -f(r)dt^2 + f(r)^{-1}dr^2 + (r^2 + \ell^2)d\Omega^2 \tag{15.6}$$

where $f(r) := (1 - 2M/\sqrt{r^2 + \ell^2})$ and, in contrast to typical spherical coordinates, $r \in (-\infty, +\infty)$. The use of these coordinates allows to cover the whole spacetime with a single chart, with the wormhole throat located at $r = 0$. In order to benefit from the properties of radially logarithmic grids, the radial coordinate above is mapped to a numerical coordinate x, such that $r = c_r \sinh(x/c_r)$.

The authors find that, after accretion starts, a portion of the plasma settles at the throat, and forms a rotating cloud roughly 10^4 times denser than the accretion disk. This cloud is hot and larger than the light ring, with its emission peaking on the far UV-X-rays for stellar-mass systems. Although no GRRT calculations where performed, this would likely result in the absence of a central brightness depression that could mimic a BH shadow.

The system does not reach equilibrium. Instead, the cloud grows linearly with time, which could result in gravitational collapse onto a BH in less than a Hubble time (also of stellar-mass systems).

Another interesting finding in Reference [34] is that matter falling into the wormhole powers a mildly relativistic thermal wind at the other side of the throat. The efficiency of this outflow (the energy output divided by mass accretion rate) appears to depend on the size of the throat, and for $\ell = 2.1\ M$, this efficiency is of $\sim$20%. Astronomical observations from this other side of the wormhole would show a BH-like system that exhibits outflows without an associated inflow, contradicting the typical phenomenology expected for BHs. All of the effects described so far are consequences of the fact that the throat is a minimum of the gravitational potential. This feature is shared by most traversable wormholes, which indicates that these results could apply to many other configurations.

The presence of the central cloud and possible absence of a central brightness depression contradicts some previous expectations on wormhole images, which highlights the importance of performing GRMHD simulations in these spacetimes. Other interesting question that could be answered by future simulations are the effects of spacetime rotation and different initial conditions for the accretion flow.

15.4.2 Non-kerr Black Holes

As noted above, the Kerr metric [86] can be used to describe stationary and axisymmetric vacuum BH spacetimes. It is obtained as a solution of the dynamical equations of GR and also in several other alternative classical gravity theories, e.g., those that admit GR as a smooth limit.

Non-Kerr BHs have broadly been considered in two contexts. First, allowing for the presence of self-gravitating matter and solving for the appropriate gravitational as well as matter field equations yields a variety of non-Kerr metrics. Examples in GR of such *solution metrics* include (a) the Reissner-Nordström (RN) spacetime (see, e.g., [100]), which contains a static electromagnetic (EM) field, (b) the Kerr-Newman (KN) spacetime [106], which contains a stationary EM field, (c) the Hayward spacetime [67], which contains a static anisotropic fluid (with no viscosity or heat fluxes), (d) the Gibbons-Maeda-Garfinkle-Horowitz-Strominger (GMGHS); [57, 59] spacetime, which contains a dilaton scalar field as well as an EM field, and (e) the Kerr-Sen (KS) spacetime [129], which contains a dilaton, an EM, and an axion field. The nonspinning RN BH and its spinning generalization, the KN BH, both admit two horizons and also possess curvature singularities at $r = 0$ (in appropriate coordinates). The Hayward BH also admits two horizons but is regular at the center, $r = 0$. The GMGHS BH has a single horizon and the same surface gravity as a Schwarzschild BH, independent of the strength of its field content. Together with the KS BH, its spinning counterpart, these arise as solutions to the low-energy effective limit of the heterotic string theory. Examples of non-Kerr BHs in alternative theories include the aforementioned Einstein-scalar-Gauss-Bonnet BHs.

Perhaps the key ambition behind ascertaining whether or not non-Kerr BH solutions can be distinguished from Kerr BHs is to probe both the various types of BH models as well as their underlying theories of gravity and fields for consistency with observations performed on horizon scales.

The second context that has generated interest in the study of non-Kerr BH spacetimes is the attempt to gather further observational evidence for the Kerr hypothesis. This is achieved by working with *parametrized metrics*, which agnostically deform away from the Kerr metric smoothly, making no reference to the field equations. Several parametrized metrics have been designed to capture qualitatively different aspects of possible metric deviations, e.g., the Johannsen-Psaltis (JP) metric [77, 80], the modified gravity bumpy Kerr metric [54, 137], Konoplya-Rezzolla-Zhidenko metric [90], etc.

The EHT observations of M87* and Sgr A* [3, 5] are particularly suited to such "tests" of the space of stationary non-Kerr metrics [4, 6, 79, 89, 101, 115]. Since the two EHT sources are supermassive ultracompact objects that are accreting at extremely low rates ($\sim 10^{-3} M_\odot \mathrm{yr}^{-1}$ and $\lesssim 10^{-7} M_\odot \mathrm{yr}^{-1}$ respectively; [3, 5]), their spacetimes are stationary to a very good approximation.

15.4.2.1 Hairy Black Hole Moving Through an Uniform Medium

As mentioned previously, an important motivation to explore the consequences of the existence of ultralight bosonic fields comes from dark matter. To understand the effects of the scalar field in a very general situation, the authors of [38] simulated the flow pattern around a BH with synchronized scalar hair, from now on hairy black hole (HBH), moving through a uniform medium. For a star or a BH, this scenario is known as Bondi-Hoyle-Lyttleton (BHL) accretion [21, 75]. BHL accretion onto a BH is relevant to model several situations of interest to Astronomy, such as a wandering BH moving through the interstellar medium, a BH accreting low angular momentum winds from a binary companion, or a BH in the so called common envelop phase of binary evolution.

The HBH model employed is that from [72], where the scalar field is subject to a quadratic potential (Eq. 15.4). As it is customary, simulations are performed in the rest frame of the BH. In this frame, the fluid is moving at Mach number $\mathcal{M} = v_\infty / c_{s,\infty} = 5$, where v_∞ and $c_{s,\infty}$ are the asymptotic fluid velocity and sound speed, respectively. Four simulations are performed, characterized by different ratios between the mass of the scalar cloud and the ADM mass ($p := M_\Phi / M_{\mathrm{ADM}}$), and between the angular momentum of the scalar cloud and the total ADM angular momentum ($q := J_\Phi / J_{\mathrm{ADM}}$). The specific values for all simulations were: $p =$(0.53, 0.749, 0.909, 0.982) and $q =$ (0.128, 0.846, 0.997, 0.998), in the same order.. The simulations are carried out in pure GRHD and in two dimensions on the equatorial plane of the BHs. They employ spherical coordinates and are performed using the code BHAC.

The simulations reveal a flow pattern that is qualitatively similar to that of BHs in BHL accretion, characterized by an upstream bow shock and downstream shock cone. However, the quantitative properties such as the opening angle of the shocks, the location of the stagnation points in units of the BH radius, and the mass accretion rate, depend strongly on the relative importance of the scalar field over the central BH. The authors fit analytical expressions for the mass accretion rate to the numerical results. These could be used to study BH formation and growth scenarios in the presence of ultralight scalar fields.

Similarly to the case of BHL accretion onto BHs [51], the authors observe quasi-periodic oscillations (QPOs) as a result of pressure modes trapped inside the shock cone. The authors study these oscillations by placing detectors close to the stagnation point and producing time series of rest-mass density. They calculate power spectral densities from these time series, and study the effect of the presence of the scalar cloud. Finally, they speculate about the possibility of linking these QPOs with those observed in the X-ray lightcurves of Sgr A*and some microquasars, and compare them with the observational data.

15.4.2.2 Thin Disk Accretion Onto Non-kerr Parameterized Black Hole

The work of [104] performs GRMHD simulations of accretion onto non-Kerr BHs. However, the spacetimes considered do not belong to any particular theory or model. They come instead from the theory agnostic parameterization known as the Johannsen metric [78]. In this parameterization, four radial functions quantify deviations from the Kerr metric, which are written as series in powers of M/r. The parameterized metric retains the separability of the Kerr metric that leads to the Carter constant.

Theory-agnostic approaches have the advantage that the metric can be easily written in a convenient form, and can provide a handle on the effect of deviations from the Kerr metric in tests where it is taken as the null hypothesis. The motivation for [104] comes precisely from the possibility of testing the Kerr hypothesis for stellar mass BHs and SMBHs using X-ray spectroscopy and breaking degeneracies in the ISCO size between the BH spin and deviations from Kerr geometry.

For the GRMHD simulations, the authors employ a horizon-penetrating version of the Johannsen metric. They also keep only the leading terms of the expansion of the deviation functions, which results in four deviation parameters. Simulations are performed using the code `HARMPI` [132], which is based on the `HARM` family [55]. The authors report seven simulations, most of which are performed in 2.5 dimensions on the meridional plane. Only one simulation is performed in three dimensions as a check for the validity of the two-dimensional results. The initial conditions for the simulations are similar to those described in Sect. 15.4.1.3. However, since X-ray reflection spectroscopy is mainly focused on thin accretion disks, a cooling function is added in order to keep the disk thin (with aspect ratio $H/R = 0.12$). Several of the spacetime metrics used share the same value of the ISCO, even when they have different non-Kerr parameteres switched on and off, and some of them are purely Kerr cases.

The simulation results show that the accretion pattern is similar to that seen in typical GRMHD simulations in the Kerr spacetime. The authors also conclude that semianalytic thin disk models for reflection spectra are a reasonable approximation to their simulations. Finally, they show that even though the total flux can be similar between Kerr and non-Kerr cases, iron line plots are visible different, strengthening the case for X-ray spectroscopy as a sensitive test for deviations with respect to Kerr metric.

To finalize this Section, we would like to remark the usefulness of theory-agnostic parameterized metrics for parameter space surveys that may allow to constrain theories of gravity. Even though currently it is impractical to perform such surveys on GRMHD simulations, due to their high computational cost, understanding the effect of changing the parameters on GRMHD simulations can be used to inform semi-analytic models (such as thin disks) for which surveying the parameter space can be made in a much more efficient way.

15.4.2.3 Magnetized Accretion Onto Dilaton-Axion BHs: Approaching Naked Singularities

As noted in Sect. 15.4.1.3, the EHT observations have been used to produce images of its two main targets, which have been compared against synthetic images produced from state-of-the-art numerical simulations. While such simulations have mostly been performed using the Kerr metric, the first simulations of non-Kerr spacetimes were reported in Ref. [101].

The goal of such work was to assess the capabilities of the EHT array to distinguish between different theories of gravity by observing the BH shadows. To this end, three different simulations of spherically symmetric non-Kerr solutions were performed. These were later postprocessed via GRRT calculations to produce images, which where compared to a simulated image of a Kerr BH (see Fig. 15.7). These solutions came from the Einstein-Maxwell-Axion-Dilaton theory [56], which introduces two hypothetical scalar fields (the axion and the dilaton). The three cases considered were selected by matching one of the relevant radii to the value of a Kerr BH, namely, the innermost stable circular orbit (ISCO), the photon ring or the event horizon.

Figure 15.7, taken from there, shows ray-traced and synthetic time-averaged images (for the Sgr A*parameters) of a Kerr BH (top row) and a (non-rotating) GMGHS or "dilaton" BH (bottom row), matched to the ISCO. The right column shows synthetic images taking into account the finite resolution of the EHT array, and demonstrates the difficulty in distinguishing the two BHs from a direct comparison of their images. The inclusion of uncertainties from the instrument and observing conditions renders the two BHs practically indistinguishable.

As a follow up of the work by [101], [118] produced new simulations in these nonspinning spacetimes. These included the MAD scenario and more sophisticated radiation models. Some interesting results included a wider jet opening angle and a higher magnetization in the case of the Kerr spacetime.

However, it was also observed that the emission model has a larger effect on the images as compared to the spacetime metric or the accretion model, leading to the

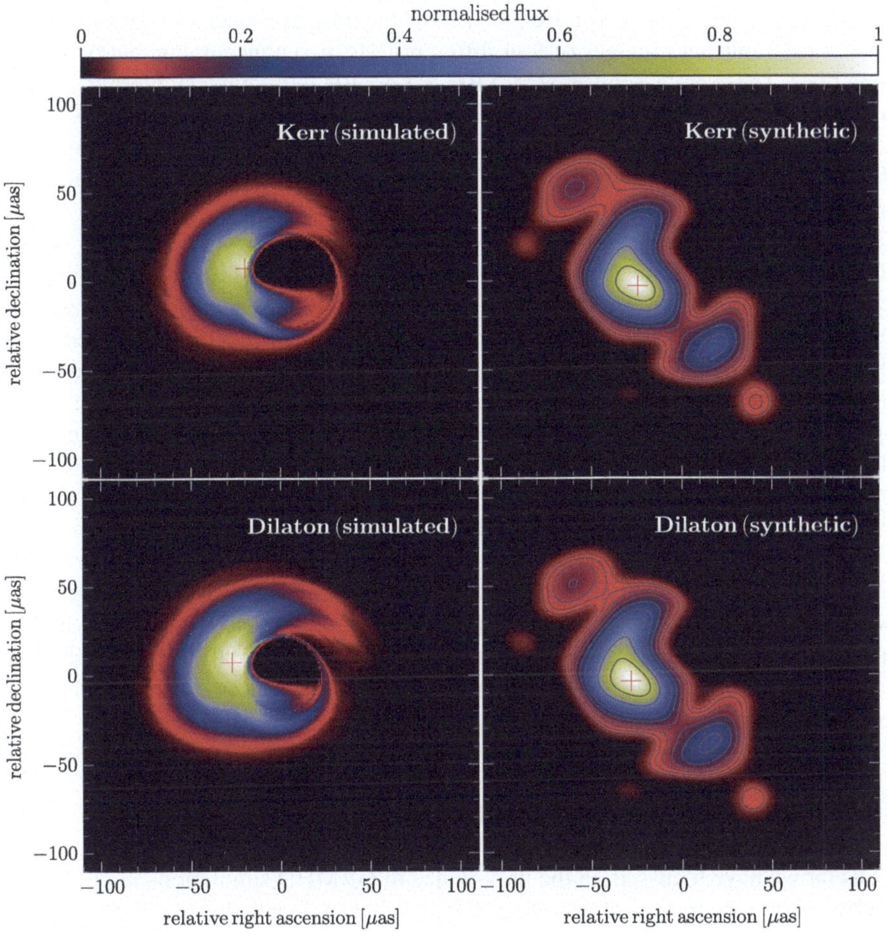

Fig. 15.7 Ray-traced (*left column*) and synthetic (*right columns*) images of a Kerr (*top row*) and a dilaton (*bottom row*) BH simulated in [101]. The two BHs are matched by the ISCO. *Credit image* Christian Fromm, Yosuke Mizuno, Ziri Younsi

conclusion that the imaging approach may not be sufficient to distinguish the two spacetimes with the current resolution of the EHT, as of 2023.

Until recently, accretion in non-Kerr spacetimes has only been explored in non-spinning spacetimes. In Ref. [104], the first 2.5D GRMHD simulation in a spinning non-Kerr spacetime was performed, which we discussed in Sect. 15.4.2.2 above. In Refs. [27, 28], a large number ($\sim$100) of 3D MAD GRMHD simulations in spinning non-Kerr spacetimes were performed. The first of these studies focussed on accretion onto Kerr-Sen (KS) or "dilaton-axion" BHs, extending the earlier work of Refs. [101, 118] discussed above. The metric for these spacetimes takes the form given in Eq. 15.1 above with $A = r(r + 2D)$ and $B = r^2 - 2(M - D)r$.

The right panels of Fig. 1 in Ref. [27] show the time-averaged 230 GHz (the EHT observing frequency) images of four different BHs, two nonspinning and two spinning ones. These are a Schwarzschild BH, a Kerr BH with dimensionless $a_* = 0.5$, a near-extremal dilaton BH with dilaton charge $D = 0.995M$, and a near-extremal dilaton-axion BH with $D = 0.5M$ and $a = 0.49M$. Since the dilaton spacetime with $D = 1M$ describes a *naked singularity* spacetime, the dilaton BH image provides hints regarding the image of such *exotic* possibilities.

The M87* BH parameters are chosen when constructing these images (mass, distance, inclination, etc.). The bright ring in each image is the photon ring [81], a collection of all higher-order images of the entire horizon-scale accretion flow. The photon ring is tightly tied to the BH shadow boundary curve, which is related to its light rings. Observable differences between the sizes of these rings are clear to see in these synthetic images. These differences may be possible to distinguish with future space-based very long baseline interferometry observations, which hope to resolve the photon rings of M87*.

Similarly, images for three classes of Johannsen-Psaltis parametrized BH spacetimes were constructed in Ref. [28]. The parameters allow one to deform smoothly away from a Kerr geometry, and control qualitatively different aspects of the spacetime. For example, the parameter α_{13} changes the size of the shadow boundary curve whereas the parameter α_{22} has a predominant effect on its shape (it changes the BH quadrupole moment) [77]. Thus, measuring the size and shape of the BH shadow from observations can be used to impose constraints on such parameters, and build confidence in the Kerr hypothesis.

15.4.2.4 Energy Extraction from Non-kerr Black Holes

Thus far we have focussed on the importance of GRMHD simulations in producing realistic synthetic images of accreting BHs, which can directly be compared with the highest angular-resolution images of accreting astrophysical BHs. Such simulations, however, provide yet another rich vein of physical insight: Spinning BHs that accrete magentized plasma generically produce "jets," which are relativistic outflows of magentically-dominated plasmas. Indeed, astrophysical jets are frequently observed across the electromagnetic spectrum (see, e.g., the Hubble Space Telescope image of the M87 jet in Ref. [18]), and much work has been devoted to explaining their production mechanism.

One of the leading proposals is the Blandford-Znajek (BZ) mechanism [19], which is an electromagnetic Penrose process involving spinning BHs [91]. The jet outflow power generated from this mechanism is heuristically understood to be given as $P \approx a^2 B^2 r_g^2 c$, where the total energy density in the outflowing plasma is effectively given by the square of the magnetic field strength $\rho \sim B^2$, the outflow speed is $\approx c$, the area of cross-section of the outflow is roughly determined by the gravitational radius $r_g = GM/c^2$, and the spin of the BH is an important modulating factor. The above can be rewritten as $P \approx \Phi^2 (a/r_g^2)$, where $\Phi = B r_g^2$ is the magnetic flux on the horizon, and depends on the specifics of the accretion flow. For Kerr BHs with finite spin, it was shown in Ref. [133] that the remainder $(a/r_g)^2$ should be replaced by Ω_H^2, i.e., the square of the horizon angular velocity, which is determined purely by the spacetime geometry, i.e., $P \approx \Phi^2 \Omega_H^2$. The outflow efficiency η defined as $\eta := P/\dot{M}c^2$ then measures the amount of outflow power per unit of rest mass energy accreted onto the BH. If $\eta > 1$, a portion of the outflow energy must be extracted from the "spin energy" of the BH itself, via an (electromagnetic) Penrose process. Assuming the BZ mechanism, the estimate of the jet power has already been used to test the Kerr hypothesis, as proposed in Ref. [14]. For further discussion, we direct the reader to an excellent review of this topic in Ref. [66].

Figure 15.8, taken from Refs. [27, 28], shows the maximum outflow efficiency η computed from MAD GRMHD simulations in $\sim$100 different Kerr and non-Kerr BH spacetimes. To compare we also show in the dashed line the analytic prediction $\eta_{BZ} \propto (\phi_H \Omega_H)^2$, where $\phi_H := \Phi/\sqrt{\dot{M}}$ is the (saturated) dimensionless horizon magnetic flux. This figure demonstrates how the BZ prediction provides an excellent universal description of the jet power in GRMHD simulations. Comparing these universal jet power predictions against actual observations may provide concrete evidence towards establishing the BZ mechanism as the key driver of astrophysical jets in active galactic nuclei. We also note that non-Kerr BHs can produce significantly higher jet powers than Kerr BHs. The application presented in this Section also showcases the insight that can be gained from libraries of simulations, that in this case were generated efficiently by the GPU-accelerated code H-AMR.

Finally, it is worth noting that, assuming the BZ mechanism, estimates of the jet power can be used to test the Kerr hypothesis, as first proposed in Ref. [15].

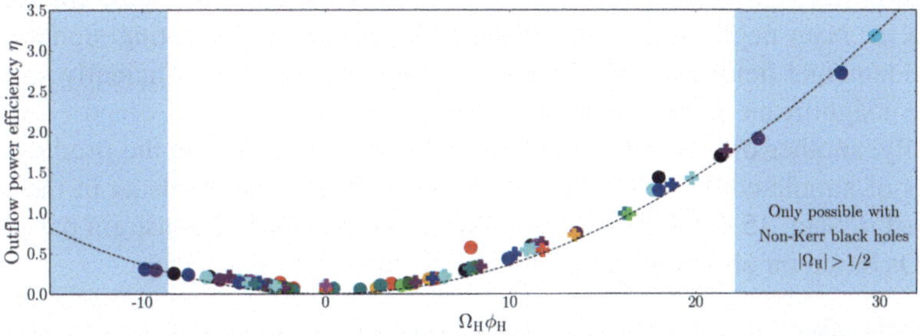

Fig. 15.8 Shown here is the outflow efficiency for Kerr-Sen BHs and Johannsen-Psaltis BHs, and their consistency the BZ prediction. This demonstrates the universality of energy extraction via an electromagnetic Penrose process from various spinning BHs, i.e., Kerr as well as non-Kerr BHs. Adapted from Refs. [27, 28]

15.5 Future Perspectives

In this chapter, we have presented a landscape of some of the most important families of ECOs, summarized some of the considerations that are in order to perform GRMHD simulations on some of these objects, and covered the existent literature on the subject while trying to provide the motivation behind each work.

As it can be noticed, the field is extremely young, and there are still many research directions to explore. For instance, regarding horizonless compact objects, it would be interesting to verify via GRMHD simulations whether it is really possible to produce central brightness depressions similar to BH shadows for objects that are dynamically stable, either by the mechanism described in Sect. 15.4.1.3, or by different mechanisms. Also, there are currently no simulations of rotating horizonless compact objects, even though such solutions exist. These could be used to study the possible formation of jets, as well as to understand what happens realistically in spacetimes that have peculiar geodesic properties, such as the non-existence of circular timelike orbits. Currently there are also no simulations of ECOs that possess surfaces. A model for the interaction between the surface and the accretion flow could yield new observational signatures to constrain the existence of these objects. A different problem, also related to boundary conditions, is that of the treatment of naked singularities within a fluid simulation. This research line would allow to extend the work on imaging such exotic objects, which so far has considered only ray-traced images with no underlying GRMHD model.

Another natural extension of existing simulations could come from taking into consideration interactions between the additional fundamental fields that are present in many ECO models. In fact, some of these fields (e.g. dilatons and axions) are expected to interact with matter or the electromagnetic field, but these interactions have so far been neglected. Such simulations could help predicting signatures of new fundamental fields even in cases where the latter are not sufficiently strong to produce a significant gravitational interaction.

Finally, another direction that we expect to be very fruitful is the production of libraries of simulated data. Besides allowing us to identify trends as in the works described in Sect. 15.4.2.4, this would allow to systematically constrain parameters for ECOs based on observations, and ultimately test their existence.

Acknowledgements We thank Z. Meliani, M. C. Teodoro, C. Fromm, Y. Mizuno and Z. Younsi for granting permission to reproduce their figures. HO is supported by the Individual CEEC program—5th edition funded by the Portuguese Foundation for Science and Technology (FCT). This work is supported by the Center for Research and Development in Mathematics and Applications (CIDMA) through the Portuguese Foundation for Science and Technology (FCT—Fundação para a Ciência e a Tecnologia), references UIDB/04106/2020, UIDP/04106/2020 (https://doi.org/10.54499/UIDB/04106/2020 and https://doi.org/10.54499/UIDP/04106/2020). The authors acknowledge support from the projects PTDC/FIS-AST/3041/2020, CERN/FIS-PAR/0024/2021 and 2022.04560.PTDC. This work has further been supported by the European Union's Horizon 2020 research and innovation (RISE) programme H2020-MSCA-RISE-2017 Grant No. FunFiCO-777740 and by the European Horizon Europe staff exchange (SE) programme HORIZON-MSCA-2021-SE-01 Grant No. NewFunFiCO-10108625.

References

1. M. Abramowicz, M. Jaroszynski, M. Sikora, Relativistic, accreting disks. A&A **63**, 221–224 (1978)
2. K. Akiyama, A. Alberdi, W. Alef et al., First M87 event horizon telescope results. V. Physical origin of the asymmetric ring. Astrophys J **875**(1):L5 (2019). Publisher: IOP Publishing
3. K. Akiyama et al., First M87 event horizon telescope results. I. The shadow of the supermassive black hole. Astrophys. J. Lett. **875**, L1 (2019)
4. K. Akiyama et al., *First M87 event horizon telescope results. VI. The shadow and mass of the central black hole*. Astrophys. J. Lett. **875**(1), L6 (2019)
5. K. Akiyama et al., First Sagittarius A* event horizon telescope results. I. The shadow of the supermassive black hole in the center of the milky way. ApJL **930**(2), L12 (2022)
6. K. Akiyama et al., First Sagittarius A* event horizon telescope results. VI. Testing the black hole metric. ApJL **930**(2), L17 (2022)
7. A. Almheiri, D. Marolf, J. Polchinski et al., Black holes: complementarity or firewalls? JHEP **02**, 062 (2013)
8. P. Anninos, P.C. Fragile, J. Wilson et al., Three-dimensional moving-mesh simulations of galactic center cloud G2. Astrophys. J. **759**(2), 132 (2012). Publisher: The American Astronomical Society
9. G. Antoniou, A. Bakopoulos, P. Kanti, Evasion of no-hair theorems and novel black-hole solutions in gauss-bonnet theories. Phys. Rev. Lett. **120**(13), 131102 (2018)
10. R. Arnowitt, S. Deser, C.W. Misner, Republication of: the dynamics of general relativity. Gen. Rel. Grav. **40**(9), 1997–2027 (2008)
11. M. Azreg-Aïnou, From static to rotating to conformal static solutions: rotating imperfect fluid wormholes with(out) electric or magnetic field. Eur. Phys. J. C **74**, 2865 (2014)
12. M. Azreg-Aïnou, Generating rotating regular black hole solutions without complexification. Phys. Rev. D **90**(6), 064041 (2014)
13. M. Azreg-Ainou, Regular and conformal regular cores for static and rotating solutions. Phys. Lett. B **730**, 95–98 (2014)
14. C. Bambi, Attempt to find a correlation between the spin of stellar-mass black hole candidates and the power of steady jets: relaxing the Kerr black hole hypothesis. Phys. Rev. D **86**, 123013 (2012). Publisher: APS ADS Bibcode: 2012PhRvD..86l3013B
15. C. Bambi, Attempt to find a correlation between the spin of stellar-mass black hole candidates and the power of steady jets: Relaxing the Kerr black hole hypothesis. Phys. Rev. D **86**(12), 123013 (2012)
16. C. Bambi, K. Freese, T. Harada et al., Accretion process onto super-spinning objects. Phys. Rev. D **80**(10), 104023 (2009). Publisher: American Physical Society
17. E. Berti, L.G. Collodel, B. Kleihaus et al., Spin-induced black-hole scalarization in Einstein-scalar-Gauss-Bonnet theory. Phys. Rev. Lett. **126**(1), 011104 (2021)
18. J.A. Biretta, W.B. Sparks, F. Macchetto, Hubble space telescope observations of superluminal motion in the M87 Jet. ApJ **520**(2), 621–626 (1999)
19. R.D. Blandford, R.L. Znajek, Electromagnetic extraction of energy from Kerr black holes. MNRAS **179**, 433–456 (1977)
20. H. Bondi, On spherically symmetrical accretion. Mon. Not. R. Astron. Soc. **112**(2), 195–204 (1952)
21. H. Bondi, F. Hoyle, On the mechanism of accretion by stars. Mon. Not. R. Astron. Soc. **104**, 273 (1944). ADS Bibcode: 1944MNRAS.104..273B
22. R.H. Boyer, R.W. Lindquist, Maximal analytic extension of the kerr metric. J. Math. Phys. **8**(2), 265–281 (1967)
23. R. Brito, V. Cardoso, C.A.R. Herdeiro et al., Proca stars: gravitating Bose-Einstein condensates of massive spin 1 particles. Phys. Lett. B **752**, 291–295 (2016)
24. R. Brito, V. Cardoso, P. Pani, in *Superradiance: New Frontiers in Black Hole Physics*, Lecture Notes in Physics, vol. 906 (2015), pp. 1–237

25. V. Cardoso, E. Franzin, P. Pani Is the gravitational-wave ringdown a probe of the event horizon? Phys. Rev. Lett. **116**(17), 171101 (2016). [Erratum: Phys.Rev.Lett. 117, 089902 (2016)]
26. V. Cardoso, P. Pani, Testing the nature of dark compact objects: a status report. Living Rev. Rel. **22**(1), 4 (2019)
27. K. Chatterjee, P. Kocherlakota, Z. Younsi et al., *Energy Extraction from Spinning Stringy Black Holes* (2023)
28. K. Chatterjee, Z. Younsi, P. Kocherlakota et al., *On the Universality of Energy Extraction from Black Hole Spacetimes* (2023)
29. Y. Choquet-Bruhat, R.P. Geroch, Global aspects of the Cauchy problem in general relativity. Commun. Math. Phys. **14**, 329–335 (1969)
30. K. Clough, T. Helfer, H. Witek et al., Ghost instabilities in self-interacting vector fields: the problem with proca fields. Phys. Rev. Lett. **129**(15), 151102 (2022)
31. A. Coates, F.M. Ramazanoğlu, Coordinate singularities of self-interacting vector field theories. Phys. Rev. Lett. **130**, 021401 (2023)
32. A. Coates, F.M. Ramazanoğlu, Intrinsic pathology of self-interacting vector fields. Phys. Rev. Lett. **129**(15), 151103 (2022)
33. E.H.T. Collaboration, K. Akiyama, A. Alberdi et al., First Sagittarius A* event horizon telescope results. VI. Testing the black hole metric. Astrophys. J. Lett. **930**(2), L17 (2022). Publisher: The American Astronomical Society,
34. L. Combi, H. Yang, E. Gutierrez et al., General relativistic magnetohydrodynamical simulations of accretion flows through traversable wormholes. Phys. Rev. D **109**(10), 103034 (2024). Publisher: American Physical Society
35. N. Comins, B.F. Schutz, On the ergoregion instability. Proc. R. Soc. Lond. Ser. A **364**(1717), 211–226 (1978)
36. A. Cruz-Osorio, S. Gimeno-Soler, J.A. Font, Non-linear evolutions of magnetized thick discs around black holes: dependence on the initial data. Mon. Not. R. Astron. Soc. **492**(4), 5730–5742 (2020)
37. A. Cruz-Osorio, F.D. Lora-Clavijo, F.S. Guzmán, Is the flip-flop behaviour of accretion shock cones on to black holes an effect of coordinates? Mon. Not. R. Astron. Soc. **426**(1), 732–738 (2012)
38. A. Cruz-Osorio, L. Rezzolla, F.D. Lora-Clavijo et al., Bondi-Hoyle-Lyttleton accretion onto a rotating black hole with ultralight scalar hair. J. Cosmol. Astroparticle Phys. **2023**(08), 057 (2023). Publisher: IOP Publishing
39. P.V.P. Cunha, E. Berti, C.A.R. Herdeiro, Light-ring stability for ultracompact objects. Phys. Rev. Lett. **119**(25), 251102 (2017)
40. P.V.P. Cunha, J. Grover, C. Herdeiro et al., Chaotic lensing around boson stars and Kerr black holes with scalar hair. Phys. Rev. D **94**(10), 104023 (2016)
41. P.V.P. Cunha, C. Herdeiro, E. Radu et al., Exotic compact objects and the fate of the light-ring instability. Phys. Rev. Lett. **130**(6), 061401 (2023)
42. P.V.P. Cunha, C.A.R. Herdeiro, Stationary black holes and light rings. Phys. Rev. Lett. **124**(18), 181101 (2020)
43. P.V.P. Cunha, C.A.R. Herdeiro, B. Kleihaus et al., Shadows of Einstein–dilaton–Gauss–Bonnet black holes. Phys. Lett. B **768**, 373–379 (2017)
44. P.V.P. Cunha, C.A.R. Herdeiro, E. Radu, Fundamental photon orbits: black hole shadows and spacetime instabilities. Phys. Rev. D **96**(2), 024039 (2017)
45. P.V.P. Cunha, C.A.R. Herdeiro, E. Radu, EHT constraint on the ultralight scalar hair of the M87 supermassive black hole. Universe **5**(12), 220 (2019)
46. P.V.P. Cunha, C.A.R. Herdeiro, E. Radu, Spontaneously scalarized kerr black holes in extended scalar-tensor–gauss-bonnet gravity. Phys. Rev. Lett. **123**(1), 011101 (2019)
47. P.V.P. Cunha, C.A.R. Herdeiro, E. Radu et al., Shadows of Kerr black holes with scalar hair. Phys. Rev. Lett. **115**(21), 211102 (2015)
48. P. Das, O. Porth, Three-dimensional GRMHD simulations of neutron star jets (2023). ArXiv:2311.05301 [astro-ph]

49. J.F.M. Delgado, C.A.R. Herdeiro, E. Radu, Equatorial timelike circular orbits around generic ultracompact objects. Phys. Rev. D **105**(6), 064026 (2022)
50. D.D. Doneva, S.S. Yazadjiev, New gauss-bonnet black holes with curvature-induced scalarization in extended scalar-tensor theories. Phys. Rev. Lett. **120**(13), 131103 (2018)
51. O. Dönmez, O. Zanotti, L. Rezzolla, On the development of quasi-periodic oscillations in Bondi-Hoyle accretion flows. Mon. Not. R. Astron. Soc. **412**(3), 1659–1668 (2011)
52. P.G.S. Fernandes, C. Burrage, A. Eichhorn et al., *Shadows and Properties of Spin-Induced Scalarized Black Holes with and without a Ricci Coupling* (2024)
53. J.A. Font, J.M. Ibanez, P. Papadopoulos, A 'Horizon adapted' approach to the study of relativistic accretion flows onto rotating black holes. Astrophys. J. Lett. **507**, L67 (1998)
54. J. Gair, N. Yunes, Approximate waveforms for extreme-mass-ratio inspirals in modified gravity spacetimes. Phys. Rev. D **84**(6), 064016 (2011)
55. C.F. Gammie, J.C. McKinney, G. Tóth, HARM: a numerical scheme for general relativistic magnetohydrodynamics. Astrophys. J. **589**(1), 444 (2003). Publisher: IOP Publishing
56. A. García, D. Galtsov, O. Kechkin, Class of stationary axisymmetric solutions of the einstein-maxwell-dilaton-axion field equations. Phys. Rev. Lett. **74**(8), 1276–1279 (1995). Publisher: American Physical Society
57. D. Garfinkle, G.T. Horowitz, A. Strominger, Charged black holes in string theory. Phys. Rev. D **43**, 3140 (1991). [Erratum: Phys.Rev.D 45, 3888 (1992)]
58. R. Ghosh, S. Sarkar, Light rings of stationary spacetimes. Phys. Rev. D **104**(4), 044019 (2021)
59. G.W. Gibbons, K.-I. Maeda, Black holes and membranes in higher dimensional theories with dilaton fields. Nucl. Phys. B **298**, 741–775 (1988)
60. S. Gillessen, R. Genzel, T.K. Fritz et al., A gas cloud on its way towards the supermassive black hole at the Galactic Centre. Nature **481**(7379), 51–54 (2012). Publisher: Nature Publishing Group
61. S. Gimeno-Soler, J.A. Font, C. Herdeiro et al., Magnetized accretion disks around Kerr black holes with scalar hair: constant angular momentum disks. Phys. Rev. D **99**(4), 043002 (2019)
62. S. Gimeno-Soler, J.A. Font, C. Herdeiro et al., Magnetized accretion disks around Kerr black holes with scalar hair: nonconstant angular momentum disks. Phys. Rev. D **104**(10), 103008 (2021)
63. E.G. Gimon, P. Hořava, Astrophysical violations of the Kerr bound as a possible signature of string theory. Phys. Lett. B **672**(3), 299–302 (2009)
64. P. Grandclément, KADATH: a spectral solver for theoretical physics. J. Comput. Phys. **229**(9), 3334–3357 (2010)
65. S.W. Hawking, Breakdown of predictability in gravitational collapse. Phys. Rev. D **14**, 2460–2473 (1976)
66. J.F. Hawley, C. Fendt, M. Hardcastle et al., Disks and jets. Gravity, rotation and magnetic fields. Space Sci. Rev. **191**(1-4):441–469 (2015)
67. S.A. Hayward, Formation and evaporation of nonsingular black holes. Phys. Rev. Lett. **96**(3), 031103 (2006)
68. C. Herdeiro, I. Perapechka, E. Radu et al., Asymptotically flat spinning scalar, dirac and proca stars. Phys. Lett. B **797**, 134845 (2019)
69. C. Herdeiro, E. Radu, H. Rúnarsson, Kerr black holes with Proca hair. Class. Quantum Gravity **33**(15), 154001 (2016)
70. C.A.R. Herdeiro, Black holes: on the universality of the kerr hypothesis. Lect. Notes Phys. **1017**, 315–331 (2023)
71. C.A.R. Herdeiro, A.M. Pombo, E. Radu et al., The imitation game: Proca stars that can mimic the Schwarzschild shadow. J. Cosmol. Astroparticle Phys. **2021**(04), 051 (2021). Publisher: IOP Publishing
72. C.A.R. Herdeiro, E. Radu, Kerr black holes with scalar hair. Phys. Rev. Lett. **112**, 221101 (2014)
73. C.A.R. Herdeiro, E. Radu, N. Sanchis-Gual et al., The non-spherical ground state of proca stars. Phys. Lett. B **852**, 138595 (2024)

74. C.A.R. Herdeiro, E. Radu, H.O. Silva et al., Spin-induced scalarized black holes. Phys. Rev. Lett. **126**(1), 011103 (2021)
75. F. Hoyle, R.A. Lyttleton, The effect of interstellar matter on climatic variation. Math. Proc. Camb. Philos. Soc. **35**(3), 405–415 (1939)
76. L. Hui, J.P. Ostriker, S. Tremaine et al., Ultralight scalars as cosmological dark matter. Phys. Rev. D **95**(4), 043541 (2017)
77. T. Johannsen, Regular black hole metric with three constants of motion. Phys. Rev. D **88**(4), 044002 (2013)
78. T. Johannsen, Regular black hole metric with three constants of motion. Phys. Rev. D **88**(4), 044002 (2013). Publisher: American Physical Society
79. T. Johannsen, D. Psaltis, Testing the no-hair theorem with observations in the electromagnetic spectrum: II. Black-hole images. Astrophys. J. **718**, 446–454 (2010)
80. T. Johannsen, D. Psaltis, A metric for rapidly spinning black holes suitable for strong-field tests of the no-hair theorem. Phys. Rev. D **83**, 124015 (2011)
81. M.D. Johnson et al., Universal interferometric signatures of a black hole's photon ring. Sci. Adv. **6**(12), eaaz1310 (2020)
82. A.B. Joshi, D. Dey, P.S. Joshi et al., Shadow of a naked singularity without photon sphere. Phys. Rev. D **102**(2), 024022 (2020)
83. P. Kanti, N.E. Mavromatos, J. Rizos et al., Dilatonic black holes in higher curvature string gravity. Phys. Rev. D **54**, 5049–5058 (1996)
84. D.J. Kaup, Klein-Gordon Geon. Phys. Rev. **172**, 1331–1342 (1968)
85. J. Keir, Slowly decaying waves on spherically symmetric spacetimes and ultracompact neutron stars. Class. Quantum Gravity **33**(13), 135009 (2016)
86. R.P. Kerr, Gravitational field of a spinning mass as an example of algebraically special metrics. Phys. Rev. Lett. **11**, 237–238 (1963)
87. B. Kleihaus, J. Kunz, E. Radu, Rotating black holes in Dilatonic Einstein-Gauss-Bonnet theory. Phys. Rev. Lett. **106**, 151104 (2011)
88. P. Kocherlakota, R. Narayan, K. Chatterjee et al., Toward general relativistic magnetohydrodynamics simulations in stationary nonvacuum spacetimes. Astrophys. J. Lett. **956**(1), L11 (2023)
89. P. Kocherlakota et al., Constraints on black-hole charges with the 2017 EHT observations of M87*. Phys. Rev. D **103**(10), 104047 (2021)
90. R. Konoplya, L. Rezzolla, A. Zhidenko, General parametrization of axisymmetric black holes in metric theories of gravity. Phys. Rev. D **93**(6), 064015 (2016)
91. J.P. Lasota, E. Gourgoulhon, M. Abramowicz et al., Extracting black-hole rotational energy: the generalized Penrose process. Phys. Rev. D **89**(2), 024041 (2014)
92. S.L. Liebling, C. Palenzuela, Dynamical boson stars. Living Rev. Rel. **26**(1), 1 (2023)
93. S.D. Mathur, *Fuzzballs and the Information Paradox: A Summary and Conjectures* (2008)
94. P.O. Mazur, E. Mottola, Gravitational vacuum condensate stars. Proc. Nat. Acad. Sci. **101**, 9545–9550 (2004)
95. J.C. McKinney, C.F. Gammie, A measurement of the electromagnetic luminosity of a Kerr black hole. Astrophys. J. **611**, 977–995 (2004)
96. Z. Meliani, F. Casse, P. Grandclément et al., On tidal disruption of clouds and disk formation near boson stars. Class. Quantum Gravity **34**(22), 225003 (2017). Publisher: IOP Publishing
97. Z. Meliani, P. Grandclément, F. Casse et al., GR-AMRVAC code applications: accretion onto compact objects, boson stars versus black holes. Class. Quantum Gravity **33**(15), 155010 (2016). Publisher: IOP Publishing
98. F.C. Michel, Accretion of matter by condensed objects. Astrophys. Space Sci. **15**(1), 153–160 (1972). Publisher: Kluwer Academic Publishers
99. A. Mignone, G. Bodo, S. Massaglia et al., PLUTO: A numerical code for computational astrophysics. Astrophys. J. Suppl. Ser. **170**, 228–242 (2007). Publisher: IOP ADS Bibcode: 2007ApJS..170..228M
100. C.W. Misner, K.S. Thorne, J.A. Wheeler, *Gravitation* (1973)

101. Y. Mizuno, Z. Younsi, C.M. Fromm et al., The current ability to test theories of gravity with black hole shadows. Nat. Astron. **2**(7):585–590 (2018). Publisher: Nature Publishing Group
102. M.S. Morris, K.S. Thorne, Wormholes in space-time and their use for interstellar travel: a tool for teaching general relativity. Am. J. Phys. **56**, 395–412 (1988)
103. Z.-G. Mou, H.-Y. Zhang, Singularity problem for interacting massive vectors. Phys. Rev. Lett. **129**(15), 151101 (2022)
104. S. Nampalliwar, A.I. Yfantis, K.D. Kokkotas, Extending GRMHD for thin disks to non-Kerr spacetimes. Phys. Rev. D **106**(6), 063009 (2022). Publisher: American Physical Society
105. R. Narayan, J.E. McClintock, *Observational Evidence for Black Holes*. arXiv:1312.6698 (1944), 21 (2013). ArXiv: 1312.6698
106. E.T. Newman, R. Couch, K. Chinnapared et al., Metric of a rotating, charged mass. J. Math. Phys. **6**, 918–919 (1965)
107. E.T. Newman, A.I. Janis, Note on the Kerr spinning particle metric. J. Math. Phys. **6**, 915–917 (1965)
108. H. Olivares, O. Porth, J. Davelaar et al., Constrained transport and adaptive mesh refinement in the black hole accretion code. Astron. Astrophys. **629**, A61 (2019). Publisher: EDP Sciences
109. H. Olivares, Z. Younsi, C.M. Fromm et al., How to tell an accreting boson star from a black hole. Mon. Not. R. Astron. Soc. **497**(1), 521–535 (2020)
110. N. Ortiz, O. Sarbach, T. Zannias, Shadow of a naked singularity. Phys. Rev. D **92**(4), 044035 (2015)
111. K. Parfrey, A. Tchekhovskoy, General-relativistic Simulations of Four States of Accretion onto Millisecond Pulsars. Astrophys. J. Lett. **851**(2):L34 (2017). Publisher: The American Astronomical Society
112. R. Penrose, Gravitational collapse and space-time singularities. Phys. Rev. Lett. **14**, 57–59 (1965)
113. E. Poisson, *A Relativist's Toolkit: The Mathematics of Black-Hole Mechanics* (2004)
114. O. Porth, H. Olivares, Y. Mizuno et al., The black hole accretion code. Comput. Astrophys. Cosmolo. **4**(1), 1 (2017). ArXiv: 1611.09720 Publisher: Springer International Publishing
115. D. Psaltis et al., Gravitational test beyond the first post-newtonian order with the shadow of the M87 black hole. Phys. Rev. Lett. **125**(14), 141104 (2020)
116. T. Regge, J.A. Wheeler, Stability of a Schwarzschild singularity. Phys. Rev. **108**, 1063–1069 (1957)
117. J.L. Ripley, F. Pretorius, Scalarized Black Hole dynamics in Einstein dilaton Gauss-Bonnet Gravity. Phys. Rev. D **101**(4), 044015 (2020)
118. J. Röder, A. Cruz-Osorio, C.M. Fromm et al., Probing the spacetime and accretion model for the Galactic Center: comparison of Kerr and dilaton black hole shadows. Astron. Astrophys. **671**, A143 (2023)
119. J.L. Rosa, P. Garcia, F.H. Vincent et al., Observational signatures of hot spots orbiting horizonless objects. Phys. Rev. D **106**(4), 044031 (2022)
120. J.L. Rosa, C.F.B. Macedo, D. Rubiera-Garcia, Imaging compact boson stars with hot spots and thin accretion disks. Phys. Rev. D **108**(4), 044021 (2023)
121. J.L. Rosa, J. Pelle, D. Pérez, *Accretion Disks and Relativistic Line Broadening in Boson Star Spacetimes* (2024)
122. J.L. Rosa, D. Rubiera-Garcia, Shadows of boson and Proca stars with thin accretion disks. Phys. Rev. D **106**(8), 084004 (2022)
123. R. Ruffini, S. Bonazzola, Systems of selfgravitating particles in general relativity and the concept of an equation of state. Phys. Rev. **187**, 1767–1783 (1969)
124. N. Sanchis-Gual, F. Di Giovanni, M. Zilhão et al., Nonlinear dynamics of spinning bosonic stars: formation and stability. Phys. Rev. Lett. **123**(22), 221101 (2019)
125. M. Schartmann, A. Burkert, C. Alig et al., Simulations of the origin and fate of the galactic center cloud G2. Astrophys. J. **755**(2), 155 (2012). Publisher: The American Astronomical Society
126. F.E. Schunck, E.W. Mielke, Rotating boson star as an effective mass torus in general relativity. Phys. Lett. A **249**, 389–394 (1998)

127. K. Schwarzschild, On the gravitational field of a mass point according to Einstein's theory. Sitzungsber. Preuss. Akad. Wiss. Berlin (Math. Phys.), 1916:189–196 (1916)
128. E. Seidel, W.-M. Suen, Formation of solitonic stars through gravitational cooling. Phys. Rev. Lett. **72**, 2516–2519 (1994)
129. A. Sen, Rotating charged black hole solution in heterotic string theory. Phys. Rev. Lett. **69**, 1006–1009 (1992)
130. I. Sengo, P.V.P. Cunha, C.A.R. Herdeiro et al., The imitation game reloaded: effective shadows of dynamically robust spinning Proca stars (2024). ArXiv:2402.14919 [astro-ph, physics:gr-qc]
131. H.O. Silva, J. Sakstein, L. Gualtieri et al., Spontaneous scalarization of black holes and compact stars from a Gauss-Bonnet coupling. Phys. Rev. Lett. **120**(13), 131104 (2018)
132. A. Tchekhovskoy, *HARMPI* (2019). https://github.com/atchekho/harmpi.git
133. A. Tchekhovskoy, R. Narayan, J.C. McKinney, Black hole spin and the radio loud/quiet dichotomy of active galactic nuclei. ApJ **711**(1), 50–63 (2010)
134. M. Teodoro, L. Collodel, D. Doneva et al., Thick toroidal configurations around scalarized Kerr black holes. Phys. Rev. D **104**(12), 124047 (2021). Publisher: American Physical Society
135. M.C. Teodoro, L.G. Collodel, J. Kunz, Retrograde polish doughnuts around boson stars. J. Cosmol. Astroparticle Phys. **2021**(03), 063 (2021). Publisher: IOP Publishing
136. M.C. Teodoro, L.G. Collodel, J. Kunz, Tidal effects in the motion of gas clouds around boson stars. Phys. Rev. D **103**(10), 104064 (2021). Publisher: American Physical Society
137. S. Vigeland, N. Yunes, L. Stein, Bumpy black holes in alternate theories of gravity. Phys. Rev. D **83**, 104027 (2011)
138. F.H. Vincent, Z. Meliani, P. Grandclement et al., Imaging a boson star at the Galactic center. Class. Quantum Gravity **33**(10), 105015 (2016)
139. F.H. Vincent, Z. Meliani, P. Grandclément et al., Imaging a boson star at the Galactic center. Class. Quantum Gravity **33**(10), 105015 (2016). Publisher: IOP Publishing
140. M. Wang, G. Guo, P. Yan et al., The ring shadow of rotating naked singularity with a complete photon sphere (2023)
141. J.A. Wheeler, Geons. Phys. Rev. **97**, 511–536 (1955)
142. B.F. Whiting, Mode stability of the kerr black hole. J. Math. Phys. **30**, 1301 (1989)
143. S. Yoshida, Y. Eriguchi, Rotating boson stars in general relativity. Phys. Rev. D **56**, 762–771 (1997)
144. Z. Younsi, O. Porth, Y. Mizuno et al., Modelling the polarised emission from black holes on event horizon-scales. Proc. Int. Astron. Union **14**(S342), 9–12 (2020)
145. Z. Younsi, K. Wu, S.V. Fuerst, General relativistic radiative transfer: formulation and emission from structured tori around black holes. Astron. Astrophys. **545**, A13 (2012). Publisher: EDP Sciences
146. Y.-F. Yuan, R. Narayan, M.J. Rees, Constraining alternate models of black holes: Type I X-ray bursts on accreting fermion-fermion and boson-fermion stars. Astrophys. J. **606**(2), 1112 (2004). Publisher: IOP Publishing
147. F.J. Zerilli, Effective potential for even parity Regge-Wheeler gravitational perturbation equations. Phys. Rev. Lett. **24**, 737–738 (1970)

Part IV
Self-Gravitating Fluids

Chapter 16
General Relativistic Magnetohydrodynamics Simulations for Binary Neutron Star Mergers

Kenta Kiuchi

Abstract Binary neutron star mergers used to be the most promising candidate for gravitational waves for ground-based gravitational wave detectors, such as advanced LIGO and advanced VIRGO. This was proved by the detection of gravitational waves from a binary neutron star merger in 2017. Numerical modeling is pivotal in predicting and interpreting binary neutron star mergers. This chapter reviews the progress of fully general relativistic magnetized binary neutron star merger simulations. From 2008 to 2024, about forty numerical relativity simulations of magnetized binary neutron star mergers were conducted with a different level of sophistication. This chapter aims to comprehensively view the magnetohydrodynamics effect in binary neutron star mergers by reviewing all the related works.

16.1 Introduction

After the observation of GW170817 associated with the electromagnetic counterparts GRB 170817A and AT 2017gfo [43, 66, 67, 73, 105], binary neutron star mergers became a leading player in the multimessenger era. Numerical relativity is a chosen way to construct a theoretical modeling of binary neutron star mergers. All fundamental interactions are equally essential in binary neutron star mergers. Thus, the numerical relativity codes should self-consistently implement all the effects of the fundamental interactions. The numerical relativity community started making an effort to build a physical model of binary neutron star mergers before the GW170817 event from two aspects. One is the magnetohydrodynamics (MHD) effect in binary neutron star mergers. It is motivated by the binary pulsars' observational fact that neutron stars generally have magnetic fields [69]. The community is trying to figure out what the MHD effect is in binary neutron star mergers. The other aspect is microphysics, i.e., the neutrino emission process (see, for example, Refs. [36, 117] as a review for the progress of this aspect). Since 2015, it has become feasible to perform a simulation that combines the two aspects, i.e., the MHD and neutrino effect [88].

K. Kiuchi (✉)
Max-Planck Institute for Gravitational Physics, Am Mühlenberg 2, 10559 Potsdam, Germany
e-mail: kenta.kiuchi@aei.mpg.de

C. Bambi et al. (eds.), *New Frontiers in GRMHD Simulations*, Springer Series in Astrophysics and Cosmology, https://doi.org/10.1007/978-981-97-8522-3_16

General relativistic magnetohydrodynamics (GRMHD) in full general relativity was initiated in 2005 [29, 113]. Subsequently, a couple of numerical relativity codes successfully implemented GRMHD [21, 26, 33, 34, 37, 53, 75, 80, 83, 85, 89], and started to explore binary neutron star mergers. The modeling of magnetized binary neutron star mergers is classified into three categories: (i) Three-dimensional binary neutron star merger simulations, particularly focusing on the merger dynamics [1–3, 6, 15, 17–19, 22, 23, 27, 32, 38–41, 46, 52, 54–59, 68, 78, 79, 81, 88, 90, 91, 97, 99–104, 123]. (ii) Simulation for binary neutron star merger remnants, in which the remnant is manually constructed by an equilibrium configuration or mapping from the three-dimensional merger simulations [16, 30, 31, 35, 53, 65, 110–112, 114, 116, 118, 119, 122]. (iii) Force-free simulation of inspiraling binary neutron star mergers, whose aim is to explore the precursor signal before the merger [76, 77, 85–87]. Since GW170817, the importance of a self-consistent modeling of binary neutron star mergers keeps growing because it is mandatory to compare numerical relativity simulations to observational data for interpreting/predicting gravitational wave events. Therefore, the numerical relativity community is responsible for conducting simulations that are quantitatively accurate enough for such a comparison. However, it is still far on the way. For this reason, in this chapter, we mainly review the category (i).

After initiation of the GRMHD simulation in binary neutron star mergers, this field is rapidly evolving, but at the same time the situation is "chaotic". Namely, there is no clear consensus among different numerical relativity groups on the role of MHD instabilities, jet launching, and neutron-rich matter ejection. Therefore, in this chapter, we intend to review all the related works with criticism as much as possible and try to seek a way to deepen our understanding of the MHD effect in binary neutron star mergers.

Readers may find a detailed discussion of the formulation, methodology, and implementation in the other chapters. Also, readers may be interested in the review of GRMHD simulations of binary neutron star mergers [20]. Despite its pioneering, we will not review Newtonian magnetized binary simulations. The notation in this section follows a standard notation in this field. For example, B^i, R, ρ, Ω, and v^i denote the magnetic field, the radius, the rest-mass density, the angular velocity, and velocity, respectively.

The structure of this chapter is as follows. In Sect. 16.2, we review a couple of crucial ingredients for the MHD effect in the context of binary neutron star mergers. Section 16.3 is devoted to the review of the magnetized binary neutron star merger simulations conducted to date. In Sect. 16.4, we summarize the current status of our understanding of the MHD effect in binary neutron star mergers and discuss prospects.

16.2 Magnetohydrodynamics Instabilities, Neutrino Emission, and Large-Scale Dynamo

This section reviews several relevant ingredients in binary neutron star mergers. The first one is magnetohydrodynamics (MHD) instabilities. We begin by estimating the magnetic-field energy in the pre-merger stage with the observed magnetic-field strength in the binary pulsars [69].

$$E_{\rm mag} \approx 4 \times 10^{41}\ {\rm erg}\ B_{12}^2 R_6^3, \tag{16.1}$$

where $B_{12} = B/10^{12}$G and $R_6 = R/10^6$cm, respectively. The magnetic field strength in the observed binary pulsars is in the range of $\sim 10^7$–10^{12} G [69]. The typical kinetic energy just before the merger is

$$E_{\rm kin} \approx 2 \times 10^{53}\ {\rm erg}\ M_{2.7} v_{10}^2, \tag{16.2}$$

where $M_{2.7} = M/2.7M_\odot$ and $v_{10} = v/10^{10}$cm s^{-1}. Therefore, the magnetic field is dynamically unimportant in the pre-merger stage unless we consider an unrealistically ultra-strong magnetic field. Several MHD instabilities are proposed, which could amplify the magnetic field efficiently in a short time scale up to the saturation level, implying that the magnetic field could become dynamically important during and after the merger.

In Ref. [95], Rasio and Shapiro first proposed the Kelvin-Helmholtz instability could amplify the magnetic field when the two neutron stars collide with each other, and the shear layer is formed at the contact interface unless the binary neutron star promptly collapses to a black hole [117]. Because the growth rate of this instability is proportional to the wavenumber, the small-scale vortices could grow in a much shorter time scale than the relevant dynamical timescale. These vortices could curl the magnetic field lines, which results in an efficient small-scale magnetic field amplification. Therefore, to confirm this picture in a grid-based simulation, the spatial grid resolution is a crucial ingredient.

The Kelvin-Helmholtz instability phase is not expected to continue during an entire post-merger stage because the shock waves generated by the colliding motion of the two neutron stars would dissipate the shear layer. Namely, the shear layer would survive only for a particular timescale. The timescale strongly depends on binary neutron star models such as an equation of state of nuclear matter, the mass of the neutron star, and neutron star spin in the pre-merger stage. For example, if we consider a "soft" equation of state, the approaching velocity becomes faster, compared to a "stiff" equation of state, when the two neutron stars collide because the neutron stars are compact [50]. Consequently, the timescale for which the shear layer survives becomes shorter. It implies that the grid resolution requirement to resolve the Kelvin-Helmholtz instability becomes challenging in such a case because we expect the instability would amplify the magnetic field up to the saturation level within the lifetime of the shear layer in reality. Pre-merger magnetic-field topology should be explored in the Kelvin-Helmholtz instability. However, a careful assessment is necessary to explore the saturation due to the Kelvin-Helmholtz instability.

After the Kelvin-Helmholtz instability phase, the merger remnant could be subject to the magneto-rotational instability (MRI) [7] because it differentially rotates in general [115]. The typical MRI wavelength of the axisymmetric fastest-growing mode is

$$\lambda_{\rm MRI} = \frac{B_p}{\sqrt{4\pi\rho}}\frac{2\pi}{\Omega} \approx 8\times 10^3 \text{ cm } B_{p,15}\rho_{15}^{-1/2}\Omega_{8000}^{-1}, \tag{16.3}$$

where $B_{p,15} = B_p/10^{15}$ G, $\rho_{15} = \rho/10^{15}$ g cm^{-3}, and $\Omega_{8000} = \Omega/8000$ rad s^{-1}. In reality, the Kelvin-Helmholtz instability saturation could initialize the poloidal magnetic-field strength B_p in the MRI phase. In a grid-based numerical relativity simulation, the spatial grid resolution is again a key ingredient in properly resolving the MRI. MRI quality factor $Q_{\rm MRI}$ quantifies the ability of the employed grid resolution to resolve the MRI:

$$Q_{\rm MRI} = \frac{\lambda_{\rm MRI}}{\Delta x}. \tag{16.4}$$

The critical MRI quality factor below which the MRI-driven turbulence cannot be sustained is ≈8–10 [47]. It should be noted that without estimating the MRI quality factor, it is hard to quantify how the MRI turbulence is sustained (see also the discussion about $Q_{\rm MRI}$ in the binary neutron star merger context below).

One caveat on this estimate based on the ideal MRI assumption is that the neutrino viscosity and drag could significantly suppress the MRI as diffusive and damping processes in reality [45]. The neutrino viscosity (drag) becomes relevant when the neutrino mean free path becomes shorter (longer) than $\lambda_{\rm MRI}$. The dispersion relation of the viscous- and damping-MRI are

$$\left[\left(\tilde{\sigma}_{\rm vis} + \tilde{k}_{\rm vis}^2\tilde{\nu}_\nu\right)\tilde{\sigma}_{\rm vis} + \tilde{k}_{\rm vis}^2\right]^2 + \tilde{\kappa}^2\left[\tilde{\sigma}_{\rm vis}^2 + \tilde{k}_{\rm vis}^2\right] - 4\tilde{k}_{\rm vis}^2 = 0, \tag{16.5}$$

and

$$\left[\left(\tilde{\sigma}_{\rm drag} + \tilde{\Gamma}_\nu\right)\tilde{\sigma}_{\rm drag} + \tilde{k}_{\rm drag}^2\right]^2 + \tilde{\kappa}^2\left[\tilde{\sigma}_{\rm drag}^2 + \tilde{k}_{\rm drag}^2\right] - 4\tilde{k}_{\rm drag}^2 = 0, \tag{16.6}$$

respectively, where $\tilde{\sigma}_{\rm vis/drag} = \sigma_{\rm vis/drag}/\Omega$, $\tilde{k}_{\rm vis/drag} = k_{\rm vis/drag}v_A/\Omega$, $\tilde{\kappa}^2 = \kappa^2/\Omega^2$, $\tilde{\nu}_\nu = \nu_\nu\Omega/v_A^2$, and $\tilde{\Gamma}_\nu = \Gamma_\nu/\Omega$. $\sigma_{\rm vis/drag}$ and $k_{\rm vis/drag}$ are the growth rate and wave number of MRI. $v_A = B^z_{\rm hyp}/\sqrt{4\pi\rho}$ is the Alfvén wave speed with the hypothetical poloidal magnetic-field strength $B^z_{\rm hyp}$. κ^2 is the epicyclic frequency. ν_ν and Γ_ν are the neutrino viscosity and drag damping rates, respectively. Reference [45] calibrated them in a one-dimensional supernova simulation with an *ab initio* neutrino radiation transport:

$$\nu_\nu = 1.2 \times 10^{10}\ \mathrm{cm^2\,s^{-1}}\ \rho_{13}^{-2} T_{10}^2, \tag{16.7}$$

$$\Gamma_\nu = 6 \times 10^3\ \mathrm{s^{-1}} T_{10}^6, \tag{16.8}$$

where $\rho_{13} = \rho/10^{13}$ g cm^{-3} and $T_{10} = T/10$ MeV. The neutrino mean free path l_ν is also calibrated by

$$l_\nu = 2 \times 10^3\ \mathrm{cm}\ \rho_{13}^{-1} \mathrm{T}_{10}^{-2}. \tag{16.9}$$

Therefore, with microphysics, i.e., finite temperature effect and neutrino emission, it is possible to estimate the growth rate of the non-ideal MRI. It suggests the importance of implementing microphysics in magnetized binary neutron star merger simulations.

It should be noted that the MRI is relevant to sustain the magneto-turbulence inside the merger remnant because the resultant turbulent viscosity facilitates the angular momentum transport and heats the matter inside the merger remnant. In other words, the MRI's linear, exponential growing phase could be irrelevant in the binary merger context unless the Kelvin-Helmholtz instability is suppressed.

The Shakura-Sunyaev parameter is the other important diagnostics to quantify the effective turbulent viscosity driven by MRI [108]:

$$\alpha_{SS} \equiv -\left\langle \frac{b^{(r)} b_{(\phi)}}{4\pi P} \right\rangle, \tag{16.10}$$

where P is the pressure, and $b_{(i)}$ is the tetrad component of the magnetic field measured in the fluid-rest frame. $\langle \cdot \rangle$ denotes the time ensemble.

Besides the turbulent viscosity, MRI-driven turbulence would play another important role. Since the Kelvin-Helmholtz instability could generate a strong, but *small-scale*, magnetic field in a short timescale, a mechanism is necessary to convert such a small-scale field to a large one for the magnetic field to be dynamically important inside the merger remnant. The mean-field dynamo theory, in which each physical quantity Q is decomposed into the mean component $\bar{Q}$ and fluctuating component q, i.e., $Q = \bar{Q} + q$, is a candidate for such a mechanism. We assume that an average of Q in the azimuthal direction gives the mean field $\bar{Q}$ in the binary neutron star merger context because the merger remnant has an approximately axisymmetric structure. Thus, in the framework of the ideal MHD, we cast the induction equations as

$$\partial_t \bar{\mathbf{B}} = \nabla \times \left(\bar{\mathbf{V}} \times \bar{\mathbf{B}} + \bar{\mathscr{E}} \right), \tag{16.11}$$

where $\bar{\mathscr{E}} = \overline{\mathbf{v} \times \mathbf{b}}$ is the electromotive force generated by the fluctuations. $\mathbf{B} = \bar{\mathbf{B}} + \mathbf{b}$ is the magnetic field and $\mathbf{V} = \bar{\mathbf{V}} + \mathbf{v}$ is the velocity field.

The simplest mean-field dynamo is $\alpha\Omega$ dynamo in the context of Solar physics, and we express the electromotive force as a function of the mean magnetic field:

$$\bar{\mathscr{E}}_i = \alpha_{ij} \bar{B}_j + \beta_{ij} \left(\overline{\nabla \times \bar{\mathbf{B}}} \right)_j, \tag{16.12}$$

where α_{ij} and β_{ij} are tensors that should not depend on the mean magnetic field [12]. The former tensor is called dynamo α, and the second term is turbulent resitivity. If we assume the diagonal component of the first term on the right-hand side is dominant and a purely cylindrical differential rotation $\bar{\mathbf{V}} = R\Omega\mathbf{e}_\phi$, we can reduce Eq. (16.11) into

$$\partial_t \bar{B}_R = -\partial_z \left[\left(\bar{\mathbf{V}}\times\bar{\mathbf{B}}\right)_\phi + \bar{\mathscr{E}}_\phi\right] \approx -\partial_z \left(\alpha_{\phi\phi}\bar{B}_\phi\right), \quad (16.13)$$

$$\partial_t \bar{B}_z = \partial_R \left[\left(\bar{\mathbf{V}}\times\bar{\mathbf{B}}\right)_\phi + \bar{\mathscr{E}}_\phi\right] \approx \partial_R \left(\alpha_{\phi\phi}\bar{B}_\phi\right), \quad (16.14)$$

$$\partial_t \bar{B}_\phi = R\bar{B}_A\partial_A\Omega, \quad (16.15)$$

where we work in the cylindrical coordinates, $\partial_\phi = 0$, and $A = R, z$.

These equations tell us that the mean poloidal magnetic field $\bar{B}_A$ is generated by the electromotive force described by the dynamo α and the mean toroidal field $\bar{B}_\phi$. It is called the α effect. The mean toroidal field $\bar{B}_\phi$ is generated by the differential rotation and the mean poloidal field $\bar{B}_A$. It is called the Ω effect or merely magnetic winding. Therefore, if it is realized inside the binary neutron star merger remnant, the dynamo cycle is closed, and the large-scale (mean-field) magnetic field is generated. In the binary neutron star merger context, the MRI could produce and sustain the turbulence, i.e., $\mathbf{v}$ and $\mathbf{b}$, which is the key to generating the electromotive force $\bar{\mathscr{E}}$. Since the turbulence is easily killed by the numerical diffusion in a simulation, the high-resolution simulation is the key to exploring the magnetic field evolution in the binary neutron star merger context.

We should mention the relevance of the convergence study for the spatial grid resolution to explore the MHD effect in binary neutron star mergers. We begin by estimating the magnetic winding timescale originating from a pre-merger large-scale magnetic field. Suppose we consider a binary neutron star merger with the highly magnetized end of 10^{12} G in the observed binary pulsars. The winding timescale t_A is

$$t_A \sim R/v_A \sim 100\text{ s } \bar{B}_{R,12}^{-1}\rho_{15}^{1/2}R_6, \quad (16.16)$$

where $\bar{B}_{R,12} = \bar{B}_R/10^{12}$ G. Therefore, the magnetic winding originating from the pre-merger large-scale magnetic field should be irrelevant during the post-merger evolution within the timescale of $O(10)$ s.

However, the existing general relativistic magnetized binary neutron star merger simulations employed the pre-merger magnetic field of 10^{15}–10^{17} G except for the simulations in Refs. [32, 38, 39, 52] and the simulations employing a sub-grid model [1–3, 41, 90, 91]. The primarily reason for using such an unrealistic magnetic field strength is to compensate the extensive computational cost to resolve the Kelvin-Helmholtz instability and the MRI (see Eq. (16.3)). However, the trade-off is to shorten the winding timescale originating from the assumed pre-merger large-scale magnetic field. t_A could become comparable to or shorter than the timescale relevant for the post-merger evolution (see Eq. (16.16)). Therefore, if we employ a *single* grid resolution with the pre-merger large-scale magnetic field of 10^{15}–10^{17} G, we can not disentangle whether the magnetic winding originating from such a large-scale magnetic field determines the subsequent evolution of the merger remnant or not. Since the magnetic winding is relatively easy to resolve numerically, the convergence study is essential to disentangle the abovementioned point. Namely, suppose the magnetic winding originating from the pre-merger large-scale field plays a primary role in the subsequent merger remnant evolution. In that case, the result should not be significantly different in the different resolution simulations. If resolving the small-scale turbulence is essential, like the MRI-driven $\alpha\Omega$ dynamo mentioned above, the convergence study would give a significantly different result.

Therefore, in the following section, we review the previous works by explicitly mentioning the grid resolution, the pre-merger large-scale magnetic field, and the level of sophistication of the microphysics.

16.3 Magnetohydrodynamics Simulations

In this section, we review the general relativistic magnetized binary neutron star merger simulations in the framework of the ideal and resistive MHD approximation. To clarify how our community has deepened our understanding of the MHD effect in binary neutron star mergers, we review all the related works chronologically.

Before starting the review, we should mention the pre-merger magnetic field topology widely used in the numerical relativity community. The field topology is called a confined field configuration:

$$A_\phi = A_b \max\left(P - P_{\rm cut}, 0\right)^{n_s}, \tag{16.17}$$

where A_ϕ is the toroidal component of the vector potential, P is the pressure, $P_{\rm cut}$ is the cut-off pressure parameterized by a fraction of the maximum pressure, and the concentration parameter n_s. A_b is chosen so that the desired magnetic field strength is achieved. Together with the assumption of no poloidal vector potential, $A_r = A_\theta = 0$, it gives a confined purely poloidal magnetic field. Since it has been revealed that the simulation result is not sensitive, at least qualitatively, to the choice of $P_{\rm cut}$ and n_s, the default setting for the pre-merger magnetic field topology is the confined configuration in this review without specifying $P_{\rm cut}$ and n_s. Otherwise, we explicitly explain the magnetic field topology.

Merger remnant of binary neutron star merger is classified into three classes; a prompt collapse case, a short-lived case, and long-lived case [117]. For the prompt collapse case, the merger remnant collapses to a black hole immediately after the merger. For the short-lived case, the remnant massive neutron star would survive within $O(0.01)$s, whose timescale would be shorter than the MRI-driven turbulent viscous timescale, the magnetic winding timescale, and the neutrino cooling timescale.[1] For the long-lived case, the merger remnant massive neutron star survives for $O(1\text{–}10)$ s.

Anderson and his collaborators initiated a general relativistic magnetized symmetric binary neutron star merger simulation [6]. They employed the grid resolution of $\Delta x = 460$ m and the pre-merger magnetic field of 9.6×10^{15} G. The neutron star is modeled by the polytropic equation of state with $\Gamma = 2$. They concluded the Kelvin-Helmholtz instability amplifies the magnetic field. They also reported the emergence of the Taylor instability with $m = 1$ mode and the magnetic buoyancy instability. They quantified how the magnetic field modifies the post-merger gravitational waveforms. However, because of the lack of a convergence study, there was room to be improved in understanding whether the Kelvin-Hemoholtz instability really activates or not.

Subsequently, Liu and his collaborators simulated magnetized symmetric and asymmetric binary neutron star mergers in full general relativity [68]. They employed the grid resolution of $\Delta x \lesssim 700$ m, and the pre-merger magnetic field of 10^{16} G.[2] The neutron star is modeled by the Γ-law with $\Gamma = 2$. Their conclusion is summarized as follows: (i) For the hypermassive neutron star formation case, the MHD effect can cause observable differences in the dynamics and gravitational waves, but they are not as significant as those reported in Ref. [6]. The difference might originate from the initial data of the binary neutron star. On the one hand, in Ref. [68], the quasi-equilibrium configuration was employed as the initial data. On the other hand, Ref. [6] set up the initial data by superposing boosted metrics of two spherical neutron stars. (ii) For the prompt black hole formation case, the MHD effect plays a minor role in the case of the symmetric binary. For the asymmetric binary, the MHD effect enhances the disk mass. However, since their employed grid resolution is not fine enough to capture the Kelvin-Helmholtz instability and MRI, as they explicitly mentioned, the role of these instabilities was not explored. Also, since their simulation time is shorter than the magnetic winding timescale t_A, the role of the magnetic winding effect needs to be more clarified.

Giacomazzo and his collaborators explored how the pre-merger magnetic field is imprinted in the inspiral gravitational waveforms [38]. They employed the grid resolution of $\Delta x \approx 350$ m, and varied the pre-merger magnetic field from 0 to 10^{17} G. The neutron star is modeled by the Γ-law equation of state with $\Gamma = 2$. Their conclusion is that the inspiral gravitational waves are essentially indistinguishable from those in the unmagnetized case unless the pre-merger magnetic field is as strong as 10^{17} G. They also investigated the magnetic field amplification via the Kelvin-Helmholtz instability in the shear layer and concluded that the toroidal field is exponentially amplified until it becomes as strong as the poloidal field. In Ref. [39], they extended

[1] The neutrino cooling timescale is estimated by $t_{\nu,\rm cool} \sim E_{\rm thr,53}/L_{\nu,53} \sim O(1)$ s where $E_{\rm thr,53} = E_{\rm thr}/10^{53}$ erg and $L_{\nu,53} = L_\nu/10^{53}$ erg/s are the thermal energy and the neutrino luminosity, respectively.

[2] Precisely speaking, their simulation result is scale-free because of the normalization with the polytropic constant K.

their work by employing the grid resolution of $\Delta x = 221$ m and varying the pre-merger magnetic field from 10^8 to 10^{12} G. The equation of state is the same as that in their previous work. They also, for the first time, performed the convergence test by changing the grid resolutions $\Delta x = 177$ m and 354 m. They reported the growth rate of the magnetic field due to the Kelvin-Helmholtz instability is $\sim$0.5ms^{-1}, and it does not significantly depend on the grid resolution. They argued the potential reason for this is the shortness of the lifetime of the shear layer. In their particular model, it is $\approx$1 ms. Therefore, the efficiency of the Kelvin-Helmholtz instability is still a riddle. Also, since the employed resolution cannot capture the MRI (16.3) and the lifetime of the remnant massive neutron star in their "high-mass binaries" is shorter than the magnetic winding timescale (16.16), the MHD effect in binary neutron star mergers needs to be more clarified.

Rezzolla and his collaborators reported the emergence of a jet-like structure in Ref. [97]. They employed the pre-merger magnetic field of 10^{12} G. The grid resolution was likely $\Delta x = 221$ m. Their model collapsed to a black hole after $\approx$7–8 ms, and the subsequent massive torus formation occurred. They argued that the magnetic field exponentially grows inside the massive torus, and the ordered-magnetic field line is generated by a non-linear dynamo. Then, the jet-like structure is produced, which can power short gamma-ray bursts. Although detailed discussions and analyses about the MRI and the dynamo mechanism were absent, the paper has the merit of showing that the binary neutron star merger with a magnetic field may explain some aspects of the gamma-ray burst phenomenology.

Giacomazzo and Perna reported a stable magnetar formation in a binary neutron star merger [40]. They employed the pre-merger magnetic field of 10^{12} G and the single grid resolution with $\Delta x = 225$ m. The NS is modeled with the Γ-law equation with $\Gamma = 2.75$. Since they chose a model in which the total mass is below the maximum mass of the TOV star, the resultant merger remnant neutron star is stable. They found the inefficient amplification via the Kelvin-Helmholtz instability, which is consistent with their previous finding [39] and subsequent amplification via the magnetic winding. Although the magnetic field did not reach the magnetar strength of $\sim 10^{14}$–10^{15} G at the end of the simulation, approximately $\approx$55 ms after the merger, they claimed the stable magnetar formation is possible because it is expected the magnetic winding and unresolved MRI would amplify the magnetic field furthermore.

Kiuchi and his collaborators reported magnetized binary neutron star merger simulations [54]. They employed the pre-merger magnetic field with $10^{14.5}$ G, 10^{15} G, and 10^{16} G. They performed an in-depth convergence study by employing the grid resolution $\Delta x = 70$ m, 110 m, and 150 m. The neutron star is modeled with the H4 equation of state [42]. They employed the piecewise polytrope prescription for the cold part [96] and Γ-law equation for the thermal part. The merger remnant is short-lived. Contrary to the previous works [39, 40, 97], they found that an efficient magnetic-field amplification via the Kelvin-Helmholtz instability. Namely, the higher grid resolution simulation exhibits a larger growth rate of the magnetic field energy. It is consistent with the property of the Kelvin-Helmholtz instability, i.e., the shorter wavelength mode has the larger growth rate. However, even with their highest grid

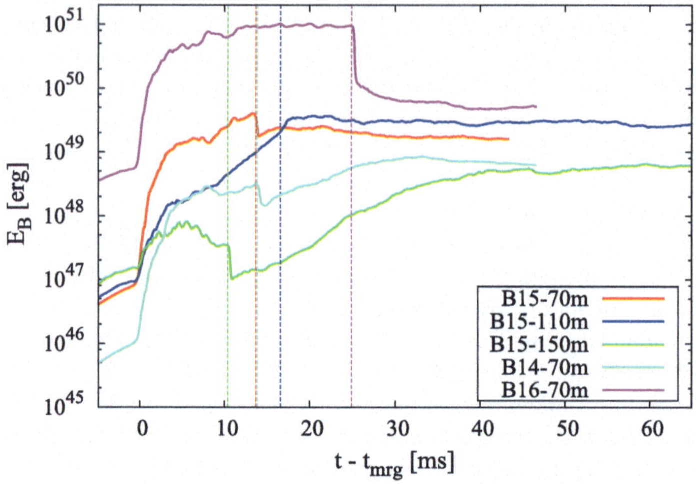

Fig. 16.1 Magnetic-field energy as a function of the post-merger time from Ref. [54]. $t - t_{\rm mrg} = 0$ corresponds to the merger. B14, B15, and B16 denote the pre-merger magnetic field of $10^{14.5}$ G, 10^{15} G, and 10^{16} G, respectively. The vertical dashed lines denote the black hole formation. The figure is taken from Ref. [54]

resolution, the saturation of the magnetic field is not achieved (see Fig. 16.1). Also, they, for the first time, estimated the MRI quality factor of the non-axisymmetric mode in the magnetized binary neutron star merger simulations. They confirmed that once the MRI quality factor exceeds the critical value of 10, the magnetic field exponentially increases inside the merger remnant (see Fig. 3 in Ref. [54]). They also reported that even with their highest grid resolution, the MRI in a high-density region with $\rho \geq 10^{13}$ g cm^{-3} cannot be captured because of the shortness of the MRI wavelength. They also reported the absence of the jet launching until the end of the simulation of $\approx$50 ms after the black hole formation.

Giacomazzo and his collaborators reported new magnetized binary neutron star merger simulations by employing a sub-grid model [41]. The pre-merger magnetic field is $\sim$2.5 $\times 10^{12}$ G. The grid resolutions are $\approx$180 m, 220 m, 300 m, and 360 m. The neutron star is modeled with the Γ-law equation of state with $\Gamma = 2$. They implemented a sub-grid model by adding a new term $\mathbf{E}_{\rm subgrid}$ in the induction equation:

$$\partial_t \mathbf{A} = -\mathbf{E}_{\rm ideal} - \mathbf{E}_{\rm subgrid}, \tag{16.18}$$

where $\mathbf{A}$ is the vector potential and $\mathbf{E}_{\rm ideal}$ is the electric field in the ideal MHD. By assuming the closure relation between $\mathbf{E}_{\rm subgrid}$ and $\mathbf{A}$,

$$\mathbf{E}_{\rm subgrid} = -S_{\rm subgrid}\mathbf{A}, \tag{16.19}$$

where S_{subgrid} is a parameter, they reproduced an exponential amplification at the merger. To mimic the Kelvin-Helmholtz instability, i.e., the energy transfer from the turbulent kinetic energy in the shear layer to the magnetic field energy, and to suppress unphysical behavior near the stellar surface, they introduced four parameters in S_{subgrid}. Two of them are proportional to the fraction of the fluid vorticity $\nabla \times \mathbf{v}$, which are calibrated by high-resolution special relativistic MHD local box simulation for MHD turbulence [129]. They performed the convergence study and concluded that the saturation of the magnetic field is $\sim 10^{51}$ erg. However, it is still an open question that the saturation is physical because it is controlled by S_{subgrid}, which is calibrated only by special relativistic MHD. Furthermore, their sub-grid model violates energy conservation. Also, their closure relation needs to be justified.

Dionysopoulou and her collaborators reported resistive MHD simulation of binary neutron star mergers for the first time [27]. They proposed a simplified model of the electrical conductivity σ_{con}[3]:

$$\sigma_{\text{con}} = \sigma_0 \max\left[1 - \frac{2}{1 + \exp\left[2D_{\text{tol}}\left(D - D_{\text{rel}}\right)/D_{\text{atm}}\right]}, 0\right], \tag{16.20}$$

where $D = \rho w$ is a conserved mass density. D_{tol}, D_{rel}, and D_{atm} are chosen such that the electric conductivity has a smooth transition from high conductivity, i.e., ideal MHD, in a high-density region to zero conductivity in a low-density region. Specifically, they chosed $D_{\text{tol}} = 0.01$, $D_{\text{rel}} = 100\rho_{\text{atm}}$, $D_{\text{atm}} = \rho_{\text{atm}}$, and $\sigma_0 = 2 \times 10^{11}\ \text{s}^{-1}$. ρ_{atm} is a tenuous atmosphere outside the neutron star and set to be $6.17 \times 10^6\ \text{g cm}^{-3}$. The pre-merger magnetic field is 1.97×10^{12} G. The employed grid resolution is $\Delta x = 148$ m. The neutron star is modeled with the Γ-law equation of state with $\Gamma = 2$. Their main finding is that the magnetic winding inside the remnant massive neutron star for the resistive MHD case becomes inefficient compared to the ideal MHD case because the magnetic field is not perfectly frozen into the fluid elements. As a result, the lifetime of the remnant massive neutron star becomes $\sim$2 ms longer in the resistive MHD case, which results in more massive torus formation after the remnant collapses into a black hole. However, these resistive effects need to be more clarified because the magnetic braking timescale with $\sim 10^{12}$ G is much longer than the lifetime of their particular model.

Palenzuela and his collaborators performed magnetized binary neutron star merger simulations incorporating neutrino physics for the first time [88]. They implemented the finite temperature nuclear density equation of states, the SFHo [121], DD2 [48], and NL3 [48], and the neutrino leakage to take into account the neutrino cooling [84]. The former results in a short-lived merger remnant, and the latter two do in a long-lived merger remnant. They also employed a sub-grid model similar to Eqs. (16.18)–(16.19) to capture the unresolved Kelvin-Helmholtz instability. The pre-merger magnetic field strength is 10^{13} G, and the grid resolution is $\Delta x = 230$ m.Their findings are summarized as follows: (i) The magnetic field is exponentially amplified up to

[3] Original notation for the electrical conductivity is σ, but to avoid confusion with the magnetization parameter σ, we use a different notation in this review.

the saturation energy of $\sim 10^{50}$ erg. (ii) The post-merger evolution of the remnant massive neutron star is not sensitive to the amplified magnetic field, at least within the timescale of ≈ 10 ms after the merger. (iii) The angular momentum transport seems to be facilitated by the magnetic braking due to the amplified magnetic field. However, there are a couple of caveats. The first one is the sub-grid term breaks the divergence-free condition since their MHD implementation is based on "**B**," not on "**A**." The second one is the saturation energy of the magnetic field is very sensitive to the sub-grid implementation since the saturation energy is one order magnitude smaller than that reported in Ref. [41]. The third point is the role of the MRI turbulence for the angular momentum transport needs to be more clarified since they did not analyze the MRI.

Kiuchi and his collaborators reported new magnetized binary neutron star merger simulations [55]. The pre-merger magnetic fields are 10^{13} G, 10^{14} G, and 10^{15} G. The employed grid resolutions are $\Delta x = 17.5$ m, 27.5 m, and 37.5 m for the convergence test. It should be noted that the high-resolution mesh refinement domains are assigned only to a region where the shear layer appears. The entire neutron stars in the pre-merger phase were covered by 70, 110, and 150 m, respectively. The NS model is the same as that in Ref. [54]. They measured the growth rate of the magnetic field amplification due to the Kelvin-Helmholtz instability, and it is approximately proportional to the inverse of the employed grid resolution (see Fig. 16.2). They also explored the saturation energy of the magnetic field by varying the pre-merger magnetic field. The back reaction due to the amplified magnetic field is likely to start once the magnetic-field energy reaches $\sim 10^{49}$ erg, and it is likely to saturate at $\sim 10^{50}$ erg (see Fig. 16.3). However, the energy spectrum analysis indicated that only a small fraction of $\sim$0.5–1% of the turbulent kinetic energy is transferred to the magnetic field energy. Also, there is no prominent sign of the angular momentum transport at ≈ 5 ms after the merger because, for efficient magnetic braking, a large-scale poloidal magnetic field is necessary. It implies that (i) the physical saturation of the magnetic field energy could still be far from that reported in the simulations, and (ii) the Kelvin-Helmholtz instability only produces the small-scale field. The MRI analysis was missed in their work because of the shortness of the simulation, i.e., 5 ms after the merger.

Ruiz and his collaborators performed magnetized binary neutron star merger simulations [99]. They employed $\Delta x = 152$ m and 227 m. The neutron star is modeled by the Γ-law equation of state with $\Gamma = 2$. The merger remnant is short-lived. The pre-merger magnetic field strength was unclear because they only mentioned the field strength at pole $B_{\rm pole} = 1.75 \times 10^{15}$ G. However, from Fig. 5 in Ref. [103] reported by the same authors with a similar set-up except for the neutron star spin, the pre-merger magnetic field is guessed to be $\approx 10^{16.5}$–10^{17} G (see also Fig. 16.4). In addition to the confined pre-merger magnetic field configuration, they employed a pulsar-like configuration:

$$A_\phi = \frac{\pi r_0^2 I_0 \varpi^2}{(r_0^2 + r^2)^{3/2}} \left(1 + \frac{15 r_0^2 (r_0^2 + \varpi^2)}{8 (r_0^2 + r^2)^2}\right), \tag{16.21}$$

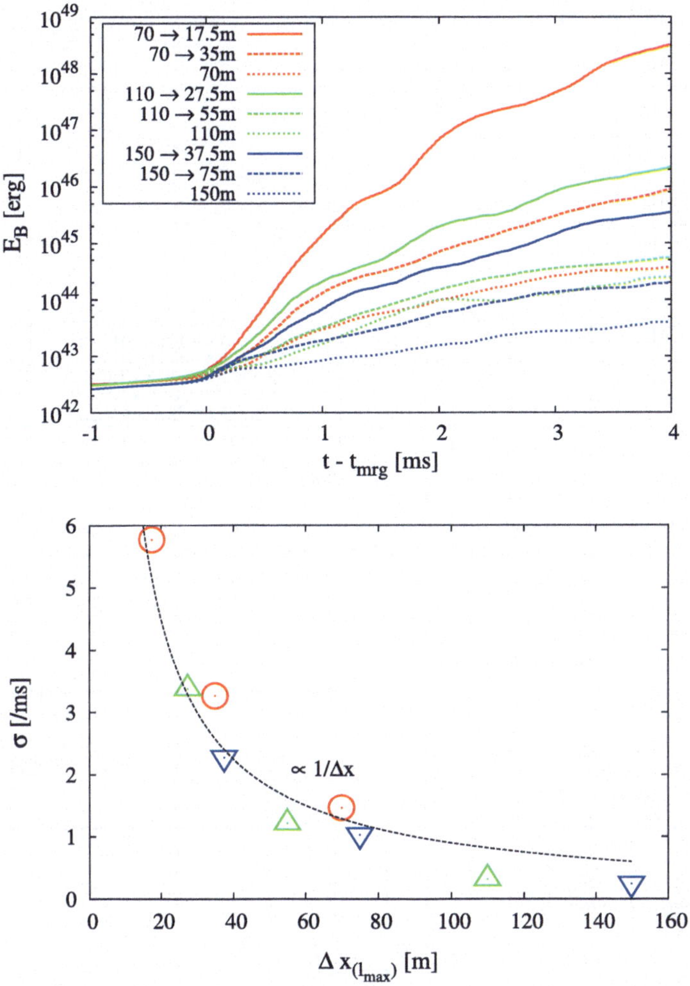

Fig. 16.2 (Top) Magnetic-field energy as a function of the post-merger time with the pre-merger magnetic field of 10^{13} G. The legend denotes the employed grid resolution. For example, the red curves correspond to the base run with $\Delta x = 70$ m in the dotted curve, the run with $\Delta x = 35$m by the increment of one mesh refinement domain from the base run in the dashed curve, and the run with $\Delta x = 17.5$ m by the increment of two mesh refinement domains from the base run in the solid curve. (Bottom) The measured growth rate of the magnetic field energy as a function of the employed grid resolution. The color scheme is the same as the upper panel. The figures are taken from Ref. [55]

where r_0 is the current loop radius and I_0 is the loop current, $r^2 = (x - x_{NS})^2 + (y - y_{NS})^2 + z^2$, and $\varpi^2 = (x - x_{NS})^2 + (y - y_{NS})^2$ where x_{NS} and y_{NS} denote the initial coordinates of the neutron star center of mass. Importantly, the neutron star exterior condition for the hydrodynamic variables is chosen such that the exterior

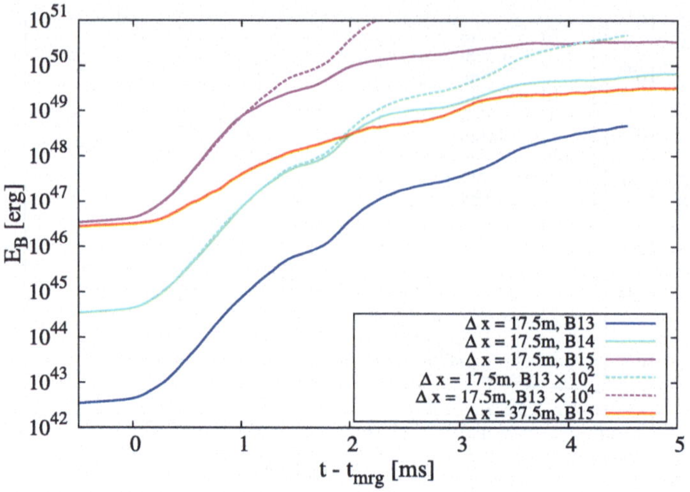

Fig. 16.3 Magnetic-field energy as a function of the post-merger time. B13, B14, and B15 denote the pre-merger magnetic field of 10^{13} G, 10^{14} G, and 10^{15} G, respectively. The cyan and purple dotted curves are the results with $\Delta x = 17.5$ m and the pre-merger magnetic field of 10^{13} G magnified by a factor of 10^2 and 10^4, respectively. The figures are taken from Ref. [55]

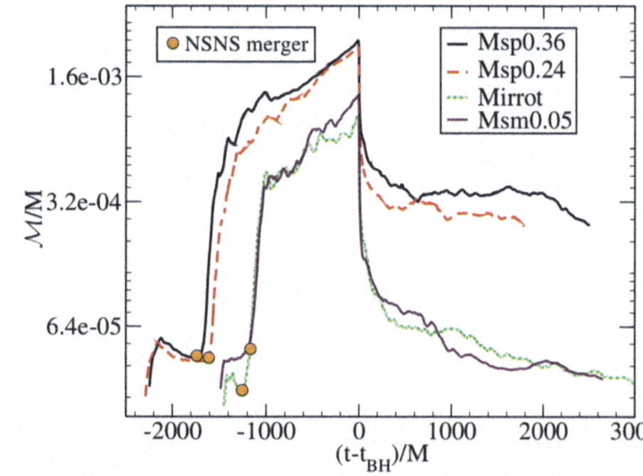

Fig. 16.4 Total magnetic energy $\mathcal{M}$ normalized by the ADM mass $M = 5.36 \times 10^{54}(M_{\rm NS}/1.625 M_\odot)$erg versus time. Dots indicate the NSNS merger time $t_{\rm mer}$. The coordinate time has been shifted to the BH formation time $t_{\rm BH}$. Msp0.36, Msp0.24, Mirrot, and Msm0.05 denote the magnetized binary with $\chi_{\rm NS} = 0.36$, 0.24, 0, and -0.05, respectively. The figure is taken from Ref. [102]

plasma beta is 0.01. They found a mildly relativistic outflow, an incipient jet, from the merger remnant composed of the rapidly spinning black hole and massive torus. They reported that this is the case irrespective of the pre-merger magnetic field geometry. Since they mentioned the MRI quality factor in the remnant massive neutron star is greater than 10, Eq. (16.3) and their grid-set up suggested $B_p \gtrsim 3 \times 10^{16}$ G. The previous works showed that it is impossible to reproduce the efficient Kelvin-Helmholtz instability with this resolution [39, 54, 55, 97] unless the sub-grid model is employed. Therefore, the poloidal magnetic field inside the merger remnant could be a consequence of the pre-merger large-scale magnetic field. After the remnant neutron star collapsed into the black hole, the pole region evacuated, and the incipient jet was launched at ≈ 40 ms after the black hole formation.

They argued the discrepancy from Ref. [54], i.e., the absence/presence of the jet launching, may originate from the difference of the neutron star model. When the two neutron stars collide, the dynamical ejecta is driven by the shock heating and tidal force [50]. A part of the dynamical ejecta falls back onto the merger remnant, and the resultant ram pressure may overcome the magnetic pressure. However, the fallback timescale may be different in different binary neutron star models. Namely, the fallback matter in their model is smaller than that in Ref. [54]. In the latter model, it may take time to launch a relativistic jet. A quantitative comparison on the fallback timescale should be necessary among the different binary neutron star merger models. Also, the large-scale magnetic field generation mechanism should be investigated in detail. The final caveat is the magnetic winding timescale originating from the pre-merger large-scale field is as short as $O(1)$ ms (see Eq. (16.16)), which is shorter than the lifetime of the remnant massive neutron star of $O(10)$ ms.

Endrizzi and his collaborators reported a new series of magnetized binary neutron star merger simulations [32]. The pre-merger magnetic field is 10^{13} G. The employed grid resolution is $\Delta x = 222$ m. The neutron star is modeled with the APR equation of state [4]. They employed a hybrid equation of state to take into account the thermal effect during the simulation. They simulated three cases: a high mass case, which results in a prompt black hole formation, and a symmetric and asymmetric low mass case, which results in a supramassive neutron star formation.[4] They reported an exponential amplification of the magnetic field energy after the merger in the two later cases (see Fig. 9 in Ref. [32].) However, the density-weighted average of the magnetic field strength did not exhibit such an exponential growth (see Fig. 11 in Ref. [32].) Also, the saturation energy of the magnetic field of 10^{48}–10^{49} erg is much smaller than the rotational kinetic energy of $\sim 10^{53}$ erg. As they reported, the Kelvin-Helmholtz instability and MRI are not resolved. Therefore, the detailed analysis is still necessary.

Kawamura and his collaborators reported a systematic study of magnetized binary neutron star merger simulations [52]. The grid resolution is $\Delta x = 222$ m for the Γ-law equation of state with $\Gamma = 2$ and $\Delta x = 186$ m for the H4 equation of state [42]. The pre-merger magnetic field is $\sim 10^{12}$ G and explored the pre-merger magnetic field topology with the field aligned to the orbital rotation axis, aligned and anti-aligned field, and both anti-aligned field. They also performed the convergence test by changing the grid resolution $\Delta x = 177$ m and 277 m for the Γ-law equation of state and $\Delta x = 150$ m for the H4 equation of state. Their results with the standard grid set-up are summarized as follows: (i) For the Γ-law equation of state, the symmetric binary results in a short-lived remnant massive neutron star formation and subsequent massive torus formation after the remnant collapses into a black hole. The Kelvin-Helmholtz instability is inefficient, and the MRI cannot be resolved during the remnant massive neutron star phase due to their set-up for the pre-merger

[4] The supramassive neutron star is defined as the mass below the uniformly rotating neutron star's maximum mass and above the TOV star's maximum mass.

magnetic field and grid (see Eq. (16.4)). Inside the massive torus, the magnetic field is amplified, but its density-weighted mean value saturated at $\approx 10^{12}$ G. $\lambda_{\rm MRI}$ in this phase is likely to be

$$\lambda_{\rm MRI} = \frac{B_{\rm mean}}{\sqrt{4\pi\rho}} 2\pi \left(\frac{GM_{\rm BH}}{R_{\rm disk}^3}\right)^{-1/2} \approx 10^3 \text{ cm } B_{\rm mean,12}\rho_{12}^{-1/2} M_{\rm BH,3}^{-1/2} R_{\rm disk,6.7}^{3/2}, \quad (16.22)$$

where $B_{\rm mean,12} = B_{\rm mean}/10^{12}$ G, $\rho_{12} = \rho/10^{12}$ g cm^{-3}, $R_{\rm disk,6.7} = R_{\rm disk}/5 \times 10^6$ cm, $M_{\rm BH,3} = M_{\rm BH}/3M_\odot$, and we assumed the Newtonian Keplarian angular velocity with the disk radius $R_{\rm disk}$. Therefore, the magnetic field amplification inside the torus is likely due to magnetic winding. However, the saturation energy of the magnetic field of $\sim 10^{44}$ erg is far from the equipartition to the rotational kinetic energy of $\sim 10^{50}$ erg. Also, they reported the magnetic field amplification in each phase is very sensitive to the grid resolution. For example, the magnetic field energy differs by six orders of magnitude at the end of the simulation with high- and low-resolution runs. (ii) For the asymmetric binary case with the Γ-law equation of state, the merger remnant almost promptly collapses into a black hole, and a torus more massive than that in the symmetric binary is formed. However, the magnetic field amplification inside the massive torus is minor, i.e., only a factor of few. It is counterintuitive because the rotational energy should be tapped into the amplification of the magnetic field energy via the MRI and magnetic winding, i.e., the more massive torus is expected to have stronger magnetization. (iii) For the symmetric binary with the H4 equation of state, the magnetic field evolution until the black hole formation is qualitatively similar to the symmetric binary with the Γ-law equation of state. However, the magnetic field is not amplified inside the massive torus and stays at $\sim 10^{44}$ erg until the end of the simulation. It is a striking difference between their result and Ref. [54] reporting the saturation energy is likely to be $\sim 10^{49}$ erg (see also Fig. 16.1) with the same binary, but different pre-merger magnetic field and grid set-up. They performed the convergence study with $\Delta x = 150$ m and reported a significant growth of the magnetic field during the massive torus phase, presumably due to the non-axisymmetric MRI, which is not observed in their standard resolution. In the high-resolution run, the magnetic field energy is amplified up to $\sim 10^{50}$ erg inside the torus at the end of the simulation. (iv) For the asymmetric binary with the H4 equation of state, they reported an exponential growth of the magnetic field during a remnant massive neutron star phase. However, given the density-weighted mean value of the magnetic field of $\sim 10^{12}$–10^{13} G (see Fig. 13 in Ref. [52]), the MRI is likely to be unresolved (16.4). As with Ref. [32], the origin of the exponential growth is a riddle. It should be noted that the MRI quality factor estimated in Ref. [54] and this work is *not* based on the poloidal field strength, i.e., $B_{\rm p}$ in Eq. (16.3), but the total magnetic field strength, implying the most likely to be the toroidal field strength. In other words, they explored the non-axisymmetric MRI, not the axisymmetric MRI [7]. Since the axisymmetric MRI has a more significant growth rate than the non-axisymmetric MRI [8], and the poloidal field is generally weaker than the toroidal field, resolving the axisymmetric MRI is more challenging.

Ciolfi and his collaborators reported new magnetized binary neutron star merger simulations, which leave a long-lived remnant [17]. They employed the grid resolution of $\Delta x = 222$ m and 177 m for the convergence test. The neutron star is modeled with the APR equation of state [4], the MS1 equation of state [82], and the H4 equation of state [42]. They explored the symmetric and asymmetric binaries. The pre-merger magnetic field is $\approx 3 \times 10^{15}$ G. They reported that after an inefficient Kelvin-Helmholtz instability amplification, the MRI is likely to play an active role in their models. However, there are a couple of caveats: (i) Since $\lambda_{\rm MRI}$ is not estimated by the poloidal field strength but by the presumably dominant toroidal field strength, it seems to be overestimated as they mentioned by themselves. (ii) Since they have not quantified the MRI quality factor with a number, it is hard to judge the ability of their simulation setup to resolve the MRI. As shown in independent simulations, it seems to be hard to resolve the axisymmetric MRI with their setup [56, 59] (see below).

Ruiz and his collaborators reported magnetized binary neutron star mergers in a prompt black hole formation scenario in Ref. [100]. The grid resolution is $\Delta x \approx 140$–150 m. The pre-merger magnetic field is the same as in their previous work [99], i.e., likely to be $10^{16.5}$–10^{17} G (see also their Table I which shows the pre-merger magnetic field energy is as high as 10^{50} erg). In additon to the confined pre-merger mangetic field configuration, they explored the dipole-like field configuration (see Eq. (16.21)). They confirmed the MRI is resolved inside the massive torus around a black hole and estimated the Shakura-Sunyaev parameter [108] for the first time. However, contrary to the remnant massive neutron star formation case [100], they did not find a launching of an incipient jet irrespective of the mass asymmetry of the binary neutron star. Since the shocked component of the dynamical ejecta and resultant fallback matter in the polar region should be negligible in the prompt black hole formation, the conclusion is puzzling compared to their previous work, i.e., the less fallback matter is favorable for the launching of the incipient jet. The absence of a large-scale magnetic field inside the disk needs to be clarified.

Kiuchi and his collaborators reported new magnetized neutron star merger simulations. The employed grid resolution is $\Delta x = 12.5$ m [56]. Similar to their previous work [55], the high-resolution mesh refinement domains are assigned to a shear layer, and the entire binary neutron star before the merger is covered by $\Delta x = 50$ m. The pre-merger magnetic field is 10^{15} G. The neutron star is modeled with the H4 equation of state [42]. They also performed the convergence study with $\Delta x = 70$ m and 110 m. They explored the less massive binary neutron star merger compared to that in Refs. [54, 55], resulting in the long-lived massive neutron star. They thoroughly assessed the MRI-driven turbulence in a non-linear phase with the axisymmetric, the non-axisymmetric MRI quality factor, the ratio of the poloidal field energy to the toroidal field energy, the Maxwell stress, and associated Shakura-Sunyaev parameters [47]. Their findings are summarized as follows: (i) The MRI quality factor, as well as the other diagnostics to quantify the ability of the simulation to sustain the MRI-driven turbulence, indicates that the resolution coarser than $\Delta x = 70$ m is insufficient in the region of $\rho \geq 10^{13}$ g cm^{-3}. Particularly, with $\Delta x = 110$ m, which is a "high" resolution in the numerical relativity community, the MRI-driven turbulence cannot be sustained in the region $\rho \geq 10^{12}$ g cm^{-3} (see Table I in Ref. [56]). It should be noted that a bulk of the rotational energy of the remnant massive neutron star is contained in such a high-density region. Unresolved MRI simulation could draw a physically *incorrect* picture unless the MRI is suppressed by a mechanism such as neutrino viscosity and drag. (ii) An ordered (large-scale) poloidal magnetic field is not established within their simulation timescale of $\approx$30 ms after the merger. It indicates the magnetic braking timescale could be $\approx$0.7–0.8 s unless the large-scale poloidal field is not established by a mechanism such as $\alpha\Omega$ dynamo. The absence of a large-scale field needs to be clarified.

Ruiz and his collaborators placed an upper limit of the maximum mass of the TOV star by combining their simulation results [99, 100] and GW170817 observation associated with GRB170817A [43, 101]. In addition to the results in Refs. [99, 100], they performed a simulation for a supramassive neutron star formation case, i.e., the lifetime of the merger remnant could be longer than the magnetic dipole radiation timescale. They concluded that the supramassive neutron star can not launch the jet. Based on these results, their scenario is that the merger remnant in GW170817 is between the supramassive limit and the prompt collapse limit. Since we know the mass of the binary neutron star in GW170817, this scenario gives

$$M_{\max}^{\text{sup}} \approx \beta M_{\max}^{\text{TOV}} \leq M_{GW170817} \approx 2.74 M_{\odot} \leq \alpha M_{\max}^{\text{TOV}}, \qquad (16.23)$$

where $\alpha = 1.3$–1.7 determines the threshold mass for the prompt black hole formation [9, 10, 49, 51, 60, 61, 127], and $M_{\max}^{\text{TOV}}$ is the maximum mass of the TOV star, and $M_{\max}^{\text{sup}}$ is the maximum mass of the uniformly rotating star. A Rhoades-Ruffini causality argument on the equation of state gives $\beta \approx 1.27$ [98]. Equilibrium configurations of uniformly rotating neutron stars with various kinds of equations of states give $\beta \approx 1.2$ [13, 25, 63, 74]. Therefore, the upper limit of the maximum mass of the TOV star is likely to be $\approx$2.16–2.28$M_{\odot}$. The caveat is that their upper limit estimation of the TOV maximum mass is based on the hypothesis that the supramassive neutron star cannot launch a relativistic jet.

Ruiz and his collaborators explored a neutron star spin effect in magnetized binary neutron star mergers [102]. The grid resolution is $\Delta x = 220$ m. The neutron star is modeled with Γ-law equation of state with $\Gamma = 2$. The pre-merger magnetic field is likely to be $10^{16.5}$–10^{17} G with pulsar-like dipole magnetic field topology (see Eq. (16.21)). Non-dimensional neutron star spin χ_{NS} is -0.05, 0.24, and 0.36[5]. It should be noted that the fastest spin observed in the binary pulsars is $\chi_{\text{NS}} \sim 0.05$ [69]. They simulated both unmagnetized and magnetized binaries. Figure 16.4 shows the magnetic field energy, and it shows the pre-merger magnetic field energy of $\sim$2–3 $\times 10^{50}$ erg is amplified by a factor of $\approx$10–15 at $\approx$2 ms after the merger. Subsequently, the magnetic field energy continues to be amplified to $\sim 10^{52}$ erg until the black hole formation. They analyzed the MRI quality factor and Shakura-Sunyaev parameter. They concluded that the magneto-turbulent state is established and, consequently,

[5] Minus sign denotes an anti-align spin with respect to the orbital angular momentum.

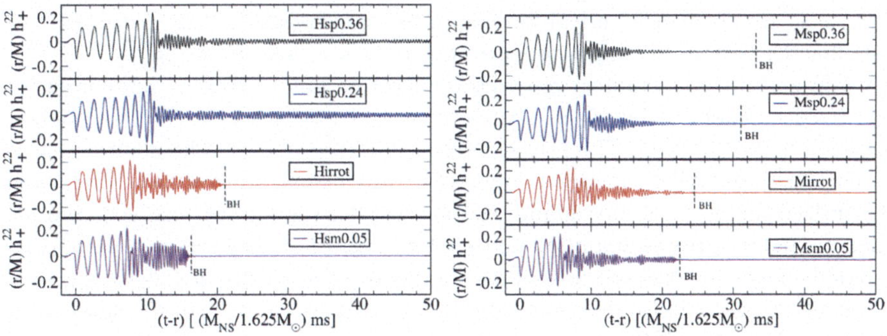

Fig. 16.5 GW strain h_+^{22} (dominant mode) as functions of retarded time. The left panel displays the GW strain in the unmagnetized cases, while the right panel displays the magnetized cases. The dashed vertical line denotes the BH formation time. Hsp0.36, Hsp0.24, Hirrot, and Hsm0.05 denote the unmagnetized binary with $\chi_{\rm NS} = 0.36$, 0.24, 0, and -0.05, respectively. The figure is taken from Ref. [102]

the angular momentum transport is facilitated. Figure 16.5 shows the gravitational waves in all the models, which summarizes the unmagnetized and magnetized binary evolution. For the aligned high spinning cases with $\chi_{\rm NS} = 0.36$ and 0.24, on the one hand, the unmagnetized binaries do not collapse into a black hole until the end of the simulation. On the other hand, the magnetized counterparts do. They argued that it indicates the MHD effect facilitates the angular momentum transport. For the non-spinning case with $\chi_{\rm NS} = 0$ or the anti-aligned low spinning case with $\chi_{\rm NS} = -0.05$, the magnetized binaries survive longer than the unmagnetized counterparts. Their interpretation for this is due to the efficient dissipation of the kinetic energy through the magneto-turbulence; the enhanced thermal pressure supports the remnant massive neutron stars. Also, they reported the incipient jet launching after the black hole formation in the magnetized binaries irrespective of the neutron star spin.

However, there are a couple of caveats: (i) With their grid set up and the figure showing $Q_{\rm MRI} \geq 10$ (see Fig. 9 in Ref. [102]), the poloidal magnetic field strength inside the remnant is estimated to be $\gtrsim 10^{16.3}$ G, which is comparable to the pre-merger magnetic field strength. The resultant winding timescale is $O(1)$ms (see Eq. (16.16)). Therefore, the MHD effect initiated from the pre-merger field is not negligible. Also, the large-scale field generation mechanism from the small-scale MRI-amplified field needs to be clarified more. (ii) The estimated MRI-turbulent viscous timescale scale of $\sim$10 ms (see Eq. (4) in Ref. [102]) is ten times shorter than that in Ref. [56] (see the second and third column in Table I in Ref. [102]). Both works estimated the Shakura-Sunyaev parameter as $\sim 10^{-2}$. Strictly speaking, the binary neutron star model is different. However, it is unlikely to make an order-of-magnitude difference in the viscous timescale. The discrepancy needs to be clarified more. (iii) Their argument on the efficient dissipation of the kinetic energy through the MRI-driven turbulence in $\chi_{\rm NS} = 0$ and -0.05 models compared to $\chi_{\rm NS} = 0.36$ and 0.24 models may not be supported by their estimated Shakura-Sunyaev parameters because it does not significantly differ among the models (see Table III in Ref. [102]). Note that the viscous heating is proportional to the viscous parameter.

Ciolfi and his collaborators reported for the first time a 100 ms simulation for long-lived magnetized binary neutron star merger remnant [18]. The pre-merger magnetic field is 10^{16} G. The grid resolution is $\Delta x = 220$ m. The neutron star is modeled with the APR equation of state [4]. They claimed that after an inefficient Kelvin-Helmholtz instability amplification, the MRI starts to be resolved, and the magnetic field energy is amplified up to $\sim 10^{51}$ erg. Also, they reported there is no sign of the jet launching at the end of the simulation of $\approx$100 ms, and they concluded the magnetar scenario for the short gamma-ray bursts is unlikely, or it could launch a jet much later than 100 ms.

However, there are a couple of caveats: (i) As in Ref. [102], the magnetic winding initiated from the pre-merger magnetic field is not negligible. (ii) With their grid setup and pre-merger magnetic field, it is hard to resolve the MRI, as they mentioned. Particularly, the increase of the MRI quality factor is likely caused by the magnetic winding, i.e., the MRI quality factor is for the toroidal field as they did previously in Refs. [17]. Also, Ref. [56] suggested the axisymmetric MRI in the high-density region with $\rho \geq 10^{13}$ g cm^{-3} can not be resolved with their setup (see Table I in Ref. [56] with the factor of five boosts, i.e., the pre-merger magnetic field ten times larger and the grid resolution twice coarser compared to the run with $\Delta x = 110$ m). (iii) The angular momentum transport due to the magnetic field needs to be clarified more as they mentioned because the angular velocity evolution, particularly in the core of the merger remnant, is similar to the unmagnetized case.

Ruiz and his collaborators explored the magnetic-field topology in binary neutron star mergers [103]. The pre-merger magnetic field is an aligned dipole-like configuration to the orbital angular momentum and a perpendicular dipole-like configuration with a field strength of $\approx 10^{16.5}$–10^{17} G (see Eq. (16.21)). The grid resolution is $\Delta x = 220$ m. The neutron star is modeled with the Γ-law equation of state with $\Gamma = 2$. The neutron star has a spin of $\chi_{\rm NS} = 0.36$. Compared to the aligned–aligned pre-merger magnetic configuration in Ref. [102], the aligned-perpendicular pre-merger magnetic configuration results in a longer lifetime of the remnant massive neutron star. They pointed out that it is caused by an inefficient angular momentum transport due to the MRI-driven turbulence because the Shakura-Sunyaev parameter is smaller than that in the aligned-aligned case. After the black hole formation, the incipient jet is launched as in the aligned-aligned case because the region with the magnetization parameter greater than 10 appears above the black hole. For the perpendicular-perpendicular case, the lifetime of the remnant massive neutron star is longer than the aligned-aligned or aligned-perpendicular case due to the inefficient angular momentum transport. Consequently, the torus mass formed after the remnant massive collapses into a black hole is larger than that in the other cases. Also, the magnetic field energy at the collapse into the black hole is ten times smaller than the aligned-perpendicular case. They mentioned that the MRI quality factor is below 6 in the perpendicular-perpendicular case, and the resultant turbulent viscosity quantified by the Shakura-Sunyaev parameter is small. They ran a higher resolution simulation, but the result is likely to be less conclusive because improving the grid resolution by only 25% is not enough to capture the MRI. After the black hole formation, they did not find an incipient jet launch in the perpendicular-perpendicular model, even though it has a larger torus mass compared to the aligned-aligned case and aligned-perpendicular case. They concluded that it is consistent with the prompt collapse case, which suggests the magnetic field energy should be more significant than 0.1% of the initial ADM mass at the formation of the black hole to launch the incipient jet.

However, in addition to the caveats raised above, there is a caveat to explore the pre-merger magnetic field topology. As demonstrated in this work, it is more challenging to reproduce the MRI-driven turbulence in the perpendicular-perpendicular case, which is likely to be natural since the aligned case is preferable to resolve the MRI, i.e., $\lambda_{\rm MRI} \propto B_z$. It implies that it may not be a fair comparison for the different pre-merger magnetic field topologies with the *same grid set up* because the inefficient development of the MRI-driven turbulence may be merely a consequence of the numerics.

Ciolfi reported a long-term magnetized binary neutron star merger simulation up to 250 ms after the merger for the first time [19]. The pre-merger magnetic field is 10^{15} G and 5×10^{15} G. The employed grid resolution is $\Delta x = 250$ m. The neutron star is modeled by the APR equation of state [4]. His finding is summarized as follows: (i) The magnetically collimated outflow appears in the strongly magnetized case but not in the weakly magnetized case. (ii) The outflow is mildly relativistic with $\lesssim 0.3$ c and the kinetic energy of the outflow within an opening angle of 15° is $\approx 3 \times 10^{49}$ erg. Therefore, except for a very optimistic scenario, the system is not likely to drive a relativistic jet compatible with GRB 170817A [43]. There are a couple of caveats: (i) The subtleness of the pre-merger magnetic field needs to be clarified more, particularly by conducting a detailed analysis of the MRI. (ii) The large-scale magnetic field generation inside the merger remnant must also be clarified more. Since the simulation set-up cannot resolve the MRI, the magnetic winding due to the pre-merger magnetic field relic could be the outflow generation mechanism.

Mösta and his collaborators performed magnetized binary neutron star remnant simulations with neutrino physics [81]. First, they performed an unmagnetized binary neutron star merger simulation. Then, they embeded a dipole-like seed poloidal magnetic field of 10^{15} G inside the remnant massive neutron star, which is short-lived. They took into account the neutrino cooling by the leakage scheme and the neutrino heating phenomenologically. They performed an in-depth convergence test with $\Delta x = 55$ m, 110 m, and 220 m. The neutron star is modeled with LS220 [64]. They also explored the binary merger remnant evolution without neutrino. Their findings are summarized as follows: (i) The highest resolution is fine enough to capture the MRI, and the toroidal magnetic field amplification is significantly enhanced compared to the low-resolution run. The enhanced toroidal field drives an MHD wind with faster velocity than the non-magnetized case, which is the neutrino-driven wind. (ii) The neutrino cooling helps mitigate the polar region's baryon pollution and aids the launch of the jet with the terminal Lorentz factor of $\sim$5. (iii) The mass ejection by the MHD effect increases with the grid resolution by a factor of three. As they mentioned, the caveat is that they assumed a dynamo activity to build up the large-scale poloidal field inside the remnant massive neutron star.

Aguilera-Miret and his collaborators performed magnetized binary neutron star merger simulations with a sub-grid model [1]. Their sub-grid model is based on the gradient sub-grid scale model in which filtering with a Gaussian kernel is applied to each physical quantity [14]. This prescription gives additional terms in the physical flux of the MHD equation:

$$\tau_N^i = -C_N \xi H_N^i \text{ (the flux in the continuity equation)}, \tag{16.24}$$

$$\tau_T^{ij} = -C_T \xi H_T^{ij} \text{ (the flux in the momentum equation)}, \tag{16.25}$$

$$\tau_M^i = -C_M \xi H_M^i \text{ (fhe flux in the induction equation)}, \tag{16.26}$$

where $\xi = \gamma^{1/3}\Delta x^2/24$. C_N, C_T, and C_M are called pre-factors, which a direct simulation should calibrate. The value of the pre-factors with $C_N = C_T = C_M = 1$ follows from the analytical calculation of large eddy simulation with the gradient sub-grid scale model. A priori tests indicate that practically, they might slightly vary with the grid resolution and other parameters, but they usually remain between 1–2 [14]. However, the numerical dissipation inherent to the employed Riemann solver with the high-order cell reconstruction at the small scale strongly attenuates the effect of the large eddy simulation at intermediate resolutions. Practically, the authors balanced this attenuation by artificially increasing the pre-factors. H_N^i, H_T^{ij}, and H_M^i are functions of the filtered fields, and the cumbersome expressions are found in Ref. [14]. The pre-merger magnetic field is 5×10^{11} G, which is approximately consistent with the highly magnetized end of the binary pulsars [69]. They employed three grid resolutions of $\Delta x = 37$ m, 74 m, and 147 m in direct simulations, i.e., without the sub-grid terms. Then, they compared them with simulations with their sub-grid model while keeping the grid resolution $\Delta x = 147$ m and changing the value of C_N, C_T, and C_M. The neutron star is modeled with the SLy equation of state [28]. Figure 16.6 summarizes their results. In the direct simulation with $\Delta x =$ 37 m, the magnetic-field energy is amplified from $\sim 10^{44}$ erg up to $\sim 10^{50}$ erg due to the Kelvin-Helmholtz instability within ≈ 10 ms after the merger. The growth rate of the amplification is determined by the employed grid resolution, which is consistent with the Kelvin-Helmholtz instability property. They also compared the energy spectrum for the turbulent kinetic energy and magnetic field energy (see Fig. 16.7). They concluded that the sub-grid model with $(C_M, C_N, C_T) = (8, 0, 0)$ can capture more efficient magnetic field amplification than the direct simulation counterpart until 5 ms after the merger. This sub-grid setup is also able to mimic the magnetic-field energy amplification of the direct simulation with $\Delta x = 74$ m at 10 ms after the merger. Also, they reported how sensitive the remnant massive neutron star evolution is to the choice of the pre-factor. If they choose $(C_M, C_N, C_T) = (8, 1, 1)$, $(8, 2, 2)$, $(8, 4, 4)$, or $(8, 8, 8)$, the Kelvin-Helmholtz amplification is less efficient than $(C_M, C_N, C_T) = (8, 0, 0)$. More importantly, it triggers the collapse of a black hole, which is not observed in the direct simulations. They concluded that the choice of the non-zero value of C_N and C_T, particularly the latter, facilitates the angular momentum transport too efficiently, which results in a spuriously early black hole formation.

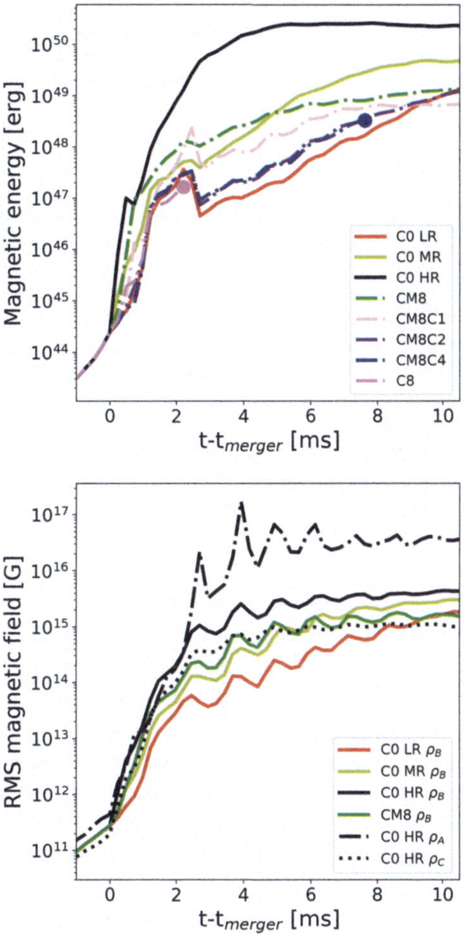

Fig. 16.6 (Top) Integrated magnetic energy as a function of the post-merger time. The circles indicate the collapse of the remnant, forming a black hole. `COLR`, `COMR`, and `COHR` in the legend denote the direct simulations with $\Delta x = 147$ m, 74 m, and 37 m, respectively. `CM8`, `CM8C1`, `CM8C2`, `CM8C4`, and `C8` denote the sub-grid simulation with $(C_M, C_N, C_T) = (8, 0, 0)$, $(8, 1, 1)$, $(8, 2, 2)$, $(8, 4, 4)$, and $(8, 8, 8)$, respectively. (Bottom) The root-mean-square value of the magnetic field for the direct simulations with different resolutions and for the most favorable sub-grid simulation case `CM8`. The r.m.s. magnetic field of the high-resolution case, `COHR`, is calculated in different regions with $\rho > \rho_X$ g cm^{-3}, being $\rho_A = 6 \times 10^9$, $\rho_B = 6 \times 10^{10}$ (value used for the top panel) and $\rho_C = 6 \times 10^{11}$. The figures are taken from Ref. [1]

Besides the subtleness of the choice of the pre-factor, there are a couple of caveats: (i) it is not clear which term(s) in H^i_M in Eq. (16.26) leads the efficient amplification during the Kelvin-Helmholtz instability phase. (ii) The magnetic field energy shown in Fig. 16.6 may not be consistent with their pre-merger magnetic field since it should be $\sim 10^{40}$ erg $B^2_{11.7} R^3_6$ where $B_{11.7} = B/10^{11.7}\mathrm{G} \approx B/(5 \times 10^{11}\mathrm{G})$ and $R_6 = R/10^6\mathrm{cm}$. (iii) The scale of the energy spectrum in Fig. 16.7 may not be consistent with their grid setup. The highest wavenumber of $k \approx 10^{-3}$ cm^{-1} is the spatial scale of 60 m, but their employed grid resolution is $\Delta x = 147$ m.

Ruiz and his collaborators reported new magnetized binary neutron star merger simulations [104]. The pre-merger magnetic field is the pulsar-like dipole field and confined dipole field with presumably $\approx 10^{16.5}$–10^{17} G. The employed grid resolution is $\Delta x = 90$ m–110 m. The neutron star is modeled with the SLy equation of sate [28] and H4 equation of state [42]. They explored the prompt black hole formation case

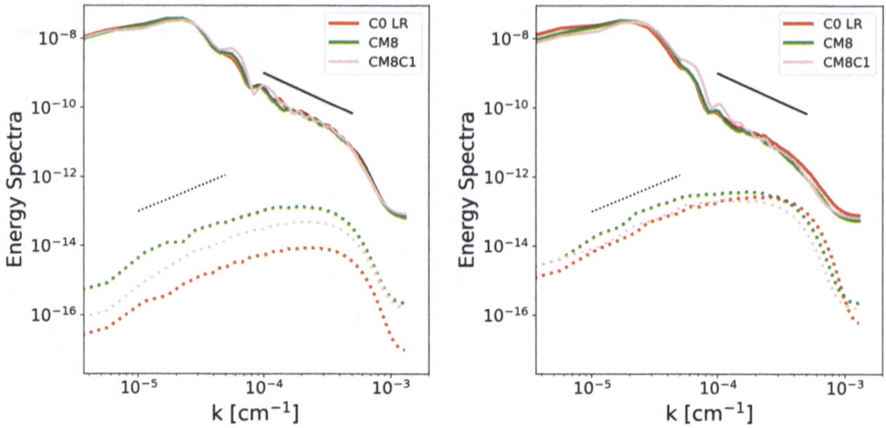

Fig. 16.7 Magnetic (dashed) and kinetic (solid) spectra of `C0LR`, `CM8` and `CM8C1` at 5 ms (left) and 10 ms (right) after the merger. Units in the vertical axis are arbitrary. Black slopes are the Kolmogorov ($\propto k^{-5/3}$) and Kazantsev ($\propto k^{3/2}$). The figure is taken from Ref. [1]

and delayed collapse case. For the latter, the collapse time is $\approx$10–40 ms after the merger. Their findings are qualitatively the same as their series works [99–103]. An incipient jet was not found for the prompt collapse case, irrespective of the model. For the delayed collapse case, the magnetic field energy is amplified to $\sim 10^{51}$ erg during the Kelvin-Helmholtz instability phase. In some models, it is furthermore amplified up to $\sim 10^{52}$ erg until the black hole formation. The amplification highly depends on the magnetic field topology. Except for one model (H4M2.8I in their terminology), which is the same as in Ref. [54] other than the pre-merger magnetic field strength and grid setup, they found a launching incipient jet. Besides the caveats mentioned above, one notable difference from the simulations in Ref. [1], in which a much weaker pre-merger magnetic field is assumed, is the saturation energy of the magnetic field $\sim 10^{50}$ erg at the end of the Kelvin-Helmholtz instability phase.

Palenzuela and his collaborators reported new magnetized binary neutron star simulations with the gradient sub-grid scale model [90]. The pre-merger magnetic field is 5×10^{11} G. They performed an in-depth resolution study with $\Delta x = 30$ m, 60 m, and 120 m with and without the sub-grid model. The neutron star is modeled with the APR equation of state [4]. The pre-factor is chosen $(C_M, C_N, C_T) = (8, 0, 0)$. Their findings are summarized as follows: (i) With the sub-grid model, the highest and middle-resolution runs indicate the convergence to $\bar{B}_{\rm tor} \approx 4$–$5 \times 10^{15}$ G and $\bar{B}_{\rm pol} \approx 5$–$6 \times 10^{15}$ G at the end of the Kelvin-Helmholtz instability phase in the core region defined by $\rho \geq 10^{13}$ g cm^{-3}. It is likely to be consistent with the result in the highest direct simulation. Such a trend is also observed as $\bar{B}_{\rm tor} \approx \bar{B}_{\rm pol} \approx 4$–$5 \times 10^{14}$ G in the envelope region defined by 10^{10} g cm$^{-3} \leq \rho \leq 10^{13}$ g cm^{-3}. (ii) The spectrum analysis suggests the kinetic turbulent energy spectrum follows the Kolmogorov power law with $\propto k^{-5/3}$ and the magnetic energy spectrum follows the Kazantsev power law with $\propto k^{3/2}$. The inverse cascade is likely to occur, which results in the increase of the coherent length scale of the magnetic field from $\approx$0.7 km to 2 km at the end of the simulation of 50 ms after the merger. (iii) During the first 50 ms at least, the efficient angular momentum redistribution is not likely to be facilitated, implying the MRI is not operating because there is no static, large-scale background field over which we can define an unstable perturbation of the MRI.

However, there are a couple of caveats: (i) They employed a cut-off density of 6×10^{13} g cm^{-3} below which the sub-grid term turns off. This cut-off density seems to be significant, and the effect seems not to be negligible for the evolution of both the remnant core and envelope. (ii) Since the highest direct simulation is likely to agree with the converged sub-grid simulation, it implies that the direct simulation looks like entering the convergent regime. It should be confirmed in a higher-resolution direct simulation. Otherwise, we cannot conclude that the saturation magnetic field energy is physical or a consequence of the sub-grid prescription. (iii) The role of the MRI should be explored in a more extended time-scale simulation because their simulation suggests the onset of the magnetic winding. Particularly, several diagnostics to explore the MRI in the non-linear phase proposed in Ref. [47] should be investigated (see also Ref. [56]). (iv) Since the high resolution of $\Delta x = 60$ m still needed for the convergence in the simulation with the sub-grid model, there might be a "double" counting for the turbulence, i.e., one developed by the direct simulation and the other by the sub-grid model which mimics the direct simulation. It is necessary to quantify how such an artifact would affect the convergent property.

Aguilera-Miret and his collaborators reported new magnetized binary neutron star merger simulations with the gradient sub-grid scale model [2]. The pre-merger magnetic field is 10^{12} G with an aligned dipole, misaligned dipole, and multipole topology. They also explored a strong magnetic field case with 10^{15} G. The employed grid resolution is $\Delta x = 60$ m. The neutron star is modeled by the APR equation of state [4]. The pre-factor is chosen $(C_M, C_N, C_T) = (8, 0, 0)$. Their findings are summarized as follows: (i) At the end of the Kelvin-Helmholtz instability, the averaged magnetic field strength in the core region converges within a factor of three for the toroidal component and two for the poloidal component irrespective of the pre-merger field topology and field strength. Consequently, the magnetic field energy saturates within a factor of three, and the strong pre-merger magnetic field results in a similar saturation level. (ii) The evolution of the energy spectrum is insensitive to the pre-merger magnetic field topology. The coherent length of the magnetic field evolves from $\sim$0.8 km to 2 km at the end of the simulation of 30 ms after the merger. Therefore, the pre-merger field topology memory is likely to be lost during the merger, implying the universality of their result.

A caveat is that they only consider the purely poloidal magnetic field, implying zero magnetic helicity. However, the magnetic field configuration after the merger remnant settles down to the quasi-equilibrium state could depend on the net magnetic field helicity because the magnetic helicity conserved in the ideal MHD framework links to the field topology [72].

Sun and his collaborators reported new magnetized binary neutron star merger simulations with neutrino physics [123]. The pre-merger magnetic field is a dipole-like field with presumably $\approx 10^{16.5}$–10^{17} G (see Eq. (16.21)). The employed grid resolution is $\Delta x = 110$ m. The neutron star is modeled with the SLy equation of states [28]. During the simulation, they employed a piecewise polytrope equation of state for the cold part combined with the analytic expression of the thermal part. They also employ the M1 scheme for the neutrino radiation field. Their finding is qualitatively similar to what they did in the past [99–104]: an incipient jet launching for the delayed collapse case irrespective of the neutrino effect. They estimated the neutrino viscous and drag effect on the MRI for the first time in the magnetized binary neutron star merger simulation and reported the effect is not significant. They also reported that neutrino cooling mitigates the baryon-loading in the funnel region above the black hole as reported in Ref. [81]. The angular momentum loss due to the neutrino emission is minor during their simulation time, $\approx$ 15 ms after the black hole formation.

The caveat, except for those in their series works, is that the luminosity for the heavy-lepton neutrino suddenly increases up to $\sim 10^{53}$ erg s^{-1} at $\approx$5 ms after the black hole formation, which seems not to be consistent with the other neutrino radiation transfer simulation of binary neutron star mergers [58, 106, 107].

Palenzuela and his collaborators reported magnetized binary neutron star merger simulations with neutrino physics and the sub-grid model [91]. They implemented a finite temperature nuclear theory-based equation of states HS [109] and LS220 [64] and neutrino leakage scheme [84]. The pre-merger magnetic field is 10^{11} G. They employed a grid resolution of $\Delta x = 187$ m. The pre-factor in the sub-grid model is set to be $(C_M, C_N, C_T) = (0.5, 0, 0)$. They concluded that although they found an efficient amplification of the magnetic field from 10^{11} G to $\sim 10^{14}$ G for $\sim$8 ms after the merger in the sub-grid model simulation the magnetic field hardly alters the neutrino emission such as the neutrino luminosity.

The caveat is that their result on the neutrino emission is not likely conclusive because (i) the previous work showed that at least the grid resolution of $\Delta x = 60$ m is necessary to obtain the saturated magnetic field of $\sim 10^{16}$ G in the sub-grid model run with $(C_M, C_N, C_T) = (8, 0, 0)$ [2] and (ii) the simulation timescale is not long enough to explore the MHD effect on the neutrino radiation.

Kiuchi and his collaborators reported a new implementation for advanced HLLD Riemann solver [70] and applied it to a magnetized binary neutron star merger with neutrino physics [57]. In the context of the accretion disk, it has been well known the HLL(E) Riemann solver[6] commonly used in the numerical relativity community is very diffusive [71, 128]. They explored how the less diffusive HLLD Riemann solver affects the post-merger evolution. They embedded a large-scale poloidal magnetic field of 10^{15} G inside a massive torus formed after a short-lived remnant massive neutron star collapses into a black hole. The employed grid resolution is $\Delta x = 150$ m. The neutron star is modeled with the SFHo equation of state [121]. Their findings are summarized as follows: (i) The small-scale magneto-turbulence due to the MRI is not able to be sustained if we employ the HLLE solver because of inherently large numerical diffusivity. (ii) As a result, the large-scale magnetic field of $\sim$ 10 km is

[6] It is known that HLL(E) and Local-Lax-Friedrich (LLF) Riemann solver [5, 62] give the essentially same result.

artificially enhanced in the simulation with the HLLE Riemann solver compared to that with the HLLD Riemann solver (see Fig. 26 in Ref. [57]). (iii) More energetic (but spurious) Poynting flux-dominated outflow is launched in the simulation with the HLLE Riemann solver compared to that with the HLLD Riemann solver (see Fig. 27 in Ref. [57]).

The caveat is that they only implemented the third-order piecewise parabolic method for the reconstruction in the Riemann problem [24]. Therefore, they did not quantify how the HLLE Riemann solver with a higher-order reconstruction such as the MP5 [124] would reduce the numerical diffusion in binary neutron star merger simulations.

Combi and Siegel reported a new magnetized binary neutron star merger simulations with neutrino physics [22]. They employed finite temperature nuclear equation of states, LS220, SFHo, and APR [4, 64, 121] and a zeroth-momentum (M0) scheme based on a ray-by-ray for the neutrino radiation transport [93]. The pre-merger magnetic field is 5×10^{15} G. The employed grid resolution is $\Delta x = 180$ m and 220 m. They reported that the magnetic field does not significantly impact the dynamical ejecta. They also reported that the MRI is fully developed inside the remnant massive neutron star and facilitates the angular momentum transport.

The caveat is as follows: (i) Many previous works reported the efficient Kelvin-Helmholtz instability and fully resolving MRI are challenging with their setup [17, 18, 56, 59, 81, 90]. (ii) Because of the shortness of their simulation timescale of $\approx$10 ms, the MHD effect, particularly the role of the angular momentum transport inside the remnant, needs to be clarified. Note that the MRI-driven viscous timescale or the magnetic winding timescale is longer than their simulation time. Also, the detailed analysis on the MRI is necessary.

De Haas and his collaborators reported magnetized binary neutron star merger simulations with neutrino physics [46], a follow-up work of Ref. [81]. They added a dipole-like large-scale magnetic field to a remnant massive neutron star:

$$A_r = A_\theta = 0, \; A_\phi = B_0 \sin\theta \frac{r_{\rm falloff}^3}{r_{\rm falloff}^3 + r^3}, \tag{16.27}$$

and they varied $B_0 = 10^{13}$ G, 10^{14} G, 10^{15} G, 5×10^{15} G and $r_{\rm falloff} = 5$ km, 10 km, 15 km, 20 km. The employed grid resolution is $\Delta x = 185$ m. The neutron star is modeled with the LS220 equation of state [64]. They employed the leakage scheme for the neutrino cooling [81] and parameterized prescription for the neutrino heating. They found in the two simulations with $(B_0, r_{\rm falloff}) = (10^{15}$ G, 20 km) and (5×10^{15} G, 10 km) the jet is launched and the velocity distribution of the ejecta has a fast component with 0.4–0.6 c.

The caveat is that (i) the role of the MRI, particularly low magnetic field cases, needs to be clarified more, and (ii), as they mentioned, the large-scale field inside the merger remnant is an assumption.

Kiuchi and his collaborators reported new magnetized binary neutron star merger simulations with neutrino physics [58]. The pre-merger magnetic field is 10^{15} G. The

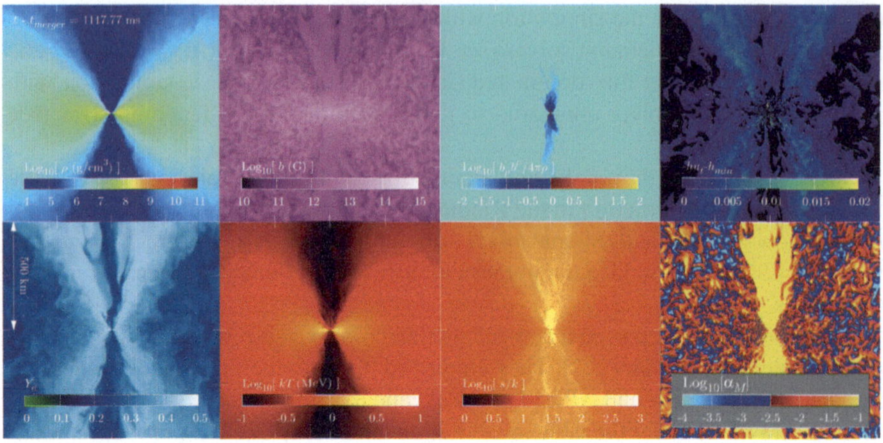

Fig. 16.8 Profiles for rest-mass density (top-left), magnetic-field strength (top-second from left), magnetization parameter (top-second from right), unboundedness defined by the Bernoulli criterion (top-right), electron fraction (bottom-left), temperature (bottom-second from left), entropy per baryon (bottom-second from right), and Shakura-Sunyaev α_{SS} parameter (bottom-right) on the $y - y_{\rm AH} = 0$ plane at $t - t_{\rm merger} \approx 1.1$ s. See also the movie: http://www2.yukawa.kyoto-u.ac.jp/~kenta.kiuchi/anime/FUGAKU/out_SFHo_12_15.mp4. The figure is taken from Ref. [58]

employed grid resolution is $\Delta x = 150$ m and 200 m for the convergence test. The neutron star is modeled with the SFHo equation of state [121]. They performed a simulation up to one second after the merger (see Fig. 16.8 for the final snapshot on a meridional slice). Their findings are summarized as follows: (i) After the short-lived massive neutron star collapses into a black hole, the magnetic field inside the massive torus is amplified by the axisymmetric MRI and magnetic winding. As a result, at ~ 0.1 s after the merger, the fully turbulent state due to the MRI is established, and the turbulent viscosity with the Shakura-Sunyaev parameter of 0.01–0.03 facilitates the angular momentum transport. The turbulent state is sustained by the MRI dynamo, proved by the butterfly diagram lasting until the end of the simulation (see Fig. 3 in Ref. [58]). (ii) Due to the angular momentum transport and turbulent viscous heating, the torus expands, and the temperature drops. As a result, the neutrino cooling becomes inefficient. (iii) Finally, the post-merger mass ejection due to the MRI-driven turbulent viscosity sets in, and the mass of $\approx 8 \times 10^{-3} M_\odot$ is ejected from the torus (see Fig. 16.9 for the detailed properties of the ejecta). (iv) The jet launching is not observed until the end of the simulation of ≈ 1 s after the merger. (v) The convergence in terms of the grid resolution is almost achieved, implying the simulation quality could be good enough to compare to the observational data such as AT 2017gfo.

The caveat is summarized as follows: (i) Since with their setup, it is impossible to resolve an efficient Kelvin-Helmholtz instability and the axisymmetric MRI inside the short-lived remnant massive neutron star whose lifetime is ≈ 17 ms, there is a

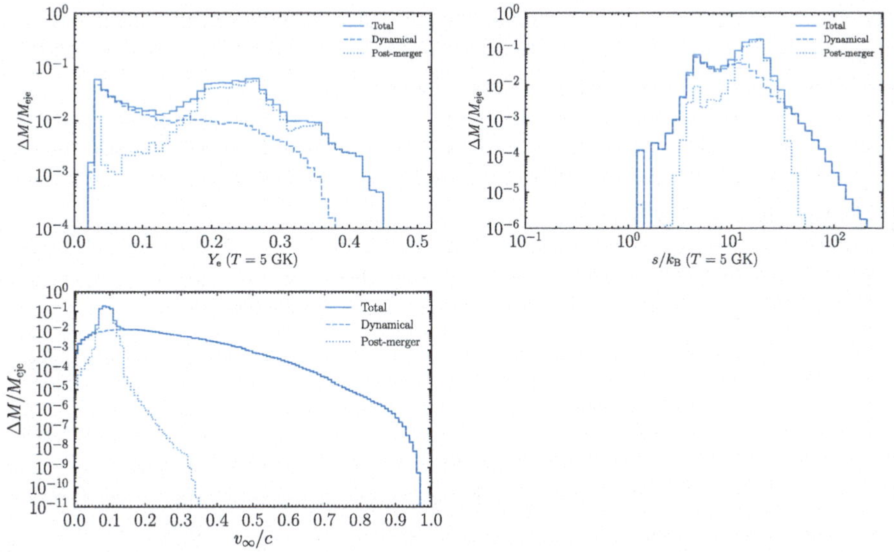

Fig. 16.9 Mass histogram of the ejecta as a function of the electron fraction (top-left), the entropy per baryon (top-right), and the terminal velocity (bottom) at $t - t_{\rm merger} \approx 1.1$ s calculated by the tracer particle. The solid, dashed, and dotted curves denote the profiles for the total, dynamical, and post-merger ejecta, respectively. The figure is taken from Ref. [58]

possibility that the large-scale magnetic field is build up before the black hole formation. (ii) Although the fallback motion lasts and resultant ram pressure overcomes the magnetic pressure until the end of the simulation of $\approx$1 s, the fallback time scale could depend on the binary neutron star model because the other works employing a different binary neutron star model reported the fallback motion ceases at $O(0.01)$ s after the black hole formation (see Ref. [104] for example). (iii) The absence of the jet launching could be caused by spurious spin down of a black hole due to an insufficient grid resolution. The resolution study with the Cowling approximation suggested it is unlikely, though.

Chabanov and his collaborators explored pre-merger magnetic field topology in magnetized binary neutron star mergers [15]. The pre-merger magnetic field is a traditional confined dipole field with $\sim 10^{14}$ G and "crustal" field with $\sim 2 \times 10^{14}$ G. The employed grid resolution is $\Delta x = 70$ m and 105 m for the convergence test. The neutron star is modeled with the TNTYST equation of state [126]. Their finding is summarized as follows: (i) The amplification process comprises the Kelvin-Helmholtz instability phase, the subsequent decay phase, and the subsequent turbulent phase. (ii) At the end of the Kelvin-Helmholtz instability phase, the magnetic field energy for the "crustal" configuration is smaller by a factor of few than the confined configuration. At the end of the turbulent phase, the former is smaller by a one-order magnitude than the latter. They conclude that the "crustal" configuration leads to inefficient Kelvin-Helmholtz instability compared to the confined configuration widely used in magnetized binary neutron star mergers.

The caveat is that the saturation energy of the magnetic field for the "crustal" configuration needs to be more extensively explored since the high-resolution simulation shows a significant growth rate, implying the physical saturation could be far from what they reported.

Most and Quataert reported new magnetized binary neutron star merger simulations with neutrino physics and a sub-grid model to reproduce a large-scale dynamo [78]. Specifically, their sub-grid model is α dynamo described by

$$e^{\mu} = \kappa b^{\mu}, \tag{16.28}$$

where e^{μ} and b^{μ} are the electric and magnetic fields in a fluid comoving frame. In the non-relativistic ideal MHD approximation, $e^i = E^i + (\mathbf{V} \times \mathbf{B})^i = 0$ (see Eqs. (16.11)–(16.12)). κ is a calibration parameter such that the saturation value of the magnetization parameter $\sigma = b^2/\rho$ is controlled. They performed two cases with $\sigma \approx 0.01$ and ≈ 0.001. The pre-merger magnetic field is 10^{15} G. The employed grid resolution is $\Delta x = 250$ m. The neutron star is modeled by the DD2 and APR equation of sates [4, 48]. The leakage scheme incorporates neutrino cooling. Figure 16.10 summarizes their result. Due to the α dynamo prescription, the strong magnetic field of $\sim 10^{17}$ G is produced, and the resultant magnetic buoyancy force pushes the fluid elements upward near the surface (left panel). A strongly magnetized loop that sticks out of the stellar surface is formed. It is twisted by a differential rotation of the merger remnant and inflate (center panel). As the twist is increased, the inflated bubble detaches from the merger remnant (right panel). Since this process repeats, a periodicity is imprinted in the Poynting flux as shown in Fig. 16.11. The luminosity and the periodicity strongly depend on the equation of state and the dynamo calibration parameter κ.

The caveat is as follows: (i) More detailed calibration κ or detailed modeling of the dynamo prescription is necessary because the result is susceptible to it. (ii) The role of the MRI needs to be more clarified.

Combi and Siegel reported new magnetized binary neutron star merger simulations with neutrino physics [23]. The pre-merger magnetic field is 3×10^{15} G. The employed grid resolution is $\Delta x = 180$ m. The neutron star is modeled with the APR equation of state [4]. The neutrino treatment is the same as the previous work [22]. Their finding is summarized as follows: (i) After the Kelvin-Helmholtz phase, the inverse turbulent cascade creates a large-scale magnetic field. (ii) The large-scale toroidal magnetic field is further amplified, and an incipient jet is launched. Also, the post-merger mass ejection due to the MHD effect sets in.

The caveat is as follows: (i) Since many previous simulations revealed that an efficient Kelvin-Helmholtz instability and the MRI are hard to be resolved by their setup, the mechanism to generate the large-scale magnetic field must be clarified more. (ii) Because of the lack of a resolution study, it is hard to quantify systematic errors in

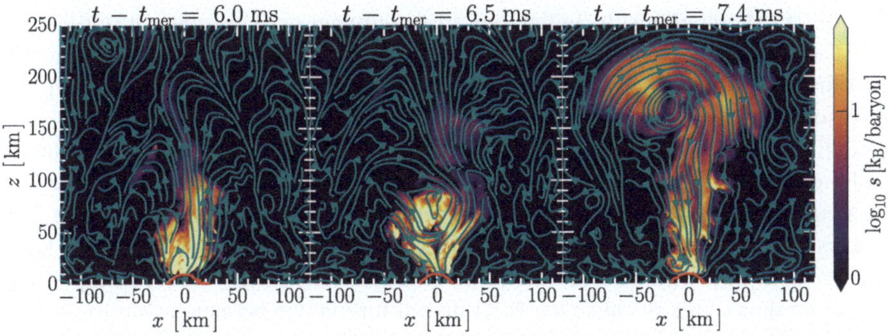

Fig. 16.10 Flaring process at the surface (red) of the neutron star merger remnant. The entropy per baryon s is shown in color (indicating heating inside the flare) with the magnetic field lines. The time is defined as the post-merger time. (Left) Built-up of magnetic stresses due to buoyancy of field lines from the star. (Center) Inflation of the connected flux tube due to strong differential rotation at the surface of the star, where both footpoints are anchored. (Right) Detachment of the flare within $\simeq$ 1ms of its creation. In the simulations, this cycle repeats quasi-periodically until a steady outflow is launched. The model is $\sigma = 0.01$. The figure is taken from Ref. [78]

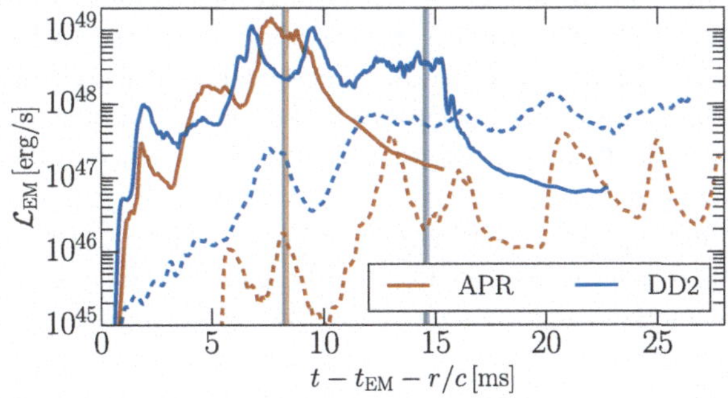

Fig. 16.11 Electromagnetic luminosity as a function of the retarded time with t_{EM}, which corresponds to the time of the first electromagnetic outburst extracted at a radius $r = 236$ km from the stellar remnant. Solid lines refer to high magnetization models, $\sigma = 0.01$, and dashed lines to low magnetization models, $\sigma = 0.001$. For the high magnetization cases, black hole formation is indicated by a vertical black line. The figure is taken from Ref. [78]

their findings, such as electromagnetic field luminosity and post-merger mass ejection, due to the grid resolution. This point is crucial to compare the simulation result with the observation data.

Kiuchi and his collaborators reported new magnetized binary neutron star merger simulations with neutrino physics [59]. The pre-merger magnetic field is 3×10^{15} G. The employed grid resolution is $\Delta x = 12.5$ m from the inspiral to the first $\approx$30 ms after the merger, subsequently $\Delta x = 50$ m up to $\approx$50 ms after the merger, and finally $\Delta x = 100$ m until the end of the simulation of 175 ms after the merger (see Method section in Ref. [59] for detail). They also performed the convergence test with $\Delta x = 200$ m. The neutron star is modeled with the DD2+Timmes equation of state [48, 125], which results in the long-lived remnant massive neutron star formation. The neutrino radiation transport is solved with the M1+GR Leakage scheme. Their findings are summarized as follows: (i) As reported previously [15, 54–56], the low resolution with $\Delta x = 200$ m is unable to capture the efficient Kelvin-Helmholtz instability and the MRI (see Extended Data Figs. 1 and 2 in Ref. [59]). The high-resolution simulation with $\Delta x = 12.5$ m can sustain the MRI-driven turbulence. (ii) The neutrino viscosity and drag are likely to be irrelevant in binary neutron star mergers because of the efficient Kelvin-Helmholtz instability (see Extended Data Fig. 3 in Ref. [59], which solved the dispersion relations (16.5)–(16.6) on top of the simulation data). (iii) Because the MRI-driven turbulence is responsible for generating the electromotive force (16.11) and the period in the butterfly diagram agrees with the prediction by the $\alpha\Omega$ dynamo theory,

$$P_{\rm theory} = 2\pi \left| \frac{1}{2}\alpha_{\phi\phi}\frac{d\Omega}{d\ln R}k_z \right|^{-1/2}, \qquad (16.29)$$

where k_z is the wavenumber corresponding to the pressure scale height (see also Eq. (16.12) for $\alpha_{\phi\phi}$), the $\alpha\Omega$ dynamo can be interpreted as a mechanism for the large-scale magnetic field generation as shown in Figs. 16.12 and 16.13.[7] (iv) The pre-merger large-scale magnetic field is harmless for the large-scale dynamo because such a relic magnetic field stays buried deep inside the merger remnant core throughout the simulation, and the dynamo wave appears from the surface of the remnant core (see Extended Data Fig. 5 in Ref. [59]). (v) The relativistic Poynting-flux dominated outflow with the luminosity of $\sim 10^{51}$ erg s^{-1} is launched by the large-scale magnetic field due to the $\alpha\Omega$ dynamo. Also, the Lorentz force due to the large-scale field drives an enormous post-merger mass ejection of $\approx 0.1 M_\odot$ as shown in Fig. 16.14. It should

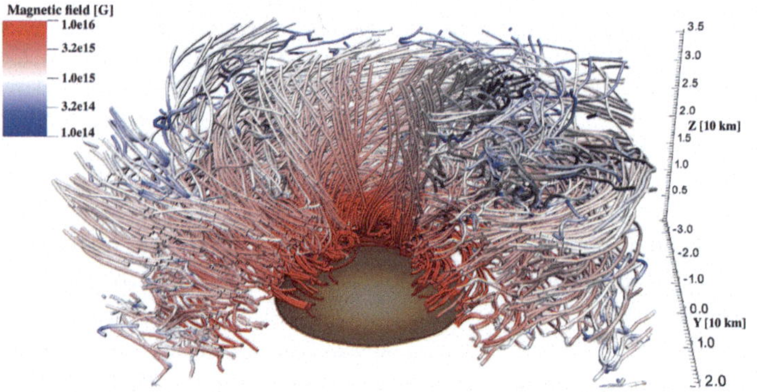

Fig. 16.12 Magnetic field lines for the density of $\rho < 10^{13}\,{\rm g\,cm^{-3}}$ at $t - t_{\rm merger} \approx 130$ ms. The core of the hypermassive neutron star is shown for the density of $\rho > 10^{13}\,{\rm g\,cm^{-3}}$. The figure is taken from Ref. [59]

[7] The working hypothesis to derive Eqs. (16.13)–(16.15) is directly verified in Ref. [59].

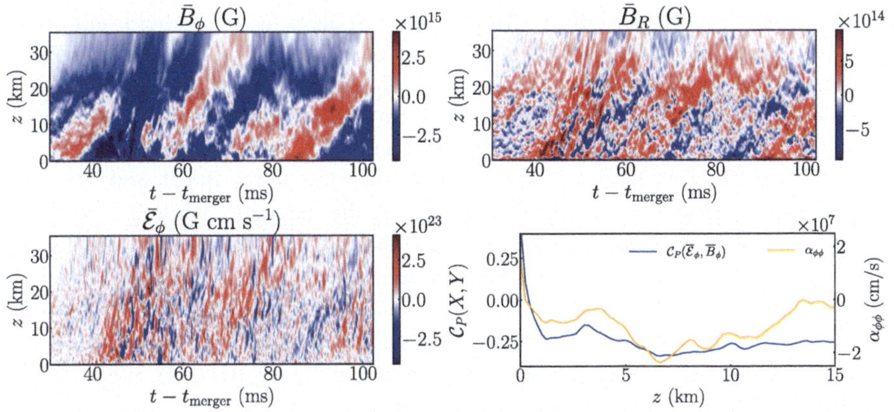

Fig. 16.13 Butterfly diagram at $R = 30$ km: (Top-left) Mean toroidal magnetic field $\bar{B}_\phi$. (Top-right) Mean radial magnetic field $\bar{B}_R$. (Bottom-left) Toroidal electromotive force $\bar{\mathscr{E}}_\phi$. (Bottom-right) $\alpha_{\phi\phi}$ parameters (orange) and correlation between $\bar{\mathscr{E}}_\phi$ and $\bar{B}_\phi$ (blue). The figure is taken from Ref. [59]

be noted that the low-resolution simulation with $\Delta x = 200$ m underestimates the luminosity of the Poynting flux by two orders of magnitude and the post-merger ejecta mass by one order of magnitude at $\approx$100 ms after the merger, which corresponds to a "longest"-term simulation among the previous simulations conducted to date for the long-lived case. Therefore, the systematic error due to the grid resolution is astonishingly large, which is crucial for the comparison to the observational data such as AT 2017gfo.

The caveat is that (i) it is necessary to confirm this picture by a simulation starting from a much weaker pre-merger magnetic field since, as claimed in Refs. [3, 90], the MRI may not be effective at least early post-merger phase because of the absence of the static and large-scale background field. (ii) The other dynamo such as the Taylor-Spruit dynamo [120] could be effective in generating the large-scale magnetic field, particularly deep inside the merger remnant core because such a region is not subject to the MRI due to the positive radial gradient of the angular velocity. (iii) Since the magnetic Prandtl number determined by the numerical viscosity and resistivity is an order of unity in their simulation, the large-scale dynamo property may change in the high magnetic Prandtl number regime.

Aguilera-Miret and his collaborators reported new magnetized binary neutron star merger simulations with the gradient sub-grid scale model [3]. The pre-merger magnetic field strength is 5×10^{11} G. They also performed a follow-up simulation with the "crustal" configuration proposed in Ref. [15]. The employed grid resolution is $\Delta x = 60$ m. The neutron star is modeled by the APR equation of state [4]. The remnant massive neutron star is long-lived. The pre-factor of the sub-grid model is $(C_M, C_N, C_T) = (8, 0, 0)$ with the cut-off density of 2×10^{11} g cm^{-3} below which the sub-grid terms are turned off. As consistent with their previous works [2, 90], the Kelvin-Helmotholz instability triggers the turbulent magnetic field amplification up to $\sim 10^{50}$ erg until $\sim$5 ms after the merger. The energy spectrum analysis suggests that isotropic turbulence results in comparable strength in the poloidal and toroidal magnetic fields. After the Kelvin-Helmholtz instability phase, the turbulent resistivity is enhanced. As a result, the small-scale magnetic field is diffused, and the magnetic field with a coherent length of a few km is developed. Because of the resultant coherent poloidal magnetic field, the magnetic winding works as a further amplification of the toroidal field, and the magnetic field energy ends up at $\sim 10^{51}$ erg at $\sim$110 ms after the merger. Although they observed a helicoidal structure of the magnetic field, they did not find a jet launching until the end of the simulation of $\approx$110 ms after the merger. The MRI potentially unstable region inside the remnant massive neutron star has a highly non-axisymmetric intensity, implying the prediction of $\lambda_{\rm MRI}$ is a non-trivial task because the classical and widespread way of evaluation of $\lambda_{\rm MRI}$ assumes a background and homogeneous field. Even starting with the "crustal" configuration, they found a similar saturation energy of $\sim 10^{50}$ erg at the end of the Kelvin-Helmholtz instability phase, which may imply the result in Ref. [15] could be merely caused by an insufficient resolution.

The caveat is that (i) there is still room for the MRI investigation because they did not explicitly show that the MRI does not emerge in their simulation, although the MRI diagnostics, such as the MRI quality factor or the Maxwell stress, indicate the emergence of the MRI. (ii) The role of the turbulent resistivity needs to be more clarified because of the inherently large numerical resistivity of their Local-Lax-Freidrich Riemann solver (see Figs. 12 and 13 in Ref. [57] for the magnetic reconnection problem with different Riemann solvers). (iii) The $\alpha\Omega$ dynamo's role needs to be clarified, particularly a correlation between the electromotive force and mean field as demonstrated in Ref. [59].

Most proposed a new sub-grid model for the $\alpha\Omega$ dynamo in magnetized binary neutron star merger context [79]. By assuming the dynamo effects grow relative to the resistive timescale in the Ohm's law, he arrived at a tensorial relation between the electric field e^μ and magnetic field b^μ in the fluid co-moving frame:

$$e^\mu = \kappa^\mu{}_\nu b^\nu, \tag{16.30}$$

where $\kappa^\mu{}_\nu = -\eta\alpha^\mu{}_\nu$ with the resistivity η and dynamo alpha (see also Eq. (16.12)). This equation is furthermore simplified by assuming that $\kappa^\mu{}_\nu = \kappa(\delta^\mu{}_\nu + u^\mu u_\nu)$ where u^μ is a fluid four velocity:

$$e^\mu = \kappa b^\mu. \tag{16.31}$$

The dynamo coefficient κ and the dynamo saturation are inspired by high-resolution magnetized binary neutron star merger simulations [56, 59]:

$$\kappa = \kappa_{\rm HMNS}\max(0, \Delta_{\rm turb}), \tag{16.32}$$

$$\Delta_{\rm turb} = 1 - \frac{\sigma}{\sigma_{\rm turb}}, \tag{16.33}$$

$$\sigma_{\rm turb} = \xi \left(l_{\rm MRI}^{\rm HMNS}\right)^2 \left(\frac{\Delta x}{12.5\ {\rm m}}\right)^2 \sigma^{\mu\nu}\sigma_{\mu\nu}, \tag{16.34}$$

$$l_{\rm MRI}^{\rm HMNS} = \max(0, a \log_{10}(\rho/\rho_*) \exp[-|b \log_{10}(\rho/\rho_*)|^{5/2}])\ {\rm m}, \tag{16.35}$$

where $\kappa_{\rm HMNS} \approx$0.025–0.035 is the dynamo parameter inferred from the ultra-high resolution simulation [59], σ is the magnetization parameter, ξ is a parameter, $\sigma_{\mu\nu}$ is the shear tensor, and $l_{\rm MRI}^{\rm HMNS}$ is the MRI wavelength inside the remnant massive neutron star. An important assumption here is that the $\alpha\Omega$ dynamo will terminate once a fraction ξ of the turbulent kinetic energy is converted into the magnetic field energy. $l_{\rm MRI}^{\rm HMNS}$ is fitted by the global simulation in Ref. [56] with $a = 22.31984$, $b = -0.425832$, and $\rho_* = 1.966769 \times 10^9$ g cm^{-3} [94]. He left ξ as a free parameter and performed a simulation with $\xi = 0.04$, 0.4, and 4.[8] The pre-merger magnetic field is 10^{15} G. The employed grid resolution is $\Delta x = 200$ m. The result is qualitatively similar to his previous work [78] (see Fig. 16.10 and its explanation). Figure 16.15 shows how the choice of ξ results in the luminosity for the Poynting flux, and it indicates $\xi = 4$ is closed to those found in Ref. [59] (see Fig. 16.14).

The caveat is that a further calibration for the sub-grid model for the $\alpha\Omega$ dynamo is necessary because the result is sensitive to the choice of ξ.

16.4 Summary and Prospect

GRMHD simulations in binary neutron star mergers have rapidly progressed in the last sixteen years. At the initial phase, it was unclear whether the Kelvin-Helmholtz instability could efficiently amplify the magnetic field at the merger, which was originally reported in the Newtonian Smoothed Particle Hydrodynamics simulation [92]. Currently, the numerical relativity community has a consensus on this picture: *the Kelvin-Helmholtz instability develops strongly magnetized turbulence in a short timescale* <5 ms *at the merger.* However, the physical saturation of the Kelvin-Helmholtz instability with $B \gtrsim 10^{15-16}$ G needs to be investigated further.

Also, the way in which the large-scale magnetic field inside the merger remnant builds up needs to be clarified more. A couple of simulations suggest *if the large-scale magnetic field is established before the remnant collapses into the black hole, jet launching could be possible.* However, the numerical relativity community keeps asking the question: Is it a relic of the large-scale pre-merger magnetic field? Or does a non-trivial physical process work to generate the large-scale magnetic field from the small-scale magnetic field? Ultimately, the community has to answer a question: does a merger simulation starting from a *strong and large-scale* pre-merger magnetic field with "standard" grid resolution lead to a physically equivalent outcome of a simulation starting from a *weak and large-scale* pre-merger magnetic field with "high" grid resolution?

---

[8] In principle, ξ should be smaller than unity.

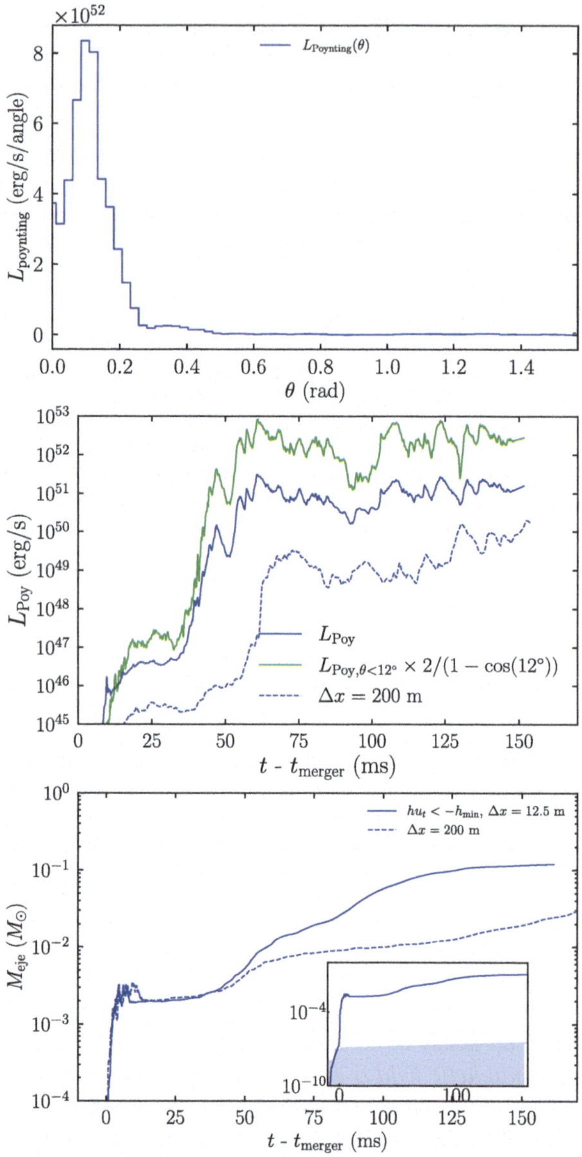

Fig. 16.14 (Top) Angular distribution of the luminosity of the Poyinting flux at the end of the simulation of $t - t_{\rm merger} \approx 150$ ms. (Middle) Luminosity for the Poynting flux as a function of the post-merger time. The green curve is the jet-opening-angle corrected luminosity. The blue-dashed curve plots the luminosity for the simulation with $\Delta x_{\rm finest} = 200$ m. (Bottom) Ejecta as a function of the post-merger time. The solid curve denotes the ejecta satisfying the Bernoulli criterion. The colored region in the inset shows the violation of the baryon mass conservation. The blue-dashed curve plots the ejecta for the simulation with $\Delta x_{\rm finest} = 200$ m. The figure is taken from Ref. [59]

A recent demonstration of the MRI-driven $\alpha\Omega$ dynamo inside the remnant massive neutron star could resolve a piece of the puzzle on this issue. The lesson from it is *that we need an in-depth resolution study and a novel analysis to disentangle the large-scale field generated by the non-trivial process from the relic large-scale field.* With them, it is possible to quantify how the assumed large-scale pre-merger field affects the simulation outcome. Otherwise, it is impossible to reject the possibility

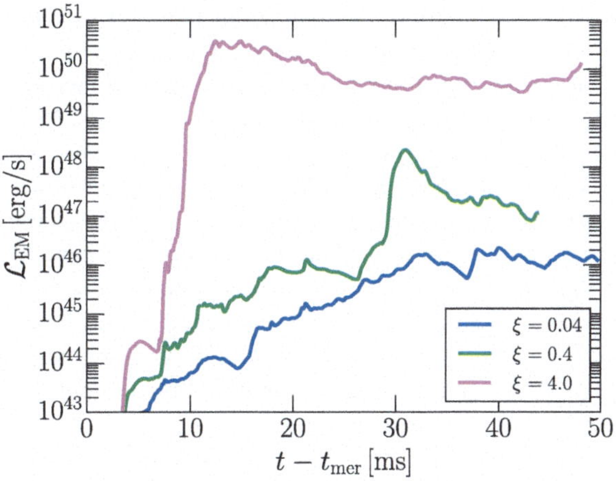

Fig. 16.15 Electromagnetic (Poynting) flux, $\mathscr{L}_{EM}$, measured at $r = 236$ km from the remnant, as a function of the post-merger time. Different curves correspond to the different dynamo parameters ξ. The figure is taken from Ref. [79]

that the outcome is merely a consequence of the unrealistically large-scale pre-merger magnetic field. It should be also emphasized that the resolution study is essential to build a physical theoretical model, which should be compared to the observables such as AT 2017gfo and GRB170817A because a higher resolution simulation makes more than one order of magnitude difference in the quantities relevant to the electromagnetic counterpart modeling as demonstrated in Ref. [59]. Also, this work brings up a new question: *What is a realistic time scale to build up the large-scale magnetic field via* $\alpha\Omega$ *dynamo? Is there any other possibilities for the large-scale dynamo such as the Taylor-Spruit dynamo?* Since Ref. [59] pointed out that the growth timescale for the large-scale magnetic field generation approximately agrees with the period of the dynamo cycle, i.e., the butterfly diagram. However, the mechanism to set the large-scale field strength just *after the merger* is an open question. A conservative estimate based on the merged poloidal magnetic flux of the pre-merger magnetic field with 10^{12} G, i.e., highly magnetized end of the binary pulsars, gives the time scale of $O(0.1)$ s for the large-scale magnetic field to build up [59]. This implies that if a remnant massive neutron star survives longer than this time scale, it could be a central engine of short gamma-ray bursts. The other possibility is the large-scale dynamo inside a torus formed after a remnant massive neutron star collapses into a black hole [44]. Once the large-scale field is established inside the torus, the Blandford-Znajek mechanism could drive a relativistic jet from the black hole [11]. However, it is an open question whether or not the large-scale field strong enough to extract the black hole rotational energy efficiently is possible via the disk dynamo. The numerical relativity community will continue exploring these possibilities by more sophisticated simulations.

The numerical relativity community does not have a consensus on the MRI inside the merger remnant massive neutron star. The gradient sub-grid scale model simulations starting from a realistic pre-merger magnetic field strength suggest that the remnant massive neutron star has a high non-axisymmetric intensity, implying the conventional way to estimate λ_{MRI} may not be valid [3, 90], which many simulations starting from the large-scale pre-merger magnetic field rely on to quantify to what extent the simulations resolve the MRI (see Ref. [56] for example). However, it does not necessarily mean whether or not the MRI sets in inside the merger remnant, but it means we need to seek a robust way to quantify the MRI's emergence or non-emergence, particularly the MRI's non-linear phase. One potential proposal is to measure the Shakura-Sunyaev parameter or the Maxwell stress. Furthermore, the diagnostics proposed in Ref. [47] could reasonably estimate how the MRI-driven turbulence is sustained. It should be noted that in the non-linear phase of the MRI, the magnetic field is strongly turbulent. It must have a high non-axisymmetric intensity and no static, large-scale background. Also, evaluating the mean poloidal field in the MRI active and inactive region could be another way to quantify the emergence of the MRI as demonstrated in Ref. [59].

The numerical relativity community has a consensus that grid resolution is essential for GRMHD simulations of binary neutron star mergers. However, the *effective* resolution could be determined by combining the quality of the Riemann solver and cell-reconstruction scheme. Most of the existing NR codes employ the HLL(E) or its variant LLF Riemann solver with a higher-order cell reconstruction scheme, such as MP5, except for Ref. [57] which employs the HLLD Riemann solver. Quantifying the effective grid resolution with the different Riemann solvers and cell-reconstruction schemes is a task that needs to be pursued in the future.

In summary, numerical modeling of binary neutron star mergers based on GRMHD simulations will continue playing a pivotal role in the multimessenger era. The numerical relativity community will keep making an effort to develop physical modeling for binary neutron star mergers, whose quality is quantitatively good enough to compare to observational data, not only the gravitational waves but also the electromagnetic signals.

Acknowledgements K. K. thanks to the Computational Relativistic Astrophysics members in AEI for a stimulating discussion. K. K. also thanks to Luciano Rezzolla, Bruno Giacomazzo, Carlos Palenzuela, Riccardo Ciolfi, Milton Ruiz, Elias Most, Ricard Aguilera-Miret, Luciano Combi and Philipp Moesta for their feedback to the draft. This work is in part supported by the Grant-in-Aid for Scientific Research (grant Nos. 23H01172) of Japan MEXT/JSPS, and by the HPCI System Research Project (Project ID: hp220174, hp220392, hp230204, hp230084, hp240039).

References

1. R. Aguilera-Miret, D. Viganò, F. Carrasco, B. Miñano, C. Palenzuela, Phys. Rev. D **102**(10), 103006 (2020). https://doi.org/10.1103/PhysRevD.102.103006, arXiv:2009.06669 [gr-qc]
2. R. Aguilera-Miret, D. Viganò, C. Palenzuela, Astrophys. J. Lett. **926**(2), L31 (2022). https://doi.org/10.3847/2041-8213/ac50a7, arXiv:2112.08406 [gr-qc]
3. R. Aguilera-Miret, C. Palenzuela, F. Carrasco, D. Viganò, Phys. Rev. D **108**(10), 103001 (2023). https://doi.org/10.1103/PhysRevD.108.103001, arXiv:2307.04837 [astro-ph.HE]
4. A. Akmal, V.R. Pandharipande, D.G. Ravenhall, Phys. Rev. C **58**, 1804–1828 (1998). https://doi.org/10.1103/PhysRevC.58.1804, arXiv:nucl-th/9804027 [nucl-th]

5. H. Amiram, L.D. Peter, L. Bram van, SIAM Rev. **25**(1), 35–61 (1983). https://doi.org/10.1137/1025002
6. M. Anderson, E.W. Hirschmann, L. Lehner, S.L. Liebling, P.M. Motl, D. Neilsen, C. Palenzuela, J.E. Tohline, Phys. Rev. Lett. **100**, 191101 (2008). https://doi.org/10.1103/PhysRevLett.100.191101, arXiv:0801.4387 [gr-qc]
7. S.A. Balbus, J.F. Hawley, Astrophys. J. **376**, 214–233 (1991). https://doi.org/10.1086/170270
8. S.A. Balbus, J.F. Hawley, Rev. Mod. Phys. **70**, 1–53 (1998). https://doi.org/10.1103/RevModPhys.70.1
9. A. Bauswein, T.W. Baumgarte, H.T. Janka, Phys. Rev. Lett. **111**(13), 131101 (2013). https://doi.org/10.1103/PhysRevLett.111.131101, arXiv:1307.5191 [astro-ph.SR]
10. A. Bauswein, O. Just, H.T. Janka, N. Stergioulas, Astrophys. J. Lett. **850**(2), L34 (2017). https://doi.org/10.3847/2041-8213/aa9994, arXiv:1710.06843 [astro-ph.HE]
11. R.D. Blandford, R.L. Znajek, Mon. Not. R. Astron. Soc. **179**, 433–456 (1977). https://doi.org/10.1093/mnras/179.3.433
12. A. Brandenburg, K. Subramanian, Phys. Rept. **417**, 1–209 (2005).https://doi.org/10.1016/j.physrep.2005.06.005, arXiv:astro-ph/0405052 [astro-ph]
13. C. Breu, L. Rezzolla, Mon. Not. R. Astron. Soc. **459**(1), 646–656 (2016). https://doi.org/10.1093/mnras/stw575, arXiv:1601.06083 [gr-qc]
14. F. Carrasco, D. Viganò, C. Palenzuela, Phys. Rev. D **101**(6), 063003 (2020). https://doi.org/10.1103/PhysRevD.101.063003, arXiv:1908.01419 [astro-ph.HE]
15. M. Chabanov, S.D. Tootle, E.R. Most, L. Rezzolla, Astrophys. J. Lett. **945**(1), L14 (2023). https://doi.org/10.3847/2041-8213/acbbc5, arXiv:2211.13661 [astro-ph.HE]
16. I.M. Christie, A. Lalakos, A. Tchekhovskoy, R. Fernández, F. Foucart, E. Quataert, D. Kasen, Mon. Not. R. Astron. Soc. **490**(4), 4811–4825 (2019). https://doi.org/10.1093/mnras/stz2552, arXiv:1907.02079 [astro-ph.HE]
17. R. Ciolfi, W. Kastaun, B. Giacomazzo, A. Endrizzi, D.M. Siegel, R. Perna, Phys. Rev. D **95**(6), 063016 (2017). https://doi.org/10.1103/PhysRevD.95.063016, arXiv:1701.08738 [astro-ph.HE]
18. R. Ciolfi, W. Kastaun, J.V. Kalinani, B. Giacomazzo, Phys. Rev. D **100**(2), 023005 (2019). https://doi.org/10.1103/PhysRevD.100.023005, arXiv:1904.10222 [astro-ph.HE]
19. R. Ciolfi, Mon. Not. R. Astron. Soc. **495**(1), L66–L70 (2020). https://doi.org/10.1093/mnrasl/slaa062, arXiv:2001.10241 [astro-ph.HE]
20. R. Ciolfi, Gen. Rel. Grav. **52**(6), 59 (2020). https://doi.org/10.1007/s10714-020-02714-x, arXiv:2003.07572 [astro-ph.HE]
21. F. Cipolletta, J.V. Kalinani, B. Giacomazzo, R. Ciolfi, Class. Quantum Gravity **37**(3), 135010 (2020). https://doi.org/10.1088/1361-6382/ab8be8, arXiv:1912.04794 [astro-ph.HE]
22. L. Combi, D.M. Siegel, Astrophys. J. **944**(1), 28 (2023). https://doi.org/10.3847/1538-4357/acac29, arXiv:2206.03618 [astro-ph.HE]
23. L. Combi, D.M. Siegel, Phys. Rev. Lett. **131**(23), 231402 (2023). https://doi.org/10.1103/PhysRevLett.131.231402, arXiv:2303.12284 [astro-ph.HE]
24. P. Colella, P.R. Woodward, J. Comput. Phys. **54**, 174 (1984)
25. G.B. Cook, S.L. Shapiro, S.A. Teukolsky, Astrophys. J. **424**, 823 (1994). https://doi.org/10.1086/173934
26. K. Dionysopoulou, D. Alic, C. Palenzuela, L. Rezzolla, B. Giacomazzo, Phys. Rev. D **88**, 044020 (2013). https://doi.org/10.1103/PhysRevD.88.044020, arXiv:1208.3487 [gr-qc]
27. K. Dionysopoulou, D. Alic, L. Rezzolla, Phys. Rev. D **92**(8), 084064 (2015). https://doi.org/10.1103/PhysRevD.92.084064, arXiv:1502.02021 [gr-qc]
28. F. Douchin, P. Haensel, Astron. Astrophys. **380**, 151 (2001). https://doi.org/10.1051/0004-6361:20011402, arXiv:astro-ph/0111092 [astro-ph]
29. M.D. Duez, Y.T. Liu, S.L. Shapiro, B.C. Stephens, Phys. Rev. D **72**, 024028 (2005). https://doi.org/10.1103/PhysRevD.72.024028, arXiv:astro-ph/0503420 [astro-ph]
30. M.D. Duez, Y.T. Liu, S.L. Shapiro, M. Shibata, B.C. Stephens, Phys. Rev. Lett. **96**, 031101 (2006). https://doi.org/10.1103/PhysRevLett.96.031101, arXiv:astro-ph/0510653 [astro-ph]

31. M.D. Duez, Y.T. Liu, S.L. Shapiro, M. Shibata, Phys. Rev. D **73**, 104015 (2006). https://doi.org/10.1103/PhysRevD.73.104015, arXiv:astro-ph/0605331 [astro-ph]
32. A. Endrizzi, R. Ciolfi, B. Giacomazzo, W. Kastaun, T. Kawamura, Class. Quantum Gravity **33**(16), 164001 (2016). https://doi.org/10.1088/0264-9381/33/16/164001, arXiv:1604.03445 [astro-ph.HE]
33. Z.B. Etienne, Y.T. Liu, S.L. Shapiro, Phys. Rev. D **82**, 084031 (2010). https://doi.org/10.1103/PhysRevD.82.084031, arXiv:1007.2848 [astro-ph.HE]
34. Z.B. Etienne, V. Paschalidis, R. Haas, P. Mösta, S.L. Shapiro, Class. Quantum Gravity **32**, 175009 (2015). https://doi.org/10.1088/0264-9381/32/17/175009, arXiv:1501.07276 [astro-ph.HE]
35. R. Fernández, A. Tchekhovskoy, E. Quataert, F. Foucart, D. Kasen, Mon. Not. R. Astron. Soc. **482**(3), 3373–3393 (2019). https://doi.org/10.1093/mnras/sty2932, arXiv:1808.00461 [astro-ph.HE]
36. F. Foucart, Liv. Rev. Comput. Astrophys. **9**(1), 1 (2023). https://doi.org/10.1007/s41115-023-00016-y, arXiv:2209.02538 [astro-ph.HE]
37. B. Giacomazzo, L. Rezzolla, Class. Quantum Gravity **24**, S235–S258 (2007). https://doi.org/10.1088/0264-9381/24/12/S16, arXiv:gr-qc/0701109 [gr-qc]
38. B. Giacomazzo, L. Rezzolla, L. Baiotti, Mon. Not. R. Astron. Soc. **399**, L164–L168 (2009). https://doi.org/10.1111/j.1745-3933.2009.00745.x, arXiv:0901.2722 [gr-qc]
39. B. Giacomazzo, L. Rezzolla, L. Baiotti, Phys. Rev. D **83**, 044014 (2011). https://doi.org/10.1103/PhysRevD.83.044014, arXiv:1009.2468 [gr-qc]
40. B. Giacomazzo, R. Perna, Astrophys. J. Lett. **771**, L26 (2013). https://doi.org/10.1088/2041-8205/771/2/L26, arXiv:1306.1608 [astro-ph.HE]
41. B. Giacomazzo, J. Zrake, P. Duffell, A.I. MacFadyen, R. Perna, Astrophys. J. **809**(1), 39 (2015). https://doi.org/10.1088/0004-637X/809/1/39, arXiv:1410.0013 [astro-ph.HE]
42. N.K. Glendenning, S.A. Moszkowski, Phys. Rev. Lett. **67**, 2414–2417 (1991). https://doi.org/10.1103/PhysRevLett.67.2414
43. A. Goldstein, P. Veres, E. Burns, M.S. Briggs, R. Hamburg, D. Kocevski, C.A. Wilson-Hodge, R.D. Preece, S. Poolakkil, O.J. Roberts et al., Astrophys. J. Lett. **848**(2), L14 (2017). https://doi.org/10.3847/2041-8213/aa8f41, arXiv:1710.05446 [astro-ph.HE]
44. O. Gottlieb, B.D. Metzger, E. Quataert, D. Issa, T. Martineau, F. Foucart, M.D. Duez, L.E. Kidder, H.P. Pfeiffer, M.A. Scheel, Astrophys. J. Lett. **958**(2), L33 (2023). https://doi.org/10.3847/2041-8213/ad096e, arXiv:2309.00038 [astro-ph.HE]
45. J. Guilet, A. Bauswein, O. Just, H.T. Janka, Mon. Not. R. Astron. Soc. **471**(2), 1879–1887 (2017). https://doi.org/10.1093/mnras/stx1739, arXiv:1610.08532 [astro-ph.HE]
46. S. de Haas, P. Bosch, P.Mösta, S. Curtis, N. Schut, Mon. Not. R. Astron. Soc. **527**(2), 2240–2250 (2023). https://doi.org/10.1093/mnras/stad2931, arXiv:2208.05330 [astro-ph.HE]
47. J.F. Hawley, X. Guan, J.H. Krolik, Astrophys. J. **738**, 84 (2011). https://doi.org/10.1088/0004-637X/738/1/84, arXiv:1103.5987 [astro-ph.HE]
48. M. Hempel, J. Schaffner-Bielich, Nucl. Phys. A **837**, 210–254 (2010). https://doi.org/10.1016/j.nuclphysa.2010.02.010, arXiv:0911.4073 [nucl-th]
49. K. Hotokezaka, K. Kyutoku, H. Okawa, M. Shibata, K. Kiuchi, Phys. Rev. D **83**, 124008 (2011). https://doi.org/10.1103/PhysRevD.83.124008, arXiv:1105.4370 [astro-ph.HE]
50. K. Hotokezaka, K. Kiuchi, K. Kyutoku, H. Okawa, Y.I. Sekiguchi, M. Shibata, K. Taniguchi, Phys. Rev. D **87**, 024001 (2013). https://doi.org/10.1103/PhysRevD.87.024001, arXiv:1212.0905 [astro-ph.HE]
51. R. Kashyap, A. Das, D. Radice, S. Padamata, A. Prakash, D. Logoteta, A. Perego, D.A. Godzieba, S. Bernuzzi, I. Bombaci et al., Phys. Rev. D **105**(10), 103022 (2022). https://doi.org/10.1103/PhysRevD.105.103022, arXiv:2111.05183 [astro-ph.HE]
52. T. Kawamura, B. Giacomazzo, W. Kastaun, R. Ciolfi, A. Endrizzi, L. Baiotti, R. Perna, Phys. Rev. D **94**(6), 064012 (2016). https://doi.org/10.1103/PhysRevD.94.064012, arXiv:1607.01791 [astro-ph.HE]
53. K. Kiuchi, K. Kyutoku, M. Shibata, Phys. Rev. D **86**, 064008 (2012). https://doi.org/10.1103/PhysRevD.86.064008, arXiv:1207.6444 [astro-ph.HE]

54. K. Kiuchi, K. Kyutoku, Y. Sekiguchi, M. Shibata, T. Wada, Phys. Rev. D **90**, 041502 (2014). https://doi.org/10.1103/PhysRevD.90.041502, arXiv:1407.2660 [astro-ph.HE]
55. K. Kiuchi, P. Cerdá-Durán, K. Kyutoku, Y. Sekiguchi, M. Shibata, Phys. Rev. D **92**(12), 124034 (2015). https://doi.org/10.1103/PhysRevD.92.124034, arXiv:1509.09205 [astro-ph.HE]
56. K. Kiuchi, K. Kyutoku, Y. Sekiguchi, M. Shibata, Phys. Rev. D **97**(12), 124039 (2018). https://doi.org/10.1103/PhysRevD.97.124039, arXiv:1710.01311 [astro-ph.HE]
57. K. Kiuchi, L.E. Held, Y. Sekiguchi, M. Shibata, Phys. Rev. D **106**(12), 124041 (2022). https://doi.org/10.1103/PhysRevD.106.124041, arXiv:2205.04487 [astro-ph.HE]
58. K. Kiuchi, S. Fujibayashi, K. Hayashi, K. Kyutoku, Y. Sekiguchi, M. Shibata, Phys. Rev. Lett. **131**(1), 011401 (2023). https://doi.org/10.1103/PhysRevLett.131.011401, arXiv:2211.07637 [astro-ph.HE]
59. K. Kiuchi, A. Reboul-Salze, M. Shibata, Y. Sekiguchi, Nat. Astron. **8**(3), 298–307 (2024). https://doi.org/10.1038/s41550-024-02194-y, arXiv:2306.15721 [astro-ph.HE]
60. M. Kölsch, T. Dietrich, M. Ujevic, B. Bruegmann, Phys. Rev. D **106**(4), 044026 (2022). https://doi.org/10.1103/PhysRevD.106.044026, arXiv:2112.11851 [gr-qc]
61. S. Köppel, L. Bovard, L. Rezzolla, Astrophys. J. Lett. **872**(1), L16 (2019). https://doi.org/10.3847/2041-8213/ab0210, arXiv:1901.09977 [gr-qc]
62. A. Kurganov, E. Tadmor, J. Comput. Phys. **160**, 241–282 (2000). https://doi.org/10.1006/jcph.2000.6459
63. J.P. Lasota, P. Haensel, M.A. Abramowicz, Astrophys. J. **456**, 300 (1996). https://doi.org/10.1086/176650
64. J.M. Lattimer, F.D. Swesty, Nucl. Phys. A **535**, 331–376 (1991). https://doi.org/10.1016/0375-9474(91)90452-C
65. L. Lehner, C. Palenzuela, S.L. Liebling, C. Thompson, C. Hanna, Phys. Rev. D **86**, 104035 (2012). https://doi.org/10.1103/PhysRevD.86.104035, arXiv:1112.2622 [astro-ph.HE]
66. B.P. Abbott et al., [LIGO Scientific, Virgo, Fermi GBM, INTEGRAL, IceCube, AstroSat Cadmium Zinc Telluride Imager Team, IPN, Insight-Hxmt, ANTARES, Swift, AGILE Team, 1M2H Team, Dark Energy Camera GW-EM, DES, DLT40, GRAWITA, Fermi-LAT, ATCA, ASKAP, Las Cumbres Observatory Group, OzGrav, DWF (Deeper Wider Faster Program), AST3, CAASTRO, VINROUGE, MASTER, J-GEM, GROWTH, JAGWAR, CaltechNRAO, TTU-NRAO, NuSTAR, Pan-STARRS, MAXI Team, TZAC Consortium, KU, Nordic Optical Telescope, ePESSTO, GROND, Texas Tech University, SALT Group, TOROS, BOOTES, MWA, CALET, IKI-GW Follow-up, H.E.S.S., LOFAR, LWA, HAWC, Pierre Auger, ALMA, Euro VLBI Team, Pi of Sky, Chandra Team at McGill University, DFN, ATLAS Telescopes, High Time Resolution Universe Survey, RIMAS, RATIR and SKA South Africa/MeerKAT]. Astrophys. J. Lett. **848**(2), L12 (2017). https://doi.org/10.3847/2041-8213/aa91c9, arXiv:1710.05833 [astro-ph.HE]
67. B.P. Abbott et al., [LIGO Scientific and Virgo]. Phys. Rev. Lett. **119**(16), 161101 (2017). https://doi.org/10.1103/PhysRevLett.119.161101, arXiv:1710.05832 [gr-qc]
68. Y.T. Liu, S.L. Shapiro, Z.B. Etienne, K. Taniguchi, Phys. Rev. D **78**, 024012 (2008). https://doi.org/10.1103/PhysRevD.78.024012, arXiv:0803.4193 [astro-ph]
69. D.R. Lorimer, Living Rev. Rel. **11**, 8 (2008). https://doi.org/10.12942/lrr-2008-8, arXiv:0811.0762 [astro-ph]
70. A. Mignone, M. Ugliano, G. Bodo, Mon. Not. R. Astron. Soc. **393**, 1141 (2009). https://doi.org/10.1111/j.1365-2966.2008.14221.x, arXiv:0811.1483 [astro-ph]
71. A. Mignone, L. Del Zanna, J. Comput. Phys. **424**, 109748 (2021). https://doi.org/10.1016/j.jcp.2020.109748, arXiv:2004.10542 [physics.comp-ph]
72. H.K. Moffatt, *Magnetic Field Generation in Electrically Conducting Fluids* (University Press, Cambridge Monographs on Mechanics and Applied Mathematics, Cambridge, 1978)
73. K.P. Mooley, A.T. Deller, O. Gottlieb, E. Nakar, G. Hallinan, S. Bourke, D.A. Frail, A. Horesh, A. Corsi, K. Hotokezaka, Nature **561**(7723), 355–359 (2018). https://doi.org/10.1038/s41586-018-0486-3, arXiv:1806.09693 [astro-ph.HE]
74. C. Musolino, C. Ecker, L. Rezzolla, Astrophys. J. **962**(1), 61 (2024). https://doi.org/10.3847/1538-4357/ad1758, arXiv:2307.03225 [gr-qc]

75. E.R. Most, L.J. Papenfort, L. Rezzolla, Mon. Not. R. Astron. Soc. **490**(3), 3588–3600 (2019). https://doi.org/10.1093/mnras/stz2809, arXiv:1907.10328 [astro-ph.HE]
76. E.R. Most, A.A. Philippov, Astrophys. J. Lett. **893**(1), L6 (2020). https://doi.org/10.3847/2041-8213/ab8196, arXiv:2001.06037 [astro-ph.HE]
77. E.R. Most, A.A. Philippov, Mon. Not. R. Astron. Soc. **515**(2), 2710–2724 (2022). https://doi.org/10.1093/mnras/stac1909, arXiv:2205.09643 [astro-ph.HE]
78. E.R. Most, E. Quataert, Astrophys. J. Lett. **947**(1), L15 (2023). https://doi.org/10.3847/2041-8213/acca84, arXiv:2303.08062 [astro-ph.HE]
79. E.R. Most, Phys. Rev. D **108**(12), 12 (2023). https://doi.org/10.1103/PhysRevD.108.123012, arXiv:2311.03333 [astro-ph.HE]
80. P. Mösta, B.C. Mundim, J.A. Faber, R. Haas, S.C. Noble, T. Bode, F. Löffler, C.D. Ott, C. Reisswig, E. Schnetter, Class. Quantum Gravity **31**, 015005 (2014). https://doi.org/10.1088/0264-9381/31/1/015005, arXiv:1304.5544 [gr-qc]
81. P. Mösta, D. Radice, R. Haas, E. Schnetter, S. Bernuzzi, Astrophys. J. Lett. **901**, L37 (2020). https://doi.org/10.3847/2041-8213/abb6ef, arXiv:2003.06043 [astro-ph.HE]
82. H. Mueller, B.D. Serot, Nucl. Phys. A **606**, 508–537 (1996). https://doi.org/10.1016/0375-9474(96)00187-X, arXiv:nucl-th/9603037 [nucl-th]
83. D. Neilsen, S.L. Liebling, M. Anderson, L. Lehner, E. O'Connor, C. Palenzuela, Phys. Rev. D **89**(10), 104029 (2014). https://doi.org/10.1103/PhysRevD.89.104029, arXiv:1403.3680 [gr-qc]
84. E. O'Connor, C.D. Ott, Class. Quantum Gravity **27**, 114103 (2010). https://doi.org/10.1088/0264-9381/27/11/114103, arXiv:0912.2393 [astro-ph.HE]
85. C. Palenzuela, Mon. Not. R. Astron. Soc. **431**, 1853–1865 (2013). https://doi.org/10.1093/mnras/stt311, arXiv:1212.0130 [astro-ph.HE]
86. C. Palenzuela, L. Lehner, M. Ponce, S.L. Liebling, M. Anderson, D. Neilsen, P. Motl, Phys. Rev. Lett. **111**(6), 061105 (2013). https://doi.org/10.1103/PhysRevLett.111.061105, arXiv:1301.7074 [gr-qc]
87. C. Palenzuela, L. Lehner, S.L. Liebling, M. Ponce, M. Anderson, D. Neilsen, P. Motl, Phys. Rev. D **88**(4), 043011 (2013). https://doi.org/10.1103/PhysRevD.88.043011, arXiv:1307.7372 [gr-qc]
88. C. Palenzuela, S.L. Liebling, D. Neilsen, L. Lehner, O.L. Caballero, E. O'Connor, M. Anderson, Phys. Rev. D **92**(4), 044045 (2015). https://doi.org/10.1103/PhysRevD.92.044045, arXiv:1505.01607 [gr-qc]
89. C. Palenzuela, B. Miñano, D. Viganò, A. Arbona, C. Bona-Casas, A. Rigo, M. Bezares, C. Bona, J. Massó, Class. Quantum Gravity **35**(18), 185007 (2018). https://doi.org/10.1088/1361-6382/aad7f6, arXiv:1806.04182 [physics.comp-ph]
90. C. Palenzuela, R. Aguilera-Miret, F. Carrasco, R. Ciolfi, J.V. Kalinani, W. Kastaun, B. Miñano, D. Viganò, Phys. Rev. D **106**(2), 023013 (2022). https://doi.org/10.1103/PhysRevD.106.023013, arXiv:2112.08413 [gr-qc]
91. C. Palenzuela, S. Liebling, B. Miñano, Phys. Rev. D **105**(10), 103020 (2022). https://doi.org/10.1103/PhysRevD.105.103020, arXiv:2204.02721 [gr-qc]
92. D. Price, S. Rosswog, Science **312**, 719 (2006). https://doi.org/10.1126/science.1125201, arXiv:astro-ph/0603845 [astro-ph]
93. D. Radice, F. Galeazzi, J. Lippuner, L.F. Roberts, C.D. Ott, L. Rezzolla, Mon. Not. R. Astron. Soc. **460**(3), 3255–3271 (2016). https://doi.org/10.1093/mnras/stw1227, arXiv:1601.02426 [astro-ph.HE]
94. D. Radice, Symmetry **12**(8), 1249 (2020). https://doi.org/10.3390/sym12081249, arXiv:2005.09002 [astro-ph.HE]
95. F.A. Rasio, S.L. Shapiro, Class. Quantum Gravity **16**, R1–R29 (1999). https://doi.org/10.1088/0264-9381/16/6/201, arXiv:gr-qc/9902019 [gr-qc]
96. J.S. Read, B.D. Lackey, B.J. Owen, J.L. Friedman, Phys. Rev. D **79**, 124032 (2009). https://doi.org/10.1103/PhysRevD.79.124032, arXiv:0812.2163 [astro-ph]
97. L. Rezzolla, B. Giacomazzo, L. Baiotti, J. Granot, C. Kouveliotou, M.A. Aloy, Astrophys. J. Lett. **732**, L6 (2011). https://doi.org/10.1088/2041-8205/732/1/L6, arXiv:1101.4298 [astro-ph.HE]

98. C.E. Rhoades, R. Ruffini, Phys. Rev. Lett. **32**, 324 (1974)
99. M. Ruiz, R.N. Lang, V. Paschalidis, S.L. Shapiro, Astrophys. J. Lett. **824**(1), L6 (2016). https://doi.org/10.3847/2041-8205/824/1/L6, arXiv:1604.02455 [astro-ph.HE]
100. M. Ruiz, S.L. Shapiro, Phys. Rev. D **96**(8), 084063 (2017). https://doi.org/10.1103/PhysRevD.96.084063, arXiv:1709.00414 [astro-ph.HE]
101. M. Ruiz, S.L. Shapiro, A. Tsokaros, Phys. Rev. D **97**(2), 021501 (2018).https://doi.org/10.1103/PhysRevD.97.021501, arXiv:1711.00473 [astro-ph.HE]
102. M. Ruiz, A. Tsokaros, V. Paschalidis, S.L. Shapiro, Phys. Rev. D **99**(8), 084032 (2019). https://doi.org/10.1103/PhysRevD.99.084032, arXiv:1902.08636 [astro-ph.HE]
103. M. Ruiz, A. Tsokaros, S.L. Shapiro, Phys. Rev. D **101**(6), 064042 (2020). https://doi.org/10.1103/PhysRevD.101.064042, arXiv:2001.09153 [astro-ph.HE]
104. M. Ruiz, A. Tsokaros, S.L. Shapiro, Phys. Rev. D **104**(12), 124049 (2021).https://doi.org/10.1103/PhysRevD.104.124049, arXiv:2110.11968 [astro-ph.HE]
105. V. Savchenko, C. Ferrigno, E. Kuulkers, A. Bazzano, E. Bozzo, S. Brandt, J. Chenevez, T.J.L. Courvoisier, R. Diehl, A. Domingo et al., Astrophys. J. Lett. **848**(2), L15 (2017).https://doi.org/10.3847/2041-8213/aa8f94, arXiv:1710.05449 [astro-ph.HE]
106. Y. Sekiguchi, K. Kiuchi, K. Kyutoku, M. Shibata, Phys. Rev. D **91**(6), 064059 (2015). https://doi.org/10.1103/PhysRevD.91.064059, arXiv:1502.06660 [astro-ph.HE]
107. Y. Sekiguchi, K. Kiuchi, K. Kyutoku, M. Shibata, K. Taniguchi, Phys. Rev. D **93**(12), 124046 (2016). https://doi.org/10.1103/PhysRevD.93.124046, arXiv:1603.01918 [astro-ph.HE]
108. N.I. Shakura, R.A. Sunyaev, Astron. Astrophys. **24**, 337 (1973)
109. H. Shen, H. Toki, K. Oyamatsu, K. Sumiyoshi, Nucl. Phys. A **637**, 435–450 (1998). https://doi.org/10.1016/S0375-9474(98)00236-X, arXiv:nucl-th/9805035 [nucl-th]
110. D.M. Siegel, R. Ciolfi, A.I. Harte, L. Rezzolla, Phys. Rev. D **87**(12), 121302 (2013). https://doi.org/10.1103/PhysRevD.87.121302, arXiv:1302.4368 [gr-qc]
111. D.M. Siegel, B.D. Metzger, Astrophys. J. **858**(1), 52 (2018). https://doi.org/10.3847/1538-4357/aabaec, arXiv:1711.00868 [astro-ph.HE]
112. D.M. Siegel, B.D. Metzger, Phys. Rev. Lett. **119**(23), 231102 (2017). https://doi.org/10.1103/PhysRevLett.119.231102, arXiv:1705.05473 [astro-ph.HE]
113. M. Shibata, Y.I. Sekiguchi, Phys. Rev. D **72**, 044014 (2005). https://doi.org/10.1103/PhysRevD.72.044014, arXiv:astro-ph/0507383 [astro-ph]
114. M. Shibata, M.D. Duez, Y.T. Liu, S.L. Shapiro, B.C. Stephens, Phys. Rev. Lett. **96**, 031102 (2006). https://doi.org/10.1103/PhysRevLett.96.031102, arXiv:astro-ph/0511142 [astro-ph]
115. M. Shibata, K. Taniguchi, Phys. Rev. D **73**, 064027 (2006). https://doi.org/10.1103/PhysRevD.73.064027, arXiv:astro-ph/0603145 [astro-ph]
116. M. Shibata, Y. Suwa, K. Kiuchi, K. Ioka, Astrophys. J. Lett. **734**, L36 (2011). https://doi.org/10.1088/2041-8205/734/2/L36, arXiv:1105.3302 [astro-ph.HE]
117. M. Shibata, K. Hotokezaka, Ann. Rev. Nucl. Part. Sci. **69**, 41–64 (2019).https://doi.org/10.1146/annurev-nucl-101918-023625, arXiv:1908.02350 [astro-ph.HE]
118. M. Shibata, S. Fujibayashi, Y. Sekiguchi, Phys. Rev. D **103**(4), 043022 (2021). https://doi.org/10.1103/PhysRevD.103.043022, arXiv:2102.01346 [astro-ph.HE]
119. M. Shibata, S. Fujibayashi, Y. Sekiguchi, Phys. Rev. D **104**(6), 063026 (2021). https://doi.org/10.1103/PhysRevD.104.063026, arXiv:2109.08732 [astro-ph.HE]
120. H.C. Spruit, Astron. Astrophys. **381**, 923 (2002). https://doi.org/10.1051/0004-6361:20011465, arXiv:astro-ph/0108207 [astro-ph]
121. A.W. Steiner, M. Hempel, T. Fischer, Astrophys. J. **774**, 17 (2013). https://doi.org/10.1088/0004-637X/774/1/17, arXiv:1207.2184 [astro-ph.SR]
122. B.C. Stephens, M.D. Duez, Y.T. Liu, S.L. Shapiro, M. Shibata, Class. Quantum Gravity **24**, S207–S220 (2007). https://doi.org/10.1088/0264-9381/24/12/S14, arXiv:gr-qc/0610103 [gr-qc]
123. L. Sun, M. Ruiz, S.L. Shapiro, A. Tsokaros, Phys. Rev. D **105**(10), 104028 (2022). https://doi.org/10.1103/PhysRevD.105.104028, arXiv:2202.12901 [astro-ph.HE]
124. A. Suresh, H.T. Huynh, J. Comput. Phys. **136**, 83 (1997)
125. F.X. Timmes, F.D. Swesty, Astrophys. J. Suppl. **126**(2), 501–516 (2000)

126. H. Togashi, K. Nakazato, Y. Takehara, S. Yamamuro, H. Suzuki, M. Takano, Nucl. Phys. A **961**, 78–105 (2017). https://doi.org/10.1016/j.nuclphysa.2017.02.010, arXiv:1702.05324 [nucl-th]
127. S.D. Tootle, L.J. Papenfort, E.R. Most, L. Rezzolla, Astrophys. J. Lett. **922**(1), L19 (2021). https://doi.org/10.3847/2041-8213/ac350d, arXiv:2109.00940 [gr-qc]
128. C.J. White, J.M. Stone, C.F. Gammie, Astrophys. J. Suppl. **225**(2), 22 (2016). https://doi.org/10.3847/0067-0049/225/2/22, arXiv:1511.00943 [astro-ph.HE]
129. J. Zrake, A.I. MacFadyen, Astrophys. J. Lett. **769**, L29 (2013). https://doi.org/10.1088/2041-8205/769/2/L29, arXiv:1303.1450 [astro-ph.HE]

Chapter 17
Black Hole-Neutron Star Binaries

Matthew D. Duez

Abstract The gravitational wave signals of black hole-neutron star (BHNS) binary systems have now been detected, and future detections might be accompanied by electromagnetic counterparts. BHNS mergers involve much of the same physics as binary neutron star mergers: strong gravity, nuclear density matter, neutrino radiation, and magnetic turbulence. They also share with binary neutron star systems the potential for bright electromagnetic signals, especially gamma ray bursts and kilonovae, and the potential to be significant sources of r-process elements. However, BHNS binaries are more asymmetric, and their mergers produce different amounts and arrangements of the various post-merger material components (e.g. disk and dynamical ejecta), together with a more massive black hole; these differences can have interesting consequences. In this chapter, we review the modeling of BHNS mergers and post-merger evolution in numerical relativistic hydrodynamics and magnetohydrodynamics. We attempt to give readers a broad understanding of the answers to the following questions. What are the main considerations that determine the merger outcome? What input physics must (or should) go into a BHNS simulation? What have the most advanced simulations to date learned?

17.1 Introduction

Black hole-neutron star (BHNS) binary inspirals and mergers involve extremes of all four fundamental forces; they involve strong, dynamical spacetime curvature, supra-nuclear matter densities, relativistic speeds, powerful neutrino emission and magnetic field production. They are strong sources of gravitational waves. In some cases, they might also produce bright electromagnetic signals via kilonovae and/or gamma ray bursts and be a significant site for the production of heavy elements. Until recently, they were known to us only by theory, their rate of occurence a matter of loosely-constrained speculation, with no observational data. Now that has changed.

M. D. Duez (✉)
Department of Physics & Astronomy, Washington State University, Pullman, WA 99164, USA
e-mail: m.duez@wsu.edu

C. Bambi et al. (eds.), *New Frontiers in GRMHD Simulations*, Springer Series in Astrophysics and Cosmology, https://doi.org/10.1007/978-981-97-8522-3_17

On Jan 5, 2020, during the LIGO O3 observing run, LIGO Livingston registered a gravitational wave signal, designated GW200105, consistent with the late inspriral of a BHNS binary. (LIGO Hanford was not operational at the time.) Ten days later, LIGO Hanford, LIGO Livingston, and Virgo coincidently detected another gravitational wave signal, GW200115, also consistent with a BHNS inspiral [1]. The two signals were reported together, although GW200115 was the far more confident detection; in fact, GW200105 was later downgraded to a "marginal candidate" [2]. The other possible BHNS detections GW190917 and GW191219 were also reported in O3 catalog papers [2–4].

The identification of these binaries as BHNS systems was based entirely on the inferred masses of the binary components. From GW200105 was inferred a primary mass of $M_1 = 8.9^{+1.2}_{-1.5}\,M_\odot$ and a secondary mass of $M_2 = 1.9^{+0.3}_{-0.2}\,M_\odot$. From GW200115 was inferred a primary mass of $M_1 = 5.7^{+1.8}_{-2.1}\,M_\odot$ and a secondary mass of $M_2 = 1.5^{+0.7}_{-0.3}\,M_\odot$. For GW191219, the masses were $M_1 = 31.1^{+2.2}_{-2.7}\,M_\odot$, $M_2 = 1.17^{+0.07}_{-0.06}\,M_\odot$; for GW190917, $M_1 = 9.7^{+3.4}_{-3.9}\,M_\odot$, $M_2 = 2.1^{+1.1}_{-0.4}\,M_\odot$. The maximum mass of a non-spinning neutron star $M_{\rm NS,max}$ is somewhere in the range 2–3.2 $M_\odot$, with the minimum value coming from the maximum inferred masses of observed pulsars [5, 6] and the maximum value derived from causality restrictions on the nuclear equation of state [7]. Thus, any compact object more massive than this is presumably a black hole. While black holes could exist with masses below $M_{\rm NS,max}$, there is no known astrophysical way to form them, so compact objects with mass less than $M_{\rm NS,max}$ are presumed to be neutron stars. The waveform also allowed LIGO-Virgo researchers to estimate a parameter related to the black hole spin, namely the effective inspiral spin parameter $\chi_{\rm eff}$. This is a mass-weighted average of the component parallel to the orbital angular momentum of the Kerr dimensionless spin parameter $\boldsymbol{\chi} \equiv \mathbf{J}/M^2$. Thus, $\chi_{\rm eff} = \left(M_1\boldsymbol{\chi}_1 + M_2\boldsymbol{\chi}_2\right) \cdot \hat{L}/(M_1 + M_2)$. This will most likely be dominated by the black hole's spin. For GW200105, $\chi_{\rm eff} = -0.01^{+0.11}_{-0.15}\,M_\odot$, consistent with zero, while for GW200115, $\chi_{\rm eff} = -0.19^{+0.23}_{-0.35}\,M_\odot$, giving a probability of 88% that the black hole has a spin *negatively* aligned with respect to the orbital angular momentum.

The detections were a matter of excitement for the astronomy community, because before there had been no confident detections of BHNS systems through any signal type. Radio surveys which have identified binary neutron star systems in our galaxy have thus far failed to find any black hole-pulsar binaries. LIGO's O1 and O2 observing runs found no BHNS signals. The first part of the O3 run turned up only two ambiguous signals: GW190426, which may have been a detector artifact, and GW190814, which had a secondary mass $M_2 = 2.59^{+0.08}_{-0.09}\,M_\odot$ and might be a binary black hole system [8, 9]. In the absence of observations, the event rate of BHNS mergers was highly uncertain. Population synthesis studies estimate it within 0.1–800 $\mathrm{Gpc^{-3}\,yr^{-1}}$, and lack of detections in O1 and O2 suggested an upper bound of $< 610\,\mathrm{Gpc^{-3}\,yr^{-1}}$. Using two candidate detections in GWTC-3, admittedly very small number statistics, allowed the LIGO-Virgo-KAGRA collaboration to constrain the rate of BHNS mergers to between $7.8\,\mathrm{Gpc^{-3}\,yr^{-1}}$ and $140\,\mathrm{Gpc^{-3}\,yr^{-1}}$ [10].

On May 29, 2023, the LIGO Livingston Observatory detected a signal, GW230529_181500, attributed to a binary with masses $3.6^{+0.8}_{-1.2}\,M_\odot$ and $1.4^{+0.6}_{-0.2}\,M_\odot$ [11]. In the most likely case that GW230529_181500 is a BHNS binary, it demonstrates the existence of BHNS with low black hole masses, which are more likely to produce electromagnetic counterparts (although none was seen in this case).

We should now be optimistic that LIGO-Virgo will detect more BHNS signals as it approaches its design sensitivity. Furthermore, a subset of BHNS mergers produce potentially detectable electromagnetic counterparts, carrying information about the fascinating postmerger dynamics. The two most-studied electromagnetic signal possibilities for a BHNS merger are a *kilonova* and a *gamma ray burst* (GRB).

Kilonovae are caused by matter ejected from the merger site. As the matter decompresses, it forms unstable nuclei whose radioactive decays power a potentially detectable thermal IR/visible/UV signal lasting days. (See Sect. 17.9 below and the review [12].)

GRBs are non-thermal high-energy emission originating from a highly-relativistic outflow. In order to achieve the needed Lorentz factors, a large energy must be released without an accompanying large load of mass. Jets from the inner region near the black hole shooting out along the polar directions are a preferred scenario. (See Sect. 17.8 on the generation of these jets.) Short duration GRBs (lasting less than two seconds) are thought to have their origin in compact binary [BHNS or neutron star-neutron star (NSNS)] mergers [13], although the detection of kilonovae following a couple of $\sim$10 s GRBs suggests that mergers may also sometimes produce longer-duration GRBs [14, 15]. Furthermore, many short GRBs show "extended emission" subsequent to the initial spike [16]. (For reviews of short GRBs, see [13, 17].) GRBs have higher isotropic equivalent luminosity than kilonovae ($\sim 10^{50}$ erg s^{-1} vs. $\sim 10^{40}$ erg s^{-1}), but their emission is highly collimated, so the observer must be somewhat close to the axis to see it. Kilonovae, on the other hand, emit fairly isotropically and for longer times. Thus, they have different strengths and weaknesses as electromagnetic counterparts to a gravitational wave signal [18, 19]. The binary neutron star gravitational wave event GW170817 was detected as both a GRB and a kilonova [20].

Other mechanisms for electromagnetic emission have been considered and will be mentioned in later sections. Interaction of the neutron star magnetosphere with the black hole might create fast radio burst and X-ray transients (see Sect. 17.6). If the jet must pierce through previously-emitted ejecta, it will produce a hot surrounding cocoon, producing its own emission and affecting observations of the jet [21–23]. Outflow interacting with the surrounding interstellar medium creates radio emission (see Sect. 17.9).

This chapter is organized as follows. In Sect. 17.2, we will provide an overview of the inspiral, merger, and postmerger process. In Sect. 17.3, we will explore the space of possible BHNS binary systems and attempts to use numerical simulations to map properties of the pre-merger binary to the post-merger outcome. Section 17.4 presents, one piece at a time, the components of a realistic simulation of a BHNS merger. Subsequent sections summarize what has been learned from numerical simulations about the various aspects and outputs of the mergers: gravitational waves (Sect. 17.5), signals from the neutron star magnetosphere (Sect. 17.6), the post-merger disk (Sect. 17.7), production of magnetic fields and jets (Sect. 17.8), and outflows (Sect. 17.9).

For other excellent reviews of BHNS systems see Kyutoku et al. [24] and Foucart [25]. Unless otherwise specified, we will use units for which $G = c = 1$ throughout this chapter.

17.2 Phases of a BHNS Merger

In this section, we outline the evolution stages of a BHNS system. We introduce the main concepts that will recur throughout the chapter and provide physical arguments for what we expect simulations to find.

The most likely formation scenario for a BHNS binary begins with a binary of high-mass main sequence stars in a region of stellar density low enough that the binary evolves in isolation (a "field binary"). If one star evolves into a neutron star and the other into a black hole–and no supernova kick disrupts the binary, there will be a BHNS binary. However, only BHNS systems that merge within the age of the Universe are interesting to us. Probably this requires the binary to undergo a common envelope phase, leaving the binary sufficiently compact that gravitational radiation can bring the neutron star and black hole to merger within a Hubble time.

Consider a binary composed of a black hole of mass $M_{\rm BH}$ and a neutron star of mass $M_{\rm NS}$ in an orbit with semimajor axis D. We expect $M_{\rm BH}$ to be significantly larger than $M_{\rm NS}$, perhaps $M_{\rm BH} \approx 5 M_{\rm NS}$ will be typical. (See Sect. 17.3.) The binary will have a Keplerian orbital angular frequency $f \approx {M_{\rm BH}}^{1/2} D^{-3/2}$. It will thus have a time-varying mass quadrupole moment and so will radiate gravitational waves. These waves carry away energy and angular momentum. The orbital energy of the binary in Newtonian gravity is $E = -M_{\rm BH} M_{\rm NS}/2D$, and this orbital energy will decrease at a rate $\dot{E} = -L_{\rm GW}$, where $L_{\rm GW}$ is the gravitational radiation luminosity. One can then infer $\dot{D} = dD/dE\,\dot{E}$. Gravitational waves cause the eccentricity to decrease, so by the time the neutron star and black hole are close, the orbit will be very close to circular. (BHNS binaries formed by multi-body gravitational interactions in dense stellar environments such as globular clusters, on the other hand, might be very eccentric at the time of merger, but such events are expected to be much rarer than field binary BHNS mergers [26].) The inspiral rate will at first be very small, but as D decreases, the binary will orbit faster, increasing $L_{\rm GW}$ and causing the inspiral rate to increase. This is the inspiral phase.

Either of two effects might terminate the inspiral phase. First, the binary separation might reach the innermost stable circular orbit $R_{\rm ISCO}$, inside of which orbital motion is unstable, and so the neutron star will plunge into the black hole on an orbital timescale. If $M_{\rm BH} \gg M_{\rm NS}$, this separation is that of test particles orbiting an isolated black hole, which depends only on $M_{\rm BH}$ and on the black hole's angular momentum

$\mathbf{J}_{\mathrm{BH}}$. In general, the orbital and black hole spin angular momentum will not be aligned, but we shall consider this first as the simplest case. Then $R_{\mathrm{ISCO}} = f(\chi)M_{\mathrm{BH}}$ where $\chi = |J_{\mathrm{BH}}|/M_{\mathrm{BH}}^2 < 1$ is the dimensionless spin, and in Boyer-Lindquist coordinates the limits are $f(0) = 6$, $f(1) = 1$. In any case, the plunge radius R_{plunge} can be taken to be of order $\sim M_{\mathrm{BH}}$.

Second, the neutron star might be torn apart by the tidal force of the black hole. This will happen when the tidal force from the black hole exceeds the neutron star's self-gravity. Both of these forces depend on the neutron star radius R_{NS}. From Newtonian physics, the gravitational force of a test body with mass m near the surface due to the neutron star's gravity is $\sim mM_{\mathrm{NS}}/R_{\mathrm{NS}}^2$ and the tidal force from the black hole (the difference in the black hole's gravitational forces for test bodies on the sides near and far from the hole) is $\sim mM_{\mathrm{BH}}R_{\mathrm{NS}}/D^3$. Setting the two forces equal we find the tidal mass transfer radius $D = R_{\mathrm{tidal}}$:

$$R_{\mathrm{tidal}} = \left(\frac{M_{\mathrm{BH}}}{M_{\mathrm{NS}}}\right)^{1/3} R_{\mathrm{NS}} = M_{\mathrm{BH}} Q^{-2/3} C^{-1} \tag{17.1}$$

where we have introduced two crucial dimensionless binary parameters: the *mass ratio* $Q \equiv M_{\mathrm{BH}}/M_{\mathrm{NS}}$ and the neutron star *compaction* $C \equiv M_{\mathrm{NS}}/R_{\mathrm{NS}}$.

Even after the neutron star begins losing mass, it persists inspiraling as a gravitationally bound compact object with most of its original mass for a short time (an orbit or so) longer before being totally disrupted. It is sometimes useful to distinguish the beginning of mass transfer off of the neutron star at $D = R_{\mathrm{tidal}}$ from tidal disruption, when the neutron star has been torn into a spiral swath. The latter does quickly follow the former, though, so we will ignore the distinction for the rest of this section.

Tidal disruption would be expected to have a profound effect on the gravitational wave signal. By spreading the neutron star matter, the variation of the quadrupole moment will drop, and the gravitational wave will quickly damp. We can estimate this cutoff frequency, using the fact that the frequency of the dominant (quadrupolar) gravitational wave mode, f_{cut} is twice the orbital frequency. Schematically,

$$f_{\mathrm{cut}} \sim {M_{\mathrm{BH}}}^{1/2} {R_{\mathrm{tidal}}}^{-3/2} \sim {M_{\mathrm{NS}}}^{1/2} {R_{\mathrm{NS}}}^{-3/2} \tag{17.2}$$

i.e. the cutoff frequency is mostly given by the dynamical timescale of the neutron star. This is one way that information about the neutron star is encoded in the waveform [27]. Unfortunately, f_{cut} comes to $\approx$kHz, which is well above LIGO's peak sensitivity.

Thus, we have three possible endpoints of the BHNS inspiral. If $R_{\mathrm{plunge}} > R_{\mathrm{tidal}}$, we have a *non-disrupting binary* whose inspiral ends with the neutron star plunging into the black hole horizon still intact. If $R_{\mathrm{tidal}} > R_{\mathrm{plunge}}$, we have a *disrupting binary* whose inspiral ends when the neutron star, still outside the black hole, is torn apart by the black hole's gravity. Finally, there could be marginal systems where $R_{\mathrm{tidal}} \approx R_{\mathrm{plunge}}$. This will be the case if

$$\zeta \equiv f(\chi) Q^{2/3} C \approx 1 \tag{17.3}$$

Disruption is expected, then, for $\zeta \lesssim 1$, which will be more likely for high black hole spin, low mass ratio (i.e. low $M_{\rm BH}$), and low compaction (in particular, for large $R_{\rm NS}$).

For non-disrupting BHNS binaries, one would not expect residual matter outside the black hole after the plunge, which eliminates the possibility of most sorts of electromagnetic signals. The infalling neutron star will disturb the spacetime near the horizon, leading to a distinctive "ringdown" segment of the waveform, and one would expect that to be the end of the story.

For disrupting binaries, the neutron star will elongate into a spiral, with some material expanding outward away from the black hole and some falling inward toward the black hole. We will find that, for non-eccentric BHNS binaries, mass transfer once begun is unstable, and the neutron star is always completely destroyed in a single tidal disruption event. Some of the outgoing material in the spiral has positive energy and is unbound. This *dynamical ejecta*, permanently expelled from the system, can be quite massive (10^{-2}–$10^{-1}\,M_\odot$) with asymptotic speeds of order 0.2–0.3 c. As this material expands and decompresses, it undergoes r-process nucleosynthesis, forming heavy nuclei. Radioactive decays heat the ejecta, causing it to radiate as a (red) kilonova.

The negative energy matter cannot escape. Most of the ingoing material falls immediately into the black hole. Some of the ingoing spiral has sufficient angular momentum to circle around the horizon, so that the inflow crashes into and shears against itself, leading to the formation of an initial *accretion disk*. (See Fig. 17.1, top.) The tidal disruption and appearance of an orbiting disk take place in of order a millisecond. The disk is initially strongly perturbed, with strong spiral waves. At the same time, the bound material that was initially outgoing begins to fall toward the black hole, and this *fallback material* incorporates itself into the disk. This *post-merger settling phase* lasts for tens of milliseconds, ending with a roughly axisymmetric black hole accretion system.

Shocks from the merger and settling quickly heat the gas in the disk to temperatures of a few to 10 MeV. The maximum density of the disk shortly after merger can be as high as $\sim 10^{12}\,\mathrm{g\,cm^{-3}}$. At these densities and temperatures, the gas is efficiently cooled by neutrino emission. Neutrinos will carry energy away from the disk and also change the composition of the gas (the ratio of protons to neutrons), both on timescales of order 10 ms. The magnetic field inside the disk is amplified by several processes, including winding, Kelvin-Helmholtz vortices at shear layers, and the magnetorotational instability (MRI). The MRI drives turbulence, which provides a means of transporting angular momentum outward, causing the material in the inner disk to accrete into the black hole, the outermost material to move farther from the hole, and the disk as a whole to spread out. The cascade of energy to small scales also provides a heating source to offset the neutrino cooling.

The subsequent late post-merger evolution is driven by the interplay of neutrino and magnetic effects. The transport and heating effects of magnetoturbulence are often conceptualized and modeled as an effective viscosity, so that it is common to refer to its effects as "viscous" (e.g. "viscous winds", "viscous timescale"), a

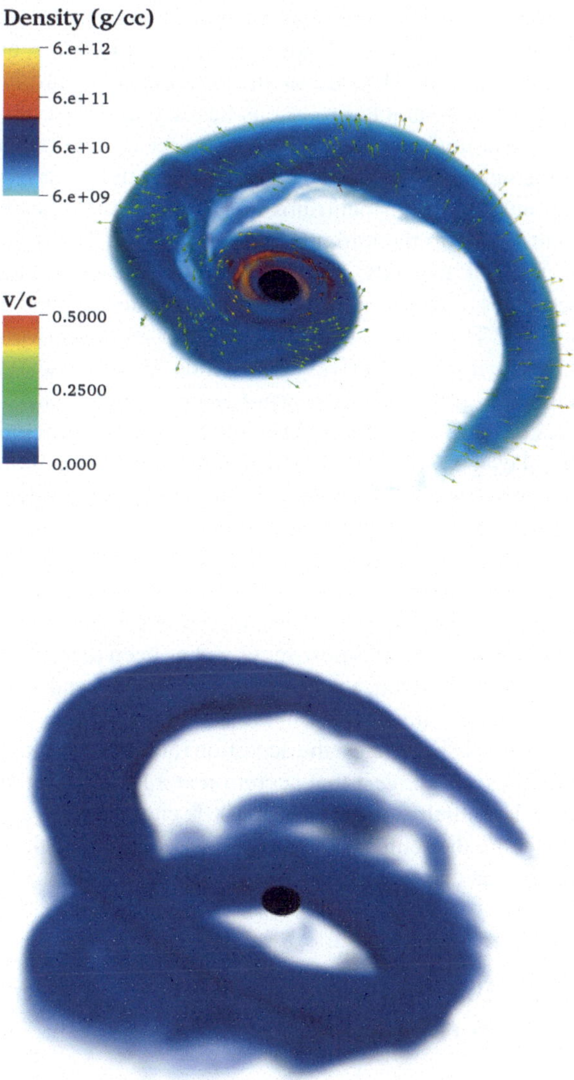

Fig. 17.1 Tidal disruption converts the neutron star into a spiral swath. **Top**: Incoming matter wraps around the black hole, crashes into itself, and forms a hot accretion disk. Color indicates density. (From Foucart et al. [28].) **Bottom**: In this system with a massive black hole with black hole spin $\chi_{\mathrm{BH}} = 0.9$ misaligned 40° from the orbital axis, there is orbital precession, and the stream avoids hitting itself. Shown is a density contour. From Foucart et al. [29]

custom we will follow, not without some reservation. To estimate the timescale of disk evolution, let us model the transport via a Shakura-Sunyaev alpha viscosity [30], in which case the effective kinematic viscosity is $\nu = \alpha_{\mathrm{SS}} c_s H$, where c_s is the sound speed, H is the disk thickness, and α_{SS} is a dimensionless constant of order 10^{-2}–10^{-1} [31, 32]. Then the characteristic timescale for accretion and expansion is

$$\tau_{\mathrm{visc}} \sim \alpha_{\mathrm{SS}}^{-1} \Omega^{-1} (H/r)^{-2} \sim 10^2\ \mathrm{ms} \tag{17.4}$$

where r is the distance from the black hole, and Ω is the Keplerian orbital velocity, and we have substituted values typical to a BHNS remnant disk.

For the first ~100 ms, neutrino luminosities are high (starting at $L_\nu \sim 10^{53}$ erg s^{-1}) and comparable to viscous heating, and the composition of the fluid is driven to an equilibrium of charged-current weak nuclear interactions. Perhaps a percent of the neutrino luminosity is deposited in the polar regions outside the disk through the annihilation of neutrinos and antineutrinos, heating matter there and possibly contributing to the initial formation of a jet. This is the *neutrino-cooled disk phase*. As the disk expands, its density and temperature drop, so that neutrino luminosity decreases, eventually becoming negligible, so that it no longer balances viscous heating. The disk is then said to be in an advective state. Viscous heating makes the disk thick and drives disk winds; these winds constitute a second source of unbound outflow, the *disk wind ejecta*, which contribute distinctly to the kilonova. This *advective disk phase* lasts longer than the previous phases as the disk depletes. One key goal of BHNS post-merger simulations is to characterize the disk wind outflows (their mass, composition, and speed) to determine the expected kilonova signals and nucleosynthesis output.

The most promising way to produce a relativistic jet, leading to a GRB, is through a poloidal magnetic field at the polar region near the horizon by the Blandford-Znajek mechanism [33]. Thus, a second key goal of magnetohydrodynamic (MHD) simulations of BHNS post-mergers has been to model the accumulation of magnetic flux on the horizon, the launching of the jet, and its eventual loss of power as the disk depletes, in order to enable comparisons with observed features of GRBs.

As the disk depletes, the accretion rate falls off with time according to a power law, which would suggest that accretion at a low level at least might continue for a long time. Eventually, after of order 100 s, energy from radioactive decays is sufficient to overcome the remnant disk's binding energy, which will quickly evaporate the disk and put an end to extended emission [34].

17.3 Binary Parameter Space and Outcome Fitting Formulae

We expect the outcome of a BHNS merger to depend on Q, χ, and C. Let us consider how we expect these parameters to be distributed among actual BHNS binaries.

Unfortunately, these distributions are very uncertain, especially regarding the black hole properties. The black hole can be characterized by its mass M_{BH} and dimensionless spin χ. Observations of black holes in low-mass X-ray binaries show a narrow mass distribution at $7.8^{+1.2}_{-1.2}\,M_\odot$ [35]. In particular, there appears to be a mass gap between black hole and neutron star masses, with few black holes with mass less than $\approx 4.5\,M_\odot$ [36]. Black holes in binary black holes detected by LIGO-Virgo have broader distribution extending to higher masses [37]. The laws of classical physics permit black holes of any mass, including within the range of neutron star masses, but no known astrophysical process would produce black holes with masses below the neutron star maximum mass. (Primordial black holes, if they exist, might have low masses, though.) From gravitational wave signals, binary components are often identified as neutron stars or black holes by their mass, the subtle effects of finite size usually being undetectable. Some numerical simulations of BHNS mergers have considered very low $Q = 1$–1.2 BHNS systems, in particular to see if they can be observationally distinguished from a binary neutron star system of equal component masses [38].

The distribution of inferred χ_{eff} in binary black holes peaks at +0.06, with a standard deviation of 0.12 [37]. It is, of course, possible that the black hole mass and spin distributions are different for BHNS binaries than they are in X-ray binaries and black hole-black hole binaries. Note that the black hole spin is a vector, and so it enlarges the dimension of our parameter space by three. Much numerical work has concentrated on the case of black hole spin aligned with (i.e. parallel to and in the same sense as) the orbital angular momentum because this turns out to be the most optimistic case for tidal disruption and electromagnetic signals.

What about neutron star masses and spins? Masses inferred from neutron stars in binaries range from about 1.2 to 2.0 $M_\odot$ [39]. The mass distribution of galactic neutron stars in binary neutron star systems is rather narrowly peaked at 1.33 $M_\odot$, but GW190425, the second detected binary neutron star merger signals, had a total mass of 3.4 $M_\odot$, implying that higher mass neutron stars in these binaries are possible [40]. Furthermore, neutron stars in binaries with white dwarfs or main sequence stars have a wider distribution of masses, and peaked at a higher value, than neutron stars in NSNS systems [39]. We have seen that candidate BHNS gravitational wave signals infer a range of M_{NS} from 1.2 to 2.1 $M_\odot$, with GW200115, the most confident case, having an inferred neutron star mass of around 1.5 $M_\odot$.

Neutron stars do spin, but the fastest observed spins are in the millisecond range (millisecond pulsars), with a corresponding spin frequency (10^2Hz) significantly smaller than the frequency of a BHNS system at merger (10^3Hz). Furthermore, neutron stars with magnetic fields spin down and will have plenty of time to do so during the long inspiral. On the other hand, tidal forces and neutron star viscosity are too weak to spin up and tidally lock the neutron star before merger [41, 42]. Therefore, neutron stars are usually approximated as being irrotational (i.e. a curl-free velocity field) when constructing initial data for merger simulations. However, the effect of a non-trivial neutron star spin has been investigated in some simulations [43, 44], where it was found that the neutron star spin can affect the disk and dynamical ejecta masses.

It should be remembered that the total neutron star mass M_{NS} is distinct from the star's "rest" or "baryonic mass" M^b_{NS}. $M_{\text{NS}} < M^b_{\text{NS}}$ because of the negative gravitational potential energy, so $M^b_{\text{NS}} - M_{\text{NS}}$ is the star's binding energy.

Given the neutron star mass, and assuming slow spin ($\ll$kHz), the radius of the neutron star, and hence the compaction, are determined by the properties of neutron star matter as codified in the equation of state, which supplies the pressure P as a function of rest mass density ρ_0. This function $P(\rho_0)$ is presumably unique–cold neutron stars have no free composition variables, unlike main sequence stars and white dwarfs. Unfortunately, this function is also unknown at high densities. One of the driving scientific interests in BHNS and NSNS binaries is to constrain it. This can be done by evolving systems with the same binary parameters (masses and spins) but different assumptions about the equation of state and determining how the observable outputs (gravitational wave and electromagnetic) depend on these assumptions. Thus, for the purposes of modeling, we must treat the equation of state as if it varied from one possible system to another. We will have much more to say about equations of state later (Sect. 17.4.2), so we defer this discussion until then. For the rest of this section, let us simply consider the compaction C to be a free parameter.

Thus, our 6D pre-merger binary parameter space has coordinates (M_{BH}, M_{NS}, χ, C). Can we define a parameter space of the merger outcome? We needn't be too concerned with the number of dimensions of this space; it is set by the number of outputs with which we are interested and does not affect the number of simulations we need to perform. Ideally, we would like to be able to express these output quantities as functions of the binary parameters. This can be done by using physical intuition to guess a form of the function, adding free parameters that can be fit using data from numerical simulations. The following outcome quantities have received the most attention.

A quantity whose dependence on binary parameters has received much attention is the post-merger baryonic mass, the material that "survives" by not falling immediately into the black hole $M^b_{\rm remnant}$. Note that this "remnant" includes disk, fallback, and dynamical ejecta. Since the disk mass is not constant, but depletes due to accretion and wind, one must specify a time at which the mass is measured–10 ms after merger, for example. From the discussion in Sect. 17.2 above, we know that the presence of tidal disruption (and thus the possibility of a disk) depends on Q C, and χ–the latter through its effect on $R_{\rm ISCO}$. One might guess that the remnant mass depends on the same parameters. In fact, this works fairly well, as Foucart and collaborators have shown by positing $M^b_{\rm remnant}/M^b_{\rm NS}$ to be a simple function of the dimensionless numbers $\eta \equiv Q/(1+Q)^2$ (the "symmetric mass ratio"), C, and $R_{\rm ISCO}(\chi)/M_{\rm BH}$. They then used the results of available numerical simulations to fit the free parameters, and are able to achieve good a reasonably good fit [45, 47].

Fits also exist for the mass and speed of dynamical ejecta. Kawaguchi et al. [46] carried out a large number of BHNS simulations, analyzed dynamical ejecta, and introduced fitting formulae. Their ejecta mass function involves C, Q, $R_{\rm ISCO}/M_{\rm BH}$, and the neutron star's specific binding energy $1 - M_{\rm NS}/M^b_{\rm NS}$. (See also [48] for a fit with better behavior for very compact neutron stars.) The ejecta mass is highest for $Q \approx 3$; it is small for very symmetric and very asymmetric binaries [49]. The ejecta velocity is adequately fit as a linear function of Q (with higher mass ratios producing faster ejecta).

Looking at the disk and dynamical ejecta masses in various regions of parameter space (see Fig. 17.2), we see that surviving matter tends to be disk-dominated at low mass ratio but ejecta-dominated at high mass ratios. Some fraction of the disk, perhaps between 5 and 30% of its mass, will be ejected in winds (see Sect. 17.9 below). Disk outflow mass will then presumably also dominate dynamical ejecta

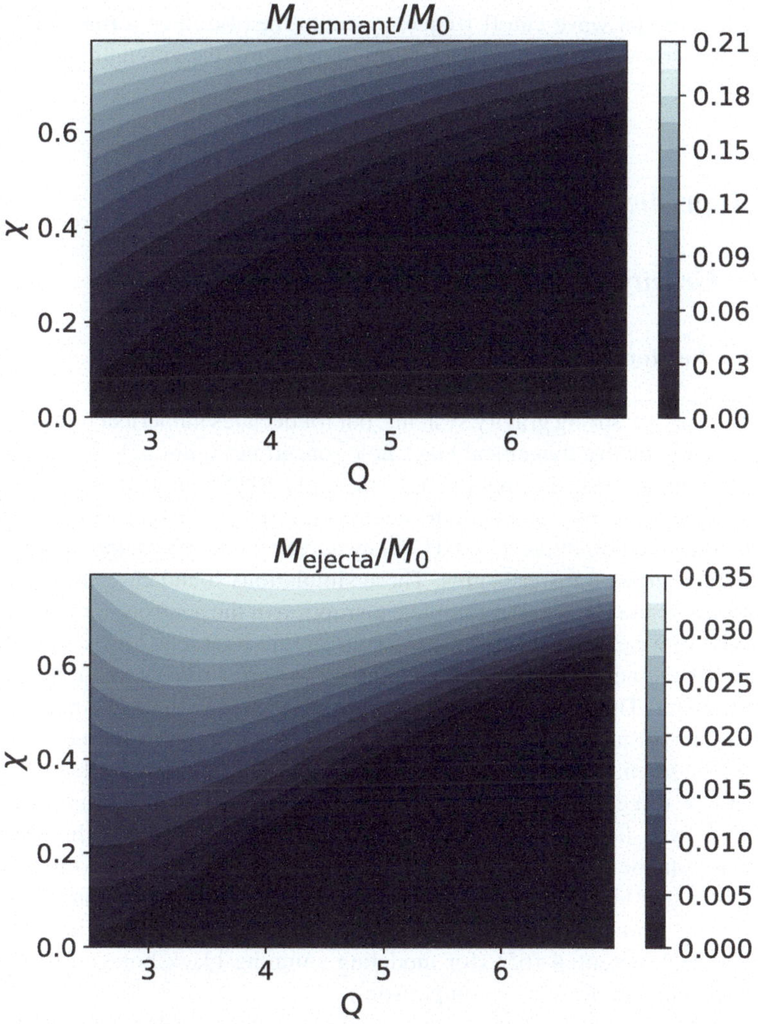

Fig. 17.2 Fraction of neutron star baryonic mass surviving in post-merger components as a function of mass ratio Q and dimensionless aligned black hole spin χ according to fitting formulae. Both plots assume a neutron star with compaction 0.161 and specific binding energy 0.094. **Top**: remnant mass after 10 ms according to the formula of Foucart et al. [45]. **Bottom**: dynamical ejecta mass according to the formula of Kawaguchi et al. [46]

mass only for very low $Q \lesssim 3$. For some range of parameters, the two components will both be important, and for high Q (and high prograde spin, so that the binary is disrupting), ejecta will mostly be dynamical. This is important, because we will see that the two ejecta components have very different properties.

Attempts have also been made to fit outcomes related to the spacetime. Pannarale fit for the post-merger black hole's mass and spin [50, 51]. (See also [52] and [53]

.) The gravitational wave cutoff frequency could then be fit as a function of these remnant properties [54, 55]. We will return to modeling the gravitational waveform in Sect. 17.5.

17.4 Ingredients of a Numerical Simulation

17.4.1 Gravity, Plus a Bit of History

17.4.1.1 Newtonian

BHNS binaries are strong gravity systems, but for decades numerical relativists were unable to stably evolve dynamical black hole spacetimes (although they could make progress on binary neutron stars [56]), so the early BHNS simulations were carried out using Newtonian physics, with the neutron star represented as an ideal fluid and the black hole as a point mass [57–61]. A key question was what happens when $R_{\rm tidal}$ is reached and mass transfer begins. These simulations found cases in which mass transfer once begun leads to the complete disruption of the neutron star but also cases where mass loss causes the neutron star to retreat into its Roche lobe so that the flow of matter off the neutron star stops and a neutron core survives to another episode of mass transfer. They found soft equations of state to favor unstable mass transfer, stiff equations of state to favor episodic mass transfer. In fact this turned out to be an artifact of Newtonian gravity; relativistic mass transfer is more unstable.

In fact, one can get qualitatively correct results, for which tidal disruption happens in one pass even for stiff realistic equations of state, by replacing the Newtonian point mass potential with "pseudo-Newtonian potentials" designed to mimic certain features of black hole orbital dynamics [62, 63]. The most common of these are the Paczynski-Wiita potential [64] (for modeling nonspinning black holes) and the Artemova et al. potential [65] (for modeling spinning black holes). (See [66] for alternatives and a performance comparison.)

It is worth pausing to gain a sense of these techniques because much of the work on late-time accretion and disk wind ejecta to be discussed later uses Newtonian physics and pseudo-Newtonian potentials. The Paczynski-Wiita potential is just $\Phi(r) = -M_{\rm BH}/(r - 2M_{\rm BH})$. The simple addition of the extra term in the denominator reproduces the innermost stable circular orbit location of a Schwarzschild black hole (for Schwarzschild radial coordinate).

17.4.1.2 Relativistic

It is beyond the scope of this chapter to explain the techniques used to stably evolve spacetime metrics (for a book-length treatment, see [67]), but perhaps we can briefly convey the flavor of it. To handle gravity relativistically, one must calculate the spacetime metric $g_{\mu\nu}$. The metric provides not only what our Newtonian intuition regards as gravitational effects but also automatically supplies the properties of the black hole and of the gravitational waves, because these three are all just different aspects of the spacetime curvature. The Einstein field equations are nonlinear second-order partial differential equations for $g_{\mu\nu}$. The field equations come in two types: *evolution* equations, which supply the time derivatives that tell how the metric changes from one time slice to the next, and *constraint* equations, which restrict how the fields must relate at each time. This is entirely analogous to the split in Maxwell's equations, e.g. Faraday's law $\partial_t \mathbf{B} = -\nabla \times \mathbf{E}$ is an evolution equation while Gauss's Law $\nabla \cdot E = 4\pi\rho$ is a constraint which must be satisfied at each time. (Analytically, if the constraints are satisfied at one slice of time, the evolution equations will keep them satisfied at later times, as with Maxwell's equations. Whether numerical-error induced constraint violations remain small depends on one's formulation of Einstein's equations.) The numerical relativist must supplement Einstein's equations with a third set of equations, the *gauge* conditions, which determine the choice of coordinates, e.g. how spacetime is to be sliced into space and time.

A key part of any numerical relativity calculation is the construction of appropriate initial data. For a BHNS simulation, this must correspond to a black hole and a neutron star in circular orbit. Numerical costs (above all the number of timesteps involved) motivate us to create binaries in the late inspiral stage, no more than roughly a dozen orbits before merger. To be a time slice of a solution of Einstein's equations, the constraint equations must be satisfied at this $t = 0$ slice. This amounts to four elliptic partial differential equations that must be solved.

In addition, one wants to impose the conditions that the star is in equilibrium and the binary is in circular orbit, which will not be true for just any choice of density distribution, orbital rotation rate, and (constraint satisfying) metric. It is impossible to impose an exact, consistent circular orbit symmetry on the spacetime, impossible because a BHNS orbiting in perfect circles forever is not a solution to Einstein's equations–actually, there must be gravitational radiation and inspiral. What is actually done, in what is called the extended conformal thin sandwich (XCTS) formulation, is to freely choose some metric variables and then use the constraints and assumed circular orbit to solve for the remaining five. For example, one factors the three-metric as $\gamma_{ij} = \Psi^4 \tilde{\gamma}_{ij}$, chooses $\tilde{\gamma}_{ij}$, e.g. to be flat or to resemble a single black hole space, and solves for Ψ with a constraint equation. Making slightly wrong assumptions makes the problem tractable, but it comes at a cost. An incorrect $\tilde{\gamma}_{ij}$ means unphysical ("junk") gravitational radiation in the initial data. Ignoring inspiral produces slightly eccentric orbits (which can be reduced by adding some initial inspiral [68]).

Equilibrium must also be imposed on the fluid. The Euler equation can be integrated to give an algebraic condition (with integration constant), the relativistic generalization of the Bernoulli integral. This turns out to be an equation for $h\Gamma$, where h is the specific enthalpy, and Γ is the Lorentz factor. Thus, given the fluid velocity, equilibrium gives h and thus (via the equation of state) ρ_0. The condition that the star is non-spinning is translated as the condition that the flow is irrotational. Recall that in Newtonian physics, this means $\nabla \times \mathbf{v} = \mathbf{0}$ so that $\mathbf{v} \equiv \nabla\phi$, and the relativistic version is similar. An elliptic equation for the velocity potential ϕ is supplied by the continuity equation.

To the extent that the inspiral timescale is much longer than the orbital timescale, one can think of the binary as simply evolving from one of these quasi-equilibrium circular orbit states to another. One can, then, track the inspiral evolution of a single binary by creating a sequence of these equilibria at different separations, holding fixed quantities like the neutron star baryonic mass and black hole irreducible mass (expected to be nearly conserved since little of the gravitational wave energy is swallowed by the black hole.) At smaller separations, the orbital angular velocity Ω increases; in fact, Ω is a better parameter along the sequence than coordinate separation, because it can be calculated in a coordinate-invariant way. As the binary inspirals to higher Ω, the total energy (ADM mass) of the binary goes down, with the reduction presumably accounted for by gravitational wave emission. Before numerical relativity evolutions were possible, these sequences provided our best understanding of the late inspiral [69–71]. The most interesting question was how the quasi-equilibrium sequences end. The ISCO of a sequence is where the binary's energy as a function of Ω reaches a minimum. The onset of tidal mass transfer is indicated by the formation of a cusp shape on the stellar surface. Unfortunately, this final and most interesting part of the sequence is precisely where inspiral is becoming rapid and the quasi-equilibrium assumption certainly breaks down. Thus, we are compelled to do full evolutions to model the end of the inspiral credibly.

For further reading on initial data construction, see [72].

Numerical relativity experienced a breakthrough in its ability to stably evolve binaries with black holes in 2005. (For a review of numerical relativity centering on this breakthrough and the work it enabled, see [73].) The next year, Loffler et al. [74] simulated a BHNS head-on collision, and Shibata and Uryu carried out the first fully relativistic simulations of BHNS merger starting from roughly circular orbit [75]. This was quickly followed by other groups [76–78].

17.4.2 Equation of State

BHNS simulations must model matter in a wide range of densities and temperatures. The high density matter inside neutron stars and the low density matter in disks and outflows each present their own challenges.

In general, the gas will be composed of a combination of photons, leptons, and baryons, with the baryons divided into free protons (p), free neutrons (n), and many different species of nuclei. At temperatures above about 0.5 MeV, we can invoke the wonderful simplification of *nuclear statistical equilibrium* (NSE). That is, strong nuclear reactions proceed so quickly that the abundances of each nuclide (including free nucleons) quickly come to their equilibrium values, and this equilibrium is maintained as fluid elements evolve. Then the composition of each isotope is fixed by a network of Saha-like equations, and the state of the fluid can be described by only three variables: the number density n (actually, the numerical relativity community prefers the equivalent number, the baryonic density ρ_0, which is just n multiplied by a standard baryonic mass), the temperature T, and the fraction of baryons that are protons $Y_e \equiv n_p/(n_p + n_n)$. The latter, Y_e, is sometimes called the reduced electron fraction and is what is being referred to when numerical relativists speak of the "composition" of our outflows. Note that Y_e of a fluid element can only change because of charged-current weak interactions (which can change protons to neutrons, and vice versa), which also create and absorb neutrinos. The timescale for these weak interactions to come to equilibrium is not always small compared to evolution timescales, so Y_e must be evolved. The equation of state thus provides pressure P and specific internal energy ϵ as functions of (ρ_0, T, Y_e). Nuclear physics equation of state models provide these to numerical relativity codes in tabulated form.

Before merger, the neutron star matter is very degenerate, so one can take $T = 0$. Furthermore, the neutron star will have had time to settle to equilibrium to weak interactions ("neutrinoless beta-equilibrium"). The latter equilibrium condition ($\mu_p + \mu_{e^-} - \mu_n = \mu_{\nu_e} = 0$, where μ_X, denoting the chemical potential of particle X, is a function of ρ_0, T, and Y_e given by the equation of state) amounts to a condition on Y_e, so the equation of state becomes one-dimensional: $\epsilon = \epsilon(\rho_0)$, $P = \rho_0^2 d\epsilon/d\rho_0$.

This 1D equation of state is all that is needed for initial data and, because most of the gravitational waveform comes from the inspiral, it is sufficient for most gravitational wave studies. Systematic studies of the effect of equation of state have been carried out assuming the piecewise polytrope family of equations of state. Here, one divides the density into intervals and takes the pressure in each interval to be a simple power law of the density. For example, in interval i covering densities between $\rho_{0,i-1}$ and $\rho_{0,i}$, one has $p(\rho_0) = \kappa_i \rho_0^{\Gamma_i}$, where κ_i and Γ_i are constants. For chosen $\rho_{0,i}$, Γ_i, and κ_0, one can infer the other κ_i by requiring continuity in $p(\rho_0)$. The range of realistic behavior can be captured with three free parameters [79]. One downside to piecewise polytropes is that they are not smooth at the boundaries of intervals, meaning the sound speed is discontinuous. There is a simple generalization to the equation of state family that fixes this [80].

An alternative flexible but smooth equation of state family is available using "spectral" equations of state [81, 82]. Here one takes the independent variable to be $x \equiv \ln(\rho_0/\rho_0^{\mathrm{ref}})$ for some reference density ρ_0^{ref}. The adiabatic index $\Gamma(x) \equiv d\ln(P)/dx$ is specified as a sum of basis functions (hence "spectral"), and then $P(x)$ and $\epsilon(x)$ can be recovered by appropriate integrals. In fact, the simple choice of a polynomial $\Gamma(x) = \Sigma_{n=0}^{N} a_n x^n$ suffices.

After merger, the densities in disks and outflows are low enough that nuclear forces are unimportant (except in the sense of holding nuclei together), and the equation of state is just a combination of classical ideal cases for free nucleons and nuclei, ideal Fermi gases for electrons/positrons, and photon radiation. As the disk expands, the NSE equilibrium changes from free nucleons to alpha particles and heavy nuclei. This recombination of nucleons releases binding energy, providing a source of thermal energy that is important for disk winds. Note that this is a reversible process; it does not generate entropy. Conservative MHD codes do not need to add a "heating" term, since the negative binding energy automatically means more thermal energy at a given total energy.

Simulations that follow disks and outflows for second timescales run into the problem of temperatures $\ll$0.5 MeV, which makes the continued assumption of NSE dubious. In particular, heating from r-process nucleosynthesis might on second timescales provide a significant source of thermal pressure, and such an effect cannot be captured with an NSE code. Unfortunately, the only adequate solution of dropping NSE and evolving isotope abundances via nuclear reaction rates would mean an explosion of evolution variables describing the composition of the fluid. Instead, a few studies have tried to add phenomenological heating terms to estimate the effects of this heating on outflows and fallback [83–85]. More recently, the necessary step of dropping NSE has been undertaken with simulations with a nuclear reaction network evolving nuclide abundances coupled to 2D ray-by-ray hydrodynamic evolution of the outflow [86].

17.4.3 Neutrino Transport

After merger, the surviving orbital material heats to $T \sim$ MeV, and neutrinos are copiously produced by weak nuclear interactions, including particularly the charged-current reactions, which alter Y_e and produce electron-flavor neutrinos and antineutrinos

$$e^- + p \rightarrow n + \nu_e \tag{17.5}$$

$$e^+ + n \rightarrow p + \overline{\nu}_e \tag{17.6}$$

and the electron-positron pair annihilations, which produce neutrinos of all flavors i

$$e^- + e^+ \rightarrow \nu_i + \overline{\nu}_i \tag{17.7}$$

Immediately post-merger, neutrino luminosities reach $L_\nu \sim 10^{53}$–10^{54} erg s^{-1}. The thermal energy of the newborn disk is $E_T \sim 10^{52}$erg, suggesting a thermal timescale of $\tau_{\rm th} \sim$ 10–100 ms [87].

Electron-type neutrinos can also be absorbed by the reverse of the processes of Eqs. 17.5 and 17.6. In addition, neutrinos of all flavors scatter off of free nucleons and nuclei. There will thus be, for neutrino species i, an opacity for absorption $\kappa_{i,a}$, an opacity for scattering $\kappa_{i,s}$, and a total opacity $\kappa_i = \kappa_{i,a} + \kappa_{i,s}$. (As with photon transport, κ_i is neutrino energy-dependent, and absorption and scattering can have somewhat different effects–details which are important in a radiation transport code but which will be ignored henceforth in this overview.) The associated mean free path is $\ell_i = 1/\kappa_i$ [or $1/(\kappa_i \rho_0)$, depending on how one defines κ_i]. How important will these scatterings and absorptions be in a post-merger disk? To answer that, one can compute the optical depth, which for a disk of height H will be $\tau_i \sim H/\ell_i$. If

$\tau_i \ll 1$, the disk is optically thin, i.e. transparent, and opacity is unimportant. Emitted neutrinos travel outward from their emission sites along almost null geodesics. If $\tau_i \gg 1$, the disk is optically thick; neutrinos are trapped, come to (Fermi-Dirac) equilibrium, escape more slowly by random walk / radiative diffusion, and can even be advected with fluid into the black hole. For BHNS disks, τ_i is at most of order $\sim$10, but after a viscous timescale of disk depletion will almost certainly be optically thin.

Because τ_i ranges from low to high, evolving the neutrino fields is difficult. Each neutrino species is described by a distribution function (the density in phase space) $f(\mathbf{x}, \mathbf{p})$. The evolution equation of f, the Boltzmann transport equation, is simple, but the fact that it is 6-dimensional makes it expensive even to store a reasonable numerical representation of it.

Many of the BHNS simulations that include neutrino effects use *neutrino leakage*, i.e. they do not evolve the neutrino fields at all, but only add a cooling term in the fluid evolution, as well as a source term for Y_e to account for lepton number changes from Eq. 17.5 and Eq. 17.6. For $\tau_i \ll 1$, one can get these from the local emission rates; for higher τ_i, one must approximate, at least to order of magnitude, the effect of diffusion. These methods decently capture neutrino cooling and were used for the earliest studies of BHNS mergers that incorporated weak interactions [28, 60, 62, 87, 88], but they have the serious drawback that neutrinos emitted in one part of the grid cannot be absorbed in another part.

A more sophisticated way to model neutrinos is provided by an *M1 moment closure scheme* [89, 90], which rather than evolving all of f only involves its lowest two moments: the neutrino energy and momentum densities. Most codes integrate the specific neutrino energy and momentum density over neutrino energies and evolve only the energy-integrated densities. Such transport schemes are called "grey". The evolution equation of each neutrino moment involves the next higher moment, so in order to obtain a closed, finite set of equations, one must assume the form of the first moment not evolved, i.e. one must provide a "closure condition". M1 does capture neutrino absorption, and it has been used in two studies of BHNS mergers [91, 92], which found that neutrino irradiation of dynamical ejecta does not significantly affect its composition, but the disk Y_e profile is noticeably different than predicted by leakage. Although a great improvement upon leakage, grey M1 remains imperfect, both because of imperfection in the closure condition and the sacrifice of spectral information.

Recently, some numerical relativity codes have evolved the full Boltzmann transport equation for the full neutrino distributions using *Monte Carlo* methods [93, 94]. Monte Carlo transport coupled to MHD was used to evolve a post-merger-like disk with the code `νbhlight` [95]. The disk had an initial optical depth of $\sim$10, so neutrino absorption is able to significantly affect the Y_e distribution of the disk and its outflows.

In addition to carrying energy, emitted neutrinos carry their momentum away with them, meaning neutrinos can provide a source of physical viscosity. Under some conditions, this viscosity could significantly slow the growth of the MRI [96]. If the MRI is suppressed, momentum transport in the disk will be greatly slowed, since the neutrino viscosity itself would be much weaker than the turbulent effective viscosity that would otherwise be present. Guilet et al. [97] have made a careful study of the effect of neutrino momentum transport on the MRI. They find that neutrino effects behave like a viscosity if the neutrino mean free path ℓ is much smaller than the wavelength of the fastest growing MRI mode: $\ell < \lambda_{MRI}$. If $\ell > \lambda_{MRI}$, neutrinos act as a drag force. For realistic (i.e. not-magnetar) initial fields, early post-merger λ_{MRI} is small enough that neutrinos will act as a drag. Simulations indicate that the drag's damping timescale is at least initially much longer than the orbital timescale on which the MRI grows [91, 98], meaning neutrinos probably will not prevent a phase of exponential field growth.

Adequate treatment of neutrino physics remains a serious challenge for BHNS simulations. Even the most sophisticated of the above schemes ignore neutrino flavor oscillations, which might affect outflow compositions [99, 100]. For a review of neutrino transport methods in numerical relativity, see [101].

17.5 Gravitational Waves

The inspiral gravitational waveform of a BHNS binary is, like the waveform from a binary black hole (BHBH) or a NSNS binary, well described by the waveform expected for a binary of two point masses. However, the fact that the neutron star is an extended object does leave a small imprint on the waveform. The gravity of the black hole tidally deforms the neutron star, giving it a quadrupole moment which affects the binding energy E and contributes to the gravitational wave luminosity L_{GW}. To lowest post-Newtonian order, the effect of tides on the gravitational wave phase depends only on one numerical property of the neutron star: its tidal deformability Λ, the ratio of the induced quadrupole moment of the star to the perturbing tidal gravitational field, in this case from the black hole [102]. This number Λ is strongly dependent on the compaction of the neutron star; it is, in fact proportional to ${R_{NS}}^5$. Thus, if this phase drift could be detected, it would provide a way of estimating the neutron star radius. Unfortunately, the tidal effect is smaller at higher Q, so BHNS systems are less promising for this sort of measurement than NSNS systems.

The effect of the end of the inspiral on the gravitational wave is dramatic, although unfortunately in the poorly accessible kHz range. The best way to understand the BHNS waveform is to compare it to a BHBH waveform. BHBH waveforms have three parts: (1) the inspiral waveform, during which the amplitude and frequency ramp up, (2) the merger waveform, at which the waveform amplitude peaks, and (3) the ringdown, a damped oscillation waveform representing the decaying perturbed modes of the settling black hole. Recall that BHNS inspiral may end, and merger may commence, in one of three ways: plunge ($R_{ISCO} > R_{tidal}$), tidal disruption ($R_{ISCO} < R_{tidal}$), or both together ($R_{ISCO} \approx R_{tidal}$). Using numerical relativity simulations for a range of Q, Shibata et al. [103] found that these three types of inspiral termination correspond to three types of BHNS merger gravitational waveforms, as illustrated in Fig. 17.3.

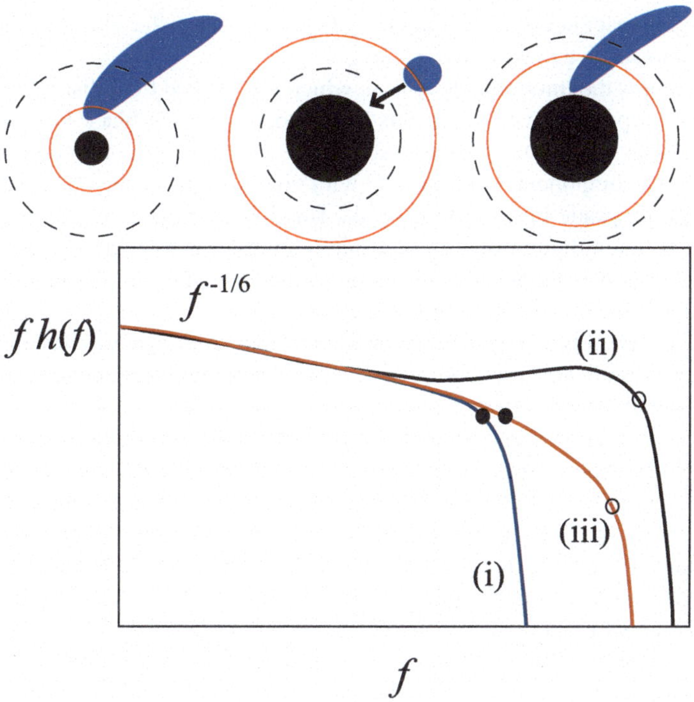

Fig. 17.3 Schematic pictures of three different merger types and their corresponding gravitational wave spectra. In the top pictures, the solid circle is the ISCO location, the dashed circle the location of tidal disruption. The types are (i) disruption with disrupted star larger than the black hole, (ii) plunge, and (iii) disruption with disrupted star smaller than the black hole. Reprinted figure with permission from Kyutoku, Okawa, and Shibata, Phys. Rev. D **84** 064018 (2011) [52]. Copyright 2011 by the American Physical Society

Systems in which the neutron star plunges into the black hole before significant mass transfer have waveforms very similar to BHBH waveforms, with inspiral, merger, and ringdown. (Indeed, the waveform is almost identical to that of a BHBH system with the same component masses [104].)

If tidal disruption happens well outside $R_{\rm ISCO}$, the waveform cuts off during the inspiral segment, and the wave amplitude decreases quickly as the disrupted neutron star spreads out and the binary's quadrupole moment drops. The matter falls through a broad, roughly axisymmetric region of the black hole and does not excite strong, coherent quasi-normal modes on the black hole, so in this case there is no ringdown waveform. One would guess that the frequency at which the waveform drops would be given by $f_{\rm cut}$ (cf. Sect. 17.2), the frequency corresponding to where mass transfer begins. Actually, the true cutoff frequency is higher, at the frequency at which the neutron star is fully tidally disrupted. The distinction has a significant effect on the dependence of the gravitational wave cutoff on the neutron star compaction [24, 52,

105]. More compact neutron stars not only begin mass transfer closer to their black hole; they also survive mass transfer longer.

There remains the intermediate case in which plunge and tidal disruption happen together. The type of waveform in this case depends on the black hole mass. For $Q \lesssim 3$, one sees an intermediate waveform type with inspiral and merger but very weak ringdown component. For higher Q with tidal disruption near the ISCO, which requires high prograde black hole spin, the ringdown component remains present. The reason is that although the neutron star is spread out by tidal disruption, if the black hole horizon is bigger than the original neutron star, the disrupted star still flows through a localized region of the horizon.

Since $R_{\rm tidal}$ depends on C, the waveform cutoff in disrupting cases has information about the neutron star and its equation of state. The information contained in the cutoff frequency in fact turns out to be nearly equivalent to Λ, presumably because both Λ and C depend (for known $M_{\rm NS}$) mostly on $R_{\rm NS}$. An ambitious numerical exploration of the BHNS parameter space was carried out with the SACRA code by Lackey et al. [106]: 134 simulations varying mass ratio, (aligned) black hole spin, and neutron star equation of state. To explore possibilities of the equation of state, a piecewise polytrope family with two free parameters was used. Although their equation of state space was 2D, a Fisher matrix analysis confirms that waveforms allow one best to measure one equation of state parameter, the tidal deformability.

It would be impossible to produce numerical relativity waveforms for every set of binary parameters that gravitational wave observatories might need as templates for detection and parameter estimation. Instead, the hope is to use waveforms from a smaller number of BHNS simulations to calibrate some simpler waveform model for which cases can be generated quickly. Thus, finite-size corrections have been added to the phenomenological ("Phenom") and effective one-body ("EOB") families of binary inspiral-merger-waveforms. The Phenom BHNS waveforms [54, 106, 107] were calibrated to the 134 SACRA waveforms. The EOB BHNS waveform models [108, 109] were calibrated to SACRA and SpEC numerical waveforms.

One way to test the adequacy of these models for LIGO-Virgo purposes is to insert numerical relativity waveforms into LIGO noise and see if the signals can be detected and the binary and equation of state parameters correctly extracted using the model waveforms as templates. Waveform models are considered good enough if the systematic errors from model inaccuracies are lower than the expected statistical errors (given whatever signal to noise ratio one is hoping for). See [110, 111] for checks of this sort.

17.6 Magnetospheres

Neutron stars observed as pulsars have magnetic fields that extend outside the neutron star surface into a magnetosphere region. In a BHNS binary, this neutron star magnetosphere could extend to and begin interacting with the black hole prior to merger. Electromagnetic signals from this magnetosphere interaction are interesting for two reasons. First of all, the signal could actually precede the merger and so perhaps appear shortly before a GRB. Second, this magnetosphere interaction and accompanying signal could still occur even for nondisrupting BHNS binaries (recall: $R_{\rm ISCO} > R_{\rm tidal}$), which otherwise seem to be such duds, electromagnetically speaking.

In the neutron star magnetosphere, the energy density and stresses of the electromagnetic field dominate over that of the fluid, i.e. $T^{\mu\nu}_{\rm EM} \gg T^{\mu\nu}_{\rm fluid}$; in this limit, the fluid's pressure and even its inertia can be ignored [112]. Energy-momentum conservation then gives

$$0 = \nabla_\nu (T^{\mu\nu}_{\rm EM} + T^{\mu\nu}_{\rm fluid}) = \nabla_\nu (T^{\mu\nu}_{\rm EM}) = F^{\mu\nu} \mathscr{J}_\nu \tag{17.8}$$

$$\mathbf{0} = \rho_e \mathbf{E} + \mathbf{J} \times \mathbf{B} \tag{17.9}$$

where $F^{\mu\nu}$ is the Faraday tensor, $\mathscr{J}_\nu$ the 4-current, ρ_e the charge density, and $\mathbf{J}$ the 3-current. This is called the "force free limit" because the above stress-energy condition is really a statement that momentum doesn't get transferred from the electromagnetic field to the fluid (i.e. no force on it). It should be emphasized that, even though $T^{\mu\nu}_{\rm fluid}$ is negligible, the force-free limit is very different from the vacuum Maxwell limit. The fluid contributes not inertia but free charges which provide conductivity. Notice from Eq. 17.9 that $\mathbf{E} \cdot \mathbf{B} = 0$. That is, there can be no electric potential difference along a magnetic field line; magnetic field lines are like *wires*.

Force-free evolutions often lead to the formation of regions in which the force-free approximation (and ideal MHD itself) breaks down. This will happen at current sheets, where $\mathbf{J}$ becomes large and the force-free condition $B^2 > E^2$ breaks down. Inside current sheets, the effects of resistivity (ignored by ideal MHD) become important, and magnetic field lines reconnect, releasing energy and producing isolated magnetic field loop regions (plasmoids).

The same unipolar inductor model that has been used to explain the Blandford-Znajek effect was used to predict electromagnetic energy release from BHNS magnetosphere interaction [114]. In the original Faraday disk, a spinning conducting disk threaded by a magnetic field has an electric field (by the MHD condition $\mathbf{E} = \mathbf{B} \times \mathbf{v}$) and so a potential difference; by touching the disk at different points with wires, the disk can be used as a battery. In the Blandford-Znajek mechanism, the black hole horizon and magnetic field lines play analogous roles. A similar unipolar inductor model of a BHNS system would predict an energy-releasing black hole battery circuit driven by the orbital motion of field lines at the horizon.

This has been modeled in numerical relativity by Paschalidis et al. [113] using an ideal MHD code able to extend to the force-free limit [115] and by East et al. [116] and Most et al. [117] using resistive MHD codes whose chosen form of Ohm's law recovers the force-free limit in the appropriate regime [118]. A snapshot from one such simulations showing magnetic field lines from the neutron star magnetosphere under the influence of the black hole is shown in Fig. 17.4. All of these studies report the electromagnetic energy outflow as a component of energy output. This

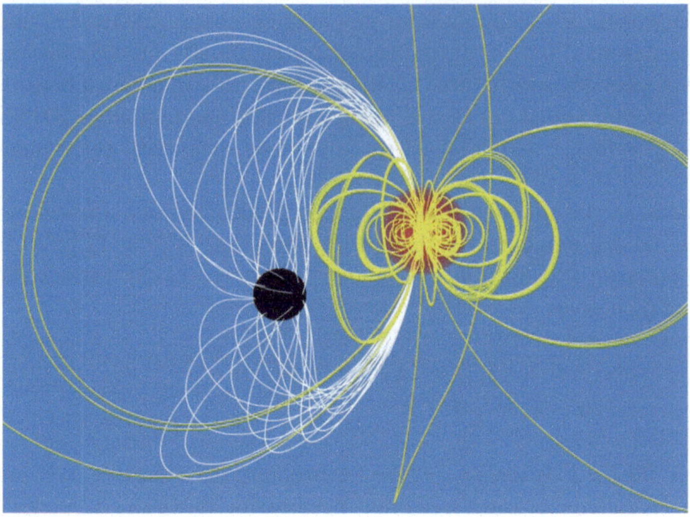

Fig. 17.4 Magnetic field lines from a neutron star magnetosphere interacting with a black hole before merger. Reprinted figure with permission from Paschalidis, Etienne, and Shapiro, Phys. Rev. D **88**, 021504 (2013) [113]. Copyright 2013 by the American Physical Society

can be calculated from the Poynting flux through an arbitrary distant sphere. One can also compute energy drained from the magnetic field via reconnection in current sheets [119].

The unipolar inductor model is found to estimate well the Poynting power output near the time of merger. In the magnetosphere of an isolated pulsar with spin angular frequency Ω, a crucial role is played by the light cylinder, the cylindrical radius $\varpi \approx c/\Omega$ at which the magnetosphere would have to orbit at the speed of light to spin at the same rate as the star, where one sees a switch from corotating closed field lines inside to open field lines and a current sheet. Similarly, in the BHNS, there is an orbital light cylinder at $\varpi \approx c/\Omega$ with Ω the orbital angular frequency, and the black hole twisting field lines close to this cylinder plays an important role in the energy release. Energy is released both by reconnection heating and by outgoing Poynting flux. The resulting radiation might be observed as a fast radio burst or X-rays.

17.7 Disk Formation and Structure

Relativistic simulations confirm that tidal disruption occurs in a single pass, even for stiff realistic equations of state [120]. Episodic mass transfer, as well as other interesting dynamics, *is* possible in general relativity for the merger of *eccentric* BHNS binaries [121]. Relativistic simulations are also the most reliable for studying the effect of black hole spin. Aligned prograde spin can make what would have been nondisrupting binaries disrupting and leads to more massive remnant disks [52, 122, 123], retrograde spin has the opposite effect, and misaligned spins have intermediate effect. This influence on disk mass is reproduced by Foucart's fitting formulae [45, 47] through the effect of black hole spin on $R_{\rm ISCO}$. For high mass ratios $Q \gtrsim 5$, corresponding to what are likely the most common black hole masses, disruption and massive disk formation are only possible for high ($\chi_{\rm BH} \gtrsim 0.7$) prograde spin [124], suggesting the depressing (to multimessenger astronomers and MHD numericists) likelihood that most BHNS mergers in nature are non-disrupting. High Q BHNS binaries that do disrupt have a larger fraction of the surviving matter outside the black hole in dynamical ejecta as opposed to disk [29, 125].

If the black hole spin is misaligned with the orbit, then there are additional effects of the spin [29, 126, 127]. Spin-orbit coupling leads to orbital precession–the neutron star and black hole "bob" above and below the initial orbital plane. For high misalignment angles ($\gtrsim 40°$) and high black hole spin, there can be notable differences in the post-merger state. If the neutron star disrupts, the swath of matter is not restricted to one plane, so it can avoid hitting itself, at least for a time, as it wraps around the black hole, delaying the formation of a disk. (See bottom panel on Fig. 17.1.) Disks from significantly misaligned systems are themselves misaligned, orbiting in a plane inclined with respect to the equator defined by the black hole's spin, although this inclination angle decreases on a timescale of tens of milliseconds. Disk-black hole spin misalignment can also lead to precession of the jet, which could produce a periodicity in an observed GRB signal as the angle between direction of jet and direction to observer oscillates. Since binary neutron star mergers do not seem to have a means of producing such misalignments, this precession-induced GRB periodicity has been suggested as a way of distinguishing GRBs from BHNS mergers [128].

In the outgoing tidal tail, some matter will be unbound and some bound. The unbound matter contributes to outflows discussed below in Sect. 17.9. The bound material eventually falls back onto the central black hole plus disk. Treating the bound outflow as following highly eccentric Keplerian orbits, one can compute the fallback time, and from that the mass inflow rate $\dot{M}_{\rm fallback}$. This is found to follow a $t^{-5/3}$ power law [78, 88, 129]. This is explained as follows [130–132]. The specific orbital energy e is related to the semimajor axis a and the period t by $e \sim a^{-1} \sim t^{-2/3}$. Then the mass falling back in time interval $dt = dP$ is $dM = (dM/de)(de/dt)dt \sim t^{-5/3}(dM/de)dt$. If the ejecta/fallback mass distribution is nonzero at $e = 0$, then zooming into a narrow enough range of e near zero (corresponding to very late fallback times), the distribution will look flat, and the fallback rate will asymptote to $\dot{M} \sim t^{-5/3}$. In fact, dM/de of bound ejecta is pretty flat, so most of the fallback obeys the $t^{-5/3}$ rate [24].

In the first tens of milliseconds, fallback onto the disk can exceed accretion into the black hole and significantly perturb and grow the disk. The power law falloff in $\dot{M}_{\rm fallback}$ would suggest that fallback can persist till late times, later than the (viscous timescale) lifetime of the postmerger disk, so that fallback has been considered as a source of late-time ($\sim$hour) extended emission from some GRBs. However, two considerations alter the expectation of an uninterrupted power law evolution of $\dot{M}_{\rm fallback}$. First, the accretion disk itself generates outgoing winds which would be expected to interfere with fallback flows. Indeed, a study by Fernández et al. [133] on the interplay between disk winds and dynamical ejecta shows that, for a $Q \approx 7$ merger, disk winds can overcome and suppress fallback for times after about 100 ms. Second, the assumption of Keplerian (ballistic) trajectories requires that pressure forces in the bound ejecta remain small. However, r-process nucleosynthesis will, on timescales of order ~second, heat the fallback, providing thermal pressure to frustrate fallback [83]. This is not entirely bad news for an extended emission model, though, since simulations of r-process heated fallback find, for sufficiently massive post-merger black holes ($>6 - 8\,M_\odot$, probably true for most BHNS post-mergers), a temporary pause in fallback rather than a permanent stop, which is closer to observed GRBs followed by extended emission than an uninterrupted power law [85]. From these two considerations, fallback and associated late-time emission may be complicated and Q-dependent.

Shocks at merger time allow the newborn disk to radiate neutrinos. The importance of weak interactions and neutrino emission depends on the mass accretion rate. If the accretion rate is above an "ignition threshold" of around $10^{-3} M_\odot\,\mathrm{s}^{-1}$, the disk will be efficently neutrino cooled [134]. If not, it will be advective. Eventually, even a disk with initial $\dot{M}$ will deplete sufficiently to fall below this level, and neutrino emission loses its thermal importance.

Weak interactions and the ignition threshold also determine the composition evolution. Initially, the disk matter is very neutron rich ($Y_e \approx 0.1$), like the neutron star it came from. Thus, $n + e^+$ reactions are more frequent than $p + e^-$ reactions (cf. Eqs. 17.5 and 17.6). The fraction of nucleons that are protons would then be expected to rise as electron antineutrino emission dominates over electron neutrino emission. This is, indeed, what simulations find in the early (~10 ms) post-merger evolution. After this, in efficiently neutrino cooled disks, Y_e begins to *decrease* again, settling back at around 0.1. The reason is that the equatorial density becomes high enough that the electron-positron gas becomes mildly degenerate: $\mu_{e^-}/k_B T \approx 1$. Equilibrium of electron-positron pair creation/annihilation forces $\mu_{e^+} = -\mu_{e^-}$. The electron and positron distribution functions differ only in the $\mu_{e^\pm}/kT$ term in the Fermi-Dirac function, so as the electron-positron gas starts to become degenerate, positrons become scarce compared to electrons, reducing the rate of $n + e^+ \rightarrow p + \overline{\nu}_e$ [87, 135]. The disk does not cool to become very degenerate ($\mu_{e^-}/k_B T \gg 1$) because strong degeneracy reduces the neutrino luminosity, causing the temperature to rise, so this process of approach to mild degeneracy with $Y_e \approx 0.1$ is self-regulating [134–136]. Below the ignition threshold, the degeneracy and regulation process does not operate. A bit later, as disk density and accretion rate decrease further, weak interactions effectively shut off to the extent that Y_e is frozen in fluid elements (as long as NSE persists). This extended post-neutrino phase will turn out to be crucial for outflows (see Sect. 17.9 below).

17.8 Magnetic Field, Jets, and Gamma Ray Bursts

MHD simulations of BHNS post-merger evolution focus both on the magnetic field's role in causing the turbulent "viscosity" and in the generation of polar jets. If the jet is generated by the Blandford-Znajek mechanism, the power output $L_{\rm BZ}$ will be proportional to Φ^2, where Φ is the magnetic flux on the black hole horizon:

$$\Phi = \int_H |B \cdot n| \, dA \tag{17.10}$$

(Without the absolute value sign, field lines going in the horizon in one hemisphere would cancel field lines coming out the other.) Thus, powerful jets require large magnetic flux on the horizon. Accretion from the disk can advect onto the horizon, and this flux can accumulate with time. The pressure of the surrounding disk is also needed to confine the near-horizon field. If the flux reaches $\Phi = \Phi_{\rm MAD} \approx 50\dot{M}^{1/2} M_{\rm BH}$, the field will be strong enough to oppose further accretion, leading to a magnetically arrested disk (MAD) state [137, 138]. MAD accretion releases energy very efficiently, but since once Φ reaches $\Phi_{\rm MAD}$ it is subsequently stuck at $\Phi \approx \Phi_{\rm MAD}$, the subsequent $L_{\rm BZ}$ is tied to the (declining) mass accretion rate.

During and shortly after merger, several processes will amplify the magnetic field inside the incipient disk. First, the field will grow due to *magnetic winding* (also called the "ω-effect" in dynamo literature): as a simple consequence of the fact that magnetic field lines are "frozen into" a conducting fluid, differential rotation/shear will "wind up" field lines. Winding will lead to a linear in time growth of the toroidal component of the field (B^ϕ).

The other sources of magnetic field growth are MHD instabilities. For example, as tidal disruption turns the star into a stream swirling around the black hole, this wound spiral arm shears against itself, triggering the Kelvin-Helmholtz instability (KHI). Vortices will form the shear interface in which magnetic fields will quickly wind up [98, 139].

The most well-known is the *MRI*, which is triggered in a rotating, conducting, weakly magnetized ($B^2 \ll P$) fluid whenever

$$\varpi \, \partial_\varpi \Omega^2 + N^2 < 0 \,, \tag{17.11}$$

where ϖ is the cylindrical radius (distance from the orbital axis), Ω is the orbital angular frequency, and N is the Brunt-Väisälä frequency [140]. In a disk, the first term will dominate, and since Ω^2 always decreases with ϖ in disks, the condition will be satisfied. The MRI is a linear instability leading to exponential growth at rate $\sim\Omega$ of the B-field. The fastest growing unstable mode has wavelength $\lambda_{\rm MRI} \sim v_A/\Omega$, where $v_A = B/\sqrt{4\pi\rho}$ is the Alfven speed. Thus, stronger fields (so long as they are not strong enough to suppress the MRI) have larger $\lambda_{\rm MRI}$, and numerical relativity simulations usually insert unphysically large initial magnetic fields of order 10^{16} G in order to be able to numerically resolve the MRI in the subsequent merger. This is justified by pointing out that the inserted magnetar-level field is still dynamically weak in the disk ($B^2/P \sim 10^{-2}$), in that there is much room for it to grow to reach equipartition, so hopefully details of the seed field will not imprint the field when it has saturated.

The MRI drives turbulence, whereby energy injected at $\lambda_{\rm MRI}$ cascades to smaller scales, and if one does not resolve a sufficient portion of the inertial range at scales beneath $\lambda_{\rm MRI}$, turbulent transport effects might be improperly modeled. The highest resolution MHD BHNS simulations, by Kiuchi et al. [98] and by Izquierdo et al. [139] (minimum grid spacing 0.12 km, the latter study including subgrid turbulence modeling, both studies omitting neutrinos) fail to demonstrate numerical convergence, with the former studying finding at higher resolution much greater near-BH disk heating and early-time outflow rate and the latter study much faster initial magnetic field growth driven by the KHI.

Early BHNS merger simulations with MHD took the initial magnetic field to consist of poloidal loops with axis aligned with the orbit and with field confined inside the neutron star. These simulations found that the field in the post-merger disk quickly wound into a toroidally dominated configuration, with no observable jets, at least on the tens of millisecond timescales investigated [78, 141, 142]. (However, MRI-generated poloidal field can dominate in parts of the low density region [143].) Simulations of black hole accretion disks have looked into the differences when one begins with a poloidal vs. a toroidal seed field. While the MRI inside the disk is found to be similar in both cases, jet launching requires poloidal magnetic field reaching the black hole [144]. Thus, it is not surprising that these toroidal field-dominated merger outcomes did not quickly produce jets.

The University of Illinois Ubana-Champaign group has investigated the effect of the geometry of the neutron star's initial magnetic field on the post-merger field. If the confined magnetic field is tilted with respect to the orbital axis, there can be a somewhat larger early post-merger poloidal fields, but still insufficient to generate an early post-merger jet [142]. On the other hand, jets can more quickly ($\approx$100 ms) emerge from a BHNS merger involving a neutron star with pre-merger poloidal field extending outside the neutron star, because in such a case more magnetic field survives the merger and can be wound above the poles into an incipient jet [145, 146]. These were, in fact, the first BHNS simulations to successfully demonstrate jet formation.

Jets are still possible for disks with toroidal initial fields, but there is the delay that the disk must first generate a poloidal field by *dynamo* action, the process whereby some of the toroidal field can be converted through small-scale turbulence into a large-scale poloidal field [148]. Magnetic winding feeds poloidal field into toroidal field, so the dynamo can be a self-sustaining cycle. Distinctive oscillatory behavior is seen in the large-scale dynamo-driven field. In black hole accretion systems, one can see a cyclic drift in the latitude of toroidal field strength reminiscent of the "butterfly diagrams" in solar sunspots [32, 149]. Disk simulations with toroidal initial field find that dynamo poloidal field growth can indeed happen [150, 151], although the jets that eventually form might be more intermittant ("striped") than those from disks with poloidal seed field [151].

A few very recent simulations have followed magnetized BHNS disks for second timescales: the work of Hayashi et al. [32, 152] using the SACRA code and the work of Gottlieb et al. [147, 153] using the H-AMR code with initial data from hydrodynamic BHNS mergers with the SpEC code. The two studies differed in several ways. Notably, the SACRA simulations included neutrino effects and recombination while the H-AMR simulations did not, but the H-AMR simulations evolved for longer times. Here, we will present the main points of the picture emerging from extended post-merger MHD simulations, encouraging readers to consult the original papers for details. Readers should note that some papers only insert a seed magnetic field after merger (e.g. [147, 153, 154]). Magnetized BHNS mergers find that the field in the initial post-merger disk is predominantly toroidal, so the early evolution of disks with poloidal seed field should not be trusted much. Also, electromagnetic luminosity can be reported differently, e.g. Poynting luminosity across the whole horizon, a measure of electromagnetic energy extraction from the black hole, versus isotropic equivalent luminosity to the Poynting flux near the axis, a measure of jet strength relevant to an on-axis observer.

Dynamo activity is clearly seen in extended MHD simulations of BHNS disks, both in the growth of poloidal field and in the distinctive "butterfly" effect drift in the toroidal field [32, 152]). The polar region after merger is not empty but contains matter ejected during tidal disruption and merger as well as early-time disk outflows. This delays the formation of a magnetosphere until the density inside a funnel region falls to $\rho < b^2/8\pi$ [32, 143], and the incipient jet must push through a cocoon of this matter [147].

The horizon magnetic flux Φ, and hence the Poynting luminosity for the Blandford-Znajek effect, accumulates early and is fairly constant for order seconds, excepting fluctuations due to turbulence and dynamo cycles. During this time, the disk depletes, and the accretion rate falls off like $\dot{M} \propto t^{-2}$. Thus the MAD criterion ratio $\Phi/(\dot{M} M_{\mathrm{BH}})$ grows. Eventually, it is inevitable that the disk will enter a MAD state. The nominal efficiency of accretion ($L_{\mathrm{BZ}}/\dot{M}$) grows ("output" holding steady while "input" decreases), reaching order unity as the disk reaches MAD state. As $\dot{M}$ drops, gas pressure at the edge of the funnel region decreases, so the gas loses ability to confine the jet. Even before the MAD criterion is satisfied, this loss of confining pressure causes the jet to broaden, leading to a corresponding decrease of flux observed by an observer in the jet's path [32]. Once the disk becomes MAD, the jet luminosity is constrained to decrease along with $\dot{M}$ at a rate $\propto t^{-2}$, which we should be able to observe in short GRBs. These features of the evolution are shown in Fig. 17.5.

If it is true that the ultimate falloff of GRB jet luminosity is caused by the growing relative strength of magnetic forces on a depleting disk, then there is a connection between the post-merger disk mass (which determines $\dot{M}$), the jet luminosity, and the lifetime after which the jet luminosity drops, as illustrated in Fig. 17.5. For a given disk mass (and hence $\dot{M}(t)$), a more luminous jet would reach this time at which decay commences sooner. Considering only the MAD criterion, this is when the L_{BZ} and $\dot{M}$ curves cross [153].

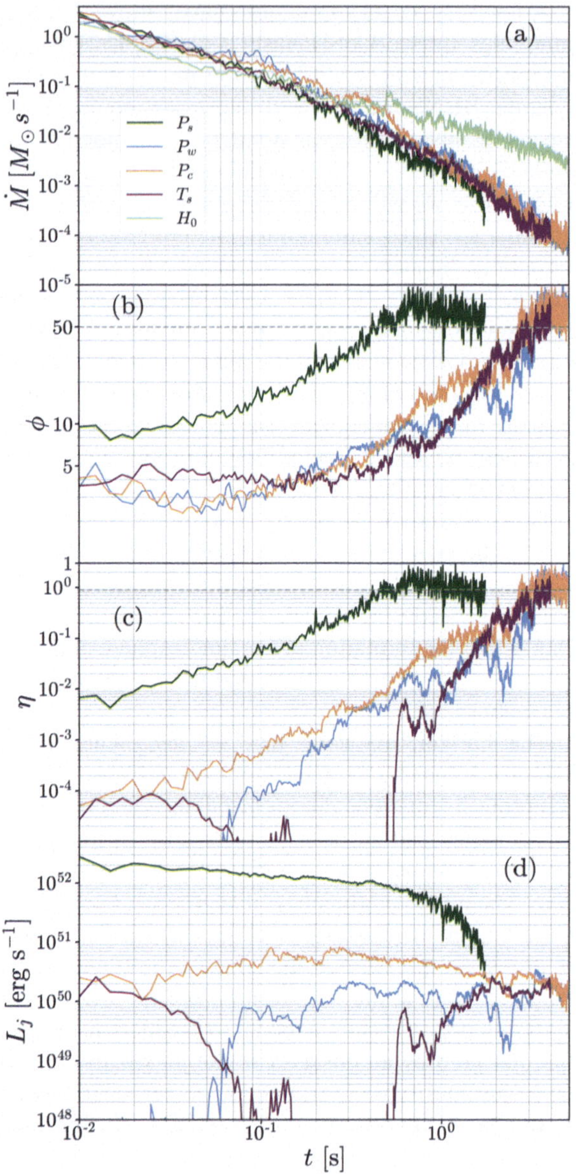

Fig. 17.5 Electromagnetic energy extraction (measured at $r = 5M_{\rm BH}$) for a $Q = 2$, $\chi_{\rm BH} = 0.6$ system with different choices of magnetic field inserted post-merger. Figure from Gottlieb *et al.* [147] under the terms of the Creative Commons Attribution 4.0 International License. **Panel (a)**: Mass accretion rate **Panel (b)**: Dimensionless magnetic flux $\phi = \Phi/\sqrt{\dot{M} M_{\rm BH}^2}$, where Φ is the magnetic flux. Ultimately, all models turn MAD when $\phi \approx 50$ (dashed line). **Panel (c)**: The jet launching efficiency **Panel (d)**: The jet luminosity, $L_j = \eta \dot{M}$

Magnetic fields are not the only mechanism that has been considered for launching jets. Another energy source comes from the annihilation of neutrinos and antineutrinos in the (neutrino) optically-thin polar region. The luminosity from neutrino annihilation for a hot, high $\dot{M}$ (but neutrino transparent) disk is roughly

$L_{\nu\overline{\nu}} \sim 10^{52}(\dot{M}/M_{\odot}s^{-1})^{2.25}$ erg s^{-1} [155]. For high accretion rates, of order a percent of the neutrino luminosity can be deposited in this way, but naturally, this source of energy shuts off when the accretion rate drops below the weak interaction ignition accretion rate, and is probably negligible for $\dot{M} < 10^{-2}M_{\odot}\,s^{-1}$ [156]. This is one problem for the $\nu\overline{\nu}$ GRB model. Another obstacle to relativistic jets is the inertia of matter ejected during merger polluting the polar regions ("baryon loading"). Radiative hydrodynamic evolutions suggest that this ejecta will prevent $\nu\overline{\nu}$ annihilation from accelerating polar outflows to relativistic speeds in NSNS remnants [157]. The same paper concludes that BHNS disks (which have somewhat cleaner polar regions after merger) can produce GRB fireballs, but the energies and durations are too low to explain most short GRBs. It remains possible that the neutrino mechanism is important early on. Since, according to high resolution MHD BHNS simulations, magnetic jets take at least tens of milliseconds to form [98], the character of the relativistic outflow might change from fireball to Poynting flux dominated (cf. [158]). It should also be remembered that, even if $\nu\overline{\nu}$ annihilation cannot drive a relativistic outflow, it might contribute to mildly relativistic winds (cf. [159]), a topic to which we now turn.

17.9 Outflows, Kilonovae, and Radio Flares

Matter from the neutron star that ends up being ejected will decompress and evolve into some kind of "normal" matter; this evolution turns out to be of great interest for electromagnetic signals and production of heavy elements. When ρ and T are low enough that outgoing gas leaves NSE, many nucleons will have collected into heavy nuclei. However, if Y_e is low (many more neutrons than protons), there will be a large number of "extra" free neutrons as well. In this case, nuclei will rapidly capture neutrons, bringing them close to the neutron drip line, i.e. the most neutrons a nucleus can hold for given number of protons–on a space of isotopes labeled by (A, Z), the maximum of A for given Z. These nuclei can eventually form the "r-process" elements. The distribution of nuclei produced by the r-process has three peaks at $A \sim$ 70–80, 120–130, and 190–200. Notably, it produces lanthanides and actinides (located past the second peak). Absent rapid neutron capture (i.e. for higher Y_e), nuclei stay closer to the valley of stable isotopes (a swath in the A-Z plane); there can be s-process nucleosynthesis in this regime. For high-Y_e outflow, the r-process does not proceed far enough to produce significant amounts of elements beyond the second or even first peak; in particular lanthanides and actinides are not efficiently produced.

The possibility that tidal disruption in BHNS mergers could produce a large mass of unbound outflow which might then undergo r-process nucleosynthesis was pointed out by Lattimer and Schramm in 1976 [160]. Li and Paczynski noticed that radioactive decays in ejecta could power a bright electromagnetic transient [161]. However, it was only sometime afterward that models were created with realistic heating rates and opacities [83, 162–164]. Opacity turns out to be crucial–a more opaque gas will trap photons longer, so that they escape on a longer timescale and at lower energy. Lanthanides, if present, introduce a thicket of absorption lines due to their partially-filled valence f shells, greatly increasing the bound-bound opacity. A lanthanide-rich ejecta glows in the near infrared on timescale of a week ("red kilonova"); a lanthanide-poor ejecta glows in the optical-UV on day timescale ("blue kilonova"). The transition between one and the other is at about $Y_e \approx 0.25$ [165–168].

As the outflow spreads and decompresses further, it interacts with the background interstellar medium (ISM); electrons are accelerated in the shock moving into the ISM and emit synchrotron radiation. This results in a distinct signal, a radio flare lasting years to decades [169–171]. This long timescale is set by the deceleration time, the time it takes the ejecta to sweep up roughly its own mass in ISM. The timescale and strength of these flares depends both on the kinetic energy of the ejecta and the density of the surrounding ISM. Thus, the properties of outflows determine multiple signals.

Numerical simulations predict the mass of outflow and its distribution in speed, entropy, and Y_e. One issue is how to identify unbound matter. In a stationary spacetime, u_t is constant on geodesics, so a $u_t < -1$ geodesic can be identified as unbound. Furthermore, $-u_t$ can be identified as the asymptotic Lorentz factor. However, the spacetime is not stationary in the early merger, and–a more serious problem–the outflow is not geodesic, at least within the grid and timeframe which most simulations can afford to maintain, because of the persistence of pressure forces (e.g. [172]). In a stationary flow (not necessarily geodesic), the Bernoulli parameter hu_t is constant on fluid elements. At asymptotically large distances and late times, $\rho_0 \to 0$, $h \to h_\infty$, so for such a flow, one can tag $(h/h_\infty)u_t < -1$ fluid elements as unbound and $-(h/h_\infty)u_t$ as the asymptotic Lorentz factor, but, unfortunately, the outflow is not stationary either. One must also worry about the effect of r-process heating, currently not accounted for in most merger and post-merger simulations. For more on the subtleties of identifying unbound matter and its asymptotic speed, see Foucart et al. [173].

Newtonian and relativistic simulation confirm that BHNS mergers produce large ejecta masses, up to $\sim 10^{-1} M_\odot$ [29, 88, 92, 123, 125, 174, 175]. Dynamical ejecta masses can thus be much larger than in NSNS systems, mainly because of the higher mass ratios Q for BHNS binaries, rather than because of something distinctive about the presence of a black hole. (Asymmetric NSNS systems also tend to produce more dynamical ejecta than equal mass systems.) Since the ejecta is produced by tidal disruption of the neutron star before merger, the ejecta is cold and little affected by neutrino (weak interaction) processes [92, 176], so its electron fraction is essentially what it was in the pre-merger neutron star, about $Y_e \approx 0.05$. Such extremely neutron-rich material will produce 2nd and 3rd peak r-process elements and produce a red kilonova. Typical asymptotic velocities are $\sim$0.2–0.3c. BHNS dynamical ejecta is extremely anisotropic, much more so than NSNS ejecta, with outflow concentrated within 20° of the equatorial plane and sweeping out only about half of this plane (180°). (See Fig. 17.6.) The ejecta thus has an overall linear momentum, so the black hole remnant can get a kick of $\sim$100 km s^{-1}, a larger effect than the gravitational wave recoil [29, 125, 174]. Further consequences of this overall direction of ejecta

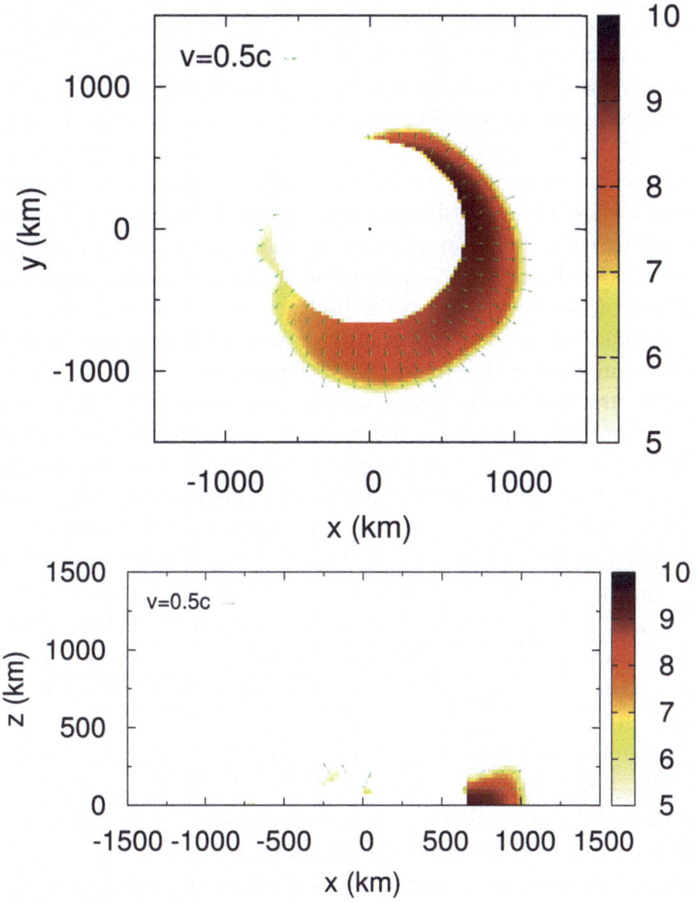

Fig. 17.6 Concentration of dynamical ejecta for a $Q = 5$, $\chi_{BH} = 0.75$, aligned spin BHNS merger. **Top:** equatorial plane. **Bottom:** a meridional plane. Color indicates density. Reprinted figure with permission from Kyutoku, Ioka, and Shibata, Phys. Rev. D **88**, 041503 (2013) [174]. Copyright 2013 by the American Physical Society

motion will be discussed below. It should be pointed out, though, that BHNS systems with significantly misaligned black hole spin can produce dynamical ejecta extending over 360° [127].

An additional component of outflow comes later from the accretion disk, especially after neutrino cooling has diminished in importance, and the disk has become advective. These disk wind outflows have been studied in Newtonian physics in 2D, with the black hole modeled as a central potential and MHD turbulence modeled as a shear viscosity, in a series of papers by Fernandez and collaborators [133, 177–180]. There have also been a small number of 2D relativistic viscous hydrodynamic studies, which find qualitatively similar results [172, 181, 182]. Outflow gas is unbound from the disk by thermal pressure, so the key to understanding disk winds is to understand how thermal energy is added to the disk. Viscosity provides the major source of heating; since viscosity is considered a model for turbulence, this viscous heating can be interpreted as dissipation of turbulent energy on small scales. Heat generated in the interior of the disk can trigger convection which transports this heat to the surface. (Thus, even simulations which don't explicitly include magnetorotational turbulence often show disk snapshots which look quite turbulent.) Thermal energy is also boosted by recombination of nucleons into nuclei, with the accompanying release of nuclear binding energy. (Recall, this is not technically "heating" since it is reversible and adiabatic so long as the fluid remains in NSE. It is automatically included in simulations that use temperature-dependent equations of state and use a conservative formulation of the fluid or MHD equations.)

A significant fraction of the disk's mass can be ejected in these winds, with 10–20% of the initial disk mass being typical. An exploration of the parameter space of disks [180] shows that the fraction of disk mass ejected decreases with the compactness of the initial disk (proportional to the black hole mass divided by the radius of the initial density peak). The dependence on disk compaction can be understood by remembering that most of the outflow launching happens after weak interaction freeze-out, when accretion becomes advective. For a more compact disk, more of the mass will accrete before this time, meaning that for a less compact disk, more mass will be remaining when strong outflows commence. Ejected mass fraction also decreases with the initial disk mass. Ejecta from disk outflows is found to be significantly slower than dynamical ejecta, with $v \approx 0.05c$ being a typical measurement. Unlike the dynamical ejecta, this ejecta undergoes significant neutrino processing from its time in the disk, and average $Y_e \approx 0.3$. Of course, within the disk outflow from a single merger, there will be gas with a range of Y_e, but more than half will be lanthanide and actinide-poor and thus lower opacity.

Viscous hydrodynamics is an approximation meant to model MHD turbulence. How well outflows from viscous evolution match those of full MHD evolution has been tested by comparing results of 3D MHD simulations to those of 2D viscous hydrodynamics, with both simulation types including neutrino cooling and nuclear recombination. One finds that MHD simulations produce late-time thermal winds that are very similar in mass, velocity, and composition to the outflows in 2D viscous hydrodynamics simulations for alpha viscosity in the range $\alpha_{SS} \approx 0.03$–0.1. Simulations with poloidal initial field (with ratio of gas pressure to magnetic pressure of $\beta \sim 10^2$) find an additional, early-time magnetically-driven ejecta, amounting to another 20% of the disk's mass, with higher velocity and lower Y_e than the later wind [31]. This early-time outflow is, however, sensitive to the choice of seed field [151]. It is essentially removed if the initial poloidal field is weak ($\beta \sim 10^3$) or if the initial field is toroidal. BHNS merger MHD simulations indicate the early post-merger field is in fact predominantly toroidal, and one does not observe the early outflow in these simulations [32]. (However, outflow might appear only at sufficiently high resolution [98], so one should be cautious with these conclusions.)

We have seen that disrupting BHNS mergers will, except for systems with low-mass black holes, have as much or more dynamical ejecta than disk ejecta. Furthermore, we have seen that the dynamical ejecta can be highly asymmetric–often concentrated near the orbital equator and to one side. Both the mass and asymmetry of dynamical ejecta distinguish typical BHNS from typical NSNS outflows. They have observable consequences.

Consider the asymmetry of ejecta. The flatness of the ejecta means photons are able to random walk to the vertical surface faster than they would be able to escape a spherical distribution. This results in a shorter time to peak emission, higher ejecta temperature at this time, higher peak luminosity, and bluer emission [174, 183]. It has even been suggested that the difference in color for a given luminosity might provide a way to distinguish BHNS from NSNS mergers from their kilonovae [183]. The asymmetry of ejecta motion, on the other hand, causes Doppler effects, so that the brightness and color depends on the viewing angle–whether the ejecta is moving toward or away from the observer [184].

The overall kilonova emission will depend on emission from and opacity of both dynamical and disk outflow components and their spatial arrangement. The dynamical ejecta will be farther out from the black hole and equatorially concentrated, while the disk outflow will be more isotropic. Thus, there may be a low-opacity disk outflow producing a blue emission, but it might only be visible from certain viewing angles, because in equatorial directions, the blue emission is blocked by the more-opaque dynamical ejecta between the disk outflow and the viewer. Some work has been done in figuring out how the different ejecta component effects combine at different viewing angles to produce overall light curves [133, 179, 185, 186].

The distinct features of BHNS dynamical ejecta also affect the later radio flare. A flat outflow takes longer than a spherical outflow of the same mass to sweep up its own mass, so the peak time might be delayed. Also, the overall motion of the ejecta in one direction results in a proper motion of the radio source which could possibly be detected [174].

17.10 Conclusions

BHNS mergers involve physics similar to that of NSNS mergers and have the potential to produce similar types of signals. They are distinguished from NSNS mergers in that higher mass ratio Q is expected, the post-merger central remnant is definitely a black hole, the black hole is likely more massive than that produced by NSNS post-merger collapse, greater dynamical ejecta masses are possible, this dynamical ejecta is colder, more neutron-rich, and more anisotropic, and the post-merger polar region is likely to be less baryon loaded. These differences may have observational consequences. The challenge of MHD simulations comes from the range of time and length scales that must be covered. However, general aspects of the merger and post-merger evolution are coming to be known with confidence, because they follow from general principles: the dependence of tidal disruption on binary parameters, the early importance but eventual shut-off of neutrino-emitting weak interactions, and the natural evolution of the disk toward a magnetically arrested state.

Nevertheless, no current simulation of the post-merger evolution should be considered definitive, because none contains all of the ingredients needed to be fully trustworthy. MHD and neutrino emission are both crucial to the post-merger emission, and while many studies include one or the other, few include them both. Those of which the author is aware include [32, 143, 152, 154]. The first two of these used neutrino leakage, which is only quantitatively reliable for optically thin disks. However, even including both magnetic fields and neutrino transport is insufficient. First, there are the difficulties of resolving turbulence induced by the MRI. Convergence studies without neutrino effects (which are thus *physically* unreliable) indicate that grid spacing of $\Delta x < 0.2\,\mathrm{km}$ is needed to capture early outflows [98]. The studies cited above that include neutrino effects do not see these outflows or this resolution sensitivity, but they also have not gone to such high resolution. Furthermore, the neutrino treatment in these studies did not include neutrino-antineutrino annihilation and flavor oscillations. Simulations continuing for seconds and tracking ejecta to large distances face the more important problem of the breakdown of nuclear statistical equilibrium, which (because of r-process heating) cannot adequately be relegated to post-processing.

In addition to these modeling issues, future simulations must more adequately explore the BHNS parameter space. Much of the work reviewed in this chapter has concentrated on low-to-moderate mass ratio systems with "canonical" neutron star masses of 1.3–$1.4 M_{\odot}$ and aligned black hole spin. Properly calibrating BHNS gravitational wave models will require more adequately sampling a fuller range of the possible masses and spins. We should be sure, for example, that the conclusions we have drawn regarding characteristically asymmetric BHNS ejecta are not too sensitive to an assumption that the black hole spin misalignment angle is small. We should be careful in extrapolating the conclusions of our few seconds-long MHD simulations [147, 152] carried out from two sets of masses and spins to the whole range of disrupting BHNS systems. As daunting as the task may be, we must carry out seconds-long radiation MHD simulations covering a wide range of BHNS systems.

Acknowledgements I am grateful to Francois Foucart, Ore Gottlieb, Koutarou Kyutoku, Nishad Muhammed, Stuart Shapiro, and Samuel Tootle for viewing a draft and providing useful comments. I also thank the authors of references [52, 113, 174] for permission to reuse Figs. 17.3, 17.4, and 17.6. I am grateful for support from the NSF through grants PHY-2110287 and PHY-2407726 and from NASA through grant 80NSSC22K0719.

References

1. R. Abbott et al., Observation of gravitational waves from two neutron star-black hole coalescences. Astrophys. J. Lett. **915**(1), L5 (2021). https://doi.org/10.3847/2041-8213/ac082e, arXiv:2106.15163 [astro-ph.HE]

2. R. Abbott et al., GWTC-3: compact binary coalescences observed by LIGO and virgo during the second part of the third observing run. Phys. Rev. X **13**(4), 041,039 (2023). https://doi.org/10.1103/PhysRevX.13.041039, arXiv:2111.03606 [gr-qc]
3. R. Abbott et al., GWTC-2.1: Deep extended catalog of compact binary coalescences observed by LIGO and Virgo during the first half of the third observing run. Phys. Rev. D **109**(2), 022,001 (2024). https://doi.org/10.1103/PhysRevD.109.022001, arXiv:2108.01045 [gr-qc]
4. J.P. Zhu, S. Wu, Y. Qin, B. Zhang, H. Gao, Z. Cao, Population properties of gravitational-wave neutron star-black hole mergers. Astrophys. J. **928**(2), 167 (2022). https://doi.org/10.3847/1538-4357/ac540c, arXiv:2112.02605 [astro-ph.HE]
5. M. Linares, T. Shahbaz, J. Casares, Peering into the dark side: magnesium lines establish a massive neutron star in PSR J2215+5135. Astrophys. J. **859**(1), 54 (2018). https://doi.org/10.3847/1538-4357/aabde6, arXiv:1805.08799 [astro-ph.HE]
6. H.T. Cromartie et al., Relativistic Shapiro delay measurements of an extremely massive millisecond pulsar. Nature Astron. **4**(1), 72–76 (2019). https://doi.org/10.1038/s41550-019-0880-2, arXiv:1904.06759 [astro-ph.HE]
7. C.E. Rhoades, R. Ruffini, Maximum mass of a neutron star. Phys. Rev. Lett. **32**(6), 324–327 (1974). https://doi.org/10.1103/PhysRevLett.32.324
8. F.J. Fattoyev, C.J. Horowitz, J. Piekarewicz, B. Reed, Gw190814: impact of a 2.6 solar mass neutron star on the nucleonic equations of state. Phys. Rev. C **102**, 065,805 (2020). https://doi.org/10.1103/PhysRevC.102.065805
9. R. Essick, P. Landry, Discriminating between neutron stars and black holes with imperfect knowledge of the maximum neutron star mass. Astrophys. J. **904**(1), 80 (2020). https://doi.org/10.3847/1538-4357/abbd3b, arXiv:2007.01372 [astro-ph.HE]
10. R. Abbott et al., Population of merging compact binaries inferred using gravitational waves through GWTC-3. Phys. Rev. X **13**(1), 011,048 (2023). https://doi.org/10.1103/PhysRevX.13.011048, arXiv:2111.03634 [astro-ph.HE]
11. A.G. Abac, et al., Observation of gravitational waves from the coalescence of a $2.5 - 4.5\ M_{\odot}$. Compact object and a neutron star. Astrophys. J. Lett. **970**(2), L34 (2024). https://doi.org/10.3847/2041-8213/ad5beb. arXiv:2404.04248 [astro-ph.HE]
12. B.D. Metzger, Kilonovae. Living Rev. Rel. **23**(1), 1 (2020). https://doi.org/10.1007/s41114-019-0024-0, arXiv:1910.01617 [astro-ph.HE]
13. E. Nakar, Short-hard gamma-ray bursts. Phys. Rep. **442**, 166–236 (2007). https://doi.org/10.1016/j.physrep.2007.02.005, arXiv:astro-ph/0701748
14. E. Troja et al., A nearby long gamma-ray burst from a merger of compact objects. Nature **612**(7939), 228–231 (2022). https://doi.org/10.1038/s41586-022-05327-3, arXiv:2209.03363 [astro-ph.HE]
15. A.J. Levan et al., Heavy-element production in a compact object merger observed by JWST. Nature **626**(8000), 737–741 (2024). https://doi.org/10.1038/s41586-023-06759-1, arXiv:2307.02098 [astro-ph.HE]
16. J.P. Norris, J.T. Bonnell, Short gamma-ray bursts with extended emission. Astrophys. J. **643**, 266–275 (2006). https://doi.org/10.1086/502796, arXiv:astro-ph/0601190
17. E. Berger, Short-duration gamma-ray bursts. Ann. Rev. Astron. Astrophys. **52**, 43–105 (2014). https://doi.org/10.1146/annurev-astro-081913-035926, arXiv:1311.2603 [astro-ph.HE]
18. S. Nissanke, M. Kasliwal, A. Georgieva, Identifying elusive electromagnetic counterparts to gravitational wave mergers: an end-to-end simulation. Astrophys. J. **767**, 124 (2013). https://doi.org/10.1088/0004-637X/767/2/124, arXiv:1210.6362 [astro-ph.HE]
19. M. Bhattacharya, P. Kumar, G. Smoot, Mergers of black hole–neutron star binaries and rates of associated electromagnetic counterparts. Mon. Not. R. Astron. Soc. **486**(4), 5289–5309 (2019). https://doi.org/10.1093/mnras/stz1147, arXiv:1809.00006 [astro-ph.HE]
20. L.S. Collaboration, V. Collaboration, F. GBM, INTEGRAL, I. Collaboration, A.C.Z.T.I. Team, I. Collaboration, I.H. Collaboration, A. Collaboration, S. Collaboration, A. Team, M. Team, D.E.C.G.E. Collaboration, D. Collaboration, D. Collaboration, G.W.I. TeAm, F.L.A.T. Collaboration, A.T.C. Array, A.S. Pathfinder, L.C.O. Group, OzGrav, DWF, AST3, C. Collaborations, V. Collaboration, M. Collaboration, J-GEM, GROWTH, JAGWAR, Caltech-NRAO,

TTU-NRAO, N. Collaborations, Pan-STARRS, M. Team, T. Consortium, K. Collaboration, N.O. Telescope, ePESSTO, GROND, T.T. University, S. Group, T.R.O. of the South Collaboration, B. Collaboration, MWA, M.W. Array, C. Collaboration, I.G.F. up Collaboration, H.E.S.S. Collaboration, L. Collaboration, L.W. Array, H. Collaboration, P.A. Collaboration, A. Collaboration, E.V. Team, P. of the Sky Collaboration, C.T. at McGill University, D.F. Network, ATLAS, H.T.R.U. Survey, RIMAS, RATIR, S.S. Africa/MeerKAT, Multi-messenger observations of a binary neutron star merger. Astrophys. J. **848**, L12 (2017). https://doi.org/10.3847/2041-8213/aa91c9, arXiv:1710.05833 [astro-ph.HE]
21. O. Gottlieb, E. Nakar, T. Piran, The cocoon emission—an electromagnetic counterpart to gravitational waves from neutron star mergers. Mon. Not. R. Astron. Soc. **473**(1), 576–584 (2018). https://doi.org/10.1093/mnras/stx2357, arXiv:1705.10797 [astro-ph.HE]
22. D. Lazzati, R. Perna, Jet-cocoon outflows from neutron star mergers: structure, light curves, and fundamental physics. Astrophys. J. **881**(2), 89 (2019). https://doi.org/10.3847/1538-4357/ab2e06, arXiv:1904.08425 [astro-ph.HE]
23. O. Gottlieb, D. Issa, J. Jacquemin-Ide, M. Liska, A. Tchekhovskoy, F. Foucart, D. Kasen, R. Perna, E. Quataert, B.D. Metzger, Hours-long near-uv/optical emission from mildly relativistic outflows in black hole-neutron star mergers. Astrophys. J. Lett. **953**(1), L11 (2023). https://doi.org/10.3847/2041-8213/acec4a, arXiv:2306.14946 [astro-ph.HE]
24. K. Kyutoku, M. Shibata, K. Taniguchi, Coalescence of black hole–neutron star binaries. Living Rev. Rel. **24**(1), 5 (2021). https://doi.org/10.1007/s41114-021-00033-4, arXiv:2110.06218 [astro-ph.HE]
25. F. Foucart, *Black Hole-Neutron Star Mergers* (Springer Nature Singapore, Singapore, 2022), pp. 611–660. https://doi.org/10.1007/978-981-16-4306-4_14
26. D. Tsang, Shattering flares during close encounters of neutron stars. Astrophys. J. **777**, 103 (2013). https://doi.org/10.1088/0004-637X/777/2/103, arXiv:1307.3554 [astro-ph.HE]
27. M. Vallisneri, Prospects for gravitational wave observations of neutron star tidal disruption in neutron star/black hole binaries. Phys. Rev. Lett. **84**, 3519 (2000). https://doi.org/10.1103/PhysRevLett.84.3519, arXiv:gr-qc/9912026
28. F. Foucart, M.B. Deaton, M.D. Duez, E. O'Connor, C.D. Ott, R. Haas, L.E. Kidder, H.P. Pfeiffer, M.A. Scheel, B. Szilagyi, Neutron star-black hole mergers with a nuclear equation of state and neutrino cooling: dependence in the binary parameters. Phys. Rev. D **90**, 024,026 (2014). https://doi.org/10.1103/PhysRevD.90.024026, arXiv:1405.1121 [astro-ph.HE]
29. F. Foucart, M.B. Deaton, M.D. Duez, L.E. Kidder, I. MacDonald, C.D. Ott, H.P. Pfeiffer, M.A. Scheel, B. Szilagyi, S.A. Teukolsky, Black hole-neutron star mergers at realistic mass ratios: equation of state and spin orientation effects. Phys. Rev. **D87**, 084,006 (2013). https://doi.org/10.1103/PhysRevD.87.084006, arXiv:1212.4810 [gr-qc]
30. N.I. Shakura, R.A. Sunyaev, in *X- and Gamma-Ray Astronomy, IAU Symposium*, vol. 55, ed. by H. Bradt, R. Giacconi (1973), p.155
31. R. Fernández, A. Tchekhovskoy, E. Quataert, F. Foucart, D. Kasen, Long-term GRMHD simulations of neutron star merger accretion discs: implications for electromagnetic counterparts. Mon. Not. Roy. Astron. Soc. **482**(3), 3373–3393 (2019). https://doi.org/10.1093/mnras/sty2932, arXiv:1808.00461 [astro-ph.HE]
32. K. Hayashi, S. Fujibayashi, K. Kiuchi, K. Kyutoku, Y. Sekiguchi, M. Shibata, General-relativistic neutrino-radiation magnetohydrodynamic simulation of seconds-long black hole-neutron star mergers. Phys. Rev. D **106**(2), 023,008 (2022). https://doi.org/10.1103/PhysRevD.106.023008, arXiv:2111.04621 [astro-ph.HE]
33. R.D. Blandford, R.L. Znajek, Electromagnetic extraction of energy from Kerr black holes. Mon. Not. R. Astron. Soc. **179**, 433–456 (1977)
34. W. Lu, E. Quataert, Late-time accretion in neutron star mergers: implications for short gamma-ray bursts and kilonovae. Mon. Not. R. Astron. Soc. **522**(4), 5848–5861 (2023). https://doi.org/10.1093/mnras/stad1336, arXiv:2208.04293 [astro-ph.HE]
35. F. Özel, D. Psaltis, R. Narayan, J.E. McClintock, The black hole mass distribution in the galaxy. Astrophys J. **725**(2), 1918–1927 (2010). https://doi.org/10.1088/0004-637X/725/2/1918, arXiv:1006.2834 [astro-ph.GA]

36. W.M. Farr, N. Sravan, A. Cantrell, L. Kreidberg, C.D. Bailyn, I. Mandel, V. Kalogera, The mass distribution of stellar-mass black holes. Astrophys. J. **741**(2), 103 (2011). https://doi.org/10.1088/0004-637X/741/2/103, arXiv:1011.1459 [astro-ph.GA]
37. R. Abbott et al., Population properties of compact objects from the second LIGO-virgo gravitational-wave transient catalog. Astrophys. J. Lett. **913**(1), L7 (2021). https://doi.org/10.3847/2041-8213/abe949, arXiv:2010.14533 [astro-ph.HE]
38. T. Hinderer et al., Distinguishing the nature of comparable-mass neutron star binary systems with multimessenger observations: GW170817 case study. Phys. Rev. D **100**(6), 06,321 (2019). https://doi.org/10.1103/PhysRevD.100.063021, arXiv:1808.03836 [astro-ph.HE]
39. F. Özel, P. Freire, masses, radii, and the equation of state of neutron stars. Ann. Rev. Astron. Astrophys. **54**, 401–440 (2016). https://doi.org/10.1146/annurev-astro-081915-023322, arXiv:1603.02698 [astro-ph.HE]
40. B.P. Abbott et al., GW190425: observation of a compact binary coalescence with total mass $\sim 3.4 M_{\odot}$. Astrophys. J. Lett. **892**(1), L3 (2020). https://doi.org/10.3847/2041-8213/ab75f5, arXiv:2001.01761 [astro-ph.HE]
41. C.S. Kochanek, Coalescing binary neutron stars. Astrophys. J. **398**, 234 (1992). https://doi.org/10.1086/171851
42. L. Bildsten, C. Cutler, Tidal interactions of inspiraling compact binaries. Astrophys. J. **400**, 175 (1992). https://doi.org/10.1086/171983
43. W.E. East, V. Paschalidis, F. Pretorius, eccentric mergers of black holes with spinning neutron stars. Astrophys. J. Lett. **807**(1), L3 (2015). https://doi.org/10.1088/2041-8205/807/1/L3, arXiv:1503.07171 [astro-ph.HE]
44. M. Ruiz, V. Paschalidis, A. Tsokaros, S.L. Shapiro, Black hole-neutron star coalescence: effects of the neutron star spin on jet launching and dynamical ejecta mass. Phys. Rev. D **102**(12), 124,077 (2020). https://doi.org/10.1103/PhysRevD.102.124077, arXiv:2011.08863 [astro-ph.HE]
45. F. Foucart, T. Hinderer, S. Nissanke, Remnant baryon mass in neutron star-black hole mergers: predictions for binary neutron star mimickers and rapidly spinning black holes. Phys. Rev. D **98**(8), 081,501 (2018). https://doi.org/10.1103/PhysRevD.98.081501, arXiv:1807.00011 [astro-ph.HE]
46. K. Kawaguchi, K. Kyutoku, M. Shibata, M. Tanaka, Models of Kilonova/macronova emission from black hole–neutron star mergers. Astrophys. J. **825**(1), 52 (2016). https://doi.org/10.3847/0004-637X/825/1/52, arXiv:1601.07711 [astro-ph.HE]
47. F. Foucart, Black hole-neutron star mergers: disk mass predictions. Phys. Rev. D **86**, 124,007 (2012). https://doi.org/10.1103/PhysRevD.86.124007, arXiv:1207.6304 [astro-ph.HE]
48. C.J. Krüger, F. Foucart, Estimates for disk and ejecta masses produced in compact binary mergers. Phys. Rev. D **101**(10), 103,002 (2020). https://doi.org/10.1103/PhysRevD.101.103002, arXiv:2002.07728 [astro-ph.HE]
49. K. Hayashi, K. Kawaguchi, K. Kiuchi, K. Kyutoku, M. Shibata, Properties of the remnant disk and the dynamical ejecta produced in low-mass black hole-neutron star mergers. Phys. Rev. D **103**(4), 043,007 (2021). https://doi.org/10.1103/PhysRevD.103.043007, arXiv:2010.02563 [astro-ph.HE]
50. F. Pannarale, Black hole remnant of black hole-neutron star coalescing binaries. Phys. Rev. D **88**(10), 104,025 (2013). https://doi.org/10.1103/PhysRevD.88.104025, arXiv:1208.5869 [gr-qc]
51. F. Pannarale, Black hole remnant of black hole-neutron star coalescing binaries with arbitrary black hole spin. Phys. Rev. D **89**(4), 044,045 (2014). https://doi.org/10.1103/PhysRevD.89.044045. arXiv:1311.5931 [gr-qc]
52. K. Kyutoku, H. Okawa, M. Shibata, K. Taniguchi, Gravitational waves from spinning black hole-neutron star binaries: dependence on black hole spins and on neutron star equations of state. Phys. Rev. D **84**, 064,018 (2011). https://doi.org/10.1103/PhysRevD.84.064018, arXiv:1108.1189 [astro-ph.HE]
53. F. Zappa, S. Bernuzzi, F. Pannarale, M. Mapelli, N. Giacobbo, Black-hole remnants from black-hole–neutron-star mergers. Phys. Rev. Lett. **123**(4), 041,102 (2019). https://doi.org/10.1103/PhysRevLett.123.041102. arXiv:1903.11622 [gr-qc]

54. F. Pannarale, E. Berti, K. Kyutoku, B.D. Lackey, M. Shibata, Aligned spin neutron star-black hole mergers: a gravitational waveform amplitude model. Phys. Rev. D **92**(8), 084,050 (2015). https://doi.org/10.1103/PhysRevD.92.084050, arXiv:1509.00512 [gr-qc]
55. F. Pannarale, E. Berti, K. Kyutoku, B.D. Lackey, M. Shibata, Gravitational-wave cutoff frequencies of tidally disruptive neutron star-black hole binary mergers. Phys. Rev. D **92**(8), 081,504 (2015). https://doi.org/10.1103/PhysRevD.92.081504, arXiv:1509.06209 [gr-qc]
56. M. Shibata, K. Uryu, Simulation of merging binary neutron stars in full general relativity: Gamma = two case. Phys. Rev. **D61**, 064,001 (2000). https://doi.org/10.1103/PhysRevD.61.064001, arXiv:gr-qc/9911058 [gr-qc]
57. W. Kluzniak, W.H. Lee, Simulations of binary coalescence of a neutron star and a black hole. Astrophys. J. **494**, L53 (1998). https://doi.org/10.1086/311151, arXiv:astro-ph/9712019 [astro-ph]
58. W.H. Lee, Newtonian hydrodynamics of the coalescence of black holes with neutron stars. 3. Irrotational binaries with a stiff equation of state. Mon. Not. Roy. Astron. Soc. **318**, 606 (2000). https://doi.org/10.1046/j.1365-8711.2000.03870.x, arXiv:astro-ph/0007206 [astro-ph]
59. W.H. Lee, Newtonian hydrodynamics of the coalescence of black holes with neutron stars IV: irrotational binaries with a soft equation of state. Mon. Not. R. Astron. Soc. **328**, 583 (2001). https://doi.org/10.1046/j.1365-8711.2001.04898.x, arXiv:astro-ph/0108236
60. H.T. Janka, T. Eberl, M. Ruffert, C.L. Fryer, Black hole: neutron star mergers as central engines of gamma-ray bursts. Astrophys. J. **527**, L39 (1999). https://doi.org/10.1086/312397, arXiv:astro-ph/9908290 [astro-ph]
61. S. Rosswog, R. Speith, G.A. Wynn, Accretion dynamics in neutron star black hole binaries. Mon. Not. R. Astron. Soc. **351**, 1121 (2004). https://doi.org/10.1111/j.1365-2966.2004.07865.x, arXiv:astro-ph/0403500 [astro-ph]
62. S. Rosswog, On the viability of neutron star black hole binaries as central engines of gamma-ray bursts (2005). arXiv:astro-ph/0505007 [astro-ph]
63. M. Ruffert, H.T. Janka, Polytropic neutron star—black hole merger simulations with a Paczyński-Wiita potential. Astron. Astrophys. **514**, A66 (2010). https://doi.org/10.1051/0004-6361/200912738, arXiv:0906.3998 [astro-ph.HE]
64. B. Paczyńsky, P.J. Wiita, Thick accretion disks and supercritical luminosities. Astron. Astrophys. **88**, 23 (1980)
65. I.V. Artemova, G. Bjoernsson, I.D. Novikov, Modified newtonian potentials for the description of relativistic effects in accretion disks around black holes. Astrophys. J. **461**, 565 (1996). https://doi.org/10.1086/177084
66. E. Tejeda, S. Rosswog, An accurate Newtonian description of particle motion around a Schwarzschild black hole. Mon. Not. Roy. Astron. Soc. **433**, 1930 (2013). https://doi.org/10.1093/mnras/stt853. arXiv:1303.4068 [astro-ph.HE]
67. T.W. Baumgarte, S.L. Shapiro, *Numerical Relativity: Solving Einstein's Equations on the Computer* (2010)
68. H.P. Pfeiffer, D.A. Brown, L.E. Kidder, L. Lindblom, G. Lovelace, M.A. Scheel, Reducing orbital eccentricity in binary black hole simulations. Class. Quantum Gravity **24**, S59–S82 (2007). https://doi.org/10.1088/0264-9381/24/12/S06, arXiv:gr-qc/0702106
69. M.A. Miller, General relativistic initial data for the binary black hole/neutron star system in quasicircular orbit (2001). arXiv:gr-qc/0106017
70. P. Grandclement, Accurate and realistic initial data for black hole-neutron star binaries. Phys. Rev. D **74**, 124,002 (2006). https://doi.org/10.1103/PhysRevD.74.124002. [Erratum: Phys.Rev.D 75, 129903 (2007)], arXiv:gr-qc/0609044
71. K. Taniguchi, T.W. Baumgarte, J.A. Faber, S.L. Shapiro, Quasiequilibrium sequences of black-hole-neutron-star binaries in general relativity. Phys. Rev. D **74**, 041,502 (2006). https://doi.org/10.1103/PhysRevD.74.041502, arXiv:gr-qc/0609053
72. W. Tichy, The initial value problem as it relates to numerical relativity. Rept. Prog. Phys. **80**(2), 026,901 (2017). https://doi.org/10.1088/1361-6633/80/2/026901, arXiv:1610.03805 [gr-qc]

73. M.D. Duez, Y. Zlochower, Numerical relativity of compact binaries in the 21st century. Rept. Prog. Phys. **82**(1), 016,902 (2019). https://doi.org/10.1088/1361-6633/aadb16, arXiv:1808.06011 [gr-qc]
74. F. Löffler, L. Rezzolla, M. Ansorg, Numerical evolutions of a black hole-neutron star system in full general relativity: head-on collision. Phys. Rev. D **74**, 104,018 (2006). https://doi.org/10.1103/PhysRevD.74.104018, https://link.aps.org/doi/10.1103/PhysRevD.74.104018
75. M. Shibata, K. Uryu, Merger of black hole-neutron star binaries in full general relativity. Class. Quantum Gravity **24**, S125–S138 (2007). https://doi.org/10.1088/0264-9381/24/12/S09, arXiv:astro-ph/0611522 [astro-ph]
76. Z.B. Etienne, J.A. Faber, Y.T. Liu, S.L. Shapiro, K. Taniguchi, T.W. Baumgarte, Fully general relativistic simulations of black hole-neutron star mergers. Phys. Rev. **D77**, 084,002 (2008). https://doi.org/10.1103/PhysRevD.77.084002, arXiv:0712.2460 [astro-ph]
77. M.D. Duez, F. Foucart, L.E. Kidder, H.P. Pfeiffer, M.A. Scheel, S.A. Teukolsky, Evolving black hole-neutron star binaries in general relativity using pseudospectral and finite difference methods. Phys. Rev. **D78**, 104,015 (2008). https://doi.org/10.1103/PhysRevD.78.104015, arXiv:0809.0002 [gr-qc]
78. S. Chawla, M. Anderson, M. Besselman, L. Lehner, S.L. Liebling, P.M. Motl, D. Neilsen, Mergers of magnetized neutron stars with spinning black holes: disruption, accretion and fallback. Phys. Rev. Lett. **105**, 111,101 (2010). https://doi.org/10.1103/PhysRevLett.105.111101, arXiv:1006.2839 [gr-qc]
79. J.S. Read, B.D. Lackey, B.J. Owen, J.L. Friedman, Constraints on a phenomenologically parameterized neutron-star equation of state. Phys. Rev. **D79**, 124,032 (2009). https://doi.org/10.1103/PhysRevD.79.124032, arXiv:0812.2163 [astro-ph]
80. M.F. O'Boyle, C. Markakis, N. Stergioulas, J.S. Read, parametrized equation of state for neutron star matter with continuous sound speed. Phys. Rev. D **102**(8), 083,027 (2020). https://doi.org/10.1103/PhysRevD.102.083027, arXiv:2008.03342 [astro-ph.HE]
81. L. Lindblom, Spectral representations of neutron-star equations of state. Phys. Rev. D **82**, 103,011 (2010). https://doi.org/10.1103/PhysRevD.82.103011, arXiv:1009.0738 [astro-ph.HE]
82. F. Foucart, M.D. Duez, A. Gudinas, F. Hebert, L.E. Kidder, H.P. Pfeiffer, M.A. Scheel, Smooth equations of state for high-accuracy simulations of neutron star binaries. Phys. Rev. D **100**(10), 104,048 (2019). https://doi.org/10.1103/PhysRevD.100.104048, arXiv:1908.05277 [gr-qc]
83. B.D. Metzger, G. Martínez-Pinedo, S. Darbha, E. Quataert, A. Arcones, D. Kasen, R. Thomas, P. Nugent, I.V. Panov, N.T. Zinner, Electromagnetic counterparts of compact object mergers powered by the radioactive decay of r-process nuclei. Mon. Not. R. Astron. Soc. **406**(4), 2650–2662 (2010). https://doi.org/10.1111/j.1365-2966.2010.16864.x, arXiv:1001.5029 [astro-ph.HE]
84. S. Rosswog, O. Korobkin, A. Arcones, F.K. Thielemann, T. Piran, The long-term evolution of neutron star merger remnants – I. The impact of r-process nucleosynthesis. Mon. Not. R. Astron. Soc. **439**(1), 744–756 (2014). https://doi.org/10.1093/mnras/stt2502, arXiv:1307.2939 [astro-ph.HE]
85. D. Desai, B.D. Metzger, F. Foucart, Imprints of r-process heating on fall-back accretion: distinguishing black hole–neutron star from double neutron star mergers. Mon. Not. R. Astron. Soc. **485**(3), 4404–4412 (2019). https://doi.org/10.1093/mnras/stz644, arXiv:1812.04641 [astro-ph.HE]
86. F. Magistrelli, S. Bernuzzi, A. Perego, D. Radice, Elements formation in radiation-hydrodynamics simulations of kilonovae (2024). arXiv:2403.13883 [astro-ph.HE]
87. M.B. Deaton, M.D. Duez, F. Foucart, E. O'Connor, C.D. Ott, L.E. Kidder, C.D. Muhlberger, M.A. Scheel, B. Szilagyi, Black hole-neutron star mergers with a hot nuclear equation of state: outflow and neutrino-cooled disk for a low-mass. High-spin case. Astrophys. J. **776**, 47 (2013). https://doi.org/10.1088/0004-637X/776/1/47, arXiv:1304.3384 [astro-ph.HE]
88. W. Brege, M.D. Duez, F. Foucart, M.B. Deaton, J. Caro, D.A. Hemberger, L.E. Kidder, E. O'Connor, H.P. Pfeiffer, M.A. Scheel, Black hole-neutron star mergers using a survey of finite-temperature equations of state. Phys. Rev. D **98**(6), 063,009 (2018). https://doi.org/10.1103/PhysRevD.98.063009, arXiv:1804.09823 [gr-qc]

89. K.S. Thorne, Relativistic radiative transfer: moment formalisms. Mon. Not. R. Astron. Soc. **194**(2), 439–473 (1981). https://doi.org/10.1093/mnras/194.2.439
90. M. Shibata, K. Kiuchi, Y.i. Sekiguchi, Y. Suwa, Truncated moment formalism for radiation hydrodynamics in numerical relativity. Prog. Theor. Phys. **125**, 1255–1287 (2011). https://doi.org/10.1143/PTP.125.1255, arXiv:1104.3937 [astro-ph.HE]
91. F. Foucart, E. O'Connor, L. Roberts, M.D. Duez, R. Haas, L.E. Kidder, C.D. Ott, H.P. Pfeiffer, M.A. Scheel, B. Szilagyi, Post-merger evolution of a neutron star-black hole binary with neutrino transport. Phys. Rev. D **91**(12), 124,021 (2015). https://doi.org/10.1103/PhysRevD.91.124021, arXiv:1502.04146 [astro-ph.HE]
92. K. Kyutoku, K. Kiuchi, Y. Sekiguchi, M. Shibata, K. Taniguchi, Neutrino transport in black hole-neutron star binaries: neutrino emission and dynamical mass ejection. Phys. Rev. D **97**(2), 023,009 (2018). https://doi.org/10.1103/PhysRevD.97.023009, arXiv:1710.00827 [astro-ph.HE]
93. J.M. Miller, B.R. Ryan, J.C. Dolence, *νbhlight*: radiation GRMHD for neutrino-driven accretion flows. Astrophys. J. Suppl. **241**(2), 30 (2019). https://doi.org/10.3847/1538-4365/ab09fc, arXiv:1903.09273 [astro-ph.IM]
94. F. Foucart, M.D. Duez, F. Hebert, L.E. Kidder, P. Kovarik, H.P. Pfeiffer, M.A. Scheel, Implementation of Monte Carlo transport in the general relativistic SpEC code. Astrophys. J. **920**(2), 82 (2021). https://doi.org/10.3847/1538-4357/ac1737, arXiv:2103.16588 [astro-ph.HE]
95. J.M. Miller, B.R. Ryan, J.C. Dolence, A. Burrows, C.J. Fontes, C.L. Fryer, O. Korobkin, J. Lippuner, M.R. Mumpower, R.T. Wollaeger, Full transport model of GW170817-like disk produces a blue kilonova. Phys. Rev. D **100**(2), 023,008 (2019). https://doi.org/10.1103/PhysRevD.100.023008, arXiv:1905.07477 [astro-ph.HE]
96. Y. Masada, T. Sano, axisymmetric magnetorotational instability in viscous accretion disks. Astrophys. J. **689**, 1234 (2008). https://doi.org/10.1086/592601, arXiv:0808.2338 [astro-ph]
97. J. Guilet, E. Mueller, H.T. Janka, Neutrino viscosity and drag: impact on the magnetorotational instability in protoneutron stars. Mon. Not. R. Astron. Soc. **447**, 3992 (2015). https://doi.org/10.1093/mnras/stu2550, arXiv:1410.1874 [astro-ph.HE]
98. K. Kiuchi, Y. Sekiguchi, K. Kyutoku, M. Shibata, K. Taniguchi, T. Wada, High resolution magnetohydrodynamic simulation of black hole-neutron star merger: mass ejection and short gamma ray bursts. Phys. Rev. D **92**(6), 064,034 (2015). https://doi.org/10.1103/PhysRevD.92.064034, arXiv:1506.06811 [astro-ph.HE]
99. A. Malkus, G.C. McLaughlin, R. Surman, Symmetric and standard matter-neutrino resonances above merging compact objects. Phys. Rev. D **93**(4), 045,021 (2016). https://doi.org/10.1103/PhysRevD.93.045021, arXiv:1507.00946 [hep-ph]
100. X. Li, D.M. Siegel, Neutrino fast flavor conversions in neutron-star postmerger accretion disks. Phys. Rev. Lett. **126**(25), 251,101 (2021). https://doi.org/10.1103/PhysRevLett.126.251101, arXiv:2103.02616 [astro-ph.HE]
101. F. Foucart, Neutrino transport in general relativistic neutron star merger simulations. Living Rev. Comput. Astrophys. **9**(1), 1 (2023). https://doi.org/10.1007/s41115-023-00016-y, (arXiv:2209.02538 [astro-ph.HE])
102. E.E. Flanagan, T. Hinderer, Constraining neutron star tidal Love numbers with gravitational wave detectors. Phys. Rev. D **77**, 021,502 (2008). https://doi.org/10.1103/PhysRevD.77.021502, arXiv:0709.1915 [astro-ph]
103. M. Shibata, K. Kyutoku, T. Yamamoto, K. Taniguchi, Gravitational waves from black hole-neutron star binaries I: classification of waveforms. Phys. Rev. D **79**, 044,030 (2009). https://doi.org/10.1103/PhysRevD.79.044030. [Erratum: Phys.Rev.D 85, 127502 (2012)], arXiv:0902.0416 [gr-qc]
104. F. Foucart, L. Buchman, M.D. Duez, M. Grudich, L.E. Kidder, I. MacDonald, A. Mroue, H.P. Pfeiffer, M.A. Scheel, B. Szilagyi, First direct comparison of nondisrupting neutron star-black hole and binary black hole merger simulations. Phys. Rev. D **88**(6), 064,017 (2013). https://doi.org/10.1103/PhysRevD.88.064017, arXiv:1307.7685 [gr-qc]
105. K. Kyutoku, M. Shibata, K. Taniguchi, Gravitational waves from nonspinning black hole-neutron star binaries: dependence on equations of state. Phys. Rev. D **82**, 044,049 (2010).

https://doi.org/10.1103/PhysRevD.82.044049. [Erratum: Phys.Rev.D 84, 049902 (2011)], arXiv:1008.1460 [astro-ph.HE]
106. B.D. Lackey, K. Kyutoku, M. Shibata, P.R. Brady, J.L. Friedman, Extracting equation of state parameters from black hole-neutron star mergers: aligned-spin black holes and a preliminary waveform model. Phys. Rev. D **89**(4), 043,009 (2014). https://doi.org/10.1103/PhysRevD.89.043009, arXiv:1303.6298 [gr-qc]
107. J.E. Thompson, E. Fauchon-Jones, S. Khan, E. Nitoglia, F. Pannarale, T. Dietrich, M. Hannam, Modeling the gravitational wave signature of neutron star black hole coalescences. Phys. Rev. D **101**(12), 124,059 (2020). https://doi.org/10.1103/PhysRevD.101.124059, arXiv:2002.08383 [gr-qc]
108. A. Matas et al., Aligned-spin neutron-star–black-hole waveform model based on the effective-one-body approach and numerical-relativity simulations. Phys. Rev. D **102**(4), 043,023 (2020). https://doi.org/10.1103/PhysRevD.102.043023, arXiv:2004.10001 [gr-qc]
109. A. Gonzalez, R. Gamba, M. Breschi, F. Zappa, G. Carullo, S. Bernuzzi, A. Nagar, Numerical-relativity-informed effective-one-body model for black-hole–neutron-star mergers with higher modes and spin precession. Phys. Rev. D **107**(8), 084,026 (2023). https://doi.org/10.1103/PhysRevD.107.084026. arXiv:2212.03909 [gr-qc]
110. K. Chakravarti et al., Systematic effects from black hole-neutron star waveform model uncertainties on the neutron star equation of state. Phys. Rev. D **99**(2), 024,049 (2019). https://doi.org/10.1103/PhysRevD.99.024049, arXiv:1809.04349 [gr-qc]
111. Y. Huang, C.J. Haster, S. Vitale, V. Varma, F. Foucart, S. Biscoveanu, Statistical and systematic uncertainties in extracting the source properties of neutron star - black hole binaries with gravitational waves. Phys. Rev. D **103**(8), 083,001 (2021). https://doi.org/10.1103/PhysRevD.103.083001, arXiv:2005.11850 [gr-qc]
112. P. Goldreich, W.H. Julian, Pulsar electrodynamics. Astrophys. J. **157**, 869 (1969). https://doi.org/10.1086/150119
113. V. Paschalidis, Z.B. Etienne, S.L. Shapiro, General relativistic simulations of binary black hole-neutron stars: precursor electromagnetic signals. Phys. Rev. **D88**(2), 021,504 (2013). https://doi.org/10.1103/PhysRevD.88.021504, arXiv:1304.1805 [astro-ph.HE]
114. S.T. McWilliams, J. Levin, Electromagnetic extraction of energy from black hole-neutron star binaries. Astrophys. J. **742**, 90 (2011). https://doi.org/10.1088/0004-637X/742/2/90, arXiv:1101.1969 [astro-ph.HE]
115. V. Paschalidis, S.L. Shapiro, A new scheme for matching general relativistic ideal magnetohydrodynamics to its force-free limit. Phys. Rev. D **88**(10), 104,031 (2013). https://doi.org/10.1103/PhysRevD.88.104031, arXiv:1310.3274 [astro-ph.HE]
116. W.E. East, L. Lehner, S.L. Liebling, C. Palenzuela, Multimessenger signals from black hole-neutron star mergers without significant tidal disruption. Astrophys. J. Lett. **912**(1), L18 (2021). https://doi.org/10.3847/2041-8213/abf566, arXiv:2101.12214 [astro-ph.HE]
117. E.R. Most, A.A. Philippov, Electromagnetic precursors to black hole-neutron star gravitational wave events: flares and reconnection-powered fast radio transients from the late inspiral. Astrophys. J. Lett. **956**(2), L33 (2023). https://doi.org/10.3847/2041-8213/acfdae, arXiv:2309.04271 [astro-ph.HE]
118. C. Palenzuela, Modeling magnetized neutron stars using resistive MHD. Mon. Not. R. Astron. Soc. **431**, 1853–1865 (2013). https://doi.org/10.1093/mnras/stt311, arXiv:1212.0130 [astro-ph.HE]
119. E.R. Most, A.A. Philippov, Electromagnetic precursor flares from the late inspiral of neutron star binaries. Mon. Not. R. Astron. Soc. **515**(2), 2710–2724 (2022). https://doi.org/10.1093/mnras/stac1909, arXiv:2205.09643 [astro-ph.HE]
120. M.D. Duez, F. Foucart, L.E. Kidder, C.D. Ott, S.A. Teukolsky, Equation of state effects in black hole-neutron star mergers. Class. Quantum Gravity **27**, 114,106 (2010). https://doi.org/10.1088/0264-9381/27/11/114106, arXiv:0912.3528 [astro-ph.HE]
121. W.E. East, F. Pretorius, B.C. Stephens, Eccentric black hole-neutron star mergers: effects of black hole spin and equation of state. Phys. Rev. **D85**, 124,009 (2012). https://doi.org/10.1103/PhysRevD.85.124009, arXiv:1111.3055 [astro-ph.HE]

122. Z.B. Etienne, Y.T. Liu, S.L. Shapiro, T.W. Baumgarte, General relativistic simulations of black-hole-neutron-star mergers: Effects of black-hole spin. Phys. Rev. **D79**, 044,024 (2009). https://doi.org/10.1103/PhysRevD.79.044024, arXiv:0812.2245 [astro-ph]
123. G. Lovelace, M.D. Duez, F. Foucart, L.E. Kidder, H.P. Pfeiffer, M.A. Scheel, B. Szilágyi, Massive disc formation in the tidal disruption of a neutron star by a nearly extremal black hole. Class. Quantum Gravity **30**, 135,004 (2013). https://doi.org/10.1088/0264-9381/30/13/135004, arXiv:1302.6297 [gr-qc]
124. F. Foucart, M.D. Duez, L.E. Kidder, M.A. Scheel, B. Szilagyi, S.A. Teukolsky, Black hole-neutron star mergers for 10 solar mass black holes. Phys. Rev. D **85**, 044,015 (2012). https://doi.org/10.1103/PhysRevD.85.044015, arXiv:1111.1677 [gr-qc]
125. K. Kyutoku, K. Ioka, H. Okawa, M. Shibata, K. Taniguchi, Dynamical mass ejection from black hole-neutron star binaries. Phys. Rev. D **92**, 044,028 (2015). https://doi.org/10.1103/PhysRevD.92.044028, arXiv:1502.05402 [astro-ph.HE]
126. F. Foucart, M.D. Duez, L.E. Kidder, S.A. Teukolsky, Black hole-neutron star mergers: effects of the orientation of the black hole spin. Phys. Rev. **D83**, 024,005 (2011). https://doi.org/10.1103/PhysRevD.83.024005, arXiv:1007.4203 [astro-ph.HE]
127. K. Kawaguchi, K. Kyutoku, H. Nakano, H. Okawa, M. Shibata, K. Taniguchi, Black hole-neutron star binary merger: Dependence on black hole spin orientation and equation of state. Phys. Rev. **D92**(2), 024,014 (2015). https://doi.org/10.1103/PhysRevD.92.024014, arXiv:1506.05473 [astro-ph.HE]
128. N. Stone, A. Loeb, E. Berger, Pulsations in short GRBs from black hole-neutron star mergers. Phys. Rev. D **87**, 084,053 (2013). https://doi.org/10.1103/PhysRevD.87.084053, arXiv:1209.4097 [astro-ph.HE]
129. S. Rosswog, Fallback accretion in the aftermath of a compact binary merger. Mon. Not. R. Astron. Soc. **376**, L48–L51 (2007). https://doi.org/10.1111/j.1745-3933.2007.00284.x, arXiv:astro-ph/0611440
130. M.J. Rees, Tidal disruption of stars by black holes of 10^6-10^8 solar masses in nearby galaxies. Nature **333**(6173), 523–528 (1988). https://doi.org/10.1038/333523a0
131. E.S. Phinney, in *The Center of the Galaxy*, vol. 136, ed. by M. Morris (1989), p. 543
132. G. Lodato, A.R. King, J.E. Pringle, Stellar disruption by a supermassive black hole: is the light curve really proportional to $t^{-5/3}$? Mon. Not. R. Astron. Soc. **392**, 332 (2009). https://doi.org/10.1111/j.1365-2966.2008.14049.x, arXiv:0810.1288 [astro-ph]
133. R. Fernández, E. Quataert, J. Schwab, D. Kasen, S. Rosswog, The interplay of disc wind and dynamical ejecta in the aftermath of neutron star–black hole mergers. Mon. Not. R. Astron. Soc. **449**(1), 390–402 (2015). https://doi.org/10.1093/mnras/stv238, arXiv:1412.5588 [astro-ph.HE]
134. S. De, D.M. Siegel, Igniting weak interactions in neutron star postmerger accretion disks. Astrophys. J. **921**(1), 94 (2021). https://doi.org/10.3847/1538-4357/ac110b, arXiv:2011.07176 [astro-ph.HE]
135. A.M. Beloborodov, Nuclear composition of gamma-ray burst fireballs. Astrophys. J. **588**, 931–944 (2003). https://doi.org/10.1086/374217, arXiv:astro-ph/0210522
136. D.M. Siegel, B.D. Metzger, Three-dimensional GRMHD simulations of neutrino-cooled accretion disks from neutron star mergers. Astrophys. J. **858**(1), 52 (2018). https://doi.org/10.3847/1538-4357/aabaec, arXiv:1711.00868 [astro-ph.HE]
137. R. Narayan, I.V. Igumenshchev, M.A. Abramowicz, Magnetically arrested disk: an energetically efficient accretion flow. Publ. Astron. Soc. Japn. **55**, L69 (2003). https://doi.org/10.1093/pasj/55.6.L69, arXiv:astro-ph/0305029
138. A. Tchekhovskoy, R. Narayan, J.C. McKinney, Efficient generation of jets from magnetically arrested accretion on a rapidly spinning black hole. Mon. Not. R. Astron. Soc. **418**(1), L79–L83 (2011). https://doi.org/10.1111/j.1745-3933.2011.01147.x, arXiv:1108.0412 [astro-ph.HE]
139. M.R. Izquierdo, M. Bezares, S. Liebling, C. Palenzuela, Large Eddy Simulations of Magnetized Mergers of Black Holes and Neutron Stars (2024). arXiv:2403.09770 [astro-ph.HE]
140. S.A. Balbus, J.F. Hawley, Instability, turbulence, and enhanced transport in accretion disks. Rev. Modern Phys. **70**(1), 1–53 (1998). https://doi.org/10.1103/RevModPhys.70.1

141. Z.B. Etienne, Y.T. Liu, V. Paschalidis, S.L. Shapiro, General relativistic simulations of black hole-neutron star mergers: effects of magnetic fields. Phys. Rev. **D85**, 064,029 (2012). https://doi.org/10.1103/PhysRevD.85.064029, arXiv:1112.0568 [astro-ph.HE]
142. Z.B. Etienne, V. Paschalidis, S.L. Shapiro, General-relativistic simulations of black-hole-neutron-star mergers: effects of tilted magnetic fields. Phys. Rev. D. **86**(8), 084026 (2012). https://doi.org/10.1103/PhysRevD.86.084026, arXiv:1209.1632 [astro-ph.HE]
143. E.R. Most, L.J. Papenfort, S.D. Tootle, L. Rezzolla, On accretion discs formed in MHD simulations of black hole–neutron star mergers with accurate microphysics. Mon. Not. R. Astron. Soc. **506**(3), 3511–3526 (2021). https://doi.org/10.1093/mnras/stab1824, arXiv:2106.06391 [astro-ph.HE]
144. K. Beckwith, J.F. Hawley, J.H. Krolik, The influence of magnetic field geometry on the evolution of black hole accretion flows: similar disks. Drastically Different Jets. Astrophys. J. **678**, 1180 (2008). https://doi.org/10.1086/533492, arXiv:0709.3833 [astro-ph]
145. V. Paschalidis, M. Ruiz, S.L. Shapiro, Relativistic Simulations of Black Hole–neutron Star Coalescence: the jet Emerges. Astrophys. J. Lett. **806**(1), L14 (2015). https://doi.org/10.1088/2041-8205/806/1/L14, arXiv:1410.7392 [astro-ph.HE]
146. M. Ruiz, S.L. Shapiro, A. Tsokaros, Jet launching from binary black hole-neutron star mergers: Dependence on black hole spin, binary mass ratio and magnetic field orientation. Phys. Rev. D **98**(12), 123,017 (2018). https://doi.org/10.1103/PhysRevD.98.123017, arXiv:1810.08618 [astro-ph.HE]
147. O. Gottlieb et al., Large-scale evolution of seconds-long relativistic jets from black hole-neutron star mergers. Astrophys. J. Lett. **954**(1), L21 (2023). https://doi.org/10.3847/2041-8213/aceeff, arXiv:2306.14947 [astro-ph.HE]
148. H.K. Moffatt, *Magnetic Field Generation in Electrically Conducting Fluids* (1978)
149. J.D. Hogg, C. Reynolds, The influence of accretion disk thickness on the large-scale magnetic dynamo. Astrophys. J. **861**(1), 24 (2018). https://doi.org/10.3847/1538-4357/aac439, arXiv:1805.05372 [astro-ph.HE]
150. M.T.P. Liska, A. Tchekhovskoy, E. Quataert, Large-scale poloidal magnetic field dynamo leads to powerful jets in GRMHD simulations of black hole accretion with toroidal field. Mon. Not. R. Astron. Soc. **494**(3), 3656–3662 (2020). https://doi.org/10.1093/mnras/staa955, arXiv:1809.04608 [astro-ph.HE]
151. I.M. Christie, A. Lalakos, A. Tchekhovskoy, R. Fernández, F. Foucart, E. Quataert, D. Kasen, The role of magnetic field geometry in the evolution of neutron star merger accretion discs. Mon. Not. R. Astron. Soc. **490**(4), 4811–4825 (2019). https://doi.org/10.1093/mnras/stz2552, arXiv:1907.02079 [astro-ph.HE]
152. K. Hayashi, K. Kiuchi, K. Kyutoku, Y. Sekiguchi, M. Shibata, General-relativistic neutrino-radiation magnetohydrodynamics simulation of seconds-long black hole-neutron star mergers: Dependence on the initial magnetic field strength, configuration, and neutron-star equation of state. Phys. Rev. D **107**(12), 123,001 (2023). https://doi.org/10.1103/PhysRevD.107.123001, arXiv:2211.07158 [astro-ph.HE]
153. O. Gottlieb, B.D. Metzger, E. Quataert, D. Issa, T. Martineau, F. Foucart, M.D. Duez, L.E. Kidder, H.P. Pfeiffer, M.A. Scheel, A unified picture of short and long gamma-ray bursts from compact binary mergers. Astrophys. J. Lett. **958**(2), L33 (2023). https://doi.org/10.3847/2041-8213/ad096e, arXiv:2309.00038 [astro-ph.HE]
154. F.H. Nouri, et al., Evolution of the magnetized, neutrino-cooled accretion disk in the aftermath of a black hole neutron star binary merger. Phys. Rev. D **97**(8), 083,014 (2018). https://doi.org/10.1103/PhysRevD.97.083014, arXiv:1710.07423 [astro-ph.HE]
155. I. Zalamea, A.M. Beloborodov, Neutrino heating near hyper-accreting black holes. Mon. Not. R. Astron. Soc. **410**(4), 2302–2308 (2011). https://doi.org/10.1111/j.1365-2966.2010.17600.x, arXiv:1003.0710 [astro-ph.HE]
156. R. Popham, S.E. Woosley, C. Fryer, Hyperaccreting black holes and gamma-ray bursts. Astrophys. J. **518**, 356–374 (1999). https://doi.org/10.1086/307259, arXiv:astro-ph/9807028
157. O. Just, M. Obergaulinger, H.T. Janka, A. Bauswein, N. Schwarz, Neutron-star merger ejecta as obstacles to neutrino-powered jets of gamma-ray bursts. Astrophys. J. **816**(2), L30 (2016). https://doi.org/10.3847/2041-8205/816/2/L30, arXiv:1510.04288 [astro-ph.HE]

158. M.V. Barkov, A.S. Pozanenko, Model of extended emission of short Gamma-ray Bursts. Mon. Not. R. Astron. Soc. **417**, 2161 (2011). https://doi.org/10.1111/j.1365-2966.2011.19398.x, arXiv:1103.4246 [astro-ph.HE]
159. S. Fujibayashi, Y. Sekiguchi, K. Kiuchi, M. Shibata, Properties of neutrino-driven ejecta from the remnant of a binary neutron star merger: pure radiation hydrodynamics case. Astrophys. J. **846**(2), 114 (2017). https://doi.org/10.3847/1538-4357/aa8039, arXiv:1703.10191 [astro-ph.HE]
160. J.M. Lattimer, D.N. Schramm, The tidal disruption of neutron stars by black holes in close binaries. Astrophys. J. **210**, 549–567 (1976). https://doi.org/10.1086/154860
161. L.X. Li, B. Paczynski, Transient events from neutron star mergers. Astrophys. J. Lett. **507**, L59 (1998). https://doi.org/10.1086/311680, arXiv:astro-ph/9807272
162. J. Barnes, D. Kasen, Effect of a high opacity on the light curves of radioactively powered transients from compact object mergers. Astrophys. J. **775**, 18 (2013). https://doi.org/10.1088/0004-637X/775/1/18, arXiv:1303.5787 [astro-ph.HE]
163. D. Kasen, N.R. Badnell, J. Barnes, Opacities and spectra of the r-process ejecta from neutron star mergers. Astrophys. J. **774**, 25 (2013). https://doi.org/10.1088/0004-637X/774/1/25, arXiv:1303.5788 [astro-ph.HE]
164. D. Kasen, B. Metzger, J. Barnes, E. Quataert, E. Ramirez-Ruiz, Origin of the heavy elements in binary neutron-star mergers from a gravitational wave event. Nature **551**, 80 (2017). https://doi.org/10.1038/nature24453, arXiv:1710.05463 [astro-ph.HE]
165. O. Korobkin, S. Rosswog, A. Arcones, C. Winteler, On the astrophysical robustness of neutron star merger r-process. Mon. Not. Roy. Astron. Soc. **426**, 1940 (2012). https://doi.org/10.1111/j.1365-2966.2012.21859.x. arXiv:1206.2379 [astro-ph.SR]
166. S. Wanajo, Y. Sekiguchi, N. Nishimura, K. Kiuchi, K. Kyutoku, M. Shibata, Production of All the r-process nuclides in the dynamical ejecta of neutron star mergers. Astrophys. J. Lett. **789**(2), L39 (2014). https://doi.org/10.1088/2041-8205/789/2/L39, arXiv:1402.7317 [astro-ph.SR]
167. J. Lippuner, L.F. Roberts, r-process lanthanide production and heating rates in kilonovae. Astrophys. J. **815**(2), 82 (2015). https://doi.org/10.1088/0004-637X/815/2/82, arXiv:1508.03133 [astro-ph.HE]
168. M. Tanaka, D. Kato, G. Gaigalas, K. Kawaguchi, Systematic opacity calculations for kilonovae. Mon. Not. R. Astron. Soc. **496**(2), 1369–1392 (2020). https://doi.org/10.1093/mnras/staa1576, arXiv:1906.08914 [astro-ph.HE]
169. E. Nakar, T. Piran, Radio remnants of compact binary mergers—the electromagnetic signal that will follow the gravitational waves. Nature **478**, 82–84 (2011). https://doi.org/10.1038/nature10365, arXiv:1102.1020 [astro-ph.HE]
170. K. Hotokezaka, T. Piran, Mass ejection from neutron star mergers: different components and expected radio signals. Mon. Not. R. Astron. Soc. **450**, 1430–1440 (2015). https://doi.org/10.1093/mnras/stv620, arXiv:1501.01986 [astro-ph.HE]
171. K. Hotokezaka, S. Nissanke, G. Hallinan, T.J.W. Lazio, E. Nakar, T. Piran, Radio counterparts of compact binary mergers detectable in gravitational waves: a simulation for an optimized survey. Astrophys. J. **831**(2), 190 (2016). https://doi.org/10.3847/0004-637X/831/2/190, arXiv:1605.09395 [astro-ph.HE]
172. M. Haddadi, M.D. Duez, F. Foucart, T. Ramirez, R. Fernández, A.L. Knight, J. Jesse, F. Hébert, L.E. Kidder, H.P. Pfeiffer, M.A. Scheel, Late-time post-merger modeling of a compact binary: effects of relativity, r-process heating, and treatment of transport. Class. Quantum Gravity **40**(8), 085,008 (2023). https://doi.org/10.1088/1361-6382/acc0c6, arXiv:2208.02367 [gr-qc]
173. F. Foucart, P. Moesta, T. Ramirez, A.J. Wright, S. Darbha, D. Kasen, Estimating outflow masses and velocities in merger simulations: impact of r-process heating and neutrino cooling. Phys. Rev. D **104**(12), 123,010 (2021). https://doi.org/10.1103/PhysRevD.104.123010, arXiv:2109.00565ZZ [astro-ph.HE]
174. K. Kyutoku, K. Ioka, M. Shibata, Anisotropic mass ejection from black hole-neutron star binaries: Diversity of electromagnetic counterparts. Phys. Rev. **D88**(4), 041,503 (2013). https://doi.org/10.1103/PhysRevD.88.041503, arXiv:1305.6309 [astro-ph.HE]

175. S. Rosswog, Mergers of neutron star black hole binaries with small mass ratios: Nucleosynthesis, gamma-ray bursts and electromagnetic transients. Astrophys. J. **634**, 1202 (2005). https://doi.org/10.1086/497062, arXiv:astro-ph/0508138 [astro-ph]
176. L.F. Roberts, J. Lippuner, M.D. Duez, J.A. Faber, F. Foucart, J.C. Lombardi, S. Ning, C.D. Ott, M. Ponce, The influence of neutrinos on r-process nucleosynthesis in the ejecta of black hole–neutron star mergers. Mon. Not. R. Astron. Soc. **464**(4), 3907–3919 (2017). https://doi.org/10.1093/mnras/stw2622, arXiv:1601.07942 [astro-ph.HE]
177. R. Fernández, B.D. Metzger, Delayed outflows from black hole accretion tori following neutron star binary coalescence. Mon. Not. R. Astron. Soc. **435**, 502 (2013). https://doi.org/10.1093/mnras/stt1312, arXiv:1304.6720 [astro-ph.HE]
178. R. Fernández, D. Kasen, B.D. Metzger, E. Quataert, Outflows from accretion discs formed in neutron star mergers: effect of black hole spin. Mon. Not. R. Astron. Soc. **446**, 750–758 (2015). https://doi.org/10.1093/mnras/stu2112, arXiv:1409.4426 [astro-ph.HE]
179. R. Fernández, F. Foucart, D. Kasen, J. Lippuner, D. Desai, L.F. Roberts, Dynamics, nucleosynthesis, and kilonova signature of black hole—neutron star merger ejecta. Class. Quantum Gravity **34**(15), 154,001 (2017). https://doi.org/10.1088/1361-6382/aa7a77, arXiv:1612.04829 [astro-ph.HE]
180. R. Fernández, F. Foucart, J. Lippuner, The landscape of disc outflows from black hole–neutron star mergers. Mon. Not. R. Astron. Soc. **497**(3), 3221–3233 (2020). https://doi.org/10.1093/mnras/staa2209, arXiv:2005.14208 [astro-ph.HE]
181. S. Fujibayashi, M. Shibata, S. Wanajo, K. Kiuchi, K. Kyutoku, Y. Sekiguchi, Viscous evolution of a massive disk surrounding stellar-mass black holes in full general relativity. Phys. Rev. D **102**, 123,014 (2020). https://doi.org/10.1103/PhysRevD.102.123014, arXiv:2009.03895 [astro-ph.HE]
182. S. Fujibayashi, M. Shibata, S. Wanajo, K. Kiuchi, K. Kyutoku, Y. Sekiguchi, Mass ejection from disks surrounding a low-mass black hole: Viscous neutrino-radiation hydrodynamics simulation in full general relativity. Phys. Rev. D **101**, 083,029 (2020). https://doi.org/10.1103/PhysRevD.101.083029
183. M. Tanaka, K. Hotokezaka, K. Kyutoku, S. Wanajo, K. Kiuchi, Y. Sekiguchi, M. Shibata, Radioactively powered emission from black hole-neutron star mergers. Astrophys. J. **780**, 31 (2014). https://doi.org/10.1088/0004-637X/780/1/31, arXiv:1310.2774 [astro-ph.HE]
184. S. Darbha, D. Kasen, F. Foucart, D.J. Price, Electromagnetic signatures from the tidal tail of a black hole-neutron star merger. Astrophys. J. **915**(1), 69 (2021). https://doi.org/10.3847/1538-4357/abff5d, arXiv:2103.03378 [astro-ph.HE]
185. S. Darbha, D. Kasen, Inclination dependence of kilonova light curves from globally aspherical geometries. Astrophys. J. **897**(2), 150 (2020). https://doi.org/10.3847/1538-4357/ab9a34, arXiv:2002.00299 [astro-ph.HE]
186. K. Kawaguchi, M. Shibata, M. Tanaka, Diversity of kilonova light curves. Astrophys. J. **889**(2), 171 (2020). https://doi.org/10.3847/1538-4357/ab61f6, arXiv:1908.05815 [astro-ph.HE]

Chapter 18
Exotic Compact Objects: A Recent Numerical-Relativity Perspective

Miguel Bezares and Nicolas Sanchis-Gual

Abstract Beyond black holes and neutron stars, new hypothetical compact objects have been proposed as potential astrophysical entities. In general, their properties have not yet been fully explored or understood, nor has it been proven whether or not they exist in nature. They are the so-called *exotic compact objects*, theoretical equilibrium configurations in the strong regime of gravity that involve new exotic physical phenomena deeply related to fundamental questions of theoretical physics (e.g., the nature of dark matter, the formation of singularities, or the presence of horizons). Among these exotic objects, there are those that require the existence of new fields and particles beyond the Standard Model, such as boson stars and ultralight bosons; those that seek to describe the dense equation of state of neutron stars as an even more extreme state of matter made up of free quarks; or those that exhibit additional properties related to extensions of General Relativity or even quantum gravity. However, in order to move from theoretical objects to astrophysical objects, their dynamics and stability must first be assessed by performing numerical-relativity simulations under the premise that solutions that are unstable on dynamical timescales will never completely form, being, at most, a transient state that will not be able to play any astrophysical role. Furthermore, numerical simulations also make it possible to extract the gravitational radiation from relevant astrophysical scenarios, such as the collapse of exotic stars or binary mergers, which could then be compared with current and upcoming LIGO-Virgo-KAGRA gravitational-wave detections.

M. Bezares (✉)
Nottingham Centre of Gravity and School of Mathematical Sciences, University of Nottingham, University Park, Nottingham NG7 2RD, UK
e-mail: Miguel.Bezaresfigueroa@nottingham.ac.uk

N. Sanchis-Gual
Departamento de Astronomía y Astrofísica, Universitat de València, Dr. Moliner 50, 46100 Burjassot, Valencia, Spain
e-mail: nicolas.sanchis@uv.es

C. Bambi et al. (eds.), *New Frontiers in GRMHD Simulations*, Springer Series in Astrophysics and Cosmology, https://doi.org/10.1007/978-981-97-8522-3_18

18.1 Introduction

Multimessenger observations involving gravitational wave detections by the LIGO-Virgo-KAGRA collaboration [2, 5–7] together with the imaging of the shadow of the supermassive black holes at the centre of Andromeda M87 [12] and the Milky Way [13] galaxies by the Event Horizon Telescope are key to unveil the underlying theory of gravity and understand the true nature of the most compact objects in the Universe: black holes and neutron stars. Although the data so far are in excellent agreement with the Kerr hypothesis, which assumes that each astrophysical black hole in the Universe is described by the Kerr metric from General Relativity, it is worth asking if alternative, non-conventional models of compact objects [153], i.e., theoretical exotic compact objects that also involve strong gravitational fields and could mimic some key observable phenomenology of neutron stars and black holes [155, 207], also populate the Cosmos.

Exotic compact objects may be linked to relevant aspects in many areas of fundamental physics [75]:

- the nature of dark matter (bosonic stars, bosonic fields, Kerr black holes with bosonic hair);
- the equation of state of dense matter in the interior of neutron stars (anisotropic stars, strange stars, fermion-boson stars, scalarized neutron stars);
- new properties of black holes related to the presence (or not) of a horizon and singularities (fuzzballs, wormholes, gravastars, baby universes in $f(R)$ theories),
- modified theories of gravity (scalarized black holes, boson stars in $f(R)$ gravity).

While many compact objects have been proposed in the literature, in order to consider them as true viable astrophysical candidates beyond standard astrophysical objects, such models would have to meet a list of requirements [153]:

1. to be described within a sound and well-motivated physical theory (for instance, boson stars in the Einstein-Klein-Gordon system);
2. to be sufficiently stable against small perturbations and long-lived;
3. to have a dynamical formation mechanism.

These requisites, and in particular the last two, allow us to define the concept of dynamical robustness as a criterion for astrophysical viability. Therefore, any exotic compact object aiming to be astrophysically relevant should satisfy these benchmarks. Compact objects that become unstable too quickly or cannot be formed from non-fine-tuned initial conditions would not survive in our unrestful Universe.

Research on highly dynamic astrophysical systems in the strong field regime of gravity relies on numerical relativity [14, 140] to solve Einstein's field equations in a hyperbolic framework coupled to the matter field or fields under study. This means that numerical simulations can be performed to assess the stability and dynamical formation of exotic compact objects. For instance, rotating objects may form in binary mergers of non-spinning configurations. Moreover, once the viability of a particular exotic compact model has been established, such solutions can be studied in more relevant astrophysical scenarios in which gravitational-wave emission is present. Obtaining gravitational waveforms from exotic compact objects and comparing them with real gravitational wave events could lead to their discovery or to discarding them [66].

This chapter will be devoted to discussing the (in)stability and formation channel of different theoretical compact objects determined using numerical-relativity simulations and their associated gravitational wave signature, if any. In Sect. 18.2, we will present recent results on the stability of isolated single exotic objects in General Relativity and comment on the implications of the findings. In Sect. 18.3, we will discuss some examples of numerical simulations of compact objects in modified theories of gravity. Finally, in Sect. 18.4, we will examine the binary merger scenarios and the gravitational radiation emitted by these objects. Throughout this chapter, Roman letters from the beginning of the alphabet $a, b, c, \ldots$ denote space-time indices ranging from 0 to 3, while letters near the middle $i, j, k, \ldots$ range from 1 to 3, denoting spatial indices.

18.2 Exotic Matter in General Relativity

Dark matter and its properties are one of the fundamental open questions in astrophysics, cosmology, and particle physics. What massive particle(s) that mainly interacts gravitationally, if there is indeed any, is responsible for the multiple dark-matter gravitational effects we see in galaxies [227]? Although there are many candidates, such as WIMPS (Weakly Interacting Massive Particle), axions and axion-like particles, new fundamental bosons, and sterile neutrinos, there is no evidence of the existence of any of them. Here, we will focus on ultralight bosons with masses in the range $10^{-22} < m < 10^{-10}$ eV, as an example of a potential dark matter particle [22]. These particles could condense into a new type of compact object known as a bosonic star, which can be described as a self-gravitating Bose-Einstein condensate, and also grow through superradiance to form clouds around rotating black holes [61].

Besides dark matter, another open question in astrophysics is related to the interior of extremely dense neutron stars: What is the equation-of-state of dense matter? Does matter suffer a phase transition under such extreme conditions and is better described as made of quarks rather than neutrons? Furthermore, these exotic objects allow us, for example, to circumvent the Buchdahl limit and the existence of anisotropies in neutron stars. They also make possible the study of open problems in theoretical physics, such as the information-loss paradox and quantum description of the horizon.

In this section we will briefly present some compelling models of isolated compact objects described by exotic matter and discuss the numerical simulation that attempt to determine their stability and dynamical formation.

18.2.1 Boson and Proca Stars

A boson star is a collection of bosons in their ground state that composes a self-gravitating macroscopic Bose-Einstein condensate described by a complex scalar field. They are horizonless, stationary, everywhere regular, solitonic solutions of the Einstein-(complex, massive) Klein-Gordon system [169, 172, 228]. Its vector cousins are known as Proca stars and are solutions of the Einstein-(complex) Proca system [58, 159]. They are described by the Einstein-Klein-Gordon and Einstein-Proca actions (with units $c = 1 = \hbar$ and $4\pi G = 1$):

$$\mathcal{S} = \int d^4x\sqrt{-g}\left[\frac{R}{4} + \mathcal{L}_{(s)}\right], \tag{18.1}$$

with the two matter Lagrangians being:

$$\mathcal{L}_{(0)} = -g^{ab}\partial_a\Phi^*\partial_b\Phi - \mu^2\Phi^*\Phi - V_{\text{int}}(\Phi, \Phi^*)\,, \tag{18.2}$$

$$\mathcal{L}_{(1)} = -\frac{1}{4}\mathcal{F}^*_{ab}\mathcal{F}^{ab} - \frac{1}{2}\mu^2\mathcal{A}^*_a\mathcal{A}^a\,. \tag{18.3}$$

Φ is a complex scalar field, V_{int} is a high-order self-interaction potential, and $\mathcal{A}$ is a complex 4-potential, with field strength $\mathcal{F} = d\mathcal{A}$. The asterisk $*$ indicates complex conjugate and μ is the field's mass. The complex bosonic field has a harmonic dependence and oscillates in time with a well-defined frequency ω, while at the level of the stress-energy tensor, the configuration is time-independent, leading to a stationary spacetime. The scalar field ansatz for boson stars is given by:

$$\Phi(t, r, \theta, \phi) = \Phi_0(r, \theta)e^{i(m\phi - wt)}. \tag{18.4}$$

where m is the azimuthal winding number, which is an integer that can take both negative and positive values $m = 0, \pm1, \pm2$. As for fermion stars, such as white dwarfs and neutron stars, bosonic stars have a maximum mass as well, which separates the stable and unstable branches, but it can vary from one to millions of solar masses, depending on the bosonic particle mass μ [44]:

$$M_{\text{max}} = \mathcal{M}_{\text{model}}\frac{M^2_{\text{Planck}}}{\mu}, \tag{18.5}$$

where $\mathcal{M}_{\text{model}}$ is a parameter that depends on the specific bosonic star model (with $\mathcal{M}_{\text{model}} = 0.633$ for non-spinning scalar mini-boson stars[1]) and M_{Planck} is the Planck mass. This feature is what makes them especially captivating, both from an astrophysical and a particle physics point of view. A lighter boson leads to more massive stars, beautifully connecting different mass and length scales. Hence, there is a need for ultralight bosons, with masses below 10^{-10} eV to reach astrophysical mass ranges. In this regard, fundamental bosonic fields minimally coupled to gravity are widely used in cosmological models to explain the nature of dark matter [197], as a consequence of String Theory [22] or to extend the Standard Model of particle physics [133], with axions or dark photons.

Therefore, bosonic stars are exotic compact objects well motivated by particle physics and General Relativity (point 1). Nonetheless, these solutions could also be used as toy models of more involved exotic compact objects or different types of dark matter. The existence and discovery of the Higgs boson gave a fresh impetus to this research. However, as stated above, to truly contemplate them as potential astrophysical objects, numerical studies need to be undertaken to fully determine their stability and find the mechanisms of their dynamical formation. Within spherical symmetry, scalar boson star models with and without self-interaction terms in their potential have been extensively studied over the last few decades both analytically [139, 182, 241, 244] and numerically in dynamical time evolutions [186]. They have been shown to be indeed stable against (sufficiently small) linear perturbations and to form in highly dynamical scenarios through the so-called gravitational cooling mechanism [106, 245] (points 2 and 3). As an illustrative example, the evolution equations for a free scalar field (without self-interactions) are shown:

$$\partial_t \Phi = -\alpha \Pi + \mathcal{L}_\beta \Phi \ , \tag{18.6}$$

$$\partial_t \Pi = \alpha \left(K \Pi - \gamma^{ij} D_i \partial_j \Phi + \mu^2 \Phi \right) - \gamma^{ij} \partial_i \alpha \partial_j \Phi + \mathcal{L}_\beta \Pi \ , \tag{18.7}$$

where α, β, γ^{ij}, and K are the lapse function, the shift vector, the three-dimensional spatial metric, and the trace of the extrinsic curvature K_{ij}, respectively, which are variables that come from the 3+1 decomposition of Einstein equations; $\mathcal{L}_\beta$ is the Lie derivative with respect to the shift vector; and Π is the conjugated momenta to the scalar field, defined as $\Pi = -(\partial_t - \mathcal{L}_\beta)\Phi/\alpha$. These equations need to be solved coupled to a 3+1 evolution formalism for the gravitational field, such as the Baumgarte-Shapiro-Shibata-Nakamura formulation [34, 246] or the conformal and covariant Z4 formulation (CCZ4) [17]. Spherically symmetric non-spinning scalar boson stars are dynamically robust and well-motivated candidates to be (part of) dark matter.

Natural extensions of the spherical case have been proposed and studied in recent years.

[1] Boson stars with only the mass term in Eq. (18.2) [$V_{\text{int}} = 0$] are usually referred to as *mini-boson stars*.

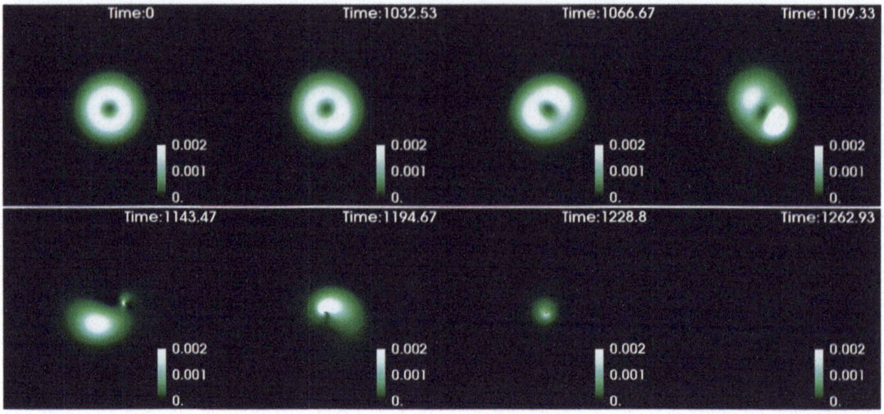

Fig. 18.1 *Spinning bosonic stars.* Time evolution of the energy density of a scalar spinning boson star. Time runs left to right, 1st to 2nd row. Extracted from [235]

- Rotating boson stars with $m \neq 0$ [268]; numerical studies showed that rotating scalar mini-boson stars are unstable against non-axisymmetric perturbations [103, 235], developing a bar-mode instability similar to that of differentially rotating neutron stars (see Fig. 18.1). These were unexpected results since rotating boson stars were believed to be stable according to catastrophe theory [175]. The decay to the spherical or lower angular momentum states is accompanied by a long-lived gravitational-wave emission [103].
- Dipolar boson stars described by a $\ell = 1, m = 0$ mode and excited states with one or more nodes in the radial profile of the field [168]; these solutions are unstable and decay dynamically to the fundamental nodeless spherically symmetric $ell = m = 0$ configurations [29, 163, 234].
- Self-interactions with high-order terms in the potential; their inclusion leads to an increase in the maximum mass of bosonic stars [90] and they can mitigate some of the dynamical instabilities of these stars (e.g. rotating models [112, 250], excited states [57, 237], and dipolar boson stars [163] can be stabilized).
- Multistate boson stars [41], ℓ-boson stars [15, 166] and multifield boson stars [141, 142, 234]; these solutions are superpositions of boson stars in different states described by several complex scalar fields coupled to gravity that can modify the stability properties of isolated stars as shown in Fig. 18.2, where the numerical evolution of multi-state stars demonstrated the quenching of instabilities.
- Proca stars were proposed in 2016 [58] as the vector counterparts of scalar boson stars. The configurations resemble very much the scalar case and share similar features. However, they have different dynamical stability properties. Spinning stars described by a $\ell = m = 1$ mode have a spheroidal morphology and contrary to the rotating scalar case, they are dynamically robust (stability and dynamical formation) [235]. On the other hand, $\ell = m = 0$ spherically symmetric stars do not constitute the fundamental ground state of solutions, according to recent numerical

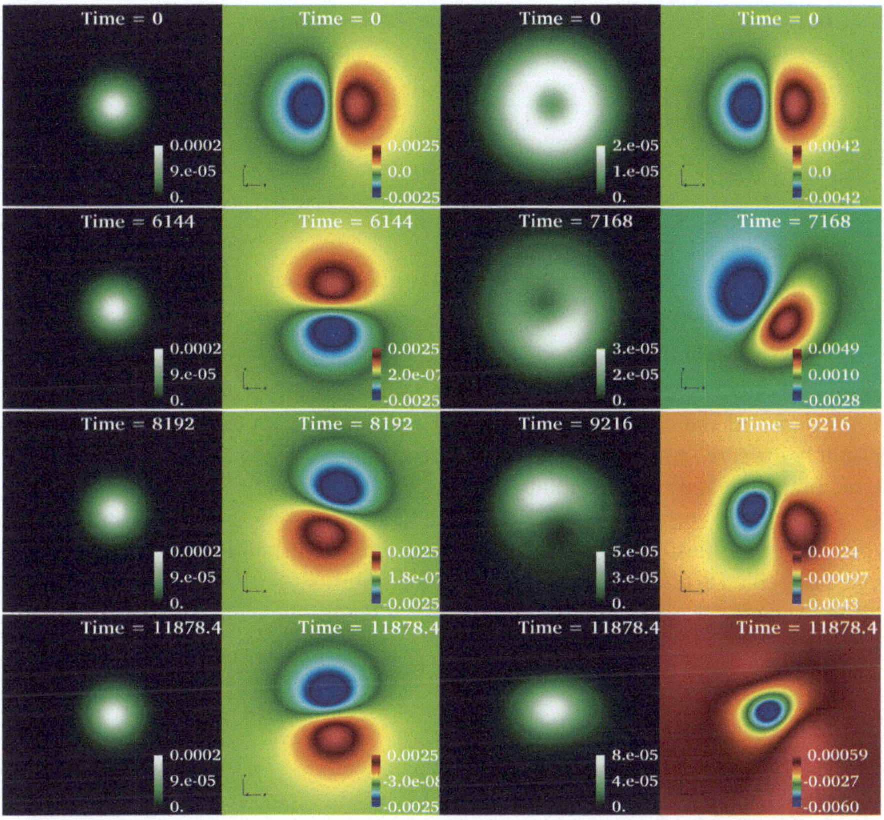

Fig. 18.2 *ℓ-bosonic stars.* Time evolution of the energy density and the real part of the $\ell = m = 1$ mode of scalar multi-field boson stars, corresponding a superposition of a static non-spinning $\ell = m = 0$ and a rotating $\ell = m = 1$ components, with $\omega_1 = 0.967$ (leftmost and left middle column), and with $\omega_1 = 0.974$ (right middle and rightmost column). Time runs from top to bottom. Extracted from [234]

simulations [159] they decay to other static configurations with axial symmetry but not spherical symmetry: these new solutions haven been dubbed as $\ell = 1, m = 0$ prolate Proca stars.

- Bosonic stars with light rings; it has been argued that horizonless ultracompact objects with light rings could mimic part of the gravitational-wave ringdown signal in binary mergers [71]. For this reason, observing gravitational wave ringdowns would not provide conclusive proof of the formation of horizons. However, such solutions must have at least two light rings, one stable and one unstable [96]. The presence of the stable light ring could act as a trapping mechanism for null perturbations that could then trigger an instability [69]. Ultracompact rotating bosonic stars (both scalar and Proca) indeed suffer a dynamical instability when light rings are present, leading to a decay to a less compact model or a collapse

Table 18.1 *Bosonic stars.* Non-radial (in)stability of mini-bosonic stars on the stable branch

Classes	Static $\ell = m = 0$	Static $\ell = 1, m = 0$	Rotating $\ell = 1, m = \pm 1$	Rotating $\ell = 2, m = \pm 2$
Scalar	Stable	Unstable	Unstable	Unstable
Proca	Unstable	Stable	Stable	Unstable

to a black hole, as shown in [97], suggesting that ultracompact objects with light rings might be unstable.

These examples illustrate why it is so important to investigate the stability of exotic compact objects, especially through numerical simulations. Rotating scalar boson stars and static, spherically symmetric Proca stars on the stable branch were definitely *expected* to be stable. And in a sense *they are stable*, since they are on the stable branch against linear radial perturbations and do not suffer the same fate as the configurations on the unstable branch. Still, these bosonic stars are prone to other types of instabilities, in particular, non-axisymmetric instabilities that lead to their decay. In Table 18.1, we give a summary of the stability of different classes of boson and Proca stars for configurations in the branch before the maximum mass model.

18.2.2 Fermion-Boson Stars

Fermion-boson stars [150, 151] are compact stars modelled as a superposition of a fermion and a boson star, respectively. The fermionic part can be described as a perfect fluid. Hence, the energy-momentum tensor for a fermion-boson star is

$$T_{ab} = T^{\Phi}_{ab} + T^{M}_{ab} , \tag{18.8}$$

where T^{Φ}_{ab} is the energy-momentum tensor associated to a (complex) scalar field and T^{M}_{ab} corresponds to a perfect fluid. Defined as follows

$$\begin{aligned} T^{\Phi}_{ab} &= \nabla_a \Phi \nabla_b \Phi^* + \nabla_a \Phi^* \nabla_b \Phi \\ &\quad - g_{ab} \left[\nabla^c \Phi \nabla_c \Phi^* + \mu^2 \Phi^* \Phi + V_{\text{int}}(\Phi, \Phi^*) \right] , \end{aligned} \tag{18.9}$$

$$T^{M}_{ab} = [\rho(1+\epsilon) + P] u_a u_b + P \, g_{ab} , \tag{18.10}$$

where Φ^* is the complex conjugate of Φ, ρ is the rest mass density, ϵ is the internal energy, P is the pressure, which is given by an equation of state, that relates the pressure with the other fluid quantities, typically the rest mass density and the internal energy, and u^a is the fluid four-velocity. The evolution equations of fermion-boson stars are then governed by the Einstein-Klein-Gordon-Euler system of equations, which are

$$G_{ab} = 8\pi \left(T^{\Phi}_{ab} + T^{M}_{ab}\right), \tag{18.11}$$

$$g^{ab}\nabla_a \nabla_b \Phi = \left(\mu^2 + \frac{dV_{\rm int}}{d\,|\Phi|^2}\right)\Phi, \tag{18.12}$$

$$\nabla^a T^{M}_{ab} = 0, \tag{18.13}$$

$$\nabla_a(\rho u^a) = 0. \tag{18.14}$$

The equilibrium configuration equations for a single fermion-boson can be obtained by combining the same techniques to build isolated non-rotating neutron and boson star solutions, respectively. The method can be summarised as follows: assuming a polar static metric (in Schwarzschild-polar coordinates), impose a harmonic time dependence for the scalar field and the static fluid condition, i.e., $v^i = 0$. Thus, the Einstein-Klein-Gordon-Euler equations become a simple set of ordinary differential equations that can be numerically integrated under suitable regularity and boundary conditions. More details can be found in [186].

Fermion-boson stars have been extensively used as a laboratory for dark matter in the context of neutron stars. Here, dark matter particles might form as a Bose-Einstein condensate, which can be represented with a single complex scalar field [59, 62]. Note that in this model, the coupling between the boson and the fermion particles is just through gravity. One of the first numerical simulations was performed by [262] considering a quadratic massive potential. The authors performed dynamical evolution using numerical relativity techniques in spherical symmetry. Due to the coupling between the fermionic and bosonic components through gravity, new oscillation modes were discovered in the fermionic sector. These new modes are significant as they provide evidence of dark matter's presence within these compact objects. Lastly, a quartic self-potential has been explored in [261]. Dynamical formation and stability of fermion-boson stars have been recently studied in Ref. [100]. The authors showed that an existing neutron star surrounded by an accreting cloud of scalar particles settles down, through gravitational cooling, dynamically into an equilibrium-stable configuration asymptotically consistent with a static solution of the Einstein-Klein-Gordon-Euler system. A follow-up of non-linear stability was conducted in [101], finding via numerical-relativity simulations a collaborative stabilisation mechanism whereby a sufficient amount of fermionic matter can stabilise an unstable excited scalar field. Interestingly, fermion-boson stars could help explain recent multi-messenger observations of compact stars [104].

The fermionic part could also be modelled by a white dwarf, with different observational consequences [80, 185, 238]. In [238], the authors explored the dynamical formation of a spherical mixed white-dwarf–boson star undergoing gravitational cooling for the first time, finding new features in various aspects of the white dwarf, including its gravitational redshift and expected electromagnetic radiation emitted from its photosphere. Additionally, differentially rotating neutron stars experiencing a bar-mode instability have been explored through numerical relativity simulations in the context of fermion-boson stars, as discussed in [105]. This study found that the accretion of dark matter onto the neutron star could alter the frequency of

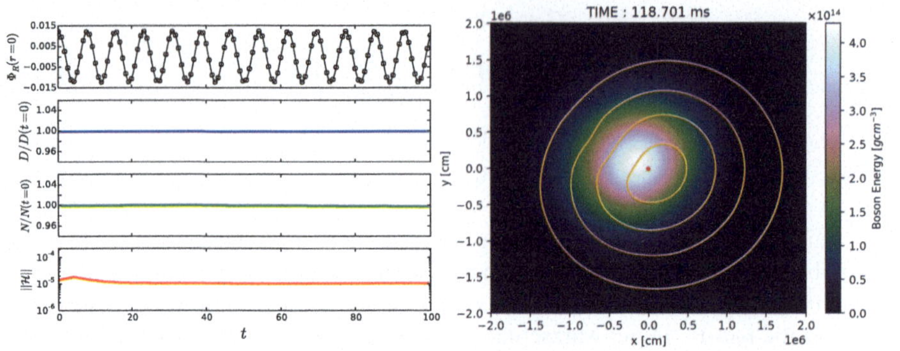

Fig. 18.3 *Fermion-Boson star.* The left panel shows the evolution of an isolated fermion-boson star. (Left panel) Evolution of the real part of Φ at $r = 0$ for a stable configuration. (Second and third row) Evolution of the integrated rest-mass density D and the Noether charge N. (Right panel) L_2 norm of the Hamiltonian $\mathscr{H}$ as a function of time. Extracted rom [52]. The right panel displays the morphology of a mixed-bar where the color represent the bosonic energy density and the isocontours represent the fermionic rest-mass density on the equatorial plane. Extracted from [105]

gravitational-wave emission expected from the bar-mode instability, thereby having an intrinsic impact on the search for continuous gravitational waves.

It should be mentioned that although most previous works have only included quadratic Φ^2 or quartic Φ^4 potentials, in Ref. [102], the authors considered a function potential that models axions and studied static solutions and non-linear stability through simulations in spherical symmetry. In the context of neutron star and dark matter gravitation-only interaction, some progress has been made in understanding non-linear evolutions in spherical symmetry. Research has been conducted on a two-fluid approach in [138] and a fermion-boson star with a realistic equation of state in [202].

Figure 18.3 shows the dynamically stable evolution of an isolated fermion-boson star configuration and the morphology of a *mixed-bar* made of ultralight bosonic field dark matter and fermionic matter on the dynamics of unstable differentially rotating neutron-boson stars.

18.2.3 Anisotropic Stars and Fuzzballs

Anisotropic stars are compact objects introduced to study mainly two different physical aspects: the first one is in the context of rescuing some equation of state that has been ruled out by astrophysical observations [28], and the second regarding Buchdahlâ€™s theorem [64] and ultra-compact objects that can violate this theorem [75]. Here, we will focus on the latter; for a complete set of references about the former topic, see the bibliography therein in [28]. Buchdahl's theorem states the maximum compactness of a self-gravitating, isotropic, spherically-symmetric object of mass M

and radius R is $M/R = 4/9$, if the object is composed of a perfect fluid. As a counter-example, Ref. [218] presents the first fully covariant and self-consistent model for spherically symmetric anisotropic fluids in General Relativity. The anisotropy was imposed using a relation through the tangential pressure and the radial one, through a parameter $\mathscr{C}$, which measures the deviation between both pressures. This model allows for the existence of stable solutions, some of which can be extremely compact, approaching $M/R = 1/2$ for different parameter values of $\mathscr{C}$. In spherical symmetry, the non-linear evolution of these anisotropic stars was performed for the first time in Ref. [218]. The authors found that the linearised stability analysis is in suitable agreement with the fully non-linear evolutions for these stars, showing that such configurations are stable below the maximum mass.

Another exotic compact object emerges when we want to evade the Buchdahl limit [65], the so-called fuzzball [188, 189, 195, 196], which can be defined as a microstate geometry with the exact mass, charge, and angular momentum of a black hole, but differing in the structure of the horizon. This object is a horizonless compact object driven by string theory with a rich multipole structure [53]. Their echoes have been computed numerically [38]. Non-linear evolutions have been performed in Ref. [162], where the authors carried out numerical simulations of small fields propagating on a large class of microstate geometries for the first time, focusing on the entire ringdown phenomenology. More details on these exotic compact objects can be found in the following reviews [75, 190].

Finally, Fig. 18.4 displays representative plots from the evolution of anisotropic neutron stars and fuzzball taken from [162, 218].

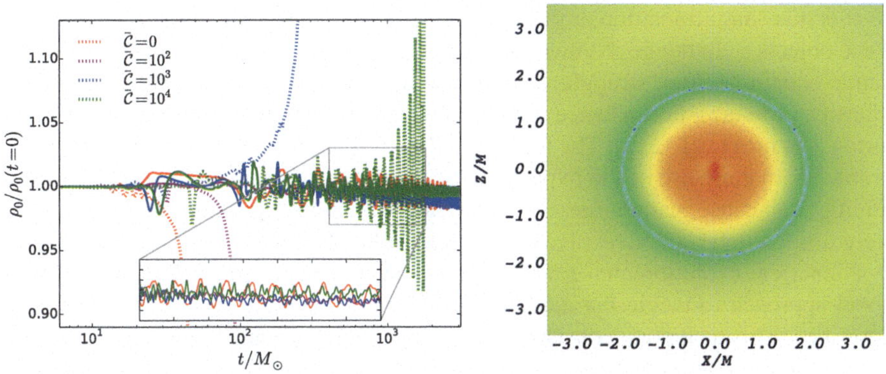

Fig. 18.4 *Anisotropic stars and fuzzball.* The left panel exhibits the normalised central density evolution for different stable(continuous curves)/unstable(dotted curves) configurations. Extracted from [218]. The right panel shows a snapshot of the evolution of a scalar field around an axisymmetric fuzzball. Extracted from [162]

18.2.4 Strange Quark Stars

The gravitational-wave event GW170817 [3] opened a new window into the interior of neutron stars and the potential identification of the cold equation of state of dense matter [4], corresponding to densities above $\sim 10^{15}$ g/cm^3, describing these objects. The different equations of state proposals from nuclear physics result in distinct properties [1, 208], such as compactness, maximum mass, and radii, that can now be tested with multi-messenger observations, ruling out some of them. In this regard, an interesting hypothesis is that strange quark matter is the true ground state of matter [266]. These exotic stars could be composed of pure strange quarks or even undergo phase transitions to quark matter during the merger phase, with neutrons and quarks coexisting in dense matter [35, 63, 79]. Rotating solutions have also been obtained in [187, 272, 273], both with uniform and differential rotation. Understanding the dynamics and stability of hybrid and strange quark stars is a necessary task to identify the viability of these models and the potential imprints of quark matter and phase transition in their gravitational wave emission. Moreover, an additional remarkable feature of strange stars is that they possess a crust, which is a discontinuity in the density between its surface and the exterior. A typical equation of state is given by the MIT bag model [84, 85]:

$$P = \frac{1}{3}(e - 4B), \tag{18.15}$$

where P is the pressure, e the energy density, and B the bag constant. The pressure goes to zero at a non-zero density, which presents a large jump at the surface.

It is this sharp transition at the stellar surface that makes the numerical study of these objects a challenge. Numerical codes solving the general relativistic hydrodynamic equations often struggle with large gradients in the hydrodynamic variables (density, pressure). For instance, shocks are dealt with High-Resolution Shock Capturing schemes [191, 223]. The idea behind such schemes is to solve local Riemann problems, making use of the hyperbolicity of the hydrodynamic equations. Discontinuities can cause many issues in numerical simulations, therefore new techniques were needed to be able to evolve even single isolated strange stars.

Numerical evolutions have been reported in [81, 270, 271, 274]. In [274], a novel approach to tackle the modelling of large discontinuities at the surface was described. This technique consists of attaching a thin hadronic "crust" described by a polytropic equation of state to the quark star, with the density going smoothly to zero. The addition of the crust only changes slightly (around $\sim 1\%$) the tidal deformability, according to the authors, and it is shown to allow the correct handling of the stellar surface. They also found that, in general, quark stars seem to require more resolution due to the sharp surface than a hadronic star of similar parameters. Binary strange star mergers are also investigated, and this will be discussed in the following Sect. 18.4.2.

Another approach implemented in [270] to evolve bare quark stars without a thin nuclear matter crust consists of modifying the recovery primitive procedure (recovering the primitive hydrodynamic variables from the conserved ones) to handle the surface discontinuity and the specific enthalpy going to unity when the pressure goes to zero. The authors successfully performed time evolutions of two triaxially deformed and differentially rotating quark star models with different masses and analyze their gravitational wave emissions. Addressing the single star and binary evolutions of triaxial quark stars is relevant since supramassive models (with masses larger than the maximum value coming from the Tolman-Oppenheimer-Volkoff equations) exist in larger parameter space than neutron stars, making them a potential outcome of supernova explosions or binary mergers. However, the evolution of isolated strange stars is found to be qualitatively similar to that of neutron stars, with a comparable gravitational-wave emission. Besides, for supramassive configurations, the emission decays faster as it settles quickly to axisymmetry. Therefore these models, and in particular the supramassive ones, are not necessarily easier to identify since they are not a better source for continuous gravitational waves. On the other hand, differentially rotating strange stars can still suffer bar-mode instability in a similar timescale than neutron stars.

Finally, in Ref. [81], stable evolutions of rotating bare quark stars close to the Keplerian limit were performed. In this case, another numerical technique is employed. The surface discontinuity is treated, making use of high-resolution shock-capturing methods together with a Riemann solver that preserves the positivity of density and pressure and a dust-like atmosphere. This allows them to study the oscillating modes and the universal relations of quark stars.

18.2.5 Black Holes with Synchronised Hair and Superradiance

An astounding property of black holes is that they can be completely characterized by a set of three parameters: mass, angular momentum, and electric charge. In the non-charged case, from the stellar to the supermassive mass ranges, all these objects are fundamentally described by the same solution (Kerr hypothesis). This is the no-hair theorem, preventing the existence of new black hole solutions in General Relativity. However, uniqueness theorems can be circumvented if we relax some of the assumptions, for instance, considering a non-electrovacuum scenario taking a complex and massive bosonic field around a Kerr black hole. Here, we will present some examples of hairy black holes in General Relativity with minimally coupled massive scalar and vector fields [152, 156, 157].

Before discussing these hairy black hole solutions, let us note that a bosonic field scattered off a rotating black hole can be amplified due to the superradiance process [61] and extract energy and angular momentum from the black hole if the oscillation frequency of the bosonic field, ω, satisfies the following condition:

$$\omega < m\Omega_{\mathrm{H}}, \tag{18.16}$$

where m is the azimuthal winding number and Ω_{H} is the angular velocity at the horizon. This phenomenon occurs for scalar, vector, and tensor fields, including electromagnetic and gravitational waves [123]. Therefore, if these superradiant-scattered waves can be trapped around a black hole by means of a cavity or a potential well due to the mass term (for massive bosonic fields), the superradiant instability can be triggered. Superradiant bosonic modes will grow exponentially and continuously extract energy in an unstoppable process leading to the black hole bomb scenario [216] when the field reaches the non-linear regime. Such a system was investigated in linear studies [70, 99, 113]. However, the endpoint of such instability remained an open question due to the long timescales involved [113, 203]. Consequently, fully non-linear numerical-relativity simulations were needed to answer this question.

A most favourable scenario for numerical simulations involving *charged* black holes instead of rotating solutions were explored in 2016. Reissner-Nordström black holes can suffer an analogue process to superradiant scattering by a charged bosonic field that extracts Coulomb energy and electric charge given rise to the charged black hole bomb [37]. In this case the superradiant condition relates ω with the electric potential at the horizon ϕ_{H} and the charge parameter of the bosonic field q:

$$\omega < q\phi_{\mathrm{H}}. \tag{18.17}$$

This problem can be tackled by solving the Einstein-(charged, complex) Klein-Gordon-Maxwell system in spherical symmetry. Moreover, the growth rates of the instability are orders of magnitude smaller than the rotating counterpart [99], significantly reducing the computational cost. A trapping mechanism must be included by imposing reflecting boundary conditions since the mass term is not enough to counter the electric repulsion between the charged black hole and the charged bosonic field. Numerical simulations showed that the development of the superradiant instability leads not to an explosive phenomenon but instead to the dynamical formation of a new static configuration at the threshold of the instability: a charged hairy black hole [54, 215, 232, 233] (see also [24]). This process consists of two phases:

1. while $\omega < q\phi_{\mathrm{H}}$: confined superradiant modes extract energy and charge from the black hole, and therefore its electric potential, ϕ_{H}, decreases;
2. until we reach $\omega = q\phi_{\mathrm{H}}$: the instability saturates and the extraction stops, the bosonic field is now in equilibrium with the black hole forming a new static solution.

This evolution is shown in Fig. 18.5. The first three panels correspond to the exponential growth of the scalar field energy, whereas the last one displays the final state of the charged hairy black hole.

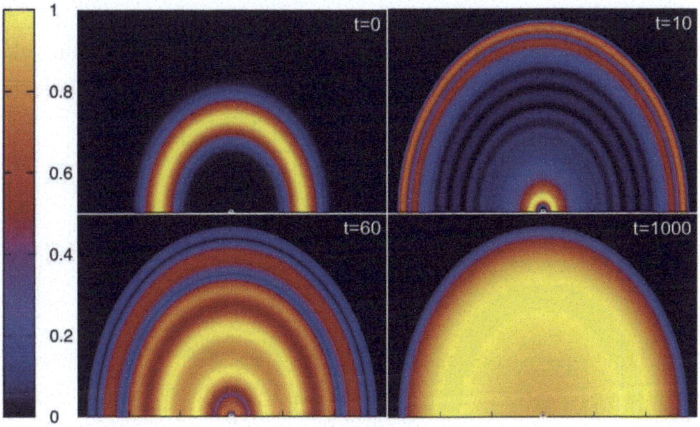

Fig. 18.5 *Black holes with synchronised hair and superradiance.* 2D snapshots of the normalized scalar field radial profile for a scalar charge $qM = 40$ at times t/M = 0, 10, 60, 1000. The apparent horizon is represented by small white circles near the origin in the 2D panels. Extracted from [233]

While the charged case is considered to be only a toy model since astrophysical black holes are not expected to be electrically charged and an artificial cavity must be placed, in 2017, pioneering numerical studies in the fully non-linear regime showed that a qualitatively identical evolution of the superradiant instability of a Proca field was also found in the Kerr case [119, 120]. They also observed the exponential growth and saturation phases, replacing the extraction of electric charge with the extraction of angular momentum. Considering Proca fields instead of scalar fields was motivated by the timescale of the superradiant instability in the scalar case ($t/M \sim 10^7$), where the current numerical relativity techniques and computational power are still insufficient to perform such simulations. However, Proca fields present shorter timescales of the order of $t/M \sim 10^4 - 10^5$ (still a computational prowess!). In order to evolve Proca fields using the standard 3+1 split, the following 3+1 quantities are used:

$$\mathscr{A}_a = \mathscr{X}_a + n_a \mathscr{X}_\phi, \quad \mathscr{X}_i = \gamma^a_i \mathscr{A}_a, \quad \mathscr{X}_\phi = -n^a \mathscr{A}_a, \tag{18.18}$$

where n^a is the timelike unit vector, $\gamma^a_b = \delta^a_b + n^a n_b$ is the operator projecting spacetime quantities onto the spatial hypersurfaces, $\mathscr{X}_i$ is the vector potential, and $\mathscr{X}_\phi$ is the scalar potential. The Proca evolution equations are then given by [275]:

$$\partial_t \mathscr{X}_i = -\alpha \left(E_i + D_i \mathscr{X}_\Phi\right) - \mathscr{X}_\Phi D_i \alpha + \mathscr{L}_\beta \mathscr{X}_i, \tag{18.19}$$

$$\partial_t E^i = \alpha \left(K E^i + D^i Z + \mu^2 \mathscr{X}^i + \epsilon^{ijk} D_j B_k\right) - \epsilon^{ijk} B_j D_k \alpha + \mathscr{L}_\beta E^i, \tag{18.20}$$

$$\partial_t \mathscr{X}_\Phi = -\mathscr{X}^i D_i \alpha + \alpha \left(K \mathscr{X}_\Phi - D_i \mathscr{X}^i - Z\right) + \mathscr{L}_\beta \mathscr{X}_\Phi, \tag{18.21}$$

$$\partial_t Z = \alpha \left(D_i E^i + \mu^2 \mathscr{X}_\Phi - \kappa Z\right) + \mathscr{L}_\beta Z\,, \tag{18.22}$$

with Z an additional variable and κ a damping parameter that help stabilize the numerical evolution, ϵ^{ijk} the Levi-Civita symbol, and the three-dimensional "electric" E^i and "magnetic" B^i fields, in analogy with Maxwell theory, are defined as

$$E_i = \gamma_i^a \mathcal{F}_{ab} n^b, \quad B_i = \gamma_i^a \star \mathcal{F}_{ab} n^b = \epsilon^{ijk} D_i \mathcal{X}_k. \tag{18.23}$$

The stress-energy tensor is given by:

$$T_{ab} = \frac{1}{2}(\mathcal{F}_{ab} \mathcal{F}^*_{bd} + \mathcal{F}^*_{ag} \mathcal{F}_{bd}) g^{gd} - \frac{1}{4} g_{ab} \mathcal{F}^{gd} \mathcal{F}^*_{gd} \\ + \frac{\mu^2}{2}(\mathcal{X}_a \mathcal{X}^*_b + \mathcal{X}^*_a \mathcal{X}_b - g_{ab} \mathcal{X}^d \mathcal{X}^*_d) . \tag{18.24}$$

Such simulations showed numerically that the growth of a Proca field perturbation around a Kerr black hole, when reaching the non-linear regime, leads to the formation of a Kerr black hole with synchronised Proca hair at the endpoint of the superradiant instability [154]. The potential formation of Kerr black holes with synchronised bosonic hair was also found in the context of boson and Proca mergers, which leads to a final bosonic cloud with a larger $m \gtrsim 4 - 6$ depending on the progenitors [47, 240]. Additionally, in 2018, the nonlinear evolution of the superradiant instability of rotating black holes in AdS_4 considering gravitational waves and solving the fully non-linear 3+1 Einstein equations [83].

Furthermore, Proca field with self-interaction has also been explored within numerical relativity. It was found that this theory can lead to significant problems, such as the loss of hyperbolicity of the field equations [32, 89] or the generation of instabilities in the context of superradiance [88].

18.3 Beyond General Relativity Theory

General Relativity faces numerous conceptual problems, such as singularity formation, horizons and the information loss paradox, its breakdown as a quantum theory at energies that exceed the Planck scale, which indicates the need for a new theory or a non-perturbative approach, or the nature of dark matter and dark energy, that seems to indicate that it is not the ultimate theory of gravity. Modified gravity theories have attempted to solve all or a least some of these issues. Within these extended theories, new aspects of well-known compact objects or even new exotic compact objects have been found.

In this section, we will discuss the numerical simulations of boson stars in modified theories of gravity and the impact of an additional scalar degree of freedom on black holes and neutron stars.

18.3.1 Boson Stars in Modified Gravity

Bosonic stars can also be studied within new theories of gravity beyond General Relativity as new compact objects with different properties or to explore new phenomena that arise in these theories, for example, related to spontaneous scalarization [98] or during gravitational collapse and the formation of black holes. Equilibrium solutions have been found in a number of theories, such as

- scalar-tensor gravity [56, 91, 158, 264],
- Brans-Dicke theory [260],
- Einstein-Dilaton-Gauss-Bonnet gravity [213],
- Einstein-Gauss-Bonnet theory [145],
- teleparallel gravity [161],
- higher-derivative gravity [27],
- Horndenski gravity [55, 224]
- tensor-multi-scalar theories [267],
- $f(T)$ extended theory [164], or
- Palatini $f(R)$ theory [192].

Despite the attention received, not many numerical simulations of boson stars in modified gravity have been carried out, generally due to the lack of well-tested evolution formalisms. Simulations have been reported in at least scalar-tensor [16, 264], Brans-Dicke [30] and more recently in Palatini $f(R)$ theories [193, 194]. The evolutions are usually performed in the so-called Einstein frame given by a conformal transformation of the metric in the physical "Jordan" frame since the former delivers equations that are similar to those in General Relativity and where the non-linearities in the gravity sector can be transferred to the matter Lagrangian. The relevant quantities can then be easily recovered by means of conformal transformations. In [193, 194], the simulations were also done in the Einstein frame instead of the $f(R)$ frame, but a mapping is needed to recover and study the true dynamics in Palatini $f(R)$ [9].

The action of a boson star in Palatini $f(R) = R + \xi R^2$ gravity, with ξ a parameter of the theory, in the Einstein frame is [193, 194]

$$S_{\rm EF} = \int d^4x\sqrt{-q}\frac{R}{2\kappa} - \frac{1}{2}\int d^4x\sqrt{-q}\left(\frac{Z - \xi\kappa Z^2 - 2V(\Phi)}{1 - 8\xi\kappa V(\Phi)}\right) , \quad (18.25)$$

where the kinetic term $Z \equiv q^{ab}\partial_a\Phi^*\partial_b\Phi$ is now contracted with the (inverse) metric q^{ab}, $V(\Phi) = -\mu^2\Phi^*\Phi/2$, $\kappa = 8\pi$ (in $G = c = 1$ units), and R is the Ricci scalar of the metric q_{ab}, i.e., $R = q^{ab}R_{ab}(q)$. In spherical symmetry, the space-time metric in the Einstein frame is given by

$$\begin{aligned} ds^2_{\rm EF} = &-(\alpha^2 - \beta^x\beta_x)dt^2 + 2\beta_x dx dt \\ &+ e^{4\chi(t,x)}\left(a(t,x)dx^2 + x^2 b(t,x)d\Omega^2\right) , \end{aligned} \quad (18.26)$$

where x is a radial coordinate, $d\Omega^2 = d\theta^2 + \sin^2\theta d\varphi^2$, $a(t,x)$ and $b(t,x)$ are the conformal metric components, and $\chi(t,x)$ is a conformal factor.

Two first-order variables are introduced to reformulate the Klein-Gordon equation

$$\Psi := \partial_x \Phi \ , \quad \Pi := n^\mu \partial_\mu \Phi = \frac{1}{\alpha}\left(\partial_t \Phi - \beta^x \Psi\right) \ . \tag{18.27}$$

Therefore, the equations of motion for the scalar field are

$$\partial_t \Phi = \beta^x \partial_x \Phi + \alpha \Pi \ , \tag{18.28}$$
$$\partial_t \Psi = \beta^x \partial_x \Psi + \Psi \partial_x \beta^x + \partial_x (\alpha \Pi) \ , \tag{18.29}$$
$$\partial_t \Pi = \frac{1 - 2\kappa\xi Z + \kappa\xi|\Pi|^2}{1 - 2\kappa\xi Z + 2\kappa\xi|\Pi|^2}\left\{\Xi - \frac{\kappa\xi\Pi^2\Xi^*}{1 - 2\kappa\xi Z + \kappa\xi|\Pi|^2}\right\} , \tag{18.30}$$

where the new variable Ξ simplifies the notation:

$$\begin{aligned}
\Xi := {} & \beta^x \partial_x \Pi + \frac{\Psi}{a e^{4\chi}} \partial_x \alpha + \frac{\alpha}{a e^{4\chi}}\left[\partial_x \Psi + \Psi\left(\frac{2}{x} - \frac{\partial_x a}{2a} + \frac{\partial_r b}{b} + 2\partial_x \chi\right)\right] + \alpha K \Pi \\
& - \frac{\alpha\mu^2\Phi}{1 - 2\kappa\xi Z} + \frac{\alpha\left(Z - \kappa\xi Z^2 + \mu^2|\Phi|^2\right) 4\xi\kappa\Phi\mu^2}{\left(1 + 4\kappa\xi\mu^2|\Phi|^2\right)(1 - 2\kappa\xi Z)} \\
& - \frac{4\kappa\xi\mu^2\alpha}{1 + 4\kappa\xi\mu^2|\Phi|^2}\left[-\frac{\Pi}{\alpha}\left(\partial_t \Phi^* \Phi + \Phi^* \partial_t \Phi\right) + \left(\frac{\Psi}{e^{4\chi} a} + \frac{\Pi\beta^x}{\alpha}\right)\left(\partial_x \Phi^* \Phi + \Phi^* \partial_x \Phi\right)\right] \\
& - \frac{\alpha\kappa\xi}{1 - 2\kappa\xi Z}\left[-\frac{\left(\partial_t \Psi^* \Psi + \Psi^* \partial_t \Psi\right) e^{4\chi} a - |\Psi|^2\left(4 e^{4\chi} a \partial_t \chi + e^{4\chi} \partial_t a\right)}{e^{8\chi} a^2} \frac{\Pi}{\alpha}\right. \\
& - \frac{\left(\partial_x \Psi^* \Psi + \Psi^* \partial_x \Psi\right) e^{4\chi} a - |\Psi|^2\left(4 e^{4\chi} a \partial_x \chi + e^{4\chi} \partial_x a\right)}{e^{8\chi} a^2}\left(\frac{\Psi}{e^{4\chi} a} + \frac{\Pi\beta^x}{\alpha}\right) \\
& \left. - \left(\frac{\Psi}{e^{4\chi} a} + \frac{\Pi\beta^x}{\alpha}\right)\left(\partial_x \Pi^* \Pi + \Pi^* \partial_x \Pi\right)\right] .
\end{aligned} \tag{18.31}$$

The main findings establish that boson stars in modified theories behave similarly to those in General Relativity in terms of stability and dynamical formation. Stable solutions oscillate around the initial configuration when perturbed, while unstable ones suffer one of the three expected outcomes: dispersion, migration to the stable branch, or collapse to black hole. Moreover, it has also been shown that boson stars can form in such theories much like they do in General Relativity, through the well-known gravitational cooling mechanism [245].

The gravitational collapse in the particular case of Palatini $f(R)$ theory was studied in [193, 194] focusing on black hole formation from unstable boson stars with a non-linear potential in spherical symmetry. It was found that, while in the Einstein frame, the endpoint of the evolution is a Schwarzschild black hole surrounded by a quasi-stationary cloud of the scalar field, when moving to the $f(R)$ frame, the collapse does not lead to the formation of a singularity. Instead, as shown in Fig. 18.6, an expanding baby universe emerges, connected to the original universe through a

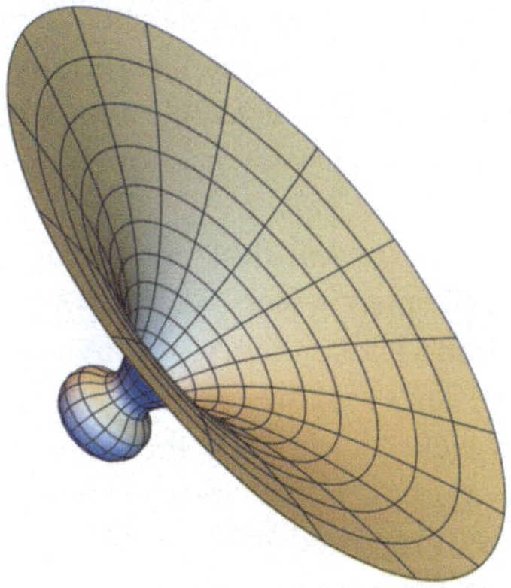

Fig. 18.6 *Boson stars in modified gravity.* Embedding diagram of the late-time spacetime geometry from the gravitational collapse of a boson star at two different times. The formation of a throat and a baby universe inside the horizon can be observed. Extracted from [194]

minimal area surface (throat or umbilical cord), supported by scalar field accretion. The throat is, however, hidden inside the horizon at all times, causally disconnecting the baby universe from external observers. This result may have implications for the understanding of singularities and bounce mechanisms in General Relativity and quantum gravity.

18.3.2 Black Hole and Neutron Star in Scalar-Tensor Theories

As explained above, one of the standard forms of modified General Relativity is adding a scalar field that is non-minimally coupled to gravity. The action gives the classic scalar-tensor theory

$$S = \frac{1}{16\pi G} \int d^4x \sqrt{-g} \left(\phi R - \frac{\omega(\phi)}{\phi} \nabla_a \phi \nabla^a \phi - V(\phi) \right) + S_m(g_{ab}, \psi) , \tag{18.32}$$

where g is the determinant of the spacetime metric g_{ab}, R is the Ricci scalar of the metric, $\omega(\phi)$ and $V(\phi)$ are functions of ϕ that must be defined to establish a specific theory, S_m is the matter action, and ψ denotes the matter fields. As explained previously, this is also known as the Jordan frame, a mathematical representation of a physical theory. Performing a conformal transformation leads to EinsteinâĿ™s theory with a minimally coupled scalar field, also known as the Einstein frame. We will not discuss details about these frames; more about this discussion can be found in Ref. [256].

With their fundamental properties and extreme conditions, black holes and neutron stars offer a unique environment for studying fundamental fields and exploring potential deviations from General Relativity. The so-called *spontaneous scalarization* is one of the most fascinating mechanisms in this context. This is a tachyonic instability that initiates the growth of a scalar field when the objectâŁ™s curvature surpasses a specific threshold value. This phenomenon is known to happen for neutron stars [98] in scalar-tensor theories, whereas black holes are protected by no-hair theorems [146]. For a complete review of this topic, we refer to the reviews Refs. [31, 116, 157, 217, 255].

One of the first numerical relativity simulations of an isolated black hole was performed in Refs. [242, 243], where the authors studied the gravitational collapse in Brans-Dicke theory of gravitation (i.e. considering $\omega = \omega_0 =$ constant and $V(\phi) = 0$). These studies focused on the emission of scalar waves (monopolar gravitational waves), confirming the no-scalar-hair theorems and proving that the dynamics of black holes in these theories are not the same as in General Relativity. Similar findings were reported in Refs. [45, 144, 173, 174, 247]. In addition, numerical relativity simulations of neutron stars collapsing to a stellar-mass black hole have also been done in scalar-tensor theories [82, 136, 198, 201, 257]. Scalarized neutron stars can shed away a large amount of scalar hair during the collapsing process, with these scalar polarization waves becoming smoking guns that could be potentially observed and measured in gravitational waves events [178, 179, 225, 226]. Furthermore, in Ref. [199], the authors carried out 1+1 numerical relativity simulations of neutrons that can acquire hair during their evolution. They studied the spectrum of these stellar oscillations, searching for some traces of those oscillations in the post-merger gravitational wave signal from a merger of two neutron stars.

A theory that has gained considerable attention is scalar-Gauss-Bonnet gravity, where the Gauss-Bonnet invariant $\mathscr{G} = R^2 - 4R_{ab}R^{ab} + R_{abcd}R^{abcd}$ is added to the action coupled to the scalar field as $\alpha f(\phi)\mathscr{G}$, where α is a coupling constant with dimensions of length square and $f(\phi)$ is a function of the scalar field. For example, if $f(\phi) = \phi$, this theory circumvents the no-hair theorem and provides hairy black holes [170]. In general, if $f'(\phi_0) = 0$, this theory admits General Relativity solutions, for some constant ϕ_0. Furthermore, theories which do not satisfy $f''(\phi_0)\mathscr{G} < 0$ exhibit the phenomenon of black hole scalarization. For more details about this phenomenon, see Refs. [19, 116, 118, 253].

One of the main challenges in fully simulating this theory is the general need for the well-posedness of the initial data (Cauchy) problem [214]. According to the criteria of Hadamard [143], the Cauchy problem for a system of partial differential equations is well-posed if a unique solution exists that continuously depends on its initial data. Since proving this statement is difficult, one usually attempts to show that the system of partial differential equations is strongly hyperbolic, a sufficient condition for well-posedness under suitable initial and boundary conditions [160].

In Refs. [220–222], the authors studied the gravitational collapse of a scalar field in the shift-symmetric Einstein dilaton-Gauss-Bonnet gravity, see as well [92, 93]. They found that higher frequency modes develop for specific couplings, leading to the breakdown of the hyperbolic structure of the system. In particular, the authors demonstrated that during the evolution, one of the characteristic speeds becomes imaginary, indicating that the evolution equations switch their character from hyperbolic to elliptic, resembling the mixed-type partial differential equation *Tricomi equation* (see Fig. 18.7). Furthermore, for sufficiently weak couplings and smooth initial data, local well-posedness of scalar-Gauss-Bonnet gravity can be established [176, 177] and successful studies have been presented in this regime [124, 125], which we will discuss in Sect. 18.4.3.

Nevertheless, a successful approach to mitigating the loss of hyperbolicity involves the *fixing-the-equations* technique [18, 76], inspired by dissipative relativistic hydrodynamics [165]. The main idea is to modify the higher-order contributions to the equations of motion *ad hoc* by replacing them with some auxiliary fields and allowing the latter to relax towards their correct values through a driver equation. In Ref. [132], in scalar-Gauss-Bonnet theory, the *fixing-the-equations* program has been employed within the gravitational collapse of a scalar field. The study demonstrates that a complete scalarized black hole solution can be achieved after passing a phase during which the original system of field equations changes character, see Fig. 18.7. This approach has been recently applied in 3+1 simulations in Ref. [181] within the same theory, and it has also been applied in the Effective Field Theory of gravity in Refs. [77, 78]. Finally, in Ref. [259] it was demonstrated that by adding a coupling between the scalar field and the Ricci scalar positively impacts the well-posedness at the nonlinear level. This finding was further confirmed in 3+1 simulations of the evolution of static and rotating black holes [115].

Another interesting modified gravity is K-essence, a scalar-tensor theory with first-order derivative self-interactions given by the action

$$S = \int \mathrm{d}^4x\sqrt{-g}\left[\frac{M_{\mathrm{Pl}}^2}{2}R + K(X)\right] + S_m\left[A(\Phi)g_{ab}, \Psi\right] ,$$

where $M_{\mathrm{Pl}} = (8\pi G)^{-1/2}$, $A(\Phi)$ is the conformal factor and $K(X)$ is a function that depends on the standard kinetic term $X \equiv \nabla_a\varphi\nabla^a\varphi$. Numerical simulations of gravitational collapse within K-essence, assuming spherical symmetry, have been conducted in [11, 43, 48, 130, 180, 184]. These works have shown that there are classes of theories for which the Cauchy problem breaks, encountering again the issue of the change of character from hyperbolic to elliptic (Tricomi) [43]. Furthermore, in Refs. [43, 48], it was found that in these theories, there may be divergences of the characteristic speeds, similar to the mixed-type Keldysh equation [43, 50]. Furthermore, k-essence can screen interactions with the scalar field on local scales while allowing for significant deviations from General Relativity on cosmological scales [25]. This screening mechanism is known as k-mouflage [26]. Simulations of non-vacuum spacetimes assuming spherical symmetry have been performed in Refs. [51, 258] and nonspherical oscillation of neutron stars in [249], showing that the screening mechanism works in the case of stellar oscillation and suppresses the monopolar scalar emission. Furthermore, in collapsing stars scenarios [51], the Cauchy problem, although locally well-posed, can lead to diverging characteristic speeds for the scalar field, which can be solved by introducing a fixing-the-equation technique [76]. A similar study about the chameleon screening mechanism has been done in Ref. [111].

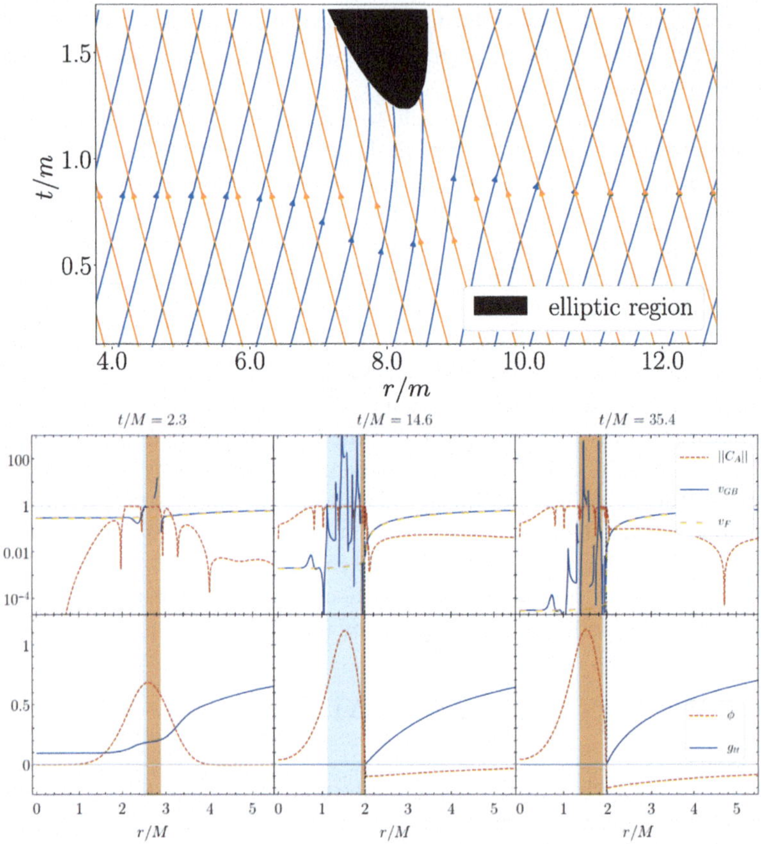

Fig. 18.7 *Black hole and neutron star in scalar-tensor theories.* The top panel shows the elliptic region (i.e., when the characteristic speeds become imaginary, indicating the character of the equations have become elliptic) forming in the simulation of gravitational collapse of Einstein dilaton Gauss-Bonnet gravity. Extracted from [221]. The bottom panel displays different time snapshots of various physical quantities, showing how the fixing-the-equation approach allows one to follow the entire simulation until the formation of hairy black holes. Extracted from [132]

Alternative theories such as action (18.32), scalar-Gauss-Bonnet gravity, Ricci-Gauss-Bonnet gravity, and k-essence are all subclasses of the Horndeski theories, which are defined as the most general scalar-tensor theory that leads to second-order field equations. For a comprehensive and specialized review of Numerical Relativity and Horndeski theories, we refer the interested reader to Ref. [219].

18.4 Exotic Compact Objects in Binaries

Once the stability of isolated configurations has been addressed, the next step is to consider a more interesting scenario from the point of view of gravitational wave astronomy. It is widely known that most astrophysical systems involving compact objects are actually binaries, for which the orbit will shrink due to the emission of gravitational radiation, leading to the eventual collision. As for black holes and neutron stars, the only way to obtain precise gravitational waveforms emitted by binaries of exotic compact objects is to perform fully non-linear numerical relativity simulations of mergers of such objects. The available numerical techniques and code infrastructures that enable the evolution of black hole and neutron star binaries can also be used to some extent for some exotic objects, facilitating the task at hand. Indeed, in many cases, binary mergers have been successfully performed, albeit under some approximations, and their gravitational wave signatures are not completely unknown. However, even though academic studies that consider only a few specific configurations are very relevant and may point out the key differences with respect to black holes and neutron stars, they might not be sufficient for any potential discovery. To reliably confirm a detection of these objects would require carrying out hundreds of simulations, systematically varying the parameters that define the models under study, as it has been done for black holes and neutron stars, with the aim of building catalogs of gravitational waves that can be then compared to the observational data of the LIGO-Virgo-KAGRA detectors.

Understanding and modelling the dynamics, endpoint, and gravitational wave emission of the different solutions is, therefore, necessary for the falsifiability of these alternative models. In this section, we will review some interesting cases of exotic binary simulations.

18.4.1 Bosonic Stars in Binaries

There are no clear mechanisms as to how bosonic stars would condense and form during the early Universe. Anisotropies in the bosonic field could lead to incomplete gravitational collapse and star formation, through gravitational cooling [245]. Therefore, as for main sequence stars, it is likely that bosonic stars could come in binaries. Binary systems would eventually merge and emit a gravitational-wave signal in frequency and amplitude ranges that directly depend on the masses of the stars and therefore on the mass of the bosonic particle μ.

18.4.1.1 Initial Data

The initial method consists in taking a superposition of the two stars as initial data [50, 200, 210, 211, 236]:

1. Construct the initial data for two non-spinning boson stars with $\{g_{ab}^{(i)}, \Phi^{(i)}\}$ and extend it to a Cartesian grid.
2. Perform a Lorentz transformation in a chosen direction.
3. Superpose the boosted solutions as:

$$\Phi(t, r) = (\Phi_0^{(1)}(r)e^{-i(\omega t + \epsilon^{(1)})})_{\text{boost}} + (\Phi_0^{(2)}(r)e^{-i(\omega t + \epsilon^{(2)})})_{\text{boost}}, \quad (18.33)$$

$$g_{ab} = g_{ab,\text{boost}}^{(1)} + g_{ab,\text{boost}}^{(2)} - \eta_{ab}\,, \quad (18.34)$$

where $\epsilon^{(i)}$ is the initial phase of the stars and η_{ab} is the Minkowski metric.

This approach is not a proper solution to Einstein's equations but only a convenient approximation. Obtaining suitable initial data requires numerically solving four elliptic equations, the Hamiltonian and momentum constraints, which is a complex process in most cases. Recent efforts have been made to solve the binary boson-star initial data. In Refs. [129, 148, 149], the authors have improved the usual superposition of the single-star spacetime for the case of head-on collisions of equal and unequal mass, non-spinning boson star considering the potential for mini-boson stars and solitonic boson stars.[2] However, the constraint violation for the Hamiltonian and momentum constraint still needs to be eliminated.

Newly, two groups have solved the initial data for boson star binaries in the last few months [23, 252]. The first has developed a method for constructing binary boson star initial data that satisfies the constraints. The authors used the conformal thin-sandwich formalism, which had previously used initial data from black holes and neutron stars [122]. They could construct initial data for various configurations considering different potentials, mass-ratios, spin magnitudes, and spin orientations. The last group that got accurate initial data was Ref. [23]. Here, the authors constructed initial data also using the conformal thin-sandwich formalism and superposed free data equations for the head-on collision of mini-boson stars. They have numerically solved conformal thin-sandwich equations using the code bamps and the hyperbolic relaxation method presented in Ref. [229]. Last but not least, one of the first papers in which the constraints have been appropriately solved was in Ref. [110], focusing on head-on collisions of boson stars with neutron stars.

18.4.1.2 Boson-Boson and Proca-Proca Mergers

In recent years, a significant number of numerical studies have been carried out on the collisions of bosonic stars and their emission of gravitational waves in General

[2] Solitonic or non-topological boson stars are characterised by an hexic potential [134, 183].

Relativity, with early simulations focusing on head-on collisions of non-spinning scalar boson stars. We will begin this section by presenting a list of the main works related to this topic:

- boson star head-on collisions [23, 42, 72, 86, 149, 167, 200, 211];
- boson star orbital mergers [47, 49, 50, 95, 129, 210, 212, 240, 251, 252];
- oscillatons (real scalar fields) [148, 265];
- ℓ-boson stars [167];
- axionic boson stars [239];
- Proca stars [66, 67, 231, 236].

Numerical simulations of bosonic star collisions have shown that their compactness $C \equiv M/R$ (M and R being the mass and radius of the boson star, the latter defined as the radius within which 99% of the total mass is contained) changes the outcome of the merger, the endpoint being a black hole or culminating in a more massive configuration [47, 50, 212, 236], as illustrated in Fig. 18.8.

Nevertheless, orbital mergers of non-spinning scalar boson stars unexpectedly failed to form rotating solutions of boson stars [47, 50, 95, 212], the final remnant losing all its angular momentum. This was later linked to the non-axisymmetric instabilities that appear when evolving isolated rotating boson stars [103, 112, 235, 250], making them quickly unstable by decaying to static spherically-symmetric configurations or collapsing to black holes. In the case of Proca stars, no rotating Proca star remnants were formed in the orbital collisions either [236], even though such rotating solutions are dynamically stable. It was recently found that the inability to retain the angular momentum is however related to the wave-like nature of bosonic fields and the quantized angular momentum given by the azimuthal winding number m (see Eq. 18.4). In particular, the interaction between the stars and the amount of energy released in the collision depend on the relative phases $\epsilon^{(i)}$ at merger between the objects [211, 231, 251]. For instance, a phase difference of $\Delta\epsilon = |\epsilon^{(1)} - \epsilon^{(2)}| = \pi$ can lead to repulsive force between the stars that will keep them apart for a longer time. Refs. [50, 211] demonstrated that two non-rotating $\ell = m = 0$ scalar boson stars with such phase difference of π will bounce back and forth in a head-on collision and could eventually settle into a $\ell = 1$, $m = 0$ dipole configuration [163] if self-interactions are considered (see also Ref. [167] for the case of ℓ-boson stars). Therefore, in mergers with orbital angular momentum, a rotating $\ell = m = 1$ boson star can only be formed if the phase difference of the stars at merger is $\Delta\epsilon \gtrsim 7\pi/8$ as found in [251]. Self-interaction must also be included in order to stabilize the rotating scalar solutions.

The phase difference $\Delta\epsilon$ plays a key role in the outcome of the final configuration in the formation scenario when the stars are not compact enough to collapse. But it also has a crucial impact on the emission of gravitational waves. The wave-like nature of bosonic stars can lead to an interference at the time of the collision, greatly changing the amount of gravitational radiation and the morphology of the waveform [231, 251]. Varying the initial phases $\epsilon^{(i)}$ does not modify the initial energy density of the stars since it is independent of the phase. However, the value of $\Delta\epsilon$ at merger

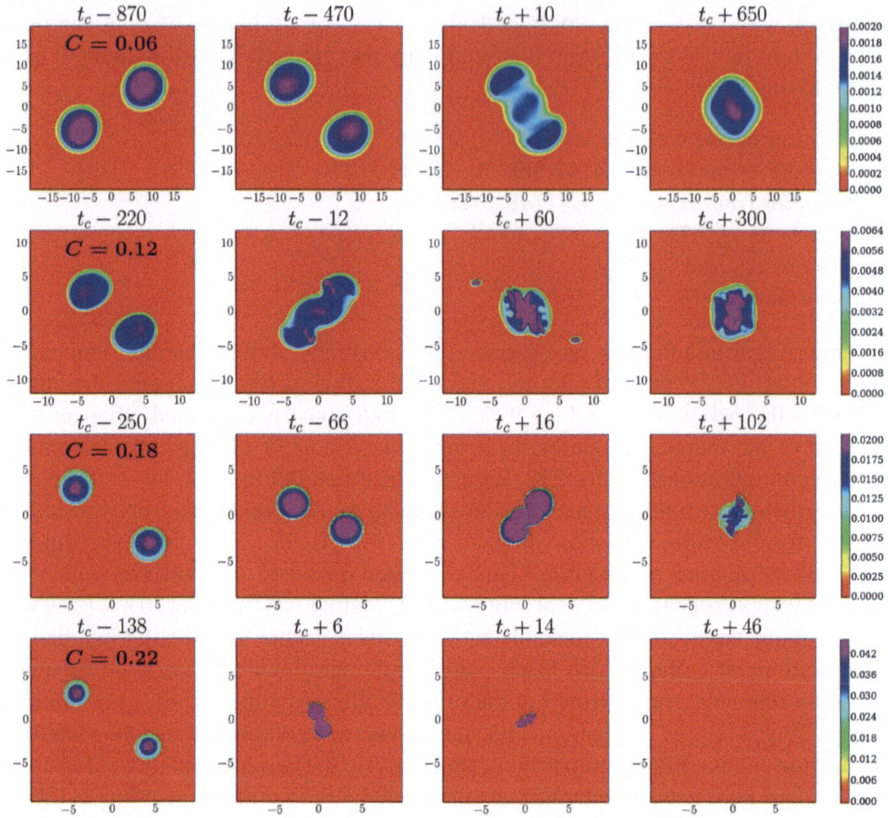

Fig. 18.8 *Boson-Boson and Proca-Proca mergers.* Coalescence of binary boson stars: Snapshots in time of the Noether charge density in the orbital plane. Each row corresponds to a different compactness (from top to bottom: 0.06, 0.12, 0.18, and 0.22). Extracted from [50]

may cause a constructive or destructive interference pattern. From the amplitude of the Proca field $\mathscr{A}$, assuming a linear superposition of both stars $\mathscr{A} = \mathscr{A}^{(1)} + \mathscr{A}^{(2)}$ oscillating at with frequencies $\omega^{(1)}$ and $\omega^{(2)}$ [231] (see [211] for the scalar case):

$$\begin{aligned} \mathrm{Re}(\mathscr{A}) &\sim \cos(\omega^{(1)} t + \epsilon^{(1)}) + \cos(\omega^{(2)} t + \epsilon^{(2)}) \\ &\sim 2\cos\left(\frac{(\omega^{(1)} + \omega^{(2)})}{2} t + \epsilon^{(1)} + \epsilon^{(2)}\right)\cos\left(\frac{(\omega^{(1)} - \omega^{(2)})}{2} t + \Delta\epsilon\right), \end{aligned} \tag{18.35}$$

$$\begin{aligned} \mathrm{Im}(\mathscr{A}) &\sim \sin(\omega^{(1)} t + \epsilon^{(1)}) + \sin(\omega^{(2)} t + \epsilon^{(2)}) \\ &\sim 2\sin\left(\frac{(\omega^{(1)} + \omega^{(2)})}{2} t + \epsilon^{(1)} + \epsilon^{(2)}\right)\cos\left(\frac{(\omega^{(1)} - \omega^{(2)})}{2} t + \Delta\epsilon\right). \end{aligned} \tag{18.36}$$

The total magnitude of the field is then given by

$$|\mathscr{A}|^2 = \mathrm{Re}(\mathscr{A})^2 + \mathrm{Im}(\mathscr{A})^2 \sim 4\cos^2\left(\frac{(\omega^{(1)} - \omega^{(2)})}{2}t + \Delta\epsilon\right)$$
$$\sim 2\big[1 + \cos\big((\omega^{(1)} - \omega^{(2)})t + \Delta\epsilon\big)\big]. \quad (18.37)$$

Equation (18.37) clearly indicates that at the time of the merger, the frequency difference $\Delta\omega$ and $\Delta\epsilon$ will change the interaction of Proca field describing both stars and induce a different gravitational wave emission depending on the value of $\cos\big((\omega^{(1)} - \omega^{(2)})t + \Delta\epsilon\big)$, in particular if $(\omega^{(1)} - \omega^{(2)})t + \Delta\epsilon = \pi$, the amplitude reaches its minimum value. Keeping fixed $\Delta\omega$, for instance in the case of equal-mass stars (for which $\omega^{(1)} = \omega^{(2)}$), the total energy emitted in gravitational waves E_{GW} in a head-on collision of rotating Proca stars greatly depends on $\Delta\epsilon$ as shown in Fig. 18.9. This result reveals the relevance of the phase difference and any study on merging bosonic stars must incorporate this parameter.

Finally, highly compact bosonic star binaries need to be highlighted in the context of the current gravitational-wave detections by the LIGO-Virgo-KAGRA network. The compactness of those stars is such that it will lead to black hole collapse quickly after merger and to the formation of an event horizon, which means that a ringdown phase in the gravitational-wave signal will occur. In consequence, these systems can imitate part of the gravitational-wave phenomenology of black hole binary mergers [50, 148, 236], which are the astrophysical scenario that best fit so far the majority of the real gravitational-wave observations. In this regard, in [66] a catalogue of around eighty numerical-relativity gravitational waveforms from head-on collision

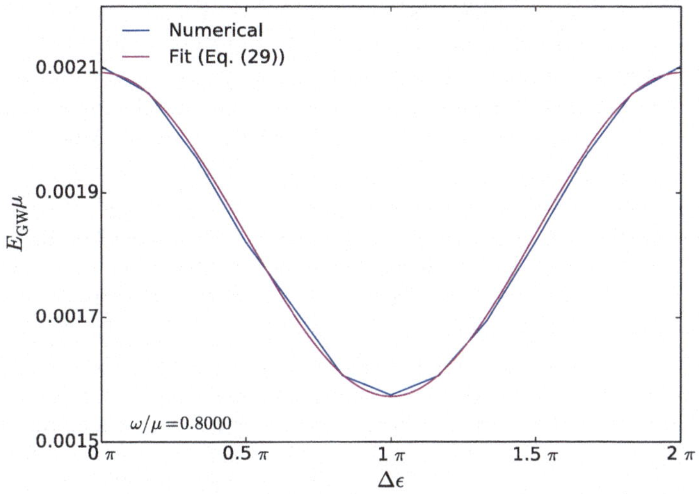

Fig. 18.9 *Boson-Boson and Proca-Proca mergers.* Coalescence of binary boson stars: Gravitational-wave energy emitted in a head-on collision of equal-mass rotating Proca stars with fixed $\omega^{(1)}/\mu = \omega^{(2)}/\mu = 0.8000$ and non-zero phase difference $\Delta\epsilon$ between the stars. The magenta line depicts the behaviour of the square of the Proca field amplitude as a function of $\Delta\epsilon$ computed from Eq. (18.37). Extracted from [231]

of rotating Proca stars (presented in [231]) was compared using a Bayesian analysis with the gravitational wave event GW190521 [8]. A slight statistical preference for the Proca model was found with respect to black hole binary mergers [66] (see Fig. 18.10), although this does not constitute any strong evidence in favour of this alternative model, especially since head-on collisions are not expected to occur in the Universe. This particular event was selected due to it is larger total mass, which led to a final black hole with 150 $M_{\odot}$, since it was observed as a short, burst-like signal which could be compatible with a head-on or eccentric collisions. Later, the same analysis was performed on this and a few other massive events with a larger set of waveforms from Proca stars including unequal-mass collisions in [67, 231]. While the preference was still there for a couple of the events, any further evidence remained inconclusive. An interesting point is that if the Proca hypothesis to explain similar events was found to be true, the mass of the hypothetical fundamental boson particle could be inferred through parameter estimation. As a proof of concept, in [66, 67] it was estimated that an ultralight vector boson of mass $\sim 9 \times 10^{-13}$ eV could be the constituent of such Proca stars and potentially be part of the dark matter in the Universe.

18.4.1.3 Fermion-Boson Fermion-Boson Mergers

As explained in the previous Sect. 18.2.2, fermion-boson stars have been used as laboratories to study dark matter using neutron stars. There are models in which dark matter can form small clusters inside neutron stars [60, 137, 171]. These dark matter clusters can change the compactness and properties of stars; see Refs. [28, 263] for a review. Therefore, collisions of these compact objects could play an important role in gravitational waves detection in binary neutron star merger events and shed some light on dark matter. Motivated by this, the first simulation of neutron star binaries with dark matter cores by numerical relativity simulations was done in Ref. [52]. The authors have considered binaries of fermion-boson stars, where the fermionic part models the neutron star and the bosonic part models the dark matter. In this case, equal-mass mergers of these stars with a fraction of dark matter in their interior form a one-arm instability during the post-merger (see Fig. 18.11), leaving an imprint on the emitted gravitational waves. In Ref. [127], a theoretical toy model of binary of neutron with dark matter in their interior has been considered, showing new frequencies in the post-merger signal. In the same context, the coalescence of binary neutron star admixed with dark matter has been studied in Ref.[36], but the dark matter is considered a test particle. The dark matter particle has been modelled and simulated using smooth particle hydrodynamic methods. The authors have shown that dark matter affects the gravitational wave, especially in the post-merger. Here, different masses and equation-of-states for the neutron stars have been considered.

Furthermore, in Ref. [128], the effects of mirror dark matter in neutron star mergers have been studied. The authors have extended the code BAM [107] to perform this simulation, considering a two-fluid approach. They found that these binaries have a long inspiral and modified the ejecta and the disk mass, see Fig. 18.11. Finally,

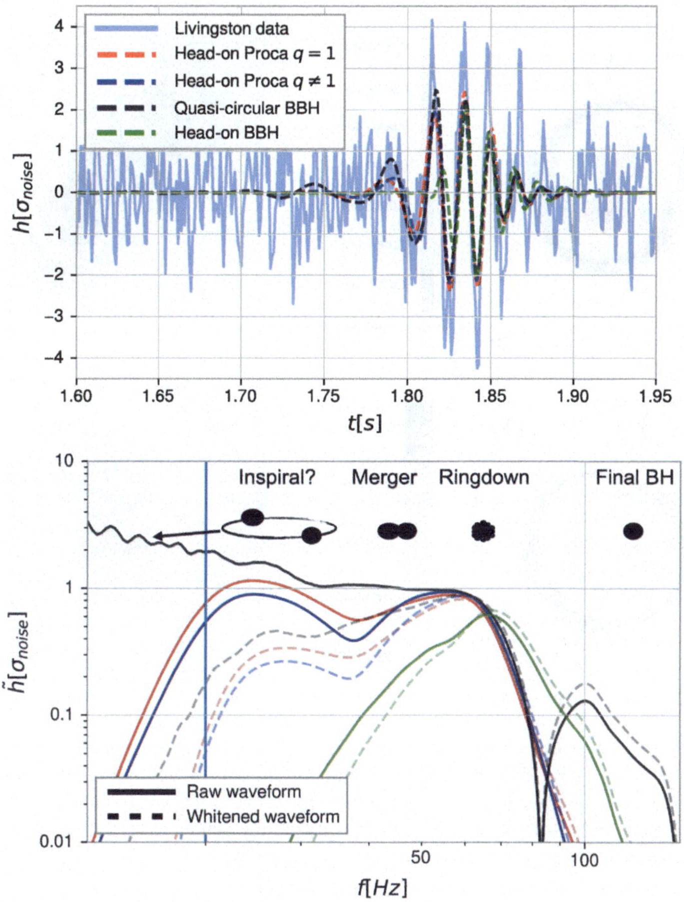

Fig. 18.10 *Boson-Boson and Proca-Proca mergers.* Time series and spectrum of GW190521. Top: Whitened strain data of the LIGO Livingston detector at the time of GW190521, together with the best fitting waveforms for a head-on merger of two black holes (green), two equal or unequal mass Proca stars (red and blue) and for a quasicircular black hole merger (black). Bottom: corresponding waveforms shown in the Fourier domain. Solid lines denote raw waveforms (scaled by a suitable, common factor) while dashed lines show the whitened versions. The vertical line denotes the 20 Hz limit, below which the detector noise increases dramatically. Extracted from [66]

consistent quasi-equilibrium configurations have been constructed for binary systems consisting of two neutron stars, each admixed with dark matter, using the SGRID code in Ref. [230].

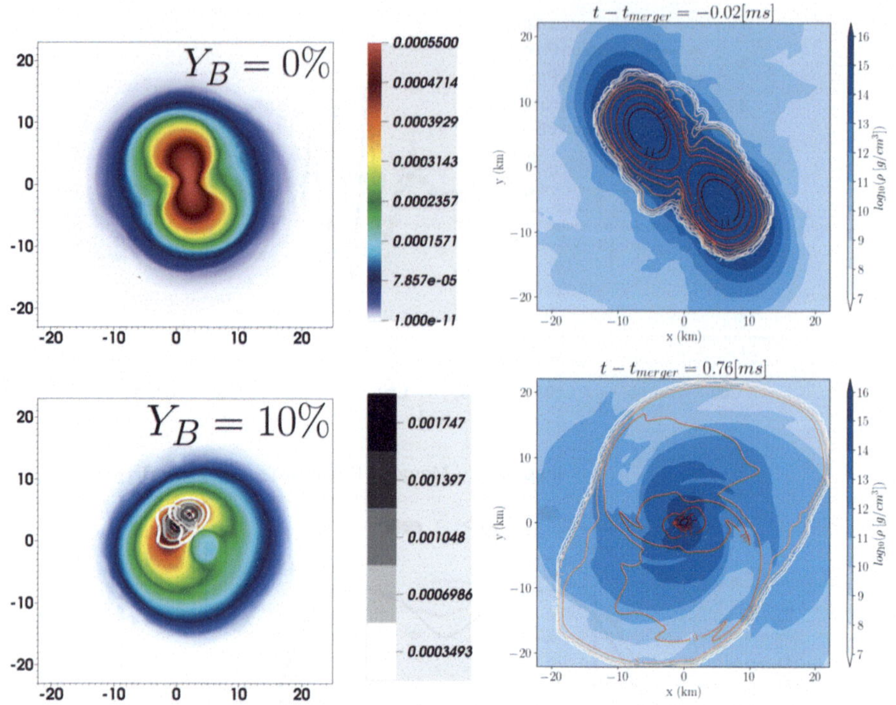

Fig. 18.11 *Fermion-boson—fermion-boson mergers.* The left panels show the symmetry breaking of the quadrupole structure when there is a significant amount of bosonic component inside the neutron star in the post-merger stage. Extracted from [52]. The right panels show time snapshots in the equatorial plane of the different phases of the coalescence taken from the simulation of equal mass neutron stars and 5% mirror dark matter. Extracted from [128]

18.4.1.4 Hybrid Binaries Mergers

Recent numerical relativity simulations have also investigated the merger of hybrid binaries of boson stars and black hole/neutron stars. One of the first three-dimensional numerical relativity simulations of neutron star - boson star collisions was performed in Ref. [110]. These collisions can result in the formation of either a black hole or a neutron star surrounded by a bosonic cloud, with gravitational wave signals offering insights into the nature of the merger remnants.

In Ref. [87], the authors have investigated the collision of axion stars with black holes and neutron stars, see Fig. 18.12. These simulations have elucidated the potential role of axion star-black hole mergers in the formation of scalar hair around black holes. Similarly, in Refs. [87, 109], studies examining neutron star-axion have shown many observable phenomena across electromagnetic and gravitational wave channels. Moreover, in Ref.[108], the authors explored if the role of nucleon-nucleon-axion bremsstrahlung has any insights into the cooling binary neutron star remnants.

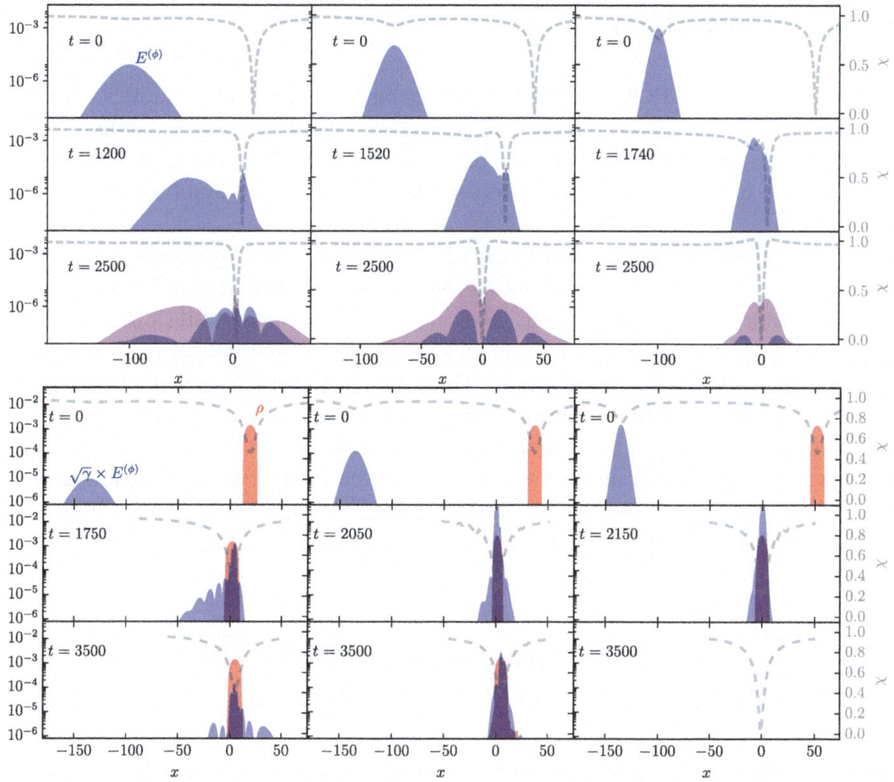

Fig. 18.12 *Hybrid binaries mergers.* Time snapshots of the evolution of the bosonic energy density and the metric's conformal factor, which roughly represents the gravitational potential (grey) for different configurations of black hole-axion star collisions considered in Ref. [87]. The bottom panel is the same as above but for neutron star-axion star collisions. The red colour represents the baryonic density. Extracted from [87]

However, they found that the imprint of axions on gravitational wave signals and ejecta mass appears to be relatively small.

Finally, for the case of the interaction between black holes and boson stars, simulations of the phenomenon of black holes piercing through massive bosonic stars have been investigated in Ref. [74]. The authors have challenging simulations involving black holes and boson stars, with length ratios as large as sixty-two. The authors have found that when the black hole pierces through the bosonic structure, the process is characterised by accretion-induced deceleration and the emission of gravitational waves. In Ref. [73], in the context of fuzzy dark matter, the accretion of a boson star by a central black hole has been investigated in spherical symmetry. Finally, the authors in Ref. [269] have extended previous works by including a hexic solitonic potential Φ^6 and exploring the piercing of a solitonic boson star by a boost black hole.

18.4.2 Strange Stars

As explained in Sect. 18.2.4, only a few studies have explored the dynamics of quark star binary systems. This is partially due to the numerical challenges that have to be faced when modelling the remarkable discontinuities on the surfaces of these stars. Here, we will only give a short overview of numerical relativity simulations of binary quark stars.

In Ref. [274], the authors have performed quark-star binaries for the first time through fully general-relativistic simulations, displaying differences between them and their hadron star counterparts. The authors have found subtle differences in merger and post-merger frequencies in these simulations, particularly regarding tidal deformability and stellar compactness. Moreover, in Ref. [271], the post-merger of the evolution of equal-mass binary quark star coalescence has unveiled the possibility of prompt collapse to a black hole for supramassive merger remnants, challenging conventional post-merger scenarios observed in binary neutron star mergers. These findings emphasise quark-star binaries' distinctive dynamics and astrophysical implications, highlighting the need for further exploration in these simulations.

18.4.3 Beyond General Relativity: Black Holes-Black Holes and Neutron Star Neutron Star Mergers

To date, null and consistency tests have primarily validated the consistency of General Relativity in binary merger dynamics and in the strong-field regime. However, these tests are insufficient to rule out specific alternative theories of gravity. This is because accurate predictions for the merger waveforms of compact objects in these alternative theories have been obtained only in a limited number of cases. Such predictions necessitate the precise numerical solution of the theory's field equations, which involve complex systems of partial differential equations. A significant challenge in this area is the general difficulty in establishing well-posed initial data (Cauchy) problems for most alternative theories of gravity, as was explained in detail in Sect. 18.3.2.

Nevertheless, recent progress in numerical relativity simulations beyond General Relativity has been made [10, 219]. Studies on binary black holes in scalar-tensor theories of gravity, such as those by [45, 147], have been among the first. Building on this, [68] examined binary black hole mergers within the $f(R)$ theory framework, finding distinct merger dynamics compared to the General Relativity counterpart. Further investigation on dynamical Chern-Simons gravity by [204–206] have provided insights into the gravitational wave signatures produced by these mergers using an order-by-order expansion scheme. Notably, this technique can lead to secular growing solutions, breaking the perturbation theory on long timescales. However, this could be solved by relying on the numerical renormalization-group-based technique presented in Ref. [135].

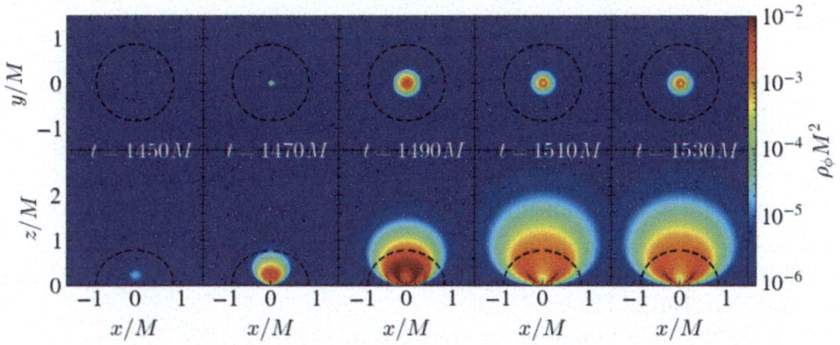

Fig. 18.13 *Beyond General Relativity: black holes—black holes and neutron star—neutron star mergers*. Time evolution of the scalar cloud after the merger of black holes binary in Einstein-scalar-Gauss-Bonnet gravity. Extracted from [21]

Additionally, progress has been made in understanding neutron star mergers within these alternative frameworks. In Refs. [33, 209] scalar-tensor theories were considered as the underlying theory of gravity, uncovering phenomena such as *dynamical scalarization*, i.e., once neutron stars come close enough to each other, they dynamically scalarize. These studies were extended in Ref. [248]. In Ref. [46], kinetic screening in simulations of merging binary neutron stars was studied, finding that the screening tends to suppress the (subdominant) dipole scalar emission but not the (dominant) quadrupolar scalar mode.

The evolution of Einstein-scalar-Gauss-Bonnet gravity using modified harmonic formulations has also been a focus of attention [94, 124, 125]. In this context, Refs. [20, 21, 114] have addressed the well-posedness of four-derivative scalar-tensor theories (including Einstein-scalar-Gauss-Bonnet gravity) by employing singularity avoiding coordinates and the CCZ4 formalism, see Fig. 18.13. Simulations in the decoupled limit aiming to study dynamical scalarization and descalarization of binary black holes have been conducted in Refs. [39, 40, 117, 126, 254]. Finally, simulations of binary neutron star mergers in scalar-Gauss-Bonnet gravity have also been performed in Ref. [121], discovering that, although there is almost no impact during the inspiral stage, in the post-merger phase; however, strong scalar effects appear. Last but not least, black hole binaries in cubic Horndeski theories with a massive potential have been investigated in Ref.[131].

Acknowledgements We thank S. Liebling and C. Herdeiro for their useful comments on the manuscript and N. Franchini for their enlightening conversations on the alternative theory of gravity and numerical relativity. NSG acknowledges support from the Spanish Ministry of Science and Innovation via the Ramón y Cajal programme (grant RYC2022-037424-I), funded by MCIN/AEI/10.13039/501100011033 and by “ESF Investing in your futureâŁž. NSG is further supported by the Spanish Agencia Estatal de Investigación (Grant PID2021-125485NB-C21) funded by MCIN/AEI/10.13039/501100011033 and ERDF A way of making Europe, and by the European Horizon Europe staff exchange (SE) programme HORIZON-MSCA2021-SE-01 Grant No. NewFunFiCO-101086251.

References

1. Compose (2018). https://compose.obspm.fr/home
2. B. Abbott et al., Phys. Rev. X **9** (2019). https://doi.org/10.1103/physrevx.9.031040, arxiv:1811.12907
3. B.P. Abbott et al., Phys. Rev. Lett. **119**, 161101 (2017). https://doi.org/10.1103/PhysRevLett.119.161101, arXiv:1710.05832
4. B.P. Abbott et al., Phys. Rev. Lett. **121**, 161101 (2018). https://doi.org/10.1103/PhysRevLett.121.161101, arXiv:1805.11581
5. R. Abbott, T. Abbott, S. Abraham et al., Phys. Rev. X **11**, 021053 (2021)
6. R. Abbott, T. Abbott, F. Acernese et al., Phys. Rev. X **13**, 011048 (2023)
7. R. Abbott, T. Abbott, F. Acernese et al., Phys. Rev. D **109**, 022001 (2024)
8. R. Abbott et al., Phys. Rev. Lett. **125**, 101102 (2020). https://doi.org/10.1103/PhysRevLett.125.101102, arXiv:2009.01075
9. V.I. Afonso, G.J. Olmo, D. Rubiera-Garcia, Phys. Rev. D **97**, 021503 (2018). https://doi.org/10.1103/PhysRevD.97.021503, arXiv:1801.10406
10. N. Afshordi et al., (2023). arXiv:2311.01300
11. R. Akhoury, D. Garfinkle, R. Saotome, JHEP **04**, 096 (2011). https://doi.org/10.1007/JHEP04(2011)096, arXiv:1103.0290
12. K. Akiyama, A. Alberdi, W. Alef et al., Astrophys. J. Lett. **875**, L4 (2019)
13. K. Akiyama, A. Alberdi, W. Alef et al., Astrophys. J. Lett. **930**, L12 (2022)
14. M. Alcubierre, vol. 140 (OUP Oxford, 2008)
15. M. Alcubierre, J. Barranco, A. Bernal et al., Class. Quantum Graviyu **35**, 19LT01 (2018). https://doi.org/10.1088/1361-6382/aadcb6, arXiv:1805.11488
16. M. Alcubierre, J.C. Degollado, D. Nunez et al., Phys. Rev. D **81**, 124018 (2010). https://doi.org/10.1103/PhysRevD.81.124018, arXiv:1003.4767
17. D. Alic, C. Bona-Casas, C. Bona et al., Phys. Rev. D **85**, 064040 (2012). https://doi.org/10.1103/PhysRevD.85.064040, arXiv:1106.2254
18. G. Allwright, L. Lehner, Class. Quantum Gravity **36**, 084001 (2019). https://doi.org/10.1088/1361-6382/ab0ee1, arXiv:1808.07897
19. G. Antoniou, A. Bakopoulos, P. Kanti, Phys. Rev. Lett. **120**, 131102 (2018). https://doi.org/10.1103/PhysRevLett.120.131102, arXiv:1711.03390
20. L. Aresté Saló, K. Clough, P. Figueras, Phys. Rev. Lett. **129**, 261104 (2022). https://doi.org/10.1103/PhysRevLett.129.261104, arXiv:2208.14470
21. L. Aresté Saló, K. Clough, P. Figueras, Phys. Rev. D **108**, 084018 (2023) . https://doi.org/10.1103/PhysRevD.108.084018, arXiv:2306.14966
22. A. Arvanitaki, S. Dimopoulos, S. Dubovsky et al., Phys. Rev. D **81**, 123530 (2010). https://doi.org/10.1103/PhysRevD.81.123530, arXiv:0905.4720
23. F. Atteneder, H.R. Rüter, D. Cors et al., Phys. Rev. D **109**, 044058 (2024). https://doi.org/10.1103/PhysRevD.109.044058, arXiv:2311.16251
24. O. Baake, O. Rinne, Phys. Rev. D **94**, 124016 (2016). https://doi.org/10.1103/PhysRevD.94.124016, arXiv:1610.08352
25. E. Babichev, C. Deffayet, Class. Quantum Gravity **30**, 184001 (2013). https://doi.org/10.1088/0264-9381/30/18/184001, arXiv:1304.7240
26. E. Babichev, C. Deffayet, R. Ziour, Int. J. Mod. Phys. D **18**, 2147–2154 (2009). https://doi.org/10.1142/S0218271809016107, arXiv:0905.2943
27. V. Baibhav, D. Maity, Phys. Rev. D **95**, 024027 (2017)
28. L. Baiotti, Prog. Part. Nucl. Phys. **109**, 103714 (2019). https://doi.org/10.1016/j.ppnp.2019.103714, arXiv:1907.08534
29. J. Balakrishna, E. Seidel, W.-M. Suen, Phys. Rev. D **58**, 104004 (1998). https://doi.org/10.1103/PhysRevD.58.104004, arXiv:gr-qc/9712064
30. J. Balakrishna, H.-A. Shinkai, Phys. Rev. D **58**, 044016 (1998). https://doi.org/10.1103/PhysRevD.58.044016, arXiv:gr-qc/9712065

31. L. Barack et al., Class. Quantum Gravity **36**, 143001 (2019). https://doi.org/10.1088/1361-6382/ab0587, arXiv:1806.05195
32. E. Barausse, M. Bezares, M. Crisostomi et al., JCAP **11** (2022) 050. https://doi.org/10.1088/1475-7516/2022/11/050, arXiv:2207.00443
33. E. Barausse, C. Palenzuela, M. Ponce et al., Phys. Rev. D **87**, 081506 (2013). https://doi.org/10.1103/PhysRevD.87.081506, arXiv:1212.5053
34. T.W. Baumgarte, S.L. Shapiro, Phys. Rev. D **59**, 024007 (1999). https://doi.org/10.1103/PhysRevD.59.024007, arXiv:gr-qc/9810065
35. A. Bauswein, N.-U.F. Bastian, D.B. Blaschke et al., Phys. Rev. Lett. **122**, 061102 (2019). https://doi.org/10.1103/PhysRevLett.122.061102, arXiv:1809.01116
36. A. Bauswein, G. Guo, J.-H. Lien et al., Phys. Rev. D **107**, 083002 (2023). https://doi.org/10.1103/PhysRevD.107.083002, arXiv:2012.11908
37. J.D. Bekenstein, Phys. Rev. D **7**, 949 (1973)
38. I. Bena, D.R. Mayerson, Phys. Rev. Lett. **125**, 221602 (2020). https://doi.org/10.1103/PhysRevLett.125.221602, arXiv:2006.10750
39. R. Benkel, T.P. Sotiriou, H. Witek, Phys. Rev. D **94**, 121503 (2016). https://doi.org/10.1103/PhysRevD.94.121503, arXiv:1612.08184
40. R. Benkel, T.P. Sotiriou, H. Witek, Class. Quant. Grav. **34**, 064001 (2017). https://doi.org/10.1088/1361-6382/aa5ce7, arXiv:1610.09168
41. A. Bernal, J. Barranco, D. Alic et al., Phys. Rev. D **81**, 044031 (2010). https://doi.org/10.1103/PhysRevD.81.044031, arXiv:0908.2435
42. A. Bernal, F.S. Guzman, Phys. Rev. D **74**, 103002 (2006)
43. L. Bernard, L. Lehner, R. Luna, Phys. Rev. D **100**, 024011 (2019). https://doi.org/10.1103/PhysRevD.100.024011, arXiv:1904.12866
44. E. Berti, V. Cardoso, Int. J. Mod. Phys. D **15**, 2209–2216 (2006). https://doi.org/10.1142/S0218271806009637, arXiv:gr-qc/0605101
45. E. Berti, V. Cardoso, L. Gualtieri et al., Phys. Rev. D **87**, 124020 (2013). https://doi.org/10.1103/PhysRevD.87.124020, arXiv:1304.2836
46. M. Bezares, R. Aguilera-Miret, L. ter Haar et al., Phys. Rev. Lett. **128**, 091103 (2022). https://doi.org/10.1103/PhysRevLett.128.091103, arXiv:2107.05648
47. M. Bezares, M. Bošković, S. Liebling et al., Phys. Rev. D **105**, 064067 (2022). https://doi.org/10.1103/PhysRevD.105.064067, arXiv:2201.06113
48. M. Bezares, M. Crisostomi, C. Palenzuela et al., JCAP **03** (2021) 072. https://doi.org/10.1088/1475-7516/2021/03/072, arXiv:2008.07546
49. M. Bezares, C. Palenzuela, Class. Quantum Gravity **35**, 234002 (2018). https://doi.org/10.1088/1361-6382/aae87c, arXiv:1808.10732
50. M. Bezares, C. Palenzuela, C. Bona, Phys. Rev. D **95**, 124005 (2017). https://doi.org/10.1103/PhysRevD.95.124005, arXiv:1705.01071
51. M. Bezares, L. ter Haar, M. Crisostomi et al., Phys. Rev. D **104**, 044022 (2021). https://doi.org/10.1103/PhysRevD.104.044022, arXiv:2105.13992
52. M. Bezares, D. Viganò, C. Palenzuela, Phys. Rev. D **100**, 044049 (2019). https://doi.org/10.1103/PhysRevD.100.044049, arXiv:1905.08551
53. M. Bianchi, D. Consoli, A. Grillo et al., Phys. Rev. Lett. **125**, 221601 (2020). https://doi.org/10.1103/PhysRevLett.125.221601, arXiv:2007.01743
54. P. Bosch, S.R. Green, L. Lehner, Phys. Rev. Lett. **116**, 141102 (2016)
55. Y. Brihaye, A. Cisterna, C. Erices, Phys. Rev. D **93**, 124057 (2016). https://doi.org/10.1103/PhysRevD.93.124057, arXiv:1604.02121
56. Y. Brihaye, B. Hartmann, JHEP **09** (2019) 049. https://doi.org/10.1007/JHEP09(2019)049, arXiv:1903.10471
57. M. Brito, C. Herdeiro, E. Radu et al., Phys. Rev. D **107**, 084022 (2023). https://doi.org/10.1103/PhysRevD.107.084022, arXiv:2302.08900
58. R. Brito, V. Cardoso, C.A.R. Herdeiro et al., Phys. Lett. B **752**, 291–295 (2016). https://doi.org/10.1016/j.physletb.2015.11.051, arXiv:1508.05395

59. R. Brito, V. Cardoso, C.F.B. Macedo et al., Phys. Rev. D **93**, 044045 (2016). https://doi.org/10.1103/PhysRevD.93.044045, arXiv:1512.00466
60. R. Brito, V. Cardoso, C.F.B. Macedo et al., Phys. Rev. D **93**, 044045 (2016). https://doi.org/10.1103/PhysRevD.93.044045, arXiv:1512.00466
61. R. Brito, V. Cardoso, P. Pani, Lect. Notes Phys. **906**, 1–237 (2015). https://doi.org/10.1007/978-3-319-19000-6, arXiv:1501.06570
62. R.P. Brito, Fundamental fields around compact objects: massive spin-2 fields, Superradiant instabilities and Stars with dark matter cores. Ph.D. thesis, Lisbon U. (2016). arXiv:1607.05146
63. M. Buballa et al., J. Phys. G **41**, 123001 (2014). https://doi.org/10.1088/0954-3899/41/12/123001, arXiv:1402.6911
64. H.A. Buchdahl, Phys. Rev. **116**, 1027 (1959). https://doi.org/10.1103/PhysRev.116.1027
65. H.A. Buchdahl, Phys. Rev. **116**, 1027–1034 (1959). https://doi.org/10.1103/PhysRev.116.1027
66. J.C. Bustillo, N. Sanchis-Gual, A. Torres-Forné et al., Phys. Rev. Lett. **126**, 081101 (2021). https://doi.org/10.1103/PhysRevLett.126.081101, arXiv:2009.05376
67. J. Calderon Bustillo, N. Sanchis-Gual, S.H.W. Leong et al., Phys. Rev. D **108**, 123020 (2023). https://doi.org/10.1103/PhysRevD.108.123020, arXiv:2206.02551
68. Z. Cao, P. Galaviz, L.-F. Li, Phys. Rev. D **87**, 104029 (2013). https://doi.org/10.1103/PhysRevD.87.104029, arXiv:1608.07816
69. V. Cardoso, L.C.B. Crispino, C.F.B. Macedo et al., Phys. Rev. D **90**, 044069 (2014). https://doi.org/10.1103/PhysRevD.90.044069, arXiv:1406.5510
70. V. Cardoso, O.J.C. Dias, J.P.S. Lemos et al., Phys. Rev. D **70**, 044039 (2004). https://doi.org/10.1103/PhysRevD.70.049903. [Erratum: Phys. Rev. D 70, 049903 (2004)], arXiv:hep-th/0404096
71. V. Cardoso, E. Franzin, P. Pani, Phys. Rev. Lett. **116**, 171101 (2016). https://doi.org/10.1103/PhysRevLett.116.171101. [Erratum: Phys. Rev. Lett. 117, 089902 (2016)], arXiv:1602.07309
72. V. Cardoso, S. Hopper, C.F.B. Macedo et al., Phys. Rev. D **94**, 084031 (2016). https://doi.org/10.1103/PhysRevD.94.084031, arXiv:1608.08637
73. V. Cardoso, T. Ikeda, R. Vicente et al., Phys. Rev. D **106**, L121302 (2022). https://doi.org/10.1103/PhysRevD.106.L121302, arXiv:2207.09469
74. V. Cardoso, T. Ikeda, Z. Zhong et al., Phys. Rev. D **106**, 044030 (2022). https://doi.org/10.1103/PhysRevD.106.044030, arXiv:2206.00021
75. V. Cardoso, P. Pani, Living Rev. Rel. **22**, 4 (2019). https://doi.org/10.1007/s41114-019-0020-4, arXiv:1904.05363
76. J. Cayuso, N. Ortiz, L. Lehner, Phys. Rev. D **96**, 084043 (2017). https://doi.org/10.1103/PhysRevD.96.084043, arXiv:1706.07421
77. R. Cayuso, P. Figueras, T. França et al., Phys. Rev. Lett. **131**, 111403 (2023). https://doi.org/10.1103/PhysRevLett.131.111403
78. R. Cayuso, L. Lehner, Phys. Rev. D **102**, 084008 (2020). https://doi.org/10.1103/PhysRevD.102.084008, arXiv:2005.13720
79. S. Chakrabarty, Phys. Rev. D **43**, 627–630 (1991). https://doi.org/10.1103/PhysRevD.43.627
80. H.-S. Chan, M.-C. Chu, S.-C. Leung, Astrophys. J. **945**, 133 (2023). https://doi.org/10.3847/1538-4357/acbc1d, arXiv:2301.06093
81. K. Chen, L.-M. Lin, Phys. Rev. D **108**, 064007 (2023). https://doi.org/10.1103/PhysRevD.108.064007, arXiv:2307.01598
82. P.C.-K. Cheong, T.G.F. Li, Phys. Rev. D **100**, 024027 (2019). https://doi.org/10.1103/PhysRevD.100.024027, arXiv:1812.04835
83. P.M. Chesler, D.A. Lowe, Phys. Rev. Lett. **122**, 181101 (2019). https://doi.org/10.1103/PhysRevLett.122.181101, arXiv:1801.09711
84. A. Chodos, R.L. Jaffe, K. Johnson et al., Phys. Rev. D **9**, 3471–3495 (1974). https://doi.org/10.1103/PhysRevD.9.3471
85. A. Chodos, R.L. Jaffe, K. Johnson et al., Phys. Rev. D **10**, 2599 (1974). https://doi.org/10.1103/PhysRevD.10.2599

86. M.W. Choptuik, F. Pretorius, Phys. Rev. Lett. **104**, 111101 (2010). https://doi.org/10.1103/PhysRevLett.104.111101, arXiv:0908.1780
87. K. Clough, T. Dietrich, J.C. Niemeyer, Phys. Rev. D **98**, 083020 (2018). https://doi.org/10.1103/PhysRevD.98.083020, arXiv:1808.04668
88. K. Clough, T. Helfer, H. Witek et al., Phys. Rev. Lett. **129**, 151102 (2022). https://doi.org/10.1103/PhysRevLett.129.151102, arXiv:2204.10868
89. A. Coates, F.M. Ramazanoğlu, Phys. Rev. Lett. **130**, 021401 (2023). https://doi.org/10.1103/PhysRevLett.130.021401, arXiv:2211.08027
90. M. Colpi, S.L. Shapiro, I. Wasserman, Phys. Rev. Lett. **57**, 2485–2488 (1986). https://doi.org/10.1103/PhysRevLett.57.2485
91. G.L. Comer, H.-A. Shinkai, Class. Quantum Gravity **15**, 669–688 (1998). https://doi.org/10.1088/0264-9381/15/3/016, arXiv:gr-qc/9708071
92. F. Corelli, M. De Amicis, T. Ikeda et al., Phys. Rev. D **107**, 044061 (2023). https://doi.org/10.1103/PhysRevD.107.044061, arXiv:2205.13007
93. F. Corelli, M. De Amicis, T. Ikeda et al., Phys. Rev. Lett. **130**, 091501 (2023). https://doi.org/10.1103/PhysRevLett.130.091501, arXiv:2205.13006
94. M. Corman, J.L. Ripley, W.E. East, Phys. Rev. D **107**, 024014 (2023). https://doi.org/10.1103/PhysRevD.107.024014, arXiv:2210.09235
95. R. Croft, T. Helfer, B.-X. Ge et al., Class. Quant. Grav. **40**, 065001 (2023). https://doi.org/10.1088/1361-6382/acace4, arXiv:2207.05690
96. P.V.P. Cunha, E. Berti, C.A.R. Herdeiro, Phys. Rev. Lett. **119**, 251102 (2017). https://doi.org/10.1103/PhysRevLett.119.251102, arXiv:1708.04211
97. P.V.P. Cunha, C. Herdeiro, E. Radu et al., Phys. Rev. Lett. **130**, 061401 (2023). https://doi.org/10.1103/PhysRevLett.130.061401, arXiv:2207.13713
98. T. Damour, G. Esposito-Farese, Phys. Rev. D **54**, 1474–1491 (1996). https://doi.org/10.1103/PhysRevD.54.1474, arXiv:gr-qc/9602056
99. J.C. Degollado, C.A. Herdeiro, H.F. Rúnarsson, Phys. Rev. D **88**, 063003 (2013)
100. F. Di Giovanni, S. Fakhry, N. Sanchis-Gual et al., Phys. Rev. D **102**, 084063 (2020). https://doi.org/10.1103/PhysRevD.102.084063, arXiv:2006.08583
101. F. Di Giovanni, S. Fakhry, N. Sanchis-Gual et al., Class. Quantum Gravity **38**, 194001 (2021). https://doi.org/10.1088/1361-6382/ac1b45, arXiv:2105.00530
102. F. Di Giovanni, D. Guerra, S. Albanesi et al., Phys. Rev. D **106**, 084013 (2022). https://doi.org/10.1103/PhysRevD.106.084013, arXiv:2207.06174
103. F. Di Giovanni, N. Sanchis-Gual, P. Cerdá-Durán et al., Phys. Rev. D **102**, 124009 (2020). https://doi.org/10.1103/PhysRevD.102.124009, arXiv:2010.05845
104. F. Di Giovanni, N. Sanchis-Gual, P. Cerdá-Durán et al., Phys. Rev. D **105**, 063005 (2022). https://doi.org/10.1103/PhysRevD.105.063005, arXiv:2110.11997
105. F. Di Giovanni, N. Sanchis-Gual, D. Guerra et al., Phys. Rev. D **106**, 044008 (2022). https://doi.org/10.1103/PhysRevD.106.044008, arXiv:2206.00977
106. F. Di Giovanni, N. Sanchis-Gual, C.A.R. Herdeiro et al., Phys. Rev. D **98**, 064044 (2018). https://doi.org/10.1103/PhysRevD.98.064044, arXiv:1803.04802
107. T. Dietrich, S. Bernuzzi, M. Ujevic et al., Phys. Rev. D **91**, 124041 (2015). https://doi.org/10.1103/PhysRevD.91.124041, arXiv:1504.01266
108. T. Dietrich, K. Clough, Phys. Rev. D **100**, 083005 (2019). https://doi.org/10.1103/PhysRevD.100.083005, arXiv:1909.01278
109. T. Dietrich, F. Day, K. Clough et al., Mon. Not. R. Astron. Soc. **483** (2019) 908–914. https://doi.org/10.1093/mnras/sty3158, arXiv:1808.04746
110. T. Dietrich, S. Ossokine, K. Clough, Class. Quantum Gravity **36**, 025002 (2019). https://doi.org/10.1088/1361-6382/aaf43e, arXiv:1807.06959
111. A. Dima, M. Bezares, E. Barausse, Phys. Rev. D **104**, 084017 (2021). https://doi.org/10.1103/PhysRevD.104.084017, arXiv:2107.04359
112. A.S. Dmitriev, D.G. Levkov, A.G. Panin et al., Phys. Rev. D **104**, 023504 (2021). https://doi.org/10.1103/PhysRevD.104.023504, arXiv:2104.00962

113. S.R. Dolan, Phys. Rev. D **87**, 124026 (2013). https://doi.org/10.1103/PhysRevD.87.124026, arXiv:1212.1477
114. D. D. Doneva, L. Aresté Saló, K. Clough et al., Phys. Rev. D **108** (2023) 084017. https://doi.org/10.1103/PhysRevD.108.084017, arXiv:2307.06474
115. D.D. Doneva, L. Aresté Saló, S.S. Yazadjiev (2024). arXiv:2404.15526
116. D.D. Doneva, F.M. Ramazanoğlu, H.O. Silva et al., Rev. Mod. Phys. **96**, 015004 (2024). https://doi.org/10.1103/RevModPhys.96.015004, arXiv:2211.01766
117. D.D. Doneva, A. Vañó Viñuales, S.S. Yazadjiev, Phys. Rev. D **106** (2022) L061502. https://doi.org/10.1103/PhysRevD.106.L061502, arXiv:2204.05333
118. D.D. Doneva, S.S. Yazadjiev, Phys. Rev. Lett. **120**, 131103 (2018). https://doi.org/10.1103/PhysRevLett.120.131103, arXiv:1711.01187
119. W.E. East, Phys. Rev. D **96**, 024004 (2017)
120. W.E. East, F. Pretorius, Phys. Rev. Lett. **119**, 041101 (2017)
121. W.E. East, F. Pretorius, Phys. Rev. D **106**, 104055 (2022). https://doi.org/10.1103/PhysRevD.106.104055, arXiv:2208.09488
122. W.E. East, F.M. Ramazanoglu, F. Pretorius, Phys. Rev. D **86**, 104053 (2012). https://doi.org/10.1103/PhysRevD.86.104053, arXiv:1208.3473
123. W.E. East, F.M. Ramazanoğlu, F. Pretorius, Phys. Rev. D **89**, 061503 (2014)
124. W.E. East, J.L. Ripley, Phys. Rev. Lett. **127**, 101102 (2021). https://doi.org/10.1103/PhysRevLett.127.101102, arXiv:2105.08571
125. W.E. East, J.L. Ripley, Phys. Rev. D **103**, 044040 (2021). https://doi.org/10.1103/PhysRevD.103.044040, arXiv:2011.03547
126. M. Elley, H.O. Silva, H. Witek et al., Phys. Rev. D **106**, 044018 (2022). https://doi.org/10.1103/PhysRevD.106.044018, arXiv:2205.06240
127. J. Ellis, A. Hektor, G. Hütsi et al., Phys. Lett. B **781**, 607–610 (2018). https://doi.org/10.1016/j.physletb.2018.04.048, arXiv:1710.05540
128. M. Emma, F. Schianchi, F. Pannarale et al., Particles **5**, 273–286 (2022). https://doi.org/10.3390/particles5030024, arXiv:2206.10887
129. T. Evstafyeva, U. Sperhake, T. Helfer et al., Class. Quantum Gravity **40**, 085009 (2023). https://doi.org/10.1088/1361-6382/acc2a8, arXiv:2212.08023
130. P. Figueras, T. França, Class. Quantm Gravity **37**, 225009 (2020). https://doi.org/10.1088/1361-6382/abb693, arXiv:2006.09414
131. P. Figueras, T. França, Phys. Rev. D **105**, 124004 (2022). https://doi.org/10.1103/PhysRevD.105.124004, arXiv:2112.15529
132. N. Franchini, M. Bezares, E. Barausse et al., Phys. Rev. D **106**, 064061 (2022). https://doi.org/10.1103/PhysRevD.106.064061, arXiv:2206.00014
133. F.F. Freitas, C.A.R. Herdeiro, A.P. Morais et al., JCAP **12**, 047 (2021). https://doi.org/10.1088/1475-7516/2021/12/047, arXiv:2107.09493
134. R. Friedberg, T.D. Lee, Y. Pang, Phys. Rev. D **35**, 3658 (1987). https://doi.org/10.1103/PhysRevD.35.3658
135. J.T. Gálvez Ghersi, L.C. Stein, Phys. Rev. E **104**, 034219 (2021). https://doi.org/10.1103/PhysRevE.104.034219, arXiv:2106.08410
136. D. Gerosa, U. Sperhake, C.D. Ott, Class. Quantum Gravity **33**, 135002 (2016). https://doi.org/10.1088/0264-9381/33/13/135002, arXiv:1602.06952
137. E. Giangrandi, V. Sagun, O. Ivanytskyi et al., Astrophys. J. **953**, 115 (2023). https://doi.org/10.3847/1538-4357/ace104, arXiv:2209.10905
138. T. Gleason, B. Brown, B. Kain, Phys. Rev. D **105**, 023010 (2022). https://doi.org/10.1103/PhysRevD.105.023010, arXiv:2201.02274
139. M. Gleiser, Phys. Rev. D **38**, 2376 (1988). https://doi.org/10.1103/PhysRevD.38.2376. [Erratum: Phys. Rev. D 39, 1257 (1989)]
140. E. Gourgoulhon (Springer, 2012)
141. F.S. Guzmán, Astron. Nachr. **342**, 398–403 (2021). https://doi.org/10.1002/asna.202113941
142. F.S. Guzmán, L.A. Ureña López, Phys. Rev. D **101**, 081302 (2020) . https://doi.org/10.1103/PhysRevD.101.081302, arXiv:1912.10585

143. J. Hadamard (Princeton University Bulletin, 1902), pp. 49–52
144. T. Harada, T. Chiba, K.-I. Nakao et al., Phys. Rev. D **55**, 2024–2037 (1997). https://doi.org/10.1103/PhysRevD.55.2024, arXiv:gr-qc/9611031
145. B. Hartmann, J. Riedel, R. Suciu, Phys. Lett. B **726**, 906–912 (2013). https://doi.org/10.1016/j.physletb.2013.09.050, arXiv:1308.3391
146. S.W. Hawking, Commun. Math. Phys. **25**, 167–171 (1972). https://doi.org/10.1007/BF01877518
147. J. Healy, T. Bode, R. Haas et al., Class. Quantum Gravity **29**, 232002 (2012). https://doi.org/10.1088/0264-9381/29/23/232002, arXiv:1112.3928
148. T. Helfer, E.A. Lim, M.A.G. Garcia et al., Phys. Rev. D **99**, 044046 (2019). https://doi.org/10.1103/PhysRevD.99.044046, arXiv:1802.06733
149. T. Helfer, U. Sperhake, R. Croft et al., Class. Quantum Gravity **39**, 074001 (2022). https://doi.org/10.1088/1361-6382/ac53b7, arXiv:2108.11995
150. A.B. Henriques, A.R. Liddle, R.G. Moorhouse, Phys. Lett. B **233**, 99 (1989). https://doi.org/10.1016/0370-2693(89)90623-0
151. A.B. Henriques, A.R. Liddle, R.G. Moorhouse, Phys. Lett. B **251**, 511–516 (1990). https://doi.org/10.1016/0370-2693(90)90789-9
152. C. Herdeiro, E. Radu, H. Rúnarsson, Class. Quantum Gravity **33**, 154001 (2016). https://doi.org/10.1088/0264-9381/33/15/154001, arXiv:1603.02687
153. C.A. Herdeiro, Black holes: on the universality of the kerr hypothesis, in *Modified and Quantum Gravity: From Theory to Experimental Searches on All Scales* (Springer, 2023), pp. 315–331
154. C.A. Herdeiro, E. Radu, Phys. Rev. Lett. **119**, 261101 (2017)
155. C.A.R. Herdeiro, A.M. Pombo, E. Radu et al., JCAP **04**, 051 (2021). https://doi.org/10.1088/1475-7516/2021/04/051, arXiv:2102.01703
156. C.A.R. Herdeiro, E. Radu, Phys. Rev. Lett. **112**, 221101 (2014). https://doi.org/10.1103/PhysRevLett.112.221101, arXiv:1403.2757
157. C.A.R. Herdeiro, E. Radu, Int. J. Mod. Phys. D **24**, 1542014 (2015). https://doi.org/10.1142/S0218271815420146, arXiv:1504.08209
158. C.A.R. Herdeiro, E. Radu, Int. J. Mod. Phys. D **27**, 1843009 (2018). https://doi.org/10.1142/S0218271818430095, arXiv:1803.08149
159. C.A.R. Herdeiro, E. Radu, N. Sanchis-Gual et al., Phys. Lett. B **852**, 138595 (2024). https://doi.org/10.1016/j.physletb.2024.138595, arXiv:2311.14800
160. D. Hilditch, Int. J. Mod. Phys. A **28**, 1340015 (2013). https://doi.org/10.1142/S0217751X13400150, arXiv:1309.2012
161. D. Horvat, S. Ilijić, A. Kirin et al., Class. Quantum Gravity **32**, 035023 (2015). https://doi.org/10.1088/0264-9381/32/3/035023, arXiv:1407.2067
162. T. Ikeda, M. Bianchi, D. Consoli et al., Phys. Rev. D **104**, 066021 (2021). https://doi.org/10.1103/PhysRevD.104.066021, arXiv:2103.10960
163. P. Ildefonso, M. Zilhão, C. Herdeiro et al., Phys. Rev. D **108**, 064011 (2023). https://doi.org/10.1103/PhysRevD.108.064011, arXiv:2307.00044
164. S. Ilijić, M. Sossich, Phys. Rev. D **102**, 084019 (2020)
165. W. Israel, J.M. Stewart, Phys. Lett. A **58**, 213–215 (1976). https://doi.org/10.1016/0375-9601(76)90075-X
166. V. Jaramillo, N. Sanchis-Gual, J. Barranco et al., Phys. Rev. D **101**, 124020 (2020). https://doi.org/10.1103/PhysRevD.101.124020, arXiv:2004.08459
167. V. Jaramillo, N. Sanchis-Gual, J. Barranco et al., Phys. Rev. D **105**, 104057 (2022). https://doi.org/10.1103/PhysRevD.105.104057, arXiv:2202.00696
168. P. Jetzer, Phys. Lett. B **222**, 447–452 (1989). https://doi.org/10.1016/0370-2693(89)90342-0
169. P. Jetzer, Phys. Rept. **220**, 163–227 (1992). https://doi.org/10.1016/0370-1573(92)90123-H
170. P. Kanti, N.E. Mavromatos, J. Rizos et al., Phys. Rev. D **54**, 5049–5058 (1996). https://doi.org/10.1103/PhysRevD.54.5049, arXiv:hep-th/9511071
171. D.R. Karkevandi, S. Shakeri, V. Sagun et al., Phys. Rev. D **105**, 023001 (2022). https://doi.org/10.1103/PhysRevD.105.023001, arXiv:2109.03801

172. D.J. Kaup, Phys. Rev. **172**, 1331–1342 (1968). https://doi.org/10.1103/PhysRev.172.1331
173. J. Kerimo, Phys. Rev. D **62**, 104005 (2000). https://doi.org/10.1103/PhysRevD.62.104005
174. J. Kerimo, D. Kalligas, Phys. Rev. D **58**, 104002 (1998). https://doi.org/10.1103/PhysRevD.58.104002
175. B. Kleihaus, J. Kunz, S. Schneider, Phys. Rev. D **85**, 024045 (2012). https://doi.org/10.1103/PhysRevD.85.024045, arXiv:1109.5858
176. A.D. Kovács, H.S. Reall, Phys. Rev. D **101**, 124003 (2020). https://doi.org/10.1103/PhysRevD.101.124003, arXiv:2003.08398
177. A.D. Kovács, H.S. Reall, Phys. Rev. Lett. **124**, 221101 (2020). https://doi.org/10.1103/PhysRevLett.124.221101, arXiv:2003.04327
178. H.-J. Kuan, J. Singh, D.D. Doneva et al., Phys. Rev. D **104**, 124013 (2021). https://doi.org/10.1103/PhysRevD.104.124013, arXiv:2105.08543
179. H.-J. Kuan, A.G. Suvorov, D.D. Doneva et al., Phys. Rev. Lett. **129**, 121104 (2022). https://doi.org/10.1103/PhysRevLett.129.121104, arXiv:2203.03672
180. G. Lara, M. Bezares, E. Barausse, Phys. Rev. D **105**, 064058 (2022). https://doi.org/10.1103/PhysRevD.105.064058, arXiv:2112.09186
181. G. Lara, H.P. Pfeiffer, N.A. Wittek et al. (2024). arXiv:2403.08705
182. T. Lee, Y. Pang, Nuclear Phys. B **315**, 477–516 (1989)
183. T.D. Lee, Y. Pang, Phys. Rep. **221**, 251–350 (1992). https://doi.org/10.1016/0370-1573(92)90064-7
184. C.D. Leonard, J. Ziprick, G. Kunstatter et al., JHEP **10**, 028 (2011). https://doi.org/10.1007/JHEP10(2011)028, arXiv:1106.2054
185. S.-C. Leung, S. Zha, M.-C. Chu et al., Astrophys. J. **884**, 9 (2019). https://doi.org/10.3847/1538-4357/ab3b5e, arXiv:1908.05102
186. S.L. Liebling, C. Palenzuela, Living Rev. Rel. **26**, 1 (2023). https://doi.org/10.1007/s41114-023-00043-4, arXiv:1202.5809
187. L.-M. Lin, J. Novak, Class. Quantum Gravity **23**, 4545 (2006). https://doi.org/10.1088/0264-9381/23/14/001, arXiv:gr-qc/0603048
188. O. Lunin, S.D. Mathur, Nucl. Phys. B **623**, 342–394 (2002). https://doi.org/10.1016/S0550-3213(01)00620-4, arXiv:hep-th/0109154
189. O. Lunin, S.D. Mathur, Phys. Rev. Lett. **88**, 211303 (2002). https://doi.org/10.1103/PhysRevLett.88.211303, arXiv:hep-th/0202072
190. E. Maggio, P. Pani, G. Raposo (2021). arXiv:2105.06410
191. J.M. Marti, J.M. Ibanez, J.A. Miralles, Phys. Rev. D **43**, 3794–3801 (1991). https://doi.org/10.1103/PhysRevD.43.3794
192. A. Masó-Ferrando, N. Sanchis-Gual, J.A. Font et al., Class. Quant. Grav. **38**, 194003 (2021). https://doi.org/10.1088/1361-6382/ac1fd0, arXiv:2103.15705
193. A. Masó-Ferrando, N. Sanchis-Gual, J.A. Font et al., JCAP **06**, 028 (2023). https://doi.org/10.1088/1475-7516/2023/06/028, arXiv:2304.12018
194. A. Masó-Ferrando, N. Sanchis-Gual, J.A. Font et al., Phys. Rev. D **109**, 044042 (2024). https://doi.org/10.1103/PhysRevD.109.044042, arXiv:2309.14912
195. S.D. Mathur, Fortsch. Phys. **53**, 793–827 (2005). https://doi.org/10.1002/prop.200410203, arXiv:hep-th/0502050
196. S.D. Mathur (2008). arXiv:0810.4525
197. T. Matos, L.A. Urena-Lopez, Class. Quantum Gravity **17**, L75–L81 (2000). https://doi.org/10.1088/0264-9381/17/13/101, arXiv:astro-ph/0004332
198. R.F.P. Mendes, N. Ortiz, Phys. Rev. D **93**, 124035 (2016). https://doi.org/10.1103/PhysRevD.93.124035, arXiv:1604.04175
199. R.F.P. Mendes, N. Ortiz, N. Stergioulas, Phys. Rev. D **104**, 104036 (2021). https://doi.org/10.1103/PhysRevD.104.104036, arXiv:2107.07036
200. B.C. Mundim. A Numerical Study of Boson Star Binaries. Ph.D. thesis (2010)
201. J. Novak, J.M. Ibanez, Astrophys. J. **533**, 392–405 (2000). https://doi.org/10.1086/308627, arXiv:astro-ph/9911298

202. J.E. Nyhan, B. Kain, Phys. Rev. D **105**, 123016 (2022). https://doi.org/10.1103/PhysRevD.105.123016, arXiv:2206.07715
203. H. Okawa, H. Witek, V. Cardoso, Phys. Rev. D **89**, 104032 (2014)
204. M. Okounkova, L.C. Stein, J. Moxon et al., Phys. Rev. D **101**, 104016 (2020). https://doi.org/10.1103/PhysRevD.101.104016, arXiv:1911.02588
205. M. Okounkova, L.C. Stein, M.A. Scheel et al., Phys. Rev. D **96**, 044020 (2017). https://doi.org/10.1103/PhysRevD.96.044020, arXiv:1705.07924
206. M. Okounkova, L.C. Stein, M.A. Scheel et al., Phys. Rev. D **100**, 104026 (2019). https://doi.org/10.1103/PhysRevD.100.104026, arXiv:1906.08789
207. H. Olivares, Z. Younsi, C. M. Fromm et al., Mon. Not. R. Astron. Soc. **497** (2020) 521–535. https://doi.org/10.1093/mnras/staa1878, arXiv:1809.08682
208. F. Özel, P. Freire, Ann. Rev. Astron. Astrophys. **54**, 401–440 (2016). https://doi.org/10.1146/annurev-astro-081915-023322, arXiv:1603.02698
209. C. Palenzuela, E. Barausse, M. Ponce et al., Phys. Rev. D **89**, 044024 (2014). https://doi.org/10.1103/PhysRevD.89.044024, arXiv:1310.4481
210. C. Palenzuela, L. Lehner, S.L. Liebling, Phys. Rev. D **77**, 044036 (2008). https://doi.org/10.1103/PhysRevD.77.044036, arXiv:0706.2435
211. C. Palenzuela, I. Olabarrieta, L. Lehner et al., Phys. Rev. D **75**, 064005 (2007). https://doi.org/10.1103/PhysRevD.75.064005, arXiv:gr-qc/0612067
212. C. Palenzuela, P. Pani, M. Bezares et al., Phys. Rev. D **96**, 104058 (2017). https://doi.org/10.1103/PhysRevD.96.104058, arXiv:1710.09432
213. P. Pani, E. Berti, V. Cardoso et al., Phys. Rev. D **84**, 104035 (2011). https://doi.org/10.1103/PhysRevD.84.104035, arXiv:1109.0928
214. G. Papallo, H.S. Reall, Phys. Rev. D **96**, 044019 (2017). https://doi.org/10.1103/PhysRevD.96.044019, arXiv:1705.04370
215. S. Ponglertsakul, E. Winstanley, S.R. Dolan, Phys. Rev. D **94**, 024031 (2016). https://doi.org/10.1103/PhysRevD.94.024031, arXiv:1604.01132
216. W.H. Press, S.A. Teukolsky, Nature **238**, 211–212 (1972)
217. I. Quiros, Int. J. Mod. Phys. D **28**, 1930012 (2019). https://doi.org/10.1142/S021827181930012X, arXiv:1901.08690
218. G. Raposo, P. Pani, M. Bezares et al., Phys. Rev. D **99**, 104072 (2019). https://doi.org/10.1103/PhysRevD.99.104072, arXiv:1811.07917
219. J.L. Ripley, Int. J. Mod. Phys. D **31**, 2230017 (2022). https://doi.org/10.1142/S0218271822300178, arXiv:2207.13074
220. J.L. Ripley, F. Pretorius, Class. Quantum Gravity **36**, 134001 (2019). https://doi.org/10.1088/1361-6382/ab2416, arXiv:1903.07543
221. J.L. Ripley, F. Pretorius, Phys. Rev. D **99**, 084014 (2019). https://doi.org/10.1103/PhysRevD.99.084014, arXiv:1902.01468
222. J.L. Ripley, F. Pretorius, Class. Quantum Gravity **37**, 155003 (2020). https://doi.org/10.1088/1361-6382/ab9bbb, arXiv:2005.05417
223. J.V. Romero, J.M. Ibanez, J.M. Marti et al., Astrophys. J. **462**, 839 (1996)
224. A.A. Roque, L.A. Ureña López, Class. Quantum Gravity **39** (2022) 044001. https://doi.org/10.1088/1361-6382/ac4614, arXiv:2109.14747
225. R. Rosca-Mead, C.J. Moore, M. Agathos et al., Class. Quantum Gravity **36**, 134003 (2019). https://doi.org/10.1088/1361-6382/ab256f, arXiv:1903.09704
226. R. Rosca-Mead, U. Sperhake, C.J. Moore et al., Phys. Rev. D **102**, 044010 (2020). https://doi.org/10.1103/PhysRevD.102.044010, arXiv:2005.09728
227. V.C. Rubin, W.K. Ford Jr., N. Thonnard, Astrophys. J. Part 2-Letters to the Editor, **225**, L107–L111 (1978)
228. R. Ruffini, S. Bonazzola, Phys. Rev. **187**, 1767–1783 (1969). https://doi.org/10.1103/PhysRev.187.1767
229. H.R. Rüter, D. Hilditch, M. Bugner et al., Phys. Rev. D **98**, 084044 (2018). https://doi.org/10.1103/PhysRevD.98.084044, arXiv:1708.07358

230. H.R. Rüter, V. Sagun, W. Tichy et al., Phys. Rev. D **108**, 124080 (2023). https://doi.org/10.1103/PhysRevD.108.124080, arXiv:2301.03568
231. N. Sanchis-Gual, J.C. Bustillo, C. Herdeiro et al., Phys. Rev. D **106**, 124011 (2022)
232. N. Sanchis-Gual, J.C. Degollado, C. Herdeiro et al., Phys. Rev. D **94**, 044061 (2016). https://doi.org/10.1103/PhysRevD.94.044061, arXiv:1607.06304
233. N. Sanchis-Gual, J.C. Degollado, P.J. Montero et al., Phys. Rev. Lett. **116**, 141101 (2016). https://doi.org/10.1103/PhysRevLett.116.141101, arXiv:1512.05358
234. N. Sanchis-Gual, F. Di Giovanni, C. Herdeiro et al., Phys. Rev. Lett. **126**, 241105 (2021). https://doi.org/10.1103/PhysRevLett.126.241105, arXiv:2103.12136
235. N. Sanchis-Gual, F. Di Giovanni, M. Zilhão et al., Phys. Rev. Lett. **123**, 221101 (2019). https://doi.org/10.1103/PhysRevLett.123.221101, arXiv:1907.12565
236. N. Sanchis-Gual, C. Herdeiro, J.A. Font et al., Phys. Rev. D **99**, 024017 (2019). https://doi.org/10.1103/PhysRevD.99.024017, arXiv:1806.07779
237. N. Sanchis-Gual, C. Herdeiro, E. Radu, Class. Quantum Gravity **39**, 064001 (2022). https://doi.org/10.1088/1361-6382/ac4b9b, arXiv:2110.03000
238. N. Sanchis-Gual, P. Izquierdo, Phys. Rev. D **105**, 084023 (2022). https://doi.org/10.1103/PhysRevD.105.084023, arXiv:2202.00434
239. N. Sanchis-Gual, M. Zilhão, V. Cardoso, Phys. Rev. D **106**, 064034 (2022). https://doi.org/10.1103/PhysRevD.106.064034, arXiv:2207.05494
240. N. Sanchis-Gual, M. Zilhão, C. Herdeiro et al., Phys. Rev. D **102**, 101504 (2020). https://doi.org/10.1103/PhysRevD.102.101504, arXiv:2007.11584
241. N.M. Santos, C.L. Benone, C.A.R. Herdeiro (2024). arXiv:2404.07257
242. M.A. Scheel, S.L. Shapiro, S.A. Teukolsky, Phys. Rev. D **51**, 4208–4235 (1995). https://doi.org/10.1103/PhysRevD.51.4208, arXiv:gr-qc/9411025
243. M.A. Scheel, S.L. Shapiro, S.A. Teukolsky, Phys. Rev. D **51**, 4236–4249 (1995). https://doi.org/10.1103/PhysRevD.51.4236, arXiv:gr-qc/9411026
244. F.E. Schunck, E.W. Mielke, Class. Quant. Grav. **20**, R301–R356 (2003). https://doi.org/10.1088/0264-9381/20/20/201, arXiv:0801.0307
245. E. Seidel, W.-M. Suen, Phys. Rev. Lett. **72**, 2516–2519 (1994). https://doi.org/10.1103/PhysRevLett.72.2516, arXiv:gr-qc/9309015
246. M. Shibata, T. Nakamura, Phys. Rev. D **52**, 5428–5444 (1995). https://doi.org/10.1103/PhysRevD.52.5428
247. M. Shibata, K.-I. Nakao, T. Nakamura, Phys. Rev. D **50**, 7304–7317 (1994). https://doi.org/10.1103/PhysRevD.50.7304
248. M. Shibata, K. Taniguchi, H. Okawa et al., Phys. Rev. D **89**, 084005 (2014). https://doi.org/10.1103/PhysRevD.89.084005, arXiv:1310.0627
249. M. Shibata, D. Traykova, Phys. Rev. D **107**, 044068 (2023). https://doi.org/10.1103/PhysRevD.107.044068, arXiv:2210.12139
250. N. Siemonsen, W.E. East, Phys. Rev. D **103**, 044022 (2021). https://doi.org/10.1103/PhysRevD.103.044022, arXiv:2011.08247
251. N. Siemonsen, W.E. East, Phys. Rev. D **107**, 124018 (2023). https://doi.org/10.1103/PhysRevD.107.124018, arXiv:2302.06627
252. N. Siemonsen, W.E. East, Phys. Rev. D **108**, 124015 (2023). https://doi.org/10.1103/PhysRevD.108.124015, arXiv:2306.17265
253. H.O. Silva, C.F.B. Macedo, T.P. Sotiriou et al., Phys. Rev. D **99**, 064011 (2019). https://doi.org/10.1103/PhysRevD.99.064011, arXiv:1812.05590
254. H.O. Silva, H. Witek, M. Elley et al., Phys. Rev. Lett. **127**, 031101 (2021). https://doi.org/10.1103/PhysRevLett.127.031101, arXiv:2012.10436
255. T.P. Sotiriou, Class. Quant. Grav. **32**, 214002 (2015). https://doi.org/10.1088/0264-9381/32/21/214002, arXiv:1505.00248
256. T.P. Sotiriou, V. Faraoni, S. Liberati, Int. J. Mod. Phys. D **17**, 399–423 (2008). https://doi.org/10.1142/S0218271808012097, arXiv:0707.2748
257. U. Sperhake, C.J. Moore, R. Rosca et al., Phys. Rev. Lett. **119**, 201103 (2017). https://doi.org/10.1103/PhysRevLett.119.201103, arXiv:1708.03651

258. L. ter Haar, M. Bezares, M. Crisostomi et al., Phys. Rev. Lett. **126**, 091102 (2021). https://doi.org/10.1103/PhysRevLett.126.091102, arXiv:2009.03354
259. F. Thaalba, M. Bezares, N. Franchini et al., Phys. Rev. D **109**, L041503 (2024). https://doi.org/10.1103/PhysRevD.109.L041503, arXiv:2306.01695
260. D.F. Torres, F.E. Schunck, A.R. Liddle, Class. Quantum Gravity **15**, 3701–3718 (1998). https://doi.org/10.1088/0264-9381/15/11/027, arXiv:gr-qc/9803094
261. S. Valdez-Alvarado, R. Becerril, L. A. Ureña López, Phys. Rev. D **102**, 064038 (2020). https://doi.org/10.1103/PhysRevD.102.064038, arXiv:2001.11009
262. S. Valdez-Alvarado, C. Palenzuela, D. Alic et al., Phys. Rev. D **87**, 084040 (2013). https://doi.org/10.1103/PhysRevD.87.084040, arXiv:1210.2299
263. L. Visinelli, Int. J. Mod. Phys. D **30**, 2130006 (2021). https://doi.org/10.1142/S0218271821300068, arXiv:2109.05481
264. A.W. Whinnett, D.F. Torres, Phys. Rev. D **60**, 104050 (1999). https://doi.org/10.1103/PhysRevD.60.104050, arXiv:gr-qc/9905017
265. J. Y. Widdicombe, T. Helfer, E.A. Lim, JCAP **01**, 027 (2020). https://doi.org/10.1088/1475-7516/2020/01/027, arXiv:1910.01950
266. E. Witten, Phys. Rev. D **30**, 272–285 (1984). https://doi.org/10.1103/PhysRevD.30.272
267. S.S. Yazadjiev, D.D. Doneva, Phys. Rev. D **99**, 084011 (2019). https://doi.org/10.1103/PhysRevD.99.084011, arXiv:1901.06379
268. S. Yoshida, Y. Eriguchi, Phys. Rev. D **56**, 762–771 (1997). https://doi.org/10.1103/PhysRevD.56.762
269. Z. Zhong, V. Cardoso, T. Ikeda et al., Phys. Rev. D **108**, 084051 (2023). https://doi.org/10.1103/PhysRevD.108.084051, arXiv:2307.02548
270. E. Zhou, K. Kiuchi, M. Shibata et al., Phys. Rev. D **103**, 123011 (2021). https://doi.org/10.1103/PhysRevD.103.123011, arXiv:2105.07498
271. E. Zhou, K. Kiuchi, M. Shibata et al., Phys. Rev. D **106**, 103030 (2022). https://doi.org/10.1103/PhysRevD.106.103030, arXiv:2111.00958
272. E. Zhou, A. Tsokaros, L. Rezzolla et al., Phys. Rev. D **97**, 023013 (2018). https://doi.org/10.1103/PhysRevD.97.023013, arXiv:1711.00198
273. E. Zhou, A. Tsokaros, K. Uryu et al., Phys. Rev. D **100**, 043015 (2019). https://doi.org/10.1103/PhysRevD.100.043015, arXiv:1902.09361
274. Z. Zhu, L. Rezzolla, Phys. Rev. D **104**, 083004 (2021). https://doi.org/10.1103/PhysRevD.104.083004, arXiv:2102.07721
275. M. Zilhão, H. Witek, V. Cardoso, Class. Quantum Gravity **32**, 234003 (2015). https://doi.org/10.1088/0264-9381/32/23/234003, arXiv:1505.00797

Chapter 19
Supernova Simulations

Bernhard Müller

Abstract Magnetohydrodynamic simulations of core-collapse supernovae have become increasingly mature and important in recent years. Magnetic fields take center stage in scenarios for explaining hypernova explosions, but are now also considered in supernova theory more broadly as an important factor even in neutrino-driven explosions, especially in the context of neutron star birth properties. Here we present an overview of simulation approaches currently used for magnetohydrodynamic supernova simulations and sketch essential physical concepts for understanding the role of magnetic fields in supernovae of slowly or rapidly rotating massive stars. We review progress on simulations of neutrino-driven supernovae, magnetorotational supernovae, and the relevant field amplification processes. Recent results on the nucleosynthesis and gravitational wave emission from magnetorotational supernovae are also discussed. We highlight efforts to provide better initial conditions for magnetohydrodynamic supernova models by simulating short phases of the progenitor evolution in 3D to address uncertainties in the treatment of rotation and magnetic fields in current stellar evolution models.

19.1 Introduction

Just like during the later lives of neutron stars and accreting black holes that were discussed in the preceding chapters, magnetic fields play an important role in the formation for these relativistic compact objects by stellar collapse. By far the most important formation channel is the collapse of massive stars that have proceeded through advanced nuclear burning stages to form an iron core that then becomes unstable to collapse due to electron captures and photodisintegration of heavy nuclei [14, 180]. In most instances, the collapse of the iron core also results in an explosion that ejects

B. Müller (✉)
School of Physics and Astronomy, Monash University, 10 College Walk,
Clayton 3800, VIC, Australia
e-mail: bernhard.mueller@monash.edu

C. Bambi et al. (eds.), *New Frontiers in GRMHD Simulations*, Springer Series
in Astrophysics and Cosmology, https://doi.org/10.1007/978-981-97-8522-3_19

the outer layers of the star as a *core-collapse supernova*.[1] While the association of observed core-collapse supernovae with compact remnants [72, 75] and with massive progenitor stars [183, 184] is now solidly established, the explosion mechanism has proved more difficult to unravel as electromagnetic observations cannot directly probe the supernova core. While neutrinos [129] and gravitational waves [2, 83] may in future supply direct information on the supernova engine, simulations remain an essential tool for understanding the inner working of core-collapse supernovae.

Magnetic fields were identified as a crucial piece of the puzzle early on in the quest for the core-collapse supernova explosion mechanism. In the early history of the field, the extraction of rotational energy from the collapsed core was seriously considered as a generic supernova mechanism [15, 16, 92, 116], but a more nuanced and tempered picture of the role of magnetic fields in core-collapse supernovae has emerged since then. Current understanding of stellar rotation points to rather slow core rotation rates in massive stars [71, 100] and is therefore not compatible with magnetorotational powering as a generic supernova mechanism. Explosions driven by rotation and magnetic fields, either in the context of rapidly spinning "milliseconds magnetars" [46, 199] or black holes with accretion disks (collapsars) [101, 102] remain the most promising scenario for hyperenergetic supernovae ("hypernovae", observationally classified as broad-lined Ic supernovae) [208], which account for about 1% of supernovae in the local Universe.

This, however, has only made the study of magnetohydrodynamics (MHD) in core-collapse supernovae much more nuanced and certainly no less relevant. Three-dimensional (3D) simulations of MHD-driven explosions are now readily available and inching towards a meaningful comparison, and there are incipient efforts to better connect them to the intricate problem of angular momentum transport and magnetic field amplification in stellar evolution. The diversity of extreme supernovae from classical hypernovae to superluminous supernovae opens up the possibility of a greater variety of scenarios for MHD-powered explosions. Furthermore, it is increasingly recognized that magnetic fields probably play an important role in supernovae with ordinary explosion energies as well, albeit not as the primary driver of the explosion. Perhaps most importantly, MHD effects in core-collapse supernovae are key to the understanding of neutron star birth spin periods and magnetic fields.

19.2 Progenitor Structure and Post-collapse Structure of the Supernova Core

It is useful to first outline the conditions in the progenitor star and in the core of the supernova after collapse to provide essential background to the reader. This will also serve to organize our discussion of MHD in core-collapse supernovae by charting

[1] Note that this term is sometimes applied more broadly to the entire compact object formation event, even when no explosion occurs.

the conditions pertinent to the evolution and impact of magnetic fields in different regions of the stellar core for different pathways of collapse and explosion.

19.2.1 Conditions in the Progenitor Star

Many of the conditions for the magnetohydrodynamic evolution during the supernova are already set prior to the collapse of the iron core. The onion shell structure of the massive progenitor star, specifically the pre-collapse density profiles, determine the accretion rate onto the young proto-neutron star after the collapse of the iron core, which in turn affects the ram pressure onto the supernova shock and the accretion power that can be converted into neutrinos and/or magnetic fields. For perfectly spherical and non-rotating stellar models, the pre-collapse density structure is therefore the primary determinant for whether or not an explosion will occur or not, and for explosion and remnant properties [50, 132]. Real stars will generally rotate, will exhibit asphericities, e.g., in convective regions [36, 52, 128, 130, 209], and will contain seed magnetic fields. For the magnetohydrodynamic evolution after collapse, the seed fields are relevant because they may jump-start the amplification of magnetic fields in the supernova core, though they likely become less relevant on longer time scales, when field amplification processes in the young neutron star and its vicinity have had sufficient time to reach saturation levels. The angular momentum distribution of the progenitor influences the rotational energy of the young neutron star after collapse, the degree of differential rotation within it and around surface, and the possibility of disk formation due to centrifugal support. It therefore impacts magnetic field amplification by differential rotation after collapse in the neutron star and its environment, and the rotational energy reservoirs that magnetic fields can tap and potentially convert into outflows.

Except for tentative ventures into 3D stellar hydrodynamics (see Sect. 19.4.2), magnetic fields and rotation currently need to be modeled using effective recipes for magnetic field amplification and angular momentum transport in one-dimensional (1D) stellar evolution models. This introduces considerable quantitative uncertainties, especially since current 1D stellar evolution models struggle to fit constraints on interior rotation stellar profiles from asteroseismology [13, 31, 47]. Naturally, the pre-collapse field geometry remains a major unknown, even though features of the field geometry can sometimes be deduced from the nature of the underlying amplification mechanisms. For example, field amplification by differential rotation via the Tayler-Spruit dynamo [185] is expected to produce predominantly toroidal fields.

The current paradigm of stellar evolution [56, 71, 100], which is informed by measurements for low-mass post-main sequence stars and white-dwarfs, suggests that core spin rates of massive stars are typically in the range of tens to hundreds of seconds [71], and that core magnetic fields are of order 10^9 G with predominantly toroidal components. Rare evolutionary scenarios that avoid strong core spin-down and achieve pre-collapse core rotation rates of $\sim$1 s can also be accommodated [3]. These would translate into millisecond-period neutron star spin periods after collapse without angular momentum redistribution and possibly provide conditions for various flavor of MHD-driven explosions for hypernovae. However, such rapid rotation rates cannot be plausibly combined with pre-collapse magnetic fields stronger than 10^9 – 10^{10} G because higher field strengths would come at the expense of more efficient core spin-down. In addition, there may be a route towards very slow pre-collapse rotation and unusually strong magnetic fields. Very strong fields can be generated in stellar mergers, and if these can be retained as fossil fields up to core collapse, central field strengths 10^{11} – 10^{12} G may be reached [176, 177]. After some inevitable reconnection, such merger-generated fields are expected to settle to a twisted-torus configuration whose decay is inhibited due approximate conservation of magnetic helicity [20].

19.2.2 Structure of the Supernova Core After Collapse

Eventually the iron core of the progenitor star becomes unstable to gravitational collapse due to deleptonisation and photo-disintegration of heavy nuclei [14, 180]. The collapse proceeds beyond nuclear saturation density and is eventually halted by the stiffening of the nuclear equation of state, which leads to a rebound ("bounce") of the inner core, and launches a shock wave into the surrounding shells that are still collapsing supersonically. Even for rapid progenitor rotation, centrifugal forces cannot stop the collapse below nuclear densities [43].

Due to the dissociation of infalling heavy nuclei in the infalling matter and rapid neutrino emission in the neutrino burst, the shock weakens and turns into an accretion shock within milliseconds [14, 95, 158]. What emerges in the supernova core is a layered structure with a high-density "proto-neutron star" at the center, an accretion shock that is initially driven to a radius about 100–200 km, and a subsonic accretion flow from the supernova shock down to the proto-neutron star (Fig. 19.1). Some mechanism for extracting energy from the proto-neutron star and/or the accretion flow is required to "revive" the shock and initiate an explosion. The two currently favored scenarios are energy deposition by neutrinos from the proto-neutron star and its surface (neutrino-driven mechanism) or the extraction of rotational energy from the proto-neutron star by magnetic fields (magnetorotational mechanism), see [14, 77, 130] for reviews.

The proto-neutron star itself consists of a low-entropy, high-density core, a hotter, roughly adiabatically stratified mantle of densities around 10^{14} g cm^{-3}, and an atmosphere (surface) with a steep positive entropy gradient and negative density gradient in the region where neutrinos decouple from matter and cool the proto-neutron star surface. The region between the proto-neutron star and the shock is roughly adiabatically stratified, and will later experience net heating by neutrinos (at which point it is termed *heating* or *gain region*). This stratification of the supernova core region is critical for magnetohydrodynamic effects because it determines the nature of instabilities and flow patterns that will dictate magnetic field amplification. The proto-neutron star mantle and the gain region tend to permit instabilities that involve radial fluid

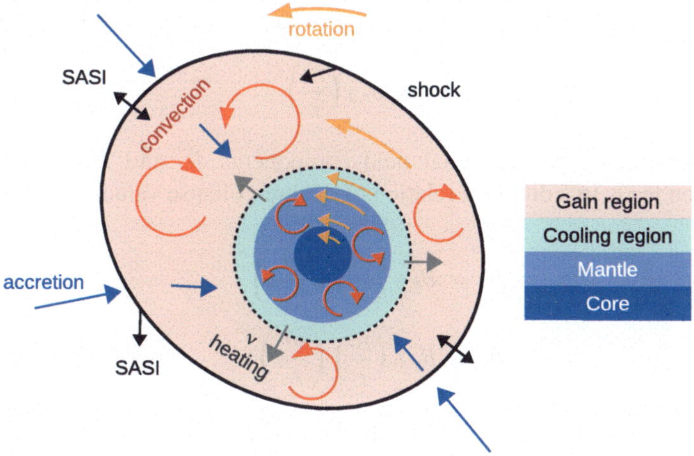

Fig. 19.1 Sketch of the structure of the supernova core during the post-bounce phase before the onset of explosion, top view along the rotation axis. The proto-neutron star consists of a non-convective core, a convective mantle, and a convectively stable cooling layer at the surface (progressively lighter shades of blue). Neutrinos emitted from the proto-neutron star (gray) deposit energy in the gain region (pink) behind the shock (black), and can also drive convective overturn there. Note also the accretion flow (blue arrows) onto the proto-neutron star, which is supersonic outside and subsonic inside the shock. In addition to convective flow in the gain region and the proto-neutron star (indicated by red circular arrows), the shock may be unstable to SASI oscillations (black arrows). Differential rotation may be present throughout the supernova core (orange arrow). Convection, SASI, differential rotation and shear flow in and near the cooling region can drive amplification of magnetic fields individually or in concert

flow (e.g., convection, [28, 74, 78]), whereas the proto-neutron star surface and the core-mantle interface are convectively stable, but may permit magnetic field amplification by shear flows. With minor modifications, this overall structure still persists initially after the shock is revived and starts to move outward. Accretion downflows and neutrino-driven outflows probably coexist in the gain region for several seconds [19, 127, 136], fueling fresh material down to the bottom of the gain region for efficient heating and then transporting the hot neutrino-heated material outward to energize the explosion. Eventually, accretion will subside, and a neutrino-driven wind outflow from the proto-neutron star will develop [45], although the onset of a "true" spherical wind is much delayed in recent simulations [19, 127, 136, 205].

With our focus on magnetic field effects in core-collapse supernovae, the rotational state of the supernova core and the magnetic field configuration after collapse are of critical importance. As a first crude approximation, one can assume that the collapse is perfectly spherical, in which case the amplification of magnetic fields during collapse by compression can be obtained easily from flux conservation. The radial magnetic field component B_r after collapse is given in terms of the initial field strength $B_{r,0}$ and the initial and final radial coordinates r_0 and r of the fluid element under consideration as

$$B_r = B_{r,0} \left(\frac{r_0}{r}\right)^2. \quad (19.1)$$

For the meridional and azimuthal field components B_θ and B_φ, magnetic flux conservation leads to a different relation to the pre-collapse values $B_{\theta,0}$ and $B_{\varphi,0}$,

$$B_\theta = B_{\theta,0} \left(\frac{r_0}{r}\right) \left(\frac{\partial r}{\partial r_0}\right)^{-1}, \quad (19.2)$$

$$B_\varphi = B_{\varphi,0} \left(\frac{r_0}{r}\right) \left(\frac{\partial r}{r_0}\right)^{-1}. \quad (19.3)$$

If the collapse is homologous ($r \propto r_0$), the result is the same as for B_r. Homologous (self-similar) collapse [58, 210] holds very well for the inner part of the iron core out to mass coordinate of $\sim 0.5 M_\odot$ for no or slow rotation [90], and the scaling of the magnetic field with r^{-2} remains a useful zeroth-order approximation in other contexts as well. Flux conservation alone would imply that for progenitor fields of 10^9 – 10^{10} G, the collapse of an iron core of $\sim$1000 km in radius to a neutron star of $\sim$12 km, fields of order $\lesssim 10^{14}$ G will be reached in the neutron star. This, however, refers to the field strength at the *center* of the *final*, cold neutron star. The warm proto-neutron star in the supernova core maintains several times its final radius for several hundreds of milliseconds after bounce so that the surface field strengths that can be reached by compression during the collapse are about two orders of magnitude lower.

For the evolution of the angular velocity ω during (spherical) collapse, angular momentum conservation also leads to a scaling with r^{-2},

$$\omega \approx \omega_0 \left(\frac{r_0^2}{r}\right)^2, \quad (19.4)$$

in terms of the pre-collapse angular velocity ω_0 of a fluid element. This leads to an interesting corollary. Differential rotation can be a key driver for magnetic field amplification, but magnetic fields in turn tend to counteract differential rotation, which has long led to the notion that the core and the surrounding burning shells in massive stars are close to rigid rotation (for an extensive discussion, see [115]), although they may rotate differentially with respect to each other. Under the assumption of spherically symmetric contraction, the rate of differential rotation after collapse is given by

$$\frac{\partial \omega}{\partial r} = \frac{\partial r^{-2} j(r)}{\partial r} = \frac{\partial r^{-2}(r_0^2 \omega_0(r))}{\partial r} = \frac{r_0^2 \partial \omega_0/\partial r_0}{r^2 \partial r/\partial r_0} + \frac{2 r_0 \omega_0}{r^2 \partial r/\partial r_0} - \frac{2 r_0^2 \omega_0}{r^3}, \quad (19.5)$$

where j denotes the specific angular momentum of a fluid element. This implies that differential rotation ($\partial\omega/\partial r \neq 0$) will emerge in regions that collapse non-homologously even from shells that rotate uniformly in the progenitor. Shear-driven instabilities such as the magnetorotational instability (MRI, [5, 7, 8]) can then occur in such regions.

19.3 Specific Requirements for Core-Collapse Supernova Simulations

Simulating the dynamics of magnetic fields in the supernova core poses different challenges than the environments discussed in previous chapters. Here we outline a few salient points and refer to in-depth reviews [118, 130] for details. Supernova explosions, especially neutrino-driven ones, are in a sense "slow" phenomena that develop and unfold over many dynamical time scales. Explosions can take hundreds of milliseconds to develop and seconds to reach their final energies [19, 29, 127, 136]. The accretion flow behind the shock is moderately subsonic, and convective flow in the proto-neutron star is very subsonic with Mach numbers of order of $\mathcal{O}(10^{-2})$. This imposes stringent requirements on the ability of codes to maintain hydrodstatic stability, conserve the total energy (ADM mass in general relativity) for a self-gravitating fluid, and to conserve angular momentum, especially in scenarios where rotation is dynamically important; all of this may require special tweaks [130]. Accurate neutrino transport is critical for the dynamics, for multi-messenger observables, and for accurate predictions of supernova nucleosynthesis. Various neutrino transport algorithms are currently in use [118], but multi-group neutrino transport, detailed interaction rates, and at least an approximate treatment of general relativistic transport effects (in particular gravitational redshift) is required. Supernova modeling is a multi-scale problem; structures of several hundred meters in the proto-neutron star need to be captured at the same time as shells in the star at radii of at least 10, 000 – 100, 000 km. The self-gravity of the stellar core and the proto-neutron star is crucial in the supernova explosion problem. This necessitates a dynamical treatment of gravity, though the approximation of a fixed space-time metric may be viable for simulating phenomena like wind outflows from the proto-neutron star or hydrodynamics instabilities in the gain region to some degree in idealized simulations.

Oidealized simulations the other hand, newly formed proto-neutron stars are less compact than cold ones, and inflow and outflow velocities are at best mildly relativistic well into the explosion phase – even in the case of magnetorotational supernovae – which often permits a pseudo-Newtonian treatment of relativistic gravity by means of a modified gravitational potential [104, 131]. However, more extreme conditions are encountered in black-hole forming collapse and during later explosion stages when the environment of the nascent compact objects becomes more dilute and relativistic outflows can develop. Special relativistic hydrodynamics with a (pseudo-)Newtonian approximation is sometimes used as an intermediate step towards general relativistic

MHD [191]. The merit of these approaches is that they can provide a better trade-off of physical and numerical model accuracy (e.g., better neutrino transport or full-wave Riemann solvers), computational cost and fidelity in treating relativistic effects.

A variety of schemes for the MHD equations have been in use for core-collapse supernova simulations. The codes 3DNSNE- IDSA [111, 192], AENUS- ALCAR [146], COCONUT- FMT [134, 137], FGR1 [88], FISH [84], FLASH- M1 [149], and GRHYDRO (part of the Einstein toolkit; [122]) are capable of 3D MHD supernova simulations. Of these, FGR1 and GRHYDRO are fully general relativistic, whereas the other codes resort to the Newtonian approximation with an effective potential; AENUS-ALCAR is also capable of special relativistic MHD. Constrained transport for enforcing the divergence-free condition [51] is used in AENUS- ALCAR, FGR1, FISH, and GRHYDRO, whereas 3DNSNE- IDSA and COCONUT- FMT implement slightly different flavors of hyperbolic divergence cleaning [39] (which is also available in GRHYDRO). Historically, constrained transport, hyperbolic divergence cleaning and projection methods have all been used in the FLASH code framework [38, 40, 93]. Older two-dimensional (2D) codes for supernova modeling have used schemes similar in spirit to constrained transport build the divergence-free condition directly into the grid discretization [6, 97].

Higher-order reconstruction schemes are used in all modern MHD supernova codes. One virtue of Newtonian codes like 3DNSNE- IDSA, AENUS- ALCAR, COCONUT- FMT, and FLASH- M1 lies in greater ease in implementing less diffusive Riemann solvers like HLLC [197] or HLLD [120] with superior performance in the subsonic and sub-Alfvénic regime, whereas general relativistic supernova simulations have so far been limited to the more diffusive HLLE [69] solver. A particular subtlety in MHD simulations concerns the difficulties of proper upwinding [57], which may make some implementations prone to (possibly mild) instabilities under certain conditions. The most ambitious and rigorous approaches for proper upwinding involve reconstruction of characteristic variables [57, 93, 187], which is not currently implemented in any MHD supernova code except FLASH [93]. For specialized purposes, e.g., for the long-term evolution of very subsonic convection in the proto-neutron star over time scales of seconds, other tailored numerical approaches are used, such as spectral methods in combination with the anelastic approximation [160].

The repercussions of these differences in the implementation of the MHD equations, and those of other approximations (pseudo-Newtonian approximation, different neutrino transport treatments) need to be gauged by code comparisons for verification and uncertainty quantification. A number of supernova code comparisons have been performed [e.g., 30, 82, 96, 133, 134, 148], but there is only one MHD code comparison in 2D between AENUS- ALCAR and COCONUT- FMT so far [202].

19.4 Supernovae of Non-rotating and Slowly Rotating Stars

While magnetic fields in core-collapse supernovae have historically been studied predominantly in the context of rapidly rotating progenitors, it is useful to first consider magnetic fields in the more generic case of non-rotating or slowly rotating progenitors. Although magnetic fields will almost invariably play at most a subsidiary role to neutrino heating in powering the explosion, the inclusion of magnetic fields is becoming important for precision modeling and indispensable for eventually understanding the origin of neutron star birth magnetic fields and spins.

The growth and potential dynamical role of magnetic fields in supernova cores in the absence of rapid rotation was already discussed before the advent of detailed multi-dimensional simulations. [195] already pointed to the possibility of turbulent dynamo action in the proto-neutron star due to convective instability and estimated saturation field strength. It has also been theorized that Alfvén waves emanating from the proto-neutron star [189] and Alfvén wave amplification within the gain region [64] may play a role the supernova explosion mechanism and for supernova nucleosynthesis during the neutrino-driven wind phase [188].

Since around 2010 the amplification of magnetic fields in supernova cores by flow instabilities not reliant on progenitor rotation has been studied extensively in multi-D simulations. Two such instabilities are known to operate in the gain region. First, neutrino heating makes the accretion flow behind the shock onto the proto-neutron star unstable to convection [28, 74, 78]. Second, the accretion shock is unstable to large-scale oscillatory motions, typically dominated by the dipole ($\ell = 1$) component, known as standing-accretion shock instability (SASI). SASI takes on the form of sloshing motions [18] or in 3D also of circular motion (spiral mode) [17], and is mediated by an amplification cycle involving vortical and acoustic perturbations in the region behind the shock and the neutron star surface [53, 63]. Idealised 2D MHD simulations of SASI without neutrino transport showed the potential for amplification of magnetic fields by about two orders of magnitude, but without a significant dynamical effect [48]. Two-dimensional simulations with neutrino transport [146] also suggested limited potential for field amplification by convection and SASI by only a factor of a few; the key limitation for field amplification in these simulations was the short time for stretching and distorting magnetic fields by convective motions before they are advected out of the gain layer onto the neutron star surface. The simulations of [146] raised the interesting prospect of a dynamical effect of magnetic fields on supernova explosions in the case of very strong pre-collapse fields of order 10^{12} G, which may be encountered for certain types of supernova progenitors, e.g., those arising from stellar mergers (see Sect. 19.2.1). A similar supporting role for very strong magnetic fields was also found in later 2D simulations with neutrino transport [79, 111].

19.4.1 Impact of Magnetic Fields on Shock Revival in 3D Simulations

A principal limitation of these early works lay in the restriction to axisymmetry, which precludes the operation of magnetic dynamos [37]. In core-collapse supernovae of non-rotating or slowly-rotating progenitors, at least small-scale turbulent dynamo amplification is expected to occur in the gain region (regardless of whether SASI or convection is the dominant instability) and in the proto-neutron star convection zone. The small-scale dynamo operates without the need for bulk differential rotation [21], driven by turbulent flow that may arise, e.g., due to convection, SASI, or shear instabilities. Turbulent dynamos are typically characterized by fast exponential growth with a growth time scale of the order of the eddy turnover time [178]. The initial exponential growth phase typically slows down and may transition to linear growth once the magnetic fields approach kinetic equipartition for higher wavenumbers [32], so that backreaction of the magnetic field on the flow becomes important. At low Reynolds number, the small-scale dynamo requires a magnetic Prandtl number (the ratio of kinematic viscosity and magnetic resistivity) $\mathrm{Pr_m} \gtrsim 1$ [70, 178], a condition which is easily fulfilled in the gain region. There are, however, still debates in dynamo theory about possible adverse effects on small-scale dynamo amplification and its saturation in the limits of high Reynolds and high-magnetic Prandtl number [175, 179, 186]. By and large, both simulations [206] and astrophysical observations [22, 33] point to a higher robustness of the small-scale dynamo, however. Small-scale dynamo amplification may in fact become possible well below $\mathrm{Pr_m} = 1$ in high-Reynolds number flow [206]. Observational phenomenology interestingly supports the relatively "naive" notion that saturation field strengths close to kinetic equipartition are reached [22, 33]. In interpreting MHD supernova simulations, it is important to consider this context from dynamo theory and astrophysical simulations in addition to raw simulation results; one should not take for granted that explosion models (which invariably employ the ideal MHD approximation) reproduce the correct physical limit in the supernova core.

The possibility of dynamo amplification in the collapse of non-rotating progenitors was first investigated by means of idealized adiabatic 3D simulations of the SASI [49]. These simulations showed rapid amplification of magnetic fields throughout the post-shock region, fed by vorticity that is generated by shear flows in the spiral mode of the SASI. Despite amplification of magnetic fields to about 10% of equipartition, the simulations of [49] did not reveal any dynamical impact of magnetic fields on the shock dynamics.

The first 3D MHD supernova simulation with multi-group neutrino transport for a "generic", slowly-rotating $15 M_\odot$ progenitor star accorded a more relevant role to turbulent dynamo amplification [137]. With a conservative choice of weak pre-collapse field strengths of $\lesssim 10^7\,\mathrm{G}$, [137] find relatively rapid small-scale dynamo amplification of the fields in the gain region to about 40% of kinetic equipartition within about 250 ms after bounce. Figure 19.2 shows the emerging structure of

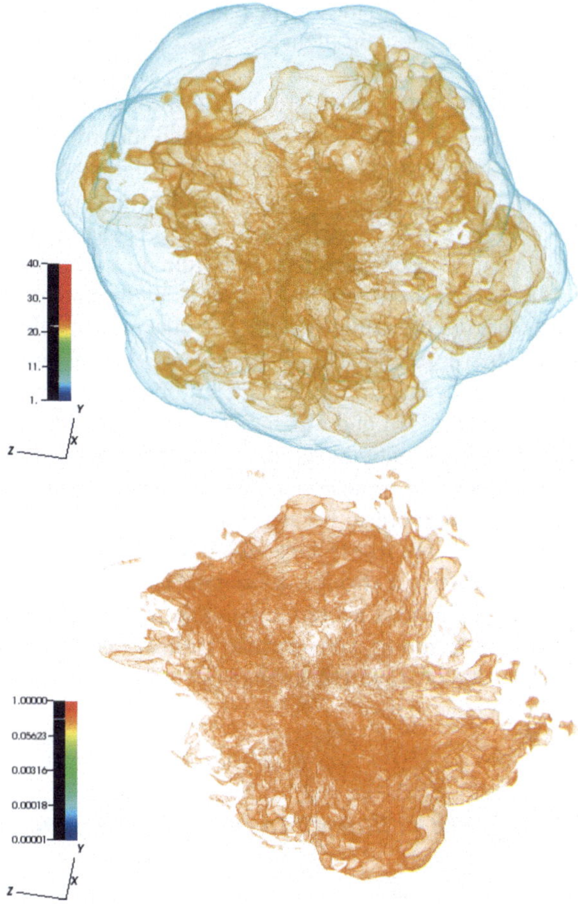

Fig. 19.2 Volume rendering of the entropy (top) and the ratio of magnetic to gas pressure (bottom) in a high-resolution (0.7° in angle) MHD supernova simulation of a non-rotating $20M_{\odot}$ stripped-envelope progenitor at a post-bounce time of 250 ms. At this time, an explosion is already underway. In addition to neutrino-heated high-entropy bubbles (orange) and regions of strong fields (orange), the shock is visible a light blue translucent surface in the top panel. The magnetic field configuration is dominated by small-scale structures. Fields of relatively homogeneous strength are present throughout the post-shock region, but they are strongest in and in between the expanding neutrino-heated bubbles. The entire structure has a diameter of about 1200 km

neutrino-heated convective bubbles and dynamo-generated magnetic fields from a recent high-resolution simulation of a $20M_{\odot}$ star with the CoCoNuT- FMT code.

The magnetic fields appear to act as an extra energy reservoir in the gain region, i.e., they do not reduce the turbulent kinetic energy compared to a non-magnetic baseline simulation. Reference [137] find that magnetic fields support shock revival, tilting the balance in favor of explosion, whereas no explosion develops in the non-magnetic baseline model. A similar behavior—higher combined kinetic and magnetic turbulent energy and more precipitous explosions with magnetic fields—was seen in more recent supernova simulations of $20M_{\odot}$ and $27M_{\odot}$ stars with the 3DnSNe- IDSA code [109, 110]. It should be noted, however, that magnetic fields play a subsidiary role to neutrino heating and aspherical mass motions in triggering an explosion. It is still neutrino energy deposition that powers the explosion; magnetic fields merely play an auxiliary, but noticeable role in pushing the stalled accretion shock out a bit to the point that neutrino heating can initiate continued shock expansion as a runaway process according to the classical neutrino-driven paradigm [76]. In fact, the ratio of magnetic to kinetic turbulent energy drops again after shock revival in the simulation of [137].

Naturally, these recent simulations prompt questions about the potential impact of resolution and the accuracy to which the implemented schemes for ideal MHD approximate the correct physical limit of high hydrodynamic and magnetic Reynolds number (see Fig. 19.2 for a snapshot from a high-resolution simulation at 0.7° with the CoCoNuT- FMT code). Further studies are clearly required to address these issues, but they may to some extent be subordinate to the question of initial conditions. One implication of these recent simulations with relatively fast turbulent field amplification is that the pre-collapse field strengths may be relevant to post-collapse dynamics in the supernova core. This is particularly interesting for the aforementioned strong fossil-field scenario in stellar merger products. Using a $16 M_\odot$ post-merger stellar evolution model [176] and a parameterized twisted-torus initial field of about 10^{12} G, [203] found a rapidly developing supernova explosion. The rapid explosion in their model appears to come about only in part due to additional magnetic stresses behind the shock, but more so because the magnetic field affects the infall of material onto the shock, giving rise to large-scale ram pressure asymmetries ahead of the shock, which facilitates a strongly unipolar explosion. Intriguingly, early shock revival at a time of high mass accretion rate and high neutrino luminosities provide the conditions for dumping considerable neutrino energy into the explosion, resulting in an explosion energy of more than 0.9×10^{51} erg, very close to canonical supernova explosion energies.

19.4.2 Pre-collapse Magnetic Fields from 3D Simulations

Both for "normal" supernova progenitors and those that presumably harbor a strong fossil-field, there is, of course, a need to put the predictions of pre-collapse fields from stellar evolution theory on a more solid footing. In recent years, 3D models of convection and nuclear burning have emerged as a means for more consistently initializing supernova simulations (see [130] for a review). This approach has also been extended to MHD, but there are as yet no full-star simulations, and reaching steady-state conditions in stellar magnetoconvection model remains challenging as this would require simulations of very long duration. Pragmatically, a major focus in 3D modeling of stellar convection has been on oxygen shell burning because this shell is typically active at collapse, has relatively high turbulent Mach numbers of order 0.1 [34], and is the shell that falls through the shock during the phase when an explosion is likely to start.

Simulations of convective oxygen shell burning provide a first glimpse tat what realistic pre-collapse fields in massive stars may look like [200]. Field strengths are predicted to saturate well below equipartition with the turbulent kinetic energy, or about 10^{10} G in absolute terms. Such fields would not have an important dynamical impact on stellar convection, but may still have an appreciably impact *after* collapse by jump-starting the turbulent dynamo in the gain region. Importantly, values of order 10^{10} G in the oxygen shell cannot be directly compared to the expected core field strengths in current 1D stellar evolution models, as the oxygen shell sits at much lower densities at radii of several thousands of kilometers; values of 10^{10} G in the oxygen shell are in fact stronger than what was conventionally expected. However, the fields are predicted to be dominated by small-scale structures, with dipole field strengths about an order of magnitude lower than isotropic field strengths. Uncertainties about the saturation of the magnetic field strength during the pre-collapse evolution exist in both direction. Due to lower magnetic Prandtl number $\mathrm{Pr_m} \lesssim 1$, high efficiency of the turbulent dynamo and high saturation field strengths are not to be taken for granted, but on the other hand, high-resolution simulations [94] suggest that saturation field strengths in turbulent convective flow may still be underestimated and actually reach a high fraction of kinetic equipartition.

19.4.3 Neutron Star Magnetic Fields and Magnetized Outflows

One of the major questions in compact object formation concerns the origin and distribution of neutron star magnetic fields. Observed neutron star magnetic fields show a broad distribution between 10^{12} G and 10^{14} G for young pulsars, and a population of *magnetars* with field strengths above 10^{14} G and up to about 10^{15} G [150]. Many evolutionary steps are interposed in the evolution of a proto-neutron star during the first few seconds of a supernova to an observable pulsar or magnetar. During the later explosion phase, ablation of material from the proto-neutron star surface by the neutrino-driven wind will also affect the neutron star surface fields. Ongoing accretion onto the neutron star early on during the explosion as well as late-time fallback may also have an impact; for example, there is the possibility of field burial below fallback material [152, 198]. Magnetic fields generated by a dynamo *inside* the warm young neutron star (see Sect. 19.5.2) may break out through the surface. These complications limit the conclusions that can be drawn on the origin of neutron star magnetic fields from current supernova simulations.

Nonetheless, some important insights can be gleaned already. Field amplification by shear motions at the bottom of the gain region is found to easily generate magnetic fields of order 10^{13} G and sometimes even 10^{14} G at the proto-neutron star surface in 3D supernova simulations [137, 203]. Dipole field strengths in simulations are somewhat lower, but are still at the upper end of the pulsar magnetic field distribution and sometimes verge into the magnetar regime. Interestingly, memory of the pre-collapse fields appears to be lost quickly, probably due to a balance between field amplification and turbulent reconnection in the neutron star surface region. Simulations starting from very weak and very strong twisted-torus fields of 10^6 G and 10^{12} G [203] showed little difference in proto-neutron surface field strengths at the end of the runs. These simulations also eventually show a decline in surface field strength in the explosion phase. Such a decline may be physical and result from turbulent reconnection and ablation of material from the fluffy surface of the warm neutron star, and could lead to more typical pulsar magnetic fields eventually. However, numerical reconnection may also play a role and calls for future high-resolution long-time MHD supernova simulations of the first seconds of the explosion until accretion onto the neutron star ceases. Over longer time scales, innovative (combinations of) numerical methods will be required to deal with the transition to the very different regime of a neutron-star with a very dilute environment and force-free magnetosphere, and potential complications like fallback.

Some of the phenomena that unfold on longer time-scales have in fact already been addressed in simulations using somewhat idealized setups and initial conditions, and simplifications to relevant physics such as neutrino heating during the neutrino-driven wind phase. The interplay of strong magnetic fields and wind outflows and the resulting effects on neutron star spin-down and nucleosynthesis in the wind are of particular interest [41, 157, 196]. In the presence of strong large-scale neutron star surface fields $\gtrsim 10^{14}$ G, the neutrino-driven wind not only becomes asymmetrical, but also unsteady with episodic mass ejection in plasmoids [196]. The altered dynamics of such strongly magnetized winds may allow for the ejection of small amounts of high-entropy material with conditions for rapid neutron capture process (r-process) nucleosynthesis up to the third r-process peak [41, 196]. Simulations of magnetized winds also indicate that the early spin-down of highly-magnetized young neutron stars by such outflows may be much more efficient than the classical dipole spin-down formula[2] would suggest [157].

19.5 Magnetorotational Explosions

The magnetorotational explosion mechanism has a long history dating back to the 1970s [15, 92, 116], and its key idea has remained current ever since. The rotational energy of rapidly rotating proto-neutron star provides a significant energy reservoir that can be tapped by magnetic fields to deliver the requisite energy to revive the shock and power an explosion. However, beyond this classical version of the magnetorotational mechanism, there is a number of closely related scenarios that involve outflows powered by rotation and magnetic fields in core-collapse supernovae. This also includes explosions powered by outflows from accretion disks after black-hole formation in stellar collapse (*collapsar scenario* [101, 102]), and covers mechanisms that power different types of outflows during different phases of a supernova. Rotation and magnetic fields may deliver the explosion energy contained in the bulk of the ejecta during the first seconds of the explosion, but they are also relevant for

[2] Using the classical dipole spin-down model, the derivative of the spin period P is given in terms for the surface dipole field B_{surf} as $\dot{P}/P \approx \left(\frac{B_{\mathrm{surf}}}{3\times 10^{19}\,\mathrm{G}}\right)^2 \mathrm{s}^{-1}$, see, e.g., [99].

explaining the relativistic jet of long gamma-ray bursts either from rapidly rotating "millisecond magnetars" [23, 24, 46, 85, 199] or collapsars [101, 102], and they may power dilute late-time wind outflows that shape the energetics and the electromagnetic emission of the supernova on longer time scales [25, 117].

Within the confines of this chapter, we shall focus mostly on the classical scenario of explosions driven by neutron star rotation within the first hundreds of milliseconds after bounce, both because it is arguably the best-explored one and because readers may find it easier to build up an end-to-end picture of MHD-powered supernovae step-by-step from the collapse phase rather than skipping ahead to later phases.

19.5.1 Basic Physical Considerations

The starting point for the classical magnetorotational mechanism is the reservoir of rotational energy that can be tapped by magnetic fields. Taking the neutron star moment of inertia to be $I \approx 0.36MR^2$ (which roughly holds for a neutron star of $1.4M_\odot$ and radius 12 km, cp. [91]), the rotational energy of a neutron star is

$$E_{\mathrm{rot}} = \frac{1}{2}I\omega^2 = \frac{2\pi^2 I}{P^2} = 2 \times 10^{52}\,\mathrm{erg}\left(\frac{M}{1.4M_\odot}\right)\left(\frac{R}{12\,\mathrm{km}}\right)^2\left(\frac{P}{1\,\mathrm{ms}}\right)^{-2}, \quad (19.6)$$

in terms of its gravitational mass M, radius R, and spin period P. For millisecond spin periods, this provides a sufficiently large energy reservoir to be compatible with the range of observed hypernova energies up to $\sim 10^{52}$ erg [113, 208] if the rotational energy can be extracted efficiently. However, one needs to bear in mind proto-neutron stars remain warm and inflated for at least a few hundred milliseconds, so for a given amount of angular momentum, the available reservoir of rotational energy will be smaller.

Furthermore, magnetic fields may not be able to tap the entire rotational energy on short time scales. The entire reservoir is available only on such time scales over which the neutron star efficiently exchanges angular momentum with its environment, i.e., initially with a dense accretion flow and later with an increasingly dilute environment. If the coupling to the environment is slow, one excepts that magnetic fields are at best able to establish uniform rotation within the neutron star. In this case, at most the free energy contained in differential rotation can be liberated [27]. Self-sustained differential rotation in the neutron star convection zone, which is known from stellar and planetary convection [80, 81, 115], could limit the available energy even further. In practice, liberation of free rotational energy can still be sufficient for reaching hypernova energies [62], especially because the neutron star can act as a transducer, accelerating accreted material into corotation and ejecting large amounts of it with little net growth in mass.

The dynamics of a supernova with a rapidly rotating proto-neutron star therefore hinges on the efficiency of energy and angular momentum extraction by magnetic fields. Extraction can in principle happen both by collimated jet outflows, or by more isotropic, wind-like outflows. Jets can emerge due to the interplay of hoop stresses and ambient pressure that leads to the formation of magnetic towers [27], but whether this occurs and whether the resulting jets are stable depends on various factors including the poloidal-to-toroidal field ratio or the initial magnetic field geometry (see Sect. 19.5.3). At this point, it is sufficient to consider order-of-magnitude estimates for the dynamical role of magnetic fields. Assuming that rotational motions dominate the flow, the angle-integrated (isotropic) radial Pointing flux F_B can be obtained from $\mathbf{B}(\mathbf{v}\cdot\mathbf{B})/4\pi$ term in the MHD energy equation and is given by

$$\begin{aligned} F_B &\sim r^2 v_\varphi B_r B_\varphi \sim r^3 \omega B_r B_\varphi \sim r^3 \omega B^2 \\ &\sim 1.1\times 10^{52}\,\mathrm{erg\,s^{-1}} \left(\frac{\omega/2\pi}{\times 10^3\,\mathrm{s^{-1}}}\right)\left(\frac{r}{12\,\mathrm{km}}\right)^3\left(\frac{B_r}{10^{15}\,\mathrm{G}}\right)\left(\frac{B_\varphi}{10^{15}\,\mathrm{G}}\right) \end{aligned} \tag{19.7}$$

in terms of the rotational velocity v_φ, and the radial and zonal magnetic field components B_r, B_φ (all of which are to be interpreted as averages or typical values at radius r). Similar estimates can be made for the angular momentum flux F_L due to magnetic stresses, which is of order

$$F_L \sim B_r B_\varphi r^3. \tag{19.8}$$

Well inside the proto-neutron star, field strengths in excess of 10^{15} G amount to only a minuscule fraction of thermal or kinetic equipartition field strengths, and fast redistribution of rotational and angular momentum towards relatively rigid rotation is therefore plausible. However, the time scale for extracting rotational energy from the neutron star is less trivial. Fast extraction on the time scale of $\sim$1 s will require rotational periods of $\sim$1 ms and magnetar field strength $\sim 10^{15}$ G at the base of accretion flow onto the proto-neutron star surface.

The principal questions in the magnetorotational paradigm are therefore how fast magnetic fields amplified, what saturation field strengths are reached, and also whether "ordered" magnetic fields are generated in the supernova core. Magnetic fields need to be ordered in the sense that B_r and B_φ in the above equations must be (strongly) correlated to achieve a net angular momentum and energy flux mediated by magnetic fields, or else the implied integrals over solid angle in Eq. (19.7) and (19.8) would fall far below order-of-magnitude estimates based on dimensional analysis. They also need to be organized in large-scale structures for the formation of jets as described above.

As already discussed for the non-rotating case, dynamically relevant magnetic fields cannot be generated simply by compression during collapse from the pre-collapse fields predicted by currently stellar evolution models. In the rotating case, however, additional dynamo mechanisms driven by rotation come into play. Furthermore, in regions with meridional flow (e.g., in the gain region or in the proto-neutron star convection zone) on top of rotational flow, substantial non-vanishing kinetic helicity of the flow,

$$H = \int \mathbf{v} \cdot (\nabla \times \mathbf{v})\, \mathrm{d}V, \tag{19.9}$$

can enable the formation of large-scale fields by an inverse magnetic helicity cascade [21, 54, 153].

In principle, differential rotation alone could lead to continuous field amplification in the collapsed cores of rapidly rotating progenitors, but the amplification would be linear with time and not sufficient to amplify weak initial fields to dynamically relevant ones on short time scales. Exponentially growing amplification mechanisms are required for this purpose. Two such mechanisms that are discussed prominently in the context of magnetorotational explosions are the magnetorotational instability, which was first discussed in the context of accretion disks [7, 8] and later identified as relevant to core-collapse supernovae [5], and the α-Ω dynamo, whose importance in proto-neutron star convection had been recognized even earlier [46].

One interesting feature of the magnetorotational instability (MRI) is that it can be driven by differential rotation alone even in convectively neutral or convectively stable regions. Entropy and lepton number gradients will naturally affect the growth and character of the MRI, however [145]. For introductory purposes, it is sufficient to review key features of the MRI in the absence of buoyancy effects as established in the theory of the MRI in accretion disks [5, 7, 8]. In the presence of a negative angular velocity gradient, the MRI leads to unstable growth of modes down to a minimum wavelength λ_{MRI} set by the Alfvén velocity v_{A} and the angular velocity ω,

$$\lambda_{\mathrm{MRI}} \sim v_{\mathrm{A}}/\omega \sim \frac{|B|}{\sqrt{4\pi\rho\omega}} \sim 450\,\mathrm{cm} \left(\frac{B}{10^{12}\,\mathrm{G}}\right) \left(\frac{\rho}{10^{10}\,\mathrm{g\,cm^{-3}}}\right)^{-1/2} \left(\frac{\omega/2\pi}{10^{3}\,\mathrm{s^{-1}}}\right)^{-1}. \tag{19.10}$$

Modes of shorter wavelength are stabilized by magnetic tension. The maximum growth rate Ω_{max} occurs for modes close to λ_{MRI}, and is of order

$$\Omega_{\mathrm{max}} = 1/2(\mathrm{d}\ln\omega/\mathrm{d}\ln r). \tag{19.11}$$

For larger wavelengths λ, the growth rate roughly scales as λ^{-1}. Thus, for moderately strong initial fields and short neutron star spin periods, numerical resolution on the meter-scale would be required to resolve the fastest growing mode. At coarser resolution, the resolved modes on the grid will grow significantly more slowly. Simple arguments for saturation can be made essentially based on dimensional arguments by requiring that the critical wavelength λ_{MRI} will grow due to increasing field strength to about the shearing length $\mathrm{d}r/\mathrm{d}\ln\omega$ before termination of growth, which results in saturation fields of order [5]

$$B^2 \sim 4\pi\rho r^2\omega^2 \frac{d\ln\omega}{d\ln r}. \tag{19.12}$$

However, such simple dimensional arguments may not hold. For example, disruption of channel models by resistive instabilities may provide a saturation mechanism for terminating MRI growth at smaller scales and lower field strength [145], which makes the question of MRI saturation in core-collapse supernovae highly non-trivial. Amplification may be limited to factors of 10–100 [166].

Field amplification by an α-Ω dynamo within the proto-neutron star convection zone [46] is interesting in that it provides a natural mechanism for generating large-scale fields due to the inverse helicity cascade. With an initial exponential growth rate of [204]

$$\Omega \sim (v_{\rm conv}/L)(\omega\tau_{\rm c}^{-1})^{1/2} \sim \sqrt{\omega\tau_{\rm conv}^{-1}}, \tag{19.13}$$

in terms of the convective velocity $v_{\rm conv}$, turnover time $\tau_{\rm conv}$, typical length scale L, and angular velocity ω, growth may initially be fast on a time scale between $\tau_{\rm conv}$ and the rotation period. However, the α-Ω-dynamo may transition to much slower growth well before reaching saturation [160]. As for the MRI, estimating the saturation field strength for the α-Ω dynamo is not trivial. Assuming complete conversion of the free energy in differential rotation into magnetic field energy [46] is likely to significantly overestimate actual field strengths. Findings from simulations will be discussed in the next section.

Recently, the Tayler-Spruit dynamo [185], which has long been studied as a field amplification mechanism in the theory of stellar evolution, has also been considered as relevant for proto-neutron stars [11]. Different from the α-Ω-dynamo, the generation of a poloidal field from a toroidal field within the dynamo loop is mediated by the Tayler instability [194], allowing the dynamo to operate in convectively stable regions as long as there is differential rotation to generate toroidal from poloidal fields.

Non-ideal effects present an important complication both for convective dynamos in the proto-neutron star [195] and for the MRI [66]. While viscosity, heat conduction, and resistivity due to nucleons, electron, positrons, and photons are extremely small, non-ideal effects due to neutrinos play a major role, which can take on the guise of a viscosity well above the typical neutrino mean free path or of drag below the neutrino mean free path. This leads to the possibility of a double cascade in proto-neutron star turbulence [195] with potentially interesting ramifications for magnetohydrodynamic instabilities. The interplay of fermion chirality with magnetic fields and its potential effect in core-collapse supernovae has also been discussed as an additional complication in the framework of chiral MHD [105].

19.5.2 Simulating Field Amplification Processes in Core-Collapse Supernovae

Due to the extreme resolution requirements for resolving the fastest growing MRI modes, studying field amplification in the context of magnetorotational supernovae is currently still a somewhat distinct endeavor from simulating magnetorotational explosions proper. Simulations focusing on field amplification processes need to sacrifice aspects that are relevant for modeling explosions, whereas explosion modeling (Sect. 19.5.3) tends to circumvent thorny issues concerning field amplification by using relatively strong parameterized initial fields. There is a significant history of local and semi-global simulations for studying field amplification by the MRI in the supernova context [65, 67, 108, 145, 167].

In the last decade, simulations have begun to bridge the gap from MRI field amplification on small scales to the global scale. A first attempt was made using the thin layer approximation as a compromise to achieve sufficiently fine resolution [106], which, however, does not permit conclusions about the emerging global field geometry. First truly global general relativistic MHD simulations of MRI and dynamo field amplification in proto-neutron stars indeed showed rapid exponential field amplification to peak field strength well above 10^{15} G, as well as the emergence of a global field due to the inverse cascade [123]. These simulations achieved resolutions as fine as 50 m and even included a basic treatment of neutrinos by means of a leakage scheme. However, they remained limited to very short simulations times of order 10 ms, and still started from relatively strong initial field strengths. Subsequent simulations have shed more light on the long-term evolution of field amplification by using pseudo-spectral methods for high accuracy and avoiding fully compressible MHD. 3D MRI simulations over more than half a second using incompressible MHD for a somewhat idealized proto-neutron star model confirmed the emergence of such high field strength, but with modest energy in the dipole, and, importantly, misalignment of the dipole from the axis [161].

More recent long-time pseudo-spectral simulations with the MAGIC code have added realism by accounting for buoyancy effects with the Boussinesq approximation, starting from PNS structures inspired by 1D core-collapse supernova simulations, and implementing physical approximations for (neutrino) viscosity, thermal conductivity and magnetic diffusivity [10, 160, 162]. Such simulations of convection in rapidly rotating proto-neutron stars have demonstrated the generation of strong dipole fields by the α-Ω dynamo of order 10^{15} G and even higher maximum toroidal field strengths of order 10^{16} G, and also established scalings with proto-neutron star rotation period and Rossby number [160]. Importantly, the initial exponential growth of the α-Ω dynamo in these simulations stagnates at a first plateau and is then followed by very slow, secular amplification until saturation is reached only after several seconds. More recently, the emergence of large-scale fields generated by an α-Ω dynamo has also been studied in compressible MHD simulations of proto-neutron star convection over shorter time scales [107], confirming a trend towards stronger large-scale fields at higher spin rate.

Subsequent simulations with MAGIC have also addressed the convectively stable "atmosphere" at the proto-neutron star surface [162], which would seem to be more directly relevant to eventual surface field strengths and hence late-time powering of magnetized outflows, barring the possibility of field breakout from the interior. The simulations of [162], which cover $\gtrsim$1 s of field evolution, showed magnetar-grade fields of $\sim$$10^{14}$ G locally, but a relatively weak (and tilted) dipole with of $\lesssim$$10^{13}$ G. Their simulations also show characteristic dynamo cycles with periods of several hundred milliseconds.

It has been theorized that field amplification in young neutron stars may also occur much later due to fallback that spins up the outer layers of the neutron star [11]. This scenario presupposes that an explosion has already occurred by other means and is thus not directly relevant for the magnetorotational mechanism proper, but such late-time field amplification is relevant for magnetar formation and the possibility of increasing the energy or brightness of a supernova by means of an "afterburner". Since this scenario would result in a positive angular velocity gradient (faster surface rotation), MRI would not act in this case, but amplification by the Tayler-Spruit dynamo will be relevant [11]. The 3D simulations of [10] investigated this scenario for various degrees of differential rotation in the proto-neutron star, identifying distinct branches of the Tayler-Spruit dynamo with scaling laws for the saturation fields conforming to the theories of [185] and [56], respectively. The dynamo-generated fields are concentrated near the rotation axis, and dipole field strengths of 10^{14} G or more can be reached.

19.5.3 Simulations of Magnetorotational Explosions

Global simulations of magnetorotational explosions date back to the 1970s. Even though many early numerical models attempted to scale the problem down to one dimension [16, 126], early work by [92] already attempted to tackle it in 2D on a 40×40 grid, even with some very crude gray treatment of neutrino transport – a pioneering approach despite the obvious limitations of the time.

19.5.3.1 Prelude: Axisymmetric Simulations

Modern 2D simulations started to revisit the problem of magnetorotational explosions about twenty years ago with mature numerical methods for MHD. In such 2D simulations, strong initial fields of order 10^{12} G need to be prescribed by hand in order for magnetic fields to influence the dynamics of the post-bounce evolution even if some amplification by field winding and 2D instabilities may occur. Early modern 2D simulations initially started with simplifications to the microphysics and neutrino transport to various degrees, but were able to model the extraction of rotational energy by magnetic fields and its impact on the shock dynamics, including the emergence of bipolar explosions [6, 121, 143, 144, 168–172, 191, 211]. Some works also investigated the repercussions of the specific dynamics of asymmetric collapse and explosion in the magnetorotational paradigm on neutrino and gravitational wave emission [86, 144, 190].

A major breakthrough came in the form of the first MHD supernova simulations [27] with multi-group neutrino transport with the VULCAN code [97], albeit using Newtonian gravity with no relativistic corrections. Even in the magnetorotational scenario, proper inclusion of neutrinos in simulations is of paramount importance as neutrino transport affects the proto-neutron star structure and contraction, will have some dynamical effect (even if it is subdominant to that of magnetic fields), and is critical for accurate predictions of nucleosynthesis from magnetorotational explosions. For initial fields of order 10^{11} - 10^{12} G, the simulations of [27] showed the emergence of bipolar magnetic tower structures that push the shock out and produce explosions with energies of up to more than 10^{51} erg, with some ejecta reaching velocities of $\sim 50,000$ km. They suggested that balance between spin-up of the proto-neutron star surface and extraction of free rotational by the jets can be maintained for time scales of seconds to eventually reach hypernova energies. Importantly, the jets remain in the non-relativistic regime, however, and are to be distinguished from relativistic long gamma-ray burst jets that could only emerge later in a more dilute environment to reduce baryon loading. A host of further 2D MHD simulations with neutrino transport has confirmed this overall picture and explored dependencies on progenitor mass, progenitor rotation profiles, and initial magnetic fields [79, 139, 140, 142].

As an interesting variation of the magnetorotational mechanism, follow-up work with the VULCAN code considered the magnetorotational explosions after the accretion-induced collapse (AIC) of white dwarfs [42], a scenario where rapid progenitor rotation can be justified naturally. For strong initial fields of $\sim 10^{12}$ G powerful explosions of $\sim 10^{51}$ erg (i.e., not in the hypernova regime) can emerge. Due to negligible production of ^{56}Ni, such explosions would be expected to be optically dark, but would be prodigious production sites of r-process material due to the ejection of very neutron-rich material with electron fraction Y_e down to ~ 0.1, which would imply strong constraints on AIC event rates.

19.5.3.2 Dynamics of Magnetorotational Explosions in 3D

First ventures into 3D MHD simulations of rotational core collapse also started in the late 2000s, and initially relied on heavily simplified physics without a microphysical equation of state and neutrino effects [89, 119]. 3D simulations with a microphysical equation of state, a parameterized treatment of deleptonization during collapse, and parameterized initial rotation profiles and magnetic fields initially focused on the gravitational wave emission during the early post-bounce phase and allowed limited conclusions on explosion dynamics and outcomes [173, 174]. Subsequent 3D simulations with neutrino leakage and very strong poloidal initial fields of 5×10^{12} G showed the emergence of jets as in 2D models and the ejection of neutron-rich material with potential for r-process nucleosynthesis [207] (for a more detailed discussion of nucleosynthesis conditions see Sect. 19.5.3.3). The first 3D simulations of magnetorotational supernovae were, however, too short to investigate explosion energetics and could only follow the growth of the explosion energy to $\sim 10^{50}$ erg, well below the energies of normal supernovae, let alone hypernovae.

Longer 3D leakage simulations [124] again with purely poloidal, but slightly weaker initial fields of 10^{12} G using the GRHYDRO code flagged an important complication in 3D in the form of the kink instability of the emerging jet (Fig. 19.3).

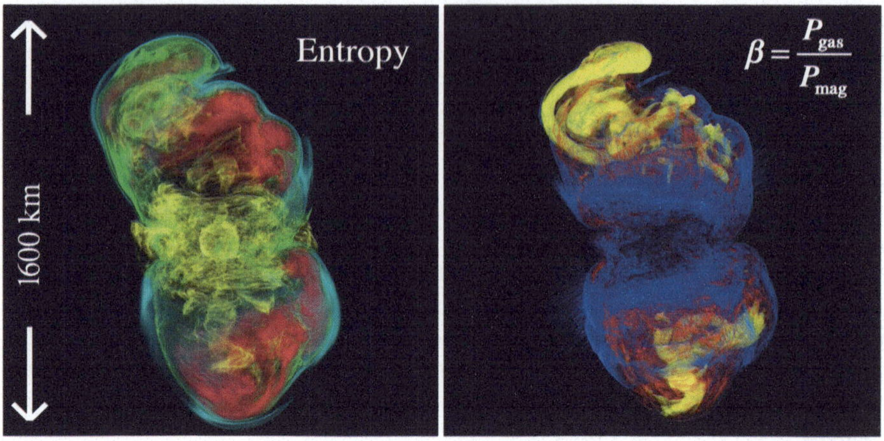

Fig. 19.3 Volume rendering of entropy (left) and plasma-β (right) in a simulation of a magnetorotational supernova [124] at a post-bounce time of 161 ms. The initial jet has been disrupted by the kink instability, but strongly magnetized lobs (yellow) still continue to expand and have pushed the shock (cyan translucent surface in the left panel) out to almost 1000 km in radius. Figure from [124]. DOI 10.1088/2041-8205/785/2/L29. ©AAS. Reproduced with permission

Kink instability for a magnetically confined plasma cylinder is expected to occur for wavelengths λ [12, 87, 124],

$$\lambda \gtrsim \frac{2\pi a}{\langle B_{tor}/B_{pol}\rangle}, \tag{19.14}$$

in terms of the toroidal and poloidal field strengths B_{tor} and B_{pol} and the cylinder radius a. Strong toroidal fields that precipitate kink instability emerge unavoidably due to field winding and are a prerequisite to jet launching in the first place. Therefore kink instability at some wavelength is expected quite generically, though the growth time scale must be sufficiently short in relation to the propagation of the jet to affect it. [124] find that an $m = 1$ mode immediately disrupts the forming jet in full 3D, different from a control simulation in octant symmetry that does not allow for an $m = 1$ instability. Magnetic lobes still push the shock out to almost 1000 km by the end of the simulation 185 ms after bounce, but accretion onto the proto-neutron star is not quenched. Reference [124] have argued that the phenomenon of jet disruption should be quite generic, especially given that jet launching is likely to occur on longer time scales if the pre-collapse magnetic fields first need to be amplified to launch jets, and hence suggested that the ultimate fate of such "fizzling" magnetorotational explosions was black hole formation and potentially a hypernova driven by a collapsar engine on longer time scales. General relativistic MHD simulations of magnetorotational explosions with multi-group neutrino transport with the FGR1 code revealed somewhat similar phenomenology where the jet is disrupted and only one big strongly magnetized lobe survives [88]. However, neutrino heating helps support the developing explosion, and a modest explosion energy of $\sim 10^{50}$ erg is obtained in [88], which still falls far short of the hypernova regime. More stable jets and an explosion energy of up to 6×10^{50} erg were found in subsequent simulations with the same code in a small parameter study with different initial rotation rates and initial fields of 10^{12} G and above [181].

Subsequent work has, however, drawn a more nuanced picture of the disruptive role of the kink instability. Reference [68] showed that with an *extremely* strong initial poloidal field of 10^{13} G, the jet can be stabilized against the kink instability, even in the case of a magnetic dipole field somewhat misaligned with the rotation axis. The 3D simulations of [141] set another optimistic counterpoint, finding the emergence of stable jets with only mild kink instability with more modest initial fields as low as $\sim 10^{11}$ G with mixed poloidal and toroidal geometry in a $39 M_\odot$ progenitor [4]. Interestingly, they found a jet-driven magnetorotational explosion even for the *predicted* field strengths in the underlying stellar evolution model (though this initial field remains far from self-consistent due to assumptions about field geometry). Due to long physical simulation times, [141] are able to reach explosion energies of $\sim 10^{51}$ erg and even $\sim 10^{52}$ erg for initial fields of $\sim 10^{12}$ G, lending more credence to the magnetorotational mechanism as a possible scenario for hypernovae. Low radial (300 zones) and angular resolution (2.8°) is a caveat, however, as this could potentially inhibit the growth of the kink instability. Higher-resolution 3D simulations of the same $39 M_\odot$ progenitor by [156] with mixed poloidal/toroidal fields of 10^{10} G and 10^{12} G also indicate that hypernova energies can be achieved in magnetorotational explosions. However, their explosion models still differ from the familiar picture of jet-driven magnetorotational explosions. In their models, most of the rotational energy of the proto-neutron star is transferred by magnetic stresses into relatively spherical outflows with quick spin-down of the neutron star by about 250 ms after bounce. Jets emerge as an epiphenomenon and carry only a small fraction of the explosion energy. The jets are subject to kink instability to varying degree with more stable jets in the 10^{10} G model and episodic disruption and reemergence of the jets in the 10^{12} G model. The explosion energies remain at the lower end of the hypernova range with 3×10^{51} erg.

The above panorama of recent simulations of magnetorotational explosions is ambiguous as there are significantly different outcomes in models by different groups. This is partly due to different choices of progenitor models and initial conditions for the magnetic fields. But there is also a clear need for controlled code comparisons to gauge to what degree the different outcomes, in particular with respect to jet disruption by the kink instability, are due to numerical uncertainties.

Furthermore, the role of the *initial* pre-collapse fields is increasingly recognized as a crucial missing piece in the magnetorotational paradigm, regardless of whether significant amplification of field *strengths* is required after collapse or not. The default assumption of poloidal or poloidal-toroidal dipole fields aligned to the rotation axis in most global simulations constitutes a strong assumption and is not in line with the phenomenology of convective stellar dynamos or expectations for the Tayler-Spruit dynamo. Simulations with inclined magnetic dipoles [26, 68] and quadrupolar pre-collapse fields [26] suggest a significant impact on the initial field geometry on the post-collapse dynamics. Thus, similar to the case of no or slow progenitor rotation, there is a need for 3D simulations of the pre-collapse phase (Sect. 19.4.2).

Compared to non-rotating or slowly rotating massive stars, MHD simulations of hypernova progenitors present significantly bigger challenges, however. Most importantly, such simulations need to rely on rotation profiles from 1D stellar evolution models, and any deviations from the "true" quasi-equilibrium angular momentum distribution will trigger significant numerical transients. Although one must bear these caveats in mind, first 3D MHD simulations of shell burning in rapidly rotating massive stars have been performed and yield some interesting insights on hypernova progenitor evolution. In an MHD simulation of a $16M_\odot$ progenitor, [201] find the development of an α-Ω dynamo that generates strong fields of more than 10^{10} G in the oxygen and neon shell, but the dipole component remains about an order of magnitude weaker. Importantly, such fields cannot be directly compared to the pre-collapse *core* field strengths quoted for typical 3D MHD supernova simulations; the usual dipole fields in 3D MHD supernova models would have significantly weaker fields in the oxygen shell than in the core. However, the work of [201] also highlights the inconsistencies of the current treatment of angular momentum transport and magnetic fields in 1D stellar evolution. In their 3D model, the dynamo-generated fields become strong enough to rapidly redistribute angular momentum between shells and suppress convection in the oxygen and neon shell. Efficient rotational coupling with the shells further out spins down the oxygen shell from an angular velocity of 0.07 rad s^{-1} initially to about 0.01 rad s^{-1} after 500 s of simulation time. While much work is needed to investigate the role of resolution, the possible role of the magnetic Prandtl number, and evolution over longer time scales, the findings of [201] highlight that our incomplete understanding of angular momentum transport and magnetic field generation in stellar evolution models is a key uncertainty for the magnetorotational explosion scenario.

19.5.3.3 Multi-Messenger Signatures from Magnetorotational Explosions

A principal challenge in determining the explosion mechanism of hypernova lies in identifying suitable direct or indirect observables that can potentially discriminate between different scenarios or at least define "litmus tests" for simulations. Unlike for normal supernovae, compact remnant populations provide little opportunities to constrain the nature of hypernova explosions, as these make up only for a small fraction of compact object formation events. This leaves the transients themselves and their contribution to the chemical evolution of galaxies as the key sources for direct and indirect clue on the hypernova mechanism. Emission from the long gamma-ray bursts often associated with hypernovae has received significant attention, but as the relativistic gamma-ray burst (GRB) only contains a minor fraction of the total explosion energy [208], it does not provide direct clues to the hypernova explosion mechanism. Furthermore, the features of the GRB and its X-ray afterglow are often ambiguous; for example, plateaus in X-ray afterglows have been taken as evidence for millisecond magnetars [35, 59], but can in fact also be accommodated in the collapsar model [44]. Lastly, there is still a gap between the core-collapse supernova simulations discussed in the this chapter in the previous subsection and models of the later GRB phase.

Among potential direct and indirect observables from hypernova explosions, nucleosynthesis outcomes have already been studied extensively on the basis of MHD supernova simulations (for a broader review of nucleosynthesis in jet-associated supernovae see [147]). The critical idea that neutron-rich matter from close to the proto-neutron star can maintain a low electron fraction Y_e if it is rapidly ejected in jets, and then provide conditions for r-process nucleosynthesis has already been formulated based on 2D simulations [42, 55, 62, 138, 165]. However, 3D modeling and neutrino transport are required for nucleosynthesis predictions to accurately capture both the explosion dynamics and the neutrino-processing of the ejecta, which tends to drive them to more proton-rich conditions as they expand to lower densities. The first generation of 3D MHD simulations still relied on neutrino leakage and therefore could not provide reliable Y_e in the ejecta directly. Post-processing to approximately capture the effects of neutrino exposure on the ejecta is possible to remedy this to some degree, however. The first nucleosynthesis calculations based on 3D MHD models with neutrino leakage suggested very neutron-rich ejecta in the jets down to $Y_e{\sim}0.15$, which is sufficiently neutron-rich for an r-process up to the third peak [207]. Subsequent simulations with leakage indicated, however, that r-process nucleosynthesis up to the third-peak is sensitive not only to assumptions about neutrino exposure, but can also be thwarted by misalignment of the (assumed) initial magnetic dipole field and jet disruption by the kink instability [68, 125]. 3D models with neutrino transport [163, 164, 212] still exhibit ejection of neutron-rich matter in the jets, but have yet to show yields consistent at best with solar r-process abundances up to the third d peak, although low-level production of elements up to the third peak has been found in some energetic explosion models. In some cases, there is little production of elements beyond $Z = 40$ even for relatively energetic explosions [212]. In the recent models of [212], the most overproduced isotopes are rather those close to the $N = 50$ shell closure (specifically ^{82}Se, ^{83}Kr, and ^{76}Ge). High overproduction factors indicate that these yield patterns can at most represent 10% of all hypernovae to avoid conflicts with constraints from chemogalactic evolution. The current state of nucleosynthesis predictions based on magnetorotational core-collapse supernova simulations is thus far from a firm identification of hypernovae as a robust r-process site, and could indicate that the dynamics of early, non-relativistic jets (provided that they are present and stable) is not yet captured correctly in current models.

Production of ^{56}Ni is a further benchmark for magnetorotational supernova models. Several $0.1\,M_\odot$ of ^{56}Ni are required to account for the luminosities of broad-lined Ic supernovae, though it is not perfectly clear whether decay of ^{56}Ni is the only power source for the light curves. 3D MHD supernova simulations are able to reach such nickel masses [156, 164, 212]. In future, comparison to supernova spectropolarimetry may provide stronger constraints for model validation. Observed broad-lined Ic supernovae appear to be characterized by a bipolar distribution of iron-group ejecta, moving at velocities up to more than $10,000\,\mathrm{km\,s^{-1}}$, and a toroidal distribution of

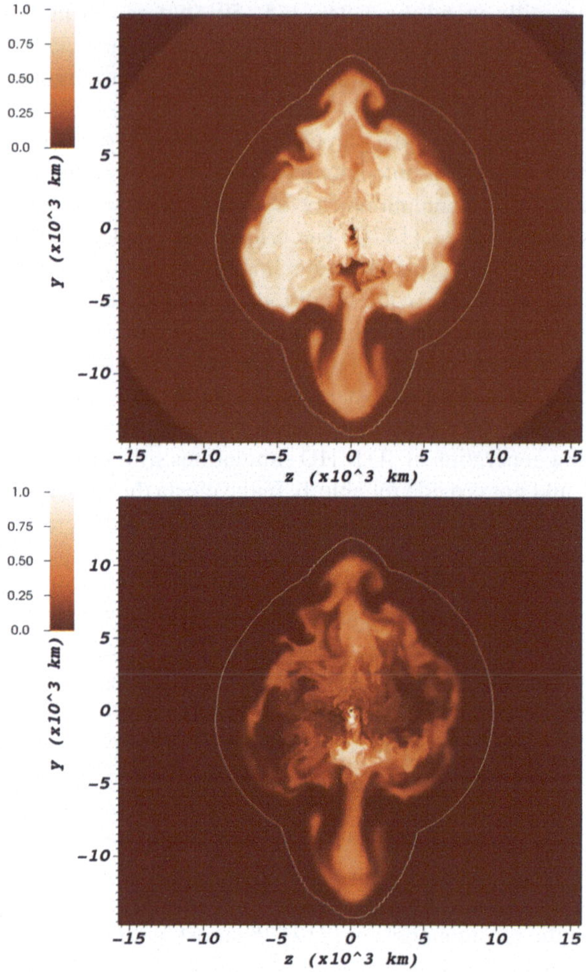

Fig. 19.4 Mass fractions of iron-group elements (top) and ^{4}He (bottom) meridional slice through the two magnerotoational explosion models of [156] with a pre-collapse core field strength of 10^{12} G by the end of the simulations. The shock is denoted by a white line. Note relatively spherical distribution of the iron-group ejecta, with only minor protrusion along the axis due to the penetrating jet. The jets have a bigger impact on shaping the distribution of ^{4}He, which is produced more abundantly in the jets due to freeze-out from nuclear statistical equilibrium at higher entropy

oxygen [103, 114, 193]. Such an ejecta structure is not yet seen in the explosion models of [156]; iron-group elements are instead distributed quite spherically and move at slower speeds (Fig. 19.4). However, the ejecta geometry can change substantially from the first second of the explosion to shock breakout due to mixing instabilities. Long-time simulations beyond shock breakout are required and need to be fed into radiative transfer calculations for meaningful comparisons with observed transients. Pipelines from multi-D core-collapse supernova simulations to synthetic light curves and spectra are now becoming available [112] and will enable more stringent validation of magnetorotational supernova models in future.

In future, gravitational waves could serve as a more direct probe of the dynamics of hypernova explosions. Observations of hypernovae with gravitational waves will, however, only be possible for nearby events even with third-generation detectors, and face the additional obstacle that the hypernova fraction is low in the local Universe. Nonetheless, gravitational wave signal predictions for magnetorotational explosions have been studied in some depth for clues about the explosion mechanism in such a fortuitous event. The bounce signal from the collapse of rapidly rotating cores is sometimes implicitly taken as a potential giveaway for magnetorotational explosions, but though it is a suitable feature for distinguishing the gravitational wave signal from potential hypernova progenitors from ordinary core-collapse supernovae [1, 73, 98, 154], it only probes the *preconditions* for magnetorotational explosions rather than the dynamics of the explosion mechanism itself. After several earlier studies had already investigated the impact of MHD effects on the bounce signal with simplified microphysics [144, 173, 190], gravitational wave signal predictions covering hundreds of milliseconds of the post-bounce and explosion phase have recently become available from MHD supernova simulations with multi-group neutrino transport [79, 155, 156]. The characteristic imprint of strong, dynamically important magnetic fields in supernova explosions remains subtle, however. The post-bounce gravitational wave signal of magnetorotational explosions does not differ fundamentally from the standard time-frequency structure of gravitational waves from core-collapse supernovae [2, 135] with a dominant rising emission band at several 100 Hz to more than 1 kHz from a low-order buoyancy-driven quadrupole oscillation (f-mode or g-mode) living near the proto-neutron star surface and other less prominent features. The gravitational wave signals from magnetorotational explosions differ from neutrino-driven ones in detail, though. Axisymmetric simulations of [79] suggested that magnetorotational explosions may be characterized by earlier subsidence of the high-frequency f/g-mode emission due to the quenching of accretion, and possibly a broader, more diffuse f/g-mode emission band. Magnetorotational explosions with strong jets should also exhibit a very strong tail signal due to the pronounced bipolarity of the expanding shock, but the detection of such a signal feature is not trivial because of its low-frequency nature. 3D simulations both with the CoCoNuT- FMT and ALCAR code [155, 156] paint a similar picture, with the exception that no strong tail signal may emerge because the jets are weaker than in axisymmetry (even if they are not disrupted by the kink instability). Nonetheless, modern signal classification methods may be able to detect the subtle differences in the time-frequency structure to distinguish magnetorotational explosions from neutrino-driven ones at moderately high overall signal-to-noise ratios of 25–45 [155]. It is also noteworthy, however, that predicted gravitational wave signals from hypernovae are generally quite strong and detectable out to 2–5 Mpc with third-generation detectors [155]. Finally, another interesting prospect is that *late-phase* gravitational wave emission may contain quite direct clues for dynamo-generated field in rapidly-rotating proto-neutron stars. In the case of rapid rotation, gravitational wave emission from the proto-neutron star convection zone is characterized by a spectrum with prominent, well-defined inertial mode frequencies, whose amplitudes and frequency structure can be related to the rotation and magnetization of the proto-neutron star [159]. The detectability of such a signal remains to be investigated, however.

19.6 Outlook

Magnetohydrodynamic simulations of core-collapse supernovae have progressed considerably in recent years. Simulations combining magnetohydrodynamics and detailed neutrino-transport have become much more common, and are becoming important beyond the traditional "niche" of the magnetorotational explosion paradigm, especially for understanding neutron star birth spins and magnetic fields. Magnetorotational explosion models are becoming sufficiently mature to deliver better predictions for observables and sufficiently widespread to enable better characterization of uncertainties in the simulations and in predicted observables. But much remains to be done. Capabilities for end-to-end modeling need to be developed to connect core-collapse simulations proper with models for GRB launching from millisecond magnetars [23, 24, 85] and collapsar-driven explosions and jets [9, 55, 60, 61, 151, 182]. Bridging the gap to neutron star birth magnetic fields also remains a challenge. Validation of MHD-driven supernova models by comparison to transient observations will become an important priority, and there is the need for a more integrated modeling of supernova progenitor evolution in 1D and 3D to address uncertainties in the interior rotation and magnetic field structure of massive stars. Magnetohydrodynamic supernova simulations are bound to develop into central hub in an interdisciplinary research network connecting computational astrophysics, stellar evolution, compact objects, and transient observations.

Acknowledgements I wish to thank B. Sykes for critical reading, P. Mösta for his kind permission to reproduce figures, and J. Powell for discussions and permission to reuse simulation data for Fig. 19.4. I acknowledge support from the Australian Research Council through Future Fellowship FT160100035, as well as computer time allocations from Astronomy Australia Limited's ASTAC scheme, the National Computational Merit Allocation Scheme (NCMAS), and from an Australasian Leadership Computing Grant. Some of this work was performed on the Gadi supercomputer with the assistance of resources and services from the National Computational Infrastructure (NCI), which is supported by the Australian Government, and through support by an Australasian Leadership Computing Grant. Some of this work was performed on the OzSTAR national facility at Swinburne University of Technology. OzSTAR is funded by Swinburne University of Technology and the National Collaborative Research Infrastructure Strategy (NCRIS).

References

1. E. Abdikamalov, S. Gossan, A.M. DeMaio, C.D. Ott, Phys. Rev. D **90**, 044001 (2014). https://doi.org/10.1103/PhysRevD.90.044001
2. E. Abdikamalov, G. Pagliaroli, D. Radice, *Handbook of Gravitational Wave Astronomy* (2022) p. 21. https://doi.org/10.1007/978-981-15-4702-721-1
3. D.R. Aguilera-Dena, N. Langer, J. Antoniadis, B. Müller, Astrophys. J. **901**, 114 (2020). https://doi.org/10.3847/1538-4357/abb138
4. D.R. Aguilera-Dena, N. Langer, T.J. Moriya, A. Schootemeijer, Astrophys. J. **858**, 115 (2018). https://doi.org/10.3847/1538-4357/aabfc1
5. S. Akiyama, J.C. Wheeler, D.L. Meier, I. Lichtenstadt, D.L. Meier, Lichtenstadt. Astrophys. J. **584**, 954 (2003). https://doi.org/10.1086/344135

6. N.V. Ardeljan, G.S. Bisnovatyi-Kogan, S.G. Moiseenko, Mon. Not. R. Astron. Soc. **359**, 333 (2005). https://doi.org/10.1111/j.1365-2966.2005.08888.x
7. S.A. Balbus, J.F. Hawley, Astrophys. J. **376**, 214 (1991). https://doi.org/10.1086/170270
8. S.A. Balbus, J.F. Hawley, Rev. Modern Phys. **70**, 1 (1998). https://doi.org/10.1103/RevModPhys.70.1
9. M.V. Barkov, S.S. Komissarov, Mon. Not. R. Astron. Soc. **385**, L28 (2008). https://doi.org/10.1111/j.1745-3933.2008.00427.x
10. P. Barrère, J. Guilet, R. Raynaud, A. Reboul-Salze, Mon. Not. R. Astron. Soc. **526**, L88 (2023). https://doi.org/10.1093/mnrasl/slad120
11. P. Barrère, J. Guilet, A. Reboul-Salze, R. Raynaud, H.T. Janka, Astron. Astrophys. **668**, A79 (2022). https://doi.org/10.1051/0004-6361/202244172
12. M.C. Begelman, Astrophys. J. **493**, 291 (1998). https://doi.org/10.1086/305119
13. K. Belkacem, J.P. Marques, M.J. Goupil, B. Mosser, T. Sonoi, R.M. Ouazzani, M.A. Dupret, S. Mathis, M. Grosjean, Astron. Astrophys. **579**, A31 (2015). https://doi.org/10.1051/0004-6361/201526043
14. H.A. Bethe, Rev. Mod. Phys. **62**, 801 (1990). https://doi.org/10.1103/RevModPhys.62.801
15. G.S. Bisnovatyi-Kogan, Astron. Zh. **47**, 813 (1970)
16. G.S. Bisnovatyi-Kogan, I.P. Popov, A.A. Samokhin, Astrophys. Space Sci. **41**, 287 (1976). https://doi.org/10.1007/BF00646184
17. J.M. Blondin, A. Mezzacappa, Nature **445**, 58 (2007). https://doi.org/10.1038/nature05428
18. J.M. Blondin, A. Mezzacappa, C. DeMarino, Astrophys. J. **584**, 971 (2003). https://doi.org/10.1086/345812
19. R. Bollig, N. Yadav, D. Kresse, H.-T. Janka, B. Müller, A. Heger, Astrophys. J. **915**, 28 (2021). https://doi.org/10.3847/1538-4357/abf82e
20. J. Braithwaite, Å. Nordlund, Astron. Astrophys. **450**, 1077 (2006). https://doi.org/10.1051/0004-6361:20041980
21. A. Brandenburg, K. Subramanian, Phys. Rep. **417**, 1 (2005). https://doi.org/10.1016/j.physrep.2005.06.005
22. A.S. Brun, M.K. Browning, Living Rev. Solar Phys. **14**, 4 (2017). https://doi.org/10.1007/s41116-017-0007-8
23. N. Bucciantini, E. Quataert, J. Arons, B.D. Metzger, T.A. Thompson, Mon. Not. R. Astron. Soc. **383**, L25 (2008). https://doi.org/10.1111/j.1745-3933.2007.00403.x
24. N. Bucciantini, E. Quataert, B.D. Metzger, T.A. Thompson, J. Arons, L. Del Zanna, Mon. Not. R. Astron. Soc. **396**, 2038 (2009). https://doi.org/10.1111/j.1365-2966.2009.14940.x
25. N. Bucciantini, T.A. Thompson, J. Arons, E. Quataert, L. Del Zanna, Mon. Not. R. Astron. Soc. **368**, 1717 (2006). https://doi.org/10.1111/j.1365-2966.2006.10217.x
26. M. Bugli, J. Guilet, M. Obergaulinger, Mon. Not. R. Astron. Soc. **507**, 443 (2021). https://doi.org/10.1093/mnras/stab2161
27. A. Burrows, L. Dessart, E. Livne, C.D. Ott, J. Murphy, Astrophys. J. **664**, 416 (2007). https://doi.org/10.1086/519161
28. A. Burrows, J. Hayes, B.A. Fryxell, Astrophys. J. **450**, 830 (1995). https://doi.org/10.1086/176188
29. A. Burrows, T. Wang, D. Vartanyan, https://doi.org/10.48550/arXiv.2401.06840 arXiv e-prints , arXiv:2401.06840 (2024)
30. R.M. Cabezón, K.-C. Pan, M. Liebendörfer, T. Kuroda, K. Ebinger, O. Heinimann, A. Perego, F.-K. Thielemann, Astron. Astrophys. **619**, A118 (2018). https://doi.org/10.1051/0004-6361/201833705
31. M. Cantiello, C. Mankovich, L. Bildsten, J. Christensen-Dalsgaard, B. Paxton, Astrophys. J. **788**, 93 (2014). https://doi.org/10.1088/0004-637X/788/1/93
32. J. Cho, E.T. Vishniac, A. Beresnyak, A. Lazarian, D. Ryu, Astrophys. J. **693**, 1449 (2009). https://doi.org/10.1088/0004-637X/693/2/1449
33. U.R. Christensen, V. Holzwarth, A. Reiners, Nature **457**, 167 (2009). https://doi.org/10.1038/nature07626

34. C. Collins, B. Müller, A. Heger, Mon. Not. R. Astron. Soc. **473**, 1695 (2018). https://doi.org/10.1093/mnras/stx2470
35. A. Corsi, P. Mészáros, Astrophys. J. **702**, 1171 (2009). https://doi.org/10.1088/0004-637X/702/2/1171
36. S.M. Couch, E. Chatzopoulos, W.D. Arnett, F.X. Timmes, Astrophys. J. **808**, L21 (2015). https://doi.org/10.1088/2041-8205/808/1/L21
37. T.G. Cowling, Mon. Not. R. Astron. Soc. **94**, 39 (1933). https://doi.org/10.1093/mnras/94.1.39
38. R.K. Crockett, P. Colella, R.T. Fisher, R.I. Klein, C.F. McKee, J. Comput. Phys. **203**, 422 (2005). https://doi.org/10.1016/j.jcp.2004.08.021
39. A. Dedner, F. Kemm, D. Kröner, C.D. Munz, T. Schnitzer, M. Wesenberg, J. Comput. Phys. **175**, 645 (2002). https://doi.org/10.1006/jcph.2001.6961
40. D. Derigs, A.R. Winters, G.J. Gassner, S. Walch, M. Bohm, J. Comput. Phys. **364**, 420 (2018). https://doi.org/10.1016/j.jcp.2018.03.002
41. D.K. Desai, D.M. Siegel, B.D. Metzger, Astrophys. J. **954**, 192 (2023). https://doi.org/10.3847/1538-4357/acea83
42. L. Dessart, A. Burrows, E. Livne, C.D. Ott, Astrophys. J. **669**, 585 (2007). https://doi.org/10.1086/521701
43. H. Dimmelmeier, C.D. Ott, H-T. Janka, A. Marek, E. Müller, Phys. Rev. Lett. **98**, 251101:1 (2007). https://doi.org/10.1103/PhysRevLett.98.251101
44. P.C. Duffell, A.I. MacFadyen, Astrophys. J. **806**, 205 (2015). https://doi.org/10.1088/0004-637X/806/2/205
45. R.C. Duncan, S.L. Shapiro, I. Wasserman, Astrophys. J. **309**, 141 (1986). https://doi.org/10.1086/164587
46. R.C. Duncan, C. Thompson, Astrophys. J. **392**, L9 (1992). https://doi.org/10.1086/186413
47. P. Eggenberger, N. Lagarde, A. Miglio, J. Montalbán, S. Ekström, C. Georgy, G. Meynet, S. Salmon, T. Ceillier, R.A. García, S. Mathis, S. Deheuvels, A. Maeder, J.W. den Hartogh, R. Hirschi, Astron. Astrophys. **599**, A18 (2017). https://doi.org/10.1051/0004-6361/201629459
48. E. Endeve, C.Y. Cardall, Budiardja, A. Mezzacappa, Astrophys. J. **713**, 1219 (2010). https://doi.org/10.1088/0004-637X/713/2/1219
49. E. Endeve, C.Y. Cardall, R.D. Budiardja, S.W. Beck, A. Bejnood, R.J. Toedte, A. Mezzacappa, J.M. Blondin, Astrophys. J. **751**, 26 (2012). https://doi.org/10.1088/0004-637X/751/1/26
50. T. Ertl, H.-T. Janka, S.E. Woosley, T. Sukhbold, M. Ugliano, Astrophys. J. **818**, 124 (2016). https://doi.org/10.3847/0004-637X/818/2/124
51. C.R. Evans, J.F. Hawley, Astrophys. J. **332**, 659 (1988). https://doi.org/10.1086/166684
52. C.E. Fields, S.M. Couch, Astrophys. J. **921**, 28 (2021). https://doi.org/10.3847/1538-4357/ac24fb
53. T. Foglizzo, P. Galletti, L. Scheck, H.-T. Janka, Astrophys. J. **654**, 1006 (2007). https://doi.org/10.1086/509612
54. U. Frisch, A. Pouquet, J. Leorat, A. Mazure, J. Fluid Mech. **68**, 769 (1975). https://doi.org/10.1017/S002211207500122X
55. S.-I. Fujimoto, N. Nishimura, M.A. Hashimoto, Astrophys. J. **680**, 1350 (2008). https://doi.org/10.1086/529416
56. J. Fuller, A.L. Piro, A.S. Jermyn, Mon. Not. R. Astron. Soc. **485**, 3661 (2019). https://doi.org/10.1093/mnras/stz514
57. T.A. Gardiner, J.M. Stone, J. Comput. Phys. **205**, 509 (2005). https://doi.org/10.1016/j.jcp.2004.11.016
58. P. Goldreich, S.V. Weber, Astrophys. J. **238**, 991 (1980). https://doi.org/10.1086/158065
59. B.P. Gompertz, P.T. O'Brien, G.A. Wynn, Mon. Not. R. Astron. Soc. **438**, 240 (2014). https://doi.org/10.1093/mnras/stt2165
60. O. Gottlieb, A. Lalakos, O. Bromberg, M. Liska, A. Tchekhovskoy, A., Mon. Not. R. Astron. Soc. **510**, 4962 (2022). https://doi.org/10.1093/mnras/stab3784
61. O. Gottlieb, M. Liska, A. Tchekhovskoy, O. Bromberg, A. Lalakos, D. Giannios, P. Mösta, Astrophys. J. **933**, L9 (2022). https://doi.org/10.3847/2041-8213/ac7530

62. J.J. Grimmett, B. Müller, A. Heger, P. Banerjee, M. Obergaulinger, Mon. Not. R. Astron. Soc. **501**, 2764 (2021). https://doi.org/10.1093/mnras/staa3819
63. J. Guilet, T. Foglizzo, Mon. Not. R. Astron. Soc. **421**, 546 (2012). https://doi.org/10.1111/j.1365-2966.2012.20333.x
64. J. Guilet, T. Foglizzo, S. Fromang, Astrophys. J. **729**, 71 (2011). https://doi.org/10.1088/0004-637X/729/1/71
65. J. Guilet, E. Müller, Mon. Not. R. Astron. Soc. **450**, 2153 (2015). https://doi.org/10.1093/mnras/stv727
66. J. Guilet, E. Müller, H.-T. Janka, Mon. Not. R. Astron. Soc. **447**, 3992 (2015). https://doi.org/10.1093/mnras/stu2550
67. J. Guilet, A. Reboul-Salze, R. Raynaud, M. Bugli, B. Gallet, Mon. Not. R. Astron. Soc. **516**, 4346 (2022). https://doi.org/10.1093/mnras/stac2499
68. G. Halevi, P. Mösta, Mon. Not. R. Astron. Soc. **477**, 2366 (2018). https://doi.org/10.1093/mnras/sty797
69. B. Harten, P.D. Lax, B. van Leer, SIAM Rev. **25**, 35 (1983)
70. N.E. Haugen, A. Brandenburg, W. Dobler, Phys. Rev. E **70**, 016308 (2004). https://doi.org/10.1103/PhysRevE.70.016308
71. A. Heger, S.E. Woosley, H.C. Spruit, Astrophys. J. **626**, 350 (2005). https://doi.org/10.1086/429868
72. D.J. Helfand, R.H. Becker, Nature **307**, 215 (1984). https://doi.org/10.1038/307215a0
73. I.S. Heng, Class. Quantum Gravity **26**, 105005 (2009). https://doi.org/10.1088/0264-9381/26/10/105005
74. M. Herant, W. Benz, W.R. Hix, C.L. Fryer, S.A. Colgate, Astrophys. J. **435**, 339 (1994). https://doi.org/10.1086/174817
75. A. Hewish, S.J. Bell, J.D.H. Pilkington, P.F. Scott, R.A. Collins, Nature **217**, 709 (1968). https://doi.org/10.1038/217709a0
76. H.-T. Janka, Astron. Astrophys. **368**, 527 (2001). https://doi.org/10.1051/0004-6361:20010012
77. H.-T. Janka, Ann. Rev. Nuclear Particle Sci **62**, 407 (2012). https://doi.org/10.1146/annurev-nucl-102711-094901
78. H.-T. Janka, E. Müller, Astron. Astrophys. **306**, 167 (1996)
79. R. Jardine, J. Powell, B. Müller, Mon. Not. R. Astron. Soc. **510**, 5535 (2022). https://doi.org/10.1093/mnras/stab3763
80. A.S. Jermyn, S.M. Chitre, P. Lesaffre, C.A. Tout, Mon. Not. R. Astron. Soc. **498**, 3758 (2020). https://doi.org/10.1093/mnras/staa2323
81. A.S. Jermyn, S.M. Chitre, P. Lesaffre, C.A. Tout, Mon. Not. R. Astron. Soc. **498**, 3782 (2020). https://doi.org/10.1093/mnras/staa2576
82. O. Just, R. Bollig, H.T. Janka, M. Obergaulinger, R. Glas, S. Nagataki, Mon. Not. R. Astron. Soc. **481**, 4786 (2018). https://doi.org/10.1093/mnras/sty2578
83. V. Kalogera, M.-A. Bizouard, A. Burrows, T. Janka, K. Kotake, B. Messer, T. Mezzacappa, B. Mueller, E. Mueller, M.A. Papa, S. Reddy, S. Rosswog, Bull. Am. Astron. Soc. **51**, 239 (2019)
84. R. Käppeli, S.C. Whitehouse, S. Scheidegger, U.L. Pen, M. Liebendörfer, Astrophys. J. Suppl. **195**, 20 (2011). https://doi.org/10.1088/0067-0049/195/2/20
85. S.S. Komissarov, M.V. Barkov, Mon. Not. R. Astron. Soc. **382**, 1029 (2007). https://doi.org/10.1111/j.1365-2966.2007.12485.x
86. K. Kotake, H. Sawai, S. Yamada, K. Sato, Astrophys. J. **608**, 391 (2004). https://doi.org/10.1086/392530
87. M. Kruskal, J.L. Tuck, Proc. R. Soc. London Ser. A **245**, 222 (1958). https://doi.org/10.1098/rspa.1958.0079
88. T. Kuroda, A. Arcones, T. Takiwaki, K. Kotake, Astrophys. J. **896**, 102 (2020). https://doi.org/10.3847/1538-4357/ab9308
89. T. Kuroda, H. Umeda, Astrophys. J. Suppl. **191**, 439 (2010). https://doi.org/10.1088/0067-0049/191/2/439

90. K. Langanke, G. Martínez-Pinedo, J.-M. Sampaio, D.J. Dean, W.R. Hix, O.E. Messer, A. Mezzacappa, M. Liebendörfer, H.-T. Janka, M. Rampp, Phys. Rev. Lett. **90**, 241102 (2003). https://doi.org/10.1103/PhysRevLett.90.241102
91. J.M. Lattimer, B.F. Schutz, Astrophys. J. **629**, 979 (2005). https://doi.org/10.1086/431543
92. J.M. LeBlanc, J.R. Wilson, Astrophys. J. **161**, 541 (1970). https://doi.org/10.1086/150558
93. D. Lee, J. Comput. Phys. **243**, 269 (2013). https://doi.org/10.1016/j.jcp.2013.02.049
94. G. Leidi, R. Andrassy, J. Higl, P.V.F. Edelmann, F.K. Röpke, Astron. Astrophys. **679**, A132 (2023). https://doi.org/10.1051/0004-6361/202347621
95. M. Liebendörfer, O.E.B. Messer, A. Mezzacappa, S.W. Bruenn, C.Y. Cardall, F.-K. Thielemann, Astrophys. J. Suppl. **150**, 263 (2004). https://doi.org/10.1086/380191
96. M. Liebendörfer, M. Rampp, H.-T. Janka, A. Mezzacappa, Astrophys. J. **620**, 840 (2005). https://doi.org/10.1086/427203
97. E. Livne, L. Dessart, A. Burrows, C.A. Meakin, Astrophys. J. Suppl. **170**, 187 (2007). https://doi.org/10.1086/513701
98. J. Logue, C.D. Ott, I.S. Heng, P. Kalmus, J.H.C. Scargill, Phys. Rev. D **86**, 044023 (2012). https://doi.org/10.1103/PhysRevD.86.044023
99. D.R. Lorimer, M. Kramer, *Handbook of Pulsar Astronomy*, vol. 4 (2004)
100. L. Ma, J. Fuller, Mon. Not. R. Astron. Soc. **488**, 4338 (2019). https://doi.org/10.1093/mnras/stz2009
101. A.I. MacFadyen, S.E. Woosley, Astrophys. J. **524**, 262 (1999). https://doi.org/10.1086/307790
102. A.I. MacFadyen, S.E. Woosley, A. Heger, Astrophys. J. **550**, 410 (2001). https://doi.org/10.1086/319698
103. K. Maeda, K. Kawabata, P.A. Mazzali, M. Tanaka, S. Valenti, K. Nomoto, T. Hattori, J. Deng, E. Pian, S. Taubenberger, M. Iye, T. Matheson, A.V. Filippenko, K. Aoki, G. Kosugi, Y. Ohyama, T. Sasaki, T. Takata, Science **319**, 1220 (2008). https://doi.org/10.1126/science.1149437
104. A. Marek, H. Dimmelmeier, H.-T. Janka, E. Müller, R. Buras, Astron. Astrophys. **445**, 273 (2006). https://doi.org/10.1051/0004-6361:20052840
105. Y. Masada, K. Kotake, T. Takiwaki, N. Yamamoto, Phys. Rev. D **98**, 083018 (2018). https://doi.org/10.1103/PhysRevD.98.083018
106. Y. Masada, T. Takiwaki, K. Kotake, Astrophys. J. **798**, L22 (2015). https://doi.org/10.1088/2041-8205/798/1/L22
107. Y. Masada, T. Takiwaki, K. Kotake, Astrophys. J. **924**, 75 (2022). https://doi.org/10.3847/1538-4357/ac34f6
108. Y. Masada, T. Takiwaki, K. Kotake, T. Sano, Astrophys. J. **759**, 110 (2012). https://doi.org/10.1088/0004-637X/759/2/110
109. J. Matsumoto, Y. Asahina, T. Takiwaki, K. Kotake, H.R. Takahashi, Mon. Not. R. Astron. Soc. **516**, 1752 (2022). https://doi.org/10.1093/mnras/stac2335
110. J. Matsumoto, T. Takiwaki, K. Kotake, Mon. Not. R. Astron. Soc. **528**, L96 (2024). https://doi.org/10.1093/mnrasl/slad173
111. J. Matsumoto, T. Takiwaki, K. Kotake, Y. Asahina, H.R. Takahashi, Mon. Not. R. Astron. Soc. **499**, 4174 (2020). https://doi.org/10.1093/mnras/staa3095
112. T. Maunder, B. Müller, F. Callan, S. Sim, A. Heger, Mon. Not. R. Astron. Soc. **527**, 2185 (2024). https://doi.org/10.1093/mnras/stad3284
113. P.A. Mazzali, A.I. McFadyen, S.E. Woosley, E. Pian, M. Tanaka, Mon. Not. R. Astron. Soc. **443**, 67 (2014). https://doi.org/10.1093/mnras/stu1124
114. P.A. Mazzali, K. Nomoto, F. Patat, K. Maeda, Astrophys. J. **559**, 1047 (2001). https://doi.org/10.1086/322420
115. L.O. McNeill, B. Müller, Mon. Not. R. Astron. Soc. **509**, 818 (2022). https://doi.org/10.1093/mnras/stab3076
116. D.L. Meier, R.I. Epstein, W.D. Arnett, D.N. Schramm, Astrophys. J. **204**, 869 (1976). https://doi.org/10.1086/154235
117. B.D. Metzger, B. Margalit, D. Kasen, E. Quataert, Mon. Not. R. Astron. Soc. **454**, 3311 (2015). https://doi.org/10.1093/mnras/stv2224

118. A. Mezzacappa, E. Endeve, O.E.B. Messer, S.W. Bruenn, Living Rev. Comput. Astrophys. **6**, 4 (2020). https://doi.org/10.1007/s41115-020-00010-8
119. H. Mikami, Y. Sato, T. Matsumoto, T. Hanawa, Astrophys. J. **683**, 357 (2008). https://doi.org/10.1086/589759
120. T. Miyoshi, K. Kusano, J. Comput. Phys. **208**, 315 (2005). https://doi.org/10.1016/j.jcp.2005.02.017
121. S.G. Moiseenko, G.S. Bisnovatyi-Kogan, N.V. Ardeljan, Mon. Not. R. Astron. Soc. **370**, 501 (2006). https://doi.org/10.1111/j.1365-2966.2006.10517.x
122. P. Mösta, B.C. Mundim, J.A. Faber, R. Haas, S.C. Noble, T. Bode, F. Löffler, C.D. Ott, C. Reisswig, E. Schnetter, Class. Quantum Gravity **31**, 015005 (2014). https://doi.org/10.1088/0264-9381/31/1/015005
123. P. Mösta, C.D. Ott, D. Radice, L.F. Roberts, E. Schnetter, R. Haas, Nature **528**, 376 (2015). https://doi.org/10.1038/nature15755
124. P. Mösta, S. Richers, C.D. Ott, R. Haas, A.L. Piro, K. Boydstun, E. Abdikamalov, C. Reisswig, E. Schnetter, Astrophys. J. **785**, L29 (2014). https://doi.org/10.1088/2041-8205/785/2/L29
125. P. Mösta, L.F. Roberts, G. Halevi, C.D. Ott, J. Lippuner, R. Haas, E. Schnetter, Astrophys. J. **864**, 171 (2018). https://doi.org/10.3847/1538-4357/aad6ec
126. E. Mueller, W. Hillebrandt, Astron. Astrophys. **80**, 147 (1979)
127. B. Müller, Mon. Not. R. Astron. Soc. **453**, 287 (2015). https://doi.org/10.1093/mnras/stv1611
128. B. Müller, Publ. Astron. Soc. Australia **33**, e048 (2016). https://doi.org/10.1017/pasa.2016.40
129. B. Müller, Ann. Rev. Nuclear Particle Sci. **69**, 253 (2019). https://doi.org/10.1146/annurev-nucl-101918-023434. (arXiv:1904.11067 [astro-ph.HE])
130. B. Müller, Living Rev. Comput. Astrophys. **6**, 3 (2020). https://doi.org/10.1007/s41115-020-0008-5
131. B. Müller, H. Dimmelmeier, E. Müller, Astron. Astrophys. **489**, 301 (2008). https://doi.org/10.1051/0004-6361:200809609
132. B. Müller, A. Heger, D. Liptai, J.B. Cameron, Mon. Not. R. Astron. Soc. **460**, 742 (2016). https://doi.org/10.1093/mnras/stw1083
133. B. Müller, H. Janka, H. Dimmelmeier, Astrophys. J. Suppl. **189**, 104 (2010). https://doi.org/10.1088/0067-0049/189/1/104
134. B. Müller, H.-T. Janka, Mon. Not. R. Astron. Soc. **448**, 2141 (2015). https://doi.org/10.1093/mnras/stv101
135. B. Müller, H.-T. Janka, A. Marek, Astrophys. J. **766**, 43 (2013). https://doi.org/10.1088/0004-637X/766/1/43
136. B. Müller, T. Melson, A. Heger, H.-T. Janka, Mon. Not. R. Astron. Soc. **472**, 491 (2017). https://doi.org/10.1093/mnras/stx1962
137. B. Müller, V. Varma, Mon. Not. R. Astron. Soc. **498**, L109 (2020). https://doi.org/10.1093/mnrasl/slaa137
138. S. Nishimura, K. Kotake, M.-A. Hashimoto, S. Yamada, N. Nishimura, S. Fujimoto, K. Sato, Astrophys. J. **642**, 410 (2006). https://doi.org/10.1086/500786
139. M. Obergaulinger, M.Á. Aloy, Mon. Not. R. Astron. Soc. **469**, L43 (2017). https://doi.org/10.1093/mnrasl/slx046
140. M. Obergaulinger, M.Á. Aloy, arXiv e-prints (2019)
141. M. Obergaulinger, M.Á. Aloy, Mon. Not. R. Astron. Soc. **503**, 4942 (2021). https://doi.org/10.1093/mnras/stab295
142. M. Obergaulinger, M.Á. Aloy, Mon. Not. R. Astron. Soc. **512**, 2489 (2022). https://doi.org/10.1093/mnras/stac613
143. M. Obergaulinger, Aloy, M. A., H. Dimmelmeier, and Müller, E., https://doi.org/10.1051/0004-6361:20064982 Astron. Astrophys. **457**, 209 (2006)
144. M. Obergaulinger, M.A. Aloy, E. Müller, Astron. Astrophys. **450**, 1107 (2006). https://doi.org/10.1051/0004-6361:20054306
145. M. Obergaulinger, P. Cerdá-Durán, E. Müller, M.A. Aloy, Astron. Astrophys. **498**, 241 (2009). https://doi.org/10.1051/0004-6361/200811323

146. M. Obergaulinger, H.-T. Janka, M.A. Aloy, Mon. Not. R. Astron. Soc. **445**, 3169 (2014). https://doi.org/10.1093/mnras/stu1969
147. M. Obergaulinger, M. Reichert, arXiv e-prints arXiv:2303.12458 (2023). https://doi.org/10.48550/arXiv.2303.12458
148. E. O'Connor, R. Bollig, A. Burrows, S. Couch, T. Fischer, H.-T. Janka, K. Kotake, E.J. Lentz, M. Liebendörfer, O.E.B. Messer, A. Mezzacappa, T. Takiwaki, D. Vartanyan, J. Phys. G Nuclear Phys. **45**, 104001 (2018). https://doi.org/10.1088/1361-6471/aadeae
149. E.P. O'Connor, S.M. Couch, Astrophys. J. **854**, 63 (2018). https://doi.org/10.3847/1538-4357/aaa893
150. S.A. Olausen, V.M. Kaspi, Astrophys. J. Suppl. **212**, 6 (2014). https://doi.org/10.1088/0067-0049/212/1/6
151. M. Ono, M. Hashimoto, S. Fujimoto, K. Kotake, S. Yamada, Progr. Theor. Phys. **128**, 741 (2012). https://doi.org/10.1143/PTP.128.741
152. D.J.B. Payne, A. Melatos, Mon. Not. R. Astron. Soc. **351**, 569 (2004). https://doi.org/10.1111/j.1365-2966.2004.07798.x
153. A. Pouquet, U. Frisch, J. Leorat, J. Fluid Mech. **77**, 321 (1976). https://doi.org/10.1017/S0022112076002140
154. J. Powell, S.E. Gossan, J. Logue, I.S. Heng, Phys. Rev. D **94**, 123012 (2016). https://doi.org/10.1103/PhysRevD.94.123012
155. J. Powell, A. Iess, M. Llorens-Monteagudo, M. Obergaulinger, B. Müller, A. Torres-Forné, E. Cuoco, J.A. Font, arXiv e-prints , arXiv:2311.18221 (2023)
156. J. Powell, B. Müller, D.R. Aguilera-Dena, N. Langer, Mon. Not. R. Astron. Soc. **522**, 6070 (2023). https://doi.org/10.1093/mnras/stad1292
157. T. Prasanna, M.S.B. Coleman, M.J. Raives, T.A. Thompson, Mon. Not. R. Astron. Soc. **517**, 3008 (2022). https://doi.org/10.1093/mnras/stac2651
158. M. Rampp, H.-T. Janka, Astron. Astrophys. **396**, 361 (2002). https://doi.org/10.1051/0004-6361:20021398
159. R. Raynaud, P. Cerdá-Durán, J. Guilet, Mon. Not. R. Astron. Soc. **509**, 3410 (2022). https://doi.org/10.1093/mnras/stab3109
160. R. Raynaud, J. Guilet, H.-T. Janka, T. Gastine, Sci. Adv. **6**, eaay2732 (2020). https://doi.org/10.1126/sciadv.aay2732
161. A. Reboul-Salze, J. Guilet, R. Raynaud, M. Bugli, Astron. Astrophys. **645**, A109 (2021). https://doi.org/10.1051/0004-6361/202038369
162. A. Reboul-Salze, J. Guilet, R. Raynaud, M. Bugli, Astron. Astrophys. **667**, A94 (2022). https://doi.org/10.1051/0004-6361/202142368
163. M. Reichert, M. Bugli, J. Guilet, M. Obergaulinger, M.Á. Aloy, A. Arcones, Mon. Not. R. Astron. Soc. (2024). https://doi.org/10.1093/mnras/stae561
164. M. Reichert, M. Obergaulinger, M.Á. Aloy, M. Gabler, A. Arcones, F.K. Thielemann, Mon. Not. R. Astron. Soc. **518**, 1557 (2023). https://doi.org/10.1093/mnras/stac3185
165. M. Reichert, M. Obergaulinger, M. Eichler, M.Á. Aloy, A. Arcones, Mon. Not. R. Astron. Soc. **501**, 5733 (2021). https://doi.org/10.1093/mnras/stab029
166. T. Rembiasz, J. Guilet, M. Obergaulinger, P. Cerdá-Durán, M.A. Aloy, E. Müller, Mon. Not. R. Astron. Soc. **460**, 3316 (2016). https://doi.org/10.1093/mnras/stw1201
167. T. Rembiasz, M. Obergaulinger, P. Cerdá-Durán, E. Müller, M.A. Aloy, Mon. Not. R. Astron. Soc. **456**, 3782 (2016). https://doi.org/10.1093/mnras/stv2917
168. H. Sawai, K. Kotake, S. Yamada, Astrophys. J. **631**, 446 (2005). https://doi.org/10.1086/432529
169. H. Sawai, K. Kotake, S. Yamada, Astrophys. J. **672**, 465 (2008). https://doi.org/10.1086/523624
170. H. Sawai, S. Yamada, Astrophys. J. **784**, L10 (2014). https://doi.org/10.1088/2041-8205/784/1/L10
171. H. Sawai, S. Yamada, Astrophys. J. **817**, 153 (2016). https://doi.org/10.3847/0004-637X/817/2/153

172. H. Sawai, S. Yamada, H. Suzuki, Astrophys. J. **770**, L19 (2013). https://doi.org/10.1088/2041-8205/770/2/L19
173. S. Scheidegger, T. Fischer, S.C. Whitehouse, M. Liebendörfer, Astron. Astrophys. **490**, 231 (2008). https://doi.org/10.1051/0004-6361:20078577
174. S. Scheidegger, S.C. Whitehouse, R. Käppeli, M. Liebendörfer, Class. Quantum Gravity **27**, 114101 (2010). https://doi.org/10.1088/0264-9381/27/11/114101
175. A.A. Schekochihin, S.C. Cowley, G.W. Hammett, J.L. Maron, J.C. McWilliams, New J. Phys. **4**, 84 (2002). https://doi.org/10.1088/1367-2630/4/1/384
176. F.R.N. Schneider, S.T. Ohlmann, P. Podsiadlowski, F.K. Röpke, S.A. Balbus, R. Pakmor, Mon. Not. R. Astron. Soc. **495**, 2796 (2020). https://doi.org/10.1093/mnras/staa1326
177. F.R.N. Schneider, S.T. Ohlmann, P. Podsiadlowski, F.K. Röpke, S.A. Balbus, R. Pakmor, V. Springel, Nature **574**, 211 (2019). https://doi.org/10.1038/s41586-019-1621-5
178. J. Schober, D. Schleicher, C. Federrath, R. Klessen, R. Banerjee, Phys. Rev. E **85**, 026303 (2012). https://doi.org/10.1103/PhysRevE.85.026303
179. J. Schober, D.R.G. Schleicher, C. Federrath, S. Bovino, R.S. Klessen, Phys. Rev. E **92**, 023010 (2015). https://doi.org/10.1103/PhysRevE.92.023010
180. S.L. Shapiro, S.A. Teukolsky, *Black Holes, White Dwarfs, and Neutron Stars: The Physics of Compact Objects* (Wiley-Interscience, New York, 1983)
181. S. Shibagaki, T. Kuroda, K. Kotake, T. Takiwaki, T. Fischer, arXiv e-prints arXiv:2309.05161 (2023)
182. D.M. Siegel, J. Barnes, B.D. Metzger, Nature **569**, 241 (2019). https://doi.org/10.1038/s41586-019-1136-0
183. S.J. Smartt, Annu. Rev. Astron. Astrophys. **47**, 63 (2009). https://doi.org/10.1146/annurev-astro-082708-101737
184. S.J. Smartt, Publ. Astron. Soc. Australia **32**, 16 (2015). https://doi.org/10.1017/pasa.2015.17
185. H.C. Spruit, Astron. Astrophys. **381**, 923 (2002). https://doi.org/10.1051/0004-6361:20011465
186. R. Stepanov, F. Plunian, Astrophys. J. **680**, 809 (2008). https://doi.org/10.1086/587795
187. J.M. Stone, K. Tomida, C.J. White, K.G. Felker, Astrophys. J. Suppl. **249**, 4 (2020). https://doi.org/10.3847/1538-4365/ab929b
188. T.K. Suzuki, S. Nagataki, Astrophys. J. **628**, 914 (2005). https://doi.org/10.1086/430847
189. T.K. Suzuki, K. Sumiyoshi, S. Yamada, Astrophys. J. **678**, 1200 (2008). https://doi.org/10.1086/533515
190. T. Takiwaki, K. Kotake, Astrophys. J. **743**, 30 (2011). https://doi.org/10.1088/0004-637X/743/1/30
191. T. Takiwaki, K. Kotake, K. Sato, Astrophys. J. **691**, 1360 (2009). https://doi.org/10.1088/0004-637X/691/2/1360
192. T. Takiwaki, K. Kotake, Y. Suwa, Mon. Not. R. Astron. Soc. **461**, L112 (2016). https://doi.org/10.1093/mnrasl/slw105
193. M. Tanaka, K. Maeda, P.A. Mazzali, K.S. Kawabata, K. Nomoto, Astrophys. J. **837**, 105 (2017). https://doi.org/10.3847/1538-4357/aa6035
194. R.J. Tayler, Mon. Not. R. Astron. Soc. **161**, 365 (1973). https://doi.org/10.1093/mnras/161.4.365
195. C. Thompson, R.C. Duncan, Astrophys. J. **408**, 194 (1993). https://doi.org/10.1086/172580
196. T.A. Thompson, A. ud-Doula, Mon. Not. R. Astron. Soc. **476**, 5502 (2018). https://doi.org/10.1093/mnras/sty480
197. E.F. Toro, M. Spruce, W. Speares, Shock Waves **4**, 25 (1994). https://doi.org/10.1007/BF01414629
198. A. Torres-Forné, P. Cerdá-Durán, J.A. Pons, J.A. Font, Mon. Not. R. Astron. Soc. **456**, 3813 (2016). https://doi.org/10.1093/mnras/stv2926
199. V.U. Usov, Nature **357**, 472 (1992). https://doi.org/10.1038/357472a0
200. V. Varma, B. Müller, Mon. Not. R. Astron. Soc. **504**, 636 (2021). https://doi.org/10.1093/mnras/stab883

201. V. Varma, B. Müller, Mon. Not. R. Astron. Soc. **526**, 5249 (2023). https://doi.org/10.1093/mnras/stad3113
202. V. Varma, B. Müller, M. Obergaulinger, Mon. Not. R. Astron. Soc. **508**, 6033 (2021). https://doi.org/10.1093/mnras/stab2983
203. V. Varma, B. Müller, F.R.N. Schneider, Mon. Not. R. Astron. Soc. **518**, 3622 (2023). https://doi.org/10.1093/mnras/stac3247
204. E.T. Vishniac, in *Magnetic Fields in the Universe: From Laboratory and Stars to Primordial Structures.*, American Institute of Physics Conference Series, vol. 784, ed. by E.M. de Gouveia dal Pino, G. Lugones, A. Lazarian (2005), pp. 3–15. https://doi.org/10.1063/1.2077167
205. T. Wang, A. Burrows, Astrophys. J. **954**, 114 (2023). https://doi.org/10.3847/1538-4357/ace7b2
206. J. Warnecke, M.J. Korpi-Lagg, F.A. Gent, M. Rheinhardt, Nat. Astron. **7**, 662 (2023). https://doi.org/10.1038/s41550-023-01975-1
207. C. Winteler, R. Käppeli, A. Perego, A. Arcones, N. Vasset, N. Nishimura, M. Liebendörfer, F.-K. Thielemann, Astrophys. J. **750**, L22 (2012). https://doi.org/10.1088/2041-8205/750/1/L22
208. S.E. Woosley, J.S. Bloom, Annu. Rev. Astron. Astrophys. **44**, 507 (2006). https://doi.org/10.1146/annurev.astro.43.072103.150558
209. N. Yadav, B. Müller, H.T. Janka, T. Melson, A. Heger, Astrophys. J. **890**, 94 (2020). https://doi.org/10.3847/1538-4357/ab66bb
210. A. Yahil, Astrophys. J. **265**, 1047 (1983). https://doi.org/10.1086/160746
211. S. Yamada, H. Sawai, Astrophys. J. **608**, 907 (2004). https://doi.org/10.1086/420760
212. S. Zha, B. Müller, J. Powell, arXiv e-prints , arXiv:2403.02072 (2024). https://doi.org/10.48550/arXiv.2403.02072

The manufacturer's authorised representative in the EU is Springer Nature Customer Service Centre GmbH, Europaplatz 3, 69115 Heidelberg, Germany. If you have any concerns regarding our products, please contact ProductSafety@springernature.com

Printed and bound by CPI Group (UK) Ltd, Croydon, CR0 4YY
01/10/2026
02225941-0019